Fluid Mechanics

Fluid Mechanics

Seventh Edition

Pijush K. Kundu

Ira M. Cohen

David R. Dowling

Jesse Capecelatro

Academic Press is an imprint of Elsevier
125 London Wall, London EC2Y 5AS, United Kingdom
525 B Street, Suite 1650, San Diego, CA 92101, United States
50 Hampshire Street, 5th Floor, Cambridge, MA 02139, United States

Front cover photo courtesy of Trevor Mahlmann.

ISBN: 978-0-12-819807-0

For information on all Academic Press publications
visit our website at https://www.elsevier.com/books-and-journals

Publisher: Peter Linsley
Acquisitions Editor: Stephen Merken
Editorial Project Manager: Sara Valentino
Production Project Manager: Haritha Dharmarajan
Cover Designer: Miles Hitchen

Typeset by VTeX

Printed in India

Last digit is the print number: 9 8 7 6 5 4 3 2 1

Dedication

The authors dedicate this textbook to their wives and families whose patience during this undertaking has been a source of strength and consolation, and to the many fine instructors and students with whom they have interacted who have all in some way highlighted the allure and importance of this subject.

Contents

About the authors xi
Preface xiii
Acknowledgments xv
Nomenclature xvii

1. Introduction
   1.1. Fluid mechanics 1
   1.2. Units of measurement 2
   1.3. Solids, liquids, and gases 2
   1.4. Continuum hypothesis 3
   1.5. Molecular transport phenomena 5
   1.6. Surface tension 7
   1.7. Fluid statics 8
   1.8. Classical thermodynamics 10
   1.9. Perfect gas 13
   1.10. Stability of stratified fluid media 15
   1.11. Dimensional analysis 19
   Exercises 24
   References 35
   Further reading 35

2. Cartesian tensors
   2.1. Scalars, vectors, tensors, notation 37
   2.2. Rotation of axes; formal definition of a vector 39
   2.3. Multiplication of matrices 40
   2.4. Second-order tensors 41
   2.5. Contraction and multiplication 43
   2.6. Force on a surface 44
   2.7. Kronecker delta and alternating tensor 46
   2.8. Vector dot and cross products 47
   2.9. Gradient, divergence, and curl 48
   2.10. Symmetric and antisymmetric tensors 50
   2.11. Eigenvalues and eigenvectors of a symmetric tensor 51
   2.12. Gauss' theorem 52
   2.13. Stokes' theorem 55
   Exercises 56
   References 58
   Further reading 58

3. Kinematics
   3.1. Introduction and coordinate systems 59
   3.2. Particle and field descriptions of fluid motion 61
   3.3. Flow lines, fluid acceleration, and Galilean transformation 64
   3.4. Strain and rotation rates 69
   3.5. Kinematics of simple plane flows 73
   3.6. Reynolds transport theorem 77
   Exercises 79
   References 83
   Further reading 83

4. Conservation laws
   4.1. Introduction 85
   4.2. Conservation of mass 86
   4.3. Stream functions 88
   4.4. Conservation of momentum 89
   4.5. Constitutive equation for a Newtonian fluid 98
   4.6. Navier-Stokes momentum equation 101
   4.7. Noninertial frame of reference 103
   4.8. Conservation of energy 106
   4.9. Special forms of the equations 109
   4.10. Boundary conditions 119
   4.11. Dimensionless forms of the equations and dynamic similarity 124
   Exercises 131
   References 148
   Further reading 148

5. Vorticity dynamics
   5.1. Introduction 151
   5.2. Kelvin's and Helmholtz's theorems 155
   5.3. Vorticity equation in an inertial frame of reference 158
   5.4. Vorticity equation in a rotating frame of reference 160
   5.5. Velocity induced by a vortex filament: law of Biot and Savart 163
   5.6. Interaction of vortices 165

5.7. Vortex sheet 168
Exercises 172
References 176
Further reading 176

6. Ideal flow

6.1. Relevance of irrotational constant-density flow theory 177
6.2. Two-dimensional stream function and velocity potential 179
6.3. Construction of elementary flows in two dimensions 181
6.4. Complex potential 191
6.5. Forces on a two-dimensional body 193
6.6. Conformal mapping 195
6.7. Numerical solution techniques in two dimensions 199
6.8. Axisymmetric ideal flow 202
6.9. Three-dimensional potential flow and apparent mass 206
6.10. Concluding remarks 211
Exercises 211
References 221
Further reading 221

7. Laminar flow

7.1. Introduction 223
7.2. Exact solutions for steady incompressible viscous flow 225
7.3. Elementary lubrication theory 230
7.4. Similarity solutions for unsteady incompressible viscous flow 236
7.5. Flows with oscillations 243
7.6. Low Reynolds number viscous flow past a sphere 245
7.7. Final remarks 253
Exercises 253
References 268
Further reading 269

8. Boundary layers and related topics

8.1. Introduction 271
8.2. Boundary-layer thickness definitions 275
8.3. Boundary layer on a flat plate: Blasius solution 277
8.4. Falkner-Skan similarity solutions of the laminar boundary-layer equations 282
8.5. von Karman momentum integral equation 283
8.6. Thwaites' method 285
8.7. Transition, pressure gradients, and boundary-layer separation 289
8.8. Flow past a circular cylinder 294
8.9. Flow past a sphere 299
8.10. Two-dimensional jets 303
8.11. Secondary flows 308
Exercises 309
References 317
Further reading 318

9. Instability

9.1. Introduction 319
9.2. Method of normal modes 321
9.3. Kelvin-Helmholtz instability 322
9.4. Rayleigh-Plateau instability 326
9.5. Thermal instability: the Bénard problem 329
9.6. Double-diffusive instability 336
9.7. Centrifugal instability: Taylor problem 339
9.8. Instability of continuously stratified parallel flows 344
9.9. Squire's theorem and the Orr-Sommerfeld equation 349
9.10. Inviscid stability of parallel flows 351
9.11. Results for parallel and nearly parallel viscous flows 354
9.12. Experimental verification of boundary-layer instability 357
9.13. Comments on nonlinear effects 359
9.14. Transition 359
9.15. Deterministic chaos 360
Exercises 365
References 372
Further reading 372

10. Turbulence

10.1. Introduction 373
10.2. Historical notes 375
10.3. Nomenclature and statistics for turbulent flow 376
10.4. Correlations and spectra 379
10.5. Averaged equations of motion 383
10.6. Homogeneous isotropic turbulence 389
10.7. Turbulent energy cascade and spectrum 393
10.8. Free turbulent shear flows 399
10.9. Wall-bounded turbulent shear flows 405
10.10. Turbulence modeling 417
10.11. Turbulence in a stratified medium 422
10.12. Taylor's theory of turbulent dispersion 426
Exercises 431
References 442
Further reading 443

11. Gravity waves

11.1. Introduction 445

11.2. Linear liquid-surface gravity waves 447
11.3. Influence of surface tension 457
11.4. Standing waves 459
11.5. Group velocity, energy flux, and dispersion 461
11.6. Nonlinear waves in shallow and deep water 466
11.7. Waves on a density interface 472
11.8. Internal waves in a continuously stratified fluid 477
Exercises 485
References 488

12. Geophysical fluid dynamics

12.1. Introduction 489
12.2. Vertical variation of density in the atmosphere and ocean 491
12.3. Equations of motion for geophysical flows 493
12.4. Geostrophic flow 496
12.5. Ekman layers 499
12.6. Shallow-water equations 505
12.7. Normal modes in a continuously stratified layer 507
12.8. High- and low-frequency regimes in shallow-water equations 510
12.9. Gravity waves with rotation 512
12.10. Kelvin wave 514
12.11. Potential vorticity conservation in shallow-water theory 517
12.12. Internal waves 520
12.13. Rossby wave 528
12.14. Barotropic instability 531
12.15. Baroclinic instability 532
12.16. Geostrophic turbulence 537
Exercises 540
References 541
Further reading 542

13. Aerodynamics

13.1. Introduction 543
13.2. Aircraft terminology 544
13.3. Characteristics of airfoil sections 547
13.4. Conformal transformation for generating airfoil shapes 552
13.5. Lift of a Zhukhovsky airfoil 555
13.6. Elementary lifting line theory for wings of finite span 558
13.7. Lift and drag characteristics of airfoils 565
13.8. Propulsive mechanisms of fishes and birds 567
13.9. Sailing against the wind 569
Exercises 570
References 574
Further reading 574

14. Compressible flow

14.1. Introduction 575
14.2. Acoustics 577
14.3. One-dimensional steady isentropic compressible flow in variable-area ducts 582
14.4. Normal shock waves 591
14.5. Operation of nozzles at different back pressures 597
14.6. Effects of friction and heating in constant-area ducts 600
14.7. One-dimensional unsteady compressible flow in constant-area ducts 603
14.8. Two-dimensional steady compressible flow 606
14.9. Thin-airfoil theory in supersonic flow 612
Exercises 614
References 619
Further reading 619

15. Computational fluid dynamics

Grétar Tryggvason

15.1. Introduction 621
15.2. The advection-diffusion equation 623
15.3. Incompressible flows in rectangular domains 637
15.4. Flow in complex domains 650
15.5. Velocity-pressure method for compressible flow 656
15.6. More to explore 662
Exercises 666
References 668
Further reading 668

Appendix A. Conversion factors, constants, and fluid properties

A.1. Conversion factors 669
A.2. Physical constants 669
A.3. Properties of pure water at atmospheric pressure 670
A.4. Properties of dry air at atmospheric pressure 670
A.5. Properties of standard atmosphere 671

Appendix B. Mathematical tools and resources

B.1. Partial and total differentiation 673
B.2. Changing independent variables 675
B.3. Basic vector calculus 676

B.4. The Dirac delta function 677
B.5. Common three-dimensional coordinate systems 677
B.6. Equations in curvilinear coordinates 679

Appendix C. Founders of modern fluid dynamics

Ludwig Prandtl (1875–1953) 683
Geoffrey Ingram Taylor (1886–1975) 683
Further reading 684

Appendix D. Visual resources

Index 687

About the authors

Pijush K. Kundu, 1941–1994, was born in Calcutta, India. He earned a BS degree in Mechanical Engineering from Calcutta University in 1963 and an MS degree in Engineering from Roorkee University in 1965. After a few years as a lecturer at the Indian Institute of Technology in Delhi, he came to the United States and earned a PhD at Pennsylvania State University in 1972. He then followed a lifelong interest in oceanography and held research and teaching positions at Oregon State University and the University de Oriente in Venezuela, finally settling at the Oceanographic Center of Nova Southeastern University, where he spent most of his career contributing to the understanding of coastal dynamics, mixed-layer physics, internal waves, and Indian Ocean dynamics. He authored the first edition of this textbook, which he dedicated to his mother, wife, daughter, and son.

Ira M. Cohen, 1937–2007, earned a BS degree from Polytechnic University in 1958 and a PhD from Princeton in 1963, both in aeronautical engineering. He taught at Brown University for three years prior to joining the faculty at the University of Pennsylvania in 1966. There he became a world-renowned scholar in the areas of continuum plasmas, electrostatic probe theories and plasma diagnostics, dynamics and heat transfer of lightly ionized gases, low current arc plasmas, laminar shear layer theory, and matched asymptotics in fluid mechanics. He served as Chair of the Department of Mechanical Engineering and Applied Mechanics from 1992 to 1997. During his 41 years as a faculty member, he distinguished himself through his integrity, candor, sense of humor, pursuit of physical fitness, unrivaled dedication to academics, fierce defense of high scholarly standards, and passionate commitment to teaching.

David R. Dowling, 1960–, grew up in southern California where early experiences with fluid mechanics included swimming, surfing, sailing, flying model aircraft, and trying to throw a curve ball. At the California Institute of Technology, he earned BS ('82), MS ('83), and PhD ('88) degrees in Applied Physics and Aeronautics. In 1992, after a year at Boeing Aerospace & Electronics and three at the Applied Physics Laboratory of the University of Washington, he joined the faculty in the Department of Mechanical Engineering at the University of Michigan (UM), where he taught and conducted research in fluid mechanics and acoustics. In 2021, he moved within the UM to become Chair of the Department of Naval Architecture and Marine Engineering. He is a fellow of the American Physical Society – Division of Fluid Dynamics, the American Society of Mechanical Engineers, and the Acoustical Society of America. Prof. Dowling is an avid swimmer, is married, and has seven children.

Jesse Capecelatro, 1986–, grew up in Brewster, New York, a suburb of New York City. He earned degrees in Mechanical Engineering from the State University of New York at Binghamton (BS in 2009), the University of Colorado Boulder (MS in 2011), and Cornell University (MS in 2012 and PhD in 2014). He then spent two years as a postdoctoral researcher at the University of Illinois Urbana-Champaign focused on high-performance computing and turbulent combustion. In 2016 he joined the faculty in the Department of Mechanical Engineering at the University of Michigan, where he currently holds a position. His research is broadly under the realm of fluid mechanics, with an emphasis on multiphase flow, turbulence, and scientific computing. Prof. Capecelatro is a rock climber, dances Argentine tango, is married, and has three children.

Preface

The *seventh edition* of *Fluid Mechanics* is the result of a team effort. Prof. Jesse Capecelatro is responsible for more than half of the work that went into it. Much of this work is unseen by the reader. The master text files for the *fifth and sixth editions* were produced and edited using MS Word. However, for the *seventh edition* the entire book was converted to LaTeX, a step that should facilitate production of future editions of the book and better intra- and inter-chapter linking and referencing. From the reader's perspective, the most noticeable change is rearrangement of Chapters 6 through 15. This reordering is based on commentary received about the *sixth edition*. The chapter on computational fluid dynamics (CFD) authored by Prof. Grétar Tryggvason of the Johns Hopkins University (Charles A. Miller, Jr. Distinguished Professor and Head of the Department of Mechanical Engineering) was revised and moved to the final position, since computational techniques can be applied in all branches of fluid mechanics.

The *seventh edition* of *Fluid Mechanics* retains nearly all topics covered by the *sixth edition*, and includes 10 new in-chapter examples and 66 new end-of-chapter exercises. Many of the new examples are numerical and provide sample code in `MATLAB`®. The programming language used was a perfunctory decision. Our goal was to develop self-contained problems that either could not be solved analytically or are greatly simplified using a computer. These problems can easily be translated to other languages (`Python`, `Julia`, `Fortran`, etc.) based on the preferences of the instructor, student, or any other interested reader. In addition, 120 figures were either revised or added compared to the *sixth edition*.

Several topics were added to the *seventh edition*. Chapter 5 (Vorticity dynamics) introduces a vortex method technique (the so-called vortex blob method) – a popular approach for solving the Biot-Savart equations numerically. Chapter 9 (Instability) now includes the Rayleigh-Plateau instability which leads to surface-tension induced breakup of a liquid jet. Lighthill's acoustic analogy was added to Chapter 14 (Compressible flow). In addition, concepts in multiphase flow involving particles, droplets, or bubbles, were added throughout the book. Many of these were used to enrich existing topics or extend previous concepts and theories. For example, fluid friction (drag) through dense particulate media was introduced to solve a problem involving pour-over coffee making. Stokes' first problem was extended to arrive at the Basset history force acting on an accelerating spherical object. The section on sports balls in Chapter 8 (Boundary layers and related topics) was replaced with a short exposition on Saffman and Magnus lift. In addition, turbulence produced by interphase momentum transfer and aeroacoustic sound generation in two-phase flows, can now be found in end-of-chapter exercises.

When we first embarked on production of the *seventh edition* of the book, one of our goals was to make the book shorter. The *sixth edition* is over 900 pages. Unfortunately, the authors found it difficult to remove content without sacrificing concepts deemed critical for an advanced-undergraduate or graduate-level survey of the topic. Production of the *seventh edition* began in earnest in September 2020, six months after the COVID-19 pandemic caused global disruptions. The current edition of the book was produced at the dawn of the machine learning (ML) era. Although in its infancy, it is becoming increasingly apparent that scientific ML will continue to play a role in fluid mechanics for years to come. The current edition of *Fluid Mechanics* does not include any application of ML but provides the groundwork needed to apply such techniques with appropriate knowledge of the underlying physics.

And finally, responsible stewardship and presentation of this material is our primary goal. Thus, we welcome the opportunity to correct any errors you find, to hear your opinion of how this book might be improved, and to include topics and exercises you might suggest; just contact us at drd@umich.edu (David R. Dowling) and/or jcaps@umich.edu (Jesse Capecelatro).

Instructor resources for this book are available by visiting https://educate.elsevier.com/book/details/9780128198070.
Freely accessible companion materials are available here: https://www.elsevier.com/books-and-journals/book-companion/9780128198070

David R. Dowling and Jesse Capecelatro
Ann Arbor, Michigan
September 2023

Acknowledgments

The current version of this textbook has benefited from the commentary, suggestions, and corrections provided by the reviewers of the revision proposal, and the many careful readers of the sixth edition of this textbook who provided feedback. The authors are grateful to Dr. Alexander Douglass and Mr. Matthew Wanink who assisted with the conversion to LaTeX. Prof. Dowling would also like to recognize and thank his technical mentors, Prof. Hans W. Liepmann (undergraduate advisor), Prof. Paul E. Dimotakis (graduate advisor), and Prof. Darrell R. Jackson (post-doctoral advisor), and his friends and colleagues who have contributed to the development of this text by discussing ideas and sharing their expertise, humor, and devotion to science and engineering. Prof. Capecelatro would similarly like to recognize Prof. Olivier Desjardins (graduate advisor) and Profs. Jonathan Freund and Daniel Bodony (post-doctoral advisors), in addition to Profs. Rodney O. Fox and Shankar Subramaniam who have shared their knowledge and enthusiasm of fluid mechanics and inspired many of topics appearing in this text.

Nomenclature

Notation (Relevant Equation Numbers Appear in Parentheses)

$\overline{f}$ = principle-axis version of f, background or quiescent-fluid value of f, or average or ensemble average of f, Darcy friction factor (10.101, 10.102)
$\hat{f}$ = complex amplitude of f
$\tilde{f}$ = full field value of f
f' = derivative of f with respect to its argument, or perturbation of f from its reference state
f^* = complex conjugate of f, or the value of f at the sonic condition
f^+ = the dimensionless, law-of-the-wall value of f
$f_\xi = \partial f/\partial \xi$ (15.105)
f_{cr} = critical value of f
f_{av} = average value of f
f_{CL} = centerline value of f
f_j = the jth component of the vector **f**, f at location j (15.14)
$f)_i^n, f_i^n$ = f at time n at horizontal x-location j (15.13)
f_{ij} = the i-j component of the second order tensor **f**
$f_{i,j}^n, f)_{i,j}^n$ = f at time n at horizontal x-location i and vertical y-location j ((15.52), Fig. 15.10)
f_R = rough-wall value of f
f_S = smooth-wall value of f
f_0 = reference, surface, or stagnation value of f
f_∞ = reference value of f or value of f far away from the point of interest
Δf = change in f

Symbols (Relevant Equation Numbers Appear in Parentheses)

α = contact angle (Fig. 1.8), thermal expansion coefficient (1.26), angle of rotation, iteration number (15.57), angle of attack (Fig. 13.6)
a = triangular area, cylinder radius, sphere radius, amplitude
a = generic vector, Lagrangian acceleration (3.1)
A = generic second-order (or higher) tensor
A, A = a constant, an amplitude, area, surface, surface of a material volume, planform area of a wing
A^* = control surface, sonic throat area
A_o = Avogadro's number
A_0 = reference area
A_{ij} = representative second-order tensor
β = angle of rotation, coefficient of density change due to salinity or other constituent, convergence acceleration parameter (15.57), variation of the Coriolis frequency with latitude (12.10), camber parameter (Fig. 13.13)
b = generic vector, control surface velocity (Fig. 3.20)
B, B = a constant, Bernoulli function (4.70), log-law intercept parameter (10.88)
B, B_{ij} = generic second-order (or higher) tensor
Bo = Bond number (4.120)
c = speed of sound ((1.25), (14.1h)), phase speed (11.4), chord length ((13.2), Figs. 13.2, 13.6)
c = phase velocity vector (11.8)
c_g, $\mathbf{c_g}$ = group velocity magnitude (11.67) and vector (11.141)
χ = scalar stream function (Fig. 4.1)
°C = degrees centigrade
C = a generic constant, hypotenuse length, closed contour
Ca = Capillary number (4.121)
C_f = skin friction coefficient ((8.15), (8.32))

C_p = pressure (coefficient) ((4.107), (6.32))
c_p = specific heat capacity at constant pressure (1.20)
C_D = coefficient of drag ((4.108), (8.33))
C_L = coefficient of lift (4.109)
c_v = specific heat capacity at constant volume (1.21)
C_{ij} = matrix of direction cosines between original and rotated coordinate system axes (2.5)
$C_\pm$ = Characteristic curves along which the $I\pm$ invariants are constant (14.62)
d = diameter, distance, fluid layer depth
$\mathbf{d}$ = dipole strength vector (6.28), displacement vector
δ = Dirac delta function (B.4.1), similarity-variable length scale (7.31), boundary-layer thickness, generic length scale, small increment, flow deflection angle (Fig. 14.22, and (14.69))
$\overline{\delta}$ = average boundary-layer thickness
δ^* = boundary-layer displacement thickness (8.16)
δ_{ij} = Kronecker delta function (2.16)
δ_{99} = 99% layer thickness
D = distance, drag force, diffusion coefficient (15.10)
$\mathbf{D}$ = drag force vector (Example 13.1)
D_i = lift-induced drag (13.15)
D/Dt = material derivative (3.4), (3.5), or (B.1.4)
D_T = turbulent diffusivity of particles (10.156)
$\mathcal{D}$ = generalized field derivative (2.31)
ϵ = volume fraction, void fraction
ε = roughness height, kinetic energy dissipation rate (4.59), a small distance, fineness ratio h/L (7.14), downwash angle (13.14)
$\overline{\varepsilon}$ = average dissipation rate of the turbulent kinetic energy (10.47)
$\overline{\varepsilon}_T$ = average dissipation rate of the variance of temperature fluctuations (10.141)
ε_{ijk} = alternating tensor (2.18)
e = internal energy per unit mass (1.16)
$\mathbf{e}_i$ = unit vector in the i-direction (2.1)
$\overline{e}$ = average kinetic energy of turbulent fluctuations (10.47)
Ec = Eckert number (4.117)
E_k = kinetic energy per unit horizontal area (11.39)
E_p = potential energy per unit horizontal area (11.41)
E = numerical error (15.21), average energy per unit horizontal area (12.18)
$\overline{E}$ = kinetic energy of the average flow (10.46)
EF = time average energy flux per unit length of wave crest (11.43)
f = generic function, Maxwell distribution function (1.1) and (1.4), Helmholtz free energy per unit mass, longitudinal correlation coefficient (10.38), Coriolis frequency (12.6), dimensionless friction parameter (14.50)
$\overline{f}$ = Darcy friction factor (10.101, 10.102)
f_i = unsteady body force distribution (14.5)
ϕ = velocity potential (6.10), an angle
$\mathbf{f}$ = surface force vector per unit area ((2.15), (4.13))
F = force magnitude, generic flow field property, generic flux, generic or profile function
F_f = perimeter friction force (14.30)
$\mathbf{F}$ = force vector, average wave energy flux vector (11.157)
Φ = body force potential (4.18), undetermined spectrum function (10.53)
$F_D, \overline{F}_D$ = drag force (4.108), average drag force
F_L = lift force (4.109)
Fr = Froude number (4.105)
γ = ratio of specific heats (1.30), velocity gradient, vortex sheet strength, generic dependent-field variable
$\dot{\gamma}$ = shear rate
$\mathbf{g}$ = body force per unit mass (4.13)
g = acceleration of gravity, undetermined function, transverse correlation coefficient (10.38)
g' = reduced gravity (11.116)
Γ = vertical temperature gradient or lapse rate, circulation (3.18)
Γ_a = adiabatic vertical temperature gradient (1.36)
Γ_{a} = circulation due to the absolute vorticity (5.24)
G = gravitational constant, profile function
G_n = Fourier series coefficient

G = center of mass, center of vorticity
h = enthalpy per unit mass (1.19), height, gap height, viscous layer thickness
$\hbar$ = Planck's constant
η = free surface shape, waveform, similarity variable (7.25) or (7.31), Kolmogorov microscale (10.50)
η_T = Batchelor microscale (10.143)
H = atmospheric scale height, water depth, step function, shape factor (10.46), profile function
i = an index, imaginary root
I = incident light intensity, bending moment of inertia
I_i = Modified Bessel function of the first kind of order i
$I_\pm$ = Invariants along the $C\pm$ characteristics (14.60)
j = an index
J = Jacobian of a transformation (15.110), momentum flux per unit span (8.60)
J_s = jet momentum flux per unit span (10.62)
J_i = Bessel function of order i
$\mathbf{J}_m$ = diffusive mass flux vector (1.7)
φ = a function, azimuthal angle in cylindrical and spherical coordinates (Fig. 3.3)
k = thermal conductivity (1.8), an index, wave number (6.12) or (11.2), wave number component
κ = thermal diffusivity, von Karman constant (12.88)
κ_s = diffusivity of salt
κ_T = turbulent thermal diffusivity (10.116)
κ_m = mass diffusivity of a passive scalar in Fick's law (1.7)
κ_{mT} = turbulent mass diffusivity (10.117)
k_B = Boltzmann's constant (1.27)
k_s = sand grain roughness height
Kn = Knudsen number
K = a generic constant, magnitude of the wave number vector (11.6), lift curve slope (13.15)
K_i = Modified Bessel function of the second kind of order i
K = degrees Kelvin
$\mathbf{K}$ = wave number vector (11.5)
l = molecular mean free path (1.6), spanwise dimension, generic length scale, wave number component ((11.5), (11.6)), shear correlation in Thwaites method (8.45), length scale in turbulent flow
l_T = mixing length (10.119)
L, L = generic length dimension, generic length scale, lift force
L_M = Monin-Obukhov length scale (10.138)
λ = wavelength ((11.1), (11.7)), laminar boundary-layer correlation parameter (8.44)
λ_m = wavelength of the minimum phase speed
λ_t = temporal Taylor microscale (10.19)
λ_f, λ_g = longitudinal and lateral spatial Taylor microscales (10.39)
Λ = lubrication-flow bearing number (7.16), Rossby radius of deformation, wing aspect ratio (13.1)
Λ_f, Λ_g = longitudinal and lateral integral spatial scales (10.39)
Λ_t = integral time scale (12.18)
μ = dynamic or shear viscosity (1.9), Mach angle (14.65)
μ_v = bulk viscosity (4.36)
m = molecular mass (1.1), generic mass, an index, moment order (10.1), wave number component ((11.5), (11.6))
M, M = generic mass dimension, mass, Mach number (4.113), apparent or added mass (6.110)
M_w = molecular weight
n = molecular density (1.1), an index, generic integer number, power law exponent (4.37)
$\mathbf{n}$ = normal unit vector
n_s = index of refraction
N = number of molecules (1.27), Brunt-Väisälä or buoyancy frequency ((1.35), (11.126)), number
N_{ij} = pressure rate of strain tensor (10.131)
ν = kinematic viscosity (1.10), cyclic frequency, Prandtl-Meyer function (14.72)
ν_T = turbulent kinematic viscosity (10.115)
O = origin
p = pressure
$\mathbf{p} = \mathbf{t} \times \mathbf{n}$, third unit vector
p_{atm} = atmospheric pressure
p_i = inside pressure

p_o = outside pressure
p_0 = reference pressure at $z = 0$
p_∞ = reference pressure far upstream or far away
$\overline{p}$ = average or quiescent pressure in a stratified fluid
P = average pressure
Π = wake strength parameter (10.95)
Pr = Prandtl number (4.118)
q = heat added to a system (1.16), volume flux per unit span, unsteady volume source (14.4), dimensionless heat addition parameter (14.50)
$\mathbf{q}$, q_i = heat flux (1.8)
$\dot{q}$ = generic acoustic source (14.8)
q_s = two-dimensional point source or sink strength in ideal flow (6.13)
q_n = generic parameter in dimensional analysis (1.42)
Q = volume flux in two or three dimensions, heat added per unit mass (14.26)
θ = potential temperature (1.37), unit of temperature, angle in polar coordinates, momentum thickness (8.17), local phase, an angle
ρ = mass density (1.7)
ρ_s = static density profile in stratified environment
ρ_m = mass density of a mixture
$\overline{\rho}$ = average or quiescent density in a stratified fluid
ρ_θ = potential density (1.39)
r = matrix rank, distance from the origin, distance from the axis
$\mathbf{r}$ = particle trajectory (3.1), (3.8)
R = distance from the cylindrical axis, radius of curvature, gas constant (1.29), generic nonlinearity parameter
R_Δ = dimensionless grid-resolution (15.42)
R_u = universal gas constant (1.28)
R_i = radius of curvature in direction i (1.11)
$\mathbf{R}$, R_{ij} = rotation tensor (3.13), correlation tensor (10.12), (10.23)
Ra = Rayleigh number (9.38)
Re = Reynolds number (4.104)
Ri = Richardson number, gradient Richardson number ((9.83), (10.136))
Rf = flux Richardson number (10.135)
Ro = Rossby number (12.13)
σ = surface tension (1.11), interfacial tension, vortex core size ((3.28), (3.29)), temporal growth rate (9.1), oblique shock angle (Fig. 14.22)
s = entropy (1.22), arc length, salinity, wingspan (13.1), dimensionless arc length
σ_i = standard deviation of molecular velocities (1.3)
S = salinity, scattered light intensity, an area, entropy
S_e = one-dimensional temporal longitudinal energy spectrum (10.20)
S_{11} = one-dimensional spatial longitudinal energy spectrum (10.45)
S_T = one-dimensional temperature fluctuation spectrum (10.142), (10.143)
$\mathbf{S}$, S_{ij} = strain rate tensor (3.12), symmetric tensor
St = Strouhal number (4.103)
t = time
$\mathbf{t}$ = tangent vector
T, T = temperature (1.1), generic time dimension, period
$\mathbf{T}$, T_{ij} = stress tensor (2.15), (Fig. 2.4)
Ta = Taylor number (9.69)
T_o = free stream temperature
T_w = wall temperature
τ = shear stress (1.9), time lag
$\boldsymbol{\tau}$, τ_{ij} = viscous stress tensor (4.27)
τ_w = wall or surface shear stress
υ = specific volume = $1/\rho$
u = horizontal component of fluid velocity (1.9)
$\mathbf{u}$ = generic vector, average molecular velocity vector (1.1), fluid velocity vector (3.1)
u_i = fluid velocity components, fluctuating velocity components
u_* = friction velocity (10.81)
$\mathbf{U}$ = generic uniform velocity vector
U_i = ensemble average velocity components
U = generic velocity, average stream-wise velocity

ΔU = characteristic velocity difference
U_e = local free-stream flow speed above a boundary layer (8.11), flow speed at the effective angle of attack
U_{CL} = centerline velocity (10.56)
U_∞ = flow speed far upstream or far away
v = molecular speed (1.4), component of fluid velocity along the y axis
$\mathbf{v}$ = molecular velocity vector (1.1), generic vector
V = volume, material volume, average stream-normal velocity, average velocity, complex velocity
V^* = control volume
w = vertical component of fluid velocity, complex potential (6.42), downwash velocity (13.13)
W = thermodynamic work per unit mass, wake function
$\dot{W}$ = rate of energy input from the average flow (10.49)
We = Weber number (4.119)
ω = temporal frequency (11.2)
$\boldsymbol{\omega}$, ω_i = vorticity vector (2.25, 3.16)
Ω = oscillation frequency, rotation rate, rotation rate of the earth
$\boldsymbol{\Omega}$ = angular velocity of a rotating frame of reference
x = first Cartesian coordinate
$\mathbf{x}$ = position vector (2.1)
x_i = components of the position vector (2.1)
ξ = generic spatial coordinate, integration variable, similarity variable (10.84)
y = second Cartesian coordinate
Y = mass fraction (1.7)
Y_{CL} = centerline mass fraction (10.69)
Y_i = Bessel function of order i
ψ = stream function (6.3), (6.71), (15.59)
Ψ = Reynolds stress scaling function (10.57), generic functional solution
$\boldsymbol{\Psi}$ = vector potential, three-dimensional stream function (4.12)
z = third Cartesian coordinate, complex variable (6.43)
ζ = interface displacement, relative vorticity
Z = altitude of the 500 mb isobar (Example 12.4)

Chapter 1

Introduction⊛

Contents

1.1 **Fluid mechanics** 1
1.2 **Units of measurement** 2
1.3 **Solids, liquids, and gases** 2
1.4 **Continuum hypothesis** 3
1.5 **Molecular transport phenomena** 5
1.6 **Surface tension** 7
1.7 **Fluid statics** 8
1.8 **Classical thermodynamics** 10
First law of thermodynamics 10
Equations of state 11
Specific heats 11
Second law of thermodynamics 11
Property relations 12
Speed of sound 12
Thermal expansion coefficient 12
1.9 **Perfect gas** 13
1.10 **Stability of stratified fluid media** 15
Potential temperature and density 17
Scale height of the atmosphere 18
1.11 **Dimensional analysis** 19
Step 1. Select variables and parameters 19
Step 2. Create the dimensional matrix 20
Step 3. Determine the rank of the dimensional matrix 20
Step 4. Determine the number of dimensionless groups 20
Step 5. Construct the dimensionless groups 21
Step 6. State the dimensionless relationship 22
Step 7. Use physical reasoning or additional knowledge to simplify the dimensionless relationship 22
Exercises 24
References 35
Further reading 35

Chapter objectives

- To properly introduce the subject of fluid mechanics and its importance
- To state the assumptions upon which the subject is based
- To review the basic background science of liquids and gases
- To present the relevant features of fluid statics
- To establish dimensional analysis as a broadly-applicable intellectual tool

1.1 Fluid mechanics

Fluid mechanics is the branch of science concerned with moving and stationary fluids. Given that the vast majority of the observable mass in the universe exists in a fluid state, that life as we know it is not possible without fluids, and that the atmosphere and oceans covering this planet are fluids, fluid mechanics has unquestioned scientific importance. Its allure crosses disciplinary boundaries, in part because it is described by a nonlinear field theory and also because fluid phenomena are readily observed. As the foundational discipline for transportation systems, power generation and conversion, materials processing and manufacturing, food production, and civil infrastructure, its practical importance is also unquestioned. For example, in the twentieth century, life expectancy in the United States approximately doubled. About half of this increase can be traced to advances in medical practice, particularly antibiotic therapies. The other half largely resulted from a steep decline in childhood mortality from water-borne diseases, a decline that occurred because of widespread delivery of clean water to nearly the entire population – a fluids-engineering and public-works achievement.

Advances in fluid mechanics, like any other branch of physical science, may arise from mathematical analyses, computer simulations, or experiments. Analytical approaches are often successful for finding solutions to idealized and simplified problems and such solutions can be of immense value for developing insight and understanding, and for comparisons with numerical and experimental results. Given that the governing field equations of fluid mechanics include three spatial dimensions and time, some fluency in multivariable calculus is an inherent requirement of this subject. In practice, drastic simplifications are frequently necessary to find analytical solutions because of the complexity of real fluid flow phenomena.

⊛. This book has a companion website hosting complementary materials. Visit this URL to access it: https://www.elsevier.com/books-and-journals/book-companion/9780128198070.

Fluid Mechanics. https://doi.org/10.1016/B978-0-12-819807-0.00010-7

Furthermore, some of the greatest theoretical contributions have come from people who depended rather strongly on their physical intuition. Ludwig Prandtl, one of the founders of modern fluid mechanics, first conceived the idea of a boundary layer based solely on physical intuition. Interestingly, the boundary layer concept has since been expanded into a general method in applied mathematics.

This book is an introduction to fluid mechanics that should appeal to anyone pursuing fluid mechanical inquiry. Its emphasis is on fully presenting fundamental concepts and illustrating them with relevant examples. Given its finite size, this book provides – at best – an incomplete description of the subject. However, the purpose of this book will have been fulfilled if the reader becomes more curious and interested in fluid mechanics as a result of its perusal.

1.2 Units of measurement

For mechanical systems, the units of all physical quantities can be expressed in terms of four basic units, namely, *length*, *mass*, *time*, and *temperature*. In this book, the international system of units (Système international d'unités) commonly referred to as SI (or MKS) units, is preferred. The basic units of this system are *meter* for length, *kilogram* for mass, *second* for time, and *Kelvin* for temperature. Some of the common variables used in fluid mechanics, and their SI units, are listed in Table 1.1. Conversion factors between different systems of units are listed in Appendix A. To avoid very large or very small numerical values, prefixes are used to indicate multiples of the units. Some of the common prefixes are listed in Table 1.2.

TABLE 1.1 SI units.

Quantity	Name of unit	Symbol	Equivalent
Length	Meter	m	
Mass	Kilogram	kg	
Time	Second	s	
Temperature	Kelvin	K	
Frequency	Hertz	Hz	s^{-1}
Force	Newton	N	$kg\ ms^{-2}$
Pressure	Pascal	Pa	$N\ m^{-2}$
Energy	Joule	J	N m
Power	Watt	W	$J\ s^{-1}$

TABLE 1.2 Common prefixes.

Prefix	Symbol	Multiple
Mega	M	10^6
Kilo	k	10^3
Deci	d	10^{-1}
Centi	c	10^{-2}
Milli	m	10^{-3}
Micro	μ	10^{-6}

The SI system is frequently augmented herein by specifying temperatures in degrees Celsius (°C), which is related to Kelvin (K) by: °C = K – 273.15. However, the United States customary system of units (foot, pound, °F) is not used.

1.3 Solids, liquids, and gases

Physical matter may be broadly categorized as being fluid or solid. A fluid is a substance that deforms continuously under an applied shear stress, or, equivalently, one that does not have a preferred shape. A solid is one that does not deform continuously under an applied shear stress, and does have a preferred shape to which it relaxes when external forces on it are withdrawn. Consider the solid rectangular element ABCD (Fig. 1.1a). Under the action of a shear force F the element assumes the shape ABC′D′. If the solid is perfectly elastic, it returns to its preferred shape ABCD when F is withdrawn. In contrast, a fluid deforms *continuously* under the action of a shear force, *however small*. Thus, the element of the fluid

ABCD confined between parallel plates (Fig. 1.1b) successively deforms to shapes such as ABC′D′ and ABC″D″, and keeps deforming (flowing), as long as the force F is maintained on the upper plate and the lower plate is held still. When F is withdrawn, the fluid element's final shape is retained; it does not return to a prior shape.

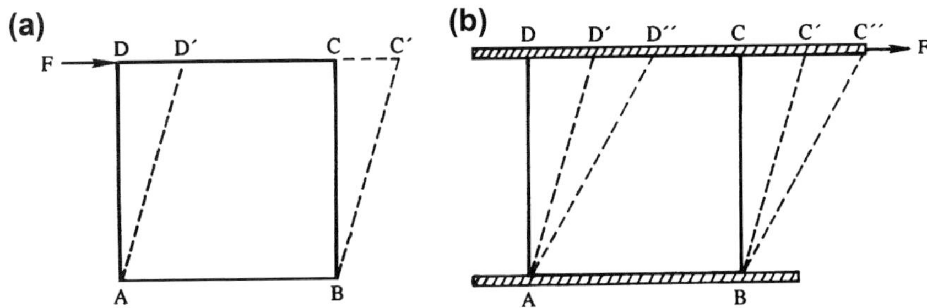

FIGURE 1.1 Deformation of solid and fluid elements under a constant shear force F. (a) Solid; the element deforms until its internal stress balances the externally applied force. (b) Fluid; the element deforms continuously as long as the shear force is applied. Hence the shape ABC′D′ in the solid (a) represents a static deformation, while the fluid (b) continues to deform with increasing time.

The qualification "however small" in the description of a fluid is significant because some solids also deform continuously if the shear stress exceeds a limiting value, the *yield point* of the solid. A solid in such a state is known as *plastic*, and plastic deformation changes the solid object's unloaded shape. Interestingly, the distinction between solids and fluids may not be well defined. Substances like paint, jelly, pitch, putty, polymer solutions, and biological substances may display both solid and fluid properties. If an elastic solid has a perfect memory and always returns to its preferred shape when unloaded, while a viscous fluid has zero memory and never springs back when unloaded, then substances like raw egg whites can be called *viscoelastic* because they partially rebound when unloaded.

Although solids and fluids behave differently when subjected to shear stresses, they behave similarly under the action of compressive stresses. However, tensile normal stresses again lead to differences in fluid and solid behavior. Solids can support both tensile and compressive normal stresses, while fluids typically expand or change phase (i.e., boil) when subjected to tensile stresses. Some liquids can support a small amount of tensile stress, the amount depending on the degree of molecular cohesion and the duration of the tensile stress.

Fluids generally fall into two classes, liquids and gases. A gas always expands to fill the entire volume of its container. In contrast, the volume of a liquid changes little, so that it may not completely fill a large container. In a gravitational field, a free surface forms that separates a liquid from its vapor. Although solids are not classified as a fluid, under certain circumstances a collection of discrete solid objects (or grains) can indeed flow with characteristics resembling a fluid state.

1.4 Continuum hypothesis

A fluid is composed of molecules in constant motion undergoing collisions with each other, and is therefore fundamentally discontinuous or discrete. However, the *average manifestation* of molecular motions is more important for macroscopic fluid mechanics. For example, forces exerted on a fluid container's walls are due to constant bombardment and rebound of the fluid's molecules; the statistical average of these collision forces per unit area is called *pressure*, a macroscopic property. When a fluid's molecular density and the smallest region of interest are large enough, such average properties are sufficient for the explanation of macroscopic phenomena and the fluid may be well described as a *continuum*. This simplifying approximation allows fluid properties like pressure, temperature, density, or velocity, to be defined continuously at every point in space because the fluid's discrete molecular structure is overlooked.

The simplest way to quantitatively assess molecular velocity variation in pure gases, and the continuum approximation's applicability limits for fluid velocity, is through use of the Maxwell distribution of molecular velocity $\mathbf{v} = (v_1, v_2, v_3)$. Here, $\mathbf{v}$ is random vector variable that represents possible molecular velocities. When the gas at the point of interest has average velocity $\mathbf{u}$, the Maxwell distribution is:

$$f(\mathbf{v})d^3v = n\left(\frac{m}{2\pi k_B T}\right)^{3/2} \exp\left\{\frac{-m}{2k_B T}|\mathbf{v}-\mathbf{u}|^2\right\} d^3v \tag{1.1}$$

(Chapman and Cowling, 1970), where n is the number of molecules per unit volume, m is the molecular mass, k_B is Boltzmann's constant, T is the absolute temperature, and $d^3v = dv_1 dv_2 dv_3$ is a small volume in velocity space centered on $\mathbf{v}$. The distribution (1.1) specifies the number of molecules at the point of interest with velocities near $\mathbf{v}$. When (1.1) is integrated over all possible molecular velocities, the molecular number density n is recovered. Thus, $f(\mathbf{v})/n$ is the probability density function for molecular velocity and the average gas velocity, $\overline{\mathbf{v}}$, and the variances of molecular velocity components, σ_i^2, can be determined from appropriate integrations:

$$\overline{\mathbf{v}} = \frac{1}{n}\int_{-\infty}^{+\infty}\int_{-\infty}^{+\infty}\int_{-\infty}^{+\infty} \mathbf{v} f(\mathbf{v}) d^3 v = \mathbf{u}, \quad \text{and} \quad \sigma_i^2 = \frac{1}{n}\int_{-\infty}^{+\infty}\int_{-\infty}^{+\infty}\int_{-\infty}^{+\infty} (v_i - u_i)^2 f(\mathbf{v}) d^3 v = \frac{k_B T}{m} \tag{1.2, 1.3}$$

(see Exercise 1.3). When $\mathbf{u} = 0$, (1.1) specifies the distribution of purely random molecular velocities in the gas and can be simplified by integrating over velocity directions to obtain the distribution of molecular speed, $v = |\mathbf{v}|$:

$$f(v) = \iint_{angles} f(\mathbf{v}) v^2 d\Omega = 4\pi n \left(\frac{m}{2\pi k_B T}\right)^{3/2} v^2 \exp\left\{\frac{-mv^2}{2k_B T}\right\}, \tag{1.4}$$

where $d\Omega$ is the differential solid-angle element. Using (1.5), the mean molecular speed can be found:

$$\overline{v} = \frac{1}{n}\int_0^\infty v f(v) dv = \left(\frac{8k_B T}{\pi m}\right)^{1/2} \tag{1.5}$$

(see Exercise 1.4). The results (1.2), (1.3), and (1.5) specify the average gas velocity, the variance of its components, and the average molecular speed. Interestingly, $\overline{v}$ (= 464 m/s for air at room temperature) may be large compared to $\mathbf{u}$. Thus, averaging over a significant number of gas molecules is necessary for the accuracy of the continuum approximation.

The continuum approximation is valid at the length scale L (a body length, a pore diameter, a turning radius, etc.) when the Knudsen number, $Kn = l/L$ where l is the molecular mean free path, is much less than unity. The molecular mean free path, l, is the average distance a gas molecule travels between collisions. It depends on the average molecular velocity $\overline{v}$, the number density of molecules n, the collision cross section of two molecules $\pi\overline{d}^2$ ($\overline{d}$ is the molecular collision diameter), and the average *relative* velocity between molecules, $\overline{v}_r = \sqrt{2}\,\overline{v}$ (see Exercises 1.5 and 1.6):

$$l = \frac{\overline{v}}{n\pi\overline{d}^2\,\overline{v}_r} = \frac{1}{\sqrt{2}n\pi\overline{d}^2}. \tag{1.6}$$

The mean free path is the average distance that a molecule travels before it communicates its presence, temperature, or momentum to other molecules. It characterizes the random molecular motions leading to the macroscopically observed phenomena of species, heat, and momentum diffusion in fluids.

For most terrestrial situations, the requirement $Kn \ll 1$ is not a great restriction since $l \approx 60$ nm for air at room temperature and pressure. Furthermore, l is more than two orders of magnitude smaller for water under the same conditions. However, a molecular-kinetic-theory approach may be necessary for analyzing flows over very small objects or in very narrow passages (where L is small), or in the tenuous gases at the upper reaches of the atmosphere (where l is large).

Example 1.1

The number density and nominal collision diameter of air molecules at 295 K and atmospheric pressure are approximately $2.5 \times 10^{25}/\text{m}^3$ and 4.0×10^{-10} m. Determine the molecular mean-free path and the Knudsen number for a 1 μm diameter aerosol particle suspended in this gas.

Solution Evaluate (1.6) to determine the mean-free path:

$$l = \frac{1}{\sqrt{2}n\pi\overline{d}^2} = \left[\sqrt{2}(2.5 \times 10^{25}\ \text{m}^{-3})\pi(4.0 \times 10^{-10}\ \text{m})^2\right]^{-1} = 5.6 \times 10^{-8}\ \text{m}.$$

For a 1 μm diameter particle, the Knudsen number is

$$Kn = \frac{l}{L} = \frac{5.6 \times 10^{-8}\ \text{m}}{1 \times 10^{-6}\ \text{m}} = 0.056.$$

Thus, the continuum theory is likely acceptable for predicting the settling velocity of fine aerosol particles in air.

1.5 Molecular transport phenomena

Random molecular motions lead to diffusive transport of molecular species, temperature, or momentum in fluids that is incorporated in the continuum description of fluid motion through the specification of transport coefficients (κ_m, k, and μ or ν in the following paragraphs).

First consider the diffusion of molecular species across a surface area AB within a mixture of two gases, say nitrogen and oxygen (Fig. 1.2), and assume that the nitrogen mass fraction Y varies in the direction perpendicular to AB. Here the mass of nitrogen per unit volume is ρY, where ρ is the overall density of the gas mixture (kg m^{-3}). Random migration of molecules across AB in both directions will result in a *net* flux of nitrogen across AB, from the region of higher Y toward the region of lower Y. To a good approximation, the diffusive flux of one constituent in a mixture is proportional to its gradient:

$$\mathbf{J}_m = -\rho\kappa_m \nabla Y. \tag{1.7}$$

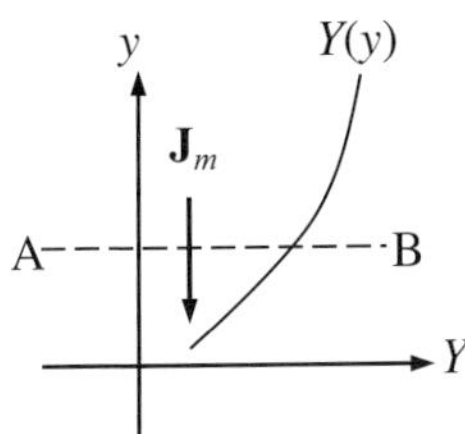

FIGURE 1.2 Mass flux $\mathbf{J}_m$ due to variation in the mass fraction $Y(y)$. Here the mass fraction profile increases with increasing y, so Fick's law of diffusion states that the diffusive mass flux that acts to smooth out mass-fraction differences is downward across AB.

Here the vector $\mathbf{J}_m$ is the mass flux (kg m^{-2} s^{-1}) of the constituent, ∇Y is the mass-fraction gradient of that constituent, and κ_m is a (positive) constant of proportionality, known as the species or mass diffusivity, that depends on the particular pair of constituents in the mixture and the local thermodynamic state. In gases, diffusivities are typically proportional to $\overline{v}l$. The linear relation (1.7) for mass diffusion is generally known as *Fick's law*, and the minus sign reflects the fact that species diffuse from higher to lower concentrations. Relations like this are based on empirical evidence, and are called *phenomenological laws*. Statistical mechanics can sometimes be used to derive such laws, but only for simple situations.

The analogous relation to (1.7) for heat transport via a temperature gradient ∇T is *Fourier's law*:

$$\mathbf{q} = -k\nabla T, \tag{1.8}$$

where $\mathbf{q}$ is the heat flux (J m^{-2} s^{-1}), and k is the material's thermal conductivity.

The analogous relationship to (1.7) for momentum transport via a velocity gradient is qualitatively similar to (1.7) and (1.8) but is more complicated because momentum and velocity are vectors. So as a first step, consider the effect of a vertical gradient, du/dy, in the horizontal velocity u (Fig. 1.3). Molecular motions and collisions cause the faster fluid above AB to pull the fluid underneath AB forward, thereby speeding it up. Molecular motions and collisions also cause the slower fluid below AB to pull the upper faster fluid backward, thereby slowing it down. Thus, without an external influence to maintain du/dy, the flow profile shown by the solid curve will evolve toward a profile shown by the dashed curve. This is analogous to saying that u, the horizontal momentum per unit mass (a momentum *concentration*), *diffuses* downward. Here, the resulting momentum flux, from high to low u, is equivalent to a shear stress, τ, existing in the fluid. Experiments show that the magnitude of τ along a surface such as AB is, to a good approximation, proportional to the local velocity gradient,

$$\tau = \mu(du/dy), \tag{1.9}$$

where the constant of proportionality μ (with units of kg m^{-1} s^{-1}) is known as the *dynamic viscosity*. This is *Newton's law* of viscous friction. It is analogous to (1.7) and (1.8) for the simple unidirectional shear flow depicted in Fig. 1.3. A more general tensor form of (1.9) that accounts for three velocity components and three possible orientations of the surface AB is presented in Chapter 4. For gases and liquids, μ depends on the local temperature T. In ideal gases, μ is nearly proportional to $\rho\overline{v}l$. At constant pressure, ρ is proportional to $1/T$ (see Section 1.8), $\overline{v}$ is proportional to $T^{1/2}$, and l is proportional to T. Consequently μ varies approximately as $T^{1/2}$. For liquids, shear stress is caused more by the intermolecular cohesive forces than by the thermal motion of the molecules. These cohesive forces decrease with increasing T so μ for a liquid decreases with increasing T.

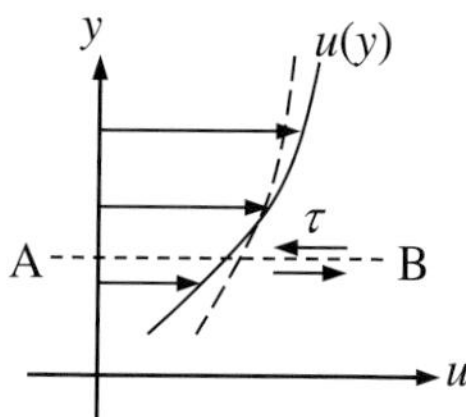

FIGURE 1.3 Shear stress τ on surface AB. The diffusive action of fluid viscosity tends to decrease velocity gradients, so that the continuous curve for $u(y)$ tends toward the dashed curve.

Although the shear stress is proportional to μ, we will see in Chapter 4 that the tendency of a fluid to diffuse momentum via velocity gradients is determined by the quantity

$$\nu \equiv \mu/\rho. \tag{1.10}$$

The units of ν ($m^2 s^{-1}$) do not involve the mass, so ν is frequently called the *kinematic viscosity*; it is a diffusivity, and for gases it is proportional to $\bar{v}l$.

The transport laws (1.7), (1.8), and (1.9) are relatively simple and linear. Only *first* derivatives appear on the right side in each case because, when the continuum approximation is valid, molecular transport is carried out at length scales too small for any influence from higher derivatives. Similarly, nonlinear terms involving higher powers of the first derivatives, for example $[\partial u/\partial y]^2$, also do not appear because experiments show that the linear relations are accurate enough for nearly all practical situations.

Example 1.2

An adult human being utilizes approximately 8.4×10^{-6} kg/s of oxygen (O_2) and the diffusivity of O_2 in air is approximately 1.8×10^{-5} m^2/s. The concentration of O_2 in room temperature air at atmospheric pressure is 0.28 kg/m^3. If human safety is imperiled when the O_2 concentration reaches 0.21 kg/m^3, estimate the opening size needed in the 10-cm-thick exterior wall of a fully-enclosed chamber, lacking forced-air ventilation, that will allow constant occupancy by one person (see Fig. 1.4).

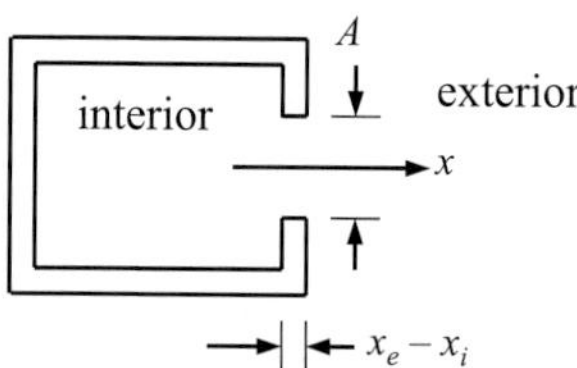

FIGURE 1.4 An enclosed chamber with a single opening of area A and wall thickness $x_e - x_i$. Sufficient diffusion of oxygen into the chamber is essential for safe occupancy.

Solution For the chamber to be safely occupied, the diffusive mass flux of O_2 through the opening must be sufficient to sustain life when the chamber-internal O_2 concentration is at 0.21 kg/m^3 and the chamber-external concentration is 0.28 kg/m^3. To make this problem simple and tractable (both appropriate for an *estimate*), assume one-dimensional diffusion through the opening. As in Fig. 1.4, let A be the required area, and define an x-axis pointing outward from the chamber perpendicular to the opening. Using subscripts 'i' and 'e' for the chamber interior and exterior, multiply (1.7) by A and evaluate as follows:

$$8.4 \times 10^{-6}\ \text{kg/s} = -A\mathbf{J}_m \cdot \mathbf{e}_x = A\rho\kappa_m \frac{\partial Y_{O2}}{\partial x} \approx A\kappa_m \frac{(\rho Y_{O2})_e - (\rho Y_{O2})_i}{x_e - x_i}.$$

The minus sign appears because the requisite O_2 flux is in the negative-x direction. The difference in O_2 concentration is 0.07 kg/m^3, and the diffusion distance is $x_e - x_i = 0.10$ m. Thus:

$$A \approx \frac{(8.4 \times 10^{-6}\ \text{kg/s})(0.1\ \text{m})}{(1.8 \times 10^{-5}\ \text{m}^2/\text{s})(0.07\ \text{kg/m}^3)} = 0.67\ \text{m}^2$$

This opening is equivalent to a square window with side length of ~0.8 m. Thus, forced-air ventilation, which can easily supply many orders of magnitude more O_2, is typically required in interior spaces that are continuously occupied by human beings.

1.6 Surface tension

A discontinuity may exist whenever two immiscible fluids are in contact, for example at the interface between water and air. Here unbalanced attractive intermolecular forces cause the interface to behave as if it were a stretched membrane under tension, like the surface of a balloon or soap bubble. The magnitude of the tensile force that acts per unit length is called *surface tension* σ; its units are N m^{-1}. Alternatively, σ can be thought of as the energy needed to create a unit of interfacial area. In general, σ depends on the pair of fluids in contact, the temperature, and the presence of surface-active chemicals (surfactants) or impurities on the interface, even at very low concentrations.

An important consequence of surface tension is that it causes a pressure difference across curved interfaces. Consider a spherical interface having a radius of curvature R (Fig. 1.5a). If p_i and p_o are the pressures on the inner and outer sides of the interface, respectively, then a static force balance gives:

$$\sigma(2\pi R) = (p_i - p_o)\pi R^2,$$

from which the pressure jump is found to be

$$p_i - p_o = 2\sigma/R$$

showing that the pressure on the concave side (the inside) is higher, regardless of whether the curved surface arises as a bubble in a liquid or a droplet in a gas.

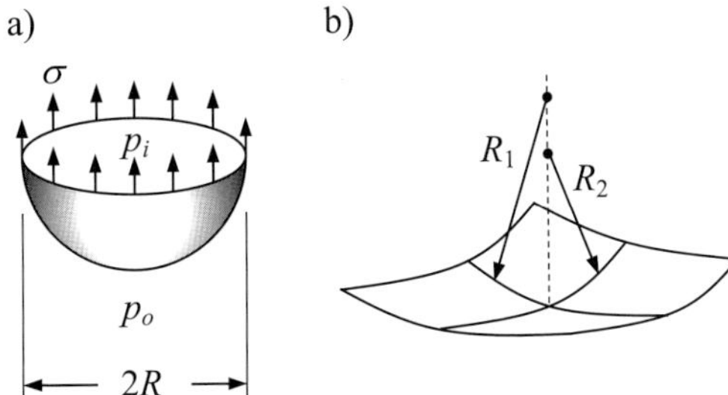

FIGURE 1.5 (a) Section of a spherical bubble or droplet, showing surface tension forces. (b) An interface with radii of curvatures R_1 and R_2 along two orthogonal directions.

The curvature of a general surface can be specified by the radii of curvature along two orthogonal directions, say, R_1 and R_2 (Fig. 1.5b). A similar analysis shows that the pressure difference across the interface is given by

$$p_i - p_o = \sigma\left(\frac{1}{R_1} + \frac{1}{R_2}\right) \tag{1.11}$$

which reduces to the spherical interface result when $R_1 = R_2$. This pressure difference is called the Laplace pressure.

Example 1.3

Determine the pressure difference between the inside and outside of a bubble in water having a radius of 1 μm using a surface tension of 0.072 N/m.

Solution The pressure difference calculation is based on the above equation:

$$p_i - p_o = 2\sigma/R = 2(0.072\ \text{N/m})/(10^{-6}\ \text{m}) = 144\ \text{kPa},$$

and this is more than atmospheric pressure. If the gas inside the bubble is soluble in water, then the extra surface-tension-induced pressure may cause more gas to dissolve into the water, which causes the bubble to shrink. In the smaller bubble, the surface-tension-induced pressure is even higher and this can cause even more gas to dissolve. Thus, small bubbles containing gases that are soluble in the surrounding liquid can be squeezed out of existence by surface tension.

In addition, the free surface of a liquid in a narrow tube rises above the surrounding level due to the influence of surface tension, as explained in Example 1.4. Narrow tubes are called *capillary tubes* (from Latin *capillus*, meaning hair); thus, the phenomena that arise from surface tension are called *capillarity*. Section 4.10 further describes surface tension, including a derivation of (1.11).

1.7 Fluid statics

The magnitude of the force per unit area in a static fluid is called the *pressure*; pressure in a moving fluid is defined in Chapter 4. Sometimes the ordinary pressure is called the *absolute pressure*, in order to distinguish it from the *gauge pressure*, which is defined as the absolute pressure minus the local atmospheric pressure:

$$p_{\text{gauge}} = p - p_{\text{atm}}.$$

The standard value for atmospheric pressure p_{atm} is 101.3 kPa = 1.013 bar where 1 bar = 10^5 Pa. An absolute pressure of zero implies vacuum while a gauge pressure of zero implies local atmospheric pressure.

In a fluid at rest, tangential viscous stresses are absent so only normal surface forces are present. In this case, the surface force per unit area (or pressure) is equal in all directions. Consider a small volume of fluid with a triangular cross section (Fig. 1.6) of unit thickness normal to the page, and let p_1, p_2, and p_3 be the pressures on the three faces. The z-axis is taken vertically upward. The only forces acting on the element are the pressure forces normal to the faces and the weight of the element. Because there is no acceleration of the element in the x direction, a balance of forces in that direction gives

$$(p_1 ds)\sin\theta - p_3\,dz = 0.$$

Because $dz = \sin\theta\, ds$, the foregoing gives $p_1 = p_3$. A balance of forces in the vertical direction gives

$$-(p_1 ds)\cos\theta + p_2 dx - \frac{1}{2}\rho g dx dz = 0.$$

As $\cos\theta\, ds = dx$, this gives

$$p_2 - p_1 - \frac{1}{2}\rho g dz = 0.$$

As the triangular element is shrunk to a point, that is, $dz \to 0$ with $\theta =$ constant, the gravity force term drops out, giving $p_1 = p_2$. Thus, at a point in a static fluid:

$$p_1 = p_2 = p_3, \tag{1.12}$$

so that surface force per unit area is independent of the surface's orientation. The pressure is therefore a scalar quantity.

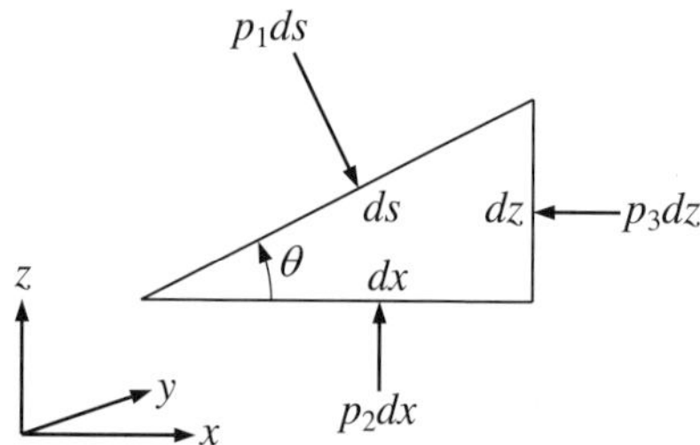

FIGURE 1.6 Demonstration that $p_1 = p_2 = p_3$ in a static fluid. Here the vector sum of the three arrows is zero when the volume of the element shrinks to zero.

The *spatial distribution* of pressure in a static fluid can be determined from a force balance. Consider an infinitesimal cube of sides dx, dy, and dz, with the z-axis vertically upward (Fig. 1.7). A balance of forces in the x direction shows that the pressures on the two sides perpendicular to the x-axis are equal. A similar result holds in the y direction, so that

$$\partial p/\partial x = \partial p/\partial y = 0. \tag{1.13}$$

This fact is expressed by *Pascal's law*, which states that all points in a resting fluid medium (and connected by the *same* fluid) are at the same pressure if they are at the same depth. For example, the pressure at points F and G in Fig. 1.8 are the same.

Vertical equilibrium of the element in Fig. 1.7 requires that

$$p\,dx\,dy - (p + dp)dx\,dy - \rho g\,dx\,dy\,dz = 0,$$

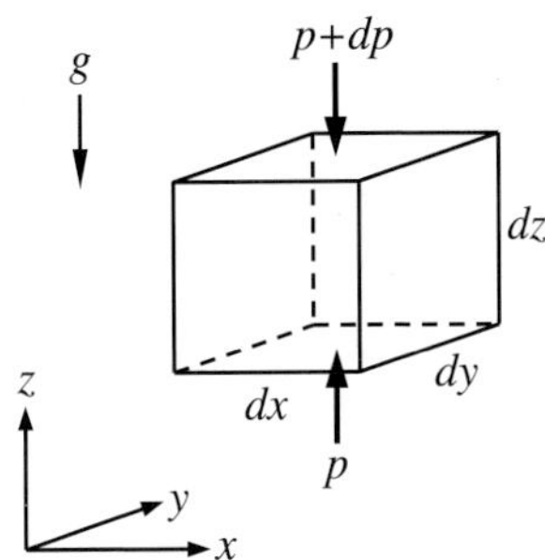

FIGURE 1.7 Fluid element at rest. Here the pressure difference between the top and bottom of the element balances the element's weight when gravity only acts in the vertical (z) direction.

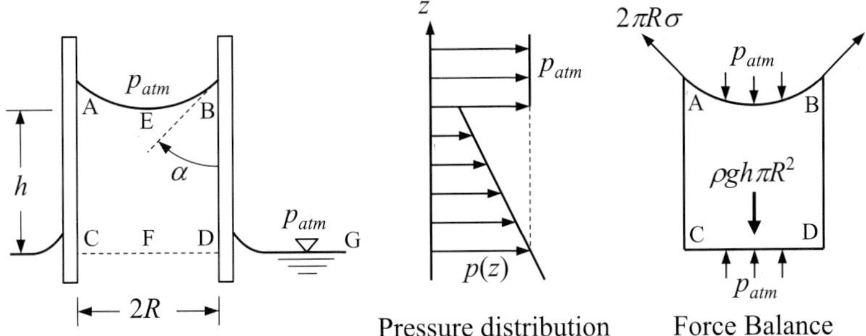

FIGURE 1.8 Rise of a liquid in a narrow tube (Example 1.4) because of the action of surface tension. The curvature of the surface and the surface tension cause a pressure difference to occur across the surface. The pressure distribution shown applies along EF, the centerline of the tube.

which simplifies to

$$dp/dz = -\rho g. \tag{1.14}$$

This shows that the pressure in a static fluid subject to a constant gravitational field decreases with height. For a fluid of uniform density, (1.14) can be integrated to give

$$p = p_0 - \rho g z, \tag{1.15}$$

where p_0 is the pressure at $z = 0$. Eq. (1.15) is the well-known result of *hydrostatics*, and shows that the pressure in a constant-density liquid decreases *linearly* with increasing height. It implies that the pressure rise at a depth h below the free surface of a liquid is equal to $\rho g h$, which is the weight of a column of liquid of height h and unit cross section.

Example 1.4

Using Fig. 1.8, show that the rise of a liquid in a narrow tube of radius R is given by

$$h = \frac{2\sigma \cos\alpha}{\rho g R},$$

where σ is the surface tension and α is the *contact angle* between the fluid and the tube's inner surface.

Solution Since the free surface is concave upward and exposed to the atmosphere, the pressure just below the interface at point E is below atmospheric. The pressure then increases linearly along EF. At F the pressure again equals the atmospheric pressure, since F is at the same level as G where the pressure is atmospheric. The pressure forces on faces AB and CD therefore balance each other. Vertical equilibrium of the element ABCD then requires that its weight be balanced by the vertical component of the surface tension force. When the raised column of water is considered to be cylindrical, this force balance is:

$$(2\pi R\sigma)\cos\alpha = \rho g h(\pi R^2),$$

which gives the required result.

1.8 Classical thermodynamics

Classical thermodynamics is the study of equilibrium states of matter, in which properties are assumed uniform in space and time. This section reviews the main ideas and most commonly used relationships. A thermodynamic *system* is a quantity of matter that exchanges heat and work, but not mass, with its surroundings. A system in equilibrium is free of fluctuations, such as those generated during heat or work exchange with its surroundings. After any such exchange, fluctuations die out or *relax*, a new equilibrium is reached, and the system's properties, such as pressure and temperature, again become well defined. A system's *relaxation time* is defined as the time after heat or work exchange for the system to reach a new equilibrium state.

This thermodynamic system concept is not directly applicable to a macroscopic volume of a moving fluid in which pressure and temperature may vary. However, experiments show that classical thermodynamics does apply to small fluid volumes commonly called *fluid particles*. A fluid particle is a small deforming volume carried by the flow that: 1) always contains the same fluid molecules, 2) is large enough so that its thermodynamic properties are well defined when it is at equilibrium, but 3) is small enough so that its *relaxation time* is short compared to the time scales of fluid-motion-induced thermodynamic changes. Under ordinary conditions (the emphasis here), molecular densities (n), average speeds ($\overline{v}$), and average collision rates ($\overline{v}/l$) are high enough so that the conditions for the existence of fluid particles are met, and classical thermodynamics can be directly applied to flowing fluids. However, there are circumstances involving rarified gases, shock waves, and high-frequency acoustic waves where one or more of the fluid particle requirements are not met and molecular-kinetic and quantum theories are needed.

The laws of classical thermodynamics are empirical, and essentially establish *definitions*, upon which the subject is built. The first law defines a system's *internal energy* while the second law defines its *entropy*.

First law of thermodynamics

The first law of thermodynamics states that the energy of a system is conserved;

$$\Delta e = \delta q + \delta w, \tag{1.16}$$

where Δe is the change of the system's *internal energy*, δq is the heat added to the system, δw is the work done on the system. All thermodynamic quantities in (1.16) are normalized by the mass of the system and have units of J kg^{-1}; thus, these are *intensive* variables and appear as lower case letters. When (1.16) is written without normalization by the system mass, $\Delta E = \delta Q + \delta W$, it portrays the same thermodynamic law but the variables have units of energy (J); these are *extensive* variables and appear as capital letters. If two identical thermodynamic systems are combined, the combination's intensive variables are unchanged while its extensive variables are doubled.

Internal energy, heat, and work, are all different. The internal energy (aka, thermal energy) e is a manifestation of the random molecular motion of the system's constituents. In fluid flows, the kinetic energy of fluid particles' average macroscopic motion is included in the e-term in (1.16). However, within classical thermodynamics, e only represents thermal energy. Heat and work are forms of *energy in transition*, which appear at the *boundary* of the system and are *not contained* within the matter. In contrast, internal energy resides within the matter. If equilibrium states 1 and 2 of a system are known, then Q and W depend on the *process* or *path* followed by the system in going from state 1 to state 2. The change $\Delta e = e_2 - e_1$, in contrast, does not depend on the path. In short, e is a thermodynamic property and is a function of the thermodynamic state of the system. Thermodynamic properties are called *state functions*, in contrast to heat and work, which are *path functions*.

Frictionless quasi-static processes, carried out at an extremely slow rate so that the system is at all times in equilibrium with the surroundings, are called *reversible processes*. For a compressible fluid, the most common type of reversible work is by the expansion or contraction of the boundaries of the fluid particle. Let $\upsilon = 1/\rho$ be the *specific volume*, that is, the volume per unit mass. The work done per unit mass by a fluid particle in an infinitesimal reversible process is $-pd\upsilon$, where $d\upsilon$ is the increase of υ. The first law (1.16) for a reversible process then becomes

$$de = \delta q - pd\upsilon, \tag{1.17}$$

provided that q is also reversible. Irreversible forms of work, such as those done against frictional stresses, are excluded from (1.17).

Equations of state

A relation defining one state function in terms of two or more others is called an *equation of state*. For a simple compressible substance composed of a single component (the applicable model for nearly all pure fluids), the specification of two independent thermodynamic properties completely determines the state of the system. We can write relations such as the *thermal* and *caloric equations of state*:

$$p = p(\upsilon, T) \quad \text{or} \quad e = e(p, T), \tag{1.18}$$

respectively. For multi-component systems, the specification of additional properties is needed to completely determine the state. For example, seawater contains dissolved salt so its density is a function of temperature, pressure, and salinity.

Specific heats

The sum of a thermodynamic system's thermal energy and pressure-volume potential energy, h, is the *enthalpy*:

$$h \equiv e + p\upsilon, \tag{1.19}$$

and it arises naturally in the study of compressible fluid flows. For single-component systems, the specific heat capacities at constant pressure and constant volume are defined as

$$c_{\mathrm{p}} \equiv (\partial h/\partial T)_{\mathrm{p}}, \quad \text{and} \quad c_{\mathrm{v}} \equiv (\partial e/\partial T)_{\mathrm{v}}, \tag{1.20, 1.21}$$

respectively. Here, (1.20) means that we regard h as a function of p and T, and find the partial derivative of h with respect to T, keeping p constant. Eq. (1.21) has an analogous interpretation. The specific heats as defined are thermodynamic properties because they are defined in terms of other properties of the system. That is, c_{p} and c_{v} can be determined when two other system properties (say, p and T) are known.

For certain processes common in fluid flows, the heat exchanged can be related to the specific heats. Consider a reversible process in which the work done is given by $pd\upsilon$, so that the first law of thermodynamics has the form of (1.17). Dividing by the change of temperature, it follows that the heat transferred per unit mass per unit temperature change in a constant volume process is

$$(\partial q/\partial T)_{\mathrm{v}} = (\partial e/\partial T)_{\mathrm{v}} = c_{\mathrm{v}}.$$

This shows that $c_{\mathrm{v}}dT$ represents the heat transfer per unit mass in a reversible constant-volume process, in which the only type of work done is of the $pd\upsilon$ type. It is misleading to define $c_{\mathrm{v}} = (dq/dT)_{\mathrm{v}}$ without any restrictions imposed, as the temperature of a constant-volume system can increase without heat transfer, such as by vigorous stirring. Similarly, the heat transferred at constant pressure during a reversible process is given by

$$(\partial q/\partial T)_{\mathrm{p}} = (\partial h/\partial T)_{\mathrm{p}} = c_{\mathrm{p}}.$$

Second law of thermodynamics

The second law of thermodynamics restricts the direction of real processes with increasing time. Its implications are discussed in Chapter 4. Some consequences of this law are the following:

(i) There must exist a thermodynamic property s, known as *entropy*, whose change between states 1 and 2 is given by

$$s_2 - s_1 = \int_1^2 \frac{dq_{\mathrm{rev}}}{T}, \tag{1.22}$$

where the integral is taken along any reversible path between the two states.

(ii) For an *arbitrary* process between states 1 and 2, the entropy change is

$$s_2 - s_1 \geq \int_1^2 \frac{dq}{T} (\text{Clausius} - \text{Duhem}),$$

which states that the entropy of an isolated system ($dq = 0$) can only increase. Such increases may be caused by friction, mixing, and other irreversible phenomena.

(iii) Molecular transport coefficients such as species diffusivity κ_m, thermal conductivity k, and viscosity μ must be positive. Otherwise, spontaneous unmixing, or thermal or momentum separation would occur and lead to a decrease of entropy of an isolated system.

Property relations

Two common relations are useful in calculating entropy changes during a process. For a reversible process, the entropy change is given by

$$Tds = \delta q. \tag{1.23}$$

Combining this with (1.17) and (1.19) leads to:

$$Tds = de + pdv, \quad \text{or} \quad Tds = dh - vdp. \text{ (Gibbs)} \tag{1.24}$$

It is interesting that these relations (1.24) are also valid for irreversible (frictional) processes, although the relations (1.17) and (1.23), from which (1.24) are derived, are true for reversible processes only. This is because (1.24) are relations between thermodynamic *state functions* alone and are therefore true for *any* process. The association of Tds with heat and $-pdv$ with work does not hold for irreversible processes. Consider stirring work done at constant volume that raises a fluid element's temperature; here $de = Tds$ is the increment of stirring work done.

Speed of sound

In a compressible fluid, infinitesimal isentropic changes in density and pressure propagate through the medium at a finite speed, c, the speed of sound. In Chapter 14, it is shown that c^2 is given by the thermodynamic derivative:

$$c^2 \equiv (\partial p/\partial \rho)_s, \tag{1.25}$$

where the subscript s signifies that the derivative is taken at constant entropy. For incompressible fluids, $\partial \rho/\partial p \to 0$ under all conditions so $c \to \infty$.

Thermal expansion coefficient

The thermal expansion coefficient, α, quantifies a fluid density's dependence on temperature at constant pressure:

$$\alpha \equiv \frac{-1}{\rho}\left(\frac{\partial \rho}{\partial T}\right)_p. \tag{1.26}$$

This coefficient appears frequently in the study of nonisothermal systems.

Example 1.5

For any equation of state relating p, ρ, and T, show that

$$(\partial p/\partial \rho)_T(\partial \rho/\partial T)_p(\partial T/\partial p)_\rho = -1.$$

Solution Assume the equation of state can be written: $p = p(\rho, T)$, and form the general differential:

$$dp = (\partial p/\partial \rho)_T d\rho + (\partial p/\partial T)_\rho dT,$$

Divide this expression by dT and consider a thermodynamic path that follows an isobar (a line or surface of constant pressure) where $dp = 0$.

$$dp/dT = (\partial p/\partial T)_p = 0 = (\partial p/\partial \rho)_T(\partial \rho/\partial T)_p + (\partial p/\partial T)_\rho, \quad \text{or} \quad (\partial p/\partial \rho)_T(\partial \rho/\partial T)_p = -(\partial p/\partial T)_\rho$$

Multiply the final equation by $(\partial T/\partial p)\rho$ to reach the desired result.

1.9 Perfect gas

A basic result from kinetic theory and statistical mechanics for the thermal equation of state for N identical noninteracting gas molecules confined within a container having volume V is:

$$pV = Nk_BT, \tag{1.27}$$

where p is the average pressure inside the container, $k_B = 1.381 \times 10^{-23}\mathrm{JK}^{-1}$ is Boltzmann's constant, and T is the absolute temperature. Eq. (1.27) is the molecule-based version of the perfect gas law. It is valid when attractive forces between molecules are negligible and when $V/N = 1/n$ is much larger than the (average) volume of an individual molecule. When used with the continuum approximation, (1.27) is commonly rearranged by noting that $\rho = mn = mN/V$, where m is the (average) mass of one gas molecule. Here m is calculated (in SI units) as M_w/A_o where M_w is the (average) molecular weight in kg (kg-mole)$^{-1}$ of the gas molecules, and A_o is the kilogram-based version of Avogadro's number, 6.023×10^{26} (kg-mole)$^{-1}$. With these replacements, (1.27) becomes:

$$p = \frac{N}{V}k_BT = \frac{Nm}{V}\left(\frac{k_B}{m}\right)T = \rho\left(\frac{k_BA_o}{M_w}\right)T = \rho\left(\frac{R_u}{M_w}\right)T = \rho RT, \tag{1.28}$$

where the product $k_BA_o = R_u = 8314\ \mathrm{J\,kmol^{-1}\ K^{-1}}$ is the *universal gas constant*, and $R = k_B/m = R_u/M_w$ is the *gas constant* for the gas under consideration. A perfect gas is one that obeys (1.28), even if it is a mixture of several different molecular species. For example, the average molecular weight of dry air is 28.966 kg kmol^{-1}, for which (1.28) gives $R = 287\ \mathrm{J\,kg^{-1}\ K^{-1}}$. At ordinary temperatures and pressures most gases can be treated as perfect gases.

The gas constant for a particular gas is related to the specific heats of the gas through the relation

$$R = c_p - c_v, \tag{1.29}$$

where c_p and c_v are defined in (1.20) and (1.21). In general, c_p and c_v increase with temperature. The ratio of specific heats

$$\gamma \equiv c_p/c_v \tag{1.30}$$

is important in compressible fluid dynamics. For air at ordinary temperatures, $\gamma = 1.40$ and $c_p = 1004\ \mathrm{J\,kg^{-1}\ K^{-1}}$. It can be shown that (1.27) or (1.28) is equivalent to $e = e(T)$ and $h = h(T)$, and conversely, so that the internal energy and enthalpy of a perfect gas are only functions of temperature (Exercise 1.14).

A process is called *adiabatic* if it takes place without the addition of heat. A process is called *isentropic* if it is adiabatic and frictionless because then the entropy of the fluid does not change. From (1.24) it can be shown (Exercise 1.15) that isentropic flow of a perfect gas with constant specific heats obeys:

$$p/\rho^\gamma = const. \tag{1.31}$$

Using (1.28) and (1.31), the temperature and density changes during an isentropic process from a reference state (subscript 0) to a current state (no subscript) are:

$$T/T_0 = (p/p_0)^{(\gamma-1)/\gamma} \text{ and } \rho/\rho_0 = (p/p_0)^{1/\gamma}. \tag{1.32}$$

In addition, simple expressions can be found for the speed of sound c and the thermal expansion coefficient α for a perfect gas:

$$c = \sqrt{\gamma RT} \quad \text{and} \quad \alpha = 1/T. \qquad (1.33, 1.34)$$

Example 1.6

A single-family residence encloses a volume $V = 700\ \mathrm{m^3}$ of air. In the winter and the summer the average interior temperatures are $T_w = 17\,^\circ\mathrm{C}$ (290 K) and $T_s = 26\,^\circ\mathrm{C}$ (299 K), respectively. If the pressure is constant at 1 atmosphere and the air has constant c_v, what is the change in the total thermal energy of the air in the residence from winter to summer?

Solution Assume air behaves as a perfect gas, and determine the winter and summer total energies from E = (air mass) (internal energy per unit mass) $= \rho Ve = \rho Vc_vT$, where T is the absolute temperature. Obtain the gas density from (1.28), $\rho = p/RT$ so:

$$E_w = \frac{p}{RT_w}Vc_vT_w = \frac{c_v}{R}pV = \frac{1}{\gamma-1}pV, \quad \text{and} \quad E_s = \frac{p}{RT_s}Vc_vT_s = \frac{c_v}{R}pV = \frac{1}{\gamma-1}pV.$$

The air's total thermal energy is the same in both seasons! Thus, the energy expended for winter heating and summer cooling in the residence only counter acts unwanted heat transfer. It does not change the total thermal energy of the interior air. This situation occurs because the residence is not sealed and does not trap air. Instead the total interior air mass increases (or decreases) as the temperature falls (or rises), so that E remains independent of temperature.

Example 1.7

Acoustic cavitation is common in a variety of ultrasonic applications. For example, microbubbles have specific acoustic properties that make them useful as a contrast agent in ultrasound imaging. The Rayleigh–Plesset equation is useful in describing the bubble dynamics, given by

$$R\frac{d^2R}{dt^2}+\frac{3}{2}\left(\frac{dR}{dt}\right)^2+\frac{4\nu}{R}\frac{dR}{dt}+\frac{2\sigma}{\rho R}+\frac{p_\infty-p_b}{\rho}=0, \tag{1.35}$$

where $R(t)$ is the bubble radius, ρ and ν are the density and viscosity of the surrounding liquid, respectively, σ is the surface tension coefficient, p_b is the liquid pressure at the surface of the bubble, and p_∞ is the far-field pressure. The driving sound field in medical ultrasound is oscillatory and pulsed, and therefore the far-field pressure can be reasonably approximated by $p_\infty(t) = p_{atm} + p_a \sin(2\pi f)$, where p_{atm} is the ambient pressure, p_a is the ultrasound pressure amplitude, and f is the driving frequency.

Numerical solutions for Eq. (1.35) are required in this case. Using MATLAB® (or another programming language), plot the temporal evolution of the bubble radius over 5.0 ms with an initial radius $R_0 = 5$ μm. Treat the gas inside the bubble as a polytripic and adiabatic, such that

$$p_b=\left(p_{atm}+\frac{2\sigma}{R_0}\right)\left(\frac{R_0}{R}\right)^{3\gamma}.$$

Using properties for water and air and driving frequency $f = 1$ kHz, compare the temporal evolution of the bubble radius, $R(t)$, under realistic conditions (viscosity and surface tension present) and idealized conditions ($\nu = \sigma = 0$). Consider a pressure amplitude of $p_a = 10$ kPa and 75 kPa.

Solution

```
global R0 rho p_atm freq sigma nu gamma pa

%% Input parameters
R0=5e-6; % initial radius [m]
rho=1000; % density [kg/m^3]
p_atm=101325; % reference pressure [Pa]
gamma=1.4; %ratio of specific heats [-]
freq=1e3; % ultrasound frequency [Hz]
pa=10e3; % ultrasound pressure amplitude [Pa]
tf=50e-4; % final time [s]

% Solve for radius including surface tension & viscosity
nu=1e-6; % kinetmatic viscosity
sigma=7.28e-2; % surface tension
[t,R] = ode15s(@RP,[0 tf],[R0 0]);

% Solve for radius neglecting surface tension & viscosity
nu=0;
sigma=0;
[t2,R2] = ode15s(@RP,[0 tf],[R0 0]);

%% Plot the bubble dynamics
figure()
plot(t*1e3,R(:,1)/R0,'-k',t2*1e3,R2(:,1)/R0,'--b','linewidth',2)
xlabel('Time [ms]','interpreter','latex','fontsize',20);
str=['$f=' num2str(freq/1000) '$ kHz, $p_a=' num2str(pa/1000) '$ kPa'];
title(str,'interpreter','latex','fontsize',20);
ylabel('$R/R_0$','interpreter','latex','fontsize',20);
l=legend('With surface tension \& viscosity','Without surface tension \& ...
    viscosity','location','NorthWest');
```

```
set(l,'Interpreter','latex','EdgeColor',[1 1 1],'fontsize',16,'color','none')
set(gca, ...
    'box', 'on',...
    'tickdir', 'in',...
    'ticklength',[.015 .015],...
    'xminortick','on',...
    'yminortick','on',...
    'linewidth',1,...
    'fontsize',20,'FontName','Times New Roman');
set(l,'Interpreter','latex','EdgeColor',[1 1 1],'fontsize',16,'color','none')

%% Rayleigh–Plesset equation
function R = RP(t,y)
global R0 rho p_atm sigma nu gamma

Pb = (p_atm+2*sigma/R0)*(R0/y(1))^(3*gamma);
Pinf = p_atm+pulse(t);
R = [y(2); ( (Pb–Pinf)/(rho*y(1)) – 1.5*y(2)^2/y(1) –...
    4*nu*y(2)/y(1)^2 – 2*sigma/(rho*y(1)^2) )];
return
end

%% Driving sound field
function z=pulse(t)
global pa freq
z = pa*sin(2*pi*freq*t);
return
end
```

The output from the MATLAB script above showing the temporal evolution of bubble radius and effect of viscosity and surface tension is given in Fig. 1.9. Note the function `ode15s` was employed due to the stiff nature of the equations. At low pressure amplitudes, the change in bubble radius is small. Viscosity and surface tension is seen to resist deformation resulting in smaller amplitudes. At higher pressure amplitudes, the variation in bubble radius is larger, and surface tension and viscosity play a larger role.

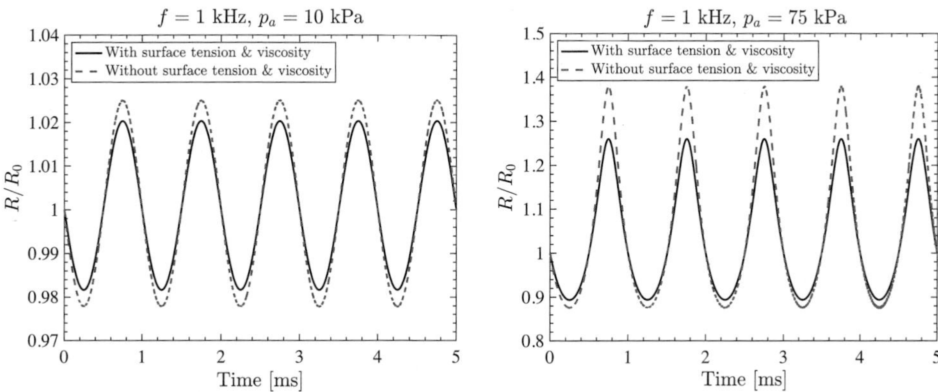

FIGURE 1.9 Solution to Example 1.7.

1.10 Stability of stratified fluid media

In a static fluid environment subject to a gravitational field, p, ρ, and T may vary with height z, but (1.14) and (1.18) provide two constraints. Thus, knowledge of the vertical profile of one thermodynamic variable allows the profiles of the others to be determined. In addition, our experience suggests that the fluid medium will be stable (i.e., not prone to growing vertical motions) if $\rho(z)$ decreases with increasing z. Interestingly, the rate at which the density decreases also plays a role in the stability of the fluid medium when the fluid is compressible, as in a planetary atmosphere.

To assess the stability of a static fluid medium, consider the situation depicted in Fig. 1.10 which shows sample density profiles that are stable (dashed curve on the left), isentropic (solid curve), and unstable (dashed curve on the right). If a fluid particle with density $\rho(z_o)$ in an atmosphere (or ocean) at equilibrium at height z_o (point o) is displaced upward a distance ζ via a frictionless adiabatic process to reach point c and then released from rest, its subsequent motion is determined by its buoyancy at its new height, $z = z_o + \zeta$. As depicted, the displaced fluid particle may have a different density, ρ_c, than that of the quiescent fluid at the same height. The profile $\rho(z)$ is *stable* if the density of the quiescent fluid at the new height, ρ_b, is less than the displaced particle's density, ρ_c. In this case the displaced fluid is heavier than the surrounding fluid, so it will fall back toward its original location and the action of viscous forces and thermal conduction will arrest any oscillatory motion. Thus, a stable density profile decreases with height *faster* than in an isentropic profile.

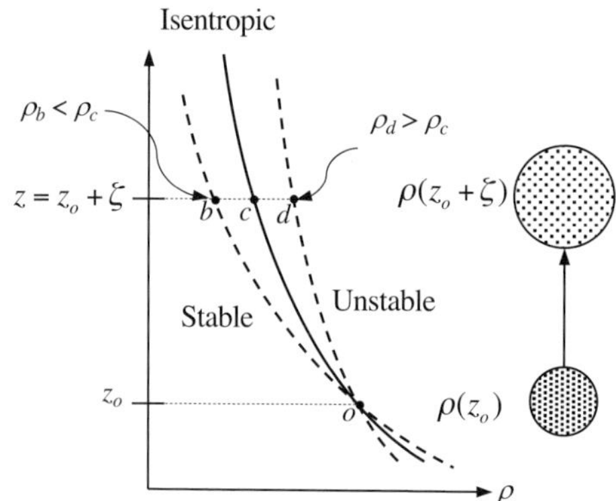

FIGURE 1.10 Isentropic expansion of a fluid particle displaced upward in a compressible medium. In a static pressure field, if the fluid particle rises adiabatically from height z_o without friction it likely encounters a lower pressure and expands along the solid curve to reach density ρ_c. When the density profile is stable (left dashed line), ρ_c is greater than the local density (ρ_b) and the particle falls back toward z_o. When the density profile is unstable (right dashed line), ρ_c is less than the local density (ρ_d) and the particle continues to rise away from z_o.

The profile $\rho(z)$ is *unstable* if the quiescent density at the new height, $\rho(z) = \rho_d$, is greater than the displaced particle's density, ρ_c. In this case the displaced fluid is lighter than the surrounding fluid, so buoyancy will push it further upward away from its original location. Thus, an unstable density profile is one that changes with height *slower* than an isentropic a profile, and vertical fluctuations in fluid particle location are amplified.

In reality, density profiles in the earth's atmosphere and ocean may be simultaneously stable and unstable over different ranges of the vertical coordinate (see Exercise 1.24). To accommodate this possibility, a local criterion for stability at height z_o can be determined from Newton's second law for the displaced element using weight and buoyancy (but not friction) forces when the initial vertical displacement ζ is small (see Exercise 1.18). The resulting second-order differential equation is

$$\frac{d^2\zeta}{dt^2} + \frac{g}{\rho(z_o)}\left(\frac{d\rho_a}{dz} - \frac{d\rho}{dz}\right)\zeta = 0$$

when first-order terms in ζ are retained. The coefficient of ζ in the second term is the square of the *Brunt-Väisälä frequency*, N,

$$N^2(z_o) = \frac{g}{\rho}\left(\frac{d\rho_a}{dz} - \frac{d\rho}{dz}\right), \tag{1.36}$$

where $\rho_a(z)$ is the local isentropic (adiabatic) density distribution near height z_o, and both derivatives and ρ are evaluated at z_o. When N^2 is positive, the fluid medium is *stable* at $z = z_o$. When N^2 is negative, the fluid medium is *unstable* at $z = z_o$. When N^2 is zero, the fluid medium is *neutrally stable* and the element will not move if released from rest after being displaced; it will have zero vertical acceleration. There are two ways to achieve neutral stability: 1) the fluid density may be independent of the vertical coordinate so that $d\rho/dz = d\rho_a/dz = 0$, or 2) the equilibrium density gradient in the fluid medium may equal the isentropic density gradient, $d\rho/dz = d\rho_a/dz$. The former case implies that constant-density fluid media are neutrally stable. The latter case requires a *neutrally stable atmosphere* to be one where p, ρ, and T decrease with increasing height in such a way that the entropy is constant.

In atmospheric science, $\Gamma \equiv dT/dz$ is the atmospheric temperature gradient or *lapse rate*. The rate of temperature decrease in an isentropic atmosphere Γ_a is

$$dT_a/dz \equiv \Gamma_a = -g\alpha T/c_p \tag{1.37}$$

(see Exercise 1.19) and is called the *adiabatic temperature gradient* or *adiabatic lapse rate*. It is the highest rate at which the temperature can decrease with increasing height without causing instability. In the earth's atmosphere, the adiabatic lapse rate is approximately $-10\,°\text{C}\,\text{km}^{-1}$.

Fig. 1.11(a) shows a typical distribution of temperature in the earth's atmosphere. The lower part has been drawn with a slope nearly equal to the adiabatic temperature gradient because mixing processes near the ground tend to form a neutral (constant entropy) atmosphere. Observations show that the neutral atmosphere ends at a layer where the temperature increases with height, a stable situation. Meteorologists call this an *inversion*, because the temperature gradient changes sign here. Atmospheric turbulence and mixing processes below such an inversion typically cannot penetrate above it. Above this inversion layer the temperature decreases again, but less rapidly than near the ground, which again corresponds to stability. An isothermal atmosphere [a vertical line in Fig. 1.11(a)] is quite stable.

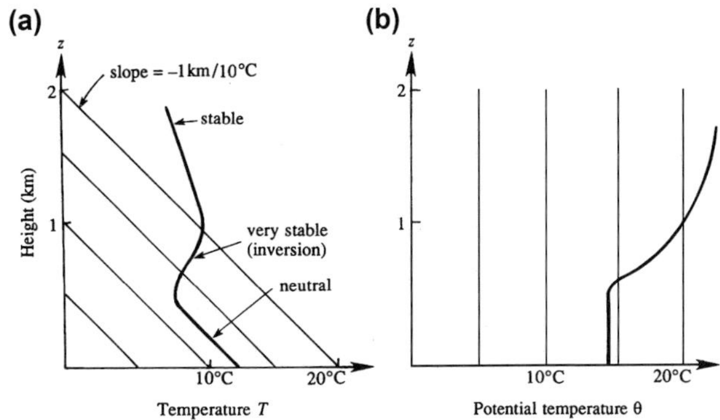

FIGURE 1.11 Vertical variation of the (a) actual and (b) potential temperature in the atmosphere. Thin straight lines represent temperatures for a neutral atmosphere. Slopes less than the neutral atmosphere lines lead to atmospheric instability. Slopes greater than the neutral atmosphere lines indicate a stable atmosphere.

Potential temperature and density

The foregoing discussion of static stability of a compressible atmosphere can be expressed in terms of the concept of *potential temperature*, θ. Suppose the pressure and temperature of a fluid particle at a height z are $p(z)$ and $T(z)$. Now if we take the particle *adiabatically* to a standard pressure $p_o = p(0)$ (say, the sea level pressure, nearly equal to 100 kPa), then the temperature θ attained by the particle is called its *potential temperature*. Using (1.32) for a perfect gas, it follows that the actual temperature T and the potential temperature θ are related by

$$T(z) = \theta(z)(p(z)/p_o)^{(\gamma-1)/\gamma}. \tag{1.38}$$

Taking the logarithm and differentiating, we obtain

$$\frac{1}{T}\frac{dT}{dz} = \frac{1}{\theta}\frac{d\theta}{dz} + \frac{(\gamma-1)}{\gamma p}\frac{dp}{dz}.$$

Substituting $dp/dz = -\rho g$, $p = \rho RT$, and $\alpha = 1/T$ produces

$$\frac{T}{\theta}\frac{d\theta}{dz} = \frac{dT}{dz} + \frac{g}{c_p} = \frac{d}{dz}(T - T_a) = \Gamma - \Gamma_a. \tag{1.39}$$

If the temperature decreases at a rate $\Gamma = \Gamma_a$, then the potential temperature θ (and therefore the entropy) is uniform with height. It follows that an atmosphere is stable, neutral, or unstable depending upon whether $d\theta/dz$ is positive, zero, or negative, respectively. This is illustrated in Fig. 1.11b. It is the gradient of *potential* temperature that determines the stability of a column of gas, not the gradient of the actual temperature. However, this difference is negligible for laboratory-scale phenomena. For example, a 1.0 m vertical change may result in an air temperature decrease of only $1.0\,\text{m} \times (10\,°\text{C}\,\text{km}^{-1}) = 0.01\,°\text{C}$.

Similarly, *potential density* ρ_θ is the density attained by a fluid particle if taken via an isentropic process to a standard pressure p_o. Using (1.32), the actual density $\rho(z)$ and potential density are related by

$$\rho(z) = \rho_\theta(z)(p(z)/p_o)^{1/\gamma}. \tag{1.40}$$

Multiplying (1.38) and (1.40), and using $p=\rho RT$, we obtain $\theta\rho_\theta = p_o/R = \text{const}$. Taking the logarithm and differentiating, we obtain

$$\frac{-1}{\rho_\theta}\frac{d\rho_\theta}{dz} = \frac{1}{\theta}\frac{d\theta}{dz}. \tag{1.41}$$

Thus, an atmosphere is stable, neutral, or unstable depending upon whether $d\rho_\theta/dz$ is negative, zero, or positive, respectively.

Interestingly, compressibility effects are also important in the deep ocean where saltwater density depends not only on the temperature and pressure, but also on the *salinity* (S) defined as kilograms of salt per kilogram of water. The average salinity of seawater is approximately 3.5%. Here, the potential density is defined as the density attained if a fluid particle is taken *at constant salinity* to a reference pressure via an isentropic process. The potential density thus defined must decrease with height for stable water column conditions. Oceanographers automatically account for the compressibility of seawater by converting their density measurements at any depth to sea level pressure, which serves as the reference pressure.

Because depth-change-induced density changes are relatively small in percentage terms (~0.5% for a 1.0 km change in depth) for seawater, the static stability of the *ocean* is readily determined from (1.36). In particular, the vertical isentropic density gradient in (1.36) may be rewritten using $dp_a/dz = -\rho_a g$ and the definition of the sound speed c (1.25) to find:

$$\frac{d\rho_a}{dz} = \left(\frac{\partial\rho_a}{\partial p}\right)_{s,S}\frac{dp_a}{dz} = -\left(\frac{\partial\rho_a}{\partial p}\right)_{s,S}\rho_a g = \frac{-\rho_a g}{c^2} \cong \frac{-\rho g}{c^2},$$

where the approximation $\rho_a \cong \rho$ produces the final result. Thus, (1.36) and its ensuing discussion imply that the ocean is stable, neutral, or unstable depending upon whether

$$\frac{d\rho_\theta}{dz} = \frac{d\rho}{dz} - \frac{d\rho_a}{dz} \cong \frac{d\rho}{dz} + \frac{\rho g}{c^2} \tag{1.42}$$

is negative, zero, or positive, respectively.

Scale height of the atmosphere

Approximate expressions for the pressure distribution and the thickness or *scale height* of the atmosphere can be obtained by assuming isothermal conditions. This is a reasonable assumption in the lower 70 km of the atmosphere, where the absolute temperature generally remains within 15% of 250 K. The hydrostatic distribution (1.14) and the perfect gas law (1.28) require

$$dp/dz = -\rho g = -pg/RT$$

When g, R, and T are constants, integration gives

$$p(z) = p_0 e^{-gz/RT},$$

where p_0 is the pressure at $z=0$. The pressure therefore falls to e^{-1} of its surface value in a height $H = RT/g$. Thus, the quantity RT/g is called the *scale height* of the atmosphere, a generic thickness of the atmosphere. For an average atmospheric temperature of $T = 250$ K, the scale height is $RT/g = 7.3$ km.

Example 1.8

Determine $\rho(z)$ and the Brunt-Väisälä frequency N at height z_o in an ideal gas atmosphere with surface pressure p_0 and constant temperature T.

Solution From (1.14) and the last equation:

$$\rho(z) = \frac{-1}{g}\frac{dp}{dz} = \frac{-1}{g}\frac{d}{dz}\left(p_0 e^{-gz/RT}\right) = \frac{p_0}{RT}e^{-gz/RT}.$$

Now determine the density derivatives that lead to N^2 at height z_o, again using (1.14)

$$\left(\frac{d\rho}{dz}\right)_{z=z_o} = \frac{d}{dz}\left(\frac{p_0}{RT}e^{-gz/RT}\right)_{z=z_0} = \frac{-gp_0 e^{-gz_o/RT}}{R^2T^2}$$

$$\left(\frac{d\rho_a}{dz}\right)_{z=z_o} = \frac{d}{dz}\left(\rho(z_o)\left(\frac{p(z)}{p(z_o)}\right)^{1/\gamma}\right)_{z=z_o} = \frac{\rho(z_o)}{p(z_o)^{1/\gamma}}\left(\frac{p(z)^{(1/\gamma)-1}}{\gamma}\frac{dp}{dz}\right)_{z=z_o} = \frac{[\rho(z_o)]^2}{\gamma p(z_o)}g = \frac{-g p_0 e^{-gz_o/RT}}{\gamma R^2 T^2}$$

From (1.36),

$$N^2(z_o) = \frac{g}{\rho}\left(\frac{d\rho_a}{dz} - \frac{d\rho}{dz}\right) = \frac{gRT}{p_0 e^{-gz_o/RT}}\left(\frac{-1}{\gamma} - (-1)\right)\frac{g p_0 e^{-gz_o/RT}}{R^2T^2} = \frac{g^2}{RT}\left(1 - \frac{1}{\gamma}\right)$$

Here N is independent of height z_o and N^2 is always positive so this atmosphere is stable.

1.11 Dimensional analysis

Interestingly, a physical quantity's units may be exploited to learn about its relationship to other physical quantities because matter and natural laws are independent of any human-defined unit system. Consider Newton's second law, generically stated as *force* = (*mass*) × (*acceleration*); it is true when stated in cgs (centimeter, gram, second), MKS (meter, kilogram, second), and even US customary (inch or foot, pound, second) units. Nature's independence from our unit systems leads to two important conclusions: 1) all correct physical relationships can be stated in dimensionless form, and 2) in any comparison, the units of the items being compared must be the same for the comparison to be valid. The first conclusion leads to the problem-simplification or scaling-law-development technique known as *dimensional analysis*. The second conclusion is known as the principle of *dimensional homogeneity*. It provides a means of error detection in derivations and formulae; if terms in an equation have different dimensions, an error has been made.

Dimensional analysis is a broadly applicable technique for developing scaling laws, interpreting experimental data, and simplifying problems. Occasionally it can even be used to solve problems. Dimensional analysis has utility throughout the physical sciences, and is presented here for subsequent use in this chapter's exercises and in the remaining chapters of this text. Of the various formal methods of dimensional analysis, the description here is based on Buckingham's method from 1914. Let $q_1, q_2, \ldots, q_n$ be n variables and parameters involved in a particular problem or situation, so that there exists a functional relationship of the form

$$f(q_1, q_2, \ldots, q_n) = 0. \tag{1.43}$$

Buckingham's theorem states that the n variables can always be combined to form exactly $(n - r)$ independent dimensionless parameter groups, where r is the number of independent dimensions. Each dimensionless parameter group is commonly called a "Π-group" or a dimensionless group. Thus, (1.43) can be written as a functional relationship:

$$\phi(\Pi_1, \Pi_2, \ldots, \Pi_{n-r}) = 0 \quad \text{or} \quad \Pi_1 = \phi(\Pi_2, \Pi_3, \ldots, \Pi_{n-r}). \tag{1.44}$$

The dimensionless groups are not unique, but $(n - r)$ of them are *independent* and form a *complete set* that spans the parametric solution space of (1.44). The power of dimensional analysis is most apparent when n and r are single-digit numbers of comparable size so (1.44), which involves $n - r$ dimensionless groups, represents a significant simplification of (1.43), which has n parameters. The process of dimensional analysis is presented here as a series of six steps that should be followed by a seventh whenever possible. Each step is described in the following paragraphs and illustrated by determining the functional dependence of the pressure difference Δp between two locations in a round pipe carrying a flowing viscous fluid.

Step 1. Select variables and parameters

Selecting variables and parameters in a dimensional analysis effort is the most important step. The list should usually contain only one unknown variable, the *solution variable*. The rest of the variables and parameters should represent the underlying physics of the situation and come from the problem's geometry, boundary conditions, initial conditions, and material parameters. Physical constants may also be included. Shorter parameter lists tend to produce the most powerful dimensional analysis results.

For the round-pipe pressure drop example, select Δp as the solution variable, and then choose as additional parameters: the distance Δx between the pressure measurement locations, the inside diameter d of the pipe, the average height ε of the pipe's wall roughness, the average flow speed U, the fluid density ρ, and the fluid viscosity μ. The resulting functional dependence between these seven parameters can be stated as:

$$f(\Delta p, \Delta X, d, \varepsilon, U, \rho, \mu) = 0. \tag{1.45}$$

Note, (1.45) does not include the fluid's thermal conductivity, heat capacities, thermal expansion coefficient, or speed of sound, so this dimensional analysis example will not account for the thermal or compressible flow effects embodied by these missing parameters. Thus, this parameter listing implies isothermal and constant density flow.

Step 2. Create the dimensional matrix

Fluid flow problems without electromagnetic forces and chemical reactions involve only mechanical variables (such as velocity and density) and thermal variables (such as temperature and specific heat). The dimensions of all these variables can be expressed in terms of four basic dimensions – mass M, length L, time T, and temperature θ. We shall denote the dimension of a variable q by $[q]$. For example, the dimensions of the velocity u are $[u] = \mathrm{L/T}$, that of pressure are $[p] = [\text{force}]/[\text{area}] = \mathrm{MLT^{-2}/L^2} = \mathrm{M/LT^2}$, and that of specific heat are $[c_p] = [\text{energy}]/[\text{mass}][\text{temperature}] = \mathrm{ML^2T^{-2}/M\theta} = \mathrm{L^2/\theta T^2}$. When thermal effects are not considered, all variables can be expressed in terms of three fundamental dimensions, namely, M, L, and T. If temperature is considered only in combination with Boltzmann's constant ($k_B\theta$), a gas constant ($R\theta$), or a specific heat ($c_p\theta$ or $c_v\theta$), then the units of the combination are simply $\mathrm{L^2/T^2}$, and only the three dimensions M, L, and T are required.

The dimensional matrix is created by listing the powers of M, L, T, and θ in a column for each parameter selected. For the pipe-flow pressure difference example, the selected variables and their dimensions produce the following dimensional matrix:

$$
\begin{array}{c|ccccccc}
 & \Delta p & \Delta x & d & \varepsilon & U & \rho & \mu \\
\hline
\mathrm{M} & 1 & 0 & 0 & 0 & 0 & 1 & 1 \\
\mathrm{L} & -1 & 1 & 1 & 1 & 1 & -3 & -1 \\
\mathrm{T} & -2 & 0 & 0 & 0 & -1 & 0 & -1
\end{array}
\tag{1.46}
$$

where the seven variables have been written above the matrix entries and the three units have been written in a column to the left of the matrix. The matrix in (1.46) portrays $[\Delta p] = \mathrm{ML^{-1}T^{-2}}$ via the first column of numeric entries.

Step 3. Determine the rank of the dimensional matrix

The *rank* r of any matrix is defined to be the size of the largest square sub-matrix that has a nonzero determinant. The determinant of the first three columns of (1.46) is zero, as is the determinant of the second through fourth, and third through fifth columns. However, the determinant of the fourth through sixth columns, and fifth through seventh columns are non-zero. For example, the determinant of the last three columns is:

$$
\begin{vmatrix} 0 & 1 & 1 \\ 1 & -3 & -1 \\ -1 & 0 & -1 \end{vmatrix} = -1.
$$

Thus, the rank of the dimensional matrix (1.46) is $r = 3$. If *all* possible third-order determinants were zero, we would have concluded that $r < 3$ and proceeded to testing second-order determinants.

For dimensional matrices, the rank is less than the number of rows only when one of the rows can be obtained by a linear combination of the other rows. For example, the matrix (not from (1.46)):

$$
\begin{vmatrix} 0 & 1 & 0 & 1 \\ -1 & 2 & 1 & -2 \\ -1 & 4 & 1 & 0 \end{vmatrix}
$$

has $r = 2$, as the last row can be obtained by adding the second row to twice the first row. A rank of less than 3 commonly occurs in static situations, in which inertia is really not relevant but the dimensions of the variables (such as force) involve M. In most fluid mechanics problems without thermal effects, $r = 3$.

Step 4. Determine the number of dimensionless groups

The number of dimensionless groups is $n - r$ where n is the number of variables and parameters, and r is the rank of the dimensional matrix. In the pipe-flow pressure difference example, the number of dimensionless groups is $4 = 7 - 3$.

Step 5. Construct the dimensionless groups

This can be done by exponent algebra or by inspection. The latter is preferred because it commonly produces dimensionless groups that are easier to interpret, but the former is sometimes required. Examples of both techniques follow. Whatever the method, the best approach is usually to create the first dimensionless group with the solution variable appearing to the first power.

When using exponent algebra, select r parameters from the dimensional matrix as *repeating parameters* that will be found in all the subsequently constructed dimensionless groups. These repeating parameters must span the appropriate r-dimensional space of M, L, and/or T, that is, the determinant of the dimensional matrix formed from these r parameters must be nonzero. For many fluid-flow problems, a characteristic velocity, a characteristic length, and a fluid property involving mass are ideal repeating parameters.

To form dimensionless groups for the pipe-flow problem, choose U, d, and ρ as the repeating parameters. The determinant of the dimensional matrix formed by these three parameters is nonzero. Other repeating parameter choices will result in a different set of dimensionless groups, but any such alternative set will still span the solution space of the problem. Thus, any satisfactory choice of the repeating parameters is equivalent to any other, so choices that simplify the work are most appropriate. Each dimensionless group is formed by combining the three repeating parameters, raised to unknown powers, with one of the nonrepeating variables or parameters from the list constructed for the first step. Here we ensure that the first dimensionless group involves the solution variable raised to the first power:

$$\Pi_1 = \boldsymbol{\Delta} p U^a d^b \rho^c.$$

The exponents a, b, and c are obtained from the requirement that Π_1 is dimensionless. Replicating this equation in terms of dimensions produces:

$$\mathrm{M}^0\mathrm{L}^0\mathrm{T}^0 = [\Pi_1] = [\boldsymbol{\Delta} p U^a d^b \rho^c] = (\mathrm{ML}^{-1}\mathrm{T}^{-2})(\mathrm{LT}^{-1})^a(\mathrm{L})^b(\mathrm{ML}^{-3})^c = \mathrm{M}^{c+1}\mathrm{L}^{a+b-3c-1}\mathrm{T}^{-a-2}.$$

Equating exponents between the two extreme ends of this extended equality produces three algebraic equations that are readily solved to find $a=-2$, $b=0$, $c=-1$, so

$$\Pi_1 = \boldsymbol{\Delta} p/\rho U^2.$$

A similar procedure with $\boldsymbol{\Delta} p$ replaced by the other unused variables ($\boldsymbol{\Delta} x$, ε, μ) produces:

$$\Pi_2 = \boldsymbol{\Delta} x/d, \qquad \Pi_3 = \varepsilon/d, \quad \text{and} \quad \Pi_4 = \mu/\rho U d.$$

The inspection method proceeds directly from the dimensional matrix, and may be less tedious than exponent algebra. It involves selecting individual parameters from the dimensional matrix and sequentially eliminating their M, L, T, and θ units by forming ratios with other parameters. For the pipe-flow pressure difference example we again start with the solution variable $[\boldsymbol{\Delta} p] = \mathrm{ML}^{-1}\mathrm{T}^{-2}$ and notice that the next entry in (1.46) that includes units of mass is $[\rho] = \mathrm{ML}^{-3}$. To eliminate M from a combination of $\boldsymbol{\Delta} p$ and ρ, we form the ratio $[\boldsymbol{\Delta} p/\rho] = \mathrm{L}^2\mathrm{T}^{-2} = [\text{velocity}^2]$. An examination of (1.46) shows that U has units of velocity, LT^{-1}. Thus, $\boldsymbol{\Delta} p/\rho$ can be made dimensionless if it is divided by U^2 to find: $[\boldsymbol{\Delta} p/\rho U^2] =$ dimensionless. Here we have the good fortune to eliminate L and T in the same step. Therefore, the first dimensionless group is $\Pi_1 = \boldsymbol{\Delta} p/\rho U^2$. To find the second dimensionless group Π_2, start with $\boldsymbol{\Delta} x$, the left-most unused parameter in (1.46), and note $[\boldsymbol{\Delta} x] = \mathrm{L}$. The first unused parameter to the right of $\boldsymbol{\Delta} x$ involving only length is d. Thus, $[\boldsymbol{\Delta} x/d] =$ dimensionless so $\Pi_2 = \boldsymbol{\Delta} x/d$. The third dimensionless group is obtained by starting with the next unused parameter, ε, to find $\Pi_3 = \varepsilon/d$. The final dimensionless group must include the last unused parameter $[\mu] = \mathrm{ML}^{-1}\mathrm{T}^{-1}$. Here it is better to eliminate the mass dimension with the density since reusing $\boldsymbol{\Delta} p$ would place the solution variable in two places in the final scaling law, an unnecessary complication. Therefore, form the ratio μ/ρ which has units $[\mu/\rho] = \mathrm{L}^2\mathrm{T}^{-1}$. These can be eliminated with d and U, $[\mu/\rho U d] =$ dimensionless, so $\Pi_4 = \mu/\rho U d$.

Forming the dimensionless groups by inspection becomes easier with experience. For example, since there are three length scales $\boldsymbol{\Delta} x$, d, and ε in (1.46), the dimensionless groups $\boldsymbol{\Delta} x/d$ and ε/d can be formed immediately. Furthermore, Bernoulli equations (see Section 4.9, "Bernoulli equations") tell us that ρU^2 has the same units as p so $\boldsymbol{\Delta} p/\rho U^2$ is readily identified as a dimensionless group. Similarly, the dimensionless group that describes viscous effects in the fluid mechanical equations of motion is found to be $\mu/\rho U d$ when these equations are cast in dimensionless form (see Section 4.11).

Other dimensionless groups can be obtained by combining established groups. For the pipe flow example, the group $\boldsymbol{\Delta} p d^2\rho/\mu^2$ can be formed from Π_1/Π_4^2, and the group $\varepsilon/\boldsymbol{\Delta} x$ can be formed as Π_3/Π_2. However, only four dimensionless groups will be independent in the pipe-flow example.

Step 6. State the dimensionless relationship

This step merely involves placing the $(n-r)\Pi$-groups in one of the forms in (1.44). For the pipe-flow example, this dimensionless relationship is:

$$\frac{\Delta p}{\rho U^2}=\phi\left(\frac{\Delta x}{d},\frac{\varepsilon}{d},\frac{\mu}{\rho U d}\right), \tag{1.47}$$

where ϕ is an undetermined function. This relationship involves only four dimensionless groups, and is therefore a clear simplification of (1.43) which lists seven independent parameters. The four dimensionless groups in (1.47) have familiar physical interpretations and have even been given special names. For example, $\Delta x/d$ is the pipe's aspect ratio, and ε/d is the pipe's roughness ratio. Common dimensionless groups in fluid mechanics are presented and discussed in Section 4.11.

Step 7. Use physical reasoning or additional knowledge to simplify the dimensionless relationship

This seventh step may not always be possible, but when it is, significant and powerful results may be achieved from dimensional analysis. For example, sometimes there are only two extensive thermodynamic variables involved and these must be proportional in the final scaling law. Besides that, an overall conservation law might be applied that restricts one or more parametric dependencies, or a phenomena may be known to be linear, quadratic, etc. in one of the parameters and this dependence must be reflected in the final scaling law.

Once the dimensionless groups have been identified, and the dimensionless law has been stated and possibly simplified, the resulting relationship can be used for similarity-scaling analysis between two or more different scenarios involving the same physical principles. For example, consider (1.47) the final dimensional analysis result for incompressible flow in a round rough-walled pipe. If $\Delta x/d=10$, $\varepsilon/d=10^{-3}$, and $\mu/\rho U d=10^{-5}$, this could represent the flow of room temperature and pressure air at 15 m/s in a 0.10-m-diameter pipe over a distance of 1.0 m with a wall roughness of 100 μm. Alternatively it could represent, the flow of room temperature water at 10 m/s in a 1.0-cm-diameter tube over a distance of 0.10 m with a wall roughness of 10 μm. If Δp is measured in the air case to be 30 Pa, then the function in (1.47) can be evaluated for this condition: $\varphi(10, 10^{-3}, 10^{-5})=(\Delta p/\rho U^2)_{air}=30Pa/(1.2\text{kgm}^{-3}\cdot(15\ \text{ms}^{-1})^2)=0.11$. This value of φ, also applies to in the water case, $\Delta p_{water}=\varphi(10, 10^{-3}, 10^{-5})\cdot(\rho U^2)_{water}=0.11(10^3\text{kgm}^{-3}\cdot(10\ \text{ms}^{-1})^2)=11$ kPa. Thus, quantitative knowledge of one situation immediately applies to the other through a relationship like (1.47). This is an illustration of the principle of dynamic similarity which is more fully discussed in Section 4.11.

Example 1.9

Use dimensional analysis to find the parametric dependence of the scale height H in a static isothermal atmosphere at temperature T_o composed of a perfect gas with average molecular weight M_w when the gravitational acceleration is g.

Solution Follow the six steps just described.

1. The parameter list must include H, T_o, M_w, and g. Here there is no velocity parameter, and there is no need for a second specification of a thermodynamic variable since a static pressure gradient prevails. However, the universal gas constant R_u must be included to help relate the thermal variable T_o to the mechanical ones.
2. The dimensional matrix is:

	H	T_o	M_w	g	R_u
M	0	0	1	0	1
L	1	0	0	1	2
T	0	0	0	−2	−2
θ	0	1	0	0	−1

   Note that the kmole^{-1} specification of M_w and R_u is lost in the matrix above since a *kmole* is a pure number.
3. The rank of this matrix is four, so $r=4$.
4. The number of dimensionless groups is: $n-r=5-4=1$.
5. Use H as the solution parameter, and the others as the repeating parameters. Proceed with exponent algebra to find the dimensionless group:

$$\text{M}^0\text{L}^0\text{T}^0\theta^0=[\Pi_1]=[HT_a^aM_w^bg^cR_v^d=(\text{L})(\theta)^a(\text{M})^b(\text{LT}^{-2})^c(\text{ML}^2\text{T}^{-2}\theta^{-1})^d=\text{M}^{b+d}\text{L}^{1+c+2}\text{T}^{-2c-2d}\theta^{a-d}.$$

Equating exponents yields four linear algebraic equations:

$$b+d=0,\ 1+c+2d=0,\ -2c-2d=0,\quad \text{and } a-d=0,$$

which are solved by: $a=-1$, $b=1$, $c=1$, $d=-1$. Thus, the lone dimensionless group is: $\Pi_1 = HgM_w/R_uT_o$.

6. Because there is only a single dimensionless group, its most general behavior is to equal a constant, so $HgM_w/R_uT = \varphi(\ldots)$, or $H = const.\,(R_uT_o/gM_w)$. Based on the finding at the end of the previous section and $R = R_u/M_w$ from (1.28), this parametric dependence is correct and the constant is unity in this case.

Example 1.10

Use dimensional analysis to determine the energy E released in an intense point blast if the blast-wave propagation distance D into an undisturbed atmosphere of density ρ is known as a function of time t following the energy release (Taylor, 1950; see Fig. 1.12).

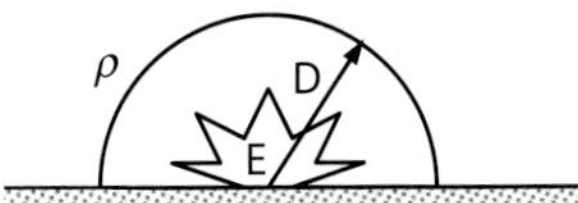

FIGURE 1.12 In an atmosphere with undisturbed density ρ, a point release of energy E produces a hemispherical blast wave that travels a distance D in time t.

Solution Again follow the six steps given earlier.

1. The parameter list $(E,\ D,\ \rho,\ t)$ is given in the problem statement so $n=4$.
2. The dimensional matrix is:

	E	D	ρ	t
M	1	0	1	0
L	2	1	−3	0
T	−2	0	0	1

3. Using MATLAB, the rank can easily be determined by

```
C = [1 0 1 0; 2 1 -3 0; -2 0 0 1];
r = rank(C);
```

which yields $r=3$.

4. The number of dimensionless groups is: $n-r=4-3=1$.
5. Construct the lone dimensionless group using MATLAB by solving a system of equations, determined by $M^0L^0T^0 = [\Pi_1] = [ED^a\rho^bt^c] = (ML^2T^{-2})(L)^a(ML^{-3})^b(T)^c$. Rewrite this in form of $Ax=b$ and solve using the '\' operator.

$$\begin{bmatrix} 0 & 1 & 0 \\ 1 & -3 & 0 \\ 0 & 0 & 1 \end{bmatrix} * \begin{bmatrix} a \\ b \\ c \end{bmatrix} = \begin{bmatrix} -1 \\ -2 \\ 2 \end{bmatrix}$$

```
A = [0 1 0; 1 -3 0; 0 0 1];
b = [-1; -2; 2];
x = A\b;
```

The output yields $x=[-5;-1;2]$, so $\Pi_1 = Et^2/\rho D^5$.

6. Here there is only a single dimensionless group, so it must be a constant (K). This produces: $Et^2/\rho D^5 = \varphi(\ldots) = K$ which implies: $E = K\rho D^5/t^2$, where K is not determined by dimensional analysis.
7. The famous physicist and fluid mechanician G.I. Taylor was able to estimate the yield of the first atomic-bomb test conducted on the White Sands Proving Grounds in New Mexico in July 1945 using: 1) the dimensional analysis shown above, 2) a

declassified movie made by J.E. Mack, and 3) timed photographs supplied by the Los Alamos National Laboratory and the Ministry of Supply. He determined the fireball radius as a function of time and then estimated E using a nominal atmospheric value for ρ. His estimate of $E = 17$ kilotons of TNT was very close to the actual yield (20 kilotons of TNT) in part because the undetermined constant K is close to unity in this case. At the time, the movie and the photographs were not classified but the yield of the bomb was entirely secret.

Example 1.11

Use dimensional analysis to determine how the average light intensity S (Watts/m^2) scattered from an isolated particle depends on the incident light intensity I (Watts/m^2), the wavelength of the light λ (m), the volume of the particle V (m^3), the index of refraction of the particle n_s (dimensionless), and the distance d (m) from the particle to the observation point. Can the resulting dimensionless relationship be simplified to better determine parametric effects when $\lambda > V^{1/3}$?

Solution Again follow the six steps given earlier knowing that the seventh step will likely be necessary to produce a useful final relationship.

1. The parameter list (S, I, λ, V, n_s, d) is given in the problem statement so $n = 6$.
2. The dimensional matrix is:

	S	I	λ	V	N_s	d
M	1	1	0	0	0	0
L	0	0	1	3	0	1
T	−3	−3	0	0	0	0

3. The rank of this matrix is 2 because all the dimensions are either intensity or length, so $r = 2$.
4. The number of dimensionless groups is: $n - r = 6 - 2 = 4$.
5. Let scattered light intensity S be the solution parameter. By inspection the four dimensionless groups are: $\Pi_1 = S/I$, $\Pi_2 = d/\lambda$, $\Pi_3 = V/\lambda^3$, and $\Pi_4 = n_s$.
6. Therefore, the dimensionless relationship is: $S/I = \varphi_1(d/\lambda, V/\lambda^3, n_s)$.
7. There are two physical features of this problem that allow refinement of this dimensional analysis result. First, light scattering from the particle must conserve energy and this implies: $4\pi d^2 S = const.$ so $S \propto 1/d^2$. Therefore, the step-6 result must simplify to: $S/I = (\lambda/d)^2 \varphi_2(V/\lambda^3, n_s)$. Second, when λ is large compared to the size of the scatterer, the scattered field amplitude will be produced from the dipole moment induced in the scatterer by the incident field, and this scattered field amplitude will be proportional to V. Thus, S, which is proportional to field amplitude squared, will be proportional to V^2. These deductions allow a further simplification of the dimensional analysis result to:

$$\frac{S}{I} = \left(\frac{\lambda}{d}\right)^2 \left(\frac{V}{\lambda^3}\right)^2 \varphi_3(n_s) = \frac{V^2}{d^2\lambda^4}\varphi_3(n_s).$$

This is Lord Rayleigh's celebrated small-particle scattering law. He derived it in the 1870s while investigating light scattering from small scatterers to understand why the cloudless daytime sky was blue while the sun appeared orange or red at dawn and sunset. At the time, he imagined that the scatterers were smoke, dust, mist, aerosols, etc. However, the atmospheric abundance of these are insufficient to entirely explain the color change phenomena but the molecules that compose the atmosphere can accomplish enough scattering to explain the observations.

Exercises

1.1 [1]Many centuries ago, a mariner poured 100 cm^3 of water into the ocean. It was subsequently mixed uniformly throughout the earth's oceans, lakes, and rivers. Ignoring salinity, estimate the probability that the next sip (5 ml) of water you drink will contain at least one water molecule from the mariner's cup. Assess your chances of ever drinking truly pristine water. (*Consider the following facts*: M_w for water is 18.0 kg per kg-mole, the radius of the earth is 6370 km, the mean depth of the oceans is approximately 3.8 km and they cover 71% of the surface of the earth.)

1.2 [1]An adult human expels approximately 500 mL of air with each breath during ordinary breathing. Imagining that two people exchanged greetings (one breath each) many centuries ago and that their breath subsequently has been

1. Based on a homework problem posed by Professor P.E. Dimotakis.

mixed uniformly throughout the atmosphere, estimate the probability that the next breath you take will contain at least one air molecule from that age-old verbal exchange. Assess your chances of ever getting a truly fresh breath of air. For this problem, assume that air is composed of identical molecules having $M_w = 29.0$ kg per kg-mole and that the average atmospheric pressure on the surface of the earth is 100 kPa. Use 6370 km for the radius of the earth and 1.20 kg/m^3 for the density of air at room temperature and pressure.

1.3 The Maxwell probability distribution, $f(\mathbf{v}) = f(v_1, v_2, v_3)$, of molecular velocities in a gas flow at a point in space with average velocity $\mathbf{u}$ is given by (1.1).

a) Verify that $\mathbf{u}$ is the average molecular velocity, and determine the standard deviations $(\sigma_1, \sigma_2, \sigma_3)$ of each component of $\mathbf{u}$ using $\sigma_i = [\frac{1}{n}\iiint_{all\,\mathbf{v}}(v_i - u_i)^2 f(v)d^3v]^{1/2}$ for $i = 1$, 2, and 3.

b) Using (1.27) or (1.28), determine $n = N/V$ at room temperature $T = 295$ K and atmospheric pressure $p = 101.3$ kPa.

c) Determine $N = nV$ = number of molecules in volumes $V = (10\ \mu\text{m})^3$, $1\ \mu\text{m}^3$, and $(0.1\ \mu\text{m})^3$.

d) For the ith velocity component, the standard deviation of the average, $\sigma_{a,i}$, over N molecules is $\sigma_{a,i} = \sigma_i/\sqrt{N}$ when $N \gg 1$. For an airflow at $\mathbf{u} = (1.0\ \text{ms}^{-1}, 0, 0)$, compute the relative uncertainty, $2\sigma_{a,1}/U_1$, at the 95% confidence level for the average velocity for the three volumes listed in part c).

e) For the conditions specified in parts b) and d), what is the smallest volume of gas that ensures a relative uncertainty in U of one percent or less?

1.4 Using the Maxwell molecular speed distribution given by (1.5),

a) Determine the most probable molecular speed,

b) Show that the average molecular speed is as given in (1.5),

c) Determine the root-mean square molecular speed $= v_{rms} = [\frac{1}{n}\int v^2 f(v)dv]^{1/2}$,

d) and Compare the results from parts a), b) and c) with c = speed of sound in a perfect gas under the same conditions.

1.5 By considering the volume swept out by a moving molecule, estimate how the mean-free path, l, depends on the average molecular cross section dimension $\overline{d}$ and the molecular number density n for nominally spherical molecules. Find a formula for $ln^{1/3}$ (the ratio of the mean-free path to the mean intermolecular spacing) in terms of the nominal *molecular volume* $(\overline{d}^3)$ and the available *volume per molecule* $(1/n)$. Is this ratio typically bigger or smaller than one?

1.6 Compute the average *relative* speed, $\overline{v}_r$, between molecules in a gas using the Maxwell speed distribution f given by (1.5) via the following steps.

a) If $\mathbf{u}$ and $\mathbf{v}$ are the velocities of two molecules then their relative velocity is: $\mathbf{v}_r = \mathbf{u} - \mathbf{v}$. If the angle between $\mathbf{u}$ and $\mathbf{v}$ is θ, show that the relative speed is: $v_r = |\mathbf{v}_r|\sqrt{u^2 + v^2 - 2uv\cos\theta}$ where $u = |\mathbf{u}|$, and $v = |\mathbf{v}|$.

b) The averaging of v_r necessary to determine $\overline{v}_r$ must include all possible values of the two speeds (u and v) and all possible angles θ. Therefore, start from:

$$\overline{v}_r = \frac{1}{2n^2}\int_{all\ u,v,\theta} v_r f(u)f(v)sin\theta\, d\theta\, dvdu,$$

and note that $\overline{v}_r$ is unchanged by exchange of u and v, to reach:

$$\overline{v}_r = \frac{1}{n^2}\int_{u=0}^{\infty}\int_{v=u}^{\infty}\int_{\theta=0}^{\pi}\sqrt{u^2 + v^2 - 2uv\,cos\theta}\,sin\theta\ f(u)f(v)d\theta dvdu$$

c) Note that v_r must always be positive and perform the integrations, starting with the angular one, to find:

$$\overline{v}_r = \frac{1}{3n^2}\int_{u=0}^{\infty}\frac{\int_{v=u}^{\infty}2u^3 + 6uv^2}{uv}f(u)f(v)dvdu = \left(\frac{16k_BT}{\pi}\right)^{1/2} = \sqrt{2}\overline{v}.$$

1.7 In a gas, the molecular momentum flux (MF_{ij}) in the j-coordinate direction that crosses a flat surface of unit area with coordinate normal direction i is: $MF_{ij} = \frac{1}{V}\iiint_{all\ v} mv_iv_j f(v)d^3v$ where $f(\mathbf{v})$ is the Maxwell velocity distribution (1.1). For a perfect gas that is not moving on average (i.e., $\mathbf{u} = 0$), show that $MF_{ij} = p$ (the pressure), when $i = j$, and that $MF_{ij} = 0$, when $i \neq j$.

1.8 Consider viscous flow in a channel of width $2b$. The channel is aligned in the x-direction, and the velocity u in the x-direction at a distance y from the channel centerline is given by the parabolic distribution $u(y) = U_o[1 - (y/b)^2]$. Calculate the shear stress τ as a function y, μ, b, and U_o. What is the shear stress at $y = 0$?

1.9 [2]Hydroplaning occurs on wet roadways when sudden braking causes a moving vehicle's tires to stop turning when the tires are separated from the road surface by a thin film of water. When hydroplaning occurs the vehicle may slide a significant distance before the film breaks down and the tires again contact the road. For simplicity, consider a hypothetical version of this scenario where the water film is somehow maintained until the vehicle comes to rest.

a) Develop a formula for the friction force delivered to a vehicle of mass M and tire-contact area A that is moving at speed u on a water film with constant thickness h and viscosity μ.

b) Using Newton's second law, derive a formula for the hypothetical sliding distance D traveled by a vehicle that started hydroplaning at speed U_o.

c) Evaluate this hypothetical distance for $M = 1200$ kg, $A = 0.1$ m^2, $U_o = 20$ m/s, $h = 0.1$ mm, and $\mu = 0.001$ kgm^{-1}s^{-1}. Compare this to the dry-pavement stopping distance assuming a tire-road coefficient of kinetic friction of 0.8.

1.10 Estimate the height to which water at 20 °C will rise in a capillary glass tube 3 mm in diameter that is exposed to the atmosphere. For water in contact with glass the contact angle is nearly 0°. At 20 °C, the surface tension of a water-air interface is $\sigma = 0.073$ N/m.

1.11 A long thin rectangular slab of glass with thickness t is dipped into a liquid and withdrawn to a height h with the liquid surface attached to the lower edges of the slab (as shown). Far from the ends of the slab, the surface's shape is two-dimensional and can be described by coordinate functions $x(s)$ and $z(s)$, where s is the arc length along the curve.

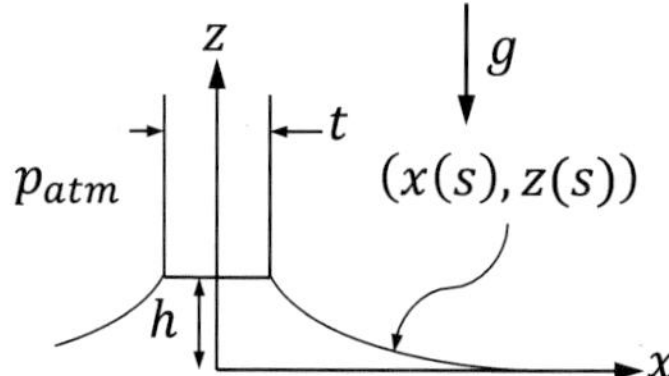

For this situation, the following equations for the hydrostatic pressure

$$p = p_{atm} - \rho g z,$$

the Laplace pressure difference across the water surface (1.11)

$$p - p_{atm} = -\sigma \frac{x''}{\sqrt{1 - x'^2}},$$

and a geometric constraint

$$x'^2 + z'^2 = 1,$$

can be used to numerically determine the shape (x, z) of the liquid surface. The primes in these equations denote differentiation with respect to s. The boundary conditions in this case are $x = t/2$ at $s = 0$, $z = h$ at $s = 0$, $x' \to 1$ as $s \to \infty$, and $z \to 0$ as $s \to \infty$.

a) Calculate and plot the surface shape when the liquid is water with $\sigma = 72$ dyne/cm, $t = 1$ mm, and $h = 2$ mm.

b) Determine the force on the slab per unit depth into the page arising from hydrostatic and surface tension forces using the parameters specified in part a).

c) Redo the part a) calculations for increasing h, but constant $t = 1$ mm, to determine the lift height, h_{max}, at which the raised liquid detaches from slab, or its minimum thickness goes to zero.

1.12 A *manometer* is a U-shaped tube containing mercury of density ρ_m. Manometers are used as pressure-measuring devices. If the fluid in tank A has a pressure p and density ρ, then show that the gauge pressure in the tank is: $p - p_{\text{atm}} = \rho_m g h - \rho g a$. Note that the last term on the right side is negligible if $\rho \ll \rho_m$. (*Hint*: Equate the pressures at X and Y.)

2. Based on a homework problem posed by Prof. Christopher Brennen.

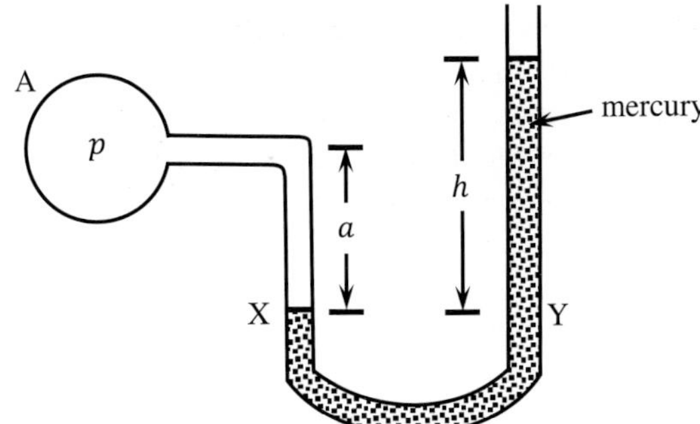

1.13 A block of concrete with density ρ_c, height H, width w, and distance into the page b is used to restrain a stagnant pool of water with density $\rho_{w)}$ and depth h formed on a hard impenetrable horizontal surface at z = 0. Gravity g acts downward, and the surrounding air pressure is p_{atm}. The block is not attached to the surface and some water seeps under it so that the hydrostatic pressure under the block is $p(x,0) - p_{atm} = (1 - x/w)[p(0,0) - p_{atm}]$. Using the (x, z)-coordinate system shown, answer the following questions in terms of the given parameters.

a) What is the horizontal hydrostatic force on the block?

b) If the coefficient of friction between the block and the hard surface is μ_f, determine a formula for the value of h above which the block will begin to slide to the right.

c) Which parameter(s) of the block should be increased to prevent block sliding?

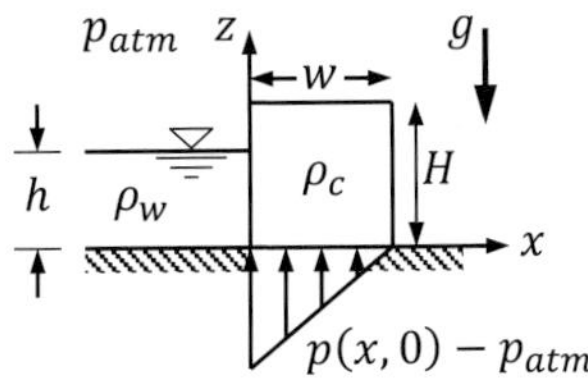

1.14 Prove that if $e(T, \upsilon) = e(T)$ only and if $h(T, p) = h(T)$ only, then the (thermal) equation of state is (1.28) or $p\upsilon = kT$, where k is a constant.

1.15 Starting from the property relationships (1.24) prove (1.31) and (1.32) for a reversible adiabatic process involving a perfect gas when the specific heats c_p and c_v are constant.

1.16 For an ideal gas $dh = c_p dT$ and $de = c_v dT$. Use one of these, one of the property relationships (1.24), the perfect gas law (1.28), and the definition of the gas constant R (1.29) to show: $\frac{s_2 - s_1}{c_v} = \ln(\frac{p_2}{p_1}(\frac{\rho_1}{\rho_2})^{\gamma})$.

1.17 A cylinder contains 2 kg of air at 50 °C and a pressure of 3 bars. The air is compressed until its pressure rises to 8 bars. What is the initial volume? Find the final volume for both isothermal compression and isentropic compression.

1.18 Derive (1.36) starting from Fig. 1.10 and the discussion at the beginning of Section 1.10.

1.19 Starting with the hydrostatic pressure law (1.14), prove (1.37) without using perfect gas relationships.

1.20 Assume that the temperature of the atmosphere varies with height z as $T = T_0 + Kz$ where K is a constant. Show that the pressure varies with height as $p = p_0(\frac{T_0}{T_0 + Kz})^{g/KR}$ where g is the acceleration of gravity and R is the gas constant for the atmospheric gas.

1.21 Suppose the atmospheric temperature varies according to: $T = 15 - 0.001z$, where T is in degrees Celsius and height z is in meters. Is this atmosphere stable?

1.22 A hemispherical bowl with inner radius r containing a liquid with density ρ is inverted on a smooth flat surface. Gravity with acceleration g acts downward. Determine the weight W of the bowl necessary to prevent the liquid from escaping. Consider two cases:

a) the pressure around the rim of the bowl where it meets the plate is atmospheric, and

b) the pressure at the highest point of the bowl's interior is atmospheric.

c) Investigate which case applies via a simple experiment. Completely fill an ordinary soup bowl with water and concentrically cover it with an ordinary dinner plate. While holding the bowl and plate together, quickly invert the water-bowl-plate combination, set it on the level surface at the bottom of a kitchen sink, and let go. Does the water escape? If no water escapes after release, hold onto the bowl only and try to lift the water-bowl-plate combination a few centimeters off the bottom of the sink. Does the plate remain in contact with the bowl? Do your answers to the first two parts of this problem help explain your observations?

1.23 Consider the case of a pure gas planet where the hydrostatic law is: $dp/dz = -\rho(z)Gm(z)/z^2$, where G is the gravitational constant and $m(z) = 4\pi \int \rho(\zeta)\zeta^2 d\zeta$ is the planetary mass up to distance z from the center of the planet. If the planetary gas is perfect with gas constant R, determine $\rho(z)$ and $p(z)$ if this atmosphere is isothermal at temperature T. Are these vertical profiles of ρ and p valid as z increases without bound?

1.24 Consider a gas atmosphere with pressure distribution $p(z) = p_0(1 - (2/\pi)\tan^{-1}(z/H))$ where z is the vertical coordinate and H is a constant length scale.

a) Determine the vertical profile of ρ from (1.14).

b) Determine N^2 from (1.36) as function of vertical distance, z.

c) Near the ground where $z \ll H$, this atmosphere is unstable, but it is stable at greater heights where $z \gg H$. Specify the value of z/H above which this atmosphere is stable when $\gamma = 1.40$.

1.25 Consider an ideal gas atmosphere subject to constant gravity g, where the pressure and density are related by a power law: $p(z) = p_0(\rho(z)/\rho_0)^k$ where p_0 and ρ_0 are the pressure and density at $z = 0$, and k is a positive constant.

a) Determine the vertical profile of ρ from (1.14).

b) If the atmosphere is isentropic and $k = \gamma = 1.40$, and what is the height where $\rho = 0$.

c) If the atmosphere is isothermal, take the limit of the answer for part a) for $k \to 1$ to recover the exponential profile given in Section 1.10.

1.26 Consider a heat-insulated enclosure that is separated into two compartments of volumes V_1 and V_2, containing perfect gases with pressures and temperatures of p_1 and p_2, and T_1 and T_2, respectively. The compartments are separated by an impermeable membrane that conducts heat (but not mass). Calculate the final steady-state temperature assuming each gas has constant specific heats.

1.27 Consider the initial state of an enclosure with two compartments as described in Exercise 1.26. At $t = 0$, the membrane is broken and the gases are mixed. Calculate the final temperature.

1.28 A heavy piston of weight W is dropped onto a thermally insulated cylinder of cross-sectional area A containing a perfect gas of constant specific heats, and initially having the external pressure p_1, temperature T_1, and volume V_1. After some oscillations, the piston reaches an equilibrium position L meters below the equilibrium position of a weightless piston. Find L. Is there an entropy increase?

1.29 Starting from 295 K and atmospheric pressure, what is the final pressure of an isentropic compression of air that raises the temperature 1, 10, and 100 K.

1.30 Compute the speed of sound in air at −40 °C (very cold winter temperature), at +45 °C (very hot summer temperature), at 400 °C (automobile exhaust temperature), and 2000 °C (nominal hydrocarbon adiabatic flame temperature).

1.31 Use dimensional analysis to prove the Pythagorean theorem based on a right triangle's area a, the radian measure β of its most acute angle, and the length C of its longest side.

1.32 The oscillation frequency Ω of a simple pendulum depends on the acceleration of gravity g, and the length L of the pendulum.

a) Using dimensional analysis, determine single dimensionless group involving Ω, g, and L.

b) Perform an experiment to see if the dimensionless group is constant. Using a piece of string slightly longer than 2 m and any small heavy object, attach the object to one end of the piece of string with tape or a knot. Mark distances of 0.25, 0.5, 1.0, and 2.0 m on the string from the center of gravity of the object. Hold the string at the marked locations, stand in front of a clock with a second hand or second readout, and count the number (N) of pendulum oscillations in 20 seconds to determine $\Omega = N/(20\text{ s})$ in Hz. Evaluate the dimensionless group for these four lengths.

c) Based only on the results of parts a) and b), what pendulum frequency do you predict when $L = 1.0$ m but g is 16.6 m/s? How confident should you be of this prediction?

1.33 The spectrum of wind waves, $S(\omega)$, on the surface of the deep sea may depend on the wave frequency ω, gravity g, the wind speed U, and the fetch distance F (the distance from the upwind shore over which the wind blows with constant velocity).

a) Using dimensional analysis, determine how $S(\omega)$ must depend on the other parameters.

b) It is observed that the mean-square wave amplitude, $\eta^2 = \int_0^\infty S(\omega)d\omega$, is proportional to F. Use this fact to revise the result of part a).

c) How must η^2 depend on U and g?

1.34 One military technology for clearing a path through a minefield is to deploy a powerfully exploding cable across the minefield that, when detonated, creates a large trench through which soldiers and vehicles may safely travel. If the expanding cylindrical blast wave from such a line-explosive has radius R at time t after detonation, use dimensional analysis to determine how R and the blast wave speed dR/dt must depend on t, ρ = air density, and E' = energy released per unit length of exploding cable.

1.35 [3]One of the triumphs of classical thermodynamics for a simple compressible substance was the identification of entropy s as a state variable along with pressure p, density ρ, and temperature T. Interestingly, this identification foreshadowed the existence of quantum physics because of the requirement that it must be possible to state all physically meaningful laws in dimensionless form. To see this foreshadowing, consider an entropic equation of state for a system of N elements each having mass m.

a) Determine which thermodynamic variables amongst s, p, ρ, and T can be made dimensionless using N, m, and the non-quantum mechanical physical constants k_B = Boltzmann's constant and c = speed of light. What do these results imply about an entropic equation of state in any of the following forms: $s = s(p, \rho)$, $s = s(\rho, T)$, or $s = s(T, p)$?

b) Repeat part a) including $\hbar$ = Planck's constant (the fundamental constant of quantum physics). Can an entropic equation of state be stated in dimensionless form with $\hbar$?

1.36 [3]The natural variables of the system enthalpy H are the system entropy S and the pressure p, which leads to an equation of state in the form: $H = H(S, p, N)$, where N is the number of system elements.

a) After creating ratios of extensive variables, use exponent algebra to independently render H/N, S/N, and p dimensionless using m = the mass of a system element, and the fundamental constants k_B = Boltzmann's constant, $\hbar$ = Planck's constant, and c = speed of light. Write a dimensionless law for H.

b) Simplify the result of part a) for non-relativistic elements by eliminating c.

c) Based on the property relationship (1.24), determine the specific volume $= \upsilon = 1/\rho = (\partial h/\partial p)_s$ from the result of part b).

d) Use the result of part c) and (1.25) to show that the sound speed in this case is $\sqrt{5p/3\rho}$, and compare this result to that for a monotonic perfect gas.

1.37 [3]A gas of noninteracting particles of mass m at temperature T has density ρ, and internal energy per unit volume ε.

a) Using dimensional analysis, determine how ε must depend on ρ, T, and m. In your formulation use k_B = Boltzmann's constant, $\hbar$ = Plank's constant, and c = speed of light to include possible quantum and relativistic effects.

b) Consider the limit of slow-moving particles without quantum effects by requiring c and $\hbar$ to drop out of your dimensionless formulation. How does ε depend on ρ and T? What type of gas follows this thermodynamic law?

c) Consider the limit of massless particles (i.e., photons) by requiring m and ρ to drop out of your dimensionless formulation of part a). How does ε depend on T in this case? What is the name of this radiation law?

1.38 A compression wave in a long gas-filled constant-area duct propagates to the left at speed U. To the left of the wave, the gas is quiescent with uniform density ρ_1 and uniform pressure p_1. To the right of the wave, the gas has uniform density $\rho_2(> \rho_1)$ and uniform pressure is $p_2(> p_1)$. Ignore the effects of viscosity in this problem. Formulate a dimensionless scaling law for U in terms of the pressures and densities.

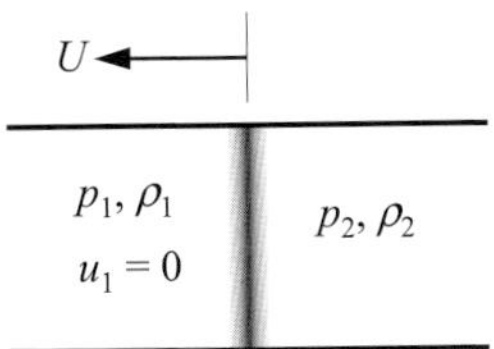

1.39 The action of viscosity causes the amplitude of a pressure waves traveling in a fluid to decrease with increasing wave-travel distance. Let D be the travel distance necessary for a pressure wave to lose half of its amplitude because of the effects of viscosity. When needed for the following items, use fluid properties at ordinary room temperature (20 °C) and pressure (1 atm.).

a) Determine a functional relationship for D in terms of the wave's pressure amplitude Δp, the wave's dominant frequency f, and the fluid's properties (density ρ, sound speed c, and viscosity μ).

b) Simplify the part a) result for low-amplitude (acoustic) pressure waves where Δp is no longer a parameter and D is proportional to μ^{-1}.

c) Using the part b) result, is D greater for low- or high-frequency acoustic waves?

3. Drawn from thermodynamics lectures of Prof. H.W. Liepmann.

d) Using the part b) result at a fixed frequency, what is the half-amplitude distance in water divided by that in air, D_{water}/D_{air}?

1.40 Many flying and swimming animals – as well as human-engineered vehicles – rely on some type of repetitive motion for propulsion through air or water. For this problem, assume the average travel speed U, depends on the repetition frequency f, the characteristic length scale of the animal or vehicle L, the acceleration of gravity g, the density of the animal or vehicle ρ_o, the density of the fluid ρ, and the viscosity of the fluid μ.

a) Formulate a dimensionless scaling law for U involving all the other parameters.

b) Simplify your answer for a) for turbulent flow where μ is no longer a parameter.

c) Fish and animals that swim at or near a water surface generate waves that move and propagate because of gravity, so g clearly plays a role in determining U. However, if fluctuations in the propulsive thrust are small, then f may not be important. Thus, eliminate f from your answer for b) while retaining L, and determine how U depends on L. Are successful competitive human swimmers likely to be shorter or taller than the average person?

d) When the propulsive fluctuations of a surface swimmer are large, the characteristic length scale may be U/f instead of L. Therefore, drop L from your answer for b). In this case, will higher speeds be achieved at lower or higher frequencies?

e) While traveling submerged, fish, marine mammals, and submarines are usually neutrally buoyant ($\rho_o \approx \rho$) or very nearly so. Thus, simplify your answer for b) so that g drops out. For this situation, how does the speed U depend on the repetition frequency f?

f) Although fully submerged, aircraft and birds are far from neutrally buoyant in air, so their travel speed is predominately set by balancing lift and weight. Ignoring frequency and viscosity, use the remaining parameters to construct dimensionally accurate surrogates for lift and weight to determine how U depends on ρ_o/ρ, L, and g.

1.41 The acoustic power W generated by a large industrial blower depends on its volume flow rate Q, the pressure rise ΔP it works against, the air density ρ, and the speed of sound c. If hired as an acoustic consultant to quiet this blower by changing its operating conditions, what is your first suggestion?

1.42 The horizontal displacement $\boldsymbol{\Delta}$ of the trajectory of a spinning ball depends on the mass m and diameter d of the ball, the air density ρ and viscosity μ, the ball's rotation rate ω, the ball's speed U, and the distance L traveled.

a) Use dimensional analysis to predict how $\boldsymbol{\Delta}$ can depend on the other parameters.

b) Simplify your result from part a) for negligible viscous forces.

c) It is experimentally observed that $\boldsymbol{\Delta}$ for a spinning sphere becomes essentially independent of the rotation rate once the surface rotation speed, $\omega d/2$, exceeds twice U. Simplify your result from part b) for this high-spin regime.

d) Based on the result in part c), how does $\boldsymbol{\Delta}$ depend on U?

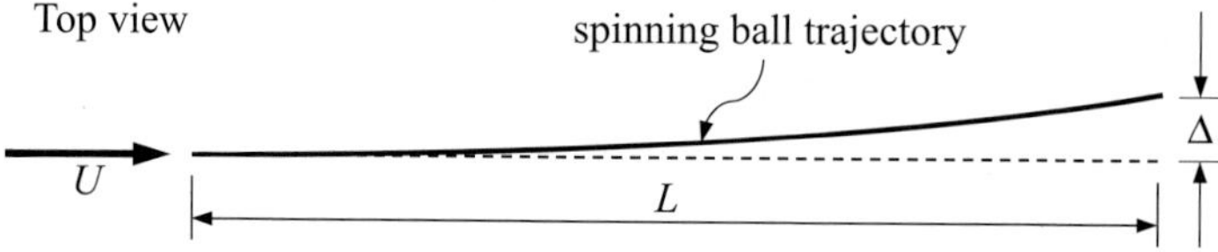

1.43 A machine that fills peanut-butter jars must be reset to accommodate larger jars. The new jars are twice as large as the old ones but they must be filled in the same amount of time by the same machine. Fortunately, the viscosity of peanut butter decreases with increasing temperature, and this property of peanut butter can be exploited to achieve the desired results since the existing machine allows for temperature control.

a) Write a dimensionless law for the jar-filling time t_f based on: the density of peanut butter ρ, the jar volume V, the viscosity of peanut butter μ, the driving pressure that forces peanut butter out of the machine P, and the diameter of the peanut butter-delivery tube d.

b) Assuming that the peanut butter flow is dominated by viscous forces, modify the relationship you have written for part a) to eliminate the effects of fluid inertia.

c) Make a reasonable assumption concerning the relationship between t_f and V when the other variables are fixed so that you can determine the viscosity ratio μ_{new}/μ_{old} necessary for proper operation of the old machine with the new jars.

d) Unfortunately, the auger mechanism that pumps the liquid peanut butter develops driving pressure through viscous forces so that P is proportional to μ. Therefore, to meet the new jar-filling requirement, what part of the machine should be changed and how much larger should it be?

1.44 As an idealization of fuel injection in a diesel engine, consider a stream of high-speed fluid (called a *jet*) that emerges into a quiescent air reservoir at $t = 0$ from a small hole in an infinite plate to form a *plume* where the fuel and air mix.

a) Develop a scaling law via dimensional analysis for the penetration distance D of the plume as a function of: Δp the pressure difference across the orifice that drives the jet, d_o the diameter of the jet orifice, ρ_o the density of the fuel, μ_∞ and ρ_∞ the viscosity and density of the air, and t the time since the jet was turned on.

b) Simplify this scaling law for turbulent flow where air viscosity is no longer a parameter.

c) For turbulent flow and $D \ll d_o$, d_o, and ρ_∞ are not parameters. Recreate the dimensionless law for D.

d) For turbulent flow and $D \gg d_o$, only the momentum flux of the jet matters, so Δp and d_o are replaced by the single parameter J_o = jet momentum flux (J_o has the units of force and is approximately equal to $\Delta p d_o^2$). Recreate the dimensionless law for D using the new parameter J_o.

1.45 [4]One of the simplest types of gasoline carburetors is a tube with small port for transverse injection of fuel. It is desirable to have the fuel uniformly mixed in the passing airstream as quickly as possible. A prediction of the mixing length L is sought. The parameters of this problem are: ρ = density of the flowing air, d = diameter of the tube, μ = viscosity of the flowing air, U = mean axial velocity of the flowing air, and J = momentum flux of the fuel stream.

a) Write a dimensionless law for L.

b) Simplify your result from part a) for turbulent flow where μ must drop out of your dimensional analysis.

c) When this flow is turbulent, it is observed that mixing is essentially complete after one rotation of the counter-rotating vortices driven by the injected-fuel momentum (see downstream-view of the drawing for this problem), and that the vortex rotation rate is directly proportional to J. Based on this information, assume that $L \propto$ (rotation time)(U) to eliminate the arbitrary function in the result of part b). The final formula for L should contain an undetermined dimensionless constant.

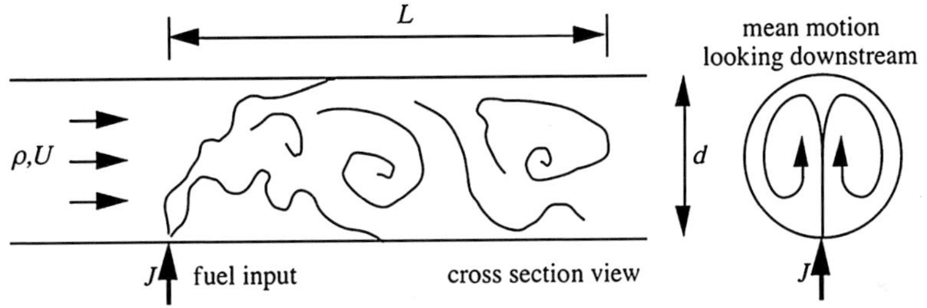

1.46 Consider dune formation in a large horizontal desert of deep sand.

a) Develop a scaling relationship that describes how the height h of the dunes depends on the average wind speed U, the length of time the wind has been blowing Δt, the average weight and diameter of a sand grain w and d, and the air's density ρ and kinematic viscosity ν.

b) Simplify the result of part a) when the sand-air interface is fully rough and ν is no longer a parameter.

c) If the sand dune height is determined to be proportional to the density of the air, how do you expect it to depend on the weight of a sand grain?

1.47 The rim-to-rim diameter D of the impact crater produced by a vertically-falling object depends on d = average diameter of the object, E = kinetic energy of the object lost on impact, ρ = density of the ground at the impact site, Σ = yield stress of the ground at the impact site, and g = acceleration of gravity.[5]

a) Using dimensional analysis, determine a scaling law for D.

b) Simplify the result of part a) when $D \gg d$, and d is no longer a parameter.

c) Further simplify the result of part b) when the ground plastically deforms to absorb the impact energy and ρ is irrelevant. In this case, does gravity influence D? And, if E is doubled how much bigger is D?

d) Alternatively, further simplify the result of part b) when the ground at the impact site is an unconsolidated material like sand where Σ is irrelevant. In this case, does gravity influence D? And, if E is doubled how much bigger is D?

e) Assume the relevant constant is unity and invert the algebraic relationship found in part d) to estimate the impact energy that formed the 1.2-km-diameter Barringer Meteor Crater in Arizona using the density of Coconino sandstone, 2.3 g/cm^3, at the impact site. The impact energy that formed this crater is likely between 10^{16} and 10^{17} J. How close to this range is your dimensional analysis estimate?

4. Developed from research discussions with Professor R. Breidenthal.
5. The scaling of impact processes to high velocities is likely not this simple (see Holsapple, 1993).

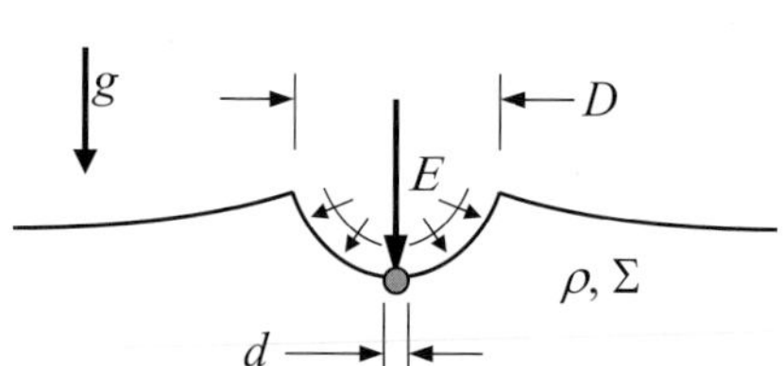

Photograph by D. Roddy of the US Geological Survey

1.48 An isolated nominally spherical bubble with radius R undergoes shape oscillations at frequency f. It is filled with air having density ρ_a and resides in water with density ρ_w and surface tension σ. What frequency ratio should be expected between two isolated bubbles with 2 cm and 4 cm diameters undergoing geometrically similar shape oscillations? If a soluble surfactant is added to the water that lowers σ by a factor of two, by what factor should air bubble oscillation frequencies increase or decrease?

1.49 Contagious diseases are often spread by airborne aerosol particles exhaled from animals and people. Thus, consider how v, the particle descent speed through air, depends on ρ_p, the density of the particle material, d, the nominal diameter of the particle, ρ_a, the density of the air, μ, the viscosity of the air, ϕ, the humidity of the air (dimensionless), and g, the acceleration of gravity.

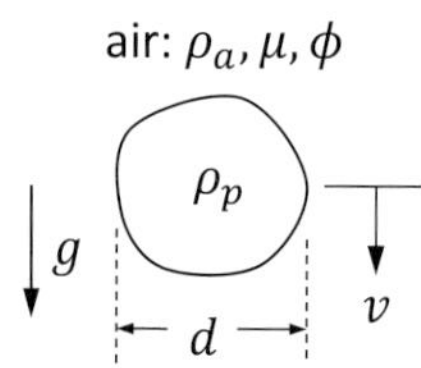

a) Find a dimensionless scaling law that indicates how v depends on the other parameters.

b) Revise your part a) result for small particles where v is proportional to g, and where the inertia of the air (as represented by ρ_a) ceases to be a parameter.

c) Assume that aerosol particles are generated at a fixed distance H above the ground and that the likelihood of disease transmission is proportional to the time a particle stays aloft. Based on this information and your part b) answer, are larger particles more likely, less likely, or equally likely to spread disease compared to smaller particles? Explain your reasoning.

d) Repeat part c) for aqueous particles. Are high humidity conditions more likely, less likely, or equally likely to spread disease compared to low humidity conditions? Explain your reasoning.

1.50 In general, boundary layer skin friction, τ_w, depends on the fluid velocity U above the boundary layer, the fluid density ρ, the fluid viscosity μ, the nominal boundary layer thickness δ, and the surface roughness length scale ε.

a) Generate a dimensionless scaling law for boundary layer skin friction.

b) For laminar boundary layers, the skin friction is proportional to μ. When this is true, how must τ_w depend on U and ρ?

c) For turbulent boundary layers, the dominant mechanisms for momentum exchange within the flow do not directly involve the viscosity μ. Reformulate your dimensional analysis without it. How must τ_w depend on U and ρ when μ is not a parameter?

d) For turbulent boundary layers on smooth surfaces, the skin friction on a solid wall occurs in a viscous sublayer that is very thin compared to δ. In fact, because the boundary layer provides a buffer between the outer flow and this viscous sub-layer, the viscous sublayer thickness l_v does not depend directly on U or δ. Determine how l_v depends on the remaining parameters.

e) Now consider nontrivial roughness. When ε is larger than l_v a surface can no longer be considered fluid-dynamically smooth. Thus, based on the results from parts a) through d) and anything you may know about the relative friction levels in laminar and turbulent boundary layers, are high- or low-speed boundary layer flows more likely to be influenced by surface roughness?

1.51 Turbulent boundary layer skin friction is one of the fluid phenomena that limit the travel speed of aircraft and ships. One means for reducing the skin friction of liquid boundary layers is to inject a gas (typically air) from the surface on which the boundary layer forms. The shear stress, τ_w, that is felt a distance L downstream of such an air injector depends on: the volumetric gas flux per unit span q (in m^2/s), the free stream flow speed U, the liquid density ρ, the liquid viscosity μ, the surface tension σ, and gravitational acceleration g.

a) Formulate a dimensionless law for τ_w in terms of the other parameters.

b) Experimental studies of air injection into liquid turbulent boundary layers on flat plates have found that the bubbles may coalesce to form an air film that provides near perfect lubrication, $\tau_w \to 0$ for $L > 0$, when q is high enough and gravity tends to push the injected gas toward the plate surface. Reformulate your answer to part a) by dropping τ_w and L to determine a dimensionless law for the minimum air injection rate, q_c, necessary to form an air layer.

c) Simplify the result of part b) when surface tension can be neglected.

d) Experimental studies (Elbing et al., 2008) find that q_c is proportional to U^2. Using this information, determine a scaling law for q_c involving the other parameters. Would an increase in g cause q_c to increase or decrease?

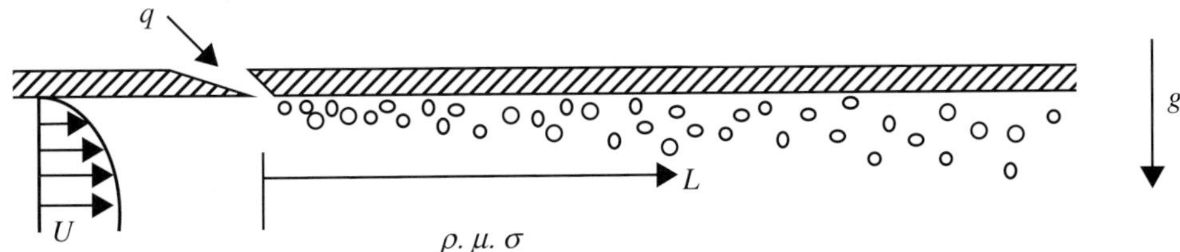

1.52 An industrial cooling system is in the design stage. The pumping requirements are known and the drive motors have been selected. For maximum efficiency the pumps will be directly driven (no gear boxes). The number N_p and type of water pumps are to be determined based on pump efficiency η (dimensionless), the total required volume flow rate Q, the required pressure rise ΔP, the motor rotation rate Ω, and the power delivered by one motor W. Use dimensional analysis and simple physical reasoning for the following items.

a) Determine a formula for the number of pumps.

b) Using Q, N_p, ΔP, Ω, and the density (ρ) and viscosity (μ) of water, create the appropriate number of dimensionless groups using ΔP as the dependent parameter.

c) Simplify the result of part b) by requiring the two extensive variables to appear as a ratio.

d) Simplify the result of part c) for high Reynolds number pumping where μ is no longer a parameter.

e) Manipulate the remaining dimensionless group until Ω appears to the first power in the numerator. This dimensionless group is known as the specific speed, and its value allows the most efficient type of pump to be chosen (see Sabersky et al., 1999).

1.53 Nearly all types of fluid filtration involve pressure driven flow through a porous material.

a) For a given volume flow rate per unit area $= Q/A$, predict how the pressure difference across the porous material $= \Delta p$, depends on the thickness of the filter material $= L$, the surface area per unit volume of the filter material $= \Psi$, and other relevant parameters using dimensional analysis.

b) Often the Reynolds number of the flow in the filter pores is very much less than unity so fluid inertia becomes unimportant. Redo the dimensional analysis for this situation.

c) To minimize pressure losses in heating, ventilating, and air-conditioning (HVAC) ductwork, should hot or cold air be filtered?

d) If the filter material is changed and Ψ is lowered to one half its previous value, estimate the change in Δp if all other parameters are constant. (Hint: make a reasonable assumption about the dependence of Δp on L; they are both *extensive* variables in this situation).

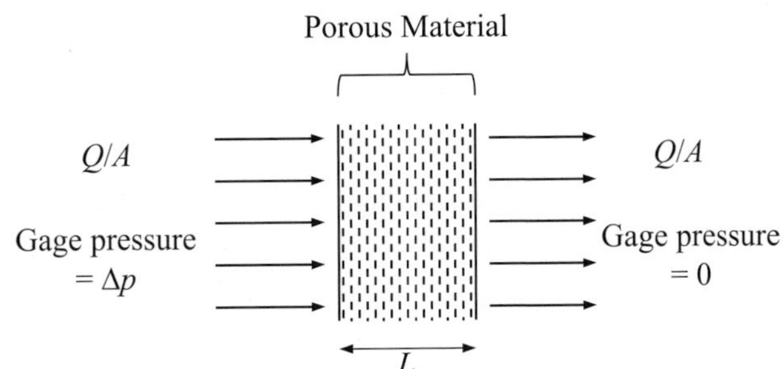

1.54 Falling precipitation particles (especially snowflakes) sometimes hit the windshield of a moving vehicle and sometimes do not. To explain this observation, consider a sparse distribution of non-interacting particles with mass m and size l that fall through air with density ρ and viscosity μ under the action of vertical body force per unit mass a.

a) Using dimensional analysis, determine a scaling law for the vertical velocity v_n of the particles in terms of m, l, ρ, μ, and a.

b) First consider hail stones that fall at high speeds; simplify the result of part a) for turbulent flow where μ is no longer a parameter.

c) Next consider fine droplets or snowflakes that fall slowly; simplify the result of part a) for laminar flow where μ is a parameter and v_n is proportional to the product ma.

Now add a nominally horizontal air flow that carries the particles at speed U over the hood and windshield of a vehicle. In this case, the near-windshield streamlines have radius of curvature R and the vertical displacement of the particles from the over-hood-and-windshield streamline while traveling over the vehicle's hood and windshield is δ (as shown). Clearly, when δ is large enough, the particles will hit the vehicle's windshield. For the items below, assume δ will be proportional to the product of v_n from part c) and the nominal particle transit time over the vehicle's hood and windshield, R/U.

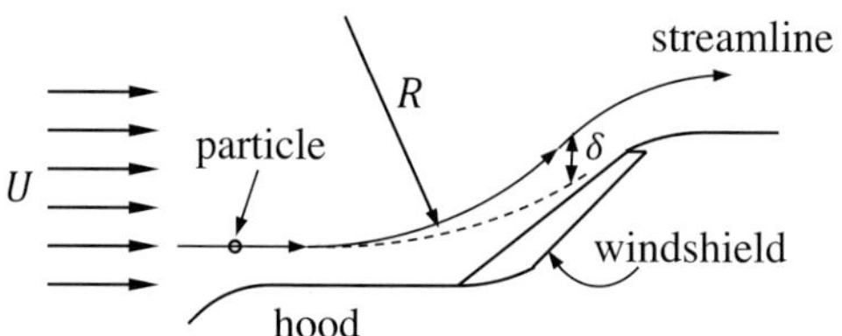

d) How does δ depend on the other parameters at low air speeds where a is equal to gravity g? In this scaling regime, are snowflakes more, less, or equally likely to hit the windshield as the air speed increases?

e) Repeat part d) for high air speeds where a is equal to U^2/R.

f) For constant m and any vehicle speed, are larger size particles more, less, or equally likely to hit the windshield compared to smaller size particles?

1.55 A new industrial process requires a volume V of hot air with initial density ρ to be moved quickly from a spherical reaction chamber to a larger evacuated chamber using a single pipe of length L and interior diameter of d. The vacuum chamber is also spherical and has a volume of V_f. If the hot air cannot be transferred fast enough, the process fails. Thus, a prediction of the transfer time t is needed based on these parameters, the air's ratio of specific heats γ, and initial values of the air's speed of sound c and viscosity μ.

a) Formulate a dimensionless scaling law for t, involving six dimensionless groups.

b) Inexpensive small-scale tests of the air-transfer process are untaken before construction of the commercial-scale reaction facility. Can all these dimensionless groups be matched if the target size for the pipe diameter in the small-scale tests is $d' = d/10$? Would lowering or raising the initial air temperature in the small-scale experiments help match the dimensionless numbers?

1.56 Create a small passive helicopter from ordinary photocopy-machine paper (as shown) and drop it from a height of 2 m or so. Note the helicopter's rotation and descent rates once it's rotating steadily. Repeat this simple experiment with different sizes of paper clips to change the helicopter's weight, and observe changes in the rotation and decent rates.

a) Using the helicopter's weight W, blade length l, and blade width (chord) c, and the air's density ρ and viscosity μ as independent parameters, formulate two independent dimensionless scaling laws for the helicopter's rotation rate Ω, and decent rate dz/dt.

b) Simplify both scaling laws for the situation where μ is no longer a parameter.

c) Do the dimensionless scaling laws correctly predict the experimental trends?

d) If a new paper helicopter is made with all dimensions smaller by a factor of two. Use the scaling laws found in part b) to predict changes in the rotation and decent rates. Make the new smaller paper helicopter and see if the predictions are correct.

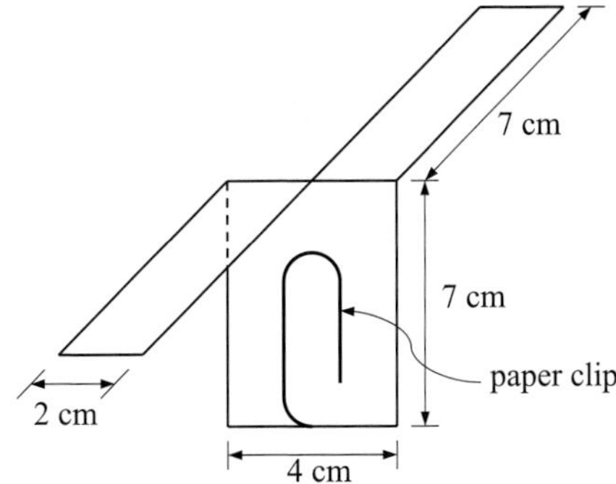

References

Chapman, S., Cowling, T.G., 1970. The Mathematical Theory of Non-Uniform Gases. University Press, Cambridge, pp. 67–73.

Elbing, B.R., Winkel, E.S., Lay, K.A., Ceccio, S.L., Dowling, D.R., Perlin, M., 2008. Bubble-induced skin friction drag reduction and the abrupt transition to air-layer drag reduction. Journal of Fluid Mechanics 612, 201–236.

Holsapple, K.A., 1993. The scaling of impact processes in planetary sciences. Annual Review of Earth and Planetary Sciences 21, 333–373.

Sabersky, R.H., Acosta, A.J., Hauptmann, E.G., Gates, E.M., 1999. Fluid Flow, 4th edition. Prentice-Hall, New Jersey, pp. 511–516.

Taylor, G.I., 1950. The formation of blast wave by a very intense explosion. Parts I and II. Proceedings of the Royal Society of London, Series A 201, 159–174. 175–186.

Further reading

Batchelor, G.K., 1967. An Introduction to Fluid Dynamics. Cambridge University Press, London (A detailed discussion of classical thermodynamics, kinetic theory of gases, surface tension effects, and transport phenomena is given).

Bridgeman, P.W., 1963. Dimensional Analysis. Yale University Press, New Haven.

Çengel, Y.A., Cimbala, J.M., 2010. Fluid Mechanics: Fundamentals and Applications, 2nd edition. McGraw-Hill, New York (A fine undergraduate fluid mechanics textbook. Chapter 7 covers dimensional analysis).

Fox, R.W., Pritchard, P.J., MacDonald, A.T., 2009. Introduction to Fluid Mechanics, 7th edition. John Wiley, New York (A fine undergraduate fluid mechanics textbook. Chapter 7 covers dimensional analysis).

Hatsopoulos, G.N., Keenan, J.H., 1981. Principles of General Thermodynamics. Krieger Publishing Co, Melbourne, FL (This is a good text on thermodynamics).

Prandtl, L., Tietjens, O.G., 1934. Fundamentals of Hydro- and Aeromechanics. Dover Publications, New York (A clear and simple discussion of potential and adiabatic temperature gradients is given).

Taylor, G.I., 1974. The interaction between experiment and theory in fluid mechanics. Annual Review of Fluid Mechanics 6, 1–16.

Tritton, D.J., 1988. Physical Fluid Dynamics, 2nd edition. Clarendon Press, Oxford (A well-written book with broad coverage of fluid mechanics).

Vincenti, W.G., Kruger, C.H., 1965. Introduction of Physical Gas Dynamics. Krieger, Malabar, Florida (A fine reference for kinetic theory and statistical mechanics).

Chapter 2

Cartesian tensors⊛

Contents

2.1 Scalars, vectors, tensors, notation 37
2.2 Rotation of axes; formal definition of a vector 39
2.3 Multiplication of matrices 40
2.4 Second-order tensors 41
2.5 Contraction and multiplication 43
2.6 Force on a surface 44
2.7 Kronecker delta and alternating tensor 46
2.8 Vector dot and cross products 47
2.9 Gradient, divergence, and curl 48
2.10 Symmetric and antisymmetric tensors 50
2.11 Eigenvalues and eigenvectors of a symmetric tensor 51
2.12 Gauss' theorem 52
2.13 Stokes' theorem 55
Exercises 56
References 58
Further reading 58

Chapter objectives

- To define the notation used in this text for scalars, vectors, and tensors
- To review the basic algebraic manipulations of vectors and matrices
- To present how vector differentiation is applied to scalars, vectors, and tensors
- To review the fundamental theorems of vector field theory

2.1 Scalars, vectors, tensors, notation

The physical quantities in fluid mechanics vary in their complexity, and may involve multiple spatial directions. The notation and proper specification of these quantities as *scalars*, *vectors*, and (second-order) *tensors* is the subject of this chapter. Here, three independent spatial dimensions are assumed to exist. The various results may be simplified to fewer, or extended to more, independent spatial dimensions.

Scalars or zero-order tensors may be defined with a magnitude and appropriate units, and may vary with spatial location, but are independent of coordinate directions. Scalars are typically denoted herein by italicized symbols. Common scalars in fluid mechanics are pressure p, temperature T, and density ρ.

Vectors or first-order tensors have both a magnitude and a direction. A vector can be completely described by its components along three orthogonal coordinate directions, and these components may change with a change in coordinate system. A vector is usually denoted herein by a boldface symbol. Common vectors in fluid mechanics are position **x**, fluid velocity **u**, and gravitational acceleration **g**. In a Cartesian coordinate system with unit vectors $\mathbf{e}_1$, $\mathbf{e}_2$, and $\mathbf{e}_3$, in the three mutually perpendicular directions, the position vector **x**, OP in Fig. 2.1, may be written:

$$\mathbf{x} = x_1\mathbf{e}_1 + x_2\mathbf{e}_2 + x_3\mathbf{e}_3, \tag{2.1}$$

where x_1, x_2, and x_3 are the components of **x** along each Cartesian axis. Here, the subscripts of **e** do *not* denote vector components but rather reference the coordinate axes 1, 2, 3; hence, each **e** is a vector itself. Sometimes, to save writing, the components of a vector are denoted with an italic symbol having one index – such as i, j, k, etc. – that implicitly is known to take on three possible values: 1, 2, or 3. For example, the components of **x** can be denoted by x_i or x_j (or x_k, etc.). A vector is often written as a column matrix; thus, (2.1) is consistent with the following vector specifications:

$$\mathbf{x} = \begin{bmatrix} x_1 \\ x_2 \\ x_3 \end{bmatrix} \text{ where } \mathbf{e}_1 = \begin{bmatrix} 1 \\ 0 \\ 0 \end{bmatrix}, \mathbf{e}_2 = \begin{bmatrix} 0 \\ 1 \\ 0 \end{bmatrix}, \text{ and } \mathbf{e}_3 = \begin{bmatrix} 0 \\ 0 \\ 1 \end{bmatrix}.$$

⊛. This book has a companion website hosting complementary materials. Visit this URL to access it: https://www.elsevier.com/books-and-journals/book-companion/9780128198070.

Fluid Mechanics. https://doi.org/10.1016/B978-0-12-819807-0.00011-9

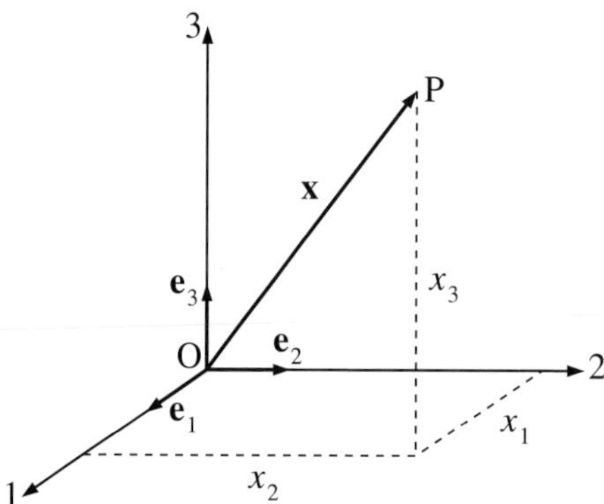

FIGURE 2.1 Position vector OP and its three Cartesian components (x_1, x_2, x_3). The three unit vectors for the coordinate directions are $\mathbf{e}_1$, $\mathbf{e}_2$, and $\mathbf{e}_3$. Once the coordinate system is chosen, the vector **x** is completely defined by its components, x_i where $i = 1, 2$, or 3.

The *transpose* of a matrix (denoted by a T-superscript) is obtained by interchanging rows and columns, so the transpose of the column matrix **x** is the row matrix: $\mathbf{x}^T = [x_1\ x_2\ x_3]$. However, to save space in the text, the square-bracket notation for vectors shown here is typically replaced by triplets (or doublets) of values separated by commas and placed inside ordinary parentheses, for example $\mathbf{x} = (x_1, x_2, x_3)$.

Second-order tensors have a component for each *pair* of coordinate directions and therefore may have as many as $3 \times 3 = 9$ separate components. A second-order tensor is sometimes denoted by a boldface symbol. A common second-order tensor in fluid mechanics is the stress **T**. Like vector components, second-order tensor components change with a change in coordinate system. Once a coordinate system is chosen, the nine components of a second-order tensor can be represented by a 3×3 matrix, or by an italic symbol having two indices, such as T_{ij} for the stress tensor. Here again the indices i and j are known implicitly to separately take on the values 1, 2, or 3. Second-order tensors are further discussed in Section 2.4.

A second implicit feature of *index-based* or *indicial* notation is the implied sum over a repeated index in terms involving multiple indices. This notational convention can be stated as follows: *whenever an index is repeated in a term, a summation over this index is implied, even though no summation sign is explicitly written.* This notational convention saves writing and increases mathematical precision when dealing with products of first- and higher-order tensors. It was introduced by Albert Einstein in 1916 and is sometimes referred to as the *Einstein summation convention*. It can be illustrated by a simple example involving the ordinary dot product of two vectors **u** and **v** having components u_i and v_j, respectively. Their dot product is the sum of component products:

$$\mathbf{u} \cdot \mathbf{v} = \mathbf{v} \cdot \mathbf{u} = u_1 v_1 + u_2 v_2 + u_3 v_3 = \sum_{i=1}^{3} u_i v_i \equiv u_i v_i, \tag{2.2}$$

where the final three-line *definition* equality (≡) follows from the repeated-index implied-sum convention. Since this notational convention is unlikely to be comfortable to the reader after a single exposure, it is repeatedly illustrated via definition equalities in this chapter before being adopted in the remainder of this text wherever indicial notation is used.

Both boldface (aka *vector* or *dyadic*) and indicial (aka *tensor*) notations are used throughout this text. Boldface notation is simpler because there are no subscripts to consider. However, algebraic manipulations may not be distinct in boldface notation since the product **uv** may not be well defined nor equal to **vu** when **u** and **v** are second-order tensors. Boldface notation has other problems too; for example, the order of a tensor is not clear if one simply denotes it **u**.

Indicial notation avoids these problems because it indicates tensor *components*, which are *scalars*. Algebraic manipulations are simpler and better defined, and special attention to the ordering of terms is unnecessary (unless differentiation is involved). In addition, the number of indices or subscripts clearly specifies the order of a tensor. However, the physical structure and meaning of terms written with index notation only become apparent after an examination of the indices. Hence, indices must be clearly written to prevent mistakes and to promote proper understanding of the terms they help define. In addition, the cross product involves the possibly cumbersome alternating tensor ε_{ijk} as described in Sections 2.7 and 2.9.

Example 2.1

If $|\mathbf{a}|^2 \equiv a_i a_i = a_i^2$ defines the magnitude of a vector, write the Cauchy–Schwartz inequality, $(\mathbf{a} \cdot \mathbf{b})^2 \le |\mathbf{a}|^2 |\mathbf{b}|^2$, in index notation.

Solution Use (2.2) for the dot product, $\mathbf{a}\cdot\mathbf{b}=a_ib_i$, and the given information to reach: $(a_ib_i)^2 \le a_j^2 b_k^2$. Here the squares on the right side cause the two indices j and k to be repeated, so they are summed over.

2.2 Rotation of axes; formal definition of a vector

A vector can be formally defined as any quantity whose components change similarly to the three components of position under rotation of the coordinate system. Let O123 be the original coordinate system, and O1′2′3′ be the rotated system that shares the same origin O (see Fig. 2.2). The position vector **x** can be written in either coordinate system:

$$\mathbf{x}=x_1\mathbf{e}_1+x_2\mathbf{e}_2+x_3\mathbf{e}_3, \quad \text{or} \quad \mathbf{x}=x_1'\mathbf{e}_1'+x_2'\mathbf{e}_2'+x_3'\mathbf{e}_3', \tag{2.1, 2.3}$$

where the components of **x** in O123 and O1′2′3′ are x_i and x_j', respectively, and the $\mathbf{e}_j'$ are the unit vectors in O1′2′3′. Forming a dot product of **x** with $\mathbf{e}_1'$, and using both (2.1) and (2.3) produces:

$$\mathbf{x}\cdot\mathbf{e}_1'=x_1\mathbf{e}_1\cdot\mathbf{e}_1'+x_2\mathbf{e}_2\cdot\mathbf{e}_1'+x_3\mathbf{e}_3\cdot\mathbf{e}_1'=x_1', \tag{2.4}$$

where the dot products between unit vectors are direction cosines; $\mathbf{e}_1\cdot\mathbf{e}_1'$ is the cosine of the angle between the 1 and 1′ axes, $\mathbf{e}_2\cdot\mathbf{e}_1'$ is the cosine of the angle between the 2 and 1′ axes, and $\mathbf{e}_3\cdot\mathbf{e}_1'$ is the cosine of the angle between the 3 and 1′ axes. These three angles are indicated in Fig. 2.2. Forming the dot products $\mathbf{x}\cdot\mathbf{e}_2'=x_2'$ and $\mathbf{x}\cdot\mathbf{e}_3'=x_3'$, and then combining these results with (2.3) produces:

$$x_j'=x_1C_{1j}+x_2C_{2j}+x_3C_{3j}=\sum_{i=1}^{3}x_iC_{ij}\equiv x_iC_{ij}, \tag{2.5}$$

where $C_{ij}=\mathbf{e}_i\cdot\mathbf{e}_j'$ is a 3×3 matrix of direction cosines and the definition equality follows from the summation convention. In (2.5), the *free* or not-summed-over index is j, while the *repeated* or summed-over index can be any letter other than j. Thus, the rightmost term in (2.5) could equally well have been written x_kC_{kj} or x_mC_{mj}. Similarly, any letter can also be used for the free index, as long as the same free index is used on *both* sides of the equation. For example, denoting the free index by i and the summed index by k allows (2.5) to be written with indicial notation as:

$$x_i'=x_kC_{ki}. \tag{2.6}$$

This index-choice flexibility exists because the three algebraic equations represented by (2.5), corresponding to the three values of j, are the same as those represented by (2.6) for the three values of i.

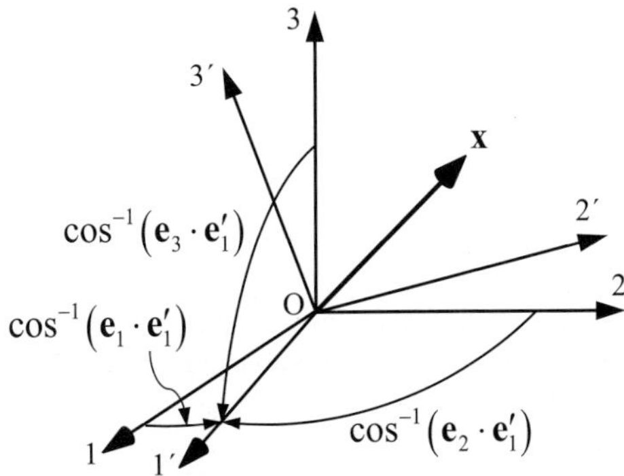

FIGURE 2.2 A rotation of the original Cartesian coordinate system O123 to a new system O1′2′3′. Here the **x** vector is unchanged, but its components in the original system x_i and in the rotated system x_i' will not be the same. The angles between the 1′ axis and each axis in the O123 coordinate system are shown and are determined from the dot products of the unit vectors.

The components of **x** in O123 are related to those in O1′2′3′ by:

$$x_j=\sum_{i=1}^{3}x_i'C_{ji}\equiv x_i'C_{ji}. \tag{2.7}$$

(see Exercise 2.2). The indicial positions on the right side of this relation are different from those in (2.5), because the first index of C_{ij} is summed in (2.5), whereas the second index of C_{ij} is summed in (2.7).

We can now formally define a Cartesian vector as any quantity that transforms like the position vector under rotation of the coordinate system. Therefore, by analogy with (2.5), **u** is a vector if its components transform as:

$$u'_j = \sum_{i=1}^{3} u_i C_{ij} \equiv u_i C_{ij}. \tag{2.8}$$

Example 2.2

Convert the two-dimensional vector $\mathbf{u} = (u_1, u_2)$ from Cartesian (x_1, x_2) to polar (r, θ) coordinates (see Fig. 3.3a).

Solution Clearly, **u** can be represented in either coordinate system: $\mathbf{u} = u_1\mathbf{e}_1 + u_2\mathbf{e}_2 = u_r\mathbf{e}_r + u_\theta\mathbf{e}_\theta$, where u_r and u_θ are the components in polar coordinates, and $\mathbf{e}_r$ and $\mathbf{e}_\theta$ are the unit vectors in polar coordinates. Here the polar coordinate system is rotated compared to the Cartesian system, as illustrated in Fig. 2.3. Forming the dot product of the above equation with $\mathbf{e}_r$ and $\mathbf{e}_\theta$ produces two algebraic equations that are equivalent to (2.5):

$$u_r = u_1\mathbf{e}_1 \cdot \mathbf{e}_r + u_2\mathbf{e}_2 \cdot \mathbf{e}_r, \text{ and } u_\theta = u_1\mathbf{e}_1 \cdot \mathbf{e}_\theta + u_2\mathbf{e}_2 \cdot \mathbf{e}_\theta,$$

with subscripts r and θ replacing $j = 1$ and 2 in (2.5). Evaluation of the unit-vector dot products leads to:

$$u_r = u_1 \cos\theta + u_2 \cos\left(\frac{\pi}{2} - \theta\right) = u_1 \cos\theta + u_2 \sin\theta, \text{ and}$$
$$u_\theta = u_1 \cos\left(\theta + \frac{\pi}{2}\right) + u_2 \cos\theta = -u_1 \sin\theta + u_2 \cos\theta.$$

Thus, in this case:

$$C_{ij} = \begin{bmatrix} \mathbf{e}_1 \cdot \mathbf{e}_r & \mathbf{e}_1 \cdot \mathbf{e}_\theta \\ \mathbf{e}_2 \cdot \mathbf{e}_r & \mathbf{e}_2 \cdot \mathbf{e}_\theta \end{bmatrix} = \begin{bmatrix} \cos\theta & -\sin\theta \\ \sin\theta & \cos\theta \end{bmatrix}.$$

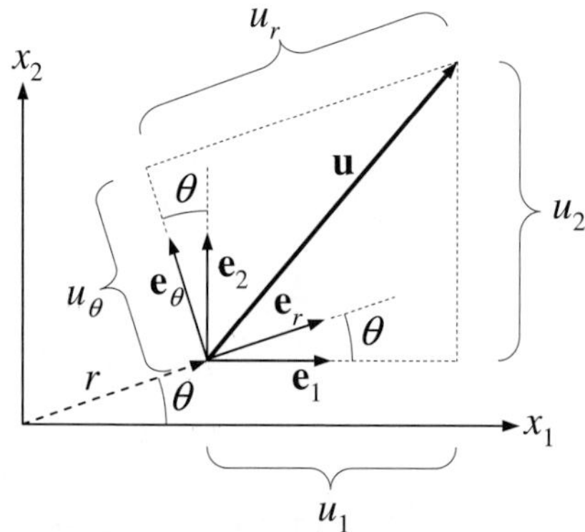

FIGURE 2.3 Resolution of a two-dimensional vector **u** in (x_1, x_2)-Cartesian and (r, θ)-polar coordinates. The angle between the $\mathbf{e}_1$ and $\mathbf{e}_r$ unit vectors, and the $\mathbf{e}_2$ and $\mathbf{e}_\theta$ unit vectors is θ. The angle between $\mathbf{e}_r$ and $\mathbf{e}_2$ unit vectors is $\pi/2 - \theta$, and the angle between the $\mathbf{e}_1$ and $\mathbf{e}_\theta$ unit vectors is $\pi/2 + \theta$. Here **u** does not emerge from the origin of coordinates (as in Fig. 2.2), but it may be well defined in either coordinate system even though its components are not the same in the (x_1, x_2)- and (r, θ)-coordinates.

2.3 Multiplication of matrices

Let **A** and **B** be two 3 × 3 matrices. The inner product of **A** and **B** is defined as the matrix **P** whose elements are related to those of **A** and **B** by:

$$P_{ij} = \sum_{k=1}^{3} A_{ik} B_{kj} \equiv A_{ik} B_{kj}, \quad \text{or} \quad \mathbf{P} = \mathbf{A} \cdot \mathbf{B}, \tag{2.9, 2.10}$$

where the definition equality in (2.9) follows from the summation convention, and the single dot between **A** and **B** in (2.10) signifies that a single index is summed to find **P**. An important feature of (2.9) is that the elements are summed over the inner or *adjacent* index k. In explicit form, (2.9) is written as:

$$\begin{bmatrix} P_{11} & P_{12} & P_{13} \\ P_{21} & P_{22} & P_{23} \\ P_{31} & P_{32} & P_{33} \end{bmatrix} = \begin{bmatrix} A_{11} & A_{12} & A_{13} \\ A_{21} & A_{22} & A_{23} \\ A_{31} & A_{32} & A_{33} \end{bmatrix} \begin{bmatrix} B_{11} & B_{12} & B_{13} \\ B_{21} & B_{22} & B_{23} \\ B_{31} & B_{32} & B_{33} \end{bmatrix}. \qquad (2.11)$$

This equation signifies that the ij-element of **P** is determined by multiplying the elements in the i-row of **A** and the j-column of **B**, and summing. For example:

$$P_{12} = A_{11}B_{12} + A_{12}B_{22} + A_{13}B_{32},$$

as indicated by the dashed-line boxes in (2.11). Naturally, the inner product $\mathbf{A}\cdot\mathbf{B}$ is only defined if the number of columns of **A** equals the number of rows of **B**.

Eq. (2.9) also applies to the inner product of a 3×3 matrix and a column vector. For example, (2.6) can be written as $x'_i = C^T_{ik}x_k$, which is now of the form of (2.9) because the summed index k is adjacent. In matrix form, (2.6) can therefore be written as:

$$\begin{bmatrix} x'_1 \\ x'_2 \\ x'_3 \end{bmatrix} = \begin{bmatrix} C_{11} & C_{12} & C_{13} \\ C_{21} & C_{22} & C_{23} \\ C_{31} & C_{32} & C_{33} \end{bmatrix}^{\mathrm{T}} \begin{bmatrix} x_1 \\ x_2 \\ x_3 \end{bmatrix}.$$

Symbolically, the preceding is $\mathbf{x}' = \mathbf{C}^{\mathrm{T}} \cdot \mathbf{x}$, whereas (2.7) is $\mathbf{x} = \mathbf{C} \cdot \mathbf{x}'$.

Example 2.3

Together (2.6) and (2.7) imply that $C^T_{ik}C_{kj}$ is the identity matrix. Show this to be true for the C_{ij} found in Example 2.2.

Solution First, rewrite $C^T_{ik}C_{kj}$ as a matrix product, and transpose the first matrix.

$$C^T_{ik}C_{kj} = \begin{bmatrix} \cos\theta & -\sin\theta \\ \sin\theta & \cos\theta \end{bmatrix}^T \begin{bmatrix} \cos\theta & -\sin\theta \\ \sin\theta & \cos\theta \end{bmatrix} = \begin{bmatrix} \cos\theta & \sin\theta \\ -\sin\theta & \cos\theta \end{bmatrix} \begin{bmatrix} \cos\theta & -\sin\theta \\ \sin\theta & \cos\theta \end{bmatrix}.$$

Perform the indicated multiplications and use the trigonometric identity, $\cos^2\theta + \sin^2\theta = 1$, to reach the desired result.

$$C^T_{ik}C_{kj} = \begin{bmatrix} \cos^2\theta + \sin^2\theta & \cos\theta(-\sin\theta) + \sin\theta\cos\theta \\ -\sin\theta\cos\theta + \cos\theta\sin\theta & (-\sin\theta)^2 + \cos^2\theta \end{bmatrix} = \begin{bmatrix} 1 & 0 \\ 0 & 1 \end{bmatrix}.$$

2.4 Second-order tensors

A simple-to-complicated hierarchical description of physically meaningful quantities starts with scalars, proceeds to vectors, and then continues to second- and higher-order tensors. A scalar can be represented by a single value. A vector can be represented by three components, one for each of three orthogonal spatial directions denoted by a single free index. A second-order tensor can be represented by nine components, one for each pair of directions and denoted by two free indices. Nearly all the tensors considered in Newtonian fluid mechanics are zero-, first-, or second-order tensors.

To better understand the structure of second-order tensors, consider the stress tensor **T** or T_{ij}. Its two free indices specify two directions; the first indicates the orientation of the *surface* on which the stress is applied while the second indicates the component of the *force per unit area* on that surface. In particular, the first (i) index of T_{ij} denotes the direction of the surface normal, and the second (j) index denotes the force component direction. This situation is illustrated in Fig. 2.4, which shows the normal and shear stresses on an infinitesimal cube having surfaces parallel to the coordinate planes. The stresses are positive if they are directed as shown in this figure. The sign convention is that, on a surface whose outward normal points in the positive direction of a coordinate axis, the normal and shear stresses are positive if they point in the

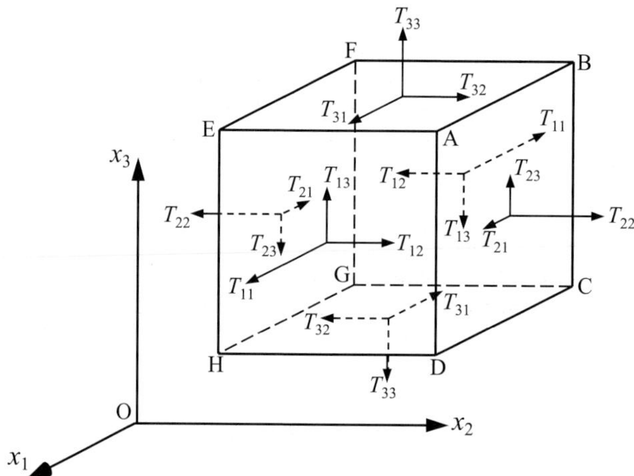

FIGURE 2.4 Illustration of the stress field at a point via stress components on a cubic volume element. Here each surface may experience one normal and two shear components of stress. The directions of positive normal and shear stresses are shown.

positive directions of the other axes. For example, on the surface ABCD, whose outward normal points in the positive x_2 direction, the positive stresses T_{21}, T_{22}, and T_{23} point in the x_1, x_2, and x_3 directions, respectively. Normal stresses are positive if they are tensile and negative if they are compressive. On the opposite face EFGH the stress components have the same value as on ABCD, but their directions are reversed. This is because Fig. 2.4 represents stresses *at a point*. The cube shown is intended to be vanishingly small, so that the faces ABCD and EFGH are just opposite sides of a plane perpendicular to the x_2-axis. Thus, stresses on the opposite faces are equal and opposite, and satisfy Newton's third law.

A vector **u** is completely specified by the three components u_i (where $i = 1, 2, 3$) because the components of **u** in any direction other than the original axes can be found from (2.8). Similarly, the state of stress at a point can be completely specified by the nine components T_{ij} (where $i, j = 1, 2, 3$) that can be written as the matrix:

$$\mathbf{T} = \begin{bmatrix} T_{11} & T_{12} & T_{13} \\ T_{21} & T_{22} & T_{23} \\ T_{31} & T_{32} & T_{33} \end{bmatrix}.$$

The specification of these nine stress components on surfaces perpendicular to the coordinate axes completely determines the state of stress at a point because the stresses on any arbitrary plane can be determined from them. To find the stresses on any arbitrary surface, we can consider a rotated coordinate system $O1'2'3'$ having one axis perpendicular to the given surface. It can be shown by a force balance on a tetrahedron element (see, e.g., Sommerfeld, 1964, page 59) that the components of **T** in the rotated coordinate system are:

$$T'_{mn} = \sum_{i=1}^{3}\sum_{j=1}^{3} C_{im}C_{jn}T_{ij} \equiv C_{im}C_{jn}T_{ij}, \tag{2.12}$$

where the definition equality follows from the summation convention. This equation may also be written as: $T'_{mn} = C^{\mathrm{T}}_{mi}T_{ij}C_{jn}$ or $\mathbf{T}' = \mathbf{C}^{\mathrm{T}} \cdot \mathbf{T} \cdot \mathbf{C}$. Note the similarity between the vector transformation rule (2.8) and (2.12). In (2.8) the first index of **C** is summed, while its second index is free. Eq. (2.12) is identical, except that **C** is used twice. A quantity that obeys (2.12) is called a *second-order tensor*.

Tensor and matrix concepts are not quite the same. A matrix is any *arrangement* of elements, written as an array. The elements of a matrix represent the components of a second-order tensor only if they obey (2.12). In general, tensors can be of any order and the number of free indices corresponds to the order of the tensor. For example, **A** is a fourth-order tensor if it has four free indices, and the associated $3^4 = 81$ components change under a rotation of the coordinate system according to:

$$A'_{mnpq} = \sum_{i=1}^{3}\sum_{j=1}^{3}\sum_{k=1}^{3}\sum_{l=1}^{3} C_{im}C_{jn}C_{kp}C_{lq}A_{ijkl} \equiv C_{im}C_{jn}C_{kp}C_{lq}A_{ijkl}, \tag{2.13}$$

where again the definition equality follows from the summation convention. Tensors of various orders arise in fluid mechanics. Common second-order tensors are the stress tensor T_{ij} and the velocity-gradient tensor $\partial u_i/\partial x_j$. The nine products $u_i v_j$ formed from the components of the two vectors **u** and **v** also transform according to (2.12), and therefore form a second-order tensor. The product of two vectors to form a tensor (aka dyadic product) is sometimes represented using boldface notation as $\mathbf{u} \otimes \mathbf{v}$. In addition, the *Kronecker-delta* and *alternating tensors* are also frequently used; these are defined and discussed in Section 2.7.

Example 2.4

Consider two orthogonal vectors in two dimensions: $\mathbf{u} = (U, 0)$ and $\mathbf{v} = (0, V)$ that are resolved in a coordinate system rotated by angle θ. Using the C_{ij}-rotation matrix from Example 2.1, show that $u'_i v'_j = (u_i v_j)'$, when the vectors u'_i and v'_j are obtained from (2.5), and the second-order tensor $(u_i v_j)'$ is obtained from (2.12).

Solution First, obtain the vectors u'_i and v'_j in the rotated coordinate system using (2.5):

$$u'_i = C_{ki}u_k = C^T_{ik}u_k = \begin{bmatrix} \cos\theta & \sin\theta \\ -\sin\theta & \cos\theta \end{bmatrix}\begin{bmatrix} U \\ 0 \end{bmatrix} = \begin{bmatrix} U\cos\theta \\ -U\sin\theta \end{bmatrix}, \text{ and}$$

$$v'_j = C_{kj}u_k = C^T_{jk}u_k = \begin{bmatrix} \cos\theta & \sin\theta \\ -\sin\theta & \cos\theta \end{bmatrix}\begin{bmatrix} 0 \\ V \end{bmatrix} = \begin{bmatrix} V\sin\theta \\ V\cos\theta \end{bmatrix}.$$

The product of these two vectors is:

$$u'_i v'_j = \begin{bmatrix} u'_1 v'_1 & u'_1 v'_2 \\ u'_2 v'_1 & u'_2 v'_2 \end{bmatrix} = UV\begin{bmatrix} \cos\theta\sin\theta & \cos^2\theta \\ -\sin^2\theta & -\sin\theta\cos\theta \end{bmatrix}.$$

Now use (2.12) to find $(u_i v_j)'$ noting that $u_i v_j$ has only one non-zero component, $u_1 v_2 = UV$:

$$(u_m v_n)' = C_{im}C_{jn}u_i v_j = C^T_{mi}u_i v_j C_{jn} = \begin{bmatrix} \cos\theta & \sin\theta \\ -\sin\theta & \cos\theta \end{bmatrix}\begin{bmatrix} 0 & UV \\ 0 & 0 \end{bmatrix}\begin{bmatrix} \cos\theta & -\sin\theta \\ \sin\theta & \cos\theta \end{bmatrix} \text{ or}$$

$$(u_m v_n)' = \begin{bmatrix} \cos\theta & \sin\theta \\ -\sin\theta & \cos\theta \end{bmatrix}\begin{bmatrix} UV\sin\theta & UV\cos\theta \\ 0 & 0 \end{bmatrix} = UV\begin{bmatrix} \cos\theta\sin\theta & \cos^2\theta \\ -\sin^2\theta & -\sin\theta\cos\theta \end{bmatrix},$$

which is the desired result.

2.5 Contraction and multiplication

When the two indices of a tensor are equated, and a summation is performed over this repeated index, the process is called *contraction*. An example is:

$$\sum_{j=1}^{3} A_{jj} \equiv A_{jj} = A_{11} + A_{22} + A_{33},$$

which is the sum of the diagonal terms of A_{ij}. The sum A_{jj} is a scalar and is independent of the coordinate system. In other words, A_{jj} is an *invariant*. (There are three independent invariants of a second-order tensor, and A_{jj} is one of them; see Exercise 2.11.)

Higher-order tensors can be formed by multiplying lower-order tensors. If **A** and **B** are two second-order tensors, then the 81 numbers defined by $P_{ijkl} \equiv A_{ij}B_{kl}$ transform according to (2.13), and therefore form a fourth-order tensor.

Lower-order tensors can be obtained by performing a contraction within a multiplied form. The four possible contractions of $A_{ij}B_{kl}$ are:

$$\sum_{i=1}^{3} A_{ij}B_{ki} \equiv A_{ij}B_{ki} = B_{ki}A_{ij}, \sum_{i=1}^{3} A_{ij}B_{ik} \equiv A_{ij}B_{ik} = A_{ji}^{T}B_{ik},$$
$$\sum_{j=1}^{3} A_{ij}B_{kj} \equiv A_{ij}B_{kj} = A_{ij}B_{jk}^{T}, \text{ and } \sum_{j=1}^{3} A_{ij}B_{jk} \equiv A_{ij}B_{jk}, \tag{2.14}$$

where all the definition equalities follow from the summation convention. All four products in (2.14) are second-order tensors. Note also in (2.14) how the terms have been rearranged until the summed index is adjacent; at this point they can be written as a product of matrices.

The contracted product of a second-order tensor **A** and a vector **u** is a vector. The two possibilities are:

$$\sum_{j=1}^{3} A_{ij}u_j \equiv A_{ij}u_j, \text{ and } \sum_{i=1}^{3} A_{ij}u_i \equiv A_{ij}u_i = A_{ji}^{T}u_i,$$

where again the definition equalities follow from the summation convention. The doubly contracted product of two second-order tensors **A** and **B** is a scalar. Using all three notations, the two possibilities are:

$$\sum_{i=1}^{3}\sum_{j=1}^{3} A_{ij}B_{ij} \equiv A_{ij}B_{ij}(=\mathbf{A{:}B}) \text{ and } \sum_{i=1}^{3}\sum_{j=1}^{3} A_{ij}B_{ij} \equiv A_{ij}B_{ij}\left(=\mathbf{A{:}B}^{\mathrm{T}}\right),$$

where the bold colon (**:**) implies a *double* contraction or double dot product.

Example 2.5

The net surface force F_j per unit volume on a fluid element is the vector derivative, $\partial/\partial x_i$, of the stress tensor T_{ij}. Determine the three components of the vector F_j.

Solution Here the vector derivative, $\partial/\partial x_i$, must appear to the left of the stress tensor T_{ij}. Thus, for F_j to be expressed in terms of matrix multiplication, it must be written as a row vector multiplied by the stress matrix.

$$F_j = \frac{\partial T_{ij}}{\partial x_i} \equiv \sum_{i=1}^{3}\frac{\partial}{\partial x_i}T_{ij} = \begin{bmatrix} \frac{\partial}{\partial x_1} & \frac{\partial}{\partial x_2} & \frac{\partial}{\partial x_3} \end{bmatrix}\begin{bmatrix} T_{11} & T_{12} & T_{13} \\ T_{21} & T_{22} & T_{23} \\ T_{31} & T_{32} & T_{33} \end{bmatrix}$$
$$= \begin{bmatrix} F_1 & F_2 & F_3 \end{bmatrix} = \begin{bmatrix} \frac{\partial T_{11}}{\partial x_1}+\frac{\partial T_{21}}{\partial x_2}+\frac{\partial T_{31}}{\partial x_3} & \frac{\partial T_{12}}{\partial x_1}+\frac{\partial T_{22}}{\partial x_2}+\frac{\partial T_{32}}{\partial x_3} & \frac{\partial T_{13}}{\partial x_1}+\frac{\partial T_{23}}{\partial x_2}+\frac{\partial T_{33}}{\partial x_3} \end{bmatrix}.$$

Since the second index of T_{ij} specifies the stress component direction, the three force components F_1, F_2, and F_3 only depend on the stress components in the '1', '2', and '3' directions, respectively, even though derivatives in all three directions are taken.

2.6 Force on a surface

A surface area element has a size (or magnitude) and an orientation, so it can be treated as a vector $d\mathbf{A}$. If dA is the surface element's size, and **n** is its normal unit vector, then $d\mathbf{A} = \mathbf{n}dA$.

Suppose the nine components, T_{ij}, of the stress tensor with respect to a given set of Cartesian coordinates 0123 are given, and we want to find the force per unit area, **f(n)** with components f_i, on an arbitrarily oriented surface element with normal **n** (see Fig. 2.5). One way of completing this task is to switch to a rotated coordinate system, and use (2.12) to find the normal and shear stresses on the surface element. An alternative method is described here. Consider the tetrahedral element shown in Fig. 2.5. The net force f_1 on the element in the first direction produced by the stresses T_{ij} is:

$$f_1 dA = T_{11}dA_1 + T_{21}dA_2 + T_{31}dA_3.$$

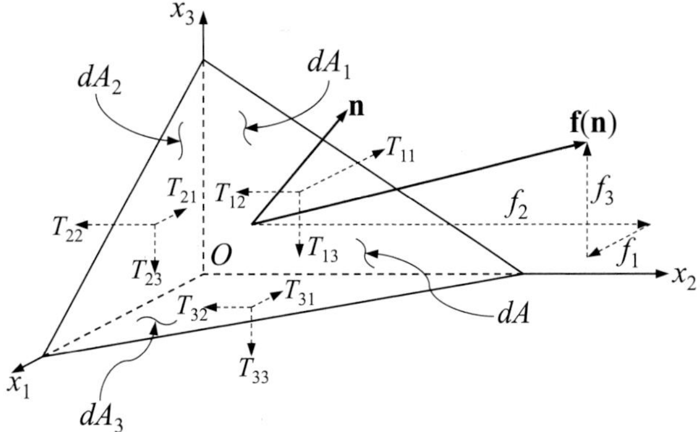

FIGURE 2.5 Force **f** per unit area on a surface element whose outward normal is **n**. The areas of the tetrahedron's faces that are perpendicular to the ith coordinate axis are dA_i. The area of the largest tetrahedron face is dA. As in Fig. 2.4, the directions of positive normal and shear stresses are shown.

The geometry of the tetrahedron requires: $dA_i = n_i dA$, where n_i are the components of the surface normal vector **n**. Thus, the net force equation can be rewritten:

$$f_1 dA = T_{11} n_1 dA + T_{21} n_2 dA + T_{31} n_3 dA.$$

Dividing by dA then produces $f_1 = T_{j1} n_j$ (with summation implied), or for any component of **f**:

$$f_i = \sum_{j=1}^{3} T_{ji} n_j \equiv T_{ji} n_j \quad \text{or} \quad \mathbf{f} = \mathbf{n} \cdot \mathbf{T}, \tag{2.15}$$

where the boldface-only version of (2.15) is unambiguous when, $T_{ij} = T_{ji}$, a claim that is proved in Chapter 4. Therefore, the contracted or inner product of the stress tensor **T** and the unit normal vector **n** gives the force per unit area on a surface perpendicular to **n**. This result is analogous to $u_n = \mathbf{u} \cdot \mathbf{n}$, where u_n is the component of the vector **u** along **n**; however, whereas u_n is a scalar, **f** in (2.15) is a vector.

Example 2.6

In two spatial dimensions, x_1 and x_2, consider parallel flow through a channel (see Fig. 2.6). Choose x_1 parallel to the flow direction. The viscous stress tensor at a point in the flow has the form:

$$\tau = \begin{bmatrix} 0 & a \\ a & 0 \end{bmatrix},$$

where a is positive in one half of the channel, and negative in the other half. Find the magnitude and direction of the force per unit area **f** on an element whose outward normal points $\phi = 30°$ from the flow direction.

Solution by using (2.15)

Start with the definition of **n** in the given coordinates:

$$\mathbf{n} = \begin{bmatrix} \cos\phi \\ \sin\phi \end{bmatrix} = \begin{bmatrix} \sqrt{3}/2 \\ 1/2 \end{bmatrix}.$$

The viscous force per unit area is therefore:

$$\mathbf{f} = \tau_{ji} n_j = \tau_{ij} n_j = \begin{bmatrix} 0 & a \\ a & 0 \end{bmatrix} \begin{bmatrix} \cos\phi \\ \sin\phi \end{bmatrix} = \begin{bmatrix} a\sin\phi \\ a\cos\phi \end{bmatrix} = \begin{bmatrix} a/2 \\ a\sqrt{3}/2 \end{bmatrix} = \begin{bmatrix} f_1 \\ f_2 \end{bmatrix}.$$

The magnitude of **f** is:

$$f = |\mathbf{f}| = (f_1^2 + f_2^2)^{1/2} = |a|.$$

If θ is the angle of **f** with respect to the x_1 axis, then:

$$\sin\theta = f_2/f = \left(\sqrt{3}/2\right)(a/|a|) \text{ and } \cos\theta = f_1/f = (1/2)(a/|a|).$$

Thus $\theta = 60°$ if a is positive (in which case both $\sin\theta$ and $\cos\theta$ are positive), and $\theta = 240°$ if a is negative (in which case both $\sin\theta$ and $\cos\theta$ are negative).

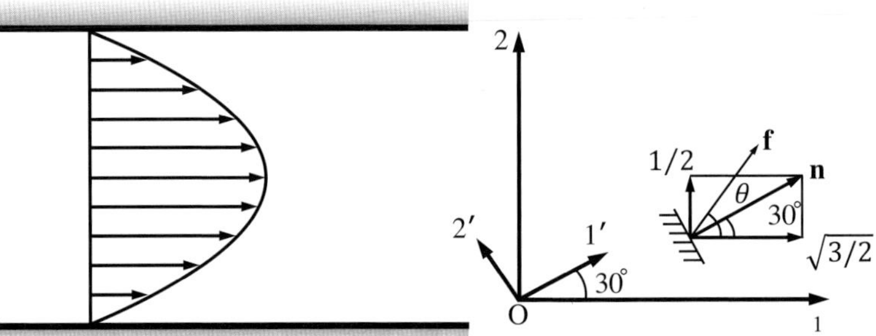

FIGURE 2.6 Determination of the viscous force per unit area on a small area element with a normal vector rotated 30° from the flow direction in a simple unidirectional shear flow parallel to the x_1-axis.

Solution by using (2.12)

Consider a rotated coordinate system O1′2′ with the x_1'-axis coinciding with **n** as shown in Fig. 2.6. Using (2.12), the components of the stress tensor in the rotated frame are:

$$\tau_{11}' = C_{11}C_{21}\tau_{12} + C_{21}C_{11}\tau_{21} = (\cos\phi\sin\phi)a + (\sin\phi\cos\phi)a = \frac{\sqrt{3}}{2}\frac{1}{2}a + \frac{1}{2}\frac{\sqrt{3}}{2}a = \frac{\sqrt{3}}{2}a, \text{ and}$$

$$\tau_{12}' = C_{11}C_{21}\tau_{12} + C_{21}C_{12}\tau_{21} = (\cos\phi)^2 a - (\sin\phi)^2 a = \frac{\sqrt{3}}{2}\frac{\sqrt{3}}{2}a - \frac{1}{2}\frac{1}{2}a = \frac{1}{2}a,$$

where $C_{ij} = \begin{bmatrix} \cos\phi & -\sin\phi \\ \sin\phi & \cos\phi \end{bmatrix}$. The normal stress is therefore $\sqrt{3}a/2$, and the shear stress is $a/2$. These results again provide the magnitude of a and a direction of 60° or 240° depending on the sign of a.

2.7 Kronecker delta and alternating tensor

The *Kronecker delta* is defined as:

$$\delta_{ij} = \left\{ \begin{array}{l} 1 \text{ if } i = j \\ 0 \text{ if } i \neq j \end{array} \right\}. \tag{2.16}$$

In three spatial dimensions it is the 3×3 identity matrix:

$$\boldsymbol{\delta} = \begin{bmatrix} 1 & 0 & 0 \\ 0 & 1 & 0 \\ 0 & 0 & 1 \end{bmatrix}.$$

In matrix multiplication operations involving the Kronecker delta, it simply replaces its summed-over index by its other index. Consider:

$$\sum_{j=1}^{3} \delta_{ij}u_j \equiv \delta_{ij}u_j = \delta_{i1}u_1 + \delta_{i2}u_2 + \delta_{i3}u_3;$$

the right-hand side is u_1 when $i = 1$, u_2 when $i = 2$, and u_3 when $i = 3$; thus:

$$\delta_{ij}u_j = u_i. \tag{2.17}$$

It can be shown that δ_{ij} is an *isotropic tensor* in the sense that its components are unchanged by a rotation of the frame of reference, that is, $\delta'_{ij} = \delta_{ij}$ (see Exercise 2.13). Isotropic tensors can be of various orders. There is no isotropic tensor of first order, and δ_{ij} is the only isotropic tensor of second order. There is also only one isotropic tensor of third order. It is called the *alternating tensor* or *permutation symbol*, and is defined as:

$$\varepsilon_{ijk} = \left\{ \begin{array}{lll} 1 & \text{if} & ijk = 123, 231, \text{or } 312 \text{ (cyclic order)}, \\ 0 & \text{if} & \text{any two indices are equal}, \\ -1 & \text{if} & ijk = 321, 213, \text{or } 132 \text{ (anti-cyclic order)} \end{array} \right\}. \tag{2.18}$$

From this definition, it is clear that *an index on ε_{ijk} can be moved two places (either to the right or to the left) without changing its value*. For example, $\varepsilon_{ijk} = \varepsilon_{jki}$ where i has been moved two places to the right, and $\varepsilon_{ijk} = \varepsilon_{kij}$ where k has been moved two places to the left. For a movement of one place, however, the sign is reversed. For example, $\varepsilon_{ijk} = -\varepsilon_{ikj}$ where j has been moved one place to the right.

A frequently used relation is the *epsilon delta relation:*

$$\sum_{k=1}^{3} \varepsilon_{ijk}\varepsilon_{klm} \equiv \varepsilon_{ijk}\varepsilon_{klm} = \delta_{il}\delta_{jm} - \delta_{im}\delta_{jl}. \tag{2.19}$$

The validity of this relationship can be verified by choosing some values for the indices *ijlm*. This relationship can be remembered by noting the following two points: 1) The adjacent index k is summed; and 2) the first two indices on the right side, namely, i and l, are the first index of ε_{ijk} and the first *free* index of ε_{klm}. The remaining indices on the right side then follow immediately.

Example 2.7

What are the singly and doubly contracted products $\varepsilon_{ijk}\delta_{kl}$ and $\varepsilon_{ijk}\delta_{jk}$?

Solution Following the mnemonic for δ_{kl} stated below (2.16), the singly contracted product $\varepsilon_{ijk}\delta_{kl}$ can be simplified by replacing the summed-over index, k, with the other index, l, of δ_{kl}. Thus, $\varepsilon_{ijk}\delta_{kl} = \varepsilon_{ijl}$. The doubly contracted product can be evaluated similarly. First consider the sum over the index j: $\varepsilon_{ijk}\delta_{jk} = \varepsilon_{ikk}$, then determine $\varepsilon_{ikk} = 0$ from the second rule in (2.18).

2.8 Vector dot and cross products

The dot product of two vectors **u** and **v** is defined in (2.2). It can also be written $\mathbf{u} \cdot \mathbf{v} = uv\cos\theta$, where u and v are the vectors' magnitudes and θ is the angle between the vectors (see Exercises 2.14 and 2.15). The dot product is therefore the magnitude of one vector times the component of the other in the direction of the first. The dot product $\mathbf{u} \cdot \mathbf{v}$ is equal to the sum of the diagonal terms of the tensor $u_i v_j$ and is zero when **u** and **v** are orthogonal.

The cross product between two vectors **u** and **v** is defined as the vector **w** whose magnitude is $uv\sin\theta$ where θ is the angle between **u** and **v**, and whose direction is perpendicular to the plane of **u** and **v** such that **u**, **v**, and **w** form a right-handed system. In this case, $\mathbf{u} \times \mathbf{v} = -\mathbf{v} \times \mathbf{u}$. Furthermore, unit vectors in right-handed coordinate systems obey the cyclic rule $\mathbf{e}_1 \times \mathbf{e}_2 = \mathbf{e}_3$. These requirements are sufficient to determine:

$$\mathbf{u} \times \mathbf{v} = (u_2v_3 - u_3v_2)\mathbf{e}_1 + (u_3v_1 - u_1v_3)\mathbf{e}_2 + (u_1v_2 - u_2v_1)\mathbf{e}_3, \tag{2.20}$$

(see Exercise 2.16). Eq. (2.20) can be written as the determinant of a matrix:

$$\mathbf{u} \times \mathbf{v} = \det \begin{bmatrix} \mathbf{e}_1 & \mathbf{e}_2 & \mathbf{e}_3 \\ u_1 & u_2 & u_3 \\ v_1 & v_2 & v_3 \end{bmatrix}.$$

In indicial notation, the k-component of $\mathbf{u} \times \mathbf{v}$ can be written as:

$$(\mathbf{u} \times \mathbf{v}) \cdot \mathbf{e}_k = \sum_{i=1}^{3}\sum_{j=1}^{3} \varepsilon_{ijk}u_iv_j \equiv \varepsilon_{ijk}u_iv_j = \varepsilon_{kij}u_iv_j. \tag{2.21}$$

As a check, for $k=1$ the nonzero terms in the double sum in (2.21) result from $i=2$, $j=3$, and from $i=3$, $j=2$. This follows from the definition (2.18) that the permutation symbol is zero if any two indices are equal. Thus, (2.21) gives:

$$(\mathbf{u}\times\mathbf{v})\cdot\mathbf{e}_1=\varepsilon_{ij1}u_iv_j=\varepsilon_{231}u_2v_3+\varepsilon_{321}u_3v_2=u_2v_3-u_3v_2,$$

which agrees with (2.20). Note that the fourth form of (2.21) is obtained from the second by moving the index k two places to the left; see the remark following (2.18).

Example 2.8

Use the definition of the cross and dot products to evaluate $(\mathbf{a}\times\mathbf{b})\cdot\mathbf{c}$ in index notation and determine its value when $\mathbf{a}=\mathbf{e}_1$, $\mathbf{b}=\mathbf{e}_2$, and $\mathbf{c}=\mathbf{e}_3$.

Solution Use (2.21) to find: $\mathbf{a}\times\mathbf{b}=\varepsilon_{ijk}a_ib_j$, which has free index is k. Form the dot product using (2.2): $(\mathbf{a}\times\mathbf{b})\cdot\mathbf{c}=[\varepsilon_{ijk}a_ib_j]\cdot\mathbf{c}=\varepsilon_{ijk}a_ib_jc_k$. When $\mathbf{a}=\mathbf{e}_1$, $\mathbf{b}=\mathbf{e}_2$, and $\mathbf{c}=\mathbf{e}_3$, then $a_1=b_2=c_3=1$; all other components of the three unit vectors are zero. Thus, the triple sum implied by $\varepsilon_{ijk}a_ib_jc_k$ reduces to one term, so that $(\mathbf{e}_1\times\mathbf{e}_2)\cdot\mathbf{e}_3=\varepsilon_{123}a_1b_2c_3=1$. Note that the placement of parentheses in the original expression is important; $\mathbf{a}\times(\mathbf{b}\cdot\mathbf{c})=\varepsilon_{ijk}a_ib_lc_l$ is a second-order tensor since $\mathbf{b}\cdot\mathbf{c}=b_lc_l$ is a scalar and this leaves two free indices (j and k).

2.9 Gradient, divergence, and curl

The vector-differentiation operator "del"[1] is defined symbolically by:

$$\nabla=\mathbf{e}_1\frac{\partial}{\partial x_1}+\mathbf{e}_2\frac{\partial}{\partial x_2}+\mathbf{e}_3\frac{\partial}{\partial x_3}=\sum_{i=1}^{3}\mathbf{e}_i\frac{\partial}{\partial x_i}\equiv\mathbf{e}_i\frac{\partial}{\partial x_i}. \tag{2.22}$$

When operating on a scalar function of position ϕ, it generates the vector:

$$\nabla\phi=\sum_{i=1}^{3}\mathbf{e}_i\frac{\partial\phi}{\partial x_i}\equiv\mathbf{e}_i\frac{\partial\phi}{\partial x_i},$$

whose i-component is $\partial\phi/\partial x_i$. The vector $\nabla\phi$ is called the *gradient* of ϕ, and $\nabla\phi$ is perpendicular to surfaces defined by $\varphi=$ constant. In addition, it specifies the magnitude and direction of the *maximum* spatial rate of change of ϕ (Fig. 2.7). The spatial rate of change of ϕ in any other direction $\mathbf{n}$ is given by:

$$\partial\phi/\partial n=\nabla\phi\cdot\mathbf{n}.$$

In Cartesian coordinates, the *divergence* of a vector field $\mathbf{u}$ is defined as the scalar

$$\nabla\cdot\mathbf{u}=\frac{\partial u_1}{\partial x_1}+\frac{\partial u_2}{\partial x_2}+\frac{\partial u_3}{\partial x_3}=\sum_{i=1}^{3}\frac{\partial u_i}{\partial x_i}\equiv\frac{\partial u_i}{\partial x_i}. \tag{2.23}$$

So far, we have defined the operations of the gradient of a scalar and the divergence of a vector. We can, however, generalize these operations. For example, the divergence, $\nabla\cdot\mathbf{T}$, of a second-order tensor $\mathbf{T}$ can be defined as the vector whose j-component is:

$$(\nabla\cdot\mathbf{T})\cdot\mathbf{e}_j=\sum_{j=1}^{3}\frac{\partial T_{ij}}{\partial x_i}\equiv\frac{\partial T_{ij}}{\partial x_i}.$$

The divergence operation *decreases* the order of the tensor by one. In contrast, the gradient operation *increases* the order of a tensor by one, changing a zero-order tensor to a first-order tensor, and a first-order tensor to a second-order tensor, i.e., $\partial u_i/\partial x_j$.

---

1. The inverted Greek delta is called a "nabla" ($\nu\alpha\beta\lambda\alpha$). The word originates from the Hebrew word for lyre, an ancient harp-like stringed instrument. It was on his instrument that the boy, David, entertained King Saul (Samuel II) and it is mentioned repeatedly in Psalms as a musical instrument to use in the praise of God.

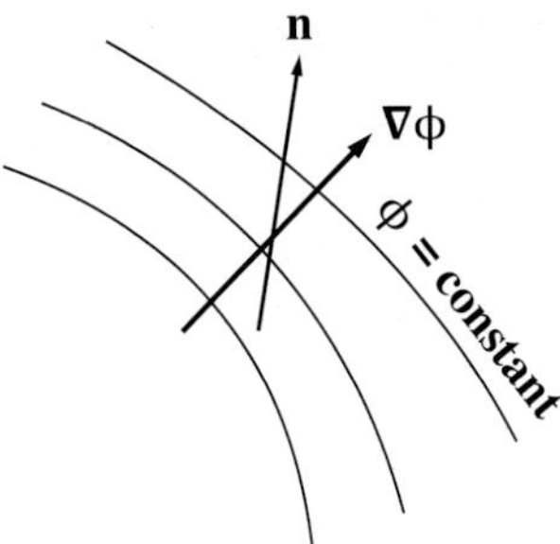

FIGURE 2.7 An illustration of the gradient, $\nabla\phi$, of a scalar function ϕ. The curves of constant ϕ and $\nabla\phi$ are perpendicular, and the spatial derivative of ϕ in the direction $\mathbf{n}$ is given by $\mathbf{n}\cdot\nabla\phi$. The most rapid change in ϕ is found when $\mathbf{n}$ and $\nabla\phi$ are parallel.

The *curl* of a vector field $\mathbf{u}$ is defined as the vector $\nabla\times\mathbf{u}$, whose i-component can be written as:

$$(\nabla\times\mathbf{u})\cdot\mathbf{e}_i=\sum_{j=1}^{3}\sum_{k=1}^{3}\varepsilon_{ijk}\frac{\partial u_k}{\partial x_j}\equiv\varepsilon_{ijk}\frac{\partial u_k}{\partial x_j},\tag{2.24}$$

using (2.21) and (2.22). The three components of the vector $\nabla\times\mathbf{u}$ can easily be found from the right-hand side of (2.24). For the $i=1$ component, the nonzero terms in the double sum result from $j=2$, $k=3$, and from $j=3$, $k=2$. The three components of $\nabla\times\mathbf{u}$ are finally found as:

$$(\nabla\times\mathbf{u})\cdot\mathbf{e}_1=\frac{\partial u_3}{\partial x_2}-\frac{\partial u_2}{\partial x_3},\ (\nabla\times\mathbf{u})\cdot\mathbf{e}_2=\frac{\partial u_1}{\partial x_3}-\frac{\partial u_3}{\partial x_1},\ \text{and}\ (\nabla\times\mathbf{u})\cdot\mathbf{e}_3=\frac{\partial u_2}{\partial x_1}-\frac{\partial u_1}{\partial x_2}.\tag{2.25}$$

A vector field $\mathbf{u}$ is called *solenoidal* or *divergence free* if $\nabla\cdot\mathbf{u}=0$, and *irrotational* or *curl free* if $\nabla\times\mathbf{u}=0$. The word solenoidal refers to the fact that the divergence of the magnetic induction is always zero because of the absence of magnetic monopoles. The reason for the word irrotational is made clear in Chapter 3.

Example 2.9

If a is a positive constant and $\mathbf{b}$ is a constant vector, determine the divergence and the curl of a vector field that diverges from the origin of coordinates, $\mathbf{u}=a\mathbf{x}$, and a vector field indicative of solid body rotation about a fixed axis, $\mathbf{u}=\mathbf{b}\times\mathbf{x}$.

Solution Using $\mathbf{u}=a\mathbf{x}=ax_1\mathbf{e}_1+ax_2\mathbf{e}_2+ax_3\mathbf{e}_3$ in (2.23) and (2.25) produces:

$$\nabla\cdot\mathbf{u}=\frac{\partial ax_1}{\partial x_1}+\frac{\partial ax_2}{\partial x_2}+\frac{\partial ax_3}{\partial x_3}=a+a+a=3a,$$

$$(\nabla\times\mathbf{u})\cdot\mathbf{e}_1=\frac{\partial ax_3}{\partial x_2}-\frac{\partial ax_2}{\partial x_3}=0,\ (\nabla\times\mathbf{u})\cdot\mathbf{e}_2=\frac{\partial ax_1}{\partial x_3}-\frac{\partial ax_3}{\partial x_1}=0,\ \text{and}$$

$$(\nabla\times\mathbf{u})\cdot\mathbf{e}_3=\frac{\partial ax_2}{\partial x_1}-\frac{\partial ax_1}{\partial x_2}=0.$$

Thus, $\mathbf{u}=a\mathbf{x}$ has a constant nonzero divergence and is irrotational. Using $\mathbf{u}=(b_2x_3-b_3x_2)\mathbf{e}_1+(b_3x_1-b_1x_3)\mathbf{e}_2+(b_1x_2-b_2x_1)\mathbf{e}_3$ in (2.23) and (2.25) produces:

$$\nabla\cdot\mathbf{u}=\frac{\partial(b_2x_3-b_3x_2)}{\partial x_1}+\frac{\partial(b_3x_1-b_1x_3)}{\partial x_2}+\frac{\partial(b_1x_2-b_2x_1)}{\partial x_3}=0,$$

$$(\nabla\times\mathbf{u})\cdot\mathbf{e}_1=\frac{\partial(b_1x_2-b_2x_1)}{\partial x_2}-\frac{\partial(b_3x_1-b_1x_3)}{\partial x_3}=2b_1,$$

$$(\nabla\times\mathbf{u})\cdot\mathbf{e}_2=\frac{\partial(b_2x_3-b_3x_2)}{\partial x_3}-\frac{\partial(b_1x_2-b_2x_1)}{\partial x_1}=2b_2,\ \text{and}$$

$$(\nabla\times\mathbf{u})\cdot\mathbf{e}_3=\frac{\partial(b_3x_1-b_1x_3)}{\partial x_1}-\frac{\partial(b_2x_3-b_3x_2)}{\partial x_2}=2b_3.$$

Thus, $\mathbf{u}=\mathbf{b}\times\mathbf{x}$ is divergence free and rotational.

2.10 Symmetric and antisymmetric tensors

A tensor **B** is called *symmetric* in the indices i and j if the components do not change when i and j are interchanged, that is, if $B_{ij} = B_{ji}$. Thus, the matrix of a symmetric second-order tensor is made up of only six distinct components (the three on the diagonal where $i = j$, and the three above or below the diagonal where $i \neq j$). On the other hand, a tensor is called *antisymmetric* if $B_{ij} = -B_{ji}$. An antisymmetric tensor's diagonal components are each zero, and it has only three distinct components (the three above or below the diagonal). Any tensor can be represented as the sum of symmetric and antisymmetric tensors. For if we write:

$$B_{ij} = \frac{1}{2}\left(B_{ij} + B_{ji}\right) + \frac{1}{2}\left(B_{ij} - B_{ji}\right) = S_{ij} + A_{ij},$$

then the operation of interchanging i and j does not change the first term, but changes the sign of the second term. Therefore, $(B_{ij} + B_{ji})/2 \equiv S_{ij}$ is called the symmetric part of B_{ij}, and $(B_{ij} - B_{ji})/2 \equiv A_{ij}$ is called the antisymmetric part of B_{ij}.

Every vector can be associated with an antisymmetric tensor, and vice versa. For example, we can associate the vector $\boldsymbol{\omega}$ having components ω_i, with an antisymmetric tensor:

$$\mathbf{R} = \begin{bmatrix} 0 & -\omega_3 & \omega_2 \\ \omega_3 & 0 & -\omega_1 \\ -\omega_2 & \omega_1 & 0 \end{bmatrix}. \tag{2.26}$$

The two are related via:

$$R_{ij} = \sum_{k=1}^{3} -\varepsilon_{ijk}\omega_k \equiv -\varepsilon_{ijk}\omega_k, \text{ and } \omega_k = \sum_{i=1}^{3}\sum_{j=1}^{3} -\frac{1}{2}\varepsilon_{ijk}R_{ij} \equiv -\frac{1}{2}\varepsilon_{ijk}R_{ij}. \tag{2.27}$$

As a check, (2.27) gives $R_{11} = 0$ and $R_{12} = -\varepsilon_{123}\omega_3 = -\omega_3$, in agreement with (2.26). (In Chapter 3, **R** is recognized as the *rotation* tensor corresponding to the *vorticity* vector $\boldsymbol{\omega}$.)

A commonly occurring operation is the doubly contracted product, P, of a *symmetric* tensor **T** and another tensor **B**:

$$P = \sum_{k=1}^{3}\sum_{l=1}^{3} T_{kl}B_{kl} \equiv T_{kl}B_{kl} = T_{kl}(S_{kl} + A_{kl}) = T_{kl}S_{kl} + T_{kl}A_{kl} = T_{ij}S_{ij} + T_{ij}A_{ij}, \tag{2.28}$$

where **S** and **A** are the symmetric and antisymmetric parts of **B**. The final equality follows from the index-summation convention; sums are completed over both k and l, so these indices can be replaced by any two distinct indices. Exchanging the indices of **A** in the final term of (2.28) produces $P = T_{ij}S_{ij} - T_{ij}A_{ji}$, but this can also be written $P = T_{ji}S_{ji} - T_{ji}A_{ji}$ because S_{ij} and T_{ij} are symmetric. Now, replace the index j by k and the index i by l to find:

$$P = T_{kl}S_{kl} - T_{kl}A_{kl}. \tag{2.29}$$

This relationship and the fourth part of the extended equality in (2.28) require that $T_{ij}A_{ij} = T_{kl}A_{kl} = 0$, and:

$$T_{ij}B_{ij} = T_{ij}S_{ij} = \frac{1}{2}T_{ij}\left(B_{ij} + B_{ji}\right).$$

Thus, the doubly contracted product of a symmetric tensor **T** with any tensor **B** equals **T** doubly contracted with the symmetric part of **B**, and the doubly contracted product of a symmetric tensor and an antisymmetric tensor is zero. The latter result is analogous to the fact that the definite integral over an even (symmetric) interval of the product of a symmetric and an antisymmetric function is zero.

Example 2.10

If $R_{ij} = -\varepsilon_{ijk}\omega_k$, what is **R:R** in terms of $\boldsymbol{\omega}$?

Solution Use the definition of the double contraction, and manipulate the indices to put the permutation-symbol product in the form of (2.19):

$$\mathbf{R{:}R} = R_{ij}R_{ij} = \left(-\varepsilon_{ijk}\omega_k\right)\left(-\varepsilon_{ijl}\omega_l\right) = \left(-\varepsilon_{kij}\right)\left(+\varepsilon_{jil}\right)\omega_k\omega_l = -(\delta_{ki}\delta_{il} - \delta_{kl}\delta_{ii})\omega_k\omega_l.$$

Inside the last parentheses, $\delta_{ki}\delta_{il} = \delta_{kl}$ and $\delta_{ii} = 3$, so:

$$\mathbf{R}{:}\mathbf{R} = -(\delta_{kl} - 3\delta_{kl})\omega_k\omega_l = 2\delta_{kl}\omega_k\omega_l = 2\omega_k^2 = 2\left(\omega_1^2 + \omega_2^2 + \omega_3^2\right) = 2|\boldsymbol{\omega}|^2.$$

2.11 Eigenvalues and eigenvectors of a symmetric tensor

The reader is assumed to be familiar with the concepts of eigenvalues and eigenvectors of a matrix, so only a brief review of the main results is provided here. Suppose $\boldsymbol{\tau}$ is a symmetric tensor with real elements. Then the following facts can be proved:

1. There are three real eigenvalues $\lambda^k (k = 1, 2, 3)$, which may or may not all be distinct. (Here, the superscript k is not an exponent, and λ^k does not denote the kth-component of a vector.) These eigenvalues (λ^1, λ^2, and λ^3) are the roots or solutions of the third-degree polynomial:

$$\det|\tau_{ij} - \lambda\delta_{ij}| = 0.$$

2. The three eigenvectors $\mathbf{b}^k$ corresponding to distinct eigenvalues λ^k are mutually orthogonal. These eigenvectors define the directions of the *principal axes* of $\boldsymbol{\tau}$. Each $\mathbf{b}$ is found by solving three algebraic equations:

$$\left(\tau_{ij} - \lambda\delta_{ij}\right) b_j = 0$$

($i = 1, 2,$ or 3), where the superscript k on λ and $\mathbf{b}$ has been omitted for clarity because there is no sum over k.

3. If the coordinate system is rotated so that its unit vectors coincide with the eigenvectors, then $\boldsymbol{\tau}$ is diagonal with elements λ^k in this rotated coordinate system:

$$\boldsymbol{\tau}' = \begin{bmatrix} \lambda^1 & 0 & 0 \\ 0 & \lambda^2 & 0 \\ 0 & 0 & \lambda^3 \end{bmatrix}.$$

4. Although the elements τ_{ij} change as the coordinate system is rotated, they cannot be larger than the largest λ or smaller than the smallest λ; the λ^k represent the extreme values of τ_{ij}.

Example 2.11

The strain rate tensor $\mathbf{S}$ is related to the velocity vector $\mathbf{u}$ by:

$$S_{ij} = \frac{1}{2}\left(\frac{\partial u_i}{\partial x_j} + \frac{\partial u_j}{\partial x_i}\right).$$

For a two-dimensional flow parallel to the 1-direction, $\mathbf{u} = (u_1(x_2), 0)$, show how $\mathbf{S}$ is diagonalized in a frame of reference rotated to coincide with the principal axes.

Solution For the given velocity profile $u_1(x_2)$, it is evident that $S_{11} = S_{22} = 0$, and $2S_{12} = 2S_{21} = du_1/dx_2 = 2\Gamma$. The strain rate tensor in the original coordinate system is therefore:

$$\mathbf{S} = \begin{bmatrix} 0 & \Gamma \\ \Gamma & 0 \end{bmatrix}.$$

The eigenvalues are determined from:

$$\det\left|S_{ij} - \lambda\delta_{ij}\right| = \det\begin{vmatrix} -\lambda & \Gamma \\ \Gamma & -\lambda \end{vmatrix} = \lambda^2 - \Gamma^2 = 0,$$

which has solutions $\lambda^1 = \Gamma$ and $\lambda^2 = -\Gamma$. The first eigenvector $\mathbf{b}^1$ is given by:

$$\begin{bmatrix} 0 & \Gamma \\ \Gamma & 0 \end{bmatrix}\begin{bmatrix} b_1^1 \\ b_2^1 \end{bmatrix} = \lambda^1\begin{bmatrix} b_1^1 \\ b_2^1 \end{bmatrix},$$

which has solution $b_1^1 = b_2^1 = 1/\sqrt{2}$, when $\mathbf{b}^1$ is normalized to have magnitude unity. The second eigenvector is similarly found so that:

$$\mathbf{b}^1 = \begin{bmatrix} 1/\sqrt{2} \\ 1/\sqrt{2} \end{bmatrix}, \text{ and } \mathbf{b}^2 = \begin{bmatrix} -1/\sqrt{2} \\ 1/\sqrt{2} \end{bmatrix}.$$

These eigenvectors are shown in Fig. 2.8. The direction cosine matrix of the original and the rotated coordinate system is therefore:

$$\mathbf{C} = \begin{bmatrix} 1/\sqrt{2} & -1/\sqrt{2} \\ 1/\sqrt{2} & 1/\sqrt{2} \end{bmatrix},$$

which represents rotation of the coordinate system by 45°. Using the transformation rule (2.12), the components of **S** in the rotated system are found as follows:

$$S'_{12} = C_{i1}C_{j2}S_{ij} = C_{11}C_{22}S_{12} + C_{21}C_{12}S_{21} = \frac{1}{\sqrt{2}}\frac{1}{\sqrt{2}}\Gamma - \frac{1}{\sqrt{2}}\frac{1}{\sqrt{2}}\Gamma = 0,\ S'_{21} = 0,$$
$$S'_{11} = C_{i1}C_{j1}S_{ij} = C_{11}C_{21}S_{12} + C_{21}C_{11}S_{21} = \Gamma, \text{ and}$$
$$S'_{22} = C_{i2}C_{j2}S_{ij} = C_{12}C_{22}S_{12} + C_{22}C_{12}S_{21} = -\Gamma.$$

(Instead of using (2.12), all the components of **S** in the rotated system can be found by carrying out the matrix product $\mathbf{C}^T \cdot \mathbf{S} \cdot \mathbf{C}$.) The matrix of **S** in the rotated frame is therefore:

$$\mathbf{S}' = \begin{bmatrix} \Gamma & 0 \\ 0 & -\Gamma \end{bmatrix}.$$

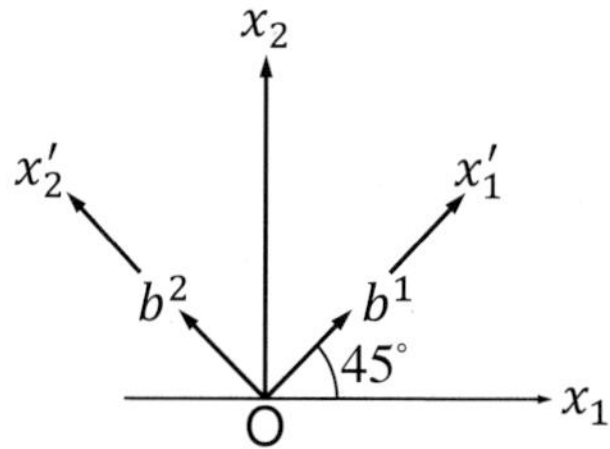

FIGURE 2.8 Original coordinate system Ox_1x_2 and the rotated coordinate system $Ox'_1x'_2$ having unit vectors that coincide with the eigenvectors of the strain-rate tensor in Example 2.6. Here the strain rate is determined from a unidirectional flow having only cross-stream variation, and the angle of rotation is determined to be 45°.

The foregoing matrix contains only diagonal terms. For positive Γ, it will be shown in the next chapter that it represents a linear stretching at a rate Γ along one principal axis, and a linear compression at a rate $-\Gamma$ along the other; the shear strains are zero in the principal-axis coordinate system of the strain rate tensor.

2.12 Gauss' theorem

This useful theorem relates volume and surface integrals. Let V be a volume bounded by a closed surface A. Consider an infinitesimal surface element dA having outward unit normal $\mathbf{n}$ with components n_i (Fig. 2.9), and let $Q(\mathbf{x})$ be a scalar, vector, or tensor field of any order. Gauss' theorem states that:

$$\iiint_V \frac{\partial Q}{\partial x_i} dV = \iint_A n_i Q dA. \tag{2.30}$$

The most common form of Gauss' theorem is when $\mathbf{Q}$ is a vector, in which case the theorem is:

$$\iiint_V \sum_{i=1}^{3} \frac{\partial Q_i}{\partial x_i} dV \equiv \iiint_V \frac{\partial Q_i}{\partial x_i} dV = \iint_A \sum_{i=1}^{3} n_i Q_i dA \equiv \iint_A n_i Q_i dA,$$
$$\text{or} \iiint_V \nabla \cdot \mathbf{Q} dV = \iint_A \mathbf{n} \cdot \mathbf{Q} dA,$$

which is commonly called the *divergence theorem*. In words, the theorem states that the volume integral of the divergence of $\mathbf{Q}$ is equal to the surface integral of the outflux of $\mathbf{Q}$.

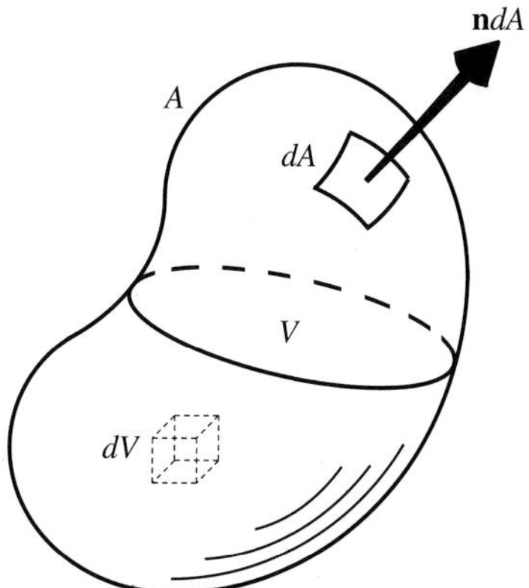

FIGURE 2.9 Illustration of Gauss' theorem for a volume V enclosed by surface area A. A small volume element, dV, and a small area element dA with outward normal $\mathbf{n}$ are shown.

Alternatively, (2.30) defines a generalized field derivative, denoted by $\mathcal{D}$, of Q when considered in its limiting form for a vanishingly small volume:

$$\mathcal{D}Q = \lim_{V\to 0}\frac{1}{V}\iint_A n_i Q dA. \tag{2.31}$$

Interestingly, this form is readily specialized to the gradient, divergence, and curl of any scalar, vector, or tensor Q. Moreover, by regarding (2.31) as a definition, the recipes for the computation of vector field derivatives may be obtained in any coordinate system. As stated, (2.31) defines the gradient of a tensor Q of any order. For a tensor of order one or higher, the divergence and curl are defined by including a dot (scalar) product or a cross (vector) product, respectively, under the integral:

$$\nabla\cdot\mathbf{Q} = \lim_{V\to 0}\frac{1}{V}\iint_A \mathbf{n}\cdot\mathbf{Q}dA, \quad \text{and} \quad \nabla\times\mathbf{Q} = \lim_{V\to 0}\frac{1}{V}\iint_A \mathbf{n}\times\mathbf{Q}dA. \tag{2.32, 2.33}$$

Example 2.12

Obtain the recipe for the divergence of a vector $\mathbf{Q}(\mathbf{x})$ in Cartesian coordinates from the integral definition (2.32).

Solution Consider an elemental rectangular volume centered on $\mathbf{x}$ with faces perpendicular to the coordinate axes (see Fig. 2.4). Denote the lengths of the sides parallel to each coordinate axis by Δx_1, Δx_2, and Δx_3, respectively. There are six faces to this rectangular volume. First consider the two that are perpendicular to the x_1 axis, EADH with $\mathbf{n} = \mathbf{e}_1$ and FBCG with $\mathbf{n} = -\mathbf{e}_1$. A Taylor expansion of $\mathbf{Q}(\mathbf{x})$ from the center of the volume to the center of each of these sides produces:

$$[\mathbf{Q}]_{EADH} = \mathbf{Q}(\mathbf{x}) + \frac{\Delta x_1}{2}\frac{\partial\mathbf{Q}(x)}{\partial x_1} + \cdots \text{ and } [\mathbf{Q}]_{FBCG} = \mathbf{Q}(\mathbf{x}) - \frac{\Delta x_1}{2}\frac{\partial\mathbf{Q}(x)}{\partial x_1} + \cdots,$$

so that the x-direction contribution to the surface integral in (2.32) is:

$$\begin{aligned}([\mathbf{n}\cdot\mathbf{Q}]_{EADH} + [\mathbf{n}\cdot\mathbf{Q}]_{FBCG})dA &= \left(\left[\mathbf{e}_1\cdot\mathbf{Q}(\mathbf{x}) + \mathbf{e}_1\cdot\frac{\Delta x_1}{2}\frac{\partial\mathbf{Q}(\mathbf{x})}{\partial x_1} + \cdots\right] + \left[-\mathbf{e}_1\cdot\mathbf{Q}(\mathbf{x}) + \mathbf{e}_1\cdot\frac{\Delta x_1}{2}\frac{\partial\mathbf{Q}(\mathbf{x})}{\partial x_1} + \cdots\right]\right)\Delta x_2\Delta x_3 \\ &= \left(\mathbf{e}_1\cdot\frac{\partial\mathbf{Q}(\mathbf{x})}{\partial x_1} + \ldots\right)\Delta x_1\Delta x_2\Delta x_3.\end{aligned}$$

Similarly for the other two directions:

$$([\mathbf{n}\cdot\mathbf{Q}]_{ABCD} + [\mathbf{n}\cdot\mathbf{Q}]_{EFGH})dA = \left(\mathbf{e}_2\cdot\frac{\partial\mathbf{Q}(\mathbf{x})}{\partial x_2} + \cdots\right)\Delta x_1\Delta x_2\Delta x_3,$$
$$([\mathbf{n}\cdot\mathbf{Q}]_{ABFE} + [\mathbf{n}\cdot\mathbf{Q}]_{DCGH})dA = \left(\mathbf{e}_3\cdot\frac{\partial\mathbf{Q}(\mathbf{x})}{\partial x_3} + \cdots\right)\Delta x_1\Delta x_2\Delta x_3.$$

Assembling the contributions from all six faces (or all three directions) to evaluate (2.32) produces:

$$\nabla \cdot \mathbf{Q} = \lim_{V\to 0} \frac{1}{V} \iint_A \mathbf{n} \cdot \mathbf{Q} dA$$
$$= \lim_{\substack{\Delta x_1 \to 0 \\ \Delta x_2 \to 0 \\ \Delta x_3 \to 0}} \frac{1}{\Delta x_1 \Delta x_2 \Delta x_3} \left(\mathbf{e}_1 \cdot \frac{\partial \mathbf{Q}(\mathbf{x})}{\partial x_1} + \mathbf{e}_2 \cdot \frac{\partial \mathbf{Q}(\mathbf{x})}{\partial x_2} + \mathbf{e}_3 \cdot \frac{\partial \mathbf{Q}(\mathbf{x})}{\partial x_3} + \dots \right) \Delta x_1 \Delta x_2 \Delta x_3,$$

and when the limit is taken, the expected Cartesian-coordinate form of the divergence emerges:

$$\nabla \cdot \mathbf{Q} = \mathbf{e}_1 \cdot \frac{\partial \mathbf{Q}(\mathbf{x})}{\partial x_1} + \mathbf{e}_2 \cdot \frac{\partial \mathbf{Q}(\mathbf{x})}{\partial x_2} + \mathbf{e}_3 \cdot \frac{\partial \mathbf{Q}(\mathbf{x})}{\partial x_3}.$$

Example 2.13

Consider a vector $\mathbf{Q} = \left(2xy^2 + 2xz^2\right)\mathbf{e}_1 + x^2y\mathbf{e}_2 + x^2z\mathbf{e}_3$. Verify Gauss' theorem numerically on a rectangular prism of size $(1 \times 1 \times 1)$ discretized using $N = 16$ grid cells in each direction (with $N + 1$ edges). Approximate the integral using the following quadrature rule: $\int f(x)dx \approx \sum_{i=1}^{N} f(x_i)\Delta x$, where x_i are discrete values of x at equidistant points from $i = 1, N$. Approximate derivatives with a forward difference scheme, i.e., $\partial f/\partial x \approx [f(x + \Delta x, y, z) - f(x, y, z)]/\Delta x$. Show that the theorem holds for any function and any value of N.

Solution First, the function handle is created for the vector $\mathbf{Q}$. Then, the domain is discretized with uniform grid spacing $\Delta x = \Delta y = \Delta z = 1/N$ using the `linspace` function. The forward difference operator used to compute the divergence was applied via the `diff` function. The difference between LHS $= \iiint_V \nabla \cdot \mathbf{Q} dV$ and RHS $= \iint_A \mathbf{n} \cdot \mathbf{Q} dA$ is found to be zero (within machine precision), regardless of the value of N.

```
% Define the vector
q = @(x,y,z) [2*x.*y.^2+2*x.*z.^2; x.^2.*y; x.^2.*z]';

% Create the grid
N=8;
x=linspace(0,1,N+1); dx=x(2)-x(1);
y=linspace(0,1,N+1); dy=y(2)-y(1);
z=linspace(0,1,N+1); dz=z(2)-z(1);

% Store the components of the vector
Qx=zeros(N+1,N+1,N+1);
Qy=zeros(N+1,N+1,N+1);
Qz=zeros(N+1,N+1,N+1);
for k=1:N+1
    for j=1:N+1
        for i=1:N+1
            Q=q(x(i),y(j),z(k));
            Qx(i,j,k)=Q(1);
            Qy(i,j,k)=Q(2);
            Qz(i,j,k)=Q(3);
        end
    end
end

% Compute the left-hand side: volume integral of div(Q)
dQdx=diff(Qx,1,1)/dx;
dQdy=diff(Qy,1,2)/dy;
dQdz=diff(Qz,1,3)/dz;
LHS=( sum(sum(sum(dQdx)))+sum(sum(sum(dQdy)))+...
    sum(sum(sum(dQdz))) )*dx*dy*dz;

% Compute the right-hand side: surface integral of Q.n
RHS=-sum(sum(Qx(1,:,:)))*dy*dz+sum(sum(Qx(N+1,:,:)))*dy*dz+...
```

```
    -sum(sum(Qy(:,1,:)))*dx*dz+sum(sum(Qy(:,N+1,:)))*dx*dz+...
    -sum(sum(Qz(:,:,1)))*dx*dy+sum(sum(Qz(:,:,N+1)))*dx*dy;

% Output the residual
disp(['Residual=',num2str(RHS-LHS)])
```

2.13 Stokes' theorem

Stokes' theorem relates the integral over an open surface A to the line integral around the surface's bounding curve C. Here, unlike Gauss' theorem, the inside and outside of A are not well defined so an arbitrary choice must be made for the direction of the outward normal $\mathbf{n}$ (here it always originates on the outside of A). Once this choice is made, the unit tangent vector to C, $\mathbf{t}$, points in the counterclockwise direction when looking at the outside of A. The final unit vector, $\mathbf{p}$, is perpendicular the curve C and tangent to the surface, so it is perpendicular to $\mathbf{n}$ and $\mathbf{t}$. Together the three unit vectors form a right-handed system: $\mathbf{t} \times \mathbf{n} = \mathbf{p}$ (see Fig. 2.10). For this geometry, Stokes' theorem states:

$$\iint_A (\nabla \times \mathbf{u}) \cdot \mathbf{n} dA = \int_C \mathbf{u} \cdot \mathbf{t} ds, \tag{2.34}$$

where s is the arc length of the closed curve C. This theorem signifies that the surface integral of the curl of a vector field $\mathbf{u}$ is equal to the line integral of $\mathbf{u}$ along the bounding curve of the surface. In fluid mechanics, the right side of (2.34) is called the *circulation* of $\mathbf{u}$ about C. In addition, (2.34) can be used to define the curl of a vector through the limit of the circulation about an infinitesimal surface as:

$$\mathbf{n} \cdot (\nabla \times \mathbf{u}) = \lim_{A \to 0} \frac{1}{A} \int_C \mathbf{u} \cdot \mathbf{t} ds. \tag{2.35}$$

The advantage of integral definitions of field derivatives is that such definitions do not depend on the coordinate system.

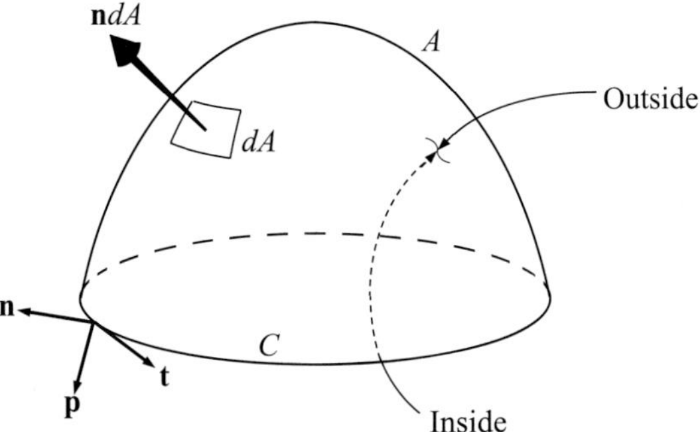

FIGURE 2.10 Illustration of Stokes' theorem for surface A bounded by the closed curve C. For the purposes of defining unit vectors, the *inside* and *outside* of A must be chosen, and one such choice is illustrated here. The unit vector $\mathbf{n}$ is perpendicular to A and originates from the outside of A. The unit vector, $\mathbf{t}$, is locally tangent to the curve C. The unit vector $\mathbf{p}$ is perpendicular to C but is locally tangent to the surface A so that it is perpendicular to both $\mathbf{n}$ and $\mathbf{t}$. The unit vectors $\mathbf{n}$, $\mathbf{t}$, and $\mathbf{p}$ define a right-handed triad of directions, $\mathbf{t} \times \mathbf{n} = \mathbf{p}$.

Example 2.14

Obtain the recipe for the curl of a vector $\mathbf{u}(\mathbf{x})$ in Cartesian coordinates from the integral definition given by (2.35).

Solution This recipe is obtained by considering rectangular contours in three perpendicular planes intersecting at the point (x, y, z) as shown in Fig. 2.11. First, consider the elemental rectangle lying in a plane defined by $x = const.$ The central point in this plane is (x, y, z) and the rectangle's area is $\Delta y \Delta z$. It may be shown by careful integration of a Taylor expansion of the integrand that the integral along each line segment may be represented by the product of the integrand at the center of the segment and the length of the segment with attention paid to the direction of integration ds. Thus we obtain:

$$(\nabla \times \mathbf{u}) \cdot \mathbf{e}_x = \lim_{\Delta y \to 0} \lim_{\Delta z \to 0} \left\{ \begin{array}{l} \dfrac{1}{\Delta y \Delta z}\left[u_z\left(x, y + \dfrac{\Delta y}{2}, z\right) - u_z\left(x, y + \dfrac{\Delta y}{2}, z\right)\right]\Delta z \\ + \dfrac{1}{\Delta y \Delta z}\left[u_y\left(x, y, z - \dfrac{\Delta z}{2}\right) - u_y\left(x, y, z + \dfrac{\Delta z}{2}\right)\right]\Delta y \end{array} \right\}.$$

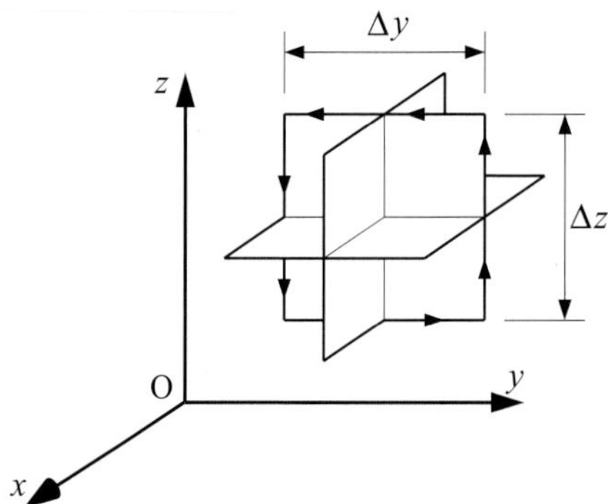

FIGURE 2.11 Drawing for Example 2.11. The point (x, y, z) is located at the center of the cruciform. The arrows show the direction of ds on the edge of the rectangular contour that lies in the $x =$ const. plane. The area inside this contour is $\Delta y \Delta z$.

Taking the limits produces:

$$(\nabla \times \mathbf{u}) \cdot \mathbf{e}_x = \frac{\partial u_z}{\partial y} - \frac{\partial u_y}{\partial z}.$$

Similarly, integrating around the elemental rectangles in the other two planes leads to:

$$(\nabla \times \mathbf{u}) \cdot \mathbf{e}_y = \frac{\partial u_x}{\partial z} - \frac{\partial u_z}{\partial x} \text{ and } (\nabla \times \mathbf{u}) \cdot \mathbf{e}_z = \frac{\partial u_y}{\partial x} - \frac{\partial u_x}{\partial y}.$$

Exercises

2.1 For three spatial dimensions, rewrite the following expressions in index notation and evaluate or simplify them using the values or parameters given, and the definitions of δ_{ij} and ε_{ijk} wherever possible. In b) through e), $\mathbf{x}$ is the position vector, with components x_i.

a. $\mathbf{b} \cdot \mathbf{c}$ where $\mathbf{b} = (1, 4, 17)$ and $\mathbf{c} = (-4, -3, 1)$.

b. $(\mathbf{u} \cdot \nabla)\mathbf{x}$ where $\mathbf{u}$ a vector with components u_i.

c. $\nabla\phi$, where $\phi = \mathbf{h} \cdot \mathbf{x}$ and $\mathbf{h}$ is a constant vector with components h_i.

d. $\nabla \times \mathbf{u}$, where $\mathbf{u} = \boldsymbol{\Omega} \times \mathbf{x}$ and $\boldsymbol{\Omega}$ is a constant vector with components Ω_i.

e. $\mathbf{C} \cdot \mathbf{x}$, where

$$\mathbf{C} = \left\{ \begin{array}{ccc} 1 & 2 & 3 \\ 0 & 1 & 2 \\ 0 & 0 & 1 \end{array} \right\}$$

2.2 Starting from (2.1) and (2.3), prove (2.7).

2.3 For two three-dimensional vectors with Cartesian components a_i and b_i, prove the Cauchy–Schwartz inequality: $(a_i b_i)^2 \le (a_i)^2 (b_i)^2$.

2.4 For two three-dimensional vectors with Cartesian components a_i and b_i, prove the triangle inequality: $|\mathbf{a}| + |\mathbf{b}| \ge |\mathbf{a} + \mathbf{b}|$.

2.5 Using Cartesian coordinates where the position vector is $\mathbf{x} = (x_1, x_2, x_3)$ and the fluid velocity is $\mathbf{u} = (u_1, u_2, u_3)$, write out the three components of the vector: $(\mathbf{u} \cdot \nabla)\mathbf{u} = u_i(\partial u_j / \partial x_i)$.

2.6 Convert $\nabla \times \nabla\rho$ to indicial notation and show that it is zero in Cartesian coordinates for any twice-differentiable scalar function ρ.

2.7 Prove the following identities using indicial notation:

a) $\mathbf{a}\times(\mathbf{b}\times\mathbf{c})=(\mathbf{a}\cdot\mathbf{c})\mathbf{b}-(\mathbf{a}\cdot\mathbf{b})\mathbf{c}$. [*Hint*: Call $\mathbf{d}\equiv\mathbf{b}\times\mathbf{c}$. Then $(\mathbf{a}\times\mathbf{d})\cdot\mathbf{e}_m=\varepsilon_{pqm}a_pd_q=\varepsilon_{pqm}a_p\varepsilon_{ijq}b_ic_j$. Using (2.19), show that $(\mathbf{a}\times\mathbf{d})\cdot\mathbf{e}_m=(\mathbf{a}\cdot\mathbf{c})\mathbf{b}\cdot\mathbf{e}_m-(\mathbf{a}\cdot\mathbf{b})\mathbf{c}\cdot\mathbf{e}_m$].]

b) $(\mathbf{a}\times\mathbf{b})\cdot(\mathbf{c}\times\mathbf{d})=(\mathbf{a}\cdot\mathbf{c})(\mathbf{b}\cdot\mathbf{d})-(\mathbf{a}\cdot\mathbf{d})(\mathbf{b}\cdot\mathbf{c})$

c) $\nabla\cdot(\nabla\times\mathbf{a})=0$

d) $\nabla\cdot\mathbf{uu}=\mathbf{u}\cdot\nabla\mathbf{u}+\mathbf{u}(\nabla\cdot\mathbf{u})$

e) $\nabla\cdot(\mathbf{a}\times\mathbf{b})=\mathbf{b}\cdot(\nabla\times\mathbf{a})-\mathbf{a}\cdot(\nabla\times\mathbf{b})$

f) $a_i\frac{\partial a_i}{\partial x_j}=\nabla\left(\frac{1}{2}\mathbf{a}\cdot\mathbf{a}\right)$

g) $\nabla\times\left[\nabla\left(\frac{1}{2}\mathbf{a}\cdot\mathbf{a}\right)-\mathbf{a}\times(\nabla\times\mathbf{a})\right]=\nabla\times(\mathbf{a}\cdot\nabla\mathbf{a})$

2.8 Show that the condition for the vectors **a**, **b**, and **c** to be coplanar is $\varepsilon_{ijk}a_ib_jc_k=0$.

2.9 Prove the following relationships: $\delta_{ij}\delta_{ij}=3$, $\varepsilon_{pqr}\varepsilon_{pqr}=6$, and $\varepsilon_{pqi}\varepsilon_{pqj}=2\delta_{ij}$.

2.10 Show that $\mathbf{C}\cdot\mathbf{C}^{\mathrm{T}}=\mathbf{C}^{\mathrm{T}}\cdot\mathbf{C}=\boldsymbol{\delta}$, where **C** is the direction cosine matrix and $\boldsymbol{\delta}$ is the matrix of the Kronecker delta. Any matrix obeying such a relationship is called an *orthogonal matrix* because it represents transformation of one set of orthogonal axes into another.

2.11 Show that for a second-order tensor **A**, the following quantities are invariant under a rotation of axes:

$$I_1=A_{ii},\ I_2=det\begin{vmatrix}A_{11}&A_{12}\\A_{21}&A_{22}\end{vmatrix}+det\begin{vmatrix}A_{22}&A_{23}\\A_{32}&A_{33}\end{vmatrix}+det\begin{vmatrix}A_{11}&A_{13}\\A_{31}&A_{33}\end{vmatrix},\ \text{and } I_3=det(A_{ij}).$$

[*Hint*: Use the result of Exercise 2.10 and the transformation rule (2.12) to show that $I_1'-A_{ii}'=A_{ii}=I_1$. Then show that $A_{ij}A_{ji}$ and $A_{ij}A_{jk}A_{ki}$ are also invariants. In fact, *all* contracted scalars of the form $A_{ij}A_{jk}\cdots A_{mi}$ are invariants. Finally, verify that $I_2=\frac{1}{2}[I_1^2-A_{ij}A_{ji}]$, and $I_3=\frac{1}{3}[A_{ij}A_{jk}A_{ki}-I_1A_{ij}A_{ji}+I_2A_{ii}]$. Because the right-hand sides are invariant, so are I_2 and I_3.]

2.12 If **u** and **v** are vectors, show that the products u_iv_j obey the transformation rule (2.12) in any number of spatial dimensions, and therefore represent a second-order tensor.

2.13 Show that δ_{ij} is an isotropic tensor. That is, show that $\delta_{ij}'=\delta_{ij}$ under rotation of the coordinate system. [*Hint*: Use the transformation rule (2.12) and the results of Exercise 2.10.]

2.14 If **u** and **v** are arbitrary vectors resolved in three-dimensional Cartesian coordinates, use the definition of vector magnitude, $|\mathbf{a}|^2=\mathbf{a}\cdot\mathbf{a}$, and the Pythagorean theorem to show that $\mathbf{u}\cdot\mathbf{v}=0$ when **u** and **v** are perpendicular.

2.15 If **u** and **v** are vectors with magnitudes u and υ, use the finding of Exercise 2.14 to show that $\mathbf{u}\cdot\mathbf{v}=u\upsilon\cos\theta$ where θ is the angle between **u** and **v**.

2.16 Determine the components of the vector **w** in three-dimensional Cartesian coordinates when **w** is defined by: $\mathbf{u}\cdot\mathbf{w}=0$, $\mathbf{v}\cdot\mathbf{w}=0$, and $\mathbf{w}\cdot\mathbf{w}=u^2\upsilon^2\sin^2\theta$, where **u** and **v** are known vectors with components u_i and υ_i and magnitudes u and υ, respectively, and θ is the angle between **u** and **v**. Choose the sign(s) of the components of **w** so that $\mathbf{w}=\mathbf{e}_3$ when $\mathbf{u}=\mathbf{e}_1$ and $\mathbf{v}=\mathbf{e}_2$.

2.17 If a is a positive constant and **b** is a constant vector, determine the divergence and the curl of $\mathbf{u}=a\mathbf{x}/x^3$ and $\mathbf{u}=\mathbf{b}\times(\mathbf{x}/x^2)$ where $x=\sqrt{x_1^2+x_2^2+x_3^2}\equiv\sqrt{x_ix_i}$ is the length of **x**.

2.18 Obtain the recipe for the gradient of a scalar function in cylindrical polar coordinates from the integral definition (2.32).

2.19 Obtain the recipe for the divergence of a vector function in cylindrical polar coordinates from the integral definition (2.32).

2.20 Obtain the recipe for the divergence of a vector function in spherical polar coordinates from the integral definition (2.32).

2.21 The hydrodynamic torque exerted on an object can be expressed as $\iint_A\mathbf{r}\times(\mathbf{n}\cdot\boldsymbol{\sigma})\,dA$, where **r** is a position vector and $\boldsymbol{\sigma}$ is a symmetric tensor. Using index notation and the divergence theorem, show the torque can be expressed as $\iiint_V\mathbf{r}\times(\nabla\cdot\boldsymbol{\sigma})\,dV$. [*Hint*: $\partial r_i/\partial x_j=\delta_{ij}$]

2.22 Use the vector integral theorems to prove that $\nabla\cdot(\nabla\times\mathbf{u})=0$ for any twice-differentiable vector function **u** regardless of the coordinate system.

2.23 Use Stokes' theorem to prove that $\nabla\times(\nabla\phi)=0$ for any single-valued twice-differentiable scalar ϕ regardless of the coordinate system.

References

Sommerfeld, A., 1964. Mechanics of Deformable Bodies. Academic Press, New York (Chapter 1 contains brief but useful coverage of Cartesian tensors).

Further reading

Aris, R., 1962. Vectors, Tensors, and the Basic Equations of Fluid Mechanics. Prentice-Hall, Englewood Cliffs, NJ (This book gives a clear and easy treatment of tensors in Cartesian and non-Cartesian coordinates, with applications to fluid mechanics).

Prager, W., 1961. Introduction to Mechanics of Continua. Dover Publications, New York (Chapters 1 and 2 contain brief but useful coverage of Cartesian tensors).

Chapter 3

Kinematics⊛

Contents

3.1 Introduction and coordinate systems 59
3.2 Particle and field descriptions of fluid motion 61
3.3 Flow lines, fluid acceleration, and Galilean transformation 64
3.4 Strain and rotation rates 69
3.5 Kinematics of simple plane flows 73
3.6 Reynolds transport theorem 77
Exercises 79
References 83
Further reading 83

Chapter objectives

- To review the basic Cartesian and curvilinear coordinate systems
- To link fluid flow kinematics with the particle kinematics
- To define the various flow lines in unsteady fluid velocity fields
- To present fluid particle acceleration in the Eulerian flow-field formulation
- To establish the fundamental meaning of the strain rate and rotation tensors
- To present the means for time differentiating general three-dimensional volume integrals

3.1 Introduction and coordinate systems

Kinematics is the study of motion without reference to the forces or stresses that produce the motion. In this chapter, fluid kinematics is presented in two and three dimensions starting with simple fluid-particle-path concepts and then proceeding to topics of greater complexity. These include: particle- and field-based descriptions for the time-dependent position, velocity, and acceleration of fluid particles; the relationship between the fluid velocity gradient tensor and the deformation and rotation of fluid elements; and the general mathematical relationships that govern arbitrary volumes that move and deform within flow fields. The forces and stresses that cause fluid motion are considered in subsequent chapters covering the *dynamics* or *kinetics* of fluid motion.

In general, three independent spatial dimensions and time are needed to fully describe fluid motion. When a flow does not depend on time, it is called *steady*; when it does depend on time it is called *unsteady*. In addition, fluid motion is studied in fewer than three dimensions whenever possible because the necessary analysis is usually simpler and relevant phenomena are more easily understood and visualized.

A truly *one-dimensional flow* is one in which the flow's characteristics can be entirely described with one independent spatial variable. Few real flows are strictly one dimensional, although flows in long, straight constant-cross-section conduits come close. Here, the independent coordinate may be aligned with the flow direction, as in the case of low-frequency pulsations in a pipe as shown in Fig. 3.1a, where z is the independent coordinate and darker gray indicates higher gas density. Alternatively, the independent coordinate may be aligned in the cross-stream direction, as in the case of viscous flow in a round tube where the radial distance, R, from the tube's centerline is the independent coordinate (Fig. 3.1b). In addition, higher dimensional flows are sometimes analyzed in one dimension by averaging the properties of the higher dimensional flow over an appropriate distance or area (Fig. 3.1c and d).

A *two-dimensional*, or *plane*, flow is one in which the variation of flow characteristics can be described by two spatial coordinates. The flow of an ideal fluid past a circular cylinder of infinite length having its axis perpendicular to the primary flow direction is an example of a plane flow (see Fig. 3.2a). This definition of two-dimensional flow officially includes the flow around bodies of revolution where flow characteristics are identical in any plane that contains the body's axis (e.g., flow past a sphere). However, such flows are customarily called *three-dimensional axisymmetric flows*.

⊛. This book has a companion website hosting complementary materials. Visit this URL to access it: https://www.elsevier.com/books-and-journals/book-companion/9780128198070.

Fluid Mechanics. https://doi.org/10.1016/B978-0-12-819807-0.00012-0

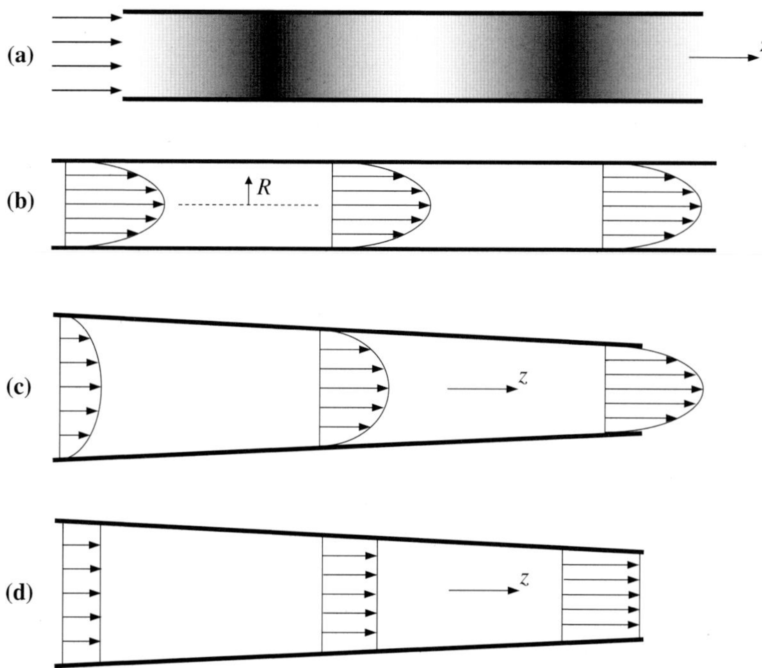

FIGURE 3.1 (a) Example of a one-dimensional fluid flow in which the gas density, shown by the grayscale, varies in the stream-wise z direction but not in the cross-stream direction. (b) Example of a one-dimensional fluid flow in which the fluid velocity varies in the cross-stream R direction but not in the stream-wise direction. (c) Example of a two-dimensional fluid flow where the fluid velocity varies in the cross-stream and stream-wise directions. (d) The one-dimensional approximation to the flow shown in part (c). Here the approximate flow field varies only in the stream-wise z direction. In (c) and (d), these arguments are for the primary velocity component in the z-direction; in such a tapered duct there must be a vertical velocity component that varies in the vertical direction, too.

A *three-dimensional* flow is one that can only be properly described with three independent spatial coordinates and is the most general case considered in this text. Sometimes curvilinear coordinates that match flow-field boundaries or symmetries greatly simplify the analysis and description of flow fields. Thus, several different coordinate systems are used in this text (see Fig. 3.3). Two-dimensional (plane) Cartesian and polar coordinates for an arbitrary point P (Fig. 3.3a) may be denoted by the coordinate pairs (x, y), (x_1, x_2), or (r, θ) with the corresponding velocity components (u, v), (u_1, u_2), or (u_r, u_θ). Here the axis perpendicular to the plane of interest will be the z-axis or x_3-axis. In three dimensions, Cartesian coordinates (Figs. 2.1 and 3.3b) may be used to locate a point P via the coordinate triplets (x, y, z) or (x_1, x_2, x_3) with corresponding velocity components (u, v, w) or (u_1, u_2, u_3). Cylindrical polar coordinates for P (Fig. 3.3c) are denoted by (R, φ, z) with corresponding velocity components (u_R, u_φ, u_z). Spherical polar coordinates for P (Fig. 3.3d) are denoted by (r, θ, φ) with the corresponding velocity components $(u_r, u_\theta, u_\varphi)$. In all cases, unit vectors are denoted by $\mathbf{e}$ with an appropriate subscript as in (2.1) and Fig. 2.1. More information about these coordinate systems is provided in Appendix B.

Example 3.1

Write the unidirectional velocity field $\mathbf{u} = V\mathbf{e}_y$ in spherical coordinates, and the uniform outflow velocity field $\mathbf{u} = U\mathbf{e}_r$ in Cartesian coordinates.

Solution The coordinate system descriptions in Fig. 3.3 and the information in Appendix B.5 are needed here. The first flow field, $\mathbf{u} = V\mathbf{e}_y$, represents a uniform velocity in the y-direction, and it is specified with a Cartesian coordinate unit vector. To find its components $(u_r, u_\theta, u_\varphi)$ in spherical coordinates, compute the appropriate dot products: $u_r = \mathbf{u} \cdot \mathbf{e}_r = V\mathbf{e}_y \cdot \mathbf{e}_r = V \sin\theta \sin\varphi$, $u_\theta = \mathbf{u} \cdot \mathbf{e}_\theta = V\mathbf{e}_y \cdot \mathbf{e}_\theta = V \cos\theta \sin\varphi$, and $u_\varphi = \mathbf{u} \cdot \mathbf{e}_\varphi = V\mathbf{e}_y \cdot \mathbf{e}_\varphi = V \cos\varphi$, so:

$$\mathbf{u} = \mathbf{e}_r V \sin\theta \sin\varphi + \mathbf{e}_\theta V \cos\theta \sin\varphi + \mathbf{e}_\varphi V \cos\varphi.$$

The second flow field $\mathbf{u} = U\mathbf{e}_r$, represents constant radial flow away from the origin of coordinates, and it is specified with a spherical coordinate unit vector. To find its components (u, v, w) in Cartesian coordinates, compute the appropriate dot products:

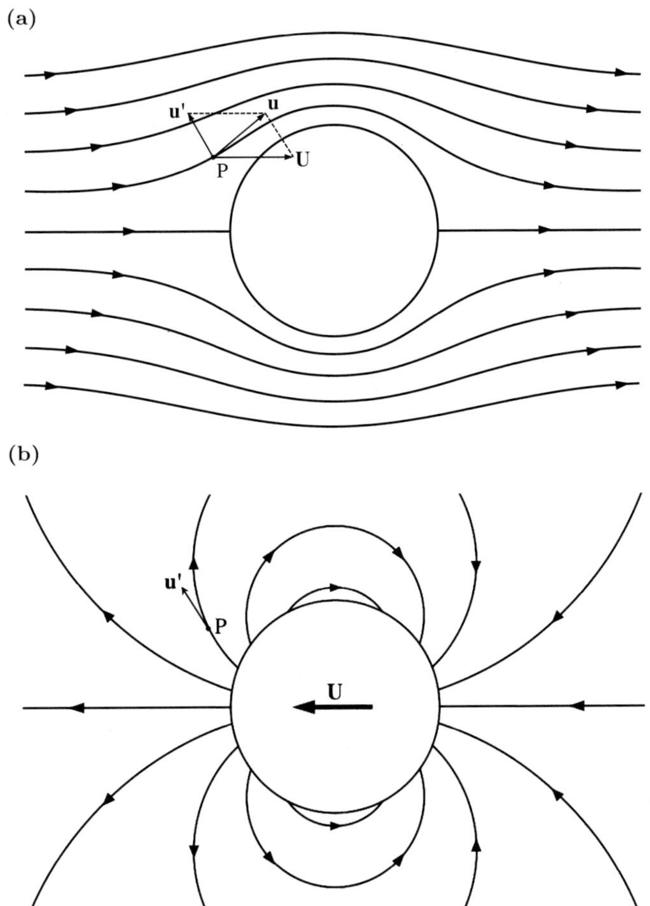

FIGURE 3.2 Sample flow fields where two spatial coordinates are needed. (a) Steady flow of an ideal incompressible fluid past a long stationary circular cylinder with its axis perpendicular to the flow. Here the total fluid velocity **u** at point P can be considered a sum of the flow velocity far from the cylinder **U**, and a velocity component **u**′ caused by the presence of the cylinder. (b) Unsteady flow of a nominally quiescent ideal incompressible fluid around a moving long circular cylinder with its axis perpendicular to the page. Here the cylinder velocity **U** is shown inside the cylinder, and the fluid velocity **u**′ at point P is caused by the presence of the moving cylinder alone. Although the two fields look quite different, they only differ by a Galilean transformation. The streamlines in (a) can be changed to those in (b) by switching to a frame of reference where the fluid far from the cylinder is motionless.

$u = \mathbf{u}\cdot\mathbf{e}_x = U\mathbf{e}_r\cdot\mathbf{e}_x = U\sin\theta\cos\varphi$, $v = \mathbf{u}\cdot\mathbf{e}_y = U\mathbf{e}_r\cdot\mathbf{e}_y = U\sin\theta\sin\phi$, and $w = \mathbf{u}\cdot\mathbf{e}_z = U\mathbf{e}_r\cdot\mathbf{e}_z = U\cos\theta$, so:

$$\mathbf{u} = \mathbf{e}_x U\sin\theta\cos\varphi + \mathbf{e}_y U\sin\theta\sin\varphi + \mathbf{e}_z U\cos\theta.$$

The flow-field specifications in the problem statement are much simpler than either answer. This fact motivates selection of a coordinate system that matches the flow geometry. Mathematical expressions for velocity fields, boundary conditions, and other aspects of fluid flows are much simpler when stated using an appropriate coordinate system.

3.2 Particle and field descriptions of fluid motion

There are two ways to describe fluid motion. In the *Lagrangian* description, fluid particles are followed as they move through a flow field. In the *Eulerian* description, a flow field's characteristics are monitored at fixed locations or in stationary regions of space. In fluid mechanics, an understanding of both descriptions is necessary because the acceleration following a fluid particle is needed for application of Newton's second law to fluid motion while observations, measurements, and

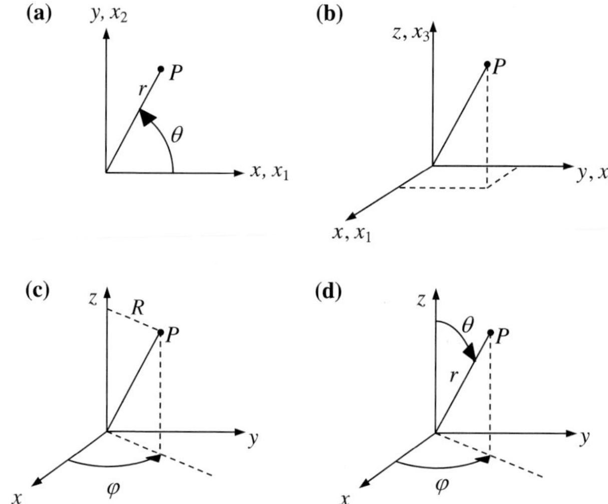

FIGURE 3.3 Coordinate systems commonly used in this text. In each case P is an arbitrary point away from the origin. (a) Plane Cartesian or polar coordinates where P is located by the coordinate pairs (x, y), (x_1, x_2), or (r, θ). (b) Three-dimensional Cartesian coordinates where P is located by the coordinate triplets (x, y, z) or (x_1, x_2, x_3). (c) Cylindrical polar coordinates where P is located by the coordinate triplet (R, φ, z). (d) Spherical polar coordinates where P is located by the coordinate triplet (r, θ, φ).

simulations of fluid flows are commonly made at fixed locations or in stationary spatial regions with the fluid moving past these locations or through the regions of interest.

The Lagrangian description is the direct extension of single particle kinematics (e.g., see Meriam and Kraige, 2007) to a whole field of fluid particles that are labeled by their location, $\mathbf{r}_o$, at a reference time, $t = t_o$. The subsequent position $\mathbf{r}$ of each fluid particle as a function of time, $\mathbf{r}(t; \mathbf{r}_o, t_o)$, specifies the flow field. Here, $\mathbf{r}_o$ and t_o are boundary or initial condition parameters that label fluid particles, and are not independent variables. Thus, the current velocity $\mathbf{u}$ and acceleration $\mathbf{a}$ of the fluid particle that was located at $\mathbf{r}_o$ at time t_o are obtained from the first and second temporal derivatives of particle position $\mathbf{r}(t; \mathbf{r}_o, t_o)$:

$$\mathbf{u} = d\mathbf{r}(t; \mathbf{r}_o, t_o)/dt \text{ and } \mathbf{a} = d^2\mathbf{r}(t; \mathbf{r}_o, t_o)/dt^2. \quad (3.1)$$

These are Lagrangian definitions for $\mathbf{u}$ and $\mathbf{a}$, and are valid for any fluid particle as it moves along its trajectory through the flow field (Fig. 3.4). In this particle-based description of fluid motion, fluid particle kinematics are identical to that in ordinary particle mechanics, and any scalar, vector, or tensor flow-field property F may depend on the path(s) followed by the relevant fluid particle(s) and time: $F = F[\mathbf{r}(t; \mathbf{r}_o, t_o), t]$. The Lagrangian description of fluid motion is used in some simulations of combustion and multiphase flows, and in cinema-graphic animations.

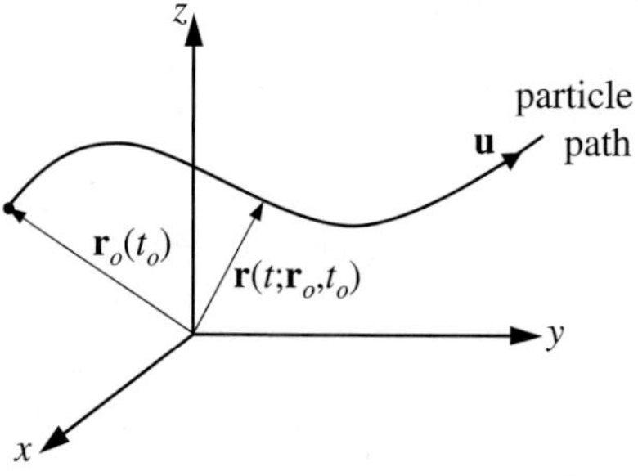

FIGURE 3.4 Lagrangian description of the motion of a fluid particle at location $\mathbf{r}_o$ at time t_o. The particle path or particle trajectory $\mathbf{r}(t; \mathbf{r}_o, t_o)$ specifies the location of the fluid particle at later times.

The Eulerian description of fluid kinematics focuses on flow field properties at locations or in regions of interest, and involves four independent variables: the three spatial coordinates represented by the position vector $\mathbf{x}$, and time t. Thus, in this field-based description, a flow-field property F depends directly on $\mathbf{x}$ and t: $F = F(\mathbf{x}, t)$, which is often convenient in describing fluid motion, as individual particles need not be followed.

Kinematic relationships between the two descriptions can be determined by requiring equality of flow-field properties when $\mathbf{r}$ and $\mathbf{x}$ define the same point in space, both are resolved in the same coordinate system, and a common clock is used to determine the time t:

$$F[\mathbf{r}(t;\mathbf{r}_o,t_o),t] = F(\mathbf{x},t) \text{ when } \mathbf{x} = \mathbf{r}(t;\mathbf{r}_o,t_o). \tag{3.2}$$

Here the second equation specifies the trajectory followed by a fluid particle. This compatibility requirement forms the basis for determining and interpreting time derivatives in the Eulerian description of fluid motion. Applying a total time derivative to the first equation in (3.2) produces:

$$\frac{d}{dt}F[\mathbf{r}(t;\mathbf{r}_o,t_o)t] = \frac{\partial F}{\partial r_1}\frac{dr_1}{dt} + \frac{\partial F}{\partial r_2}\frac{dr_2}{dt} + \frac{\partial F}{\partial r_3}\frac{dr_3}{dt} + \frac{\partial F}{\partial t} = \frac{d}{dt}F(\mathbf{x},t) \text{ when } \mathbf{x} = \mathbf{r}(t;\mathbf{r}_o,t_o), \tag{3.3}$$

where the components of $\mathbf{r}$ are r_i. In (3.3), the time derivatives of r_i are the components u_i of the fluid particle's velocity $\mathbf{u}$ from (3.1). In addition, $\partial F/\partial r_i = \partial F/\partial x_i$ when $\mathbf{x} = \mathbf{r}$, so the right side of (3.3) can be rewritten entirely in the Eulerian description:

$$\frac{d}{dt}F[\mathbf{r}(t;\mathbf{r}_o,t_o),t] = \frac{\partial F}{\partial x_1}u_1 + \frac{\partial F}{\partial x_2}u_2 + \frac{\partial F}{\partial x_3}u_3 + \frac{\partial F}{\partial t} = (\nabla F)\cdot\mathbf{u} + \frac{\partial F}{\partial t} \equiv \frac{D}{Dt}F(\mathbf{x},t), \tag{3.4}$$

where the final equality defines D/Dt as the total time derivative in the Eulerian description of fluid motion. It is the equivalent of the total time derivative d/dt in the Lagrangian description and is known as the *material derivative*, *substantial derivative*, or *particle derivative*, where the final attribution emphasizes the fact that it provides time derivative information following a fluid particle.

The material derivative D/Dt defined in (3.4) is composed of unsteady and advective parts. (1) The *unsteady* part of DF/Dt is $\partial F/\partial t$ and represents the *local* temporal rate of change of F at the location $\mathbf{x}$. It is zero when F is independent of time. (2) The *advective* (or *convective*) part of DF/Dt is $\mathbf{u}\cdot\nabla F$ and represents the rate of change of F that occurs as fluid particles move from one location to another. It is zero where F is spatially uniform, the fluid is not moving, or $\mathbf{u}$ and ∇F are perpendicular. For clarity and consistency in this book, the movement of fluid particles from place to place is referred to as *advection* with the term *convection* being reserved for the special circumstance of heat transport by fluid movement. In vector and index notations, (3.4) is commonly rearranged slightly and written as:

$$\frac{DF}{Dt} \equiv \frac{\partial F}{\partial t} + \mathbf{u}\cdot\nabla F, \text{ or } \frac{DF}{Dt} \equiv \frac{\partial F}{\partial t} + u_i\frac{\partial F}{\partial x_i}. \tag{3.5}$$

The scalar product $\mathbf{u}\cdot\nabla F$ is the magnitude of $\mathbf{u}$ times the component of ∇F in the direction of $\mathbf{u}$ so (3.5) can then be written in scalar notation as:

$$\frac{DF}{Dt} \equiv \frac{\partial F}{\partial t} + |\mathbf{u}|\frac{\partial F}{\partial s}, \tag{3.6}$$

where s is a path-length coordinate along the fluid particle trajectory $\mathbf{x} = \mathbf{r}(t;\mathbf{r}_o,t_o)$, that is, $d\mathbf{r} = \mathbf{e}_\mathbf{u}ds$ with $\mathbf{e}_\mathbf{u} = \mathbf{u}/|\mathbf{u}|$.

Example 3.2

A fluid particle in a steady flow moves along the x-axis. Its distance from the origin is r_o at time t_o and its trajectory is $r(t) = [K(t-t_o) + r_o^3]^{1/3}$, where K is a positive constant with units of volume/time. Determine this flow's Eulerian velocity and acceleration, $u(x)$ and $a(x)$, and show that $a(x)$ may also be obtained from $Du(x)/Dt$.

Solution First, perform the differentiations indicated in (3.1) to determine the particle's Lagrangian velocity and acceleration:

$$u(t) = \frac{dr}{dt} = \frac{K}{3}[K(t-t_o) + r_o^3]^{-2/3} \quad \text{and}$$
$$a(t) = \frac{d^2r}{dt^2} = -\frac{2K^2}{9}[K(t-t_o) + r_o^3]^{-5/3}.$$

To find the Eulerian fluid velocity and acceleration (functions of position), require $u(x)$ and $a(x)$ to be equal to $u(t)$ and $a(t)$ when $x = r(t)$. This step can be completed by substituting the given form of $r(t)$ into the equations for $u(t)$ and $a(t)$ to eliminate the particle-specification information (r_o and t_o) as follows:

$$u(x) = [u(t)]_{x=r(t)} = \left[\frac{K}{3r^2}\right]_{x=r} = \frac{K}{3x^2}, \quad \text{and}$$
$$a(x) = [a(t)]_{x=r(t)} = \left[-\frac{2K^2}{9r^5}\right]_{x=r} = -\frac{2K}{9x^5}.$$

To show that this $a(x)$ may also be obtained from $Du(x)/Dt$, use (3.5) with $F = u(x)$. Here, the Eulerian velocity does not depend on time and it is unidirectional, so the partial time derivative in $Du(x)/Dt$ drops out, and the dot product in $Du(x)/Dt$ simplifies to a single term. This lone dot-product term yields the desired result:

$$\frac{Du}{Dt} = \frac{\partial u(x)}{\partial t} + u(x)\frac{\partial u(x)}{\partial x} = 0 + \frac{K}{3x^2}\frac{\partial}{\partial x}\left(\frac{K}{3x^2}\right) = \frac{K}{3x^2}\left(\frac{-2K}{3x^3}\right) = -\frac{2K}{9x^5}.$$

3.3 Flow lines, fluid acceleration, and Galilean transformation

In the Eulerian description, three types of curves are commonly used to describe fluid motion – streamlines, path lines, and streak lines. These are defined and described here assuming that the fluid velocity vector, $\mathbf{u}$, is known at every point of space and instant of time throughout the region of interest. These curves coincide when the flow is steady, and are often valuable for understanding fluid motion. They form the basis for experimental techniques that track seed particles or dye filaments. Pictorial and photographic examples of flow lines can be found in specialty volumes devoted to flow visualization (Van Dyke, 1982; Samimy et al., 2003).

A *streamline* is a curve that is instantaneously tangent to the fluid velocity throughout the flow field. In unsteady flows the streamline pattern changes with time. In Cartesian coordinates, if $d\mathbf{s} = (dx, dy, dz)$ is an element of arc length along a streamline (Fig. 3.5) and $\mathbf{u} = (u, v, w)$ is the local fluid velocity vector, then the tangency requirement on $d\mathbf{s}$ and $\mathbf{u}$ leads to:

$$dx/u = dy/v = dz/w, \tag{3.7}$$

(see Exercise 3.5), and $\mathbf{u} \times d\mathbf{s} = 0$ because $d\mathbf{s}$ and $\mathbf{u}$ are locally parallel. Integrating (3.7) in both the upstream and downstream directions from a variety of reference locations allows streamlines to be determined throughout the flow field. If these reference locations lie on a closed curve C, the resulting stream surface is called a *stream tube* (Fig. 3.6). No fluid crosses a stream tube's surface because the fluid velocity vector is everywhere tangent to it. Streamlines are useful in the depiction of flow fields and important for calculations involving simplifications (Bernoulli equations) of the full equations of fluid motion. In experiments, streamlines may be visualized by particle streak photography or by integrating (3.7) using measured velocity fields.

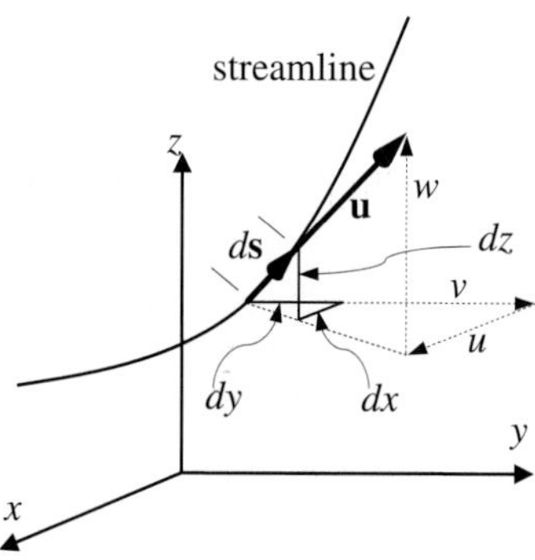

FIGURE 3.5 Streamline geometry. The arc-length element of a streamline, ds, is locally tangent to the fluid velocity $\mathbf{u}$ so its components and the components of the velocity must follow (3.7).

A *path line* is the trajectory of a fluid particle of fixed identity. It is defined in (3.2) and (3.3) as $\mathbf{x} = \mathbf{r}(t; \mathbf{r}_o, t_o)$. The equation of the path line for the fluid particle launched from $\mathbf{r}_o$ at t_o is obtained from the fluid velocity $\mathbf{u}$ by integrating:

$$d\mathbf{r}/dt = [\mathbf{u}(\mathbf{x}, t)]_{x=r} = \mathbf{u}(\mathbf{r}, t) \tag{3.8}$$

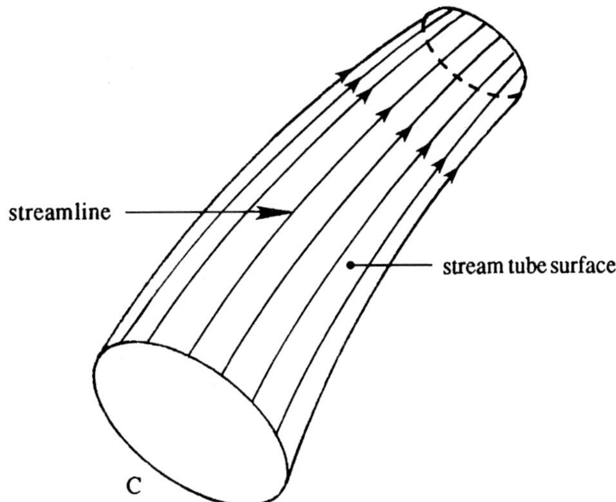

FIGURE 3.6 Stream tube geometry for the closed curve C.

subject to the requirement $\mathbf{r}(t_o) = \mathbf{r}_o$. Other path lines are obtained by integrating (3.8) from different values of $\mathbf{r}_o$ or t_o. A discretized version of (3.8) is the basis for particle image velocimetry (PIV), a popular and powerful flow field measurement technique (Raffel et al., 1998).

A *streak line* is the curve obtained by connecting all the fluid particles that will pass, or have passed through, a fixed point in space. The streak line through the point $\mathbf{x}_o$ at time t is found by integrating (3.8) for all relevant reference times, t_o, subject to the requirement $\mathbf{r}(t_o) = \mathbf{x}_o$. When completed, this integration provides a path line, $\mathbf{x} = \mathbf{r}(t; \mathbf{x}_o, t_o)$, for each value of t_o. At a fixed time t, the components of these path-line equations, $x_i = r_i(t; \mathbf{x}_o, t_o)$, provide a parametric specification of the streak line with t_o as the parameter. Alternatively, these path-line component equations can sometimes be combined to eliminate t_o and thereby produce an equation that directly specifies the streak line through the point $\mathbf{x}_o$ at time t. Streak lines may be visualized in experiments by injecting a passive marker, like dye or smoke, from a small port and observing where it goes as it is carried through the flow field.

Example 3.3

In two-dimensional (x, y)-Cartesian coordinates, determine the streamline, path line, and streak line that pass through the origin of coordinates at $t = t'$ in the unsteady two-dimensional near-surface flow field typical of long-wavelength water waves with amplitude ξ_o: $u = \omega\xi_o \cos(\omega t)$ and $v = \omega\xi_o \sin(\omega t)$.

Streamline solution

Utilize the first equality in (3.7) to find:

$$\frac{dy}{dx} = \frac{v}{u} = \frac{\omega\xi_o \sin(\omega t')}{\omega\xi_o \cos(\omega t')} = \tan(\omega t').$$

Integrating once produces: $y = x\tan(\omega t') + \text{const}$. For the streamline to pass through the origin $(x = y = 0)$, the constant must equal zero, so the streamline equation is: $y = x\tan(\omega t')$.

Path-line solution

Set $\mathbf{r} = [(x(t), y(t)]$, and use both components of (3.8) to find:

$$dx/dt = u = \omega\xi_o \cos(\omega t), \text{ and } dy/dt = v = \omega\xi_o \sin(\omega t).$$

Integrate each of these equations once to find: $x = \xi_o \sin(\omega t) + x_o$, and $y = -\xi_o \cos(\omega t) + y_o$, where x_o and y_o are integration constants. The path-line requirement at $x = y = 0$ and $t = t'$ implies $x_o = -\xi_o \sin(\omega t')$, and $y_o = \xi_o \cos(\omega t')$, so the path-line component equations are:

$$x = \xi_o[\sin(\omega t) - \sin(\omega t')] \text{ and } y = \xi_o[-\cos(\omega t) + \cos(\omega t')].$$

Here, the time variable t can be eliminated via a little algebra to find:

$$(x + \xi_o \sin(\omega t'))^2 + (y - \xi_o \cos(\omega t'))^2 = \xi_o^2,$$

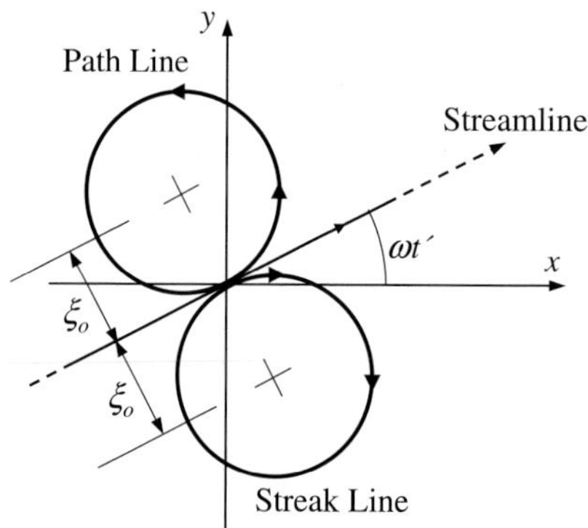

FIGURE 3.7 Streamline, path line, and streak line for Example 3.3. All three are distinct because the flow field is unsteady.

which is the equation of a circle of radius ξ_o centered on the location $[-\xi_o \sin(\omega t'), \xi_o \cos(\omega t')]$.

Streak-line solution

To determine the streak line that passes through the origin of coordinates at $t = t'$, the location of the fluid particle that passed through $x = y = 0$ at $t = t_o$ must be found. Use the path-line results above but evaluate at t_o instead of t' to find different constants. Thus the parametric streak-line component equations are:

$$x = \xi_o[\sin(\omega t) - \sin(\omega t_o)] \text{ and } y = \xi_o[-\cos(\omega t) + \cos(\omega t_o)].$$

Combine these equations to eliminate t_o and evaluate the result at $t = t'$ to find the required streak line:

$$(x - \xi_o \sin(\omega t'))^2 + (y + \xi_o \cos\cos(\omega t'))^2 = \xi_o^2.$$

This is the equation of a circle of radius ξ_o centered on the location $[\xi_o \sin(\omega t'), -\xi_o \cos(\omega t')]$. The three flow lines in this example are shown in Fig. 3.7. In this case, the streamline, path line, and streak line are all tangent to each other at the origin of coordinates.

From this example, it should be clear that streamlines, path lines, and streak lines differ in an unsteady flow field. This situation is also illustrated in Fig. 3.2, which shows streamlines when there is relative motion of a circular cylinder and an ideal fluid. Fig. 3.2a shows streamlines for a stationary cylinder with the fluid moving past it, a steady flow. Here, fluid particles that approach the cylinder are forced to move up or down to go around it. Fig. 3.2b shows streamlines for a moving cylinder in a nominally quiescent fluid, an unsteady flow. Streamlines originate on the left side of the advancing cylinder where fluid particles are pushed to the left to make room for the cylinder. These streamlines curve backward and fluid particles move rightward at the cylinder's widest point. These streamlines terminate on the right side of the cylinder where fluid particles again move to the left to fill in the region behind the moving cylinder. Although their streamline patterns appear dissimilar, these flow fields only differ by a Galilean transformation. Consider the fluid velocity at a point P that lies at the same location relative to the cylinder in both fields. If $\mathbf{u}'$ is the fluid velocity at P in Fig. 3.2b where the cylinder is moving at speed $\mathbf{U}$, then the fluid velocity $\mathbf{u}$ at P in Fig. 3.2a is $\mathbf{u} = \mathbf{U} + \mathbf{u}'$. If $\mathbf{U}$ is constant, the fluid acceleration in both fields must be the same at the same location relative to the cylinder.

This expectation can be verified in general using (3.5) with F replaced by the fluid velocity observed in different coordinate frames. Consider a Cartesian coordinate system $O'x'y'z'$ that moves at a constant velocity $\mathbf{U}$ with respect to a stationary system $Oxyz$ having parallel axes (Fig. 3.8). The fluid velocity $\mathbf{u}'(\mathbf{x}', t')$ observed in $O'x'y'z'$ will be related to the fluid velocity $\mathbf{u}(\mathbf{x}, t)$ observed in $Oxyz$ by $\mathbf{u}(\mathbf{x}, t) = \mathbf{U} + \mathbf{u}'(\mathbf{x}', t')$ when $t = t'$ and $\mathbf{x} = \mathbf{x}' + \mathbf{U}t + \mathbf{x}'_o$, where $\mathbf{x}'_o$ is the vector distance from O to O′ at $t = 0$. Under these conditions it can be shown that:

$$\frac{\partial \mathbf{u}}{\partial t} + (\mathbf{u} \cdot \nabla)\mathbf{u} = \left(\frac{D\mathbf{u}}{Dt}\right)_{in\ Oxyz} = \left(\frac{D\mathbf{u}'}{Dt'}\right)_{in\ O'x'y'z'} = \frac{\partial \mathbf{u}'}{\partial t'} + (\mathbf{u}' \cdot \nabla')\mathbf{u}' \tag{3.9}$$

(Exercise 3.14), where ∇' operates on the primed coordinates. The first and second terms of the left most part of (3.9) are the unsteady and advective acceleration terms in $Oxyz$. The unsteady acceleration term, $\partial \mathbf{u}/\partial t$, is nonzero at $\mathbf{x}$ when $\mathbf{u}$ varies with time at $\mathbf{x}$. It is zero everywhere when the flow is steady. The advective acceleration term, $(\mathbf{u} \cdot \nabla)\mathbf{u}$, is nonzero when fluid particles move between locations where the fluid velocity is different. It is zero when the fluid velocity is zero, the fluid velocity is uniform in space, or when the fluid velocity only varies in the cross-stream direction. In addition, the unsteady term is linear in $\mathbf{u}$ while the advective term is nonlinear (quadratic) in $\mathbf{u}$. This nonlinearity is a primary feature of fluid mechanics. When $\mathbf{u}$ is small enough for this nonlinearity to be ignored, fluid mechanics reduces to acoustics or, when $\mathbf{u} = 0$, to fluid statics.

When examined together, the sample flow fields in Fig. 3.2 and the Galilean invariance of the Eulerian fluid acceleration, (3.9), show that the relative importance of the steady and advective fluid-acceleration terms depends on the frame of reference of the observer. Fig. 3.2a depicts a steady flow where the streamlines do not depend on time. Thus, the unsteady acceleration term, $\partial \mathbf{u}/\partial t$, is zero. However, the streamlines do bend in the vicinity of the cylinder so fluid particles must feel some acceleration because the absence of fluid-particle acceleration in a flow field corresponds to constant fluid-particle velocity and straight streamlines. Therefore, the advective acceleration term, $(\mathbf{u} \cdot \nabla)\mathbf{u}$, is nonzero for the flow in Fig. 3.2a. In Fig. 3.2b, the flow is unsteady and the streamlines are curved, so both acceleration terms in the right-most part of (3.9) are nonzero. These observations imply that a Galilean transformation can alter the relative importance of the unsteady and advective fluid acceleration terms without changing the overall fluid-particle acceleration. Thus, an astutely chosen steadily moving coordinate system can be used to enhance (or reduce) the relative importance of either the unsteady or advective fluid-acceleration term.

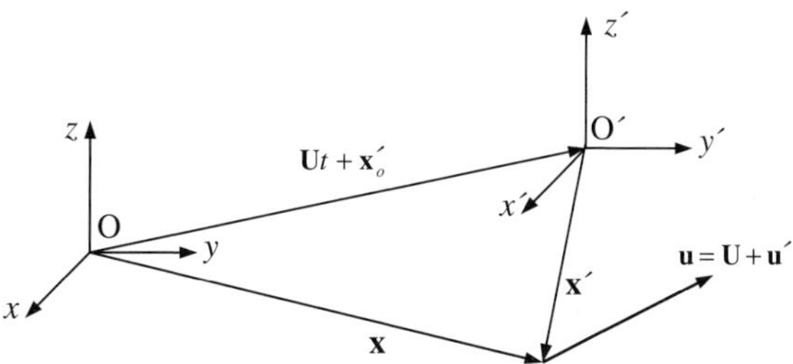

FIGURE 3.8 Geometry for showing that the fluid particle acceleration as determined by (3.9) is independent of the frame of reference when the frames differ by a Galilean transformation. Here $Oxyz$ is stationary and $O'x'y'z'$ moves with respect to it at a constant speed $\mathbf{U}$, the axes of the two frames are parallel, and $\mathbf{x}$ and $\mathbf{x}'$ represent the same location. The fluid velocity observed at $\mathbf{x}$ in frame $Oxyz$ is $\mathbf{u}$. The fluid velocity observed at $\mathbf{x}'$ in frame $O'x'y'z'$ is $\mathbf{u}'$.

Additional insights into the character of the unsteady and advective acceleration terms might also be obtained from the reader's observations and experiences. For example, a nonzero unsteady acceleration is readily observed at any street intersection regulated by a traffic light with the moving or stationary vehicles playing the role of fluid particles. Here, a change in the traffic light may halt east-west vehicle flow and allow north-south vehicle flow to begin, thereby producing a time-dependent 90° rotation of the traffic-flow streamlines at the intersection location. Similarly, a nonzero advective acceleration is readily felt by rollercoaster riders when an analogy is made between the rollercoaster track and a streamline. While stationary and waiting in line, soon-to-be rollercoaster riders can observe that the track's shape involves hills, curves, and bends, and that this shape does not depend on time. This situation is analogous to the stationary observer of a nontrivial steady fluid flow – like that depicted in Fig. 3.2a – who readily notes that streamlines curve and bend but do not depend on time. Thus, the unsteady acceleration term is zero for both the rollercoaster and a steady flow because both the rollercoaster cars and fluid particles travel through space on fixed-shape trajectories and achieve consistent (time-independent) velocities at any point along the track or streamline. However, anyone who has ever ridden a rollercoaster will know that significant acceleration is possible while following a rollercoaster's fixed-shape track because a rollercoaster car's velocity varies in magnitude and direction as it traverses the track. These velocity variations result from the advective acceleration, and fluid particles that follow curved fixed-shape streamlines experience it as well. Within this rollercoaster-streamline analogy a nonzero unsteady acceleration would correspond to rollercoaster cars and fluid particles following time-dependent paths. Such a possibility is certainly unusual for rollercoaster riders; rollercoaster tracks are nearly rigid, seldom fall down (thankfully), and are typically designed to produce consistent car velocities at each point along the track.

Example 3.4

In some cases, the local velocity field may be known, but its derivatives may not be easily obtained analytically, for example, if only discrete values are available from a numerical simulation. Consider an inviscid, constant-density flow past a two-dimensional half-body formed from a horizontal free stream and a point source at the origin. In Chapter 6, a solution is obtained from irrotational constant-density flow theory, where the two components of velocity in Cartesian coordinates are

$$u = U + \frac{q_s}{2\pi}\frac{x}{x^2+y^2} \quad \text{and} \quad v = \frac{q_s}{2\pi}\frac{y}{x^2+y^2},$$

where q_s is a constant (with units of length^2/time) that sets the point source's strength. Determine the streamlines numerically that pass through several points above and below the origin and plot the solution.

Solution The first equality in (3.7) is solved numerically using a first-order Euler integration scheme (see Chapter 15), according to $y_{j+1} = y_j + v/u\Delta x$, where the subscript $j = 0 \ldots N$ denotes the discrete point y is evaluated at, with y_0 specifying the point the streamline passes through, and the velocity fields are defined discretely on a grid evenly space by Δx. The code is provided below and the solution is plotted in Fig. 3.9.

```
% Input parameters
a=0.5;
U=1;
qs=2*pi*U*a;

% Define the x-coordinate
x=linspace(-8*a,8*a,400);
dx=x(2)-x(1);

% Points the streamlines pass through (note y0=0 is challenging numerically)
y0=[(-2:.4:-.4) -.01 .01 (.4:.4:2)]*a;
y=zeros(1,length(x));

figure()
hold all
for j=1:length(y0)
    y(1)=y0(j);
    for i=1:length(x)-1
        u=U+qs/(2*pi)*x(i)/(x(i)^2+y(i)^2);
        v=qs/(2*pi)*y(i)/(x(i)^2+y(i)^2);
        y(i+1)=y(i)+v/u*dx;
    end

    % Plot
    plot(x/a,y/a,'-k','linewidth', 2)
end
hold off
xlabel('$x/a$','interpreter','latex')
ylabel('$y/a$','interpreter','latex')
axis equal
```

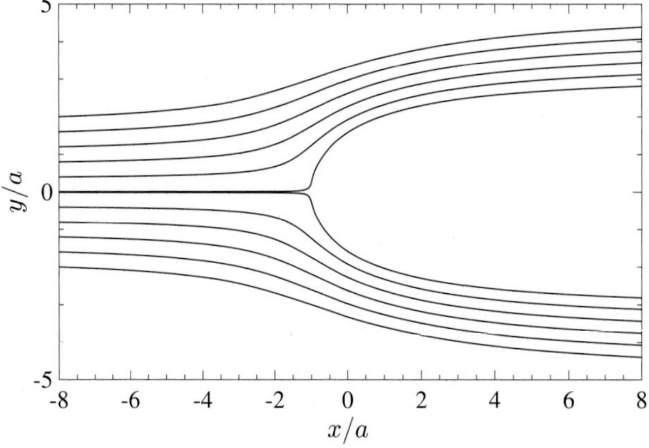

FIGURE 3.9 Solution to Example 3.4.

3.4 Strain and rotation rates

Given the definition of a fluid as a material that deforms continuously under the action of a shear stress, the basic constitutive law for fluids relates fluid element *deformation rates* to the stresses (surface forces per unit area) applied to a fluid element. This section describes fluid-element deformation and rotation rates in terms of the fluid *velocity gradient tensor*, $\partial u_i/\partial x_j$. The constitutive law for Newtonian fluids is covered in the next chapter. The various illustrations and interpretations provided here are analogous to their counterparts in solid mechanics when the fluid-appropriate *strain rate* (based on velocity $\mathbf{u}$) is replaced by the solid-appropriate *strain* (based on displacement).

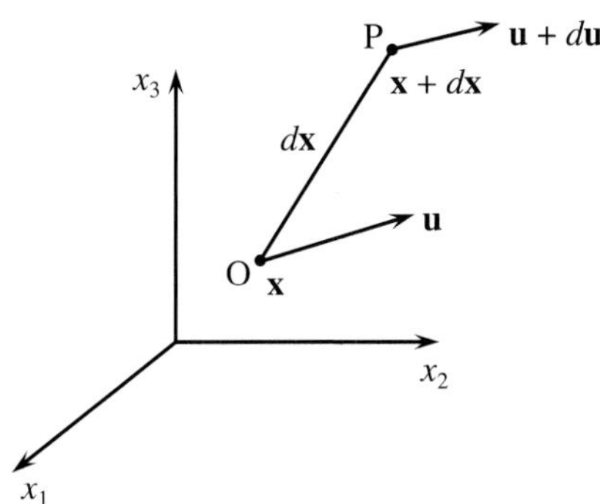

FIGURE 3.10 Velocity vectors $\mathbf{u}$ and $\mathbf{u} + d\mathbf{u}$ at two neighboring points O and P, respectively, that are separated by the short distance $d\mathbf{x}$.

The relative motion between two neighboring points can be written as the sum of the motion due to local rotation and deformation. Consider the situation depicted in Fig. 3.10, and let $\mathbf{u}(\mathbf{x}, t)$ be the velocity at point O (position vector $\mathbf{x}$), and let $\mathbf{u} + d\mathbf{u}$ be the velocity at the same time at a nearby neighboring point P (position vector $\mathbf{x} + d\mathbf{x}$). A three-dimensional first-order Taylor expansion of $\mathbf{u}$ about $\mathbf{x}$ leads to the following relationship between the components of $d\mathbf{u}$ and $d\mathbf{x}$:

$$du_i = \left(\partial u_i/\partial x_j\right) dx_j. \tag{3.10}$$

The term in parentheses in (3.10), $\partial u_i/\partial x_j$, specifies the components of the velocity gradient tensor, and it can be decomposed into symmetric, S_{ij}, and antisymmetric, R_{ij}, matrices:

$$\frac{\partial u_i}{\partial x_j} = S_{ij} + \frac{1}{2}R_{ij}, \quad \text{where } S_{ij} = \frac{1}{2}\left(\frac{\partial u_i}{\partial x_j} + \frac{\partial u_j}{\partial x_i}\right), \text{ and } R_{ij} = \frac{\partial u_i}{\partial x_j} - \frac{\partial u_j}{\partial x_i}. \tag{3.11, 3.12, 3.13}$$

Here, S_{ij} is the *strain rate tensor*, and R_{ij} is the *rotation tensor*. The decomposition of $\partial u_i/\partial x_j$ provided by (3.11) is important when formulating the conservation equations for fluid motion because S_{ij}, which embodies fluid element deformation, is related to the stress field in a moving fluid while R_{ij}, which embodies fluid element rotation, is not.

The strain rate tensor has on- and off-diagonal terms. The diagonal terms of S_{ij} represent elongation and contraction per unit length in the various coordinate directions, and are sometimes called *linear* strain rates. A geometrical interpretation of S_{ij}'s first component, S_{11}, is provided in Fig. 3.11. The rate-of-change of fluid element length in the x_1-direction per unit length is:

$$\frac{1}{\delta x_1}\frac{D}{Dt}(\delta x_1) = \lim_{dt\to 0}\frac{1}{dt}\left(\frac{A'B' - AB}{AB}\right) = \lim_{dt\to 0}\frac{1}{\delta x_1 dt}\left(\delta x_1 + \frac{\partial u_1}{\partial x_1}\delta x_1 dt - \delta x_1\right) = \frac{\partial u_1}{\partial x_1},$$

where D/Dt indicates that the fluid element is followed as extension takes place. This simple construction is readily extended to the other two Cartesian directions, and in general the linear strain rate in the η direction is $\partial u_\eta/\partial x_\eta$ where *no summation* over the repeated η-index is implied. (Greek subscripts are commonly used when the summation convention is not followed.)

The off-diagonal terms of S_{ij} represent shear deformations that change the relative orientations of material line segments initially parallel to the i- and j-directions in the flow. A geometrical interpretation of S_{ij}'s first off-diagonal component, $S_{12} = S_{21}$, is provided in Fig. 3.12. The average rate at which the initially perpendicular segments δx_1 and δx_2 rotate toward each other is:

$$\frac{1}{2}\frac{D(\alpha+\beta)}{Dt} = \lim_{dt\to 0}\frac{1}{2dt}\left(\frac{1}{\delta x_2}\left(\frac{\partial u_1}{\partial x_2}\delta x_2 dt\right) + \frac{1}{\delta x_1}\left(\frac{\partial u_2}{\partial x_1}\delta x_1 dt\right)\right) = \frac{1}{2}\left(\frac{\partial u_1}{\partial x_2} + \frac{\partial u_2}{\partial x_1}\right) = S_{12} = S_{21},$$

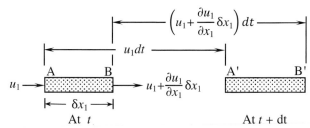

FIGURE 3.11 Illustration of positive linear strain rate in the first coordinate direction. Here, $A'B' = AB + BB' - AA'$, and a positive $S_{11} = \partial u_1/\partial x_1$ corresponds to a lengthening of the fluid element.

where again D/Dt indicates that the fluid element is followed as shear deformation takes place, and again this simple construction is readily extended to the other two Cartesian direction pairs. Thus, the off-diagonal terms of S_{ij} represent the average rate at which material line segments initially parallel to the i- and j-directions rotate *toward* each other.

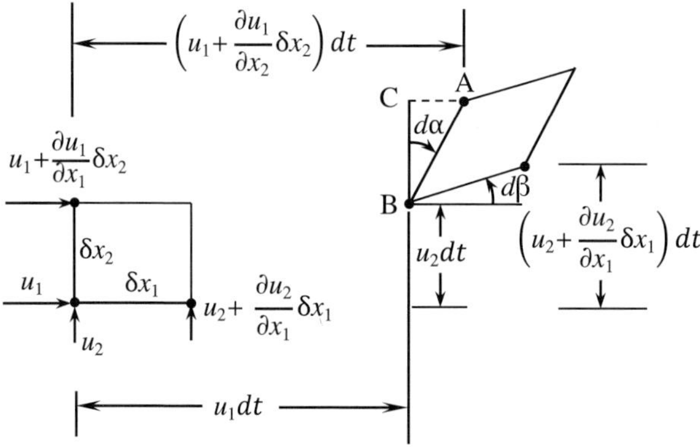

FIGURE 3.12 Illustration of positive deformation of a fluid element in the plane defined by the first and second coordinate directions. Here, both $\partial u_1/\partial x_2$ and $\partial u_2/\partial x_1$ are shown as positive, so $S_{12} = S_{21}$ from (3.12) is also positive. The deformation angle $d\alpha = \angle CBA$ is proportional to $\partial u_1/\partial x_2$ while $d\beta$ is proportional to $\partial u_2/\partial x_1$.

Here we also note that S_{ij} is zero for any rigid body motion composed of translation at a spatially uniform velocity **U** and rotation at a constant rate **Ω** (see Exercise 3.21). Thus, S_{ij} is independent of the frame of reference in which it is observed, even if **U** depends on time and the frame of reference is rotating.

The first invariant of S_{ij} (the sum of its diagonal terms) is the *volumetric strain rate* or *bulk strain rate*. For a small volume $\delta V = \delta x_1 \delta x_2 \delta x_3$, it can be shown (Exercise 3.22) that:

$$\frac{1}{\delta V}\frac{D}{Dt}(\delta V) = \frac{\partial u_1}{\partial x_1} + \frac{\partial u_2}{\partial x_2} + \frac{\partial u_3}{\partial x_3} = \frac{\partial u_i}{\partial x_i} = S_{ii}. \tag{3.14}$$

Thus, S_{ii} specifies the rate of volume change per unit volume and it does not depend on the orientation of the coordinate system.

The second member of the strain-rate decomposition (3.11) is the rotation tensor, R_{ij}. It is antisymmetric so its diagonal elements are zero and its off-diagonal elements are equal and opposite. Furthermore, its three independent elements can be put in correspondence with a vector. From (2.26), (2.27), or (3.13), this vector is the *vorticity*, $\boldsymbol{\omega} = \nabla \times \mathbf{u}$, and the correspondence is:

$$R_{ij} = -\varepsilon_{ijk}\omega_k = \begin{bmatrix} 0 & -\omega_3 & \omega_2 \\ \omega_3 & 0 & -\omega_1 \\ -\omega_2 & \omega_1 & 0 \end{bmatrix}, \tag{2.26, 2.27, 3.15}$$

where

$$\omega_1 = \frac{\partial u_3}{\partial x_2} - \frac{\partial u_2}{\partial x_3},\quad \omega_2 = \frac{\partial u_1}{\partial x_3} - \frac{\partial u_3}{\partial x_1},\quad \text{and } \omega_3 = \frac{\partial u_2}{\partial x_1} - \frac{\partial u_1}{\partial x_2}. \tag{2.25, 3.16}$$

Fig. 3.12 illustrates the motion of an initially square fluid element in the (x_1, x_2)-plane when $\partial u_1/\partial x_2$ and $\partial u_2/\partial x_1$ are nonzero and unequal so that $-\omega_3 = R_{12} = -R_{21} \neq 0$. In this situation, the fluid element translates and deforms in the (x_1, x_2)-plane, and rotates about the third coordinate axis. The average rotation rate is:

$$\frac{1}{2}\frac{D(-\alpha+\beta)}{Dt} = \lim_{dt\to 0}\frac{1}{2dt}\left(-\frac{1}{\delta x_2}\left(\frac{\partial u_1}{\partial u_2}\delta x_2 dt\right) + \frac{1}{\delta x_1}\left(\frac{\partial u_2}{\partial x_1}\delta x_1 dt\right)\right) = \frac{1}{2}\left(-\frac{\partial u_1}{\partial x_2} + \frac{\partial u_2}{\partial x_1}\right) = -\frac{R_{12}}{2} = \frac{R_{21}}{2},$$

where again D/Dt indicates that the fluid element is followed as rotation takes place, and again this simple construction is readily extended to the other two Cartesian direction pairs. Thus, $\boldsymbol{\omega}$ and R_{ij} represent twice the fluid element rotation rate (see also Exercise 3.21). This means that $\boldsymbol{\omega}$ and R_{ij} depend on the frame of reference in which they are determined since it is possible to choose a frame of reference that rotates with the fluid particle of interest at the time of interest. In such a co-rotating frame, $\boldsymbol{\omega}$ and R_{ij} will be zero but they will be nonzero if they are determined in a frame of reference that rotates at a different rate (see Exercise 3.23).

Interestingly, the presence or absence of fluid rotation often determines the character of a flow, and this dependence leads to two additional kinematic concepts related to fluid rotation. First, fluid motion is called *irrotational* if:

$$\boldsymbol{\omega} = 0, \text{ or equivalently } R_{ij} = \partial u_i/\partial x_j - \partial u_j/\partial x_i = 0. \tag{3.17}$$

When (3.17) is true, the fluid velocity $\mathbf{u}$ can be written as the gradient of a scalar function $\phi(\mathbf{x}, t)$ because $u_i = \partial\phi/\partial x_i$ satisfies the condition of irrotationality (see Exercises 2.6 and 2.23). Although this may seem to be an unnecessary mathematical complication, finding a scalar function $\phi(\mathbf{x}, t)$ such that $\nabla\phi$ solves the irrotational equations of fluid motion is sometimes easier than solving these equations directly for the vector velocity $\mathbf{u}(\mathbf{x}, t)$ in the same circumstance.

The second concept related to fluid rotation is the extension of the vorticity, twice the fluid rotation rate at a point, to the *circulation* Γ, the amount of fluid rotation within a closed contour (or *circuit*) C. Here the circulation Γ is defined by:

$$\Gamma \equiv \oint_C \mathbf{u}\cdot d\mathbf{s} = \int_A \boldsymbol{\omega}\cdot\mathbf{n}dA, \tag{3.18}$$

where $d\mathbf{s}$ is an element of C, and the geometry is shown in Fig. 3.13. The loop through the first integral sign signifies that C is a closed circuit and is often omitted. The second equality in (3.18) follows from Stokes' theorem (Section 2.13) and the definition of the vorticity $\boldsymbol{\omega} = \nabla\times\mathbf{u}$. The second equality requires the line integral of $\mathbf{u}$ around a closed curve C to be equal to the *flux of vorticity* through the arbitrary surface A bounded by C. Here, and elsewhere in this text, the term *flux* is used for the integral of a vector field normal to a surface. Eq. (3.18) allows $\boldsymbol{\omega}$ to be identified as *the circulation per unit area*. This identification also follows directly from the definition of the curl as the limit of the circulation integral (see (2.35)).

Returning to the situation in Fig. 3.10, Eqs. (3.11) through (3.14) allow (3.10) to be rewritten:

$$du_i = \left(S_{ij} - \frac{1}{2}\varepsilon_{ijk}\omega_k\right)dx_j, \tag{3.19}$$

where $\varepsilon_{ijk}\omega_k dx_j$ is the i-component of the cross product $-\omega\times d\mathbf{x}$ (see (2.21)). Thus, the meaning of the second term in (3.19) can be deduced as follows. The velocity at a distance $\mathbf{x}$ from the axis of rotation of a rigid body rotating at angular velocity Ω is $\boldsymbol{\Omega}\times\mathbf{x}$. The second term in (3.19) therefore represents the velocity of point P relative to point O because of an angular velocity of $\omega/2$.

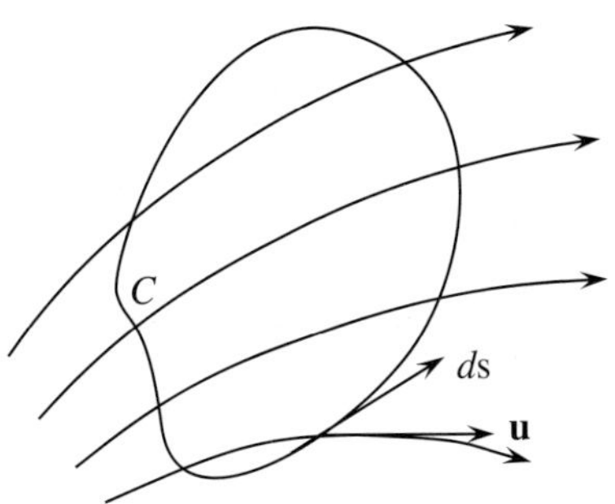

FIGURE 3.13 The circulation around the closed contour C is the line integral of the dot product of the velocity $\mathbf{u}$ and the contour element $d\mathbf{s}$.

The first term in (3.19) is the relative velocity between point P and point O caused by deformation of the fluid element defined by $d\mathbf{x}$. This deformation becomes particularly simple in a coordinate system coinciding with the principal axes of the strain-rate tensor. The components S_{ij} change as the coordinate system is rotated, and for one particular orientation of the coordinate system, a symmetric tensor has only diagonal components; these are called the *principal axes* of the tensor (see Section 2.12 and Example 2.11). Denoting the variables in this principal coordinate system by an over bar (Fig. 3.14), the *first* part of (3.19) can be written as:

$$d\overline{\mathbf{u}} = \overline{\mathbf{S}} \cdot d\overline{\mathbf{x}} = \begin{bmatrix} \overline{S}_{11} & 0 & 0 \\ 0 & \overline{S}_{22} & 0 \\ 0 & 0 & \overline{S}_{33} \end{bmatrix} \begin{bmatrix} d\overline{x}_1 \\ d\overline{x}_2 \\ d\overline{x}_3 \end{bmatrix}. \tag{3.20}$$

Here, $\overline{S}_{11}$, $\overline{S}_{22}$, and $\overline{S}_{33}$ are the diagonal components of $\mathbf{S}$ in the principal-axis coordinate system and are called the eigenvalues of $\mathbf{S}$. The three components of (3.20) can be written:

$$d\overline{u}_1 = \overline{S}_{11} d\overline{x}_1, \quad d\overline{u}_2 = \overline{S}_{22} d\overline{x}_2, \quad \text{and} \quad d\overline{u}_3 = \overline{S}_{33} d\overline{x}_3. \tag{3.21}$$

Consider the significance of $d\overline{u}_1 = \overline{S}_{11} d\overline{x}_1$ when $\overline{S}_{11}$ is positive. This equation implies that point P in Fig. 3.10 is moving *away* from point O in the $\overline{x}_1$-direction at a rate proportional to the distance $d\overline{x}_1$. Considering all points on the surface of a sphere centered on O and having radius $|d\mathbf{x}|$ (see Fig. 3.14), the movement of P in the $\overline{x}_1$-direction is maximum when P coincides with point M (where $d\overline{x}_1 = |d\mathbf{x}|$) and is zero when P coincides with point N (where $d\overline{x}_1 = 0$). Fig. 3.14 illustrates the intersection of this sphere with the $(\overline{x}_1, \overline{x}_2)$-plane for the case where $\overline{S}_{11} > 0$ and $\overline{S}_{22} < 0$; the deformation in the x_3 direction is not shown in this figure. In a small interval of time, *a spherical fluid element around* O *therefore becomes an ellipsoid whose axes are the principal axes of the strain-rate tensor* $\mathbf{S}$.

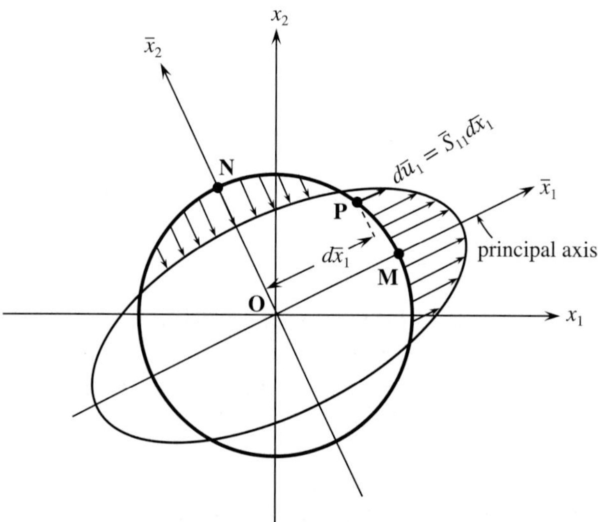

FIGURE 3.14 Deformation of a spherical fluid element into an ellipsoid. Here only the intersection of the element with the plane defined by the first and second coordinate directions is shown.

Example 3.5

A steady two-dimensional flow field incorporating fluid element rotation and strain is given in (x, y)-Cartesian coordinates by $\mathbf{u} = (u, v) = (qy/x^2, qy^2/x^3)$ where q is a positive constant with units of area/time. Sample profiles of $u(x, y)$ vs. y are shown at two x-locations in Fig. 3.15. Determine the streamlines, vorticity, and strain rate tensor in this flow away from $x = 0$. What are the equations of the streamlines along which the x- and y-axes are aligned with the principal axes of the flow?

Solution Utilize the first equality in (3.7) to find:

$$\frac{dy}{dx} = \frac{v}{u} = \frac{qy^2/x^3}{qy/x^2} = \frac{y}{x}.$$

Integrating once produces: $\ln(y) = \ln(x) + \text{const}$. Exponentiate both sides to find $y = mx$, where m is another constant. Thus the flow's streamlines are straight lines with slope m that pass through the origin of coordinates (such as the dashed line in Fig. 3.15).

For two-dimensional flow in the (x, y)-plane, there is only a z-component of vorticity. For the given flow field it is:

$$\omega_z = \frac{\partial v}{\partial x} - \frac{\partial u}{\partial y} = -\frac{q}{x^2}\left(1 + \frac{3y^2}{x^2}\right).$$

Thus, all fluid elements rotate clockwise at a position dependent rate.

The strain rate tensor for this flow is:

$$S_{ij} = \begin{bmatrix} \dfrac{\partial u}{\partial x} & \dfrac{1}{2}\left(\dfrac{\partial v}{\partial x} + \dfrac{\partial u}{\partial y}\right) \\ \dfrac{1}{2}\left(\dfrac{\partial u}{\partial y} + \dfrac{\partial v}{\partial x}\right) & \dfrac{\partial v}{\partial y} \end{bmatrix} = \begin{bmatrix} \dfrac{-2qy}{x^3} & \left(\dfrac{q}{2x^2}\right)\left(1 - \dfrac{3y^2}{x^2}\right) \\ \left(\dfrac{q}{2x^2}\right)\left(1 - \dfrac{3y^2}{x^2}\right) & \dfrac{+2qy}{x^3} \end{bmatrix}.$$

In the first quadrant where x and y are both positive, fluid elements contract in the x-direction and expand in the y-direction. For the streamlines $y = \pm x/\sqrt{3}$, the off-diagonal terms of the strain rate tensor are zero, so the x- and y-axes are aligned with the flow's principal axes along these streamlines.

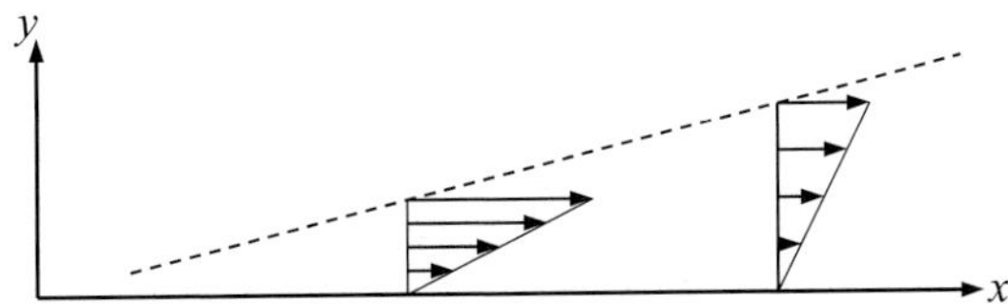

FIGURE 3.15 Sample horizontal velocity profiles at two downstream locations for the flow field specified in Example 3.5. The dashed line intersects the origin.

Summary

The relative velocity in the neighborhood of a point can be divided into two parts. One part comes from rotation of the element, and the other part comes from deformation of the element. When element rotation is absent, a spherical element deforms into an ellipsoid whose axes coincide with the principal axes of the local strain-rate tensor.

3.5 Kinematics of simple plane flows

In this section, the rotation and deformation of fluid elements in two simple steady flows with straight and circular streamlines are considered in two-dimensional (x_1, x_2)-Cartesian and (r, θ)-polar coordinates, respectively. In both cases, the flows can be described with a single independent spatial coordinate that increases perpendicular to the flow direction.

First consider parallel shear flow where $\mathbf{u} = (u_1(x_2), 0)$ as shown in Fig. 3.16. The lone nonzero velocity gradient is $\gamma(x_2) \equiv du_1/dx_2$, and, from (3.16), the only nonzero component of vorticity is $\omega_3 = -\gamma$. In Fig. 3.16, the angular velocity of line element AB is $-\gamma$, and that of BC is zero, giving $-\gamma/2$ as the overall angular velocity (half the vorticity). The average value does not depend on *which* two mutually perpendicular elements in the (x_1, x_2)-plane are chosen to compute it.

In contrast, the components of the strain rate do depend on the orientation of the element. From (3.11), S_{ij} for a fluid element such as ABCD, with sides parallel to the x_1, x_2-axes, is:

$$S_{ij} = \begin{bmatrix} 0 & \gamma/2 \\ \gamma/2 & 0 \end{bmatrix},$$

which shows that there are only off-diagonal elements of $\mathbf{S}$. Therefore, the element ABCD undergoes shear, but no normal strain. As discussed in Section 2.11 and Example 2.11, a symmetric tensor with zero diagonal elements can be diagonalized by rotating the coordinate system through 45°. It is shown there that, along these *principal axes* (denoted by an overbar in Fig. 3.14), the strain rate tensor is:

$$\overline{S}_{ij} = \begin{bmatrix} \gamma/2 & 0 \\ 0 & -\gamma/2 \end{bmatrix},$$

so that along the first principle axis there is a linear extension rate of $\gamma/2$, along the second principle axis there is a linear compression rate of $-\gamma/2$, and no shear. This can be seen geometrically in Fig. 3.16 by examining the deformation of an element PQRS oriented at 45°, which deforms to P′Q′R′S′. The side PS elongates and the side PQ contracts, but the angles between the sides of the element remain 90°. In a small time interval, a small spherical element in this flow would become an ellipsoid oriented at 45° to the x_1, x_2-coordinate system. However, elements in this shear flow still rotate, so the orientation of this ellipsoid changes as time progresses.

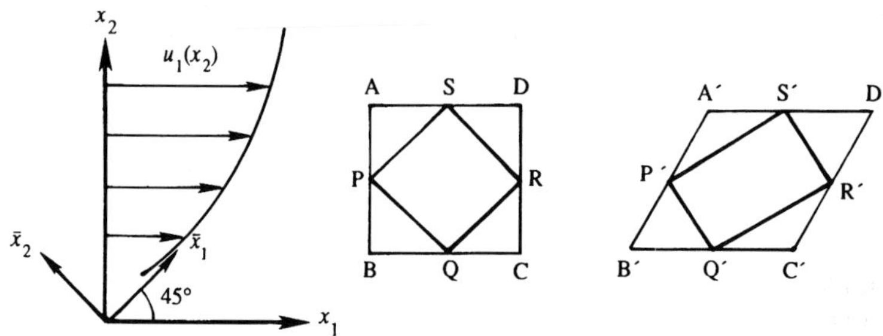

FIGURE 3.16 Deformation of elements in a parallel shear flow. The element is stretched along the principal axis $\bar{x}_1$ and compressed along the principal axis $\bar{x}_2$. The lengths of the sides of ADCB remain unchanged while the corner angles of SRQP also remain unchanged.

In summary, the element ABCD in a parallel shear flow deforms via shear without normal strain, whereas the element PQRS deforms via normal strain without shear strain. And, both elements rotate at the same angular velocity.

Now consider two steady vortex flows having circular streamlines. In (r, θ)-polar coordinates, both flows are defined by $u_r = 0$ and $u_\theta = u_\theta(r)$, with the first one being *solid body rotation*:

$$u_r = 0 \quad \text{and} \quad u_\theta = \omega_0 r, \tag{3.22}$$

where ω_0 is a constant equal to the angular velocity of each particle about the origin (Fig. 3.17). Such a flow can be generated by steadily rotating a cylindrical tank containing a viscous fluid about its axis and waiting until the transients die out. From Appendix B, the vorticity component in the z-direction perpendicular to the (r, θ)-plane is:

$$\omega_z = \frac{1}{r}\frac{\partial}{\partial r}(ru_\theta) - \frac{1}{r}\frac{\partial u_r}{\partial \theta} = 2\omega_0, \tag{3.23}$$

which is independent of location. Thus, each fluid element is rotating about its own center at the same rate that it rotates about the origin of coordinates. This is evident in Fig. 3.17, which shows the location of element ABCD at two successive times. The two mutually perpendicular fluid lines AD and AB both rotate counterclockwise (about the center of the element) with speed ω_0. The time period for one *rotation* of the particle about its own center equals the time period for one *revolution* around the origin of coordinates. In addition, $\mathbf{S} = 0$ for this flow so fluid elements do not deform and each retains its location relative to other elements, as is expected for solid body rotation.

The circulation around a circuit of radius r in this flow is:

$$\Gamma = \oint_C \mathbf{u} \cdot d\mathbf{s} = \int_0^{2\pi} u_\theta r d\theta = 2\pi r u_\theta = 2\pi r^2 \omega_0, \tag{3.24}$$

which shows that circulation equals the vorticity, $2\omega_0$, times the area contained by C. This result is true for *any* circuit C, regardless of whether or not it contains the origin (see Exercise 3.27).

Another flow with circular streamlines is that from an ideal vortex line oriented perpendicular to the (r, θ)-plane. Here, the θ-component of fluid velocity is inversely proportional to the radius of the streamline and the radial velocity is again zero:

$$u_r = 0 \text{ and } u_\theta = B/r, \tag{3.25}$$

where B is a constant. From (3.23), the vorticity in this flow at any point away from the origin is $\omega_z = 0$, but the circulation around a circuit of radius r centered on the origin is a nonzero constant:

$$\Gamma = \int_0^{2\pi} u_\theta r d\theta = 2\pi r u_\theta = 2\pi B \tag{3.26}$$

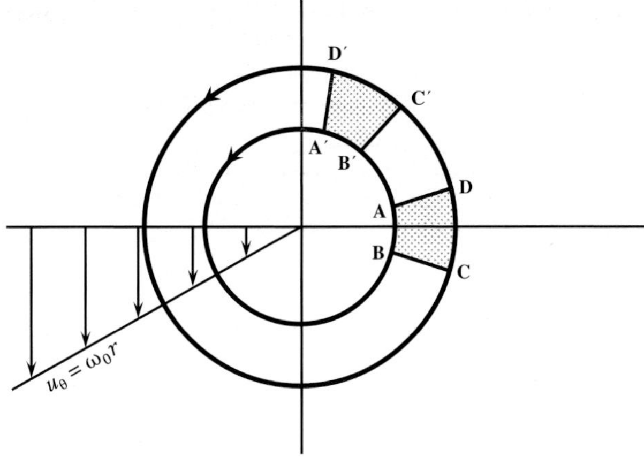

FIGURE 3.17 Solid-body rotation. The streamlines are circular and fluid elements spin about their own centers at the same rate that they revolve around the origin. There is no deformation of the elements, only rotation.

independent of r. Thus, considering vorticity to be the circulation per unit area, as in (3.18) when $\mathbf{n} = \mathbf{e}_z$, then (3.26) implies that the flow specified by (3.25) is *irrotational everywhere except at* $r = 0$ *where the vorticity is infinite with a finite area integral*:

$$[\omega_z]_{r\to 0} = \lim_{r\to 0} \frac{1}{A}\int_A \omega_z dA = \lim_{r\to 0} \frac{1}{\pi r^2}\int_C \mathbf{u}\cdot d\mathbf{s} = \lim_{r\to 0} \frac{2B}{r^2}. \tag{3.27}$$

Although the circulation around a circuit containing the origin in an irrotational vortex flow is nonzero, that around a circuit *not* containing the origin is zero. The circulation around the contour ABCD (Fig. 3.18) is:

$$\Gamma_{\text{ABCD}} \left\{ \int_{\text{AB}} + \int_{\text{BC}} + \int_{\text{CD}} + \int_{\text{DA}} \right\} \mathbf{u}\cdot d\mathbf{s}.$$

The line integrals of $\mathbf{u}\cdot d\mathbf{s}$ on BC and DA are zero because $\mathbf{u}$ and $d\mathbf{s}$ are perpendicular, and the remaining parts of the circuit ABCD produce:

$$\Gamma_{\text{ABCD}} = -[u_\theta r]_r \Delta\theta + [u_\theta r]_{r+\Delta r} \Delta\theta = 0,$$

where the line integral along AB is negative because $\mathbf{u}$ and $d\mathbf{s}$ are oppositely directed, and the final equality is obtained by noting that the product $u_\theta r = B$ is a constant. In addition, zero circulation around ABCD is expected because of Stokes' theorem and the fact that the vorticity vanishes everywhere within ABCD.

Real vortices, such as a bathtub vortex, a wing-tip vortex, or a tornado, do not mimic solid body rotation over large regions of space, nor do they produce unbounded fluid velocity magnitudes near their axes of rotation. Instead, real vortices combine elements of the ideal vortex flows described by (3.22) and (3.25). Near the center of rotation, a real vortex's core flow is nearly solid-body rotation, but far from this core, real-vortex-induced flow is nearly irrotational. Two common idealizations of this behavior are the Rankine vortex defined by:

$$\omega_z(r) = \left\{ \begin{array}{ll} \Gamma/\pi\sigma^2 = \text{const. for } r \le \sigma \\ 0 \qquad\qquad\quad\ \text{for } r > \sigma \end{array} \right\} \quad \text{and}$$
$$u_\theta(r) = \left\{ \begin{array}{lll} (\Gamma/2\pi\sigma^2)r & \text{for} & r \le \sigma \\ \Gamma/2\pi r & \text{for} & r > \sigma \end{array} \right\}, \tag{3.28}$$

and the Gaussian vortex defined by:

$$\omega_z(r) = \frac{\Gamma}{\pi\sigma^2}\exp(-r^2/\sigma^2), \quad \text{and} \quad u_\theta(r) = \frac{\Gamma}{2\pi r}(1 - \exp(-r^2/\sigma^2)). \tag{3.29}$$

In both cases, σ is a core-size parameter that determines the radial distance where real vortex behavior transitions from solid-body rotation to irrotational-vortex flow. For the Rankine vortex, this transition is abrupt and occurs at $r = \sigma$ where

u_θ reaches its maximum. For the Gaussian vortex, this transition is gradual and the maximum value of u_θ is reached at $r/\sigma \approx 1.12091$ (see Exercise 3.31).

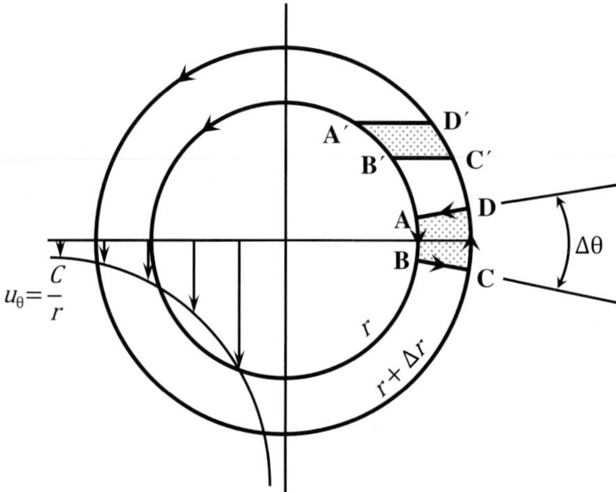

FIGURE 3.18 Irrotational vortex. The streamlines are circular, as for solid body rotation, but the fluid velocity varies with distance from the origin so that fluid elements only deform; they do not spin. The vorticity of fluid elements is zero everywhere, except at the origin where it is infinite.

Example 3.6

The two-dimensional flow described in Cartesian coordinates by $\mathbf{u} = (Ax, -Ay)$, where A is a constant, produces pure straining motion. Determine the streamlines, vorticity, and strain rate tensor in this flow, and sketch fluid element shapes in the first quadrant when A is positive.

Solution Again, utilize the first equality in (3.7) to find:

$$\frac{dy}{dx} = \frac{v}{u} = \frac{-Ay}{Ax} = -\frac{y}{x}.$$

Integrating once produces: $\ln(y) = -\ln(x) + \text{const.}$, or $\ln(y) + \ln(x) = \text{const.}$ Exponentiate both sides to find $xy = c$, where c is another constant. Thus, the flow's streamlines are hyperbolae that asymptote to the x- and y-axes.

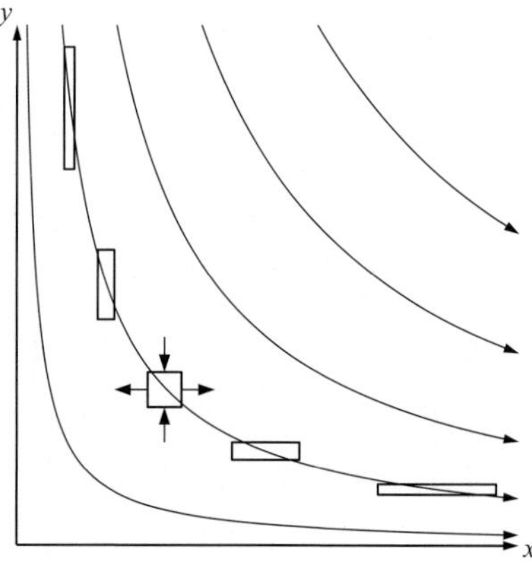

FIGURE 3.19 Sketch of the flow for Example 3.6 in x-y coordinates. The streamlines are hyperbolae. Arrowheads at the right side show the flow direction. A rectangular fluid element changes shape as it passes through the flow field following the second streamline. It does not rotate but it lengthens in the x-direction and contracts in the y-direction. Arrows indicating this strain are shown when the element is square.

For two-dimensional flow in the (x, y)-plane, there is only a z-component of vorticity. For the given flow field it is:

$$\omega_z = \frac{\partial v}{\partial x} - \frac{\partial u}{\partial y} = 0 - 0 = 0.$$

Thus, fluid elements do not rotate in this flow.

The strain rate tensor for this flow is:

$$S_{ij} = \begin{bmatrix} \dfrac{\partial u}{\partial x} & \dfrac{1}{2}\left(\dfrac{\partial v}{\partial x} + \dfrac{\partial u}{\partial y}\right) \\ \dfrac{1}{2}\left(\dfrac{\partial u}{\partial y} + \dfrac{\partial v}{\partial x}\right) & \dfrac{\partial v}{\partial y} \end{bmatrix} = \begin{bmatrix} A & 0 \\ 0 & -A \end{bmatrix}.$$

Thus, the x- and y-axes are principle axes, and, when A is positive, fluid elements expand in the x-direction and contract in the y-direction. A sketch of these flow results appears in Fig. 3.19.

3.6 Reynolds transport theorem

The final kinematic result needed for developing the differential and the control-volume versions of the conservation equations for fluid motion is the Reynolds transport theorem for time differentiation of integrals over arbitrarily moving and deforming volumes. Reynolds transport theorem is the three-dimensional extension of *Leibniz's theorem* for differentiating a single-variable integral having a time-dependent integrand and time-dependent limits (see Riley et al., 1998).

Consider a function F that depends on one independent spatial variable, x, and time t. In addition assume that the time derivative of its integral is of interest when the limits of integration, a and b, are themselves functions of time. Leibniz's theorem states the time derivative of the integral of $F(x,t)$ between $x = a(t)$ and $x = b(t)$ is:

$$\frac{d}{dt}\int_{x=a(t)}^{x=b(t)} F(x,t)dx = \int_a^b \frac{\partial F}{\partial t}dx + \frac{db}{dt}F(b,t) - \frac{da}{dt}F(a,t), \quad (3.30)$$

where a, b, F, and their derivatives appearing on the right side of (3.30) are all evaluated at time t. This situation is depicted in Fig. 3.20, where the three contributions are shown by dots and cross-hatches. The continuous line shows the integral $\int F dx$ at time t, and the dashed line shows the integral at time $t + dt$. The first term on the right side of (3.30) is the integral of $\partial F/\partial t$ between $x = a$ and b, the second term is the gain of F at the upper limit which is moving at rate db/dt, and the third term is the loss of F at the lower limit which is moving at rate da/dt. The essential features of (3.30) are the total time derivative on the left, an integral over the partial time derivative of the integrand on the right, and terms that account for the time-dependence of the limits of integration on the right. These features persist when (3.30) is generalized to three dimensions.

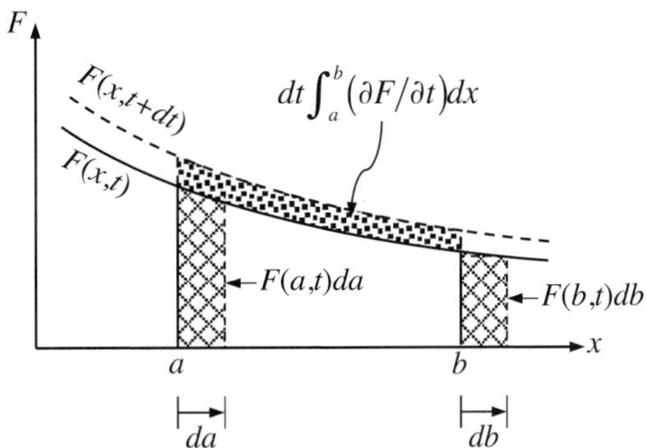

FIGURE 3.20 Graphical illustration of the Liebniz theorem. The three marked areas correspond to the three contributions shown on the right in (3.30). Here da, db, and $\partial F/\partial t$ are all shown as positive.

A largely geometrical development of this generalization is presented here using notation drawn from Thompson (1972). Consider a moving volume $V^*(t)$ having a (closed) surface $A^*(t)$ with outward normal **n** and let **b** denote the local velocity of A^* (Fig. 3.21). The volume V^* and its surface A^* are commonly called a *control volume* and its *control surface*, respectively. The situation is quite general. The volume and its surface need not coincide with any particular boundary, interface, or surface. The velocity **b** need not be steady or uniform over $A^*(t)$. No specific coordinate system or origin of coordinates is needed. The goal of this effort is to determine the time derivative of the integral of a single-valued continuous function $F(\mathbf{x}, t)$ in the volume $V^*(t)$. The starting point for this effort is the definition of a time derivative:

$$\frac{d}{dt}\int_{V^*(t)} F(\mathbf{x},t)dV = \lim_{\Delta t\to 0}\frac{1}{\Delta t}\left\{\int_{V^*(t+\Delta t)} F(\mathbf{x},t+\Delta t)dV - \int_{V^*(t)} F(\mathbf{x},t)dV\right\}. \tag{3.31}$$

The geometry for the two integrals inside the {,}-braces is shown in Fig. 3.21 where solid lines are for time t while the dashed lines are for time $t + \Delta t$. The time derivative of the integral on the left is properly written as a total time derivative since the volume integration subsumes the possible spatial dependence of F. The first term inside the {,}-braces can be expanded to four terms by defining the volume increment $\Delta V \equiv V^*(t + \Delta t) - V^*(t)$ and Taylor expanding the integrand function $F(\mathbf{x}, t + \Delta t) \cong F(\mathbf{x}, t) + \Delta t(\partial F/\partial t)$ for $\Delta t \to 0$:

$$\int_{V^*(t+\Delta t)} F(\mathbf{x},t+\Delta t)dV \cong \int_{V^*(t)} F(\mathbf{x},t)dV + \int_{V^*(t)} \Delta t\frac{\partial F(\mathbf{x},t)}{\partial t}dV + \int_{\Delta V} F(\mathbf{x},t)dV + \int_{\Delta V} \Delta t\frac{\partial F(\mathbf{x},t)}{\partial t}dV. \tag{3.32}$$

The first term on the right in (3.32) will cancel with the final term in (3.31), and, when the limit in (3.31) is taken, both Δt and ΔV go to zero so the final term in (3.32) will not contribute because it is second order. Thus, when (3.32) is substituted into (3.31), the result is:

$$\frac{d}{dt}\int_{V^*(t)} F(\mathbf{x},t)dV = \lim_{\Delta t\to 0}\frac{1}{\Delta t}\left\{\int_{V^*(t)} \Delta t\frac{\partial F(\mathbf{x},t)}{\partial t}dV + \int_{\Delta V} F(\mathbf{x},t)dV\right\}, \tag{3.33}$$

and this limit may be taken once the relationship between ΔV and Δt is known.

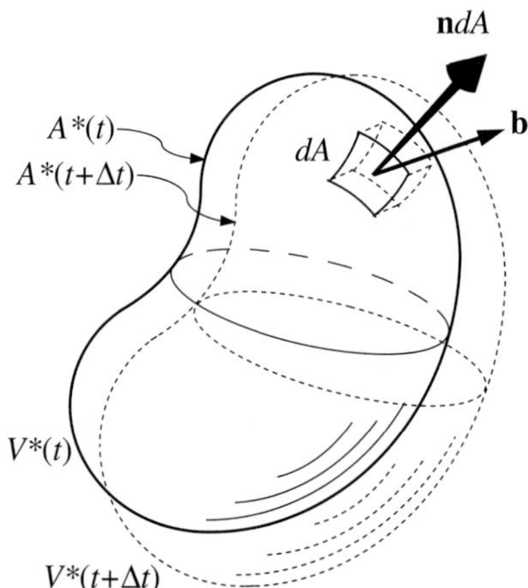

FIGURE 3.21 Geometrical depiction of a control volume $V^*(t)$ having a surface $A^*(t)$ that moves at a nonuniform velocity $\mathbf{b}$ during a small time increment Δt. When Δt is small enough, the volume increment $\Delta V = V^*(t + \Delta t) - V^*(t)$ will lie very near $A^*(t)$, so the volume-increment element adjacent to dA will be $(\mathbf{b}\Delta t) \cdot \mathbf{n}dA$ where $\mathbf{n}$ is the outward normal on $A^*(t)$.

To find this relationship consider the motion of the small area element dA shown in Fig. 3.21. In time Δt, dA sweeps out an elemental volume $(\mathbf{b}\Delta t) \cdot \mathbf{n}dA$ of the volume increment ΔV. Furthermore, this small element of ΔV is located adjacent to the surface $A^*(t)$. All these elemental contributions to ΔV may be summed together via a surface integral, and, as Δt goes to zero, the integrand value of $F(\mathbf{x}, t)$ within these elemental volumes may be taken as that of F on the surface $A^*(t)$, thus:

$$\int_{\Delta V} F(\mathbf{x},t)dV \cong \int_{A^*(t)} F(\mathbf{x},t)(\mathbf{b}\Delta t\cdot\mathbf{n})dA \text{ as } \Delta t \to 0. \tag{3.34}$$

Substituting (3.34) into (3.33), and taking the limit, produces the following statement of Reynolds transport theorem:

$$\frac{d}{dt}\int_{V^*(t)} F(\mathbf{x},t)dV = \int_{V^*(t)} \frac{\partial F(\mathbf{x},t)}{\partial t}dV + \int_{A^*(t)} F(\mathbf{x},t)\mathbf{b}\cdot\mathbf{n}dA. \tag{3.35}$$

This final result follows the pattern set by Liebniz's theorem that the total time derivative of an integral with time-dependent limits equals the integral of the partial time derivative of the integrand plus a term that accounts for the motion of the integration boundary. In (3.35), both inflows and outflows of F are accounted for through the dot product in the surface-integral term that monitors whether $A^*(t)$ is locally advancing ($\mathbf{b}\cdot\mathbf{n} > 0$) or retreating ($\mathbf{b}\cdot\mathbf{n} < 0$) along $\mathbf{n}$, so separate terms as in (3.30) are unnecessary. In addition, the $(\mathbf{x}, t)$-space-time dependence of the control volume's surface velocity $\mathbf{b}$ and unit normal $\mathbf{n}$ are not explicitly shown in (3.35) because $\mathbf{b}$ and $\mathbf{n}$ are only defined on $A^*(t)$; neither is a field quantity like $F(\mathbf{x}, t)$. Eq. (3.35) is an entirely kinematic result, and it shows that d/dt may be moved inside a volume integral and replaced by $\partial/\partial t$ only when the integration volume, $V^*(t)$, is fixed in space so that $\mathbf{b} = 0$ (or more precisely $\mathbf{b}\cdot\mathbf{n} = 0$).

There are two physical interpretations of (3.35). The first, obtained when $F = 1$, is that volume is conserved as $V^*(t)$ moves through three-dimensional space, and under these conditions (3.35) is equivalent to (3.14) for small volumes (see Exercise 3.36). The second is that (3.35) is the extension of (3.5) to finite-size volumes (see Exercise 3.38). Nevertheless, (3.35) and judicious choices of F and $\mathbf{b}$ are the starting points in the next chapter for deriving the field equations of fluid motion from the principles of mass, momentum, and energy conservation.

Example 3.7

The base radius r of a fixed-height right circular cone is increasing at the rate $\dot{r}$. Use Reynolds transport theorem to determine the rate at which the cone's volume is increasing when the cone's base radius is r_o if its height is h.

Solution At any time, the volume V of the right circular cone is: $V = (1/3)\pi h r^2$, which can be differentiated directly and evaluated at $r = r_o$ to find: $dV/dt = (2/3)\pi h r_o \dot{r}$. However, the task is to obtain this answer using (3.35). Choose V^* to perfectly enclose the cone so that $V^* = V$, and set $F = 1$ in (3.35) so that the time derivative of the cone's volume appears on the left. In this case, $\partial F/\partial t = 0$ so (3.35) reduces to:

$$dV/dt = \int_{A^*(t)} \mathbf{b}\cdot\mathbf{n}\,dA.$$

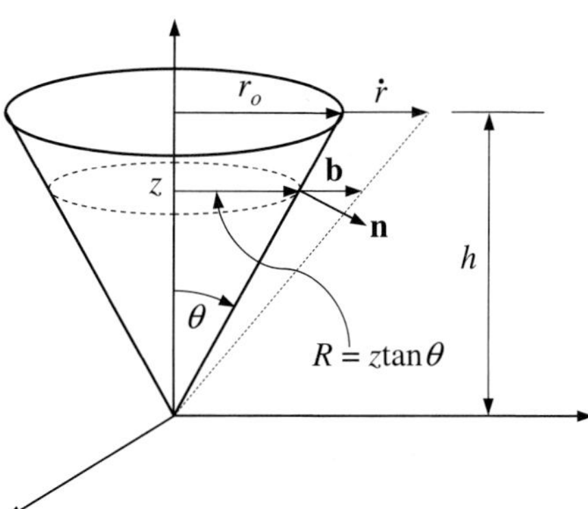

FIGURE 3.22 Conical geometry for Example 3.7. The cone's height is fixed but the radius of its circular surface (base) is increasing.

Use the cylindrical coordinate system shown in Fig. 3.22 with the cone's apex at the origin. Here, $\mathbf{b} = 0$ on the cone's base while $\mathbf{b} = (z/h)\dot{r}\mathbf{e}_R$ on its conical sides. The normal vector on the cone's sides is $\mathbf{n} = \mathbf{e}_R\cos\theta - \mathbf{e}_z\sin\theta$ where $r_o/h = \tan\theta$. Here, at the height z, the cone's surface area element is $dA = z\tan\theta d\varphi(dz/\cos\theta)$, where φ is the azimuthal angle, and the extra cosine factor enters because the conical surface is sloped. Thus, the volumetric rate of change becomes:

$$\frac{dV}{dt} = \int_{z=0}^{h}\int_{\varphi=0}^{2\pi}\frac{z\dot{r}}{h}\mathbf{e}_R\cdot(\mathbf{e}_R\cos\theta - \mathbf{e}_z\sin\theta)z\tan\theta d\varphi\left(\frac{dz}{\cos\theta}\right) = 2\pi\frac{\dot{r}\tan\theta}{h}\int_{z=0}^{h} z^2 dz = \frac{2}{3}\pi h^2\dot{r}\tan\theta = \frac{2}{3}\pi h r_o\dot{r},$$

which recovers the answer obtained by direct differentiation.

Exercises

3.1 The gradient operator in Cartesian coordinates (x, y, z) is: $\nabla = \mathbf{e}_x(\partial/\partial x) + \mathbf{e}_y(\partial/\partial y) + \mathbf{e}_z(\partial/\partial z)$ where $\mathbf{e}_x$, $\mathbf{e}_y$, and $\mathbf{e}_z$ are the unit vectors. In cylindrical polar coordinates (R, φ, z) having the same origin, (see Fig. 3.3c), coordinates and unit vectors are related by: $R = \sqrt{x^2 + y^2}$, $\varphi = \tan^{-1}(y/x)$, and $z = z$; and $\mathbf{e}_R = \mathbf{e}_x\cos\varphi + \mathbf{e}_y\sin\varphi$, $\mathbf{e}_\varphi = -\mathbf{e}_x\sin\varphi + \mathbf{e}_y\cos\varphi$, and $\mathbf{e}_z = \mathbf{e}_z$. Determine the following in the cylindrical polar coordinate system. [This exercise requires some mathematical patience.]

a. $\partial \mathbf{e}_r/\partial \varphi$ and $\partial \mathbf{e}_\varphi/\partial \varphi$
b. the gradient operator ∇
c. the divergence of the velocity field $\nabla \cdot \mathbf{u}$
d. the Laplacian operator $\nabla \cdot \nabla \equiv \nabla^2$
e. the advective acceleration term $(\mathbf{u} \cdot \nabla)\mathbf{u}$
[See Appendix B for answers].

3.2 Consider Cartesian coordinates (as given in Exercise 3.1) and spherical polar coordinates (r, θ, φ) having the same origin (see Fig. 3.3d). Here coordinates and unit vectors are related by: $r = \sqrt{x^2 + y^2 + z^2}$, $\theta = \tan^{-1}(\sqrt{x^2 + y^2}/z)$, and $\varphi = \tan^{-1}(y/x)$; and $\mathbf{e}_r = \mathbf{e}_x \cos\varphi \sin\theta + \mathbf{e}_y \sin\varphi \sin\theta + \mathbf{e}_z \cos\theta$, $\mathbf{e}_\theta = \mathbf{e}_x \cos\varphi \cos\theta + \mathbf{e}_y \sin\varphi \cos\theta - \mathbf{e}_z \sin\theta$, and $\mathbf{e}_\varphi = -\mathbf{e}_x \sin\varphi + \mathbf{e}_y \cos\varphi$. In the spherical polar coordinate system, determine the following items. [This exercise requires mathematical patience.]
a. $\partial \mathbf{e}_r/\partial \theta$, $\partial \mathbf{e}_r/\partial \varphi$, $\partial \mathbf{e}_\theta/\partial \theta$, $\partial \mathbf{e}_\theta/\partial \varphi$, and $\partial \mathbf{e}_\varphi/\partial \varphi$
b. the gradient operator ∇
c. the divergence of the velocity field $\nabla \cdot \mathbf{u}$
d. the Laplacian operator $\nabla \cdot \nabla \equiv \nabla^2$
e. the advective acceleration term $(\mathbf{u} \cdot \nabla)\mathbf{u}$
[See Appendix B for answers].

3.3 In a steady two-dimensional flow, Cartesian-component particle trajectories are given by: $x(t) = r_o \cos(\gamma(t - t_o) + \theta_o)$ and $y(t) = r_o \sin(\gamma(t - t_o) + \theta_o)$ where $r_o = \sqrt{x_o^2 + y_o^2}$ and $\theta_o = \tan^{-1}(y_o/x_o)$.
a. From these trajectories determine the Lagrangian particle velocity components $u(t) = dx/dt$ and $v(t) = dy/dt$, and convert these to Eulerian velocity components $u(x, y)$ and $v(x, y)$.
b. Compute Cartesian particle acceleration components, $a_x = d^2x/dt^2$ and $a_y = d^2y/dt^2$, and show that they are equal to D/Dt of the Eulerian velocity components $u(x, y)$ and $v(x, y)$.

3.4 In a steady two-dimensional flow, polar coordinate particle trajectories are given by: $r(t) = r_o$ and $\theta(t) = \gamma(t - t_o) + \theta_o$.
a. From these trajectories determine the Lagrangian particle velocity components $u_r(t) = dr/dt$ and, $u_\theta(t) = r d\theta/dt$, and convert these to Eulerian velocity components $u_r(r, \theta)$ and $u_\theta(r, \theta)$.
b. Compute polar-coordinate particle acceleration components, $a_r = d^2r/dt^2 - r(d\theta/dt)^2$ and $a_\theta = r d^2\theta/dt^2 + 2(dr/dt)(d\theta/dt)$, and show that they are equal to D/Dt of the Eulerian velocity with components $u_r(r, \theta)$ and $u_\theta(r, \theta)$.

3.5 If $d\mathbf{s} = (dx, dy, dz)$ is an element of arc length along a streamline (Fig. 3.5) and $\mathbf{u} = (u, v, w)$ is the local fluid velocity vector, show that if $d\mathbf{s}$ is everywhere tangent to $\mathbf{u}$ then $dx/u = dy/v = dz/w$.

3.6 For the two-dimensional steady flow having velocity components $u = Sy$ and $v = Sx$, determine the following when S is a positive real constant having units of 1/time.
a. equations for the streamlines with a sketch of the flow pattern
b. the components of the strain-rate tensor
c. the components of the rotation tensor
d. the coordinate rotation that diagonalizes the strain-rate tensor, and the principal strain rates
e. How is this flow field related to that in Example 3.6.

3.7 At the instant shown in Fig. 3.2b, the (u, v)-velocity field in Cartesian coordinates is $u = A(y^2 - x^2)/(x^2 + y^2)^2$, and $v = -2Axy/(x^2 + y^2)^2$ where A is a positive constant. Determine the equations for the streamlines by rearranging the first equality in (3.7) to read $udy - vdx = 0 = (\partial\psi/\partial y)dy + (\partial\psi/\partial x)dx$ and then looking for a solution in the form $\psi(x, y) = \text{const}$.

3.8 Determine the equivalent of the first equality in (3.7) for two-dimensional (r, θ)-polar coordinates, and then find the equation for the streamline that passes through (r_o, θ_o) when $\mathbf{u} = (u_r, u_\theta) = (A/r, B/r)$ where A and B are constants.

3.9 Air flows downward toward an infinitely wide horizontal flat plate. A small time-varying perturbation is superposed onto the mean flow. The velocity field is given by $u = Ax(1 + \epsilon \cos\omega t)$ and $v = -Ay(1 + \epsilon \cos\omega t)$, where A, ϵ and ω are constants. Plot four representative streamlines. Plot four representative path lines in the x-y plane. Determine the acceleration vector along the wall $(y = 0)$.

3.10 Determine the streamline, path line, and streak line that pass through the origin of coordinates at $t = t'$ when $u = U_o + \omega\xi_o \cos(\omega t)$ and $v = \omega\xi_o \sin(\omega t)$ in two-dimensional Cartesian coordinates where U_o is a constant horizontal velocity. Compare your results to those in Example 3.3 for $U_o \to 0$.

3.11 Compute and compare the streamline, path line, and streak line that pass through (1,1,0) at $t = 0$ for the following Cartesian velocity field $\mathbf{u} = (x, -yt, 0)$.

3.12 Consider a time-dependent flow field in two-dimensional Cartesian coordinates where $u = \ell\tau/t^2$, $v = xy/\ell\tau$, and ℓ and τ are constant length and time scales, respectively.

a. Use dimensional analysis to determine the functional form of the streamline through $\mathbf{x}'$ at time t'.

b. Find the equation for the streamline through $\mathbf{x}'$ at time t' and put your answer in dimensionless form.

c. Repeat b) for the path line through $\mathbf{x}'$ at time t'.

d. Repeat b) for the streak line through $\mathbf{x}'$ at time t'.

3.13 The velocity components in an unsteady plane flow are given by $u = x/(1+t)$ and $v = 2y/(2+t)$. Determine equations for the streamlines and path lines subject to $\mathbf{x} = \mathbf{x}_0$ at $t = 0$.

3.14 Using the geometry and notation of Fig. 3.8, prove (3.9).

3.15 Determine the unsteady, $\partial\mathbf{u}/\partial t$, and advective, $(\mathbf{u}\cdot\nabla)\mathbf{u}$, fluid acceleration terms for the following flow fields specified in Cartesian coordinates.

a. $\mathbf{u} = (u(y, z, t), 0, 0)$

b. $\mathbf{u} = \Omega \times \mathbf{x}$ where $\Omega = (0, 0, \Omega_z(t))$

c. $\mathbf{u} = A(t)(x, -y, 0)$

d. $\mathbf{u} = (U_o + u_o \sin(kx - \Omega t), 0, 0)$ where U_o, u_o, k, and Ω are positive constants

3.16 Consider the following Cartesian velocity field $\mathbf{u} = A(t)(f(x), g(y), h(z))$ where A, f, g, and h are nonconstant functions of only one independent variable.

a. Determine $\partial\mathbf{u}/\partial t$, and $(\mathbf{u}\cdot\nabla)\mathbf{u}$ in terms of A, f, g, and h, and their derivatives.

b. Determine A, f, g, and h when $D\mathbf{u}/Dt = 0$, $\mathbf{u} = 0$ at $\mathbf{x} = 0$, and $\mathbf{u}$ is finite for $t > 0$.

c. For the conditions in b), determine the equation for the path line that passes through $\mathbf{x}_o$ at time t_o, and show directly that the acceleration $\mathbf{a}$ of the fluid particle that follows this path is zero.

3.17 If a velocity field is given by $u = ay$ and $v = 0$, compute the circulation around a circle of radius r_o that is centered on at the origin. Check the result by using Stokes' theorem.

3.18 Consider a plane Couette flow of a viscous fluid confined between two flat plates a distance b apart. At steady state the velocity distribution is $u = Uy/b$ and $v = w = 0$, where the upper plate at $y = b$ is moving parallel to itself at speed U, and the lower plate is held stationary. Find the rates of linear strain, the rate of shear strain, and vorticity in this flow.

3.19 The steady two-dimensional flow field inside a sloping passage is given in (x, y)-Cartesian coordinates by $\mathbf{u} = (u, v) = (3q/4h)(1 - (y/h)^2)(1, (y/h)(dh/dx))$ where q is the volume flow rate per unit length into the page, and h is the passage's half thickness. Determine the streamlines, vorticity, and strain rate tensor in this flow away from $x = 0$ when $h = \alpha x$ where α is a positive constant. Sample profiles of $u(x, y)$ vs. y are shown at two x-locations in the figure. What are the equations of the streamlines along which the x- and y-axes are aligned with the principal axes of the flow? What is the fluid particle rotation rate along these streamlines?

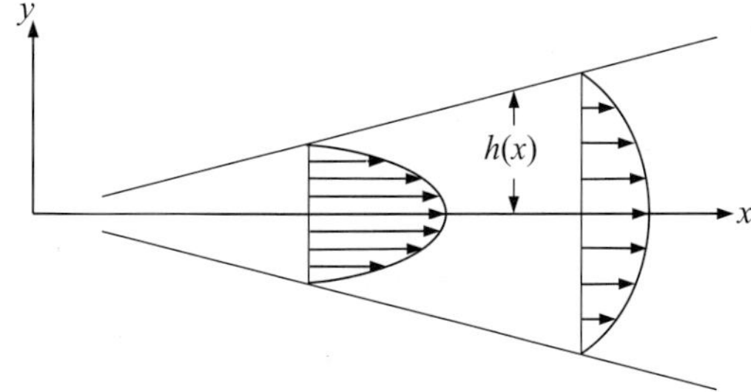

3.20 While the concept of vorticity is well defined mathematically, detection and identification of vortex structures in a flow field is less straightforward. A popular method for the visualization of coherent vortical structures is based on the second invariant of the velocity gradient tensor–the so-called Q-criterion (Hunt et al., 1988). Assuming an incompressible flow ($\partial u_i/\partial x_i = 0$), express the second invariant of $\partial u_i/\partial x_j$ (Q) in terms of S_{ij} and R_{ij} and comment on what positive and negative values correspond to. Next, consider the classical 'ABC' flow, where the three-dimensional Cartesian flow field is given by $u = A\sin z + C\cos y$, $v = B\sin x + A\cos z$, and $w = C\sin y + B\cos x$, with $A = \sqrt{3}$, $B = \sqrt{2}$, and $C = 1$. Plot the streamlines and regions where $Q > 0$.

3.21 For the flow field $\mathbf{u} = \mathbf{U} + \mathbf{\Omega} \times \mathbf{x}$, where $\mathbf{U}$ and $\mathbf{\Omega}$ are constant linear- and angular-velocity vectors, use Cartesian coordinates to a) show that S_{ij} is zero, and b) determine R_{ij}.

3.22 Starting with a small rectangular volume element $\delta V = \delta x_1 \delta x_2 \delta x_3$, prove (3.14).

3.23 Let Oxyz be a stationary frame of reference, and let the z-axis be parallel with the fluid vorticity vector in the vicinity of O so that $\boldsymbol{\omega} = \nabla \times \mathbf{u} = \omega_z \mathbf{e}_z$ in this frame of reference. Now consider a second rotating frame of reference O$x'y'z'$ having the same origin that rotates about the z-axis at angular rate $\Omega \mathbf{e}_z$. Starting from the kinematic relationship, $\mathbf{u} = (\Omega \mathbf{e}_z) \times \mathbf{x} + \mathbf{u}'$, show that in the vicinity of O the vorticity $\boldsymbol{\omega}' = \nabla' \times \mathbf{u}'$ in the rotating frame of reference can only be zero when $2\Omega = \omega_z$, where ∇' is the gradient operator in the primed coordinates. The following unit vector transformation rules may be of use: $\mathbf{e}' = \mathbf{e}_x \sin(\Omega t) + \mathbf{e}_y \cos(\Omega t)$, $\mathbf{e}'_y = -\mathbf{e}_x \sin(\Omega t) + \mathbf{e}_y \cos(\Omega t)$, and $\mathbf{e}'_z = \mathbf{e}_z$.

3.24 Consider a plane-polar area element having dimensions dr and $rd\theta$. For two-dimensional flow in this plane, evaluate the right-hand side of Stokes' theorem $\int \boldsymbol{\omega} \cdot \mathbf{n} dA = \int \mathbf{u} \cdot d\mathbf{s}$ and thereby show that the expression for vorticity in plane-polar coordinates is: $\omega_z = \frac{1}{r}\frac{\partial}{\partial r}(ru_\theta) - \frac{1}{r}\frac{\partial u_r}{\partial \theta}$.

3.25 The velocity field of a certain flow is given by $u = 2xy^2 + 2xz^2$, $v = x^2 y$, and $w = x^2 z$. Consider the fluid region inside a spherical volume $x^2 + y^2 + z^2 = a^2$. Verify the validity of Gauss' theorem $\iiint_V \nabla \cdot \mathbf{u} dV = \iint_A \mathbf{u} \cdot \mathbf{n} dA$ by integrating over the sphere.

3.26 A flow field on the xy-plane has the velocity components $u = 3x + y$ and $v = 2x - 3y$. Show that the circulation around the circle $(x-1)^2 + (y-6)^2 = 4$ is 4π.

3.27 Consider solid-body rotation about the origin in two dimensions: $u_r = 0$ and $u_\theta = \omega_0 r$. Use a polar-coordinate element of dimension $rd\theta$ and dr, and verify that the circulation is vorticity times area. (In Section 3.5 this was verified for a circular element surrounding the origin.)

3.28 Consider the following steady Cartesian velocity field $\mathbf{u} = \left(\frac{-Ay}{(x^2+y^2)^\beta}, \frac{+Ax}{(x^2+y^2)^\beta}, 0\right)$.

a. Determine the streamline that passes through $\mathbf{x} = (x_o, y_o, 0)$

b. Compute R_{ij} for this velocity field.

c. For $A > 0$, explain the sense of rotation (i.e., clockwise or counterclockwise) for fluid elements for $\beta < 1$, $\beta = 1$, and $\beta > 1$.

3.29 Compute the streamlines and path lines for an idealized vortex pair of circulation Γ convected at a free-stream velocity U along the x-direction and centered at $y = 0$, where the two-dimensional velocity field is given by

$$u(x, y, t) = U + \frac{\Gamma}{2\pi}\frac{y+h}{(x-Ut)^2 + (y+h)^2} - \frac{\Gamma}{2\pi}\frac{y-h}{(x-Ut)^2 + (y-h)^2}, \tag{3.36}$$

$$v(x, y, t) = -\frac{\Gamma}{2\pi}\frac{x-Ut}{(x-Ut)^2 + (y+h)^2} - \frac{\Gamma}{2\pi}\frac{x-Ut}{(x-Ut)^2 + (y-h)^2}. \tag{3.37}$$

3.30 Using indicial notation (and no vector identities), show that the acceleration $\mathbf{a}$ of a fluid particle is given by: $\mathbf{a} = \partial \mathbf{u}/\partial t + \nabla\left(\frac{1}{2}|\mathbf{u}|^2\right) + \boldsymbol{\omega} \times \mathbf{u}$ where ω is the vorticity.

3.31 Starting from (3.29), show that the maximum u_θ in a Gaussian vortex occurs when $1 + 2(r^2/\sigma^2) = \exp(r^2/\sigma^2)$. Verify that this implies $r \approx 1.12091\sigma$.

3.32 Using (3.35) in two dimensions with $F = 1$, show that the time-rate-of-change of the area of the parallelogram shown is $hl(d\theta/dt)\cos\theta$ when θ depends on time while h and l are constants.

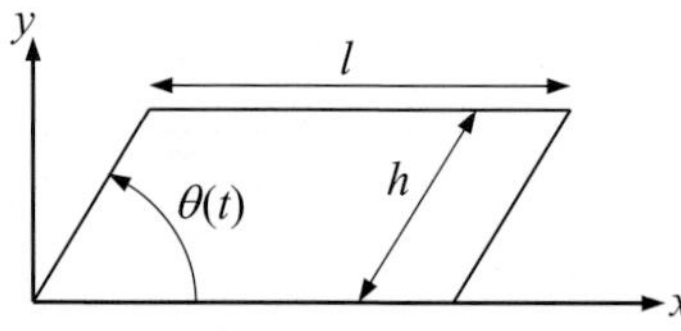

3.33 Using (3.35) in two dimensions with $F = 1$, show that the time-rate-of-change of the area of the triangle shown is $\frac{1}{2}b(dh/dt)$ when h depends on time and b is constant.

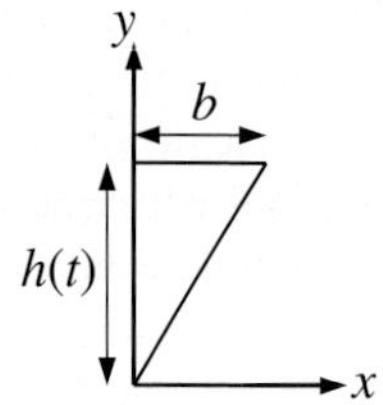

3.34 Using (3.35) in two dimensions with $F = 1$, show that the time-rate-of-change of the area of the ellipse shown is $\pi b(da/dt)$ when a depends on time and b is constant.

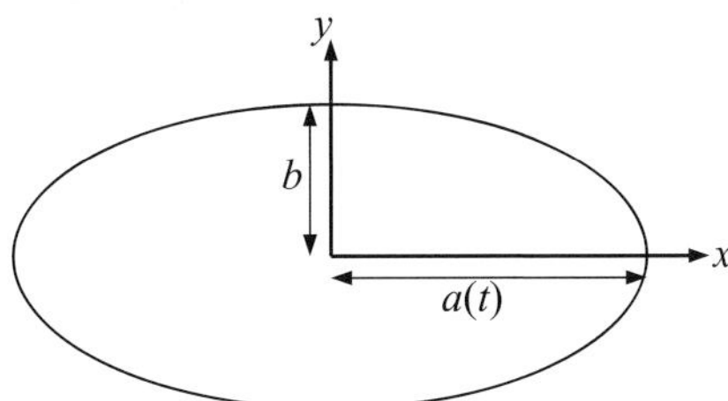

3.35 [1]For the following time-dependent volumes $V^*(t)$ and smooth single-valued integrand functions F, choose an appropriate coordinate system and show that $(d/dt)\int_{V*t} F dV$ obtained from (3.30) is equal to that obtained from (3.35).

a. $V^*(t) = L_1(t)L_2L_3$ is a rectangular solid defined by $0 \le x_i \le L_i$, where L_1 depends on time while L_2 and L_3 are constants, and the integrand function $F(x_1, t)$ depends only on the first coordinate and time.

b. $V^*(t) = (\pi/4)d^2(t)L$ is a cylinder defined by $0 \le R \le d(t)/2$ and $0 \le z \le L$, where the cylinder's diameter d depends on time while its length L is constant, and the integrand function $F(R, t)$ depends only on the distance from the cylinder's axis and time.

c. $V^*(t) = (\pi/6)D^3(t)$ is a sphere defined by $0 \le r \le D(t)/2$ where the sphere's diameter D depends on time, and the integrand function $F(r, t)$ depends only on the radial distance from the center of the sphere and time.

3.36 Starting from (3.35), set $F = 1$ and derive (3.14) when $\mathbf{b} = \mathbf{u}$ and $V^*(t) = \delta V \to 0$.

3.37 For a smooth, single-valued function $F(\mathbf{x})$ that only depends on space and an arbitrarily shaped control volume that moves with velocity $\mathbf{b}(t)$ that only depends on time, show that $(d/dt)\int_{V*(t)} F(\mathbf{x})dV = \mathbf{b} \cdot (\int_{V*(t)} \nabla F(\mathbf{x})dV)$.

3.38 Show that (3.35) reduces to (3.5) when $V^*(t) = \delta V \to 0$ and the control surface velocity $\mathbf{b}$ is equal to the fluid velocity $\mathbf{u}(\mathbf{x}, t)$.

References

Hunt, J.C.R., Wray, A.A., Moin, P., 1988. Eddies, stream, and convergence zones in turbulent flows. Report CTR-S88. Center for Turbulence Research.

Meriam, J.L., Kraige, L.G., 2007. Engineering Mechanics, Dynamics, 6th ed. John Wiley and Sons, Inc., Hoboken, NJ.

Raffel, M., Willert, C., Kompenhans, J., 1998. Particle Image Velocimetry. Springer, Berlin.

Riley, K.F., Hobson, M.P., Bence, S.J., 1998. Mathematical Methods for Physics and Engineering, 3rd ed. Cambridge University Press, Cambridge, UK.

Samimy, M., Breuer, K.S., Leal, L.G., Steen, P.H., 2003. A Gallery of Fluid Motion. Cambridge University Press, Cambridge, UK.

Thompson, P.A., 1972. Compressible-Fluid Dynamics. McGraw-Hill, New York.

Van Dyke, M., 1982. An Album of Fluid Motion. Stanford, California. Parabolic Press.

Further reading

Aris, R., 1962. Vectors, Tensors, and the Basic Equations of Fluid Mechanics. Prentice-Hall, Englewood Cliffs, NJ (The distinctions among streamlines, path lines, and streak lines in unsteady flows are explained; with examples).

Prandtl, L., Tietjens, O.C., 1934a. Fundamentals of Hydro- and Aeromechanics. Dover Publications, New York (Chapter V contains a simple but useful treatment of kinematics).

Prandtl, L., Tietjens, O.G., 1934b. Applied Hydro- and Aeromechanics. Dover Publications, New York (This volume contains classic photographs from Prandtl's laboratory).

1. Developed from Problem 1.9 on page 48 in Thompson (1972).

Chapter 4

Conservation laws⊛

Contents

4.1 Introduction 85
4.2 Conservation of mass 86
4.3 Stream functions 88
4.4 Conservation of momentum 89
4.5 Constitutive equation for a Newtonian fluid 98
4.6 Navier-Stokes momentum equation 101
4.7 Noninertial frame of reference 103
4.8 Conservation of energy 106
4.9 Special forms of the equations 109
Angular momentum principle for a stationary control volume 110
Bernoulli equations 111
Neglect of gravity in constant density flows 117
The Boussinesq approximation 117
Summary 119
4.10 Boundary conditions 119
Conservation of mass boundary condition 120
Constant surface tension boundary condition 120
Conservation of momentum boundary conditions 122
Conservation of energy boundary conditions 123
4.11 Dimensionless forms of the equations and dynamic similarity 124
Exercises 131
References 148
Further reading 148

Chapter objectives

- To present a derivation of the governing equations for a material volume moving in a fluid starting from the principles of mass, momentum, and energy conservation
- To illustrate the application of the integral forms of the mass and momentum conservation equations to stationary, steadily moving, and accelerating control volumes
- To develop the constitutive equation for a Newtonian fluid and provide the Navier-Stokes differential momentum equations
- To show how the differential momentum equations are modified in noninertial frames of reference
- To develop the differential energy equation and highlight its internal coupling between mechanical and thermal energies
- To present several common extensions and simplified forms of the equations of fluid motion
- To derive and describe the dimensionless numbers that appear naturally when the equations of motion are put in dimensionless form

4.1 Introduction

The governing principles of fluid mechanics are the conservation laws for mass, momentum, and energy. These laws are presented in this order in this chapter and can be stated in *integral* form, applicable to an extended region, or in *differential* form, applicable at a point or to a fluid particle. Both forms are equally valid and may be derived from each other. The integral forms of the equations of motion are stated in terms of the evolution of a control volume and the fluxes of mass, momentum, and energy that cross its control surface. The integral forms are typically useful when potentially complicated flow phenomena can be isolated within a control volume and an average or integral flow property, such as a mass flux, a surface pressure force, or an overall velocity or acceleration, is sought. The integral forms are commonly taught in first courses on fluid mechanics for a variety of different control volume conditions. Here they are presented in a unified way without such conditions.

The differential forms of the equations of fluid motion are coupled nonlinear partial differential equations for the dependent flow-field variables of density, velocity, pressure, temperature, etc. Thus, the differential forms are often more appropriate when detailed field information is needed instead of average or integrated quantities. However, both approaches can be used for either scenario when appropriately refined. In the development of the differential equations of fluid motion, attention is given to determining when a solvable system of equations has been found by comparing the number of equations

⊛. This book has a companion website hosting complementary materials. Visit this URL to access it: https://www.elsevier.com/books-and-journals/book-companion/9780128198070.

Fluid Mechanics. https://doi.org/10.1016/B978-0-12-819807-0.00013-2

with the number of unknown dependent field variables. At the outset of this monitoring effort, the fluid's thermodynamic characteristics are assumed to provide as many as two equations, the thermal and caloric equations of state (1.18).

The development of the integral and differential equations of fluid motion presented in this chapter is not unique, and alternatives are readily found in other references. The version provided here is primarily based on that in Thompson (1972).

Example 4.1

In isothermal liquid flows, the fluid density is typically a known constant. What are the dependent field variables in this case? How many equations are needed for a successful mathematical description of such flows? What physical principles supply these equations?

Solution When the fluid's temperature is constant and its density is a known constant, the thermal energy of fluid elements cannot be changed by heat transfer or work because $dT = dv = 0$, so the thermodynamic characterization of the flow is complete from knowledge of the density. Thus, the dependent field variables are $\mathbf{u}$, the fluid's velocity (momentum per unit mass), and the pressure, p. Here, p is not a thermodynamic variable; instead it is a normal force (per unit area) developed between neighboring fluid particles that either causes or results from fluid-particle acceleration, or arises from body forces. Thus, four equations are needed; one for each component of $\mathbf{u}$, and one for p. These equations are supplied by the principle of mass conservation, and three components of Newton's second law for fluid motion (conservation of momentum).

4.2 Conservation of mass

Setting aside nuclear reactions and relativistic effects, mass is neither created nor destroyed. Thus, individual mass elements – molecules, grains, fluid particles, etc. – may be tracked within a flow field because they will not disappear and new elements will not spontaneously appear. The equations representing conservation of mass in a flowing fluid are based on the principle that the mass of a specific collection of neighboring fluid particles is constant. The volume occupied by a specific collection of fluid particles is called a *material volume* $V(t)$. Such a volume moves and deforms within a fluid flow so that it always contains the same mass elements; none enter the volume and none leave it. This implies that a material volume's surface $A(t)$, a material surface, must move at the local fluid velocity $\mathbf{u}$ so that fluid particles inside $V(t)$ remain inside and fluid particles outside $V(t)$ remain outside. Thus, a statement of conservation of mass for a material volume in a flowing fluid is:

$$\frac{d}{dt}\int_{V(t)} \rho(\mathbf{x}, t)dV = 0, \tag{4.1}$$

where ρ is the fluid density. Fig. 3.21 depicts a material volume when the control surface velocity $\mathbf{b}$ is equal to $\mathbf{u}$. The primary concept here is equivalent to an infinitely flexible, perfectly sealed thin-walled balloon containing fluid. The balloon's contents play the role of the material volume $V(t)$ with the balloon itself defining the material surface $A(t)$. And, because the balloon is sealed, the total mass of fluid inside the balloon remains constant as the balloon moves, expands, contracts, or deforms.

From (4.1), it can be seen that the principle of mass conservation constrains the fluid density. The implications of (4.1) for the fluid velocity field may be better displayed by using the Reynolds transport theorem (3.35) with $F = \rho$ and $\mathbf{b} = \mathbf{u}$ to expand the time derivative in (4.1):

$$\int_{V(t)} \frac{\partial \rho(\mathbf{x}, t)}{\partial t} dV + \int_{A(t)} \rho(\mathbf{x}, t)\mathbf{u}(\mathbf{x}, t) \cdot \mathbf{n}\, dA = 0. \tag{4.2}$$

This is a mass-balance statement between integrated density changes within $V(t)$ and integrated motion of its surface $A(t)$. Although general and correct, (4.2) may be hard to utilize in practice because the motion and evolution of $V(t)$ and $A(t)$ are determined by the flow, which may be unknown.

To develop the integral equation that represents mass conservation for an *arbitrarily moving* control volume $V^*(t)$ with surface $A^*(t)$, (4.2) must be modified to involve integrations over $V^*(t)$ and $A^*(t)$. This modification is motivated by the frequent need to conserve mass within a volume that is not a material volume, for example a stationary control volume. The first step in this modification is to set $F = \rho$ in (3.35) to obtain:

$$\frac{d}{dt}\int_{V^*(t)} \rho(\mathbf{x}, t)dV - \int_{V^*(t)} \frac{\partial \rho(\mathbf{x}, t)}{\partial t} dV - \int_{A^*(t)} \rho(\mathbf{x}, t)\mathbf{b} \cdot \mathbf{n}\, dA = 0. \tag{4.3}$$

The second step is to choose the arbitrary control volume $V^*(t)$ to be instantaneously coincident with material volume $V(t)$ so that *at the moment of interest* $V(t) = V^*(t)$ and $A(t) = A^*(t)$. At this coincidence moment, the $(d/dt)\int \rho dV$-terms in (4.1) and (4.3) are not equal; however, the volume integration of $\partial\rho/\partial t$ in (4.2) is equal to that in (4.3) and the surface integral of $\rho\mathbf{u}\cdot\mathbf{n}$ over $A(t)$ is equal to that over $A^*(t)$:

$$\int_{V^*(t)} \frac{\partial\rho(\mathbf{x},t)}{\partial t} dV = \int_{V(t)} \frac{\partial\rho(\mathbf{x},t)}{\partial t} dV = -\int_{A(t)} \rho(\mathbf{x},t)\mathbf{u}(\mathbf{x},t)\cdot\mathbf{n}\,dA = -\int_{A^*(t)} \rho(\mathbf{x},t)\mathbf{u}(\mathbf{x},t)\cdot\mathbf{n}\,dA, \tag{4.4}$$

where the middle equality follows from (4.2). The two ends of (4.4) allow the central volume-integral term in (4.3) to be replaced by a surface integral to find:

$$\frac{d}{dt}\int_{V^*(t)} \rho(\mathbf{x},t)dV + \int_{A^*(t)} \rho(\mathbf{x},t)(\mathbf{u}(\mathbf{x},t)-\mathbf{b})\cdot\mathbf{n}\,dA = 0, \tag{4.5}$$

where $\mathbf{u}$ and $\mathbf{b}$ must both be observed in the same frame of reference; they are not otherwise restricted. This is the general integral statement of conservation of mass for an arbitrarily moving control volume. It can be specialized to stationary, steadily moving, accelerating, or deforming control volumes by appropriate choice of $\mathbf{b}$. In particular, when $\mathbf{b} = \mathbf{u}$, the arbitrary control volume becomes a material volume and (4.5) reduces to (4.1).

The differential equation that represents mass conservation is obtained by applying Gauss' divergence theorem (2.30) to the surface integration in (4.2):

$$\int_{V(t)} \frac{\partial\rho(\mathbf{x},t)}{\partial t} dV + \int_{A(t)} \rho(\mathbf{x},t)\mathbf{u}(\mathbf{x},t)\cdot\mathbf{n}\,dA = \int_{V(t)} \left\{\frac{\partial\rho(\mathbf{x},t)}{\partial t} + \nabla\cdot(\rho(\mathbf{x},t)\mathbf{u}(\mathbf{x},t))\right\} dV = 0. \tag{4.6}$$

The final equality can only be possible if the integrand vanishes at every point in space. If the integrand did not vanish at every point in space, then integrating (4.6) in a small volume around a point where the integrand is nonzero would produce a nonzero integral. Thus, (4.6) requires:

$$\frac{\partial\rho(\mathbf{x},t)}{\partial t} + \nabla\cdot(\rho(\mathbf{x},t)\mathbf{u}(\mathbf{x},t)) = 0 \quad \text{or,} \quad \text{in index notation:} \quad \frac{\partial\rho}{\partial t} + \frac{\partial}{\partial x_i}(\rho u_i) = 0. \tag{4.7}$$

This relationship is called the *continuity equation*. It expresses the principle of conservation of mass in differential form, but is insufficient for fully determining flow fields because it is a single equation that involves two field quantities, ρ and $\mathbf{u}$, and $\mathbf{u}$ is a vector with three components.

The second term in (4.7) is the divergence of the mass-density flux $\rho\mathbf{u}$. Such *flux divergence* terms frequently arise in conservation statements and can be interpreted as the net loss at a point due to divergence of a flux. For example, the local ρ will decrease with time if $\nabla\cdot(\rho\mathbf{u})$ is positive. Flux divergence terms are also called *transport* terms because they transfer quantities from one region to another without making a net contribution over the entire field. When integrated over the entire domain of interest, their contribution vanishes if there are no sources at the boundaries.

The continuity equation may alternatively be written using the definition of D/Dt (3.5) and $\partial(\rho u_i)/\partial x_i = u_i\partial\rho/\partial x_i + \rho\partial u_i/\partial x_i$ [see (B.3.6)]:

$$\frac{1}{\rho(\mathbf{x},t)}\frac{D}{Dt}\rho(\mathbf{x},t) + \nabla\cdot\mathbf{u}(\mathbf{x},t) = 0. \tag{4.8}$$

The derivative $D\rho/Dt$ is the time rate of change of fluid density following a fluid particle. It will be zero for *constant density* flow, and for *incompressible* flow where the density of fluid particles does not change but different fluid particles may have different density:

$$\frac{D\rho}{Dt} \equiv \frac{\partial\rho}{\partial t} + \mathbf{u}\cdot\nabla\rho = 0. \tag{4.9}$$

Taken together, (4.8) and (4.9) imply:

$$\nabla\cdot\mathbf{u} = 0 \tag{4.10}$$

for incompressible flows. Constant density flows are a subset of incompressible flows; $\rho =$ constant is a solution of (4.9) but it is not a general solution. A fluid is usually called *incompressible* if its density does not change with *pressure*. Density stratification in the ocean is an example where the flow is incompressible but the density is non-constant. Liquids are almost incompressible. Gases are compressible, but for flow speeds less than ~100 m/s (that is, for Mach numbers <0.3) the fractional change of absolute pressure in a room temperature airflow is small. In this and several other situations, density changes in the flow are also small and (4.9) and (4.10) are valid.

The general form of the continuity equation (4.7) is typically required when the derivative $D\rho/Dt$ is nonzero because of changes in the pressure, temperature, or molecular composition of fluid particles.

Example 4.2

The density in a horizontal flow $\mathbf{u} = U(y,z)\mathbf{e}_x$ is given by $\rho(\mathbf{x},t) = f(x - Ut, y, z)$, where $f(x,y,z)$ is the density distribution at $t = 0$. Is this flow incompressible?

Solution There are two ways to answer this question. First, consider (4.9) and evaluate $D\rho/Dt$, letting $\xi = x - Ut$:

$$\frac{D\rho}{Dt} = \frac{\partial\rho}{\partial t} + \mathbf{u}\cdot\nabla\rho = \frac{\partial\rho}{\partial t} + U\frac{\partial\rho}{\partial x} = \frac{\partial\rho}{\partial\xi}\frac{\partial\xi}{\partial t} + U\frac{\partial\rho}{\partial\xi}\frac{\partial\xi}{\partial x} = \frac{\partial\rho}{\partial\xi}(-U) + U\frac{\partial\rho}{\partial\xi}(1) = 0.$$

Second, consider (4.10) and evaluate $\nabla\cdot\mathbf{u}$:

$$\nabla\cdot\mathbf{u} = \frac{\partial U(y,z)}{\partial x} + 0 + 0 = 0.$$

In both cases, the result is zero. This is an incompressible flow, but the density may vary when f is not constant.

4.3 Stream functions

Consider the steady form of the continuity equation (4.7):

$$\nabla\cdot(\rho\mathbf{u}) = 0. \tag{4.11}$$

The divergence of the curl of any vector field is identically zero (see Exercise 2.22), so $\rho\mathbf{u}$ will satisfy (4.11) when written as the curl of a vector potential $\boldsymbol{\Psi}$:

$$\rho\mathbf{u} = \nabla\times\boldsymbol{\Psi}, \tag{4.12}$$

which can be specified in terms of χ and ψ, two scalar (stream) functions: $\boldsymbol{\Psi} = \chi\nabla\psi$. Putting this specification for $\boldsymbol{\Psi}$ into (4.12) produces $\rho\mathbf{u} = \nabla\chi\times\nabla\psi$, because the curl of any gradient is identically zero (see Exercise 2.23). Furthermore, $\nabla\chi$ is perpendicular to surfaces of constant χ, and $\nabla\psi$ is perpendicular to surfaces of constant ψ, so the mass flux $\rho\mathbf{u} = \nabla\chi\times\nabla\psi$ will be parallel to surfaces of constant χ and constant ψ. Therefore, three-dimensional streamlines are the intersections of such stream surfaces in a three-dimensional flow.

The situation is illustrated in Fig. 4.1. Consider two stream surfaces from each stream function: $\chi = a$, $\chi = b$, $\psi = c$, $\psi = d$. The intersections, shown as darkened lines in Fig. 4.1, are the streamlines. The mass flux $\dot{m}$ through the surface A bounded by the four stream surfaces (shown in gray in Fig. 4.1) is calculated with area element dA, normal $\mathbf{n}$ (as shown), and Stokes' theorem.

Defining the mass flux $\dot{m}$ through A, and using Stokes' theorem produces:

$$\begin{aligned}\dot{m} &= \int_A \rho\mathbf{u}\cdot\mathbf{n}\,dA = \int_A (\nabla\times\boldsymbol{\Psi})\cdot\mathbf{n}\,dA = \int_C \boldsymbol{\Psi}\cdot d\mathbf{s} = \int_C \chi\nabla\psi\cdot d\mathbf{s} = \int_C \chi\,d\psi \\ &= b(d-c) + a(c-d) = (b-a)(d-c).\end{aligned}$$

Here we have used the vector identity $\nabla\psi\cdot d\mathbf{s} = d\psi$. The mass flow rate of the stream tube defined by the adjacent stream surfaces is just the product of the differences of the numerical values of the respective stream functions.

As a special case, consider two-dimensional flow in (x, y)-Cartesian coordinates where all the streamlines lie in $z =$ constant planes. In this situation, z is one of the three-dimensional stream functions, so we can set $\chi = -z$, where the sign is chosen to obey the usual convention. This produces $\nabla\chi = -\mathbf{e}_z$, so $\rho\mathbf{u} = -\mathbf{e}_z\times\nabla\psi$, or:

$$\rho u = \partial\psi/\partial y, \quad \text{and} \quad \rho v = -\partial\psi/\partial x,$$

in conformity with Exercise 4.8.

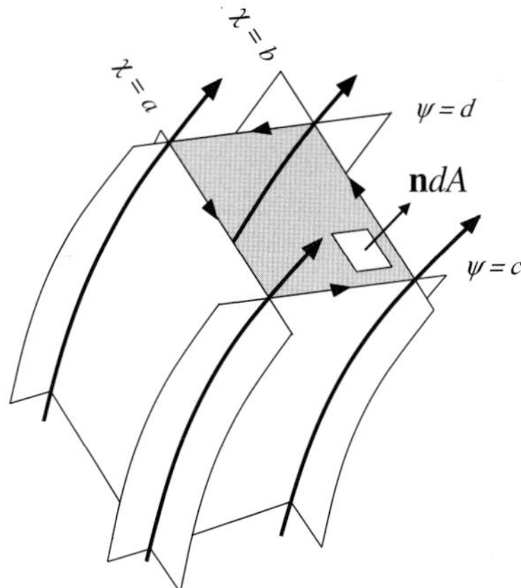

FIGURE 4.1 Isometric view of neighboring stream surfaces defined by $\chi = a$, $\chi = b$, $\psi = c$, and $\psi = d$ where a, b, c, and d are constants. The solid curves are streamlines and these lie at the intersections of the surfaces. The unit vector **n** points in the stream direction and is perpendicular to the gray surface A that is bordered by the nearly rectangular curve C The arrows on this boarder indicate the integration direction for Stokes' theorem.

Similarly, for axisymmetric three-dimensional flow in cylindrical polar coordinates (Fig. 3.3c), all the streamlines lie in $\varphi =$ constant planes that contain the z-axis so $\chi = -\varphi$ is one of the stream functions. This produces $\nabla\chi = -R^{-1}\mathbf{e}_\varphi$ and $\rho\mathbf{u} = \rho(u_R, u_z) = -R^{-1}\,\mathbf{e}_\varphi \times \nabla\psi$, or:

$$\rho u_R = -R^{-1}(\partial\psi/\partial z), \quad \text{and} \quad \rho u_z = R^{-1}(\partial\psi/\partial R).$$

We note here that if the density is constant, mass conservation reduces to $\nabla \cdot \mathbf{u} = 0$ (steady or not) and the entire preceding discussion follows for **u** rather than $\rho\mathbf{u}$ with the interpretation of stream function values in terms of volumetric flux rather than mass flux.

Example 4.3

In a planar constant-density flow, the stream function ψ is defined by $\mathbf{u} = (u, v) = -\mathbf{e}_z \times \nabla\psi$. Show that a curve of $\psi(x, y) = \text{const.}$ satisfies the stream function tangency requirement, the first equality of (3.7).

Solution On a curve of $\psi(x, y) = \text{const.}$, the following will be true:

$$d\psi = \frac{\partial\psi}{\partial x}dx + \frac{\partial\psi}{\partial y}dy = 0, \quad \text{or} \quad \frac{dx}{\partial\psi/dy} = \frac{dy}{-\partial\psi/\partial x},$$

where the second equation is a rearrangement of the first. Evaluate the cross product in the problem statement to find $u = \partial\psi/\partial y$ and $v = -\partial\psi/\partial x$, and substitute these into the second equality to reach: $dx/u = dy/v$, which is the first equality of (3.7).

4.4 Conservation of momentum

In this section, the momentum-conservation equivalent of (4.5) is developed from Newton's second law, the fundamental principle governing fluid momentum. When applied to a material volume $V(t)$ with surface area $A(t)$, Newton's second law can be stated directly as:

$$\frac{d}{dt}\int_{V(t)} \rho(\mathbf{x}, t)\mathbf{u}(\mathbf{x}, t)dV = \int_{V(t)} \rho(\mathbf{x}, t)\mathbf{g}dV + \int_{A(t)} \mathbf{f}(\mathbf{n}, \mathbf{x}, t)dA, \tag{4.13}$$

where $\rho\mathbf{u}$ is the momentum per unit volume of the flowing fluid, **g** is the body force per unit mass acting on the fluid within $V(t)$, **f** is the surface force per unit area acting on $A(t)$, and **n** is the outward normal on $A(t)$. The implications of (4.13) are better displayed when the time derivative is expanded using Reynolds transport theorem (3.35) with $F = \rho\mathbf{u}$ and $\mathbf{b} = \mathbf{u}$:

$$\int_{V(t)} \frac{\partial}{\partial t}(\rho(\mathbf{x}, t)\mathbf{u}(\mathbf{x}, t))\,dV + \int_{A(t)} \rho(\mathbf{x}, t)\mathbf{u}(\mathbf{x}, t)(\mathbf{u}(\mathbf{x}, t)\cdot\mathbf{n})\,dA = \int_{V(t)} \rho(\mathbf{x}, t)\mathbf{g}dV + \int_{A(t)} \mathbf{f}(\mathbf{n}, \mathbf{x}, t)\,dA. \tag{4.14}$$

This is a momentum-balance statement between integrated momentum changes within $V(t)$, integrated momentum contributions from the motion of $A(t)$, and integrated volume and surface forces. It is the momentum conservation equivalent of (4.2).

To develop an integral equation that represents momentum conservation for an arbitrarily moving control volume $V^*(t)$ with surface $A^*(t)$, (4.14) must be modified to involve integrations over $V^*(t)$ and $A^*(t)$. The steps in this process are entirely analogous to those taken between (4.2) and (4.5) for conservation of mass. First, set $F = \rho \boldsymbol{u}$ in (3.35) and rearrange it to obtain:

$$\int_{V^*(t)} \frac{\partial}{\partial t}(\rho(\mathbf{x},t)\mathbf{u}(\mathbf{x},t))\,dV = \frac{d}{dt}\int_{V^*(t)} \rho(\mathbf{x},t)\mathbf{u}(\mathbf{x},t)\,dV - \int_{A^*(t)} \rho(\mathbf{x},t)\mathbf{u}(\mathbf{x},t)\mathbf{b}\cdot\mathbf{n}\,dA, \tag{4.15}$$

then choose $V^*(t)$ to be instantaneously coincident with $V(t)$ so that at the moment of interest:

$$\int_{V(t)} \frac{\partial}{\partial t}(\rho(\mathbf{x},t)\mathbf{u}(\mathbf{x},t))dV = \int_{V^*(t)} \frac{\partial}{\partial t}(\rho(\mathbf{x},t)\mathbf{u}(\mathbf{x},t))dV,$$

$$\int_{A(t)} \rho(\mathbf{x},t)\mathbf{u}(\mathbf{x},t)(\mathbf{u}(\mathbf{x},t)\cdot\mathbf{n})dA = \int_{A^*(t)} \rho(\mathbf{x},t)\mathbf{u}(\mathbf{x},t)(\mathbf{u}(\mathbf{x},t)\cdot\mathbf{n})dA, \qquad \text{(4.16a, 4.16b, 4.16c, 4.16d)}$$

$$\int_{V(t)} \rho(\mathbf{x},t)\mathbf{g}dV = \int_{V^*(t)} \rho(\mathbf{x},t)\mathbf{g}dV, \quad \text{and} \quad \int_{A(t)} \mathbf{f}(\mathbf{n},\mathbf{x},t)dA = \int_{A^*(t)} \mathbf{f}(\mathbf{n},\mathbf{x},t)dA.$$

Now substitute (4.16a) into (4.15) and use this result plus (4.16b, 4.16c, 4.16d) to convert (4.14) to:

$$\frac{d}{dt}\int_{V^*(t)} \rho(\mathbf{x},t)\mathbf{u}(\mathbf{x},t)dV + \int_{A^*(t)} \rho(\mathbf{x},t)\mathbf{u}(\mathbf{x},t)(\mathbf{u}(\mathbf{x},t)-\mathbf{b})\cdot\mathbf{n}\,dA = \int_{V^*(t)} \rho(\mathbf{x},t)\mathbf{g}\,dV + \int_{A^*(t)} \mathbf{f}(\mathbf{n},\mathbf{x},t)dA. \tag{4.17}$$

This is the general integral statement of momentum conservation for an arbitrarily moving control volume. Just like (4.5), it can be specialized to stationary, steadily moving, accelerating, or deforming control volumes by appropriate choice of **b**. For example, when $\mathbf{b} = \mathbf{u}$, the arbitrary control volume becomes a material volume and (4.17) reduces to (4.13).

At this point, the forces in (4.13), (4.14), and (4.17) merit some additional description that facilitates the derivation of the differential equation representing momentum conservation and allows its simplification under certain circumstances.

The body force, $\rho\mathbf{g}dV$, acting on the fluid element dV does so without physical contact. Body forces commonly arise from gravitational or electromagnetic *force fields*. In addition, in accelerating or rotating frames of reference, *fictitious* body forces arise from the frame's noninertial motion (see Section 4.7). By definition, body forces are distributed through the fluid and are proportional to mass (or electric charge, electric current, etc.). In this book, body forces are specified per unit mass and carry the units of acceleration.

Body forces may be conservative or nonconservative. *Conservative body forces* are those that can be expressed as the gradient of a potential function:

$$\mathbf{g} = -\nabla\Phi \quad \text{or} \quad g_j = -\partial\Phi/\partial x_j, \tag{4.18}$$

where Φ is called the *force potential*; it has units of energy per unit mass. When the z-axis points vertically upward, the force potential for gravity is $\Phi = gz$, where g is the acceleration of gravity, and (4.18) produces $\mathbf{g} = -g\mathbf{e}_z$. Forces satisfying (4.18) are called *conservative* because the work done by conservative forces is independent of the path, and the sum of fluid-particle kinetic and potential energies is conserved when friction is absent.

Surface forces, **f**, act on fluid elements through direct contact with the surface of the element. They are proportional to the contact area and carry units of stress (force per unit area). Surface forces are commonly resolved into components normal and tangential to the contact area. Consider an arbitrarily oriented element of area dA in a fluid (Fig. 2.5). If **n** is the surface normal with components n_i, then from (2.15) the components f_j of the surface force per unit area $\mathbf{f}(\mathbf{n},\mathbf{x},t)$ on this element are $f_j = n_i T_{ij}$ where T_{ij} is the stress tensor. Thus, the normal component of **f** is $\mathbf{n}\cdot\mathbf{f} = n_i f_i$, while the tangential component is the vector $\mathbf{f} - (\mathbf{n}\cdot\mathbf{f})\mathbf{n}$ which has components $f_k - (n_i f_i)n_k$.

Other forces that influence fluid motion are surface- and interfacial-tension forces that act on lines or curves embedded within interfaces between liquids and gases or between immiscible liquids (see Fig. 1.5). Although these forces are commonly important in flows with such interfaces, they do not commonly appear in the equations of motion, entering instead through the boundary conditions.

Before proceeding to the differential equation representing momentum conservation, the use of (4.5) and (4.17) for stationary, moving, and accelerating control volumes having a variety of sizes and shapes is illustrated through a few examples. In all four examples, equations representing mass and momentum conservation must be solved simultaneously.

Example 4.4

A long bar with constant cross section is held perpendicular to a uniform horizontal flow of speed U_∞, as shown in Fig. 4.2. The flowing fluid has density ρ and viscosity μ (both constant). The bar's cross section has characteristic transverse dimension d, and the span of the bar is l with $l \gg d$. The average horizontal velocity profile measured downstream of the bar is $U(y)$, which is less than U_∞ due to the presence of the bar. Determine the required force per unit span, $-F_D/l$, applied to the ends of the bar to hold it in place. Assume the flow is steady and two dimensional in the plane shown. Ignore body forces.

Solution Before beginning, it is important to explain the sign convention for fluid dynamic drag forces. The drag force on an inanimate object is *the force applied to the object by the fluid*. Thus, for stationary objects, drag forces are positive in the downstream direction, the direction the object would accelerate if released. However, the control volume laws are written for *forces applied to the contents of the volume*. Thus, from Newton's third law, a positive drag force on an object implies a negative force on the fluid. Therefore, the F_D appearing in Fig. 4.2 is a positive number and this will be borne out by the final results. Here we also note that since the horizontal velocity downstream of the bar, the wake velocity $U(y)$, is less than U_∞, the fluid has been decelerated inside the control volume and this is consistent with a force from the body opposing the motion of the fluid as shown.

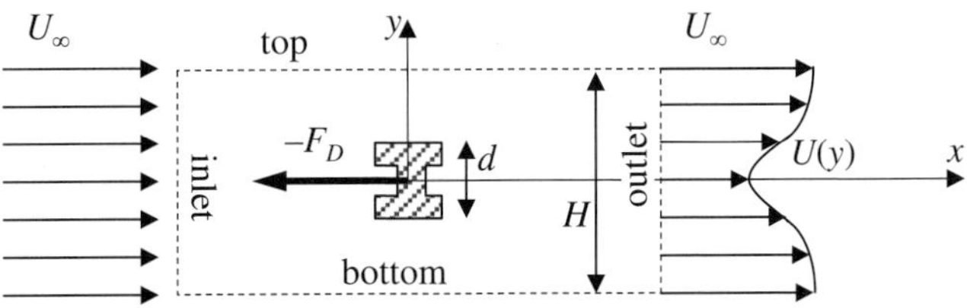

FIGURE 4.2 Momentum and mass balance for flow past long bar of constant cross section placed perpendicular to the flow. The intersection of the recommended stationary control volume with the x-y plane is shown with dashed lines. The force $-F_D$ holds the bar in place and slows the fluid that enters the control volume.

The basic strategy is to select a stationary control volume, and then use (4.5) and (4.17) to determine the force F_D that the body exerts on the fluid per unit span in terms of ρ, U_∞, and $U(y)$. The first quantitative step in the solution is to select a rectangular control volume with flat control surfaces aligned with the coordinate directions. The inlet, outlet, top, and bottom sides of such a control volume are shown in Fig. 4.2. The vertical sides parallel to the x-y plane are not shown. However, the flow does not vary in the third direction and is everywhere parallel to these surfaces so these merely need be selected a comfortable distance l apart. The inlet control surface should be far enough upstream of the bar so that the inlet fluid velocity is $U_\infty \mathbf{e}_x$, the pressure is p_∞, and both are uniform. The top and bottom control surfaces should be separated by a distance H that is large enough so that these boundaries are free from shear stresses, and the horizontal velocity and pressure are so close to U_∞ and p_∞ that any difference can be ignored. And finally, the outlet surface should be far enough downstream so that streamlines are nearly horizontal there, the pressure can again be treated as equal to p_∞, and viscous normal stresses can be ignored.

For steady flow and the chosen stationary volume, the control surface velocity is $\mathbf{b} = 0$ and the time derivative terms in (4.5) and (4.17) are both zero. In addition, the surface force integral contributes $-F_D\mathbf{e}_x$ where the beam crosses the control volume's vertical sides parallel to the x-y plane. The remainder of the surface force integral contains only pressure terms since the shear stress is zero on the control surface boundaries. After setting the pressure to p_∞ on all control surfaces, (4.5) and (4.17) simplify to:

$$\int_{A^*(t)} \rho\mathbf{u}(\mathbf{x}) \cdot \mathbf{n}\, dA = 0, \quad \text{and} \quad \int_{A^*} \rho\mathbf{u}(\mathbf{x})\mathbf{u}(\mathbf{x}) \cdot \mathbf{n}\, dA = -\int_{A^*} p_\infty \mathbf{n}\, dA - F_D\mathbf{e}_x.$$

In this case the pressure integral may be evaluated immediately using Gauss' divergence theorem:

$$\int_{A^*} p_\infty \mathbf{n}\, dA = \int_{V^*} \nabla p_\infty dV = 0,$$

with the final value (zero) occurring because p_∞ is a constant. After this simplification, denote the fluid velocity components by $(u, v) = \mathbf{u}$, and evaluate the mass and x-momentum conservation equations:

$$-\int_{inlet} \rho U_\infty l\, dy + \int_{top} \rho v l\, dx - \int_{bottom} \rho v l\, dx + \int_{outlet} \rho U(y) l\, dy = 0, \text{ and}$$

$$-\int_{inlet} \rho U_\infty^2 l\, dy + \int_{top} \rho U_\infty v l\, dx - \int_{bottom} \rho U_\infty v l\, dx + \int_{outlet} \rho U^2(y) l\, dy = -F_D,$$

where $\mathbf{u}\cdot\mathbf{n}dA$ is: $-U_\infty l dy$ on the inlet surface, $+vldx$ on the top surface, $-vldx$ on the bottom surface, and $+U(y)ldy$ on the outlet surface where l is the span of the flow into the page. Dividing both equations by ρl, and combining like integrals produces:

$$\int_{top} v\,dx - \int_{bottom} v\,dx = \int_{-H/2}^{+H/2} (U_\infty - U(y))dy \quad \text{and}$$

$$U_\infty\left(\int_{top} v\,dx - \int_{bottom} v\,dx\right) + \int_{-H/2}^{+H/2} (U^2(y) - U_\infty^2)dy = -F_D/\rho l.$$

Eliminating the top and bottom control surface integrals between these two equations leads to:

$$F_D/l = \rho\int_{-H/2}^{+H/2} U(y)(U_\infty - U(y))dy,$$

which produces a positive value of F_D when $U(y)$ is less than U_∞. An essential feature of this analysis is that there are nonzero mass fluxes through the top and bottom control surfaces. The final formula here is genuinely useful in experimental fluid mechanics since it allows F_D/l to be determined from single-component velocity measurements made in the wake of an object.

Example 4.5

Using a stream-tube control volume of differential length ds, derive the Bernoulli equation, $\rho U^2/2 + gz + p/\rho = \text{constant}$ along a streamline, for steady, inviscid, constant density flow where U is the local flow speed.

Solution The basic strategy is to use a stationary stream-tube-element control volume, (4.5), and (4.17) to determine a simple differential relationship that can be integrated along a streamline. The geometry is shown in Fig. 4.3. For steady inviscid flow and a stationary control volume ($\mathbf{b} = 0$, V^* and A^* time independent), the surface friction forces are zero, and the time derivative terms in (4.5) and (4.17) are both zero. Thus, these two equations simplify to:

$$\int_{A^*} \rho\mathbf{u}(\mathbf{x})\cdot\mathbf{n}dA = 0, \quad \text{and} \quad \int_{A^*} \rho\mathbf{u}(\mathbf{x})\mathbf{u}(\mathbf{x})\cdot\mathbf{n}dA = \int_{V^*} \rho\mathbf{g}dV - \int_{A^*} p\mathbf{n}dA.$$

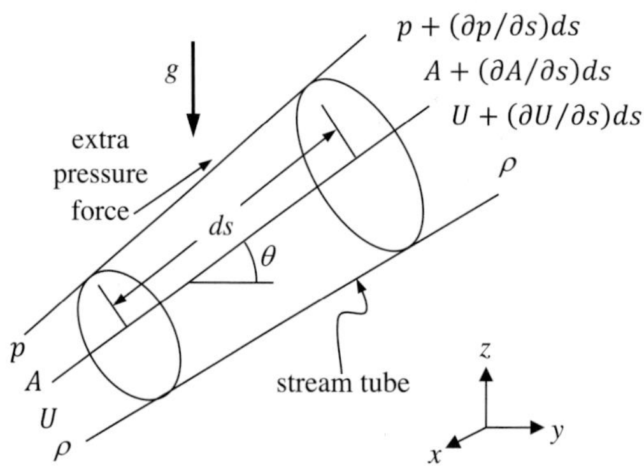

FIGURE 4.3 Momentum and mass balance for a short segment of a stream tube in steady inviscid constant-density flow. Here, the inlet and outlet areas are perpendicular to the flow direction, and they are small enough so that only first order corrections in the stream tube need to be considered. The alignment of gravity and the stream tube leads to a vertical change of $\sin\theta ds = dz$ between its two ends. The area difference between the two ends of the stream-tube leads to an extra pressure force.

The geometry of the volume plays an important role here. The nearly conical curved surface is tangent to the velocity while the inlet and outlet areas are perpendicular to it. Thus, $\mathbf{u}\cdot\mathbf{n}dA$ is: $-UdA$ on the inlet surface, zero on the nearly conical curved surface, and $+[U + (\partial U/\partial s)ds]dA$ on the outlet surface. Therefore, conservation of mass with constant density leads to:

$$-\rho UA + \rho\left(U + \frac{\partial U}{\partial s}ds\right)\left(A + \frac{\partial A}{\partial s}ds\right) = 0,$$

where first-order variations in U and A in the stream-wise direction are accounted for. Now consider the stream-wise component of the momentum equation recalling that $\mathbf{u} = U\mathbf{e}_u$ and setting $\mathbf{g} = -g\mathbf{e}_z$. For inviscid flow, the only surface force is pressure, so the simplified version of (4.17) becomes:

$$-\rho U^2 A+\rho\left(U+\frac{\partial U}{\partial s}ds\right)^2\left(A+\frac{\partial A}{\partial s}ds\right)$$
$$=-\rho g\sin\theta\left(A+\frac{\partial A}{\partial s}\frac{ds}{2}\right)ds+pA+\left(p+\frac{\partial p}{\partial s}\frac{ds}{2}\right)\frac{\partial A}{\partial s}ds-\left(p+\frac{\partial p}{\partial s}ds\right)\left(A+\frac{\partial A}{\partial s}ds\right).$$

Here, the middle pressure term comes from the extra pressure force on the nearly conical surface of the stream tube.

To reach the final equation, use the conservation of mass result to simplify the flux terms on the left side of the stream-wise momentum equation. Then, simplify the pressure contributions by canceling common terms, and note that $\sin\theta ds = dz$ to find:

$$-\rho U^2 A+\rho U\left(U+\frac{\partial U}{\partial s}ds\right)A=\rho UA\frac{\partial U}{\partial s}ds=-\rho g\left(A+\frac{\partial A}{\partial s}\frac{ds}{2}\right)dz+\frac{\partial p}{\partial s}\frac{\partial A}{\partial s}\frac{(ds)^2}{2}-A\frac{\partial p}{\partial s}ds-\frac{\partial p}{\partial s}\frac{\partial A}{\partial s}(ds)^2.$$

Drop the second-order terms containing $(ds)^2$ or $dsdz$, and divide by ρA to reach:

$$U\frac{\partial U}{\partial s}ds=-gdz-\frac{1}{\rho}\frac{\partial p}{\partial s}ds,\quad\text{or}\quad [d(U^2/2)+gdz+(1/\rho)dp=0]_{\text{along a streamline}}.$$

Integrate the final differential expression along the streamline to find:

$$U^2/2+gz+p/\rho=\text{a constant along a streamline.} \tag{4.19}$$

Example 4.6

Consider a small solitary wave that moves from right to left on the surface of a water channel of undisturbed depth h (Fig. 4.4). Denote the acceleration of gravity by g. Assuming a small change in the surface elevation across the wave, derive an expression for its propagation speed, U, when the channel bed is flat and frictionless.

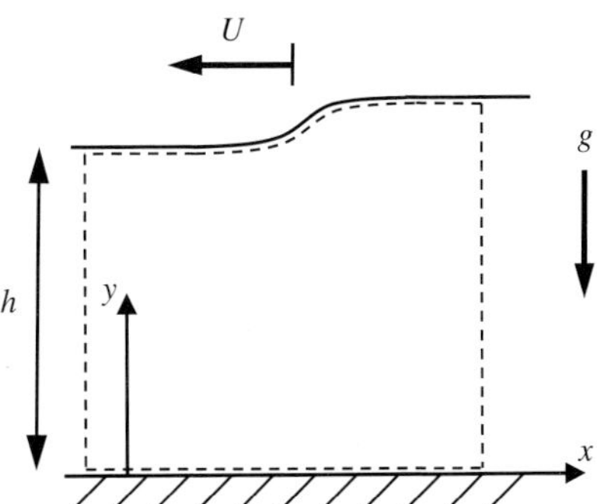

FIGURE 4.4 Momentum and mass balance for a small amplitude water wave moving into quiescent water of depth h. The recommended moving control volume is shown with dashed lines. The wave is driven by the imbalance of static pressure forces on the vertical inlet (left) and outlet (right) control surfaces.

Solution Before starting the control volume analysis, dimensional analysis goes a *long* way toward the final solution. The problem statement has only three parameters, U, g, and h, and there are two independent units (length and time). Thus, there is only one dimensionless group, U^2/gh, so it must be a constant. Therefore, the final answer must be in the form: $U=\text{const.}\cdot\sqrt{gh}$, so the value of the following control volume analysis lies merely in determining the constant.

Choose the control volume shown and assume it is moving at speed $\mathbf{b}=-U\mathbf{e}_x$. Here we assume that the upper and lower control surfaces coincide with the water surface and the channel's frictionless bed. They are shown close to these boundaries in Fig. 4.4 for clarity. Apply the integral conservation laws for mass and momentum, (4.5) and (4.17).

With this choice of a moving control volume, its contents and shape are constant so the d/dt terms in both equations are zero, and V^* and A^* are again time independent, leaving:

$$\int_{A^*}\rho(\mathbf{u}+U\mathbf{e}_x)\cdot\mathbf{n}dA=0\quad\text{and}\quad\int_{A^*}\rho\mathbf{u}(\mathbf{u}+U\mathbf{e}_x)\cdot\mathbf{n}dA=\int_{V^*}\rho\mathbf{g}dV+\int_{A^*}\mathbf{f}dA.$$

Here, all velocities are referred to a stationary coordinate frame, so that $\mathbf{u}=0$ on the inlet side of the control volume in the undisturbed fluid layer. In addition, label the inlet (left) and outlet (right) water depths as h_{in} and h_{out}, respectively, and save consideration of the simplifications that occur when $(h_{out} - h_{in}) \ll (h_{out} + h_{in})/2$ for the end of the analysis. Let U_{out} be the horizontal flow speed on the outlet side of the control volume and assume its profile is uniform. Therefore $(\mathbf{u} + U\mathbf{e}_x) \cdot \mathbf{n}dA$ is $-Uldy$ on the inlet surface, and $+(U_{out} + U)ldy$ on the outlet surface, where l is (again) the width of the flow into the page. With these replacements, the conservation of mass equation becomes:

$$-\rho U h_{in} l + \rho(U_{out} + U)h_{out} l = 0, \quad \text{or} \quad U h_{in} = (U_{out} + U)h_{out},$$

and the horizontal momentum equation becomes:

$$-\rho(0)(0+U)h_{in}l + \rho U_{out}(U_{out}+U)h_{out}l = -\int_{inlet} p\mathbf{n}\cdot\mathbf{e}_x\,dA - \int_{outlet} p\mathbf{n}\cdot\mathbf{e}_x\,dA - \int_{top} p\mathbf{n}\cdot\mathbf{e}_x\,dA.$$

Here, no friction terms are included, and the body force term does not appear because it has no horizontal component. First, consider the pressure integral on the top of the control volume, and let $y = h(x)$ define the shape of the water surface:

$$-p_o\int \mathbf{n}\cdot\mathbf{e}_x\,dA = -p_o\int \frac{(-dh/dx, 1)}{\sqrt{1+(dh/dx)^2}}\cdot(1,0)l\sqrt{1+(dh/dx)^2}dx$$
$$= -p_o\int\left(-\frac{dh}{dx}\right)l\,dx = p_o\int_{h_{in}}^{h_{out}} l\,dh = p_o l(h_{out} - h_{in}),$$

where the various square-root factors arise from the surface geometry; p_o is the (constant) atmospheric pressure on the water surface. The pressure on the inlet and outlet sides of the control volume is hydrostatic. Using the coordinate system shown, integrating (1.14), and evaluating the constant on the water surface produces $p = p_o + \rho g(h - y)$. Thus, the integrated inlet and outlet pressure forces are:

$$\int_{inlet} p\,dA - \int_{outlet} p\,dA - \int_{top} p_o\mathbf{n}\cdot\mathbf{e}_x\,dA = \int_0^{h_{in}}(p_o + \rho g(h_{in} - y))l\,dy - \int_0^{h_{out}}(p_o + \rho g(h_{out} - y))l\,dy + p_o(h_{out} - h_{in})l$$
$$= \int_0^{h_{in}} \rho g(h_{in} - y)l\,dy - \int_0^{h_{out}} \rho g(h_{out} - y)l\,dy = \rho g\left(\frac{h_{in}^2}{2} - \frac{h_{out}^2}{2}\right)l,$$

where the signs of the inlet and outlet integrals have been determined by evaluating the dot products and we again note that the constant reference pressure p_o does not contribute to the net pressure force. Substituting this pressure force result into the horizontal momentum equation produces:

$$-\rho(0)(0+U)h_{in}l + \rho U_{out}(U_{out}+U)h_{out}l = \frac{\rho g}{2}(h_{in}^2 - h_{out}^2)l.$$

Dividing by the common factors of ρ and l to reach:

$$U_{out}(U_{out}+U)h_{out} = \frac{g}{2}(h_{in}^2 - h_{out}^2),$$

and eliminating U_{out} via the conservation of mass relationship, $U_{out} = (h_{in} - h_{out})U/h_{out}$, leads to:

$$U\frac{(h_{in} - h_{out})}{h_{out}}\left(U\frac{(h_{in} - h_{out})}{h_{out}} + U\right)h_{out} = \frac{g}{2}(h_{in}^2 - h_{out}^2).$$

Dividing by the common factor of $(h_{in} - h_{out})$ and simplifying the left side of the equation produces:

$$U^2\frac{h_{in}}{h_{out}} = \frac{g}{2}(h_{in} + h_{out}), \quad \text{or} \quad U = \sqrt{\frac{g h_{out}}{2h_{in}}(h_{in} + h_{out})} \approx \sqrt{gh},$$

where the final approximate equality holds when the inlet and outlet heights differ by only a small amount with both nearly equal to h.

Example 4.7

Derive the differential equation for the vertical motion for a simple rocket having nozzle area A_e that points downward, exhaust discharge speed V_e, and exhaust density ρ_e, without considering the internal flow within the rocket (Fig. 4.5). Denote the mass of the rocket by $M(t)$ and assume the discharge flow is uniform.

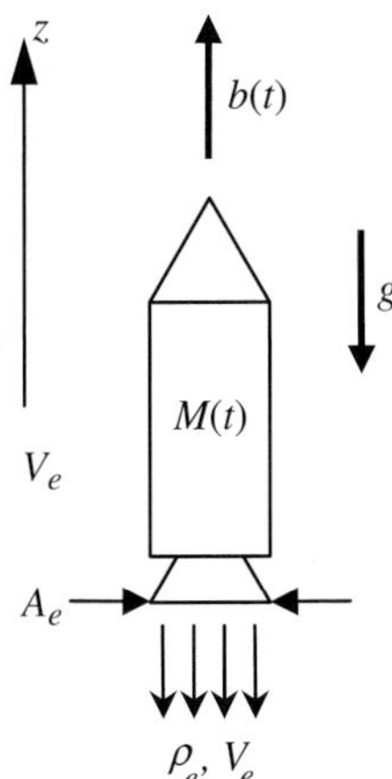

FIGURE 4.5 Geometry and parameters for a simple rocket having mass $M(t)$ that is moving vertically at speed $b(t)$. The rocket's exhaust area, density, and velocity (or specific impulse) are A_e, ρ_e, and V_e, respectively.

Solution Select a fixed-shape control volume (not shown) that contains the rocket and travels with it. This will be a time-independent accelerating control volume and its velocity $\mathbf{b} = b(t)\mathbf{e}_z$ will be the rocket's vertical velocity. In addition, the discharge velocity is specified with respect to the rocket, so in a stationary frame of reference, the absolute velocity of the rocket's exhaust is $\mathbf{u} = u_z\mathbf{e}_z = (-V_e + b)\mathbf{e}_z$.

The conservation of mass and vertical-momentum equations are:

$$\frac{d}{dt}\int_{V^*}\rho dV + \int_{A^*}\rho(\mathbf{u}-\mathbf{b})\cdot\mathbf{n}dA = 0$$

$$\frac{d}{dt}\int_{V^*}\rho u_z dV + \int_{A^*}\rho u_z(\mathbf{u}-\mathbf{b})\cdot\mathbf{n}dA = -g\int_{V^*}\rho dV + \int_{A^*} f_z dA.$$

Here we recognize the first term in each equation as the time derivative of the rocket's mass M, and the rocket's vertical momentum Mb, respectively. (The second of these identifications is altered when the rocket's internal flows are considered; see Thompson, 1972, pp. 43–47.) For ordinary rocketry, the rocket exhaust exit will be the only place that mass and momentum cross A^* and here $\mathbf{n} = -\mathbf{e}_z$; thus $(\mathbf{u}-\mathbf{b})\cdot\mathbf{n}dA = (-V_e\mathbf{e}_z)\cdot(-\mathbf{e}_z)dA = V_e dA$ over the nozzle exit. In addition, denote the integral of vertical surface stresses by F_S, a force that includes the aerodynamic drag on the rocket and the pressure thrust produced when the rocket nozzle's outlet pressure exceeds the local ambient pressure. With these replacements, the above equations become:

$$\frac{dM}{dt} + \rho_e V_e A_e = 0 \quad \text{and} \quad \frac{d}{dt}(Mb) + \rho_e(-V_e + b)V_e A_e = -Mg + F_S.$$

Eliminating $\rho_e V_e A_e$ between the two equations produces:

$$\frac{d}{dt}(Mb) + (-V_e + b)\left(-\frac{dM}{dt}\right) = -Mg + F_S,$$

which reduces to:

$$M\frac{d^2 z_R}{dt^2} = -V_e\frac{dM}{dt} - Mg + F_S,$$

where z_R is the rocket's vertical location and $dz_R/dt = b$. From this equation it is clear that negative dM/dt (mass loss) may produce upward acceleration of the rocket when its exhaust discharge velocity V_e is high enough. In fact, V_e is the crucial figure of merit in rocket propulsion and is commonly referred to as the *specific impulse*, the thrust produced per unit rate of mass discharged.

Example 4.8

Pour-over (or drip) coffee is a popular brewing method that involves pouring hot water through coffee grounds (see Fig. 4.6). The flow resistance through the coffee can be expressed by a drag force (per unit volume) as $-\beta(\mathbf{u}-\mathbf{u}_s)$, where $\mathbf{u}$ and $\mathbf{u}_s$ are the velocities of the water and solid phase, respectively. Assuming the coffee grounds are spherical with an average diameter d and closely packed such that the void fraction (i.e., the volume fraction of the liquid phase) is $\epsilon < 0.8$, β can be modeled as (Gidaspow, 1994)

$$\beta = \frac{150\,(1-\epsilon)^2\,\mu}{\epsilon d^2} + \frac{1.75\rho|\mathbf{u}-\mathbf{u}_s|\,(1-\epsilon)}{d}.$$

Consider the pour over coffee dripper to be a truncated cone with height $H = 0.08$ m and half-angle at the cone vertex $\theta = 20^o$ as depicted in Fig. 4.6. The coffee is packed to a height $H_c = 0.06$ m with radius $R_0 = 0.05$ m at $z = 0$. Assume (i) the coffee is sufficiently packed ($\epsilon = 0.25$) so there is no grain movement ($\mathbf{u}_s = 0$), (ii) absorption is negligible (ϵ and d are constant), (iii) the fluid has constant density, and (iv) velocity and pressure depend only on z. Determine the maximum volumetric flow rate Q without causing the flow to back-up above the top of the coffee dripper. Plot the corresponding vertical velocity and pressure distributions, $w(z)$ and $p(z)$, and report the residence time of a fluid particle within the bed of coffee. Compare results for finely ground ($d = 50$ µm) and coarsely-ground ($d = 100$ µm) coffee. Use properties of water at $T = 90^o$ C.

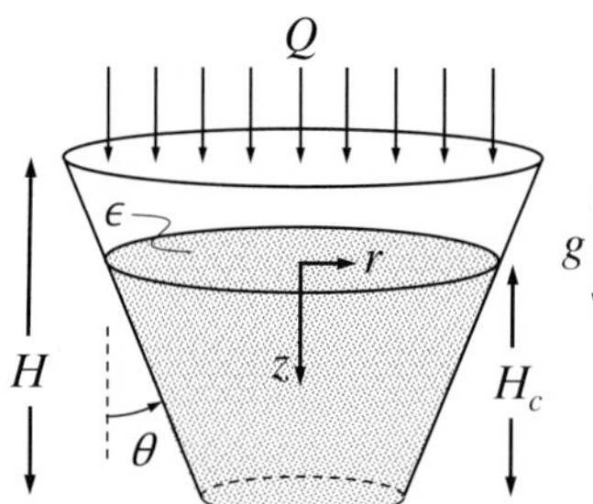

FIGURE 4.6 A pour-over coffee dripper is modeled as a truncated cone. Water is poured over coffee grounds with void fraction ϵ at a steady flow rate Q and exits at the bottom ($z = H$) into ambient pressure.

Solution The radius and cross-sectional area at depth z into the coffee are given by

$$R(z) = (H^* - z)\tan\theta \quad \text{and} \quad A(z) = \pi(H^* - z)^2\tan^2\theta,$$

where $H^* = R_0/\tan\theta$ is the virtual height of the cone if it was not truncated. To solve for the vertical velocity and pressure, a control volume is defined as a thin disk at a depth z with radius $R(z)$ and thickness Δz. Assuming steady and uniform flow, conservation of mass inside the bed ($z > 0$) implies $Q = w(z)\epsilon A(z)$, where the inclusion of ϵ accounts for the flow in the interstitial space between the coffee grounds. Thus, the vertical fluid velocity at any depth in the coffee is:

$$w(z) = \frac{Q}{\epsilon\pi(H^* - z)^2\tan^2\theta} \quad \text{for} \quad z > 0.$$

Without fluid acceleration, the integral conservation law for momentum (4.17) becomes:

$$0 = -\int_{A^*}\epsilon p\mathbf{n}\,dA + \int_{V^*}\epsilon\rho\mathbf{g}\,dV - \int_{V^*}\beta\mathbf{u}\,dV,$$

which for unidirectional vertical flow reduces to:

$$p_{out} - p_{in} = -\left(\frac{w(z)\beta}{\epsilon} - \rho g\right)\Delta z \quad \text{for} \quad z > 0,$$

where $p_{in} = p(z - \Delta z/2)$ and $p_{out} = p(z + \Delta z/2)$ are the pressures incoming and outgoing of the control volume, respectively. The pressure can be solved for numerically by discretizing the depth into N segments each with thickness Δz and enforcing the boundary condition $p(z = 0) = p_0 = \rho g h$, where $h = H - H_c$ is the maximum liquid height above the coffee bed. With an initial guess for Q, the problem can be solved iteratively until ambient pressure at the exit, $p(z = H) = p_\infty$, is obtained. An example code is provided below.

```
global rho mu eps

% Inputs
rho=965; mu=3.15e-4; % (@ T=90 C)
g=9.81; eps=.25; R0=.05; Hc=.06; H=.08; theta=20;

% Fine & coarse parameters
```

```
df=50e-6; dc=100e-6;

% Guess flowrates
Qf=8.17e-8; Qc=3.26e-7;

% Store coordinate midpoint (zm)
N=128; dz=Hc/N;
zm=dz/2:dz:Hc-dz/2;

% Compute velocity
Hs=R0/tand(theta);
A=pi*(Hs-zm).^2*tand(theta)^2;
wf=Qf./(eps*A); wc=Qc./(eps*A);

% Compute residence time (in minutes)
tf=sum(dz./wf)/60
tc=sum(dz./wc)/60

% Hydrostatic pressure above the coffee
p0=rho*g*(H-Hc);

% Compute pressure (apply BC p(z=0)=p0)
pf=zeros(1,N); pc=zeros(1,N);
for i=1:N
    if i==1
        pf_in=p0;
        pc_in=p0;
    else
        pf_in=pf(i-1);
        pc_in=pc(i-1);
    end
    beta=drag(wf(i),df);
    pf(i)=pf_in-(beta*wf(i)/eps-rho*g)*dz;
    beta=drag(wc(i),dc);
    pc(i)=pc_in-(beta*wc(i)/eps-rho*g)*dz;
end

%% Drag (Gidaspow 1994)
function beta=drag(w,d)
global rho mu eps
beta=150*(1-eps)^2*mu/(eps*d^2)+1.75*rho*abs(w)*(1-eps)/d;
end
```

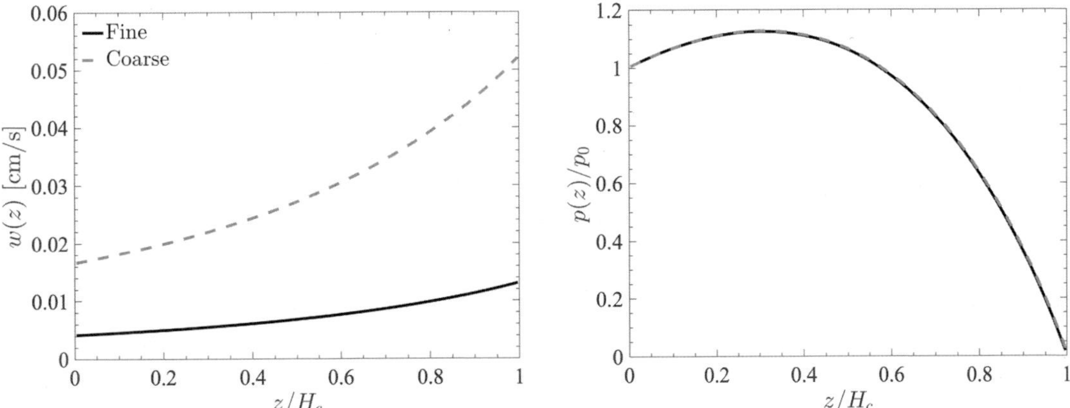

FIGURE 4.7 Solution to Example 4.8.

A flow rate of $Q = 8.17 \times 10^{-2}$ cm^3/s and $Q = 3.26 \times 10^{-1}$ cm^3/s for finely-ground and coarsely-ground coffee, respectively, is needed to satisfy the boundary conditions when the water above the bed is at its maximum height. As shown in Fig. 4.7, the liquid accelerates as it travels through the coffee due to the reduction in area. The increased diameter of the coarse coffee grounds results in a higher velocity compared to the finely-ground coffee, yet the pressure profiles for both are identical. The trajectory of a fluid particle is given by $dz_p/dt = w(z)$. Thus, the residence time can be determined by $\Delta t = \int_0^{H_c} dz/w(z)$. Solving this numerically yields $\Delta t \approx 15$ min for the fine coffee and $\Delta t \approx 3.8$ min for the coarse coffee. It should be noted that discretizing the problem's domain into many small control volumes is the premise behind finite volume methods used in computational fluid dynamics (Chapter 15).

The differential equation that represents momentum conservation is obtained from (4.14) after collecting all four terms into the same volume integral. The first step is to convert the two surface integrals in (4.14) to volume integrals using Gauss' theorem (2.30):

$$\int_{A(t)} \rho(\mathbf{x},t)\mathbf{u}(\mathbf{x},t)(\mathbf{u}(\mathbf{x},t)\cdot\mathbf{n})dA = \int_{V(t)} \nabla\cdot(\rho(\mathbf{x},t)\mathbf{u}(\mathbf{x},t)\mathbf{u}(\mathbf{x},t))dV = \int_{V(t)} \frac{\partial}{\partial x_i}(\rho u_i u_j)dV,$$
$$\text{and} \quad \int_{A(t)} \mathbf{f}(\mathbf{n},\mathbf{x},t)dA = \int_{A(t)} n_i T_{ij} dA = \int_{V(t)} \frac{\partial}{\partial x_i}(T_{ij})dV, \tag{4.20a, 4.20b}$$

where the explicit listing of the independent variables has been dropped upon moving to index notation, and T_{ij} is the stress tensor in the fluid. Substituting (4.20a, 4.20b) into (4.14) and collecting all the terms on one side of the equation into the same volume integration produces:

$$\int_{V(t)} \left\{ \frac{\partial}{\partial t}(\rho u_j) + \frac{\partial}{\partial x_i}(\rho u_i u_j) - \rho g_j - \frac{\partial}{\partial x_i}(T_{ij}) \right\} dV = 0. \tag{4.21}$$

Similarly to (4.6), the integral in (4.21) can only be zero for any material volume if the integrand vanishes at every point in space; thus (4.21) requires:

$$\frac{\partial}{\partial t}(\rho u_j) + \frac{\partial}{\partial x_i}(\rho u_i u_j) = \rho g_j + \frac{\partial}{\partial x_i}(T_{ij}). \tag{4.22}$$

This equation can be put into a more standard form by expanding the leading two terms,

$$\frac{\partial}{\partial t}(\rho u_j) + \frac{\partial}{\partial x_i}(\rho u_i u_j) = \rho\frac{\partial u_j}{\partial t} + u_j\left[\frac{\partial \rho}{\partial t} + \frac{\partial}{\partial x_i}(\rho u_i)\right] + \rho u_i \frac{\partial u_j}{\partial x_i} = \rho\frac{Du_j}{Dt}, \tag{4.23}$$

recognizing that the contents of the [,]-brackets are zero because of (4.7), and using the definition of D/Dt from (3.5). The final result is:

$$\rho\frac{Du_j}{Dt} = \rho g_j + \frac{\partial}{\partial x_i}(T_{ij}), \tag{4.24}$$

which is sometimes called *Cauchy's equation of motion*. It relates fluid-particle acceleration to the net body (ρg_i) and surface ($\partial T_{ij}/\partial x_j$) forces on the particle. It is true in any continuum, solid or fluid, no matter how the stress tensor T_{ij} is related to the velocity field. However, (4.24) does not provide a complete description of fluid dynamics, even when combined with (4.7) because the number of dependent field variables exceeds the number of equations. Taken together, (4.7), (4.24), and two thermodynamic equations provide at most $1+3+2=6$ scalar equations but (4.7) and (4.24) contain ρ, u_j, and T_{ij} for a total of $1+3+9=13$ unknowns. Thus, the number of unknowns must be decreased to produce a solvable system. The fluid's stress-strain rate relationship(s) or constitutive equation provides much of the requisite reduction.

4.5 Constitutive equation for a Newtonian fluid

As previously described in Section 2.4, the stress at a point can be completely specified by the nine components of the stress tensor $\mathbf{T}$; these components are illustrated in Figs. 2.4 and 2.5, which show the directions of *positive* stresses on the various faces of small cubical and tetrahedral fluid elements. The first index of T_{ij} indicates the direction of the normal to the surface on which the stress is considered, and the second index indicates the direction in which the stress acts. The diagonal elements T_{11}, T_{22}, and T_{33} of the stress matrix are the normal stresses, and the off-diagonal elements are the tangential or shear stresses. Although finite size elements are shown in these figures, the stresses apply on the various planes when the elements shrink to a point and have vanishingly small mass. Denoting the cubical volume in Fig. 2.4 by

$dV = dx_1 dx_2 dx_3$ and considering the torque produced on it by the various stress components, it can be shown that the stress tensor is symmetric,

$$T_{ij} = T_{ji}, \tag{4.25}$$

by considering the element's rotational dynamics in the limit $dV \to 0$ (see Exercise 4.41). Therefore, the stress tensor has only six independent components. However, this symmetry is violated if there are body-force couples proportional to the mass of the fluid element, such as those exerted by an electric field on polarized fluid molecules. Antisymmetric stresses must be included in such circumstances.

The relationship between the stress and deformation in a continuum is called a *constitutive equation*, and a linear constitutive equation between stress T_{ij} and $\partial u_i/\partial x_j$ is examined here. A fluid that follows the simplest possible linear constitutive equation is known as a *Newtonian* fluid. Non-Newtonian fluids are briefly discussed near the end of this section.

In a fluid at rest, there are only normal components of stress on a surface, which do not depend on the orientation of the surface; stress is *isotropic*. The only second-order isotropic tensor is the Kronecker delta, δ_{ij}, from (2.16). Therefore, stress in a static fluid must be of the form

$$T_{ij} = -p\delta_{ij}, \tag{4.26}$$

where p is the *thermodynamic pressure* related to ρ and T by an equation of state such as that for a perfect gas $p = \rho RT$ (1.28). The negative sign in (4.26) occurs because the normal components of $\mathbf{T}$ are regarded as positive if they indicate tension rather than compression of fluid elements (see Fig. 2.4).

A moving fluid develops additional stress components, τ_{ij}, because of its viscosity, and these stress components appear as both diagonal and off-diagonal components of $\mathbf{T}$. A simple extension of (4.26) that captures this phenomenon and reduces to (4.26) when fluid motion ceases is:

$$T_{ij} = -p\delta_{ij} + \tau_{ij}. \tag{4.27}$$

This decomposition of the stress into fluid-static (p) and fluid-dynamic (τ_{ij}) contributions is approximate, because p is only well defined for equilibrium conditions. However, molecular densities, speeds, and collision rates are typically high enough, so that fluid particles (as defined in Section 1.8) reach local thermodynamic equilibrium conditions in nearly all fluid flows so that p in (4.27) is still the thermodynamic pressure.

The fluid-dynamic contribution, τ_{ij}, to the stress tensor is called the *deviatoric stress tensor*. For it to be invariant under Galilean transformations, it cannot depend on the absolute fluid velocity, but instead on the velocity gradient tensor $\partial u_i/\partial x_j$. However, by definition, stresses only develop in fluid elements that change shape. Therefore, only the symmetric part of $\partial u_i/\partial x_j$, S_{ij} from (3.12), should be considered in the fluid constitutive equation because the antisymmetric part of $\partial u_i/\partial x_j$, R_{ij} from (3.13), corresponds to pure rotation of fluid elements. The most general linear relationship between τ_{ij} and S_{ij} that produces $\tau_{ij} = 0$ when $S_{ij} = 0$ is

$$\tau_{ij} = K_{ijmn} S_{mn}, \tag{4.28}$$

where K_{ijmn} is a fourth-order tensor having 81 components that may depend on the local thermodynamic state of the fluid. Eq. (4.28) allows *each* of the nine components of τ_{ij} to be linearly related to *all* nine components of S_{ij}. However, this level of generality is unnecessary when the stress tensor is symmetric, and the fluid is isotropic.

In an isotropic fluid medium, the stress-strain rate relationship is independent of the orientation of the coordinate system. This is only possible if K_{ijmn} is an isotropic tensor. All fourth-order isotropic tensors must be of the form:

$$K_{ijmn} = \lambda\delta_{ij}\delta_{mn} + \mu\delta_{im}\delta_{jn} + \gamma\delta_{in}\delta_{jm} \tag{4.29}$$

(see Aris, 1962, pp. 30–33), where λ, μ, and γ are scalars that depend on the local thermodynamic state. In addition, τ_{ij} is symmetric in i and j, so (4.28) requires that K_{ijmn} also be symmetric in i and j. This requirement is consistent with (4.29) only if:

$$\gamma = \mu. \tag{4.30}$$

Therefore, only two constants, μ and λ, of the original 81, remain after the imposition of material-isotropy and stress-symmetry restrictions. Substitution of (4.29) into the constitutive equation (4.28) yields:

$$\tau_{ij} = 2\mu S_{ij} + \lambda S_{mm}\delta_{ij},$$

where $S_{mm} = \nabla \cdot \mathbf{u}$ is the volumetric strain rate (see Section 3.6). The complete stress tensor (4.27) then becomes:

$$T_{ij} = -p\delta_{ij} + 2\mu S_{ij} + \lambda S_{mm}\delta_{ij}, \tag{4.31}$$

which represents the appropriate multidimensional extension of (1.9).

The two scalar constants μ and λ can be further related as follows. Setting $i = j$, summing over the repeated index, and noting that $\delta_{ii} = 3$, we obtain:

$$T_{ii} = -3p + (2\mu + 3\lambda)S_{mm},$$

from which the pressure is found to be:

$$p = -\frac{1}{3}T_{ii} + \left(\frac{2}{3}\mu + \lambda\right)\nabla \cdot \mathbf{u}. \tag{4.32}$$

The diagonal terms of T_{ij} in a flow may be unequal because of the term proportional to μ in (4.31) and the fact that the diagonal terms of S_{ij} in a flow may be unequal. We can therefore take the average of the diagonal terms of $\mathbf{T}$ and define a *mean pressure* (as opposed to thermodynamic pressure p) as:

$$\overline{p} \equiv -\frac{1}{3}T_{ii}. \tag{4.33}$$

Substitution into (4.32) gives:

$$p - \overline{p} = \left(\frac{2}{3}\mu + \lambda\right)\nabla \cdot \mathbf{u}. \tag{4.34}$$

For a completely incompressible fluid we can only define a mechanical or mean pressure, because there is no equation of state to determine a thermodynamic pressure. (In fact, the *absolute pressure in an incompressible fluid is indeterminate,* and only its *gradients* can be determined from the equations of motion.) The λ-term in the constitutive equation (4.31) drops out when $S_{mm} = \nabla \cdot \mathbf{u} = 0$, and no consideration of (4.34) is necessary. So, for *incompressible fluids*, the constitutive equation (4.31) takes the simple form:

$$T_{ij} = -p\delta_{ij} + 2\mu S_{ij} \text{ (incompressible)}, \tag{4.35}$$

where p can only be interpreted as the mean pressure experienced by a fluid particle. For a compressible fluid, on the other hand, a thermodynamic pressure can be defined, and it seems that p and $\overline{p}$ can be different. In fact, (4.34) relates this difference to the rate of expansion through the proportionality constant $\mu_v = \lambda + 2\mu/3$, which is called the *coefficient of bulk viscosity*. It has an appreciable effect on sound absorption and shock-wave structure. It is generally found to be nonzero in polyatomic gases because of relaxation effects associated with molecular translation and rotation. However, the *Stokes assumption*,

$$\lambda + \frac{2}{3}\mu = 0,$$

is found to be accurate in many situations because either the fluid's μ_v or the flow's dilatation rate is small. Interesting historical aspects of the Stokes assumption can be found in Truesdell (1952).

Without using the Stokes assumption, the stress tensor (4.31) is:

$$T_{ij} = -p\delta_{ij} + \tau_{ij} = -p\delta_{ij} + 2\mu\left(S_{ij} - \frac{1}{3}S_{mm}\delta_{ij}\right) + \mu_v S_{mm}\delta_{ij}. \tag{4.36}$$

This linear relation between $\mathbf{T}$ and $\mathbf{S}$ is consistent with Newton's definition of the viscosity coefficient μ in a simple parallel flow $u(y)$, for which (4.36) gives a viscous shear stress of $\tau = \mu(du/dy)$. Consequently, a fluid obeying Eq. (4.36) is called a *Newtonian fluid* where μ and μ_v may only depend on the local thermodynamic state. The off-diagonal terms of T_{ij} are merely $2\mu S_{ij}$, while the on-diagonal terms also include pressure and velocity-divergence terms.

Example 4.9

Write out all the components of the stress tensor $\mathbf{T}$ in (x, y, z)-coordinates in terms of $\mathbf{u} = (u, v, w)$, and its derivatives.

Solution Evaluate each component of (4.36) and abbreviate $S_{mm} = \partial u/\partial x + \partial v/\partial y + \partial w/\partial z = \nabla \cdot \mathbf{u}$ to find:

$$T = \begin{bmatrix} -p + 2\mu \frac{\partial u}{\partial x} + \left(\mu_v - \frac{2}{3}\mu\right)\nabla \cdot \mathbf{u} & \mu\left(\frac{\partial u}{\partial y} + \frac{\partial v}{\partial x}\right) & \mu\left(\frac{\partial u}{\partial z} + \frac{\partial w}{\partial x}\right) \\ \mu\left(\frac{\partial v}{\partial x} + \frac{\partial u}{\partial y}\right) & -p + 2\mu \frac{\partial v}{\partial y} + \left(\mu_v - \frac{2}{3}\mu\right)\nabla \cdot \mathbf{u} & \mu\left(\frac{\partial v}{\partial z} + \frac{\partial w}{\partial y}\right) \\ \mu\left(\frac{\partial w}{\partial x} + \frac{\partial u}{\partial z}\right) & \mu\left(\frac{\partial w}{\partial y} + \frac{\partial v}{\partial z}\right) & -p + 2\mu \frac{\partial w}{\partial z} + \left(\mu_v - \frac{2}{3}\mu\right)\nabla \cdot \mathbf{u} \end{bmatrix}$$

The linear Newtonian friction law (4.36) might only be expected to hold for small strain rates since it is essentially a first-order expansion of the stress in terms of S_{ij} around $T_{ij} = 0$. However, the linear relationship is surprisingly accurate for many common fluids such as air, water, gasoline, and oils.

Yet, other liquids display non-Newtonian behavior at moderate rates of strain. These include solutions containing long-chain polymer molecules, concentrated soaps, melted plastics, emulsions and slurries containing suspended particles, and many liquids of biological origin. Non-Newtonian fluid mechanics, an important and fascinating sub-field of fluid mechanics, is largely beyond the scope of this text, and its fundamentals are well covered elsewhere (see Bird et al., 1987). Thus, the remainder of this section merely describes three common non-Newtonian flow phenomena.

First, shear stress in a non-Newtonian flow may be a *nonlinear* function of the local strain rate. The simplest model of this behavior for a unidirectional shear flow $\mathbf{u} = (u_1(x_2), 0, 0)$ (see Section 3.5), is a power law:

$$\tau_{12} = \eta \frac{\partial u_1}{\partial x_2} = (m\dot{\gamma}^{n-1}) \frac{\partial u_1}{\partial x_2} = m\dot{\gamma}^n, \tag{4.37}$$

where $\eta = m\dot{\gamma}^{n-1}$ is the non-Newtonian viscosity, m is the power law coefficient, $\dot{\gamma} = \partial u_1/\partial x_2$ is the shear rate, and n is the power law exponent. For a Newtonian fluid, $n = 1$ and m is the fluid's viscosity. Departures from Newtonian behavior are characterized by the value of n. When $n < 1$, the fluid is known as *shear thinning* or *pseudoplastic* and its viscosity drops with increasing strain rate. Most liquid plastics and polymeric solutions are shear thinning. When $n > 1$, the fluid is known as *shear thickening* or *dilatant*. Solutions of water and starch and concentrated suspensions of small particles can be shear thickening.

Second, the current stress on a non-Newtonian fluid particle may depend on the local strain rate and on its *history*. Such memory effects give the fluid some elastic properties that may allow it to mimic solid behavior over sufficiently short periods of time. In fact, there is a whole class of *viscoelastic* substances that are neither fully fluid nor fully solid. For linear viscoelastic materials, the general linear stress-strain rate law (4.28) is replaced by:

$$\tau_{ij}(t) = \int_{-\infty}^{t} K_{ijmn}(t - t')S_{mn}(t')dt',$$

where K_{ijmn} is the tensorial relaxation modulus of the material, and the temporal argument of each factor is shown. The integral in this constitutive relationship allows a fluid-element's strain-rate history to influence its current stress state.

And finally, flowing non-Newtonian fluids may develop normal stresses that do not occur in Newtonian fluids. For example, a non-Newtonian fluid undergoing the simple shear flow mentioned above, $\mathbf{u} = (u_1(x_2), 0, 0)$, may develop a nonzero *first normal-stress difference*, $T_{11} - T_{22}$, and a nonzero *second normal-stress difference*, $T_{22} - T_{33}$. For polymeric fluids, the first normal stress difference is typically negative and it corresponds to the development of tensile stress along streamlines, while the second normal stress difference is typically positive and smaller in absolute magnitude. The existence of a negative first normal stress difference largely explains the phenomena of extrudate swell and rotating rod climbing exhibited by some polymeric fluids.

4.6 Navier-Stokes momentum equation

The momentum conservation equation for a Newtonian fluid is obtained by substituting (4.36) into Cauchy's equation (4.24) to obtain:

$$\rho\left(\frac{\partial u_j}{\partial t} + u_i \frac{\partial u_j}{\partial x_i}\right) = -\frac{\partial p}{\partial x_j} + \rho g_j + \frac{\partial}{\partial x_i}\left[\mu\left(\frac{\partial u_j}{\partial x_i} + \frac{\partial u_i}{\partial x_j}\right) + \left(\mu_v - \frac{2}{3}\mu\right)\frac{\partial u_m}{\partial x_m}\delta_{ij}\right], \tag{4.38}$$

where we have used $(\partial p/\partial x_i)\delta_{ij} = \partial p/\partial x_j$, (3.5) with $F = u_j$, and (3.12). This is the *Navier-Stokes momentum equation*. The viscosities, μ and μ_v, in this equation can depend on the thermodynamic state and indeed μ, for most fluids, displays a rather strong dependence on temperature, decreasing with T for liquids and increasing with T for gases. Together, (4.7) and (4.38) provide $1+3=4$ scalar equations, and they contain ρ, p, and u_j for $1+1+3=5$ dependent variables. Therefore, when combined with suitable boundary conditions, (4.7) and (4.38) provide a complete description of fluid dynamics when ρ is a known constant or when a single known relationship exists between p and ρ. In the latter case, the fluid or the flow is said to be *barotropic*. When the relationship between p and ρ also includes the temperature T, the internal (or thermal) energy e of the fluid must also be considered. These additions allow a caloric equation of state to be added to the equation listing, but introduce two more dependent variables, T and e. Thus, in general, a third field equation representing conservation of energy is needed to fully describe fluid dynamics.

When temperature differences are small within the flow, μ and μ_v can be taken outside the spatial derivative operating on the contents of the [,]-brackets in (4.38), which then reduces to

$$\rho\frac{Du_j}{Dt} = -\frac{\partial p}{\partial x_j} + \rho g_j + \mu\frac{\partial^2 u_j}{\partial x_i^2} + \left(\mu_v + \frac{1}{3}\mu\right)\frac{\partial}{\partial x_j}\frac{\partial u_m}{\partial x_m} \text{ (compressible).} \tag{4.39}$$

For incompressible fluids $\nabla \cdot \mathbf{u} = \partial u_m/\partial x_m = 0$, so (4.39) in vector notation reduces to:

$$\rho\frac{D\mathbf{u}}{Dt} = -\nabla p + \rho\mathbf{g} + \mu\nabla^2\mathbf{u} \text{ (incompressible).} \tag{4.40}$$

Interestingly, the net viscous force per unit volume in incompressible flow, the last term on the right in this equation $= \mu(\partial^2 u_j/\partial x_i^2)$, can be obtained from the divergence of the strain rate tensor or from the curl of the vorticity (see Exercise 4.51):

$$\mu\frac{\partial^2 u_j}{\partial x_i^2} = 2\mu\frac{\partial S_{ij}}{\partial x_i} = \mu\frac{\partial}{\partial x_i}\left(\frac{\partial u_j}{\partial x_i} + \frac{\partial u_i}{\partial x_j}\right) = -\mu\varepsilon_{jik}\frac{\partial \omega_k}{\partial x_i}, \quad \text{or} \quad \mu\nabla^2\mathbf{u} = -\mu\nabla\times\boldsymbol{\omega}. \tag{4.41}$$

This result would seem to pose a paradox since it shows that the net viscous force depends on the vorticity even though rotation of fluid elements was explicitly excluded from entering (4.36), the precursor of (4.41). This paradox is resolved by realizing that the net viscous force is given by either a spatial *derivative* of the vorticity or a spatial *derivative* of the deformation rate. The net viscous force vanishes when $\boldsymbol{\omega}$ is uniform in space (as in solid-body rotation), in which case the incompressibility condition requires that the deformation rate is zero everywhere as well.

If viscous effects are negligible, which is commonly true in exterior flows away from solid boundaries, (4.39) further simplifies to the *Euler equation*:

$$\rho\frac{D\mathbf{u}}{Dt} = -\nabla p + \rho\mathbf{g}. \tag{4.42}$$

Example 4.10

Write out the Navier-Stokes equations in two-dimensional (x, y)-coordinates when $\mathbf{u} = (u, v)$, and simplify these to the one-dimensional flow case where $u = u(x, t)$ and $v = 0$.

Solution Evaluate each component of (4.38) to find:

$$\rho\left(\frac{\partial u}{\partial t} + u\frac{\partial u}{\partial x} + v\frac{\partial u}{\partial y}\right) = -\frac{\partial p}{\partial x} + \rho g_x + \frac{\partial}{\partial x}\left[2\mu\frac{\partial u}{\partial x} + \left(\mu_v - \frac{2}{3}\mu\right)\left(\frac{\partial u}{\partial x} + \frac{\partial v}{\partial y}\right)\right] + \frac{\partial}{\partial y}\left[\mu\left(\frac{\partial v}{\partial x} + \frac{\partial u}{\partial y}\right)\right] \quad \text{and}$$

$$\rho\left(\frac{\partial v}{\partial t} + u\frac{\partial v}{\partial x} + v\frac{\partial v}{\partial y}\right) = -\frac{\partial p}{\partial y} + \rho g_y + \frac{\partial}{\partial x}\left[\mu\left(\frac{\partial u}{\partial y} + \frac{\partial v}{\partial x}\right)\right] + \frac{\partial}{\partial y}\left[2\mu\frac{\partial v}{\partial y} + \left(\mu_v - \frac{2}{3}\mu\right)\left(\frac{\partial u}{\partial x} + \frac{\partial v}{\partial y}\right)\right],$$

where $\mathbf{g} = (g_x, g_y)$. To reach the appropriate one-dimensional form, work from the x-component equation above, and drop the terms containing v and y-derivatives of u. The result is:

$$\rho\left(\frac{\partial u}{\partial t} + u\frac{\partial u}{\partial x}\right) = -\frac{\partial p}{\partial x} + \rho g_x + \frac{\partial}{\partial x}\left[\left(\mu_v + \frac{4}{3}\mu\right)\left(\frac{\partial u}{\partial x}\right)\right].$$

In this case, the y-component equation reduces to a hydrostatic balance: $0 = -\partial p/\partial y + \rho g_y$.

4.7 Noninertial frame of reference

The equations of motion given in Sections 4.4 through 4.6 are valid in an inertial frame of reference, one that is stationary or that is moving at a constant speed with respect to a stationary frame of reference. Although a stationary frame of reference cannot be defined precisely, a frame of reference that is not moving with respect to distant stars is adequate for present purposes. Thus, noninertial-frame effects may be found in other frames of reference known to undergo nonuniform translation and rotation. For example, the fluid mechanics of rotating machinery and large-scale atmospheric, oceanic, and geophysical flows are often best analyzed in a rotating frame of reference. Plus, modern life commonly places us in the noninertial frame of reference of a moving and maneuvering vehicle. Fortunately, in many laboratory situations, the relevant distances and time scales are short enough so that a frame of reference attached to the earth (sometimes referred to as the *laboratory frame* of reference) is a suitable inertial frame of reference.

In a noninertial frame of reference, the continuity equation (4.7) is unchanged but the momentum equation (4.38) must be modified. Consider a frame of reference O′1′2′3′ that translates at velocity $d\mathbf{X}(t)/dt = \mathbf{U}(t)$ and rotates at angular velocity $\boldsymbol{\Omega}(t)$ with respect to a stationary frame of reference O123 (see Fig. 4.8). The vectors $\mathbf{U}$ and $\boldsymbol{\Omega}$ may be resolved in either frame. The same clock is used in both frames so $t = t'$. The spatial coordinates of a fluid particle P in the rotating and stationary frames are $\mathbf{x}' = (x_1', x_2', x_3')$ and $\mathbf{x} = (x_1, x_2, x_3)$, respectively, and these are related by vector addition: $\mathbf{x} = \mathbf{X} + \mathbf{x}'$. The velocity $\mathbf{u}$ of the fluid particle is obtained by time differentiation:

$$\begin{aligned}\mathbf{u} &= \frac{d\mathbf{x}}{dt} = \frac{d\mathbf{X}}{dt} + \frac{d\mathbf{x}'}{dt} = \mathbf{U} + \frac{d}{dt}(x_1'\mathbf{e}_1' + x_2'\mathbf{e}_2' + x_3'\mathbf{e}_3') \\ &= \mathbf{U} + \frac{dx_1'}{dt}\mathbf{e}_1' + \frac{dx_2'}{dt}\mathbf{e}_2' + \frac{dx_3'}{dt}\mathbf{e}_3' + x_1'\frac{d\mathbf{e}_1'}{dt} + x_2'\frac{d\mathbf{e}_2'}{dt} + x_3'\frac{d\mathbf{e}_3'}{dt} = \mathbf{U} + \mathbf{u}' + \boldsymbol{\Omega} \times \mathbf{x}',\end{aligned} \tag{4.43}$$

where the final equality is based on the geometric construction of the cross product shown in Fig. 4.9 for $\mathbf{e}_1'$, one of the unit vectors in the rotating frame. In a small time dt, the rotation of O′1′2′3′ causes $\mathbf{e}_1'$ to trace a small portion of a cone with radius $\sin\alpha$ as shown. The magnitude of the change in $\mathbf{e}_1'$ is $d|\mathbf{e}_1'| = (\sin\alpha)|\boldsymbol{\Omega}|dt$, so $d|\mathbf{e}_1'|/dt = (\sin\alpha)|\boldsymbol{\Omega}|$, which is equal to the magnitude of $\boldsymbol{\Omega} \times \mathbf{e}_1'$. The direction of the rate of change of $\mathbf{e}_1'$ is perpendicular to $\boldsymbol{\Omega}$ and $\mathbf{e}_1'$, which is the direction of $\boldsymbol{\Omega} \times \mathbf{e}_1'$. Thus, by geometric construction, $d\mathbf{e}_1'/dt = \boldsymbol{\Omega} \times \mathbf{e}_1'$, and by direct extension to the other unit vectors, $d\mathbf{e}_i'/dt = \boldsymbol{\Omega} \times \mathbf{e}_i'$ (in mixed notation).

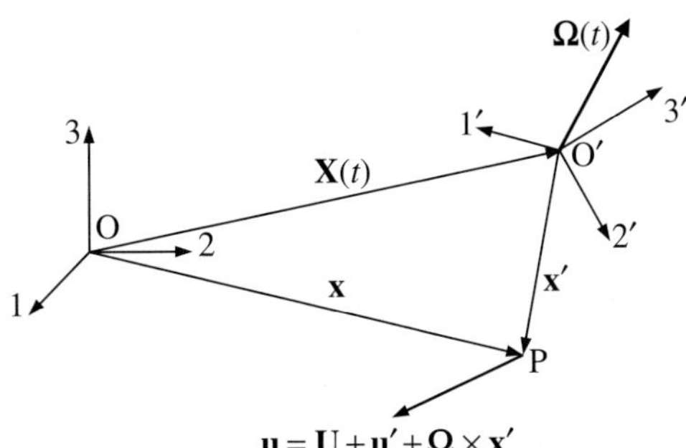

FIGURE 4.8 Geometry showing the relationship between a stationary coordinate system O123 and a noninertial coordinate system O′1′2′3′ that is moving, accelerating, and rotating with respect to O123. In particular, the vector connecting O and O′ is $\mathbf{X}(t)$ and the rotational velocity of O′1′2′3′ is $\boldsymbol{\Omega}(t)$. The vector velocity $\mathbf{u}$ at point P in O123 is shown. The vector velocity $\mathbf{u}'$ at point P in O′1′2′3′ differs from $\mathbf{u}$ because of the motion of O′1′2′3′.

To find the acceleration $\mathbf{a}$ of a fluid particle at P, take the time derivative of the final version of (4.43) to find:

$$\mathbf{a} = \frac{d\mathbf{u}}{dt} = \frac{d}{dt}(\mathbf{U} + \mathbf{u}' + \boldsymbol{\Omega} \times \mathbf{x}') = \frac{d\mathbf{U}}{dt} + \mathbf{a}' + 2\boldsymbol{\Omega} \times \mathbf{u}' + \frac{d\boldsymbol{\Omega}}{dt} \times \mathbf{x}' + \boldsymbol{\Omega} \times (\boldsymbol{\Omega} \times \mathbf{x}') \tag{4.44}$$

(see Exercise 4.57), where $d\mathbf{U}/dt$ is the acceleration of O′ with respect to O, $\mathbf{a}'$ is the fluid particle acceleration viewed in the noninertial frame, $2\boldsymbol{\Omega} \times \mathbf{u}'$ is the *Coriolis* acceleration, $(d\boldsymbol{\Omega}/dt) \times \mathbf{x}'$ is the acceleration caused by angular acceleration of the noninertial frame, and the final term is the *centripetal* acceleration.

In fluid mechanics, the acceleration $\mathbf{a}$ of fluid particles is denoted $D\mathbf{u}/Dt$, so (4.44) is rewritten:

$$\left(\frac{D\mathbf{u}}{Dt}\right)_{O123} = \left(\frac{D'\mathbf{u}'}{Dt}\right)_{O1'2'3'} + \frac{d\mathbf{U}}{dt} + 2\boldsymbol{\Omega} \times \mathbf{u}' + \frac{d\boldsymbol{\Omega}}{dt} \times \mathbf{x}' + \boldsymbol{\Omega} \times (\boldsymbol{\Omega} \times \mathbf{x}'). \tag{4.45}$$

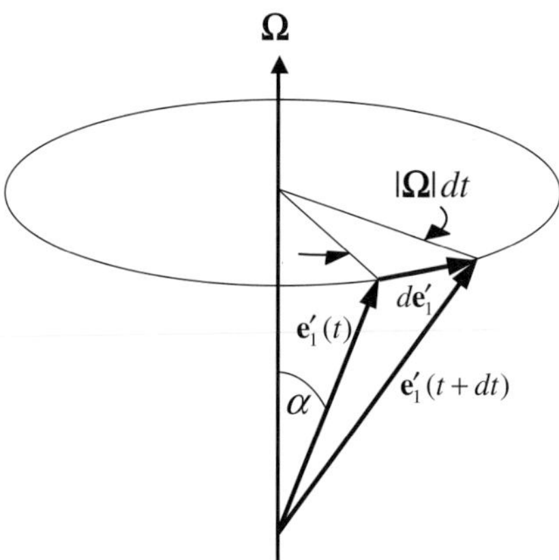

FIGURE 4.9 Geometry showing the relationship between $\mathbf{\Omega}$, the rotational velocity vector of O′1′2′3′, and the first coordinate unit vector $\mathbf{e}_1'$ in O′1′2′3′. Here, the increment $d\mathbf{e}_1'$ is perpendicular to $\mathbf{\Omega}$ and $\mathbf{e}_1'$.

This equation states that fluid particle acceleration in an inertial frame is equal to the sum of: the particle's acceleration in the noninertial frame, the acceleration of the noninertial frame, the Coriolis acceleration, the particle's apparent acceleration from the noninertial-frame's angular acceleration, and the particle's centripetal acceleration. Substituting (4.45) into (4.40) produces

$$\rho\left(\frac{D'\mathbf{u}'}{Dt}\right)_{O1'2'3'} = -\nabla' p + \rho\left[\mathbf{g} - \frac{d\mathbf{U}}{dt} - 2\mathbf{\Omega}\times\mathbf{u}' - \frac{d\mathbf{\Omega}}{dt}\times\mathbf{x}' - \mathbf{\Omega}\times(\mathbf{\Omega}\times\mathbf{x}')\right] + \mu\nabla'^2\mathbf{u}' \tag{4.46}$$

as the incompressible-flow momentum conservation equation in a noninertial frame of reference where the primes denote differentiation, velocity, and position in the noninertial frame. Thermodynamic variables and the net viscous stress are independent of the frame of reference. Eq. (4.46) makes it clear that the primary effect of a noninertial frame is the addition of extra body force terms that arise from the motion of the noninertial frame. The terms in [,]-brackets reduce to $\mathbf{g}$ alone when O′1′2′3′ is an inertial frame ($\mathbf{U}$ = constant and $\mathbf{\Omega} = 0$).

The four new terms in (4.46) may each be significant. The first new term $d\mathbf{U}/dt$ accounts for the acceleration of O′ relative to O. It provides the apparent force that pushes occupants back into their seats or makes them tighten their grip on a handrail when a vehicle accelerates. An aircraft that is flown on a parabolic trajectory produces weightlessness in its interior when its acceleration $d\mathbf{U}/dt$ equals $\mathbf{g}$.

The second new term, the Coriolis term, depends on the fluid particle's velocity, not on its position. Thus, even at the earth's rotation rate of one cycle per day, it has important consequences for the accuracy of artillery and for navigation during air and sea travel. The earth's angular velocity vector $\mathbf{\Omega}$ points out of the ground in the northern hemisphere. The Coriolis acceleration $-2\mathbf{\Omega}\times\mathbf{u}$ therefore tends to deflect a particle to the right of its direction of travel in the northern hemisphere and to the left in the southern hemisphere. Imagine a low-drag projectile shot horizontally from the north pole with speed u (Fig. 4.10). The Coriolis acceleration $2\Omega u$ constantly acts perpendicular to its path and therefore does not change the speed u of the projectile. The forward distance traveled in time t is ut, and the deflection is Ωut^2. The angular deflection is $\Omega ut^2/ut = \Omega t$, which is the earth's rotation in time t. This demonstrates that the projectile in fact travels in a straight line if observed from outer space (an inertial frame); its apparent deflection is merely due to the rotation of the earth underneath it. Observers on earth invoke the Coriolis acceleration to account for this deflection. Further description of the Coriolis acceleration is provided in Stommel and Moore (1989).

In the atmosphere, the Coriolis acceleration is responsible for wind circulation patterns around centers of high and low pressure in the earth's atmosphere. In an inertial frame, a nonzero pressure gradient accelerates fluid from regions of higher pressure to regions of lower pressure, as the first term on the right of (4.38) and (4.46) indicates. Imagine a cylindrical polar coordinate system (Fig. 3.3c), with the z-axis normal to the earth's surface and the origin at the center of a high- or low-pressure region in the atmosphere. If it is a high pressure zone, u_R would be outward (positive) away from the z-axis in the absence of rotation since fluid will leave a center of high pressure. In this situation when there is rotation, the Coriolis acceleration $-2\mathbf{\Omega}\times\mathbf{u} = -2\Omega_z u_R \mathbf{e}_\varphi$ is in the $-\varphi$ direction (in the Northern hemisphere), or clockwise as viewed from above. On the other hand, if the flow is inward toward the center of a low-pressure zone, which reverses the direction of u_R, the Coriolis acceleration is counterclockwise. In the southern hemisphere, the direction of Ω_z is reversed so that the circulation patterns described above are reversed. The effects of Earth's rotation appear prominently throughout geophysical fluid mechanics (see Chapter 12).

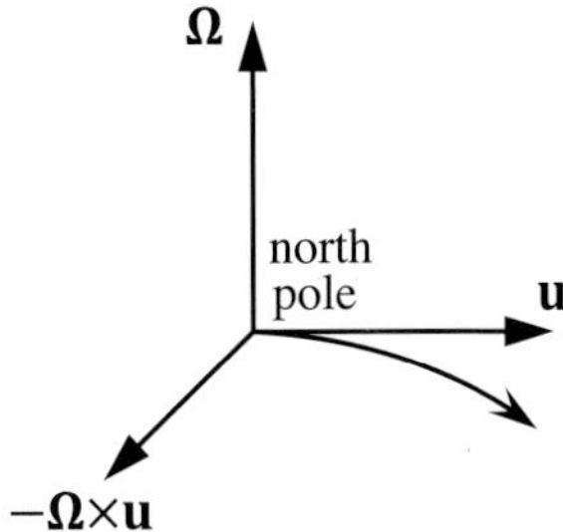

FIGURE 4.10 Particle trajectory deflection caused by the Coriolis acceleration when observed in a rotating frame of reference. If observed from a stationary frame of reference, the particle trajectory would be straight.

The third new acceleration term in [,]-brackets in (4.46) is caused by changes in the rotation rate of the frame of reference so it unimportant for geophysical flows and flows in machinery that rotate at a constant rate about a fixed axis. However, it does play a role when rotation speed or the direction of rotation varies with time.

The final new acceleration term in (4.46), the centrifugal acceleration, depends strongly on the rotation rate and the distance of the fluid particle from the axis of rotation. If the rotation rate is steady and the axis of rotation coincides with the z-axis of a cylindrical polar coordinate system so that $\mathbf{\Omega} = (0, 0, \Omega)$ and $\mathbf{x}' = (R, \varphi, z)$, then $-\mathbf{\Omega} \times (\mathbf{\Omega} \times \mathbf{x}') = +\Omega^2 R\mathbf{e}_R$. This additional apparent acceleration can be added to the Newtonian gravitational acceleration $\mathbf{g}_\mathbf{n}$ to define an *effective gravity* $\mathbf{g} = \mathbf{g}_\mathbf{n} + \Omega^2 R\mathbf{e}_R$ (Fig. 4.11). Interestingly, a body-force potential for the new term can be found, but its impact might only be felt for relatively large atmospheric- or oceanic-scale flows (Exercise 4.74). The effective gravity is not precisely directed at the center of the earth and its acceleration value varies slightly over the surface of the earth. The equipotential surfaces (shown by the dashed lines in Fig. 4.11) are perpendicular to the effective gravity, and the average sea level is one of these equipotential surfaces. Thus, at least locally on the earth's surface, we can write $\Phi = gz$, where z is measured perpendicular to an equipotential surface, and g is the local acceleration caused by the effective gravity. Use of the locally correct acceleration and direction for the earth's gravitational body force in the equations of fluid motion accounts for the centrifugal acceleration and the fact that the earth is really an ellipsoid with equatorial diameter 42 km larger than the polar diameter.

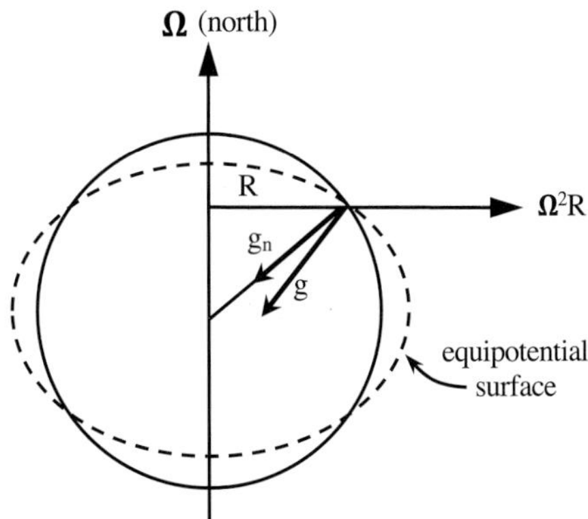

FIGURE 4.11 The earth's rotation causes it to budge near the equator and this leads to a mild distortion of equipotential surfaces from perfect spherical symmetry. The total gravitational acceleration is a sum of a centrally directed acceleration $\mathbf{g}_\mathbf{n}$ (the Newtonian gravitation) and a rotational correction $\Omega^2 R$ that points away from the axis of rotation.

Example 4.11

Find the radial and angular fluid momentum equations for viscous flow in the gaps between plates of a von Karman viscous-impeller pump (see Fig. 4.12) that rotates at a constant angular speed Ω_z. Assume steady constant-density constant-viscosity flow, neglect the body force for simplicity, and use cylindrical coordinates (Fig. 3.3c).

Solution A von Karman viscous impeller pump uses rotating plates to pump viscous fluids via a combination of viscous and centrifugal forces. Although such pumps may be inefficient, they are wear-tolerant and may be used to pump abrasive fluids that would damage the vanes or blades of other pumps. Plus, their pumping action is entirely steady so they can be exceptionally quiet, a feature occasionally exploited for air-moving applications for interior spaces occupied by human beings.

For steady, constant-density, constant-viscosity flow without a body force in a steadily rotating frame of reference, the momentum equation is a simplified version of (4.46):

$$\rho(\mathbf{u}'\cdot\nabla')\mathbf{u}' = -\nabla' p + \rho[-2\boldsymbol{\Omega}\times\mathbf{u}' - \boldsymbol{\Omega}\times(\boldsymbol{\Omega}\times\mathbf{x}')] + \mu\nabla'^2\mathbf{u}'.$$

Here we are not concerned with the axial inflow or the flow beyond the outer edges of the disks. Now choose the z-axis of the coordinate system to be coincident with the axis of rotation. For this choice, the flow between the disks should be axisymmetric, so we can presume that u'_R, u'_φ, u'_z and p only depend on R and z. To further simplify the momentum equation, drop the primes, evaluate the cross products:

$$\boldsymbol{\Omega}\times\mathbf{u} = \Omega_z\mathbf{e}_z\times(u_R\mathbf{e}_R + u_\varphi\mathbf{e}_\varphi) = +\Omega_z u_R\mathbf{e}_\varphi - \Omega_z u_\varphi\mathbf{e}_R, \quad \text{and} \quad \boldsymbol{\Omega}\times(\boldsymbol{\Omega}\times\mathbf{x}') = -\Omega_z^2 R\mathbf{e}_R,$$

and separate the radial, angular, and axial components to find:

$$\rho\left(u_R\frac{\partial u_R}{\partial R} + u_z\frac{\partial u_R}{\partial z} - \frac{u_\varphi^2}{R}\right) = -\frac{\partial p}{\partial R} + \rho[2\Omega_z u_\varphi + \Omega_z^2 R] + \mu\left(\frac{1}{R}\frac{\partial}{\partial R}\left(R\frac{\partial u_R}{\partial R}\right) + \frac{\partial^2 u_R}{\partial z^2} - \frac{u_R}{R^2}\right)$$

$$\rho\left(u_R\frac{\partial u_\varphi}{\partial R} + u_z\frac{\partial u_\varphi}{\partial z} + \frac{u_R u_\varphi}{R}\right) = \rho[-2\Omega_z u_R] + \mu\left(\frac{1}{R}\frac{\partial}{\partial R}\left(R\frac{\partial u_\varphi}{\partial R}\right) + \frac{\partial^2 u_\varphi}{\partial z^2} - \frac{u_\varphi}{R^2}\right)$$

$$\rho\left(u_R\frac{\partial u_z}{\partial R} + u_z\frac{\partial u_z}{\partial z}\right) = -\frac{\partial p}{\partial z} + \mu\left(\frac{1}{R}\frac{\partial}{\partial R}\left(R\frac{\partial u_z}{\partial R}\right) + \frac{\partial^2 u_z}{\partial z^2}\right).$$

Here we have used the results found in the Appendix B for cylindrical coordinates. In the first two momentum equations, the terms in [,]-brackets result from rotation of the coordinate system.

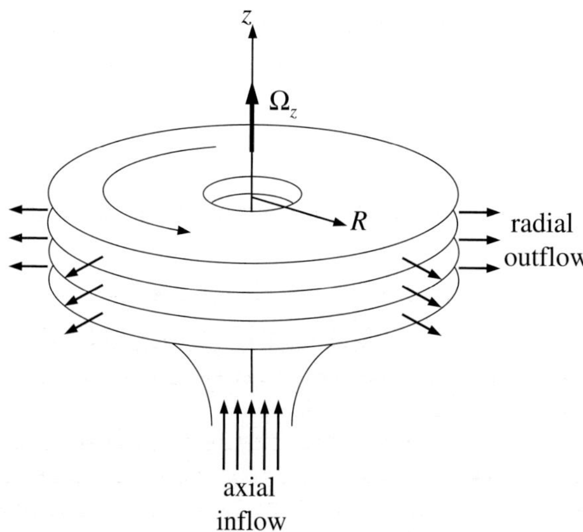

FIGURE 4.12 Schematic drawing of the impeller of a von Karman pump (Example 4.11).

4.8 Conservation of energy

In this section, the integral energy-conservation equivalent of (4.5) and (4.17) is developed from a mathematical statement of conservation of energy for a fluid particle in an inertial frame of reference. The subsequent steps that lead to a differential energy-conservation equivalent of (4.7) and (4.24) follow the pattern set in Sections 4.2 and 4.4 of this chapter. For clarity and conciseness, the explicit listing of independent variables is dropped from the equations in this section.

When applied to a material volume $V(t)$ with surface area $A(t)$, conservation of internal energy per unit mass e and the kinetic energy per unit mass $|\mathbf{u}|^2/2$ for a single-component fluid can be stated:

$$\frac{d}{dt}\int_{V(t)}\rho\left(e + \frac{1}{2}|\mathbf{u}|^2\right)dV = \int_{V(t)}\rho\mathbf{g}\cdot\mathbf{u}\,dV + \int_{A(t)}\mathbf{f}\cdot\mathbf{u}\,dA - \int_{A(t)}\mathbf{q}\cdot\mathbf{n}\,dA, \tag{4.47}$$

where the terms on the right are: work done on the fluid in $V(t)$ by body forces, work done on the fluid in $V(t)$ by surface forces, and heat transferred out of $V(t)$. Here, $\mathbf{q}$ is the heat flux vector and in general includes thermal conduction and radiation. The final term in (4.47) has a negative sign because the energy in $V(t)$ decreases when heat leaves $V(t)$ and this occurs when $\mathbf{q}\cdot\mathbf{n}$ is positive. In general, terms associated with differing molecular enthalpies and chemical reactions must be added to (4.47) in multicomponent fluid flows (see Kuo, 1986). The implications of (4.47) are better displayed when the time derivative is expanded using Reynolds transport theorem (3.35),

$$\int_{V(t)}\frac{\partial}{\partial t}\left(\rho e+\frac{\rho}{2}|\mathbf{u}|^2\right)dV+\int_{A(t)}\left(\rho e+\frac{\rho}{2}|\mathbf{u}|^2\right)(\mathbf{u}\cdot\mathbf{n})dA=\int_{V(t)}\rho\mathbf{g}\cdot\mathbf{u}dV+\int_{A(t)}\mathbf{f}\cdot\mathbf{u}dA-\int_{A(t)}\mathbf{q}\cdot\mathbf{n}dA. \quad (4.48)$$

Similar to the prior developments for mass and momentum conservation, this result can be generalized to an arbitrarily moving control volume $V^*(t)$ with surface $A^*(t)$:

$$\frac{d}{dt}\int_{V^*(t)}\left(\rho e+\frac{\rho}{2}|\mathbf{u}|^2\right)dV+\int_{A^*(t)}\left(\rho e+\frac{\rho}{2}|\mathbf{u}|^2\right)(\mathbf{u}-\mathbf{b})\cdot\mathbf{n}dA$$
$$=\int_{V^*(t)}\rho\mathbf{g}\cdot\mathbf{u}dV+\int_{A^*(t)}\mathbf{f}\cdot\mathbf{u}dA-\int_{A^*(t)}\mathbf{q}\cdot\mathbf{n}dA, \quad (4.49)$$

when $V^*(t)$ is instantaneously coincident with $V(t)$. And, just like (4.5) and (4.17), (4.49) can be specialized to stationary, steadily moving, accelerating, or deforming control volumes by appropriate choice of the control surface velocity $\mathbf{b}$.

The differential equation that represents energy conservation is obtained from (4.48) after collecting all four terms under the same volume integration. The first step is to convert the three surface integrals in (4.48) to volume integrals using Gauss' theorem (2.30):

$$\int_{A(t)}\left(\rho e+\frac{\rho}{2}|\mathbf{u}|^2\right)(\mathbf{u}\cdot\mathbf{n})dA=\int_{V(t)}\nabla\cdot\left(\rho e\mathbf{u}+\frac{\rho}{2}|\mathbf{u}|^2\mathbf{u}\right)dV=\int_{V(t)}\frac{\partial}{\partial x_i}\left(\rho\left(e+\frac{1}{2}u_j^2\right)u_i\right)dV, \quad (4.50)$$

$$\int_{A(t)}\mathbf{f}\cdot\mathbf{u}dA=\int_{A(t)}n_iT_{ij}u_jdA=\int_{V(t)}\frac{\partial}{\partial x_i}(T_{ij}u_j)dV, \quad (4.51)$$

and $$\int_{A(t)}\mathbf{q}\cdot\mathbf{n}dA=\int_{A(t)}q_in_idA=\int_{V(t)}\nabla\cdot\mathbf{q}dV=\int_{V(t)}\frac{\partial q_i}{\partial x_i}dV, \quad (4.52)$$

where in (4.50) $u_j^2=u_1^2+u_2^2+u_3^2$ because the index j is implicitly repeated. Substituting (4.50) through (4.52) into (4.48) and putting all the terms together into the same volume integration produces:

$$\int_{V(t)}\left\{\frac{\partial}{\partial t}\left(\rho\left[e+\frac{1}{2}u_j^2\right]\right)+\frac{\partial}{\partial x_i}\left(\rho\left[e+\frac{1}{2}u_j^2\right]u_i\right)-\rho g_iu_i-\frac{\partial}{\partial x_i}(T_{ij}u_j)+\frac{\partial q_i}{\partial x_i}\right\}dV=0. \quad (4.53)$$

Similar to (4.6) and (4.21), the integral in (4.53) can only be zero for any material volume if its integrand vanishes at every point in space; thus (4.53) requires:

$$\frac{\partial}{\partial t}\left(\rho\left[e+\frac{1}{2}u_j^2\right]\right)+\frac{\partial}{\partial x_i}\left(\rho\left[e+\frac{1}{2}u_j^2\right]u_i\right)=\rho g_iu_i+\frac{\partial}{\partial x_i}(T_{ij}u_j)-\frac{\partial q_i}{\partial x_i}. \quad (4.54)$$

This differential equation is a statement of conservation of energy containing terms for fluid particle internal energy, fluid particle kinetic energy, work, energy exchange, and heat transfer. It is commonly revised and simplified so that its terms are more readily interpreted. The second term on the right side of (4.54) represents the total rate of work done on a fluid particle by surface stresses. By performing the differentiation, and then using (4.27) to separate out pressure and viscous surface-stress terms, it can be decomposed as follows:

$$\frac{\partial}{\partial x_i}(T_{ij}u_j)=T_{ij}\frac{\partial u_j}{\partial x_i}+u_j\frac{\partial T_{ij}}{\partial x_i}=\left(-p\frac{\partial u_j}{\partial x_j}+\tau_{ij}\frac{\partial u_j}{\partial x_i}\right)+\left(-u_j\frac{\partial p}{\partial x_j}+u_j\frac{\partial\tau_{ij}}{\partial x_i}\right). \quad (4.55)$$

In the final equality, the terms in the first set of (,)-parentheses are the pressure and viscous-stress work terms that lead to the deformation of fluid particles while the terms in the second set of (,)-parentheses are the product of the local fluid velocity with the net pressure force ($\partial p/\partial x_j$) and the net viscous force ($\partial\tau_{ij}/\partial x_i$) on a fluid particle that lead to either an increase or decrease in its kinetic energy. Substituting (4.55) into (4.54), expanding the differentiations on the left in (4.54), and using the continuity equation (4.7) to drop terms produces:

$$\rho\frac{D}{Dt}\left(e+\frac{1}{2}u_j^2\right)=\rho g_i u_i+\left(-p\frac{\partial u_j}{\partial x_j}+\tau_{ij}\frac{\partial u_j}{\partial x_i}\right)+\left(-u_j\frac{\partial p}{\partial x_j}+u_j\frac{\partial \tau_{ij}}{\partial x_i}\right)-\frac{\partial q_i}{\partial x_i} \tag{4.56}$$

(see Exercise 4.60). This equation contains both mechanical and thermal energy terms. A separate equation for the mechanical energy can be constructed by multiplying (4.22) by u_j and summing over j. After some manipulation, the result is:

$$\rho\frac{D}{Dt}\left(\frac{1}{2}u_j^2\right)=\rho g_j u_j-u_j\frac{\partial p}{\partial x_j}+u_j\frac{\partial \tau_{ij}}{\partial x_i}=u_j\left(\rho g_j-\frac{\partial p}{\partial x_j}+\frac{\partial \tau_{ij}}{\partial x_i}\right) \tag{4.57}$$

(see Exercise 4.61), where (4.27) has been used for T_{ij}. The final product of u_j with the terms in parentheses is a dot product. Thus, this equation shows that a fluid element's kinetic energy increases whenever the sum of body, pressure, and viscous forces has a net positive component in the fluid element's direction of travel. Subtracting (4.57) from (4.56), dividing by $\rho = 1/\upsilon$, and using (4.8) produces:

$$\frac{De}{Dt}=-p\frac{D\upsilon}{Dt}+\frac{1}{\rho}\tau_{ij}S_{ij}-\frac{1}{\rho}\frac{\partial q_i}{\partial x_i}, \tag{4.58}$$

where the fact that τ_{ij} is symmetric has been exploited so $\tau_{ij}(\partial u_j/\partial x_i)=\tau_{ij}(S_{ji}+R_{ji})=\tau_{ij}S_{ij}$ with S_{ij} given by (3.12). Eq. (4.58) is entirely equivalent to the first law of thermodynamics (1.16) – the change in energy of a system equals the work put into the system minus the heat lost by the system. The difference is that in (4.58), all the terms have units of power per unit mass instead of energy. The first two terms on the right in (4.58) are the rates at which pressure and viscous work are done on a fluid particle while the final term represents the heat transfer rate from the fluid particle.

The viscous work term in (4.58) is the kinetic energy dissipation rate per unit mass, and it is commonly denoted by $\varepsilon = (1/\rho)\tau_{ij}S_{ij}$. It is the product of the viscous stress acting on a fluid element and the deformation rate of that fluid element, and represents the viscous work put into fluid element deformation. This work is irreversible because deformed fluid elements do not return to their prior shape when a viscous stress is relieved. Thus, ε represents the irreversible conversion of mechanical energy to thermal energy through the action of viscosity. It is always positive and can be written in terms of the viscosities and squares of velocity field derivatives (see (4.56)):

$$\varepsilon\equiv\frac{1}{\rho}\tau_{ij}S_{ij}=\frac{1}{\rho}\left(2\mu S_{ij}+\left(\mu_\upsilon-\frac{2}{3}\mu\right)\frac{\partial u_m}{\partial x_m}\delta_{ij}\right)S_{ij}=2\nu\left(S_{ij}-\frac{1}{3}\frac{\partial u_m}{\partial x_m}\delta_{ij}\right)^2+\frac{\mu_\upsilon}{\rho}\left(\frac{\partial u_m}{\partial x_m}\right)^2, \tag{4.59}$$

where $\nu \equiv \mu/\rho$ is the kinematic viscosity (1.10), and together (4.27) and (4.36) imply

$$\tau_{ij}=+\mu\left(\frac{\partial u_i}{\partial x_j}+\frac{\partial u_j}{\partial x_i}\right)+\left(\mu_\upsilon-\frac{2}{3}\mu\right)\frac{\partial u_m}{\partial x_m}\delta_{ij} \tag{4.60}$$

for a Newtonian fluid. As described in Chapter 10, ε plays an important role in the physics and description of turbulent flow. It is proportional to μ (and μ_υ) and the square of velocity gradients, so it is more important in regions of high shear. The internal energy increase resulting from high ε can appear as hot lubricant in a bearing, or as burning of the surface of a spacecraft on reentry into the atmosphere.

The final energy-equation manipulation is to express q_i in terms of the other dependent field variables. For nearly all the circumstances considered in this text, heat transfer is caused by thermal conduction alone, so using (4.59) and Fourier's law of heat conduction (1.8), (4.58) can be rewritten:

$$\rho\frac{De}{Dt}=-p\frac{\partial u_m}{\partial x_m}+2\mu\left(S_{ij}-\frac{1}{3}\frac{\partial u_m}{\partial x_m}\delta_{ij}\right)^2+\mu_\upsilon\left(\frac{\partial u_m}{\partial x_m}\right)^2+\frac{\partial}{\partial x_i}\left(k\frac{\partial T}{\partial x_i}\right), \tag{4.61}$$

where k is the fluid's thermal conductivity. It is presumed to only depend on thermodynamic conditions, as is the case for μ and μ_υ.

The development of the differential equations of fluid motion is now complete. The field equations (4.7), (4.38), and (4.61) are general for a Newtonian fluid that follows Fourier's law of heat conduction. These field equations and two thermodynamic equations provide: $1+3+1+2=7$ scalar equations. The dependent variables in these equations are ρ,

e, p, T, and u_j, a total of $1+1+1+1+3=7$ unknowns. Thus, solutions are in principle possible for suitable boundary conditions.

Interestingly, the evolution of the entropy s in fluid flows can be deduced from (4.58) by using Gibb's property relation (1.24) for the internal energy $de = Tds - pd(1/\rho)$. When made specific to time variations following a fluid particle, it becomes:

$$\frac{De}{Dt} = T\frac{Ds}{Dt} - p\frac{D(1/\rho)}{Dt}. \tag{4.62}$$

Combining (4.58), (4.59), and (4.62) produces:

$$\frac{Ds}{Dt} = -\frac{1}{\rho T}\frac{\partial q_i}{\partial x_i} + \frac{\varepsilon}{T} = -\frac{1}{\rho}\frac{\partial}{\partial x_i}\left(\frac{q_i}{T}\right) - \frac{q_i}{\rho T^2}\left(\frac{\partial T}{\partial x_i}\right) + \frac{\varepsilon}{T}, \tag{4.63}$$

and using Fourier's law of heat conduction, this becomes:

$$\frac{Ds}{Dt} = +\frac{1}{\rho}\frac{\partial}{\partial x_i}\left(\frac{k}{T}\frac{\partial T}{\partial x_i}\right) + \frac{k}{\rho T^2}\left(\frac{\partial T}{\partial x_i}\right)^2 + \frac{\varepsilon}{T}. \tag{4.64}$$

The first term on the right side is the entropy gain or loss from heat conduction. The last two terms, which are proportional to the square of temperature and velocity gradients (see (4.59)), represent the *entropy production* caused by heat conduction and viscous generation of heat. The second law of thermodynamics requires that the entropy production due to these irreversible phenomena should be positive; thus, a fluid's transport coefficients (μ, μ_v, k) must all be positive.

With this requirement in place, explicit appeal to the second law of thermodynamics is not required in most analyses of fluid flows because the second law has already been satisfied by use of positive values for μ, μ_v, and k in (4.38) and (4.61). In addition (4.64) requires that fluid particle entropy be preserved along particle trajectories when the flow is inviscid and nonheat-conducting, i.e., when $Ds/Dt = 0$.

Example 4.12

Starting from (4.57) and (4.61), determine the mechanical and thermal energy equations when the flow is incompressible, and μ and k are constants.

Solution When the flow is incompressible, $\partial u_m/\partial x_m = 0$ and $\tau_{ij} = \mu(\partial u_i/\partial x_j + \partial u_j/\partial x_i)$ from (4.60). For the mechanical energy equation (4.57), the only impact is on the last term:

$$u_j\frac{\partial}{\partial x_i}(\tau_{ij}) = u_j\frac{\partial}{\partial x_i}\left(\mu\left(\frac{\partial u_i}{\partial x_j} + \frac{\partial u_j}{\partial x_i}\right)\right) = \mu u_j\left(\frac{\partial}{\partial x_j}\left(\frac{\partial u_i}{\partial x_i}\right) + \frac{\partial^2 u_j}{\partial x_i^2}\right) = \mu u_j\frac{\partial^2 u_j}{\partial x_i^2}.$$

where $\mu =$ constant allows the second equality, and $\partial u_i/\partial x_i = 0$ allows the third. Thus, the incompressible constant viscosity form of (4.57) is:

$$\rho\frac{D}{Dt}\left(\frac{1}{2}u_j^2\right) = \rho g_j u_j - u_j\frac{\partial p}{\partial x_j} + u_j\mu\frac{\partial^2 u_j}{\partial x_i^2} = \mathbf{u}\cdot\left(\rho\mathbf{g} - \nabla p + \mu\nabla^2\mathbf{u}\right).$$

For the thermal energy equation (4.61), the incompressibility condition allows three terms to be dropped. And, when k is constant it may be brought outside the divergence in the final term. Thus, when (3.12) is used for S_{ij}, (4.61) simplifies to:

$$\rho\frac{De}{Dt} = +2\mu\left(S_{ij}\right)^2 + \frac{\partial}{\partial x_i}\left(k\frac{\partial T}{\partial x_i}\right) = \frac{\mu}{2}\left(\frac{\partial u_i}{\partial x_j} + \frac{\partial u_j}{\partial x_i}\right)^2 + k\frac{\partial^2 T}{\partial x_i^2}.$$

When the first term on the right is small and e is proportional to T (both common occurrences), this becomes a linear equation for the fluid temperature T.

4.9 Special forms of the equations

The general equations of motion for a fluid may be put into a variety of special forms when certain symmetries or approximations are valid. Several special forms are presented in this section. The first applies to the integral form of the momentum equation and corresponds to the classical mechanics principle of conservation of angular momentum. The second through fifth special forms arise from manipulations of the differential equations to generate Bernoulli equations. The sixth special form applies when the flow has constant density and the gravitational body force and hydrostatic pressure cancel. The final special form for the equations of motion presented here, known as the Boussinesq approximation, is for low-speed incompressible flows with constant transport coefficients and small changes in density.

Angular momentum principle for a stationary control volume

In the mechanics of solid bodies, it is shown that

$$d\mathbf{H}/dt = \mathbf{M}, \tag{4.65}$$

where $\mathbf{M}$ is the torque of all external forces on the body about any chosen axis, and $d\mathbf{H}/dt$ is the rate of change of angular momentum of the body about the same axis. For the fluid in a material control volume, the angular momentum is

$$\mathbf{H} = \int_{V(t)} (\mathbf{r} \times \rho\mathbf{u}) dV,$$

where $\mathbf{r}$ is the position vector from the chosen axis (Fig. 4.13). Inserting this in (4.65) produces:

$$\frac{d}{dt}\int_{V(t)} (\mathbf{r} \times \rho\mathbf{u}) dV = \int_{V(t)} (\mathbf{r} \times \rho\mathbf{g}) dV + \int_{A(t)} (\mathbf{r} \times \mathbf{f}) dA,$$

where the two terms on the right are the torques produced by body forces and surface stresses, respectively. As before, the left-hand term can be expanded via Reynolds transport theorem to find:

$$\begin{aligned}
\frac{d}{dt}\int_{V(t)} (\mathbf{r} \times \rho\mathbf{u}) dV &= \int_{V(t)} \frac{\partial}{\partial t}(\mathbf{r} \times \rho\mathbf{u}) dV + \int_{A(t)} (\mathbf{r} \times \rho\mathbf{u})(\mathbf{u}\cdot\mathbf{n}) dA \\
&= \int_{V_o} \frac{\partial}{\partial t}(\mathbf{r} \times \rho\mathbf{u}) dV + \int_{A_o} (\mathbf{r} \times \rho\mathbf{u})(\mathbf{u}\cdot\mathbf{n}) dA \\
&= \frac{d}{dt}\int_{V_o} (\mathbf{r} \times \rho\mathbf{u}) dV + \int_{A_o} (\mathbf{r} \times \rho\mathbf{u})(\mathbf{u}\cdot\mathbf{n}) dA,
\end{aligned}$$

where V_o and A_o are the volume and surface of a stationary control volume that is instantaneously coincident with the material volume, and the final equality holds because V_o does not vary with time. Thus, the stationary volume angular momentum principle is:

$$\frac{d}{dt}\int_{V_o} (\mathbf{r} \times \rho\mathbf{u}) dV + \int_{A_o} (\mathbf{r} \times \rho\mathbf{u})(\mathbf{u}\cdot\mathbf{n}) dA = \int_{V_o} (\mathbf{r} \times \rho\mathbf{g}) dV + \int_{A_o} (\mathbf{r} \times \mathbf{f}) dA. \tag{4.66}$$

The angular momentum principle (4.66) is analogous to the linear momentum principle (4.17) when $\mathbf{b} = 0$, and is useful in investigating rotating fluid systems such as turbomachines, fluid couplings, dishwashing-machine spray rotors, and even lawn sprinklers.

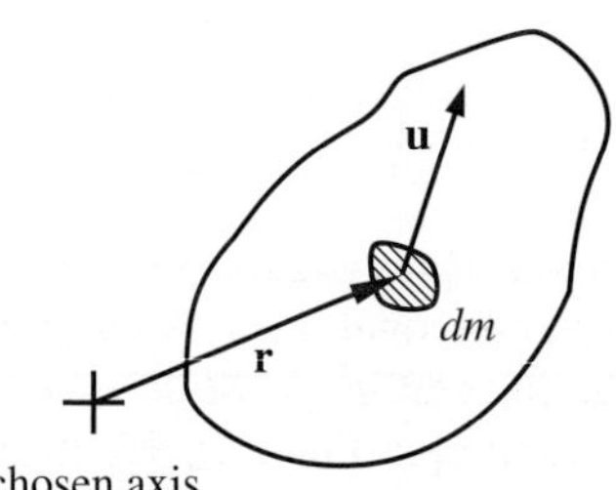

FIGURE 4.13 Definition sketch for the angular momentum theorem where $dm = \rho dV$. Here the chosen axis points out of the page, and elemental contributions to the angular momentum about this axis are $\mathbf{r} \times \rho\mathbf{u} dV$.

Example 4.13

Consider a lawn sprinkler as shown in Fig. 4.14. The area of each nozzle exit is A, and the jet velocity is U. Find the torque required to hold the rotor stationary.

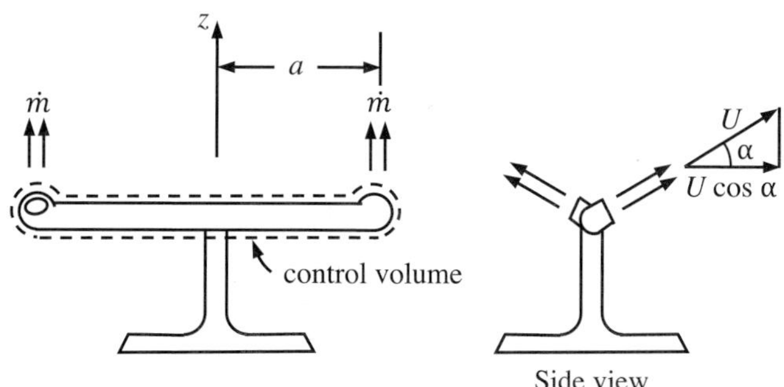

FIGURE 4.14 Lawn sprinkler.

Solution Select a stationary volume V_o with area A_o as shown by the dashed lines. Pressure everywhere on the control surface is atmospheric so there is no net moment due to pressure forces. The control surface cuts through the vertical support and the torque M exerted by the support on the sprinkler arm is the only torque acting on V_o. Apply the angular momentum balance

$$\int_{A_o} (\mathbf{r} \times \rho \mathbf{u})(\mathbf{u} \cdot \mathbf{n}) dA = \int_{A_o} (\mathbf{r} \times \mathbf{f}) dA = M,$$

where the time derivative term must be zero for a stationary rotor. Evaluating the surface flux terms produces:

$$\int_{A_o} (\mathbf{r} \times \rho \mathbf{u})(\mathbf{u} \cdot \mathbf{n}) dA = (a\rho U \cos\alpha) UA + (a\rho U \cos\alpha) UA = 2a\rho AU^2 \cos\alpha.$$

Therefore, the torque required to hold the rotor stationary is $M = 2a\rho AU^2 \cos\alpha$. When the sprinkler is rotating at a steady rate, the fluid flux torque is balanced by air resistance and mechanical friction.

Bernoulli equations

Various conservation laws for mass, momentum, energy, and entropy were presented in the preceding sections. Bernoulli equations are not separate laws, but are instead derived from the Navier-Stokes momentum equation (4.38) or the energy equation (4.61) under various sets of conditions.

First consider inviscid flow ($\mu = \mu_v = 0$) where gravity is the only body force so that (4.38) reduces to the Euler equation (4.42):

$$\frac{\partial u_j}{\partial t} + u_i \frac{\partial u_j}{\partial x_i} = -\frac{1}{\rho}\frac{\partial p}{\partial x_j} - \frac{\partial}{\partial x_j}\Phi, \tag{4.67}$$

where $\Phi = gz$ is the body force potential, g is the acceleration of gravity, and the z-axis is vertical. If the flow is also barotropic, then $\rho = \rho(p)$, and the pressure gradient term can be rewritten in terms of an integral:

$$\frac{\partial}{\partial x_j}\int_{p_o}^{p} \frac{dp'}{\rho(p')} = \frac{\partial}{\partial p}\left(\int_{p_o}^{p} \frac{dp'}{\rho(p')}\right)\frac{\partial p}{\partial x_j} = \frac{1}{\rho}\frac{\partial p}{\partial x_j}, \tag{4.68}$$

where dp/ρ is a perfect differential, p_o is a reference pressure, p' is the integration variable, the middle expression follows from the chain-rule for partial differentiation, and the final one follows from the rules for differentiating an integral with respect to its upper limit. When $\rho = \rho(p)$, the integral in (4.68) depends only on its endpoints, and not on the path of integration. Constant density, isothermal, and isentropic flows are barotropic. In addition, the advective acceleration in (4.67) may be rewritten in terms of the velocity-vorticity cross product, and the gradient of the kinetic energy per unit mass:

$$u_i \frac{\partial u_j}{\partial x_i} = -\varepsilon_{ikj} u_i \omega_k + \frac{\partial}{\partial x_j}\left(\frac{1}{2}u_i^2\right), \quad \text{or} \quad (\mathbf{u} \cdot \nabla)\mathbf{u} = -\mathbf{u} \times \boldsymbol{\omega} + \nabla\left(|\mathbf{u}|^2/2\right) \tag{4.69}$$

(see Exercise 4.65). Substituting (4.68) and (4.69) into (4.67) produces:

$$\frac{\partial \mathbf{u}}{\partial t} + \nabla\left[\frac{1}{2}|\mathbf{u}|^2 + \int_{p_o}^{p} \frac{dp'}{\rho(p')} + gz\right] = \mathbf{u} \times \boldsymbol{\omega}, \tag{4.70}$$

where all the gradient terms have been collected together to form the Bernoulli function B = the contents of the [,]-brackets.

Eq. (4.70) can be used to deduce the evolution of the Bernoulli function in inviscid barotropic flow. First consider steady flow ($\partial \mathbf{u}/\partial t = 0$) so that (4.70) reduces to

$$\nabla B = \mathbf{u} \times \boldsymbol{\omega}. \tag{4.71}$$

The left-hand side is a vector normal to the surface defined by B = constant, whereas the right-hand side is a vector perpendicular to both $\mathbf{u}$ and $\boldsymbol{\omega}$ (Fig. 4.15). It follows that surfaces of constant B must contain the streamlines and vortex lines. Thus, an inviscid, steady, barotropic flow satisfies:

$$\frac{1}{2}|\mathbf{u}|^2 + \int_{p_o}^{p} \frac{dp'}{\rho(p')} + gz = \text{constant along streamlines and vortex lines.} \tag{4.72}$$

This is the first of several possible *Bernoulli equations*. If, in addition, the flow is irrotational ($\boldsymbol{\omega} = 0$), then (4.71) implies that

$$\frac{1}{2}|\mathbf{u}|^2 + \int_{p_o}^{p} \frac{dp'}{\rho(p')} + gz = \text{constant everywhere.} \tag{4.73}$$

It may be shown that a sufficient condition for the existence of the surfaces containing streamlines and vortex lines is that the flow be barotropic. Incidentally, these are called Lamb surfaces in honor of the distinguished English applied mathematician and hydrodynamicist, Horace Lamb. In a general nonbarotropic flow, a path composed of streamline and vortex line segments can be drawn between any two points in a flow field. Then (4.72) is valid with the proviso that the integral be evaluated on the specific path chosen. As written, (4.72) requires that the flow be steady, inviscid, and have only gravity (or other conservative) body forces acting upon it. Irrotational flow is presented in Chapter 6. We shall note only the important point here that, in a nonrotating frame of reference, barotropic irrotational flows remain irrotational if viscous effects are negligible. Consider the flow around a solid object, say an airfoil (Fig. 4.16). The flow is irrotational at all points outside the thin viscous layer close to the surface of the object. This is because a particle P on a streamline outside the viscous layer started from some point S, where the flow is uniform and consequently irrotational. The Bernoulli equation (4.73) is therefore satisfied everywhere outside the viscous layer in this example.

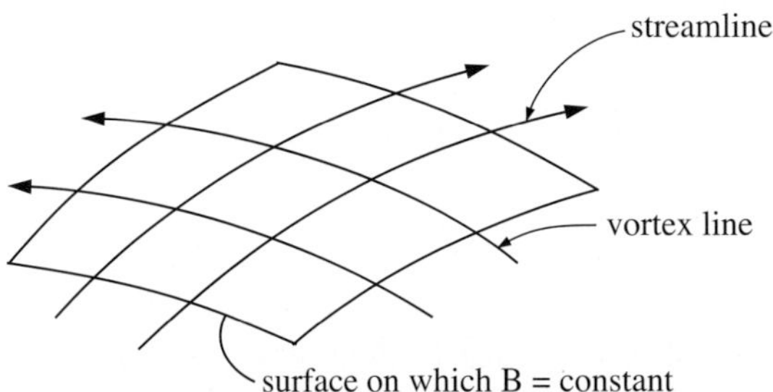

FIGURE 4.15 A surface defined by streamlines and vortex lines. Within this surface the Bernoulli function defined as the contents of the [,]-brackets in (4.70) is constant in steady flow. Note that the streamlines and vortex lines can be at an arbitrary angle.

An unsteady form of Bernoulli's equation can be derived only if the flow is irrotational. In this case, the velocity vector can be written as the gradient of a scalar potential ϕ (called the velocity potential):

$$\mathbf{u} \equiv \nabla \phi. \tag{4.74}$$

Putting (4.74) into (4.70) with $\boldsymbol{\omega} = 0$ produces:

$$\nabla\left[\frac{\partial \phi}{\partial t} + \frac{1}{2}|\nabla \phi|^2 + \int_{p_o}^{p} \frac{dp'}{\rho(p')} + gz\right] = 0, \quad \text{or} \quad \frac{\partial \phi}{\partial t} + \frac{1}{2}|\nabla \phi|^2 + \int_{p_o}^{p} \frac{dp'}{\rho(p')} + gz = B(t), \tag{4.75}$$

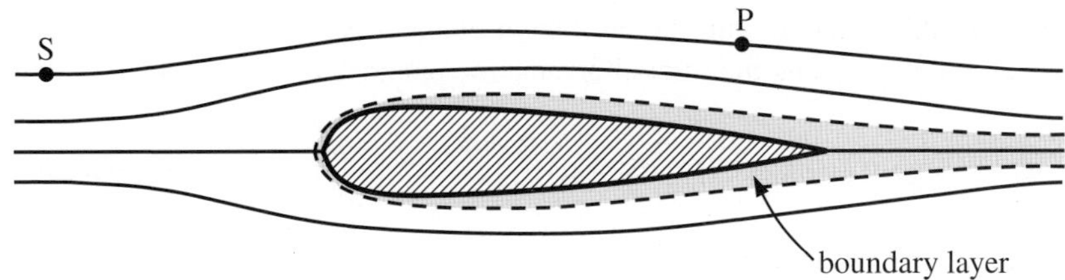

FIGURE 4.16 Flow over a solid object. Viscous shear stresses are usually confined to a thin layer near the body called a boundary layer. Flow outside the boundary layer is irrotational, so if a fluid particle at S is initially irrotational it will remain irrotational at P because the streamline it travels on does not enter the boundary layer.

where the integration function $B(t)$ is independent of location. Here ϕ can be redefined to include B,

$$\phi \rightarrow \phi + \int_{t_o}^{t} B(t')dt',$$

without changing its use in (4.74). Then, the second part of (4.75) provides a second Bernoulli equation for unsteady, inviscid, irrotational, barotropic flow:

$$\frac{\partial \phi}{\partial t} + \frac{1}{2}|\nabla\phi|^2 + \int_{p_o}^{p} \frac{dp'}{\rho(p')} + gz = \text{constant}. \tag{4.76}$$

This form of the Bernoulli equation will be used in studying irrotational wave motions in Chapter 11.

A third Bernoulli equation can be obtained for steady flow ($\partial/\partial t = 0$) from the energy equation (4.56) in the absence of viscous stresses and heat transfer ($\tau_{ij} = q_i = 0$):

$$\rho u_i \frac{\partial}{\partial x_i}\left(e + \frac{1}{2}u_j^2\right) = \rho u_i g_i - \frac{\partial}{\partial x_j}\left(\rho u_j p/\rho\right). \tag{4.77}$$

When the body force is conservative with potential gz, and the steady continuity equation, $\partial(\rho u_i)/\partial x_i = 0$, is used to simplify (4.77), it becomes:

$$\rho u_i \frac{\partial}{\partial x_i}\left(e + \frac{p}{\rho} + \frac{1}{2}u_j^2 + gz\right) = 0. \tag{4.78}$$

From (1.19) $h = e + p/\rho$, so (4.78) states that gradients of the sum $h + |\mathbf{u}|^2/2 + gz$ must be normal to the local streamline direction u_i. Therefore, a third Bernoulli equation is:

$$h + \frac{1}{2}|\mathbf{u}|^2 + gz = \text{constant on streamlines.} \tag{4.79}$$

This result is consistent with the momentum Bernoulli equations (4.72), (4.73), and (4.76). Eq. (4.64) requires that inviscid, nonheat-conducting flows are isentropic (s does not change along particle paths), and (1.24) implies $dp/\rho = dh$ when $s = \text{constant}$. Thus the path integral $\int dp/\rho$ in (4.72), (4.73), and (4.76) becomes a function h of the endpoints only if both heat conduction and viscous stresses may be neglected. Eq. (4.79) is useful for high-speed gas flows where there is significant interplay between kinetic and thermal energies along a streamline. It is nearly the same as (4.72), but does not include the other barotropic and vortex-line-evaluation possibilities allowed by (4.72).

Interestingly, there is also a Bernoulli equation for constant-viscosity constant-density irrotational flow. It can be obtained by starting from (4.40), using (4.69) for the advective acceleration, and noting from (4.41) that $\nabla^2\mathbf{u} = -\nabla \times \boldsymbol{\omega}$ in incompressible flow:

$$\rho\frac{D\mathbf{u}}{Dt} = \rho\frac{\partial \mathbf{u}}{\partial t} + \rho\nabla\left(\frac{1}{2}|\mathbf{u}|^2\right) - \rho\mathbf{u} \times \boldsymbol{\omega} = -\nabla p + \rho\mathbf{g} + \mu\nabla^2\mathbf{u} = -\nabla p + \rho\mathbf{g} - \mu\nabla \times \boldsymbol{\omega}. \tag{4.80}$$

When $\rho = \text{constant}$, $\mathbf{g} = -\nabla(gz)$, and $\boldsymbol{\omega} = 0$, the second and final parts of this extended equality require:

$$\rho\frac{\partial \mathbf{u}}{\partial t} + \nabla\left(\frac{1}{2}\rho|\mathbf{u}|^2 + \rho gz + p\right) = 0. \tag{4.81}$$

Now, form the dot product of this equation with the arc-length element $\mathbf{e}_u ds = d\mathbf{s}$ directed along a chosen streamline, integrate from location 1 to location 2 along this streamline, and recognize that $\mathbf{e}_u \cdot \nabla = \partial/\partial s$ to find:

$$\rho \int_1^2 \frac{\partial \mathbf{u}}{\partial t} \cdot d\mathbf{s} + \int_1^2 \mathbf{e}_u \cdot \nabla \left(\frac{1}{2}\rho|\mathbf{u}|^2 + \rho g z + p \right) ds = \rho \int_1^2 \frac{\partial \mathbf{u}}{\partial t} \cdot d\mathbf{s} + \int_1^2 \frac{\partial}{\partial s} \left(\frac{1}{2}\rho|\mathbf{u}|^2 + \rho g z + p \right) ds = 0. \tag{4.82}$$

The integration in the second term is elementary, so a fourth Bernoulli equation for constant-viscosity constant-density irrotational flow is:

$$\int_1^2 \frac{\partial \mathbf{u}}{\partial t} \cdot d\mathbf{s} + \left(\frac{1}{2}|\mathbf{u}|^2 + gz + \frac{p}{\rho} \right)_2 = \left(\frac{1}{2}|u|^2 + gz + \frac{p}{\rho} \right)_1, \tag{4.83}$$

where 1 and 2 denote upstream and downstream locations on the same streamline at a single instant in time. Alternatively, (4.81) can be written using (4.74) as:

$$\frac{\partial \phi}{\partial t} + \frac{1}{2}|\nabla \phi|^2 + gz + \frac{p}{\rho} = \text{constant}. \tag{4.84}$$

To summarize, there are (at least) four Bernoulli equations: (4.72) is for inviscid, steady, barotropic flow; (4.76) is for inviscid, irrotational unsteady, barotropic flow, (4.79) is for inviscid, isentropic, steady flow, and (4.83) or (4.84) are for constant-viscosity, irrotational, unsteady, constant density flow. Perhaps the simplest form of these is (4.19).

There are many useful and important applications of Bernoulli equations. A few of these are described in the following paragraphs.

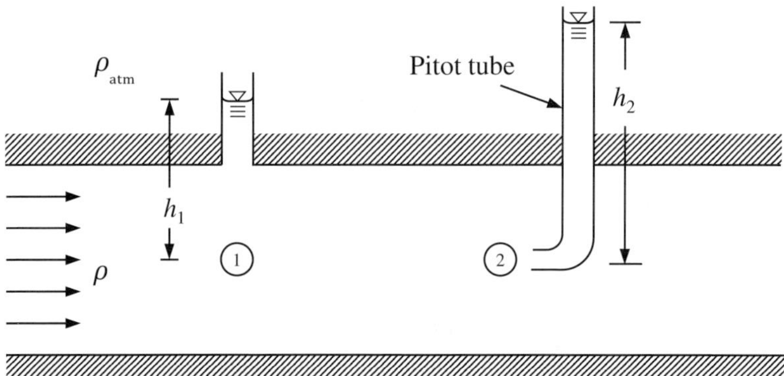

FIGURE 4.17 Pitot tube for measuring velocity in a duct. The first port measures the static pressure while the second port measures the static and dynamic pressure. Using the steady Bernoulli equation for incompressible flow, the height difference $h_2 - h_1$ can be related to the flow speed.

Consider first a simple device to measure the local velocity in a fluid stream by inserting a narrow bent tube (Fig. 4.17), called a *pitot tube* after the French mathematician Henri Pitot (1695–1771), who used a bent glass tube to measure the velocity of the river Seine. Consider two points (1 and 2) at the same level, point 1 being away from the tube and point 2 being immediately in front of the open end where the fluid velocity $\mathbf{u}_2$ is zero. If the flow is steady and irrotational with constant density along the streamline that connects 1 and 2, then (4.19) gives:

$$\frac{p_1}{\rho} + \frac{1}{2}|\mathbf{u}_1|^2 = \frac{p_2}{\rho} + \frac{1}{2}|\mathbf{u}_2|^2 = \frac{p_2}{\rho}$$

from which the magnitude of $\mathbf{u}_1$ is found to be:

$$|\mathbf{u}_1| = \sqrt{2(p_2 - p_1)/\rho}.$$

Pressures at the two points are found from the hydrostatic balance:

$$p_1 = \rho g h_1 \quad \text{and} \quad p_2 = \rho g h_2,$$

so that the magnitude of $\mathbf{u}_1$ can be found from:

$$|\mathbf{u}_1| = \sqrt{2g(h_2 - h_1)}.$$

Because it is assumed that the fluid density is much greater than that of the atmosphere to which the tubes are exposed, the pressures at the tops of the two fluid columns are assumed to be the same. They actually differ by $\rho_{atm}g(h_2 - h_1)$. Use of the hydrostatic approximation above station 1 is valid when the streamlines are straight and parallel between station 1 and the upper wall.

Here, the pressure measured by the Pitot tube, $p_2 = p_1 + \rho|\mathbf{u}_1|^2/2$, is the *stagnation pressure* or *total pressure* in this flow. In general, *stagnation pressure* is defined as the pressure reached when the local flow is slowed to zero velocity without friction. For incompressible flows, local stagnation pressure is the sum of the local static pressure p and the local *dynamic pressure* $\rho|\mathbf{u}|^2/2$.

As another application of Bernoulli's equation, consider the flow through an orifice or opening in a tank (Fig. 4.18). The flow is slightly unsteady due to lowering of the water level in the tank, but this effect is small if the tank area is large compared to the orifice area. Viscous effects are negligible everywhere away from the walls of the tank. All streamlines can be traced back to the free surface in the tank, where they have the same value of the Bernoulli constant $B = |\mathbf{u}|^2/2 + p/\rho + gz$. It follows that the flow is irrotational, and that B is constant *throughout* the flow.

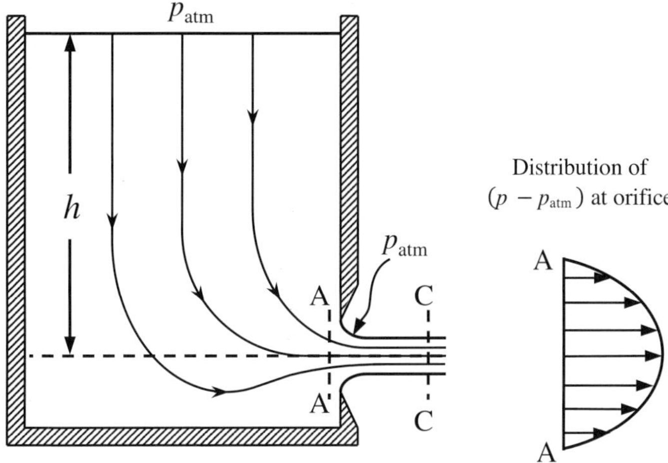

FIGURE 4.18 Flow through a sharp-edged orifice. The pressure is atmospheric everywhere across section CC; its distribution across orifice AA is indicated. The basic finding here is that the hole from which the jet emerges is larger than the width of the jet that crosses CC.

Application of the Bernoulli equation (4.19) for steady constant-density flow between a point on the free surface in the tank and a point in the jet downstream of CC gives:

$$\frac{p_{atm}}{\rho} + gh = \frac{p_{atm}}{\rho} + \frac{u^2}{2},$$

from which the average jet velocity magnitude u is found as:

$$u = \sqrt{2gh},$$

which simply states that the loss of potential energy equals the gain of kinetic energy.

To recover the jet's mass flow, its cross sectional area is needed. However, the conditions right at the opening (section AA in Fig. 4.18) are not simple because the pressure is *not* uniform across the jet outlet and streamlines are curved. Although atmospheric pressure acts everywhere on the free surface of the jet (neglecting small surface tension effects), it is not equal to the atmospheric pressure *inside* the jet at the section AA. The curved streamlines at the orifice indicate that pressure must increase toward the centerline of the jet to balance the centrifugal force. A sketch of the pressure distribution across the orifice (section AA) is shown in Fig. 4.18. However, the streamlines in the jet become parallel a short distance away from the orifice (section CC in Fig. 4.18), where the jet area is smaller than the orifice area. The pressure across section CC is uniform and equal to the atmospheric value (p_{atm}). Thus, the mass flow rate in the jet is approximately: $\dot{m} = \rho A_c u = \rho A_c\sqrt{2gh}$, where A_c is the area of the jet at CC. For round orifices with sharp edges, A_c has been found to be $\approx 62\%$ of the orifice area because the jet contracts downstream of the orifice opening.

If the orifice has a well-rounded opening (Fig. 4.19), then the jet does not contract, the streamlines right at the exit are parallel, and the pressure at the exit is uniform and equal to the atmospheric pressure. Consequently the mass flow rate is simply $\rho A\sqrt{2gh}$, where A is the orifice area. Thus, a simple way to increase the flow rate from such an orifice is to provide a well-rounded opening.

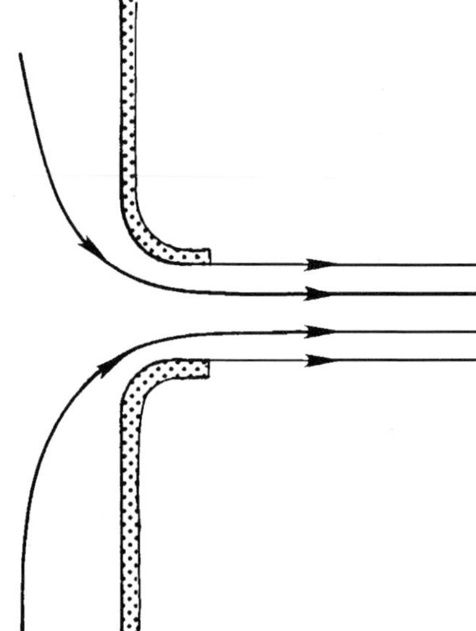

FIGURE 4.19 Flow through a rounded orifice. Here the pressure and velocity can achieve parallel outflow inside the tank, so the width of the jet does not change outside the tank.

Example 4.14

A perfect gas with specific heat ratio γ at temperature T_1 escapes horizontally at the speed of sound ($u_1 = c$) from a small leak in a pressure vessel (Fig. 4.20). What is the temperature T_2 of the gas as it leaves the pressure vessel if its enthalpy is proportional to T (i.e., $h = c_p T$) and the flow is steady and frictionless?

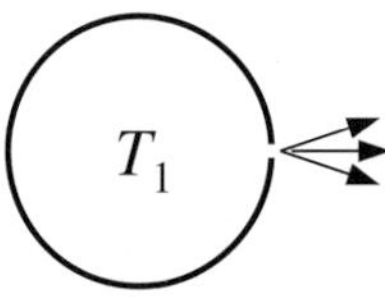

FIGURE 4.20 Pressure vessel with a small leak.

Solution Start with the steady compressible-flow Bernoulli equation (4.79) without the gravity term and use (1.33), $c^2 = \gamma RT$, for the speed of sound c:

$$h_1 + \frac{1}{2}u_1^2 = h_2 + \frac{1}{2}c^2 \quad \text{or} \quad c_p T_1 + 0 = c_p T_2 + \frac{1}{2}\gamma R T_2.$$

Here, $u_1 \approx 0$ is the speed of the quiescent gas in the pressure vessel, and the sound speed is evaluated at the gas temperature as it passes through the orifice. The second equation is readily solved for T_2:

$$T_2 = c_p T_1 \bigg/ \left(c_p + \frac{1}{2}\gamma R\right) = 2T_1/(\gamma + 1),$$

where (1.29) and (1.30) have been used to reach the final form. Interestingly, this answer does not depend on the pressure in the vessel, and it leads to a noticeable temperature change: $T_1 - T_2 = [(\gamma - 1)/(\gamma + 1)]T_1$, which is nearly 50 °C for pressurized air starting at room temperature.

Example 4.15

The density and flow speed in the intake manifold of a reciprocating engine are approximately ρ_o (a constant) and $u(t) = U_o(1 + \sin(2\pi ft))$. If the throttle-plate-to-cylinder-intake-valve runner is a straight horizontal tube of length L, (see Fig. 4.21) determine a formula for the pressure difference required between the ends of this tube to sustain this fluid motion, assuming frictionless incompressible flow.

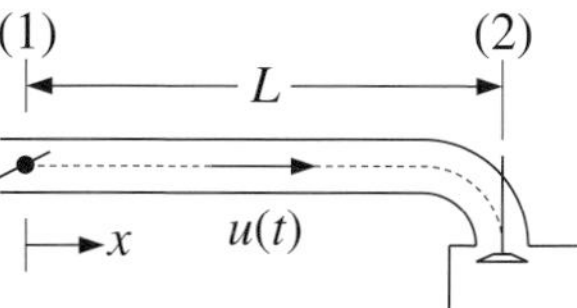

FIGURE 4.21 Simple intake manifold for a reciprocating internal combustion engine.

Solution Start with the unsteady Bernoulli equation (4.83) and use the tube-centerline streamline between the throttle plate (1) and the intake valve (2). This streamline is the dashed line in Fig. 4.21. Several simplifications to (4.83) can be made immediately: the flow is horizontal so the gravity terms do not enter; the fluid velocity is the same at both ends of the tube so the kinetic energy terms cancel; and the dot-product in the acceleration term is readily evaluated $(\partial \mathbf{u}/\partial t) \cdot d\mathbf{s} = (\partial u/\partial t)dx$. Thus, (4.83) becomes:

$$\int_1^2 \frac{\partial u}{\partial t} dx + \frac{p_2}{\rho} = \frac{p_1}{\rho}, \quad \text{or} \quad p_1 - p_2 = \rho \int_1^2 \frac{\partial u}{\partial t} dx = \rho \frac{\partial u}{\partial t}(x_2 - x_1) = 2\pi \rho f U_o L \cos(2\pi ft),$$

where $x_2 - x_1 = L$, and the integral is elementary because $\partial u/\partial t = 2\pi f U_o \cos(2\pi ft)$ does not depend on x.

Interestingly, even for a low air density (0.5 kg/m^3), a low frequency (50 Hz), a low flow speed (10 m/s), and a short runner length (0.3 m), this estimate produces pressure fluctuations that are enormous from a sound amplitude standpoint: $|p_1 - p_2| = 2\pi(0.5\,\text{kg/m}^3)(50\,\text{Hz})(10\,\text{m/s})(0.3\,\text{m}) \approx 470\,\text{Pa}$. Although this pressure is a tiny fraction of atmospheric pressure, it corresponds to a sound pressure level of more than 140 dB (where 20 μPa defines 0 dB), a level that quickly causes hearing damage or loss.

Neglect of gravity in constant density flows

When the flow velocity is zero, the Navier-Stokes momentum equation for incompressible flow (4.40) reduces to a balance between the hydrostatic pressure p_s, and the steady body force acting on the hydrostatic density ρ_s:

$$0 = -\nabla p_s + \rho_s \mathbf{g},$$

which is equivalent to (1.14). When this hydrostatic balance is subtracted from (4.40), the pressure difference from hydrostatic, $p' = p - p_s$, and the density difference from hydrostatic, $\rho' = \rho - \rho_s$, appear:

$$\rho \frac{D\mathbf{u}}{Dt} = -\nabla p' + \rho' \mathbf{g} + \mu \nabla^2 \mathbf{u}. \tag{4.85}$$

When the fluid density is constant, $\rho' = 0$ and the gravitational-body-force term disappears leaving:

$$\rho \frac{D\mathbf{u}}{Dt} = -\nabla p' + \mu \nabla^2 \mathbf{u}. \tag{4.86}$$

Because of this, steady body forces (like gravity) in constant density flow are commonly omitted from the momentum equation, and pressure is measured relative to its local hydrostatic value. Furthermore, the prime on p in (4.86) is typically dropped in this situation. However, when the flow includes a free surface, a fluid-fluid interface across which the density changes, or other variations in density, the gravitational-body-force term should reappear.

The Boussinesq approximation

For flows satisfying certain conditions, Boussinesq in 1903 suggested that density changes in the fluid can be neglected except where ρ is multiplied by g. This approximation also treats the other properties of the fluid (such as μ, k, c_p) as constants. It is commonly useful for analyzing oceanic and atmospheric flows. Here the basis for the approximation is presented in a somewhat intuitive manner, and the resulting simplifications of the equations of motion are examined. A formal justification, and the conditions under which the Boussinesq approximation holds, is given in Spiegel and Veronis (1960).

The Boussinesq approximation replaces the full continuity equation (4.7) by its incompressible form (4.10), $\nabla \cdot \mathbf{u} = 0$, to indicate that the fractional density changes following a fluid particle, $\rho^{-1}(D\rho/Dt)$, are small compared to the velocity gradients that compose $\nabla \cdot \mathbf{u}$. Thus, the Boussinesq approximation cannot be applied to high-speed gas flows where density variations induced by velocity divergence cannot be neglected (see Section 4.11). Similarly, it cannot be applied when the vertical scale of the flow is so large that hydrostatic pressure variations cause significant changes in density. In a hydrostatic field, the vertical distance over which the density changes become important is of order $c^2/g \sim 10$ km for air where c is the speed of sound. (This vertical distance estimate is consistent with the scale height of the atmosphere; see Section 1.10.) The Boussinesq approximation therefore requires that the vertical scale of the flow be $L \ll c^2/g$.

In both cases just mentioned, density variations are caused by pressure variations. Now suppose that such pressure-compressibility effects are small and that density changes are caused by temperature variations alone, as in a thermal convection problem. In this case, the Boussinesq approximation applies when the temperature variations in the flow are small. Assume that ρ changes with T according to $\delta\rho/\rho = -\alpha\delta T$, where $\alpha = -\rho^{-1}(\partial\rho/\partial T)_p$ is the thermal expansion coefficient (1.26). For a perfect gas at room temperature $\alpha = 1/T \sim 3 \times 10^{-3}$ K^{-1} but for typical liquids $\alpha \sim 5 \times 10^{-4}$ K^{-1}. Thus, for a temperature difference in the fluid of 10 °C, density variations can be at most a few percent, and it turns out that $\rho^{-1}(D\rho/Dt)$ can also be no larger than a few percent of the velocity gradients in $\nabla \cdot \mathbf{u}$, such as $\partial u/\partial x$. To see this, assume that the flow field is characterized by a length scale L, a velocity scale U, and a temperature scale δT. By this we mean that the velocity varies by U and the temperature varies by δT between locations separated by a distance of order L. The magnitude ratio of the two representative terms in the continuity equation (4.8) is:

$$\frac{\rho^{-1}(u\partial\rho/\partial x)}{\partial u/\partial x} \sim \frac{(U/\rho)\delta\rho/L}{U/L} = \frac{\delta\rho}{\rho} = \alpha\delta T \ll 1,$$

which allows (4.8) to be replaced by its incompressible form (4.10).

The Boussinesq approximation for the momentum equation is based on its form for incompressible flow, and proceeds from (4.85) divided by ρ_s:

$$\frac{\rho}{\rho_s}\frac{D\mathbf{u}}{Dt} = -\frac{1}{\rho_s}\nabla p' + \frac{\rho'}{\rho_s}\mathbf{g} + \frac{\mu}{\rho_s}\nabla^2\mathbf{u}.$$

When the density fluctuations are small $\rho/\rho_s \cong 1$ and $\mu/\rho_s \cong \nu$ (= the kinematic viscosity), so this equation implies:

$$\frac{D\mathbf{u}}{Dt} = -\frac{1}{\rho_0}\nabla p' + \frac{\rho'}{\rho_0}\mathbf{g} + \nu\nabla^2\mathbf{u}, \tag{4.87}$$

where ρ_0 is a constant reference value of ρ_s. This equation states that density changes are negligible when conserving momentum, except when ρ' is multiplied by g. In flows involving buoyant convection, the magnitude of $\rho' g/\rho_s$ is of the same order as the vertical acceleration $\partial w/\partial t$ or the viscous term $\nu\nabla^2 w$.

The Boussinesq approximation to the energy equation starts from (4.61), written in vector notation:

$$\rho\frac{De}{Dt} = -p\nabla \cdot \mathbf{u} + \rho\varepsilon - \nabla \cdot \mathbf{q}, \tag{4.88}$$

where (4.59) has been used to insert ε, the kinetic energy dissipation rate per unit mass. Although the continuity equation is approximately $\nabla \cdot \mathbf{u} = 0$, an important point is that the volume expansion term $p(\nabla \cdot \mathbf{u})$ is *not* negligible compared to other dominant terms of (4.88); only for incompressible liquids is $p(\nabla \cdot \mathbf{u})$ negligible in (4.88). We have

$$-p\nabla \cdot \mathbf{u} = \frac{p}{\rho}\frac{D\rho}{Dt} \cong \frac{p}{\rho}\left(\frac{\partial\rho}{\partial T}\right)_p \frac{DT}{Dt} = -p\alpha\frac{DT}{Dt}.$$

Assuming a perfect gas, for which $p = \rho RT$, $c_p - c_v = R$, and $\alpha = 1/T$, the foregoing estimate becomes:

$$-p\nabla \cdot \mathbf{u} = -\rho RT\alpha\frac{DT}{Dt} = -\rho\left(c_p - c_v\right)\frac{DT}{Dt}.$$

Eq. (4.88) then becomes:

$$\rho c_{\mathrm{p}} \frac{DT}{Dt} = \rho \varepsilon - \nabla \cdot \mathbf{q}, \tag{4.89}$$

where $e = c_{\mathrm{v}} T$ for a perfect gas. Note that c_{v} (instead of c_{p}) would have appeared on the left side of (4.89) if $\nabla \cdot \mathbf{u}$ had been dropped from (4.88).

The heating due to viscous dissipation of energy is negligible under the restrictions underlying the Boussinesq approximation. Comparing the magnitude of $\rho\varepsilon$ with the left-hand side of (4.89), we obtain:

$$\frac{\rho \varepsilon}{\rho c_{\mathrm{p}} (DT/Dt)} \sim \frac{2\mu S_{ij} S_{ij}}{\rho c_{\mathrm{p}} u_i (\partial T/\partial x_i)} \sim \frac{\mu U^2/L^2}{\rho c_{\mathrm{p}} U (\delta T/L)} = \frac{\nu U}{(c_{\mathrm{p}} \delta T) L}.$$

In typical situations, the final ratio is extremely small ($\sim 10^{-7}$). Neglecting $\rho\varepsilon$, and assuming Fourier's law of heat conduction (1.8) with constant k, (4.89) finally reduces the energy equation to:

$$\frac{DT}{Dt} = \kappa \nabla^2 T, \tag{4.90}$$

where $\kappa \equiv k/\rho c_{\mathrm{p}}$ is the thermal diffusivity.

Summary

The Boussinesq approximation applies if the Mach number of the flow is small, propagation of sound or shock waves is not considered, the vertical scale of the flow is not too large, and the temperature differences in the fluid are small. Then the density can be treated as a constant in both the continuity and the momentum equations, except in the gravity term. Properties of the fluid such as μ, k, and c_{p} are also assumed constant. Omitting Coriolis accelerations, the set of equations corresponding to the Boussinesq approximation is: (4.9) and/or (4.10), (4.87) with $\mathbf{g} = -g\mathbf{e}_z$, (4.90), and $\rho = \rho_0[1 - \alpha(T - T_0)]$, where the z-axis points upward. The constant ρ_0 is a reference density corresponding to a reference temperature T_0, which can be taken to be the mean temperature in the flow or the temperature at an appropriate boundary. Applications of the Boussinesq set can be found in several places in this book, for example, in the analysis of wave propagation in a density-stratified medium, thermal instability, turbulence in a stratified medium, and geophysical fluid dynamics.

4.10 Boundary conditions

The differential equations for the conservation laws require boundary conditions for proper solution. Specifically, the Navier-Stokes momentum equation (4.38) requires the specification of the velocity vector on all surfaces bounding the flow domain. For an external flow, one that is not contained by walls or surfaces at specified locations, the fluid's velocity vector and the thermodynamic state must be specified on a closed distant surface.

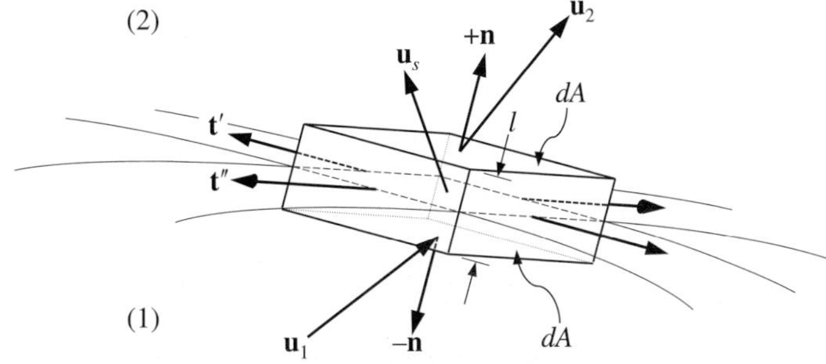

FIGURE 4.22 Interface between two media for evaluation of boundary conditions. Here medium 2 is a fluid, and medium 1 is a solid or a second fluid that is immiscible with the fluid above it. Boundary conditions can be determined by evaluating the equations of motion in the small rectangular control volume shown and then letting l go to zero with the upper and lower square areas straddling the interface.

On a solid boundary or at the interface between two immiscible fluids, some of the necessary boundary conditions may be derived from the conservation laws by examining a small thin control volume that spans the interface. A suitable control volume is shown in Fig. 4.22 for the interface separating medium 2 (a fluid) from medium 1 (a solid or a fluid immiscible with fluid 2). This control volume moves with the interface's velocity $\mathbf{u}_s$. When necessary, $\mathbf{u}_s$ can be developed from a specification of the shape of the interface (see Section 11.2). Here $+\mathbf{n}$ and $-\mathbf{n}$ are the unit normal vectors pointing into medium 2 and medium 1, respectively. The square surfaces of the control volume have area dA, are locally parallel to the interface, and are separated from each other by a small distance l. The two tangent vectors to the surface are $\mathbf{t}'$ and $\mathbf{t}''$, and are chosen so that $\mathbf{t}' \times \mathbf{t}'' = \mathbf{n}$. Application of the conservation laws to the rectangular volume ldA as $l \to 0$, keeping the two square area elements in the two different media, produces five boundary conditions. As $l \to 0$, all volume integrals $\to 0$ and the surface integrals over the four rectangular side areas, which are proportional to l, tend to zero unless there is interfacial (surface) tension.

Conservation of mass boundary condition

The mass conservation result, obtained from Fig. 4.22 control volume and (4.5) with $\mathbf{b} = \mathbf{u}_s$, is:

$$\dot{m}_s = \rho_1(\mathbf{u}_1 - \mathbf{u}_s)\cdot\mathbf{n} = \rho_2(\mathbf{u}_2 - \mathbf{u}_s)\cdot\mathbf{n}, \tag{4.91}$$

where $\dot{m}_s$ is the surface mass flux per unit area. Importantly, only the normal component $(\mathbf{u}_s \cdot \mathbf{n})$ of $\mathbf{u}_s$ enters this boundary condition formulation. Thus, $\mathbf{u}_s$ can be chosen with arbitrary or convenient tangential components, and this feature of (4.91) allows simplifications when curved moving interfaces are studied using Cartesian coordinates. If medium 1 is a fluid that is immiscible with fluid 2, then no mass flows across the boundary, $\dot{m}_s = 0$, and (4.91) reduces to $\mathbf{u}_1 \cdot \mathbf{n} = \mathbf{u}_s \cdot \mathbf{n}$ and $\mathbf{u}_2 \cdot \mathbf{n} = \mathbf{u}_s \cdot \mathbf{n}$. If medium 1 is a solid that is not dissolving, subliming, ablating, or otherwise emitting material at the interface, then $\mathbf{u}_1 = \mathbf{u}_s$ so $\dot{m}_s$ is again zero, and the conservation of mass boundary condition reduces to $\mathbf{u}_2 \cdot \mathbf{n} = \mathbf{u}_s \cdot \mathbf{n}$. In addition, if the solid is stationary then $\mathbf{u}_2 \cdot \mathbf{n} = 0$. In general, (4.91) must be used without simplification when there is mass-flow through a moving surface, as in the case of a moving shockwave observed from a stationary vantage point.

Constant surface tension boundary condition

Before proceeding to the conservation of momentum boundary condition, the effects of surface tension forces and their relationship to the geometry of the interface within the thin rectangular control volume shown in Fig. 4.22 must be identified. Surface tension and interfacial tension arise because of the differences in attractive intermolecular forces at gas-liquid and liquid-liquid interfaces, respectively. For clarity, the following discussion emphasizes surface tension at gas-liquid interfaces; however, the results are equally applicable to interfacial tension at liquid-liquid interfaces.

In general, attractive intermolecular forces dominate in liquids, whereas repulsive forces dominate in gases. However, as a gas-liquid interface is approached from the liquid side, the attractive forces are not felt equally because there are many fewer liquid-phase molecules near the interface. Thus there tends to be an unbalanced attraction toward the interior of the liquid on the molecules near the gas-liquid boundary. This unbalanced attraction leads to *surface tension* and a pressure increment across a curved gas-liquid interface that must be properly accounted for when conserving fluid momentum. A somewhat more detailed description is provided in texts on physicochemical hydrodynamics. Two excellent sources are Probstein (1994, Chapter 8) and Levich (1962, Chapter VII).

The thermodynamic basis for surface tension starts from consideration of the Helmholtz free energy (per unit mass) f, defined by:

$$f = e - Ts,$$

where the notation is consistent with that used in Section 1.8. H. Lamb, in *Hydrodynamics* (1945, 6th Edition, p. 456) writes, "Since the condition of stable equilibrium is that the free energy be a minimum, the surface tends to contract as much as is consistent with the other conditions of the problem." If the free energy is a minimum, then the system is in a state of stable equilibrium, and f is called the thermodynamic potential at constant volume (Fermi, 1956, *Thermodynamics*, p. 80). For a reversible isothermal change, the work done on the system increases the free energy f:

$$df = de - Tds - sdT,$$

where the last term is zero for an isothermal change. Then, from (1.24), $df = -pdv =$ work done on the system. (These relations suggest that surface tension decreases with increasing temperature.)

For an interface of area $= A$, separating two fluids of densities ρ_1 and ρ_2, with volumes V_1 and V_2, respectively, and with a surface tension coefficient σ (corresponding to free energy per unit area), the total (Helmholtz) free energy F of the system can be written as:

$$F = \rho_1 V_1 f_1 + \rho_2 V_2 f_2 + A\sigma.$$

If $\sigma > 0$, then the two media (fluids) are immiscible and A will reach a local minimum value at equilibrium. On the other hand, if $\sigma < 0$, corresponding to surface compression, then the two fluids mix freely since the minimum free energy will occur when A has expanded to the point that the spacing between its folds reaches molecular dimensions and the two-fluid system has uniform composition.

When $\sigma > 0$, minimum interface area is achieved by pressure forces that cause fluid elements to move. These pressure forces are determined by σ and the local curvature of the interface. Consider the situation depicted in Fig. 4.23 where the pressure above a curved interface is higher than that below it by an increment $\boldsymbol{\Delta} p$, and the shape of the fluid interface is given by $\eta(x, y, z) = z - h(x, y) = 0$. Here σ is assumed constant. The influence of surface tension gradients on fluid boundary conditions is considered in the next subsection. (Flows driven by surface tension gradients are called Marangoni flows and are largely beyond the scope of this text.) The origin of coordinates and the direction of the z-axis are chosen so that h, $\partial h/\partial x$, and $\partial h/\partial y$ are all zero at $\mathbf{x} = (0, 0, 0)$. Plus, the directions of the x- and y-axes are chosen so that the surface's principal radii of curvature, R_1 and R_2, are found in the x-z and y-z planes, respectively. Thus, the surface's shape is given by:

$$\eta(x, y, z) = z - (x^2/2R_1) - (y^2/2R_2) = 0$$

in the vicinity of the origin. A closed curve C is defined by the intersection of the curved surface and the plane $z = \zeta$. The goal here is to determine how the pressure increment $\boldsymbol{\Delta} p$ depends on R_1 and R_2 when pressure and surface tension forces are balanced as the area enclosed by C approaches zero.

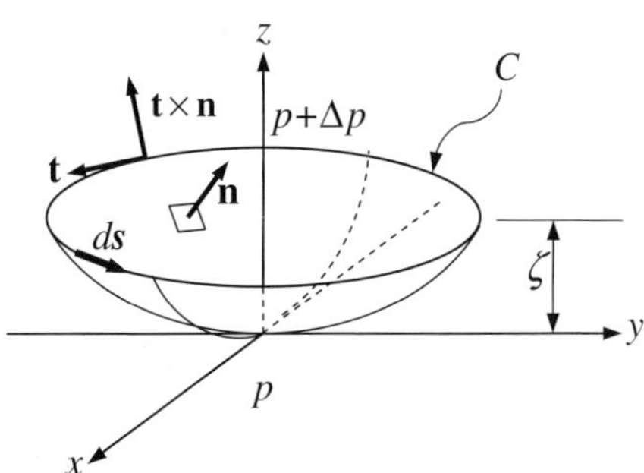

FIGURE 4.23 The curved surface shown is tangent to the x-y plane at the origin of coordinates. The pressure above the surface is $\boldsymbol{\Delta} p$ higher than the pressure below the surface, creating a downward force. Surface tension forces pull in the local direction of $\mathbf{t} \times \mathbf{n}$, which is slightly upward, all around the curve C and thereby balance the downward pressure force.

First, determine the net pressure force $\mathbf{F}_p$ on the surface A bounded by C. The unit normal $\mathbf{n}$ to the surface η is:

$$\mathbf{n} = \frac{\nabla\eta}{|\nabla\eta|} = \frac{(-x/R_1, -y/R_2, 1)}{\sqrt{(x/R_1)^2 + (y/R_2)^2 + 1}},$$

and the area element is:

$$dA = \sqrt{1 + (\partial\eta/\partial x)^2 + (\partial\eta/\partial y)^2}dxdy = \sqrt{1 + (x/R_1)^2 + (y/R_2)^2}dxdy, \text{ so}$$

$$\mathbf{F}_p = -\int\!\!\int_A \boldsymbol{\Delta} p\mathbf{n}dA = -\boldsymbol{\Delta} p\int_{-\sqrt{2R_1\zeta}}^{+\sqrt{2R_1\zeta}}\left[\int_{-\sqrt{2R_2\zeta - x^2R_2/R_1}}^{+\sqrt{2R_2\zeta - x^2R_2/R_1}}(-x/R_1, -y/R_2, 1)dy\right]dx. \tag{4.92}$$

The minus sign appears here because greater pressure above the surface (positive $\boldsymbol{\Delta} p$) must lead to a downward force and the vertical component of $\mathbf{n}$ is positive. The x- and y-components of $\mathbf{F}_p$ are zero because of the symmetry of the situation (odd integrand with even limits). The remaining double integration for the z-component of $\mathbf{F}_p$ produces:

$$\left(\mathbf{F}_p\right)_z = \mathbf{e}_z \cdot \mathbf{F}_p = -\pi\boldsymbol{\Delta} p\sqrt{2R_1\zeta}\sqrt{2R_2\zeta}.$$

This result could have been anticipated from the given geometry; it is merely the negative of the pressure increment, $-\boldsymbol{\Delta} p$, times the area of the curved surface projected onto the plane $z = \zeta$, $\pi\sqrt{2R_1\zeta}\sqrt{2R_2\zeta}$.

The net surface tension force $\mathbf{F}_{st}$ on bounding curve C can be determined from the integral:

$$\mathbf{F}_{st} = \sigma \oint_C \mathbf{t} \times \mathbf{n} ds, \tag{4.93}$$

where $ds = dx\sqrt{1 + (dy/dx)^2}$ is an arc length element of C, and $\mathbf{t}$ is the unit tangent to C so:

$$\mathbf{t} = -\frac{(1, dy/dx, 0)}{\sqrt{1 + (dy/dx)^2}} = \frac{(-y/R_2, x/R_1, 0)}{\sqrt{(y/R_2)^2 + (x/R_1)^2}},$$

and dy/dx is found by differentiating the equation for C, $\zeta = (x^2/2R_1) - (y^2/2R_2)$, with ζ regarded as constant. On each element of C, the surface tension force acts perpendicular to $\mathbf{t}$ and tangent to the curved interface. This direction is given by $\mathbf{t} \times \mathbf{n}$ so the integrand in (4.93) is:

$$\mathbf{t} \times \mathbf{n} ds = \frac{(R_2/y)dx}{\sqrt{1 + (x/R_1)^2 + (y/R_2)^2}} \left(\frac{x}{R_1}, \frac{y}{R_2}, \frac{x^2}{R_1^2} + \frac{y^2}{R_2^2} \right) \cong \frac{R_2}{y} \left(\frac{x}{R_1}, \frac{y}{R_2}, \frac{x^2}{R_1^2} + \frac{y^2}{R_2^2} \right) dx,$$

where the approximate equality holds when x/R_1 and $y/R_2 \ll 1$ and the area enclosed by C approaches zero. The symmetry of the integration path will cause the x- and y-components of $\mathbf{F}_{st}$ to be zero, leaving:

$$(\mathbf{F}_{st})_z = \mathbf{e}_z \cdot \mathbf{F}_{st} = 4\sigma \int_0^{\sqrt{2R_1\zeta}} \frac{R_2}{\sqrt{2R_2\zeta - (R_2/R_1)x^2}} \left[\frac{2\zeta}{R_2} + \frac{x^2}{R_1} \left(\frac{1}{R_1} - \frac{1}{R_2} \right) \right] dx,$$

where y has been eliminated from the integrand using the equation for C, and the factor of four appears because the integral shown only covers one-quarter of the path defined by C. An integration variable substitution in the form $\sin\xi = x/\sqrt{2R_1\zeta}$ allows the integral to be evaluated:

$$(\mathbf{F}_{st})_z = \mathbf{e}_z \cdot \mathbf{F}_{st} = \pi\sigma\sqrt{2R_1\zeta}\sqrt{2R_2\zeta} \left(\frac{1}{R_1} + \frac{1}{R_2} \right).$$

For static equilibrium, $\mathbf{F}_p + \mathbf{F}_{st} = 0$, so the evaluated results of (4.92) and (4.93) require:

$$\Delta p = \sigma(1/R_1 + 1/R_2), \tag{1.11}$$

where the pressure is greater on the side of the surface with the centers of curvature of the interface. Thus, in the absence of buoyant forces and fluid motion, a bubble in water will assume a spherical shape since that shape minimizes its radii of curvature, or equivalently, its surface area (see Rayleigh, 1890, or Batchelor, 1967).

For air bubbles in water, gravity is an important factor for bubbles of millimeter size. The hydrostatic pressure in a liquid is obtained from $p_L = p_o - \rho g z$, where z is measured positively upwards from the free surface and gravity acts downwards and p_o is the pressure at $z = 0$. Thus, for a gas bubble beneath the free surface:

$$p_G = p_L + \sigma(1/R_1 + 1/R_2) = p_o - \rho g z + \sigma(1/R_1 + 1/R_2).$$

Gravity and surface tension forces are of the same order over a length scale $(\sigma/\rho g)^{1/2}$. For an air bubble in water at 288 K, this length scale is: $(\sigma/\rho g)^{1/2} = [7.35 \times 10^{-2}\,\mathrm{N/m}/(9.81\,\mathrm{m/s^2} \times 10^3\,\mathrm{kg/m^3})]^{1/2} = 2.74 \times 10^{-3}\,\mathrm{m}$. Analysis of surface tension effects results in the appearance of additional dimensionless parameters in which surface tension is compared with other effects such as viscous stresses, body forces such as gravity, and inertia. These are defined in Section 4.11.

Conservation of momentum boundary conditions

Now return to the development of boundary conditions from Fig. 4.22 and consider the normal direction. In this case, the momentum conservation result, obtained from (4.18) and (4.20b) with $\mathbf{b} = \mathbf{u}_s$, is:

$$\dot{m}_s(\mathbf{u}_2 - \mathbf{u}_1) \cdot \mathbf{n} = -(p_2 - p_1) + ((n_i\tau_{ij})_2 - (n_i\tau_{ij})_1)n_j + \sigma(1/R' + 1/R''), \tag{4.94}$$

where τ_{ij} is the viscous stress tensor given by (4.60), and the tangent unit vectors $\mathbf{t}'$ and $\mathbf{t}''$ lie along the principal directions of interface curvature (with radii of curvature R' and R''). This condition sets the normal velocity difference at an interface. When both fluids are not moving, or when $\dot{m}_s = 0$ and the fluids are inviscid, (4.94) reduces to (1.11).

Interestingly, a requirement on the tangential fluid velocity components at an interface cannot be developed from the equations of motion. Fortunately, the *no-slip condition* provides a simple experimentally verified result that addresses this analytical insufficiency. The simplest statement of the no-slip condition is that tangential velocity components must match at the interface:

$$\mathbf{u}_1 \cdot \mathbf{t}' = \mathbf{u}_2 \cdot \mathbf{t}', \quad \text{and} \quad \mathbf{u}_1 \cdot \mathbf{t}'' = \mathbf{u}_2 \cdot \mathbf{t}''. \tag{4.95a,b}$$

This condition has been under discussion for centuries, and kinetic theory does provide insights into its validity for gases. The no-slip condition is widely accepted as an experimental fact in macroscopic flows of ordinary fluids (Panton, 2005; White, 2006). For the simplest case of a viscous fluid (medium 2) moving with respect to an impermeable solid (medium 1), (4.91) and (4.95a,b) all together reduce to the interface condition:

$$\mathbf{u}_2 = \mathbf{u}_1, \tag{4.96}$$

and this viscous-flow boundary condition is used throughout this text. Known violations of the no-slip boundary condition occur in rarefied gases and for superfluid helium at or below 2.17 K, where it has an immeasurably small (essentially zero) viscosity. Slip has also been observed in microscopic flows, and on micro-patterned and super-hydrophobic surfaces (Tretheway and Meinhart, 2002; Gogte et al., 2005).

For the control volume in Fig. 4.22, the tangential momentum conservation results from (4.18) and (4.95) are:

$$0 = +((n_i\tau_{ij})_2 - (n_i\tau_{ij})_1)t'_j + (\partial\sigma/\partial x_j)t'_j \quad \text{and} \quad 0 = +((n_i\tau_{ij})_2 - (n_i\tau_{ij})_1)t''_j + (\partial\sigma/\partial x_j)t''_j, \tag{4.97a,b}$$

where τ_{ij} is given by (4.60). These conditions include surface tension gradients and are statements of tangential stress matching at fluid-fluid interfaces. In general, (4.91), (4.94), (4.95), and (4.97) are required for analyzing multiphase flows with phase change.

Conservation of energy boundary conditions

When Fig. 4.22 control volume is used with (4.49) and $\mathbf{b} = \mathbf{u}_s$, the following energy-conservation boundary condition can be developed:

$$\dot{m}_s\left[\left(h + \frac{1}{2}|\mathbf{u}|^2\right)_2 - \left(h + \frac{1}{2}|\mathbf{u}|^2\right)_1\right] = -(k\nabla T)_2 \cdot \mathbf{n} + (k\nabla T)_1 \cdot \mathbf{n}. \tag{4.98}$$

When $\dot{m}_s = 0$, the conductive heat flux must be continuous at the interface.

For a complete set of boundary conditions, the thermal equivalent of the no-slip condition is needed:

$$T_1 = T_2 \tag{4.99}$$

on the interface; no temperature jump is permitted. This condition is also widely accepted and is applied in the remainder of the text. However, it is known to be violated in rarefied gases and is related to viscous slip in such flows. Both topics are reviewed by McCormick (2005).

Example 4.16

Calculate the shape of the free surface of a liquid adjoining an infinite vertical plane wall.

Solution Here let $z = \zeta(x)$ define the free surface shape. With reference to Fig. 4.24 where the y axis points into the page, $1/R_1 = [\partial^2\zeta/\partial x^2][1 + (\partial\zeta/\partial x)^2]^{-3/2}$, and $1/R_2 = [\partial^2\zeta/\partial y^2][1 + (\partial\zeta/\partial y)^2]^{-3/2} = 0$. At the free surface, $\rho g\zeta - \sigma/R_1 = \text{const}$. As $x \to \infty$, $\zeta \to 0$, and $R_1 \to \infty$, so const $= 0$. Then $\rho g\zeta/\sigma - \zeta''/(1 + \zeta'^2)^{3/2} = 0$.

Multiply by the integrating factor ζ' and integrate to obtain $(\rho g/2\sigma)\zeta^2 + (1 + \zeta'^2)^{-1/2} = C$. Evaluate C as $x \to \infty$, $\zeta \to 0$, $\zeta' \to 0$. Then $C = 1$. Consider $x = 0$, $z = \zeta(0) = h$ to find h. The surface slope at the wall is $\zeta' = \tan(\theta + \pi/2) = -\cot\theta$. Then $1 + \zeta'^2 = 1 + \cot^2\theta = \csc^2\theta$. Thus we now have $(\rho g/2\sigma)\, h^2 = 1 - 1/\csc\theta = 1 - \sin\theta$, so that $h^2 = (2\sigma/\rho g)(1 - \sin\theta)$. Finally we seek to integrate to obtain the shape of the interface. Squaring and rearranging the result above, the differential equation we must

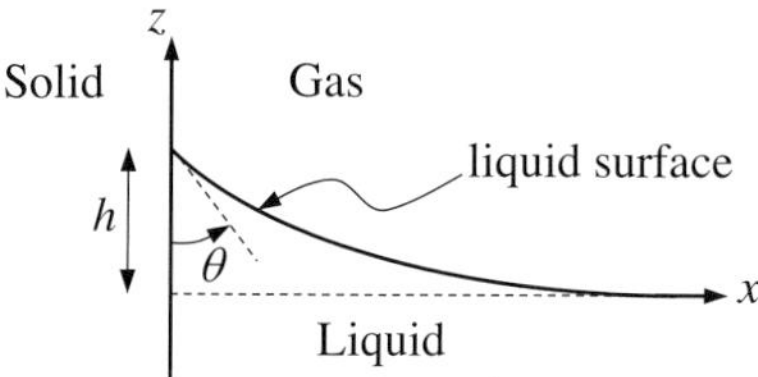

FIGURE 4.24 Free surface of a liquid adjoining a vertical plane wall. Here the contact angle is θ and the liquid rises to $z = h$ at the solid wall.

solve may be written as $1 + (d\zeta/dx)^2 = [1 - (\rho g/2\sigma)\zeta^2]^{-2}$. Solving for the slope and taking the negative square root (since the slope is negative for positive x):

$$d\zeta/dx = -\left\{1 - [1 - (\rho g/2\sigma)\zeta^2]^2\right\}^{1/2}[1 - (\rho g/2\sigma)\zeta^2]^{-1}.$$

Define $\sigma/\rho g = \delta^2$, $\zeta/\delta = \gamma$. Rewriting the equation in terms of x/δ and γ, and separating variables leads to:

$$2(1 - \gamma^2/2)\gamma^{-1}(4 - \gamma^2)^{-1/2}d\gamma = d(x/\delta).$$

The integrand on the left is simplified by partial fractions and the constant of integration is evaluated at $x = 0$ when $\gamma = h/\delta$. Finally:

$$\cosh^{-1}(2\delta/\zeta) - (4 - \zeta^2/\delta^2)^{1/2} - \cosh^{-1}(2\delta/h) + (4 - h^2/\delta^2)^{1/2} = x/\delta$$

gives the shape of the interface in terms of $x(\zeta)$.

4.11 Dimensionless forms of the equations and dynamic similarity

For a properly specified fluid flow problem or situation, the differential equations of fluid motion, the fluid's constitutive and thermodynamic properties, and the boundary conditions may be used to determine the dimensionless parameters that govern the situation of interest even before a solution of the equations is attempted. The dimensionless parameters indicate which phenomena will be important in the resulting flow. This section describes and presents the primary dimensionless parameters or numbers required in the remainder of the text, though many others exist in the broad realm of fluid mechanics.

The dimensionless parameters for any particular problem can be determined in two ways. They can be deduced directly from the governing differential equations if these equations are known; this method is illustrated here. However, if the governing differential equations are unknown or the parameter of interest does not appear in the known equations, dimensionless parameters can be determined from dimensional analysis (see Section 1.11). The advantage of adopting the former strategy is that dimensionless parameters determined from the equations of motion are more readily interpreted and linked to the physical phenomena occurring in the flow. Thus, knowledge of the relevant dimensionless parameters frequently aids the solution process, especially when assumptions and approximations are necessary to reach a solution.

In addition, the dimensionless parameters obtained from the equations of fluid motion set the conditions under which scale model testing will prove useful for predicting the performance of smaller or larger devices. In particular, two flow fields are considered to be dynamically similar when: (1) their geometries are scale similar; and (2) their dimensionless parameters match. The first requirement implies that any length scale in the first flow field may be mapped to its counterpart in the second flow field by multiplication with a single scale ratio. The second requirement allows predictions for the larger- or smaller-scale flow to be made from quantitative knowledge of the model scale flow when the scale ratio is accounted for. Moreover, use of standard dimensionless parameters typically reduces the parameters that must be varied in an experiment or calculation, and greatly facilitates the comparison of measured or computed results with prior work conducted under potentially different conditions.

To illustrate these advantages, consider the drag force, F_D, on a sphere of diameter d moving at a speed U through a fluid of density ρ and viscosity μ. Dimensional analysis (Section 1.11) using these five parameters produces the following possible dimensionless scaling laws:

$$\frac{F_D}{\rho U^2 d^2} = \Psi\left(\frac{\rho U d}{\mu}\right), \quad \text{or} \quad \frac{F_D \rho}{\mu^2} = \Phi\left(\frac{\mu}{\rho U d}\right). \tag{4.100}$$

Both are valid, but the first is preferred because it contains dimensionless groups that either come from the equations of motion or are traditionally defined in the study of fluid dynamic drag. If dimensionless groups were not used, experiments would have to be conducted to determine F_D as a function of d, keeping U, ρ, and μ fixed. Then, experiments would have to be conducted to determine F_D as a function of U, keeping d, ρ, and μ fixed, and so on. However, such a duplication of effort is unnecessary when dimensionless groups are used. In fact, use of the first dimensionless law above allows experimental results from a wide range of conditions to be simply plotted with two axes (see Fig. 4.25) even though the full complement of experiments may have involved variations in all five dimensional parameters.

The idea behind dimensional analysis is intimately associated with the concept of similarity. In fact, a collapse of all the data on a single graph such as the one in Fig. 4.25 is possible only because in this problem all flows having the same value of the dimensionless group known as the *Reynolds number* $\text{Re} = \rho Ud/\mu$ are dynamically similar. This dynamic similarity is assured because the Reynolds number appears when the equations of motion are cast in dimensionless form.

The use of dimensionless parameters pervades fluid mechanics to such a degree that this chapter and this text would be considered incomplete without this section, even though this topic is well covered in first-course fluid mechanics texts where the content of this section is commonly combined with that in Section 1.11. For clarity, the following discussion first covers the dimensionless groups associated with the momentum equation, and then proceeds to the continuity and energy equations.

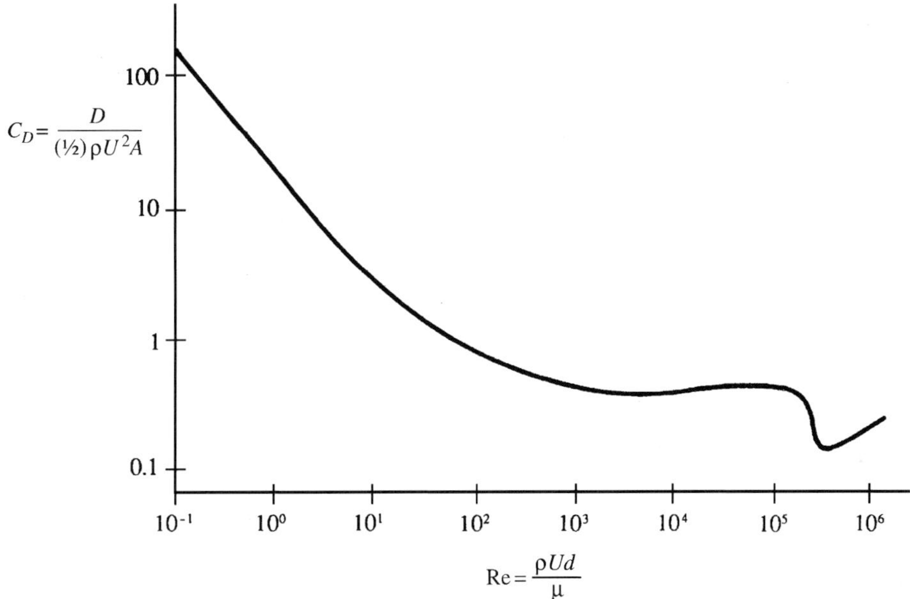

FIGURE 4.25 Coefficient of drag C_D for a sphere vs. the Reynolds number Re based on sphere diameter. At low Reynolds number $C_D \sim 1/\text{Re}$, and above $\text{Re} \sim 10^3$, $C_D \sim$ constant (except for the dip between $\text{Re} = 10^5$ and 10^6). These behaviors (except for the dip) can be explained by simple dimensional reasoning. The reason for the dip is the transition of the laminar boundary layer to a turbulent one, as explained in Chapter 8.

Consider the flow of a fluid having nominal density ρ and viscosity μ with a length scale l, a velocity scale U, and a rotation or oscillation frequency Ω. The situation here is intended to be general so that the dimensional parameters obtained from this effort will be broadly applicable. Particular situations that would involve all five parameters include pulsating flow through a tube, flow past an undulating self-propelled body, or flow through a turbomachine.

The starting point is the Navier-Stokes momentum equation (4.40) simplified for incompressible flow. (The effect of compressibility is deduced from the continuity equation in the next subsection.)

$$\rho\left(\frac{\partial \mathbf{u}}{\partial t} + (\mathbf{u}\cdot\nabla)\mathbf{u}\right) = -\nabla p + \rho\mathbf{g} + \mu\nabla^2\mathbf{u} \tag{4.40}$$

This equation can be rendered dimensionless by defining dimensionless variables:

$$x_i^* = x_i/l, \quad t^* = \Omega t, \quad u_j^* = u_j/U, \quad p^* = (p - p_\infty)/\rho U^2, \quad \text{and} \quad g_j^* = g_j/g, \tag{4.101}$$

where g is the acceleration of gravity. When these dimensionless variables are used, the boundary conditions can be stated in terms of pure numbers and are independent of l, U, and Ω. For example, consider the viscous flow over a circular cylinder of radius R. When the velocity scale U is the free-stream velocity and the length scale is the radius R, then, in terms of the dimensionless velocity $u^* = u/U$ and the dimensionless coordinate $r^* = r/R$, the boundary condition at infinity is $u^* \to 1$ as $r^* \to \infty$, and the condition at the surface of the cylinder is $u^* = 0$ at $r^* = 1$. In addition, because pressure enters (4.40) only as a gradient, the pressure itself is not of consequence; only pressure differences are important. The conventional practice is to render $p - p_\infty$ dimensionless, where p_∞ is a suitably chosen reference pressure. Depending on the nature of the flow, $p - p_\infty$ could be made dimensionless with a generic viscous stress $\mu U/l$, a hydrostatic pressure $\rho g l$, or as in (4.101), a dynamic pressure ρU^2. Substitution of (4.101) into (4.40) produces:

$$\left[\frac{\Omega l}{U}\right]\frac{\partial \mathbf{u}^*}{\partial t^*} + (\mathbf{u}^* \cdot \nabla^*)\mathbf{u}^* = -\nabla^* p^* + \left[\frac{gl}{U^2}\right]\mathbf{g}^* + \left[\frac{\mu}{\rho U l}\right]\nabla^{*2}\mathbf{u}^*, \tag{4.102}$$

where $\nabla^* = l\nabla$. The form of this equation implies that two flows having different values of Ω, U, l, g, or μ, will obey the exactly the same differential momentum equation if the values of the dimensionless groups $\Omega l/U$, gl/U^2, and $\mu/\rho U l$ are identical. Because the dimensionless boundary conditions are also identical in the two flows, it follows that *they will have the same dimensionless solutions*. Products of these dimensionless groups appear as coefficients in front of different terms when the pressure is presumed to have alternative scalings (see Exercise 4.79).

The parameter groupings shown in [,]-brackets in (4.101) have the following names and interpretations:

$$\text{St} = \text{Strouhal number} \equiv \frac{\text{unsteady acceleration}}{\text{advective acceleration}} \propto \frac{\partial u/\partial t}{u(\partial u/\partial x)} \propto \frac{\Omega U}{U^2/l} = \frac{\Omega l}{U}, \tag{4.103}$$

$$\text{Re} = \text{Reynolds number} \equiv \frac{\text{inertia force}}{\text{viscous force}} \propto \frac{\rho u(\partial u/\partial x)}{\mu(\partial^2 u/\partial x^2)} \propto \frac{\rho U^2/l}{\mu U/l^2} = \frac{\rho U l}{\mu}, \quad \text{and} \tag{4.104}$$

$$\text{Fr} = \text{Froude number} \equiv \left[\frac{\text{inertia force}}{\text{gravity force}}\right]^{1/2} \propto \left[\frac{\rho u(\partial u/\partial x)}{\rho g}\right]^{1/2} \propto \left[\frac{\rho U^2/l}{\rho g}\right]^{1/2} = \frac{U}{\sqrt{gl}}. \tag{4.105}$$

The Strouhal number sets the importance of unsteady fluid acceleration in flows with oscillations. It is relevant when flow unsteadiness arises naturally or because of an imposed frequency. The Reynolds number is the most common dimensionless number in fluid mechanics. Low Re flows involve small sizes, low speeds, and high kinematic viscosity such as bacteria swimming through mucous. High Re flows involve large sizes, high speeds, and low kinematic viscosity such as an ocean liner steaming at full speed.

St, Re, and Fr have to be equal for dynamic similarity between two flows in which unsteadiness, and viscous and gravitational effects are important. Note that the mere presence of gravity does not make the gravitational effects dynamically important. For flow around an object in a homogeneous incompressible fluid, gravity is important only if surface waves are generated. Otherwise, the effect of gravity is simply to add a hydrostatic pressure to the entire system that changes the local pressure reference (see "Neglect of Gravity in Constant Density Flows" in Section 4.9).

Interestingly, in a density-stratified fluid, gravity can play a significant role without the presence of a free surface. The effective gravity force per unit volume in a two-fluid-layer situation is $(\rho_2 - \rho_1)g$, where ρ_1 and ρ_2 are fluid densities in the two layers. In such a case, an internal Froude number is defined as:

$$\text{Fr}' \equiv \left[\frac{\text{inertia force}}{\text{buoyancy force}}\right]^{1/2} \propto \left[\frac{\rho_1 U^2/l}{(\rho_2 - \rho_1)g}\right]^{1/2} = \frac{U}{\sqrt{g'l}}, \tag{4.106}$$

where $g' \equiv g\ (\rho_2 - \rho_1)/\rho_1$ is the *reduced gravity*. For a continuously stratified fluid having a maximum buoyancy frequency N (see (1.36)), the equivalent of (4.106) is $\text{Fr}' \equiv U/Nl$. Alternatively, the internal Froude number may be replaced by the Richardson Number $= \text{Ri} \equiv 1/\text{Fr}'^2 = g'l/U^2$, which can also be refined to a gradient Richardson number $\equiv N^2(z)/(dU/dz)^2$ that is important in studies of instability and turbulence in stratified fluids.

Under dynamic similarity, the dimensionless numbers in the model-scale flow are matched to their counterparts in the larger- or smaller-scale flow, and this ensures that the dimensionless solutions are identical. Furthermore, the dimensional consistency of the equations of motion ensures that all flow quantities may be set in dimensionless form. For example, the local pressure at point $\mathbf{x} = (x, y, z)$ can be made dimensionless in the form:

$$\frac{p(\mathbf{x}, t) - p_\infty}{\frac{1}{2}\rho U^2} \equiv C_p = \Psi\left(\text{St, Fr, Re}; \frac{\mathbf{x}}{l}, \Omega t\right), \tag{4.107}$$

where $C_p = (p - p_\infty)/(1/2)\rho U^2$ is called the *pressure coefficient* (or the Euler number $=$ Eu), and Ψ represents the dimensionless solution for the pressure coefficient in terms of dimensionless parameters and variables. The factor of 1/2 in (4.107) is conventional but not necessary. Similar relations also hold for any other dimensionless flow variable such as velocity $\mathbf{u}/U$. It follows that in dynamically similar flows, dimensionless flow variables are identical at *corresponding points and times* (that is, for identical values of $\mathbf{x}/l$, and Ωt). Of course there are many instances where the flow geometry may require two or more length scales: $l, l_1, l_2, \ldots l_n$. When this is the case, the aspect ratios $l_1/l, l_2/l, \ldots l_n/l$ provide a dimensionless description of the geometry, and would also appear as arguments of the function Ψ in a relationship like (4.107). Here a difference between relations (4.100) and (4.107) should be noted. Eq. (4.100) is a relation between *overall* flow parameters, whereas (4.107) holds *locally* at a point.

Incidentally, in liquid flows, when p_∞ in (4.107) is replaced by the liquid's vapor pressure, p_v, the dimensionless ratio is known as the *cavitation number*. Zero or negative cavitation number at any point in a flow indicates likely vapor-bubble formation at that location, and the presence of such bubbles may completely change the character of the flow, often in detrimental ways. For example, cavitation often sets performance limits and/or dictates the operational lifetime of hydrodynamic machinery such as water pumps, hydroelectric power turbines, and ship propellers.

In the foregoing analysis, the imposed unsteadiness in boundary conditions was assumed important. However, time may also be made dimensionless via $t^* = Ut/l$, as would be appropriate for a flow with steady boundary conditions. In this case, the time derivative in (4.40) should still be retained because the resulting flow may still be naturally unsteady since flow oscillations can arise spontaneously even if the boundary conditions are steady. But, from dimensional considerations, such unsteadiness must have a time scale proportional to l/U.

In the foregoing analysis we have also assumed that an imposed velocity U is relevant. Consider now a situation in which the imposed boundary conditions are purely unsteady. To be specific, consider an object having a characteristic length scale l oscillating with a frequency Ω in a fluid at rest at infinity. This is a problem having an imposed length scale and an *imposed time scale* $1/\Omega$. In such a case a velocity scale $U = l\Omega$ can be constructed. The preceding analysis then goes through, leading to the conclusion that $\mathrm{St} = 1$, $\mathrm{Re} = Ul/\nu = \Omega l^2/\nu$, and $\mathrm{Fr} = U/(gl)^{1/2} = \Omega(l/g)^{1/2}$ have to be duplicated for dynamic similarity.

All dimensionless quantities are identical in dynamically similar flows. For flow around an immersed body, like a sphere, we can define the (dimensionless) drag and lift coefficients:

$$C_D \equiv \frac{F_D}{\frac{1}{2}\rho U^2 A} \quad \text{and} \quad C_L \equiv \frac{F_L}{\frac{1}{2}\rho U^2 A}, \tag{4.108, 4.109}$$

where A is a reference area, and F_D and F_L are the drag and lift forces, respectively, experienced by the body; as in (4.107) the factor of 1/2 in (4.108) and (4.109) is conventional but not necessary. For blunt bodies such as spheres and cylinders, A is taken to be the maximum cross section perpendicular to the flow. Therefore, $A = \pi d^2/4$ for a sphere of diameter d, and $A = bd$ for a cylinder of diameter d and length b, with its axis perpendicular to the flow. For flows over flat plates, and airfoils, on the other hand, A is taken to be the *planform area*, that is, $A = sl$; here, l is the average length of the plate or chord of the airfoil in the direction of flow and s is the width perpendicular to the flow, sometimes called the *span*.

The values of the drag and lift coefficients are identical for dynamically similar flows. For flow about a steadily moving ship, the drag is caused both by gravitational and viscous effects so we must have a functional relation of the form $C_D = C_D(\mathrm{Fr}, \mathrm{Re})$. However, in many flows gravitational effects are unimportant. An example is flow around a body that is far from a free surface and does not generate gravity waves. In this case, Fr is irrelevant, so $C_D = C_D(\mathrm{Re})$ is all that is needed when the effects of compressibility are unimportant. This is the situation portrayed by the first member of (4.100) and illustrated in Fig. 4.25.

A recurring limitation in scale-model testing is the inability to match Reynolds numbers to achieve full dynamic similarity between a model and a larger and/or faster full-scale device. This situation is commonly known as *incomplete* similarity, and it can be managed in a variety of ways. First of all, if model- and full-scale Reynolds numbers cannot be matched with the same fluid, then use of a special fluid with a desirable density or viscosity for the model tests may be possible. Thus, hydrodynamic tests are sometimes performed on aerodynamic devices because the kinematic viscosity of water is approximately 1/15 that of air, so smaller devices tested at lower speeds in water can achieve the same Reynolds number as larger ones tested in faster moving air. Compressed air and liquid helium are other fluids that allow high-Reynolds number testing of model-scale devices. Second, the scale-model tests may show that the important performance metrics (C_L and C_D perhaps) are independent of *Re* above a threshold Reynolds number. In this case, model-to-full-scale extrapolation of performance results can be successful, but such extrapolation is inherently risky. However, such extrapolation uncertainty and risk from incomplete similarity in scale-model tests can be reduced if the model- and full-scale Reynolds numbers are as close as possible. In practice, this means that scale-model tests are typically conducted with the largest possible models at the highest possible speeds the available resources will allow. The two examples at the end of this subsection both describe performance predictions based on incomplete similarity.

Now return to the development of the dimensionless groups that naturally arise from the equations of motion. A dimensionless form of the continuity equation should indicate when flow-induced pressure differences induce significant departures from incompressible flow. However, the simplest possible scaling fails to provide any insights because the continuity equation itself does not contain the pressure. Thus, a more fruitful starting point for determining the relative size of $\nabla \cdot \mathbf{u}$ is a rearranged version of (4.9):

$$\nabla \cdot \mathbf{u} = -\frac{1}{\rho}\frac{D\rho}{Dt} = -\frac{1}{\rho c^2}\frac{Dp}{Dt}, \tag{4.110}$$

along with the assumption that pressure-induced density changes will be isentropic, $dp = c^2 d\rho$ where c is the sound speed, see (1.25). Using the following dimensionless variables:

$$x_i^* = x_i/l, \quad t^* = Ut/l, \quad u_j^* = u_j/U, \quad p^* = (p - p_\infty)/\rho_o U^2, \quad \text{and} \quad \rho^* = \rho/\rho_o, \tag{4.111}$$

where ρ_o is a reference density, the outside members of (4.110) can be rewritten:

$$\nabla^* \cdot \mathbf{u}^* = -\left[\frac{U^2}{c^2}\right]\frac{1}{\rho^*}\frac{Dp^*}{Dt^*}, \tag{4.112}$$

which specifically shows that the square of:

$$M = \text{Mach number} \equiv \left[\frac{\text{inertia force}}{\text{compressibility force}}\right]^{1/2} \propto \left[\frac{\rho U^2/l}{\rho c^2/l}\right]^{1/2} = \frac{U}{c} \tag{4.113}$$

sets the size of isentropic departures from incompressible flow. In engineering practice, gas flows are considered incompressible when $M < 0.3$, and from (4.112) this corresponds to ~10% departure from ideal incompressible behavior when $(1/\rho^*)(Dp^*/Dt^*)$ is unity. Of course, there may be nonisentropic changes in density too and these are considered in Thompson (1972, pp. 137–146). Flows in which $M < 1$ are called *subsonic*, whereas flows in which $M > 1$ are called *supersonic*. At high subsonic and supersonic speeds, matching Mach number between flows is required for dynamic similarity.

There are many possible thermal boundary conditions for the energy equation, so a fully general scaling of (4.61) is not possible. Instead, a simple scaling is provided based on constant specific heats, neglect of μ_v, and constant free-stream and wall temperatures, T_o and T_w, respectively. In addition, for simplicity, an imposed flow oscillation frequency is not considered. The starting point of the scaling provided here is a mild revision of (4.61) that involves the enthalpy h per unit mass:

$$\rho\frac{Dh}{Dt} = \frac{Dp}{Dt} + \rho\varepsilon + \frac{\partial}{\partial x_i}\left(k\frac{\partial T}{\partial x_i}\right), \tag{4.114}$$

where ε is given by (4.59). Using $dh \cong c_p dT$, the following dimensionless variables:

$$\varepsilon^* = \rho_o l^2 \varepsilon/\mu_o U^2, \quad \mu^* = \mu/\mu_o, \quad k^* = k/k_o, \quad T^* = (T - T_o)/(T_w - T_o), \tag{4.115}$$

and those defined in (4.107), (4.108) become:

$$\rho^*\frac{DT^*}{Dt^*} = \left[\frac{U^2}{c_p(T_w - T_o)}\right]\frac{Dp^*}{Dt^*} + \left[\frac{U^2}{c_p(T_w - T_o)}\frac{\mu_o}{\rho_o U l}\right]\rho^*\varepsilon^* + \left[\frac{k_o}{c_p\mu_o}\frac{\mu_o}{\rho_o U l}\right]\nabla^*(k^*\nabla^* T^*). \tag{4.116}$$

Here the relevant dimensionless parameters are:

$$\text{Ec} = \text{Eckert number} \equiv \frac{\text{kinetic energy}}{\text{thermal energy}} = \frac{U^2}{c_p(T_w - T_o)}, \tag{4.117}$$

$$\text{Pr} = \text{Prandtl number} \equiv \frac{\text{momentum diffusivity}}{\text{thermal diffusivity}} = \frac{\nu}{\kappa} = \frac{\mu_o/\rho_o}{k_o/\rho_o c_p} = \frac{\mu_o c_p}{k_o}, \tag{4.118}$$

and we recognize $\rho_o U l/\mu_o$ as the Reynolds number in (4.116) as well. In low speed flows where the Eckert number is small, the middle terms drop out of (4.116), and the full energy equation (4.114) reduces to (4.90). Thus, low Ec is needed for the Boussinesq approximation.

The Prandtl number is a ratio of two molecular transport properties. It is therefore a fluid property and independent of flow geometry. For air at ordinary temperatures and pressures, $\mathrm{Pr} = 0.72$, which is close to the value of 0.67 predicted from a simplified kinetic theory model assuming hard spheres and monatomic molecules (Hirschfelder et al., 1954, pp. 9–16). For water at 20 °C, $\mathrm{Pr} = 7.1$. Dynamic similarity between flows involving thermal effects requires equality of the Eckert, Prandtl, and Reynolds numbers.

And finally, for flows involving surface tension σ, there are several relevant dimensionless numbers:

$$\mathrm{We} = \text{Weber number} \equiv \frac{\text{inertia force}}{\text{surface tension force}} \propto \frac{\rho U^2 l^2}{\sigma l} = \frac{\rho U^2 l}{\sigma}, \tag{4.119}$$

$$\mathrm{Bo} = \text{Bond number} \equiv \frac{\text{gravity force}}{\text{surface tension force}} \propto \frac{\rho l^3 g}{\sigma l} = \frac{\rho l^2 g}{\sigma}, \tag{4.120}$$

$$\mathrm{Ca} = \text{Capillary number} \equiv \frac{\text{viscous stress}}{\text{surface tension stress}} \propto \frac{\mu U/l}{\sigma/l} = \frac{\mu U}{\sigma}. \tag{4.121}$$

Here, for the Weber and Bond numbers, the ratio is constructed based on a ratio of forces as in (4.108) and (4.109), and not forces per unit volume as in (4.104), (4.105), and (4.113). At high Weber number, droplets and bubbles are easily deformed by fluid acceleration or deceleration, for example during impact with a solid surface. At high Bond numbers surface tension effects are relatively unimportant compared to gravity, as is the case for long-wavelength, ocean surface waves. At high capillary numbers viscous forces dominate those from surface tension; this is commonly the case in machinery lubrication flows. However, for slow bubbly flow through porous media or narrow tubes (low Ca) the opposite is true.

Example 4.17

A ship 100 m long (l) is expected to cruise at 10 m/s (U). It has a submerged surface of 300 m^2 (A). Find the model speed for a 1/25 scale model, neglecting frictional effects. The drag force F_D is measured to be 60 N when the model is tested in a towing tank at this model speed. Estimate the full scale drag when the skin-friction drag coefficient for the model is 0.003 and that for the full-scale ship is 0.0015.

Solution The ship's hull will interact with the water's surface and directly with the water, so both wave drag and friction drag will occur. In dimensionless form, this requires a dimensionless drag force to depend on the Froude number, the Reynolds number, and an aspect ratio:

$$F_D \Big/ \left(\frac{1}{2}\rho U^2 A\right) = \Psi\left(U \Big/ \sqrt{gl}, \rho U l/\mu, \sqrt{A} \Big/ l\right),$$

where Ψ is an undetermined function. Here it will not be possible to match both Fr and Re between the model and the full-scale device. However, the aspect ratio is matched automatically for a true scale-model test so it's not considered further.

To find the model test speed, equate the model and ship Froude numbers:

$$\left(U \Big/ \sqrt{gl}\right)_m = \left(U \Big/ \sqrt{gl}\right)_s, \text{ which implies: } U_m = U_s\sqrt{gl_m/gl_s} = (10\,\mathrm{m/s})\sqrt{1/25} = 2\,\mathrm{m/s}.$$

Here subscripts "m" and "s" denote *model* and *ship* parameters, respectively.

The total drag on the model is measured to be 60 N at this model speed, and part of this is friction drag. Here we can use Froude's hypothesis that the unknown function Ψ is a sum of a frictional drag term Ψ_f that only depends on the Reynolds number (and surface roughness ratio), and a wave drag term Ψ_w that only depends on the Froude number.

$$F_D \Big/ \left(\frac{1}{2}\rho U^2 A\right) = \Psi_w\left(U \Big/ \sqrt{gl}\right) + \Psi_f(\rho U l/\mu).$$

Furthermore, treat the submerged portion of the hull as a flat plate for which the friction drag coefficient C_D is a function of the Reynolds number, i.e., set $C_D = \Psi_f$. The problem statement sets the frictional drag coefficients as $C_{D,m} = 0.003$ and $C_{D,s} = 0.0015$, and these are consistent with the length-based Re values for the model and the ship, 8×10^6 and 10^9, respectively. Using a value of $\rho = 1000$ kg/m^3 for water, the model's friction drag can be estimated:

$$\left(\frac{1}{2}\rho U^2 A C_D\right)_m = (0.5)(10^3\,\mathrm{kg/m^3})(2\,\mathrm{m/s})^2(300\,\mathrm{m^2}/25^2)(0.003) = 2.88\,\mathrm{N}.$$

Thus, out of the total model drag of 60 N, the model's wave drag is $(F_{WD})_m = 60 - 2.88 = 57.12\,\text{N}$. And this *wave drag* obeys the scaling law above, which means that:

$$(\Psi_w)_m = (\Psi_w)_s = \left(\frac{F_{WD}}{\frac{1}{2}\rho U^2 A}\right)_m = \left(\frac{F_{WD}}{\frac{1}{2}\rho U^2 A}\right)_s.$$

Thus, the wave drag on the ship $(F_{WD})_s$ can be estimated as follows:

$$(F_{WD})_s = (F_{WD})_m \frac{\rho_s}{\rho_m}\frac{U_s^2}{U_m^2}\frac{A_s}{A_m} = (57.12\,\text{N})(1)\left(\frac{10\,\text{m/s}}{2\,\text{m/s}}\right)^2 (25)^2 = 8.925 \times 10^5\,\text{N},$$

where the area ratio is the square of the length-scale ratio. This must be added the ship's frictional drag:

$$\left(\frac{1}{2}\rho U^2 A C_D\right)_s = (0.5)(10^3\,\text{kg/m}^3)(10\,\text{m/s})^2(300\,\text{m}^2)(0.0015) = 0.225 \times 10^5\,\text{N}.$$

Therefore, total drag on the ship is predicted to be: $(8.925 + 0.225) \times 10^5 = 9.15 \times 10^5$ N. If no correction was made for friction, and all the measured model drag was assumed due to wave effects, then the predicted ship drag would be:

$$(F_D)_s = (F_D)_m \frac{\rho_s}{\rho_m}\frac{U_s^2}{U_m^2}\frac{A_s}{A_m} = (60\,\text{N})(1)\left(\frac{10\,\text{m/s}}{2\,\text{m/s}}\right)^2 (25)^2 = 9.37 \times 10^5\,\text{N},$$

which is a few percent higher than the friction-corrected estimate.

Example 4.18

A table top centrifugal blower with diameter of d_1 is tested at rotational speed Ω_1 and generates a volume flow rate of Q_1 against a pressure difference of Δp_1 when moving air with density ρ_1 and viscosity μ_1. What are the three relevant dimensionless groups? A scale-similar centrifugal water pump with diameter $d_2 = 2d_1$ is operated at rotational speed $\Omega_2 = \Omega_1/10$. If turbomachine performance is independent of Re in the operational range of these devices, what are the pump's volume flow rate Q_2 and pressure rise Δp_1 if Re is not matched but the other two dimensionless parameters are?

Solution Dimensional analysis using the six parameters yields the following dimensionless groups:

$$\text{The flow coefficient} = \frac{Q}{\Omega d^3}, \quad \text{the head coefficient} = \frac{\Delta p}{\rho \Omega^2 d^2}, \quad \text{and} \quad \text{Re} = \frac{\rho \Omega d^2}{\mu}.$$

These three, along with the device's efficiency, are routinely used when scaling turbomachine performance. Matching flow coefficients produces:

$$\frac{Q_1}{\Omega_1 d_1^3} = \frac{Q_2}{\Omega_2 d_2^3} = \frac{Q_2}{(\Omega_1/10)8d_1^3}, \quad \text{or} \quad Q_2 = \frac{4}{5}Q_1.$$

Matching head coefficients produces:

$$\frac{\Delta p_1}{\rho_1 \Omega_1^2 d_1^2} = \frac{\Delta p_2}{\rho_2 \Omega_2^2 d_2^2} = \frac{\Delta p_2}{825\rho_1(\Omega_1/10)^2 4d_1^2}, \quad \text{or} \quad \Delta p_2 = \frac{825 \cdot 4}{10^2}\Delta p_1 = 33\Delta p_1,$$

where the water/air-density ratio is 825 at room temperature and pressure. For these conditions, the water pump power is $Q_2 \Delta p_2 = 26.4 Q_1 \Delta p_1$, a substantial increase above the blower's power. And, the Reynolds number ratio is:

$$\frac{\text{Re}_2}{\text{Re}_1} = \frac{\Omega_2 d_2^2/\nu_2}{\Omega_1 d_1^2/\nu_1} \cong \frac{(\Omega_1/10)4d_1^2/(\nu_1/15)}{\Omega_1 d_1^2/\nu_1} = 6,$$

where ν is the kinematic viscosity and $\nu_2/\nu_1 \approx 1/15$ at room temperature and pressure.

Exercises

4.1 Let a one-dimensional velocity field be $u = u(x, t)$, with $v = 0$ and $w = 0$. The density varies as $\rho = \rho_0(2 - \cos\omega t)$. Find an expression for $u(x, t)$ if $u(0, t) = U$.

4.2 Consider the one-dimensional Cartesian velocity field: $\mathbf{u} = (\alpha x/t, 0, 0)$ where α is a constant.

a) Find a spatially uniform, time-dependent density field, $\rho = \rho(t)$, that renders this flow field mass conserving when $\rho = \rho_o$ at $t = t_o$.

b) What are the unsteady $(\partial\mathbf{u}/\partial t)$, advective $([\mathbf{u}\cdot\nabla]\mathbf{u})$, and particle $(D\mathbf{u}/Dt)$ accelerations in this flow field? What does $\alpha = 1$ imply?

4.3 Find a nonzero density field $\rho(x, y, z, t)$ that renders the following Cartesian velocity fields mass conserving. Comment on the physical significance and uniqueness of your solutions.

a) $\mathbf{u} = (U\sin(\omega t - kx), 0, 0)$ where U, ω, k are positive constants [*Hint*: exchange the independent variables x, t for a single independent variable $\xi = \omega t - kx$]

b) $\mathbf{u} = (-\Omega y, +\Omega x, 0)$ with $\Omega =$ constant [*Hint*: switch to cylindrical coordinates.]

c) $\mathbf{u} = (A/x, B/y, C/z)$ where A, B, C are constants

4.4 A proposed conservation law for ξ, a new fluid property, takes the following form: $\frac{d}{dt}\int_{V(t)}\rho\xi dV + \int_{A(t)}\mathbf{Q}\cdot\mathbf{n}dS = 0$, where $V(t)$ is a material volume that moves with the fluid velocity $\mathbf{u}$, $A(t)$ is the surface of $V(t)$, ρ is the fluid density, and $\mathbf{Q} = -\rho\gamma\nabla\xi$.

a) What partial differential equation is implied by the above conservation statement?

b) Use the part a) result and the continuity equation to show: $(\partial\xi/\partial t) + \mathbf{u}\cdot\nabla\xi = (1/\rho)\nabla\cdot(\rho\gamma\nabla\xi)$.

4.5 The components of a mass flow vector $\rho\mathbf{u}$ are $\rho u = 4x^2y$, $\rho v = xyz$, $\rho w = yz^2$.

a) Compute the net mass outflow through the closed surface formed by the planes $x = 0$, $x = 1$, $y = 0$, $y = 1$, $z = 0$, $z = 1$.

b) Compute $\nabla\cdot(\rho\mathbf{u})$ and integrate over the volume bounded by the surface defined in part a).

c) Explain why the results for parts a) and b) should be equal or unequal.

4.6 Consider a simple fluid mechanical model for the atmosphere of an ideal spherical star that has a surface gas density of ρ_o and a radius r_o. The escape velocity from the surface of the star is v_e. Assume that a tenuous gas leaves the star's surface radially at speed v_o uniformly over the star's surface. Use the steady continuity equation for the gas density ρ and fluid velocity $\mathbf{u} = (u_r, u_\theta, u_\varphi)$ in spherical coordinates:

$$\frac{1}{r^2}\frac{\partial}{\partial r}(r^2\rho u_r) + \frac{1}{r\sin\theta}\frac{\partial}{\partial\theta}(\rho u_\theta\sin\theta) + \frac{1}{r\sin\theta}\frac{\partial}{\partial\varphi}(\rho u_\varphi) = 0$$

for the following items.

a) Determine ρ when $v_o \geq v_e$ so that $\mathbf{u} = (u_r, u_\theta, u_\varphi) = (v_o\sqrt{1 - (v_e^2/v_o^2)(1 - (r_o/r))}, 0, 0)$.

b) Simplify the result from part a) when $v_o \gg v_e$ so that: $\mathbf{u} = (u_r, u_\theta, u_\varphi) = (v_o, 0, 0)$.

c) Simplify the result from part a) when $v_o = v_e$.

d) Use words, sketches, or equations to describe what happens when $v_o < v_e$. State any assumptions that you make.

4.7 Consider the three-dimensional flow field $u_i = \beta x_i$ or equivalently $\mathbf{u} = \beta r\mathbf{e}_r$, where β is a constant with units of inverse time, x_i is the position vector from the origin, r is the distance from the origin, and $\mathbf{e}_r$ is the radial unit vector. Find a density field ρ that conserves mass when:

a) $\rho(t)$ depends only on time t and $\rho = \rho_o$ at $t = 0$, and

b) $\rho(r)$ depends only on the distance r and $\rho = \rho_1$ at $r = 1$ m.

c) Does the sum $\rho(t) + \rho(r)$ also conserve mass in this flow field? Explain your answer.

4.8 The definition of the stream function for two-dimensional, constant-density flow in the x-y plane is: $\mathbf{u} = -\mathbf{e}_z \times \nabla\psi$, where $\mathbf{e}_z$ is the unit vector perpendicular to the x-y plane that determines a right-handed coordinate system.

a) Verify that this vector definition is equivalent to $u = \partial\psi/\partial y$ and $v = -\partial\psi/\partial x$ in Cartesian coordinates.

b) Determine the velocity components in r-θ polar coordinates in terms of r-θ derivatives of ψ.

c) Determine an equation for the z-component of the vorticity in terms of ψ.

4.9 A curve of $\psi(x, y) = C_1$ (= a constant) specifies a streamline in steady two-dimensional, constant-density flow. If a neighboring streamline is specified by $\psi(x, y) = C_2$, show that the volume flux per unit depth into the page between the streamlines equals $C_2 - C_1$ when $C_2 > C_1$.

4.10 Consider steady two-dimensional incompressible flow in r-θ polar coordinates where $\mathbf{u} = (u_r, u_\theta)$, $u_r = +(\Lambda/r^2)\cos\theta$, and Λ is positive constant. Ignore gravity.

a) Determine the simplest possible u_θ.

b) Show that the simplest stream function for this flow is $\psi = (\Lambda/r)\sin\theta$.
c) Sketch the streamline pattern. Include arrowheads to show stream direction(s).
d) If the flow is frictionless and the pressure far from the origin is p_∞, evaluate the pressure $p(r, \theta)$ on $\theta = 0$ for $r > 0$ when the fluid density is ρ. Does the pressure increase or decrease as r increases?

4.11 The well-known undergraduate fluid mechanics textbook by Fox et al. (2009) provides the following statement of conservation of momentum for a constant-shape (nonrotating) control volume moving at a nonconstant velocity $\mathbf{U} = \mathbf{U}(t)$:

$$\frac{d}{dt}\int_{V^*(t)} \rho \mathbf{u}_{rel} dV + \int_{A^*(t)} \rho \mathbf{u}_{rel}(\mathbf{u}_{rel}\cdot\mathbf{n})dA = \int_{V^*(t)} \rho \mathbf{g} dV + \int_{A^*(t)} \mathbf{f} dA - \int_{V^*(t)} \rho \frac{d\mathbf{U}}{dt} dV.$$

Here $\mathbf{u}_{rel} = \mathbf{u} - \mathbf{U}(t)$ is the fluid velocity observed in a frame of reference moving with the control volume while $\mathbf{u}$ and $\mathbf{U}$ are observed in a nonmoving frame. Meanwhile, (4.17) states this law as

$$\frac{d}{dt}\int_{V^*(t)} \rho \mathbf{u} dV + \int_{A^*(t)} \rho \mathbf{u}(\mathbf{u} - \mathbf{U})\cdot\mathbf{n} dA = \int_{V^*(t)} \rho \mathbf{g} dV + \int_{A^*(t)} \mathbf{f} dA$$

where the replacement $\mathbf{b} = \mathbf{U}$ has been made for the velocity of the accelerating control surface $A^*(t)$. Given that the two equations above are not identical, determine if these two statements of conservation of fluid momentum are contradictory or consistent.

4.12 A jet of water with a diameter of 8 cm and a speed of 25 m/s impinges normally on a large stationary flat plate. Find the force required to hold the plate stationary. Compare the average pressure on the plate with the stagnation pressure if the plate is 20 times the area of the jet.

4.13 Show that the thrust developed by a stationary rocket motor is $F = \rho A U^2 + A(p - p_{atm})$, where p_{atm} is the atmospheric pressure, and p, ρ, A, and U are, respectively, the pressure, density, area, and velocity of the fluid at the nozzle exit.

4.14 Consider the propeller of an airplane moving with a velocity U_1. Take a reference frame in which the air is moving and the propeller [disk] is stationary. Then the effect of the propeller is to accelerate the fluid from the upstream value U_1 to the downstream value $U_2 > U_1$. Assuming incompressibility, show that the thrust developed by the propeller is given by $F = \rho A(U_2^2 - U_1^2)/2$, where A is the projected area of the propeller and ρ is the density (assumed constant). Show also that the velocity of the fluid at the plane of the propeller is the average value $U = (U_1 + U_2)/2$. [*Hint*: The flow can be idealized by a pressure jump of magnitude $\Delta p = F/A$ right at the location of the propeller. Also apply Bernoulli's equation between a section far upstream and a section immediately upstream of the propeller. Also apply the Bernoulli equation between a section immediately downstream of the propeller and a section far downstream. This will show that $\Delta p = \rho(U_2^2 - U_1^2)/2$.]

4.15 Consider the control volume depicted in Fig. 4.2 from Example 4.4. As described in the example, a long bar with constant cross section is held perpendicular to a uniform horizontal flow of speed U_∞. The fluid has constant density ρ and viscosity μ. The bar's cross section has characteristic transverse dimension d, and the span of the bar is l with $l \gg d$. In this exercise, the average horizontal velocity profile measured downstream of the bar, $U(y)$, is given by

$$U(y) = U_\infty\left(1 - \frac{1}{2}\left[\tanh\left(\frac{y + W/2}{\delta}\right) - \tanh\left(\frac{y - W/2}{\delta}\right)\right]\right),$$

where δ is a wake-profile smoothness parameter and W is the width of the wake flow's velocity deficit, given by $W = (0.6 - \delta)/0.6$. Determine the required force per unit span, $-F_D/l$, applied to the ends of the bar to hold it in place for $\delta \in [0, 0.6]$. Plot $U(y)/U_\infty$ for several values of δ spanning this range. Solve for the force numerically and plot it as a function of δ. Assume the flow is steady, two dimensional, and ignore body forces. Comment on the resulting force in the limit $\delta = 0$ (i.e., the velocity deficit resembles a 'top-hat' profile) and when $W = 0$ (no wake). Provide physical reasoning for these results.

4.16 Generalize the control volume analysis of Example 4.4 by considering the control volume geometry shown for steady two-dimensional flow past an arbitrary body in the absence of body forces. Show that the force the fluid exerts on the body is: $F_j = -\int_{A_1}(\rho u_i u_j - T_{ij})n_i dA$ and that $0 = \int_{A_1} \rho u_i n_i dA$.

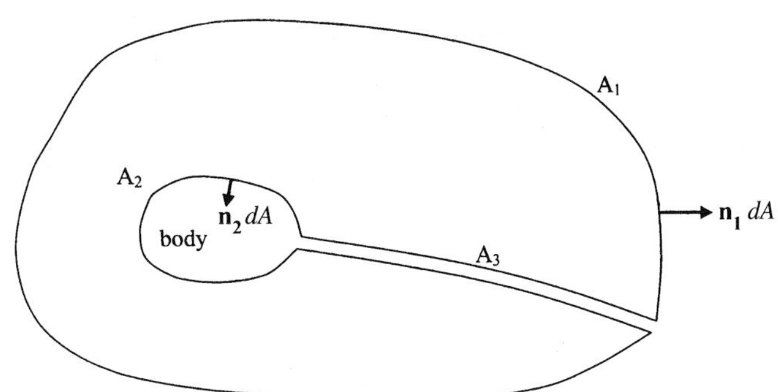

4.17 The pressure rise $\Delta p = p_2 - p_1$ that occurs for flow through a sudden pipe-cross-sectional-area expansion can depend on the average upstream flow speed U_{ave}, the upstream pipe diameter d_1, the downstream pipe diameter d_2, and the fluid density ρ and viscosity μ. Here p_2 is the pressure downstream of the expansion where the flow is first fully adjusted to the larger pipe diameter.

a) Find a dimensionless scaling law for Δp in terms of U_{ave}, d_1, d_2, ρ, and μ.

b) Simplify the result of part a) for high-Reynolds-number turbulent flow where μ does not matter.

c) Use a control volume analysis to determine Δp in terms of U_{ave}, d_1, d_2, and ρ for the high Reynolds number limit. [*Hints*: 1) a streamline drawing might help in determining or estimating the pressure on the vertical surfaces of the area transition, and 2) assume uniform flow profiles wherever possible.]

d) Compute the ideal flow value for Δp and compare this to the result from part c) for a diameter ratio of $d_1/d_2 = 1/2$. What fraction of the maximum possible pressure rise does the sudden expansion achieve?

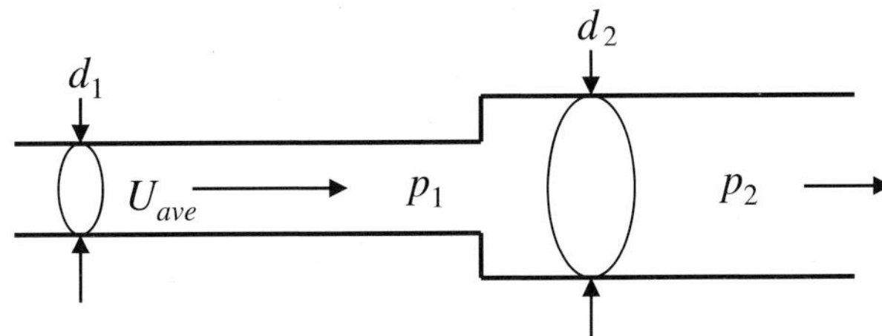

4.18 Consider how pressure gradients and skin friction develop in an empty wind tunnel or water tunnel test section when the flow is incompressible. Here the fluid has viscosity μ and density ρ, and flows into a horizontal cylindrical pipe of length L with radius R at a uniform horizontal velocity U_o. The inlet of the pipe lies at $x = 0$. Boundary layer growth on the pipe's walls induces the horizontal velocity on the pipe's centerline to be U_L at $x = L$; however, the pipe-wall boundary layer thickness remains much smaller than R. Here, L/R is of order 10, and $\rho U_o R/\mu \gg 1$. The radial coordinate from the pipe centerline is r.

a) Determine the displacement thickness, δ_L^*, of the boundary layer at $x = L$ in terms of U_o, U_L, and R. Assume that the boundary layer displacement thickness is zero at $x = 0$. [The boundary layer displacement thickness, δ^*, is the thickness of the zero-flow-speed layer that displaces the outer flow by the same amount as the actual boundary layer. For a boundary layer velocity profile $u(y)$ with $y =$ wall-normal coordinate and $U =$ outer flow velocity, δ^* is defined by: $\delta^* = \int_0^\infty (1 - (u/U))dy$.]

b) Determine the pressure difference, $\Delta P = P_L - P_o$, between the ends of the pipe in terms of ρ, U_o, and U_L.

c) Assume the horizontal velocity profile at the outlet of the pipe can be approximated by: $u(r, x = L) = U_L(1 - (r/R)^n)$ and estimate average skin friction, $\overline{\tau}_w$, on the inside of the pipe between $x = 0$ and $x = L$ in terms of ρ, U_o, U_L, R, L, and n.

d) Calculate the skin friction coefficient, $c_f = \overline{\tau}_w/\frac{1}{2}\rho U_o^2$, when $U_o = 20.0$ m/s, $U_L = 20.5$ m/s, $R = 1.5$ m, $L = 12$ m, $n = 80$, and the fluid is water, i.e., $\rho = 10^3$ kg/m^3.

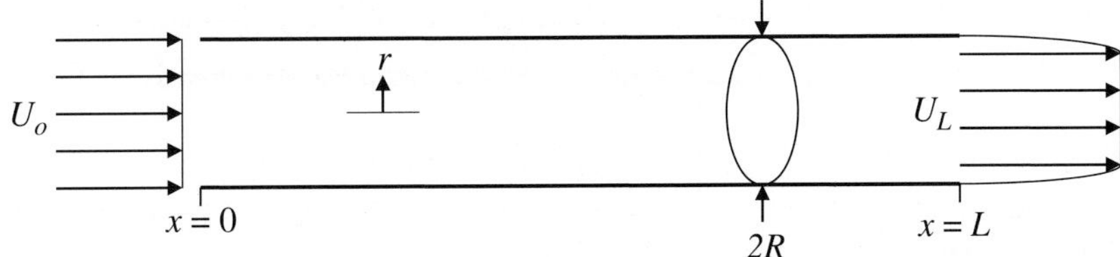

4.19 An acid solution with density ρ flows horizontally into a mixing chamber at speed V_1 at $x = 0$ where it meets a buffer solution with the same density moving at speed V_2. The inlet flow layer thicknesses are h_1 and h_2 as shown, the mixer chamber height is constant at $h_1 + h_2$, and the chamber width into the page is b. Assume steady uniform flow across the two inlets and the outlet. Ignore fluid friction on the interior surfaces of the mixing chamber for parts a) and b).

a) By conserving mass and momentum in a suitable control volume, determine the pressure difference, $\Delta p = p(L) - p(0)$, between the outlet ($x = L$) and inlet ($x = 0$) of the mixing chamber in terms of V_1, V_2, h_1, h_2, and ρ. Do not use the Bernoulli equation.

b) Is the pressure at the outlet higher or lower than that at the inlet when $V_1 \neq V_2$?

c) Explain how your answer to a) would be modified by friction on the interior surfaces of the mixing chamber.

4.20 Consider the situation depicted below. Wind strikes the side of a simple residential structure and is deflected up over the top of the structure. Assume the following: two-dimensional steady inviscid constant-density flow, uniform upstream velocity profile, linear gradient in the downstream velocity profile (velocity U at the upper boundary and zero velocity at the lower boundary as shown), no flow through the upper boundary of the control volume, and constant pressure on the upper boundary of the control volume. Using the control volume shown:

a) determine h_2 in terms of U and h_1, and

b) determine the *direction* and *magnitude* of the horizontal force on the house per unit depth into the page in terms of the fluid density ρ, the upstream velocity U, and the height of the house h_1.

c) Evaluate the magnitude of the force for a house that is 10 m tall and 20 m long in wind of 22 m/sec (approximately 80 km per hour).

4.21 For room ventilation, air with constant density ρ flows horizontally at uniform speed U_o out of a register of height h at the lower junction of a vertical interior wall (at $x = 0$) and the floor (at $y = 0$) of a room as shown. The flow interacts with the room air above and with the impenetrable floor surface below so that the horizontal velocity at $x = L$ of the ventilation flow is $u(L, y) = U_L f(y/H)$, where $f(\eta)$ is a smooth function with the following properties: $f(0) = 1$, $f(1) = 0$, $\int_0^1 f(\eta)d\eta = 3/5$, and $\int_0^1 f^2(\eta)d\eta = 2/5$. For the items below, utilize the following 5 assumptions: (*i*) two-dimensional flow with dimension B into the page, (*ii*) the airflow above $y = H$ is perfectly downward with uniform speed V, (*iii*) there is no friction between the moving air and the floor, (*iv*) the average pressure on the floor for $0 \leq x \leq L$ is P_f, and (*v*) atmospheric pressure P_a acts on the vertical wall and elsewhere within the room. By conserving mass, horizontal momentum, and vertical momentum, determine a formula for $P_f - P_a$ in terms of ρ, U_o, L, h, and H.

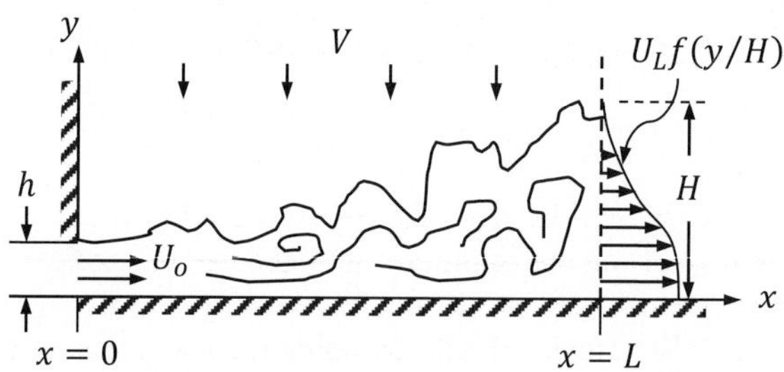

4.22 A large wind turbine with diameter D extracts a fraction η of the kinetic energy from the airstream (density $= \rho =$ constant) that impinges on it with velocity U.

a) What is the diameter of the wake zone, E, downstream of the wind turbine?
b) Determine the magnitude and direction of the force on the wind turbine in terms of ρ, U, D, and η.
c) Does your answer approach reasonable limits as $\eta \to 0$ and $\eta \to 1$?

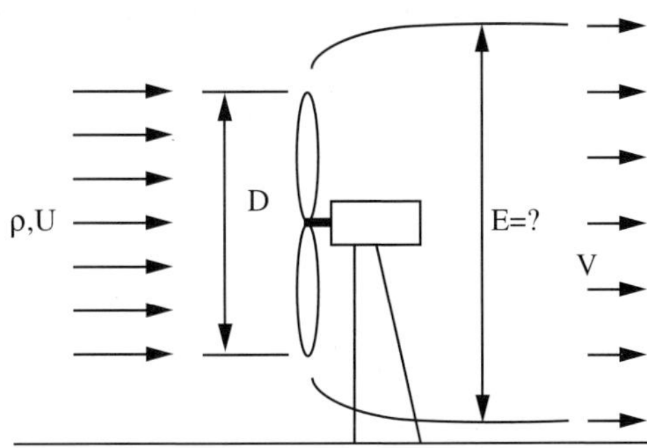

4.23 Consider the implications of the integral law for conservation of energy for steady viscous constant-density flow through a variable-cross-sectional-area pipe without input or output of work or heat.
a) Start with (4.48), choose a stationary control volume that encloses the pipe, and show:

$$\left(\frac{U^2}{2} + gz + \frac{p}{\rho}\right)_{in} - \left(\frac{U^2}{2} + gz + \frac{p}{\rho}\right)_{out} = loss,$$

where U is the local cross-sectional-average downstream flow speed in the pipe, g is the acceleration of gravity, z is the vertical coordinate, p is the pressure, ρ is the fluid density, $loss = e_{in} - e_{out}$, e is the fluid's internal (thermal) energy per unit mass, and the in and out subscripts denote pipe inlet and outlet conditions. For this derivation, assume that: (i) the only forces on the fluid at the pipe's inlet and outlet arise from pressure and gravity, (ii) the flow's properties are uniform across the pipe's cross section, and (iii) the fluid velocity, $\mathbf{u} = U\mathbf{e}_d$, at any location is adequately characterized by its cross-sectional average, U, and the pipe-centerline downstream-pointing unit vector, $\mathbf{e}_d$.
b) Water at 20 °C enters a constant cross-section horizontal pipe at a gage pressure of 100 kPa. If the water flows without heat transfer and emerges from the pipe at atmospheric pressure, what is its temperature? Is this inlet-to-outlet thermal change likely to be significant?

4.24 Pumped-storage hydroelectricity is a type of hydroelectric energy storage used by electric power systems for load balancing. The idea is to pump water from a lower elevation reservoir to a higher elevation reservoir when electricity demand is low and inexpensive, such as at night. During periods of high electrical demand, such as during the day, the stored water is released through turbines to produce power.

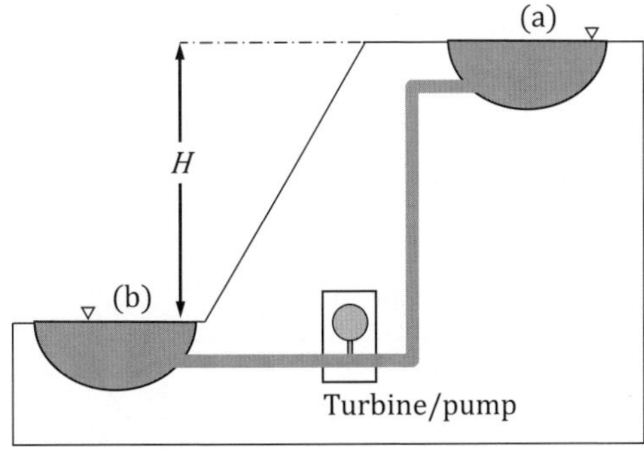

a) If water of density ρ flows by gravity from the higher elevation reservoir (a) to the lower one (b) at a steady flow rate Q over a height H, using the solution from Exercise 4.23 what is the energy loss per unit mass of fluid associated with this flow?
b) If this same amount of loss is associated with pumping the water from the lower reservoir to the higher one at the same flow rate, determine the power required to run the pump.

4.25 An incompressible fluid of density ρ flows through a horizontal rectangular duct of height h and width b. A uniform flat plate of length L and width b attached to the top of the duct at point A is deflected to an angle θ as shown.

a) Estimate the pressure difference between the upper and lower sides of the plate in terms of x, ρ, U_o, h, L, and θ when the flow separates cleanly from the tip of the plate.

b) If the plate has mass M and can rotate freely about the hinge at A, determine a formula for the angle θ in terms of the other parameters. You may leave your answer in terms of an integral.

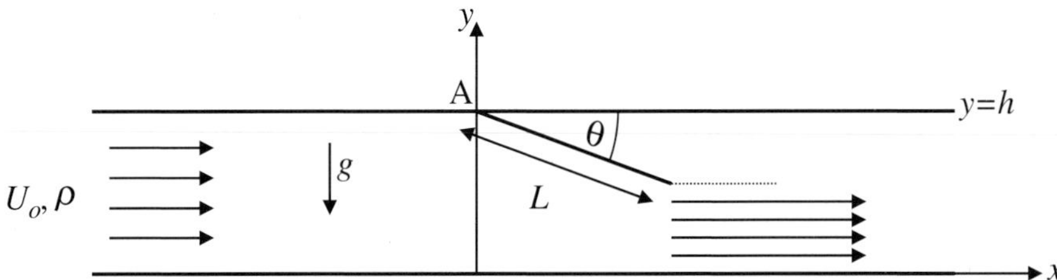

4.26 A pipe of length L and cross sectional area A is to be used as a fluid-distribution manifold that expels a steady uniform volume flux per unit length of an incompressible liquid from $x = 0$ to $x = L$. The liquid has density ρ, and is to be expelled from the pipe through a slot of varying width, $w(x)$. The goal of this problem is to determine $w(x)$ in terms of the other parameters of the problem. The pipe-inlet pressure and liquid velocity at $x = 0$ are P_o and U_o, respectively, and the pressure outside the pipe is P_e. If $P(x)$ denotes the pressure on the inside of the pipe, then the liquid velocity through the slot U_e is determined from: $P(x) - P_e = \frac{1}{2}\rho U_e^2$. For this problem assume that the expelled liquid exits the pipe perpendicular to the pipe's axis, and note that $wU_e = \text{const.} = U_oA/L$, even though w and U_e both depend on x.

a) Formulate a dimensionless scaling law for w in terms of x, L, A, ρ, U_o, P_o, and P_e.

b) Ignore the effects of viscosity, assume all profiles through the cross section of the pipe are uniform, and use a suitable differential-control-volume analysis to show that:

$$A\frac{dU}{dx} + wU_e = 0, \quad \text{and} \quad \rho\frac{d}{dx}U^2 = -\frac{dP}{dx}.$$

c) Use these equations and the relationships stated above to determine $w(x)$ in terms of x, L, A, ρ, U_o, P_o, and P_e. Is the slot wider at $x = 0$ or at $x = L$?

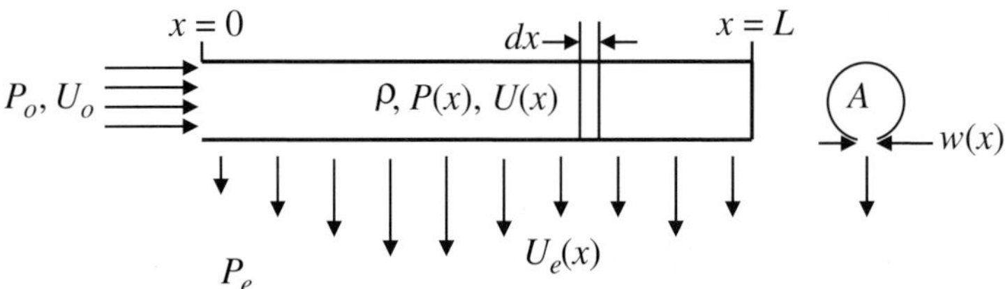

4.27 The take-off mass of a Boeing 747-400 may be as high as 400,000 kg. An Airbus A380 may be even heavier. Using a control volume (CV) that comfortably encloses the aircraft, explain why such large aircraft do not crush houses or people when they fly low overhead. Of course, the aircraft's wings generate lift but they are entirely contained within the CV and do not coincide with any of the CV's surfaces; thus merely stating the lift balances weight is not a satisfactory explanation. Given that the CV's vertical body-force term, $-g\int_{CV}\rho dV$, will exceed 4×10^6 N when the airplane and air in the CV's interior are included, your answer should instead specify which of the CV's surface forces or surface fluxes carries the signature of a large aircraft's impressive weight.

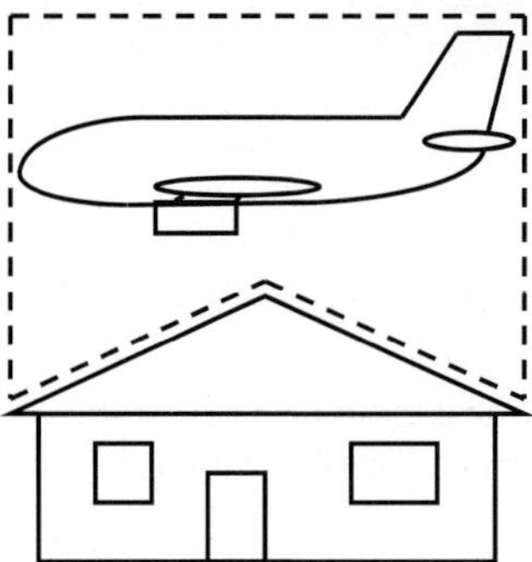

4.28 [1]An inviscid incompressible liquid with density ρ flows in a wide conduit of height H and width B into the page. The inlet stream travels at a uniform speed U and fills the conduit. The depth of the outlet stream is less than H. Air with negligible density fills the gap above the outlet stream. Gravity acts downward with acceleration g. Assume the flow is steady for the following items.

a) Find a dimensionless scaling law for U in terms of ρ, H, and g.

b) Denote the outlet stream depth and speed by h and u, respectively, and write down a set of equations that will allow U, u, and h to be determined in terms of ρ, H, and g.

c) Solve for U, u, and h in terms of ρ, H, and g. [*Hint*: solve for h first.]

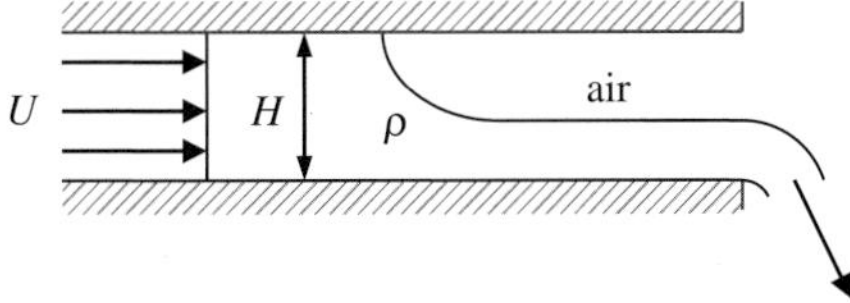

4.29 A hydraulic jump is the shallow-water-wave equivalent of a gas-dynamic shock wave. A steady radial hydraulic jump can be observed safely in one's kitchen, bathroom, or backyard where a falling stream of water impacts a shallow pool of water on a flat surface. The radial location R of the jump will depend on gravity g, the depth of the water behind the jump H, the volume flow rate of the falling stream Q, and stream's speed, U, where it impacts the plate. In your work, assume $\sqrt{2gh} \ll U$ where r is the radial coordinate from the point where the falling stream impacts the surface.

a) Formulate a dimensionless law for R in terms of the other parameters.

b) Use the Bernoulli equation and a control volume with narrow angular and negligible radial extents that contains the hydraulic jump to show that:

$$R \cong \frac{Q}{2\pi U H^2}\left(\frac{2U^2}{g} - H\right).$$

c) Rewrite the results of part b) in terms of the dimensionless parameters found for part a).

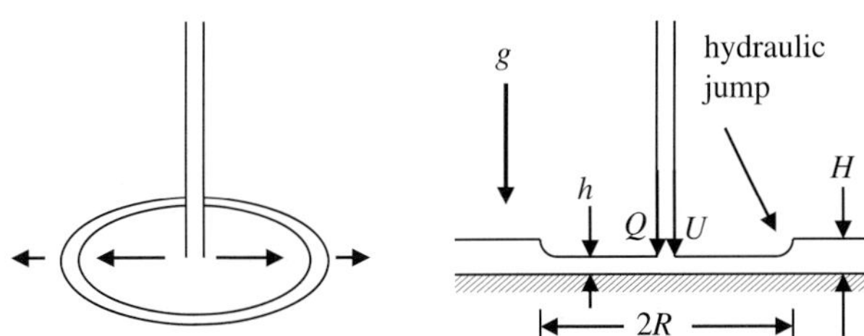

4.30 A jet of water with thickness a and density ρ traveling at uniform speed U and angle θ with respect to the horizontal strikes a right-angle corner with horizontal and vertical sides (see figure). After splashing against the solid surfaces, the water flows to the right in a layer of depth h at $x = L$. The flow is two-dimensional with dimension B into the page. Gravity g acts downward. Use the (x, y)-coordinate system shown.

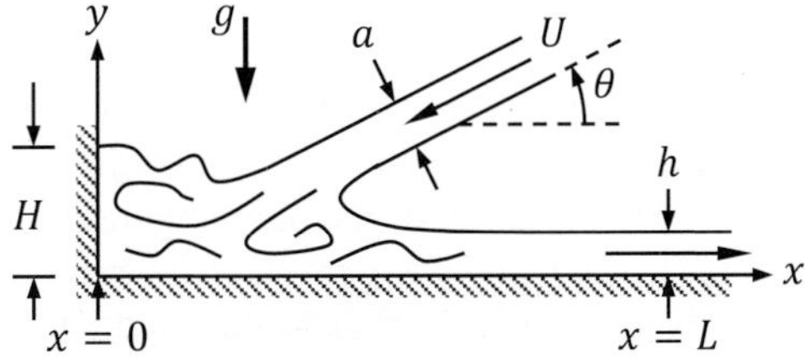

a) Use dimensional analysis to formulate a functional relationship for H = the water depth at $x = 0$ in terms of the other parameters in the problem.

b) Neglect friction forces on the surfaces (but do not use the Bernoulli equation), assume a uniform outflow, and use a control volume to find a formula for H in terms of U, g, a, h, and θ.

1. Based on a lecture example of Professor P.E. Dimotakis.

c) Revise the solution to part b) when shear stress $\tau_w(x)$ acts at $y = 0$ from $x = 0$ to $x = L$. This revised solution should include an integral.

d) Considering $0° \leq \theta \leq 90°$, what values of the angle θ lead to the smallest and largest values of H?

4.31 A fine uniform mist of an inviscid incompressible fluid with density ρ settles steadily at a total volume flow rate (per unit depth into the page) of $2q$ onto a flat horizontal surface of width $2s$ to form a liquid layer of thickness $h(x)$ as shown. The geometry is two dimensional.

a) Formulate a dimensionless scaling law for h in terms of x, s, q, ρ, and g.

b) Use a suitable control volume analysis, assuming $u(x)$ does not depend on y, to find a single cubic equation for $h(x)$ in terms of $h(0)$, s, q, x, and g.

c) Determine $h(0)$.

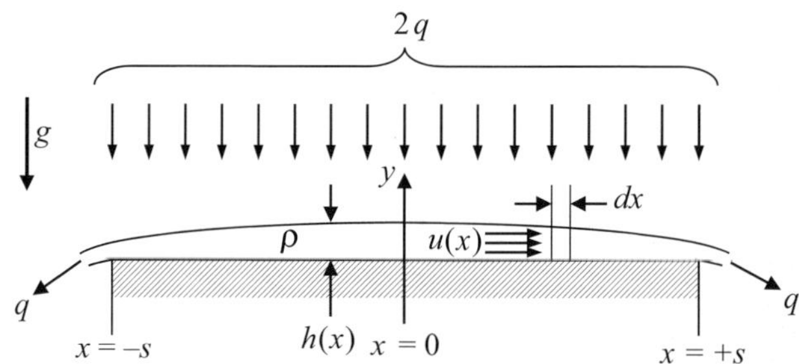

4.32 A thin-walled pipe of mass m_O, length L, and cross-sectional area A is free to swing in the x-y plane from a frictionless pivot at point O. Water with density ρ fills the pipe, flows into it at O perpendicular to the x-y plane, and is expelled at a right angle from the pipe's end as shown. The pipe's opening also has area A and gravity g acts downward. For a steady mass flow rate of $\dot{m}$, the pipe hangs at angle θ with respect to the vertical as shown. Ignore fluid viscosity.

a) Develop a dimensionless scaling law for θ in terms of m_O, L, A, ρ, $\dot{m}$, and g.

b) Use a control volume analysis to determine the force components, F_x and F_y, applied to the pipe at the pivot point in terms of θ, m_O, L, A, ρ, $\dot{m}$, and g.

c) Determine θ in terms of m_O, L, A, ρ, $\dot{m}$, and g.

d) Above what value of $\dot{m}$ will the pipe rotate without stopping?

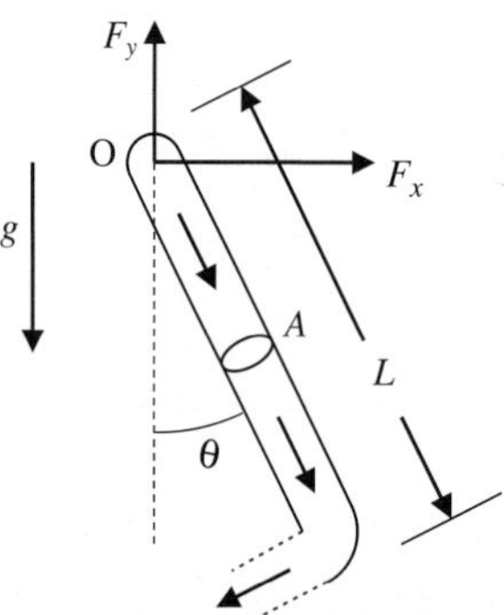

4.33 Construct a house of cards, or light a candle. Get the cardboard tube from the center of a roll of paper towels and back away from the cards or candle a meter or two so that by blowing you cannot knock down the cards or blow out the candle unaided. Now use the tube in two slightly different configurations. First, place the tube snugly against your face encircling your mouth, and try to blow down the house of cards or blow out the candle. Repeat the experiment while moving closer until the cards are knocked down or the candle is blown out (you may need to get closer to your target than might be expected; do not hyperventilate; do not start the cardboard tube on fire). Note the distance between the end of the tube and the card house or candle at this point. Rebuild the card house or relight the candle and repeat the experiment, except this time hold the tube a few centimeters away from your face and mouth, and blow through it. Again, determine the greatest distance from which you can knock down the cards or blow out the candle.

a) Which configuration is more effective at knocking the cards down or blowing the candle out?

b) Explain your findings with a suitable control-volume analysis.

c) List some practical applications where this effect might be useful.

4.34 [2]Attach a drinking straw to a 15-cm-diameter cardboard disk with a hole at the center using tape or glue. Loosely fold the corners of a standard piece of paper upward so that the paper mildly cups the cardboard disk (see drawing). Place the cardboard disk in the central section of the folded paper. Attempt to lift the loosely folded paper off a flat surface by blowing or sucking air through the straw.

a) Experimentally determine if blowing or suction is more effective in lifting the folded paper.

b) Explain your findings with a control volume analysis.

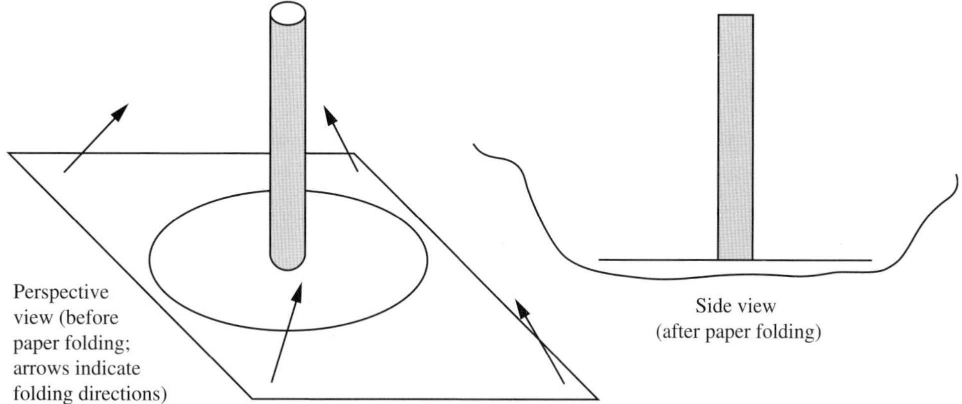

4.35 A compression wave in a long gas-filled constant-area duct propagates to the left at speed U. To the left of the wave, the gas is quiescent with uniform density ρ_1 and uniform pressure p_1. To the right of the wave, the gas has uniform density ρ_2 ($> \rho_1$) and uniform pressure is p_2 ($> p_1$). Ignore the effects of viscosity in this problem.

a) Formulate a dimensionless scaling law for U in terms of the pressures and densities.

b) Determine U in terms of ρ_1, ρ_2, p_1, and p_2 using a control volume.

c) Put your answer to part b) in dimensionless form and thereby determine the unknown function from part a).

d) When the density and pressure changes are small, they are proportional:

$$p_2 - p_1 = c^2(\rho_2 - \rho_1) \quad \text{for} \quad (\rho_2 - \rho_1)/\rho_2 \ll 1,$$

where $c^2 = (\partial p/\partial \rho)_s$. Under these conditions, U is associated with what common property of the gas?

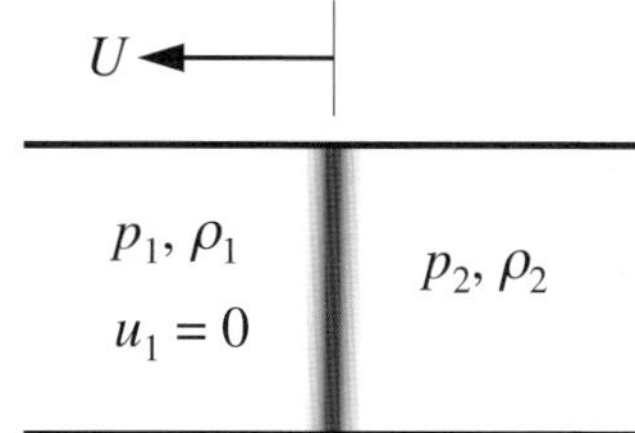

4.36 A rectangular tank is placed on wheels and is given a constant horizontal acceleration a. Show that, at steady state, the angle made by the free surface with the horizontal is given by $\tan\theta = a/g$.

4.37 Starting from rest at $t = 0$, an airliner of mass M accelerates at a constant rate $\mathbf{a} = a\mathbf{e}_x$ into a headwind, $\mathbf{u} = -u_i\mathbf{e}_x$. For the following items, assume that: 1) the x-component of the fluid velocity is $-u_i$ on the front, sides, and back upper half of the control volume (CV), 2) the x-component of the fluid velocity is $-u_o$ on the back lower half of the CV, 3) changes in M can be neglected, 4) changes of air momentum inside the CV can be neglected, and 5) frictionless wheels. In addition, assume constant air density ρ and uniform flow conditions exist on the various control surfaces. In your work, denote the CV's front and back area by A. (This approximate model is appropriate for real commercial airliners that have the engines hung under the wings).

a) Find a dimensionless scaling law for u_o at $t = 0$ in terms of u_i, ρ, a, M, and A.

b) Using a CV that encloses the airliner (as shown) determine a formula for $u_o(t)$, the time-dependent air velocity on the lower half of the CV's back surface.

2. Based on a demonstration done by Prof. G. Tryggvason.

c) Evaluate u_o at $t = 0$, when $M = 4 \times 10^5$ kg, $a = 2.0$ m/s^2, $u_i = 5$ m/s, $\rho = 1.2$ kg/m^3, and $A = 1200$ m^2. Would you be able to walk comfortably behind the airliner?

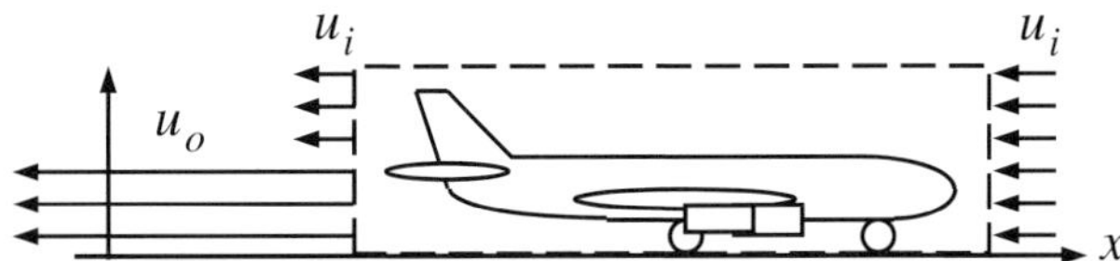

4.38 [3]A cart that can roll freely in the x-direction deflects a horizontal water jet into its tank with a vane inclined to the vertical at an angle θ. The jet issues steadily at velocity U with density ρ, and has cross-sectional area A. The cart is initially at rest with a mass of m_o. Ignore the effects of surface tension, the cart's rolling friction, and wind resistance in your answers.

a) Formulate dimensionless law for the mass, $m(t)$, in the cart at time t in terms of t, θ, U, ρ, A, and m_o.
b) Formulate a differential equation for $m(t)$.
c) Find a solution for $m(t)$ and put it in dimensionless form.

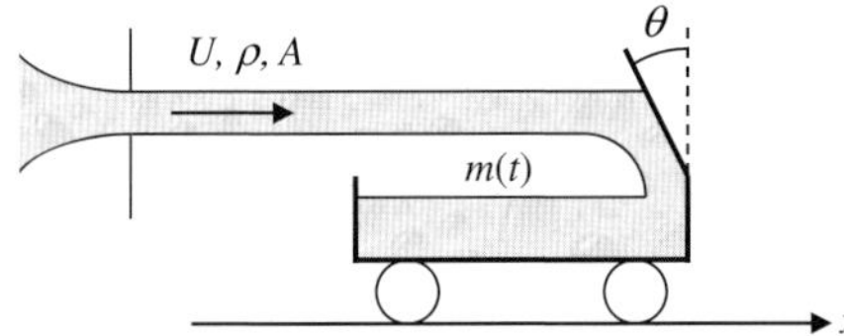

4.39 A spherical balloon of mass M_b is filled with air of density ρ and is initially stationary at $x = 0$ with diameter D_o. At $t = 0$, an opening of area A is created and the balloon travels horizontally along the x-axis. The aerodynamic drag force on the balloon is given by $(1/2)\rho U^2 \pi (D/2)^2 C_D$, where: C_D is a constant, $U(t)$ is the velocity of the balloon, and $D(t)$ is the current diameter of the balloon. Assume incompressible air flow and that $D(t)$ is known.

a) Find a differential equation for U that includes: M_b, ρ, C_D, D, and A.
b) Solve the part a) equation when $C_D = 0$, and the mass flow rate of air out of the balloon, $\dot{m}$, is constant, so that the mass of the balloon and its contents are $M_b + \rho(\pi/6)D_o^3 - \dot{m}t$ at time t.
c) What is the maximum value of U under the conditions of part b)?
d) Relax the assumption that C_D is constant. From Fig. 4.25, it can be seen that drag over a sphere varies with Reynolds numbers. A common formula for the drag coefficient used for spheres valid for Re < 800 is given by

$$C_D = \frac{24}{\mathrm{Re}}\left(1 + 0.15\mathrm{Re}^{0.687}\right).$$

Solve the equation from part a) numerically using the variable drag coefficient given above. Plot C_D as a function of time and compare U when C_D is constant.

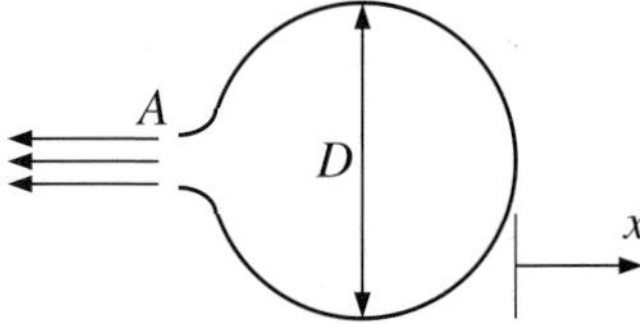

4.40 For time $t < 0$, a rolling water tank with frictionless wheels, horizontal cross-sectional area A, and empty mass M sits stationary while filled to a depth h_o with water of density ρ. At $t = 0$, the outlet of the tank is opened and the tank starts moving to the right. The outlet tube has cross sectional area a and contains a narrow-passage honeycomb so that the flow speed through the tube is $U_e = gh/R$, where R is the specific flow resistivity of the honeycomb material, g is the acceleration of gravity, and $h(t)$ is the average water depth in the rolling tank for $t > 0$. Here, U_e is the leftward speed of the water with respect to the outlet tube; it is independent of the speed $b(t)$ of the rolling tank. Assume uniform flow at the pipe outlet and use an appropriate control volume analysis for the following items.

a) By conserving mass, develop a single equation for $h(t)$ in terms of a, A, g, R, and t.
b) Solve the part a) equation for $h(t)$.

3. Similar to problem 4.170 on page 157 in Fox et al. (2009).

c) By conserving horizontal momentum, develop a single equation for $b(t)$ in terms of a, A, M, h, ρ, g, and R.
d) Determine for $b(t)$ in terms of a, A, M, h_o, ρ, g, R, and t. [*Hint:* use $db/dt = (db/dh)(dh/dt)$]

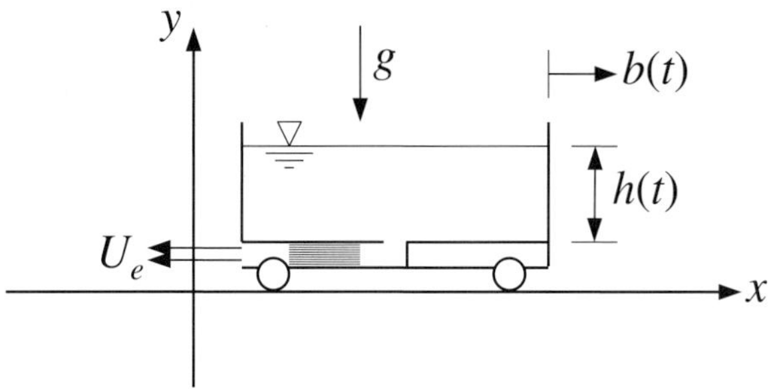

4.41 Prove that the stress tensor is symmetric by considering first-order changes in surface forces on a vanishingly small cube in rotational equilibrium. Work with rotation about the number 3 coordinate axis to show $T_{12} = T_{21}$. Cyclic permutation of the indices will suffice for showing the symmetry of the other two shear stresses.

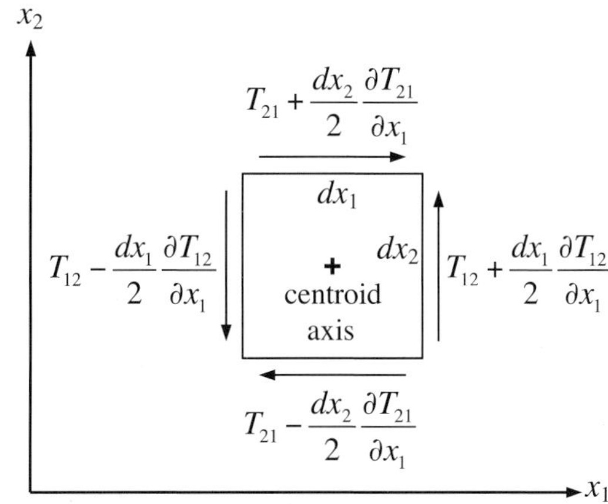

4.42 Obtain an empty plastic milk jug with a cap *that seals tightly*, and a frying pan. Fill both the pan and jug with water to a depth of approximately 1 cm. Place the jug in the pan with the cap off. Place the pan on a stove and turn up the heat until the water in the frying pan boils vigorously for a few minutes. Turn the stove off, and quickly put the cap tightly on the jug. *Avoid spilling or splashing hot water on yourself.* Remove the capped jug from the frying pan and let it cool to room temperature. Does anything interesting happen? If something does happen, explain your observations in terms of surface forces. What is the origin of these surface forces? Can you make any quantitative predictions about what happens?

4.43 In cylindrical coordinates (R, φ, z), two components of a steady incompressible viscous flow field are known: $u_\varphi = 0$, and $u_z = -Az$ where A is a constant, and body force is zero.
a) Determine u_R so that the flow field is smooth and conserves mass.
b) If the pressure, p, at the origin of coordinates is P_o, determine $p(R, \varphi, z)$ when the density is constant.

4.44 Consider solid-body rotation of an isothermal perfect gas (with constant R) at temperature T in (r, θ)-plane-polar coordinates: $u_r = 0$, and $u_\theta = \Omega_z r$, where Ω_z is a constant rotation rate and the body force is zero. What is the pressure distribution $p(r)$ if $p(0) = p_{atm}$? If the gas is air at 295 K and the container has a radius of $r_o = 10$ cm, what Ω_z is needed to produce $p(r_o) = 2p_{atm}$?

4.45 Solid body rotation with a constant angular velocity, $\mathbf{\Omega}$, is described by the following Cartesian velocity field: $\mathbf{u} = \mathbf{\Omega} \times \mathbf{x}$. For this velocity field:
a) Compute the components of:

$$T_{ij} = -p\delta_{ij} + \mu\left[\left(\frac{\partial u_i}{\partial x_j} + \frac{\partial u_j}{\partial x_i}\right) - \frac{2}{3}\delta_{ij}\frac{\partial u_k}{\partial x_k}\right] + \mu_v\delta_{ij}\frac{\partial u_k}{\partial x_k}.$$

b) Consider the case of $\Omega_1 = \Omega_2 = 0$, $\Omega_3 \neq 0$, $\rho =$ constant, with $p = p_o$ at $x_1 = x_2 = 0$. Use the differential momentum equation in Cartesian coordinates to determine $p(r)$, where $r^2 = x_1^2 + x_2^2$, when there is no body force and $\rho =$ constant. Does your answer make sense? Can you check it with a simple experiment?

4.46 Using only (4.7), (4.22), (4.36), and (3.12) show that $\rho(D\mathbf{u}/Dt) + \nabla p = \rho\mathbf{g} + \mu\nabla^2\mathbf{u} + (\mu_\upsilon + \frac{1}{3}\mu)\nabla(\nabla\cdot\mathbf{u})$ when the dynamic (μ) and bulk (μ_υ) viscosities are constants.

4.47 [4]Air, water, and petroleum products are important engineering fluids and can usually be treated as Newtonian fluids. Consider the following materials and try to classify them as: Newtonian fluid, non-Newtonian fluid, or solid. State the reasons for your choices and note the temperature range where you believe your answers are correct. Simple impact, tensile, and shear experiments in your kitchen or bathroom are recommended. Test and discuss at least five items.

a) toothpaste
b) peanut butter
c) shampoo
d) glass
e) honey
f) mozzarella cheese
g) hot oatmeal
h) creamy salad dressing
i) ice cream
j) silly putty

4.48 The equations for conservation of mass and momentum for a viscous Newtonian fluid are (4.7) and (4.39) when the viscosities are constant.

a) Simplify these equations and write them out in primitive form for steady constant-density flow in two dimensions where $u_i = (u_1(x_1, x_2), u_2(x_1, x_2), 0)$, $p = p(x_1, x_2)$, and $g_j = 0$.

b) Determine $p = p(x_1, x_2)$ when $u_1 = Cx_1$ and $u_2 = -Cx_2$, where C is a positive constant.

4.49 [5]Simplify the planar Navier-Stokes momentum equations (given in Example 4.10) for incompressible flow, constant viscosity, and conservative body forces. Cross differentiate these equations and eliminate the pressure to find a single equation for $\omega_z = \partial v/\partial x - \partial u/\partial y$. What process(es) might lead to the changes in ω_z for fluid elements in this flow?

4.50 Starting from (4.7) and (4.40), derive a Poisson equation for the pressure, p, by taking the divergence of the constant-density momentum equation. [In other words, find an equation where $\partial^2 p/\partial x_j^2$ appears by itself on the left side and other terms not involving p appear on the right side]. What role does the viscosity μ play in determining the pressure in constant density flow?

4.51 Prove the equality of the two ends of index notation version of (4.41) without leaving index notation or using vector identities.

4.52 The viscous compressible fluid conservation equations for mass and momentum are (4.7) and (4.38). Simplify these equations for constant-density, constant-viscosity flow and where the body force has a potential, $g_j = -\partial\Phi/\partial x_j$. Assume the velocity field can be found from $u_j = \partial\phi/\partial x_j$, where the scalar function ϕ depends on space and time. What are the simplified conservation of mass and momentum equations for ϕ?

4.53 Consider the near-surface flow over a stationary locally-flat surface defined by $x_2 = 0$ using flow-aligned Cartesian coordinates where the x_1-axis points downstream, the x_2-axis is normal to the surface, and the x_3-axis points in the cross-stream direction defined by $\hat{e}_3 = \hat{e}_1 \times \hat{e}_2$. The general forms for the continuity and Navier-Stokes momentum equations are (4.4) and (4.38).

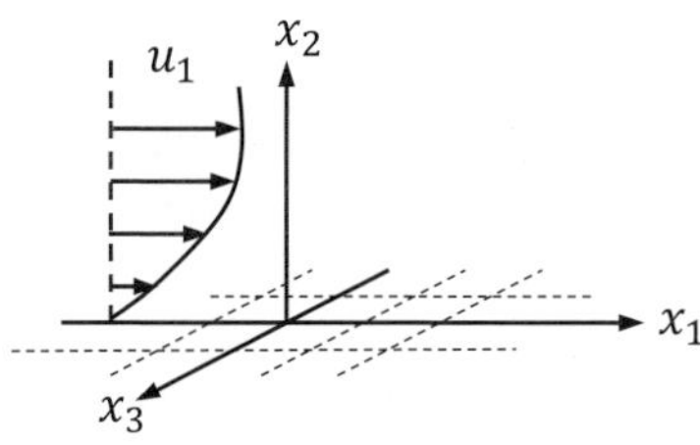

a) Simplify these equations for steady, two-dimensional, constant-density and constant-viscosity flow in the (x_1, x_2)-plane when there is no body force, and write them out term-by-term without using index notation in terms of x_1, x_2, u_1, u_2, ρ, p, and μ.

b) Near to the surface, the remaining (dependent) field variables can be written in terms of a two-variable near-surface expansion in powers of x_2:

4. Based on a suggestion from Professor W.W. Schultz.
5. Based on a homework problem posed by Professor C.E. Brennan.

$$u_1(x_1, x_2) = a_0(x_1) + a_1(x_1)x_2 + a_2(x_1)x_2^2 + \ldots,$$
$$u_2(x_1, x_2) = b_0(x_1) + b_1(x_1)x_2 + b_2(x_1)x_2^2 + \ldots,$$
$$\text{and } p(x_1, x_2) = p(x_1, 0) + (\partial p/\partial x_2)_{x_2=0}x_2 + \ldots.$$

Here, x_2 is presumed to be near zero, and the 'a's, 'b's, $p(x_1, 0)$, and $(\partial p/\partial x_2)_{x_2=0}$ are functions of x_1 alone. Using the definition of the shear stress, τ_w, on the solid surface:

$$\tau_w \equiv \mu(\partial u_1/\partial x_2)_{x_2=0},$$

the other boundary conditions on the solid surface, and the equations produced for part a), determine: a_0, a_1, a_2, b_0, b_1, and b_2 in terms of τ_w, μ, $\partial p(x_1, 0)/\partial x_1$, and $(\partial p/\partial x_2)_{x_2=0}$.

c) What relationship is required between τ_w and $(\partial p/\partial x_2)_{x_2=0}$? When τ_w is constant, does p tend to increase, decrease, or stay the same with increasing x_2?

4.54 The viscous compressible fluid conservation equations for mass and momentum are (4.7) and (4.38).

a) In Cartesian coordinates (x, y, z) with $\mathbf{g} = (g_x, 0, 0)$, simplify these equations for unsteady one-dimensional unidirectional flow where: $\rho = \rho(x, t)$ and $\mathbf{u} = (u(x, t), 0, 0)$.

b) If the flow is also incompressible, show that the fluid velocity depends only on time, i.e., $u(x, t) = U(t)$, and show that the equations found for part a) reduce to

$$\frac{\partial \rho}{\partial t} + u\frac{\partial \rho}{\partial x} = 0, \quad \text{and} \quad \rho\frac{\partial u}{\partial t} = -\frac{\partial p}{\partial x} + \rho g_x.$$

c) If $\rho = \rho_o(x)$ at $t = 0$, and $u = U(0) = U_o$ at $t = 0$, determine an implicit solution for $\rho = \rho(x, t)$ and $U(t)$ in terms of x, t, $\rho_o(x)$, U_o, $\partial p/\partial x$, and g_x.

4.55 The general forms for the continuity and Navier-Stokes momentum equations are (4.4) and (4.38). Simplify these equations for two-dimensional, constant-density, constant-viscosity flow and show that they can be written in terms of two dependent functions: the two-dimensional stream function $\psi(x, y, t)$, and the pressure, $p(x, y, t)$:

$$\rho\frac{\partial^2\psi}{\partial t\partial y} + \rho\frac{\partial\psi}{\partial y}\frac{\partial^2\psi}{\partial x\partial y} - \rho\frac{\partial\psi}{\partial x}\frac{\partial^2\psi}{\partial y^2} = -\frac{\partial p}{\partial x} + \mu\left(\frac{\partial^3\psi}{\partial x^2\partial y} + \frac{\partial^3\psi}{\partial y^3}\right), \text{ and}$$
$$-\rho\frac{\partial^2\psi}{\partial t\partial x} - \rho\frac{\partial\psi}{\partial y}\frac{\partial^2\psi}{\partial x^2} + \rho\frac{\partial\psi}{\partial x}\frac{\partial^2\psi}{\partial y\partial x} = -\frac{\partial p}{\partial y} + \mu\left(\frac{\partial^3\psi}{\partial x^3} + \frac{\partial^3\psi}{\partial y^2\partial x}\right)$$

Determine $p(x, y)$ when the flow is steady and the stream function is given by $\psi(x, y) = (x^2 + y^2)/\tau$, where τ is a constant with units of time.

4.56 a) [6]Derive the following equation for the velocity potential for irrotational inviscid compressible flow in the absence of a body force:

$$\frac{\partial^2\phi}{\partial t^2} + \frac{\partial}{\partial t}\left(|\nabla\phi|^2\right) + \frac{1}{2}\nabla\phi\cdot\nabla\left(|\nabla\phi|^2\right) - c^2\nabla^2\phi = 0$$

where $\nabla\phi = \mathbf{u}$, as usual. Start from the Euler equation (4.42), use the continuity equation, assume that the flow is isentropic so that p depends only on ρ, and denote $(\partial p/\partial\rho)_s = c^2$.

b) What limit does $c \to \infty$ imply?

c) What limit does $|\nabla\phi| \to 0$ imply?

4.57 Derive (4.44) from (4.43).

4.58 Observations of the velocity $\mathbf{u}'$ of an incompressible viscous fluid are made in a frame of reference rotating steadily at rate $\boldsymbol{\Omega} = (0, 0, \Omega_z)$. The pressure at the origin is p_o and $\mathbf{g} = -g\mathbf{e}_z$.

a) In Cartesian coordinates with $\mathbf{u}' = (U, V, W) =$ a constant, find $p(x, y, z)$.

b) In cylindrical coordinates with $\mathbf{u}' = -\Omega_z R\mathbf{e}_\varphi$, determine $p(R, \varphi, z)$. Guess the result if you can.

4.59 For many atmospheric flows, rotation of the earth is important. The momentum equation for inviscid flow in a frame of reference rotating at a constant rate $\boldsymbol{\Omega}$ is:

$$\partial\mathbf{u}/\partial t + (\mathbf{u}\cdot\nabla)\mathbf{u} = -\nabla\Phi - (1/\rho)\nabla p - 2\boldsymbol{\Omega}\times\mathbf{u} - \boldsymbol{\Omega}\times(\boldsymbol{\Omega}\times\mathbf{x})$$

6. Obtained from Professor Paul Dimotakis.

For steady two-dimensional horizontal flow, $\mathbf{u} = (u, v, 0)$, with $\Phi = gz$ and $\rho = \rho(z)$, show that the streamlines are parallel to constant pressure lines when the fluid particle acceleration is dominated by the Coriolis acceleration $|(\mathbf{u} \cdot \nabla)\mathbf{u}| \ll |2\boldsymbol{\Omega} \times \mathbf{u}|$, and when the local pressure gradient dominates the centripetal acceleration $|\boldsymbol{\Omega} \times (\boldsymbol{\Omega} \times \mathbf{x})| \ll |\nabla p|/\rho$. [This seemingly strange result governs just about all large-scale weather phenomena like hurricanes and other storms, and it allows weather forecasts to be made based on surface pressure measurements alone. *Hints*:

1. If $Y(x)$ defines a streamline contour, then $dY/dx = v/u$ is the streamline slope.
2. Write out all three components of the momentum equation and build the ratio v/u.
3. Using hint 1, the pressure increment along a streamline is: $dp = (\partial p/\partial x)dx + (\partial p/\partial y)dY$.]

4.60 Show that (4.56) can be derived from (4.7), (4.54), and (4.55).

4.61 Multiply (4.22) by u_j and sum over j to derive (4.57).

4.62 Starting from $\varepsilon = (1/\rho)\tau_{ij}S_{ij}$, derive the right most expression in (4.59).

4.63 For many gases and liquids (and solids too!), the following equations are valid: $\mathbf{q} = -k\nabla T$ (Fourier's law of heat conduction, k = thermal conductivity, T = temperature), $e = e_o + c_v T$ (e = internal energy per unit mass, c_v = specific heat at constant volume), and $h = h_o + c_p T$ (h = enthalpy per unit mass, c_p = specific heat at constant pressure), where e_o and h_o are constants, and c_v and c_p are also constants. Start with the energy equation

$$\rho\frac{\partial e}{\partial t} + \rho u_i\frac{\partial e}{\partial x_i} = -p\frac{\partial u_i}{\partial x_i} + \tau_{ij}S_{ij} - \frac{\partial q_i}{\partial x_i}$$

for each of the following items.

a) Derive an equation for T involving u_j, k, ρ, and c_v for incompressible flow when $\tau_{ij} = 0$.
b) Derive an equation for T involving u_j, k, ρ, and c_p for flow with p = const. and $\tau_{ij} = 0$.
c) Provide a physical explanation why the answers to a) and b) are different.

4.64 Derive the following alternative form of (4.61): $\rho c_p(DT/Dt) = \alpha T(Dp/Dt) + \rho\varepsilon + (\partial/\partial x_i)(k(\partial T/\partial x_i))$, where ε is given by (4.59) and α is the thermal expansion coefficient defined in (1.26). [*Hint*: $dh = (\partial h/\partial T)_p dT + (\partial h/\partial p)_T dp$]

4.65 Show that the first version of (4.69) is true without abandoning index notation or using vector identities.

4.66 Consider an incompressible planar Couette flow, which is the flow between two parallel plates separated by a distance b. The upper plate is moving parallel to itself at speed U, and the lower plate is stationary. Let the x-axis lie on the lower plate. The pressure and velocity fields are independent of x, and fluid has uniform density and viscosity.

a) Show that the pressure distribution is hydrostatic and that the solution of the Navier-Stokes equation is $u(y) = Uy/b$.
b) Write the expressions for the stress and strain rate tensors, and show that the viscous kinetic-energy dissipation per unit volume is $\mu U^2/b^2$.
c) Evaluate the kinetic energy equation (4.57) within a rectangular control volume for which the two horizontal surfaces coincide with the walls and the two vertical surfaces are perpendicular to the flow and show that the viscous dissipation and the work done in moving the upper surface are equal.

4.67 Determine the outlet speed, U_2, of a chimney in terms of ρ_o, ρ_2, g, H, A_1, and A_2. For simplicity, assume the fire merely decreases the density of the air from ρ_o to ρ_2 ($\rho_o > \rho_2$) and does not add any mass to the airflow. (This mass flow assumption isn't true, but it serves to keep the algebra under control in this problem.) The relevant parameters are shown in the figure. Use the steady Bernoulli equation into the inlet and from the outlet of the fire, but perform a control volume analysis across the fire. Ignore the vertical extent of A_1 compared to H and the effects of viscosity.

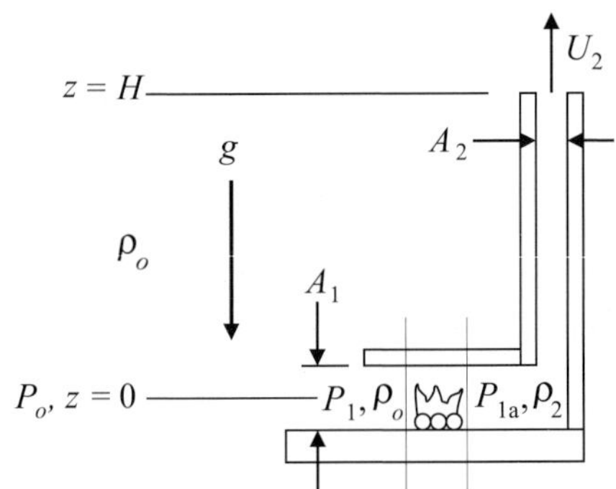

4.68 A hemispherical vessel of radius R has a small rounded orifice of area A at the bottom. Show that the time required to lower the level from h_1 to h_2 is given by

$$t = \frac{2\pi}{A\sqrt{2g}}\left[\frac{2}{3}R\left(h_1^{3/2} - h_2^{3/2}\right) - \frac{1}{5}\left(h_1^{5/2} - h_2^{5/2}\right)\right].$$

4.69 Water flows through a pipe in a gravitational field as shown in the accompanying figure. Neglect the effects of viscosity and surface tension. Solve the appropriate conservation equations for the variation of the cross-sectional area of the fluid column $A(z)$ after the water has left the pipe at $z = 0$. The velocity of the fluid at $z = 0$ is uniform at v_0 and the cross-sectional area is A_0.

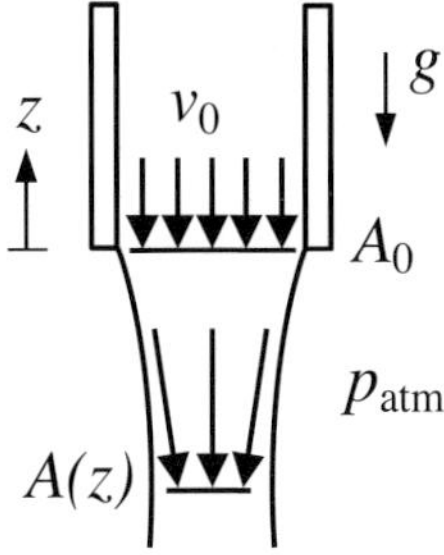

4.70 Redo the solution for the orifice-in-a-tank problem allowing for the fact that in Fig. 4.18, $h = h(t)$, but ignoring fluid acceleration. Estimate how long it takes for the tank take to empty.

4.71 Consider the planar flow of Example 3.6, $\mathbf{u} = (Ax, -Ay)$, but allow $A = A(t)$ to depend on time. Here the fluid density is ρ, the pressure at the origin or coordinates is p_o, and there are no body forces.

a) If the fluid is inviscid, determine the pressure on the x-axis, $p(x, 0, t)$ as a function of time from the unsteady Bernoulli equation.

b) If the fluid has constant viscosities μ and μ_v, determine the pressure throughout the flow field, $p(x, y, t)$, from the x-direction and y-direction differential momentum equations.

c) Are the results for parts a) and b) consistent with each other? Explain your findings.

4.72 A circular plate is forced down at a steady velocity U_o against a flat surface. Frictionless incompressible fluid of density ρ fills the gap $h(t)$. Assume that $h \ll r_o$ = the plate radius, and that the radial velocity $u_r(r, t)$ is constant across the gap.

a) Obtain a formula for $u_r(r, t)$ in terms of r, U_o, and h.

b) Determine $\partial u_r(r, t)/\partial t$.

c) Calculate the pressure distribution under the plate assuming that $p(r = r_o) = 0$.

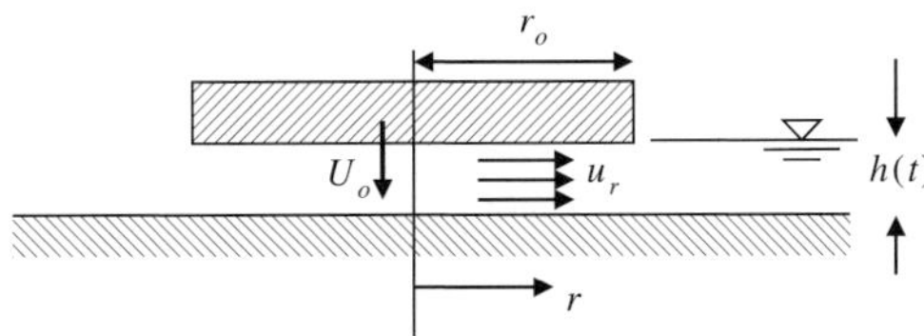

4.73 A frictionless, incompressible fluid with density ρ resides in a horizontal nozzle of length L having a cross-sectional area that varies smoothly between A_i and A_o via: $A(x) = A_i + (A_o - A_i)f(x/L)$, where f is a function that goes from 0 to 1 as x/L goes from 0 to 1. Here the x-axis lies on the nozzle's centerline, and $x = 0$ and $x = L$ are the horizontal locations of the nozzle's inlet and outlet, respectively. At $t = 0$, the pressure at the inlet of the nozzle is raised to $p_i > p_o$, where p_o is the (atmospheric) outlet pressure of the nozzle, and the fluid begins to flow horizontally through the nozzle.

a) Derive the following equation for the time-dependent volume flow rate $Q(t)$ through the nozzle from the unsteady Bernoulli equation and an appropriate conservation-of-mass relationship.

$$\frac{\dot{Q}(t)}{A_i}\int_{x=0}^{x=L}\frac{A_i}{A(x)}dx + \frac{Q^2(t)}{2}\left(\frac{1}{A_o^2} - \frac{1}{A_i^2}\right) = \left(\frac{p_i - p_o}{\rho}\right)$$

b) Solve the equation of part a) when $f(x/L) = x/L$.

c) If $\rho = 10^3\,\text{kg/m}^3$, $L = 25\,\text{cm}$, $A_i = 100\,\text{cm}^2$, $A_o = 30\,\text{cm}^2$, and $p_i - p_o = 100\,\text{kPa}$ for $t \geq 0$, how long does it take for the flow rate to reach 99% of its steady-state value?

4.74 For steady constant-density inviscid flow with body force per unit mass $\mathbf{g} = -\nabla\Phi$, it is possible to derive the following Bernoulli equation: $p + \frac{1}{2}\rho|\mathbf{u}|^2 + \rho\Phi = \text{constant along a streamline}$.

a) What is the equivalent form of the Bernoulli equation for constant-density inviscid flow that appears steady when viewed in a frame of reference that rotates at a constant rate about the z-axis, i.e., when $\boldsymbol{\Omega} = (0, 0, \Omega_z)$ with Ω_z constant?

b) If the extra term found in the Bernoulli equation is considered a pressure correction: Where on the surface of the earth (i.e., at what latitude) will this pressure correction be the largest? What is the absolute size of the maximum pressure correction when changes in R on a streamline are 1 m, 1 km, and 10^3 km.

4.75 Starting from (4.46) derive the following unsteady Bernoulli equation for inviscid incompressible irrotational fluid flow observed in a nonrotating frame of reference undergoing acceleration $d\mathbf{U}/dt$ with its z-axis vertical.

$$\int_1^2 \frac{\partial \mathbf{u}}{\partial t} \cdot d\mathbf{s} + \left(\frac{|\mathbf{u}|^2}{2} + \frac{p}{\rho} + gz + \mathbf{x} \cdot \frac{d\mathbf{U}}{dt} \right)_2 = \left(\frac{|\mathbf{u}|^2}{2} + \frac{p}{\rho} + gz + \mathbf{x} \cdot \frac{d\mathbf{U}}{dt} \right)_1$$

4.76 The Challenger Deep, located beneath the Pacific Ocean in the southern end of the Mariana Trench, represents the deepest known part of the ocean with a depth of $H = 10.924$ km. Although water can typically be treated as incompressible, at this depth the water density does vary due to compressibility.

a) Consider a linear expansion of density $\rho = \rho(p_0) + (\partial\rho/\partial p)_s (p - p_0)$, where $\rho(p_0) = \rho_0$ and p_0 are the density and pressure at the surface ($z = 0$), respectively, and $(\partial\rho/\partial p)_s = 1/c^2$ with c the speed of sound. Starting from (4.73), show that the (steady) Bernoulli equation for such a compressible fluid can be expressed as

$$\frac{1}{2}|\mathbf{u}|^2 + c^2 \ln\left(1 + \frac{p - p_0}{\rho_0 c^2}\right) + gz = \text{constant}.$$

b) Using the solution from **a)** and assuming constant speed of sound $c = 1500$ m/s and $\mathbf{u} = 0$, compute the pressure of the water at the bottom of the Challenger Deep. What is the density variation compared to ρ_0?

4.77 Using the small slope version of the surface curvature $1/R_1 \approx d^2\zeta/dx^2$, redo Example 4.16 to find h and $\zeta(x)$ in terms of x, σ, ρ, g, and θ. Show that the two answers are consistent when θ approaches $\pi/2$.

4.78 A spherical bubble with radius $R(t)$, containing gas with negligible density, creates purely radial flow, $\mathbf{u} = (u_r(r,t), 0, 0)$, in an unbounded bath of a quiescent incompressible liquid with density ρ and viscosity μ. Determine $u_r(t)$ in terms of $R(t)$, its derivatives. Ignoring body forces, and assuming a pressure of p_∞ far from the bubble, find (and solve) an equation for the pressure distribution, $p(r,t)$, outside the bubble. Integrate this equation from $r = R$ to $r \to \infty$, and apply an appropriate boundary condition at the bubble's surface to find the Rayleigh-Plesset equation for the pressure $p_B(t)$ inside the bubble:

$$\frac{p_B(t) - p_\infty}{\rho} = R\frac{d^2R}{dt^2} + \frac{3}{2}\left(\frac{dR}{dt}\right)^2 + \frac{4\mu}{\rho R}\frac{dR}{dt} + \frac{2\sigma}{\rho R},$$

where μ is the fluid's viscosity and σ is the surface tension.

4.79 Redo the dimensionless scaling leading to (4.102) by choosing a generic viscous stress, $\mu U/l$, and then a generic hydrostatic pressure, $\rho g l$, to make $p - p_\infty$ dimensionless. Interpret the revised dimensionless coefficients that appear in the scaled momentum equation, and relate them to St, Re, and Fr.

4.80 A solid sphere of mass m and diameter D is released from rest and falls through an incompressible viscous fluid with density ρ and viscosity μ under the action of gravity g. When the z coordinate increases downward, the vertical component of Newton's second law for the sphere is: $m(du_z/dt) = +mg - F_B - F_D$, where u_z is positive downward, F_B is the buoyancy force on the sphere, and F_D is the fluid-dynamic drag force on the sphere. Here, with $u_z > 0$, F_D opposes the sphere's downward motion. At first the sphere is moving slowly so its Reynolds number is low, but $\text{Re}_D = \rho u_z D/\mu$ increases with time as the sphere's velocity increases. To account for this variation in Re_D, the sphere's coefficient of drag may be approximated as: $C_D \cong \frac{1}{2} + 24/\text{Re}_D$. For the following items, provide answers in terms of m, ρ, μ, g and D; do not use z, u_z, F_B, or F_D.

a) Assume the sphere's vertical equation of motion will be solved by a computer after being put into dimensionless form. Therefore, use the information provided and the definition $t^* \equiv \rho g t D/\mu$ to show that this equation may be rewritten: $(d\text{Re}_D/dt^*) = A\text{Re}_D^2 + B\text{Re}_D + C$, and determine the coefficients A, B, and C.

b) Solve the part a) equation for Re_D analytically in terms of A, B, and C for a sphere that is initially at rest.
c) Undo the dimensionless scaling to determine the terminal velocity of the sphere from the part c) answer as $t \to \infty$.

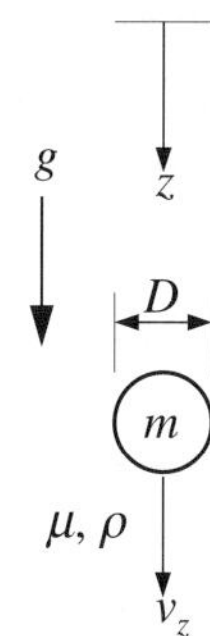

4.81 a) From (4.102), what is the dimensional differential momentum equation for steady incompressible viscous flow as $\text{Re} \to \infty$ when $\mathbf{g} = 0$.
b) Repeat part a) for $\text{Re} \to 0$. Does this equation include the pressure gradient?
c) Given that pressure gradients are important for fluid mechanics at low Re, revise the pressure scaling in (4.101) to obtain a more satisfactory low-Re limit for (4.40) with $\mathbf{g} = 0$.

4.82 a) Simplify (4.46) for motion of a constant-density inviscid fluid observed in a frame of reference that does not translate but does rotate at a constant rate $\mathbf{\Omega} = \Omega \mathbf{e}_z$.
b) Use length, velocity, acceleration, rotation, and density scales of L, U, g, Ω, and ρ to determine the dimensionless parameters for this flow when $\mathbf{g} = -g\mathbf{e}_z$ and $\mathbf{x} = (x, y, z)$. [*Hint:* substract out the static pressure distribution.]
c) The Rossby number, Ro, in this situation is $U/\Omega L$. What are the simplified equations of motion for a steady horizontal flow, $\mathbf{u} = (u, v, 0)$, observed in the rotating frame of reference when $\text{Ro} \ll 1$.

4.83 From Fig. 4.25, it can be seen that $C_D \propto 1/\text{Re}$ at small Reynolds numbers and that C_D is approximately constant at large Reynolds numbers. Redo the dimensional analysis leading to (4.100) to verify these observations when:
a) Re is low and fluid inertia is unimportant so ρ is no longer a parameter.
b) Re is high and the drag force is dominated by fore-aft pressure differences on the sphere and μ is no longer a parameter.

4.84 Suppose that the power to drive a propeller of an airplane depends on d (diameter of the propeller), U (free-stream velocity), ω (angular velocity of propeller), c (velocity of sound), ρ (density of fluid), and μ (viscosity). Find the dimensionless groups. In your opinion, which of these are the most important and should be duplicated in model testing?

4.85 A 1/25 scale model of a submarine is being tested in a wind tunnel in which $p = 200$ kPa and $T = 300$ K. If the speed of the full-size submarine is 30 km/hr, what should be the free-stream velocity in the wind tunnel? What is the drag ratio? Assume that the submarine would not operate near the free surface of the ocean.

4.86 The volume flow rate Q from a centrifugal blower depends on its rotation rate Ω, its diameter d, the pressure rise it works against Δp, and the density ρ and viscosity μ of the working fluid.
a) Develop a dimensionless scaling law for Q in terms of the other parameters.
b) Simplify the result of part c) for high Reynolds number pumping where μ is no longer a parameter.
c) For $d = 0.10$ m and $\rho = 1.2$ kg/m^3, plot the measured centrifugal blower performance data from the table in dimensionless form to determine if your result for part b) is a useful simplification. Here RPM is revolutions per minute, Q is in liter/s, and Δp is in kPa.

RPM = 5000		8000		11,000	
Q	ΔP	Q	ΔP	Q	ΔP
0.3	0.54	0.5	1.4	0.9	2.6
1.1	0.51	2.0	1.3	4.1	2.2
1.5	0.48	3.3	1.1	6.2	1.8
2.8	0.37	5.0	0.84	7.7	1.3
3.8	0.24	6.5	0.49	9.5	0.81

d) What maximum pressure rise would you predict for a geometrically similar blower having twice the diameter if it were spun at 6500 RPM?

4.87 A set of small-scale tank-draining experiments are performed to predict the liquid depth, h, as a function of time t for the draining process of a large cylindrical tank that is geometrically similar to the small-experiment tanks. The relevant parameters are gravity g, initial tank depth H, tank diameter D, orifice diameter d, and the density and viscosity of the liquid, ρ and μ, respectively.

a) Determine a general relationship between h and the other parameters.

b) Using the following small-scale experiment results, determine whether or not the liquid's viscosity is an important parameter.

$H = 8$ cm, $D = 24$ cm, $d = 8$ mm		$H = 16$ cm, $D = 48$ cm, $d = 1.6$ cm	
h (cm)	t (s)	h (cm)	t (s)
8.0	0.00	16.0	0.00
6.8	1.00	13.3	1.50
5.0	2.00	9.5	3.00
3.0	3.00	5.3	4.50
1.2	4.00	1.8	6.00
0.0	5.30	0.0	7.50

c) Using the small-scale-experiment results above, predict how long it takes to completely drain the liquid from a large tank having $H = 10$ m, $D = 30$ m, and $d = 1.0$ m.

References

Aris, R., 1962. Vectors, Tensors, and the Basic Equations of Fluid Mechanics. Prentice-Hall, Englewood Cliffs, NJ (The basic equations of motion and the various forms of the Reynolds transport theorem are derived and discussed).

Batchelor, G.K., 1967. An Introduction to Fluid Dynamics. Cambridge University Press, London (This contains an excellent and authoritative treatment of the basic equations).

Bird, R.B., Armstrong, R.C., Hassager, O., 1987. Dynamics of polymeric liquids. In: Fluid Mechanics, vol. 1, 2nd ed. John Wiley & Sons, New York.

Fermi, E., 1956. Thermodynamics. Dover Publications, Inc., New York.

Gidaspow, D., 1994. Multiphase Flow and Fluidization: Continuum and Kinetic Theory Descriptions. Academic Press.

Gogte, S., Vorobieff, P., Truesdell, R., Mammoli, A., van Swol, F., Shah, P., Brinker, C.J., 2005. Effective slip on textured superhydrophobic surfaces. Physics of Fluids 17, 051701.

Hirschfelder, J.O., Curtiss, C.F., Bird, R.B., 1954. Molecular Theory of Gases and Liquids. John Wiley and Sons, New York.

Kuo, K.K., 1986. Principles of Combustion. John Wiley & Sons, New York.

Lamb, H., 1945. Hydrodynamics, 6th ed. Dover Publications, Inc., New York.

Levich, V.G., 1962. Physicochemical Hydrodynamics, 2nd ed. Prentice-Hall, Englewood Cliffs, NJ.

Lord Rayleigh (J.W. Strutt), 1890. On the theory of surface forces. Philosophical Magazine Series 5 30, 285–298. 456–475.

McCormick, N.J., 2005. Gas-surface accommodation coefficients from viscous slip and temperature jump coefficients. Physics of Fluids 17, 107104.

Probstein, R.F., 1994. Physicochemical Hydrodynamics, 2nd ed. John Wiley & Sons, New York.

Panton, R.L., 2005. Incompressible Flow, 3rd ed. John Wiley and Sons, New York.

Spiegel, E.A., Veronis, G., 1960. On the Boussinesq approximation for a compressible fluid. The Astrophysical Journal 131, 442–447.

Stommel, H.M., Moore, D.W., 1989. An Introduction to the Coriolis Force. Columbia University Press, New York.

Thompson, P.A., 1972. Compressible-Fluid Dynamics. McGraw-Hill, New York.

Tretheway, D.C., Meinhart, C.D., 2002. Apparent fluid slip at hydrophobic microchannel walls. Physics of Fluids 14, L9–L12.

Truesdell, C.A., 1952. Stokes' principle of viscosity. Journal of Rational Mechanics and Analysis 1, 228–231.

White, F.M., 2006. Viscous Fluid Flow, 3rd ed. McGraw-Hill, New York.

Further reading

Çengel, Y.A., Cimbala, J.M., 2010. Fluid Mechanics: Fundamentals and Applications, 2nd ed. McGraw-Hill, New York (A fine undergraduate-level fluid mechanics textbook).

Chandrasekhar, S., 1961. Hydrodynamic and Hydromagnetic Stability. Oxford University Press, London (This is a good source to learn the basic equations in a brief and simple way).

Dussan, V.E.B., 1979. On the spreading of liquids on solid surfaces: static and dynamic contact lines. Annual Review of Fluid Mechanics 11, 371–400.

Fox, R.W., Pritchard, P.J., MacDonald, A.T., 2009. Introduction to Fluid Mechanics, 7th ed. John Wiley, New York (Another fine undergraduate-level fluid mechanics textbook).

Levich, V.G., Krylov, V.S., 1969. Surface tension driven phenomena. Annual Review of Fluid Mechanics 1, 293–316.

Pedlosky, J., 1987. Geophysical Fluid Dynamics. Springer-Verlag, New York.

Sabersky, R.H., Acosta, A.J., Hauptmann, E.G., Gates, E.M., 1999. Fluid Flow: A First Course in Fluid Mechanics, 4th ed. Prentice Hall, New Jersey (Another fine undergraduate-level fluid mechanics textbook).

White, F.M., 2008. Fluid Mechanics, 6th ed. McGraw-Hill, New York (A fine undergraduate-level fluid mechanics textbook).

Chapter 5

Vorticity dynamics⊛

Contents

5.1 Introduction 151
5.2 Kelvin's and Helmholtz's theorems 155
5.3 Vorticity equation in an inertial frame of reference 158
5.4 Vorticity equation in a rotating frame of reference 160
5.5 Velocity induced by a vortex filament: law of Biot and Savart 163
5.6 Interaction of vortices 165
5.7 Vortex sheet 168
Exercises 172
References 176
Further reading 176

Chapter objectives

- To introduce the basic concepts and phenomena associated with vortex lines, tubes, and sheets in viscous and inviscid flows
- To derive and state classical theorems and equations for vorticity production and transport in inertial and rotating frames of reference
- To develop the relationship that describes how vorticity at one location induces fluid velocity at another
- To present some of the intriguing phenomena of vortex dynamics

5.1 Introduction

Vorticity is a vector field that is twice the angular velocity of fluid particles in the field. A concentration of codirectional or nearly codirectional vorticity is called a vortex. Fluid motion leading to circular or nearly circular streamlines is called vortex motion. In two dimensions (r, θ), a uniform distribution of plane-normal vorticity with magnitude ω, produces solid body rotation:

$$u_\theta = \omega r/2, \tag{5.1}$$

while a perfect concentration of plane-normal vorticity located at $r = 0$ with circulation Γ produces irrotational flow for $r > 0$:

$$u_\theta = \Gamma/2\pi r. \tag{5.2}$$

Both of these flow fields are steady and produce closed (circular) streamlines. However, in the first, fluid particles rotate, but in the second, for $r \neq 0$, they do not. In the second flow field, the vorticity is infinite on a line perpendicular to the r-θ plane that intersects it at $r = 0$, but is zero elsewhere. Thus, such an *ideal line vortex* is also known as an *irrotational vortex*. It is a useful idealization that will be exploited in this chapter, and in Chapters 6 and 13.

In general, vorticity in a flowing fluid is neither unidirectional nor steady, and it can be thought of as permeating fluid elements. Thus, a fluid element's vorticity may be reoriented or concentrated or diffused depending on the motion and deformation of that element, and on the torques applied to that element by surrounding fluid elements and surfaces. This conjecture is based on the fact that the dynamics of three-dimensional time-dependent vorticity fields can often be interpreted in terms of a few fundamental principles. This chapter presents these principles and some aspects of flows with vorticity, starting with fundamental vortex concepts.

A vortex line is a curve in the fluid that is everywhere tangent to the local vorticity vector. Here, of course, we recognize that a *vortex line* is not strictly linear; it may be curved just as a streamline may be curved. A vortex line is related to the vorticity vector the same way a streamline is related to the velocity vector. Thus, if ω_x, ω_y, and ω_z are the Cartesian components of the vorticity vector $\boldsymbol{\omega}$, then the components of an element $d\mathbf{s} = (dx, dy, dz)$ of a vortex line satisfy:

$$dx/\omega_x = dy/\omega_y = dz/\omega_z, \tag{5.3}$$

⊛. This book has a companion website hosting complementary materials. Visit this URL to access it: https://www.elsevier.com/books-and-journals/book-companion/9780128198070.

Fluid Mechanics. https://doi.org/10.1016/B978-0-12-819807-0.00014-4

which is analogous to (3.7) for a streamline. As a further similarity, vortex lines do not exist in irrotational flow just as streamlines do not exist in stationary fluid. Elementary examples of vortex lines are supplied by the flow fields (5.1) and (5.2). For solid-body rotation (5.1), all lines perpendicular to the r-θ plane are vortex lines, while in the flow field of an irrotational vortex (5.2) the lone vortex line is perpendicular to the r-θ plane and passes through it at $r = 0$.

In a region of flow with nontrivial vorticity, the vortex lines passing through any closed curve form a tubular surface called a *vortex tube* (Fig. 5.1), which is akin to a stream tube (Fig. 3.6). The circulation around a narrow vortex tube is $d\Gamma = \boldsymbol{\omega} \cdot \mathbf{n} dA$ just as the volume flow rate in a narrow stream tube is $dQ = \mathbf{u} \cdot \mathbf{n} dA$. The *strength of a vortex tube* is defined as the circulation computed on a closed circuit lying on the surface of the tube that encircles it just once. From Stokes' theorem it follows that the strength of a vortex tube, Γ, is equal to the vorticity in the tube integrated over its cross-sectional area. Thus, when Gauss' theorem is applied to the volume V defined by a section of a vortex of tube, such as that shown in Fig. 5.1, we find that:

$$\begin{aligned} \int_V \nabla \cdot \boldsymbol{\omega} dV = \int_A \boldsymbol{\omega} \cdot \mathbf{n} dA = \left\{ \int_{lower\ end} + \int_{curved\ side} + \int_{upper\ end} \right\} \boldsymbol{\omega} \cdot \mathbf{n} dA \\ = \Gamma_{lower\ end} + \Gamma_{upper\ end} = 0, \end{aligned} \tag{5.4}$$

where $\boldsymbol{\omega} \cdot \mathbf{n}$ is zero on the curved sides of the tube, and the final equality follows from $\nabla \cdot \boldsymbol{\omega} = \nabla \cdot (\nabla \times \mathbf{u}) = 0$. Eq. (5.4) states that a vortex tube's strength is independent of where it is measured, $\Gamma_{lower\ end} = \Gamma_{upper\ end}$, and this implies that vortex tubes cannot end within the fluid, a concept that can be extended to vortex lines in the limit as a vortex tube's cross-sectional area goes to zero. However, vortex lines and tubes can terminate on solid surfaces or free surfaces, or they can form loops. This kinematic constraint is often useful for determining the topology of vortical flows.

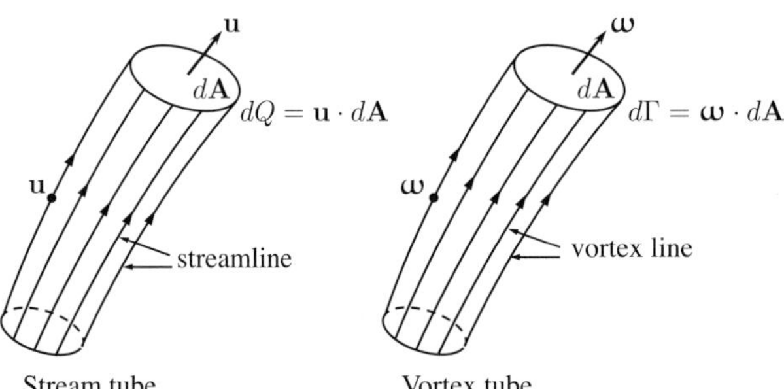

FIGURE 5.1 Analogy between stream tubes and vortex tubes. The lateral sides of stream and vortex tubes are locally tangent to the flow's velocity and vorticity fields, respectively. Stream and vortex tubes with cross-sectional area $d\mathbf{A}$ carry constant volume flux $\mathbf{u} \cdot d\mathbf{A}$ and constant circulation $\boldsymbol{\omega} \cdot d\mathbf{A}$, respectively.

As we will see in this and other chapters, fluid viscosity plays an essential role in the diffusion of vorticity, and in the reconnection of vortex lines. However, before considering these effects, the role of viscosity in the two basic vortex flows (5.1) and (5.2) is examined. Assuming incompressible flow, we shall see that in one of these flows the viscous terms in the momentum equation drop out, although the viscous stress and dissipation of energy are nonzero.

As discussed in Chapter 3, fluid elements undergoing solid-body rotation (5.1) do not deform ($S_{ij} = 0$), so the Newtonian viscous stress tensor (4.36) reduces to $T_{ij} = -p\delta_{ij}$, and Cauchy's equation (4.24) reduces to Euler's equation (4.42). When the solid-body rotation field, $u_r = 0$ and $u_\theta = \omega r/2$, is substituted into (4.42), it simplifies to:

$$-\rho u_\theta^2/r = -\partial p/\partial r, \quad \text{and} \quad 0 = -\partial p/\partial z - \rho g. \tag{5.5a, 5.5b}$$

Integrating (5.5a) produces $p(r, z) = \rho\omega^2 r^2/8 + f_1(z)$, where f_1 is an undetermined function. Integrating (5.5b) produces $p(r, z) = -\rho g z + f_2(r)$, where f_2 is an undetermined function. These two equations are consistent when:

$$p(r, z) - p_o = \frac{1}{8}\rho\omega^2 r^2 - \rho g z, \tag{5.6}$$

where p_o is the pressure at $r = 0$ and $z = 0$. To determine the shape of constant pressure surfaces, solve (5.6) for z to find:

$$z = \frac{\omega^2 r^2}{8g} - \frac{p(r, z) - p_o}{\rho g}.$$

Hence, surfaces of constant pressure are paraboloids of revolution (Fig. 5.2).

The important point to note is that viscous stresses are absent in steady solid-body rotation. (The viscous stresses, however, are important during the transient period of *initiating* solid body rotation, say by steadily rotating a tank containing a viscous fluid initially at rest.) In terms of velocity, (5.6) can be written as:

$$-\frac{1}{2}u_\theta^2 + gz + \frac{p(r,z)}{\rho} = \text{const.},$$

and, when compared to (4.19), this shows that the Bernoulli function $B = u_\theta^2/2 + gz + p/\rho$ is *not* constant for points on different streamlines. This outcome is expected because the flow is rotational.

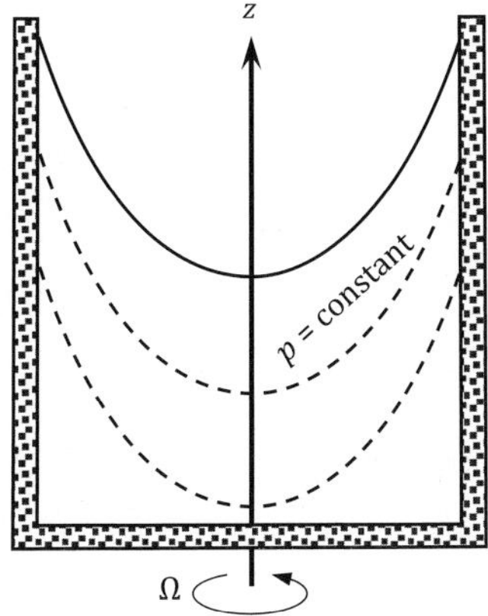

FIGURE 5.2 The steady flow field of a viscous liquid in a steadily rotating tank is solid body rotation. When the axis of rotation is parallel to the (downward) gravitational acceleration, surfaces of constant pressure in the liquid are paraboloids of revolution.

For the flow induced by an irrotational vortex (5.2), the viscous stress is:

$$\tau_{r\theta} = \mu\left[\frac{1}{r}\frac{\partial u_r}{\partial \theta} + r\frac{\partial}{\partial r}\left(\frac{u_\theta}{r}\right)\right] = -\frac{\mu\Gamma}{\pi r^2},$$

which is nonzero everywhere because fluid elements deform (see Fig. 3.18). However, the interesting point is that the *net viscous force* on an element is zero for $r > 0$ (see Exercise 5.4) because the viscous forces on the surfaces of an element cancel out, leaving a zero resultant. Thus, momentum conservation is again represented by the Euler equation. Substitution of (5.2) into (5.5), followed by integration yields:

$$p(r,z) - p_\infty = -\frac{\rho\Gamma^2}{8\pi^2 r^2} - \rho g z, \tag{5.7}$$

where p_∞ is the pressure far from the line vortex at $z = 0$. This can be rewritten:

$$z = -\frac{\Gamma^2}{8\pi^2 r^2 g} - \frac{p(r,z) - p_\infty}{\rho g},$$

which shows that surfaces of constant pressure are hyperboloids of revolution of the second degree (Fig. 5.3). Eq. (5.7) can also be rewritten:

$$\frac{1}{2}u_\theta^2 + gz + \frac{p(r,z)}{\rho} = \text{const.},$$

which shows that Bernoulli's equation is applicable between any two points in the flow field, as is expected for steady incompressible irrotational flow.

One way of generating the flow field of an irrotational vortex is by rotating a solid circular cylinder with radius a in an infinite viscous fluid (see Fig. 7.7). It is shown in Section 7.2 that the steady solution of the Navier-Stokes equations satisfying the no-slip boundary condition ($u_\theta = \Omega a$ at $r = a$) is:

$$u_\theta = \Omega a^2/r \quad \text{for} \quad r \geq a,$$

where Ω is the cylinder's constant rotation rate; see (7.11). When the motions inside and outside the cylinder are considered, this flow field precisely matches that of a Rankine vortex with core size a, see (3.28) with $\Gamma = 2\pi a^2\Omega$. The presence of the nonzero-radius cylinder leads to a flow field without a singularity that is irrotational for $r > a$. Viscous stresses are present, and the resulting viscous dissipation of kinetic energy is exactly compensated by the work done at the surface of the cylinder. However, there is no *net* viscous force at any point in the steady state. Interestingly, the application of the moment of momentum principle (see Section 4.9) to a large-radius cylindrical control volume centered on the rotating solid cylinder shows that the torque that rotates the solid cylinder is transmitted to an arbitrarily large distance from the axis of rotation. Thus, any attempt to produce this flow in a stationary container would require the application of a counteracting torque on the container.

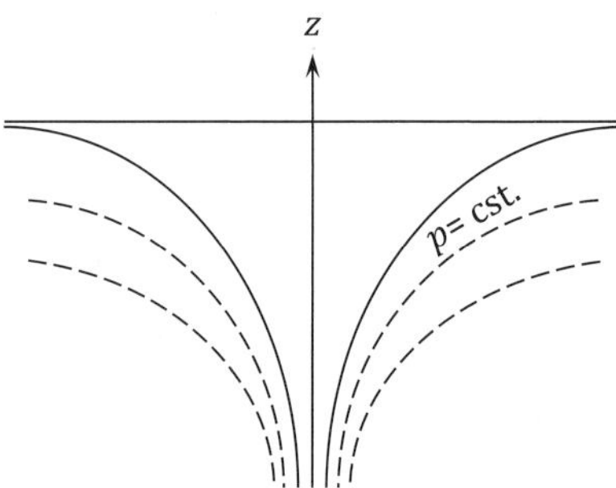

FIGURE 5.3 Surfaces of constant pressure in the flow induced by an ideal linear vortex that coincides with the z-axis and is parallel to the (downward) gravitational acceleration.

These examples suggest that *irrotationality does not imply the absence of viscous stresses*. Instead, it implies the *absence of net viscous forces*. Viscous stresses will be present whenever fluid elements deform. Yet, when $\boldsymbol{\omega}$ is uniform and nonzero (solid body rotation), there is no viscous stress at all. However, solid-body rotation is unique in this regard, and this uniqueness is built into the Newtonian-fluid viscous stress tensor (4.60). In general, fluid element rotation is accomplished and accompanied by viscous effects. Indeed, viscosity is a primary agent for vorticity generation and diffusion.

Example 5.1

The surface of a nominally quiescent pool of water is deflected symmetrically downward near $R = 0$ because of vortical flow within the pool as shown in Fig. 5.4. In cylindrical coordinates, the water surface profile is $z = h(R)$. Assume the velocity field in the water has only an angular component, $\mathbf{u} = u_\varphi(R)\mathbf{e}_\varphi$, and determine $u_\varphi(R)$ when the pressure on the water surface is p_o.

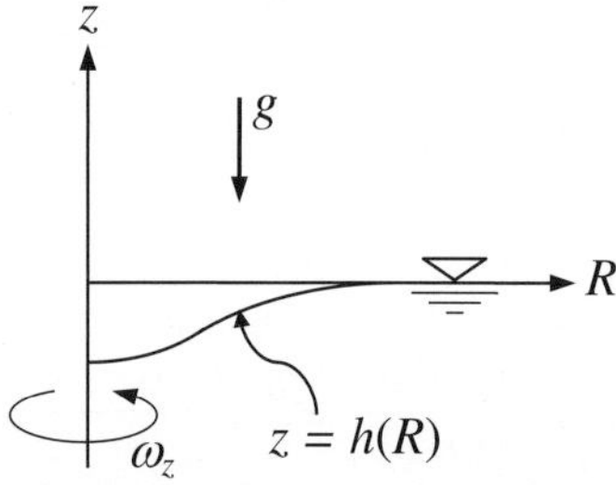

FIGURE 5.4 Swirling water flow causes a water surface to deflect downward near $R = 0$. The surface shape is $z = h(R)$. Here $h(R)$ is negative, but its radial derivative, dh/dR, is positive.

Solution Integrating the equivalents of (5.5a) and (5.5b) in cylindrical coordinates leads to:

$$p(R,z) - p_o = \int_0^R \rho \frac{u_\varphi^2}{R} dR + f_1(z), \quad \text{and} \quad p(R,z) - p_o = -\rho g z + f_2(R),$$

where f_1 and f_2 are single-variable functions of integration. These two results are consistent when:

$$p(R,z) - p_o = \int_0^R \rho \frac{u_\varphi^2}{R} dR - \rho g z.$$

On the water surface the pressure is p_o, that is $p(R, h) = p_o$, so this pressure equation implies:

$$\int_0^R \rho \frac{u_\varphi^2}{R} dR = \rho g h, \quad \text{or} \quad \rho \frac{u_\varphi^2}{R} = \rho g \frac{dh}{dR}, \text{ which leads to: } u_\varphi(R) = \pm\sqrt{gR(dh/dR)}.$$

In this case, the vortical flow in the water may swirl with either sense of rotation to produce the surface shape $h(R)$.

5.2 Kelvin's and Helmholtz's theorems

By considering the analogy with electrodynamics, Helmholtz published several theorems for vortex motion in an inviscid fluid in 1858. Ten years later, Kelvin introduced the idea of circulation and proved the following theorem: *In an inviscid, barotropic flow with conservative body forces, the circulation around a closed curve moving with the fluid remains constant with time*, if the motion is observed from a nonrotating frame. Kelvin's theorem is considered first and can be stated mathematically as:

$$D\Gamma/Dt = 0, \tag{5.8}$$

where D/Dt is defined by (3.5) and represents the total time rate of change following the fluid elements that define the closed curve, C (a material contour), used to compute the circulation Γ. Such a material contour is shown in Fig. 5.5.

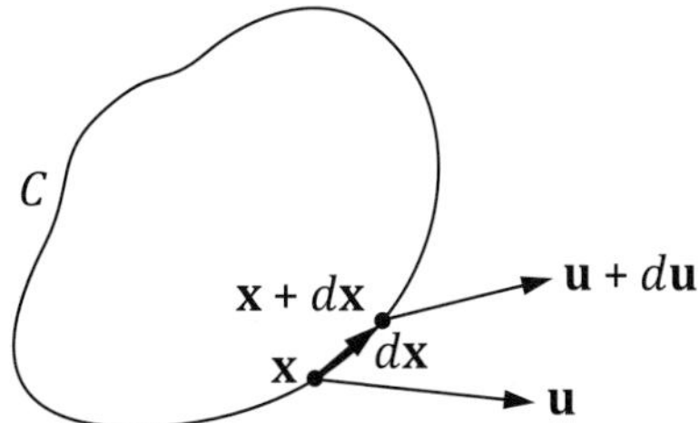

FIGURE 5.5 Contour geometry for the Proof of Kelvin's circulation theorem. Here the short segment $d\mathbf{x}$ of the contour C moves with the fluid so that $D(d\mathbf{x})/Dt = d\mathbf{u}$.

Kelvin's theorem can be proved by time differentiating the definition of the circulation (3.18):

$$\frac{D\Gamma}{Dt} = \frac{D}{Dt}\int_C u_i dx_i = \int_C \frac{Du_i}{Dt} dx_i + \int_C u_i \frac{D}{Dt}(dx_i), \tag{5.9}$$

where dx_i are the components of the vector arc length element $d\mathbf{x}$ of C. Using (4.39) and (4.60), the first term on the right side of (5.9) may be rewritten:

$$\int_C \frac{Du_i}{Dt} dx_i = \int_C \left(-\frac{1}{\rho}\frac{\partial p}{\partial x_i} + g_i + \frac{1}{\rho}\frac{\partial \tau_{ij}}{\partial x_j} \right) dx_i = -\int_C \frac{1}{\rho} dp - \int_C d\Phi + \int_C \left(\frac{1}{\rho}\frac{\partial \tau_{ij}}{\partial x_j} \right) dx_i, \tag{5.10}$$

where the replacements $(\partial p/\partial x_i)dx_i = dp$ and $(\partial \Phi/\partial x_i)dx_i = d\Phi$ have been made, and Φ is the body force potential (4.18). For a barotropic fluid, the first term on the right side of (5.10) is zero because C is a closed contour, and ρ and p are single valued at each point in space. Similarly, the second integral on the right side of (5.10) is zero since Φ is also single valued at each point in space.

Now consider the second term on the right side of (5.9). The velocity at point $\mathbf{x} + d\mathbf{x}$ on C is:

$$\mathbf{u} + d\mathbf{u} = \frac{D}{Dt}(\mathbf{x} + d\mathbf{x}) = \frac{D\mathbf{x}}{Dt} + \frac{D}{Dt}(d\mathbf{x}), \quad \text{so} \quad du_i = \frac{D}{Dt}(dx_i).$$

Thus, the last term in (5.9) then becomes:

$$\int_C u_i \frac{D}{Dt}(dx_i) = \int_C u_i du_i = \int_C d\left(\frac{1}{2}u_i^2\right) = 0,$$

where the final equality again follows because C is a closed contour and $\mathbf{u}$ is a single-valued vector function. Hence, (5.9) simplifies to:

$$\frac{D\Gamma}{Dt} = \int_C \left(\frac{1}{\rho} \frac{\partial \tau_{ij}}{\partial x_j} \right) dx_i, \tag{5.11}$$

and Kelvin's theorem (5.8) is proved when the fluid is inviscid ($\mu = \mu_v = 0$) or when the integrated viscous force ($\partial \tau_{ij}/\partial x_j$) is zero around the contour C. This latter condition can occur when C lies entirely in irrotational fluid.

This short proof indicates the three ways to create or destroy vorticity in a flow are: nonconservative body forces, a nonbarotropic pressure-density relationship, and nonzero net viscous torques. Examples of each follow. The Coriolis acceleration leads to a nonconservative body force that occurs in rotating frames of reference, and it, together with vortex stretching (see Example 5.3), can generate a drain or *bathtub* vortex when a fully quiescent water tank on the earth's surface is drained. Nonbarotropic effects can lead to vorticity generation when a vertical barrier is removed between two side-by-side initially motionless fluids having different densities in the same container and subject to a gravitational field. The two fluids will tumble as the heavier one slumps to the container's bottom and the lighter one surges to the container's top (see Fig. 5.6, and Exercise 5.5). Nonzero net viscous torques create vorticity. The conditions for vorticity creation via viscous torques often occur at solid boundaries where the no-slip condition (4.95) is maintained. A short distance above a solid boundary, the velocity parallel to the boundary may be nearly uniform. The resulting shear-flow velocity profile that links the surface with the above-surface uniform flow commonly leads to a net viscous torque and vorticity creation (see Example 5.2).

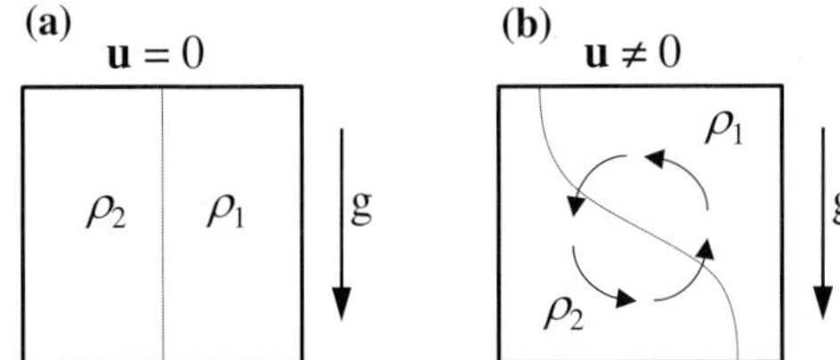

FIGURE 5.6 Schematic drawings of two fluids with differing density that are initially stationary and separated within a rectangular container. Gravity acts downward as shown. Here the density difference is baroclinic because it depends on fluid composition and pressure, not on pressure alone. (a) This drawing shows the initial condition immediately before the barrier between the two fluids is removed. (b) This drawing shows the resulting fluid motion a short time after barrier removal. The deflection of the fluid interface clearly indicates that vorticity has been created.

Kelvin's theorem implies that irrotational flow will remain irrotational if the following four restrictions are satisfied:

1. There are no net viscous forces along C. If C moves into regions where there are net viscous forces such as within a boundary layer that forms on a solid surface, then the circulation changes. The presence of viscous effects causes *diffusion* of vorticity into or out of a fluid circuit and consequently changes the circulation.
2. The body forces are conservative. Conservative body forces such as gravity act through the center of mass of a fluid particle and therefore do not generate torques that cause fluid particle rotation.
3. The fluid density must depend on pressure only (barotropic flow). A flow will be barotropic if the fluid is homogeneous and one of the two independent thermodynamic variables is constant. Isentropic, isothermal, and constant density conditions lead to barotropic flow. Flows that are not barotropic are called *baroclinic.* Here fluid density depends on the pressure *and* the temperature, composition, salinity, and/or concentration of dissolved constituents. Consider fluid elements in barotropic and baroclinic flows (Fig. 5.7). For the barotropic element, lines of constant p are parallel to lines of constant ρ, which implies that the resultant pressure forces pass through the center of mass of the element. For the baroclinic element, the lines of constant p and ρ are not parallel. The net pressure force does not pass through the center of mass, and the resulting torque changes the vorticity and circulation. As described above, Fig. 5.7 depicts a situation where vorticity is generated in a baroclinic flow.
4. The frame of reference must be an inertial frame. As described in Section 4.7, the conservation of momentum equation includes extra terms when the frame of reference rotates and accelerates, and these extra terms were not considered in the short proof given above.

Under the same four restrictions, Helmholtz proved the following theorems for vortex motion:

1. Vortex lines move with the fluid.
2. The strength of a vortex tube (its circulation) is constant along its length.
3. A vortex tube cannot end within the fluid. It must either end at a boundary or form a closed loop – a *vortex ring* or *loop*.

4. The strength of a vortex tube remains constant in time.

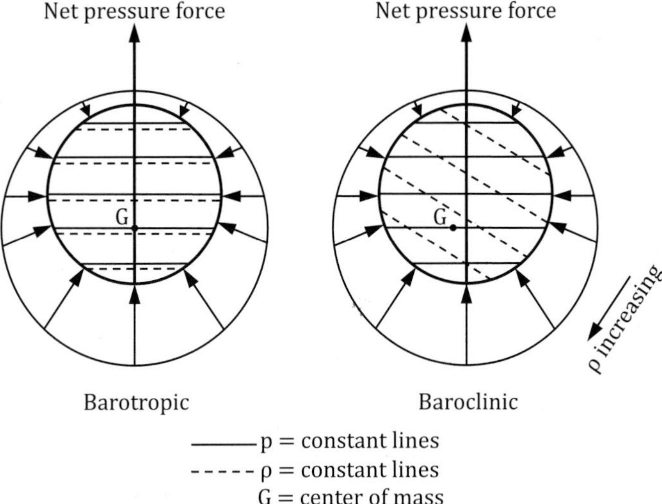

FIGURE 5.7 Mechanism of vorticity generation in baroclinic flow, showing that the net pressure force does not pass through the center of mass G of the fluid element. The radially inward arrows indicate pressure forces on an element.

Here, we only highlight the proof of the first theorem, which essentially says that fluid particles that at any time are part of a vortex line always belong to the same vortex line. To prove this result, consider an area S, bounded by a curve, lying on the surface of a vortex tube without embracing it (Fig. 5.8). Since the vorticity vectors are everywhere lying parallel to S (none are normal to S), it follows that the circulation around the edge of S is zero. After an interval of time, the same fluid particles form a new surface, S'. According to Kelvin's theorem, the circulation around S' must also be zero. As this is true for any S, the component of vorticity normal to every element of S' must vanish, demonstrating that S' must lie on the surface of the vortex tube. Thus, vortex tubes move with the fluid, a result we will also be able to attain from the field equation for vorticity. Applying this result to an infinitesimally thin vortex tube, we get the Helmholtz vortex theorem that vortex lines move with the fluid. A different proof may be found in Sommerfeld (1964, pp. 130–132).

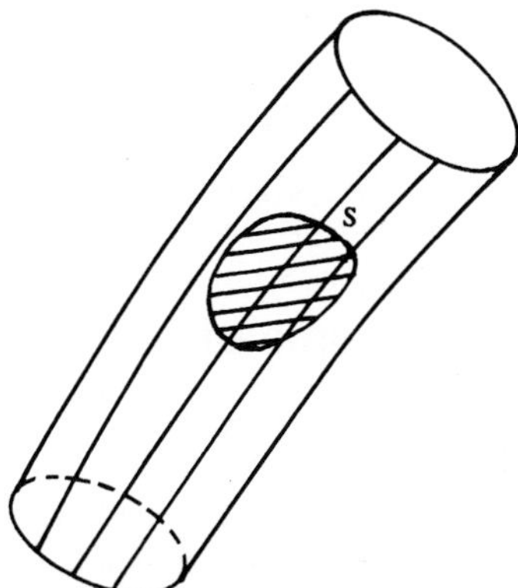

FIGURE 5.8 Vortex tube and surface geometry for Helmholtz's first vortex theorem. The surface S lies within a closed contour on the surface of a vortex tube.

Example 5.2

For a time dependent version of the planar shear flow in Fig. 3.16, $\mathbf{u} = (u_1(x_2, t), 0)$, what is $D\Gamma/Dt$ when Γ is computed from a small rectangular contour, ABCD – centered on (x_1, x_2) with sides Δx_1 and Δx_2 – when ρ and μ are constants?

Solution Evaluate (5.11) to determine $D\Gamma/Dt$. The first step is to determine the viscous stress tensor τ_{ij} from the given velocity field. This is an incompressible flow ($\nabla \cdot \mathbf{u} = 0$), so $\tau_{ij} = 2\mu S_{ij}$ and can be written in as a 2-by-2 matrix in terms of the one nonzero velocity gradient. Similarly the net viscous force on a fluid element, $\partial\tau_{ij}/\partial x_j$, can be written as a 2-by-1 row vector:

$$\tau_{ij} = 2\mu S_{ij} = \mu\begin{bmatrix} 0 & \partial u_1/\partial x_2 \\ \partial u_1/\partial x_2 & 0 \end{bmatrix} \quad \text{and} \quad \frac{\partial\tau_{ij}}{\partial x_j} = \mu\begin{bmatrix} \partial^2 u_1/\partial x_2^2 & 0 \end{bmatrix}.$$

Here, a net horizontal viscous force is present when the velocity profile has a nonzero second derivative. For the rectangular contour ABCD, the integral in (5.11) can be written:

$$\frac{D\Gamma}{Dt} = \int_C \left(\frac{1}{\rho}\frac{\partial\tau_{ij}}{\partial x_j}\right) dx_i = \frac{1}{\rho}\left[\int_{AB} + \int_{BC} + \int_{CD} + \int_{DA}\right]\left(\frac{\partial\tau_{ij}}{\partial x_j}\right) dx_i.$$

On the vertical AB and CD portions of the contour $d\mathbf{x} = (0, dx_2)$, so the dot product with $\partial\tau_{ij}/\partial x_j$ will be zero. This leaves the horizontal BC and DA portions of the contour:

$$\frac{D\Gamma}{Dt} = \frac{1}{\rho}\left[\int_{BC} + \int_{DA}\right]\left(\frac{\partial\tau_{ij}}{\partial x_j}\right) dx_i = \frac{1}{\rho}\left[\int_{x_1-\Delta x_1/2}^{x_1+\Delta x_1/2}\left(\mu\frac{\partial^2 u_1}{\partial u_2^2}\right)_{x_2-\Delta x_2/2} dx_1' - \int_{x_1-\Delta x_1/2}^{x_1+\Delta x_1/2}\left(\mu\frac{\partial^2 u_1}{\partial u_2^2}\right)_{x_2+\Delta x_2/2} dx_1'\right],$$

where x_1' is an integration variable, and the minus sign arises in the DA integration because the direction of integration opposes the direction of increasing x_1. Both integrands are independent of x_1, and, when Δx_2 is small enough, they may be Taylor-expanded to find:

$$\frac{D\Gamma}{Dt} = -\frac{\mu}{\rho}\frac{\partial^3 u_1}{\partial x_2^3}\Delta x_1\Delta x_2.$$

Thus, the circulation around fluid elements in a simple shear flow will only change when the third derivative of the velocity profile is nonzero. This finding can be understood as follows. When the velocity profile is linear, $\partial u_1/\partial x_2 = \text{const.}$, fluid elements have vorticity and Γ computed on ABCD is not zero, but there is no net viscous force or torque on fluid elements so Γ does not change. When the velocity profile is quadratic, $\partial^2 u_1/\partial x_2^2 = \text{const.}$, fluid elements again have vorticity, Γ computed on ABCD is not zero, and there is a net viscous force on fluid elements, but no net viscous torque. In this case, fluid elements may accelerate, or pressure or body forces may balance the net viscous force, but fluid elements feel no net torque so Γ computed on ABCD does not change. However, when the velocity profile is cubic (or higher) so that $\partial^3 u_1/\partial x_2^3 \neq 0$, then there is a net viscous torque on fluid elements and Γ computed on ABCD will increase or decrease appropriately.

Interestingly, the final result of this example is a precursor to (5.13), the main result of the next section, since a limit allows it to be rewritten in terms $\omega_3 = -\partial u_1/\partial x_2$, the plane-normal vorticity (circulation per unit area) in the simple shear flow:

$$\lim_{\Delta x_1,\Delta x_2\to 0}\frac{1}{\Delta x_1\Delta x_2}\frac{D\Gamma}{Dt} = \frac{D\omega_3}{Dt} = -\frac{\mu}{\rho}\frac{\partial^3 u_1}{\partial x_2^3} = \nu\frac{\partial^2\omega_3}{\partial x_2^2},$$

where $\Delta x_1\Delta x_2$ is the area inside the square ABCD contour, and $\nu = \mu/\rho$ is the kinematic viscosity.

5.3 Vorticity equation in an inertial frame of reference

An equation governing the vorticity in an inertial frame of reference is derived in this section. The *fluid density* ρ *is assumed to be constant*, so that the flow is barotropic. Viscous effects are retained but the viscosity is assumed to be constant. Baroclinic effects and a rotating frame of reference are considered in Section 5.4.

An equation for the vorticity can be obtained from the curl of the momentum conservation equation (4.40):

$$\nabla\times\left\{\frac{D\mathbf{u}}{Dt} = -\frac{1}{\rho}\nabla p + \mathbf{g} + \nu\nabla^2\mathbf{u}\right\}. \tag{5.12}$$

When $\mathbf{g}$ is conservative and (4.18) applies, the curl of the first two terms on the right side of (5.12) will be zero because they are gradients of scalar functions. The acceleration term on the left side of (5.12) becomes:

$$\nabla\times\left\{\frac{\partial\mathbf{u}}{\partial t} + (\mathbf{u}\cdot\nabla)\mathbf{u}\right\} = \frac{\partial\boldsymbol{\omega}}{\partial t} + \nabla\times\{(\mathbf{u}\cdot\nabla)\mathbf{u}\} = \frac{\partial\boldsymbol{\omega}}{\partial t} + \nabla\times\left\{\nabla\left(\frac{\mathbf{u}\cdot\mathbf{u}}{2}\right) + \boldsymbol{\omega}\times\mathbf{u}\right\}$$
$$= \frac{\partial\boldsymbol{\omega}}{\partial t} + \nabla\times(\boldsymbol{\omega}\times\mathbf{u}),$$

so (5.12) reduces to:

$$\frac{\partial \boldsymbol{\omega}}{\partial t} + \nabla \times (\boldsymbol{\omega} \times \mathbf{u}) = \nu \nabla^2 \boldsymbol{\omega},$$

where we have also used the identity $\nabla \times \nabla^2 \mathbf{u} = \nabla^2(\nabla \times \mathbf{u})$ in rewriting the viscous term. The second term in the above equation can be written as $\nabla \times (\boldsymbol{\omega} \times \mathbf{u}) = (\mathbf{u} \cdot \nabla)\boldsymbol{\omega} - (\boldsymbol{\omega} \cdot \nabla)\mathbf{u}$, based on the vector identity (B.3.10), and the fact that $\nabla \cdot \mathbf{u} = 0$ and $\nabla \cdot \boldsymbol{\omega} = \nabla \cdot (\nabla \times \mathbf{u}) = 0$. Thus, (5.12) becomes:

$$\frac{D\boldsymbol{\omega}}{Dt} = (\boldsymbol{\omega} \cdot \nabla)\mathbf{u} + \nu \nabla^2 \boldsymbol{\omega}. \tag{5.13}$$

This is the field equation governing vorticity in a fluid with constant ρ and conservative body forces. The term $\nu \nabla^2 \boldsymbol{\omega}$ represents the rate of change of $\boldsymbol{\omega}$ caused by diffusion of vorticity in the same way that $\nu \nabla^2 \mathbf{u}$ represents acceleration caused by diffusion of momentum. The term $(\boldsymbol{\omega} \cdot \nabla)\mathbf{u}$ represents the rate of change of vorticity caused by the stretching and tilting of vortex lines. Note that pressure and gravity terms do not appear in (5.13) since these forces act through the center of mass of an element and therefore generate no torque. In addition, note that (5.13) might appear upon first glance to be a linear equation for $\boldsymbol{\omega}$. However, the vorticity is the curl of the velocity so both the advective part of the $D\boldsymbol{\omega}/Dt$ term and the $(\boldsymbol{\omega} \cdot \nabla)\mathbf{u}$ term represent nonlinearities.

To better understand the vortex-stretching term, consider the natural coordinate system where s is the arc length along a vortex line, n points away from the center of vortex-line curvature, and m lies along the second normal to s (Fig. 5.9). In this coordinate system, the vortex stretching term becomes

$$(\boldsymbol{\omega} \cdot \nabla)\mathbf{u} = \left[\boldsymbol{\omega} \cdot \left(\mathbf{e}_s \frac{\partial}{\partial s} + \mathbf{e}_n \frac{\partial}{\partial n} + \mathbf{e}_m \frac{\partial}{\partial m}\right)\right]\mathbf{u} = \omega \frac{\partial \mathbf{u}}{\partial s}, \tag{5.14}$$

where, by definition, $\boldsymbol{\omega} \cdot \mathbf{e}_n = \boldsymbol{\omega} \cdot \mathbf{e}_m = 0$, and $\boldsymbol{\omega} \cdot \mathbf{e}_s = \omega = |\boldsymbol{\omega}|$. Thus, (5.14) shows that $(\boldsymbol{\omega} \cdot \nabla)\mathbf{u}$ equals the magnitude of $\boldsymbol{\omega}$ times the derivative of $\mathbf{u}$ in the direction of $\boldsymbol{\omega}$. The quantity $\omega(\partial \mathbf{u}/\partial s)$ is a vector and has the components $\omega(\partial u_s/\partial s)$, $\omega(\partial u_n/\partial s)$, and $\omega(\partial u_m/\partial s)$. Among these, $\partial u_s/\partial s$ represents the increase of u_s along the vortex line s, that is, the stretching of a vortex line. On the other hand, $\partial u_n/\partial s$ and $\partial u_m/\partial s$ represent the change of the normal velocity components along s and, therefore, the rate of turning or tilting of vortex lines about the m and n axes, respectively.

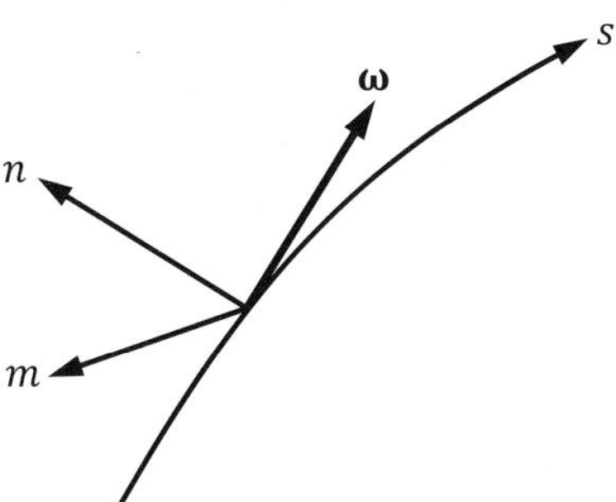

FIGURE 5.9 Natural coordinate system aligned with the vorticity vector.

To see the effect of these terms more clearly, write out the components of (5.13) for an inviscid flow:

$$\frac{D\omega_s}{Dt} = \omega \frac{\partial u_s}{\partial s}, \quad \frac{D\omega_n}{Dt} = \omega \frac{\partial u_n}{\partial s}, \quad \text{and} \quad \frac{D\omega_m}{Dt} = \omega \frac{\partial u_m}{\partial s}. \tag{5.15}$$

The first equation of (5.15) shows that the vorticity along s changes due to stretching of vortex lines, reflecting the principle of conservation of angular momentum. Stretching decreases the moment of inertia of fluid elements that constitute a vortex line, resulting in an increase of their angular rotation speed (as in Example 5.3). Vortex stretching plays an especially crucial role in the dynamics of turbulent and geophysical flows. The second and third equations of (5.15) show how vorticity along n and m is created by the tilting of vortex lines. For example, in Fig. 5.12, the turning of the vorticity vector $\boldsymbol{\omega}$ toward the n-axis will generate a vorticity component along n. *The vortex stretching and tilting term* $(\boldsymbol{\omega} \cdot \nabla)\mathbf{u}$ *is zero in two-dimensional flows where* $\boldsymbol{\omega}$ *is perpendicular to the plane of flow*. This has important consequences on turbulent flows, as a reduction in spatial dimensions will fail to capture important dynamics associated with this term in three dimensions.

Example 5.3

In cylindrical coordinates, pure-strain extensional (stretching) flow along the z-axis is given by: $u_R = -(\gamma/2)R$ and $u_z = \gamma z$, where γ is the strain rate. A vortex aligned with the z-axis has vorticity $\boldsymbol{\omega} = \omega_z \mathbf{e}_z = \omega_o(R)\mathbf{e}_z$ at $t = 0$. What is $\omega_z(R, t)$ in this flow when the fluid is inviscid? When γ is positive, does the vorticity at $R = 0$ strengthen or weaken as t increases? When $\gamma \neq 0$, does the vortex's circulation change?

Solution To determine $\omega_z(R, t)$, (5.13) must be solved for the given flow field and initial condition. For the given velocity field with $\nu = 0$, the z-direction component of (5.13) is:

$$\frac{\partial \omega_z}{\partial t} + u_R \frac{\partial \omega_z}{\partial R} + \frac{u_\varphi}{R}\frac{\partial \omega_z}{\partial \varphi} + u_z \frac{\partial \omega_z}{\partial z} = \omega_z \frac{\partial u_z}{\partial z}.$$

When ω_z depends only on R and t only, the third and fourth terms on the left side are zero. In addition, $u_R = -(\gamma/2)R$ and $\partial u_z/\partial z = \gamma$, so this vorticity component equation becomes:

$$\frac{\partial \omega_z}{\partial t} - \frac{\gamma R}{2}\frac{\partial \omega_z}{\partial R} = \gamma \omega_z, \quad \text{or} \quad \frac{\partial \omega_z}{\partial t} - \frac{1}{R}\frac{\partial}{\partial R}\left(\frac{1}{2}\gamma R^2 \omega_z\right) = 0.$$

To solve the second equation, rewrite it in terms of a new dependent function $\phi(R, t) = (1/2)\gamma R^2 \omega_z(R, t)$:

$$\frac{\partial \phi}{\partial t} - \frac{\gamma R}{2}\frac{\partial \phi}{\partial R} = 0,$$

and assume that there is a trajectory $R(t)$ in R-t space where $dR/dt = -(1/2)\gamma R$ so that the last equation represents the total derivative $d\phi/dt = 0$ (see Appendix B.1). Here, the equation for the trajectories $R(t)$ is readily separated and integrated to find: $R(t) = R_o \exp\{-\gamma t/2\}$ where R_o is the radial distance of a trajectory at $t = 0$.

In this case ϕ will be constant along the curves $R(t)$, so the solution for ϕ is given by:

$$\phi(R(t), t) = \text{const.} = \frac{\gamma R^2(t)}{2}\omega_z(R, t) = \frac{\gamma R_o^2}{2}\omega_o(R_o),$$

where the final equality follows from the initial condition. Algebraic rearrangement of the last equality using $R_o/R = \exp\{+\gamma t/2\}$ produces the final solution for ω_z:

$$\omega_z(R, t) = e^{+\gamma t}\omega_o(Re^{+\gamma t/2}).$$

When γ is positive, the vorticity at $R = 0$ increases exponentially as t increases. The circulation of the vortex at time t, is:

$$\Gamma = \int \omega_z dA = 2\pi \int_0^\infty \omega_z(R, t)RdR = 2\pi \int_0^\infty e^{+\gamma t}\omega_o(Re^{+\gamma t/2})RdR = 2\pi \int_0^\infty \omega_o(\xi)\xi d\xi,$$

where an integration variable substitution $\xi = R\exp\{+\gamma t/2\}$ has been made, and the final form is the circulation of the vortex at $t = 0$. Thus, the vortex's initial circulation is not changed by axial stretching, and this makes sense; axial stretching does not apply a torque to fluid elements.

This example illustrates how vortex stretching can concentrate vorticity, and is analogous to the figure skater who starts by moving in a wide circle but ends up in a rapid vertical spin. It also suggests two means for developing a strong vortex. First, start with a large horizontally-spread vorticity-containing fluid mass so that Γ is substantial. And second, allow vertical stretching to take place over a sufficiently long period of time ($\gamma t \gg 1$) so that the exponential factors dominate the solution for ω_z. Although both also involve three-dimensional effects, drain vortices and tornadoes are primarily formed by the vorticity-intensification mechanism illustrated in this example.

5.4 Vorticity equation in a rotating frame of reference

The vorticity equation (5.13) is valid for a fluid of uniform density and viscosity observed from an *inertial* frame of reference. Here, this equation is generalized to a steadily rotating frame of reference and a variable density fluid. The flow, however, will be assumed incompressible (or nearly so). The resulting vorticity equation is applicable to flows in rotating machinery and to oceanic and atmospheric fluid flows of sufficient size and duration to necessitate inclusion of the earth's rotation rate when conserving momentum.

The continuity and momentum equations for flow of a variable-density incompressible fluid observed in a steadily rotating frame of reference are:

$$\frac{\partial u_i}{\partial x_i}=0, \quad \text{and} \quad \frac{\partial u_i}{\partial t}+u_j\frac{\partial u_i}{\partial x_j}+2\varepsilon_{ijk}\Omega_j u_k=-\frac{1}{\rho}\frac{\partial p}{\partial x_i}+g_i+\nu\frac{\partial^2 u_i}{\partial x_j^2}, \tag{4.10, 5.16}$$

where $\mathbf{\Omega}$ is the angular velocity of the coordinate system and g_i is the effective gravity (including centrifugal acceleration); see Section 4.7. In particular, (5.16) is merely a simplified version of (4.46) where $\mathbf{U}$ is zero, $\mathbf{\Omega}$ is constant, and the primes have been dropped. The advective acceleration and viscous diffusion terms of (5.16) can be rewritten:

$$u_j\frac{\partial u_i}{\partial x_j}=-\varepsilon_{ijk}u_j\omega_k+\frac{\partial}{\partial x_i}\left(\frac{1}{2}u_j^2\right) \quad \text{and} \quad \nu\frac{\partial^2 u_i}{\partial x_j^2}=-\nu\varepsilon_{ijk}\frac{\partial \omega_k}{\partial x_j}. \tag{4.69, 4.41}$$

In addition, the Coriolis acceleration term in (5.16) can be rewritten:

$$2\varepsilon_{ijk}\Omega_j u_k=-2\varepsilon_{ijk}u_j\Omega_k. \tag{5.17}$$

Substituting (4.69), (4.41), and (5.17) into (5.16) produces:

$$\frac{\partial u_i}{\partial t}+\frac{\partial}{\partial x_i}\left(\frac{1}{2}u_j^2+\Phi\right)-\varepsilon_{ijk}u_j(\omega_k+2\Omega_k)=-\frac{1}{\rho}\frac{\partial p}{\partial x_i}-\nu\varepsilon_{ijk}\frac{\partial \omega_k}{\partial x_j}, \tag{5.18}$$

where $g_i=-\partial\Phi/\partial x_i$ (4.18) has also been used.

Eq. (5.18) is another form of the Navier-Stokes momentum equation, so the rotating-frame-of-reference vorticity equation is obtained by taking its curl. This task is accomplished by applying $\varepsilon_{nqi}\partial/\partial x_q$ on the left side of each term in (5.18):

$$\varepsilon_{nqi}\frac{\partial}{\partial x_q}\left(\frac{\partial u_i}{\partial t}\right)+0-\varepsilon_{nqi}\varepsilon_{ijk}\frac{\partial}{\partial x_q}\left[u_j\left(\omega_k+2\Omega_k\right)\right]=-\varepsilon_{nqi}\frac{\partial}{\partial x_q}\left(\frac{1}{\rho}\frac{\partial p}{\partial x_i}\right)-\nu\varepsilon_{nqi}\varepsilon_{ijk}\frac{\partial^2 \omega_k}{\partial x_q\partial x_j}. \tag{5.19}$$

The second term on the left side of (5.18) vanished because the curl of a gradient is zero (see Exercise 2.23). The second nontrivial term on the left side of (5.19) can be rewritten using the identity (2.19), the continuity equation (4.10), $\partial\omega_k/\partial x_k=0$ (the divergence of a curl is zero; see Exercise 2.22), and the fact that the derivatives of Ω are zero:

$$\begin{aligned}
-\varepsilon_{nqi}\varepsilon_{ijk}\frac{\partial}{\partial x_q}[u_j(\omega_k+2\Omega_k)] &= -(\delta_{nj}\delta_{qk}-\delta_{nk}\delta_{qj})\frac{\partial}{\partial x_q}[u_j(\omega_k+2\Omega_k)] \\
&= -\frac{\partial}{\partial x_k}[u_n(\omega_k+2\Omega_k)]+\frac{\partial}{\partial x_j}[u_j(\omega_n+2\Omega_n)] \\
&= -(\omega_k+2\Omega_k)\frac{\partial u_n}{\partial x_k}+0+0+u_j\frac{\partial}{\partial x_j}(\omega_n+2\Omega_n) \\
&\quad -(\omega_j+2\Omega_j)\frac{\partial u_n}{\partial x_j}+u_j\frac{\partial \omega_n}{\partial x_j}+0.
\end{aligned} \tag{5.20}$$

The first term on the right side of (5.19) can be simplified to a cross product of gradients:

$$-\varepsilon_{nqi}\frac{\partial}{\partial x_q}\left(\frac{1}{\rho}\frac{\partial p}{\partial x_i}\right)=0+\frac{1}{\rho^2}\varepsilon_{nqi}\frac{\partial \rho}{\partial x_q}\frac{\partial p}{\partial x_i}, \tag{5.21}$$

because the curl of the pressure gradient is zero. The viscous term in (5.19) can be rewritten using (2.19):

$$-\nu\varepsilon_{nqi}\varepsilon_{ijk}\frac{\partial^2\omega_k}{\partial x_q\partial x_j}=-\nu(\delta_{nj}\delta_{qk}-\delta_{nk}\delta_{qj})\frac{\partial^2\omega_k}{\partial x_q\partial x_j}=-\nu\frac{\partial^2\omega_k}{\partial x_k\partial x_n}+\nu\frac{\partial^2\omega_n}{\partial x_j^2}=0+\nu\frac{\partial^2\omega_n}{\partial x_j^2}, \tag{5.22}$$

where the final equality follows because $\partial\omega_k/\partial x_k=0$. Substituting (5.20) through (5.22) into (5.19) leads to:

$$\frac{\partial \omega_n}{\partial t}=\frac{\partial u_n}{\partial x_j}(\omega_j+2\Omega_j)-u_j\frac{\partial \omega_n}{\partial x_j}+\frac{1}{\rho^2}\varepsilon_{nqi}\frac{\partial \rho}{\partial x_q}\frac{\partial p}{\partial x_i}+\nu\frac{\partial^2\omega_n}{\partial x_j^2}.$$

Using (3.5), the definition of D/Dt, produces

$$\frac{D\omega_n}{Dt} = (\omega_j + 2\Omega_j)\frac{\partial u_n}{\partial x_j} + \frac{\varepsilon_{nqi}}{\rho^2}\frac{\partial \rho}{\partial x_q}\frac{\partial p}{\partial x_i} + \nu\frac{\partial^2 \omega_n}{\partial x_j^2}, \quad \text{or}$$
$$\frac{D\boldsymbol{\omega}}{Dt} = (\boldsymbol{\omega} + 2\boldsymbol{\Omega})\cdot\nabla\mathbf{u} + \frac{1}{\rho^2}\nabla\rho\times\nabla p + \nu\nabla^2\boldsymbol{\omega}, \tag{5.23}$$

where the second equation is merely the first rewritten in vector notation. The terms on the right-hand side are known as the *vortex stretching* term, the *baroclinic torque*, and the *viscous torque*. The vortex stretching term is discussed in Section 5.3. The baroclinic torque is illustrated in Fig. 5.6. The viscous torque is considered in Example 5.2.

This is the *vorticity equation* for variable-density incompressible flow with constant viscosity observed from a frame of reference rotating at a constant rate $\boldsymbol{\Omega}$. Here $\mathbf{u}$ and $\boldsymbol{\omega}$ are, respectively, the velocity and vorticity observed in this rotating frame of reference. Given that vorticity is defined as twice the angular velocity, $2\boldsymbol{\Omega}$ is known as the *planetary vorticity* in atmospheric flows, and $(\boldsymbol{\omega} + 2\boldsymbol{\Omega})$ is the *absolute vorticity* of the fluid measured in an inertial frame. The variable-density incompressible flow vorticity equation in a nonrotating frame is obtained from (5.23) by setting $\boldsymbol{\Omega}$ to zero and interpreting $\mathbf{u}$ and $\boldsymbol{\omega}$ as the absolute velocity and vorticity, respectively.

Similar to (5.12), the left side of (5.23) represents the rate of change of vorticity following a fluid particle and the last term $\nu\nabla^2\boldsymbol{\omega}$ represents the rate of change due to viscous diffusion of vorticity. The first term on the right side of (5.23) represents vortex stretching and plays a crucial role in vorticity dynamics even when $\boldsymbol{\Omega} = 0$. The second term on the right-hand side is the rate of generation of vorticity due to baroclinicity (varying density) in the flow, as discussed in Section 5.2. In a barotropic flow, density is a function of pressure alone, so $\nabla\rho$ and ∇p are parallel vectors and this term disappears.

To better understand how rotation of the frame of reference influences vorticity, consider $\boldsymbol{\Omega} = \Omega\mathbf{e}_z$ so that $2(\boldsymbol{\Omega}\cdot\nabla)\mathbf{u} = 2\Omega(\partial\mathbf{u}/\partial z)$ and suppress all other terms on the right side of (5.23) to obtain the component equations:

$$\frac{D\omega_z}{Dt} = 2\Omega\frac{\partial w}{\partial z}, \quad \frac{D\omega_x}{Dt} = 2\Omega\frac{\partial u}{\partial z}, \quad \text{and} \quad \frac{D\omega_y}{Dt} = 2\Omega\frac{\partial v}{\partial z}.$$

This shows that stretching of fluid lines in the z direction increases ω_z, whereas a tilting of vertical lines changes the relative vorticity along the x and y directions. Note that merely stretching or turning of vertical *material lines* is required for this mechanism to operate, in contrast to $(\boldsymbol{\omega}\cdot\nabla)\mathbf{u}$ where a stretching or turning of *vortex lines* is needed. This is because vertical material lines contain the planetary vorticity $2\Omega\mathbf{e}_z$. Thus, a vertically stretching fluid column tends to acquire positive ω_z, and a vertically shrinking fluid column tends to acquire negative ω_z (Fig. 5.10). For this reason large-scale geophysical flows are almost always influenced by vorticity, and the change of $\boldsymbol{\omega}$ due to the presence of planetary vorticity $2\boldsymbol{\Omega}$ is a central feature of geophysical fluid dynamics.

Kelvin's circulation theorem for inviscid flow in a rotating frame of reference is modified to:

$$\frac{D\Gamma_a}{Dt} = 0 \quad \text{where} \quad \Gamma_a \equiv \int_A (\boldsymbol{\omega} + 2\boldsymbol{\Omega})\cdot\mathbf{n}\,dA = \Gamma + 2\int_A \boldsymbol{\Omega}\cdot\mathbf{n}\,dA \tag{5.24}$$

(see Exercise 5.11). Here, Γ_a is circulation due to the absolute vorticity, $\boldsymbol{\omega} + 2\boldsymbol{\Omega}$, and differs from Γ by the amount of planetary vorticity intersected by the area A.

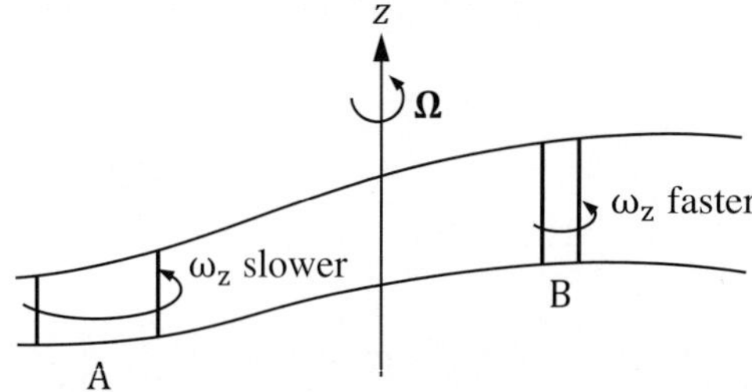

FIGURE 5.10 Generation of vorticity in a steadily rotating frame of reference due to stretching of fluid columns parallel to the planetary vorticity $2\boldsymbol{\Omega}$. A fluid column acquires $\boldsymbol{\omega}_z$ (in the same sense as $\boldsymbol{\Omega}$) by moving from location A to location B.

Example 5.4

Consider constant-density pure-strain extensional (stretching) flow along the z-axis, $u_R = -(\gamma/2)R$ and $u_z = \gamma z$, where γ is the strain rate, in a steadily rotating frame of reference with $\mathbf{\Omega} = \Omega \mathbf{e}_z$. If $\boldsymbol{\omega} = \omega_z \mathbf{e}_z = 0$ at $t = 0$, what is $\omega_z(t)$?

Solution To determine $\omega_z(t)$, (5.23) must be solved for the given flow field and initial condition. Here, the solution $\omega_z(t)$ is presumed to be independent of the spatial coordinates. So, for the given velocity field with $\rho =$ constant, the z-direction component of (5.23) is:

$$\frac{\partial \omega_z}{\partial t} = (\omega_z + 2\Omega)\frac{\partial u_z}{\partial z} = (\omega_z + 2\Omega)\gamma.$$

The two ends of this equality form a simple linear differential equation. Integrating after separating variables yields:

$$\ln(2\Omega + \omega_z) = \gamma t + \text{const.}$$

Here, the initial condition requires const. $= \ln(2\Omega)$. A little algebra leads to the final answer:

$$\omega_z(t) = 2\Omega(e^{\gamma t} - 1).$$

Therefore, when both Ω and γ are both positive, vorticity can arise from an apparently irrotational background. This example illustrates how vertical stretching can concentrate planetary vorticity.

5.5 Velocity induced by a vortex filament: law of Biot and Savart

For variety of applications in aero- and hydrodynamics, the flow induced by a concentrated distribution of vorticity (a vortex) with arbitrary orientation must be calculated. Here we consider the simple case of incompressible flow where $\nabla \cdot \mathbf{u} = 0$. Taking the curl of the vorticity produces:

$$\nabla \times \boldsymbol{\omega} = \nabla \times (\nabla \times \mathbf{u}) = \nabla(\nabla \cdot \mathbf{u}) - \nabla^2 \mathbf{u} = -\nabla^2 \mathbf{u},$$

where the second equality follows from an identity of vector calculus (B.3.13). The two ends of this extended equality form a Poisson equation, and its solution is the vorticity-induced portion of the fluid velocity:

$$\mathbf{u}(\mathbf{x}, t) = -\frac{1}{4\pi}\int_{V'} \frac{1}{|\mathbf{x} - \mathbf{x}'|}(\nabla' \times \boldsymbol{\omega}(\mathbf{x}', t))d^3x', \tag{5.25}$$

where V' encloses the vorticity of interest and ∇' operates on the $\mathbf{x}'$ coordinates (see Exercise 5.10). This result can be further simplified by rewriting the integrand in (5.25):

$$\begin{aligned}\frac{1}{|\mathbf{x} - \mathbf{x}'|}(\nabla' \times \boldsymbol{\omega}(\mathbf{x}', t)) &= \nabla' \times \left(\frac{\boldsymbol{\omega}(\mathbf{x}', t)}{|\mathbf{x} - \mathbf{x}'|}\right) - \nabla'\left(\frac{1}{|\mathbf{x} - \mathbf{x}'|}\right) \times \boldsymbol{\omega}(\mathbf{x}', t)\\ &= \nabla' \times \left(\frac{\boldsymbol{\omega}(\mathbf{x}', t)}{|\mathbf{x} - \mathbf{x}'|}\right) + \left(\frac{\mathbf{x} - \mathbf{x}'}{|\mathbf{x} - \mathbf{x}'|^3}\right) \times \boldsymbol{\omega}(\mathbf{x}', t),\end{aligned}$$

to obtain:

$$\mathbf{u}(\mathbf{x}, t) = -\frac{1}{4\pi}\int_{V'} \nabla' \times \left(\frac{\boldsymbol{\omega}(\mathbf{x}', t)}{|\mathbf{x} - \mathbf{x}'|}\right)d^3x' + \frac{1}{4\pi}\int_{V'} \frac{\boldsymbol{\omega}(\mathbf{x}', t) \times (\mathbf{x} - \mathbf{x}')}{|\mathbf{x} - \mathbf{x}'|^3}d^3x'.$$

Here the first integral is zero when V' is chosen to capture a segment of the vortex, but it takes several steps to deduce this. First, rewrite the curl operation in index notation and apply Gauss' divergence theorem:

$$\begin{aligned}\int_{V'} \nabla' \times \left(\frac{\boldsymbol{\omega}(\mathbf{x}', t)}{|\mathbf{x} - \mathbf{x}'|}\right)d^3x' &= \int_{V'} \varepsilon_{kij}\frac{\partial}{\partial x_i'}\left(\frac{\omega_j(\mathbf{x}', t)}{|\mathbf{x} - \mathbf{x}'|}\right)d^3x' = \int_{A'} \varepsilon_{kij}\left(\frac{\omega_j(\mathbf{x}', t)}{|\mathbf{x} - \mathbf{x}'|}\right)n_i d^2x'\\ &= \int_{A'} \frac{\mathbf{n} \times \boldsymbol{\omega}(\mathbf{x}', t)}{|\mathbf{x} - \mathbf{x}'|}d^2x',\end{aligned} \tag{5.26}$$

where A' is the surface of V' and $\mathbf{n}$ is the outward normal on A'. Now choose V' to be a volume aligned so that its end surfaces are locally normal to $\boldsymbol{\omega}(\mathbf{x}', t)$ while its curved lateral surface lies outside the concentration of vorticity as shown in Fig. 5.11. For this volume, the final integral in (5.26) is zero because $\mathbf{n} \times \boldsymbol{\omega} = 0$ on its end surfaces since $\boldsymbol{\omega}(\mathbf{x}', t)$ and $\mathbf{n}$ are parallel there, and because $\boldsymbol{\omega}(\mathbf{x}', t) = 0$ on its lateral surface. Alternatively, the same result may be achieved when $\boldsymbol{\omega}(\mathbf{x}', t)$ is nonzero on the volume's lateral surfaces when the integrand is self-canceling on the volume's lateral surfaces. In either case, (5.25) reduces to:

$$\mathbf{u}(\mathbf{x}, t) = \frac{1}{4\pi} \int_{V'} \frac{\boldsymbol{\omega}(\mathbf{x}', t) \times (\mathbf{x} - \mathbf{x}')}{|\mathbf{x} - \mathbf{x}'|^3} d^3x'. \tag{5.27}$$

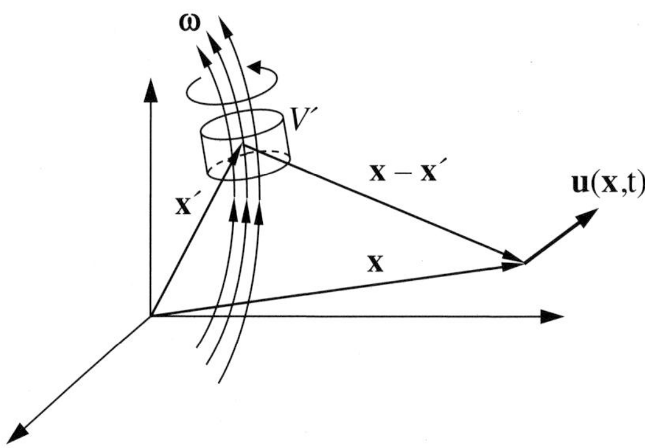

FIGURE 5.11 Geometry for derivation of Law of Biot and Savart. The location of the vorticity concentration or vortex is $\mathbf{x}'$. The location of the vortex-induced velocity $\mathbf{u}$ is $\mathbf{x}$. The volume V' contains an element of the vortex. Its flat ends are perpendicular to the vorticity in the vortex, while its curved lateral sides lie outside the vortex.

This result can be further specialized to a slender vortex element of length dl by choosing $V' = \Delta A' dl$, and assuming the observation location, $\mathbf{x}$, is sufficiently distant from the vortex location $\mathbf{x}'$ so that $(\mathbf{x} - \mathbf{x}')/|\mathbf{x} - \mathbf{x}'|^3$ is effectively constant over the vorticity concentration. In this situation, (5.27) may be simplified to:

$$\begin{aligned} d\mathbf{u}(\mathbf{x}, t) &\cong \frac{1}{4\pi} \int_{\Delta A'} |\boldsymbol{\omega}(\mathbf{x}', t)| \mathbf{e}_\omega d^2x' \times \frac{(\mathbf{x} - \mathbf{x}')}{|\mathbf{x} - \mathbf{x}'|^3} dl = \frac{\Gamma dl}{4\pi} \mathbf{e}_\omega \times \frac{(\mathbf{x} - \mathbf{x}')}{|\mathbf{x} - \mathbf{x}'|^3}, \text{ or} \\ \mathbf{u}(\mathbf{x}, t) &= \int_{vortex} d\mathbf{u} \cong \frac{\Gamma}{4\pi} \int_{vortex} \mathbf{e}_\omega \times \frac{(\mathbf{x} - \mathbf{x}')}{|\mathbf{x} - \mathbf{x}'|^3} dl, \end{aligned} \tag{5.28}$$

where $d\mathbf{u}$ is the velocity induced by an element of the slender vortex having length dl, $\mathbf{u}$ is the velocity induced by the entire slender vortex, and Γ and $\mathbf{e}_\omega$ are the strength and direction, respectively, of the vortex element at $\mathbf{x}'$. These are expressions of the Biot–Savart vortex induction law. They are useful for determining vortex-induced velocities, and are the basis for some fluid-flow simulation techniques (see Example 5.8).

Example 5.5

Consider a thin ideal vortex segment of uniform strength Γ that lies along the z-axis between z_1 and z_2, and has a sense of rotation that points along the z-axis. Use the Biot–Savart law to show that the induced velocity $\mathbf{u}$ at the location (R, φ, z) will be $\mathbf{u}(\mathbf{x}, t) = (\Gamma/4\pi R)(\cos\theta_1 - \cos\theta_2)\mathbf{e}_\varphi$ where the polar angles θ_1 and θ_2 are as shown in Fig. 5.12.

Solution Apply (5.28) to the geometry shown with $\mathbf{x}' = (0, 0, z')$:

$$\frac{(\mathbf{x} - \mathbf{x}')}{|\mathbf{x} - \mathbf{x}'|^3} = \frac{(x, y, z - z')}{[x^2 + y^2 + (z - z')^2]^{3/2}} \text{ and } \mathbf{e}_z \times \frac{(\mathbf{x} - \mathbf{x}')}{|\mathbf{x} - \mathbf{x}'|^3} = \frac{(-y, x, 0)}{[x^2 + y^2 + (z - z')^2]^{3/2}}.$$

Switch to cylindrical coordinates where $R^2 = x^2 + y^2$, and $\mathbf{e}_\varphi = -\mathbf{e}_x \sin_\varphi + \mathbf{e}_y \cos_\varphi$. This allows the geometrical results to be mildly simplified:

$$\mathbf{e}_z \times \frac{(\mathbf{x} - \mathbf{x}')}{|\mathbf{x} - \mathbf{x}'|^3} = \frac{R\mathbf{e}_\varphi}{\left[R^2 + (z - z')^2\right]^{3/2}}.$$

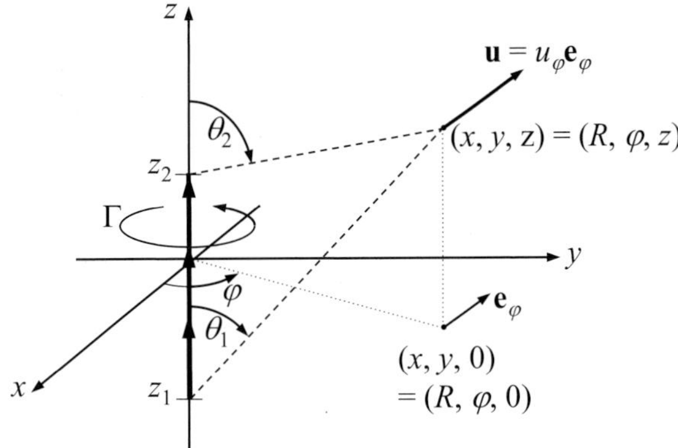

FIGURE 5.12 Geometry for determination of the induced velocity $\mathbf{u}$ from a straight ideal vortex segment with strength Γ that lies along the z-axis between $z = z_1$ and $z = z_2$. Here the induced velocity will be in the $\mathbf{e}_\varphi$ direction and will depend on the polar angles θ_1 and θ_2.

With $dl = dz$, the total induced velocity from the vortex segment is given by (5.28) integrated between z_1 and z_2:

$$\mathbf{u}(\mathbf{x},t) = \frac{\Gamma \mathbf{e}_\varphi}{4\pi} \int_{z_1}^{z_2} \frac{R\,dz}{\left[R^2 + (z - z')^2\right]^{3/2}}.$$

The integral can be evaluated by switching to a polar-angle integration variable:

$$\cos\theta = \frac{z - z'}{\left[R^2 + (z - z')^2\right]^{1/2}},$$

having a convenient differential:

$$\begin{aligned} d(\cos\theta) &= \left\{ \frac{-1}{\left[R^2 + (z - z')^2\right]^{1/2}} - \frac{1}{2} \frac{(z - z')}{\left[R^2 + (z - z')^2\right]^{3/2}} \cdot 2(z - z')(-1) \right\} dz' \\ &= -\left\{ \frac{R^2}{\left[R^2 + (z - z')^2\right]^{3/2}} \right\} dz'. \end{aligned}$$

After making this variable change, the remaining integral is elementary:

$$\mathbf{u}(\mathbf{x},t) = -\frac{\Gamma \mathbf{e}_\varphi}{4\pi R} \int_{\theta_1}^{\theta_2} d(\cos\theta) = \frac{\Gamma \mathbf{e}_\varphi}{4\pi R} (\cos\theta_1 - \cos\theta_2), \tag{5.29}$$

where θ_1 and θ_2 are the polar angles from the vortex segment's lower and upper ends, respectively:

$$\cos\theta_1 = \frac{z - z_1}{\left[R^2 + (z - z_1)^2\right]^{1/2}} \quad \text{and} \quad \cos\theta_2 = \frac{z - z_2}{\left[R^2 + (z - z_2)^2\right]^{1/2}}.$$

For an infinite line vortex ($\theta_1 = 0$, and $\theta_2 = \pi$), (5.29) produces $\mathbf{u}(\mathbf{x},t) = \Gamma \mathbf{e}_\varphi / 2\pi R$.

5.6 Interaction of vortices

Vortices placed close to one another can mutually interact through their induced velocities and generate interesting motions. To examine such interactions, consider ideal concentrated line vortices. A real vortex, with a vorticity distribution in its core, can be idealized at distances larger than several core diameters as a concentrated vortex line with circulation equal to that of the real vortex. In this idealization, motion outside the vortex core is irrotational and therefore inviscid. Furthermore, the Biot–Savart law for concentrated vortices embodies the superposition principle since it states that the induced velocity at a point of interest is a sum (an integral) of the induced velocity contributions from all vortex elements. To determine the mutual interaction of line vortices, the important principle to keep in mind is the first Helmholtz vortex theorem – vortex lines move with the flow.

Consider the interaction of two ideal line vortices of strengths Γ_1 and Γ_2, where both Γ_1 and Γ_2 are positive (i.e., counterclockwise rotation). Let $h = h_1 + h_2$ be the distance between the vortices (Fig. 5.13). Then the velocity at point 2 due to vortex Γ_1 is directed upward and equals:

$$V_1 = \Gamma_1/2\pi h. \tag{5.30}$$

Similarly, the velocity at point 1 due to vortex Γ_2 is downward and equals:

$$V_2 = \Gamma_2/2\pi h. \tag{5.31}$$

When both vortices can move, they travel counterclockwise on circular trajectories centered on point G, the center of vorticity, which remains stationary. In terms of Γ_1, Γ_2, and h, G is found from:

$$h_1 = \frac{\Gamma_2 h}{\Gamma_1 + \Gamma_2}, \quad \text{and} \quad h_2 = \frac{\Gamma_1 h}{\Gamma_1 + \Gamma_2} \tag{5.32a,b}$$

(see Exercise 5.22).

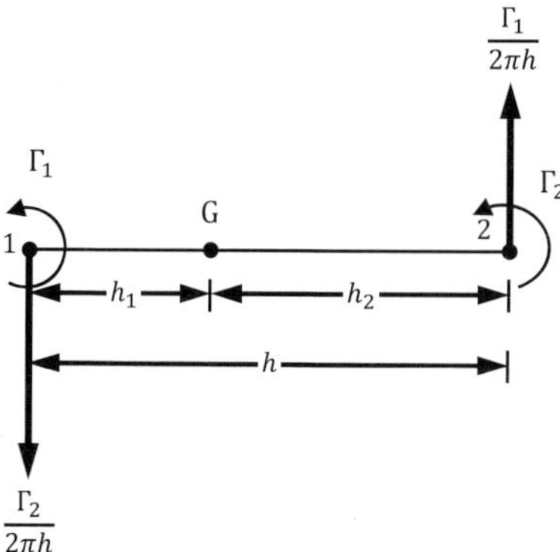

FIGURE 5.13 Interaction of two line vortices of the same sign. Here the induced velocities are in opposite directions and perpendicular to the line connecting the vortices. Thus, if free to move, the two vortices will travel on circular paths centered on the point G, the center of circulation.

Now suppose that the two vortices have the same circulation magnitude Γ, but an opposite sense of rotation (Fig. 5.14). Then the velocity of each vortex at the location of the other is $\Gamma/(2\pi h)$ so the dual-vortex system translates at a speed $\Gamma/(2\pi h)$ relative to the fluid. A pair of counter-rotating vortices can be set up by stroking a rectangular canoe paddle horizontally, or by briefly moving the blade of a knife in a bucket of water (Fig. 5.15). After the paddle or knife is withdrawn, the vortices do not remain stationary but continue to move.

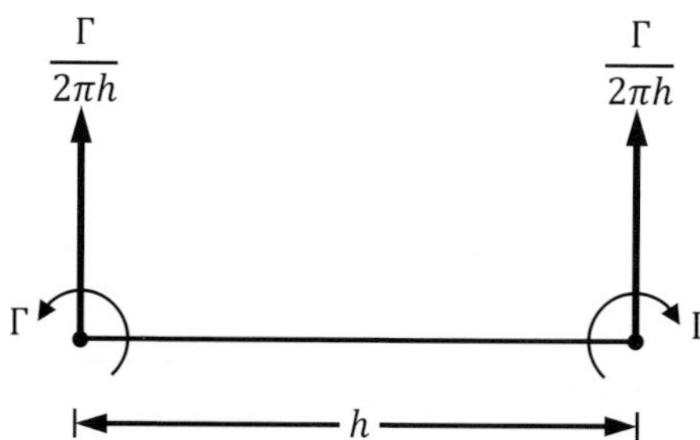

FIGURE 5.14 Interaction of line vortices of opposite spin, but of the same magnitude. Here Γ refers to the *magnitude* of circulation, and the induced velocities are in same direction and perpendicular to the line connecting the vortices. Thus, if free to move, the two vortices will travel along straight lines in the direction shown at speed $\Gamma/2\pi h$.

The behavior of a single ideal vortex near a wall can be found by superposing two vortices of equal and opposite strength. The technique involved is called the *method of images* and has wide application in irrotational flow, heat conduction, acoustics, and electromagnetism. The inviscid flow pattern due to vortex A at distance h from a wall can be obtained by eliminating the wall and introducing instead a vortex of equal and opposite strength at the image point B (Fig. 5.16). The fluid velocity at any point P on the wall, made up of V_A due to the real vortex and V_B due to the image vortex, is parallel to the wall. The wall is therefore a streamline, and the inviscid boundary condition of zero normal velocity across a solid wall

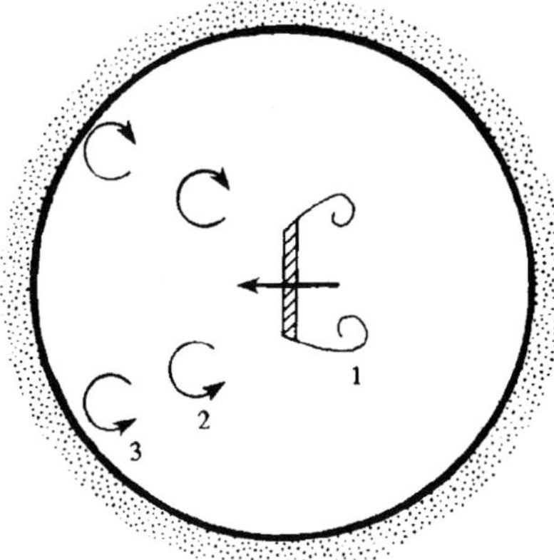

FIGURE 5.15 Top view of a vortex pair generated by moving the blade of a knife in a bucket of water. Positions at three instances of time 1, 2, and 3 are shown (see Lighthill, 1986).

is satisfied. Because of the flow induced by the image vortex, vortex A moves with speed $\Gamma/(4\pi h)$ parallel to the wall. For this reason, the vortices in Fig. 5.15 move apart along the boundary on reaching the side of the vessel.

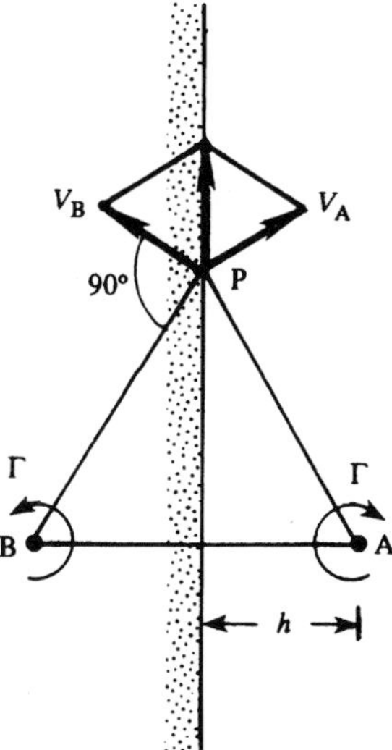

FIGURE 5.16 Line vortex A near a wall and its image B. The sum of the induced velocities is parallel to the wall at all points P on the wall when the two vortices have equal and opposite strengths and they are equidistant from the wall.

Now consider the interaction of two doughnut-shaped vortex rings (such as smoke rings) of equal and opposite circulation (Fig. 5.17a). According to the method of images, the flow field for a single ring near a wall is identical to the flow of two rings of opposite circulations. The translational motion of each element of the ring is caused by the induced velocity from each element of the same ring, plus the induced velocity from each element of the other vortex ring. In the figure, the motion at A is the resultant of V_B, V_C, and V_D, and this resultant has components parallel to and toward the wall. Consequently, the vortex ring increases in diameter and moves toward the wall with a speed that decreases monotonically (Fig. 5.17b).

Finally, consider the interaction of two vortex rings of equal magnitude and similar sense of rotation. It is left to the reader (Exercise 5.19) to show that they should both translate in the same direction, but the one in front increases in radius and therefore slows down in its translational speed, while the rear vortex contracts and translates faster. This continues until the smaller ring passes through the larger one, at which point the roles of the two vortices are reversed. The two vortices can pass through (or leapfrog) each other forever in an ideal fluid. Further discussion of this intriguing problem can be found in Sommerfeld (1964, p. 161).

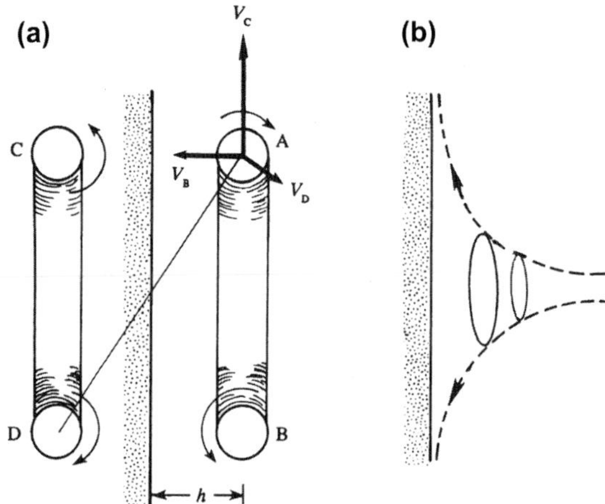

FIGURE 5.17 (a) Torus or doughnut-shaped vortex ring near a wall and its image. A section through the middle of the ring is shown along with primary induced velocities at A from the vortex segments located at B, C, and D. (b) Trajectory of a vortex ring, showing that it widens while its translational velocity toward the wall decreases.

Example 5.6

Two ideal vortices with circulation Γ are located at $(-b, b)$ and (b, b) at the time of interest in a semi-infinite inviscid fluid (see Fig. 5.18). A flat impenetrable surface lies at $y = 0$. Determine the velocities of the point vortices presuming they are free to move.

Solution Each vortex will feel induced velocities from its own image vortex, the other vortex, and the other vortex's image. For the vortex on the left (no. 1), these contributions lead to:

$$\mathbf{u}(-b, b) = \frac{\Gamma}{2\pi \cdot 2b}\mathbf{e}_x - \frac{\Gamma}{2\pi \cdot 2b}\mathbf{e}_y + \frac{\Gamma}{2\pi \cdot 2\sqrt{2}b}\left(\frac{\mathbf{e}_x + \mathbf{e}_y}{\sqrt{2}}\right) = \frac{\Gamma}{4\pi b}\left(\frac{3}{2}\mathbf{e}_x - \frac{1}{2}\mathbf{e}_y\right).$$

For the vortex on the right (no. 2), these contributions are:

$$\mathbf{u}(b, b) = \frac{\Gamma}{2\pi \cdot 2b}\mathbf{e}_x + \frac{\Gamma}{2\pi \cdot 2b}\mathbf{e}_y + \frac{\Gamma}{2\pi \cdot 2\sqrt{2}b}\left(\frac{\mathbf{e}_x - \mathbf{e}_y}{\sqrt{2}}\right) = \frac{\Gamma}{4\pi b}\left(\frac{3}{2}\mathbf{e}_x + \frac{1}{2}\mathbf{e}_y\right).$$

These results show that the vortex pair near a flat wall has a higher horizontal velocity than a single vortex at the same distance from the wall. Plus, vortex no. 1 is moving toward the wall while vortex no. 2 is moving away from the wall at the instant shown.

5.7 Vortex sheet

Consider an infinite number of ideal line vortices, placed side by side on a surface AB (Fig. 5.19). Such a surface is called a *vortex sheet.* If the vortex filaments all rotate clockwise, then the tangential velocity immediately above AB is to the right, while that immediately below AB is to the left. Thus, a discontinuity of tangential velocity exists across a vortex sheet. If the vortex filaments are not infinitesimally thin, then the vortex sheet has a finite thickness, and the velocity change is spread out over this thickness.

In Fig. 5.19, consider the circulation around a circuit of dimensions dn and ds. The normal velocity component v is continuous across the sheet ($v = 0$ if the sheet does not move normal to itself), while the tangential component u experiences a sudden jump. If u_1 and u_2 are the tangential velocities on the two sides, then:

$$d\Gamma = u_2 ds + v dn - u_1 ds - v dn = (u_2 - u_1)ds.$$

Therefore the circulation per unit length, called the *strength of a vortex sheet*, equals the jump in tangential velocity:

$$\gamma \equiv \frac{d\Gamma}{ds} = u_2 - u_1. \tag{5.33}$$

The concept of a vortex sheet is especially useful in discussing the instability of shear flows (Chapter 9), and the flow over aircraft wings (Chapter 13).

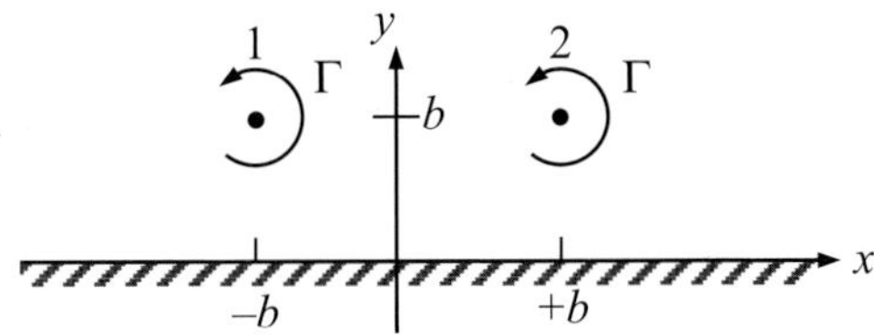

FIGURE 5.18 Two ideal vortices of strength Γ a distance b from a flat surface and a distance $2b$ from each other. Each vortex will move in the velocity field induced by the other vortex, the other vortex's image, and its own image vortex.

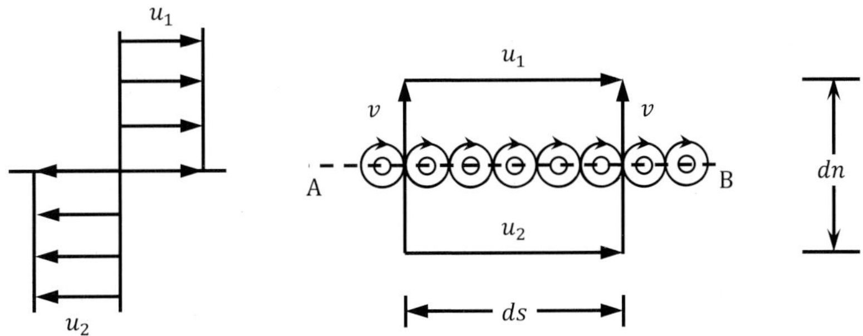

FIGURE 5.19 A vortex sheet produces a change in the velocity that is tangent to it. Vortex sheets may be formed by placing many parallel ideal line vortices next to each other. The strength of a vortex sheet, $d\Gamma/ds = u_1 - u_2$, can be determined by computing the circulation on the rectangular contour shown and this strength may depend on the sheet-tangent coordinate.

Example 5.7

Show that the velocity induced by an infinite stationary vortex sheet of strength γ lying in the plane defined by $y = 0$ (see Fig. 5.20), is $\mathbf{u} = +(\gamma/2)\mathbf{e}_x$ for $y < 0$, and $\mathbf{u} = -(\gamma/2)\mathbf{e}_x$ for $y > 0$.

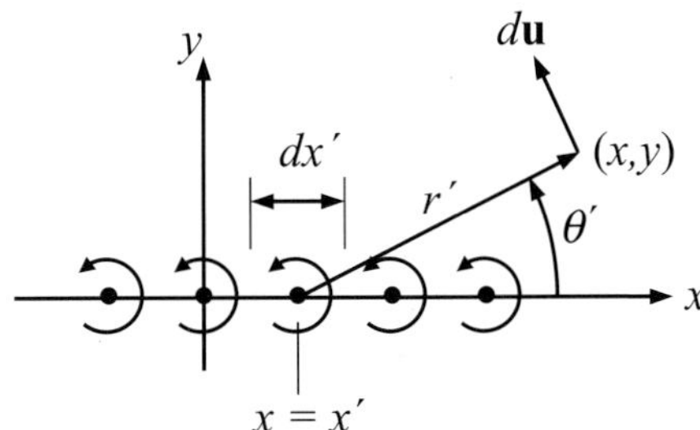

FIGURE 5.20 An ideal vortex sheet lies in the $y = 0$ plane. The fluid velocity $\mathbf{u}$ induced by the whole sheet at the point (x, y) is of interest. If the vortex sheet strength is γ, the circulation of an element of the sheet is $\gamma dx'$.

Solution Consider the infinite sheet composed of many line vortices perpendicular to the x-y plane as shown. As depicted in Fig. 5.18, the circulation of each vortex line element is $\gamma dx'$. Thus, the increment of induced velocity $d\mathbf{u}$ at location (x, y) from the vortex at $x = x'$ is:

$$d\mathbf{u} = \frac{\gamma dx'}{2\pi r'}(-\mathbf{e}_x \sin\theta' + \mathbf{e}_y \cos\theta').$$

Therefore, the induced velocity from all such vortices is:

$$\mathbf{u} = \int_{whole\ sheet} d\mathbf{u} = u\mathbf{e}_x + v\mathbf{e}_y = \frac{\gamma}{2\pi}\int_{-\infty}^{+\infty} \frac{(-\mathbf{e}_x \sin\theta' + \mathbf{e}_y \cos\theta')}{r'} dx',$$

where $r' = \sqrt{(x - x')^2 + y^2}$, $\sin\theta' = y/r'$, and $\cos\theta' = (x - x')/r'$. Using these geometrical relationships and first considering the vertical velocity component leads to:

$$v = \frac{\gamma}{2\pi}\int_{-\infty}^{+\infty} \frac{(x - x')}{(x - x')^2 + y^2} dx' = \frac{\gamma}{2\pi}\int_{-\infty}^{+\infty} \frac{\xi}{\xi^2 + y^2} dx' = 0.$$

This result occurs because the integrand is an odd function of $\xi = x - x'$, but the integration interval is even. The integral for the horizontal velocity component leads to:

$$u = -\frac{\gamma}{2\pi}\int_{-\infty}^{+\infty}\frac{y\,dx'}{(x-x')^2+y^2} = -\frac{\gamma}{2\pi}\left[\mathrm{sgn}(y)\tan^{-1}\left(\frac{x-x'}{|y|}\right)\right]_{-\infty}^{+\infty} = -\frac{\gamma\,\mathrm{sgn}(y)}{2\pi}\left[\frac{\pi}{2}-\left(-\frac{\pi}{2}\right)\right] = -\mathrm{sgn}(y)\frac{\gamma}{2},$$

and this is the desired result.

Example 5.8

Consider a wake represented by two parallel vortex sheets of opposite sign separated by a distance h. In this example, the Biot–Savart vortex induction law will be solved numerically to simulate the dynamics of the two vortex sheets. The vorticity distribution can be approximated by N points vortices each at a location $\mathbf{x}_i$ according to

$$\omega(\mathbf{x}, t) \approx \sum_{i=1}^{N}\Gamma_i\delta(\mathbf{x}-\mathbf{x}_i(t)),$$

where δ is the two-dimensional Dirac delta function. Recasting (5.28) for the vorticity-induced fluid velocity in two-dimensions (see Exercise 5.14), the path of each point vortex $\mathbf{x}_i(t)$ can be obtained by solving the following system of ordinary differential equations

$$\frac{d\mathbf{x}_i}{dt} \equiv \mathbf{u}_i = \frac{1}{2\pi}\sum_{\substack{j=1\\ j\neq i}}^{N}\Gamma_j\frac{\mathbf{e}_z\times(\mathbf{x}_i-\mathbf{x}_j)}{|\mathbf{x}_i-\mathbf{x}_j|^2}.$$

A direct solution to the expression above is numerically unstable (i.e., it is known to result in unbounded velocities) when two vortices come very close to each other. The so-called "vortex blob" method was introduced to alleviate this issue (Leonard, 1980). The key idea is to replace the Dirac delta function with a regularized (smooth) function of finite size ϵ that approaches δ when $\epsilon \to 0$. A common approach is to replace δ with a Gaussian function by multiplying the term in the summation by $1-\exp(-|\mathbf{x}_i-\mathbf{x}_j|^2/2\epsilon^2)$.

Use MATLAB® or another programming language to simulate the behavior of 2 rows of $N = 100$ point vortices in a periodic domain of length $L = 1$. The speed of the wake is $U_\infty = 1$. Add sinusoidal perturbations to each sheet with amplitude $A = 0.05L$ and wavelength $\lambda = L$. Set the width of the vortex blob $\epsilon = 4L/N$. Show a time series of the vortex distribution for $0 \le tU_\infty/L \le 1.6$ with $h = 0.25L$ and $h = 0.1L$.

Solution To solve for the path of each vortex, the expression above for the induced velocity is discretized using a second-order Runge–Kutta method. The circulation of each point vortex Γ_i can be determined from the tangential velocity of the wake U_∞ according to (5.33), resulting in $\Gamma_i = \pm U_\infty\Delta x$, where $\Delta x = L/N$ is the initial spacing between the vortex pairs. The upper sheet has positive circulation while the lower sheet has opposite sign. A MATLAB code is provided below.

```
% Parameters
global L U sig
U=1; L=1;
lam=L; A=0.15*L; h=0.5*L;

% Initialize point vorticies
N=100; dx=L/N;
x=linspace(dx/2,L-dx/2,N); x = [x,x];
y(1:N)=A*sin(2*pi.*x(1:N)/lam)+h/2;
y(N+1:2*N)=A*sin(2*pi.*x(N+1:2*N)/lam)-h/2;
gamma(1:N)=U*dx;
gamma(N+1:2*N)=-U*dx;
sig=4*dx;

% Loop through time
t=linspace(0,1.6,1000); dt=t(2)-t(1);
for i=1:length(t)

    % Second-order Runge-Kutta
    xold=x; yold=y;
```

```
    [u,v] = biotsavart(x,y,gamma);
    x = xold + 0.5*dt*u;
    y = yold + 0.5*dt*v;

    % Correct – Midpoint rule
    [u,v] = biotsavart(x,y,gamma);
    x = xold + dt*u;
    y = yold + dt*v;

    % Correct for periodicity
    for ni=1:2*N
        if x(ni)>L
            x(ni)=x(ni)-L;
        elseif x(ni)<0
            x(ni)=x(ni)+L;
        end
    end

% ------------------------- %
% Biot-Savart function %
% ------------------------- %
function [u,v] = biotsavart(x,y,gamma)
global L U sig

N=length(x);
u=zeros(1,N);
v=zeros(1,N);

% Effect of vortex j on vortex i
for i=1:N
    for j=1:N
        if i≠j
            % Get separation distance (account for periodicity)
            if x(i)<x(j)
                xjp=x(j)-L;
                x12p=x(i)-xjp;
            else
                xjp=x(j)+L;
                x12p=x(i)-xjp;
            end
            x12=x(i)-x(j);
            if abs(x12)>abs(x12p)
                x12=x12p;
            end
            y12=y(i)-y(j);
            r=sqrt(x12^2+y12^2);

            % Induced velocity
            K=1/(2*pi*r^2)*(1-exp(-r^2/(2*sig^2)));
            u(i)=u(i)-gamma(j)*K*y12;
            v(i)=v(i)+gamma(j)*K*x12;
        end
    end
end

end
```

The evolution of vorticity distribution is shown in Fig. 5.21. For large separation h, the two vortex sheets roll up without interacting significantly. As the distance between the sheets is decreased, the dynamics of each sheet becomes increasingly affected by the other.

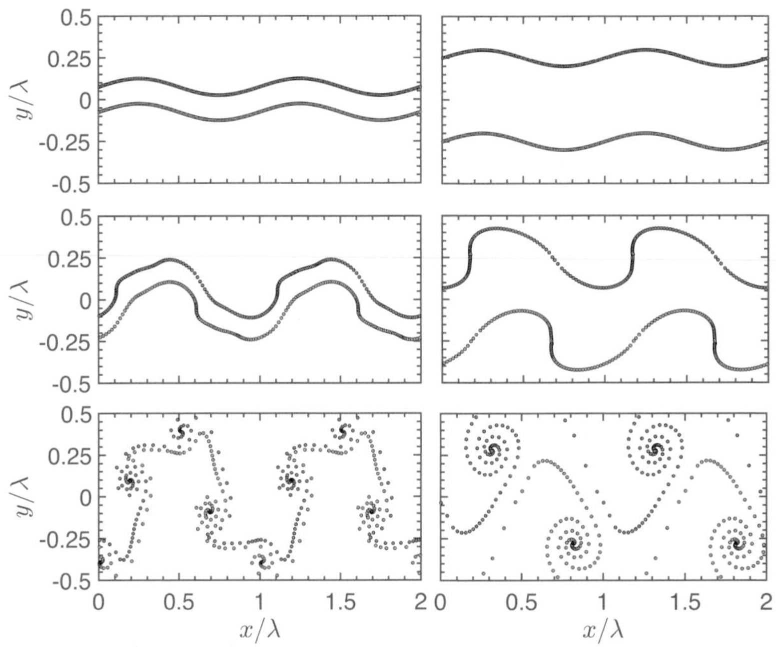

FIGURE 5.21 Solution to Example 5.8 showing point vortices with counterclockwise [blue (dark gray in print version)] and clockwise [red (mid gray in print version)] rotation. $h = 0.15L$ (left) and $h = 0.5L$ (right) at $t = 0$ (top), $t = 0.8L/U_\infty$ (middle), and $t = 1.6L/U_\infty$ (bottom).

Exercises

5.1 A closed cylindrical tank 4 m high and 2 m in diameter contains water to a depth of 3 m. When the cylinder is rotated at a constant angular velocity of 40 rad/s, show that nearly 0.71 m^2 of the bottom surface of the tank is uncovered. [*Hint*: The free surface is in the form of a paraboloid of revolution. For a point on the free surface, let h be the height above the (imaginary) vertex of the paraboloid and r be the local radius of the paraboloid. From Section 5.1, $h = \omega_0^2 r^2/2g$, where ω_0 is the angular velocity of the tank. Apply this equation to the two points where the paraboloid cuts the top and bottom surfaces of the tank.]

5.2 A tornado can be idealized as a Rankine vortex with a core of diameter 30 m. The gauge pressure at a radius of 15 m is -2000 N/m^2 (i.e., the absolute pressure is 2000 N/m^2 below atmospheric).

a) Show that the circulation around any circuit surrounding the core is 5485 m^2/s. [*Hint*: Apply the Bernoulli equation between infinity and the edge of the core.]

b) Such a tornado is moving at a linear speed of 25 m/s relative to the ground. Find the time required for the gauge pressure to drop from -500 to -2000 N/m^2. Neglect compressibility effects and assume an air temperature of 25 °C. (Note that the tornado causes a sudden decrease of the local atmospheric pressure. The damage to structures is often caused by the resulting excess pressure on the inside of the walls, which can cause a house to explode.)

5.3 The velocity field of a flow in cylindrical coordinates (R, φ, z) is $\mathbf{u} = (u_R, u_\varphi, u_z) = (0, aRz, 0)$ where a is a constant.

a) Show that the vorticity components are $\boldsymbol{\omega} = (\omega_R, \omega_\varphi, \omega_z) = (-aR, 0, 2az)$.

b) Verify that $\nabla \cdot \boldsymbol{\omega} = 0$.

c) Sketch the streamlines and vortex lines in an (R, z)-plane. Show that the vortex lines are given by $zR^2 = \text{constant}$.

5.4 Starting from the flow field of an ideal vortex (5.2), compute the viscous stresses σ_{rr}, $\sigma_{r\theta}$, and $\sigma_{\theta\theta}$, and show that the net viscous force on a fluid element, $(\partial\tau_{ij}/\partial x_i)$, is zero.

5.5 Consider the situation depicted in Fig. 5.6. Use a Cartesian coordinate system with a horizontal x-axis that puts the barrier at $x = 0$, a vertical y-axis that puts the bottom of the container at $y = 0$ and the top of the container at $y = H$, and a z-axis that points out of the page. Show that, at the instant the barrier is removed, the rate of baroclinic vorticity production at the interface between the two fluids is:

$$\frac{D\omega_z}{Dt} = \frac{2(\rho_2 - \rho_1)g}{(\rho_2 + \rho_1)\delta},$$

where the thickness of the density transition layer just after barrier removal is $\delta \ll H$, and the density in this thin interface layer is assumed to be $(\rho_1 + \rho_2)/2$. If necessary, also assume that fluid pressures match at $y = H/2$ just after barrier removal, and that the width of the container into the page is b. State any additional assumptions that you make.

5.6 At $t = 0$ a constant-strength z-directed vortex sheet is created in the x-z plane ($y = 0$) in an infinite pool of a fluid with kinematic viscosity ν, that is, $\boldsymbol{\omega}(y, 0) = \mathbf{e}_z \gamma \delta(y)$. The symmetry of the initial condition suggests that $\boldsymbol{\omega} = \omega_z \mathbf{e}_z$ and that ω_z will only depend on y and t. Determine $\boldsymbol{\omega}(y, t)$ for $t > 0$ via the following steps.

a) Determine a dimensionless scaling law for ω_z in terms of γ, ν, y, and t.

b) Simplify the general vorticity equation (5.13) to a linear field equation for ω_z for this situation.

c) Based on the fact that the field equation is linear, simplify the result of part a) by requiring ω_z to be proportional to γ, plug the simplified dimensionless scaling law into the equation determined for part b), and solve this equation to find the undetermined function to reach:

$$\omega_z(y, t) = \frac{\gamma}{2\sqrt{\pi \nu t}} \exp\left\{-\frac{y^2}{4\nu t}\right\}$$

5.7 [1]**a)** Starting from the continuity and Euler equations for an inviscid compressible fluid, $\partial\rho/\partial t + \nabla \cdot (\rho \mathbf{u}) = 0$ and $\rho(D\mathbf{u}/Dt) = -\nabla p + \rho \mathbf{g}$, derive the Vazsonyi equation:

$$\frac{D}{Dt}\left(\frac{\boldsymbol{\omega}}{\rho}\right) = \left(\frac{\boldsymbol{\omega}}{\rho}\right) \cdot \nabla u + \frac{1}{\rho^3} \nabla \rho \times \nabla p,$$

when the body force is conservative: $\mathbf{g} = -\nabla \Phi$. This equation shows that $\boldsymbol{\omega}/\rho$ in a compressible flow plays nearly the same dynamic role as $\boldsymbol{\omega}$ in an incompressible flow [see (5.23) with $\boldsymbol{\Omega} = 0$ and $\nu = 0$].

b) Show that the final term in the Vazsonyi equation may also be written: $(1/\rho)\nabla T \times \nabla s$.

c) Simplify the Vazsonyi equation for barotropic flow.

5.8 Consider rotational and variable-density flow of an inviscid fluid with density ρ.

a) Use the continuity, Euler, and/or Vazsonyi equations, to derive the following conservation equation for $\rho\boldsymbol{\omega}$, where $\boldsymbol{\omega} = \nabla \times \mathbf{u}$ is the vorticity.

$$\frac{D\rho\boldsymbol{\omega}}{Dt} = [(\rho\boldsymbol{\omega} \cdot \nabla]\mathbf{u} - 2(\rho\boldsymbol{\omega})\nabla \cdot \mathbf{u} + \frac{1}{\rho}\nabla\rho \times \nabla p.$$

b) Consider uniform expansion of a uniformly rotating flow field in two dimensional Cartesian coordinates: $\mathbf{u} = (u, v, 0) = (-\Omega y + Ax/2, \Omega x + Ay/2, 0)$, where A is a constant. If ρ and Ω depend on t = time, and are equal to ρ_o and Ω_o at $t = 0$, determine $\rho(t)$ and $\Omega(t)$.

c) Does the baroclinic torque act in the part b) flow field?

d) Consider a fluid disk centered on $x = y = 0$ of radius $a(t)$ and height h = constant in the part-b) flow field that expands (or contracts) so it has constant mass and always contains the same fluid particles. At any instant in time, the angular momentum (H) of the fluid inside this disk about the z-axis is $H(t) = (\pi/2)\rho a^4 h\Omega$. Determine $H(t)$ in terms of the initial radius of the disk $= a_o$, and the other parameters in this problem. Is there anything unusual or interesting about your answer? Explain. [Hint: $da/dt = (u_R)_{R=a}$ where R is the radial coordinate from the z-axis.]

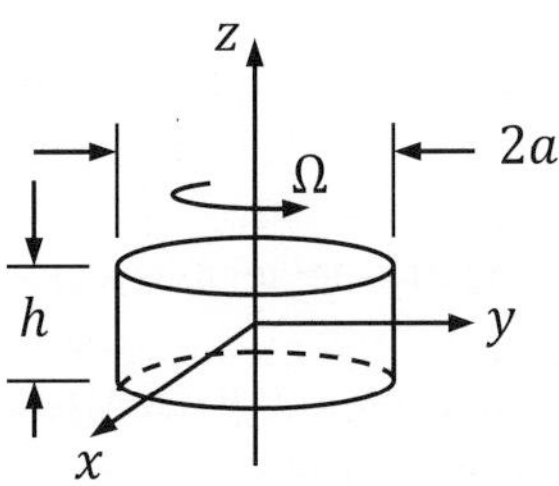

1. Obtained from Prof. Paul Dimotakis.

5.9 Starting from the unsteady momentum equation for a compressible fluid with constant viscosities, $\rho(D\mathbf{u}/Dt) + \nabla p = \rho\mathbf{g} + \mu\nabla^2\mathbf{u} + \left(\mu_v + \frac{1}{3}\mu\right)\nabla(\nabla\cdot\mathbf{u})$, show that:

$$\frac{\partial\mathbf{u}}{\partial t} + \boldsymbol{\omega}\times\mathbf{u} = T\nabla s - \nabla\left(h + \frac{1}{2}|\mathbf{u}|^2 + \Phi\right) - \frac{\mu}{\rho}\nabla\times\boldsymbol{\omega} + \frac{1}{\rho}\left(\mu_v + \frac{4}{3}\mu\right)\nabla(\nabla\cdot\mathbf{u})$$

where T = temperature, h = enthalpy per unit mass, s = entropy per unit mass, and the body force is conservative: $\mathbf{g} = -\nabla\Phi$. This is the viscous Crocco-Vazsonyi equation. Simplify this equation for steady inviscid nonheat-conducting flow to find the Bernoulli equation (4.79), $h + \frac{1}{2}|\mathbf{u}|^2 + \Phi$ = constant along a streamline, which is valid when the flow is rotational and nonisothermal.

5.10 **a)** Solve $\nabla^2 G(\mathbf{x},\mathbf{x}') = \delta(\mathbf{x}-\mathbf{x}')$ for $G(\mathbf{x},\mathbf{x}')$ in a uniform, unbounded three-dimensional domain, where $\delta(\mathbf{x}-\mathbf{x}') = \delta(x-x')\delta(y-y')\delta(z-z')$ is the three-dimensional Dirac delta function.

b) Use the result of part a) to show that: $\phi(\mathbf{x}) = -\frac{1}{4\pi}\int_{all\ x'}\frac{q(\mathbf{x}')}{|\mathbf{x}-\mathbf{x}'|}d^3x'$ is the solution of Poisson equation $\nabla^2\phi(\mathbf{x}) = q(\mathbf{x})$ in a uniform, unbounded three-dimensional domain.

5.11 Start with the equations of motion in the rotating coordinates, and prove (5.24) Kelvin's circulation theorem for the absolute vorticity. Assume that the flow is inviscid and barotropic and that the body forces are conservative. Explain the result physically.

5.12 In (R,φ,z) cylindrical coordinates, consider the radial velocity $u_R = -R^{-1}(\partial\psi/\partial z)$, and the axial velocity $u_z = R^{-1}(\partial\psi/\partial R)$ determined from the axisymmetric stream function $\psi(R,z) = \frac{Aa^4}{10}\left(\frac{R^2}{a^2}\right)\left(1 - \frac{R^2}{a^2} - \frac{z^2}{a^2}\right)$ where A is a constant. This flow is known as Hill's spherical vortex.

a) For $R^2 + z^2 \le a^2$, sketch the streamlines of this flow in a plane that contains the z-axis. What does a represent?

b) Determine $\mathbf{u} = u_R(R,z)\mathbf{e}_R + u_z(R,z)\mathbf{e}_z$

c) Given $\omega_\varphi = (\partial u_R/\partial u_z) - (\partial u_z/\partial u_R)$, show that $\boldsymbol{\omega} = AR\mathbf{e}_\varphi$ in this flow and that this vorticity field is a solution of the vorticity equation, (5.13).

d) Does this flow include stretching of vortex lines?

5.13 In (R,φ,z) cylindrical coordinates, consider the flow field $u_R = -\alpha R/2$, $u_\varphi = 0$, and $u_z = \alpha z$.

a) Compute the strain rate components S_{RR}, S_{zz}, and S_{Rz}. What sign of α causes fluid elements to elongate in the z-direction? Is this flow incompressible?

b) Show that it is possible for a steady vortex (a Burgers' vortex) to exist in this flow field by adding $u_\varphi = (\Gamma/2\pi R)[1 - \exp(-\alpha R^2/\nu)]$ to u_R and u_z from part a) and then determining a pressure field $p(R,\varphi)$ that together with $\mathbf{u} = (u_R, u_\varphi, u_z)$ solves the Navier-Stokes momentum equation for a fluid with constant density ρ and kinematic viscosity ν.

c) Determine the vorticity in the Burgers' vortex flow of part b).

d) Explain how the vorticity distribution can be steady when $\alpha \neq 0$ and fluid elements are stretched or compressed.

e) Interpret what is happening in this flow when $\alpha > 0$ and when $\alpha < 0$.

5.14 The vorticity distribution for long straight concentrated vortex that coincides with the z-axis of an (x,y,z)-coordinate system is:

$$\boldsymbol{\omega}(\mathbf{x},t) = \Gamma\delta(x)\delta(y)\mathbf{e}_z.$$

Start with (5.27), the most general version of the Biot–Savart law, and use this vorticity distribution to show that fluid velocity induced by this vortex is:

$$\mathbf{u}(\mathbf{x},t) = \frac{\Gamma}{2\pi\sqrt{x^2+y^2}}\mathbf{e}_\varphi,$$

where $\mathbf{e}_\varphi = \mathbf{e}_x\cos\varphi + \mathbf{e}_y\sin\varphi$ is the azimuthal-angle unit vector in the x-y plane. [Hint use the integration variable substitution suggested in Example 5.5.]

5.15 A vortex ring of radius a and strength Γ lies in the x-y plane as shown in the figure.

a) Use the Biot–Savart law (5.28) to reach the following formula for the induced velocity along the x-axis:

$$\mathbf{u}(\mathbf{x}) = -\frac{\Gamma\mathbf{e}_z}{4\pi}\int_{\varphi=0}^{2\pi}\frac{(x\cos\varphi - a)a\,d\varphi}{[x^2 - 2ax\cos\varphi + a^2]^{3/2}}.$$

b) What is $\mathbf{u}(0)$, the induced velocity at the origin of coordinates?
c) What is $\mathbf{u}(x)$ to leading order in a/x when $x \gg a$?

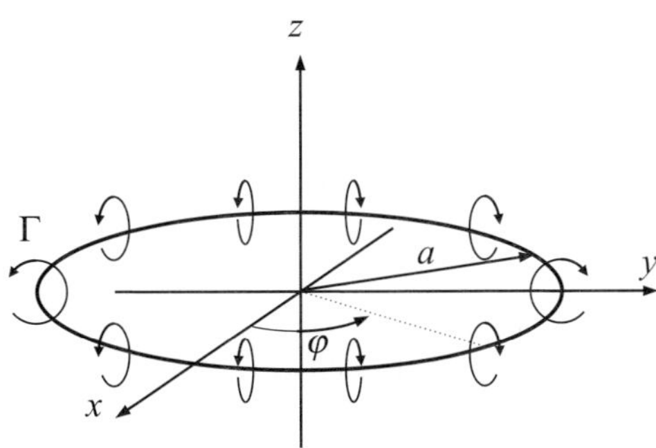

5.16 Consider the same configuration from Exercise 5.15 (an ideal circular vortex of radius a and strength Γ that lies in the x-y plane). Because there is no vortex core thickness, the speed of the vortex ring, $w = \mathbf{u} \cdot \mathbf{e}_z$, is ill-defined, preventing an analytic solution from being obtained.

a) Discretize the vortex ring into N equidistant vortex line segments. The figure below shows an example with $N = 8$. Compute the velocity at the center of one segment induced by all other $N - 1$ segments by solving the Biot–Savart law (5.28) numerically. Use geometric arguments to obtain the length and direction of each vortex element, dl and $\mathbf{e}_\omega$. Plot the translational speed of the vortex ring, w, as a function of N. As N increases, does w converge? If not, does it diverge as $\mathcal{O}(N)$, $\mathcal{O}(\ln N)$, or some other scaling?

b) Kelvin showed the speed of an ideal vortex ring with a finite but small core thickness $a' \ll a$ is

$$w = \frac{\Gamma}{4\pi a}\left(\ln\frac{8a}{a'} - \frac{1}{4}\right).$$

Relate this to your findings from part a.

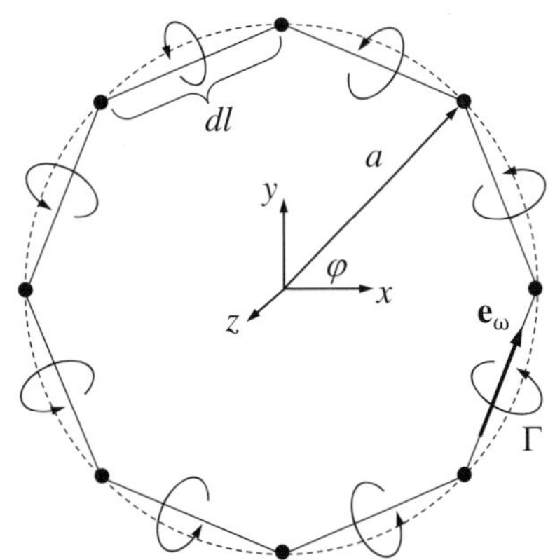

5.17 An ideal line vortex parallel to the z-axis of strength Γ intersects the x-y plane at $x = 0$ and $y = h$. Two solid walls are located at $y = 0$ and $y = H > 0$. Use the method of images for the following.

a) Based on symmetry arguments, determine the horizontal velocity u of the vortex when $h = H/2$.

b) Show that for $0 < h < H$ the horizontal velocity of the vortex is:

$$u(0, h) = \frac{\Gamma}{4\pi h}\left(1 - 2\sum_{n=1}^{\infty}\frac{1}{(nH/h)^2 - 1}\right),$$

and evaluate the sum when $h = H/2$ to verify your answer to part a).

5.18 The axis of an infinite solid circular cylinder with radius a coincides with the z-axis. The cylinder is stationary and immersed in an incompressible inviscid fluid, and the net circulation around it is zero. An ideal line vortex parallel to the cylinder with circulation Γ passes through the x-y plane at $x = L > a$ and $y = 0$. Here two image vortices are needed to satisfy the boundary condition on the cylinder's surface. If one of these is located at $x = y = 0$ and has strength Γ, determine the strength and location of the second image vortex.

5.19 Consider the interaction of two vortex rings of equal strength and similar sense of rotation. Argue that they go through each other, as described near the end of Section 5.6.

5.20 A constant-density irrotational flow in a rectangular torus has a circulation Γ and volumetric flow rate Q. The inner radius is r_1, the outer radius is r_2, and the height is h. Compute the total kinetic energy of this flow in terms of only ρ, Γ, and Q.

5.21 Consider a cylindrical tank of radius R filled with a viscous fluid spinning steadily about its axis with constant angular velocity $\boldsymbol{\Omega}$. Assume that the flow is in a steady state.

a) Find $\int_A \boldsymbol{\omega} \cdot \mathbf{n} dA$ where A is a horizontal plane surface through the fluid normal to the axis of rotation and bounded by the wall of the tank.

b) The tank then stops spinning. Find again the value of $\int_A \boldsymbol{\omega} \cdot \mathbf{n} dA$.

5.22 Using Fig. 5.13, prove (5.32) assuming that: (i) the two vortices travel in circles, (ii) each vortex's speed along its circular trajectory is constant, and (iii) the period of the motion is the same for both vortices.

5.23 Consider two-dimensional steady flow in the x-y plane outside of a long circular cylinder of radius a that is centered on and rotating about the z-axis at a constant angular rate of Ω_z. Show that the fluid velocity on the x-axis is $\mathbf{u}(x, 0) = (\Omega_z a^2/x)\mathbf{e}_y$ for $x > a$ when the cylinder is replaced by:

a) A circular vortex sheet of radius a with strength $\gamma = \Omega_z a$

b) A circular region of uniform vorticity $\boldsymbol{\omega} = 2\Omega_z \mathbf{e}_z$ with radius a

c) Describe the flow for $x^2 + y^2 < a^2$ for parts a) and b).

5.24 An ideal line vortex in a half space filled with an inviscid constant-density fluid has circulation Γ, lies parallel to the z-axis, and passes through the x-y plane at $x = 0$ and $y = h$. The plane defined by $y = 0$ is a solid surface.

a) Use the method of images to find $\mathbf{u}(x, y)$ for $y > 0$ and show that the fluid velocity on $y = 0$ is $\mathbf{u}(x, 0) = \Gamma h \mathbf{e}_x/[\pi(x^2 + h^2)]$.

b) Show that $\mathbf{u}(0, y)$ is unchanged for $y > 0$ if the image vortex is replaced by a vortex sheet of strength $\gamma(x) = -u(x, 0)$ on $y = 0$.

c) (If you have the patience) Repeat part b) for $\mathbf{u}(x, y)$ when $y > 0$.

5.25 Consider a two-dimensional x-y plane rotating about the z-axis at a constant angular rate Ω_z. Using the two-dimensional vortex method outlined in Example 5.8 with $\epsilon = 0$, plot the trajectories of two point vortices initially placed $h = 0.1$ apart with equal and opposite circulation $\Gamma = \pm 1$, as depicted in Fig. 5.14. The expression for the induced velocity must be modified to account for solid-body rotation by the rotation rate $\boldsymbol{\Omega} = (0, 0, \Omega_z)$ via $\mathbf{x} \times \boldsymbol{\Omega}$.

a) Show that the induced velocity V reproduces Eqs. (5.30) and (5.31) when $\Omega_z = 0$. [*Hint*: Choose a timestep size $\Delta t \ll h/V$.]

b) Plot the trajectories for different values of Ω_z. Comment on the relative contributions from vorticity-induced velocity and solid-body rotation.

c) Redo the calculation with $N \gg 1$ vortices randomly placed within a circular patch of radius $R = 1$ each with random circulation $\Gamma \in (0, 1)$ for different values of Ω_z. Comment on any coherent motion observed and the relative contribution of the rotation rate.

References

Leonard, A., 1980. Vortex methods for flow simulation. Journal of Computational Physics 37 (3), 289–335.

Lighthill, M.J., 1986. An Informal Introduction to Theoretical Fluid Mechanics. Clarendon Press, Oxford, England.

Sommerfeld, A., 1964. Mechanics of Deformable Bodies. Academic Press, New York.

Further reading

Batchelor, G.K., 1967. An Introduction to Fluid Dynamics. Cambridge University Press, London.

Pedlosky, J., 1987. Geophysical Fluid Dynamics. Springer-Verlag, New York (This book discusses the vorticity dynamics in rotating coordinates, with application to geophysical systems).

Prandtl, L., Tietjens, O.G., 1934. Fundamentals of Hydro- and Aeromechanics. Dover Publications, New York (This book contains a good discussion of the interaction of vortices).

Saffman, P.G., 1992. Vortex Dynamics. Cambridge University Press, Cambridge (This book presents a wide variety vortex topics from an applied mathematics perspective).

Chapter 6

Ideal flow⍟

Contents

6.1 Relevance of irrotational constant-density flow theory 177
6.2 Two-dimensional stream function and velocity potential 179
6.3 Construction of elementary flows in two dimensions 181
6.4 Complex potential 191
6.5 Forces on a two-dimensional body 193
6.6 Conformal mapping 195
6.7 Numerical solution techniques in two dimensions 199
6.8 Axisymmetric ideal flow 202
6.9 Three-dimensional potential flow and apparent mass 206
6.10 Concluding remarks 211
Exercises 211
References 221
Further reading 221

Chapter objectives

- To describe the formulation and limitations of ideal flow theory
- To illustrate the use of the stream function and the velocity potential in two-dimensional, axisymmetric, and three-dimensional flows
- To derive and present classical ideal flow results for flows past simple objects

6.1 Relevance of irrotational constant-density flow theory

When a constant-density fluid flows without rotation, and pressure is measured relative to its local hydrostatic value (see Section 4.9), the equations of fluid motion in an inertial frame of reference, (4.7) and (4.38), simplify to:

$$\nabla \cdot \mathbf{u} = 0 \quad \text{and} \quad \rho(D\mathbf{u}/Dt) = -\nabla p, \tag{4.10, 6.1}$$

even though the fluid's viscosity μ may be non-zero. These are the equations of *ideal* flow. They are useful for developing a first-cut understanding of nearly any macroscopic fluid flow, and are directly applicable to low-Mach-number irrotational flows of homogeneous fluids away from solid boundaries. Ideal flow theory has abundant applications in the exterior aero- and hydrodynamics of moderate- to large-scale objects at non-trivial subsonic speeds. Here, moderate size (L) and non-trivial speed (U) are determined jointly by the requirement that the Reynolds number, $\mathrm{Re} = \rho UL/\mu$ (4.104), be large enough (typically $\mathrm{Re} \sim 10^3$ or greater) so that the combined influence of fluid viscosity and fluid element rotation is confined to thin layers on solid surfaces, commonly known as *boundary layers*.

The conditions necessary for the application of ideal flow theory are commonly present on the upstream side of many ordinary objects, and may even persist to the downstream side of some. Ideal flow analysis can predict fluid velocity away from solid surfaces, surface-normal pressure forces (when the boundary layer on the surface is thin and attached), acoustic streamlines, flow patterns that minimize form drag, and unsteady-flow fluid-inertia effects. Ideal flow theory does not predict viscous effects like skin friction or energy dissipation, so it is not directly applicable to interior flows in pipes and ducts, to boundary-layer flows, or to any rotational flow region. This final specification excludes low-Re flows and regions of turbulence.

Because (6.1) involves only first-order spatial derivatives, ideal flows only satisfy the no-through-flow boundary condition on solid surfaces. The no-slip boundary condition (4.95) is not applied in ideal flows, so non-zero tangential velocity at a solid surface may exist (Fig. 6.1a). In contrast, a real fluid with a non-zero shear viscosity satisfy the no-slip boundary condition (4.95) because (4.38) contains second-order spatial derivatives. At sufficiently high Re, there are two primary differences between ideal and real flows over the same object. First, viscous boundary layers containing rotational fluid form on solid surfaces in the real flow, and the thickness of such boundary layers, within which viscous diffusion of vorticity is important, approaches zero as $\mathrm{Re} \to \infty$ (Fig. 6.1b). The second difference is the possible formation in the real

⍟. This book has a companion website hosting complementary materials. Visit this URL to access it: https://www.elsevier.com/books-and-journals/book-companion/9780128198070.

Fluid Mechanics. https://doi.org/10.1016/B978-0-12-819807-0.00015-6

flow of *separated flow* or *wake* regions that occur when boundary layers leave the surface on which they have developed to create a wider zone of rotational flow (Fig. 6.2). Ideal flow theory is not directly applicable to such layers or regions of rotational flow. However, rotational flow regions may be easy to anticipate or identify, and may represent a small fraction of a total flow field so that predictions from ideal flow theory may remain worthwhile even when viscous flow phenomena are present. Further discussion of viscous-flow phenomena is provided in Chapters 7 and 8.

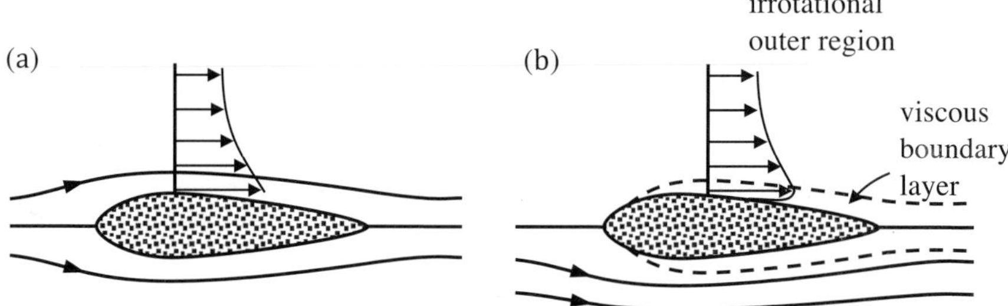

FIGURE 6.1 Comparison of a completely irrotational constant-density (ideal) flow (a) and a comparable high Reynolds number flow (b). In both cases the no-through-flow boundary condition is applied. However, the ideal flow is effectively inviscid and the fluid velocity is tangent to and non-zero on the body surface. The high-Re flow includes thin boundary layers where fluid rotation and viscous effects are prevalent, and the no-slip boundary condition is enforced, but the velocity above the thin boundary layer is almost identical to that in the ideal flow.

For (4.10) and (6.1) to apply, fluid density ρ must be constant and the flow must be irrotational. If the flow is merely incompressible and contains baroclinic density variations, (4.10) will still be satisfied but (6.1) will not; it will need a body-force term like that in (4.85) and the reference pressure would have to be redefined. If the fluid is a homogeneous compressible gas with sound speed c, the constant density requirement will be satisfied when the Mach number, $M = U/c$ (4.113), of the flow is much less than unity. The irrotationality condition is satisfied when fluid elements enter the flow field of interest without rotation and do not acquire any while they reside in it. Based on Kelvin's circulation theorem (5.11) for constant density flow, this is possible when the body force is conservative and the net viscous torque on a fluid element is zero. Thus, a fluid element that is initially irrotational will stay that way unless it enters a boundary layer, wake, or separated flow region where it acquires rotation via viscous diffusion. So, when initially irrotational fluid flows over a solid object, ideal flow theory most-readily applies to the *outer* region of the flow away from the object's surface(s). Viscous flow theory is needed in the *inner* region where viscous diffusion of vorticity is important. Often, at high Re, the outer flow can be approximately predicted by ignoring the existence of viscous boundary layers. With this outer flow prediction, viscous flow equations can be solved for the boundary-layer flow and, under the right conditions, the two solutions can be adjusted until they match in a suitable region of overlap. This approach works well for objects like thin airfoils at low angles of attack when boundary layers remain thin and stay attached all the way to the foil's trailing edge (Fig. 6.1b). However, it is not satisfactory when the solid object has such a shape that one or more boundary layers separate from its surface before reaching its downstream edge (Fig. 6.2), giving rise to a rotational wake flow or region of separated flow (sometimes called a *separation bubble*) that is not necessarily thin, no matter how high the Reynolds number. In this case, the limit of a real flow as $\mu \to 0$ does not approach that of an ideal flow ($\mu = 0$). Yet, upstream of boundary-layer separation, ideal flow theory may still provide a good approximation of the real flow.

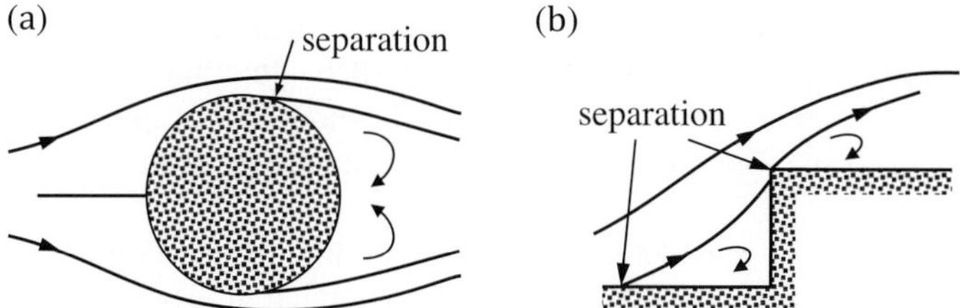

FIGURE 6.2 Schematic drawings of flows with boundary-layer separation. (a) Real flow past a cylinder where the boundary layers on the top and bottom of the cylinder leave the surface near its widest point. (b) Real flow past a sharp-cornered obstacle where the boundary layer leaves the surface at the corner. Upstream of the point of separation, ideal flow theory is usually a good approximation of the real flow.

In summary, the theory presented here does not apply to inhomogeneous fluids, high subsonic or supersonic flow speeds, boundary-layer flows, wake flows, interior flows, or any flow region where fluid elements rotate. However, the remaining flow possibilities are abundant, and include those that are commercially valuable (flight), naturally important (water waves), or readily encountered in our everyday lives (flow around vehicles, also see Chapters 11 and 13 for further examples).

Steady and unsteady irrotational constant-density flow fields around simple objects and through simple geometries in two and three dimensions are the subjects of this chapter. All the coordinate systems presented in Fig. 3.3 are utilized herein.

Example 6.1

If the pressure is constant in an ideal flow, what does this imply about the fluid velocity?

Solution When the pressure is constant, ∇p is zero so (6.1) simplifies to $D\mathbf{u}/Dt = 0$. Thus, the fluid velocity is constant following fluid particles, and each fluid particle retains the velocity $\mathbf{U}(\mathbf{x}_o, t_o)$ it had at a reference time t_o and reference location $\mathbf{x}_o$. If a moving coordinate system is attached to any fluid particle, then that coordinate system will be inertial and the fluid particle of interest will remain stationary at the origin of these moving coordinates. The flow about such a stationary point can only be composed of straining motion and rotation (3.10, 3.19), but there can be no rotation in an ideal flow. Thus, the most general possible ideal flow field about the origin in the moving coordinate system is straining flow. However, in this moving frame, the origin of coordinates is a stagnation point, a pressure maximum (see Example 6.9). Thus, to achieve constant pressure, the fluid velocity in the moving frame must be zero everywhere, which implies constant velocity in the original stationary frame of reference. Thus, constant pressure in a steady ideal flow implies constant fluid velocity.

6.2 Two-dimensional stream function and velocity potential

The two-dimensional incompressible continuity equation:

$$\partial u/\partial x + \partial v/\partial y = 0 \tag{6.2}$$

is identically satisfied when u, v-velocity components are determined from a scalar stream function $\psi(x, y)$:

$$u \equiv \partial\psi/\partial y, \quad \text{and} \quad v \equiv -\partial\psi/\partial x. \tag{6.3}$$

Along a curve of $\psi =$ constant, $d\psi = 0$, and this implies:

$$0 = d\psi = \frac{\partial\psi}{\partial x}dx + \frac{\partial\psi}{\partial y}dy = -v dx + u dy, \quad \text{or} \quad \left(\frac{dy}{dx}\right)_{\psi=\text{const}} = \frac{v}{u},$$

which is the definition of a streamline in two dimensions. The vorticity ω_z in a flow described by ψ is:

$$\frac{\partial v}{\partial x} - \frac{\partial u}{\partial y} = \omega_z = \frac{\partial}{\partial x}\left(-\frac{\partial\psi}{\partial x}\right) - \frac{\partial}{\partial y}\left(\frac{\partial\psi}{\partial y}\right) = -\nabla^2\psi. \tag{6.4}$$

In constant-density irrotational flow, ω_z will be zero everywhere except at the locations of ideal irrotational vortices. Thus, we are interested in solutions of:

$$\nabla^2\psi = 0, \quad \text{and} \quad \nabla^2\psi = -\Gamma\delta(x - x')\delta(y - y'), \tag{6.5, 6.6}$$

where δ is the Dirac delta-function (see Appendix B.4), and $\mathbf{x}' = (x', y')$ is the location of an ideal irrotational vortex of strength Γ.

In an unbounded domain, the most elementary non-trivial solutions of (6.5) and (6.6) are:

$$\psi = -Vx + Uy, \quad \text{and} \quad \psi = -\frac{\Gamma}{2\pi}\ln\sqrt{(x - x')^2 + (y - y')^2}, \tag{6.7, 6.8}$$

respectively. These correspond to uniform fluid velocity with horizontal component U and vertical component V, and to the flow induced by an irrotational vortex located at $\mathbf{x}'$ (see Exercise 6.1).

These two-dimensional stream function results have been obtained by considering incompressibility first, and irrotationality second. An equivalent formulation of two-dimensional ideal flow based on a different scalar function is possible when incompressibility and irrotationality are considered in the other order. The condition of irrotationality in two dimensions is:

$$\partial v/\partial x - \partial u/\partial y = 0, \tag{6.9}$$

and it is identically satisfied when u, v-velocity components are determined from a scalar *velocity potential* function $\phi(x, y)$:

$$u \equiv \partial\phi/\partial x, \quad \text{and} \quad v \equiv \partial\phi/\partial y, \tag{6.10}$$

or $\nabla\phi = \mathbf{u}$. The velocity potential must exist in all irrotational flows, so such flows are frequently called *potential flows*. Curves of $\phi =$ constant are defined by:

$$0 = d\phi = \frac{\partial\phi}{\partial x}dx + \frac{\partial\phi}{\partial y}dy = udx + vdy \quad \text{or} \quad \left(\frac{dy}{dx}\right)_{\phi=\text{const}} = -\frac{u}{v},$$

and are perpendicular to streamlines. When using $\phi(x, y)$, the condition for incompressibility becomes:

$$\frac{\partial u}{\partial x} + \frac{\partial v}{\partial y} = \frac{\partial}{\partial x}\left(\frac{\partial\phi}{\partial x}\right) + \frac{\partial}{\partial y}\left(\frac{\partial\phi}{\partial y}\right) = \nabla^2\phi = q(x, y), \tag{6.11}$$

where $q(x, y)$ is the spatial distribution of the source strength in the flow field. Of course, in real incompressible flows, $q(x, y) = 0$; however, ideal point sources and sinks of fluid are useful idealizations that allow the flow around objects of various shapes to be determined. These point sources and sinks are the ϕ-field equivalents of positive- and negative-circulation ideal vortices in flow fields described by ψ. Thus, we are interested in solutions of:

$$\nabla^2\phi = 0, \quad \text{and} \quad \nabla^2\phi = q_s\delta(x - x')\delta(y - y'), \tag{6.12, 6.13}$$

where q_s is a constant (with units of length^2/time) that sets the strength of the singularity of at $\mathbf{x}'$.

In an unbounded domain, the most elementary solutions of (6.12) and (6.13) are:

$$\phi = Ux + Vy, \quad \text{and} \quad \phi = \frac{q_s}{2\pi}\ln\sqrt{(x - x')^2 + (y - y')^2}, \tag{6.14, 6.15}$$

respectively. These correspond to uniform fluid velocity with horizontal component U and vertical component V, and to the flow induced by an ideal point source of strength q_s located at $\mathbf{x}'$ (see Exercise 6.3). Here, q_s is the source's volume flow rate per unit length perpendicular to the plane of the flow.

Either ψ or ϕ can provide a complete description of a two-dimensional ideal flow, and they can be combined to form a complex potential that follows the theory of *harmonic* functions (see Section 6.4). In addition, ψ is readily extended to rotational flows while ϕ is readily extended to unsteady and three-dimensional flows.

Boundary conditions must be considered to extend the elementary ideal flow solutions (6.7), (6.8), (6.14), and (6.15) to more interesting geometries. The boundary conditions commonly encountered in irrotational flows are as follows.

(1) *No flow through a solid surface.* As required for conservation of mass, the component of fluid velocity normal to a solid surface must equal the velocity of the boundary normal to itself, see (4.91) and its ensuing discussion. This can be stated as $\mathbf{n}\cdot\mathbf{U}_s = (\mathbf{n}\cdot\mathbf{u})_{on\ the\ surface}$, where $\mathbf{n}$ is the surface's normal and $\mathbf{U}_s$ is the velocity of the surface at the point of interest. For a stationary body, this condition reduces to $(\mathbf{n}\cdot\mathbf{u})_{on\ the\ surface} = 0$, which implies:

$$\partial\phi/\partial n = 0 \quad \text{or} \quad \partial\psi/\partial s = 0 \textit{ on the surface}, \tag{6.16}$$

where s is the arc-length along the surface, and n is the surface-normal coordinate. However, $\partial\psi/\partial s$ is also zero along a streamline. Thus, a stationary solid boundary in an ideal flow must also be a streamline. Therefore, if any ideal-flow streamline is replaced by a stationary solid boundary having the same shape, then the remainder of the flow is not changed.

(2) *Recovery of conditions at infinity.* For the typical case of a body immersed in a uniform fluid flowing in the x-direction with speed U, the condition far from the body is:

$$\partial\phi/\partial x = U, \quad \text{or} \quad \partial\psi/\partial y = U. \tag{6.17}$$

When $U = 0$, the fluid far from the body is said to be quiescent.

Historically, irrotational flow theory was developed by finding functions that satisfy the Laplace, (6.5) or (6.12), or Poisson, (6.6) or (6.13), equations and then determining the boundary conditions met by those functions. Directly solving these equations with geometrically complicated boundary shapes using the conditions (6.16) and (6.17) typically requires numerical techniques. However, since the Laplace and Poisson equations are linear, any superposition of known solutions provides another solution, but the superposition of two or more solutions may satisfy different boundary conditions than either of the constituents of the superposition. Thus, through collection and combination, a rich variety of interesting ideal-flow solutions has emerged. This solution-construction approach to ideal flow theory is adopted in this chapter. When appropriate, numerical solutions of these equations can be obtained through the techniques described in Section 6.7 and Chapter 15.

After a solution of (6.5), (6.6), (6.12), or (6.13) has been obtained, fluid velocity components are determined by taking derivatives of ϕ or ψ as defined by (6.3) and (6.10). The pressure in the flow is determined by conserving momentum via (6.1) in the form of a Bernoulli equation:

$$p+\frac{1}{2}\rho|\mathbf{u}|^2 = p+\frac{1}{2}\rho(u^2+v^2) = p+\frac{1}{2}\rho|\nabla\phi|^2 = p+\frac{1}{2}\rho|\nabla\psi|^2 = \text{const.}, \tag{6.18}$$

when the flow is steady. For unsteady flow, the term $\rho(\partial\phi/\partial t)$ must be added (see Exercise 6.4). With this procedure, solutions of (4.10) and (6.1) for $\mathbf{u}$ and p are obtained for ideal flows in a simple manner, even though (6.1) is a non-linear partial differential equation.

For quick reference, the important equations in planar polar coordinates are:

$$\frac{1}{r}\frac{\partial}{\partial r}(ru_r)+\frac{1}{r}\frac{\partial u_\theta}{\partial \theta}=0 \text{ (continuity)} \quad \text{and} \quad \frac{1}{r}\frac{\partial}{\partial r}(ru_\theta)-\frac{1}{r}\frac{\partial u_r}{\partial \theta}=0 \text{ (irrotationality)}, \tag{6.19, 6.20}$$

$$u_r=\frac{\partial\phi}{\partial r}=\frac{1}{r}\frac{\partial\psi}{\partial\theta} \quad \text{and} \quad u_\theta=\frac{1}{r}\frac{\partial\phi}{\partial\theta}=-\frac{\partial\psi}{\partial r}, \tag{6.21, 6.22}$$

$$\nabla^2\psi=\frac{1}{r}\frac{\partial}{\partial r}\left(r\frac{\partial\psi}{\partial r}\right)+\frac{1}{r^2}\frac{\partial^2\psi}{\partial\theta^2}=0 \quad \text{and} \quad \nabla^2\phi=\frac{1}{r}\frac{\partial}{\partial r}\left(r\frac{\partial\phi}{\partial r}\right)+\frac{1}{r^2}\frac{\partial^2\phi}{\partial\theta^2}=0. \tag{6.23a, 6.23b}$$

These are the counterparts of (6.2), (6.3), (6.5), (6.9), (6.10), and (6.12).

Example 6.2

For steady ideal flow, show that (6.1) is equivalent to (6.18).

Solution For steady flow, $\partial\mathbf{u}/\partial t=0$, so (6.1) reduces to $(\mathbf{u}\cdot\nabla)\mathbf{u}+(1/\rho)\nabla p=0$. The vector identity (B.3.9), $(\mathbf{u}\cdot\nabla)\mathbf{u}=\nabla\left(\frac{1}{2}|\mathbf{u}|^2\right)-\mathbf{u}\times(\nabla\times\mathbf{u})$, allows the advective acceleration to be replaced with gradient and cross-product terms:

$$\nabla\left(\frac{1}{2}|\mathbf{u}|^2\right)-\mathbf{u}\times(\nabla\times\mathbf{u})+\frac{1}{\rho}\nabla p=0.$$

The second term in this equation is zero because ideal flow is irrotational so $\boldsymbol{\omega}=\nabla\times\mathbf{u}=0$. Multiplying through by ρ (= constant), and collecting both terms under a single gradient produces:

$$\nabla\left(\frac{1}{2}\rho|\mathbf{u}|^2+p\right)=0, \quad \text{or} \quad p+\frac{1}{2}\rho|\mathbf{u}|^2=\text{const.},$$

which is (6.18). Thus, once a steady ideal flow velocity field has been found, conservation of momentum merely determines the requisite pressure variations.

6.3 Construction of elementary flows in two dimensions

In this section, elementary solutions of the Laplace equation are developed and then superimposed to produce a variety of geometrically simple ideal flows in two dimensions.

First, consider polynomial solutions of the Laplace equation in Cartesian coordinates. A zero-order polynomial, ψ or $\phi=$ constant, is not interesting since this yields $\mathbf{u}=0$. First-order polynomial solutions are given by (6.7) and (6.14), and these solutions represent spatially uniform velocity fields, $\mathbf{u}=(U,V)$. Quadratic functions in x and y are the next possibilities, and there are two of these:

$$\psi=2Axy \quad \text{or} \quad \phi=2Axy, \quad \text{and} \quad \psi=A(x^2-y^2) \quad \text{or} \quad \phi=A(x^2-y^2), \tag{6.24–6.27}$$

where A is a constant. (The reason for the two in (6.24) and (6.25) will be clear in the next section.) Here, only the flow field for $2Axy$ is constructed. Production of flow-field results for ψ and $\phi = A(x^2 - y^2)$ is left as an exercise.

Examine $\psi = 2Axy$ first, and by direct differentiation find $u = 2Ax$, and $v = -2Ay$. Thus, for $A > 0$, the flow is toward the origin along the y-axis, away from it along the x-axis, and the streamlines are hyperbolae given by $xy = \psi/2A$ (Fig. 6.3). Considering the first quadrant only, this is ideal flow in a 90° corner. Now consider $\phi = 2Axy$, and by direct differentiation find $u = 2Ay$, and $v = 2Ax$. The equipotential lines are hyperbolae given by $xy = \phi/2A$. The flow is away from the origin along the line $y = x$ and toward it along the line $y = -x$. Thus, $\phi = 2Axy$ produces a flow that is equivalent to that of $\psi = 2Axy$ after a 45° rotation. Interestingly, higher-order polynomial solutions lead to flows in smaller-angle corners, while fractional powers lead to flows in larger-angle corners (see Section 6.4 and Exercises 6.8 and 6.9).

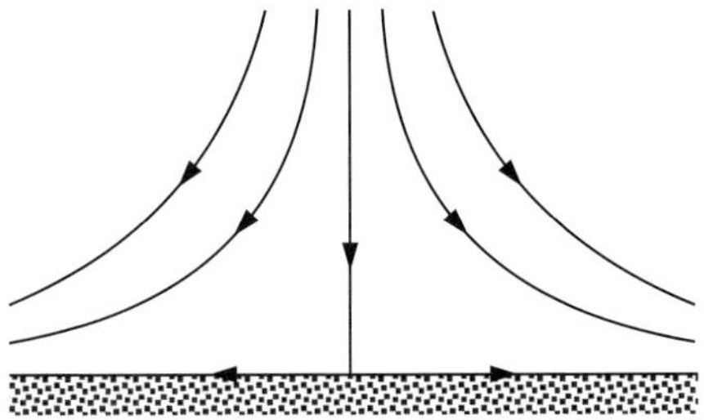

FIGURE 6.3 Stagnation point flow represented by $\psi = 2Axy$. Here, the flow impinges on the flat horizontal surface from above. The stagnation point is located where the single vertical streamline touches the horizontal surface.

The next set of solutions to consider are (6.8) and (6.15) with $x'=y'=0$. In this case, curves of $\psi=-(\Gamma/2\pi)\ln\sqrt{x^2+y^2}=$ const. are circles centered on the origin of coordinates (Fig. 6.4), and direct differentiation of (6.8) produces:

$$u = \frac{\partial}{\partial y}\left(-\frac{\Gamma}{2\pi}\ln\sqrt{x^2+y^2}\right) = -\frac{\Gamma}{2\pi}\frac{y}{x^2+y^2} = -\frac{\Gamma}{2\pi r}\sin\theta, \quad \text{and}$$

$$v = -\frac{\partial}{\partial x}\left(-\frac{\Gamma}{2\pi}\ln\sqrt{x^2+y^2}\right) = +\frac{\Gamma}{2\pi}\frac{x}{x^2+y^2} = \frac{\Gamma}{2\pi r}\cos\theta,$$

where r and θ are defined in Fig. 3.3a. These results may be rewritten using the outcome of Example 2.2 as $u_r = 0$ and $u_\theta = \Gamma/2\pi r$, which is the flow field of the ideal irrotational vortex (5.2). Similarly, curves of $\phi = (q_s/2\pi)\ln\sqrt{x^2+y^2} =$ const. are circles centered on the origin of coordinates (Fig. 6.5), and direct differentiation of (6.15) produces:

$$u = \frac{\partial}{\partial x}\left(\frac{q_s}{2\pi}\ln\sqrt{x^2+y^2}\right) = \frac{q_s}{2\pi}\frac{x}{x^2+y^2} = \frac{q_s}{2\pi r}\cos\theta, \quad \text{and}$$

$$v = \frac{\partial}{\partial y}\left(\frac{q_s}{2\pi}\ln\sqrt{x^2+y^2}\right) = \frac{q_s}{2\pi}\frac{y}{x^2+y^2} = \frac{q_s}{2\pi r}\sin\theta.$$

These results may be rewritten as $u_r = q_s/2\pi r$ and $u_\theta = 0$, which is purely radial flow away from the origin (Fig. 6.5). Here, $\nabla\cdot\mathbf{u}$ is zero everywhere except at the origin. Thus, this potential represents flow from an ideal incompressible point source for $q_s > 0$, or sink for $q_s < 0$, that is located at $r = 0$ in two dimensions.

A source of strength $+q_s$ at $(-\varepsilon, 0)$ and sink of strength $-q_s$ at $(+\varepsilon, 0)$, can be considered together

$$\phi = \frac{q_s}{2\pi}\ln\sqrt{(x+\varepsilon)^2+y^2} - \frac{q_s}{2\pi}\ln\sqrt{(x-\varepsilon)^2+y^2}$$

to obtain the potential for a *doublet* in the limit that $\varepsilon \to 0$ and $q_s \to \infty$, so that the dipole strength vector:

$$\mathbf{d} = \sum_{sources}\mathbf{x}_i q_{s,i} = -\varepsilon\mathbf{e}_x q_s + \varepsilon\mathbf{e}_x(-q_s) = -2q_s\varepsilon\mathbf{e}_x \tag{6.28}$$

remains constant. Here, the dipole strength points from the sink toward the source. As $\varepsilon \to 0$, the logarithm of the square roots can be simplified:

$$\ln\sqrt{(x\pm\varepsilon)^2+y^2} = \ln r + \ln\sqrt{1\pm 2\varepsilon x/r^2 + \varepsilon^2/r^2} = \ln r + \ln(1\pm\varepsilon x/r^2 + \ldots) \cong \ln(r) \pm \varepsilon x/r^2,$$

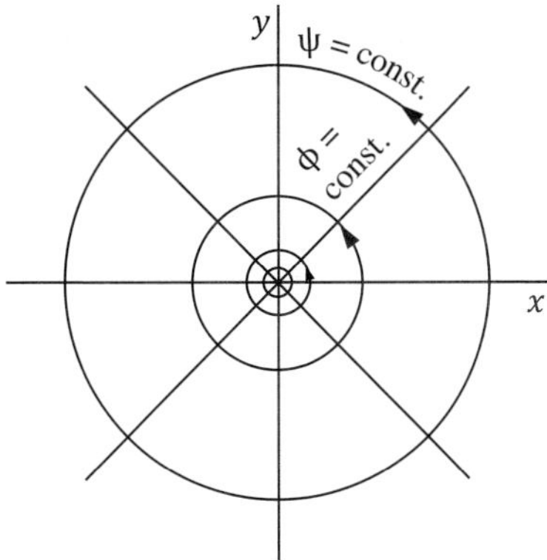

FIGURE 6.4 The flow field of an ideal vortex located at the origin of coordinates in two dimensions. The streamlines are circles and the potential lines are radials. Here, the vortex line is perpendicular to the x-y plane.

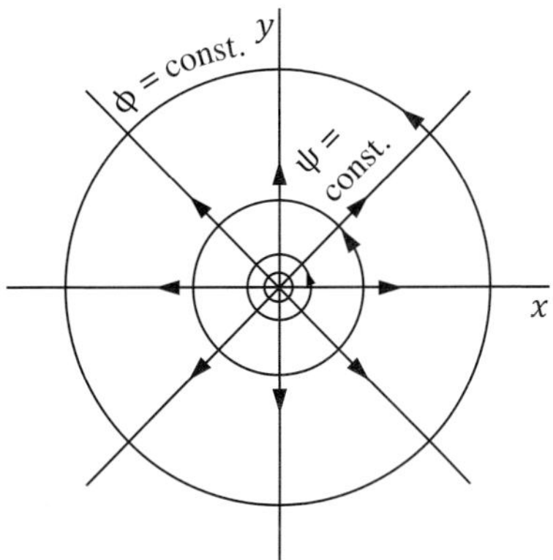

FIGURE 6.5 The flow field of an ideal source located at the origin of coordinates in two dimensions. The streamlines are radials and the potential lines are circles.

where $r^2 = x^2 + y^2$, so

$$\lim_{\varepsilon\to 0}\lim_{m\to\infty}\phi \cong \frac{q_s}{2\pi}\left(\ln r + \frac{\varepsilon x}{r^2} + \ldots - \ln r + \frac{\varepsilon x}{r^2} + \ldots\right) = \frac{q_s\varepsilon}{\pi}\frac{x}{r^2} = -\frac{\mathbf{d}\cdot\mathbf{x}}{2\pi r^2} = \frac{|\mathbf{d}|}{2\pi}\frac{\cos\theta}{r}. \tag{6.29}$$

The doublet flow field is illustrated in Fig. 6.6. The stream function for the doublet can be derived from (6.29) (Exercise 6.11).

The flows described by (6.7), (6.14), and (6.24) through (6.27) are solutions of the Laplace equation. The flows described by (6.8), (6.15), and (6.29) are singular at the origin and satisfy the Laplace equation for $r > 0$. Perhaps the most common and useful superposition of these solutions involves combining a uniform stream parallel to the x-axis, $\psi = Uy$ or $\phi = Ux$, and one or more of the singular solutions. The simplest example is the combination of a source and a uniform stream, which can be written in Cartesian and polar coordinates as:

$$\phi = Ux + \frac{q_s}{2\pi}\ln\sqrt{x^2+y^2} = Ur\cos\theta + \frac{q_s}{2\pi}\ln r, \quad \text{or}$$
$$\psi = Uy + \frac{q_s}{2\pi}\tan^{-1}\left(\frac{y}{x}\right) = Ur\sin\theta + \frac{q_s}{2\pi}\theta. \tag{6.30, 6.31}$$

Here the velocity field components are:

$$u = U + \frac{q_s}{2\pi}\frac{x}{x^2+y^2} \quad \text{and} \quad v = \frac{q_s}{2\pi}\frac{y}{x^2+y^2},$$

and streamlines are shown in Fig. 6.7. The stagnation point is located at $x = -a = -q_s/2\pi U$, and $y = 0$, and the value of the stream function on the stagnation streamline is $\psi = q_s/2$.

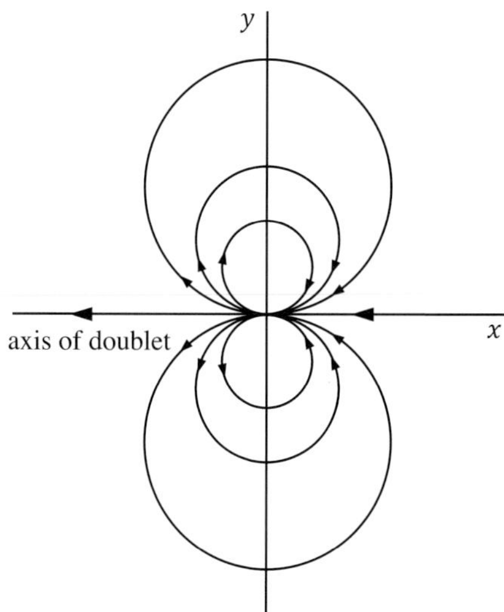

FIGURE 6.6 The flow field of an ideal two-dimensional doublet that points along the negative x-axis. The net source strength is zero so all streamlines begin and end at the origin. In this flow, the streamlines are circles tangent to the x-axis at the origin.

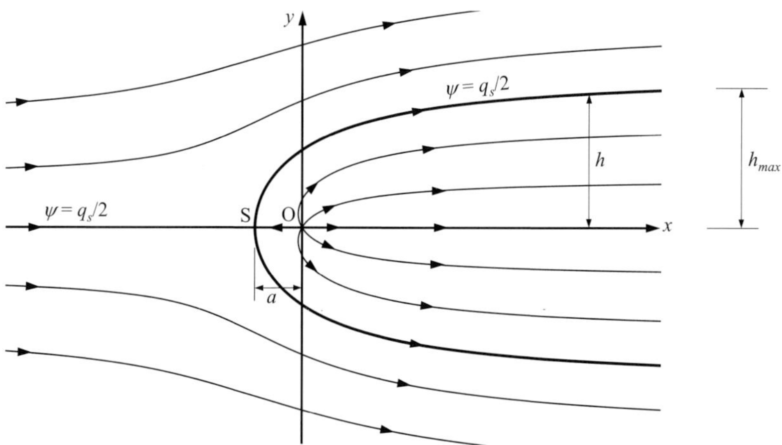

FIGURE 6.7 Ideal flow past a two-dimensional half-body formed from a horizontal free stream and a point source at the origin. The boundary streamline, shown as a darker curve, is given by $\psi = q_s/2$.

The streamlines that emerge vertically from the stagnation point (the darker curves in Fig. 6.7) form a semi-infinite body with a smooth nose, generally called a *half-body*. These stagnation streamlines divide the field into regions external and internal to the half body. The internal flow consists entirely of fluid emanating from the source, and the external region contains fluid from upstream of the source. The half-body resembles several practical shapes, such as the leading edge of an airfoil or the front part of a bridge pier. The half-width of the body, h, can be found from (6.31) with $\psi = q_s/2$:

$$h = q_s(\pi - \theta)/2\pi U.$$

Far downstream ($\theta \to 0$), the half-width tends to $h_{max} = q_s/2U$ (Fig. 6.7).

The pressure distribution on the half body can be found from Bernoulli's equation, (6.18) with const. $= p_\infty + \rho U^2/2$, and is commonly reported as a dimensionless excess pressure via the *pressure coefficient* C_p or Euler number (4.107):

$$C_p = \frac{p - p_\infty}{\frac{1}{2}\rho U^2} = 1 - \frac{|\mathbf{u}|^2}{U^2}. \tag{6.32}$$

A plot of C_p on the surface of the half-body is given in Fig. 6.8, which shows that there is pressure excess near the nose of the body and a pressure deficit beyond it. Interestingly, integrating p over the surface leads to no net pressure force (see Exercise 6.15).

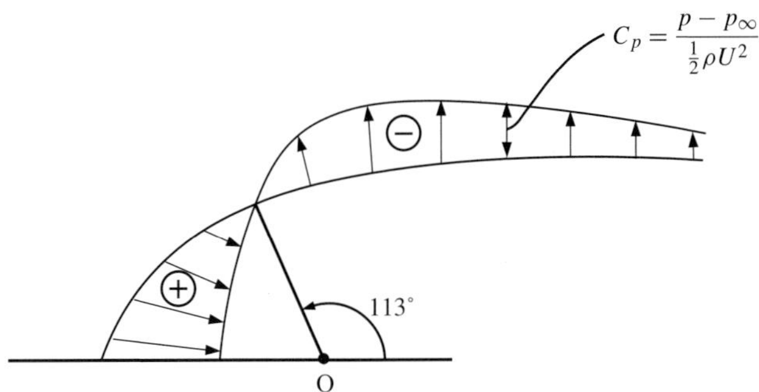

FIGURE 6.8 Pressure distribution in ideal flow over the half-body shown in Fig. 6.7. Pressure excess near the nose is indicated by the circled "+" and pressure deficit elsewhere is indicated by the circled "−".

As a second example of flow construction via superposition, consider a horizontal free stream U and a doublet with strength $\mathbf{d} = -2\pi U a^2 \mathbf{e}_x$:

$$\phi = Ux + \frac{Ua^2 x}{x^2 + y^2} = U\left(r + \frac{a^2}{r}\right)\cos\theta, \quad \text{or} \quad \psi = Uy - \frac{Ua^2 y}{x^2 + y^2} = U\left(r - \frac{a^2}{r}\right)\sin\theta. \tag{6.33}$$

Here, $\psi = 0$ at $r = a$ for all values of θ, showing that the streamline $\psi = 0$ represents a circular cylinder of radius a. The streamline pattern is shown in Fig. 6.9 (and Fig. 3.2a). In this flow, the net source strength is zero, so the cylindrical body is closed and does not extend downstream. The velocity field is:

$$u_r = U\left(1 - \frac{a^2}{r^2}\right)\cos\theta, \quad \text{and} \quad u_\theta = -U\left(1 + \frac{a^2}{r^2}\right)\sin\theta. \tag{6.34}$$

The velocity components on the surface of the cylinder are $u_r = 0$ and $u_\theta = -2U\sin\theta$, so the cylinder-surface pressure coefficient is:

$$C_p(r = a, \theta) = 1 - 4\sin^2\theta, \tag{6.35}$$

and this is shown by the continuous line in Fig. 6.10. There are stagnation points on the cylinder's surface at r-θ coordinates, $(a, 0)$ and (a, π). The cylinder-surface pressure minima occur at r-θ coordinates $(a, \pm\pi/2)$ where the surface flow speed is maximum. The fore-aft symmetry of the pressure distribution implies that there is no net pressure force on the cylinder. In fact, a general result of two-dimensional ideal flow theory is that a steadily moving body experiences no drag. This result is at variance with observations and is sometimes known as *d'Alembert's paradox*. The existence of real-flow tangential stress on a solid surface, commonly known as *skin friction*, is not the only reason for the discrepancy. For blunt bodies such as a cylinder, most of the drag comes from flow separation and the formation of a wake, which is likely to be unsteady or even turbulent. When a wake is present, the flow loses fore-aft symmetry, and the surface pressure on the downstream side of the object is smaller than that predicted by ideal flow theory (Fig. 6.10), resulting in pressure drag. These facts will be discussed in further in Chapter 8.

As discussed in Section 3.3, the flow due to a cylinder moving steadily through a fluid appears unsteady to an observer at rest with respect to the fluid at infinity. This flow is shown in Fig. 3.2b and can be obtained by superposing a uniform stream in the negative x-direction with the flow shown in Fig. 6.9. The resulting instantaneous streamline pattern is simply that of a doublet, as depicted by the decomposition shown in Fig. 6.11.

Although there is no net drag force on a circular cylinder in steady irrotational flow, there may be a lateral or lift force perpendicular to the free stream when circulation is added. Consider the flow field (6.33) with the addition of a point vortex of circulation $-\Gamma$[1] at the origin that induces a *clockwise* velocity:

$$\psi = U\left(r - \frac{a^2}{r}\right)\sin\theta + \frac{\Gamma}{2\pi}\ln\left(\frac{r}{a}\right). \tag{6.36}$$

Here, a has been added to the logarithm's argument to make it dimensionless.

1. This minus sign is necessary to achieve the usual fluid dynamic result given by (6.40).

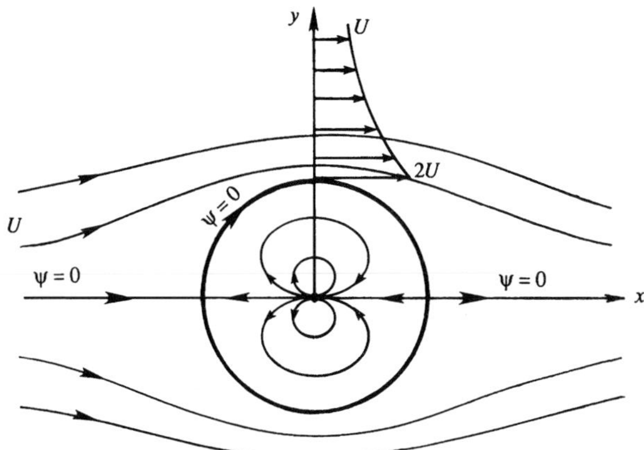

FIGURE 6.9 Ideal flow past a circular cylinder without circulation. This flow field is formed by combining a horizontal uniform stream flowing in the $+x$-direction with a doublet pointing in the $-x$-direction. The streamline that passes through the two stagnation points and forms the body surface is given by $\psi = 0$.

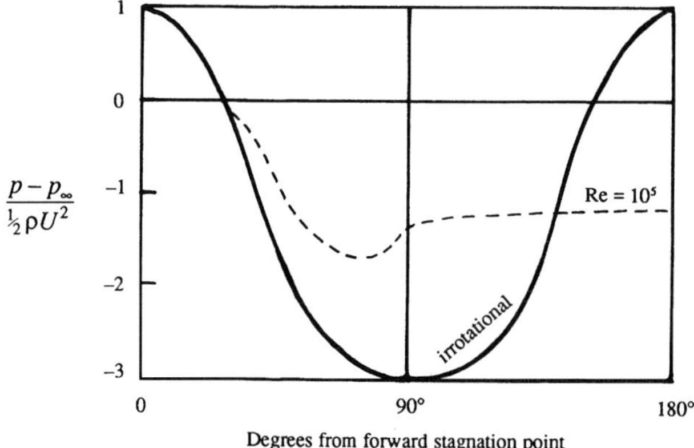

FIGURE 6.10 Comparison of irrotational and observed pressure distributions over a circular cylinder. Here 0° is the most upstream point of the cylinder and 180° is the most downstream point. The observed distribution changes with the Reynolds number; a typical behavior at high Re is indicated by the dashed line.

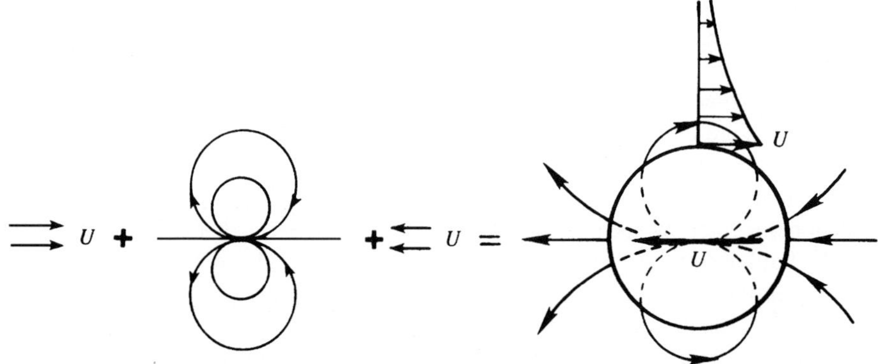

FIGURE 6.11 Decomposition of the irrotational flow pattern due to a moving cylinder. Here a horizontal free stream of $+U$ and doublet form a cylinder. When a uniform stream of $-U$ is added, the flow field of a moving cylinder is obtained.

Fig. 6.12 shows the resulting streamline pattern for various values of Γ. The close streamline spacing and higher velocity on top of the cylinder is due to the addition of velocities from the clockwise vortex and the uniform stream. In contrast, the smaller velocities at the bottom of the cylinder are a result of the vortex field counteracting the uniform stream. Bernoulli's equation consequently implies a higher pressure below the cylinder than above it, and this pressure difference leads to an upward lift force on the cylinder.

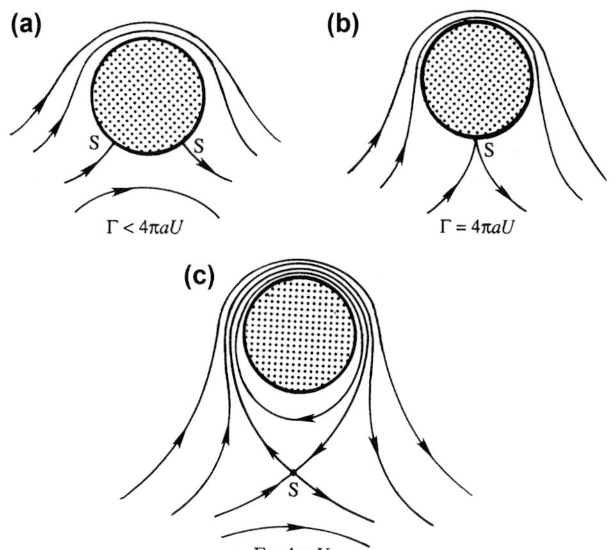

FIGURE 6.12 Irrotational flow past a circular cylinder for different circulation values. Here S represents the stagnation point(s) in the flow. (a) At low values of the circulation, there are two stagnation points on the surface of the cylinder. (b) When the circulation is equal to $4\pi aU$, there is one stagnation point on the surface of the cylinder. (c) When the circulation is even greater, there is one stagnation point below the cylinder.

The tangential velocity component at any point in the flow is:

$$u_\theta = -\frac{\partial \psi}{\partial r} = -U\left(1+\frac{a^2}{r^2}\right)\sin\theta - \frac{\Gamma}{2\pi r}.$$

At the surface of the cylinder, the fluid velocity is entirely tangential and is given by:

$$u_\theta(r=a,\theta) = -2U\sin\theta - \Gamma/2\pi a, \tag{6.37}$$

which vanishes if:

$$\sin\theta = -\Gamma/4\pi aU. \tag{6.38}$$

For $\Gamma < 4\pi aU$, two values of θ satisfy (6.38), implying that there are two stagnation points on the cylinder's surface. The stagnation points progressively move down as Γ increases (Fig. 6.12) and coalesce when $\Gamma = 4\pi aU$. For $\Gamma > 4\pi aU$, the stagnation point moves out into the flow along the negative y-axis. The radial distance of the stagnation point in this case is found from:

$$u_\theta(r,\theta=-\pi/2) = U\left(1+\frac{a^2}{r^2}\right) - \frac{\Gamma}{2\pi r} = 0, \quad \text{or} \quad r = \frac{1}{4\pi U}\left[\Gamma \pm \sqrt{\Gamma^2-(4\pi aU)^2}\right],$$

one root of which has $r > a$; the other root corresponds to a stagnation point inside the cylinder.

The cylinder surface pressure is found from (6.18) with const. $= p_\infty + \rho U^2/2$, and (6.37) to be:

$$p(r=a,\theta) = p_\infty + \frac{1}{2}\rho\left[U^2 - \left(-2U\sin\theta - \frac{\Gamma}{2\pi a}\right)^2\right]. \tag{6.39}$$

The upstream-downstream symmetry of the flow implies that the pressure force on the cylinder has no stream-wise component. The vertical pressure force (per unit length perpendicular to the flow plane) is:

$$L = -\int_0^{2\pi} p(r=a,\theta)\mathbf{n}\,dl\cdot\mathbf{e}_y = -\int_0^{2\pi} p(r=a,\theta)\sin\theta a\,d\theta,$$

where $\mathbf{n} = \mathbf{e}_r$ is the outward normal from the cylinder, and $dl = ad\theta$ is a surface element of the cylinder's cross section; L is known as the *lift* force in aerodynamics (Fig. 6.13). Evaluating the integral using (6.39) produces:

$$L = \rho U \Gamma. \tag{6.40}$$

It is shown in Section 6.5 that (6.40) holds for irrotational flow around *any* two-dimensional object; it is not just for circular cylinders. The result that L is proportional to Γ is of fundamental importance in aerodynamics. Eq. (6.40) was proved independently by the German mathematician, Wilhelm Kutta, and the Russian aerodynamist, Nikolai Zhukhovsky just after 1900; it is called the *Kutta-Zhukhovsky lift theorem*. (Older Western texts transliterated Zhukhovsky's name as Joukowsky.) The interesting question of how certain two-dimensional shapes, such as an airfoil, develop circulation when placed in a moving fluid is discussed in Chapter 13. It is described there how fluid viscosity is responsible for the development of circulation. The magnitude of circulation, however, is independent of viscosity but does depend on the flow speed U, and the shape and orientation of the object.

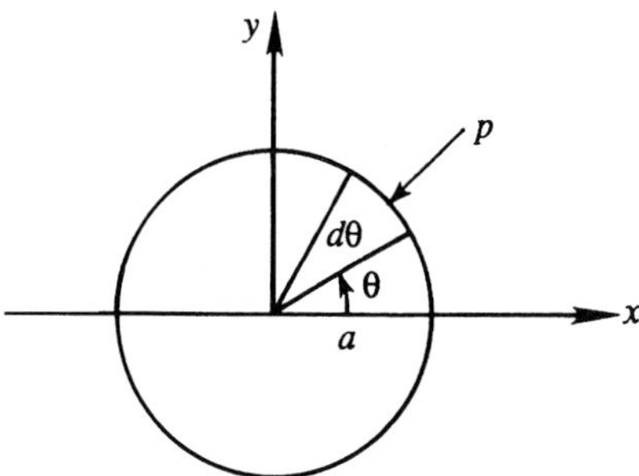

FIGURE 6.13 Calculation of pressure force on a circular cylinder. Surface pressure forces on top of the cylinder, where **n** has a positive vertical component, push the cylinder down. Thus the surface integral for the lift force applied by the fluid to the cylinder contains a minus sign.

For a circular cylinder, the only way to develop circulation is by rotating it. Although viscous effects are important in this case, the observed flow pattern for *large* values of cylinder rotation displays a striking similarity to the ideal flow pattern for $\Gamma > 4\pi aU$; see Fig. 3.25 in the book by Prandtl (1952). For lower rates of cylinder rotation, the retarded flow in the boundary layer is not able to overcome the adverse pressure gradient behind the cylinder, leading to separation; the real flow is therefore rather unlike the irrotational pattern. However, even in the presence of separation, observed flow speeds are higher on the upper surface of the cylinder, implying a lift force.

A second reason for the presence of lift on a rotating cylinder is the flow asymmetry generated by a delay of boundary-layer separation on the upper surface of the cylinder. The contribution of this mechanism is small for two-dimensional objects such as the circular cylinder, but it is the only mechanism for side forces experienced by spinning three-dimensional objects like sports balls. The lateral force experienced by rotating bodies is called the *Magnus effect*.

Two-dimensional ideal flow solutions are commonly not unique and the topology of the flow domain determines uniqueness. In short, a two-dimensional ideal flow solution is unique when any closed contour lying entirely within the fluid can be reduced to a point by continuous deformation without ever cutting through a flow-field boundary. Such fluid domains are *singly connected*. Thus, fluid domains, like those shown in Figs. 6.9 and 6.12, that entirely encircle an object may not provide unique ideal flow solutions based on boundary conditions alone. In particular, consider the ideal flow (6.36) depicted in Fig. 6.12 for various values of Γ. All satisfy the *same* boundary condition on the solid surface ($u_r = 0$) and at infinity ($\mathbf{u} = U\mathbf{e}_x$). The ambiguity occurs in these domains because there exist closed contours lying entirely within the fluid that cannot be reduced to a point, and on these contours a non-zero circulation can be computed. Fortunately, this ambiguity may often be resolved by considering real-flow effects. For example, the circulation strength that should be assigned to a streamlined object in two-dimensional ideal flow can be determined by applying the viscous-flow-based *Kutta* condition at the object's trailing edge. This point is further explained in Chapter 13.

Another important consequence of the superposition principle for ideal flow is that it allows boundaries to be built into ideal flows through the method of images. For example, if the flow of interest in an unbounded domain is the solution of $\nabla^2\psi_1 = -\omega_1(x, y)$, then $\nabla^2\psi_2 = -\omega_1(x, y) + \omega_1(x, -y)$ will determine the solution for the same vorticity distribution with a solid wall along the x-axis. Here, $\psi_2 = \psi_1(x, y) - \psi_1(x, -y)$, so that the *zero* streamline, $\psi_2 = 0$, occurs on $y = 0$ (Fig. 6.14). Similarly, if the flow of interest in an unbounded domain is the solution of $\nabla^2\phi_1 = q_1(x, y)$, then $\nabla^2\phi_2 = q_1(x, y) + q_1(x, -y)$ will determine the solution for the same source distribution with a solid wall along the x-axis. Here, $\phi_2 = \phi_1(x, y) + \phi_1(x, -y)$ so that $v = \partial\phi_2/\partial y = 0$ on $y = 0$ (Fig. 6.15).

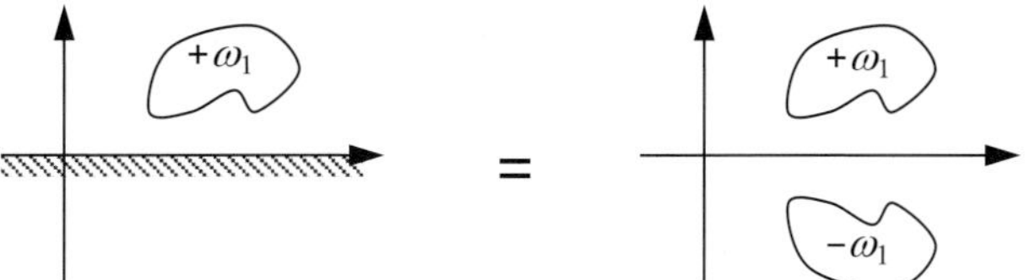

FIGURE 6.14 Illustration of the method of images for the flow generated by a vorticity distribution near a horizontal wall. An image distribution of equal strength and opposite sign mimics the effect of the solid wall.

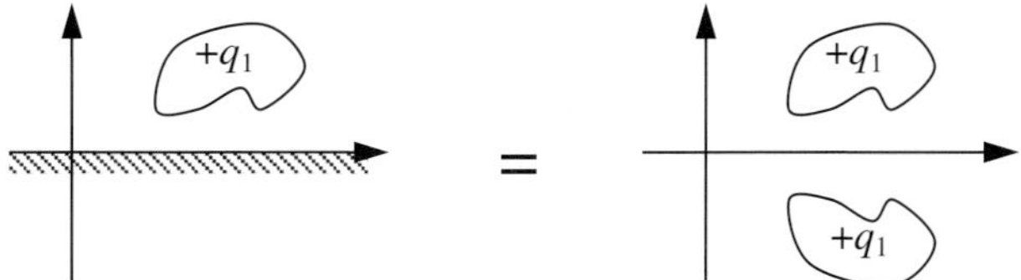

FIGURE 6.15 Illustration of the method of images for the flow generated by a source distribution near a horizontal wall. An image distribution of equal strength mimics the effect of the solid wall.

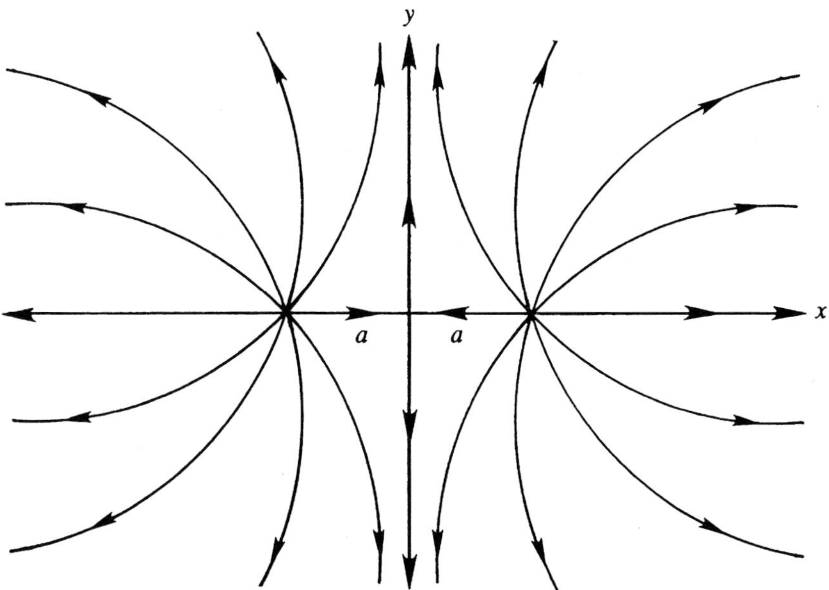

FIGURE 6.16 Ideal flow from two equal sources placed at $x = \pm a$. The origin is a stagnation point. The vertical axis is a streamline and may be replaced by a solid surface. This flow field further illustrates the method of images.

As an example of the method of images, consider the flow induced by an ideal source of strength q_s a distance a from a straight vertical wall (Fig. 6.16). Here an image source of the same strength and sign is needed a distance a on the other side of the wall. The stream function and potential for this flow are:

$$\psi = \frac{q_s}{2\pi}\left[\tan^{-1}\left(\frac{y}{x+a}\right) + \tan^{-1}\left(\frac{y}{x-a}\right)\right] \quad \text{and}$$
$$\phi = \frac{q_s}{2\pi}\left[\ln\sqrt{(x+a)^2+y^2} + \ln\sqrt{(x-a)^2+y^2}\right]. \tag{6.41}$$

After some rearranging and use of the two-angle formula for the tangent function, the equation for the streamlines may be found:

$$x^2 - y^2 - 2xy\cot(2\pi\psi/q_s) = a^2.$$

The x and y axes form part of the streamline pattern, with the origin as a stagnation point. This flow represents three interesting situations: flow from two equal sources (all of Fig. 6.16), flow from a source near a flat vertical wall (right half of Fig. 6.16), and flow through a narrow slit at $x = a$ into a right-angled corner (first quadrant of Fig. 6.16).

The method of images can sometimes be extended to circular boundaries by allowing more than one image vortex or source (Exercises 5.18, 6.28), and to unsteady flows as long as the image distributions of vorticity or source strength move appropriately. Unsteady two-dimensional ideal flow merely involves the inclusion of time as an independent variable in ψ or ϕ, and the addition of $\rho(\partial\phi/\partial t)$ in the Bernoulli equation (see Exercise 6.4). The following example, which validates the free-vortex results stated in Section 5.6, illustrates these changes.

Example 6.3

At $t = 0$, an ideal free vortex with strength $-\Gamma$ is located at point A near a flat solid vertical wall as shown in Fig. 5.16. If the x, y-coordinates of A are $(h,0)$ and the fluid far from the vortex is quiescent at pressure p_∞, determine the trajectory $\boldsymbol{\xi}(t) = (\xi_x, \xi_y)$ of the vortex, and the pressure at the origin of coordinates as a function of time.

Solution From the method of images, the stream function for this flow field will be:

$$\psi(x, y, t) = \frac{\Gamma}{2\pi}\left[-\ln\sqrt{(x+\xi_x(t))^2 + (y-\xi_y(t))^2} + \ln\sqrt{(x-\xi_x(t))^2 + (y-\xi_y(t))^2}\right].$$

The first term is for the image vortex and second term is for the original vortex. The horizontal and vertical components of the induced velocity for both vortices are:

$$u(x, y, t) = \frac{\partial\psi}{\partial y} = \frac{\Gamma}{2\pi}\left[-\frac{y-\xi_y(t)}{(x+\xi_x(t))^2 + (y-\xi_y(t))^2} + \frac{y-\xi_y(t)}{(x-\xi_x(t))^2 + (y-\xi_y(t))^2}\right] = \frac{\partial\phi}{\partial x},$$

$$v(x, y, t) = -\frac{\partial\psi}{\partial x} = -\frac{\Gamma}{2\pi}\left[-\frac{x+\xi_x(t)}{(x+\xi_x(t))^2 + (y-\xi_y(t))^2} + \frac{x-\xi_x(t)}{(x-\xi_x(t))^2 + (y-\xi_y(t))^2}\right] = \frac{\partial\phi}{\partial y}.$$

As expected, the use of an opposite sign image vortex produces $u(0, y) = 0$. Free vortices move with fluid elements and follow path lines, thus:

$$\frac{d\xi_x(t)}{dt} = \lim_{x\to\xi}(u(x, y, t)), \quad \text{and} \quad \frac{d\xi_y(t)}{dt} = \lim_{x\to\xi}(v(x, y, t)).$$

The equations for the velocity components given above include contributions from the image and original vortices. The limit of the image vortex's induced velocity at the location of the original vortex is well defined. The velocity induced on the original vortex by itself is not well defined, but this ambiguity is a mathematical artifact of ideal vortices. Any real vortex has a finite core size, and the self-induced velocity is well defined and equal to zero on the vortex axis when the core is axisymmetric. Thus, the self-induced velocity of an ideal vortex is taken to be zero, so the vortex's path-line equations above become:

$$\frac{d\xi_x}{dt} = 0, \quad \text{and} \quad \frac{d\xi_y}{dt} = \frac{\Gamma}{2\pi}\frac{1}{2\xi_x}.$$

The solution of the first path-line equation is $\xi_x = \text{const.} = h$, where the second equality follows from the initial condition. The solution of the second equation is: $\xi_y = \Gamma t/4\pi h$, where the initial condition requires the constant of integration to be zero. Therefore, the vortex's trajectory is: $\boldsymbol{\xi}(t) = (h, \Gamma t/4\pi h)$.

To determine the pressure, integrate the velocity components to determine the potential:

$$\phi(x, y, t) = +\frac{\Gamma}{2\pi}\tan^{-1}\left(\frac{y - \Gamma t/4\pi h}{x+h}\right) - \frac{\Gamma}{2\pi}\tan^{-1}\left(\frac{y - \Gamma t/4\pi h}{x-h}\right).$$

The fluid far from the vortex is quiescent so the appropriate Bernoulli equation is:

$$\frac{\partial\phi}{\partial t} + \frac{1}{2}|\nabla\phi|^2 + \frac{p}{\rho} = \text{const.} = \frac{p_\infty}{\rho}$$

[(4.84) with $g = 0$], where the second equality follows from evaluating the constant far from the vortex. Thus, at $x = y = 0$:

$$\frac{p(0,0,t) - p_\infty}{\rho} = -\left(\frac{\partial\phi}{\partial t} + \frac{1}{2}(u^2 + v^2)\right)_{x=y=0}.$$

Here,

$$\left(\frac{\partial\phi}{\partial t}\right)_{x=y=0} = \frac{-\Gamma^2}{4\pi^2}\left(\frac{1}{h^2+(\Gamma t/4\pi h)^2}\right),\ u(0,0,t)=0,\quad \text{and}\quad v(0,0,t)=\frac{\Gamma h}{\pi}\left(\frac{1}{h^2+(\Gamma t/4\pi h)^2}\right),\ \text{so}$$

$$\frac{p(0,0,t)-p_\infty}{\rho} = \frac{\Gamma^2}{4\pi^2}\left(\frac{1}{h^2+(\Gamma t/4\pi h)^2}\right) - \frac{\Gamma^2 h^2}{2\pi^2}\left(\frac{1}{h^2+(\Gamma t/4\pi h)^2}\right)^2 = \frac{\Gamma^2}{4\pi^2}\frac{(\Gamma t/4\pi h)^2-h^2}{\left((\Gamma t/4\pi h)^2+h^2\right)^2}.$$

6.4 Complex potential

Using complex variables, the two-dimensional ideal flow developments for ψ and ϕ provided in the prior two sections can be recast in terms of a single complex function $w(z)$:

$$w(z)=\phi(x,y)+i\psi(x,y),\quad \text{where}\quad z\equiv x+iy=re^{i\theta} \tag{6.42, 6.43}$$

is a complex variable, $i=\sqrt{-1}$ is the imaginary root, (x, y) are plane Cartesian coordinates, and (r,θ) are plane polar coordinates. There are many fine texts, such as Churchill et al. (1974) or Carrier et al. (1966), that cover the relevant mathematics of complex analysis, so it is merely alluded to here. In its Cartesian form, the complex number z represents a point in the (x, y)-plane with x increasing on the real axis and y increasing on the imaginary axis (Fig. 6.17). In its polar form, z represents the position vector Oz, with magnitude $r=(x^2+y^2)^{1/2}$ and angle with respect to the x-axis of $\theta=\tan^{-1}(y/x)$.

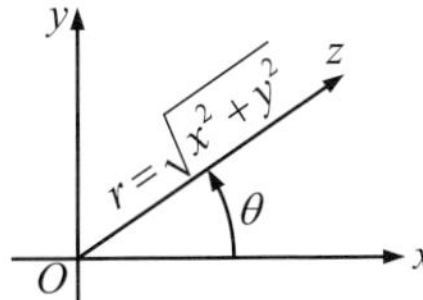

FIGURE 6.17 The complex plane where $z=x+iy=re^{i\theta}$ is the independent complex variable, $i=\sqrt{-1}$, $r=(x^2+y^2)^{1/2}$, and $\tan\theta=y/x$.

The complex function $w(z)$ is *analytic* and has a unique derivative dw/dz independent of the direction of differentiation within the complex z-plane. This condition leads to the *Cauchy-Riemann conditions*:

$$\frac{\partial\phi}{\partial x}=\frac{\partial\psi}{\partial y},\quad \text{and}\quad \frac{\partial\phi}{\partial y}=-\frac{\partial\psi}{\partial x}, \tag{6.44}$$

where derivatives of $w(z)$ in the x- and iy-directions are computed separately and then equated. These equations imply that lines of constant ϕ and ψ are orthogonal. Points in the z-plane where w or dw/dz is zero or infinite are called *singularities* and at these points this orthogonality is lost.

If ϕ is interpreted as the velocity potential and ψ as the stream function, then w is the *complex potential* for the flow and (6.44) ensures the equality of the (u, v)-velocity components. Here, the *complex velocity* can be determined from:

$$dw/dz=u-iv. \tag{6.45}$$

Applying the *Cauchy-Riemann conditions* to the complex velocity in (6.45) leads to the conditions for incompressible (6.2) and irrotational (6.9) flow, and Laplace equations for ϕ and ψ, (6.5) and (6.12), respectively. Thus, any twice-differentiable complex function of $z=x+iy$ produces solutions to Laplace's equation in the (x, y)-plane, a genuinely remarkable result! In general, a function of the two variables (x, y) may be written as $f(z, z^*)$, where $z^*=x-iy$ is the complex conjugate of z. Thus, it is the special case when $f(z, z^*)=w(z)$ alone that is considered here.

With these formal mathematical results, the correspondence between $w(z)$ and the prior results for ψ and ϕ are summarized in this and the following short paragraphs. The complex potential for flow in a corner of angle $\alpha=\pi/n$ is obtained from a power law in z:

$$w(z)=Az^n=A(re^{i\theta})^n=Ar^n(\cos n\theta+i\sin n\theta)\ \text{for}\ n\geq 1/2, \tag{6.46}$$

where A is a real constant. When $n = 2$, the streamline pattern, $\psi = \text{Im}\{w\} = Ar^2 \sin 2\theta$, represents flow in a region bounded by perpendicular walls, and (6.24) and (6.27) are readily recovered from (6.46). By including the field within the second quadrant of the z-plane, it is clear that $n = 2$ also represents the flow impinging against a flat wall (Fig. 6.3). The streamlines and equipotential lines are all rectangular hyperbolas. This is called a *stagnation flow* because it includes a stagnation point. For comparison, the streamline pattern for $n = 1/2$ corresponds to flow around a semi-infinite plate. In general, the complex velocity computed from (6.46) is:

$$dw/dz = nAz^{n-1} = (A\pi/\alpha)z^{(\pi-\alpha)/\alpha},$$

which shows that $dw/dz = 0$ at the origin for $\alpha < \pi$ while $dw/dz \to \infty$ at the origin for $\alpha > \pi$. Thus, in this flow the origin is a stagnation point for flow with wall angle smaller than 180°; in contrast, it is a point of infinite velocity for wall angle larger than 180°. In both cases the origin is a singular point.

The complex potential for an irrotational vortex of strength Γ at (x', y'), the equivalent of (6.8), is:

$$w(z) = -\frac{i\Gamma}{2\pi}\ln(z - z') = \frac{\Gamma}{2\pi}\theta' - i\frac{\Gamma}{2\pi}\ln r', \tag{6.47}$$

where $z' = x' + iy'$, $r' = \sqrt{(x - x')^2 + (y - y')^2}$, and $\theta' = \tan^{-1}((y - y')/(x - x'))$.

The complex potential for a source or sink of volume flow rate q_s per unit depth located at (x', y'), the equivalent of (6.15), is:

$$w(z) = \frac{q_s}{2\pi}\ln(z - z') = \frac{q_s}{2\pi}\ln\left(r'e^{i\theta'}\right) = \frac{q_s}{2\pi}\ln r' + i\frac{q_s\theta'}{2\pi}. \tag{6.48}$$

The complex potential for a doublet with dipole strength $-d\mathbf{e}_x$ located at (x', y'), the equivalent of (6.29), is:

$$w = \frac{d}{2\pi(z - z')}. \tag{6.49}$$

The complex potential for uniform flow at speed U past a half body – see (6.29), (6.30), and Fig. 6.7 – is the combination of a source of strength q_s at the origin and a uniform horizontal stream:

$$w(z) = Uz + \frac{q_s}{2\pi}\ln z. \tag{6.50}$$

The complex potential for uniform flow at speed U past a circular cylinder, see (6.33) and Fig. 6.9, is the combination of a doublet with dipole strength $\mathbf{d} = -2\pi Ua^2\mathbf{e}_x$ and a uniform stream:

$$w = U\left(z + a^2/z\right). \tag{6.51}$$

When *clockwise* circulation $-\Gamma$ is added to the cylinder, the complex potential becomes:

$$w = U\left(z + \frac{a^2}{z}\right) + \frac{i\Gamma}{2\pi}\ln(z/a), \tag{6.52}$$

the flow field is altered (see Fig. 6.12), and the cylinder experiences a lift force. Here, the imaginary part of (6.52) reproduces (6.36).

The method of images also applies to the complex potential. For example, the complex potential for the flow described by (6.41) is:

$$w = \frac{q_s}{2\pi}\ln\left(\frac{z-a}{a}\right) + \frac{q_s}{2\pi}\ln\left(\frac{z+a}{a}\right) = \frac{q_s}{2\pi}\ln\left(\frac{z^2 - a^2}{a^2}\right) = \frac{q_s}{2\pi}\ln\left(\frac{x^2 - y^2 - a^2 + 2ixy}{a^2}\right). \tag{6.53}$$

As described in the next section, the complex variable description of ideal flow also allows some general results to be obtained for pressure forces (per unit depth perpendicular to the plane of the flow) that act on two-dimensional bodies.

Example 6.4

Separate real and imaginary parts of (6.52) to determine ϕ and recover ψ as given by (6.36).

Solution Start from (6.52), and insert $x+iy$ for z:

$$w=U\left(z+\frac{a^2}{z}\right)+\frac{i\Gamma}{2\pi}\ln\left(\frac{z}{a}\right)=U(x+iy)+\frac{Ua^2}{x+iy}+\frac{i\Gamma}{2\pi}\ln\left(\frac{x+iy}{a}\right).$$

Remove i from the denominator of the second term and the argument of the logarithm:

$$w=U(x+iy)+\frac{Ua^2(x-iy)}{x^2+y^2}+\frac{i\Gamma}{2\pi}\ln\left(\frac{\sqrt{x^2+y^2}}{a}\right)+\frac{i\Gamma}{2\pi}\tan^{-1}\left(\frac{y}{x}\right).$$

Group terms without and with i together:

$$w=\phi+i\psi=Ux\left(1+\frac{a^2}{x^2+y^2}\right)-\frac{\Gamma}{2\pi}\tan^{-1}\left(\frac{y}{x}\right)+iUy\left(1-\frac{a^2}{x^2+y^2}\right)+\frac{i\Gamma}{2\pi}\ln\left(\frac{\sqrt{x^2+y^2}}{a}\right).$$

Switch to (r,θ)-polar coordinates using $x=r\cos\theta$ and $y=r\sin\theta$, and separate real and imaginary parts in the last equation to reach the final results for ϕ and ψ:

$$\phi=U\left(r+\frac{a^2}{r}\right)\cos\theta-\frac{\Gamma\theta}{2\pi},\quad\text{and}\quad\psi=U\left(r-\frac{a^2}{r}\right)\sin\theta+\frac{\Gamma}{2\pi}\ln\left(\frac{r}{a}\right).$$

6.5 Forces on a two-dimensional body

In Section 6.3 the drag and lift forces per unit length on a circular cylinder in steady ideal flow were found to be zero and $\rho U\Gamma$, respectively, when the circulation is clockwise. These results are also valid for any object with an *arbitrary* non-circular cross section that does not vary perpendicular to the x-y plane.

Blasius theorem

Consider a stationary object of this type with extent B perpendicular to the plane of the flow, and let D (drag) be the stream-wise (x) force component and L (lift) be cross-stream or lateral (y) force (per unit depth) exerted on the object by the surrounding fluid. Thus, from Newton's third law, the total force applied to the fluid by the object is $\mathbf{F}=-B(D\mathbf{e}_x+L\mathbf{e}_y)$. For steady irrotational constant-density flow, conservation of momentum (4.17) within a stationary control volume implies:

$$\int_{A^*}\rho\mathbf{u}(\mathbf{u}\cdot\mathbf{n})dA=-\int_{A^*}p\mathbf{n}dA+\mathbf{F}. \tag{6.54}$$

If the control surface A^* is chosen to coincide with the body surface and the body is not moving, then $\mathbf{u}\cdot\mathbf{n}=0$ and the flux integral on the left in (6.54) is zero, so:

$$D\mathbf{e}_x+L\mathbf{e}_y=-\frac{1}{B}\int_{A^*}p\mathbf{n}dA. \tag{6.55}$$

If C is the contour of the body's cross section, then $dA=Bds$ where $d\mathbf{s}=\mathbf{e}_xdx+\mathbf{e}_ydy$ is an element of C and $ds=[(dx)^2+(dy)^2]^{1/2}$. By definition, $\mathbf{n}$ must have unit magnitude, must be perpendicular to $d\mathbf{s}$, and must point outward from the control volume, so $\mathbf{n}=(\mathbf{e}_xdy-\mathbf{e}_ydx)/ds$. Using these relationships for $\mathbf{n}$ and dA, (6.55) can be separated into force components:

$$D\mathbf{e}_x+L\mathbf{e}_y=-\frac{1}{B}\oint_C p\frac{(\mathbf{e}_xdy-\mathbf{e}_ydx)}{ds}B\,ds=\left(-\oint_C pdy\right)\mathbf{e}_x+\left(\oint_C pdx\right)\mathbf{e}_y, \tag{6.56}$$

to identify the contour integrals leading to D and L. Here, C must be traversed in the counterclockwise direction.

Now switch from the physical domain to the complex z-plane to make use of the complex potential. This switch is accomplished here by replacing $d\mathbf{s}$ with $dz=dx+idy$ and exploiting the dichotomy between real and imaginary parts to keep track of horizontal and vertical components (see Fig. 6.18). To achieve the desired final result, construct the complex force:

$$D-iL=\left(-\oint_C pdy\right)-i\left(\oint_C pdx\right)=-i\oint_C p(dx-idy)=-i\oint_C pdz^*, \tag{6.57}$$

where * denotes a complex conjugate. The pressure is found from the Bernoulli equation (6.18):

$$p_\infty + \tfrac{1}{2}\rho U^2 = p + \tfrac{1}{2}\rho(u^2 + v^2) = p + \tfrac{1}{2}\rho(u - iv)(u + iv),$$

where p_∞ and U are the pressure and horizontal flow speed far from the body. Inserting this into (6.57) produces:

$$D - iL = -i\oint_C \left[p_\infty + \tfrac{1}{2}\rho U^2 - \tfrac{1}{2}\rho(u - iv)(u + iv)\right]dz^*. \tag{6.58}$$

The integral of the constant terms, $p_\infty + \rho U^2/2$, around a closed contour is zero. The body-surface velocity vector and the surface element $dz = |dz|e^{i\theta}$ are parallel, so $(u + iv)dz^*$ can be rewritten:

$$(u + iv)dz^* = [u^2 + v^2]^{1/2}e^{i\theta}|dz|e^{-i\theta} = [u^2 + v^2]^{1/2}e^{-i\theta}|dz|e^{i\theta} = (u - iv)dz = (dw/dz)dz, \tag{6.59}$$

where (6.45) has been used for the final equality. Thus, (6.58) reduces to:

$$D - iL = \frac{i\rho}{2}\oint_C \left(\frac{dw}{dz}\right)^2 dz, \tag{6.60}$$

a result known as the *Blasius theorem*. It applies to any steady planar ideal flow. Interestingly, the integral need not be carried out along the contour of the body because the theory of complex variables allows *any contour surrounding the body to be chosen* provided there are no singularities in $(dw/dz)^2$ between the body and the contour chosen.

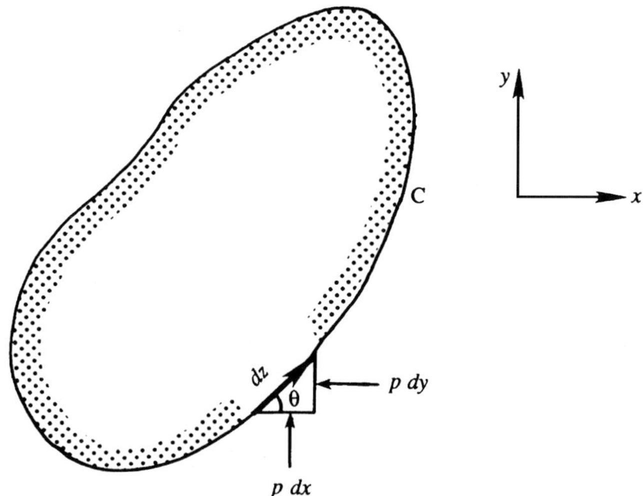

FIGURE 6.18 Elemental forces in a plane on a two-dimensional object. Here the elemental horizontal and vertical force components (per unit depth) are $-pdy$ and $+pdx$, respectively.

Kutta-Zhukhovsky lift theorem

The Blasius theorem can be readily applied to an arbitrary cross-section object around which there is circulation $-\Gamma$. The flow can be considered a superposition of a uniform stream and a set of singularities such as vortex, doublet, source, and sink.

As there are no singularities outside the body, we shall take the contour C in the Blasius theorem at a large distance from the body. From large distances, all singularities appear to be located near the origin $z = 0$, so the complex potential on the contour C will be of the form:

$$w = Uz + \frac{q_s}{2\pi}\ln z + \frac{i\Gamma}{2\pi}\ln z + \frac{d}{2\pi z} + \ldots$$

When U, q_s, Γ, and d are positive and real, the first term represents a uniform flow in the x-direction, the second term represents a net source of fluid, the third term represents a clockwise vortex, and the fourth term represents a doublet.

Because the body contour is closed, there can be no net flux of fluid into the domain. The sinks must scavenge all the flow introduced by the sources, so $q_s = 0$. The Blasius theorem, (6.60), then becomes:

$$D - iL = \frac{i\rho}{2} \oint_C \left(U + \frac{i\Gamma}{2\pi z} - \frac{d}{2\pi z^2} + \ldots \right)^2 dz = \frac{i\rho}{2} \oint_C \left(U^2 + \frac{iU\Gamma}{\pi} \frac{1}{z} + \left(\frac{Ud}{\pi} - \frac{\Gamma^2}{4\pi^2} \right) \frac{1}{z^2} + \ldots \right) dz. \quad (6.61)$$

To evaluate the contour integral in (6.61), we simply have to find the coefficient of the term proportional to $1/z$ in the integrand. This coefficient is known as the *residue* at $z = 0$ and the residue theorem of complex variable theory states that the value of a contour integral like (6.61) is $2\pi i$ times the sum of the residues at all singularities inside C. Here, the only singularity is at $z = 0$, and its residue is $iU\Gamma/\pi$, so:

$$D - iL = \frac{i\rho}{2} 2\pi i \left(\frac{iU\Gamma}{\pi} \right) = -i\rho U\Gamma, \quad \text{or} \quad D = 0 \quad \text{and} \quad L = \rho U\Gamma. \quad (6.62)$$

Thus, there is no drag on an arbitrary-cross-section object in steady two-dimensional, irrotational constant-density flow, a more general statement of d'Alembert's paradox. Given that non-zero drag forces are an omnipresent fact of everyday life, this might seem to eliminate any practical utility for ideal flow. However, there are at least three reasons to avoid this presumption. First, ideal flow streamlines indicate what a real flow should look like to achieve minimum pressure drag. Lower drag on real objects is often realized when object-geometry changes are made or boundary-layer separation-control strategies are implemented that allow real-flow streamlines to better match their ideal-flow counterparts. Second, the predicted circulation-dependent force on the object perpendicular to the oncoming stream – the lift force, $L = \rho U\Gamma$ – is basically correct. The result (6.62) is called the *Kutta-Zhukhovsky lift theorem*, and it plays a fundamental role in aero- and hydrodynamics. As described in Chapter 13, the circulation developed by an air- or hydrofoil is nearly proportional to U, so L is nearly proportional to U^2. And third, the influence of viscosity in real fluid flows takes some time to develop, so impulsively started flows and rapidly oscillating flows (i.e., acoustic fluctuations) often follow ideal flow streamlines.

Example 6.5

Revise the Kutta-Zhukhovsky Lift Theorem results for the case when $q_s \neq 0$.

Solution Start from (6.60) and consider a closed contour far from the origin as was done to reach (6.62), but include the source-or-sink term in the complex potential:

$$\frac{dw}{dz} = U + \frac{q_s}{2\pi z} + \frac{i\Gamma}{2\pi z} - \frac{d}{2\pi z^2} + \ldots, \text{ so}$$

$$D - iL = \frac{i\rho}{2} \oint_C \left(U^2 + \frac{U(q_s + i\Gamma)}{\pi z} + \left(\frac{\mathrm{Ud}}{\pi} + \frac{(q_s + i\Gamma)^2}{4\pi^2} \right) \frac{1}{z^2} + \ldots \right) dz.$$

Here, the residue is $U(q_s + i\Gamma)/\pi$, so:

$$D - iL = \frac{i\rho}{2} 2\pi i \left(\frac{U(q_s + i\Gamma)}{\pi} \right) = -\rho U q_s - i\rho U\Gamma, \quad \text{or} \quad D = -\rho U q_s \quad \text{and} \quad L = \rho U\Gamma.$$

The lift force is the same as in (6.62), but there is thrust (negative drag) on a source ($q_s > 0$) and drag on a sink ($q_s < 0$). The sign of this drag result can be understood as follows. The streamlines for a uniform stream and a positive source are shown in Fig. 6.7. Imagine a control volume that surrounds the fluid that comes from the source, extends downstream of the source, and terminates via a vertical segment between upper and lower branches of the boundary streamline (approximately the shape of a backwards letter "D"). The streamlines exterior to this CV are essentially the same as those surrounding a rocket heading into the on-coming stream, which implies a thrust force on the CV. Similarly, when a sink replaces the source, the sink removes a portion of the on-coming stream, along with its momentum, from flow field. Thus, the sink feels a drag force that pushes in the direction (downstream) of the fluid momentum the sink removes.

6.6 Conformal mapping

The technique known as *conformal mapping* in complex variable theory can be used to transform simple flow patterns into more complicated ones and vice versa. Consider the functional relationship $w = f(z)$, which maps a point in the w-plane to a point in the z-plane. If $w = f(z)$ is an analytic transformation, then infinitesimal figures in the two planes preserve their geometric similarity. Let lines C_z and C'_z in the z-plane be transformations of the curves C_w and C'_w in the w-plane, respectively (Fig. 6.19). Let δz, $\delta' z$, δw, and $\delta' w$ be infinitesimal elements along the curves as shown. The four elements are related by:

$$\delta w = \frac{dw}{dz}\delta z, \quad \text{and} \quad \delta' w = \frac{dw}{dz}\delta' z. \tag{6.63, 6.64}$$

If $w = f(z)$ is analytic, then dw/dz is independent of orientation of the elements, and therefore has the same value in (6.63) and (6.64) These two equations then imply that the elements δz and $\delta' z$ are rotated by the *same amount* (equal to the argument of dw/dz) to obtain the elements δw and $\delta' w$. Thus, the angles in Fig. 6.19 are equal, $\alpha = \beta$, which demonstrates that infinitesimal figures in the two planes are geometrically similar, but they are not necessarily the same size or at the same orientation. This demonstration fails at singular points at which dw/dz is either zero or infinite. Away from singular points, the amount of magnification and rotation that an element δz undergoes during transformation from the z-plane to the w-plane varies because dw/dz is a function of z. Consequently, *large* figures become distorted during the transformation.

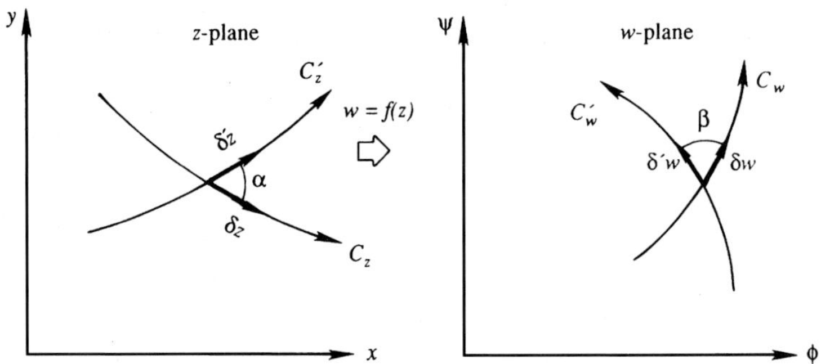

FIGURE 6.19 Preservation of geometric similarity of small elements in conformal mapping between the complex z and w planes.

In applications of conformal mapping, a rectangular grid consisting of constant ϕ and ψ lines is commonly chosen in the w-plane (Fig. 6.20). In other words, ϕ and ψ are chosen to be the real and imaginary parts of w:

$$w = \phi + i\psi. \tag{6.42}$$

The rectangular mesh in the w-plane represents a uniform flow in this plane. The constant ϕ and ψ lines are transformed into certain curves in the z-plane through the transformation $w = f(z)$ or its inverse $f^{-1}(w) = z$. *The flow pattern in the z-plane is typically the physical pattern under investigation*, and the images of constant ϕ and ψ lines in the z-plane form the equipotential lines and streamlines, respectively, of the flow of interest. The transformation $w = f(z)$ maps a uniform flow in the w-plane into the desired flow in the z-plane, and all the preceding flow patterns presented in Section 6.4 can be interpreted this way.

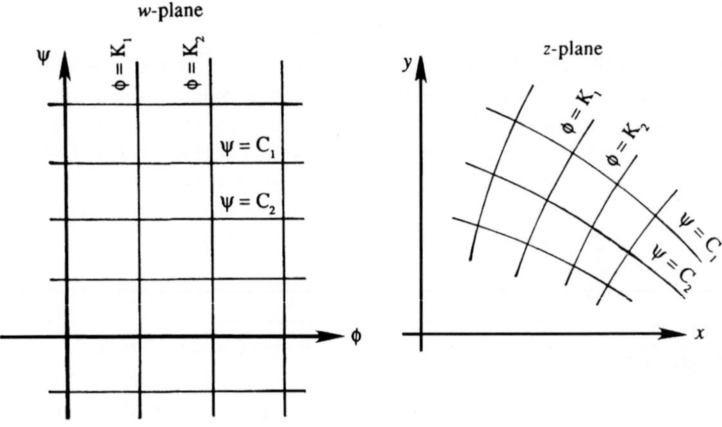

FIGURE 6.20 Flow patterns in the complex w-plane and the z-plane. The w-plane represents uniform flow with straight potential lines and streamlines. In the z-plane, these lines curve to represent the flow of interest.

If the physical pattern under investigation is too complicated, intermediate transformations in going from the w-plane to the z-plane may be introduced. For example, the transformation $w = \ln(\sin z)$ can be broken into: $w = \ln \zeta$, and $\zeta = \sin z$. In this particular case, velocity components in the z-plane are given by:

$$u - iv = \frac{dw}{dz} = \frac{dw}{d\zeta}\frac{d\zeta}{dz} = \frac{1}{\zeta}\cos z = \cot z.$$

As an example of conformal mapping, consider the transformation, $w = \phi + i\psi = z^2 = x^2 + y^2 + 2ixy$. Streamlines are given by $\psi = \text{const.} = 2xy$, rectangular hyperbolae (see Fig. 6.3). Here uniform flow in the w-plane has been mapped onto flow in a 90° corner in the z-plane by this transformation. A more involved example follows. Additional applications are discussed in Chapter 13.

The Zhukhovsky transformation relates two complex variables z and ζ, and has applications in airfoil theory:

$$z = \zeta + b^2/\zeta. \tag{6.65}$$

When $|\zeta|$ or $|z|$ is large compared to b, this transformation becomes an identity, so it does not change the flow condition far from the origin when moving between the z and ζ planes. However, close to the origin, (6.65) transforms a circle of radius b centered at the origin of the ζ-plane into a line segment on the real axis of the z-plane. To establish this, let $\zeta = b\exp(i\theta)$ on the circle (Fig. 6.21) so that (6.65) provides the corresponding point in the z-plane as:

$$z = be^{i\theta} + be^{-i\theta} = 2b\cos\theta.$$

As θ varies from 0 to π, z goes along the x-axis from $2b$ to $-2b$. As θ varies from π to 2π, z goes from $-2b$ to $2b$. The circle of radius b in the ζ-plane is thus transformed into a line segment of length $4b$ in the z-plane. The region *outside* the circle in the ζ-plane is mapped into the *entire* z-plane. It can be shown that the region inside the circle is also transformed into the entire z-plane. This, however, is not a concern because the interior of the circle in the ζ-plane is not important.

Now consider a circle of radius $a > b$ in the ζ-plane (Fig. 6.21). A point $\zeta = a\exp(i\theta)$ on this circle is transformed to:

$$z = ae^{i\theta} + \left(\frac{b^2}{a}\right)e^{-i\theta}, \tag{6.66}$$

which traces out an ellipse for various values of θ; the geometry becomes clear by separating real and imaginary parts of (6.66) and eliminating θ:

$$\frac{x^2}{(a + b^2/a)^2} + \frac{y^2}{(a - b^2/a)^2} = 1. \tag{6.67}$$

For $a > b$, (6.67) represents a family of ellipses in the z-plane with foci at $x = \pm 2b$.

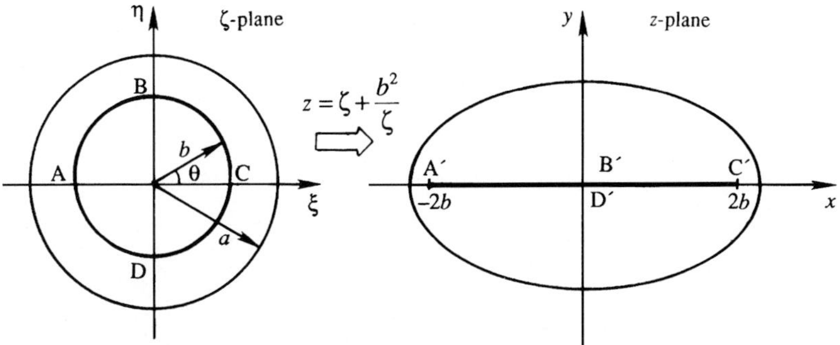

FIGURE 6.21 Transformation of a circle of radius a in the ζ-plane into an ellipse in the z-plane by means of the Zhukhovsky transformation $z = \zeta + b^2/\zeta$. A circle of radius b in the ζ-plane transforms into a line segment between $z = \pm 2b$ in the z-plane.

The flow around one of these ellipses (in the z-plane) can be determined by first finding the flow around a circle of radius a in the ζ-plane, and then using (6.65) to go to the z-plane. To be specific, suppose the desired flow in the z-plane is that of flow around an elliptic cylinder with clockwise circulation Γ placed in a stream moving at U. The corresponding flow in the ζ-plane is that of flow with the same circulation around a circular cylinder of radius a placed in a horizontal stream of speed U. The complex potential for this flow is (6.52) with z replaced by ζ:

$$w = U\left(\zeta + \frac{a^2}{\zeta}\right) + \frac{i\Gamma}{2\pi}\ln(\zeta/a). \tag{6.68}$$

The complex potential $w(z)$ in the z-plane can be found by substituting the inverse of (6.65):

$$\zeta = \frac{1}{2}z + \frac{1}{2}\sqrt{z^2 - 4b^2}, \tag{6.69}$$

into (6.68). Here, the negative root, which falls inside the cylinder, has been excluded from (6.69). Instead of finding the complex velocity in the z-plane by directly differentiating $w(z)$, it is easier to find it using the chain rule, $u - iv = dw/dz = (dw/d\zeta)(d\zeta/dz)$. The resulting flow around an elliptic cylinder with circulation is qualitatively quite similar to that around a circular cylinder as shown in Fig. 6.12.

Example 6.6

The Schwarz-Christoffel transformation is a means for obtaining ideal flow patterns past objects with sharp corners (see Milne-Thompson, 1962, or Currie, 1993). It can be used to obtain the following complex potential for ideal horizontal flow at speed U past a forward facing step of height h:

$$w = \frac{2hU}{\pi}\zeta, \quad \text{where} \quad z = \frac{2h}{\pi}\left\{\sqrt{\zeta^2 + \zeta} - \ln\left[\sqrt{\zeta} + \sqrt{\zeta + 1}\right] + i\pi/2\right\}.$$

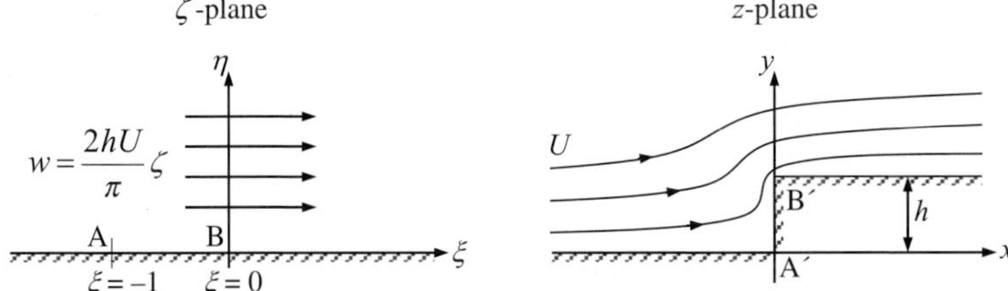

FIGURE 6.22 Complex plane-flow patterns for Example 6.6. The horizontal ξ-axis in the ζ-plane is transformed into a forward facing step in the z-plane. Here the point A, $\xi = -1$ and $\eta = 0$ in the ζ-plane, is mapped into the point A′, the origin in the z-plane. And, the point B, the origin in the ζ-plane, is mapped into the point B′, $x = 0$ and $y = h$, in the z-plane.

The flows in both planes are shown in Fig. 6.22. Determine the complex velocity, dw/dz, and the pressure coefficient, C_p from (6.32), on the front face of the step. Does ideal flow predict a drag force on the step?

Solution First, compute the complex velocity using the chain rule:

$$u - iv = \frac{dw}{dz} = \frac{dw}{d\zeta}\frac{d\zeta}{dz} = \frac{2hU}{\pi}\left[\frac{dz}{d\zeta}\right]^{-1} = \frac{2hU}{\pi}\left[\frac{2h}{\pi}\sqrt{\frac{\zeta}{\zeta + 1}}\right]^{-1} = U\sqrt{\frac{\zeta + 1}{\zeta}},$$

where the second to last equality requires some algebraic effort. From (6.32), the coefficient of pressure is:

$$C_p = \frac{p - p_\infty}{\frac{1}{2}\rho U^2} = 1 - \frac{u^2 + v^2}{U^2} = 1 - \frac{|dw/dz|^2}{U^2} = 1 - \left|\sqrt{\frac{\zeta + 1}{\zeta}}\right|^2 = 1 - \sqrt{\frac{(\xi + 1)^2 + \eta^2}{\xi^2 + \eta^2}},$$

where p_∞ is the pressure far from the step, and $\zeta = \xi + i\eta$. For the given transformation, the face of the step occurs where $-1 < \xi < 0$ and $\eta = 0$, so:

$$(C_p)_{face} = 1 - \frac{\xi + 1}{(-\xi)} = \frac{2\xi + 1}{\xi},$$

where the minus sign in the denominator comes from taking the positive square root of ξ when $\xi < 0$. This pressure coefficient is singular at the upper corner of the step (B′) where $\xi = 0$.

The drag force (per unit length into the page) on the step will be:

$$D=\frac{\rho U^2}{2}\int_{y=0}^{h}(C_p)_{face}dy=\frac{\rho U^2}{2}\int_{\xi=-1}^{0}\frac{2\xi+1}{\xi}\left(\frac{dy}{d\xi}\right)d\xi=\frac{\rho U^2}{2}\int_{-1}^{0}\frac{2\xi+1}{\xi}\left(\frac{-2ih}{\pi}\sqrt{\frac{\xi}{\xi+1}}\right)d\xi,$$

and the integration variable is changed from y to ξ. Interestingly, the final integral can be evaluated exactly:

$$D=\frac{-ih\rho U^2}{\pi}\int_{-1}^{0}\frac{2\xi+1}{\sqrt{\xi^2+\xi}}d\xi=\frac{-ih\rho U^2}{\pi}\left[\frac{1}{2}\sqrt{\xi^2+\xi}\right]_{-1}^{0}=0.$$

Thus, ideal flow predicts no drag force on the step even though there is a stagnation point ($C_p=1$) at the base of the step (point A′). This overpressure is balanced by suction at the upper corner of the step (point B′) where $C_p\to-\infty$, an (unrealistic) ideal flow phenomena sometimes known as corner or edge suction. Actual flow past a forward facing step separates upstream of the step and at the upper step corner as depicted in Fig. 6.2b when $\rho Uh/\mu \gg 1$.

6.7 Numerical solution techniques in two dimensions

Exact solutions for ideal flows can be obtained only for relatively simple geometries, so approximate methods become necessary for most problems of practical significance. One of these approximate methods is that of building up a flow by superposing a distribution of sources and sinks; this method is illustrated in Section 6.8 for axisymmetric flows. Another method is to apply perturbation techniques by assuming the body is thin. A third method is to solve the Laplace equation numerically. In this section, we illustrate the numerical method as a means of introducing computational techniques in their simplest form without considerations of computational efficiency. The main idea is to replace the continuous partial differential equation with a discrete system of algebraic equations that can be solved numerically. For ideal flows, this results in a (potentially large) system of linear equations, a common situation in modern computational fluid dynamics (CFD). Although applied to ideal flow in this section, the same methodology can be applied to other types of flows. A more comprehensive introduction to CFD is provided in Chapter 15.

Numerical techniques for solving the Laplace equation typically rely on discretizing the spatial domain into a system of *grid points*, then using finite difference techniques between adjacent grid points to determine field values. Let the coordinates of a point be represented by

$$x=i\Delta x\;(i=0,1,2,\ldots),\quad\text{and}\quad y=j\Delta y\;(j=0,1,2,\ldots).$$

Here, Δx and Δy are the lengths of each side of a grid cell, and the integers i and j are the indices associated with a grid point (Fig. 6.23); the complex root is not needed here. The two-dimensional stream function $\psi(x,y)$ can be represented at these discrete locations by

$$\psi(x,y)=\psi(i\Delta x,j\Delta y)\equiv\psi_{i,j},$$

where $\psi_{i,j}$ is the value of ψ at the grid point (i,j). In finite difference form, the second derivatives of ψ are approximated using central differences (see Chapter 15):

$$\left(\frac{\partial^2\psi}{\partial x^2}\right)_{ij}\approx\frac{1}{\Delta x^2}\left[\psi_{i+1,j}-2\psi_{i,j}+\psi_{i-1,j}\right]\quad\text{and}$$
$$\left(\frac{\partial^2\psi}{\partial y^2}\right)_{ij}\approx\frac{1}{\Delta y^2}\left[\psi_{i,j+1}-2\psi_{i,j}+\psi_{i,j-1}\right].\tag{6.70}$$

Here, the derivatives are approximated by the expressions on the right-hand side as they neglect high-order terms. These expressions approach the left-hand side as $\Delta x,\ \Delta y\to 0$ but with the trade off of computational cost associated with more grid points. Using (6.70), the Laplace equation (6.5) for the stream function in a planar two-dimensional flow has the following finite difference representation:

$$\frac{1}{\Delta x^2}\left[\psi_{i+1,j}-2\psi_{i,j}+\psi_{i-1,j}\right]+\frac{1}{\Delta y^2}\left[\psi_{i,j+1}-2\psi_{i,j}+\psi_{i,j-1}\right]=0.$$

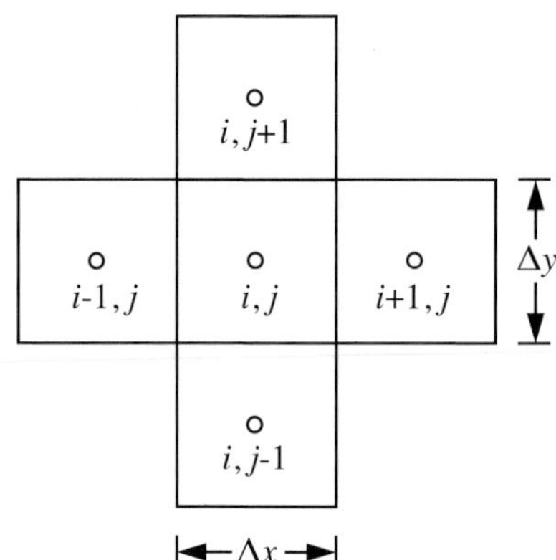

FIGURE 6.23 Adjacent grid cells in a numerical calculation showing how the indices i and j change.

Taking $\Delta x = \Delta y$ for simplicity, this reduces to

$$\psi_{ij} = \frac{1}{4}\left[\psi_{i-1,j} + \psi_{i+1,j} + \psi_{i,j-1} + \psi_{i,j+1}\right], \tag{6.71}$$

which shows that ψ numerically satisfies the Laplace equation if its value at each grid point equals the average of its values at the four surrounding points.

Eq. (6.71) can be solved by a simple iteration technique when the values of ψ are given on the flow's boundaries. A common approach is to start with an initial guess for ψ at all interior points (excluding the boundaries where values are known), solve (6.71), then improve the guess and repeat until the solution converges within a desired threshold. For example, starting with an initial guess of $\psi = 0$, the entries of ψ_{ij} in (6.71) are replaced with updated values as the calculation proceeds. In this manner, we can sweep over the entire flow domain in a systematic manner, *always using the latest available value at each grid point*. Once the first estimate at every point has been obtained, the domain sweep can be repeated to obtain second estimates and this process can be repeated again and again until the values of ψ_{ij} do not change appreciably between two successive steps. At this point, the iteration process has *converged*.

Such an approach is ideally suited for a computer where it is easy to replace old values at a point as soon as a new value is available. In practice, more sophisticated and efficient numerical techniques are used for large calculations. However, the purpose here is to present the simplest numerical solution technique, which is illustrated in the following example.

Example 6.7

Fig. 6.24 shows a contraction in a channel through which the flow rate per unit depth is 5 m^2/s. The velocity is uniform and parallel across the inlet and outlet sections. Find the stream function values within this flow field.

Solution In this example, the stream function values are solved following the numerical solution technique outlined in this Section. The difference in ψ values is equal to the flow rate between two streamlines. Thus, setting $\psi = 0$ at the bottom wall requires $\psi = 5$ m^2/2 at the top wall. The stream function is discretized into a system of grid points as shown, with $\Delta x = \Delta y = 1$ m. Because $\Delta\psi/\Delta y (= u)$ is given to be uniform across the inlet and the outlet, we must have $\Delta\psi = 1$ m^2/s at the inlet and $\Delta\psi = 5/3 = 1.67$ m^2/s at the outlet. The resulting values of ψ at the boundary points are indicated in Fig. 6.24. A MATLAB® code for solving the problem is given below. The resulting contour lines are provided in Fig. 6.25. Note that better resolution can be obtained by reducing Δx and Δy.

```
% Initialize the stream function
psi=zeros(10,6);

% Set boundary conditions at top and bottom walls
psi(1:6,1)=0;
psi(6,1:3)=0;
psi(6:10,3)=0;
psi(1:10,6)=5;

```

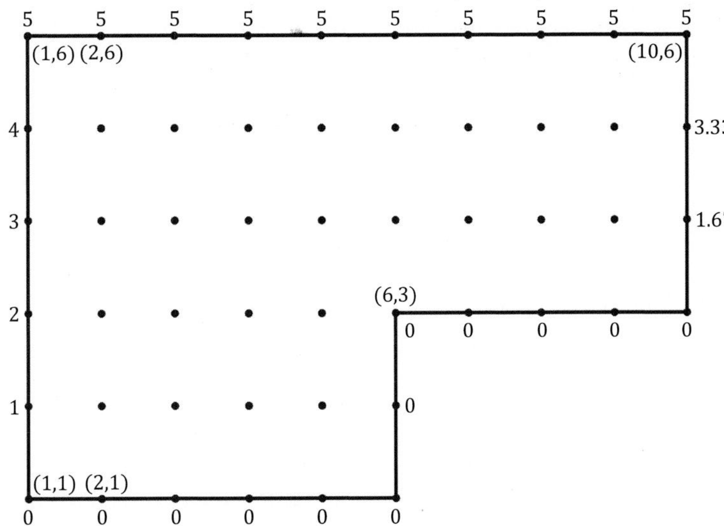

FIGURE 6.24 Grid pattern for irrotational flow through an abrupt, sharp-cornered contraction (Example 6.6). The flow enters on the left and exits on the right. The boundary values of ψ are indicated on the outside. The values of i, j for some grid points are indicated on the inside.

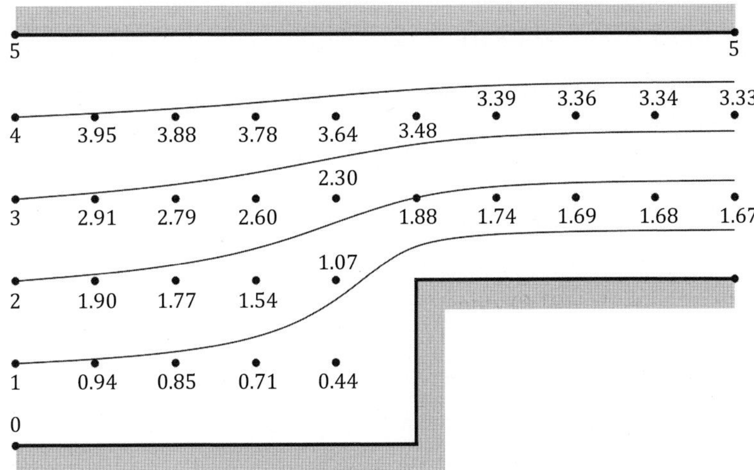

FIGURE 6.25 Numerical solution of Example 6.7 for the boundary conditions shown in Fig. 6.24. The numerical values listed are for the corresponding grid points shown as black dots. Note that there is some streamline curvature after the abrupt contraction.

```
% Set boundary conditions at inlet
for j=2:6
    psi(1,j)=j-1;
end

% Set boundary conditions at outlet
for j=4:6
    psi(10,j)=(j-3)*5/3;
end

% Iterate
for n=1:20
    for i=2:5
        for j=2:5
            psi(i,j)=0.25*(psi(i,j+1)+psi(i,j-1)+...
            psi(i+1,j)+psi(i-1,j));
        end
    end
    for i=6:9
```

```
        for j=4:5
            psi(i,j)=0.25*(psi(i,j+1)+psi(i,j-1)+...
            psi(i+1,j)+psi(i-1,j));
        end
    end
end

% Display values
for j=1:6
    sprintf('%10.2f', psi(:,j))
end
```

6.8 Axisymmetric ideal flow

Two stream functions are required to describe a fully three-dimensional flow (see Section 4.3). However, when the flow is axisymmetric, a single stream function can again be used. Thus, the development presented here parallels that in Section 6.2 for plane ideal flow.

In (R, φ, z) cylindrical coordinates, the axisymmetric incompressible continuity equation is:

$$\frac{1}{R}\frac{\partial}{\partial R}(Ru_R) + \frac{\partial u_z}{\partial z} = 0. \tag{6.72}$$

As discussed near the end of Section 4.3, this equation may be solved by choosing the first three-dimensional stream function $\chi = -\varphi$, so that $\mathbf{u} = (u_R, 0, u_z) = \nabla\chi \times \nabla\psi = -(1/R)\mathbf{e}_\varphi \times \nabla\psi$, which implies

$$u_R = -\frac{1}{R}\frac{\partial\psi}{\partial z}, \quad \text{and} \quad u_z = \frac{1}{R}\frac{\partial\psi}{\partial R}. \tag{6.73}$$

Substituting these into the equation for the φ-component of vorticity,

$$\omega_\varphi = \frac{\partial u_R}{\partial z} - \frac{\partial u_z}{\partial R}, \tag{6.74}$$

produces the field equation for the axisymmetric stream function in irrotational flow:

$$\frac{\partial}{\partial R}\left(\frac{1}{R}\frac{\partial\psi}{\partial R}\right) + \frac{1}{R}\frac{\partial^2\psi}{\partial z^2} = -\omega_\varphi = 0. \tag{6.75}$$

This *is not* the two-dimensional Laplace equation. Therefore, the complex variable formulation for plane ideal flows does not apply to axisymmetric ideal flow.

The axisymmetric stream function is sometimes called the *Stokes stream function*. It has units of volume/time, in contrast to the plane-flow stream function, which has units of area/time. Surfaces of ψ = constant in axisymmetric flow are surfaces of revolution. The volume flow rate dQ between two axisymmetric stream surfaces described by constant values ψ and $\psi + d\psi$ (Fig. 6.26) is:

$$dQ = 2\pi R(\mathbf{u}\cdot\mathbf{n})ds = 2\pi R(-u_R dz + u_z dR) = 2\pi\left(\frac{\partial\psi}{\partial z}dz + \frac{\partial\psi}{\partial R}dR\right) = 2\pi d\psi, \tag{6.76}$$

where $\mathbf{u} = u_R\mathbf{e}_R + u_z\mathbf{e}_z$, $\mathbf{n} = (-\mathbf{e}_R dz + \mathbf{e}_z dR)/ds$, and (6.73) has been used, and dz is negative as drawn in Fig. 6.26. The form $d\psi = dQ/2\pi$ shows that the difference in ψ values is the flow rate between two concentric stream surfaces per unit radian angle around the axis. This is consistent with the discussion of stream functions in Section 4.3. The factor of 2π is absent in plane flows, where $d\psi = dQ$ is the flow rate per unit depth perpendicular to the plane of the flow.

An axisymmetric potential function ϕ can also be defined via $\mathbf{u} = \nabla\phi$ or:

$$u_R = \partial\phi/\partial R \quad \text{and} \quad u_z = \partial\phi/\partial z, \tag{6.77}$$

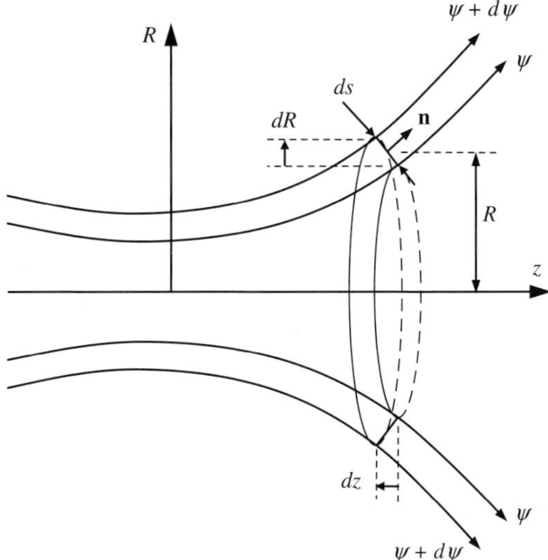

FIGURE 6.26 Geometry for calculating the volume flow rate between axisymmetric-flow stream surfaces with values of ψ and $\psi + d\psi$. The z-axis is the axis of symmetry, R, is the radial distance from the z-axis, $d\mathbf{s} = \mathbf{e}_R dR + \mathbf{e}_z dz$ is the distance between the two surfaces, and $\mathbf{n}$ is a unit vector perpendicular to $d\mathbf{s}$. The volume flow rate between the two surfaces is $2\pi d\psi$.

so that the flow identically satisfies $\omega_\varphi = 0$. Substituting (6.77) into the incompressible continuity equation produces the field equation for the axisymmetric potential function:

$$\frac{1}{R}\frac{\partial}{\partial R}\left(R\frac{\partial \phi}{\partial R}\right) + \frac{\partial^2 \phi}{\partial z^2} = 0. \tag{6.78}$$

While this *is* the axisymmetric Laplace equation, it is not the same as the two-dimensional version (6.12).

In axisymmetric flow problems, both (R, φ, z)-cylindrical and (r, θ, φ)-spherical polar coordinates are commonly used. These are illustrated in Fig. 3.3 with the z-axis and polar-axis vertical. The angle φ is the same in both systems. Axisymmetric flows are independent of φ, and their velocity component, u_φ, in the φ-direction is zero. In this section, we will commonly point the z-axis horizontal. Note that R is the radial distance from the axis of symmetry (the z-axis or polar axis) in cylindrical coordinates, whereas r is the distance from the origin in spherical coordinates. Important expressions for these curvilinear coordinates are listed in Appendix B. Several relevant expressions are provided here for easy reference:

Cylindrical	Spherical	
$x = R\cos\varphi$	$x = r\sin\theta\cos\varphi$	
$y = R\sin\varphi$	$y = r\sin\theta\sin\varphi$	(6.79)
$z = z$	$z = r\cos\theta$	
Continuity equations		
(6.72)	$\frac{1}{r^2}\frac{\partial}{\partial r}(r^2 u_r) + \frac{1}{r\sin\theta}\frac{\partial}{\partial \theta}(u_\theta \sin\theta) = 0$	(6.80)
Velocity components		
(6.73), (6.77)	$u_r = \frac{1}{r^2\sin\theta}\frac{\partial \psi}{\partial \theta} = \frac{\partial \phi}{\partial r}, u_\theta = -\frac{1}{r\sin\theta}\frac{\partial \psi}{\partial r} = \frac{1}{r}\frac{\partial \phi}{\partial \theta}$	(6.81)
Vorticity		
(6.74)	$\omega_\varphi = \frac{1}{r}\left[\frac{\partial}{\partial r}(ru_\theta) - \frac{\partial u_r}{\partial \theta}\right]$	(6.82)
Laplace equation		
(6.78)	$\frac{1}{r^2}\frac{\partial}{\partial r}(r^2\frac{\partial \phi}{\partial r}) + \frac{1}{r^2\sin\theta}\frac{\partial}{\partial \theta}\left(\sin\theta\frac{\partial \phi}{\partial \theta}\right) = 0$	(6.83)

Some simple examples of axisymmetric irrotational flows around bodies of revolution, such as spheres and airships, are provided in the rest of this section.

Axisymmetric ideal flows can be constructed from elementary solutions in the same manner as plane flows, except that complex variables cannot be used. Several elementary flows are tabulated here:

Cylindrical	Spherical	
Uniform flow in the z-direction		
$\phi = Uz,\quad \psi = \frac{1}{2}UR^2$	$\phi = Ur\cos\theta,\quad \psi = \frac{1}{2}Ur^2\sin^2\theta$	(6.84)
Point source of strength Q (volume/time) at the origin of coordinates		
$\phi = \frac{-Q}{4\pi\sqrt{R^2+z^2}},\ \psi = \frac{-Qz}{4\pi\sqrt{R^2+z^2}}$	$\phi = -\frac{Q}{4\pi r},\quad \psi = -\frac{Q}{4\pi}\cos\theta$	(6.85)
Doublet with dipole strength $-d\mathbf{e}_z$ at the origin of coordinates		
$\phi = \frac{d}{4\pi}\frac{z}{(R^2+z^2)^{3/2}},\quad \psi = -\frac{d}{4\pi}\frac{R^2}{(R^2+z^2)^{3/2}}$	$\phi = \frac{d}{4\pi r^2}\cos\theta,\ \psi = -\frac{d}{4\pi r}\sin^2\theta$	(6.86)

For these three flows, streamlines in any plane containing the axis of symmetry will be qualitatively similar to those of their two-dimensional counterparts.

Potential flow around a sphere can be generated by the superposition of a uniform stream $U\mathbf{e}_z$ and an axisymmetric doublet opposing the stream of strength $d = 2\pi a^3 U$. In spherical coordinates, the stream and potential functions are:

$$\psi = \frac{1}{2}Ur^2\sin^2\theta - \frac{d}{4\pi r}\sin^2\theta = \frac{1}{2}Ur^2\left(1 - \frac{a^3}{r^3}\right)\sin^2\theta;$$
$$\phi = Ur\cos\theta + \frac{d}{4\pi r^2}\cos\theta = Ur\left(1 + \frac{a^3}{2r^3}\right)\cos\theta. \tag{6.87}$$

This shows that $\psi = 0$ for $\theta = 0$ or π (for any $r \neq 0$), or for $r = a$ (for any θ). Thus the entire z-axis and the spherical surface of radius a form the stream surface $\psi = 0$. Streamlines for this flow are shown in Fig. 6.27. The velocity components are:

$$u_r = \frac{1}{r^2\sin\theta}\frac{\partial\psi}{\partial\theta} = U\left[1 - \left(\frac{a}{r}\right)^3\right]\cos\theta, \quad \text{and} \quad u_\theta = -\frac{1}{r\sin\theta}\frac{\partial\psi}{\partial r} = -U\left[1 + \frac{1}{2}\left(\frac{a}{r}\right)^3\right]\sin\theta. \tag{6.88}$$

The pressure coefficient on the sphere's surface is:

$$[C_p]_{r=a} = \frac{p - p_\infty}{\frac{1}{2}\rho U^2} = 1 - \left(\frac{u_\theta}{U}\right)^2 = 1 - \frac{9}{4}\sin^2\theta, \tag{6.89}$$

which is fore-aft symmetrical, again demonstrating zero drag in steady ideal flow.

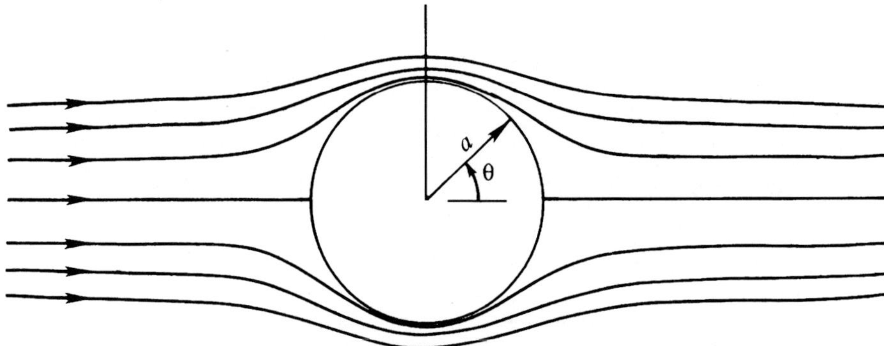

FIGURE 6.27 Axisymmetric streamlines for ideal flow past a sphere in a plane containing the axis of symmetry. The flow is fastest and the streamlines are closest together at $\theta = 90°$. The streamlines upstream and downstream of the sphere are the same, so there is no drag on the sphere.

Interestingly, the potential for this flow can be rewritten in terms of dot products to eliminate explicit dependence on the orientation of the coordinate system. Start from the first equality for the potential in (6.87) and use $\mathbf{x} = r\mathbf{e}_r$, $|\mathbf{x}| = r$,

$\cos\theta = \mathbf{e}_z \cdot \mathbf{e}_r$, $\mathbf{U} = U\mathbf{e}_z$, and $\mathbf{d} = -d\mathbf{e}_z$, to find:

$$\phi = Ur\cos\theta + \frac{\mathbf{d}}{4\pi r^2}\cos\theta = U\mathbf{e}_z \cdot r\mathbf{e}_r - \frac{\mathbf{d}}{4\pi r^3}\cdot r\mathbf{e}_r = \mathbf{U}\cdot\mathbf{x} - \frac{\mathbf{d}}{4\pi|\mathbf{x}|^3}\cdot\mathbf{x} = \left(\mathbf{U} - \frac{\mathbf{d}}{4\pi|\mathbf{x}|^3}\right)\cdot\mathbf{x}, \tag{6.90}$$

a result that is useful in the next section.

As in plane flows, fluid motion around a closed body of revolution can be generated by superposition of a uniform stream and a collection of sources and sinks whose net strength is zero. The closed surface becomes *streamlined* (that is, has a gradually tapering tail) if, for example, the sinks are *distributed* over a finite length. Consider Fig. 6.28, where there is a point source Q at the origin O, and a continuously distributed line sink on the z-axis from O to A (distance $= a$). Let the volume extraction rate per unit length of the line sink be k. An elemental length $d\xi$ of the distributed sink can be regarded as a point sink of strength $kd\xi$, for which the stream function at any point P is [see (6.85)]:

$$d\psi_{sink} = \frac{kd\xi}{4\pi}\cos\alpha.$$

The total stream function at P due to the entire line sink from O to A is:

$$\psi_{sink} = \int d\psi_{sink} = \frac{k}{4\pi}\int_0^a \cos\alpha\, d\xi. \tag{6.91}$$

The integral can be evaluated by noting that $z - \xi = R\cot\alpha$. This gives $d\xi = Rd\alpha/\sin^2\alpha$ because z and R remain constant as we go along the sink. The stream function of the line sink is therefore:

$$\psi_{sink} = \frac{k}{4\pi}\int_\theta^{\alpha_1}\cos\alpha\frac{R}{\sin^2\alpha}d\alpha = \frac{kR}{4\pi}\int_\theta^{\alpha_1}\frac{d(\sin\alpha)}{\sin^2\alpha} = \frac{kR}{4\pi}\left[\frac{1}{\sin\theta} - \frac{1}{\sin\alpha_1}\right] = \frac{k}{4\pi}(r - r_1). \tag{6.92}$$

To obtain a closed body, we must adjust the strengths so that the volume flow from the source (Q) is absorbed by the sink, that is, $Q = ak$. Then the stream function at any point P due to the superposition of a point source of strength Q, a distributed line sink of strength $k = Q/a$, and a uniform stream of velocity U along the z-axis, is:

$$\psi = -\frac{Q}{4\pi}\cos\theta + \frac{Q}{4\pi a}(r - r_1) + \frac{1}{2}Ur^2\sin^2\theta. \tag{6.93}$$

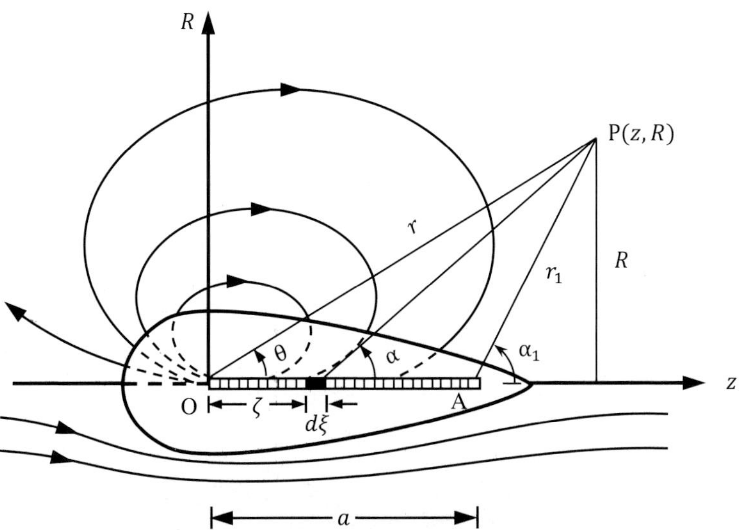

FIGURE 6.28 Ideal flow past an axisymmetric streamlined body generated by a point source at O and a distributed line sink from O to A. The upper half of the figure shows the streamlines induced by the source and the line-segment sink alone. The lower half of the figure shows streamlines when a uniform stream along the axis of symmetry is added to the flow in the upper half of the figure.

A plot of the steady streamline pattern is shown in the bottom half of Fig. 6.28, in which the top half shows instantaneous streamlines in a frame of reference at rest with respect to the fluid at infinity.

Here, we have assumed that the strength of the line sink is uniform along its length. Other interesting streamline patterns can be generated by assuming that the strength $k(\xi)$ is non-uniform.

So far, we have assumed certain distributions of singularities, and then determined the resulting body shape when the distribution is superposed on a uniform stream. The flow around a body with a given shape, $R = B(z)$, can be determined by superposing a uniform stream on source distribution of unknown strength lying on the z-axis between the nose ($z = -L/2$) and tail ($z = +L/2$) of the body:

$$\psi(R,z) = \frac{1}{2}UR^2 + \frac{1}{4\pi}\int_{-L/2}^{+L/2} k(\xi)\frac{z-\xi}{\sqrt{(z-\xi)^2+R^2}}d\xi. \tag{6.94}$$

The radial and axial velocities can be found from (6.94) using (6.73):

$$u_R = -\frac{1}{R}\frac{\partial\psi}{\partial z} = -\frac{R}{4\pi}\int_{-L/2}^{+L/2}\frac{k(\xi)d\xi}{((z-\xi)^2+R^2)^{3/2}}, \quad \text{and}$$
$$u_z = \frac{1}{R}\frac{\partial\psi}{\partial R} = U - \frac{1}{4\pi}\int_{-L/2}^{+L/2}\frac{k(\xi)(z-\xi)d\xi}{((z-\xi)^2+R^2)^{3/2}}. \tag{6.95}$$

The source strength distribution $k(\xi)$ must be adjusted so that the fluid velocity at $R = B(z)$ is tangent to the body surface:

$$(u_R/u_z)_{R=B(z)} = dB/dz. \tag{6.96}$$

Together (6.95) and (6.96) lead to an integral equation for $k(\xi)$:

$$-\frac{1}{4\pi}\int_{-L/2}^{+L/2} k(\xi)\frac{[B(z)-(z-\xi)(dB(z)/dz)]}{\left((z-\xi)^2+R^2\right)^{3/2}}d\xi = U\frac{dB(z)}{dz}, \tag{6.97}$$

which can be solved numerically by discretizing the integrand and the specified body shape, and solving the resulting algebraic system of equations.

Example 6.8

What is the shape of the axisymmetric body in the flow defined by (6.94) when $k(\xi) = -2\pi a^3 d\delta(\xi)/d\xi$, where $\delta(\xi)$ is a Dirac delta-function?

Solution Insert the given source distribution into (6.94) and evaluate the integral by parts:

$$\psi(R,z) = \frac{1}{2}UR^2 - \frac{a^3U}{2}\int_{-L/2}^{+L/2}\frac{d\delta(\xi)}{d\xi}\frac{z-\xi}{\sqrt{(z-\xi)^2+R^2}}d\xi$$
$$= \frac{1}{2}UR^2 - \frac{a^3U}{2}\left(\left[\delta(\xi)\frac{z-\xi}{\sqrt{(z-\xi)^2+R^2}}\right]_{-L/2}^{+L/2} - \int_{-L/2}^{+L/2}\delta(\xi)\frac{d}{d\xi}\left(\frac{z-\xi}{\sqrt{(z-\xi)^2+R^2}}\right)d\xi\right)$$
$$= \frac{1}{2}UR^2 - \frac{a^3U}{2}\left(-\int_{-L/2}^{+L/2}\delta(\xi)\frac{-R^2d\xi}{\left[(z-\xi)^2+R^2\right]^{3/2}}\right) = \frac{1}{2}UR^2 - \frac{a^3U}{2}\frac{R^2}{[z^2+R^2]^{3/2}}$$
$$= \frac{1}{2}Ur^2\left(1-\frac{a^3}{r^3}\right)\sin^2\theta.$$

Here, the boundary term from integrating by parts is zero because of the properties of the Dirac delta-function, and the final form is obtained by switching to spherical coordinates using: $R = r\sin\theta$ and $z = r\cos\theta$. This final form matches (6.87) and represents ideal flow past a sphere with radius a. Thus, the given $k(\xi)$ is the source-distribution of a doublet.

6.9 Three-dimensional potential flow and apparent mass

In three dimensions, ideal flow concepts can be used effectively for a variety of problems in aerodynamics and hydrodynamics. However, d'Alembert's paradox persists and it can be shown that steady ideal flow in three dimensions cannot predict fluid mechanical drag on closed bodies (Exercise 6.46). However, non-zero drag forces can be predicted on submerged three-dimensional objects when the flow is unsteady or some vorticity is present. This section concentrates on the former while leaving the latter to Chapter 13. The objective here is to establish the origin of the *apparent mass* or *added mass* of an accelerating object immersed in a fluid. In general terms, apparent mass is the enhanced and/or altered inertia of an object that arises from unsteady motion of the fluid around the object. Knowledge of apparent mass is essential for predicting the performance of underwater vehicles, lighter-than-air airships, and ultra-light aircraft. It is important for describing and understanding the maneuverability of fish, the dynamics of kites and bubbles, and the differences between sailboat and motorboat motions on the surface of a wavy sea. However, to minimize complexity and to emphasize the core concepts, the focus here is on the simplest possible three-dimensional object, a sphere.

In general, the velocity potential ϕ is extended to three dimensions merely by considering all three components of its definition $\mathbf{u} \equiv \nabla\phi$ to be non-zero. In Cartesian coordinates, this means augmenting (6.10) to include $w \equiv \partial\phi/\partial z$, where (in this section) w is the z-component of the fluid velocity and z is the third spatial coordinate.

Example 6.9

Three-dimensional ideal straining flow is described by $\phi = \frac{1}{2}(\overline{S}_{11}x_1^2 + \overline{S}_{22}x_2^2 + \overline{S}_{33}x_3^2)$ when the coordinate axes coincide with the principal axes of the strain-rate tensor. Determine the fluid velocity, sketch the flow field when $\overline{S}_{11}$ and $\overline{S}_{22}$ are greater than zero, and determine the pressure $p(\mathbf{x})$ when p_o is the pressure at the origin of coordinates.

Solution This is a three-dimensional potential flow, so the fluid velocity $\mathbf{u}$ is obtained from all three components of the gradient of ϕ:

$$\mathbf{u} = \nabla\phi = \mathbf{e}_1\overline{S}_{11}x_1 + \mathbf{e}_2\overline{S}_{22}x_2 + \mathbf{e}_3\overline{S}_{33}x_3.$$

For this flow field, there is a stagnation point ($\mathbf{u} = 0$) at the origin.

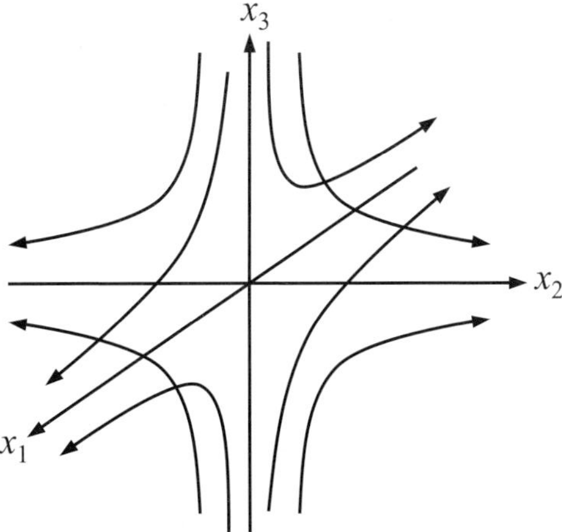

FIGURE 6.29 Three-dimensional straining flow when the coordinate axes coincide with the principal axes of the strain-rate tensor and $\overline{S}_{11} > 0$ and $\overline{S}_{22} > 0$. Here, fluid elements are stretched in the x_1- and x_2-directions, but compressed in the x_3-direction. The origin of coordinates is a stagnation point.

When $\overline{S}_{11} > 0$ and $\overline{S}_{22} > 0$, fluid elements will be stretched in the x_1- and x_2-directions, but compressed in the x_3-direction, because $\nabla \cdot \mathbf{u} = \overline{S}_{11} + \overline{S}_{22} + \overline{S}_{33} = 0$ in ideal flow so $\overline{S}_{33}$ must be negative when $\overline{S}_{11}$ and $\overline{S}_{22}$ are positive. Fig. 6.29 is sketch of this flow.

The pressure field can be obtained from the Bernoulli equation (6.18) and the velocity field given above:

$$p(\mathbf{x}) + \frac{1}{2}\rho|\nabla\phi|^2 = p_o, \quad \text{or} \quad p_o - p(x) = \frac{1}{2}\rho\left(\overline{S}_{11}^2x_1^2 + \overline{S}_{22}^2x_2^2 + \overline{S}_{33}^2x_{33}^2\right).$$

This final result shows that p_o is a local maximum since the right side of the last equation is a sum of positive terms away from the origin, regardless of the location.

The situation of interest is depicted in Fig. 6.30, where a sphere with radius a moves in a quiescent fluid with undisturbed pressure p_∞ via an external force, $\mathbf{F}_E$, that acts only on the sphere.[2] The location $\mathbf{x}_s(t)$, velocity $\mathbf{u}_s(t) = d\mathbf{x}_s/dt$, and

2. The development provided here is based on a lecture given by Prof. P. Dimotakis in 1984.

acceleration $d\mathbf{u}_s/dt$ of the sphere are presumed known, and the fluid dynamic force $\mathbf{F}_s$ on the sphere is to be determined. This is an idealization of the situation for a maneuvering fish, submarine, or airship; and is directly relevant to the motion of non-deforming bubbles.

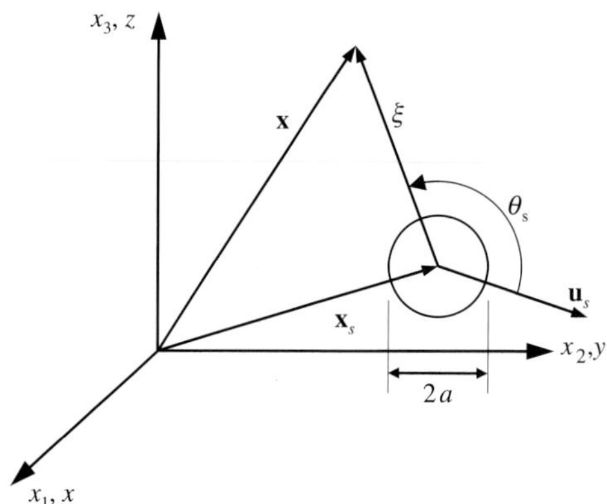

FIGURE 6.30 Three-dimensional geometry for calculating the fluid dynamic force on an arbitrarily moving submerged sphere of radius a centered at $\mathbf{x}_S$. The angle between the sphere's velocity, $\mathbf{u}_S$, and the observation point $\mathbf{x}$ is θ_S. The vector distance from the center of the sphere to the observation point is $\boldsymbol{\xi}$.

The potential for an arbitrarily moving sphere is a modified version of (6.94) with the sphere centered at $\mathbf{x}_s(t)$ and the fluid far from the sphere at rest. These changes are implemented by replacing $\mathbf{x}$ in (6.94) with $\mathbf{x} - \mathbf{x}_s(t)$, and by setting $\mathbf{U} = 0$, which leaves:

$$\phi = -\frac{1}{4\pi|\mathbf{x} - \mathbf{x}_s(t)|^3}\mathbf{d}\cdot(\mathbf{x} - \mathbf{x}_s(t)). \tag{6.98}$$

For this potential to represent a constant-size moving sphere at each instant in time, the dipole strength must continuously change direction and magnitude to point into the flow impinging on the sphere. If the sphere's velocity is $\mathbf{u}_s(t)$, then, to an observer on the sphere, the oncoming flow velocity is $-\mathbf{u}_s(t)$. Thus, at any instant in time the dipole strength must be $\mathbf{d}(t) = -2\pi a^3[-\mathbf{u}_s(t)]$, a direct extension of the steady flow result. Substitution of this $\mathbf{d}(t)$ into (6.98) produces:

$$\phi(\mathbf{x}, \mathbf{x}_s, \mathbf{u}_s) = -\frac{a^3}{2|\mathbf{x} - \mathbf{x}_s|^3}\mathbf{u}_s\cdot(\mathbf{x} - \mathbf{x}_s) = -\frac{a^3}{2|\boldsymbol{\xi}|^3}\mathbf{u}_s\cdot\boldsymbol{\xi}. \tag{6.99}$$

Here, explicit listing of the time argument of $\mathbf{x}_s$ and $\mathbf{u}_s$ has been dropped for clarity, and $\boldsymbol{\xi} = \mathbf{x} - \mathbf{x}_s$ is the vector distance from the center of the sphere to the location $\mathbf{x}$.

For ideal flow, an integral of the pressure forces over the surface of the sphere determines $\mathbf{F}_s$:

$$\mathbf{F}_s = -\int_{sphere's\ surface}(p - p_\infty)\mathbf{n}\,dA. \tag{6.100}$$

This is the three-dimensional equivalent of (6.55) since $\int p_\infty \mathbf{n}\,dA = 0$ for a closed surface and constant p_∞. The pressure difference in (6.100) can be obtained from the unsteady Bernoulli equation evaluated on the sphere's surface and far from the sphere where the pressure is p_∞, $\mathbf{u} = 0$, and $\partial\phi/\partial t = 0$:

$$\left[\frac{\partial\phi}{\partial t} + \frac{1}{2}|\nabla\phi|^2 + \frac{p}{\rho}\right]_{sphere's\ surface} = \frac{p_\infty}{\rho}. \tag{6.101}$$

For the geometry shown in Fig. 6.30, the sphere's surface is defined by $|\mathbf{x} - \mathbf{x}_s| = |\boldsymbol{\xi}| = a$, so for notational convenience the subscript "a" will denote quantities evaluated on the sphere's surface. Thus, (6.101) can be rewritten:

$$\frac{p_a - p_\infty}{\rho} = -\left(\frac{\partial\phi}{\partial t}\right)_a - \frac{1}{2}|\nabla\phi|_a^2. \tag{6.102}$$

The time derivative of ϕ can be evaluated as follows:

$$\frac{\partial}{\partial t}\phi(\mathbf{x},\mathbf{x}_s,\mathbf{u}_s)=\frac{\partial\phi}{\partial(x_s)_i}\frac{d(x_s)_i}{dt}+\frac{\partial\phi}{\partial(u_s)_i}\frac{d(u_s)_i}{dt}=-\mathbf{u}\cdot\mathbf{u}_s-\frac{a^3}{2|\boldsymbol{\xi}|^3}\boldsymbol{\xi}\cdot\frac{d\mathbf{u}_s}{dt}, \tag{6.103}$$

where the middle of this extended equality presents a temporary switch to index notation. The final form in (6.103) is obtained from the definition $d(x_s)_i/dt=\mathbf{u}_s$, and the fact that $\partial\phi/\partial(x_s)_i=-\partial\phi/\partial x_i=-\mathbf{u}$ for the potential (6.99) since it only depends on $\mathbf{x}-\mathbf{x}_s$. When evaluated on the sphere's surface, this becomes:

$$\left(\frac{\partial\phi}{\partial t}\right)_a=-\mathbf{u}_a\cdot\mathbf{u}_s-\frac{a}{2}\mathbf{e}_\xi\cdot\frac{d\mathbf{u}_s}{dt}, \tag{6.104}$$

where $\mathbf{e}_\xi=\boldsymbol{\xi}/|\boldsymbol{\xi}|$. The independent spatial variable $\mathbf{x}$ appears twice in (6.99) so the gradient of ϕ involves two terms:

$$\nabla\phi=-\frac{a^3}{2}\left[-\frac{3(\mathbf{x}-\mathbf{x}_s)}{|\mathbf{x}-\mathbf{x}_s|^5}\mathbf{u}_s\cdot(\mathbf{x}-\mathbf{x}_s)+\frac{1}{|\mathbf{x}-\mathbf{x}_s|^3}\mathbf{u}_s\right], \tag{6.105}$$

which are readily evaluated on the surface of the sphere where $(\mathbf{x}-\mathbf{x}_s)_a=a\mathbf{e}_\xi$:

$$\mathbf{u}_a=(\nabla\phi)_a=-\frac{a^3}{2}\left[-\frac{3a\mathbf{e}_\xi}{a^5}\mathbf{u}_s\cdot a\mathbf{e}_\xi-\frac{1}{a^3}\mathbf{u}_s\right]=\frac{3}{2}(\mathbf{u}_s\cdot\mathbf{e}_\xi)e_\xi-\frac{1}{2}\mathbf{u}_s. \tag{6.106}$$

Combining (6.102), (6.104), and (6.106) produces a final relationship for the surface pressure p_a on the sphere in terms of the orientation $\mathbf{e}_\xi$, and the sphere's velocity $\mathbf{u}_s$ and acceleration $d\mathbf{u}_s/dt$:

$$\begin{aligned}\frac{p_a-p_\infty}{\rho}&=\left(\frac{3}{2}(\mathbf{u}_s\cdot\mathbf{e}_\xi)\mathbf{e}_\xi-\frac{1}{2}\mathbf{u}_s\right)\cdot\mathbf{u}_s+\frac{a}{2}\mathbf{e}_\xi\cdot\frac{d\mathbf{u}_s}{dt}-\frac{1}{2}\left|\frac{3}{2}(\mathbf{u}_s\cdot\mathbf{e}_\xi)\mathbf{e}_\xi-\frac{1}{2}\mathbf{u}_s\right|^2\\&=\left(\frac{3}{2}(\mathbf{u}_s\cdot\mathbf{e}_\xi)^2-\frac{1}{2}|\mathbf{u}_s|^2\right)+\frac{a}{2}\mathbf{e}_\xi\cdot\frac{d\mathbf{u}_s}{dt}-\frac{1}{8}\left(9(\mathbf{u}_s\cdot\mathbf{e}_\xi)^2-6(\mathbf{u}_s\cdot\mathbf{e}_\xi)^2+|\mathbf{u}_s|^2\right)\\&=\frac{1}{2}|\mathbf{u}_s|^2\left(\frac{9}{4}\frac{(\mathbf{u}_s\cdot\mathbf{e}_\xi)^2}{|\mathbf{u}_s|^2}-\frac{5}{4}\right)+\frac{a}{2}\mathbf{e}_\xi\cdot\frac{d\mathbf{u}_s}{dt}.\end{aligned} \tag{6.107}$$

When the sphere is not accelerating and θ_s is the angle between $\mathbf{e}_\xi$ and $\mathbf{u}_s$, then:

$$\left(\frac{p_a-p_\infty}{\frac{1}{2}\rho|\mathbf{u}_s|^2}\right)_{steady}=\frac{9}{4}\cos^2\theta_s-\frac{5}{4}=1-\frac{9}{4}\sin^2\theta_s, \tag{6.108}$$

which is identical to (6.89). Thus, as expected from the Galilean invariance of Newtonian mechanics, steady flow past a stationary sphere and steady motion of a sphere through an otherwise quiescent fluid lead to the same pressure distribution on the sphere. And, once again, no drag on the sphere is predicted.

However, (6.107) includes a second term that depends on the direction and magnitude of the sphere's acceleration. To understand the effects of this term, reorient the coordinate system in Fig. 6.30 so that at the time of interest the sphere is at the origin of coordinates and its acceleration is parallel to the polar z- or x_3-axis: $d\mathbf{u}_s/dt=|d\mathbf{u}_s/dt|\mathbf{e}_z$. With this revised geometry, $\mathbf{e}_\xi\cdot d\mathbf{u}_s/dt=|d\mathbf{u}_s/dt|\cos\theta$, and the fluid dynamic force on the sphere can be obtained from (6.100) in spherical polar coordinates:

$$\mathbf{F}_s=-\rho\frac{a}{2}\left|\frac{d\mathbf{u}_s}{dt}\right|\int_{\theta=0}^{\pi}\int_{\varphi=0}^{2\pi}\cos\theta(\mathbf{e}_x\sin\theta\cos\varphi+\mathbf{e}_y\sin\theta\sin\varphi+\mathbf{e}_z\cos\theta)a^2\sin\theta\,d\varphi d\theta. \tag{6.109}$$

The φ-integration causes the x- and y-force components to be zero, leaving:

$$\mathbf{F}_s=-\pi\rho a^3\left|\frac{d\mathbf{u}_s}{dt}\right|\mathbf{e}_z\int_{\theta=0}^{\pi}\cos^2\theta\,\sin\theta\,d\varphi d\theta=-\frac{2}{3}\pi\rho a^3\frac{d\mathbf{u}_s}{dt}=-M\frac{d\mathbf{u}_s}{dt}, \tag{6.110}$$

where $M = 2\pi a^3 \rho/3$ is the *apparent* or *added mass* of the sphere. Thus, the ideal-flow fluid-dynamic force on an accelerating sphere opposes the acceleration, and its magnitude is proportional to the sphere's acceleration and one-half of the mass of the fluid displaced by the sphere.

This fluid-inertia-based loading is the apparent mass or added mass of the sphere. It occurs because fluid must move more rapidly out of the way, in front of, and more rapidly fill in behind, an accelerating sphere. To illustrate its influence, consider an elementary dynamics problem involving a rigid sphere of mass m and radius a that is subject to an external force $\mathbf{F}_E$ while submerged in a large bath of nominally quiescent inviscid fluid with density ρ. In this case, Newton's second law (sum of forces = mass times acceleration) implies:

$$\mathbf{F}_E + \mathbf{F}_s = \mathbf{F}_E - M\frac{d\mathbf{u}_s}{dt} = m\frac{d\mathbf{u}_s}{dt}, \quad \text{or} \quad \mathbf{F}_E = \left(m + \frac{2\pi}{3}\rho a^3\right)\frac{d\mathbf{u}_s}{dt}. \tag{6.111}$$

The expression above can be rewritten as $\mathbf{F}_E = m\,(1 + C_m\rho/\rho_s)\,d\mathbf{u}_s/dt$, where ρ_s is the density of the object and C_m is the added mass coefficient that depends on the object's shape ($C_m = 0.5$ for a sphere). Thus, a submerged sphere will behave as if its inertia is larger by one-half of the mass of the fluid it displaces compared to its behavior in vacuum. It can also be seen that the relative contribution of the apparent mass scales like ρ/ρ_s, thus it plays a larger role for bubbles submerged in liquid than for liquid droplets or solid particles surrounded by gas. For a sphere, the apparent mass is a scalar because of its rotational symmetry. In general, apparent mass is a tensor and the final equality in (6.110) is properly stated $(F_s)_i = M_{ij}\,d(u_s)_j/dt$.

Example 6.10

Consider a tiny spherical bubble with radius a and density ρ_s in an oscillating flow with density ρ and velocity $u(t) = U\cos(2\pi t/T)$, where U is the amplitude of the velocity oscillations, T is the period, and t is time. Assume the Reynolds number is small such that the flow is in the Stokes regime. For an air bubble in water. Find the bubble trajectory for four periods $t \in [0, 4T]$ with $a = 0.5mm$, $U = 1.0cm/s$ and $T = 10ms$. Compare the results with and without the added mass contribution.

Solution In the Stokes regime, the external force in (6.111) is given by $F_E = 6\pi\mu a(u - u_s)$ (see Chapter 7), where u_s is the velocity of the spherical bubble in the flow direction. The equations of motion for the bubble can thus be expressed as

$$\frac{dx_s}{dt} = u_s \quad \text{and} \quad m\left(1 + C_m\frac{\rho}{\rho_s}\right)\frac{du_s}{dt} = 6\pi\mu a(u - u_s).$$

The equations are solved numerically in the MATLAB code provided below.

```
global a mu U T m rho rhos Cm

T=.01; % Period of oscillation in seconds
U=.01; % Carrier phase velocity amplitude in m/s
mu=10e-3; % Carrier phase viscosity
rho=998; % Carrier phase density
rhos=1.2; % Sphere density
a=.5e-3; % Sphere radius in m
m=4*pi*a^3*rhos/3; % Sphere mass
Cm=0.5; % Added mass coefficient

[t,x] = ode45(@RHS,[0 4*T],[0 0]); % Solve w/ out added mass
[t2,x2] = ode45(@RHS2,[0 4*T],[0 0]); % Solve w/ added mass

% Right-hand side (w/ out added mass)
function x = RHS(t,y)
global a mu U T m
u=U*cos(2*pi*t/T);
x = [y(2); 6*pi*mu*a*(u-y(2))/m];
return
end

% Right-hand side (w/ added mass)
function x = RHS2(t,y)
```

```
global a mu U T m rho rhos Cm
u=U*cos(2*pi*t/T);
x = [y(2); 6*pi*mu*a*(u-y(2))/(m*(1+Cm*rho/rhos))];
return
end
```

A comparison of the bubble trajectory with and without added mass is given in Fig. 6.31. Without added mass, the bubble responds to the oscillating flow according to Stokes drag. Added mass can be seen to delay the bubble's response to the fluid oscillations and reduce the amplitude of its motion.

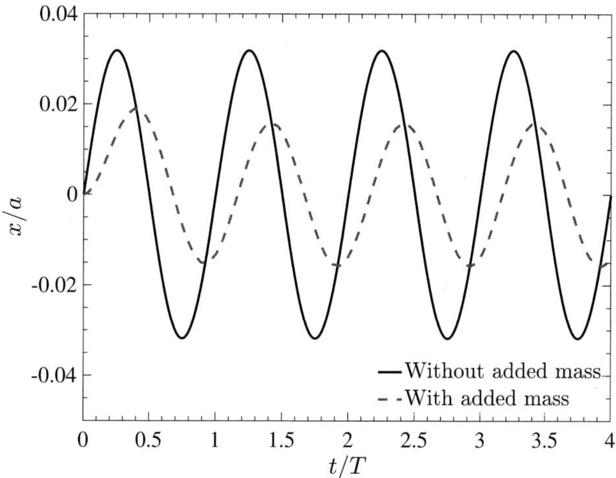

FIGURE 6.31 Solution to Example 6.10.

6.10 Concluding remarks

The theory of irrotational constant-density (ideal) flow has reached a highly developed stage during the last 250 years because of the efforts of theoretical physicists such as Euler, Bernoulli, D'Alembert, Lagrange, Stokes, Helmholtz, Kirchhoff, and Kelvin. Interest in the subject has resulted from the applicability of potential theory to other fields such as heat conduction, elasticity, and electromagnetism. When applied to fluid flows, however, the theory predicts zero fluid dynamic drag on a steadily moving body, a result that is at variance with observations. Meanwhile, the theory of viscous flow was developed during the middle of the nineteenth century, after the Navier-Stokes equations were formulated. Early analytical viscous-flow solutions generally applied either to very slow flows where the nonlinear advection terms in the equations of motion were negligible, or to flows in which the advective terms were identically zero (such as the viscous flow through a straight pipe). Such viscous-flow solutions are highly rotational, and it was not clear where irrotational flow theory was applicable and why, a quandary left for Prandtl to explain (see Chapter 8).

Ideal flow theory no longer occupies center stage in fluid mechanics, although it did so in the past. However, the subject is still useful for aerodynamics and hydrodynamics. We shall see in Chapter 8 that the pressure distribution around streamlined bodies can still be predicted with a fair degree of accuracy from ideal flow theory. In Chapter 13, we shall see that the lift of an airfoil is due to the development of circulation around it, and the magnitude of the lift agrees with the Kutta-Zhukhovsky lift theorem. And, the technique of conformal mapping remains useful for studying flow around airfoil shapes.

Exercises

6.1 **a)** Show that (6.7) solves (6.5) and leads to $\mathbf{u} = (U, V)$.
b) Integrate (6.6) within circular area centered on (x', y') of radius $r' = \sqrt{(x - x')^2 + (y - y')^2}$ to show that (6.8) is a solution of (6.6).

6.2 For two-dimensional ideal flow, show separately that:
a) $\nabla\psi \cdot \nabla\phi = 0$

b) $-\nabla\psi \times \nabla\phi = |u|^2\mathbf{e}_z$
c) $|\nabla\psi|^2 = |\nabla\phi|^2$
d) $\nabla\phi = -\mathbf{e}_z \times \nabla\psi$

6.3 **a)** Show that (6.14) solves (6.12) and leads to $\mathbf{u} = (U, V)$.
b) Integrate (6.13) within circular area centered on (x', y') of radius $r' = \sqrt{(x-x')^2 + (y-y')^2}$ to show that (6.15) is a solution of (6.13).
c) For the flow described by (6.15), show that the volume flux (per unit depth into the page) $\oint_C \mathbf{u}\cdot\mathbf{n}ds$ computed from a closed contour C that encircles the point (x', y') is q_s. Here $\mathbf{n}$ is the outward normal on C and ds is a differential element of C.

6.4 Show that (6.1) reduces to $\dfrac{\partial\phi}{\partial t} + \dfrac{1}{2}|\nabla\phi|^2 + \dfrac{p}{\rho} = \text{const.}$ when the flow is described by the velocity potential ϕ.

6.5 Consider the following two-dimensional Cartesian flow fields: (*i*) solid body rotation (SBR) at angular rate Ω about the origin: $(u, v) = (-\Omega y, \Omega x)$; and (*ii*) uniform expansion (UE) at linear expansion rate Θ: $(u, v) = (\Theta x, \Theta y)$. Here Ω, Θ, and the fluid density ρ are positive real constants and there is no body force.
a) What is the stream function $\psi_{\text{SBR}}(x, y)$ for solid body rotation?
b) Is there a potential function $\phi_{\text{SBR}}(x, y)$ for solid body rotation? Specify it if it exists.
c) What is the pressure, $p_{\text{SBR}}(x, y)$, in the solid body rotation flow when $p_{\text{SBR}}(0, 0) = p_o$?
e) What is the potential function $\phi_{\text{UE}}(x, y)$ for uniform expansion?
e) Is there a stream function $\psi_{\text{UE}}(x, y)$ for uniform expansion? Specify it if it exists.
f) Determine the pressure, $p_{\text{UE}}(x, y)$, in the uniform expansion flow when $p_{\text{UE}}(0, 0) = p_o$.

6.6 Determine u and v, and sketch streamlines for:
a) $\psi = A(x^2 - y^2)$
b) $\phi = A(x^2 - y^2)$

6.7 For the following two-dimensional stream and potential functions, find the fluid velocity $\mathbf{u} = (u, v)$, and determine why these are or are not ideal flows.
a) $\psi = A(x^2 + y^2)$
b) $\phi = A(x^2 + y^2)$

6.8 Assume $\psi = ax^3 + bx^2y + cxy^2 + dy^3$ where a, b, c, and d are constants; and determine two independent solutions to the Laplace equation. Sketch the streamlines for both flow fields.

6.9 Repeat Exercise 6.8 for $\psi = ax^4 + bx^3y + cx^2y^2 + dxy^3 + ey^4$ where a, b, c, d, and e are constants.

6.10 Without using complex variables, determine:
a) The potential ϕ for an ideal vortex of strength Γ starting from (6.8)
b) The stream function for an ideal point source of strength q_s starting from (6.15)
c) Is there any ambiguity in your answers to parts a) and b)? If so, does this ambiguity influence the fluid velocity?

6.11 Determine the stream function of a doublet starting from (6.29) and show that the streamlines are circles having centers on the y-axis that are tangent to the x-axis at the origin.

6.12 Consider steady horizontal flow at speed U past a stationary source of strength q_s located at the origin of coordinates in two dimensions, (6.30) or (6.31). To hold it in place, an external force per unit depth into the page, $\mathbf{F}$, is applied to the source.
a) Develop a dimensionless scaling law for $F = |\mathbf{F}|$.
b) Use a cylindrical control volume centered on the source with radius R and having depth B into the page, the steady ideal-flow momentum conservation equation for a control volume:

$$\int_{A^*} \rho\mathbf{u}(\mathbf{u}\cdot\mathbf{n})dA = -\int_{A^*} p\mathbf{n}dA + \mathbf{F},$$

and an appropriate Bernoulli equation to determine the magnitude and direction of $\mathbf{F}$ without using Blasius Theorem.
c) Is the direction of $\mathbf{F}$ unusual in anyway? Explain it physically.

6.13 Repeat all three parts of Exercise 6.12 for steady ideal flow past a stationary irrotational vortex located at the origin when the control volume is centered on the vortex. The stream function for this flow is: $\psi = Ur\sin\theta - (\Gamma/2\pi)\ln(r)$.

6.14 Use the principle of conservation of mass (4.5) and an appropriate control volume to show that maximum half thickness of the half body described by (6.30) or (6.31) is $h_{max} = q_s/2U$.

6.15 By integrating the surface pressure, show that the drag on a plane half-body (Fig. 6.7) is zero.

6.16 Ideal flow past a cylinder (6.33) is perturbed by adding a small vertical velocity without changing the orientation of the doublet:

$$\psi = -U\gamma x + Uy - \frac{Ua^2 y}{x^2 + y^2} = -U\gamma r\cos\theta + U\left(r - \frac{a^2}{r}\right)\sin\theta.$$

a) Show that the stagnation point locations are $r_s = a$ and $\theta_s = \gamma/2$, or $\pi + \gamma/2$, when $\gamma \ll 1$.
b) Does this flow include a closed body?

6.17 For the following flow fields (b, U, Q, and Γ are positive real constants), sketch streamlines and compute the pressure on the x-axis. Be sure to specify the reference pressure and location.
a) $\psi = b\sqrt{r}\cos(\theta/2)$ for $|\theta| < 180°$
b) $\psi = Uy + (\Gamma/2\pi)\left[\ln\left(\sqrt{x^2 + (y-b)^2}\right) - \ln\left(\sqrt{x^2 + (y+b)^2}\right)\right]$
c) $\phi = \sum_{n=-\infty}^{n=+\infty} (q_s/2\pi)\ln\left(\sqrt{x^2 + (y-2na)^2}\right)$ for $|y| < a$

6.18 [3]Take a standard sheet of paper and cut it in half. Make a simple airfoil with one half and a cylinder with the other half that are approximately the same size as shown.
a) If the cylinder and the airfoil are dropped from the same height at the same time with the airfoil pointed toward the ground in its most streamlined configuration, predict which one reaches the ground first.
b) Stand on a chair and perform this experiment. What happens? Are your results repeatable?
c) Can you explain what you observe?

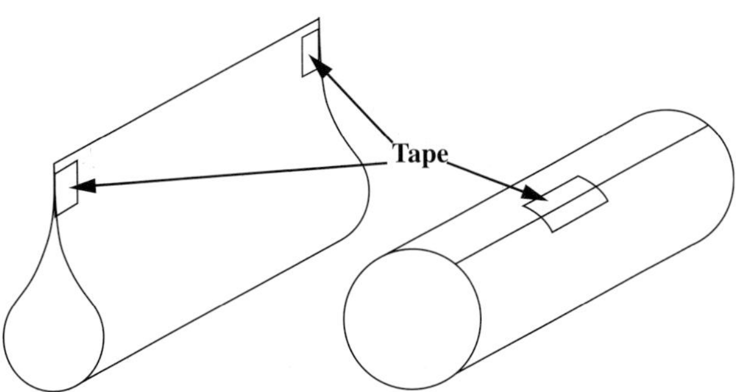

6.19 Consider the following two-dimensional stream function composed of a uniform horizontal stream of speed U and two vortices of equal and opposite strength in (x, y)-Cartesian coordinates:

$$\psi(x, y) = Uy + (\Gamma/2\pi)\ln\sqrt{x^2 + (y-b)^2} - (\Gamma/2\pi)\ln\sqrt{x^2 + (y+b)^2}$$

a) Simplify this stream function for the combined limit of $b \to 0$ and $\Gamma \to \infty$ with $2b\Gamma = C =$ constant to find: $\psi(x, y) = Uy(1 - (C/2\pi U)(x^2 + y^2)^{-1})$
b) Switch to (r, θ)-polar coordinates and find both components of the velocity using the simplified stream function.
c) For the simplified stream function, determine where $u_r = 0$.
d) Sketch the streamlines for the simplified stream function, and describe this flow.

6.20 Graphically generate the streamline pattern for a plane half-body in the following manner. Take a source of strength $q_s = 200$ m^2/s and a uniform stream $U = 10$ m/s. Draw radial streamlines from the source at equal intervals of $\Delta\theta = \pi/10$, with the corresponding stream function interval: $\Delta\psi_{source} = (q_s/2\pi)\Delta\theta = 10$ m^2/s. Now draw streamlines of the uniform flow with the same interval, that is, $\Delta\psi_{stream} = U\Delta\psi = 10$ m^2/s. This requires $\Delta y = 1$ m, which can be plotted assuming a linear scale of 1 cm = 1 m. Connect points of equal $\psi = \psi_{\text{source}} + \psi_{\text{stream}}$ to display the flow pattern.

6.21 Consider the two-dimensional steady flow formed by combining a uniform stream of speed U in the positive x-direction, a source of strength $q_s > 0$ at $(x, y) = (-a, 0)$, and a sink of strength q_s at $(x, y) = (+a, 0)$ where $a > 0$. The pressure far upstream of the origin is p_∞.
a) Write down the velocity potential and the stream function for this flow field.
b) What are the coordinates of the stagnation points, marked by s in the figure?

---

3. Based on a suggestion from Professor William Schultz.

c) Determine the pressure in this flow field along the y-axis.

d) There is a closed streamline in this flow that defines a Rankine body. Obtain a transcendental algebraic equation for this streamline, and show that the half-width, h, of the body in the y-direction is given by: $h/a = \cot(\pi Uh/q_s)$. (The introduction of angles may be useful here.)

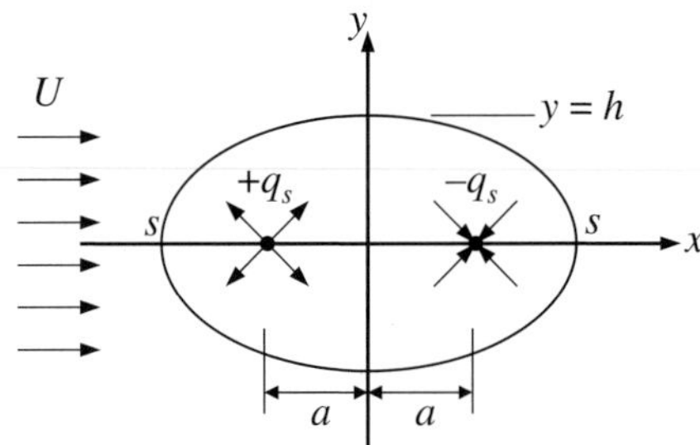

6.22 A stationary ideal two-dimensional vortex with clockwise circulation Γ is located at point $(0, a)$, above a flat plate. The plate coincides with the x-axis. A uniform stream U directed along the x-axis flows past the vortex.

a) Sketch the flow pattern and show that it represents the flow over an oval-shaped body when $\Gamma/\pi a > U$. [*Hint*: Introduce the image vortex and locate the two stagnation points on the x-axis.]

b) If the pressure far from the origin just *above* the x-axis is p_∞ show that the pressure p at any location on the plate is: $p_\infty - p = \frac{\rho\Gamma^2 a^2}{2\pi^2(x^2+a^2)^2} - \frac{\rho U\Gamma a}{\pi(x^2+a^2)}$.

c) Using the result of part b), show that the total upward force F on the plate per unit depth into the page is $F = -\rho U\Gamma + \rho\Gamma^2/4\pi a$ when the pressure everywhere below the plate is p_∞.

6.23 Consider plane flow around a circular cylinder with clockwise circulation Γ. Use the complex potential and Blasius theorem (6.60) to show that the drag is zero and the lift is $L = \rho U\Gamma$. (In Section 6.3, these results were obtained by integrating the surface pressure distribution.)

6.24 For the doublet flow described by (6.29) and sketched in Fig. 6.6, show $u < 0$ for $y < x$ and $u > 0$ for $y > x$. Also, show that $v < 0$ in the first quadrant and $v > 0$ in the second quadrant.

6.25 Hurricane winds blow over a *Quonset hut*, that is, a long half-circular cylindrical cross-section building, 6 m in diameter. If the velocity far upstream is $U_\infty = 40$ m/s and $p_\infty = 1.003 \times 10^5$ N/m, $\rho_\infty = 1.23$ kg/m^3, find the force per unit depth on the building, assuming the pressure inside the hut is a) p_∞, and b) stagnation pressure, $p_\infty + \frac{1}{2}\rho_\infty U_\infty^2$.

6.26 In a two-dimensional ideal flow, a source of strength q_s is located a meters above an infinite plane. Find the fluid velocity on the plane, the pressure on the plane, and the reaction force on the plane assuming constant pressure p_∞ below the plane.

6.27 Consider two-dimensional ideal flow over a circular cylinder of radius $r = a$ with axis coincident with a right-angle corner, as shown in the figure below. Solve for the stream function and velocity components.

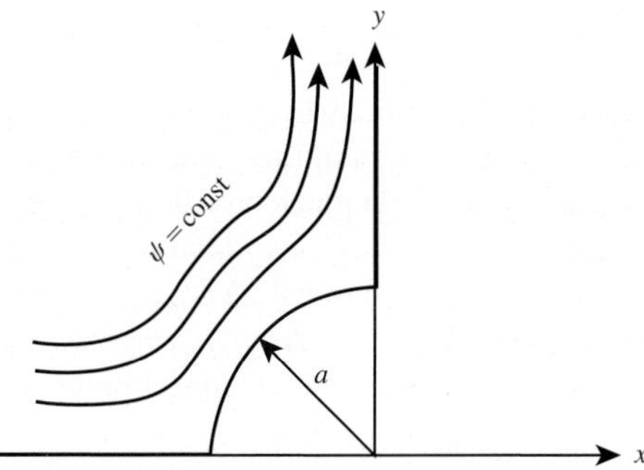

6.28 Consider the following two-dimensional velocity potential consisting of two sources and one sink, all of strength q_s:

$$\phi(x, y) = (q_s/2\pi)\left(\ln\sqrt{(x-b)^2+y^2} + \ln\sqrt{(x-a^2/b)^2+y^2} - \ln\sqrt{x^2+y^2}\right).$$

Here a and b are positive constants and $b > a$.

a) Determine the locations of the two stagnation points in this flow field.
b) Sketch the streamlines in this flow field.
c) Show that the closed streamline in this flow is given by $x^2 + y^2 = a^2$.

6.29 In two-dimensional ideal flow, the potential $\phi(x, y)$ for the flow around a symmetrical body with upper (+) and lower (−) surfaces defined by $y = \pm B(x)$ can be determined from:

$$\phi(x, y) = Ux + \frac{1}{2\pi}\int_{c/2}^{c/2} q(\xi)\ln\left(\sqrt{(x-\xi)^2 + y^2}\right)d\xi,$$

where U is the speed of a uniform horizontal stream, c is the downstream length of the body, and $q(\xi)$ is a source strength distribution located at $|x| < c/2$ and $y = 0$.

a) Use the exact body surface boundary condition: $\left(\frac{\partial\phi/\partial y}{\partial\phi/\partial x}\right)_{y=B(x)} = \frac{dB(x)}{dx}$ to determine a single linear integral equation for the source strength distribution q.
b) If $\delta(x)$ is the Dirac δ-function, what body shape is represented by $q(x) = -2\pi a^2 U \frac{d\delta(x)}{dx}$?

6.30 Show that $\nabla^2\phi = 0$ and $\nabla^2\psi = 0$ are satisfied for any complex function $w(z) = z^n$ when n is a positive integer value, where z is given by (6.43). Plot the corresponding streamlines for different values of n.

6.31 Without using complex variables, derive the results of the Kutta-Zhukhovsky lift theorem (6.62) for steady two-dimensional irrotational constant-density flow past an arbitrary-cross-section object by considering the *clam-shell* control volume (shown as a dashed line) in the limit as $r \to \infty$. Here A_1 is a large circular contour, A_2 follows the object's cross section contour, and A_3 connects A_1 and A_2. Let p_∞ and $U\mathbf{e}_x$ be the pressure and flow velocity far from the origin of coordinates, and denote the flow extent perpendicular to the x-y plane by B.

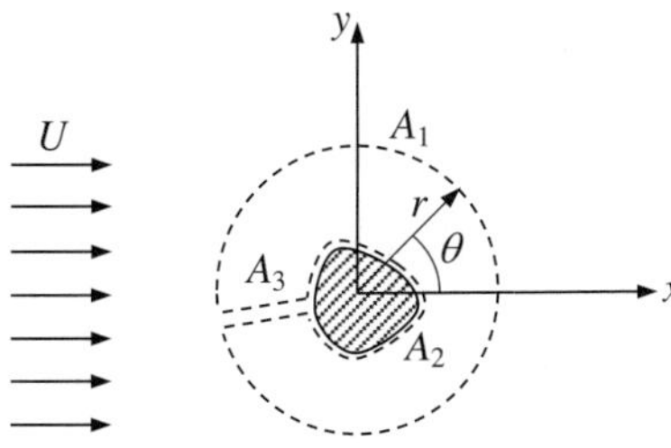

6.32 Pressure fluctuations in wall-bounded turbulent flows are a common source of flow noise. Such fluctuations are caused by turbulent eddies as they move over the bounding surface. A simple ideal-flow model that captures some of the important phenomena involves a two-dimensional vortex that moves above a flat surface in a fluid of density ρ. Thus, for the following items, use the potential:

$$\phi(x, y, t) = -\frac{\Gamma}{2\pi}\tan^{-1}\left(\frac{y-h}{x-Ut}\right) + \frac{\Gamma}{2\pi}\tan^{-1}\left(\frac{y+h}{x-Ut}\right),$$

where h is the distance of the vortex above the flat surface, Γ is the vortex strength, and U is the convection speed of the vortex.

a) Compute the horizontal u and vertical v velocity components and verify that $v = 0$ on $y = 0$.
b) Determine the pressure at $x = y = 0$ in terms of ρ, t, Γ, h, and U.
c) Based on your results from part b), is it possible for a fast-moving, high-strength vortex far from the surface to have the same pressure signature as a slow-moving, low-strength vortex closer to the surface?

6.33 A pair of equal strength ideal line vortices having axes perpendicular to the x-y plane are located at $\mathbf{x}_a(t) = (x_a(t), y_a(t))$, and $\mathbf{x}_b(t) = (x_b(t), y_b(t))$, and move in their mutually induced velocity fields. The stream function for this flow is given by: $\psi(x, y, t) = -(\Gamma/2\pi)(\ln|\mathbf{x} - \mathbf{x}_a(t)| + \ln|\mathbf{x} - \mathbf{x}_b(t)|)$. Explicitly determine $\mathbf{x}_a(t)$ and $\mathbf{x}_b(t)$, given $\mathbf{x}_a(0) = (-r_o, 0)$ and $\mathbf{x}_b(0) = (r_o, 0)$. Switching to polar coordinates at some point in your solution may be useful.

6.34 Two unequal strength ideal line vortices having axes perpendicular to the x-y plane are located at $\mathbf{x}_1(t) = (x_1(t), y_1(t))$ with circulation Γ_1, and $\mathbf{x}_2(t) = (x_2(t), y_2(t))$ with circulation Γ_2, and move in their mutually induced velocity fields. The stream function for this flow is given by: $\psi(x, y, t) = -(\Gamma_1/2\pi)\ln|\mathbf{x} - \mathbf{x}_1(t)| - (\Gamma_2/2\pi)\ln|\mathbf{x} - \mathbf{x}_2(t)|$. Explicitly determine $\mathbf{x}_1(t)$ and $\mathbf{x}_2(t)$ in terms of Γ_1, Γ_2, h_1, h_2, and h, given $\mathbf{x}_1(0) = (h_1, 0)$, $\mathbf{x}_2(0) = (h_2, 0)$, and $h_2 - h_1 = h > 0$ [Hint: choose a convenient origin of coordinates, and switch to polar coordinates after finding the shape of the trajectories.]

6.35 Consider the unsteady potential flow of two ideal sinks located at $\mathbf{x}_a(t) = (x_a(t), 0)$ and $\mathbf{x}_b(t) = (x_b(t), 0)$ that are free to move along the x-axis in an ideal fluid that is stationary far from the origin. Assume that each sink will move in the velocity field induced by the other:

$$\phi(x, y, t) = -\frac{q_s}{2\pi}\left[\ln\sqrt{(x - x_a(t))^2 + y^2} + \ln\sqrt{(x - x_b(t))^2 + y^2}\right] \text{ with } q_s > 0.$$

a) Determine $x_a(t)$ and $x_b(t)$ when $\mathbf{x}_a(0) = (-L, 0)$ and $\mathbf{x}_b(0) = (+L, 0)$.

b) If the pressure far from the origin is p_∞ and the fluid density is ρ, determine the pressure p at $x = y = 0$ as function of p_∞, ρ, q_s, and $x_a(t)$.

6.36 Consider the unsteady potential flow of an ideal source and sink located at $\mathbf{x}_1(t) = (x_1(t), 0)$ and $\mathbf{x}_2(t) = (x_2(t), 0)$ that are free to move along the x-axis in an ideal fluid that is stationary far from the origin. Assume that the source and sink will move in the velocity field induced by the other:

$$\phi(x, y, t) = \frac{q_s}{2\pi}\left[\ln\sqrt{(x - x_1(t))^2 + y^2} - \ln\sqrt{(x - x_2(t))^2 + y^2}\right], \text{ with } q_s > 0.$$

a) Determine $x_1(t)$ and $x_2(t)$ when $\mathbf{x}_1(0) = (-\ell, 0)$ and $\mathbf{x}_2(0) = (+\ell, 0)$.

b) If the pressure far from the origin is p_∞ and the fluid density is ρ, determine the pressure p at $x = y = 0$ as function of p_∞, ρ, q_s, ℓ, and t.

6.37 Consider a free ideal line vortex oriented parallel to the z-axis in a 90° corner defined by the solid walls at $\theta = 0$ and $\theta = 90°$. If the vortex passes the through the plane of the flow at (x, y), show that the vortex path is given by: $x^{-2} + y^{-2} =$ constant. [*Hint*: Three image vortices are needed at points $(-x, -y)$, $(-x, y)$, and $(x, -y)$. Carefully choose the directions of rotation of these image vortices, show that $dy/dx = v/u = -y^3/x^3$, and integrate to produce the desired result.]

6.38 In ideal flow, streamlines are defined by $d\psi = 0$, and potential lines are defined by $d\phi = 0$. Starting from these relationships, show that streamlines and potential lines are perpendicular

a) in plane flow where x and y are the independent spatial coordinates, and

b) in axisymmetric flow where R and z are the independent spatial coordinates.

[*Hint*: For any two independent orthogonal coordinates x_1 and x_2, the unit tangent to the curve $x_2 = f(x_1)$ is $\mathbf{t} = (\mathbf{e}_1 + (df/dx_1)\mathbf{e}_2) \big/ \sqrt{1 + (df/dx_1)^2}$; thus, for a) and b) it is sufficient to show $(\mathbf{t})_{\psi=\text{const}} \cdot (\mathbf{t})_{\phi=\text{const}} = 0$.]

6.39 Following the numerical method outlined in Section 6.7, recompute the stream function for the channel flow provided in Example 6.7 using a finer grid with $\Delta x = \Delta y = 0.1$ m. Note that a finer grid will require more iterations for the solution to converge.

a) Compute the local velocity components u and v at each grid point. [*Hint*: a first order finite difference approximation of ψ in the x-direction is $\partial\psi/\partial x \approx [\psi_{i+1,j} - \psi_{i,j}]/\Delta x$.]

b) Solve for the pathline of a fluid particle that passes through $x = 0$ and $y = 1$. Repeat for a particle passing through $x = 0$ and $y = 4$. Plot the velocity magnitude obtained from a) in addition to the trajectories of each particle. [*Hint*: this requires interpolating the field values (u and v) from the grid points to the particle location. Different approaches should be tested, such as using the value closest to the particle in addition to bilinear interpolation.]

c) Determine the time it takes each particle to traverse the channel.

d) Assuming the density of the fluid is $\rho = 1$, determine the pressure distribution along the streamline passing through $x = 0$ and $y = 1$ and along the streamline passing through $x = 0$ and $y = 4$. Plot $p - p_o$ as a function of x, where p_o is the pressure of the point at $x = 0$.

6.40 Consider the ideal-flow force, F_x, per unit depth into the page that the fluid feels because of the vertical face of the abrupt contraction shown in Fig. 6.25.

a) Ignore gravity, use Cartesian coordinates with x increasing to the right, and assume uniform inflow and outflow from a suitable control volume, to show that:

$$F_x = \rho U^2 \frac{(H_1 - H_2)^2}{2H_2} - P(H_1 - H_2),$$

where ρ is the fluid density, U is the upstream flow speed in the larger duct, P is the upstream pressure in the larger duct, and H_1 and H_2 are the heights of the larger and smaller ducts, respectively.

b) With $H_1 - H_2$ held constant, does this formula reach the correct limit as $H_2 \to \infty$? Explain your answer in light of the discussion in Example 6.6.

c) What changes in the part a) formula if the downstream duct is larger?

d) If you have completed the previous exercise, determine the pressure on the face of the abrupt contraction from your numerical solution, numerically integrate it, and compare the result with an appropriate evaluation of the part a) formula using parameter values from Example 6.7.

6.41 Consider a three-dimensional point source of strength Q (m^3/s). Use a spherical control volume and the principle of conservation of mass to argue that the velocity components in spherical coordinates are $u_\theta = 0$ and $u_r = Q/4\pi r^2$ and that the velocity potential and stream function must be of the form $\phi = \phi(r)$ and $\psi = \psi(\theta)$. Integrate the velocity, to show that $\phi = -Q/4\pi r$ and $\psi = -Q\cos\theta/4\pi$.

6.42 Solve the Poisson equation $\nabla^2\phi = Q\delta(\mathbf{x} - \mathbf{x}')$ in a uniform, unbounded three-dimensional domain to obtain the velocity potential $\phi = -Q/4\pi|\mathbf{x} - \mathbf{x}'|$ for an ideal point source located at $\mathbf{x}'$.

6.43 Using (R, φ, z)-cylindrical coordinates, consider steady three-dimensional potential flow for a point source of strength Q at the origin in a free stream flowing along the z-axis at speed U:

$$\phi(R, \varphi, z) = Uz - \frac{Q}{4\pi\sqrt{R^2 + z^2}}.$$

a) Sketch the streamlines for this flow in any R-z half-plane.

b) Find the coordinates of the stagnation point that occurs in this flow.

c) Determine the pressure gradient, ∇p, at the stagnation point found in part b).

d) If $R = a(z)$ defines the stream surface that encloses the fluid that emerges from the source, determine $a(z)$ for $z \to +\infty$.

e) Use Stokes' stream function to determine an implicit equation for $a(z)$ that is valid for any value of z.

f) Use the control-volume momentum equation, $\int_A \rho\mathbf{u}(\mathbf{u}\cdot\mathbf{n})dA = -\int_A p\mathbf{n}\,dS + \mathbf{F}$ where $\mathbf{n}$ is the outward normal from the control volume, to determine the force $\mathbf{F}$ applied to the point source to hold it stationary.

g) If the fluid expelled from the source is replaced by a solid body having the same shape, what is the drag on the front of this body?

6.44 In (R, φ, z) cylindrical coordinates, the three-dimensional potential for a point source at $(0, 0, s)$ is given by: $\phi = -(Q/4\pi)[R^2 + (z-s)^2]^{-1/2}$.

a) By combining a source of strength $+Q$ at $(0, 0, -b)$, a sink of strength $-Q$ at $(0, 0, +b)$, and a uniform stream with velocity $U\mathbf{e}_z$, derive the potential (6.87) for flow around a sphere of radius a by taking the limit as $Q \to \infty$ and $b \to 0$, such that $\mathbf{d} = -2bQ\mathbf{e}_x = -2\pi a^3 U\mathbf{e}_x =$ constant. Put your final answer in spherical coordinates in terms of U, r, θ, and a.

b) Repeat part a) for the Stokes stream function starting from

$$\psi = -(Q/4\pi)(z-s)\left[R^2 + (z-s)^2\right]^{-1/2}.$$

6.45 a) Determine the locus of points in uniform ideal flow past a circular cylinder of radius a without circulation where the velocity perturbation produced by the presence of the cylinder is 1% of the free-stream value.

b) Repeat for uniform ideal flow past a sphere.

c) Explain the physical reason(s) for the differences between the answers for a) and b).

6.46 Using the figure for Exercise 6.31 with $A_3 \to 0$ and $r \to \infty$, expand the three-dimensional potential for a stationary arbitrary-shape closed body – formed by a superposition of sources and sinks – in inverse powers of the distance r and prove that ideal flow theory predicts zero drag on the body.

6.47 Consider steady ideal flow over a hemisphere of constant radius a lying on the y-z plane. For the spherical coordinate system shown, the potential for this flow is:

$$\phi(r, \theta, \varphi) = Ur(1 + a^3/2r^3)\cos\theta,$$

where U is the flow velocity far from the hemisphere. Assume gravity acts downward along the x-axis. Ignore fluid viscosity in this problem.

a) Determine all three components of the fluid velocity on the surface of the hemisphere, $r = a$, in spherical polar coordinates: $(u_r, u_\theta, u_\varphi) = \nabla\phi = (\partial\phi/\partial r, (1/r)\partial\phi/\partial\theta, (1/r\sin\theta)\partial\phi/\partial\varphi)$.

b) Determine the pressure, p, on $r = a$.

c) Determine the hydrodynamic force, R_x, on the hemisphere assuming stagnation pressure is felt everywhere underneath the hemisphere. [*Hints*: $\mathbf{e}_r \cdot \mathbf{e}_x = \sin\theta\cos\varphi$, $\int_0^\pi \sin^2\theta\,d\theta = \pi/2$, and $\int_0^\pi \sin^4\theta\,d\theta = 3\pi/8$].

d) For the conditions of part c) what density ρ_h must the hemisphere have to remain on the surface.

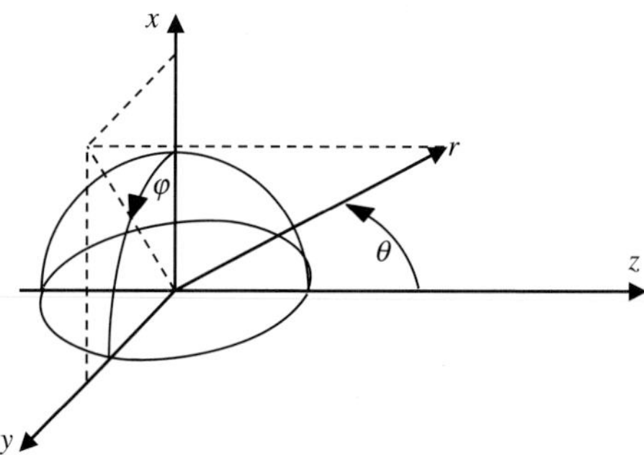

6.48 The flow-field produced by suction flow into a round vacuum cleaner nozzle held above a large flat surface can be easily investigated with a simple experiment, and analyzed via potential flow in (R, φ, z)-cylindrical coordinates with the method of images.

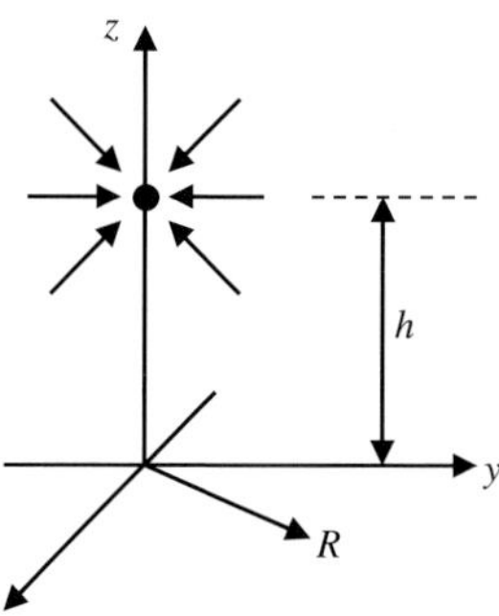

a) Do the experiment first. Obtain a vacuum cleaner that has a hose for attachments. Remove any cleaning attachments (brush, wand, etc.) or unplug the hose from the cleaning head, and attach an extension hose or something with a round opening (~4 cm diameter is recommended). Find a smooth dry flat horizontal surface that is a ~0.5 meter or more in diameter. Sprinkle the central third of the surface with a light granular material that is easy to see (granulated sugar, dry coffee grounds, salt, flour, talcum powder, etc., should work well). The grains should be 0.5 to 1 mm apart on average. Turn on the vacuum cleaner and lower the vacuum hose opening from ~0.25 meter above the surface toward the surface with the vacuum opening facing toward the surface. When the hose gets to within about one opening diameter of the surface or so, the granular material should start to move. Once the granular material starts moving, hold the hose opening at the same height or lift the hose slightly so that grains are not sucked into it. If many grains are vacuumed up, distribute new ones in the bare spot(s) and start over. Once the correct hose-opening-to-surface distance is achieved, hold the hose steady and let the suction airflow of the vacuum cleaner scour a pattern into the distributed granular material. Describe the shape of the final pattern, and measure any relevant dimensions.

Now see if ideal flow theory can explain the pattern observed in part a). As a first approximation, the flow field near the hose inlet can be modeled as a sink (a source with strength $-Q$) above an infinite flat boundary since the vacuum cleaner outlet (a source with strength $+Q$) is likely to be far enough away to be ignored. Denote the fluid density by ρ, the pressure far away by p_∞, and the pressure on the flat surface by $p(R)$. The potential for this flow field will be the sum of two terms:

$$\phi(R, z) = \frac{+Q}{4\pi\sqrt{R^2 + (z-h)^2}} + K(R, z)$$

b) Sketch the streamlines in the y-z plane for $z > 0$.

d) Determine $K(R, z)$.

d) Use dimensional analysis to determine how $p(R) - p_\infty$ must depend on ρ, Q, R, and h.

e) Compute $p(R) - p_\infty$ from the steady Bernoulli equation. Is this pressure distribution consistent with the results of part a)? Where is the lowest pressure? (This is also the location of the highest speed surface flow). Is a grain at the origin of coordinates the one most likely to be picked up by the vacuum cleaner?

6.49 An ideal point source with time-dependent strength $Q(t)$ is located at $\mathbf{x} = (0, 0, h)$ in an infinite three-dimensional bath of quiescent inviscid fluid with density ρ. A second source with strength $\gamma Q(t)$ is located at $\mathbf{x} = (0, 0, -h)$, where γ is a weighting constant $(-1 \leq \gamma \leq 1)$. Use the cylindrical coordinates shown, $\mathbf{x} = (R, \phi, z)$, for your work.

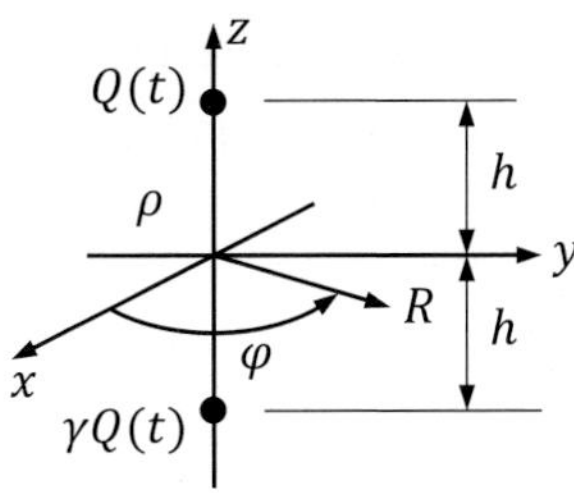

a) Write down the velocity potential, $\phi(R, z, t)$, for this situation in terms of $Q(t)$, h, γ, R, and z.
b) Determine $p(R) - p_\infty$, where $p(R)$ is the pressure on the $x - y$ plane ($z = 0$) and p_∞ is the pressure far from the origin.
c) What value of γ causes $p(R) - p_\infty$ to decrease most quickly as R increases?

6.50 There is a point source of strength Q (m^3/s) at the origin, and a uniform line sink of strength $k = Q/a$ extending from $z = 0$ to $z = a$. The two are combined with a uniform stream U parallel to the z-axis. Show that the combination represents the flow past a closed surface of revolution of airship shape, whose total length is the difference of the roots of:

$$\frac{z^2}{a^2}\left(\frac{z}{a} \pm 1\right) = \frac{Q}{4\pi U a^2}.$$

6.51 Using a computer, determine the surface contour of an axisymmetric half-body formed by a line source of strength k (m^2/s) distributed uniformly along the z-axis from $z = 0$ to $z = a$ and a uniform stream. The nose of this body is more pointed than that formed by the combination of a point source and a uniform stream. From a mass balance, show that the radius of the half-body far downstream of the origin is $r = [ak/\pi U]^{1/2}$.

6.52 [4]Consider the radial flow induced by the collapse of a spherical cavitation bubble of radius $R(t)$ in a large quiescent bath of incompressible inviscid fluid of density ρ. The pressure far from the bubble is p_∞. Ignore gravity.
a) Determine the velocity potential $\phi(r, t)$ for the radial flow outside the bubble.
b) Determine the pressure $p(R(t), t)$ on the surface of the bubble.
c) Suppose that at $t = 0$ the pressure on the surface of the bubble is p_∞, the bubble radius is R_o, and its initial velocity is $-\dot{R}_o$ (i.e., the bubble is shrinking), how long will it take for the bubble to completely collapse if its surface pressure remains constant?

6.53 A moving point source of constant strength Q travels below the x-axis at constant speed U and constant depth D in an ideal liquid with density ρ as depicted in the figure below. In three-dimensional coordinates, the time-dependent source location is $x_s = (Ut, 0, -D)$. Gravity g acts downward parallel to the z-axis.

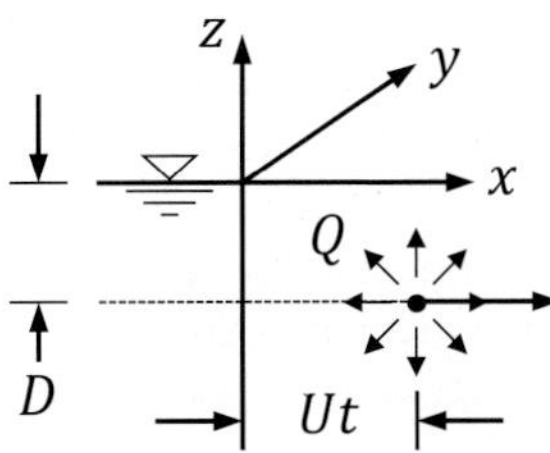

a) Assuming the vertical deflection of the liquid surface at $z = 0$ is negligible, write down the velocity potential for this situation.
b) Using this velocity potential, determine the pressure, $p(x, y, 0, t)$, on $z = 0$.
c) Now consider a deep source ($Q \ll UD^2$) that barely perturbs the fluid surface, and use the linearized free surface boundary condition, $\rho[\partial\phi/\partial t]_{(z=0)} + \rho g\zeta \cong 0$, to obtain an approximate relationship for the vertical surface deflection $\zeta(x, y, t)$ caused by the moving source.

4. Based on problem 5.7 in Currie (1993).

d) Sketch $\zeta(x, y, t)$ at the instant $t = 0$ in the plane defined by $y = 0$.

6.54 Derive the apparent mass per unit depth into the page of a cylinder of radius a that travels at speed $U_c(t) = dx_c/dt$ along the x-axis in a large reservoir of an ideal quiescent fluid with density ρ. Use an appropriate Bernoulli equation and the following time-dependent two-dimensional potential: $\phi(x, y, t) = -\dfrac{a^2 U_c(x - x_c)}{(x - x_c)^2 + y^2}$, where $x_c(t)$ is location of the center of the cylinder, and the Cartesian coordinates are x and y. [*Hint*: steady cylinder motion does *not* contribute to the cylinder's apparent mass; keep only the term (or terms) from the Bernoulli equation necessary to determine apparent mass].

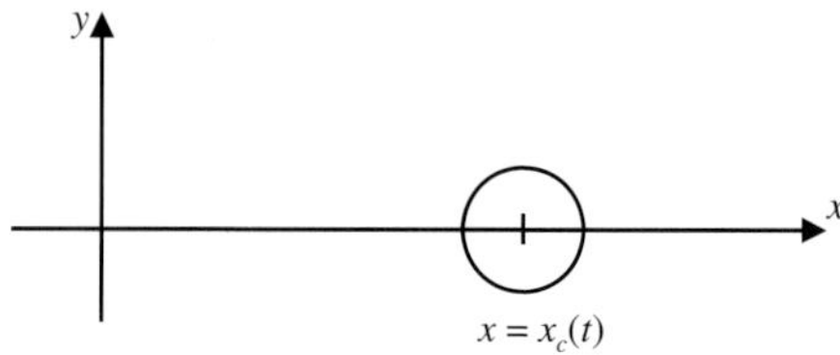

6.55 A stationary sphere of radius a and mass m resides in inviscid fluid with constant density ρ.

a) Determine the buoyancy force on the sphere when gravity g acts downward.

b) At $t = 0$, the sphere is released from rest. What is its initial acceleration?

c) What is the sphere's initial acceleration if it is a bubble in a heavy fluid (i.e., when $m \to 0$)?

6.56 A sphere of mass m and volume V is attached to the end of a light, thin flexible cable of length L. In a vacuum, with gravity g acting, the natural frequencies for small longitudinal (bouncing) and transverse (pendulum) oscillations of the sphere are ω_b and ω_p. Ignore the effects of viscosity and estimate these natural frequencies when the same sphere and cable are submerged in water with density ρ_w. What is ω_p when $m \ll \rho_w V$?

6.57 Determine the ideal-flow force on a stationary sphere for the following unsteady flow conditions ignoring the sphere-internal potential flow.

a) The free stream of velocity $U\mathbf{e}_z$ is constant but the sphere's radius $a(t)$ varies.

b) The free stream velocity magnitude changes, $U(t)\mathbf{e}_z$, but the sphere's radius a is constant.

c) The free stream velocity changes direction $U(\mathbf{e}_x \cos\Omega t + \mathbf{e}_y \sin\Omega t)$, but its magnitude U and the sphere's radius a are constant.

6.58 Consider the flow field produced by a sphere of radius a that moves in the x-direction at constant speed U along the x-axis in an unbounded environment of a quiescent ideal fluid with density ρ. The pressure far from the sphere is p_∞ and there is no body force. The velocity potential for this flow field is:

$$\phi(x, y, z, t) = -\frac{a^3 U}{2} \frac{x - Ut}{\left[(x - Ut)^2 + y^2 + z^2\right]^{3/2}}.$$

a) If $u = (u, v, w)$, what is $\mathbf{u}(x, 0, 0, t)$, the velocity along the x-axis as a function of time? [Hint: consider the symmetry of the situation before differentiating in all directions.]

b) What is $p(x, 0, 0, t)$, the pressure along the x-axis as a function of time?

c) What is the pressure on the x-axis at $x = Ut \pm a$?

d) If the plane defined by $z = h$ is an impenetrable flat surface and the sphere executes the same motion, what additional term should be added to the given potential?

e) Compared to the sphere's apparent mass in an unbounded environment, is the sphere's apparent mass larger, the same, or smaller when the impenetrable flat surface is present?

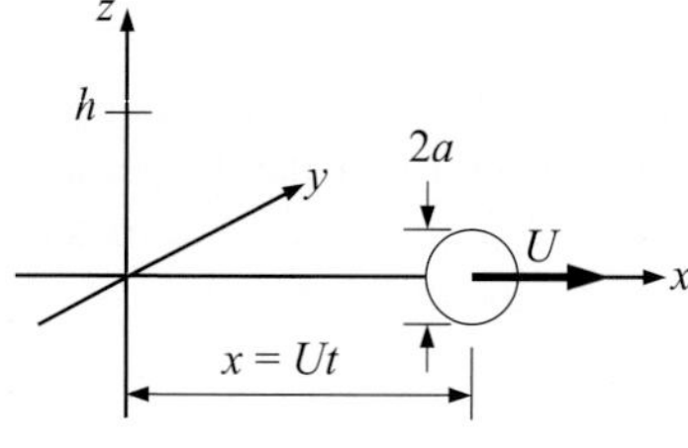

6.59 In three dimensions, consider a solid object moving with velocity **U** near the origin of coordinates in an unbounded quiescent bath of inviscid incompressible fluid with density ρ. The kinetic energy of the moving fluid in this situation is:

$$KE = \frac{\rho}{2}\int_V |\nabla\phi|^2 dV,$$

where ϕ is the velocity potential and V is a control volume that contains all of the moving fluid but excludes the object. (Such a control volume is shown in the figure for Exercise 6.31 when $A_3 \to 0$ and $U = 0$.)

a) Show that $KE = -\frac{\rho}{2}\int_A \phi(\nabla\phi \cdot \mathbf{n})dA$ where A encloses the body and is coincident with its surface, and **n** is the outward normal on A.

b) The apparent mass, M, of the moving body may be defined by $KE = \frac{1}{2}M|\mathbf{U}|^2$. Using this definition, the result of a), and (6.99) with $\mathbf{x}_s = 0$, show that $M = 2\pi a^3\rho/3$ for a sphere.

6.60 Using the code provided in Example 6.10, investigate the effect of density ratio (ρ/ρ_s) on a spherical particle's trajectory in the same oscillating flow given in the example. What is the relative effect of added mass for bubbly flows $(\rho/\rho_s = \mathcal{O}(10^3))$, liquid-solid flows $(\rho/\rho_s = \mathcal{O}(1))$, and droplet-laden flows $(\rho/\rho_s = \mathcal{O}(10^{-3}))$.

References

Carrier, G.F., Krook, M., Pearson, C.E., 1966. Functions of a Complex Variable. McGraw-Hill, New York.
Churchill, R.V., Brown, J.W., Verhey, R.F., 1974. Complex Variables and Applications, 3rd ed. McGraw-Hill, New York.
Currie, I.G., 1993. Fundamentals of Fluid Mechanics, 2nd ed. McGraw-Hill, New York.
Milne-Thompson, L.M., 1962. Theoretical Hydrodynamics. Macmillan & Company Ltd., London.
Prandtl, L., 1952. Essentials of Fluid Dynamics. Hafner Publishing, New York.

Further reading

Batchelor, G.K., 1967. An Introduction to Fluid Dynamics. Cambridge University Press, London.
Shames, I.H., 1962. Mechanics of Fluids. McGraw-Hill, New York.
Vallentine, H.R., 1967. Applied Hydrodynamics. Plenum Press, New York.

Chapter 7

Laminar flow⊛

Contents

7.1 **Introduction** 223
7.2 **Exact solutions for steady incompressible viscous flow** 225
Steady flow between parallel plates 226
Steady flow in a round tube 227
Steady flow between concentric rotating cylinders 228
7.3 **Elementary lubrication theory** 230
7.4 **Similarity solutions for unsteady incompressible viscous flow** 236
7.5 **Flows with oscillations** 243
7.6 **Low Reynolds number viscous flow past a sphere** 245
7.7 **Final remarks** 253
Exercises 253
References 268
Further reading 269

Chapter objectives

- To present a variety of exact and approximate solutions to the viscous equations of fluid motion in confined and unconfined geometries
- To introduce lubrication theory and indicate its utility
- To define and present similarity solutions to exact and approximate viscous flow field equations
- To develop the equations for creeping flow and illustrate their use

7.1 Introduction

Chapters 6 and 11 cover flows in which the viscous terms in the Navier-Stokes equations are dropped because the flow is ideal (irrotational and constant density) or the effects of viscosity are small. For these situations, the underlying assumptions are either that 1) viscous forces and rotational flow are spatially confined to a small portion of the flow domain (thin boundary layers near solid surfaces), or that 2) fluid particle accelerations associated with fluid inertia $\sim U^2/L$ are much larger than those caused by viscosity $\sim \mu U/\rho L^2$, where U is a characteristic velocity, L is a characteristic length, ρ is the fluid's density, and μ is the fluid's viscosity. Both of these assumptions are valid if the Reynolds number is large and boundary layers stay attached to the surface on which they have formed.

However, for low values of the Reynolds number, the *entire flow* may be influenced by viscosity, and inviscid flow theory is no longer even approximately correct. The purpose of this chapter is to present some exact and approximate solutions of the Navier-Stokes equations for simple geometries and situations, retaining the viscous terms in (4.38) everywhere in the flow and applying the no-slip boundary condition at solid surfaces (see Section 4.10).

Viscous flows generically fall into three categories, *laminar*, *transitional*, and *turbulent*, but the boundaries between them are imperfectly defined. The differences between the categories are phenomenological and were first demonstrated by Osborne Reynolds in 1883, who injected a thin stream of dye into the flow of water through a tube (Fig. 7.1). At low flow rates, the dye stream was observed to follow a well-defined straight path, indicating that the fluid moved in parallel layers (laminae) with no unsteady mixing or overturning motion of the layers. If the flow rate was increased above a certain critical value, the dye stream broke up into irregular filaments and spread throughout the cross-section of the tube, indicating the presence of unsteady, apparently chaotic, three-dimensional mixing motions, referred to as *turbulence*. Reynolds demonstrated that the transition from laminar to turbulent flow always occurred at or near a fixed value of the ratio that bears his name, the Reynolds number, $\mathrm{Re} = Ud/\nu \sim 2000$ to 3000 where U is the velocity averaged over the tube's cross-section, d is the tube diameter, and $\nu = \mu/\rho$ is the kinematic viscosity. At or near the critical Reynolds number, the flow is transitional and may vary between laminar and turbulent characteristics.

⊛. This book has a companion website hosting complementary materials. Visit this URL to access it: https://www.elsevier.com/books-and-journals/book-companion/9780128198070.

Fluid Mechanics. https://doi.org/10.1016/B978-0-12-819807-0.00016-8

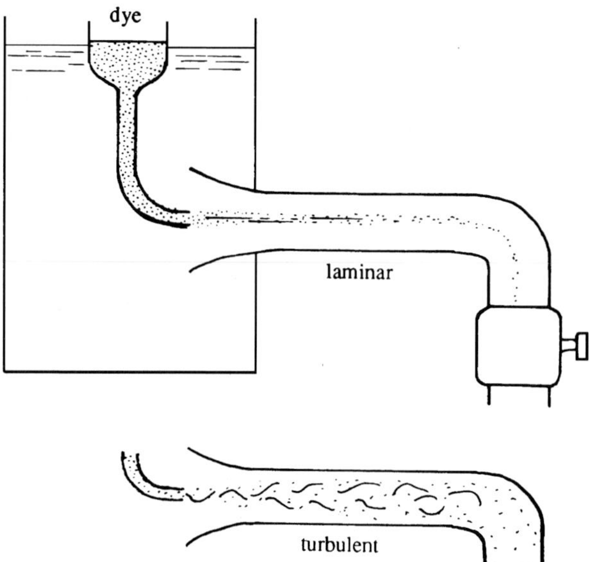

FIGURE 7.1 Reynolds' experiment to distinguish between laminar and turbulent flows. At low flow rates (the upper drawing), the pipe flow was laminar and the dye filament moved smoothly through the pipe. At high flow rates (the lower drawing), the flow became turbulent and the dye filament was mixed throughout the cross-section of the pipe.

The fluid's kinematic viscosity, $\nu = \mu/\rho$, specifies the propensity for vorticity to diffuse through a fluid. Consider (5.13) for the z-component of vorticity in a two-dimensional flow confined to the x-y plane so that $\boldsymbol{\omega} \cdot \nabla \mathbf{u} = 0$:

$$D\omega_z/Dt = \nu \nabla^2 \omega_z.$$

This equation states that the rate of change of ω_z following a fluid particle is caused by diffusion of vorticity. For the same initial vorticity distribution, a fluid with larger ν will produce a larger diffusion term, $\nu \nabla^2 \omega$, and more rapid changes in the vorticity. This equation is similar to the Boussinesq heat equation:

$$DT/Dt = \kappa \nabla^2 T, \tag{4.90}$$

where $\kappa \equiv k/\rho c_p$ is the *thermal diffusivity*, and this similarity suggests that vorticity diffuses in a manner analogous to heat. At a coarse level, this suggestion is correct since both ν and κ arise from molecular processes in real fluids and both have the same units (length^2/time). The similarity emphasizes that the diffusive effects are controlled by ν and κ, and not by μ (viscosity) and k (thermal conductivity). In fact, the constant-density, constant-viscosity momentum equation:

$$D\mathbf{u}/Dt = -(1/\rho)\nabla p + \nu \nabla^2 \mathbf{u}, \tag{4.86, 7.1}$$

also shows that the fluid particle acceleration due to viscous diffusion is proportional to ν. Thus, at room temperature and pressure, air ($\nu = 15 \times 10^{-6}$ m^2/s) is 15 times more diffusive than water ($\nu = 1 \times 10^{-6}$ m^2/s), although μ for water is larger. Since ν and κ have the units of (length)^2/time, the kinematic viscosity ν is sometimes called the *momentum diffusivity*, in analogy with κ, the *thermal diffusivity*. However, velocity cannot be simply regarded as being diffused and advected in a flow because of the presence of the pressure gradient in (7.1).

Laminar flows in which viscous effects are important throughout the flow are the subject of the present chapter. The primary field equations are $\nabla \cdot \mathbf{u} = 0$ (4.10) and (7.1) or the version that includes a body force (4.40). The velocity boundary conditions on a solid surface are simplified versions of (4.91) and (4.95):

$$\mathbf{n} \cdot \mathbf{u}_s = (\mathbf{n} \cdot \mathbf{u})_{on\ the\ surface} \quad \text{and} \quad \mathbf{t} \cdot \mathbf{u}_s = (\mathbf{t} \cdot \mathbf{u})_{on\ the\ surface}, \tag{7.2, 7.3}$$

where $\mathbf{u}_s$ is the velocity of the surface, $\mathbf{n}$ is the normal to the surface, and $\mathbf{t}$ is the tangent to the surface in the plane of interest. Here fluid density will be assumed constant, and the frame of reference will be inertial. Thus, gravity can be dropped from the momentum equation as long as no free surface is present (see Section 4.9 "Neglect of gravity in constant density flows"). Laminar flows in which frictional effects are confined to boundary layers near solid surfaces are discussed in the next chapter. Chapter 9 considers the stability of laminar flows and their transition to turbulence; fully turbulent flows are discussed in Chapter 10. Some viscous flow solutions in rotating coordinates, such as the Ekman layers, are presented in Chapter 12.

Example 7.1

Consider incompressible viscous flow near a smooth flat surface coincident with the x-y plane. If ∇p is not zero at $z=0$, what is the simplified momentum equation for $z \to 0$?

Solution For the geometric situation described, the boundary conditions (7.2) and (7.3) require all three components of $\mathbf{u}$ to go to zero as $z \to 0$. However, if $\mathbf{u}$ is simply set to 0 in (7.1) this requires the pressure gradient to be zero, a contradiction of the problem statement. Instead, consider a Taylor expansion of $\mathbf{u}$ about $z=0$:

$$\mathbf{u} \cong 0 + z\left(\frac{\partial \mathbf{u}}{\partial z}\right)_{z=0} + \frac{z^2}{2}\left(\frac{\partial^2 \mathbf{u}}{\partial z^2}\right)_{z=0} + \ldots$$

This expansion satisfies (7.2) and (7.3), and it allows the following limits to be found:

$$\lim_{z\to 0}\frac{D\mathbf{u}}{Dt} = \lim_{z\to 0}\left\{\frac{\partial \mathbf{u}}{\partial t} + (\mathbf{u}\cdot\nabla)\mathbf{u}\right\} = 0 \quad \text{and} \quad \lim_{z\to 0}\nabla^2\mathbf{u} = \lim_{z\to 0}\left\{\frac{\partial^2 \mathbf{u}}{\partial x^2} + \frac{\partial^2 \mathbf{u}}{\partial y^2} + \frac{\partial^2 \mathbf{u}}{\partial z^2}\right\} = \left(\frac{\partial^2 \mathbf{u}}{\partial z^2}\right)_{z=0},$$

since all the terms in both sets of {,}-braces, except $\partial^2\mathbf{u}/\partial z^2$, are proportional to z, z^2, or higher powers of z. Thus, the limiting form of (7.1) as $z \to 0$ is:

$$0 = -\frac{1}{\rho}(\nabla p)_{z=0} + \frac{\mu}{\rho}\left(\frac{\partial^2 \mathbf{u}}{\partial z^2}\right)_{z=0}, \quad \text{or} \quad 0 = -(\nabla p)_{z=0} + \mu\left(\frac{\partial^2 \mathbf{u}}{\partial z^2}\right)_{z=0}.$$

Hence, the surface pressure gradient sets the near-wall curvature of the velocity profile. This result represents a balance of pressure forces and viscous forces on near-wall fluid elements. The fluid density ρ does not appear in the final equation because fluid inertia becomes unimportant as $\mathbf{u} \to 0$ when $z \to 0$.

7.2 Exact solutions for steady incompressible viscous flow

Because of the presence of the nonlinear acceleration term $\mathbf{u}\cdot\nabla\mathbf{u}$ in (7.1), few exact solutions of the Navier-Stokes equations are known in closed form. An example of an exact solution is that for steady laminar flow between infinite parallel plates (Fig. 7.2). Such a flow is said to be *fully developed* when its velocity profile $u(x, y)$ becomes independent of the downstream coordinate x so that $u = u(y)$ alone. The entrance length of the flow, where the velocity profile depends on the downstream distance, may be several or even many times longer than the spacing between the plates. Within the entrance length, the derivative $\partial u/\partial x$ is not zero so the continuity equation $\partial u/\partial x + \partial v/\partial y = 0$ requires that $v \neq 0$, so that the flow is *not* parallel to the walls within the entrance length. Laminar flow development is the subject of Section 7.4, and the next chapter. Here, the topic is fully developed flows.

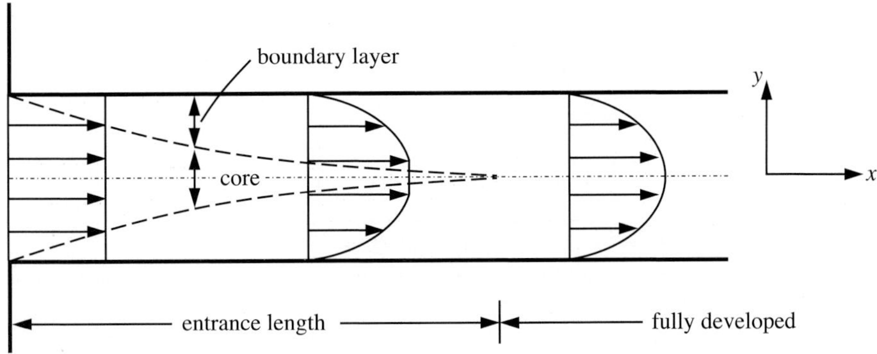

FIGURE 7.2 Developing and fully developed flows in a channel. Within the entrance length, the viscous boundary layers on the upper and lower walls are distinct and the flow profile $u(x, y)$ depends on both spatial coordinates. Downstream of the point where the boundary layers have completely merged, the flow is fully developed and its profile $u(y)$ is independent of the stream-wise coordinate x.

Steady flow between parallel plates

Consider the situation depicted in Fig. 7.3 where a viscous fluid flows between plates parallel to the x-axis with lower and upper plates at $y = 0$ and $y = h$, respectively. The flow is sustained by an externally applied pressure gradient $(\partial p/\partial x \neq 0)$ in the x-direction, and horizontal motion of the upper plate at speed U in the x-direction. For this situation, the flow should be independent of the z-direction so $w = 0$ and $\partial/\partial z = 0$ can be used in the equations of motion. A steady, fully developed flow will have a horizontal velocity $u(y)$ that does not depend on x so $\partial u/\partial x = 0$. Thus, the continuity equation, $\partial u/\partial x + \partial v/\partial y = 0$, requires $\partial v/\partial y = 0$, and since $v = 0$ at $y = 0$ and h, it follows that $v = 0$ everywhere, which reflects the fact that the flow is parallel to the walls. Under these circumstances, $\mathbf{u} = (u(y), 0, 0)$, and the x- and y-momentum equations reduce to:

$$0 = -\frac{1}{\rho}\frac{\partial p}{\partial x} + \nu\frac{d^2u}{dy^2}, \quad \text{and} \quad 0 = -\frac{1}{\rho}\frac{\partial p}{\partial y}. \tag{7.4a,b}$$

The y-momentum equation shows that p is not a function of y, so $p = p(x)$. Thus, the first term in the x-momentum equation must be a function of x alone, while the second term must be a function of y alone. The only way the equation can be satisfied throughout x-y space is if both terms are constant. The *pressure gradient is therefore a constant*, which implies that the pressure varies linearly along the channel. To find the velocity profile, integrate the x-momentum equation twice to obtain:

$$0 = -\frac{y^2}{2}\frac{dp}{dx} + \mu u + Ay + B,$$

where A and B are constants and dp/dx has replaced $\partial p/\partial x$ because p is a function of x alone. The constants are determined from the boundary conditions: $u = 0$ at $y = 0$, and $u = U$ at $y = h$. The results are $B = 0$ and $A = (h/2)(dp/dx) - \mu U/h$, so the velocity profile becomes:

$$u(y) = \frac{U}{h}y - \frac{1}{2\mu}\frac{dp}{dx}y(h - y), \tag{7.5}$$

which is illustrated in Fig. 7.4 for various cases. The volume flow rate q per unit width of the channel is:

$$q = \int_0^h u\,dy = U\frac{h}{2}\left[1 - \frac{h^2}{6\mu U}\frac{dp}{dx}\right],$$

so that the average velocity is:

$$V \equiv \frac{q}{h} = \frac{1}{h}\int_0^h u\,dy = \frac{U}{2}\left[1 - \frac{h^2}{6\mu U}\frac{dp}{dx}\right].$$

Here, negative and positive pressure gradients increase and decrease the flow rate, respectively.

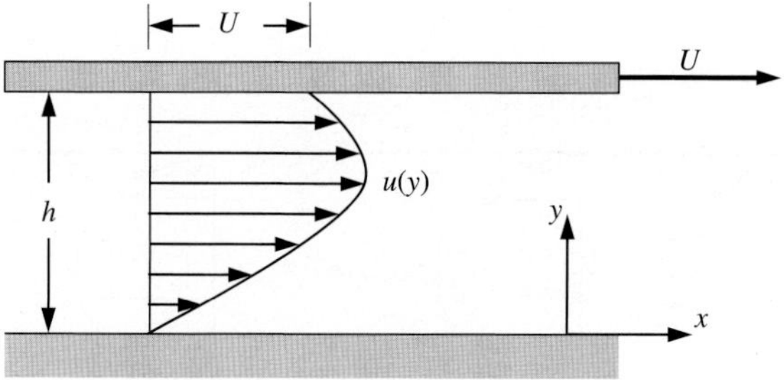

FIGURE 7.3 Flow between parallel plates when the lower plate at $y = 0$ is stationary, the upper plate at $y = h$ is moving in the positive-x direction at speed U, and a non-zero $dp/dx < 0$ leads to velocity profile curvature.

When the flow is driven by motion of the upper plate alone, without any externally imposed pressure gradient, it is called a *plane Couette flow* (Fig. 7.4(c)). In this case (7.5) reduces to $u(y) = Uy/h$, and the magnitude of the shear stress is $\tau = \mu(du/dy) = \mu U/h$, which is uniform across the channel. When the flow is driven by an externally imposed pressure gradient without motion of either plate, it is called a *plane Poiseuille flow* (Fig. 7.4(d)). In this case, (7.5) reduces to the

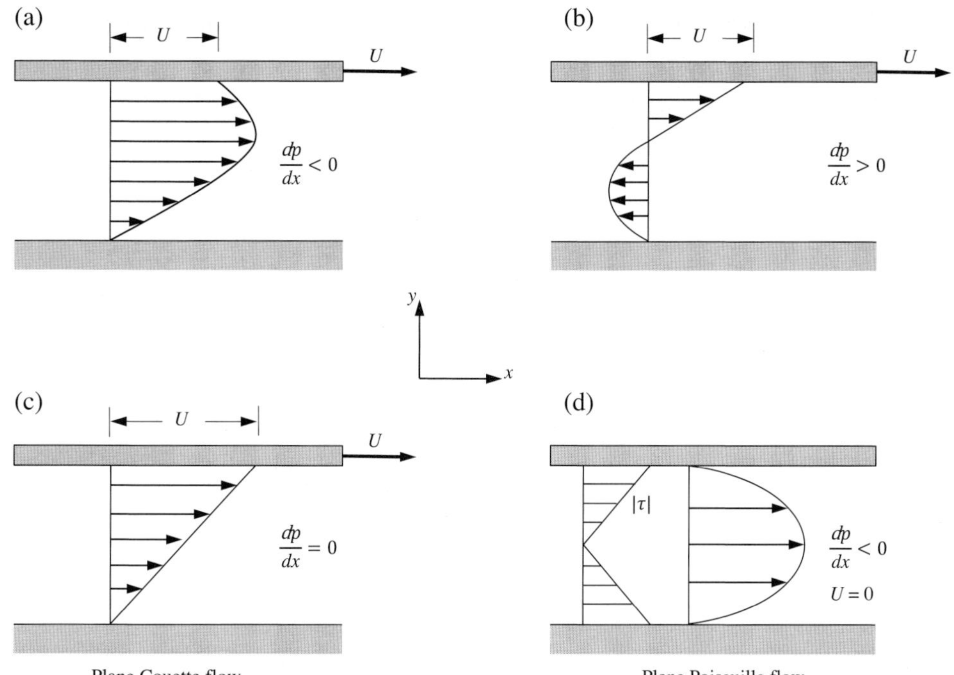

FIGURE 7.4 Various cases of parallel flow in a channel: (a) positive U and favorable $dp/dx < 0$, (b) positive U and adverse $dp/dx > 0$, (c) positive U and $dp/dx = 0$, and (d) $U = 0$ and favorable $dp/dx < 0$.

parabolic profile:

$$u(y) = -\frac{1}{2\mu}\frac{dp}{dx}y(h-y).$$

The shear stress is:

$$\tau = \mu\frac{du}{dy} = -\left(\frac{h}{2} - y\right)\frac{dp}{dx},$$

which is linear with a magnitude of $(h/2)(dp/dx)$ at the walls (Fig. 7.4(d)).

Interestingly, the constancy of the pressure gradient and the linearity of the shear stress distribution are general results for a fully developed channel flow and persist for non-Newtonian fluids, and appropriate averages of these quantities when the flow is turbulent.

Steady flow in a round tube

A second geometry for which there is an exact solution of Eqs. (4.10) and (7.1) is steady, fully developed laminar flow through a round tube of constant radius a, frequently called *circular Poiseuille flow*. We employ cylindrical coordinates (R, φ, z), with the z-axis coinciding with the axis of the tube (Fig. 7.5). The equations of motion in cylindrical coordinates are given in Appendix B. The only non-zero component of velocity is the axial velocity $u_z(R)$, and $\mathbf{u} = (0, 0, u_z(R))$ automatically satisfies the continuity equation. The radial and angular equations of motion reduce to:

$$0 = \partial p/\partial\varphi \quad \text{and} \quad 0 = \partial p/\partial R,$$

so p is a function of z alone. The z-momentum equation gives:

$$0 = -\frac{dp}{dz} + \frac{\mu}{R}\frac{d}{dR}\left(R\frac{du_z}{dR}\right).$$

As for flow between parallel plates, the first term must be a function of the stream-wise coordinate, z, alone and the second term must be a function of the cross-stream coordinate, R, alone, so both terms must be constant. The pressure therefore falls linearly along the length of the tube. Integrating the stream-wise momentum equation twice produces:

$$u_z(R) = \frac{R^2}{4\mu}\frac{dp}{dz} + A\ln R + B.$$

To keep u_z bounded at $R = 0$, the constant A must be zero. The no-slip condition $u_z = 0$ at $R = a$ gives $B = -(a^2/4\mu)(dp/dz)$. The velocity distribution therefore takes the parabolic shape:

$$u_z(R) = \frac{R^2 - a^2}{4\mu}\frac{dp}{dz}. \tag{7.6}$$

From Appendix B, the shear stress at any point is:

$$\tau_{zR} = \mu\left(\frac{\partial u_R}{\partial z} + \frac{\partial u_z}{\partial R}\right).$$

In this case the radial velocity u_R is zero. Dropping the subscripts on τ and differentiating (7.6) yields:

$$\tau = \mu\frac{\partial u_z}{\partial R} = \frac{R}{2}\frac{dp}{dz}, \tag{7.7}$$

which shows that the stress distribution is linear, having a maximum value at the wall of:

$$\tau_w = \frac{a}{2}\frac{dp}{dz}. \tag{7.8}$$

Here again, (7.8) is also valid for appropriate averages of τ_w and p for turbulent flow in a round pipe.

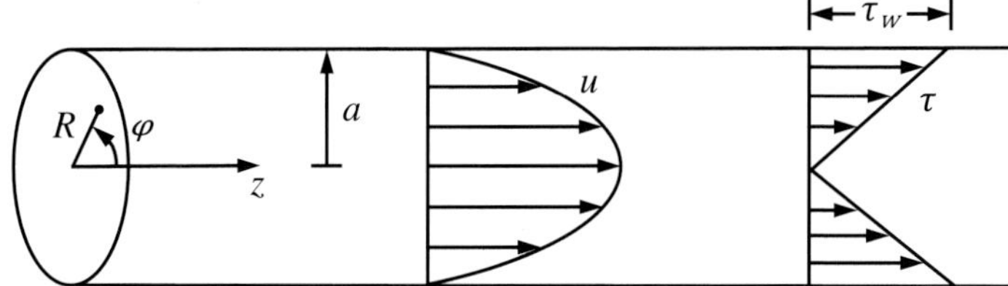

FIGURE 7.5 Laminar flow through a round tube. The flow profile is parabolic, similar to pressure-driven flow between stationary parallel plates (Fig. 7.4(d)).

The volume flow rate in the tube is:

$$Q = \int_0^a u(R)2\pi R\,dR = -\frac{\pi a^4}{8\mu}\frac{dp}{dz},$$

where the negative sign offsets the negative value of dp/dz. The average velocity over the cross section is:

$$V = \frac{Q}{\pi a^2} = -\frac{a^2}{8\mu}\frac{dp}{dz}.$$

Steady flow between concentric rotating cylinders

A third example in which the nonlinear advection terms drop out of the equations of motion is steady flow between two concentric, rotating cylinders, also known as *circular Couette flow* (or *Taylor-Couette flow* for sufficiently high Reynolds numbers). Let the radius and angular velocity of the inner cylinder be R_1 and Ω_1 and those for the outer cylinder be R_2 and Ω_2 (Fig. 7.6). Using cylindrical coordinates and assuming that $\mathbf{u} = (0, u_\varphi(R), 0)$, the continuity equation is automatically satisfied, and the momentum equations for the radial and tangential directions are:

$$-\frac{u_\varphi^2}{R} = -\frac{1}{\rho}\frac{dp}{dR}, \quad \text{and} \quad 0 = \mu\frac{d}{dR}\left[\frac{1}{R}\frac{d}{dR}(Ru_\varphi)\right].$$

The R-momentum equation shows that the pressure increases radially outward due to the centrifugal acceleration. The pressure distribution can therefore be determined once $u_\varphi(R)$ has been found. Integrating the φ-momentum equation twice produces:

$$u_\varphi(R) = AR + B/R. \tag{7.9}$$

Using the boundary conditions $u_\varphi = \Omega_1 R_1$ at $R = R_1$, and $u_\varphi = \Omega_2 R_2$ at $R = R_2$, A and B are found to be:

$$A = \frac{\Omega_2 R_2^2 - \Omega_1 R_1^2}{R_2^2 - R_1^2}, \quad \text{and} \quad B = -\frac{(\Omega_2 - \Omega_1) R_1^2 R_2^2}{R_2^2 - R_1^2}.$$

Substitution of these into (7.9) produces the velocity distribution:

$$u_\varphi(R) = \frac{1}{R_2^2 - R_1^2}\left\{[\Omega_2 R_2^2 - \Omega_1 R_1^2]R - [\Omega_2 - \Omega_1]\frac{R_1^2 R_2^2}{R}\right\}, \tag{7.10}$$

which has interesting limiting cases when $R_2 \to \infty$ with $\Omega_2 = 0$, and when $R_1 \to 0$ with $\Omega_1 = 0$.

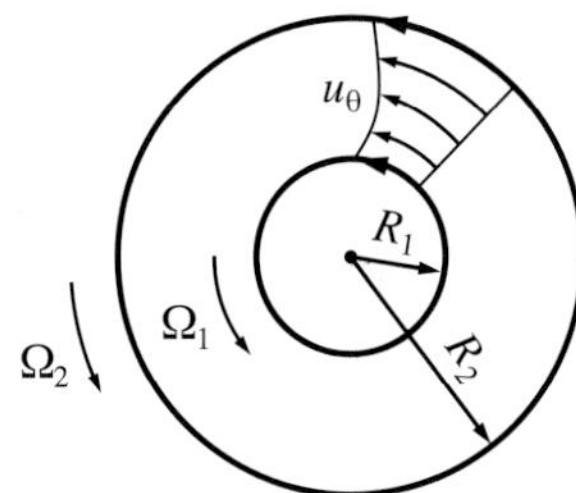

FIGURE 7.6 Circular Couette flow. Viscous fluid flows in the gap between an inner cylinder with radius R_1 that rotates at angular speed Ω_1 and an outer cylinder with radius R_2 that rotates at angular speed Ω_2.

The first limiting case produces the flow outside a long circular cylinder with radius R_1 rotating with angular velocity Ω_1 in an infinite bath of viscous fluid (Fig. 7.7). By direct simplification of (7.10), the velocity distribution is:

$$u_\varphi(R) = \frac{\Omega_1 R_1^2}{R}, \tag{7.11}$$

which is identical to that of an ideal vortex, see (5.2), for $R > R_1$ when $\Gamma = 2\pi\Omega_1 R_1^2$. This is an example of a viscous solution that is completely irrotational. As described in Section 5.1, shear stresses do exist in this flow, but there is no *net* viscous force on a fluid element. The viscous shear stress at any point is given by:

$$\sigma_{R\varphi} = \mu\left[\frac{1}{R}\frac{\partial u_R}{\partial \varphi} + R\frac{\partial}{\partial R}\left(\frac{u_\varphi}{R}\right)\right] = -\frac{2\mu\Omega_1 R_1^2}{R^2}.$$

The mechanical power supplied to the fluid (per unit length of cylinder) is $(2\pi R_1)\tau_{R\varphi}u_\phi$, and it can be shown that this power equals the integrated viscous dissipation of the flow field (Exercise 7.17).

The second limiting case of (7.10) produces steady viscous flow within a cylindrical tank of radius R_2 rotating at rate Ω_2. Setting R_1 and Ω_1 equal to zero in (7.10) leads to:

$$u_\varphi(R) = \Omega_2 R, \tag{7.12}$$

which is the velocity field of solid body rotation, see (5.1) and Section 5.1.

The three exact solutions of the incompressible viscous flow equations (4.10) and (7.1) described in this section are all known as internal or confined flows. In each case, the velocity field was confined between solid walls and the symmetry of each situation eliminated the nonlinear advective acceleration term from the equations. Other exact solutions of the incompressible viscous flow equations for confined and unconfined flows are described in other fine texts (Sherman, 1990; White, 2006), in Section 7.4, and in the Exercises of this chapter. However, before proceeding to these, a short diversion into elementary lubrication theory is provided in the next section.

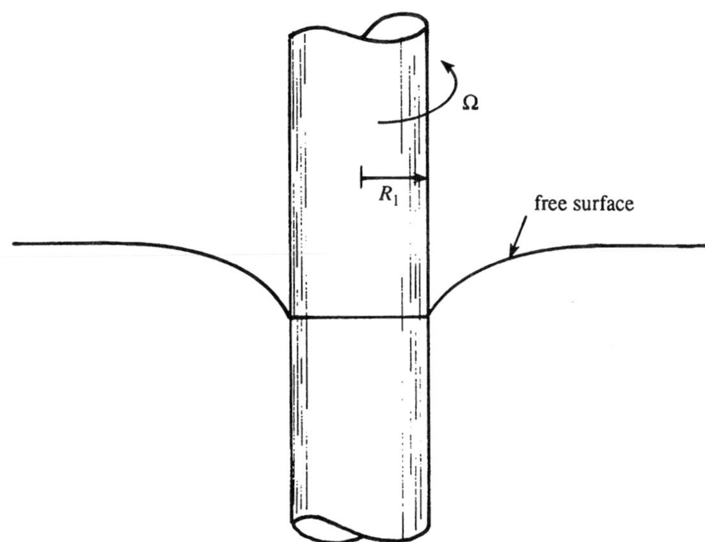

FIGURE 7.7 Rotation of a solid cylinder of radius R in an infinite body of viscous fluid. If gravity points downward along the cylinder's axis, the shape of a free surface pierced by the cylinder is also indicated. The flow field is viscous but irrotational.

Example 7.2

Using the circular Poiseuille flow results given above, determine the time necessary for a round tube of radius a and length L to convey a fluid volume equal to its internal volume when the fluid viscosity is μ and the pressure difference between the tube ends is Δp. Evaluate this conveyance time when the fluid is water at 20°C, $L = 1.0$ cm, $\Delta p = 1$ kPa, and $a = 1.0$ mm, 100 μm, and 10 μm.

Solution First, determine the flow time. The volume of the tube is $\pi a^2 L$ and the volume flow rate of fluid it conveys is $(\pi a^4/8\mu)(-dp/dz)$, so the time to convey one tube volume is:

$$\frac{Volume}{Q} = \pi a^2 L \left[\frac{\pi a^4}{8\mu} \left| \frac{dp}{dz} \right| \right]^{-1} = \pi a^2 L \frac{8\mu L}{\pi a^4 \Delta p} = 8 \frac{\mu}{\Delta p} \left(\frac{L}{a} \right)^2,$$

where $|dp/dz| = \Delta p/L$. Here the leading factor is the same for all three geometries: $8\mu/\Delta p = 8$ μs, but the tube's conveyance time depends on its aspect ratio, L/a, squared. Thus, for $L = 1.0$ cm, and $a = 1.0$ mm, 100 μm, and 10 μm, the conveyance times are: 0.8 ms, 80 ms, and 8 s, respectively. These numbers indicate that fluid motion may be significantly retarded in small confined passages. This is one of the challenges of micro- and nano-fluidics that may be overcome with electro-chemical hydrodynamic effects (see Kirby, 2010 or Conlisk, 2013).

7.3 Elementary lubrication theory

The exact viscous flow solutions for ideal geometries presented in the prior section indicate that a linear, quadratic, or simply varying velocity profile is a robust solution for flow within a confined space. This observation has been developed into the theory of lubrication, which provides approximate solutions to the viscous flow equations when the geometry is not ideal but at least one transverse flow dimension is small. The elementary features of lubrication theory are presented here because of its connection to the exact solutions described in Section 7.2, especially the Couette and Poiseuille flow solutions. Plus, the development of approximate equations in this section parallels that necessary for the boundary layer approximation (see Section 8.1).

The economic importance of lubrication with viscous fluids is hard to overestimate, and lubrication theory covers the mathematical formulation and analysis of such flows. The purpose of this section is to develop the most elementary equations of lubrication theory and illustrate some interesting phenomena that occur in viscous constant-density flows where the flow's boundaries or confining walls are close together, but not precisely parallel, and their motion is mildly unsteady. For simplicity, consider two spatial dimensions, x and y, where the primary flow direction, x, lies along a narrow passage with gap height $h(x, t)$ (see Fig. 7.8). The length L of this passage is presumed to be large compared to h so that

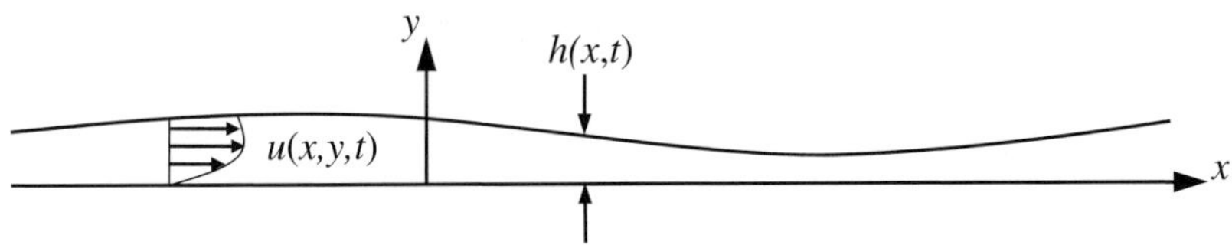

FIGURE 7.8 Nearly parallel flow of a viscous fluid having a film thickness of $h(x,t)$ above a flat stationary surface.

viscous and pressure forces are the primary terms in any fluid-momentum balance. If the passage is curved, this will not influence the analysis as long as the radius of curvature is much larger than the gap height h. The field equations are (4.10) and (7.1) for the horizontal u, and vertical v velocity components, and the pressure p in the fluid:

$$\frac{\partial u}{\partial x}+\frac{\partial v}{\partial y}=0, \tag{6.2}$$

$$\frac{\partial u}{\partial t}+u\frac{\partial u}{\partial x}+v\frac{\partial u}{\partial y}=-\frac{1}{\rho}\frac{\partial p}{\partial x}+\frac{\mu}{\rho}\left(\frac{\partial^2 u}{\partial x^2}+\frac{\partial^2 u}{\partial y^2}\right),\quad\text{and}$$

$$\frac{\partial v}{\partial t}+u\frac{\partial v}{\partial x}+v\frac{\partial v}{\partial y}=-\frac{1}{\rho}\frac{\partial p}{\partial y}+\frac{\mu}{\rho}\left(\frac{\partial^2 v}{\partial x^2}+\frac{\partial^2 v}{\partial y^2}\right). \tag{7.13a, 7.13b}$$

Here, the boundary conditions are $u=U_0(t)$ on $y=0$ and $u=U_h(t)$ on $y=h(x,t)$, and the pressure is presumed to be time dependent as well.

To determine which terms are important and which may be neglected when the passage is narrow, recast these equations in terms of dimensionless variables:

$$x^*=x/L,\quad y^*=y/h=y/\varepsilon L,\quad t^*=Ut/L,\quad u^*=u/U,\quad v^*=v/\varepsilon U,\quad\text{and}\quad p^*=p/P_a, \tag{7.14}$$

where U is a characteristic velocity of the flow, P_a is atmospheric pressure, and $\varepsilon=h/L$ is the passage's *fineness ratio* (the inverse of its aspect ratio). The goal of this effort is to find a set of approximate equations that are valid for common lubrication geometries where $\varepsilon\ll 1$. Because of the passage geometry, the magnitude of v is expected to be much less than the magnitude of u and gradients along the passage, $\partial/\partial x\sim 1/L$, are expected to be much smaller than gradients across it, $\partial/\partial y\sim 1/h$. These expectations have been incorporated into the dimensionless scaling (7.14). Combining (6.2), (7.13), and (7.14) leads to the following dimensionless equations:

$$\frac{\partial u^*}{\partial x^*}+\frac{\partial v^*}{\partial y^*}=0, \tag{7.15}$$

$$\varepsilon^2\mathrm{Re}_L\left(\frac{\partial u^*}{\partial t^*}+u^*\frac{\partial u^*}{\partial x^*}+v^*\frac{\partial u^*}{\partial y^*}\right)=-\frac{1}{\Lambda}\frac{\partial p^*}{\partial x^*}+\varepsilon^2\frac{\partial^2 u^*}{\partial x^{*2}}+\frac{\partial^2 u^*}{\partial y^{*2}},\quad\text{and} \tag{7.16a}$$

$$\varepsilon^4\mathrm{Re}_L\left(\frac{\partial v^*}{\partial t^*}+u^*\frac{\partial v^*}{\partial x^*}+v^*\frac{\partial v^*}{\partial y^*}\right)=-\frac{1}{\Lambda}\frac{\partial p^*}{\partial y^*}+\varepsilon^4\frac{\partial^2 v^*}{\partial x^{*2}}+\varepsilon^{*2}\frac{\partial^2 v^*}{\partial y^2}, \tag{7.16b}$$

where $\mathrm{Re}_L=\rho UL/\mu$, and $\Lambda=\mu UL/P_a h^2$ is the ratio of the viscous and pressure forces on a fluid element; it is sometimes called the *bearing number*. All the dimensionless derivative terms should be of order unity when the scaling (7.14) is correct. Thus, possible simplifying approximations are based on the size of the dimensionless coefficients of the various terms. The scaled two-dimensional continuity equation (7.15) does not contain any dimensionless coefficients so mass must be conserved without approximation. The two scaled momentum equations (7.16) contain ε, Re_L, and Λ. For the present purposes, Λ must be considered to be near unity, Re_L must be finite, and $\varepsilon\ll 1$. When $\varepsilon^2\mathrm{Re}_L\ll 1$, the left side and the middle term on the right side of (7.16a) may be ignored. In (7.16b) the pressure derivative is the only term not multiplied by ε. Therefore, the momentum equations can be approximately simplified to:

$$0\cong-\frac{\partial p}{\partial x}+\mu\frac{\partial^2 u}{\partial y^2}\quad\text{and}\quad 0\cong-\frac{\partial p}{\partial y}, \tag{7.17a, 7.17b}$$

when $\varepsilon^2\mathrm{Re}_L\to 0$. As a numerical example of this approximation, $\varepsilon^2\mathrm{Re}_L=0.001$ for room temperature flow of common 30-weight oil with $v\approx 4\times 10^{-4}\ \mathrm{m^2/s}$ within a 0.1 mm gap between two 25-cm-long surfaces moving with a differential speed of 10 m/s. When combined with a statement of conservation of mass, Eqs. (7.17) are the simplest form of the lubrication approximation (the *zeroth*-order approximation), and these equations are readily extended to two-dimensional gap-thickness variations (see Exercise 7.25). Interestingly, the approximations leading to (7.17) eliminated both the unsteady and the advective fluid acceleration terms from (7.16) This occurs because the flow's time scale was assumed to be L/U. When there is an externally imposed time scale, τ (the period of a mechanical oscillation for example), the additional requirement $\rho h^2/\mu\tau \ll 1$ is needed for the validity of (7.17) (see Exercise 7.18). Thus, time is typically still an independent variable in lubrication-flow analysis even though it does not explicitly appear in either Eq. (7.17a) or (7.17b).

A generic solution to (7.17) is readily produced by repeating the steps leading to (7.5). Eq. (7.17b) implies that p is not a function of y, so (7.17a) can be integrated twice to produce:

$$u(x, y, t) \cong \frac{1}{\mu}\frac{\partial p(x,t)}{\partial x}\frac{y^2}{2} + Ay + B, \tag{7.18}$$

where A and B might be functions of x and t but not y. Applying the boundary conditions mentioned earlier allows A and B to be evaluated, and the fluid velocity within the gap is found to be:

$$u(x, y, t) \cong -\frac{h^2(x,t)}{2\mu}\frac{\partial p(x,t)}{\partial x}\frac{y}{h(x,t)}\left(1 - \frac{y}{h(x,t)}\right) + (U_h(t) - U_0(t))\frac{y}{h(x,t)} + U_0(t). \tag{7.19}$$

The basic result here is that balancing viscous and pressure forces leads to a velocity profile that is parabolic in the cross-stream direction. While (7.19) represents a significant simplification of the two momentum equations (7.13), it is not a complete solution because the pressure $p(x,t)$ within the gap has not yet been determined. The complete solution to an elementary lubrication flow problem is typically obtained by combining (7.19), or an appropriate equivalent, with a differential or integral form of (4.10) or (6.2), and pressure boundary conditions. Such solutions are illustrated in the following examples.

Example 7.3

A sloped bearing pad of width B into the page moves horizontally at a steady speed U on a thin layer of oil with density ρ and viscosity μ. The gap between the bearing pad and a stationary hard, flat surface located at $y = 0$ is $h(x) = h_o(1 + \alpha x/L)$ where $\alpha \ll 1$. If p_e is the exterior pressure and $p(x)$ is the pressure in the oil under the bearing pad, determine the load W (per unit width into the page) that the bearing can support.

Solution The solution plan is to conserve mass exactly using (4.5), a control volume (CV) that is attached to the bearing pad, and the generic velocity profile (7.19). Then, pressure boundary conditions at the ends of the bearing pad should allow the pressure distribution under the pad to be found. Finally, W can be determined by integrating this pressure distribution.

Use the fixed-shape, but moving CV shown in Fig. 7.9 that lies between x_1 and x_2 at the moment of interest. The mass of fluid in the CV is constant so the unsteady term in (4.5) is zero, and the control surface velocity is $\mathbf{b} = U\mathbf{e}_x$. Denote the fluid velocity as $\mathbf{u} = u(x, y)\mathbf{e}_x$, and recognize $\mathbf{n}dA = -\mathbf{e}_x B dy$ on the vertical CV surface at x_1 and $\mathbf{n}dA = +\mathbf{e}_x B dy$ on the vertical CV surface at x_2. Thus, (4.5) simplifies to:

$$\rho B\left[-\int_0^{h(x_1)}(u(x_1, y) - U)dy + \int_0^{h(x_2)}(u(x_2, y) - U)dy\right] = 0.$$

Dividing this equation by $\rho B(x_2 - x_1)$ and taking the limit as $(x_2 - x_1) \to 0$ produces:

$$\frac{d}{dx}\left[\int_0^{h(x)}(u(x, y) - U)dy\right] = 0, \quad \text{or} \quad \int_0^{h(x)}(u(x, y) - U)dy = C_1,$$

where C_1 is a constant. For the flow geometry and situation in Fig. 7.9, (7.19) simplifies to:

$$u(x, y, t) \cong -\frac{h^2(x)}{2\mu}\frac{dp(x)}{dx}\frac{y}{h(x)}\left(1 - \frac{y}{h(x)}\right) + U\frac{y}{h(x)},$$

which can be substituted into the conservation of mass result and integrated to determine C_1 in terms of dp/dx and $h(x)$:

$$C_1 = -\frac{h^3(x)}{12\mu}\frac{dp(x)}{dx} - \frac{Uh(x)}{2}, \quad \text{or} \quad \frac{dp(x)}{dx} = -\frac{12\mu}{h^3(x)}C_1 - \frac{6\mu U}{h^2(x)}.$$

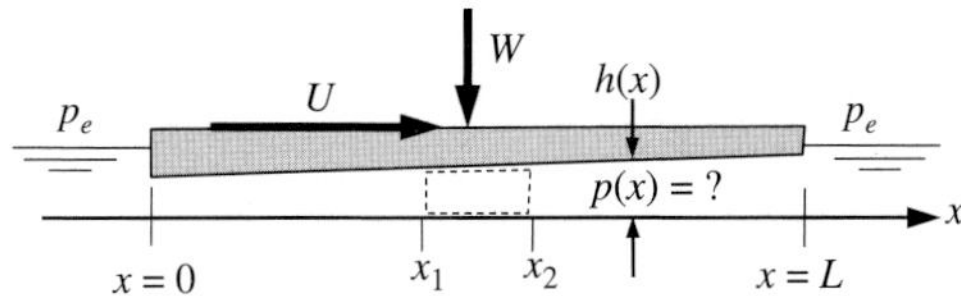

FIGURE 7.9 Schematic of a bearing pad with load W moving above a stationary flat surface coated with viscous oil. The gap below the pad has a mild slope and the pressure ahead, behind, and on top of the bearing pad is p_e.

The second equation is just an algebraic rearrangement of the first, and is a simple first-order differential equation for the pressure that can be integrated using the known $h(x)$ from the problem statement:

$$p(x) = -\frac{12\mu C_1}{h_o^3}\int \frac{dx}{(1+\alpha x/L)^3} - \frac{6\mu U}{h_o^2}\int \frac{dx}{(1+\alpha x/L)^2} + C_2 = \frac{6\mu L}{h_o^2\alpha}\left[\frac{C_1/h_o}{(1+\alpha x/L)^2} + \frac{U}{1+\alpha x/L}\right] + C_2.$$

Using the two ends of this extended equality and the pressure conditions $p(x=0) = p(x=L) = p_e$ produces two algebraic equations that can be solved simultaneously for the constants C_1 and C_2:

$$C_1 = -\left(\frac{1+\alpha}{2+\alpha}\right)Uh_o, \quad \text{and} \quad C_2 = p_e - \frac{6\mu LU}{h_o^2\alpha}\left(\frac{1}{2+\alpha}\right).$$

Thus, after some algebra the following pressure distribution is found:

$$p(x) - p_e = \frac{6\mu LU}{h_o^2}\left[\frac{\alpha(x/L)(1-x/L)}{(2+\alpha)(1+\alpha x/L)}\right].$$

However, this distribution may contain some superfluous dependence on α and x, because no approximations have been made regarding the size of α while (7.19) is only valid when $\alpha \ll 1$. Thus, keeping only the linear term in α produces:

$$p(x) - p_e \cong \frac{3\alpha\mu LU}{h_o^2}\left(\frac{x}{L}\right)\left(1 - \frac{x}{L}\right) \quad \text{for} \quad \alpha \ll 1.$$

The bearing load per unit depth into the page is:

$$W = \int_0^L (p(x) - p_e)dx = \frac{3\alpha\mu LU}{h_o^2}\int_0^L \left(\frac{x}{L}\right)\left(1 - \frac{x}{L}\right)dx = \frac{\alpha\mu L^2 U}{2h_o^2}.$$

This result shows that larger loads may be carried when the bearing slope, the fluid viscosity, the bearing size, and/or the bearing speed are larger, or the oil passage is smaller. Thus, the lubrication action of this bearing pad as described in this example is stable to load perturbations when the other parameters are held constant; an increase in load will lead to a smaller h_o where the bearing's load-carrying capacity is higher. However, the load-carrying capacity of this bearing goes to zero when α, μ, L, or U go to zero, and this bearing design fails (i.e., W becomes negative so the pad and surface are drawn into contact) when either α or U are negative. Thus, the bearing only works when it moves in the proper direction. A more detailed analysis of this bearing flow is provided in Sherman (1990).

Example 7.4 Viscous Flow Between Parallel Plates (Hele-Shaw, 1898)

A viscous fluid flows with velocity $\mathbf{u} = (u, v, w)$ in a narrow gap between stationary parallel plates lying at $z = 0$ and $z = h$ as shown in Fig. 7.10. Non-zero x- and y-directed pressure gradients are maintained at the plates' edges, and obstacles or objects of various sizes may be placed between the plates. Using the continuity equation (4.10) and the two horizontal (x, y) and one vertical (z) momentum equations (deduced in Exercise 7.25):

$$0 \cong -\frac{\partial p}{\partial x} + \mu\frac{\partial^2 u}{\partial z^2}, \quad 0 \cong -\frac{\partial p}{\partial y} + \mu\frac{\partial^2 v}{\partial z^2}, \quad \text{and } 0 \cong \frac{\partial p}{\partial z}$$

shows that the two in-plane velocity components parallel to the plates, u and v, can be determined from the equations for two-dimensional potential flow:

$$u = \frac{\partial \phi}{\partial x} \quad \text{and} \quad v = \frac{\partial \phi}{\partial y} \quad \text{with} \quad \frac{\partial^2 \phi}{\partial x^2} + \frac{\partial^2 \phi}{\partial y^2} = 0, \qquad (6.10, 6.12)$$

for an appropriate choice of ϕ.

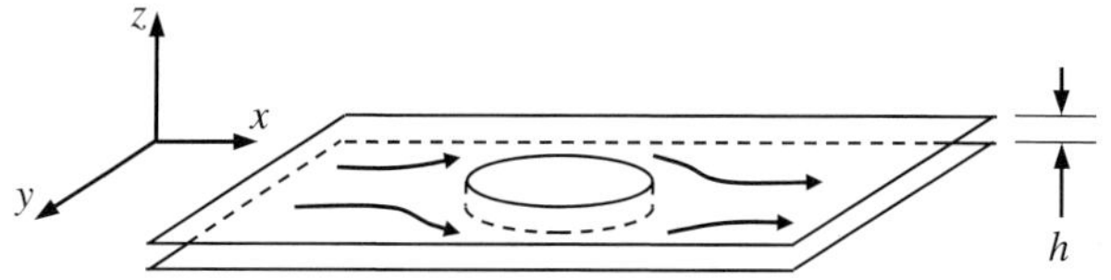

FIGURE 7.10 Pressure-driven viscous flow between parallel plates that trap an obstacle. The gap height h is small compared to the extent of the plates and the extent of the obstacle, shown here as a round disk.

Solution The solution plan is to use the two horizontal momentum equations given above to determine the functional forms of u and v. Then ϕ can be determined via integration of (6.10). Combining these results into (4.10) should produce (6.12), the two-dimensional Laplace equation for ϕ. Integrating the two horizontal momentum equations twice in the z-direction produces:

$$u \cong \frac{1}{\mu}\frac{\partial p}{\partial x}\frac{z^2}{2} + Az + B, \quad \text{and} \quad v \cong \frac{1}{\mu}\frac{\partial p}{\partial y}\frac{z^2}{2} + Cz + D,$$

where A, B, C, and D are constants that can be determined from the boundary conditions on $z = 0$, $u = v = 0$ which produces $B = D = 0$, and on $z = h$, $u = v = 0$, which produces $A = -(h/2\mu)(\partial p/\partial x)$ and $C = -(h/2\mu)(\partial p/\partial y)$. Thus, the two in-plane velocity components are:

$$u \cong -\frac{1}{2\mu}\frac{\partial p}{\partial x}z(h-z) = \frac{\partial \phi}{\partial x} \quad \text{and} \quad v \cong -\frac{1}{2\mu}\frac{\partial p}{\partial y}z(h-z) = \frac{\partial \phi}{\partial y}.$$

Integrating the second equality in each case produces:

$$\phi = -\frac{z(h-z)}{2\mu}p + E(y) \quad \text{and} \quad \phi = -\frac{z(h-z)}{2\mu}p + F(x).$$

These equations are consistent when $E = F = const.$, and this constant can be set to zero without loss of generality because it does not influence u and v, which are determined from derivatives of ϕ. Therefore, the velocity field requires a potential ϕ of the form:

$$\phi = -\frac{z(h-z)}{2\mu}p.$$

To determine the equation satisfied by p or ϕ, place the results for u and v into the continuity equation (4.10) and integrate in the z-direction from $z = 0$ to h to find:

$$\int_0^h \left(\frac{\partial u}{\partial x} + \frac{\partial v}{\partial y}\right)dz = -\int_0^h \frac{\partial w}{\partial z}dz = -(w)_{z=0}^{z=h} = 0 \to -\frac{1}{2\mu}\left(\frac{\partial^2 p}{\partial x^2} + \frac{\partial^2 p}{\partial y^2}\right)\int_0^h z(h-z)dz = 0,$$

where the no-through-flow boundary condition ensures $w = 0$ on $z = 0$ and h. The vertical momentum equation given above requires p to be independent of z, that is, $p = p(x, y, t)$, so p may be taken outside the z integration. The integral of $z(h - z)$ from $z = 0$ to h is not zero, so it and $-1/2\mu$ can be divided out of the last equation to achieve:

$$\frac{\partial^2 p}{\partial x^2} + \frac{\partial^2 p}{\partial y^2} = 0, \quad \text{or} \quad \frac{\partial^2 \phi}{\partial x^2} + \frac{\partial^2 \phi}{\partial y^2} = 0,$$

where the final equation follows from the form of ϕ determined from the velocity field.

This is a rather unusual and unexpected result because it requires viscous flow between closely spaced parallel plates to produce the same potential-line and streamline patterns as two-dimensional ideal flow. Interestingly, this suggestion is correct, except in thin layers having a thickness of order h near the surface of obstacles where the no-slip boundary condition on the obstacle prevents the tangential-flow slip that occurs in ideal flow. (Hele-Shaw flow near the surface of an obstacle is considered in Exercise 7.51). Thus, two-dimensional, ideal-flow streamlines past an object or obstacle may be visualized by injecting dye into pressure-driven viscous flow between closely spaced glass plates that trap a cross-sectional slice of the object or obstacle. Hele-Shaw flow has practical applications, too. Much of the manufacturing design analysis done to create molds and tooling for plastic-forming operations is based on Hele-Shaw flow.

The basic balance of pressure and viscous stresses underlying lubrication theory can be extended to gravity-driven viscous flows by appropriately revising the meaning of the pressure gradient and evaluating the constants A and B in (7.18) for different boundary conditions. Such an extension is illustrated in the next example in two dimensions for gravity-driven flow of magma, paint, or viscous oil over a horizontal surface. Gravity re-enters the formulation here because there is a large density change across the free surface of the viscous fluid (see Section 4.9 "Neglect of gravity in constant density flows").

Example 7.5

A two-dimensional bead of a viscous fluid with density ρ and viscosity μ spreads slowly on a smooth horizontal surface under the action of gravity. Ignoring surface tension and fluid acceleration, determine a differential equation for the thickness $h(x,t)$ of the spreading bead as a function of time.

Solution The solution plan is to conserve mass exactly using (4.5), a stationary control volume (CV) of thickness dx, height h, and unit depth into the page (see Fig. 7.11), and the generic velocity profile (7.18) when the pressure gradient is recast in terms of the thickness gradient $\partial h/\partial x$. The constants A and B in (7.18) can be determined from the no-slip condition at $y=0$, and a stress-free condition on $y=h$. When this refined version of (7.18) is put into the conservation of mass statement, the result is the differential equation that is sought.

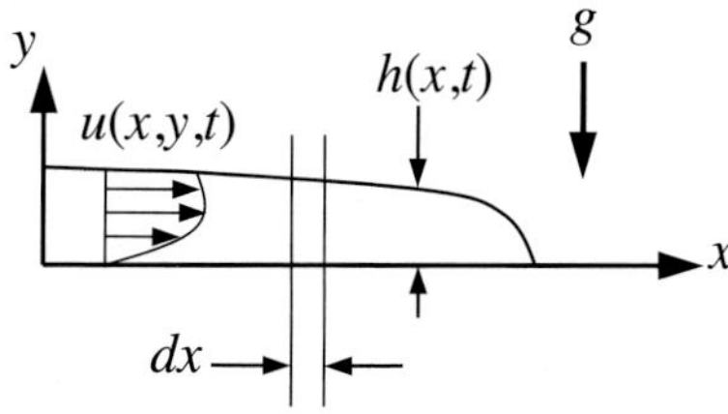

FIGURE 7.11 Gravity-driven spreading of a two-dimensional drop or bead on a flat, stationary surface. The fluid is not confined from above. Hydrostatic pressure forces cause the fluid to move but it is impeded by the viscous shear stress at $y=0$. The flow is assumed to be symmetric about $x=0$ so only half of it is shown.

By conserving mass between the two vertical lines in Fig. 7.11, (4.5) becomes:

$$\rho\frac{\partial h}{\partial t}dx - \int_0^{h(x,t)} \rho u(x,y,t)dy + \int_0^{h(x+dx,t)} \rho u(x+dx,y,t)dy = 0.$$

When rearranged and the limit $dx \to 0$ is taken, this becomes:

$$\frac{\partial h}{\partial t} + \frac{\partial}{\partial x}\left(\int_0^{h(x,t)} u(x,y,t)dy\right) = 0.$$

At any location within the spreading bead, the pressure p is hydrostatic: $p(x,y,t) = \rho g(h(x,t)-y)$ when fluid acceleration is ignored. Thus, the horizontal pressure gradient in the viscous fluid is:

$$\frac{\partial p}{\partial x} = \frac{\partial}{\partial x}[\rho g(h(x,t)-y)] = \rho g\frac{\partial h}{\partial x},$$

which is independent of y, so (7.18) becomes:

$$u(x,y,t) \cong \frac{\rho g}{2\mu}\frac{\partial h(x,t)}{\partial x}y^2 + Ay + B.$$

The no-slip condition at $y=0$ implies that $B=0$, and the no-stress condition at $y=h$ implies:

$$0 = \mu\left(\frac{\partial u}{\partial y}\right)_{y=h(x,t)} = \rho g\frac{\partial h(x,t)}{\partial x}h(x,t) + \mu A, \quad \text{so} \quad A = -\frac{\rho g}{\mu}h\frac{\partial h}{\partial x}.$$

So, the velocity profile within the bead is:

$$u \cong -\frac{\rho g}{2\mu}\frac{\partial h}{\partial x}y(2h-y),$$

and its integral is:

$$\int_0^h u(x,y,t)dy \cong -\frac{\rho g}{2\mu}\frac{\partial h}{\partial x}\int_0^h y(2h-y)dy = -\frac{\rho g}{3\mu}h^3\frac{\partial h}{\partial x}.$$

When this result is combined with the integrated conservation of mass statement, the final equation is:

$$\frac{\partial h}{\partial t}=\frac{\rho g}{3\mu}\frac{\partial}{\partial x}\left(h^3\frac{\partial h}{\partial x}\right).$$

This is a single nonlinear partial differential equation for $h(x,t)$ that in principle can be solved if a bead's initial thickness, $h(x,0)$, is known. Although this completes the effort for this example, a *similarity solution* to this equation does exist.

7.4 Similarity solutions for unsteady incompressible viscous flow

Similarity solutions to partial differential equations are possible when a variable transformation exists that allows the partial differential equation to be rewritten as an ordinary differential equation. Several such solutions for the Navier-Stokes equations are presented in this section.

So far, steady flows with parallel, or nearly parallel, streamlines have been considered. In this situation, the nonlinear advective acceleration is zero, or small, and the stream-wise velocity reduces to a function of one spatial coordinate, and time. When a viscous flow with parallel or nearly parallel streamlines is impulsively started from rest, the flow depends on the spatial coordinate(s) and time. For such unsteady flows, exact solutions still exist because the nonlinear advective acceleration drops out again (see Exercise 7.48). In this section, several simple and physically revealing unsteady flow problems are presented and solved. The first is the flow due to impulsive motion of a flat plate parallel to itself, commonly known as *Stokes' first problem*. (The flow is sometimes unfairly associated with the name of Rayleigh, who used Stokes' solution to predict the thickness of a developing boundary layer on a semi-infinite plate.)

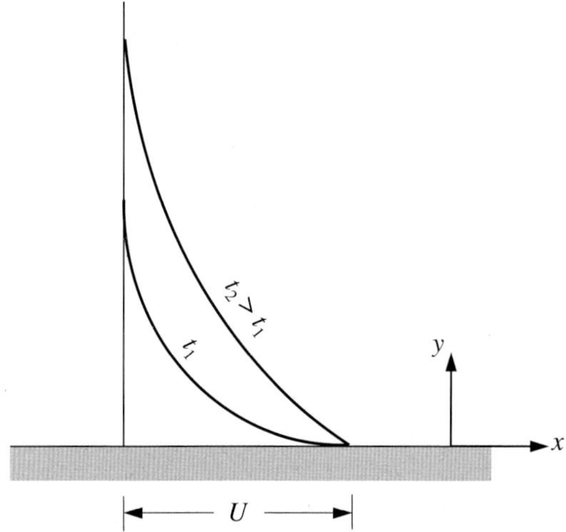

FIGURE 7.12 Laminar flow profiles resulting from a flat plate that starts moving parallel to itself at speed U at $t=0$. Before $t=0$, the entire fluid half-space ($y>0$) was quiescent. As time progresses, more and more of the viscous fluid above the plate is drawn into motion. Thus, the flow profile with greater vertical extent corresponds to the later time.

A similarity solution is one of several ways to solve Stokes' first problem. The geometry of this problem is shown in Fig. 7.12. An infinite flat plate lies along $y=0$, surrounded by an initially quiescent fluid (with constant ρ and μ) for $y>0$. The plate is impulsively given a velocity U at $t=0$ and constant pressure is maintained at $x=\pm\infty$. At first, only the fluid near the plate will be drawn into motion, but as time progresses the thickness of this moving region will increase. Since the resulting flow at any time is invariant in the x direction ($\partial/\partial x=0$), the continuity equation $\partial u/\partial x+\partial v/\partial y=0$ requires $\partial v/\partial y=0$. Thus, it follows that $v=0$ everywhere since it is zero at $y=0$. Therefore, the simplified horizontal and vertical momentum equations are:

$$\rho\frac{\partial u}{\partial t}=-\frac{\partial p}{\partial x}+\mu\frac{\partial^2 u}{\partial y^2},\quad\text{and}\quad 0=-\frac{\partial p}{\partial y}.$$

Just before $t=0$, all the fluid is at rest so $p=$ constant. For $t>0$, the vertical momentum equation only allows the fluid pressure to depend on x and t. However, at any finite time, there will be a vertical distance from the plate where the fluid velocity is still zero, and, at this vertical distance from the plate, $\partial p/\partial x$ is zero. However, if $\partial p/\partial x = 0$ far from the plate, then $\partial p/\partial x = 0$ on the plate because $\partial p/\partial y = 0$. Thus, the horizontal momentum equation reduces to:

$$\frac{\partial u}{\partial t} = \nu \frac{\partial^2 u}{\partial y^2}, \tag{7.20}$$

subject to the boundary and initial conditions:

$$u(y, t=0) = 0, \quad u(y=0, t) = \left\{ \begin{array}{l} 0 \text{ for } t < 0 \\ U \text{ for } t \geq 0 \end{array} \right\}, \quad \text{and} \quad u(y \to \infty, t) = 0. \qquad (7.21, 7.22, 7.23)$$

The problem is well posed because (7.22) and (7.23) are conditions at two values of y, and (7.21) is a condition at one value of t; this is consistent with (7.20), which involves a first derivative in t and a second derivative in y.

The partial differential equation (7.20) can be transformed into an ordinary differential equation by switching to a similarity variable. The reason for this is the absence of enough other parameters in this problem to render y and t dimensionless without combining them. Based on dimensional analysis (see Section 1.11), the functional form of the solution to (7.20) can be written:

$$u/U = f\left(y/\sqrt{\nu t}, y/Ut\right), \tag{7.24}$$

where f is an undetermined function. However, (7.20) is a linear equation, so u must be proportional to U. This means that the final dimensionless group in (7.24) must be dropped, leaving:

$$u/U = F\left(y/\sqrt{\nu t}\right) \equiv F(\eta), \tag{7.25}$$

where F is an undetermined function, but this time it is a function of only one dimensionless group and this dimensionless group $\eta = y/(\nu t)^{1/2}$ combines both independent variables. This reduces the dimensionality of the solution space from two to one, an enormous simplification.

Eq. (7.25) is the similarity form for the fluid velocity in Stokes' first problem. The similarity variable η could have been defined differently, such as $\nu t/y^2$, but different choices for η merely change F, not the final answer. The chosen η allows F to be interpreted as a velocity profile function with y appearing to the first power in the numerator of η. At any fixed $t > 0$, y and η are proportional.

Using (7.25) to form the derivatives in (7.20), leads to:

$$\frac{\partial u}{\partial t} = U\frac{dF}{d\eta}\frac{\partial \eta}{\partial t} = -\frac{Uy}{2\sqrt{\nu t^3}}\frac{dF}{d\eta} = -\frac{U\eta}{2t}\frac{dF}{d\eta} \quad \text{and}$$

$$U\frac{\partial^2 F}{\partial y^2} = U\frac{\partial}{\partial y}\left(\frac{dF}{d\eta}\frac{\partial \eta}{\partial y}\right) = U\frac{\partial}{\partial y}\left(\frac{1}{\sqrt{\nu t}}\frac{dF}{d\eta}\right) = \frac{U}{\sqrt{\nu t}}\frac{d}{d\eta}\left(\frac{dF}{d\eta}\right)\frac{\partial \eta}{\partial y} = \frac{U}{\nu t}\frac{d}{d\eta}\left(\frac{dF}{d\eta}\right)$$

and these can be combined to provide the equivalent of (7.20) in similarity form:

$$-\frac{\eta}{2}\frac{dF}{d\eta} = \frac{d}{d\eta}\left(\frac{dF}{d\eta}\right) \tag{7.26}$$

The initial and boundary conditions (7.21) through (7.23) for F reduce to:

$$F(\eta = 0) = 1, \quad \text{and} \quad F(\eta \to \infty) = 0, \qquad (7.27, 7.28)$$

because (7.21) and (7.23) reduce to the same condition in terms of η. This reduction is expected because (7.20) is a partial differential equation and needs two conditions in y and one condition in t to be solved, while (7.26) is a second-order ordinary differential equation and needs only two boundary conditions to be solved. Eq. (7.26) is readily separated:

$$-\frac{\eta}{2}d\eta = \frac{d(dF/d\eta)}{dF/d\eta},$$

and integrated to reach:

$$-\frac{\eta^2}{4} = \ln(dF/d\eta) + \text{const.}, \quad \text{or} \quad dF/d\eta = A\exp(-\eta^2/4),$$

where A is a constant. Integrating again leads to:

$$F(\eta) = A\int_0^{\eta} \exp(-\xi^2/4)d\xi + B,$$

where ξ is just an integration variable and B is another constant. The condition (7.27) sets $B = 1$, while condition (7.28) gives:

$$0 = A\int_0^{\infty} \exp(-\xi^2/4)d\xi + 1, \quad \text{or} \quad 0 = 2A\int_0^{\infty} \exp(-\zeta^2)d\zeta + 1, \text{ so } 0 = 2A\left(\sqrt{\pi}/2\right) + 1,$$

thus $A = -1/\sqrt{\pi}$, where the tabulated integral $\int_{-\infty}^{+\infty} \exp(-\zeta^2)d\zeta = \sqrt{\pi}$ has been used. The final solution for u then becomes:

$$\frac{u(y,t)}{U} = 1 - \operatorname{erf}\left(\frac{y}{2\sqrt{\nu t}}\right), \quad \text{where} \quad \operatorname{erf}(\zeta) = \frac{2}{\sqrt{\pi}}\int_0^{\zeta} \exp(-\xi^2)d\xi, \tag{7.29}$$

is the error function and again ξ is just an integration variable. The error function is a standard tabulated function (see Abramowitz and Stegun, 1972). Eq. (7.29) is the solution to the problem and the form of (7.29) makes it is apparent that *the velocity profile at different times will collapse into a single curve of u/U vs.* η, as shown in Fig. 7.13.

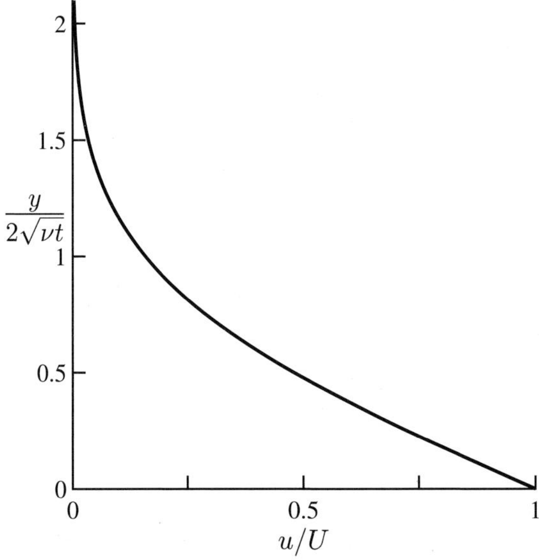

FIGURE 7.13 Similarity solution of laminar flow due to an impulsively started flat plate. Using these scaled coordinates, all flow profiles like those shown on Fig. 7.12 will collapse to the same curve. The factor of two in the scaling of the vertical axis follows from (7.29).

The nature of the variation of u/U with y for various values of t is sketched in Fig. 7.12, and this solution has a diffusive nature. At $t = 0$, a vortex sheet (that is, a velocity discontinuity) is created at the plate surface. The initial vorticity is in the form of a delta function, which is infinite at the plate surface and zero elsewhere. The integral $\int_0^{\infty} \omega dy = \int_0^{\infty} (-\partial u/\partial y)dy = -U$ is independent of time, so *no new vorticity is generated after the initial time*. The flow given by (7.29) occurs as the initial vorticity diffuses away from the wall. The situation is analogous to a heat conduction problem in a semi-infinite solid extending from $y = 0$ to $y = \infty$. Initially, the solid has a uniform temperature, and at $t = 0$ the face at $y = 0$ is suddenly brought to a different temperature. The temperature distribution for this heat conduction problem is given by an equation similar to (7.29).

We may arbitrarily define the thickness of the diffusive layer as the distance at which u falls to 1% of U. From Fig. 7.13, $u/U = 0.01$ corresponds to $y/(\nu t)^{1/2} = 3.64$. Therefore, in time t the diffusive effects propagate to a distance of:

$$\delta_{99} \sim 3.64\sqrt{\nu t}, \tag{7.30}$$

which defines the 99% thickness of the layer of moving fluid and this layer's thickness increases as $t^{1/2}$. Obviously, the factor of 3.64 is somewhat arbitrary and can be changed by choosing a different ratio of u/U as the definition for the edge of the diffusive layer. However, 99% thicknesses are commonly considered in boundary layer theory (see Chapter 8).

Stokes' first problem illustrates an important class of fluid mechanical problems that have *similarity solutions*. Because of the absence of suitable scales to render the independent variables dimensionless, the only possibility was a combination of variables that resulted in a reduction in the number of independent variables required to describe the problem. In this case the reduction was from two (y, t) to one (η) so that the formulation reduced a partial differential equation in y and t to an ordinary differential equation in η.

The solution (7.29) for $u(y, t)$ is *self-similar* in the sense that at different times $t_1, t_2, t_3, \ldots$ the various velocity profiles $u(y, t_1), u(y, t_2), u(y, t_3), \ldots$ fall on a single curve if u is scaled by U and y is scaled by the diffusive thickness $(\nu t)^{1/2}$. Moreover, such a collapse will occur for different values of U and for fluids having different ν.

Similarity solutions arise in situations in which there are no imposed length or time scales provided by the initial or boundary conditions (or the field equation). A similarity solution would not be possible if, for example, the boundary conditions were changed after a certain time t_1 since this introduces a time scale into the problem (see Exercise 7.47). Likewise, if the flow in Stokes' first problem was bounded above by a second parallel plate, there could be no similarity solution because the distance to the second plate introduces a length scale into the problem.

Similarity solutions are often ideal for developing an understanding of flow phenomena, so they are sought wherever possible. A method for finding similarity solutions starts from a presumed form for the solution:

$$\gamma = At^{-n}F(\xi/\delta(t)) \equiv At^{-n}F(\eta) \quad \text{or} \quad \gamma = A\xi^{-n}F(\xi/\delta(t)) \equiv A\xi^{-n}F(\eta), \tag{7.31a,b}$$

where γ is the dependent field variable of interest, (a velocity component, for example), A is a constant (units $= [\gamma] \times [\text{time}]^n$ or $[\gamma] \times [\text{length}]^n$), ξ is the independent spatial coordinate, t is time, $\eta = \xi/\delta$ is the similarity variable, and $\delta(t)$ is a time-dependent length scale (not the Dirac delta-function). The factor of At^{-n} or $A\xi^{-n}$ that multiplies F in (7.31) is sometimes needed for similarity solutions that are infinite (or zero) at $t = 0$ or $\xi = 0$. Use of (7.31) is illustrated in the following examples.

Example 7.6

Use (7.31a) to find the similarity solution to Stokes' first problem.

Solution The solution plan is to populate (7.31a) with the appropriate variables, substitute it into the field equation (7.20), and then require that the coefficients all have the same time dependence. For Stokes' first problem $\gamma = u/U$, and the independent spatial variable is y. For this flow, the coefficient At^{-n} is not needed since $u/U = 1$ at $\eta = 0$ for all $t > 0$ and this can only happen when $A = 1$ and $n = 0$. Thus, the dimensional analysis result (7.25) may be replaced by (7.31a) with $A = 1$, $n = 0$, and $\xi = y$:

$$u/U = F(y/\delta(t)) \equiv F(\eta).$$

Partial derivatives in time and space produce:

$$\frac{\partial u}{\partial t} = U\frac{dF}{d\eta}\left(-\frac{y}{\delta^2}\right)\frac{d\delta}{dt}, \quad \text{and} \quad \frac{\partial^2 u}{\partial y^2} = U\frac{d^2F}{d\eta^2}\frac{1}{\delta^2}.$$

Reconstructing (7.20) with these replacements yields:

$$\frac{\partial u}{\partial t} = \nu\frac{\partial^2 u}{\partial y^2} \rightarrow U\frac{dF}{d\eta}\left(-\frac{y}{\delta^2}\right)\frac{d\delta}{dt} = \nu U\frac{d^2F}{d\eta^2}\frac{1}{\delta^2},$$

which can be rearranged to find:

$$-\left[\frac{1}{\delta}\frac{d\delta}{dt}\right]\eta\frac{dF}{d\eta} = \left[\frac{\nu}{\delta^2}\right]\frac{d^2F}{d\eta^2}.$$

For this equation to be in similarity form, the coefficients in [,]-brackets must both have the same time dependence so that division by this common time dependence will leave an ordinary differential equation for $F(\eta)$ and t will no longer appear. Thus, we require the two coefficients to be proportional:

$$\frac{1}{\delta}\frac{d\delta}{dt} = C_1\frac{\nu}{\delta^2},$$

where C_1 is the constant of proportionality. This is a simple differential equation for $\delta(t)$ that is readily rearranged and solved:

$$\delta\frac{d\delta}{dt} = C_1\nu \rightarrow \frac{\delta^2}{2} = C_1\nu t + C_2 \rightarrow \delta = \sqrt{2C_1\nu t},$$

where the condition $\delta(0) = 0$ has been used to determine that $C_2 = 0$. When $C_1 = 1/2$, the prior definition of η in (7.25) is recovered, and the solution for u proceeds as before (see (7.26) through (7.29)).

Example 7.7

At $t = 0$ an infinitely thin vortex sheet in a fluid with density ρ and viscosity μ coincides with the plane defined by $y = 0$, so that the fluid velocity is U for $y > 0$ and $-U$ for $y < 0$. The coordinate axes are aligned so that only the z-component of vorticity is non-zero. Determine the similarity solution for $\omega_z(y, t)$ for $t > 0$.

Solution The solution plan is the same as for Example 7.6, except here the coefficient At^{-n} must be included. In this circumstance, there will be only one component of the fluid velocity, $\mathbf{u} = u(y, t)\mathbf{e}_x$, so $\omega_z(y, t) = -\partial u/\partial y$. The independent coordinate y does not appear in the initial condition, so (7.31a) is the preferred choice. Its appropriate form is:

$$\omega_z(y, t) = At^{-n}F(y/\delta(t)) \equiv At^{-n}F(\eta),$$

and the field equation:

$$\frac{\partial \omega_z}{\partial t} = \nu \frac{\partial^2 \omega_z}{\partial y^2},$$

is obtained by applying $\partial/\partial y$ to (7.20). Here, the derivatives of the similarity solution are:

$$\frac{\partial \omega_z}{\partial t} = -nAt^{-n-1}F(\eta) + At^{-n}\frac{dF}{d\eta}\left(-\frac{y}{\delta^2}\right)\frac{d\delta}{dt}, \quad \text{and} \quad \frac{\partial^2 \omega_z}{\partial y^2} = At^{-n}\frac{d^2F}{d\eta^2}\frac{1}{\delta^2}.$$

Reassembling the field equation and canceling common factors produces:

$$-\left[\frac{n}{t}\right]F(\eta) - \left[\frac{1}{\delta}\frac{d\delta}{dt}\right]\eta\frac{dF}{d\eta} = \left[\frac{\nu}{\delta^2}\right]\frac{d^2F}{d\eta^2}.$$

From Example 7.6, we know that requiring the second and third coefficients in [,]-brackets to be proportional with a proportionality constant of 1/2 produces $\delta = (\nu t)^{1/2}$. With this choice for δ, each of the coefficients in [,]-brackets is proportional to $1/t$ so, the similarity equation becomes:

$$-nF(\eta) - \frac{1}{2}\eta\frac{dF}{d\eta} = \frac{d^2F}{d\eta^2}.$$

The boundary conditions are: 1) at any finite time the vorticity must go to zero infinitely far from the initial location of the vortex sheet, $F(\eta) \to 0$ for $\eta \to \infty$, and 2) the velocity difference across the diffusing vortex sheet is constant and equal to $2U$:

$$-\int_{-\infty}^{+\infty} \omega_z dy = \int_{-\infty}^{+\infty} \frac{\partial u}{\partial y} dy = [u(y, t)]_{-\infty}^{+\infty} = U - (-U) = 2U.$$

Substituting the similarity solution into this second requirement leads to:

$$-\int_{-\infty}^{+\infty} \omega_z dy = -\int_{-\infty}^{+\infty} At^{-n}F(\eta)dy = -At^{-n}\delta\int_{-\infty}^{+\infty} F(\eta)d(y/\delta) = -At^{-n}\delta\int_{-\infty}^{+\infty} F(\eta)d\eta = 2U.$$

The final integral is just a number so $t^{-n}\delta(t)$ must be constant, and this implies $n = 1/2$ so the similarity equation may be rewritten, and integrated:

$$-\frac{1}{2}\left(F(\eta) + \eta\frac{dF}{d\eta}\right) = -\frac{1}{2}\frac{d}{d\eta}(\eta F) = \frac{d}{d\eta}\left(\frac{dF}{d\eta}\right) \to \frac{dF}{d\eta} + \frac{1}{2}\eta F = C.$$

The first boundary condition implies that both F and $dF/d\eta \to 0$ when η is large enough. Therefore, assume that $\eta F \to 0$ when $\eta \to \infty$ so that the constant of integration C can be set to zero (this assumption can be checked once F is found). When $C = 0$, the last equation can be separated and integrated to find:

$$F(\eta) = D\exp(-\eta^2/4),$$

where D is a constant, and the assumed limit, $\eta F \to 0$ when $\eta \to \infty$, is verified so C is indeed zero. The velocity-difference constraint and the tabulated integral used to reach (7.29) allow the product AD to be evaluated. Thus, the similarity solutions for the vorticity $\omega_z = -\partial u/\partial y$ and velocity u are:

$$\omega_z(y,t) = -\frac{U}{\sqrt{\pi \nu t}}\exp\left\{-\frac{y^2}{4\nu t}\right\}, \quad \text{and} \quad u(y,t) = U\,\text{erf}\left\{\frac{y}{2\sqrt{\nu t}}\right\}.$$

Schematic plots of the vorticity and velocity distributions are shown in Fig. 7.14. If we define the width of the velocity transition layer as the distance between the points where $u = \pm 0.95U$, then the corresponding values of η are ± 2.77 and consequently the width of the transition layer is $5.54(\nu t)^{1/2}$.

The results of this example are closely related to Stokes' first problem, and to the laminar boundary layer flows discussed in the next chapter, for several reasons. First of all, this flow is essentially the same as that in Stokes' first problem. The velocity field in the upper half of Fig. 7.14 is identical to that in Fig. 7.13 after a Galilean transformation to a coordinate system moving at speed $+U$ followed by a sign change. In addition, the flow for $y > 0$ represents a *temporally developing* boundary layer that begins at $t = 0$. The velocity far from the surface is irrotational and uniform at speed U while the no-slip condition ($u = 0$) is satisfied at $y = 0$. Here, the wall shear stress, τ_w, and skin friction coefficient C_f are time dependent:

$$\tau_w = \mu\left(\frac{\partial u}{\partial y}\right)_{y=0} = \frac{\mu U}{\sqrt{\pi \nu t}}, \quad \text{or} \quad C_f = \frac{\tau_w}{\frac{1}{2}\rho U^2} = \frac{2}{\sqrt{\pi}}\sqrt{\frac{\nu}{U^2 t}}.$$

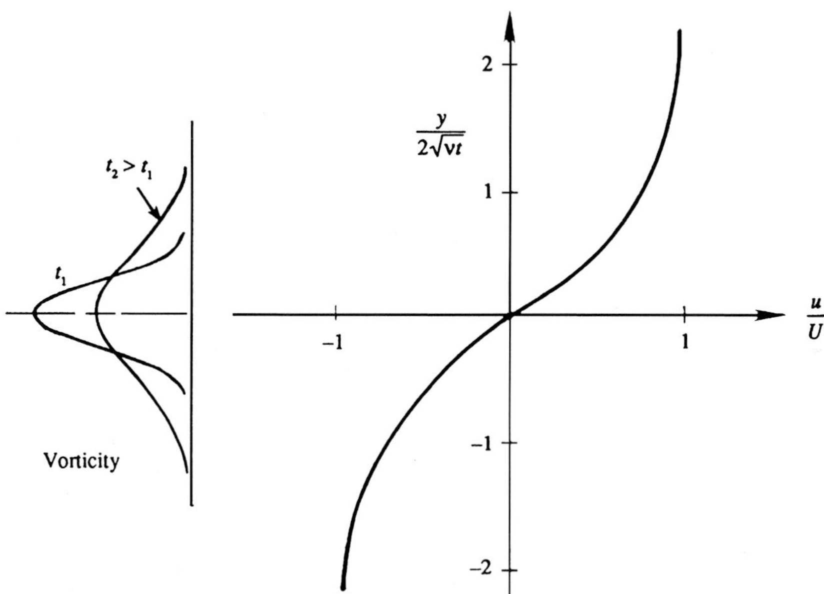

FIGURE 7.14 Viscous thickening of a vortex sheet. The left panel indicates the vorticity distribution at two times, while the right panel shows the velocity field solution in similarity coordinates. The upper half of this flow is equivalent to a temporally developing boundary layer.

When Ut is interpreted as a surrogate for the downstream distance, x, in a *spatially developing* boundary layer, the last square-root factor above becomes $(\nu/Ux)^{1/2} = \text{Re}_x^{-1/2}$, and this is the correct parametric dependence for C_f in a laminar boundary layer that develops on a smooth, flat surface below a steady uniform flow.

Example 7.8

A thin, rapidly spinning cylinder produces the two-dimensional flow field, $u_\theta = \Gamma/2\pi r$, of an ideal vortex of strength Γ located at $r = 0$. At $t = 0$, the cylinder stops spinning. Use (7.31) to determine $u_\theta(r,t)$ for $t > 0$.

Solution Follow the approach specified for Example 7.6 but this time use (7.31b) because r appears in the initial condition. Here u_θ is the dependent field variable and r is the independent spatial variable, so the appropriate form of (7.31b) is:

$$u_\theta(r,t) = Ar^{-n}F(r/\delta(t)) \equiv Ar^{-n}F(\eta).$$

The initial and boundary conditions are: $u_\theta(r, 0) = \Gamma/2\pi r = u_\theta(r \to \infty, t)$, $u_\theta(0, t) = 0$ for $t > 0$, which are simplified to $F(\eta \to \infty) = 1$ and $F(0) = 0$ when Ar^{-n} is set equal to $\Gamma/2\pi r$. In this case, the field equation for u_θ (see Appendix B) is:

$$\frac{\partial u_\theta}{\partial t} = \nu \frac{\partial}{\partial r}\left(\frac{1}{r}\frac{\partial}{\partial r}(ru_\theta)\right).$$

Inserting $u_\theta = (\Gamma/2\pi r)F(\eta)$ produces:

$$-\frac{\Gamma}{2\pi r}\left(\frac{1}{\delta}\frac{d\delta}{dt}\right)\eta\frac{dF}{d\eta} = \nu\frac{\Gamma}{2\pi}\left(-\frac{1}{\delta r^2}\frac{dF}{d\eta} + \frac{1}{\delta^2 r}\frac{d^2F}{d\eta^2}\right) \to -\left[\frac{r^2}{\nu\delta}\frac{d\delta}{dt}\right]\eta\frac{dF}{d\eta} = -\eta\frac{dF}{d\eta} + \eta^2\frac{d^2F}{d\eta^2}.$$

For a similarity solution, the coefficient in [,]-brackets must depend on η alone, not on r or t. Here, this coefficient reduces to $\eta^2/2$ when $\delta = (\nu t)^{1/2}$ (as in the prior examples). With this replacement, the similarity equation can be integrated twice:

$$\left(\frac{1}{\eta} - \frac{\eta}{2}\right)\frac{dF}{d\eta} = \frac{d}{d\eta}\left(\frac{dF}{d\eta}\right) \to \ln\eta - \frac{\eta^2}{4} + \text{const.} = \ln\left(\frac{dF}{d\eta}\right)$$
$$\to C\int \eta\exp\left\{-\eta^2/4\right\}d\eta + D = F(\eta).$$

The remaining integral is elementary, and the boundary conditions given above for F allow the constants C and D to be evaluated. The final result is $F(\eta) = 1 - \exp\{-\eta^2/4\}$, so the velocity distribution is:

$$u_\theta(r, t) = \frac{\Gamma}{2\pi r}\left[1 - \exp\left\{-\frac{r^2}{4\nu t}\right\}\right],$$

which is identical to the Gaussian vortex of (3.29) when $\sigma^2 = 4\nu t$. A sketch of the velocity distribution for various values of t is given in Fig. 7.15. Near the center, $r \ll (\nu t)^{1/2}$, the flow has the form of a rigid-body rotation, while in the outer region, $r \gg (\nu t)^{1/2}$, the motion has the form of an ideal vortex.

The foregoing presentation applies to the *decay* of a line vortex. The case where a line vortex is suddenly *introduced* into a fluid at rest leads to the velocity distribution:

$$u_\theta(r, t) = \frac{\Gamma}{2\pi r}\exp\left\{-\frac{r^2}{4\nu t}\right\}$$

(see Exercise 7.39). This situation is equivalent to the impulsive rotational start of an infinitely thin and quickly rotating cylinder located at $r = 0$.

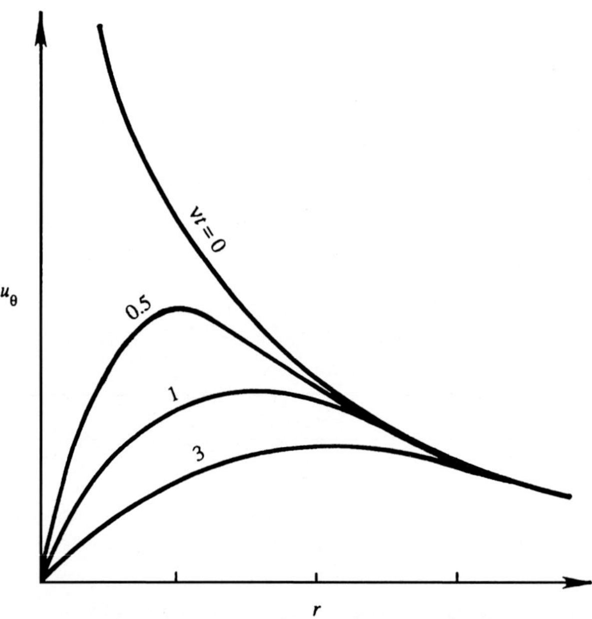

FIGURE 7.15 Viscous decay of a line vortex showing the tangential velocity u_θ at different times. The velocity field nearest to the axis of rotation changes the most quickly. At large radii, flow alterations occur more slowly.

After reviewing these examples, it should be clear that diffusive length scales in unsteady viscous flow are proportional to $(\nu t)^{1/2}$. The viscous bead-spreading example produces a length scale with a different power, but this is not a diffusion time scale. Instead it is an advection time scale that specifies how far fluid elements travel in the direction of the flow.

7.5 Flows with oscillations

The unsteady flows discussed in the preceding sections have similarity solutions because there were no imposed or specified length or time scales. The two flows discussed here are unsteady viscous flows that include an imposed time scale.

First, consider an infinite flat plate lying at $y = 0$ that executes sinusoidal oscillations parallel to itself. (This is sometimes called *Stokes' second problem.*) Here, only the steady periodic solution after the starting transients have died out is considered, thus there are no initial conditions to satisfy. The governing equation (7.20) is the same as that for Stokes' first problem. The boundary conditions are:

$$u(y=0,t) = U\cos(\omega t), \quad \text{and} \quad u(y\to\infty,t) = \text{bounded}, \tag{7.32, 7.33}$$

where ω is the oscillation frequency (rad/s). In the steady state, the flow variables must have a periodicity equal to the periodicity of the boundary motion. Consequently, a complex separable solution of the form:

$$u(y,t) = \text{Re}\left\{f(y)e^{i\omega t}\right\}, \tag{7.34}$$

is used here, and the specification of the real part is dropped until the final equation for u is reached. Substitution of (7.34) into (7.20) produces:

$$i\omega f = \nu\left(d^2f/dy^2\right), \tag{7.35}$$

which is an ordinary differential equation with constant coefficients. It has exponential solutions of the form: $f = \exp(ky)$ where $k = (i\omega/\nu)^{1/2} = \pm(i+1)(\omega/2\nu)^{1/2}$. Thus, the solution of (7.35) is:

$$f(y) = A\exp\left\{-(i+1)y\sqrt{\omega/2\nu}\right\} + B\exp\left\{+(i+1)y\sqrt{\omega/2\nu}\right\}. \tag{7.36}$$

The condition (7.33) requires that the solution must remain bounded as $y\to\infty$, so $B = 0$ and the complex solution only involves the first term in (7.36). The surface boundary condition (7.32) requires $A = U$. Thus, after taking the real part as in (7.34), the final velocity distribution for Stokes' second problem is:

$$u(y,t) = U\exp\left\{-y\sqrt{\frac{\omega}{2\nu}}\right\}\cos\left(\omega t - y\sqrt{\frac{\omega}{2\nu}}\right). \tag{7.37}$$

The cosine factor in (7.37) represents a dispersive wave traveling in the positive-y direction, while the exponential term represents amplitude decay with increasing y. The flow therefore resembles a highly damped transverse wave (Fig. 7.16). However, this is a diffusion problem and *not* a wave-propagation problem because there are no restoring forces involved here. The apparent propagation is merely a result of the oscillating boundary condition. For $y = 4(\nu/\omega)^{1/2}$, the amplitude of u is $U\exp\left\{-4/\sqrt{2}\right\} = 0.06U$, which means that the influence of the wall is confined within a distance of order $\delta\sim 4(\nu/\omega)^{1/2}$, which decreases with increasing frequency.

The solution (7.37) has several interesting features. First of all, it cannot be represented by a single curve in terms of dimensionless variables. A dimensional analysis of Stokes' second problem produces three dimensionless groups: u/U, ωt, and $y(\omega/\nu)^{1/2}$. Here the independent spatial variable y can be fully separated from the independent time variable t. Self-similar solutions exist only when the independent spatial and temporal variables must be combined in the absence of imposed time or length scales. However, the fundamental concept associated with viscous diffusion holds true, the spatial extent of the solution is parameterized by $(\nu/\omega)^{1/2}$, the square root of the product of the kinematic viscosity, and the imposed time scale $1/\omega$. In addition, (7.37) can be used to predict the weak absorption of sound at solid flat surfaces.

A second oscillating viscous flow with an exact solution occurs in a straight round tube with an oscillating pressure gradient (Womersley, 1955a,b). The problem specification and the flow geometry are the same as that in Section 7.2 for steady laminar flow in a round tube with two exceptions. First, the unsteady term, $\partial u_z/\partial t$, in the axial momentum equation is retained:

$$\frac{\partial u_z}{\partial t} = -\frac{1}{\rho}\frac{dp}{dz} + \frac{\nu}{R}\frac{d}{dR}\left(R\frac{du_z}{dR}\right), \tag{7.38}$$

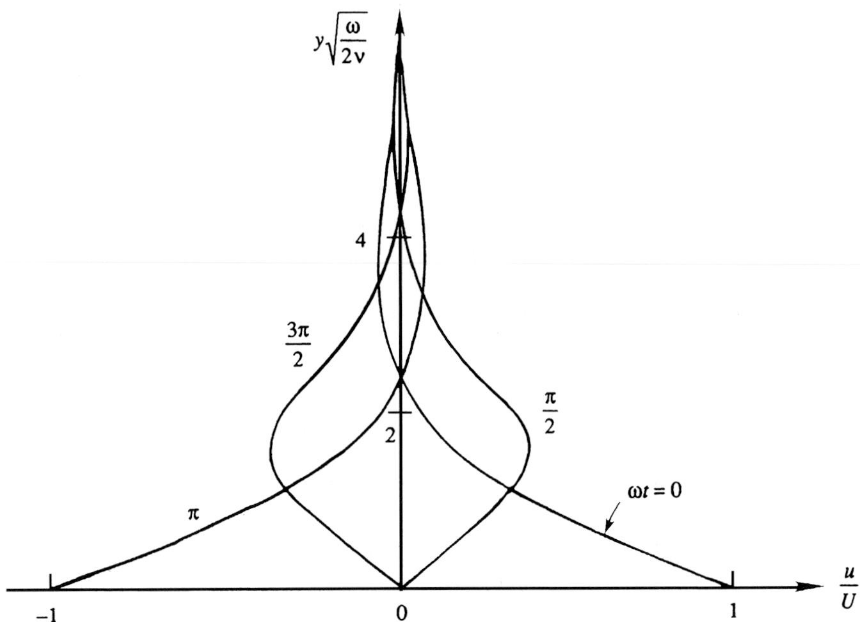

FIGURE 7.16 Velocity distribution in laminar flow near an oscillating plate. The distributions at $\omega t = 0$, $\pi/2$, π, and $3\pi/2$ are shown. The diffusive distance is of order $\delta \sim 4(\nu/\omega)^{1/2}$.

because $\mathbf{u} = (0, 0, u_z(R, t))$, and second, the pressure gradient oscillates at radian frequency ω:

$$dp/dz = \text{Re}\left\{(\Delta p/L)e^{i\omega t}\right\}, \tag{7.39}$$

where Δp is the pressure fluctuation amplitude between the ends of the pipe, and L is the pipe length (both are constants here). Substituting the equivalent of (7.34) for this geometry:

$$u_z(R, t) = \text{Re}\left\{f(R)e^{i\omega t}\right\}, \tag{7.40}$$

into (7.38) leads to:

$$\frac{d^2 f}{dR^2} + \frac{1}{R}\frac{df}{dR} - i\frac{\omega}{\nu}f = \frac{\Delta p}{\mu L}. \tag{7.41}$$

This is a form of the zeroth-order Bessel equation. The solutions for $f(R)$ are zeroth-order Bessel functions of complex argument with the radial coordinate R scaled by the diffusion distance $(\nu/\omega)^{1/2}$:

$$f(R) = AJ_o\left(\frac{i^{3/2}R}{\sqrt{\nu/\omega}}\right) + BY_o\left(\frac{i^{3/2}R}{\sqrt{\nu/\omega}}\right) + i\frac{\Delta p}{\omega\rho L},$$

where A and B are constants, and J_o and Y_o are zeroth-order Bessel functions of the first kind. Here $f(0)$ must be finite, so B must be zero since Y_o grows without bound as its argument goes to zero. The remaining constant A can be determined from the no slip condition at $r = a$:

$$f(R = a) = 0 = AJ_o\left(\frac{i^{3/2}a}{\sqrt{\nu/\omega}}\right) + i\frac{\Delta p}{\omega\rho L}, \quad \text{or} \quad A = -i\frac{\Delta p}{\omega\rho L}\Big/ J_o\left(\frac{i^{3/2}a}{\sqrt{\nu/\omega}}\right),$$

which leads to:

$$u_z(R, t) = \text{Re}\left\{i\frac{\Delta p}{\omega\rho L}\left[1 - J_o\left(\frac{i^{3/2}R}{\sqrt{\nu/\omega}}\right)\Big/ J_o\left(\frac{i^{3/2}a}{\sqrt{\nu/\omega}}\right)\right]e^{i\omega t}\right\}. \tag{7.42}$$

Even though evaluation of Bessel functions at arbitrary points in the complex plane requires special techniques (see Kurup and Koithyar, 2013), (7.42) does represent an exact solution of (4.10) and (7.1). It simplifies to the profile (7.6) in the limit as $\omega \to 0$, and to an appropriate profile, similar to (7.37), in the limit as $\omega \to \infty$ (see Exercise 7.37).

Example 7.9

Show that $\dot{w}$ = the rate of work (per unit area) done on the fluid by the oscillating plate is balanced by $\dot{e}$ = the viscous dissipation of energy (per unit area) in the fluid above the plate.

Solution The rate of work (per unit area) done on the fluid by the moving plate is the product of the shear stress on the fluid, τ_w, and the plate velocity, $U\cos(\omega t)$:

$$\dot{w} = \tau_w U\cos(\omega t) = -\mu(\partial u/\partial y)_{y=0} \cdot U\cos(\omega t).$$

The negative sign appears because the outward normal from the fluid points downward on the surface of the plate. Differentiating (7.37) with respect to y, leads to:

$$\frac{\partial u}{\partial y} = U\sqrt{\frac{\omega}{2\nu}}\exp\left\{-y\sqrt{\frac{\omega}{2\nu}}\right\}\left[-\cos\left(\omega t - y\sqrt{\frac{\omega}{2\nu}}\right) + \sin\left(\omega t - y\sqrt{\frac{\omega}{2\nu}}\right)\right],$$

and evaluating the result at $y = 0$ produces:

$$\left(\frac{\partial u}{\partial y}\right)_{y=0} = U\sqrt{\frac{\omega}{2\nu}}[-\cos(\omega t) + \sin(\omega t)].$$

Thus, the time-average rate of work (per unit area) done by the plate on the fluid is:

$$\overline{\dot{w}} = \frac{\omega}{2\pi}\int_0^{2\pi/\omega} \tau_w U\cos(\omega t)dt = \frac{\omega}{2\pi}\int_0^{2\pi/\omega} -\mu U\sqrt{\frac{\omega}{2\nu}}[-\cos(\omega t) + \sin(\omega t)]U\cos(\omega t)dt = \mu\frac{U^2}{2}\sqrt{\frac{\omega}{2\nu}},$$

where $2\pi/\omega$ is the period of the plate's oscillations.

From (4.59), the rate of dissipation of fluid kinetic energy per unit volume is $\tau_{ij}S_{ij}$, which reduces to $2\mu S_{ij}S_{ij}$ for an incompressible viscous fluid. Thus, the time-average energy dissipation rate (per unit area) above the plate will be:

$$\overline{\dot{e}} = \frac{\omega}{2\pi}\int_0^{2\pi/\omega}\int_0^{\infty} 2\mu S_{ij}S_{ij}dydt = \frac{\omega}{\pi}\mu\int_0^{2\pi/\omega}\int_0^{\infty}(S_{xy}^2 + S_{yx}^2)dydt = \frac{\omega}{2\pi}\mu\int_0^{2\pi/\omega}\int_0^{\infty}\left(\frac{\partial u}{\partial y}\right)^2 dydt,$$

since the only strain-rate component in this flow is $S_{xy} = S_{xy} = (1/2)(\partial u/\partial y)$. The final result is easiest to obtain by performing the time average first:

$$\frac{\omega}{2\pi}\mu\int_0^{2\pi/\omega}\left(\frac{\partial u}{\partial y}\right)^2 dt = \mu U^2\frac{\omega}{2\nu}\exp\left\{-2y\sqrt{\frac{\omega}{2\nu}}\right\}.$$

This leaves the vertical integral:

$$\overline{\dot{e}} = \int_0^{\infty}\mu U^2\frac{\omega}{2\nu}\exp\left\{-y\sqrt{\frac{2\omega}{\nu}}\right\}dy = \mu U^2\frac{\omega}{2\nu}\sqrt{\frac{\nu}{2\omega}} = \mu\frac{U^2}{2}\sqrt{\frac{\omega}{2\nu}},$$

and this matches the time-averaged result for $\dot{w}$. Thus, the average rates of work input and energy dissipation are equal. They are not instantaneously equal, so the fluid's kinetic energy (per unit area) fluctuates, but it does not grow without bound.

7.6 Low Reynolds number viscous flow past a sphere

Many physical problems can be described by the behavior of a system when a certain parameter is either very small or very large. Consider the problem of steady constant-density flow of a viscous fluid at speed U around an object of size L. The governing equations will be (4.10) and the steady flow version of (7.1):

$$\rho\mathbf{u}\cdot\nabla\mathbf{u} + \nabla p = \mu\nabla^2\mathbf{u}. \tag{7.43}$$

As described in Section 4.11, this equation can be scaled to determine which terms are most important. The purpose of such a scaling is to generate dimensionless terms that are of order unity in the flow field. For example, when the flow speeds are high and the viscosity is small, the pressure and inertia forces dominate the momentum balance, showing that pressure changes are of order ρU^2. Consequently, for high Reynolds number, the scaling (4.101) is appropriate for non-dimensionalizing (7.43) to obtain:

$$\mathbf{u}^* \cdot \nabla^* \mathbf{u}^* + \nabla^* p^* = \frac{1}{\mathrm{Re}} \nabla^{*2} \mathbf{u}^*, \tag{7.44}$$

where $\mathrm{Re} = \rho UL/\mu$ is the Reynolds number. For $\mathrm{Re} \gg 1$, (7.44) may be solved by treating 1/Re as a small parameter, and as a first approximation, 1/Re may be set to zero everywhere in the flow, which reduces (7.44) to the inviscid Euler equation without a body force.

However, viscous effects may still be felt at high Re because a single length scale is typically inadequate to describe all regions of high-Re flows. For example, complete omission of the viscous term cannot be valid near a solid surface because the inviscid flow cannot satisfy the no-slip condition at the body surface. Viscous forces are important near solid surfaces because of the high shear rate in the boundary layer near the body surface. The scaling (4.101), which assumes that velocity gradients are proportional to U/L, is invalid in such boundary layers. Thus, there is a region of *non-uniformity* near the body where a perturbation expansion in terms of the small parameter 1/Re becomes *singular*. The proper scaling in the *boundary layer* and a procedure for analyzing wall-bounded high Reynolds number flows will be discussed in Chapter 8. A hint of what is to come is provided by the scaling (7.14), which leads to the lubrication approximation and involves different length scales for the stream-wise and cross-stream flow directions.

Now consider flows in the opposite limit of very low Reynolds numbers, $\mathrm{Re} \to 0$. Such flows should have negligible inertia forces, with pressure and viscous forces providing the dominant balance. Therefore, multiply (7.44) by Re to obtain:

$$\mathrm{Re}(\mathbf{u}^* \cdot \nabla^* \mathbf{u}^* + \nabla^* p^*) = \nabla^{*2} \mathbf{u}^*. \tag{7.45}$$

Although this equation does have negligible inertia terms as $\mathrm{Re} \to 0$, it does not lead to a balance of pressure and viscous forces as $\mathrm{Re} \to 0$ since it reduces to $0 = \mu \nabla^2 \mathbf{u}$, which is not the proper governing equation for low Reynolds number flows. The source of the inadequacy is the scaling of the pressure term specified by (4.101). For low Reynolds number flows, pressure is *not* of order ρU^2. Instead, at low Re, pressure differences should be scaled with a generic viscous stress such as $\mu \partial u/\partial y \sim \mu U/L$. Thus, the pressure scaling $p^* = (p - p_\infty)/\rho U^2$ in (4.101) should be replaced by $p^* = (p - p_\infty)L/\mu U$, and this leads to a correctly revised version of (7.45):

$$\mathrm{Re}(\mathbf{u}^* \cdot \nabla^* \mathbf{u}^*) = -\nabla^* p^* + \nabla^{*2} \mathbf{u}^*, \tag{7.46}$$

which does exhibit the proper balance of terms as $\mathrm{Re} \to 0$ and becomes the linear (dimensional) equation:

$$\nabla p = \mu \nabla^2 \mathbf{u}, \tag{7.47}$$

when this limit is taken.

Flows with $\mathrm{Re} \to 0$ are called *creeping* flows, and they occur at low flow speeds of viscous fluids past small objects or through narrow passages. Examples of such flows are the motion of a thin film of oil in the bearing of a shaft, the settling of sediment particles in nominally quiescent water, the fall of mist droplets in the atmosphere, or the flow of molten plastic during a molding process.

From this discussion of scaling, we conclude that the proper length and time scales depend on the nature and the region of the flow, and are obtained by balancing the terms that are most important in the region of the flow field under consideration. Identifying the proper length and time scales is commonly the goal of experimental and numerical investigations of viscous flows, so that the most appropriate simplified versions of the full equations for fluid motion may be analyzed. The remainder of this section presents a solution for the creeping flow past a sphere, first given by Stokes in 1851. This is a flow where different field equations should be used in regions close to and far from the sphere.

We begin by considering the near-field flow around a stationary sphere of radius a placed in a uniform stream of speed U (Fig. 7.17) with $\mathrm{Re} \to 0$. The problem is axisymmetric, that is, the flow patterns are identical in all planes parallel to U and passing through the center of the sphere. Since $\mathrm{Re} \to 0$, as a first approximation, neglect the inertia forces altogether and seek a solution to (7.47). Taking the curl of (7.47) produces an equation for the vorticity alone:[1]

$$\nabla^2 \boldsymbol{\omega} = 0,$$

1. In spherical polar coordinates, the operator in the footnoted equations is actually $-\nabla \times \nabla \times ()$ or $-$curl curl (), which is different from the Laplace operator defined in Appendix B.

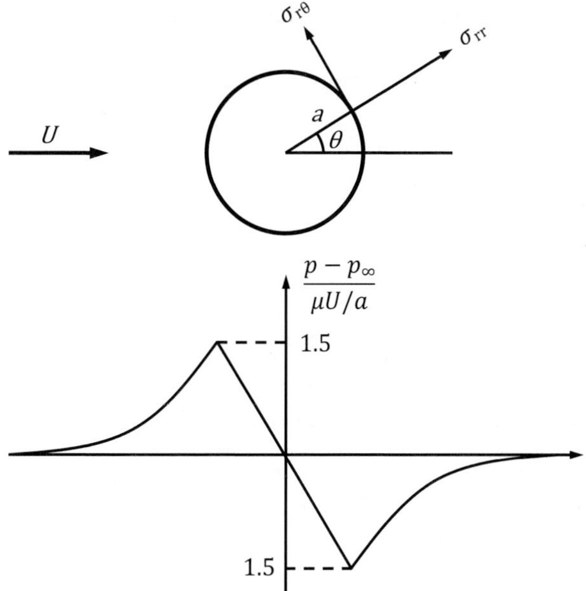

FIGURE 7.17 Creeping flow over a sphere. The upper panel shows the viscous stress components at the surface. The lower panel shows the pressure distribution in an axial ($\varphi =$ const.) plane.

because $\nabla \times \nabla p = 0$ and the order of the operators curl and ∇^2 can be interchanged. (The reader may verify this using indicial notation.) The only component of vorticity in this axisymmetric problem is ω_φ, the component perpendicular to $\varphi =$ constant planes in Fig. 7.17, and it is given by:

$$\omega_\varphi = \frac{1}{r}\left[\frac{\partial(ru_\theta)}{\partial r} - \frac{\partial u_r}{\partial \theta}\right].$$

This is an axisymmetric flow, so the r and θ velocity components can be found from an axisymmetric stream function ψ:

$$u_r = \frac{1}{r^2 \sin\theta}\frac{\partial \psi}{\partial \theta}, \quad \text{and} \quad u_\theta = -\frac{1}{r \sin\theta}\frac{\partial \psi}{\partial r}. \tag{6.86}$$

In terms of this stream function, the vorticity becomes:

$$\omega_\varphi = -\frac{1}{r}\left[\frac{1}{\sin\theta}\frac{\partial^2 \psi}{\partial r^2} + \frac{1}{r^2}\frac{\partial}{\partial \theta}\left(\frac{1}{\sin\theta}\frac{\partial \psi}{\partial \theta}\right)\right],$$

which is governed by[1]:

$$\nabla^2 \omega_\varphi = 0.$$

Combining the last two equations, we obtain[2]:

$$\left[\frac{\partial^2}{\partial r^2} + \frac{\sin\theta}{r^2}\frac{\partial}{\partial \theta}\left(\frac{1}{\sin\theta}\frac{\partial}{\partial \theta}\right)\right]^2 \psi = 0. \tag{7.48}$$

The boundary conditions on the preceding equation are:

$$\psi(r = a, \theta) = 0 \ [u_r = 0 \text{ at surface}], \tag{7.49}$$

$$\partial\psi(r = a, \theta)/\partial r = 0 \ [u_\theta = 0 \text{ at surface}], \text{ and} \tag{7.50}$$

$$\psi(r \to \infty, \theta) = \frac{1}{2}Ur^2 \sin^2\theta \ [\text{uniform flow far from the sphere}]. \tag{7.51}$$

2. Eq. (7.48) is the square of the operator, and not the biharmonic.

The last condition follows from the fact that the stream function for a uniform flow is (1/2) $Ur^2 \sin^2\theta$ in spherical coordinates (see (6.84)).

The far-field condition (7.51) suggests a separable solution of the form:

$$\psi(r,\theta) = f(r)\sin^2\theta.$$

Substitution of this into the governing equation (7.48) gives:

$$f^{iv} - \frac{4f''}{r^2} + \frac{8f'}{r^3} - \frac{8f}{r^4} = 0,$$

which is an equi-dimensional equation with power-law solutions $\sim r^n$. Here, the possible values for n are -1, 1, 2, and 4, so:

$$f = Ar^4 + Br^2 + Cr + D/r.$$

The far-field boundary condition (7.51) requires that $A = 0$ and $B = U/2$ while the surface boundary conditions require $C = -3Ua/4$ and $D = Ua^3/4$. The solution then reduces to:

$$\psi = Ur^2\sin^2\theta\left(\frac{1}{2} - \frac{3a}{4r} + \frac{a^3}{4r^3}\right). \tag{7.52}$$

The velocity components are found from (7.52) using (6.86):

$$u_r = U\cos\theta\left(1 - \frac{3a}{2r} + \frac{a^3}{2r^3}\right), \quad \text{and} \quad u_\theta = -U\sin\theta\left(1 - \frac{3a}{4r} - \frac{a^3}{4r^3}\right). \tag{7.53}$$

The pressure is found by integrating the momentum equation $\nabla p = \mu\nabla^2\mathbf{u}$. The result is:

$$p - p_\infty = -\frac{3\mu a U\cos\theta}{2r^2}, \tag{7.54}$$

which is sketched in Fig. 7.17. The maximum $p - p_\infty = 3\mu U/2a$ occurs at the forward stagnation point ($\theta = \pi$), while the minimum $p - p_\infty = -3\mu U/2a$ occurs at the rear stagnation point ($\theta = 0$).

The drag force D on the sphere can be determined by integrating its surface pressure and shear stress distributions (see Exercise 7.52) to find:

$$D = 6\pi\mu aU, \tag{7.55}$$

of which one-third is pressure drag and two-thirds is skin friction drag. It follows that drag in a creeping flow is proportional to the velocity; this is known as *Stokes' law of resistance*.

In a well-known experiment to measure the charge of an electron, Millikan (1911) used (7.55) to estimate the radius of an oil droplet falling through air. Suppose ρ' is the density of a falling spherical particle and ρ is the density of the surrounding fluid. Then the effective weight of the sphere is $4\pi a^3 g(\rho' - \rho)/3$, which is the weight of the sphere minus the weight of the displaced fluid. The falling body reaches its terminal velocity when it no longer accelerates, at which point the viscous drag equals the effective weight. Then:

$$(4/3)\pi a^3 g(\rho' - \rho) = 6\pi\mu aU,$$

from which the radius a can be estimated.

Millikan (1911) was able to deduce the charge on an electron (and win a Nobel prize) making use of Stokes' drag formula by the following experiment. Two horizontal parallel plates can be charged by a battery (see Fig. 7.18). Oil is sprayed through a very fine hole in the upper plate and develops static charge (+) by losing a few (n) electrons in passing through the small hole. If the plates are charged, then an electric force neE will act on each of the drops. Now n is not known but the electric field E is known provided that the charge density in the gap is very low: $E = -V_b/L$ where V_b is the battery voltage and L is the gap between the plates. With the plates uncharged, measurement of the downward terminal velocity allowed the radius of a drop to be calculated assuming that the viscosity of the drop is much larger than the viscosity of the air. The switch is thrown to charge the upper plate negatively. The same droplet then reverses direction and is forced upward. It quickly achieves its terminal velocity U_u by virtue of the balance of upward forces (electric + buoyancy) and downward forces (weight + drag). This gives:

$$6\pi\mu U_u a + (4/3)\pi a^3 g(\rho' - \rho) = neE,$$

where U_u is measured by the observation telescope and the radius of the particle is now known. The data then allow for the calculation of ne. As n must be an integer, data from many droplets may be differenced to identify the minimum difference that must be e, the charge of a single electron.

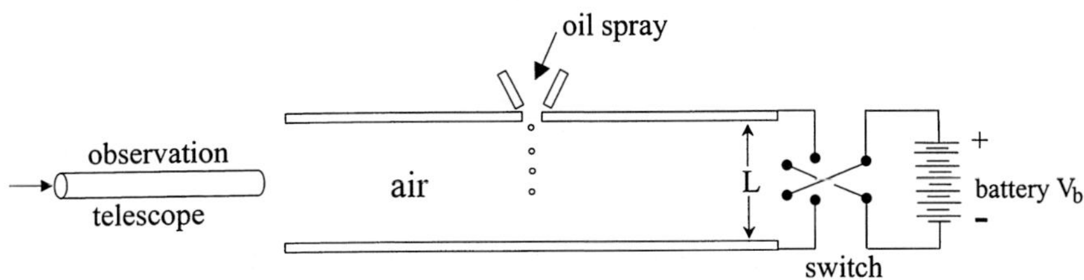

FIGURE 7.18 Simplified schematic of the Millikan oil drop experiment where observations of charged droplet motion and Stokes' drag law were used to determine the charge on an electron.

The drag coefficient, C_D, defined by (4.108) with $A = \pi a^2$, for Stokes' sphere is:

$$C_D = \frac{D}{\frac{1}{2}\rho U^2 \pi a^2} = \frac{24}{\text{Re}}, \tag{7.56}$$

where $\text{Re} = 2aU/\nu$ is the Reynolds number based on the diameter of the sphere. This dependence on the Reynolds number can be predicted from dimensional analysis when fluid inertia, represented by ρ, is not a parameter (see Exercise 4.83). Without fluid density, the drag force on a slowly moving sphere may only depend on the other parameters of the problem:

$$D = f(\mu, U, a).$$

Here there are four variables and the three basic dimensions of mass, length, and time. Therefore, only one dimensionless parameter, $D/\mu Ua$, can be formed. Hence, it must be a constant, and this leads to $C_D \propto 1/\text{Re}$.

The flow pattern in a reference frame fixed to the fluid at infinity can be found by superposing a uniform velocity U to the left. This cancels out the first term in (7.52), giving:

$$\psi = Ur^2 \sin^2\theta \left(-\frac{3a}{4r} + \frac{a^3}{4r^3} \right),$$

which gives the streamline pattern for a sphere moving from right to left in front of an observer (Fig. 7.19). The pattern is symmetric between the upstream and the downstream directions, which is a result of the linearity of the governing equation (7.47); reversing the direction of the free-stream velocity merely changes $\mathbf{u}$ to $-\mathbf{u}$ and $p - p_\infty$ to $-p + p_\infty$. The flow therefore does not leave a velocity-field wake behind the sphere.

In spite of its fame and success, the Stokes solution is not valid at large distances from the sphere because the advective terms are not negligible compared to the viscous terms at these distances. At large distances, the viscous terms are of the order:

$$\text{viscous force/volume} = \text{stress gradient} \sim \frac{\mu Ua}{r^3} \quad \text{as} \quad r \to \infty,$$

while from (7.53), the largest inertia term is:

$$\text{inertia force/volume} \sim \rho u_r \frac{\partial u_\theta}{\partial r} \sim \frac{\rho U^2 a}{r^2} \quad \text{as} \quad r \to \infty;$$

therefore:

$$\text{inertia force/viscous force} \sim \frac{\rho Ua}{\mu}\frac{r}{a} \sim \text{Re}\frac{r}{a} \quad \text{as} \quad r \to \infty,$$

which shows that the inertia forces are not negligible for distances larger than $r/a \sim 1/\text{Re}$.

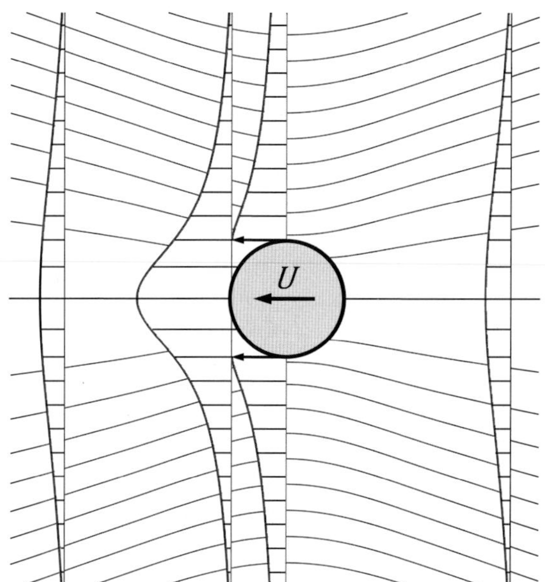

FIGURE 7.19 Streamlines and velocity distributions in Stokes' solution of creeping flow due to a moving sphere. Note the upstream and downstream symmetry, which is a result of complete neglect of nonlinearity.

Solutions of problems involving a small parameter can be developed in terms of a perturbation series in which the higher-order terms act as corrections on the lower-order terms. If we regard the Stokes solution as the first term of a series expansion in the small parameter Re, then the expansion is not uniformly valid because it breaks down as $r \to \infty$. If we tried to calculate the next term (to order Re) of the perturbation series, we would find that the velocity corresponding to the higher-order term becomes unbounded compared to that of the first term as $r \to \infty$.

The situation becomes worse for two-dimensional objects such as the circular cylinder. In this case, the Stokes balance, $\nabla p = \mu \nabla^2 \mathbf{u}$, has *no solution at all* that can satisfy the uniform-flow boundary condition at infinity. From this, Stokes concluded that steady, slow flows around cylinders cannot exist in nature. It has now been realized that the non-existence of a first approximation of the Stokes flow around a cylinder is due to the *singular* nature of low Reynolds number flows in which there is a region of *non-uniformity* for $r \to \infty$. The non-existence of the second approximation for flow around a sphere is due to the same reason. In a different (and more familiar) class of singular perturbation problems, the region of non-uniformity is a thin layer (the boundary layer) near the surface of an object. This is the class of flows with $\mathrm{Re} \to \infty$, that are discussed in the next chapter. For these high Reynolds number flows the small parameter 1/Re multiplies the *highest*-order derivative in the governing equations, so that the solution with 1/Re identically set to zero cannot satisfy all the boundary conditions. In low Reynolds number flows, this classic symptom of the loss of the highest derivative is absent, but it is a singular perturbation problem nevertheless.

Oseen (1910) provided an improvement to Stokes' solution by partly accounting for the inertia terms at large distances. He made the substitutions:

$$u = U + u', \quad v = v', \quad \text{and} \quad w = w',$$

where u', v', and w' are the Cartesian components of the perturbation velocity, and are small at large distances. Substituting these, the advection term of the x-momentum equation becomes:

$$u\frac{\partial u}{\partial x} + v\frac{\partial u}{\partial y} + w\frac{\partial u}{\partial z} = U\frac{\partial u'}{\partial x} + \left[u'\frac{\partial u'}{\partial x} + v'\frac{\partial u'}{\partial y} + w'\frac{\partial u'}{\partial z}\right].$$

Neglecting the quadratic terms, a revised version of the equation of motion (7.47) becomes:

$$\rho U\frac{\partial u_i'}{\partial x} = -\frac{\partial p}{\partial x_i} + \mu \nabla^2 u_i',$$

where u_i' represents u', v', or w'. This is called *Oseen's equation*, and the approximation involved is called *Oseen's approximation*. In essence, the Oseen approximation linearizes the advective acceleration term $\mathbf{u} \cdot \nabla \mathbf{u}$ to $U(\partial \mathbf{u}/\partial x)$, whereas the Stokes approximation drops advection altogether. Near the body both approximations have the same order of accuracy. However, the Oseen approximation is better in the far field where the velocity is only slightly different from U. The Oseen equations provide a lowest-order solution that is uniformly valid everywhere in the flow field.

The boundary conditions for a stationary sphere with the fluid moving past it at velocity $U\mathbf{e}_x$ are:

$$u', v', w' \to 0 \quad \text{as} \quad r \to \infty, \quad \text{and} \quad u' = -U \quad \text{and} \quad v', w' = 0$$

on the sphere's surface. The solution found by Oseen is:

$$\frac{\psi}{Ua^2} = \left[\frac{r^2}{2a^2} + \frac{a}{4r}\right]\sin^2\theta - \frac{3}{\text{Re}}(1+\cos\theta)\left\{1 - \exp\left[-\frac{\text{Re}}{4}\frac{r}{a}(1-\cos\theta)\right]\right\}, \tag{7.57}$$

where $\text{Re} = 2aU/\nu$. Near the surface $r/a \approx 1$, a series expansion of the exponential term shows that Oseen's solution is identical to the Stokes solution (7.52) to the lowest order. The Oseen approximation predicts that the drag coefficient is:

$$C_D = \frac{24}{\text{Re}}\left(1 + \frac{3}{16}\text{Re}\right),$$

which should be compared with the Stokes formula (7.56). Experimental results show that the Oseen and the Stokes formulas for C_D are both fairly accurate for $\text{Re} < 5$ (experimental results fall between them), an impressive range of validity for a theory developed for $\text{Re} \to 0$.

The streamlines corresponding to the Oseen solution (7.57) are shown in Fig. 7.20, where a uniform flow of U is added to the left to generate the pattern of flow due to a sphere moving in front of a stationary observer. It is seen that the flow is no longer symmetric, but has a wake where the streamlines are closer together than in the Stokes flow. The velocities in the wake are larger than in front of the sphere. Relative to the sphere, the flow is slower in the wake than in front of the sphere.

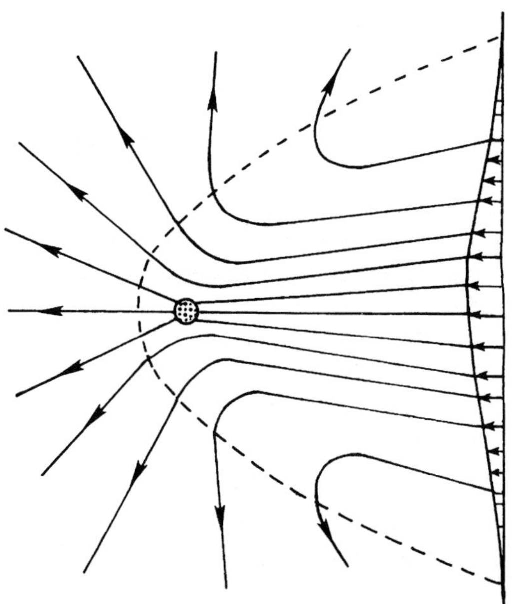

FIGURE 7.20 Streamlines and velocity distribution in Oseen's solution of creeping flow due to a moving sphere. Note the upstream and downstream asymmetry, which is a result of partial accounting for advection in the far field.

Stokes' solution for the drag force on a sphere (7.55) and Oseen's correction (7.57) were derived for steady flows. Just as the added mass force (6.111) accounts for drag on an accelerating sphere in an ideal fluid (see Section 6.9), Basset (1888) showed that an additional force arises when a sphere accelerates in a viscous fluid. Following Crowe et al. (2011), the most direct approach to understanding this additional force is to consider an impulsively accelerated infinite flat plate (Fig. 7.12). As shown in Section 7.4, the resulting wall shear stress is $\tau_w = \mu U/(\pi \nu t)^{1/2}$, or equivalently $\tau_w = (\rho\mu/(\pi t))^{1/2}U$, where U is the speed of the plate. Now, suppose the plate velocity varies in time according to a series of small incremental step changes as shown in Fig. 7.21, such that at time $t = 0$ there is a change ΔU_0, at time t_1 a change ΔU_1, and so on. The resulting shear stress is

$$\tau_w = \sqrt{\frac{\rho\mu}{\pi}}\left[\frac{\Delta U_0}{\sqrt{t}} + \frac{\Delta U_1}{\sqrt{t-t_1}} + \frac{\Delta U_2}{\sqrt{t-t_2}}\cdots\right].$$

For a time increment $\Delta t'$ the change in velocity is $(dU/dt')\Delta t'$ so the summation can be generalized to

$$\tau_w = \sqrt{\frac{\rho\mu}{\pi}}\sum_{n=0}^{N}\frac{dU/dt'}{\sqrt{t-n\Delta t'}}\Delta t',$$

where $N\Delta t'$ is the time interval from the initiation of the acceleration to the present time t. Taking the limit $\Delta t' \to 0$ and $N\Delta t' \to t'$, this becomes

$$\tau_w = \sqrt{\frac{\rho\mu}{\pi}}\int_0^t \frac{dU'/dt'}{\sqrt{t-t'}}dt'. \tag{7.58}$$

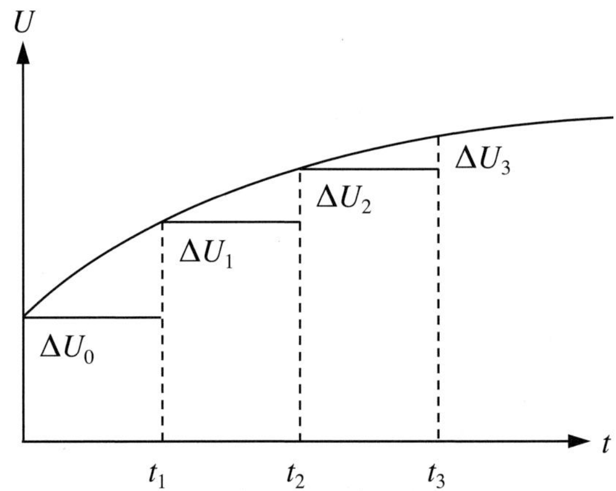

FIGURE 7.21 Stepwise impulsive acceleration of a flat plate. Adapted from Crowe et al. (2011) with permission.

Applying a similar approach to an impulsively started sphere at low Reynolds number, Basset found the drag force was equal to

$$D = 6a^2\sqrt{\pi\rho\mu}\int_0^t \frac{1}{\sqrt{t-t'}}\frac{d}{dt'}(U-V)dt', \tag{7.59}$$

where U is the fluid velocity and V is the velocity of the sphere. This result is sometimes called the *history term* and can be combined with Stokes' solution (7.55) to capture drag acting on a moving sphere at low Reynolds number. In practice, (7.59) is difficult to evaluate since it requires knowledge of the acceleration history up to the present time. Scaling arguments show that (7.55) is insignificant when the density of the sphere is much greater than the fluid density, such as the case of solid particles or liquid droplets in air (see Exercise 7.57).

Example 7.10

By treating small water droplets in air as a solid objects, determine the settling velocities and Reynolds numbers of a naturally occurring cloud and falling-mist droplets with diameters $d = 10$ μm and $d = 100$ μm at 20 °C.

Solution The settling (or terminal) velocity, U, occurs when the droplet's weight (minus buoyancy) is balanced by its viscous drag.

$$3\pi\mu_a U d = \frac{\pi}{6}d^3 g(\rho_w - \rho_a),$$

where μ_a is the air's viscosity, and ρ_a and ρ_w are the air and water densities, respectively. Thus, the fog droplet's settling velocity is:

$$U = \frac{d^2 g}{18\mu_a}(\rho_w - \rho_a) = \frac{(10^{-5}\text{ m})^2(9.81\text{ ms}^{-2})}{18(1.8\times10^{-5}\text{ kgm}^{-1}\text{s}^{-1})}(998 - 1.2)\text{ kgm}^{-3} = 3.0\times10^{-3}\text{ ms}^{-1},$$

and this is low enough so that weak vertical air currents can keep the droplet suspended. Its Reynolds number is $\text{Re} = \rho_a U d/\mu_a = (1.2\text{ kgm}^{-3})(0.003\text{ ms}^{-1})(10^{-5}\text{ m})/(1.8\times10^{-5}\text{ kgm}^{-1}s^{-1}) = 0.002$.

The falling mist droplet is 10 times larger so it's settling velocity is 100 times larger, $U = 0.30\text{ ms}^{-1}$, and its Reynolds number is 1000 times larger, Re = 2. Droplets of this size typically descend through the atmosphere and grow to become raindrops or shrink to become cloud droplets, depending on the local thermodynamic conditions.

7.7 Final remarks

As in other fields of physical science, analytical methods in fluid mechanics are useful in understanding the physics of fluid flows and in making generalizations. However, it is probably fair to say that most, if not all, analytically tractable problems in ordinary laminar flow have already been solved, so approximate and numerical methods are necessary for further advancing our knowledge. A successful approximate technique is the perturbation method, where the flow is assumed to deviate slightly from a basic linear state. But, as will be discussed in Chapter 15, the most common modern alternative to analytic solution techniques is, of course, to solve the Navier-Stokes equations numerically using a computer. Interestingly, this situation has not relegated exact analytical solutions to the annals of scientific history. Instead, exact and asymptotically-exact laminar flow solutions are commonly used to verify numerical solution schemes, and new measurement techniques as well. Thus, this chapter's coverage of classical laminar flow solutions remains essential for a complete appreciation of modern fluid mechanics.

Exercises

7.1 **a)** Write out the three components of (7.1) in x-y-z Cartesian coordinates.
b) Set $\mathbf{u} = (u(y), 0, 0)$, and show that the x- and y-momentum equations reduce to:

$$0 = -\frac{1}{\rho}\frac{\partial p}{\partial x} + \nu\frac{d^2u}{dy^2}, \quad \text{and} \quad 0 = -\frac{1}{\rho}\frac{\partial p}{\partial y}.$$

7.2 For steady pressure-driven flow between parallel plates (see Fig. 7.3), there are 7 parameters: $u(y)$, U, y, h, ρ, μ, and dp/dx. Determine a dimensionless scaling law for $u(y)$, and rewrite the flow-field solution (7.5) in dimensionless form.

7.3 An incompressible viscous liquid with density ρ fills the gap between two large, smooth parallel walls that are both stationary. The upper and lower walls are located at $x_2 = \pm h$, respectively. An additive in the liquid causes its viscosity to vary in the x_2 direction. Here the flow is driven by a constant non-zero pressure gradient: $\partial p/\partial x_1 = const.$

a) Assume steady flow, ignore the body force, set $\mathbf{u} = (u_1(x_2), 0, 0)$ and use:

$$\frac{\partial\rho}{\partial t} + \frac{\partial}{\partial x_i}(\rho u_i) = 0, \text{ and}$$

$$\rho\frac{\partial u_j}{\partial t} + \rho u_i\frac{\partial u_j}{\partial x_i} = -\frac{\partial p}{\partial x_j} + \rho F_j + \frac{\partial}{\partial x_i}\left[\mu\left(\frac{\partial u_i}{\partial x_j} + \frac{\partial u_j}{\partial x_i}\right)\right] + \frac{\partial}{\partial x_j}\left[\left(\mu_v - \frac{2}{3}\mu\right)\frac{\partial u_i}{\partial x_i}\right]$$

to determine $u_1(x_2)$ when $\mu = \mu_o(1 + \gamma(x_2/h)^2)$.

b) What shear stress is felt on the lower wall?

c) What is the volume flow rate (per unit depth into the page) in the gap when $\gamma = 0$?

d) If $-1 < \gamma < 0$, will the volume flux be higher or lower than the case when $\gamma = 0$?

7.4 An incompressible viscous liquid with density ρ fills the gap between two large, smooth parallel plates. The upper plate at $x_2 = h$ moves in the positive x_1-direction at speed U. The lower plate at $x_2 = 0$ is stationary. An additive in the liquid causes its viscosity to vary in the x_2 direction.

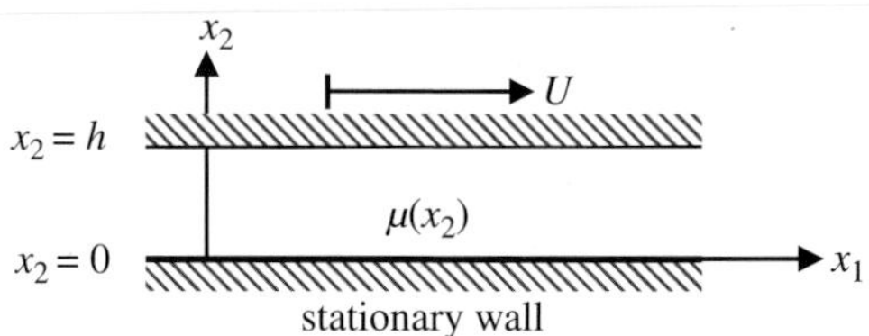

a) Assume steady flow, ignore the body force, set $\mathbf{u} = (u_1(x_2), 0, 0)$ and $\partial p/\partial x_1 = 0$, and use the equations specified in Exercise 7.3 to determine $u_1(x_2)$ when $\mu = \mu_o(1 + \gamma x_2/h)$.

b) What shear stress is felt on the lower plate?

c) Are there any physical limits on γ? If, so specify them.

7.5 Oil with density ρ and viscosity μ fills the gap between two large parallel plates separated by a small distance h. The oil flows with velocity $\mathbf{u} = (u, v, 0)$ under the action of a constant pressure gradient $\nabla p = (dp/dx, 0, 0)$ and a constant velocity $\mathbf{V} = (0, V, 0)$ of the upper plate. The lower plate, at $z = 0$, is stationary. Use the (x, y, z)-Cartesian coordinate system shown for your work.

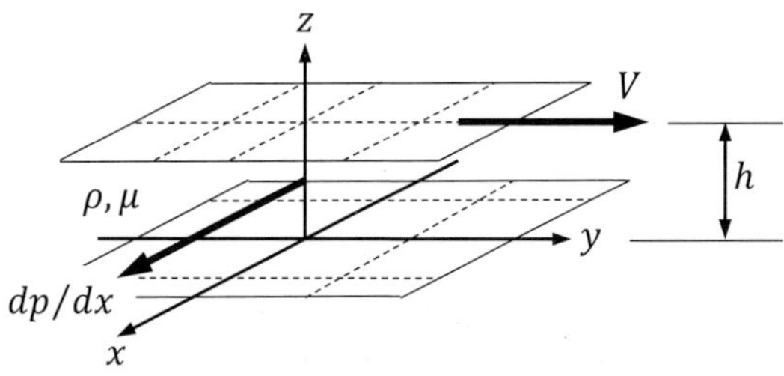

a) Develop formulae for u and v, the two horizontal components of the oil's velocity as functions of the given parameters and the vertical coordinate, z.

b) What are the two components of the shear stress, τ_{xz} and τ_{yz}, on the lower plate.

7.6 Planar Couette flow is generated by placing a viscous fluid between two infinite parallel plates and moving one plate (say, the upper one) at a velocity U with respect to the other one. The plates are a distance h apart. Two immiscible viscous liquids are placed between the plates as shown in the diagram. The lower fluid layer has thickness d. Solve for the velocity distributions in the two fluids.

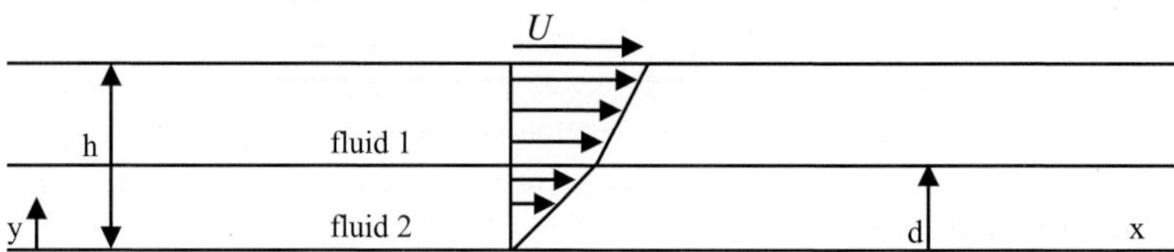

7.7 Consider plane Poiseuille flow of a non-Newtonian power-law fluid using the coordinate system and geometry shown in Fig. 7.3. Here the fluid's constitutive relationship is given by (4.37): $\tau_{xy} = m(\partial u/\partial y)^n$, where m is the power law coefficient, and n is the power law exponent.

a) Determine the velocity profile $u(y)$ in the lower half of the channel, $0 \le y \le h/2$, using the boundary condition $u(0) = 0$.

b) Given that the maximum velocity occurs at $y = h/2$ and that the flow profile is symmetric about this location, plot $u(y)/u(h/2)$ vs. $y/(h/2)$ for $n = 2$ (a shear thickening fluid), $n = 1$ (a Newtonian fluid), and $n = 0.4$ (a shear thinning fluid).

c) Explain in physical terms why the shear-thinning velocity profile is the bluntest.

7.8 Consider the laminar flow of a fluid layer falling down a plane inclined at an angle θ with respect to the horizontal. If h is the thickness of the layer in the fully developed stage, show that the velocity distribution is $u(y) = (g/2\nu)(h^2 - y^2)\sin\theta$, where the x-axis points in the direction of flow along the free surface, and the y-axis points toward the plane. Show that the volume flow rate per unit width is $Q = (gh^3/3\nu)\sin\theta$, and that the frictional stress on the wall is $\tau_w = \rho g h \sin\theta$.

7.9 In two-dimensional (x, y)-coordinates, the Navier-Stokes equations for the fluid velocity, $\mathbf{u} = (u, v)$, in a constant-viscosity constant-density flow are: $\partial u/\partial x + \partial v/\partial y = 0$:

$$\frac{\partial u}{\partial t} + u\frac{\partial u}{\partial x} + v\frac{\partial u}{\partial y} = -\frac{1}{\rho}\frac{\partial p}{\partial x} + \nu\left(\frac{\partial^2 u}{\partial x^2} + \frac{\partial^2 u}{\partial y^2}\right), \quad \text{and}$$

$$\frac{\partial v}{\partial t} + u\frac{\partial v}{\partial x} + v\frac{\partial v}{\partial y} = -\frac{1}{\rho}\frac{\partial p}{\partial y} + \nu\left(\frac{\partial^2 v}{\partial x^2} + \frac{\partial^2 v}{\partial y^2}\right).$$

a) Cross-differentiate and sum the two momentum equations to reach the following equation for $\omega_z = \partial v/\partial x - \partial u/\partial y$, the vorticity normal to the x-y plane:

$$\frac{\partial \omega_z}{\partial t} + u\frac{\partial \omega_z}{\partial x} + v\frac{\partial \omega_z}{\partial y} = \nu\left(\frac{\partial^2 \omega_z}{\partial x^2} + \frac{\partial^2 \omega_z}{\partial y^2}\right).$$

b) The simplest non-trivial solution of this equation is uniform shear or solid body rotation (ω_z = constant). The next simplest solution is a linear function of the independent coordinates: $\omega_z = ax + by$, where a and b are constants. Starting from this vorticity field, derive the following velocity field:

$$u = -\frac{b}{2}\left[\frac{(ax+by)^2}{a^2+b^2} + c\right] \quad \text{and} \quad v = \frac{a}{2}\left[\frac{(ax+by)^2}{a^2+b^2} + c\right],$$

where c is an undetermined constant.

c) For the part b) flow, sketch the streamlines. State any assumptions you make about a, b, and c.

d) For the part b) flow when $a = 0$, $b > 0$, and $\mathbf{u} = (U_o, 0)$ at the origin of coordinates with $U_o > 0$, sketch the velocity profile along a line x = constant, and determine ∇p.

7.10 Consider circular Couette flow as described by (7.10) in the limit of a thin gap between the cylinders. Use the definitions: $\overline{R} = (R_1 + R_2)/2$, $h = R_2 - R_1$, $\overline{\Omega} = (\Omega_1 + \Omega_2)/2$, $\Delta\Omega = \Omega_2 - \Omega_1$, and $R = \overline{R} + y$ to complete the following items.

a) Show that $u_\varphi(y) \cong \overline{\Omega}\,\overline{R} + \Delta\Omega\overline{R}(y/h)$ when $y \ll \overline{R}$, $h \ll \overline{R}$, and all terms of order $y/\overline{R}$ and $h/\overline{R}$ or higher have been dropped. This is the lubrication approximation for circular Couette flow.

b) Compute the shear stress $\mu(du_\varphi/dy)$ from the part a) flow field and put it terms of μ, R_1, R_2, Ω_1, and Ω_2.

c) Compute the exact shear stress from (7.10) using: $\tau_{R\varphi} = 2\mu\left[\frac{R}{2}\frac{\partial}{\partial R}\left(\frac{u_\varphi}{R}\right) + \frac{1}{2R}\frac{\partial u_R}{\partial \varphi}\right]$, and evaluate it at $R = R_1$, and $R = R_2$.

d) What values of $h/\overline{R}$ lead to shear stress errors of 1%, 3% and 10%.

7.11 Room temperature water drains through a round vertical tube with diameter d. The length of the tube is L. The pressure at the tube's inlet and outlet is atmospheric, the flow is steady, and $L \gg d$.

a) Using dimensional analysis write a physical law for the mass flow rate $\dot{m}$ through the tube.

b) Assume that the velocity profile in the tube is independent of the vertical coordinate, determine a formula for $\dot{m}$, and put it in dimensionless form.

c) What is the change in $\dot{m}$ if the temperature is raised and the water's viscosity drops by a factor of two?

7.12 Consider steady laminar flow through the annular space formed by two coaxial tubes aligned with the z-axis. The flow is along the axis of the tubes and is maintained by a pressure gradient dp/dz. Show that the axial velocity at any radius R is:

$$u_z(R) = \frac{1}{4\mu}\frac{dp}{dz}\left[R^2 - a^2 - \frac{b^2 - a^2}{\ln(b/a)}\ln\frac{R}{a}\right],$$

where a is the radius of the inner tube and b is the radius of the outer tube. Find the radius at which the maximum velocity is reached, the volume flow rate, and the stress distribution.

7.13 A long, round wire with radius a is pulled at a steady speed U, along the axis of a long round tube of radius b that is filled with a viscous fluid. Assuming laminar, fully developed axial flow with $\partial p/\partial z = 0$ in cylindrical coordinates (R, φ, z) with $\mathbf{u} = (0, 0, w(R))$, determine $w(R)$ assuming constant fluid density ρ and viscosity μ with no body force.

a) What force per unit length of the wire is needed to maintain the motion of the wire?

b) Explain what happens to $w(R)$ when $b \to \infty$. Is this situation physically meaningful? What additional term(s) from the equations of motion need to be retained to correct this situation?

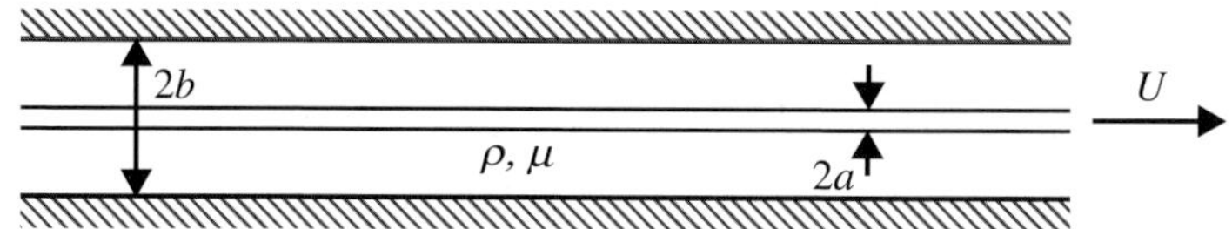

7.14 [3]Consider steady unidirectional incompressible viscous flow in Cartesian coordinates, $u = v = 0$ with $w = w(x, y)$ without body forces.

a) Starting from the steady version of (7.1), derive a single equation for w assuming that $\partial p/\partial z$ is non-zero and constant.

b) Guess $w(x, y)$ for a tube with elliptical cross-section: $(x/a)^2 + (y/b)^2 = 1$.

c) Determine $w(x, y)$ for a tube of rectangular cross-section: $-a/2 \le x \le +a/2$, $-b/2 \le y \le +b/2$. [*Hint*: find particular (a polynomial) and homogeneous (a Fourier series) solutions for w.]

7.15 Consider the steady unidirectional viscous flow in a tube of rectangular cross-section $-a/2 \le x \le +a/2$, $-b/2 \le y \le +b/2$ from Exercise 7.14.

a) Using the finite difference approximation given in Section 6.7 with uniform grid spacing Δ in x and y, show that the z-component of velocity at index (i, j) can be expressed as

$$w_{i,j} = \frac{\Delta^2}{4\mu}\frac{dp}{dz} + \frac{1}{4}\left[w_{i-1,j} + w_{i+1,j} + w_{i,j-1} + w_{i,j+1}\right]. \tag{7.60}$$

b) Write a code to determine w numerically using the iterative procedure outlined in Section 6.7 with no-slip boundary conditions applied at the channel walls. Take $dp/dz = 0.1$, $\mu = 10^{-2}$, $\Delta = .01$, and vary $a/b = 1$, 2, and 4.

c) Compute the average velocity $w_{av} = \int w(x, y)dxdy/A$ for each case in part b where $A = ab$ is the cross-sectional area. Compare the solutions with a velocity profile given for a Poiseuille flow in a round tube, given by Eq. (7.6), using a hydraulic diameter $d_h = 4A/l$ where $l = 2a + 2b$ is the perimeter of the channel cross section.

7.16 A long vertical cylinder of radius b rotates with angular velocity Ω concentrically outside a smaller stationary cylinder of radius a. The annular space is filled with fluid of viscosity μ. Show that the steady velocity distribution is: $u_\varphi = \dfrac{R^2 - a^2}{b^2 - a^2}\dfrac{b^2\Omega}{R}$, and that the torque exerted on either cylinder, per unit length, equals $4\pi\mu\Omega a^2 b^2/(b^2 - a^2)$.

7.17 Consider a solid cylinder of radius a, steadily rotating at angular speed Ω in an infinite viscous fluid. The steady solution is irrotational: $u\theta = \Omega a^2/R$. Show that the work done by the external agent in maintaining the flow (namely, the value of $2\pi R u\theta \tau_{r\theta}$ at $R = a$) equals the viscous dissipation rate of fluid kinetic energy in the flow field.

7.18 Redo the lubrication-theory scaling provided in Section 7.3 for the situation when there is an imposed external time scale τ so that the appropriate dimensionless time is $t^* = t/\tau$, instead of that shown in (7.14). Show that this leads to the additional requirement $\rho h^2/\mu\tau \ll 1$ for the validity of (7.17a). Interpret this new requirement in terms of the viscous diffusion length $\sqrt{\nu\tau}$. Is the unsteady acceleration term needed to analyze the effect of a 100 Hz oscillation imposed on a 0.1-mm film of 30-weight oil ($\nu \approx 4 \times 10^{-4}$ m^2/s)?

7.19 For lubrication flow under the sloped bearing of Example 7.3, the assumed velocity profile was $u(x, y) = -(1/2\mu)(dP/dx)y(h(x) - y) + Uy/h(x)$, the derived pressure was $P(x) = P_e + (3\mu U\alpha/h_o^2 L)x(L - x)$, and the load (per unit depth) carried by the bearing was $W = \mu U\alpha L^2/2h_o^2$. Use these equations to determine the frictional

3. Developed from problem 2 on page 383 in Yih (1979).

force (per unit depth), F_f, applied to the lower (flat) stationary surface in terms of W, h_o/L, and α. What is the spatially averaged coefficient of friction under the bearing?

7.20 A bearing pad of total length $2L$ moves to the right at constant speed U above a thin film of incompressible oil with viscosity μ and density ρ. There is a step change in the gap thickness (from h_1 to h_2) below the bearing as shown. Assume that the oil flow under the bearing pad follows: $u(y) = -\dfrac{y(h_j - y)}{2\mu}\dfrac{dP(x)}{dx} + \dfrac{Uy}{h_j}$, where $j = 1$ or 2. The pad is instantaneously aligned above the coordinate system shown. The pressure in the oil ahead and behind the bearing is P_e.

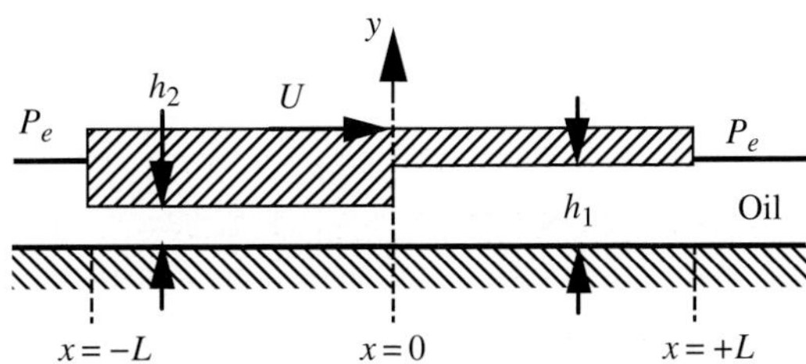

a) By conserving mass for the oil flow, find a relationship between μ, U, h_j, dP/dx, and an unknown constant C.

b) Use the result of part a) and continuity of the pressure at $x = 0$, to determine

$$P(0) - P_e = \frac{6\mu UL(h_1 - h_2)}{h_2^3 + h_1^3},$$

c) Can this bearing support an externally applied downward load when $h_1 < h_2$?

7.21 A flat disk of radius a rotates above a solid boundary at a steady rotational speed of Ω. The gap, $h(\ll a)$, between the disk and the boundary is filled with an incompressible Newtonian fluid with viscosity μ and density ρ. The pressure at the edge of the disk is $p(a)$.

a) Using cylindrical coordinates and assuming that the only non-zero velocity component is $u_\varphi(R, z)$, determine the torque necessary to keep the disk turning.

b) If $p(a)$ acts on the exposed (upper) surface of the disk, will the pressure distribution on the disk's wetted surface tend to pull the disk *toward* or *away from* the solid boundary?

c) If the gap is increased, eventually the assumption of part a) breaks down. What happens? Explain why and where u_r and u_z might be non-zero when the gap is no longer narrow.

7.22 A circular block with radius a and weight W is released at $t = 0$ on a thin layer of an incompressible fluid with viscosity μ that is supported by a smooth horizontal motionless surface. The fluid layer's initial thickness is h_o. Assume that flow in the gap between the block and the surface is quasi-steady with a parabolic velocity profile:

$$u_R(R, z, t) = -(dP(R)/dR)z(h(t) - z))/2\mu,$$

where R is the distance from the center of the block, $P(R)$ is the pressure at R, z is the vertical coordinate from the smooth surface, $h(t)$ is the gap thickness, and t is time.

a) By considering conservation of mass, show that: $dh/dt = (h^3/6\mu R)(dP/dR)$.

b) If W is known, determine $h(t)$ and note how long it takes for $h(t)$ to reach zero.

7.23 Consider the inverse of the previous exercise. A block and a smooth surface are separated by a thin layer of a viscous fluid with thickness h_o. At $t = 0$, a force, F, is applied to separate them. If h_o is arbitrarily small, can the block and plate be separated easily? Perform some tests in your kitchen. Use maple syrup, peanut butter, liquid soap, pudding, etc., for the viscous liquid. The flat top side of a metal jar lid or the flat bottom of a drinking glass makes a good circular block. (Lids with raised edges and cups and glasses with ridges or sloped bottoms do not work well.) A flat countertop or the flat portion of a dinner plate can be the motionless smooth surface. Can the item used for the block be more easily separated from the surface when tilted relative to the surface? Describe your experiments and try to explain your results.

7.24 A rectangular slab of width $2L$ (and depth B into the page) moves *vertically* on a thin layer of oil that flows horizontally as shown. Assume $u(y,t) = -(h^2/2\mu)(dP/dx)(y/h)(1 - y/h)$, where $h(x,t)$ is the instantaneous gap between the slab and the surface, μ is the oil's viscosity, and $P(x,t)$ is the pressure in the oil below the slab. The slab is slightly misaligned with the surface so that $h(x,t) = h_o(1 + \alpha x/L) + \dot{h}_o t$ where $\alpha \ll 1$ and $\dot{h}_o$ is the vertical velocity of the slab. The pressure in the oil outside the slab is P_o. Consider the instant $t = 0$ in your work below.

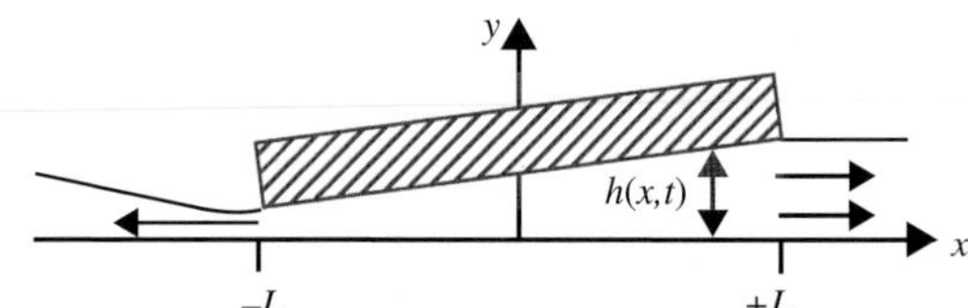

a) Conserve mass in an appropriate CV to show that: $\dfrac{\partial h}{\partial t} + \frac{\partial}{\partial x}\left(\int_0^{h(x,t)} u(y,t)dy\right) = 0.$

b) Keeping only linear terms in α, and noting that C and D are constants, show that:

$$P(x,t=0) = \frac{12\mu}{h_o^3}\left(\frac{\dot{h}_o x^2}{2}\left(1 - \frac{2\alpha x}{L}\right) + Cx\left(1 - \frac{3\alpha x}{2L}\right)\right) + D.$$

c) State the boundary conditions necessary to evaluate the constants C and D.

d) Evaluate the constants to show that the pressure distribution below the slab is:

$$P(x,t) - P_o = -\frac{6\mu \dot{h}_o L^2}{h_o^3(t)}(1 - (x/L)^2)\left(1 - 2\alpha\frac{x}{L}\right).$$

e) Does this pressure distribution act to increase or decrease alignment between the slab and surface when the slab is moving downward? Answer the same question for upward slab motion.

7.25 Show that the lubrication approximation can be extended to viscous flow within narrow gaps $h(x,y,t)$ that depend on two spatial coordinates. Start from (4.10) and (7.1), and use Cartesian coordinates oriented so that x-y plane is locally tangent to the center-plane of the gap. Scale the equations using a direct extension of (7.14):

$$x^* = x/L, \quad y^* = y/L, \quad z^* = z/h = z/\varepsilon L, \quad t^* = Ut/L, \quad u^* = u/U,$$
$$v^* = v/U, \quad w^* = w/\varepsilon U, \quad \text{and} \quad p^* = p/P_a,$$

where L is the characteristic distance for the gap thickness to change in either the x or y direction, and $\varepsilon = h/L$. Simplify these equations when $\varepsilon^2 \text{Re}_L \to 0$, but $\mu UL/P_a h^2$ remains of order unity to find:

$$0 \cong -\frac{\partial p}{\partial x} + \mu\frac{\partial^2 u}{\partial z^2}, \quad 0 \cong -\frac{\partial p}{\partial y} + \mu\frac{\partial^2 v}{\partial z^2}, \quad \text{and} \quad 0 \cong \frac{\partial p}{\partial z}.$$

7.26 A squeegee is pulled across a smooth flat stationary surface at a constant speed U. The gap between the squeegee and surface is $h(x) = h_o \exp\{+\alpha x/L\}$ and this gap is filled with a fluid of constant density ρ and viscosity μ. If the squeegee is wetted by the fluid for $0 \le x \le L$, and the pressure in the surrounding air is p_e, what is the pressure $p(x)$ distribution under the squeegee and what force W perpendicular to the surface is needed to hold the squeegee in place? Ignore gravity in your work.

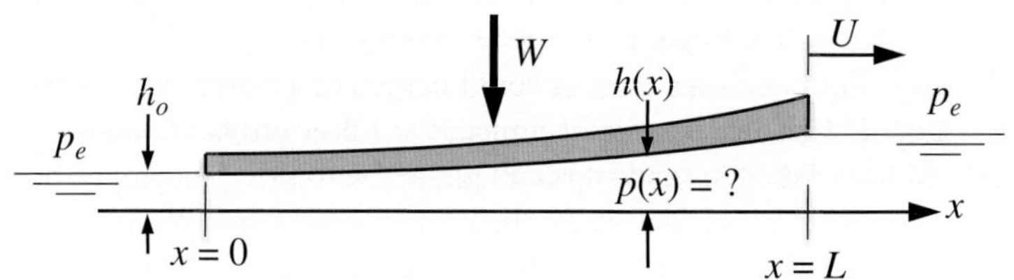

7.27 A close-fitting solid cylinder with net weight W (= actual weight − buoyancy), length L, and radius a is centered in and may slide along the axis of a long vertical tube with radius $a + h$, where $h \ll a$. The tube is filled with oil having constant viscosity μ that is pumped slowly upward at a volume flow rate Q.

a) Use dimensional analysis to find a scaling law for the value of Q that holds the cylinder stationary when fluid inertia is unimportant.

b) Use the lubrication approximation and assume that the pressure is uniform above and below the cylinder to determine a formula for the value of Q that holds the cylinder stationary.

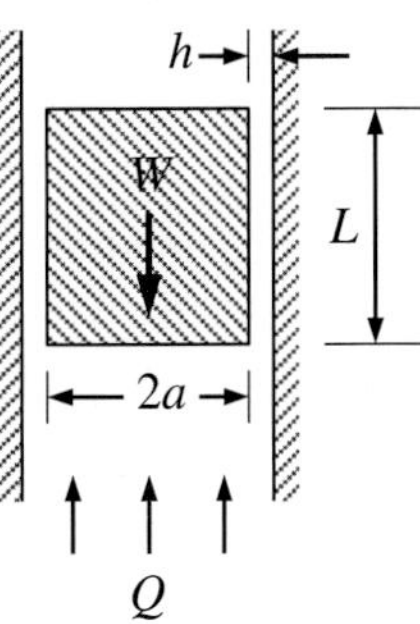

7.28 A thin film of viscous fluid is bounded below by a flat stationary plate at $z = 0$. If the in-plane velocity at the upper film surface, $z = h(x, y, t)$, is $\mathbf{U} = U(x, y, t)\mathbf{e}_x + V(x, y, t)\mathbf{e}_y$, use the equations derived in Exercise 7.25 to produce the Reynolds equation for constant-density, thin-film lubrication:

$$\nabla \cdot \left[\left(\frac{h^3}{\mu} \right) \nabla p \right] = 12 \frac{\partial h}{\partial t} + 6\nabla \cdot (h\mathbf{U}),$$

where $\nabla = \mathbf{e}_x(\partial/\partial x) + \mathbf{e}_y(\partial/\partial y)$ merely involves the two in-plane dimensions.

7.29 Fluid of density ρ and viscosity μ flows inside a long tapered tube of length L and radius $R(x) = (1 - \alpha x/L)R_o$, where $\alpha < 1$ and $R_o \ll L$.

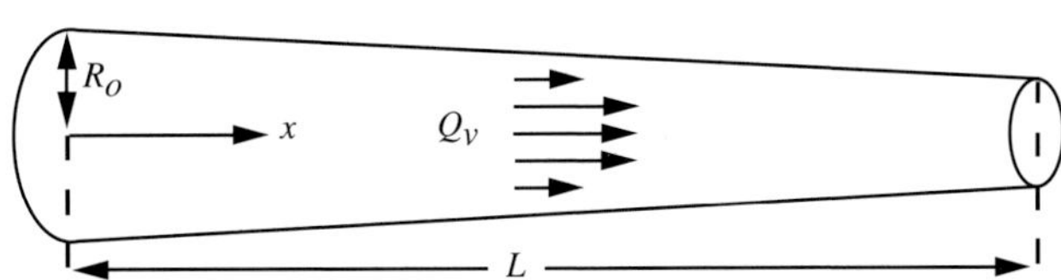

a) Estimate the volume discharge rate Q_v through the tube, for a given pressure difference Δp sustained between the inlet and the outlet.

b) Discuss the range of validity of your solution in terms of the parameters of the problem.

7.30 A circular lubricated bearing of radius a holds a stationary round shaft. The bearing hub rotates at angular rate Ω as shown. A load per unit depth on the shaft, $\mathbf{W}$, causes the center of the shaft to be displaced from the center of the rotating hub by a distance εh_o, where h_o is the average gap thickness and $h_o \ll a$. The gap is filled with an incompressible oil of viscosity μ. Neglect the shear stress contribution to $\mathbf{W}$.

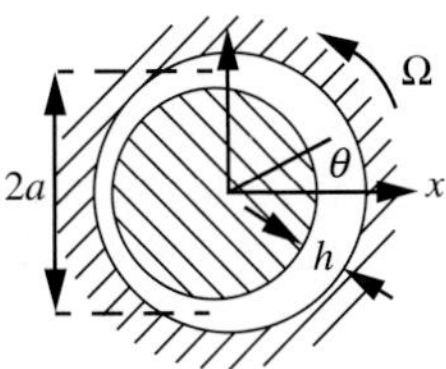

a) Determine a dimensionless scaling law for $|\mathbf{W}|$.

b) Determine $\mathbf{W}$ by assuming a lubrication flow profile in the gap and $h(\theta) = h_o(1 + \varepsilon \cos\theta)$ with $\varepsilon \ll 1$.

c) If $\mathbf{W}$ is increased a little bit, is the lubrication action stabilizing?

7.31 As a simple model of small-artery blood flow, consider slowly varying viscous flow through a round flexible tube with inlet at $z = 0$ and outlet at $z = L$. At $z = 0$, the volume flux entering the tube is $Q_o(t)$. At $z = L$, the pressure equals the exterior pressure p_e. The radius of the tube, $a(z,t)$, expands and contracts in proportion to pressure variations within the tube so that: 1) $a - a_e = \gamma(p - p_e)$, where a_e is the tube radius when the pressure, $p(z,t)$, in the tube is equal to p_e, and γ is a positive constant. Assume the local volume flux, $Q(z,t)$, is related to $\partial p/\partial z$ by 2) $Q = -(\pi a^4/8\mu)(\partial p/\partial z)$.

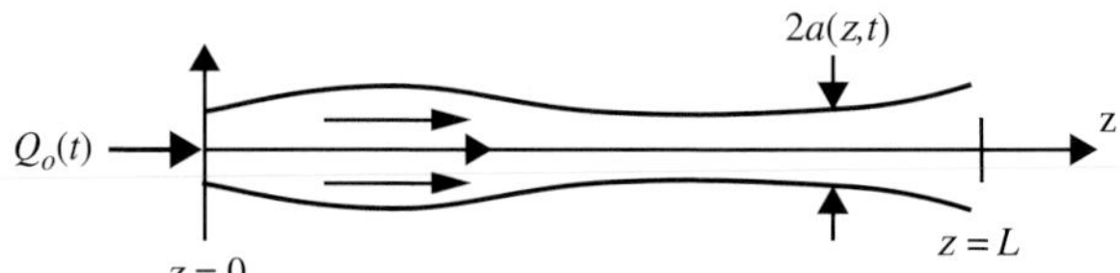

a) By conserving mass, find a partial differential equation that relates Q and a.
b) Combine 1), 2) and the result of part a) into one partial differential equation for $a(z,t)$.
c) Determine $R(z)$ when Q_o is a constant and the flow is perfectly steady.

7.32 Consider a simple model of flow from a tube of toothpaste. A liquid with viscosity μ and density ρ is squeezed out of a round horizontal tube having radius $a(t)$. In your work, assume that a is decreasing and use cylindrical coordinates with the z-axis coincident with the centerline of the tube. The tube is closed at $z = 0$, but is open to the atmosphere at $z = L$. Ignore gravity.

a) If w is the fluid velocity along the z-axis, show that: $za\dfrac{da}{dt} + \displaystyle\int_0^a w(z,R,t)R\,dR = 0$.
b) Determine the pressure distribution, $p(z) - p(L)$, by assuming the flow in the tube can be treated within the lubrication approximation by setting $w(z,R,t) = -\dfrac{1}{4\mu}\dfrac{dp}{dz}(a^2(t) - R^2)$.
c) Find the cross-section-average flow velocity $w_{ave}(z,t)$ in terms of z, a, and da/dt.
d) If the pressure difference between $z = 0$ and $z = L$ is ΔP, what is the volume flux exiting the tube as a function of time. Docs this answer partially explain why fully emptying a toothpaste tube by squeezing it is essentially impossible?

7.33 A uniform rectangular slab of length l (and depth b into the page) is hinged at $x = 0$ to a horizontal flat surface at $y = 0$. The slab is tilted by a small angle θ with respect to the surface and an incompressible fluid with density ρ and viscosity μ fills the narrow triangular gap. Air at pressure p_{atm} acts on the slab and the fluid outside the gap. For the following items, assume: (i) the gap height h is given by $h(x,t) = x\theta(t)$ when θ is measured in radians, (ii) fluid flow in the gap is two dimensional (no flow perpendicular to the page), (iii) the hinge angle is changing slowly at the rate $d\theta/dt$, and (iv) hydrostatic pressure within the gap can be ignored.

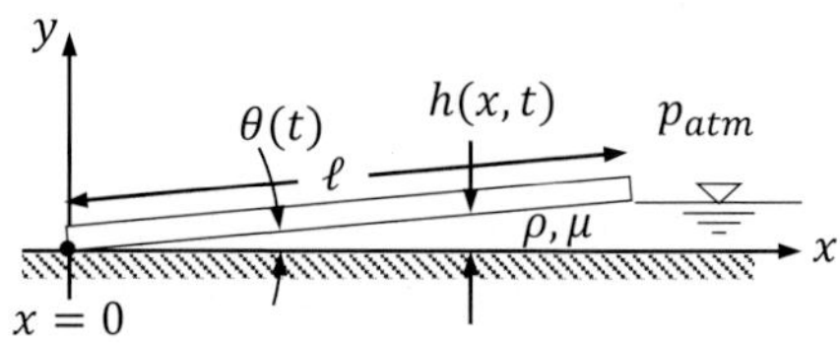

a) Conserve fluid mass using a suitable control volume to deduce:

$$\int_0^{h(x,t)} u(x,y,t)dy = -\frac{1}{2}x^2\frac{d\theta}{dt},$$

where $u(x,y,t)$ is the horizontal fluid velocity in the gap at time t.

b) Assume that $u(x,y,t)$ at any x location can be approximated by the equivalent plane Poiseuille flow profile for that gap thickness,

$$u(x,y,t) = -\frac{h^2}{2\mu}\left(\frac{\partial p}{\partial x}\right)\left(\frac{y}{h}\right)\left(1 - \frac{y}{h}\right),$$

and show that:

$$p(x) - p_{atm} = +\left(6\mu/\theta^3\right)(d\theta/dt)\ln(x/l).$$

c) If $d\theta/dt$ is positive, where under the slab is the highest pressure in the viscous fluid?

d) I f the block's weight per unit depth into the page is W and $\theta = \theta_o$ at $t = 0$, balance moments about the origin, assume the slab's rotary acceleration is negligible, and use the result of part b) to determine $\theta(t)$.

7.34 When a road surface is covered with water, hydroplaning is a significant moving-vehicle safety concern. Hydroplaning occurs when a thin film of water develops between a vehicle's tires and the road surface. When formed, this thin water layer prevents effective steering and braking, and significantly increases a vehicle's stopping distance, even when the vehicle's brakes are locked and its wheels are not rotating.
Consider a steady two-dimensional idealization of this situation in the sliding vehicle's frame of reference where a round (slick) tire with radius R is stationary and the smooth roadway moves to the left below it at speed U, as shown.

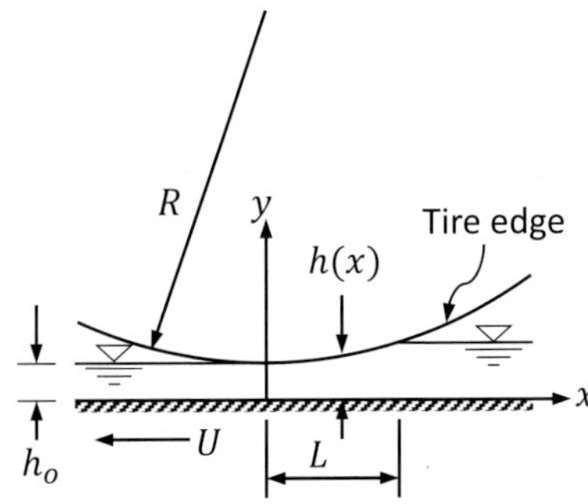

Here the clearance distance between the road and the tire is: $h(x) = h_o(1 + x^2/(2h_o R))$, and this gap is filled with water (density ρ, viscosity μ, both constant) for $0 \le x \le L$. The pressure in the thin water layer upstream and downstream of the tire is p_o. Ignore gravity in the water layer.

a) Assume $h(x) \ll R$ and $h(x) \ll L$, and use the lubrication approximation to determine a formula for the horizontal fluid velocity, $u(x, y)$, in the gap in terms of y, $h(x)$, μ, U, and dp/dx.

b) Use the part a) result and the principal of conservation of mass to reach the following equation for the pressure in the water layer trapped between the tire and the road:

$$\frac{dp}{dx} = -\frac{12\mu}{h^3}\left(C + \frac{Uh}{2}\right),$$

where C is a constant.

c) Define new functions: $\Psi(x) \equiv h_o^2 \int_0^x [1/h^3(\xi)]d\xi$ and $\Phi(x) \equiv h_o \int_0^x [1/h^2(\xi)]d\xi$, and evaluate constants to show that:

$$p(x) - p_o = \frac{6\mu U}{h_o}(\Phi(L) - \Phi(0))\left(\frac{\Psi(x) - \Psi(0)}{\Psi(L) - \Psi(0)} - \frac{\Phi(x) - \Phi(0)}{\Phi(L) - \Phi(0)}\right).$$

d) Evaluate the integrals to explicitly determine $\Psi(x) = \Psi(0)$ and $\Phi(x) = \Phi(0)$.

e) If you are the driver of a vehicle that encounters a long puddle, starts hydroplaning, and then slides uncontrollably. Based on the results given here, what should you do to regain control of the vehicle? Explain your reasoning.

7.35 A large flat plate below an infinite stationary incompressible viscous fluid is set in motion with a constant acceleration, $\dot{u}$, at $t = 0$. A prediction for the subsequent fluid motion, $u(y, t)$, is sought.

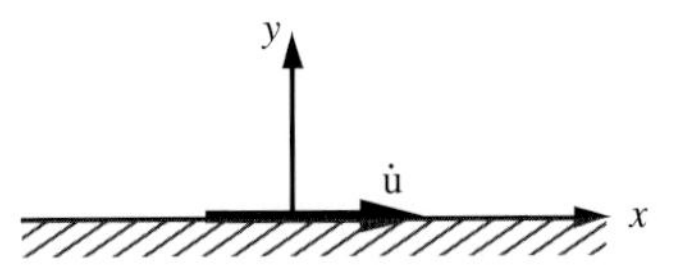

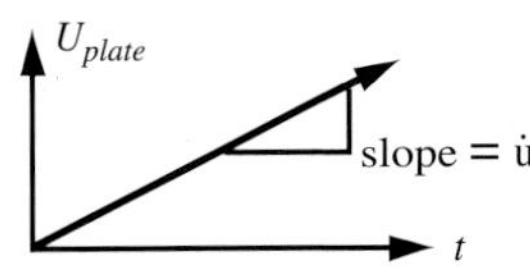

a) Use dimensional analysis to write a physical law for $u(y, t)$ in this flow.
b) Starting from the x-component of (7.1) determine a linear partial differential equation for $u(y, t)$.
c) The linearity of the equation obtained for part c) suggests that $u(y, t)$ must be directly proportional to $\dot{u}$. Simplify your dimensional analysis to incorporate this requirement.
d) Let $\eta = y/(\nu t)^{1/2}$ be the independent variable, and derive a second-order ordinary linear differential equation for the unknown function $f(\eta)$ left from the dimensional analysis.
e) From an analogy between fluid acceleration in this problem and fluid velocity in Stokes' first problem, deduce the solution $u(y, t) = \dot{u} \int_0^t [1 - \text{erf}(y/2\sqrt{\nu t'})]dt'$ and show that it solves the equation of part b).
f) Determine $f(\eta)$ and – if your patience holds out – show that it solves the equation found in part d).
g) Sketch the expected velocity profile shapes for several different times. Note the direction of increasing time on your sketch.

7.36 A quiescent fluid with constant density and constant kinematic viscosity ν fills the x-y half-space above a large horizontal flat plate located at $y = 0$. The plate is stationary until $t = 0$ at which time the plate impulsively moves a distance Δ along the x axis. Assume the resulting time-dependent fluid velocity is entirely horizontal, $\mathbf{u} = (u, 0)$, and is described by:

$$u(y, t) = (\Delta/t)F(\eta)$$

where $\eta = y/\sqrt{4\nu t}$ and F is an undetermined function.

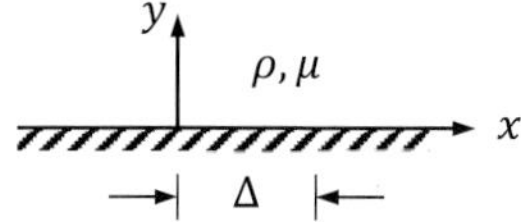

a) Simplify the two-dimensional Navier-Stokes momentum equations, and show that:

$$d^2F(\eta)/d\eta^2 + 2\eta\, dF(\eta)/d\eta + 4F(\eta) = 0.$$

b) What are the boundary conditions on F?
c) Solve the part a) equation and show that $F(\eta) = (\eta\sqrt{\pi})\exp(-\eta^2)$ by considering how viscosity must impact the displacement of fluid particles near the plate.

7.37 **a)** When z is complex, the small-argument expansion of the zeroth-order Bessel function $J_o(z) = 1 - \frac{1}{4}z^2 + \ldots$ remains valid. Use this to show that (7.42) reduces to (7.6) as $\omega \to 0$ when $dp/dz = \Delta p/L$. The next term in the series is $\frac{1}{64}z^4$. At what value of $a/\sqrt{\upsilon/\omega}$ is the magnitude of this term equal to 5% of the second term.
b) When z is complex, the large-argument expansion of the zeroth-order Bessel function $J_o(z) \cong (2/\pi z)^{1/2}\cos\left[z - \frac{1}{4}\pi\right]$ remains valid for $|\arg(z)| < \pi$. Use this to show that (7.42) reduces to the velocity profile of a viscous boundary layer on a plane wall beneath an oscillating flow as $\omega \to \infty$:

$$u_z(y, t) = -\frac{\Delta p}{\rho\omega L}\left[\sin(\omega t) - \exp\left\{-y\sqrt{\frac{\omega}{2\nu}}\right\}\sin\left(\omega t - y\sqrt{\frac{\omega}{2\nu}}\right)\right],$$

where y is the distance from the tube wall, $R = a - y$, $y \ll$ a, and $dp/dz = \Delta p/L$.

7.38 A round tube bent into a U-shape having inner diameter d holds a column of liquid with overall length L. Initially the column of liquid is pushed upward on the right side of the U-tube and downward on the left side of the U-tube so that the two liquid surfaces are a vertical distance $2h_o$ apart. If the liquid has density ρ and viscosity μ, and the column is released from rest, find and solve an approximate ordinary differential equation that describes the subsequent damped oscillations of $h(t)$, the liquid height above equilibrium in the right side of the U-tube, assuming that the flow profile at any time throughout the tube is parabolic. Under what condition(s) is this approximate solution valid? Will oscillations occur in this parameter regime?

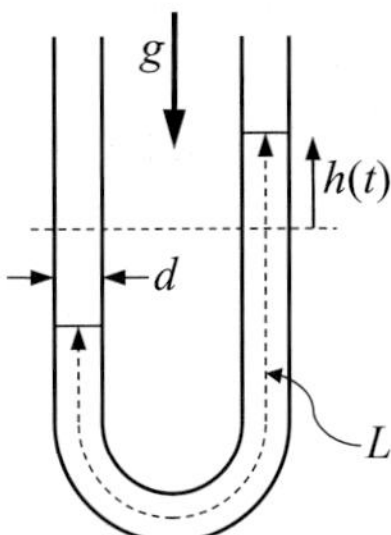

7.39 Suppose a line vortex of circulation Γ is suddenly introduced into a fluid at rest at $t = 0$. Show that the solution is $u_\theta(r, t) = (\Gamma/2\pi r)\exp\{-r^2/4\nu t\}$. Sketch the velocity distribution at different times. Calculate and plot the vorticity, and observe how it diffuses outward.

7.40 Use (7.31a) and an appropriate constraint on the total volume of fluid to determine the form of the similarity solution to the two-dimensional, viscous, drop-spreading equation of Example 7.5.

7.41 Following Taylor and Green (1937), consider the two-dimensional vortex flow field with constant density ρ:

$$\mathbf{u} = (u, v) = (A\sin(kx)\cos(ky),\ B\cos(kx)\sin(ky)).$$

a) If the flow is steady and inviscid, and A and B are constants, explicitly determine the pressure, $p(x, y)$, in terms of x, y, A, ρ, and k from (4.10) and (7.1) in two dimensions.

b) If the flow field is unsteady and viscous (with viscosity μ), A and B are functions of time t, and $A = A_o$ at $t = 0$, determine $A(t)$, $B(t)$, and $p(x, y, t)$ so that the given $\mathbf{u}$ is an exact solution of (4.10) and (7.1) in two dimensions.

c) How long does it take for $A(t)$ to fall to $A_o/2$? Does the parametric dependence of this decay time follow typical diffusion scaling?

7.42 Consider a collection of solid, spherical particles placed in the steady two-dimensional vortex flow field from part a) of Exercise 7.41 with $k = 2\pi$ and $A = 1$ as shown in the figure. Each particle has radius a and density ρ_p.

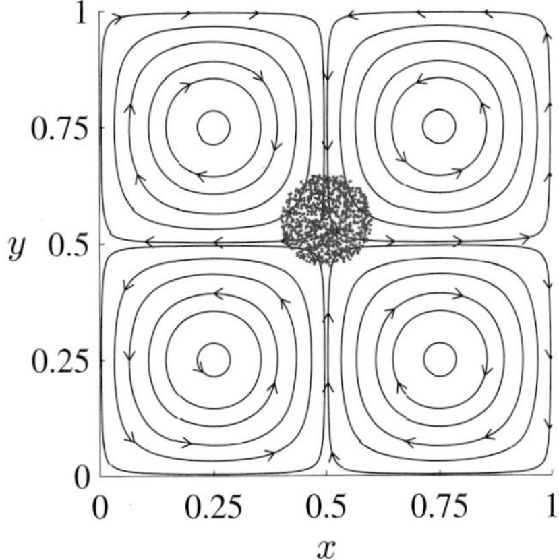

a) Neglecting body forces such as gravity and assuming the particles are sufficiently small so their motion can be described by Stokes' drag, the equation of motion for each particle is given by

$$m_p\frac{d\mathbf{v}}{dt} = 6\pi\mu a(\mathbf{u} - \mathbf{v}),$$

where μ is the fluid viscosity and $m_p = 4/3\rho_p\pi a^3$ is the particle's mass. Using a reference velocity scale U and reference fluid time scale τ_f, show that the non-dimensional particle acceleration can be expressed in terms of a dimensionless Stokes number, $\text{St} = \tau_p/\tau_f$, where τ_p is the particle response time. Provide a physical explanation of the Stokes number.

b) Write a code that numerically integrates 1000 particles in a $[0, 1] \times [0, 1]$ domain. Randomly distribute the particles within a small circle just above the center of the Taylor-Green vortex with radius $r = 0.1$ and center $x_0 = 0.5$ and $y_0 = 0.55$, as shown in the figure above. Placing the particles off center breaks the symmetry of their motion, and thus makes the effect of the Stokes number easier to observe. Set the initial particle velocity to the fluid velocity at its location. Apply periodic boundary conditions to the particle positions when they cross the domain boundary. [*Hint*: the particle acceleration can be discretized using a simple first-order Euler method according to $d\mathbf{v}^*/dt^* \approx (\mathbf{v}^{*n+1} - \mathbf{v}^{*n})/\Delta t$, where Δt is the time step size and n is the iteration number.]

c) Using the code generated in a), solve for the particle positions until a non-dimensional time of $t^* = 2$ is reached with a time step size $\Delta t = 0.005$. The particle behavior depends on a critical Stokes number $\mathrm{St}_c = 1/(8\pi)$. What do you observe as you vary St below and above St_c? below and above this value? Are these results expected?

7.43 Obtain several liquids of differing viscosity (water, cooking oil, pancake syrup, shampoo, etc.). Using an eyedropper, a small spoon, or your finger, place a drop of each on a smooth vertical surface (a bathroom mirror perhaps) and measure how far the drops have moved or extended in a known period of time (perhaps a minute or two). Try to make the mass of all the drops equal. Using dimensional analysis, determine how the drop-sliding distance depends on the other parameters. Does this match your experimental results?

7.44 A drop of an incompressible viscous liquid is allowed to spread on a flat horizontal surface under the action of gravity. Assume the drop spreads in an axisymmetric fashion and use cylindrical coordinates (R, φ, z). Ignore the effects of surface tension.

a) Show that conservation of mass implies: $\dfrac{\partial h}{\partial t} + \dfrac{1}{R}\dfrac{\partial}{\partial R}\left(R\int_o^h u\,dz\right) = 0$, where $u = u(R, z, t)$ is the horizontal velocity within the drop, and $h = h(R, t)$ is the thickness of the spreading drop.

b) Assume that the lubrication approximation applies to the horizontal velocity profile, that is, $u(R, z, t) = a(R, t) + b(R, t)z + c(R, t)z^2$, apply the appropriate boundary conditions on the upper and lower drop surfaces, and require a pressure and shear-stress force balance within a differential control volume $h(R, t)RdRd\theta$ to show that: $u(R, z, t) = -\dfrac{g}{2\nu}\dfrac{\partial h}{\partial R}z(2h - z)$.

c) Combine the results of a) and b) to find $\dfrac{\partial h}{\partial t} = \dfrac{g}{3\nu R}\dfrac{\partial}{\partial R}\left(Rh^3\dfrac{\partial h}{\partial R}\right)$.

d) Assume a similarity solution: $h(R, t) = \dfrac{A}{t^n}f(\eta)$ with $\eta = \dfrac{BR}{t^m}$, use the result of part c) and $2\pi\int_o^{R_{\max}(t)} h(R, t)RdR = V$, where $R_{max}(t)$ is the radius of the spreading drop and V is the initial volume of the drop to determine $m = 1/8$, $n = 1/44$, and a single nonlinear ordinary differential equation for $f(\eta)$ involving only A, B, g/ν, and η. You need not solve this equation for f. [Given that $f \to 0$ as $\eta \to \infty$, there will be a finite value of η for which f is effectively zero. If this value of η is $\eta_{\max}$ then the radius of the spreading drop, $R(t)$, will be: $R_{\max}(t) = \eta_{\max}t^m/B$.]

7.45 Obtain a clean, flat glass plate, a watch, a ruler, and some non-volatile oil that is more viscous than water. The plate and oil should be at room temperature. Dip the tip of one of your fingers in the oil and smear the oil over the center of the plate so that a thin bubble-free oil film covers a circular area ~10 to 15 cm in diameter. Set the plate on a horizontal surface and place a single drop of oil at the center of the oil-film area and observe how the drop spreads. Measure the spreading drop's diameter 1, 10, 10^2, 10^3, and 10^4 seconds after the drop is placed on the plate. Plot your results and determine if the spreading drop diameter grows as $t^{1/8}$ (the predicted drop-diameter time dependence from the prior problem) to within experimental error.

7.46 A fine stream of a viscous fluid with density ρ and viscosity μ falls slowly at a constant volume flow rate Q onto the center of a flat horizontal circular disk of radius R. The fluid flows steadily under the action of gravity g from the center of the disk to its edge in a layer of thickness $h(r)$, where r is the radial coordinate. For the following items, assume Q is constant, and apply the approximate boundary condition $h(R^+) = 0$, where R^+ is a radial location just beyond the edge of the disk:

a) Determine a scaling law for h from dimensional analysis.

b) Using the lubrication approximation determine a formula for $h(r)$ that is valid for $0 < r < R$.

c) Increasing which parameters increase the thickness of the fluid layer on the disk?

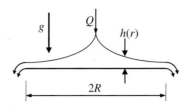

7.47 An infinite flat plate located at $y = 0$ is stationary until $t = 0$ when it begins moving horizontally in the positive x-direction at a constant speed U. This motion continues until $t = T$ when the plate suddenly stops moving.

a) Determine the fluid velocity field, $u(y, t)$ for $t > T$. At what height above the plate does the peak velocity occur for $t > T$? [*Hint*: the governing equation is linear so superposition of solutions is possible.]

b) Determine the mechanical impulse I (per unit depth and length) imparted to the fluid while the plate is moving: $I = \int_0^T \tau_w dt$.

c) As $t \to \infty$, the fluid slows down and eventually stops moving. How and where was the mechanical impulse dissipated? What is t/T when 99% of the initial impulse has been lost?

7.48 Consider the development from rest of plane Couette flow. The flow is bounded by two rigid boundaries at $y = 0$ and $y = h$, and the motion is started from rest by suddenly accelerating the lower plate to a steady velocity U. The upper plate is held stationary. Here a similarity solution cannot exist because of the appearance of the parameter h. Show that the velocity distribution is given by:

$$u(y, t) = U\left(1 - \frac{y}{h}\right) - \frac{2U}{\pi}\sum_{n=1}^{\infty}\frac{1}{n}\exp\left(-n^2\pi^2\frac{\nu t}{h^2}\right)\sin\left(\frac{n\pi y}{h}\right).$$

Sketch the flow pattern at various times, and observe how the velocity reaches the linear distribution for large times.

7.49 [4]Two-dimensional flow between flat non-parallel plates can be formulated in terms of a normalized angular coordinate, $\eta = \theta/\alpha$, where α is the half angle between the plates, and a normalized radial velocity, $u_r(r, \theta) = u_{\max}(r)f(\eta)$, where $\eta = \theta/\alpha$ for $|\theta| \le \alpha$. Here, $u_\theta = 0$, the Reynolds number is $\mathrm{Re} = u_{\max} r\alpha/\nu$, and Q is the volume flux (per unit width perpendicular to the page).

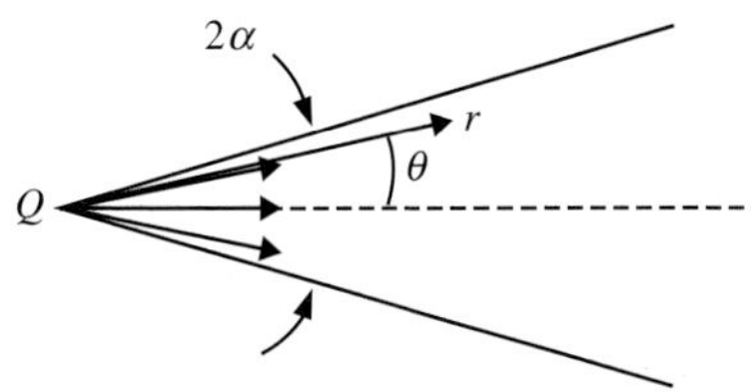

a) Using the appropriate versions of (4.10) and (7.1), show that $f'' + \mathrm{Re}\,\alpha f^2 + 4\alpha^2 f = const.$

b) Find $f(\eta)$ for symmetric creeping flow, that is, $\mathrm{Re} = 0 = f(+1) = f(-1)$, and $f(0) = 1$.

c) Above what value of the channel half-angle will backflow always occur?

7.50 Consider steady viscous flow inside a cone of constant angle θ_o. The flow has constant volume flux $= Q$, and the fluid has constant density $= \rho$ and constant kinematic viscosity $= \nu$. Use spherical coordinates, and assume that the flow only has a radial component, $\mathbf{u} = (u_r(r, \theta), 0, 0)$, which is independent of the azimuthal angle φ, so that the equations of motion are:

- Conservation of mass:

$$\frac{1}{r^2}\frac{\partial}{\partial r}(r^2 u_r) = 0,$$

- Conservation of radial momentum:

$$u_r\frac{\partial u_r}{\partial r} = -\frac{1}{\rho}\frac{\partial p}{\partial r} + \nu\left(\frac{1}{r^2}\frac{\partial}{\partial r}\left(r^2\frac{\partial u_r}{\partial r}\right) + \frac{1}{r^2 \sin\theta}\frac{\partial}{\partial\theta}\left(\sin\theta\frac{\partial u_r}{\partial\theta}\right) - \frac{2}{r^2}u_r\right),$$

4. Rephrased from White (2006) p. 211, problem 3.32.

- Conservation of θ-momentum:

$$0 = -\frac{1}{\rho r}\frac{\partial p}{\partial \theta} + \nu\left(\frac{2}{r^2}\frac{\partial u_r}{\partial \theta}\right).$$

For the following items, assume the radial velocity can be determined using: $u_r(r,\theta) = QR(r)\Theta(\theta)$. Define the Reynolds number of this flow as: $\mathrm{Re} = Q/(\pi \nu r)$.

a) Use the continuity equation to determine $R(r)$.

b) Integrate the θ-momentum equation, assume the constant of integration is zero, and combine the result with the radial momentum equation to determine a single differential equation for $\Theta(\theta)$ in terms of θ and Re.

c) State the matching and/or boundary conditions that $\Theta(\theta)$ must satisfy.

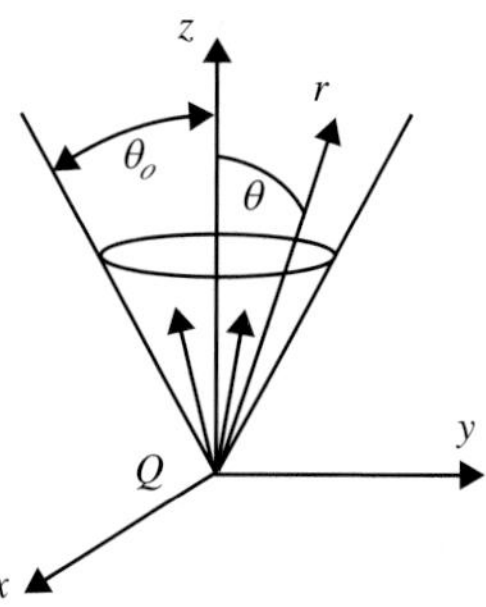

7.51 The boundary conditions on obstacles in Hele-Shaw flow were not considered in Example 7.4. Therefore, consider them here by examining Hele-Shaw flow parallel to a flat obstacle surface at $y = 0$. The Hele-Shaw potential in this case is:

$$\phi = Ux\frac{z}{h}\left(1 - \frac{z}{h}\right),$$

where (x, y, z) are Cartesian coordinates and the flow is confined to $0 < z < h$ and $y > 0$.

a) Show that this potential leads to a slip velocity of $u(x, y \to 0) = U(z/h)(1 - z/h)$, and determine the pressure distribution implied by this potential.

b) Since this is a viscous flow, the slip velocity must be corrected to match the genuine no-slip condition on the obstacle's surface at $y = 0$. The analysis of Example 7.4 did not contain the correct scaling for this situation near $y = 0$. Therefore, rescale the x-component of (7.1) using:

$$x^* = x/L,\ y^* = y/h = y/\varepsilon L,\ z^* = z/h = z/\varepsilon L,\ t^* = Ut/L,$$
$$u^* = u/U,\ v^* = v/\varepsilon U,\ w^* = w/\varepsilon U,\quad \text{and}\quad p^* = p/P_a,$$

and then take the limit as $\varepsilon^2 \mathrm{Re}_L \to 0$, with $\mu UL/P_a h^2$ remaining of order unity, to simplify the resulting dimensionless equation that has:

$$0 \cong \frac{dp}{dx} + \mu\left(\frac{\partial^2 u}{\partial y^2} + \frac{\partial^2 u}{\partial z^2}\right)$$

as its dimensional counterpart.

c) Using boundary conditions of $u = 0$ on $y = 0$, and $u = U(z/h)(1 - z/h)$ for $y \gg h$. Show that

$$u(x, y, z) = U\frac{z}{h}\left(1 - \frac{z}{h}\right) + \sum_{n=1}^{\infty} A_n \sin\left(\frac{n\pi}{h}z\right)\exp\left(-\frac{n\pi}{h}y\right), \quad \text{where}$$

$$A_n = -\frac{2U}{h}\int_0^h \frac{z}{h}\left(1 - \frac{z}{h}\right)\sin\left(\frac{n\pi}{h}z\right)dz,$$

which implies: $A_n = -8U/n^3\pi^3$ for $n =$ odd, and $A_n = 0$ for $n =$ even. [The results here are directly applicable to the surfaces of curved obstacles in Hele-Shaw flow when the obstacle's radius of curvature is much greater than h.]

7.52 Using the velocity field (7.48), determine the drag on Stokes' sphere from the surface pressure and the viscous surface stresses σ_{rr} and $\sigma_{r\theta}$. [*Hint*: the appropriate sum of surface force components is independent of θ.]

7.53 Compare ideal (inviscid) flow past a sphere, Eq. (6.88), with Stokes' flow past a sphere, Eq. (7.53). In the spherical polar coordinates shown in Fig. 7.17, where the z-axis is horizontal, a uniform stream that flows parallel to the z-axis is $U\,\mathbf{e}_z = U\cos\theta\,\mathbf{e}_r - U\sin\theta\,\mathbf{e}_\theta$.

a) Using these same polar coordinates centered on a stationary sphere with radius a, determine the flow perturbation for $r > a$ to the uniform stream,

$$\mathbf{u}' = [\mathbf{u}]_{\text{with sphere}} - U\,\mathbf{e}_z,$$

caused by the presence of the sphere in each case.

b) Where are the radial flow velocity perturbations, $\mathbf{u}'\cdot\mathbf{e}_r$, for Stokes' and ideal flow equal? Beyond what radial distance is $\mathbf{u}'\cdot\mathbf{e}_r$ for Stokes' flow always greater than $\mathbf{u}'\cdot\mathbf{e}_r$ for ideal flow?

c) For what values of θ is the angular flow velocity perturbation, $\mathbf{u}'\cdot\mathbf{e}_\theta$, positive in each case?

d) To leading order, what is the ratio $|\mathbf{u}'|_{\text{Stokes flow}}/|\mathbf{u}'|_{\text{Ideal flow}}$ when $r \gg a$?

e) If spherical particles are added to a flowing liquid, are fluid-mediated particle-particle interactions more important in viscous or inviscid flow? Explain your answer. Does it have any practical implications?

7.54 Calculate the drag on a spherical droplet of radius $r = a$, density ρ' and viscosity μ' moving with velocity U in an infinite fluid of density ρ and viscosity μ. Assume $\mathrm{Re} = \rho U a/\mu \ll 1$. Neglect surface tension.

7.55 Consider a very low Reynolds number flow over a circular cylinder of radius $r = a$. For $r/a = O(1)$ in the $\mathrm{Re} = Ua/\nu \to 0$ limit, find the equation governing the stream function $\psi(r,\theta)$ and solve for ψ with the least singular behavior for large r. There will be one remaining constant of integration to be determined by asymptotic matching with the large r solution (which is not part of this problem). Find the domain of validity of your solution.

7.56 A small, neutrally buoyant sphere is centered at the origin of coordinates in a deep bath of a quiescent viscous fluid with density ρ and viscosity μ. The sphere has radius a and is initially at rest. It begins rotating about the z-axis with a constant angular velocity Ω at $t = 0$. The relevant equations for the fluid velocity, $\mathbf{u} = (u_r, u_\theta, u_\varphi)$, in spherical coordinates (r, θ, φ) are:

$$\frac{1}{r^2}\frac{\partial}{\partial r}(r^2 u_r) + \frac{1}{r\sin\theta}\frac{\partial}{\partial\theta}(u_\theta\sin\theta) + \frac{1}{r\sin\theta}\frac{\partial}{\partial\phi}(u_\varphi) = 0, \quad \text{and}$$

$$\begin{aligned}
\frac{\partial u_\varphi}{\partial t} + u_r\frac{\partial u_\varphi}{\partial r} + \frac{u_\theta}{r}\frac{\partial u_\varphi}{\partial\theta} + \frac{u_\varphi}{r\sin\theta}\frac{\partial u_\varphi}{\partial\varphi} + \frac{1}{r}(u_r u_\varphi + u_\theta u_\varphi\cot\theta) = -\frac{1}{\rho r\sin\theta}\frac{\partial p}{\partial\phi} \\
+\nu\left(\frac{1}{r^2}\frac{\partial}{\partial r}\left(r^2\frac{\partial u_\phi}{\partial r}\right) + \frac{1}{r^2\sin\theta}\frac{\partial}{\partial\theta}\left(\sin\theta\frac{\partial u_\varphi}{\partial\theta}\right) + \frac{1}{r^2\sin^2\theta}\frac{\partial^2 u_\varphi}{\partial\varphi^2} - \frac{u_\varphi}{r^2\sin^2\theta}\right. \\
\left. + \frac{2}{r^2\sin^2\theta}\frac{\partial u_r}{\partial\varphi} + \frac{2\cos\theta}{r^2\sin^2\theta}\frac{\partial u_\theta}{\partial\varphi}\right).
\end{aligned}$$

a) Assume $\mathbf{u} = (0, 0, u_\varphi)$ and reduce these equations to:

$$\frac{\partial u_\varphi}{\partial t} = -\frac{1}{\rho r\sin\theta}\frac{\partial p}{\partial\phi} + \nu\left(\frac{1}{r^2}\frac{\partial}{\partial r}\left(r^2\frac{\partial u_\varphi}{\partial r}\right) + \frac{1}{r^2\sin\theta}\frac{\partial}{\partial\theta}\left(\sin\theta\frac{\partial u_\varphi}{\partial\theta}\right) - \frac{u_\varphi}{r^2\sin^2\theta}\right).$$

b) Set $u_\varphi(r,\theta,t) = \Omega a F(r,t)\sin\theta$, make an appropriate assumption about the pressure field, and derive the following equation for F: $\dfrac{\partial F}{\partial t} = \nu\left(\dfrac{1}{r^2}\dfrac{\partial}{\partial r}\left(r^2\dfrac{\partial F}{\partial r}\right) - 2\dfrac{F}{r^2}\right)$.

c) Determine F for $t \to \infty$ for boundary conditions $F = 1$ at $r = a$, and $F \to 0$ as $r \to \infty$.

d) Find the surface shear stress and torque on the sphere.

7.57 Consider a sphere with density ρ_s and radius a submerged in a fluid with density ρ and viscosity μ.

a) Neglecting initial transient effects, show that the response time of the sphere to changes in fluid motion is $\tau_s = 2\rho_s a^2/(9\mu)$.

b) Consider the fluid to be at rest and the sphere is accelerating with characteristic velocity U_0 and timescale τ_s. Show that the ratio of the history force (7.59) compared to Stokes' drag (7.55) is

$$\frac{\text{Basset}}{\text{Stokes}} = \left(\frac{18}{\pi}\frac{\rho}{\rho_s}\right)^{1/2}.$$

Comment on the relative importance of the history term compared to Stokes drag for an air bubble in water, solid particle in water, and solid particle in air.

c) Repeat part b) for the case when the fluid is accelerating and the sphere is initially at rest. Assume the ambient flow has a characteristic length scale L and velocity scale U_0. Show that the ratio of the history force to Stokes' drag is independent of density ratio, and instead scales like

$$\frac{\text{Basset}}{\text{Stokes}} = \left(\frac{2}{\pi}\frac{a}{L}\text{Re}\right)^{1/2}.$$

7.58 Consider the geometry of a cone and plate rheometer. A flat cone with radius R and apex angle of θ_1, slightly less than $\pi/2$, touches a large stationary horizontal flat plate at the origin and rotates at a constant rate Ω about the vertical z-axis. A viscous fluid with density ρ and viscosity μ fills the gap between the cone and the plate.

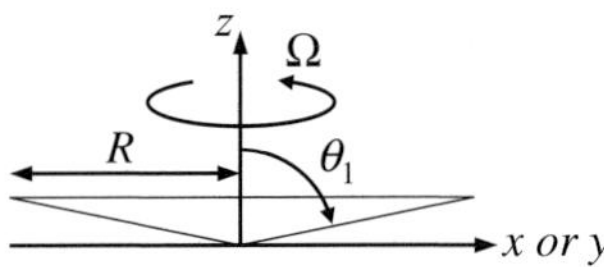

a) Assuming the fluid's velocity is steady with a single component: $\mathbf{u} = \mathbf{e}_\varphi u_\varphi(r,\theta)$, use the continuity equation and the azimuthal momentum equation for creeping flow in spherical coordinates, where $r = \sqrt{x^2+y^2+z^2}$, to find:

$$u_\varphi(r,\theta) = \Omega r \sin\theta_1 \left[\frac{\frac{1}{2}\sin\theta \cdot \ln\left(\frac{1+\cos\theta}{1-\cos\theta}\right) + \cot\theta}{\frac{1}{2}\sin\theta_1 \cdot \ln\left(\frac{1+\cos\theta_1}{1-\cos\theta_1}\right) + \cot\theta_1}\right].$$

The boundary conditions here are $u_\varphi(r,\pi/2) = 0$, and $u_\phi(r,\theta_1) = r\Omega\sin\theta_1$.

b) Use this to determine the polar-azimuthal shear stress: $\tau_{\theta\varphi} = \mu\left[\frac{\sin\theta}{r}\frac{\partial}{\partial\theta}\left(\frac{u_\varphi}{\sin\theta}\right) + \frac{1}{r\sin\theta}\frac{\partial u_\theta}{\partial\phi}\right]$.

c) Simplify the velocity field and shear stress results when θ and θ_1 both approach $\pi/2$.

d) A torque of 3.0 N-m causes the cone to rotate with an angular velocity of 1.5 rad/s. If the radius of the cone is $R = 6.35$ cm and $90° - \theta_1 = 0.30°$, what is the viscosity of the fluid?

e) For the conditions in part d) with $\rho = 10^3$ kg/m^3, compare $\rho\Omega^2R^2$ to $\tau_{\theta\varphi}$. Is neglect of fluid inertia justified here?

References

Abramowitz, M., Stegun, I.A., 1972. Handbook of Mathematical Functions. U.S. Department of Commerce, National Bureau of Standards, Washington, DC.

Basset, A.B., 1888. III. On the motion of a sphere in a viscous liquid. Philosophical Transactions of the Royal Society of London 179, 43–63.

Conlisk, A.T., 2013. Essentials of Micro- and Nanofluidics with Applications to the Biological and Chemical Sciences. Cambridge University Press.

Crowe, C., Schwarzkopf, J., Sommerfeld, M., Tsuji, Y., 2011. Multiphase Flows with Droplets and Particles, 2nd edition. CRC Press.

Hele-Shaw, H.S., 1898. Investigations of the nature of surface resistance of water and of stream line motion under certain experimental conditions. Transactions of the Royal Institution of Naval Architects 40, 21–46.

Kirby, B.J., 2010. Micro- and Nanoscale Fluid Mechanics. Cambridge University Press.

Kurup, D.G., Koithyar, A., 2013. New expansions of Bessel functions of first kind of complex argument. IEEE Transactions on Antennas and Propagation 61, 2708–2713.

Millikan, R.A., 1911. The isolation of an ion, a precision measurement of its charge, and the correction of Stokes' law. Physical Review 32, 349–397.

Oseen, C.W., 1910. Über die Stokes'sche Formel, und über eine verwandte Aufgabe in der Hydrodynamik. Arkiv för Matematik, Astronomi och Fysik 6 (29).

Sherman, F.S., 1990. Viscous Flow. McGraw-Hill, New York.

Taylor, G.I., Green, A.E., 1937. Mechanism of the production of small eddies from large ones. Proceedings of the Royal Society of London A 158, 499–521.

White, F.M., 2006. Viscous Fluid Flow. McGraw-Hill, New York.

Womersley, J.R., 1955a. Method for the calculation of velocity, rate of flow and viscous drag in arteries when the pressure gradient is known. Journal of Physiology 127, 553–563.

Womersley, J.R., 1955b. Oscillatory motion of a viscous liquid in a thin-walled elastic tube. I. The linear approximation for long waves. Philosophical Magazine 46, 199–221.

Yih, C.-S., 1979. Fluid Mechanics. West River Press, Ann Arbor.

Further reading

Batchelor, G.K., 1967. An Introduction to Fluid Dynamics. Cambridge University Press, London.

Lighthill, M.J., 1986. An Informal Introduction to Theoretical Fluid Mechanics. Clarendon Press, Oxford, England.

Schlichting, H., 1979. Boundary Layer Theory. McGraw-Hill, New York.

Chapter 8

Boundary layers and related topics⍟

Contents

8.1 **Introduction** 271
8.2 **Boundary-layer thickness definitions** 275
8.3 **Boundary layer on a flat plate: Blasius solution** 277
8.4 **Falkner-Skan similarity solutions of the laminar boundary-layer equations** 282
8.5 **von Karman momentum integral equation** 283
8.6 **Thwaites' method** 285
8.7 **Transition, pressure gradients, and boundary-layer separation** 289
8.8 **Flow past a circular cylinder** 294
Low Reynolds numbers 294
Moderate Reynolds numbers 294
High Reynolds numbers 296
8.9 **Flow past a sphere** 299
Saffman lift 300
Magnus effect 300
8.10 **Two-dimensional jets** 303
8.11 **Secondary flows** 308
Exercises 309
References 317
Further reading 318

Chapter objectives

- To describe the boundary-layer concept and the mathematical simplifications it allows in the complete equations of motion for a viscous fluid
- To present the equations of fluid motion for attached laminar boundary layers
- To provide a variety of exact and approximate steady laminar boundary-layer solutions
- To describe the basic phenomenology of boundary-layer transition and separation
- To discuss the Reynolds number dependent phenomena associated with flow past bluff bodies
- To illustrate the use of a boundary-layer solution methodology for free and wall-bounded jet flows

8.1 Introduction

Until the beginning of the twentieth century, analytical solutions of steady fluid flows were generally known for two typical situations. One of these was that of parallel viscous flows at sufficiently low Reynolds numbers, in which the nonlinear advective terms were zero (or small), and pressure and viscous forces balanced. The second type of solution was that of inviscid flows around bodies of various shapes, in which inertia and pressure forces balanced. Although the equations of motion are nonlinear in this second situation, the velocity field can be determined by solving the linear Laplace equation. These irrotational solutions predicted pressure forces on a streamlined body that agreed surprisingly well with experimental data for flow of fluids of small viscosity. However, these solutions also predicted zero drag force and a non-zero tangential velocity at the body surface, features that did not agree with the experiments.

In 1905, Ludwig Prandtl, an engineer by profession and therefore motivated to find realistic fields near bodies of various shapes, first hypothesized that, for small viscosity, the viscous forces are negligible everywhere except close to solid boundaries where the no-slip condition had to be satisfied. The thickness of these boundary layers approaches zero as the viscosity goes to zero. Prandtl's hypothesis reconciled two rather contradictory facts. It supported the intuitive idea that the effects of viscosity are indeed negligible in most of the flow field if ν is small, but it also accounted for drag by insisting that the no-slip condition must be satisfied at a solid surface, no matter how small the viscosity. This reconciliation was Prandtl's aim, which he achieved brilliantly. Prandtl also showed how the full equations of fluid motion within the boundary layer can be simplified. Since the time of Prandtl, the concept of the boundary layer has been generalized, and the mathematical techniques involved have been formalized, extended, and applied in other branches of physical science (see van Dyke, 1975; Bender and Orszag, 1978; Kevorkian and Cole, 1981; Nayfeh, 1981). The boundary-layer concept is considered a cornerstone in the intellectual foundation of fluid mechanics.

⍟. This book has a companion website hosting complementary materials. Visit this URL to access it: https://www.elsevier.com/books-and-journals/book-companion/9780128198070.

Fluid Mechanics. https://doi.org/10.1016/B978-0-12-819807-0.00017-X

This chapter presents the boundary-layer hypothesis and examines its consequences. The equations of fluid motion within the boundary layer can be simplified because of the layer's thinness, and exact or approximate solutions can be obtained in many cases. In addition, boundary-layer phenomena provide explanations for the lift and drag characteristics of bodies of various shapes in high Reynolds number flows, including turbulent flows.

The fundamental assumption of boundary-layer theory is that the layer is thin compared to other length scales such as the length or radius of curvature of the surface on which the boundary layer develops. Across this thin layer, which can exist only in high Reynolds number flows, the velocity varies rapidly enough for viscous effects to be important. This is depicted in Fig. 8.1, where the boundary-layer thickness is greatly exaggerated. (On a commercial aircraft wing, which may have a chord of several meters, the boundary-layer thickness is of order one centimeter.) However, thin viscous layers exist not only next to solid walls but also in the form of jets, wakes, and shear layers if the Reynolds number is sufficiently high. So, to be specific, we shall first consider the boundary layer contiguous to a solid surface, adopting a curving coordinate system that conforms to the surface where x increases along the surface and y increases normal to it. Here the surface may be curved but its radius of curvature is assumed to be much larger than the boundary-layer thickness. We shall refer to the solution of the irrotational flow outside the boundary layer as the *outer* problem and that of the boundary-layer flow as the *inner* problem.

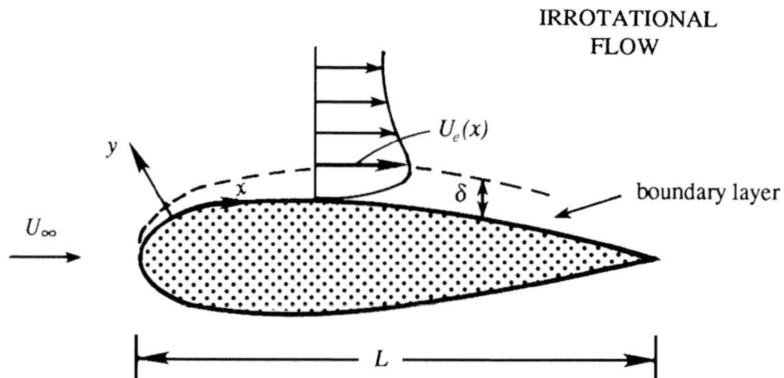

FIGURE 8.1 A boundary layer forms when a viscous fluid moves over a solid surface. Only the boundary layer on the top surface of the foil is depicted in the figure and its thickness, δ, is greatly exaggerated. Here, U_∞ is the oncoming free-stream velocity and U_e is the velocity at the edge of the boundary layer. The usual boundary layer coordinate system allows the x-axis to coincide with a mildly curved surface so that the y-axis lies in the surface-normal direction.

For a thin boundary layer that is contiguous to the solid surface on which it has formed, the full equations of motion for a constant-density constant-viscosity fluid, (4.10) and (7.1), may be simplified. Let $\overline{\delta}(x)$ be the average thickness of the boundary layer at downstream location x on the surface of a body having a local radius of curvature R. The steady-flow momentum equation for the surface-parallel velocity component, u, is:

$$u\frac{\partial u}{\partial x}+v\frac{\partial u}{\partial y}=-\frac{1}{\rho}\frac{\partial p}{\partial x}+\nu\left(\frac{\partial^2 u}{\partial x^2}+\frac{\partial^2 u}{\partial y^2}\right), \tag{8.1}$$

which is valid when $\overline{\delta}/R \ll 1$. The more general curvilinear form for arbitrary $R(x)$ is given in Goldstein (1938) and Schlichting (1979), but the essential features of viscous boundary layers can be illustrated via (8.1) without additional complications.

Let the characteristic magnitude of u be U_∞, the velocity at a large distance upstream of the body, and let L be the stream-wise distance over which u changes appreciably. The longitudinal length of the body can serve as L, because u within the boundary layer may change in the stream-wise direction by a large fraction of U_∞ over a distance L (Fig. 8.2). A measure of $\partial u/\partial x$ is therefore U_∞/L, so that the approximate size of the first advective term in (7.1) is:

$$u(\partial u/\partial x) \sim U_\infty^2/L, \tag{8.2}$$

where $\sim$ is to be interpreted as "of order." We shall see shortly that the other advective term in (8.1) is of the same order. The approximate size of the viscous stress term in (8.1) is:

$$(1/\rho)(\partial \tau/\partial y) = \nu(\partial^2 u/\partial y^2) \sim \nu U_\infty/\overline{\delta}^2. \tag{8.3}$$

The magnitude of $\overline{\delta}$ can now be estimated by noting that the advective and viscous terms should be of the same order within the boundary layer. Equating the magnitudes of advective and viscous terms in (8.2) and (8.3) leads to:

$$\overline{\delta} \sim \sqrt{\nu L/U_\infty} \quad \text{or} \quad \overline{\delta}/L \sim 1/\sqrt{\mathrm{Re}}. \tag{8.4}$$

This estimate of $\overline{\delta}$ can also be obtained by noting that viscous effects diffuse to a distance of order $[\nu t]^{1/2}$ in time t and that the time-of-flight for a fluid element along a body of length L is of order L/U_∞. Substituting L/U_∞ for t in $[\nu t]^{1/2}$ suggests the viscous layer's diffusive thickness at $x = L$ is of order $[\nu L/U_\infty]^{1/2}$.

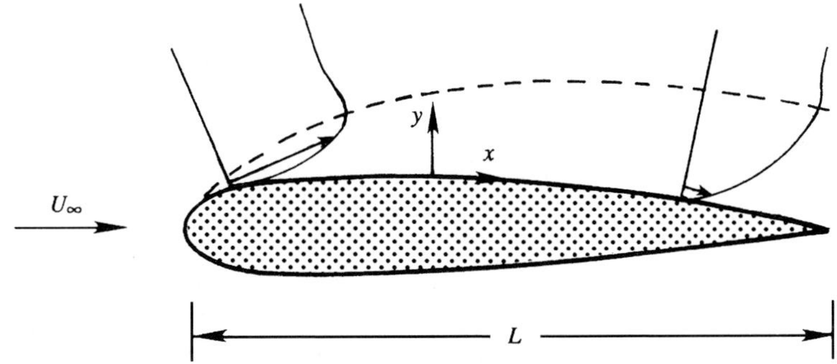

FIGURE 8.2 Velocity profiles at two positions within the boundary layer. Here again, the boundary-layer thickness is greatly exaggerated. The two velocity vectors are drawn at the same distance y from the surface, showing that the variation of u over a distance $x \sim L$ is of the order of the free-stream velocity U_∞.

A formal simplification of the equations of motion within the boundary layer can now be performed. The basic idea is that variations across the boundary layer occur over a much shorter length scale than variations along the layer, that is:

$$\partial/\partial x \sim 1/L \quad \text{and} \quad \partial/\partial y \sim 1/\overline{\delta}, \tag{8.5}$$

where $\overline{\delta} \ll L$ when $\mathrm{Re} \gg 1$ from (8.4). When applied to the continuity equation, $\partial u/\partial x + \partial v/\partial y = 0$, this derivative scaling requires $U_\infty/L \sim v/\overline{\delta}$, so the proper velocity scale for v is $\overline{\delta}U_\infty/L = U_\infty \mathrm{Re}^{-1/2}$. At high Re, experimental data show that the pressure distribution on the body is nearly that in an irrotational flow over the body, implying that the pressure variations scale with the fluid inertia: $p - p_\infty \sim \rho U_\infty^2$. Thus, the proper dimensionless variables for boundary-layer flow are:

$$\begin{gathered} x^* = x/L, \quad y^* = y/\overline{\delta} = (y/L)\mathrm{Re}^{1/2}, \quad u^* = u/U_\infty, \\ v^* = (v/U_\infty)\mathrm{Re}^{1/2}, \quad \text{and} \quad p^* = (p - p_\infty)/\rho U_\infty^2. \end{gathered} \tag{8.6}$$

For the coordinates and the velocities, this scaling is similar to that of (7.14) with $\varepsilon = \mathrm{Re}^{-1/2}$. The primary effect of (8.6) is a magnification of the surface-normal coordinate y and velocity v by a factor of $\mathrm{Re}^{1/2}$ compared to the stream-wise coordinate x and velocity u. In terms of these dimensionless variables, the steady two-dimensional equations of motion are:

$$\frac{\partial u^*}{\partial x^*} + \frac{\partial v^*}{\partial y^*} = 0, \tag{7.15}$$

$$u^* \frac{\partial u^*}{\partial x^*} + v^* \frac{\partial u^*}{\partial y^*} = -\frac{\partial p^*}{\partial x^*} + \frac{1}{\mathrm{Re}} \frac{\partial^2 u^*}{\partial x^{*2}} + \frac{\partial^2 u^*}{\partial y^{*2}}, \quad \text{and} \tag{8.7}$$

$$\frac{1}{\mathrm{Re}} \left(u^* \frac{\partial v^*}{\partial x^*} + v^* \frac{\partial v^*}{\partial y^*} \right) = -\frac{\partial p^*}{\partial y^*} + \frac{1}{\mathrm{Re}^2} \frac{\partial^2 v^*}{\partial x^2} + \frac{1}{\mathrm{Re}} \frac{\partial^2 v^*}{\partial y^{*2}}, \tag{8.8}$$

where $\mathrm{Re} \equiv U_\infty L/\nu$ is an overall Reynolds number. In these equations, each of the dimensionless variables and their derivatives should be of order unity when the scaling assumptions embodied in (8.6) are valid. Thus, it follows that the importance of each term in (7.15), (8.7), and (8.8) is determined by its coefficient. So, as $\mathrm{Re} \to \infty$, the terms with coefficients of 1/Re or 1/Re^2 drop out asymptotically. Thus, the relevant equations for laminar boundary-layer flow, in dimensional form, are:

$$\frac{\partial u}{\partial x} + \frac{\partial v}{\partial y} = 0, \quad u\frac{\partial u}{\partial x} + v\frac{\partial u}{\partial y} = -\frac{1}{\rho}\frac{\partial p}{\partial x} + \nu\frac{\partial^2 u}{\partial y^2}, \quad \text{and} \quad 0 = -\frac{\partial p}{\partial y}. \tag{6.2, 8.9, 8.10}$$

This scaling exercise has shown which terms must be kept and which terms may be dropped under the boundary-layer assumption. It differs from the scalings that produced (4.10) and (7.46) because the x and y directions are scaled differently in (8.6) which causes a second derivative term to be retained in (8.9).

Eq. (8.10) implies that the pressure is approximately uniform through the thickness of the boundary layer, an important result. The pressure at the surface is therefore equal to that at the edge of the boundary layer, so it can be found from an ideal outer-flow solution for flow above the surface. Thus, the outer flow imposes the pressure on the boundary layer. This justifies the experimental fact that the observed surface pressures underneath attached boundary layers are approximately equal to that calculated from the ideal flow theory. A vanishing $\partial p/\partial y$, however, is not valid if the boundary layer separates from the surface or if the radius of curvature of the surface is not large compared with the boundary-layer thickness.

Although the steady boundary-layer equations (6.2), (8.9), and (8.10) do represent a significant simplification of the full equations, they are still nonlinear second-order partial differential equations that can only be solved when appropriate boundary and matching conditions are specified. If the exterior flow is presumed to be known and irrotational (and the fluid density is constant), the pressure gradient at the edge of the boundary layer can be found by differentiating the steady constant-density Bernoulli equation (without the body force term), $p + \frac{1}{2}\rho U_e^2 = \text{const.}$, to find:

$$-\frac{1}{\rho}\frac{dp}{dx} = U_e \frac{dU_e}{dx}, \tag{8.11}$$

where $U_e(x)$ is the velocity at the *edge* of the boundary layer. Eq. (8.11) represents a matching condition between the outer ideal-flow solution and the inner boundary-layer solution in the region where both solutions must be valid. The (usual) remaining boundary conditions on the fluid velocities of the inner solution are:

$$u(x,0) = v(x,0) = 0 \quad \text{(no slip and no through flow at the wall)}, \tag{8.12a,b}$$

$$u(x, y \to \infty) = U_e(x) \quad \text{(matching of inner and outer solutions)}, \quad \text{and} \tag{8.13}$$

$$u(x_0, y) = u_{in}(y) \quad \text{(inlet boundary condition at } x_0). \tag{8.14}$$

For two-dimensional flow, (6.2), (8.9), and the conditions (8.11) through (8.14) completely specify the inner solution as long as the boundary layer remains thin and contiguous to the surface on which it develops. Condition (8.13) merely means that the boundary layer must join smoothly with the outer flow; for the inner solution, points outside the boundary layer are represented by $y \to \infty$, although we mean this strictly in terms of the dimensionless distance $y/\overline{\delta} = (y/L)\text{Re}^{1/2} \to \infty$. Condition (8.14) implies that an initial velocity profile $u_{\text{in}}(y)$ at some location x_0 is required for solving the problem. Such a condition is needed because the terms $u\partial u/\partial x$ and $\nu\partial^2 u/\partial y^2$ give the boundary-layer equations a parabolic character, with x playing the role of a time-like variable. Recall the Stokes problem of a suddenly accelerated plate, discussed in the preceding chapter, where the simplified field equation is $\partial u/\partial t = \nu\partial^2 u/\partial y^2$. In such problems governed by parabolic equations, the field at a certain time or place depends only on its *past* or *upstream history*. Boundary layers therefore transfer viscous effects only in the *downstream* direction. In contrast, the complete Navier-Stokes equations are elliptic and thus require boundary conditions on the velocity (or its derivative normal to the boundary) upstream, downstream, and on the top and bottom boundaries, that is, all around. (The upstream influence of the downstream boundary condition is a common concern in fluid dynamic computations.)

Considering two dimensions, an ideal outer flow solution from (6.5) or (6.12) and (6.18), and a viscous inner flow solution as described here would seem to fully solve the problem of uniform flow of a viscous fluid past a solid object. The solution procedure could be a two-step process. First, the outer flow is determined, neglecting the existence of the boundary layer, an error that gets smaller when the boundary layer becomes thinner. Then, (8.11) could be used to determine the surface pressure, and (6.2) and (8.9) could be solved for the boundary-layer flow using the surface-pressure gradient determined from the outer flow solution. If necessary this process might even be iterated to achieve higher accuracy by re-solving for the outer flow with the first-pass-solution boundary-layer characteristics included, and then proceeding to a second solution of the boundary-layer equations using the corrected outer-flow solution. In practice, such an approach can be successful and it converges when the boundary layer stays thin and attached. However, it does not converge when the boundary layer thickens or departs (separates) from the surface on which it has developed. Boundary-layer separation occurs when the surface shear stress, τ_w, produced by the boundary layer vanishes and reverse (or upstream-directed) flow occurs near the surface. Boundary-layer separation is discussed further in Sections 8.6–8.7. Here it is sufficient to point out that ideal flow around non-slender or *bluff* bodies typically produces surface pressure gradients that lead to boundary-layer separation.

In summary, the simplifications of the boundary-layer assumption are as follows. First, diffusion in the stream-wise direction is negligible compared to that in the wall normal direction. Second, the pressure in the boundary layer can be found from the outer flow, so that it is regarded as a known quantity within the boundary layer that does not vary perpendicular to the surface. Furthermore, a crude estimate of τ_w, the wall shear stress, can be made from the various scalings employed earlier: $\tau_w \sim \mu U/\overline{\delta} \sim (\mu U/L)\mathrm{Re}^{1/2}$. This implies a *skin friction coefficient* of:

$$C_f \equiv \frac{\tau_w}{\frac{1}{2}\rho U^2} \sim \frac{(\mu U/L)\sqrt{\mathrm{Re}}}{\frac{1}{2}\rho U^2} = \frac{2}{\sqrt{\mathrm{Re}}}. \tag{8.15}$$

The skin friction coefficient is an important dimensionless parameter in boundary-layer flows. It specifies the fraction of the local dynamic pressure, $\frac{1}{2}\rho U^2$, that is felt as shear stress on the surface. Here for laminar boundary layers, (8.15) provides the correct order of magnitude and parametric dependence on Reynolds number. However, the numerical factor differs for different laminar boundary-layer flows.

Example 8.1

For time-averaged turbulent boundary-layer flow, the advective acceleration scaling (8.2) is still appropriate. However, the laminar shear stress relationship (8.3) should be replaced with $\partial\tau/\partial y \sim \tau_w/\overline{\delta}$. What is the scaling for the skin friction coefficient in this case?

Solution As was done to reach (8.4), equate the advective and shear-stress accelerations:

$$\frac{U_\infty^2}{L} \sim \frac{1}{\rho}\frac{\tau_w}{\overline{\delta}} \sim \frac{1}{\rho\overline{\delta}}\left(\frac{1}{2}\rho U_\infty^2\right)C_f,$$

where the second scaling follows from (8.15), the definition of the skin friction. Canceling common terms between the two ends of this relationship then produces:

$$C_f \sim 2\overline{\delta}/L.$$

Although the Reynolds number dependence of C_f is not revealed by this simple relationship, it does suggest C_f will be much less than unity for attached turbulent boundary-layer flows. Measurements in flat-plate turbulent boundary-layer flows on smooth walls typically produce $C_f \sim 0.001$ to 0.004 with the lower values occurring at higher Reynolds number; see Section 8.7.

8.2 Boundary-layer thickness definitions

Since the fluid velocity in the boundary layer smoothly joins that of the outer flow, there is no obvious demarcation of the boundary layer's edge. Thus, a variety of thickness definitions are used to define a boundary layer's character. The three most common thickness definitions are described here.

The first, δ_{99}, is an overall boundary-layer thickness that specifies the distance from the wall where the stream-wise velocity in the boundary layer is $0.99U_e$, where U_e is the local free-stream speed. For a known boundary-layer stream-wise velocity profile, $u(x, y)$, at downstream distance x, this thickness is defined by: $u(x, \delta_{99}) = 0.99U_e(x)$. This thickness primarily plays a conceptual role in boundary-layer research. In practice it is difficult to measure accurately, and its physical importance is subjective since the choice of 99% instead of 95%, 98%, 99.5%, or another percentage is arbitrary.

A second measure of the boundary-layer thickness, and one in which there is no arbitrariness, is the *displacement thickness,* which is commonly denoted $\delta^*(x)$ or δ_1. It is defined as the thickness of a layer of zero-velocity fluid that has the same velocity deficit as the actual boundary layer. The velocity deficit in a boundary layer is $U_e - u$, so this definition implies:

$$\int_{y=0}^{h}(U_e - u)dy = \int_{y=0}^{\delta^*}(U_e - 0)dy = U_e\delta^*, \quad \text{or} \quad \delta^* = \int_{y=0}^{\infty}\left(1 - \frac{u}{U_e}\right)dy, \tag{8.16}$$

where h is a surface-normal distance that lies far outside the boundary layer (Fig. 8.3). Here the extension of $h \to \infty$ in the upper limit in the last integration is not problematic because $U_e - u \to 0$ exponentially fast as $y \to \infty$. Alternatively, the displacement thickness is the distance by which the wall would have to be displaced outward in a hypothetical frictionless flow to maintain the same mass flux as that in the actual flow. This means that the displacement thickness can be interpreted as the distance by which streamlines outside the boundary layer are displaced due to the presence of the boundary layer.

Fig. 8.4 shows the displacement of streamlines over a flat plate. Equating mass flux across two sections A and B, we obtain:

$$U_e h = \int_{y=0}^{h+\delta^*} u dy = \int_{y=0}^{h} u dy + U_e \delta^*, \quad \text{or} \quad U_e \delta^* = \int_{y=0}^{h} (U_e - u) dy,$$

where h is the wall-normal distance defined above. Here again, it can be replaced by ∞ without changing the integral in the final equation, which then reduces to (8.16).

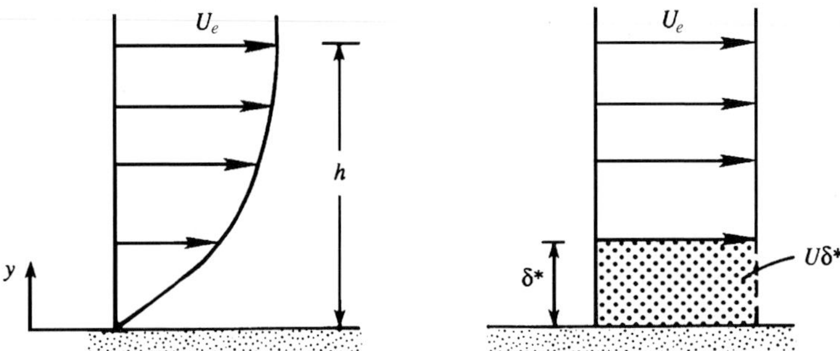

FIGURE 8.3 Schematic depiction of the displacement thickness. The panel on the left shows a typical laminar boundary-layer profile. The panel on the right shows an equivalent ideal-flow velocity profile with a zero-velocity layer having the same volume-flux deficit as the actual boundary layer. The thickness of this zero-velocity layer is the displacement thickness δ^*.

The displacement thickness is used in the design of airfoils, ducts, nozzles, intakes of air-breathing engines, wind tunnels, etc. by first assuming a frictionless flow and then revising the device's geometry to produce the desired flow condition with the boundary layer present. Here, the method for the geometric revisions involves using δ^* to correct the outer flow solution for the presence of the boundary layer. As mentioned in Section 8.1, the first approximation is to neglect the existence of thc boundary layer, and calculate the ideal-flow dp/dx over the surface of interest. A solution of the boundary-layer equations gives $u(x, y)$ and this can be integrated using (8.16) to find $\delta^*(x)$, the displacement thickness. The flow device's surface is then displaced outward by this amount and a next approximation of dp/dx is found from a new ideal flow solution over the mildly revised geometry (see Exercise 8.26). Thus, $\delta^*(x)$ is a critical ingredient in such an iterative solution procedure that alternates between the outer- and inner-flow solutions.

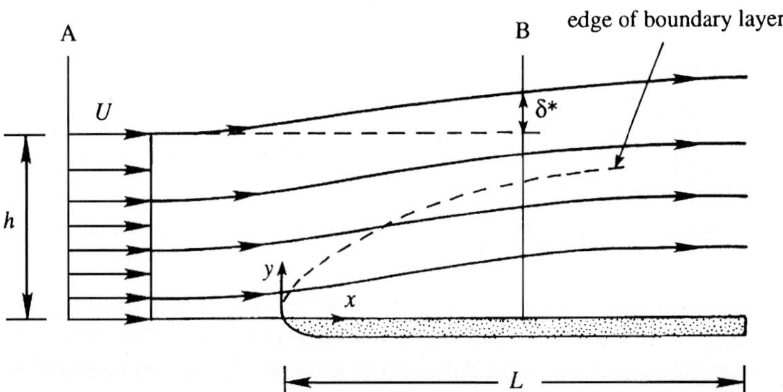

FIGURE 8.4 Displacement thickness and streamline displacement. Within the boundary layer, fluid motion in the downstream direction is retarded, that is, $\partial u/\partial x$ is negative. Thus, the continuity equation (6.2) requires $\partial v/\partial y$ to be positive, so the boundary layer produces a surface-normal velocity that deflects streamlines away from the surface. Above the boundary layer, the extent of this deflection is the displacement thickness δ^*.

A third measure of the boundary-layer thickness is the momentum thickness θ or δ_2. It is defined such that $\rho U^2\theta$ is the momentum loss in the actual flow because of the presence of the boundary layer. A control volume calculation (see Exercise 8.7) leads to the following definition:

$$\theta = \int_{y=0}^{\infty} \frac{u}{U_e}\left(1 - \frac{u}{U_e}\right) dy. \tag{8.17}$$

The momentum thickness embodies the integrated influence of the wall shear stress from the beginning of the boundary layer to the stream-wise location of interest.

Example 8.2

The shape factor, δ^*/θ, is important in boundary-layer flows as it indicates that a boundary layer is headed toward separation when it increases. This is further discussed in Section 8.6. Compute the shape factor for the following approximate-laminar (u_l) and approximate-turbulent (u_t) boundary-layer profiles:

$$\frac{u_l(y)}{U_e} = \begin{Bmatrix} 2\frac{y}{\bar{\delta}} - \left(\frac{y}{\bar{\delta}}\right)^2 & \text{for } y < \bar{\delta} \\ 1 & \text{for } y > \bar{\delta} \end{Bmatrix}, \quad \text{and} \quad \frac{u_t(y)}{U_e} = \begin{Bmatrix} \left(\frac{y}{\bar{\delta}}\right)^{1/7} & \text{for } y < \bar{\delta} \\ 1 & \text{for } y > \bar{\delta} \end{Bmatrix},$$

where $\bar{\delta}$ is a profile constant in each case. Which boundary layer is closer to separation?

Solution For the laminar profile, use (8.16) and (8.17) to find:

$$\delta_l^* = \int_0^{\bar{\delta}} \left(1 - 2\frac{y}{\bar{\delta}} + \left(\frac{y}{\bar{\delta}}\right)^2\right) dy = \frac{\bar{\delta}}{3} \quad \text{and} \quad \theta_l = \int_0^{\bar{\delta}} \left(2\frac{y}{\bar{\delta}} - \left(\frac{y}{\bar{\delta}}\right)^2\right)\left(1 - 2\frac{y}{\bar{\delta}} + \left(\frac{y}{\bar{\delta}}\right)^2\right) dy = \frac{2\bar{\delta}}{15}, \quad \text{so} \quad \frac{\delta^*}{\theta} = \frac{5}{2}.$$

Repeat for the turbulent profile to find:

$$\delta_t^* = \int_0^{\bar{\delta}} \left(1 - \left(\frac{y}{\bar{\delta}}\right)^{1/7}\right) dy = \frac{\bar{\delta}}{8} \quad \text{and} \quad \theta_t = \int_0^{\bar{\delta}} \left(\frac{y}{\bar{\delta}}\right)^{1/7}\left(1 - \left(\frac{y}{\bar{\delta}}\right)^{1/7}\right) dy = \frac{7\bar{\delta}}{72}, \quad \text{so} \quad \frac{\delta^*}{\theta} = \frac{9}{7}.$$

For the given profiles, the laminar boundary layer has a larger shape factor and is closer to separation. In general, turbulent boundary layers resist separation better than laminar ones.

8.3 Boundary layer on a flat plate: Blasius solution

The simplest-possible boundary layer forms on a semi-infinite flat plate with a constant free-stream flow speed, $U_e = U =$ constant. In this case, the boundary-layer equations simplify to:

$$\frac{\partial u}{\partial x} + \frac{\partial v}{\partial y} = 0 \quad \text{and} \quad u\frac{\partial u}{\partial x} + v\frac{\partial u}{\partial y} = \nu\frac{\partial^2 u}{\partial y^2}, \qquad (6.2, 8.18)$$

where (8.11) requires $dp/dx = 0$ because $dU_e/dx = 0$. Here, the independent variables are x and y, and the dependent field quantities are u and v. The flow is incompressible but rotational, so a guaranteed solution of (6.2) may be sought in terms of a stream function, ψ, with the two velocity components determined via derivatives of ψ (see (6.3)). Here, the flow is steady and there is no imposed length scale, so a similarity solution for ψ can be proposed based on (7.31):

$$\psi = U\delta(x)f(\eta) \quad \text{where} \quad \eta = y/\delta(x), \qquad (8.19)$$

where x is the time-like independent variable, f is an unknown dimensionless function, and $\delta(x)$ is a boundary-layer thickness that is to be determined as part of the solution (it is not a Dirac delta-function). Here the coefficient $U\delta$ in (8.19) has replaced $UA\xi^{-n}$ in (7.31) based on dimensional considerations; the stream function must have dimensions of length^2/time. A more general form of (8.19) that uses UAx^{-n} as the coefficient of $f(\eta)$ produces the same results when combined with (8.18).

The solution to (6.2) and (8.18) should be valid for $x > 0$, so the boundary conditions are:

$$u = v = 0 \quad \text{on} \quad y = 0, \qquad (8.20)$$

$$u(x, y) \to U \quad \text{as} \quad y/\delta \to \infty, \quad \text{and} \qquad (8.21)$$

$$\delta \to 0 \quad \text{as} \quad x \to 0. \qquad (8.22)$$

Here, we note that the boundary-layer approximation will not be valid near $x = 0$ (the leading edge of the plate) where the high Reynolds number approximation, $\text{Re}_x = Ux/\nu \gg 1$, used to reach (8.18) is not valid. Ideally, the exact equations of motion would be solved from $x = 0$ up some location, x_0, where $Ux_0/\nu \gg 1$. Then, the stream-wise velocity profile at this location would be used in the inlet boundary condition (8.14), and (8.18) could be solved for $x > x_0$ to determine the boundary-layer flow. However, for this similarity solution, we are effectively assuming that the distance x_0 is small compared to x and can be ignored. Thus, the boundary condition (8.22), which replaces (8.14), is really an assumption that must be shown to produce self-consistent results when $\mathrm{Re}_x \gg 1$.

The prior discussion touches on the question of a boundary layer's downstream dependence on, or memory of, its initial state. If the external stream $U_e(x)$ admits a similarity solution, is the initial condition forgotten? And, if so, how soon? Serrin (1967) and Peletier (1972) showed that for $U_e dU_e/dx > 0$ (*favorable* pressure gradients) when considering similarity solutions, the initial condition is forgotten and that the larger the free-stream acceleration the sooner similarity is achieved. However, a decelerating flow will accentuate details of the boundary layer's initial state and similarity will never be found even if it is mathematically possible. This is consistent with the experimental findings of Gallo et al. (1970). Interestingly, a flat plate for which $U_e(x) = U = \text{const.}$ is the borderline case; similarity is eventually achieved. Thus, a solution in the form (8.19) is pursued here.

The first solution steps involve performing derivatives of ψ to find u and v:

$$u = \frac{\partial \psi}{\partial y} = U\delta \frac{df}{d\eta}\frac{1}{\delta} = Uf', \qquad v = -\frac{\partial \psi}{\partial x} = -U\left(\frac{d\delta}{dx} f + \delta \frac{df}{d\eta}\left(-\frac{y}{\delta^2}\right)\frac{d\delta}{dx}\right) = U\delta'(-f + \eta f'), \tag{8.23, 8.24}$$

where a prime denotes differentiation of a function with respect to its argument. When substituted into (8.18), these relations for u and v produce:

$$Uf'Uf''\left(-\frac{y}{\delta^2}\right)\delta' + U\delta'(-f + \eta f')\frac{U}{\delta}f'' = \nu\frac{U}{\delta^2}f''', \quad \text{or} \quad -\left[\frac{U^2}{\delta}\delta'\right]ff'' = \left[\nu\frac{U}{\delta^2}\right]f''', \tag{8.25}$$

since two terms on the left side of the first equality are equal and opposite. For a similarity solution, the coefficients in [,]-braces in (8.25) must be proportional:

$$C\frac{U^2}{\delta}\delta' = \nu\frac{U}{\delta^2}, \quad \text{or} \quad C\delta\frac{d\delta}{dx} = \frac{\nu}{U}, \quad \text{which implies:} \quad C\frac{\delta^2}{2} = \frac{\nu x}{U} + D,$$

where C and D are constants. Here (8.22) requires $D = 0$, and C can be chosen equal to 2 to simplify the resulting expression for δ:

$$\delta(x) = [\nu x/U]^{1/2}. \tag{8.26}$$

As described above, this result will be imperfect as $x \to 0$ since it is based on equations that are only valid when $\mathrm{Re}_x \gg 1$. However, it is self-consistent since it produces a boundary layer that thins with decreasing distance so that $u \to U$ at any finite y as $x \to 0$. When (8.26) is substituted into (8.25), the final equation for f is found:

$$\frac{d^3 f}{d\eta^3} + \frac{1}{2}f\frac{d^2 f}{d\eta^2} = 0, \quad \text{or} \quad f''' + \frac{1}{2}ff'' = 0. \tag{8.27}$$

The boundary conditions for (8.27) are:

$$df/d\eta = f = 0 \quad \text{at} \quad \eta = 0, \quad \text{and} \quad df/d\eta \to 1 \quad \text{as} \quad \eta \to \infty, \tag{8.28, 8.29}$$

which replace (8.20) and (8.21), respectively.

A series solution of (8.27), subject to (8.28) and (8.29), was given by Blasius; today it is much easier to numerically determine $f(\eta)$ (see Exercise 8.2), and Table 8.1 provides numerical results for f, $f' = df/d\eta$, and $f'' = d^2 f/d\eta^2$ vs. η for $0 < \eta < 7.0$. The resulting profile of $u/U = f'(\eta)$ is shown in Fig. 8.5. For $\eta > 7.0$, the table may be continued via: $f = \eta - 1.7208$, $df/d\eta = 1$, and $d^2 f/d\eta^2 = 0$. The solution makes the profiles at various downstream distances collapse into a single curve of u/U vs. $y[U/\nu x]^{1/2}$, and is in excellent agreement with experimental data for laminar flow at high Reynolds numbers. The profile has a point of inflection (i.e., zero curvature) at the wall, where $\partial^2 u/\partial y^2 = 0$. This is a result of the absence of a pressure gradient in the flow (see Section 8.7).

The Blasius boundary-layer profile has a variety of noteworthy properties. First of all, an asymptotic analysis of the solution to (8.27) shows that $(df/d\eta - 1) \sim (1/\eta)\exp(-\eta^2/4)$ as $\eta \to \infty$ so u approaches U very smoothly with increasing wall-normal distance. Second, the wall-normal velocity is:

$$v = -\frac{\partial \psi}{\partial x} = \frac{1}{2}\sqrt{\frac{\nu U}{x}}\left(-f + \eta\frac{df}{d\eta}\right), \quad \text{or} \quad \frac{v}{U} = \frac{1}{2\mathrm{Re}_x^{1/2}}\left(-f + \eta\frac{df}{d\eta}\right) \sim \frac{0.86}{\mathrm{Re}_x^{1/2}} \quad \text{as} \quad \eta \to \infty,$$

TABLE 8.1 Blasius boundary-layer profile functions.

η	f	$df/d\eta$	$d^2f/d\eta^2$
0.0	0.0	0.0	0.3321
0.2	0.0068	0.0664	0.3320
0.4	0.0268	0.1328	0.3314
0.6	0.0611	0.1989	0.3299
0.8	0.1074	0.2646	0.3272
1.0	0.1667	0.3297	0.3228
1.2	0.2390	0.3937	0.3164
1.4	0.3252	0.4559	0.3074
1.6	0.4225	0.5163	0.2962
1.8	0.5310	0.5743	0.2826
2.0	0.6502	0.6297	0.2667
2.2	0.7823	0.6809	0.2481
2.4	0.9240	0.7282	0.2278
2.6	1.0741	0.7716	0.2063
2.8	1.2319	0.8109	0.1840
3.0	1.3969	0.8459	0.1614
3.2	1.5697	0.8756	0.1392
3.4	1.7478	0.9010	0.1181
3.6	1.9302	0.9226	0.0984
3.8	2.1164	0.9407	0.0804
4.0	2.3058	0.9555	0.0643
4.2	2.4983	0.9666	0.0508
4.4	2.6924	0.9758	0.0391
4.6	2.8883	0.9826	0.0296
4.8	3.0853	0.9878	0.0219
5.0	3.2833	0.9915	0.0160
5.2	3.4819	0.9942	0.0114
5.4	3.6809	0.9961	0.0080
5.6	3.8803	0.9975	0.0055
5.8	4.0799	0.9984	0.0037
6.0	4.2796	0.9990	0.0024
6.2	4.4795	0.9994	0.0016
6.4	4.6794	0.9996	0.0010
6.6	4.8793	0.9998	0.0006
6.8	5.0793	0.9999	0.0004
7.0	5.2792	0.9999	0.0002

a plot of which is shown in Fig. 8.6. The wall-normal velocity increases from zero at the wall to a maximum value at the edge of the boundary layer, a pattern that is in agreement with the streamline shapes sketched in Fig. 8.4.

The various thicknesses for the Blasius boundary layer are as follows. From Table 8.1, the distance where $u = 0.99U$ is $\eta = 4.92$, so:

$$\delta_{99} = 4.92\sqrt{\nu x/U} \quad \text{or} \quad \delta_{99}/x = 4.92/\mathrm{Re}_x^{1/2}. \tag{8.30a}$$

For air at ordinary temperatures flowing at $U = 1$ m/s, the Reynolds number at a distance of 1 m from the leading edge of a flat plate is $\mathrm{Re}_x = 6 \times 10^4$, and (8.30a) gives $\delta_{99} = 2$ cm, showing that the boundary layer is indeed thin. The displacement and momentum thicknesses, (8.16) and (8.17), of the Blasius boundary layer are:

$$\delta^* = 1.72\sqrt{\nu x/U}, \quad \text{and} \quad \theta = 0.664\sqrt{\nu x/U}. \tag{8.30b,c}$$

These thicknesses are indicated along the abscissa of Fig. 8.5.

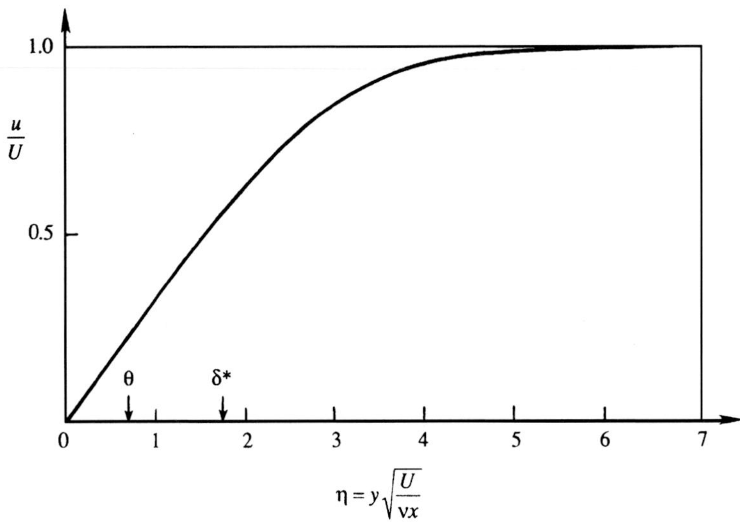

FIGURE 8.5 The Blasius similarity solution for the horizontal velocity distribution in a laminar boundary layer on a flat plate with zero-pressure gradient, $U_e = U =$ constant. The finite slope at $\eta = 0$ implies a non-zero wall shear stress τ_w. The boundary layer's velocity profile smoothly asymptotes to U as $\eta \to \infty$. The momentum θ and displacement δ^* thicknesses are indicated by arrows on the horizontal axis.

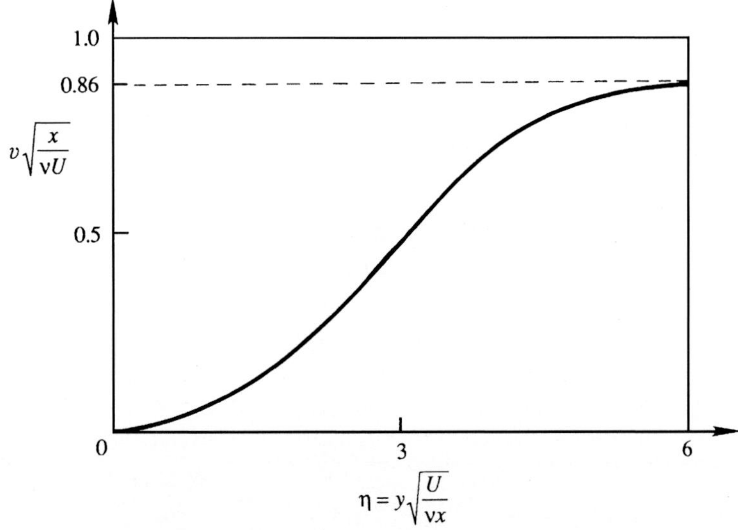

FIGURE 8.6 Surface-normal velocity component, v, in a laminar boundary layer on a flat plate with constant free-stream speed U. Here the scaling on vertical axis, $(v/U)\sqrt{\mathrm{Re}_x}$, causes it to be expanded compared to that in Fig. 8.5.

The local wall shear stress, is $\tau_w = \mu(du/dy)_{y=0} = (\mu U/\delta)(d^2 f/d\eta^2)_{\eta=0}$, so it and the skin friction coefficient are:

$$\tau_w = 0.332\rho U^2/\sqrt{\mathrm{Re}_x}, \quad \text{and} \quad C_f \equiv \frac{\tau_w}{\frac{1}{2}\rho U^2} = \frac{0.664}{\sqrt{\mathrm{Re}_x}}. \tag{8.31, 8.32}$$

The wall shear stress therefore decreases as $x^{-1/2}$, a result of the thickening of the boundary layer and the associated decrease of the velocity gradient at the surface. Note that the wall shear stress at the plate's leading edge has an integrable singularity. This is a manifestation of the fact that boundary-layer theory breaks down near the leading edge where the assumptions $\mathrm{Re}_x \gg 1$, and $\partial/\partial x \ll \partial/\partial y$ are invalid. The drag force per unit width on one side of a plate of length L is:

$$F_D = \int_0^L \tau_w dx = \frac{0.664\rho U^2 L}{\sqrt{\mathrm{Re}_L}},$$

where $\mathrm{Re}_L \equiv UL/\upsilon$ is the Reynolds number based on the plate length. This equation shows that the drag force is proportional to the 3/2-power of the velocity. This is a higher power than that in low Reynolds number flows where drag is proportional to the first power of velocity. But, it is a lower power than that in high Reynolds number flow past a *blunt* body where drag is typically proportional to the *square* of velocity.

The overall *drag coefficient* for one side of the plate, defined in the usual manner, is:

$$C_D = \frac{F_D}{\frac{1}{2}\rho U^2 L} = \frac{1.33}{\sqrt{\mathrm{Re}_L}}. \quad (8.33)$$

It is clear from (8.32) and (8.33) that:

$$C_D = \frac{1}{L}\int_0^L C_f dx,$$

which says that the overall drag coefficient is the spatial average of the local skin friction coefficient. However, carrying out an integration from $x = 0$ may be of questionable validity because the equations and solutions are valid only for $\mathrm{Re}_x \gg 1$. Nevertheless, (8.33) is found to be in good agreement with laminar flow experiments for $\mathrm{Re}_L > 10^3$.

Example 8.3

Using the information in Table 8.1 plot streamlines and δ_{99} in the Blasius boundary layer for a 1.0 m/s airflow over a 3.0-m-long surface.

Solution Start with (8.19) and insert (8.26) to reach:

$$\psi = \sqrt{\upsilon U x} f\left(y\sqrt{U/\upsilon x}\right).$$

The goal is to use this formula and Table 8.1 to plot an x-y curve that represents $\psi = \text{constant}$. To get started denote the first two entries on the ith row of Table 8.1 by η_i and f_i, and look for an algebraic parameterization of the streamline's coordinates at discrete locations: $x_i = x(\eta_i, f_i)$ and $y_i = y(\eta_i, f_i)$. The first parameterization can be found directly from the above equation:

$$\psi = \sqrt{\upsilon U x_i} f_i \quad \text{or} \quad x_i = \psi^2/(f_i^2 \upsilon U).$$

The second comes from (8.26):

$$\eta_i = y_i\sqrt{U/\upsilon x_i} \quad \text{or} \quad y_i = \eta_i\psi/(f_i U).$$

Thus, once a value of ψ is selected, x_i-y_i coordinate pairs on this streamline can be obtained by evaluating the equations for x_i and y_i using the η_i and f_i entries in Table 8.1. For the conditions given, such a streamline plot is provided in Fig. 8.7. Here, a few additional η and f values from high in the boundary ($\eta > 7.0$) were needed to plot streamlines starting from $x = 0$. And the darker line is δ_{99} from (8.30a). For $x < 0$, the streamlines are straight and horizontal.

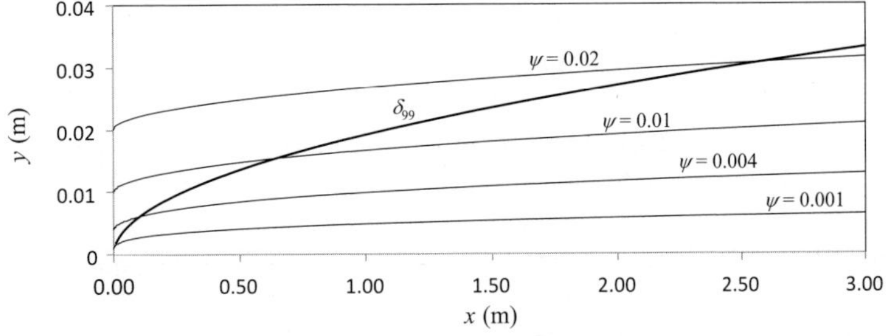

FIGURE 8.7 Blasius boundary-layer streamlines and 99% thickness for a 1.0 m/s air flow over a 3.0-m-long surface. At $x = 3$ m, the Reynolds number based on downstream distance, Re_x, is 200,000. The lighter curves are streamlines from the solution of (8.27), and the stream function values are in m^2/s. The heavier curve is the overall boundary-layer thickness, δ_{99} from (8.30a); it reaches 3.3 cm at $x = 3.0$ m.

This figure shows several important phenomena. First, even at this modest size and flow speed the boundary layer's thickness (centimeters) is much less than the corresponding development length (meters). Second, there is a kink in the streamlines at $x = 0$. This occurs because the boundary-layer equations are parabolic so the plate has no upstream influence. This kink would be absent if the full equations of fluid motion were used near the plate's leading edge. And third, streamlines that originate in the outer irrotational flow continually enter the boundary layer with increasing downstream distance.

8.4 Falkner-Skan similarity solutions of the laminar boundary-layer equations

The Blasius boundary-layer solution is one of a whole class of similarity solutions to the boundary-layer equations that were investigated by Falkner and Skan (1931). In particular, similarity solutions of (6.2), (8.9), and (8.10) are possible when $U_e(x) = ax^n$, and $\mathrm{Re}_x = ax^{(n+1)}/\nu$ is sufficiently large so that the boundary-layer approximation is valid and any dependence on an initial velocity profile has been forgotten. In this case, the initial location x_0 again disappears from the problem and a similarity solution may be sought in the form:

$$\psi(x, y) = \sqrt{\nu x U_e(x)}\, f(\eta), \quad \text{where} \quad \eta = \frac{y}{\delta(x)} = \frac{y}{x}\sqrt{\mathrm{Re}_x} = y\sqrt{\frac{a}{\nu}}x^{(n-1)/2}. \tag{8.34}$$

This is a direct extension of (8.19) to boundary-layer flow with a spatially varying free-stream speed $U_e(x)$. Here, $u/U_e = f'(\eta)$ as in the Blasius solution, but now the pressure gradient is non-trivial:

$$-dp/dx = U_e(dU_e/dx) = na^2x^{2n-1}, \tag{8.35}$$

and the generic boundary-layer thickness is:

$$\delta(x) = \sqrt{\nu x/U_e(x)} = \sqrt{\nu x^{1-n}/a},$$

which increases in size when $n < 1$ and decreases in size when $n > 1$ as x increases. When $n = 1$, then $\delta(x)$ is constant. Substituting (8.34) and (8.35) into (8.9) allows it to be reduced to the similarity form:

$$\frac{d^3 f}{d\eta^3} + \frac{n+1}{2} f \frac{d^2 f}{d\eta^2} - n\left(\frac{df}{d\eta}\right)^2 + n = 0, \quad \text{or} \quad f''' + \frac{n+1}{2}ff'' - nf'^2 + n = 0, \tag{8.36}$$

where f is subject to the boundary conditions (8.28) and (8.29) The Blasius equation (8.27) is a special case of (8.36) for $n = 0$, that is, $U_e(x) = U = \text{constant}$.

Solutions to (8.36) are displayed in Fig. 5.9.1 of Batchelor (1967) and are reproduced here in Fig. 8.8. They are parameterized by the power law exponent, n, which also sets the pressure gradient. The shapes of the various profiles can be understood by comparing them to the stream-wise velocity profiles obtained for flow between parallel plates when the upper plate moves with a positive horizontal velocity. They show a monotonically increasing shear stress $[f''(0)]$ as n increases. When n is positive, the flow accelerates as it moves to higher x, the pressure gradient is *favorable* ($dp/dx < 0$), the wall shear stress is non-zero and positive, and $(\partial^2 u/\partial y^2)_{y=0} < 0$. Thus, the profiles for $n > 0$ in Fig. 8.8 are similar to the lower half of the profiles shown in Figs. 7.4a or 7.4d. When $n = 0$, there is no flow acceleration or pressure gradient and $(\partial^2 u/\partial y^2)_{y=0} = 0$. This case corresponds to Fig. 7.4c. When n is negative, the flow decelerates as it moves downstream, the pressure gradient is *adverse* ($dp/dx > 0$), the wall shear stress may approach zero, and $(\partial^2 u/\partial y^2)_{y=0} > 0$. Thus, the profiles for $n < 0$ in Fig. 8.8 approach that shown in Fig. 7.4b. For $n = -0.0904$, $f''(0) = 0$, so $\tau_w = 0$, and boundary-layer separation is imminent all along the surface. Solutions of (8.36) exist for $n < -0.0904$ but these solutions involve reverse flow, like that shown in Fig. 7.4b, and do not necessarily represent boundary layers because the stream-wise velocity scaling in (8.6) used to reach (8.9), $u \sim U$, is invalid when $u = 0$ away from the wall.

In many real flows, boundary or initial conditions prevent similarity solutions from being directly applicable. However, after a variety of empirical and analytical advances made in the middle of the twentieth century, useful approximate methods were found to predict the properties of laminar boundary layers. These approximate techniques are based on the von Karman boundary-layer integral equation, which is derived in the next section. Then, in Section 8.6, the Thwaites method for estimating the surface shear stress, the displacement thickness, and the momentum thickness for attached laminar boundary layers is presented. In the most general cases or when greater accuracy is required, the full set of equations for fluid motion must be solved numerically by procedures discussed Chapter 15.

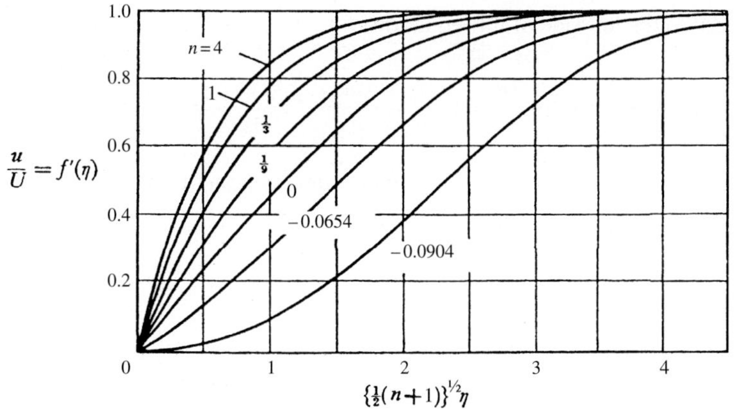

FIGURE 8.8 Falkner-Skan profiles of stream-wise velocity in a laminar boundary layer when the external stream is $U_e = ax^n$. The horizontal axis is the scaled surface-normal coordinate. The various curves are labeled by their associated value of n. When $n > 0$, the free-stream speed increases with increasing x, and $\partial^2 u/\partial y^2$ is negative throughout the boundary layer. When $n = 0$ (the Blasius boundary layer), the free-stream speed is constant, and $\partial^2 u/\partial y^2 = 0$ at the wall and is negative throughout the boundary layer. When $n < 0$, the free-stream speed decreases with increasing x, and $\partial^2 u/\partial y^2$ is positive near the wall but negative higher up in the boundary layer so there is an inflection point in the stream-wise velocity profile at a finite distance from the surface. *Reprinted with the permission of Cambridge University Press, from: G.K. Batchelor*, An Introduction to Fluid Dynamics, *1st ed. (1967).*

Example 8.4

Determine a formula for the surface-normal velocity v in a Falkner-Skan boundary-layer flow. Is v positive or negative when the exterior flow is accelerating and n is positive?

Solution Start with (8.34) and differentiate:

$$v = -\frac{\partial \psi}{\partial x} = -\frac{\partial}{\partial x}\left[\sqrt{\nu a x^{n+1}}\, f\left(y\sqrt{a x^{n-1}/\nu}\right)\right] = -\left[\frac{n+1}{2}\sqrt{\nu a x^{n-1}}\, f + \sqrt{\nu a x^{n-1}}\,\frac{n-1}{2}\eta\frac{df}{d\eta}\right].$$

Divide this result by $U(x) = ax^n$ and simplify to find:

$$\frac{v}{U_e(x)} = -\frac{1}{\mathrm{Re}_x^{1/2}}\left[\frac{n+1}{2} f + \frac{n-1}{2}\eta\frac{df}{d\eta}\right].$$

The plotted results in Fig. 8.8 show that η, f, and $df/d\eta$ are all positive for all n in the range of interest. Thus, this equation suggests that v is negative when n is positive, and this is certainly so when $n \geq 1$. Thus, an accelerating outer flow is mildly drawn *towards* the wall, and this prevents boundary-layer separation. However, the opposite is also true; a decelerating flow ($n < 0$) will produce a mild positive wall normal velocity (recall that $f < \eta(df/d\eta)$ in the Blasius boundary layer where $n = 0$). Thus, a decelerating outer flow is pushed *away* from the wall, and this may lead to boundary-layer separation.

8.5 von Karman momentum integral equation

Exact solutions of the boundary-layer equations are possible only in simple cases. In more complicated problems, approximate methods satisfy only an *integral* of the boundary-layer equations across the layer thickness. When this integration is performed, the resulting ordinary differential equation involves the boundary layer's displacement and momentum thicknesses, and its wall shear stress. This simple differential equation was derived by von Karman (1921) and applied to several situations by Pohlhausen (1921).

The common emphasis of an integral formulation is to obtain critical information with minimum effort. The important results of boundary-layer calculations are the wall shear stress, displacement thickness, momentum thickness, and separation point (when one exists). The von Karman boundary-layer momentum integral equation explicitly links the first three of these, and can be used to estimate, or at least determine the existence of, the fourth. The starting points are (6.2) and (8.9), with the pressure gradient specified in terms of $U_e(x)$ from (8.11) and the shear stress $\tau = \mu(\partial u/\partial y)$:

$$u\frac{\partial u}{\partial x} + v\frac{\partial u}{\partial y} = U_e\frac{dU_e}{dx} + \frac{1}{\rho}\frac{\partial \tau}{\partial y}. \tag{8.37}$$

Multiply (6.2) by u and add it to the left side of this equation:

$$u\left(\frac{\partial u}{\partial x}+\frac{\partial v}{\partial y}\right)+u\frac{\partial u}{\partial x}+v\frac{\partial u}{\partial y}=\frac{\partial(u^2)}{\partial x}+\frac{\partial(vu)}{\partial y}=U_e\frac{dU_e}{dx}+\frac{1}{\rho}\frac{\partial\tau}{\partial y}. \tag{8.38}$$

Move the term involving U_e to the other side of the last equality, and integrate (6.2) and (8.38) from $y=0$ where $u=v=0$ to $y=\infty$ where $u=U_e$ and $v=v_\infty$:

$$\int_0^\infty\left(\frac{\partial u}{\partial x}+\frac{\partial v}{\partial y}\right)dy=0\rightarrow\int_0^\infty\frac{\partial u}{\partial x}dy=-\int_0^\infty\frac{\partial v}{\partial y}dy=-[v]_0^\infty=-v_\infty, \tag{8.39}$$

$$\int_0^\infty\left(\frac{\partial(u^2)}{\partial x}+\frac{\partial(vu)}{\partial y}-U_e\frac{dU_e}{dx}\right)dy=+\frac{1}{\rho}\int_0^\infty\frac{\partial\tau}{\partial y}dy\rightarrow\int_0^\infty\left(\frac{\partial(u^2)}{\partial x}-U_e\frac{dU_e}{dx}\right)dy+U_ev_\infty=-\frac{1}{\rho}\tau_w, \tag{8.40}$$

where τ_w is the shear stress at $y=0$ and $\tau=0$ at $y=\infty$. Use the final form of (8.39) to eliminate v_∞ from (8.40), and exchange the order of integration and differentiation in the first term of (8.40):

$$\frac{d}{dx}\int_0^\infty u^2dy-\int_0^\infty U_e\frac{dU_e}{dx}dy-U_e\int_0^\infty\frac{\partial u}{\partial x}dy=-\frac{1}{\rho}\tau_w. \tag{8.41}$$

Now, note that:

$$U_e\int_0^\infty\frac{\partial u}{\partial x}dy=U_e\frac{d}{dx}\int_0^\infty u\,dy=\frac{d}{dx}\left(U_e\int_0^\infty udy\right)-\frac{dU_e}{dx}\int_0^\infty u\,dy,$$

and use this to rewrite the third term on the left side of (8.41) to find:

$$\frac{d}{dx}\int_0^\infty(u^2-U_eu)dy+\frac{dU_e}{dx}\int_0^\infty(u-U_e)dy=-\frac{1}{\rho}\tau_w. \tag{8.42}$$

A few final algebraic rearrangements produce:

$$\frac{1}{\rho}\tau_w=\frac{d}{dx}\left[U_e^2\int_0^\infty\frac{u}{U_e}\left(1-\frac{u}{U_e}\right)dy\right]+\frac{dU_e}{dx}U_e\int_0^\infty\left(1-\frac{u}{U_e}\right)dy,$$
$$\text{or}\quad\frac{1}{\rho}\tau_w=\frac{d}{dx}[U_e^2\theta]+U_e\delta^*\frac{dU_e}{dx}. \tag{8.43}$$

Throughout these manipulations, U_e and dU_e/dx may be moved inside or taken outside the vertical-direction integrations because they only depend on x.

Eq. (8.43) is known as the *von Karman boundary-layer momentum integral equation*, and it is valid for steady laminar boundary layers and for time-averaged flow in turbulent boundary layers. It is a single ordinary differential equation that relates three unknowns θ, δ^*, and τ_w, so additional assumptions must be made or correlations provided to obtain solutions for these parameters. The search for appropriate assumptions and empirical correlations was actively pursued by many researchers in the middle of the twentieth century starting with Pohlhausen (1921) and ending with Thwaites (1949) who combined analysis and inspired guesswork with the laminar boundary-layer measurements and equation solutions known at that time to develop the approximate empirical laminar-boundary-layer solution procedure for (8.43) described in the next section.

Example 8.5

Use the von Karman boundary-layer momentum integral equation to determine how the wall shear stress depends on downstream distance in an accelerating flow where $U_e(x)=(U_o/L)x$.

Solution The given exterior flow velocity follows a power law with $n=1$, so the generic Falkner-Skan boundary-layer thickness is:

$$\delta(x)=\sqrt{\nu x/U_e(x)}=\sqrt{\nu L/U_o}=const.$$

From this, we can deduce that θ and δ^* are constant as well since are both defined as integrals of the velocity profile and are therefore proportional to the generic thickness, δ. For example:

$$\theta \equiv \int_{y=0}^{\infty} \frac{u}{U_e}\left(1-\frac{u}{U_e}\right)dy = \delta \int_{y=0}^{\infty} f'(\eta)(1-f'(\eta))d\eta = \delta \cdot \text{const.}$$

where η and $f(\eta)$ are defined by (8.34). Now use (8.43) and $U_e(x) = (U_o/L)x$ to find:

$$\frac{1}{\rho}\tau_w = \frac{d}{dx}\left[\frac{U_o^2 x^2}{L^2}\theta\right] + \frac{U_o x}{L}\delta^* \frac{d}{dx}\left[\frac{U_o x}{L}\right] = \frac{2U_o^2 x}{L^2}\theta + \frac{U_o^2 x}{L^2}\delta^*,$$

and this can be mildly simplified to:

$$\frac{\tau_w}{\frac{1}{2}\rho U_o^2} = \left(\frac{4\theta + 2\delta^*}{L}\right)\frac{x}{L}.$$

Thus, the skin friction increases linearly with downstream distance in this case. However, the three-unknowns-and-one-equation problem persists since values for θ and δ^* are needed to determine τ_w. Thwaites method provides an approximate remedy for this problem.

8.6 Thwaites' method

To solve (8.43) at least two additional equations are needed. Using the correlation parameter:

$$\lambda \equiv \frac{\theta^2}{\nu}\frac{dU_e}{dx}, \tag{8.44}$$

introduced by Holstein and Bohlen (1940), Thwaites (1949) developed an approximate solution to (8.43) that involves two empirical dimensionless functions $l(\lambda)$ and $H(\lambda)$:

$$\tau_w \equiv \mu \frac{U_e}{\theta} l(\lambda) \quad \text{and} \quad \frac{\delta^*}{\theta} \equiv H(\lambda) \tag{8.45, 8.46}$$

that are listed in Table 8.2. This tabulation is identical to Thwaites' original for $\lambda \geq -0.060$ but includes the improvements recommended by Curle and Skan a few years later (see Curle, 1962) for $\lambda < -0.060$. The function $l(\lambda)$ is sometimes known as the shear correlation while $H(\lambda)$ is commonly called the shape factor.

Thwaites' method is developed from (8.43) by multiplying it with $\rho\theta/\mu U_e$:

$$\frac{\theta\tau_w}{\mu U_e} = \frac{\rho\theta}{\mu U_e}\frac{d}{dx}(U_e^2\theta) + \frac{\rho\theta}{\mu U_e}U_e\delta^*\frac{dU_e}{dx}, \quad \text{or} \quad \frac{\theta\tau_w}{\mu U_e} = 2\frac{\theta^2}{\nu}\frac{dU_e}{dx} + \frac{U_e\theta}{\nu}\frac{d\theta}{dx} + \frac{\theta^2}{\nu}\frac{\delta^*}{\theta}\frac{dU_e}{dx}. \tag{8.47}$$

The definitions of l and H allow the second version to be simplified:

$$l(\lambda) = (2 + H(\lambda))\frac{\theta^2}{\nu}\frac{dU_e}{dx} + \frac{U_e}{2}\frac{d}{dx}\left(\frac{\theta^2}{\nu}\right).$$

The momentum thickness θ can be eliminated from this equation using (8.44), to find:

$$U_e \frac{d}{dx}\left(\frac{\lambda}{dU_e/dx}\right) = 2l(\lambda) - 2(2 + H(\lambda))\lambda \equiv L(\lambda). \tag{8.48}$$

Fortunately, $L(\lambda) \approx 0.45 - 6.0\lambda = 0.45 + 6.0m$, is approximately linear as shown in Fig. 8.9 which is taken from Thwaites' (1949) original paper where $m = -\lambda$. With this linear fit, (8.48) can be integrated:

$$U_e \frac{d}{dx}\left(\frac{\lambda}{dU_e/dx}\right) = 0.45 - 6.0\lambda \rightarrow \frac{d}{dx}\left(\frac{\theta^2}{\nu}\right) + \frac{6.0}{U_e}\frac{dU_e}{dx}\frac{\theta^2}{\nu} = \frac{0.45}{U_e}. \tag{8.49}$$

The second version of (8.49) is a first-order linear inhomogeneous differential equation for θ^2/ν, and its integrating factor is U_e^6. The resulting solution for θ^2 involves a simple integral of the fifth power of the free-stream velocity at the edge of

TABLE 8.2 Universal functions for Thwaites' method.

λ	$l(\lambda)$	$H(\lambda)$
0.25	0.500	2.00
0.20	0.463	2.07
0.14	0.404	2.18
0.12	0.382	2.23
0.10	0.359	2.28
0.08	0.333	2.34
0.064	0.313	2.39
0.048	0.291	2.44
0.032	0.268	2.49
0.016	0.244	2.55
0.0	0.220	2.61
−0.008	0.208	2.64
−0.016	0.195	2.67
−0.024	0.182	2.71
−0.032	0.168	2.75
−0.040	0.153	2.81
−0.048	0.138	2.87
−0.052	0.130	2.90
−0.056	0.122	2.94
−0.060	0.113	2.99
−0.064	0.104	3.04
−0.068	0.095	3.09
−0.072	0.085	3.15
−0.076	0.072	3.22
−0.080	0.056	3.30
−0.084	0.038	3.39
−0.086	0.027	3.44
−0.088	0.015	3.49
−0.090	0.0	3.55

the boundary layer:

$$\frac{\theta^2 U_e^6(x)}{\nu} = 0.45\int_0^x U_e^5(x')dx' + \frac{\theta_0^2 U_0^6}{\nu} \quad \text{or} \quad \theta^2 = \frac{0.45\nu}{U_e^6(x)}\int_0^x U_e^5(x')dx' + \frac{\theta_0^2 U_0^6}{U_e^6(x)}, \tag{8.50}$$

where x' is an integration variable, and $\theta = \theta_0$ and $U_e = U_0$ at $x = 0$. If $x = 0$ is a stagnation point ($U_e = 0$), then it is safe to set $\theta_0 = 0$ since the exterior flow must accelerate away from a stagnation point and accelerating external flow leads to boundary-layer initial-condition memory loss. Once the integration specified by (8.50) is complete, the surface shear stress and displacement thickness can be recovered by computing λ and then using (8.45), (8.46), and Table 8.2.

Overall, the accuracy of Thwaites' method is ±3% or so for favorable pressure gradients, and ±10% for adverse pressure gradients but perhaps slightly worse near boundary-layer separation. The great strength of Thwaites' method is that it involves only one parameter (λ) and requires only a single integration. This simplicity makes it ideal for preliminary engineering calculations that are likely to be followed by more formal computations or experiments.

Before proceeding to example calculations, an important limitation of boundary-layer calculations that start from a steady presumed surface pressure distribution (such as Thwaites' method) must be stated. Such techniques can only predict the *existence* of boundary-layer separation; they do not reliably predict the location of boundary-layer separation. As will be further discussed in the next section, once a boundary layer separates from the surface on which it has formed, the

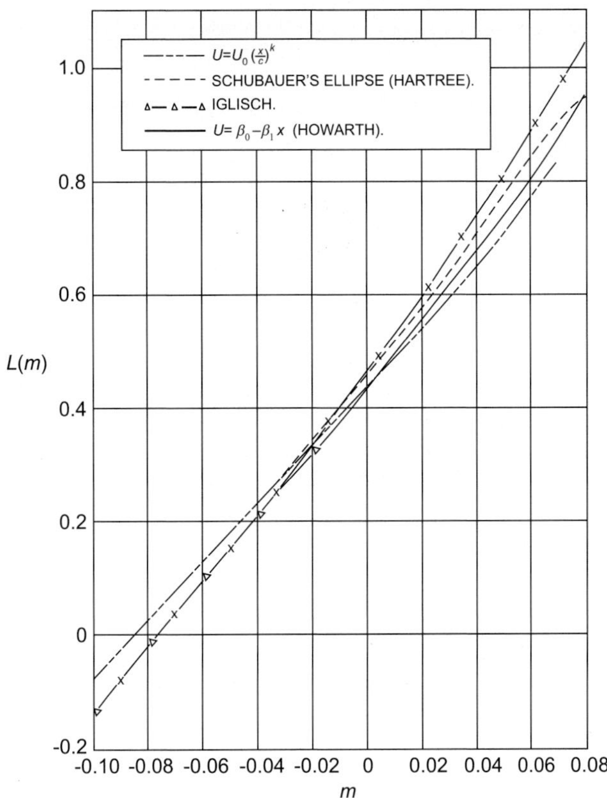

FIGURE 8.9 Plot of $L(m)$ from (8.48) vs. $m=-\lambda$ from Thwaites' (1949) paper. Here a suitable empirical fit to the four sources of laminar boundary-layer data is provided by $L(m)=0.45+6.0m=0.45-6.0\lambda$. *Reprinted with the permission of The Royal Aeronautical Society.*

fluid mechanics of the situation are entirely changed. First of all, the boundary-layer approximation is invalid downstream of the separation point because the layer is no longer thin and contiguous to the surface; thus, the scaling (8.6) is no longer valid. Second, separation commonly leads to unsteadiness because separated boundary layers are unstable and may produce fluctuations even if all boundary conditions are steady. And third, a separated boundary layer commonly has an enormous flow-displacement effect that drastically changes the outer flow so that it no longer imposes the presumed attached boundary-layer surface pressure distribution. Thus, any boundary-layer calculation that starts from a presumed surface pressure distribution should be abandoned once that calculation predicts the occurrence of boundary-layer separation.

The following two examples illustrate the use of Thwaites' method with and without a prediction of the occurrence of boundary-layer separation.

Example 8.6

Use Thwaites' method to estimate the momentum thickness, displacement thickness, and wall shear stress of the Blasius boundary layer with $\theta_0=0$ at $x=0$.

Solution The solution plan is to use (8.50) to obtain θ. Then, because $dU_e/dx=0$ for the Blasius boundary layer, $\lambda=0$ at all downstream locations and the remaining boundary-layer parameters can be determined from the θ results, (8.45), (8.46), and Table 8.2. The first step is setting $U_e=U=$ constant in (8.50) with $\theta_0=0$:

$$\theta^2=\frac{0.45\nu}{U^6}\int_0^x U^5 dx=\frac{0.45\nu}{U}x, \quad \text{or} \quad \theta=0.671\sqrt{\frac{\nu x}{U}}.$$

This approximate answer is 1% higher than the Blasius-solution value. For $\lambda=0$, the tabulated shape factor is $H(0)=2.61$, so:

$$\delta^*=\theta\left(\frac{\delta^*}{\theta}\right)=\theta H(0)=0.671\sqrt{\frac{\nu x}{U}}(2.61)=1.75\sqrt{\frac{\nu x}{U}}.$$

This approximate answer is also 1% higher than the Blasius-solution value. For $\lambda = 0$, the shear correlation value is $l(0) = 0.220$, so:

$$\tau_w = \mu \frac{U}{\theta} l(0) = \frac{\mu U}{0.671\sqrt{\nu x/U}}(0.220) = \frac{1}{2}\rho U^2 (0.656)\sqrt{\frac{\nu}{Ux}},$$

which implies a skin friction coefficient of:

$$C_f = \frac{\tau_w}{\frac{1}{2}\rho U^2} = \frac{0.656}{\sqrt{\mathrm{Re}_x}},$$

which is 1.2% below the Blasius-solution value.

Example 8.7

A shallow-angle, two-dimensional diffuser of length L is designed for installation downstream of a blower in a ventilation system to slow the blower-outlet airflow via an increase in duct cross-sectional area (see Fig. 8.10). If the diffuser should reduce the flow speed by half by doubling the flow area and the boundary layer is laminar, is boundary-layer separation likely to occur in this diffuser?

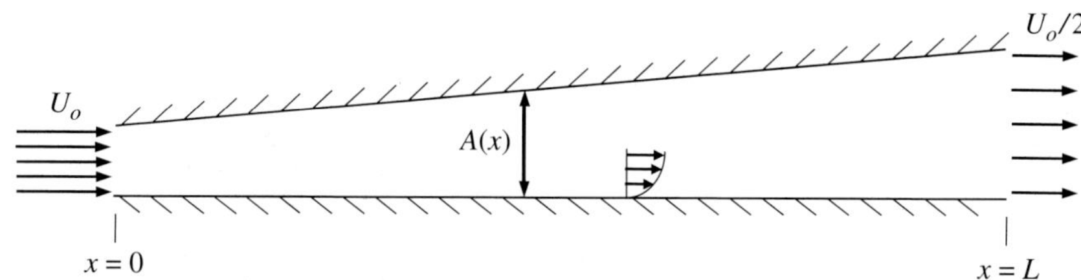

FIGURE 8.10 A simple two-dimensional diffuser of length L intended to slow the incoming flow to half its speed by doubling the flow area. The resulting adverse pressure gradient in the diffuser influences the character of the boundary layers that develop on the diffuser's inner surfaces, especially when these boundary layers are laminar.

Solution The first step is to determine the outer flow $U_e(x)$ by assuming uniform (ideal) flow within the diffuser. Then, (8.50) can be used to estimate $\theta^2(x)$ and $\lambda(x)$. Boundary-layer separation will occur if λ falls below -0.090.

For uniform incompressible flow within the diffuser: $U_1A_1 = U_e(x)A(x)$, where (1) denotes the diffuser inlet, $U_e(x)$ is the flow speed, A(x) is the diffuser's cross sectional area. For flat diffuser sides, a doubling of the flow area in a distance L, requires $A(x) = A_1(1 + x/L)$, so the ideal outer flow velocity is $U_e(x) = U_1(1 + x/L)^{-1}$. With this exterior velocity the Thwaites' integral becomes:

$$\theta^2 = \frac{0.45\nu}{U_e^6(x)}\int_0^x U_e^5(x')dx' + \frac{\theta_0^2 U_0^6}{U_e^6(x)} = \frac{0.45\nu}{U_1}\left(1+\frac{x}{L}\right)^6 \int_0^x \left(1+\frac{x}{L}\right)^{-5} dx + \theta_0^2\left(1+\frac{x}{L}\right)^6,$$

where $U_0 = U_1$ in this case. The 0-to-x integration is readily completed and this produces:

$$\theta^2 = \frac{0.45\nu}{U_1}\left(1+\frac{x}{L}\right)^6 \frac{L}{4}\left[1-\left(1+\frac{x}{L}\right)^{-4}\right] + \theta_0^2\left(1+\frac{x}{L}\right)^6.$$

From this equation it is clear that θ grows with increasing x. This relationship can be converted to λ by multiplying it with $(1/\nu)dU_e/dx = -(U_1/\nu L)(1 + x/L)^{-2}$:

$$\lambda = \frac{\theta^2}{\nu}\frac{dU_e}{dx} = -\frac{0.45}{4}\left[\left(1+\frac{x}{L}\right)^4 - 1\right] - \frac{\theta_0^2 U_1}{\nu L}\left(1+\frac{x}{L}\right)^4.$$

In this case, even when $\theta_0 = 0$, λ will (at best) start at zero and become increasingly negative with increasing x. At this point, a determination of whether or not boundary-layer separation will occur involves calculating λ as function of x/L. The following table comes from evaluating the last equation with $\theta_0 = 0$.

x/L	λ
0	0
0.05	−0.02424
0.10	−0.05221
0.15	−0.08426
0.20	−0.12078

Here, Thwaites' method predicts that boundary-layer separation will occur, since λ will fall below -0.090 at $x/L \approx 0.16$, a location that is far short of the end of the diffuser at $x = L$. While it is tempting to consider this a prediction of the location of boundary-layer separation, such a temptation should be avoided. In addition, if θ_0 was non-zero, then λ would decrease even more quickly than shown in the table, making the positive prediction of boundary-layer separation even firmer. Thus, successful prediction of the flow in this diffuser requires simultaneous assessment of the whole flow field. Partitioning the equation-solving effort into an ideal outer flow and a steady laminar inner flow is not successful in this case. (In reality, diffusers in duct work and flow systems are common but they typically operate with turbulent boundary layers that more effectively resist separation.)

8.7 Transition, pressure gradients, and boundary-layer separation

The analytical and empirical results provided in the prior sections are altered when a boundary layer transitions from laminar to turbulent flow, and when a boundary layer separates from the surface on which it has developed. Both of these phenomena, especially the second, are influenced by the pressure gradient felt by the boundary layer.

The process of changing from laminar to turbulent flow is called *transition*, and it occurs in a wide variety of flows as the Reynolds number increases. For the present purposes, the complicated phenomenon known as boundary-layer transition is described in general terms. Interestingly for a high Reynolds number theory, the agreement of solutions to the laminar boundary equations with experimental data breaks down when the downstream-distance-based Reynolds number Re_x is larger than some critical value, say Re_{cr}, that depends on fluctuations in the free stream above the boundary layer and on the surface shape, curvature, roughness, vibrations, and pressure gradient. Above Re_{cr}, a laminar boundary-layer flow becomes unstable and transitions to turbulence. Typically, the critical Reynolds number decreases when the surface roughness or free-stream fluctuation levels increase. In general, Re_{cr} varies greatly and detailed predictions of transition are often a difficult task or a research endeavor. Within a factor of five or so, the transition Reynolds number for a smooth, flat-plate boundary layer is found to be:

$$\mathrm{Re}_{\mathrm{cr}} \sim 10^6 \quad \text{(flat plate)}.$$

Fig. 8.11 schematically depicts the flow regimes on a semi-infinite flat plate (with the vertical direction greatly exaggerated). In the leading-edge region, where $\mathrm{Re}_x = Ux/\nu \sim 1$, the full Navier-Stokes equations are required to properly describe the flow. As Re_x increases toward the downstream limit of the leading-edge region, we can locate x_0 as the maximal upstream location where the laminar boundary-layer equations are valid (perhaps $\mathrm{Re}_{x0} \sim 10^3$). For some distance $x > x_0$, the boundary layer's condition at $x = x_0$ is remembered. Eventually, the influence of the initial condition may be neglected and the solution becomes of similarity form. For somewhat larger Re_x, a bit farther downstream, an initial instability appears and fluctuations of a specific wavelength or frequency may be amplified. With increasing downstream distance, a wider spatial or temporal frequency range of fluctuations may be amplified and these fluctuations interact with each other nonlinearly through the advective acceleration terms in the momentum equation. As Re_x increases further, the fluctuations may increase in strength and the flow becomes increasingly chaotic and irregular with increasing downstream distance. When the fluctuations cease their rapid growth, the flow is said to be fully turbulent and transition is complete (see also Section 9.13).

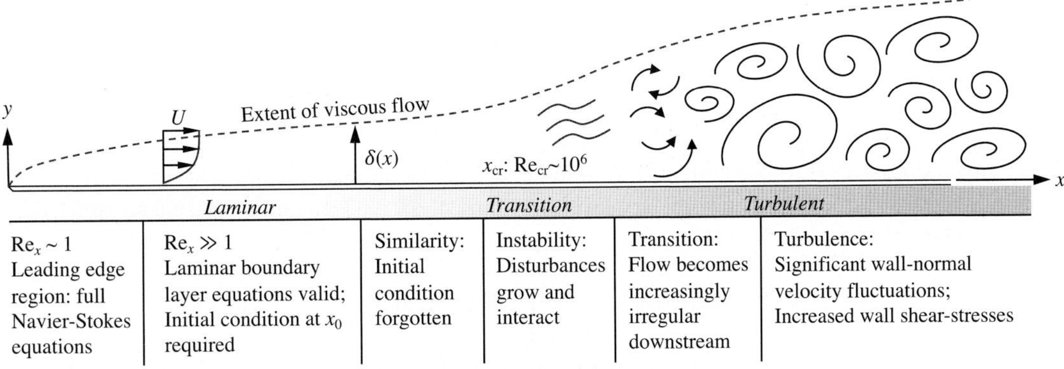

FIGURE 8.11 Schematic depiction of flow over a semi-infinite flat plate. Here, increasing x is synonymous with increasing Reynolds number.

Laminar and turbulent boundary layers differ in many important ways. A fully turbulent boundary layer produces significantly more average surface shear stress τ_w than an equivalent laminar boundary layer, and a fully turbulent boundary-layer velocity profile has a different shape and different parametric dependencies than an equivalent laminar one. For example, the thickness of a zero-pressure-gradient turbulent boundary layer grows faster than $x^{1/2}$ (Fig. 8.11), and the wall shear stress increases faster with U than in a laminar boundary layer where $\tau_w \propto U^{3/2}$. This increase in friction occurs because turbulent fluctuations produce more wall-normal transport of momentum than that possible from steady viscous diffusion alone. However, both types of boundary layers respond similarly to pressure gradients but with different sensitivities.

Fig. 8.12 sketches the nature of the observed variation of the drag coefficient in a flow over a flat plate, as a function of the Reynolds number. The lower curve applies if the boundary layer is laminar over the entire length of the plate, and the upper curve applies if the boundary layer is turbulent over the entire length. The curve joining the two applies to a boundary layer that is laminar over the initial part of the plate, begins transition at $\mathrm{Re}_L \sim 5 \times 10^5$, and is fully turbulent for $\mathrm{Re}_L > 10^7$. The exact point at which the observed drag deviates from the wholly laminar behavior depends on flow conditions, flow geometry, and surface conditions.

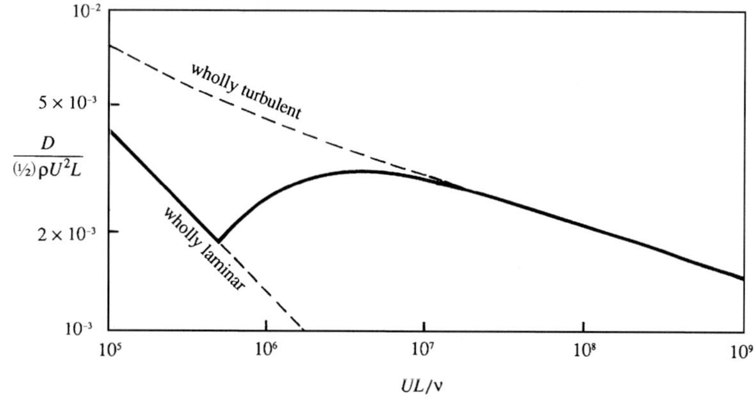

FIGURE 8.12 Measured drag coefficient for a boundary-layer flow over a flat plate. The continuous line shows the drag coefficient for a plate on which the flow is partly laminar and partly turbulent, with the transition taking place at a position where the local Reynolds number is 5×10^5. The dashed lines show the behavior if the boundary layer was either completely laminar or completely turbulent over the entire length of the plate.

Although surface pressure gradients do affect transition, it may be argued that their most important influence is on boundary-layer separation. A fundamental discussion of boundary-layer separation begins with the steady stream-wise boundary-layer-flow momentum equation, (8.9), where the pressure gradient is found from the external velocity field via (8.11) and with x taken in the stream-wise direction along the surface of interest. At the surface, both velocity components are zero so (8.9) reduces to:

$$\mu(\partial^2 u/\partial y^2)_{wall} = dp/dx$$

(see Example 7.1). In an accelerating stream $dp/dx < 0$, so:

$$(\partial^2 u/\partial y^2)_{wall} < 0 \quad \text{(accelerating)}. \tag{8.51}$$

Given that the velocity profile has to blend smoothly with the external profile, the gradient $\partial u/\partial y$ slightly below the edge of the boundary layer decreases with increasing y from a positive value to zero; therefore, $\partial^2 u/\partial y^2$ slightly below the boundary-layer edge is negative. Eq. (8.51) then shows that $\partial^2 u/\partial y^2$ has the same sign at the wall and at the boundary-layer edge, and presumably throughout the boundary layer. In contrast, for a decelerating external stream, $dp/dx > 0$, the curvature of the velocity profile at the wall is:

$$(\partial^2 u/\partial y^2)_{wall} > 0 \quad \text{(decelerating)}, \tag{8.52}$$

so that the profile curvature changes sign somewhere within the boundary layer. In other words, the boundary-layer profile in a decelerating flow has a *point of inflection* where $\partial^2 u/\partial y^2 = 0$, an important fact for boundary-layer stability and transition (see Chapter 9). In the special case of the Blasius boundary layer, the profile's inflection point is at the wall.

The shape of the velocity profiles in Fig. 8.13 and the finding in Example 8.4 suggest that a decelerating exterior flow tends to increase the thickness of the boundary layer. This can also be seen from integrating the two-dimensional continuity

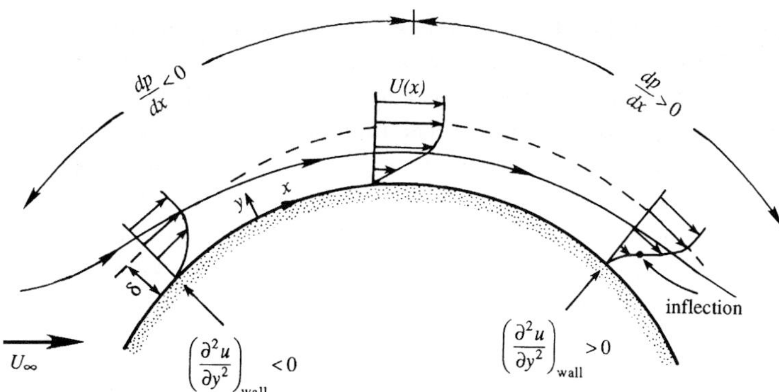

FIGURE 8.13 Velocity profiles across boundary layers with favorable ($dp/dx < 0$) and adverse ($dp/dx > 0$) pressure gradients, as indicated above the flow. The surface shear stress and stream-wise fluid velocity near the surface are highest and lowest in the favorable and adverse pressure gradients, respectively, with the $dp/dx = 0$ case falling between these limits.

equation:

$$v(y) = -\int_0^y (\partial u/\partial x)dy.$$

Compared to flow over a flat plate, a decelerating external stream causes a larger $-(\partial u/\partial x)$ within the boundary layer because the deceleration of the outer flow adds to the viscous deceleration within the boundary layer. It follows from the foregoing equation that the wall-normal velocity component (v) is larger for a decelerating flow. The boundary layer therefore thickens not only by viscous diffusion but also by advection away from the surface, resulting in a more rapid increase in the boundary-layer thickness with x than when the exterior flow is constant or accelerating.

If p falls with increasing x, $dp/dx < 0$, the pressure gradient is said to be *favorable*. If p rises with increasing x, $dp/dx > 0$, the pressure gradient is said to be *adverse*. In an adverse pressure gradient, the boundary-layer flow decelerates, thickens, and develops a point of inflection. When the adverse pressure gradient is strong enough or acts over a long enough distance, the flow next to the wall reverses direction (Fig. 8.14). The point S at which the reverse flow meets the forward flow is a local stagnation point and is known as the *separation point*. Fluid elements approach S (from either side) and are then transported away from the wall. Thus, a separation streamline emerges from the surface at S. Furthermore, the surface shear stress changes sign across S because the surface flow changes direction. Thus, the surface shear stress at S is zero, which implies:

$$(\partial u/\partial y)_{wall} = 0 \quad \text{(seperation)}.$$

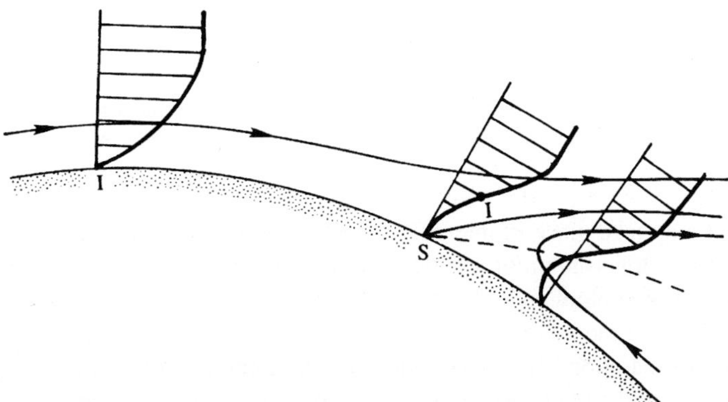

FIGURE 8.14 Streamlines and velocity profiles near a separation point S where a streamline emerges from the surface. The usual boundary-layer equations are not valid downstream of S. The inflection point in the stream-wise velocity profile is indicated by I. The dashed line is the locus of $u = 0$.

Once a boundary layer separates from the surface on which it has formed, the surface-normal displacement effect produced by divergence of the body contour and the separation streamline may be enormous. Additionally, at high Reynolds numbers, a separated boundary layer commonly acquires the properties of a vortex sheet and may rapidly become unstable and transition to a thick zone of turbulence. Thus, boundary-layer separation typically requires the presumed geometry-based inner-outer and rotational-irrotational flow dichotomies to be reconsidered or even abandoned. In such cases, recourse to experiments or multidimensional numerical calculations may be the only choices for flow investigation.

At Reynolds numbers that are not too large, flow separation may not lead to unsteadiness. For flow past a circular cylinder for $4 < \mathrm{Re} < 40$ the reversed flow downstream of a separation point may form part of a steady vortex behind the cylinder (see Fig. 8.18 in Section 8.8). At higher Reynolds numbers, when the flow on the upstream side of the cylinder develops genuine boundary-layer characteristics, the flow downstream of separation is unsteady and frequently turbulent.

The adverse-pressure gradient strength that a boundary layer can withstand without separating depends on the geometry of the flow and whether the boundary layer is laminar or turbulent. However, a severe adverse-pressure gradient, such as that on the aft side of a rounded blunt body, invariably leads to separation. In contrast, the boundary layer on the trailing surface of a slender body may overcome the weak pressure gradients involved. Therefore, to avoid separation and the resulting form drag penalty, the trailing section of a submerged body should be *gradually* reduced in size, giving it a *streamlined* (or teardrop) shape.

Experimental evidence indicates that the point of separation is relatively insensitive to the Reynolds number as long as the boundary layer is laminar. However, a *transition to turbulence delays boundary-layer separation*; that is, a turbulent boundary layer is more capable of withstanding an adverse pressure gradient. This is because the velocity profile in a turbulent boundary layer places more high-speed fluid near the surface (Fig. 8.15). For example, the laminar boundary layer over a circular cylinder separates at ~82° from the forward stagnation point, whereas a turbulent layer over the same body separates at 125° (shown later in Fig. 8.18). Experiments show that the surface pressure remains fairly uniform downstream of separation and has a lower value than the pressures on the forward face of the body. The resulting drag due to such fore-aft pressure differences is called *form drag*, as it depends crucially on the shape of the body (and the location of boundary-layer separation). For a blunt body like a sphere, the form drag is larger than the skin friction drag because of the occurrence of separation. For a streamlined body like a rowing shell for crew races, skin friction is generally larger than the form drag. As long as the separation point is located at the same place on the body, the drag coefficient of a blunt body is nearly constant at high Reynolds numbers. However, the drag coefficient may drop suddenly when the boundary layer undergoes transition to turbulence, the separation point moves aft, and the body's wake becomes narrower (see Fig. 8.24 in Section 8.8).

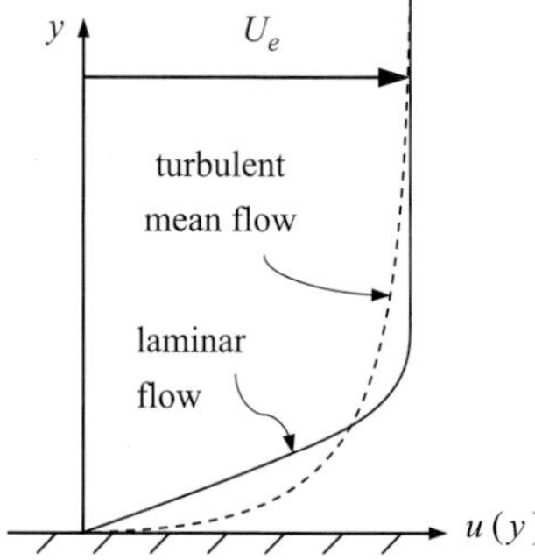

FIGURE 8.15 Nominal comparison of laminar and turbulent-mean-flow stream-wise velocity profiles for boundary layers with nominally equal displacement thickness. Here the primary differences are the presence of higher speed fluid closer to the surface and greater surface shear stress in the turbulent boundary layer.

Boundary-layer separation may take place in internal as well as external flows. An example is a divergent channel or diffuser (Example 8.7, Fig. 8.16). Downstream of a narrow point in a ducted flow, an adverse-pressure gradient can cause separation. Elbows, tees, and valves in pipes and tubes commonly lead to regions of internal flow separation, too.

Again it must be emphasized that the boundary-layer equations are valid only as far downstream as the point of separation, if it is known. Beyond separation, the basic underlying assumptions of boundary-layer theory become invalid. Moreover, the parabolic character of the boundary-layer equations requires that a numerical integration is possible only in the direction of advection (along which information is propagated). In a region of reversed flow, this integration direction is opposite the flow direction (upstream). Thus, a forward-directed (downstream) integration of the boundary-layer equations breaks down after separation. Furthermore, ideal-flow theory may not be used to determine the pressure in a separated flow region, since the flow there is rotational and the interface between irrotational and rotational flow regions no longer follows the body's solid surface. Instead, the irrotational-rotational flow interface may be some unknown shape encompassing part of the body's contour, the separation streamline, and, possibly, a wake-zone contour.

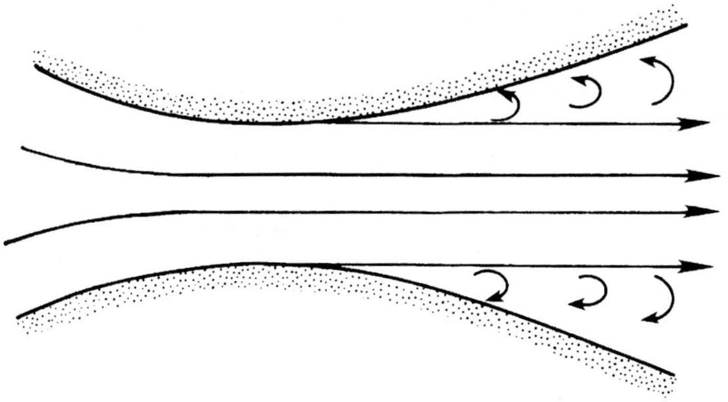

FIGURE 8.16 Separation of flow in a divergent channel. Here, an adverse pressure gradient has lead to boundary-layer separation just downstream of the narrowest part of the channel. Such separated flows are unstable and are exceedingly likely to be unsteady, even if all the boundary conditions are time independent.

Example 8.8

Using a third-order two-dimensional power-series expansion near a flat-plate boundary layer's separation point, $x = x_s$ and $y = 0$, determine how the stream function $\psi(x, y)$ depends on $\partial p/\partial x$ and β_s, the angle the separating streamline makes with the horizontal surface as shown in Fig. 8.17.

Solution A third-order power series expansion for $\psi(x, y)$ is:

$$\psi(x, y) = a_0 + a_1 x' + a_2 y + a_3 x'^2 + a_4 x' y + A y^2 + a_5 x'^3 + a_6 x'^2 + B x' y^2 + C y^3,$$

where $x' = x - x_s$, and a_0 through a_6, A, B, and C are undetermined constants. This stream function must satisfy the no-slip boundary condition, $u = v = 0$ on $y = 0$, so $\partial\psi/\partial y = -\partial\psi/\partial x = 0$ on $y = 0$. These two conditions cause a_1 through a_6 to be zero, and if $\psi = 0$ defines the plate surface, then the stream function reduces to $\psi(x, y) = Ay^2 + B(x - x_s)y^2 + Cy^3$. In addition, the surface shear stress, τ_w, is zero at the separation point, so:

$$\tau_w = \mu\left(\frac{\partial u}{\partial y}\right)_{y=0, x=x_s} = \mu\left(\frac{\partial^2\psi}{\partial y^2}\right)_{y=0, x=x_s} = (2A + 2B(x - x_s) + 6Cy)_{y=0, x=x_s} = 2A = 0,$$

and this leaves:

$$\psi(x, y) = B(x - x_s)y^2 + Cy^3.$$

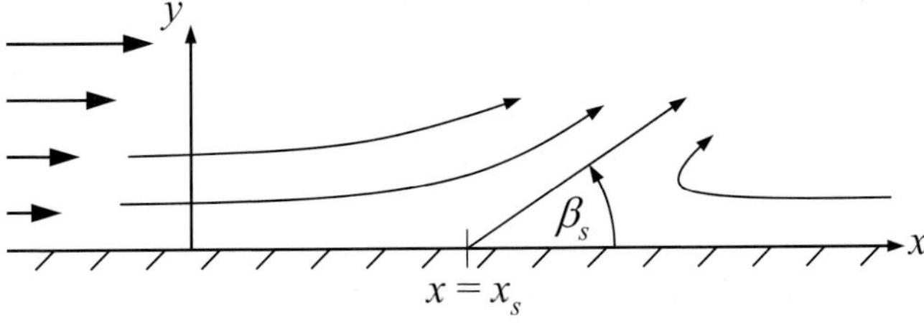

FIGURE 8.17 Streamline pattern near the separation point ($x = x_s$, $y = 0$) on a flat surface.

In the vicinity of the separation point, this stream function $\psi(x, y)$ must satisfy two additional conditions. The first comes from the limiting form of (7.1) as $y \to 0$ (see Example 7.1), which for the present coordinate system and stream function is:

$$(\partial p/\partial x)_{y=0} = \mu(\partial^2 u/\partial y^2)_{y=0} = \mu(\partial^3\psi/\partial y^3)_{y=0},$$

and this implies $C = (1/6\mu)(\partial p/\partial x)$. The second condition is that the zero-streamline must leave the surface at an angle β_s with respect to the downstream direction. The zero-streamline is given by $\psi(x, y) = 0$, which implies:

$$0 = B(x - x_s)y^2 + \frac{1}{6\mu}\left(\frac{\partial p}{\partial x}\right)y^3, \quad \text{or} \quad -\frac{1}{6\mu}\left(\frac{\partial p}{\partial x}\right)y = B(x - x_s), \quad \text{or} \quad -\frac{1}{6\mu}\left(\frac{\partial p}{\partial x}\right)\frac{dy}{dx} = B.$$

So, with $dy/dx = \tan\beta_s$, the final form for the stream function expansion is:

$$\psi(x, y) = \frac{y^2}{6\mu}\left(\frac{\partial p}{\partial x}\right)(y - (x - x_s)\tan\beta_s).$$

Thus for boundary layer separation from a flat surface, the angle of the separating streamline may be independent of the local pressure gradient. And, when the flow is in the positive x-direction upstream of the separation point (i.e. $\psi > 0$ for $y > 0$), this stream function only makes sense when $\partial p/\partial x$ is locally positive, an adverse pressure gradient.

8.8 Flow past a circular cylinder

In general, analytical solutions of viscous flows can be found (possibly in terms of perturbation series) only in two limiting cases, namely Re $\ll 1$ and Re $\gg 1$. In the Re $\ll 1$ limit the inertia forces are negligible over most of the flow field; the Stokes-Oseen solutions discussed in the preceding chapter are of this type. In the opposite limit of Re $\gg 1$, the viscous forces are negligible everywhere except close to the surface, and a solution may be attempted by matching an irrotational outer flow with a boundary layer near the surface. In the intermediate range of Reynolds numbers, analytical solutions are elusive or do not exist, and one has to depend on experimentation and numerical solutions. Some of these experimental flow patterns are described in this section, taking the flow over a circular cylinder as an example. Instead of discussing only the intermediate Reynolds number range, the experimental-observed phenomena for the entire range from small to very high Reynolds numbers are presented.

Low Reynolds numbers

Consider creeping flow around a circular cylinder, characterized by $\text{Re} = U_\infty d/\nu < 1$, where U_∞ is the upstream flow speed and d is the cylinder's diameter. Vorticity is generated close to the surface because of the no-slip boundary condition. In the Stokes approximation, this vorticity is simply diffused, not advected, which results in fore and aft symmetry of streamlines. The Oseen approximation partially takes into account the advection of vorticity, and results in an asymmetric velocity distribution *far* from the body (which was shown for a sphere in Fig. 7.20). The vorticity distribution is qualitatively analogous to the dye distribution caused by a source of colored fluid at the position of the body. The color diffuses symmetrically in very slow flows, but at higher flow speeds the dye is confined behind a parabolic boundary with the dye source at the parabola's focus.

For increasing Re above unity, the Oseen approximation breaks down, and the vorticity is increasingly confined behind the cylinder because of advection. For Re > 4, two small steady eddies appear behind the cylinder and form a closed separation zone contained with a separation streamline. This zone is sometimes called a *separation bubble*. The cylinder's wake is completely laminar and the vortices rotate in a manner that is consistent with the exterior flow (Fig. 8.18). These eddies grow in length and width as Re increases.

Moderate Reynolds numbers

An interesting sequence of events begins to develop when Re reaches 40, the point at which the wake behind the cylinder becomes unstable. Experiments show that for Re $\sim 10^2$ the wake develops a slow oscillation in which the velocity is periodic in time and downstream distance, with the amplitude of the oscillation increasing downstream. The oscillating wake rolls up into two staggered rows of vortices with opposite sense of rotation (Fig. 8.19). von Karman investigated the phenomenon as a problem of superposition of irrotational vortices; he concluded that a non-staggered row of vortices is unstable, and a staggered row is stable only if the ratio of lateral distance between the vortices to their longitudinal distance is 0.28. Because of the similarity of the wake with footprints on a street, the staggered row of vortices behind a blunt body is called a *von Karman vortex street*. The vortices move downstream at a speed smaller than U_∞. This means that the vortex pattern slowly follows the cylinder if it is pulled through a stationary fluid.

In the range $40 < \text{Re} < 80$, the vortex street does not interact with the pair of attached vortices. As Re increases above 80, the vortex street forms closer to the cylinder, and the attached eddies (whose downstream length has now grown to be about twice the diameter of the cylinder) themselves begin to oscillate. Finally the attached eddies periodically break off alternately from the two sides of the cylinder. While an eddy on one side is shed, that on the other side forms, resulting in an unsteady flow near the cylinder. As vortices of opposite circulations are shed off alternately from the two sides, the circulation around the cylinder changes sign, resulting in an oscillating lift or lateral force perpendicular to the upstream flow direction. If the frequency of vortex shedding is close to the natural frequency of some structural mode of vibration

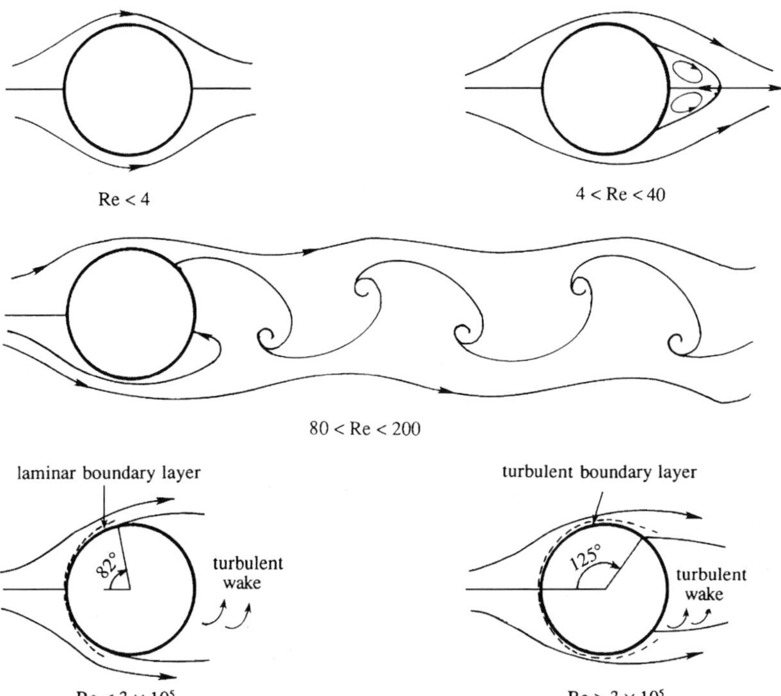

FIGURE 8.18 Depiction of some of the flow regimes for a circular cylinder in a steady uniform cross flow. Here, $\text{Re} = U_\infty d/\nu$ is the Reynolds number based on free-stream speed U_∞ and cylinder diameter d. At the lowest Re, the streamlines approach perfect fore-aft symmetry. As Re increases, asymmetry increases and steady wake vortices form. With further increase in Re, the wake becomes unsteady and forms the alternating-vortex von Karman vortex street. For Re up to $\text{Re}_{cr} \sim 3 \times 10^5$, the laminar boundary layer separates approximately 82° from the forward separation point. Above this Re value, the boundary-layer transitions to turbulence, and separation is delayed to 125° from the forward separation point.

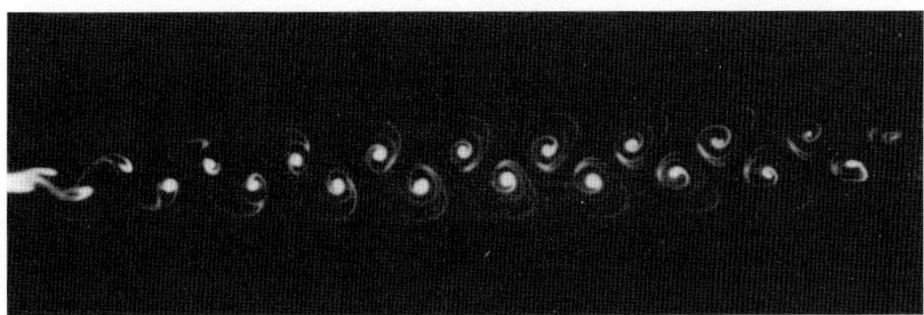

FIGURE 8.19 von Karman vortex street downstream of a circular cylinder at Re = 55. *Flow visualized by condensed milk. Taneda,* Jour. Phys. Soc., Japan ***20****: 1714–1721, 1965, and reprinted with the permission of The Physical Society of Japan and Dr. Sadatoshi Taneda.*

of the cylinder and its supports, then an appreciable lateral vibration may be observed. Engineered structures such as suspension bridges, oil drilling platforms, and even automobile components are designed to prevent coherent shedding of vortices from cylindrical structures. This is done by including spiral blades protruding out of the cylinder's surface, which break up the spanwise coherence of vortex shedding, forcing the vortices to detach at different times along the length of these structures (Fig. 8.20).

The passage of regular vortices causes velocity measurements in the cylinder's wake to have a dominant periodicity, and this frequency Ω is commonly expressed as a *Strouhal number* (4.103), $\text{St} = \Omega d/U_\infty$. Experiments show that for a circular cylinder the value of St remains close to 0.2 for a large range of Reynolds numbers. For small values of cylinder diameter and moderate values of U_∞, the resulting frequencies of the vortex shedding and oscillating lift lie in the acoustic range. For example, at $U_\infty = 10$ m/s and a wire diameter of 2 mm, the frequency corresponding to a Strouhal number of 0.2 is 1000 cycles per second. The *singing* of telephone and electrical transmission lines and automobile radio antennae have been attributed to this phenomenon. The value of St given here is that observed in three-dimensional flows with nominally two-dimensional boundary conditions. Moving soap-film experiments and calculations suggest a somewhat higher value of $\text{St} = 0.24$ in perfectly two-dimensional flow (see Wen and Lin, 2001).

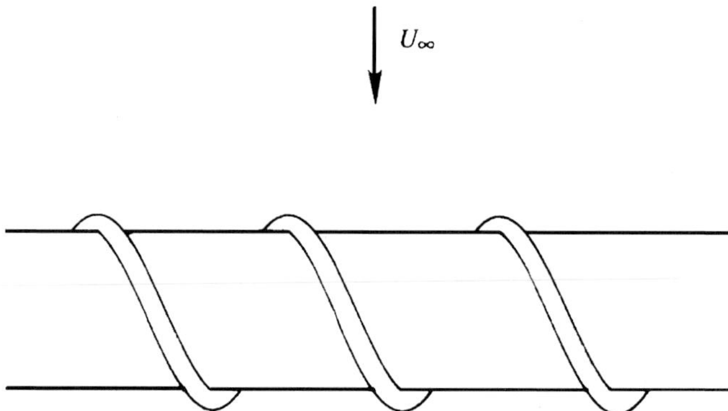

FIGURE 8.20 Spiral blades used for breaking up the span-wise coherence of vortex shedding from a cylindrical rod. Coherent vortex shedding can produce tonal noise and potentially large (and undesired) structural loads on engineered devices that encounter wind or water currents.

Below Re = 200, the vortices in the wake are laminar and continue to be so for large distances downstream. Above 200, the vortex street becomes unstable and irregular, and the flow within the vortices themselves becomes chaotic. However, the flow in the wake continues to have a strong frequency component corresponding to a Strouhal number of St = 0.2. However, above a Reynolds number of several thousand, periodicity in the wake is only perceptible near the cylinder, and the wake may be described as fully turbulent beyond several cylinder diameters downstream.

Striking examples of vortex streets have also been observed in stratified atmospheric flows. Fig. 8.21 shows a satellite photograph of the wake behind an isolated mountain peak when the wind is blowing toward the upper right of picture. The mountain pierces through the cloud level, and the flow pattern becomes visible in the clouds. The wake behind the mountain peak displays the characteristics of a von Karman vortex street. The strong density stratification in this flow has prevented vertical motions, giving the flow the two-dimensional character necessary for the formation of vortex streets.

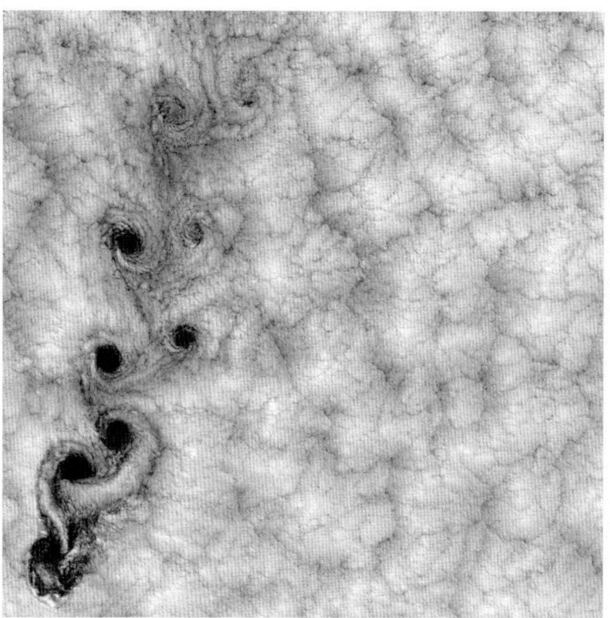

FIGURE 8.21 A von Karman vortex street downstream of a mountain peak in a strongly stratified atmosphere off the Chilean coast near the Juan Fernandez Islands. North is upward, and the wind is blowing toward the norththeast. Photographed by the Landsat 7 satellite on September 15, 1999. Image credit: NASA Earth Observatory.

High Reynolds numbers

At high Reynolds numbers the frictional effects upstream of separation are confined near the surface of the cylinder, and the boundary-layer approximation is valid as far downstream as the point of separation. For a smooth cylinder up to $\text{Re} < 3 \times 10^5$, the boundary layer remains laminar, although the wake formed behind the cylinder may be completely turbulent. The laminar boundary layer separates at $\approx 82°$ from the forward stagnation point (Fig. 8.18). The pressure in the wake downstream of the point of separation is nearly constant and lower than the upstream pressure (Fig. 8.22). The drag on the cylinder in this Re range is primarily due to the asymmetry in the pressure distribution caused by boundary-layer separation, and, since the point of separation remains fairly stationary in this Re range, the cylinder's drag coefficient C_D also stays constant at a value near unity (see Fig. 8.23).

Important changes take place beyond the critical Reynolds number of $\mathrm{Re}_{cr} \sim 3 \times 10^5$. When $3 \times 10^5 < \mathrm{Re} < 3 \times 10^6$, the laminar boundary layer becomes unstable and transitions to turbulence. Because of its greater average near-surface flow speed, a turbulent boundary layer is able to overcome a larger adverse-pressure gradient. In the case of a circular cylinder the turbulent boundary layer separates at 125° from the forward stagnation point, resulting in a thinner wake and a pressure distribution more similar to that of potential flow. Fig. 8.22 compares the pressure distributions around the cylinder for two values of Re, one with a laminar and the other with a turbulent boundary layer. It is apparent that the pressures within the wake are higher when the boundary layer is turbulent, resulting in a drop in the drag coefficient from 1.2 to 0.33 at the point of transition. For values of $\mathrm{Re} > 3 \times 10^6$, the separation point slowly moves upstream as the Reynolds number increases, resulting in a mild increase of the drag coefficient (Fig. 8.23).

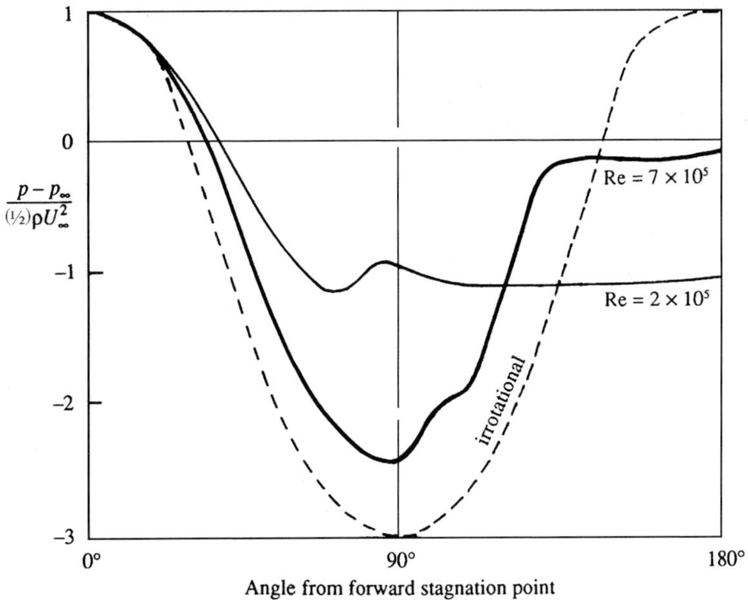

FIGURE 8.22 Surface pressure distribution around a circular cylinder at subcritical and supercritical Reynolds numbers. Note that the pressure is nearly constant within the wake and that the wake is narrower for flow at supercritical Re. The change in the top- and bottom-side, boundary-layer separation points near Re_{cr} is responsible for the change in C_p shown.

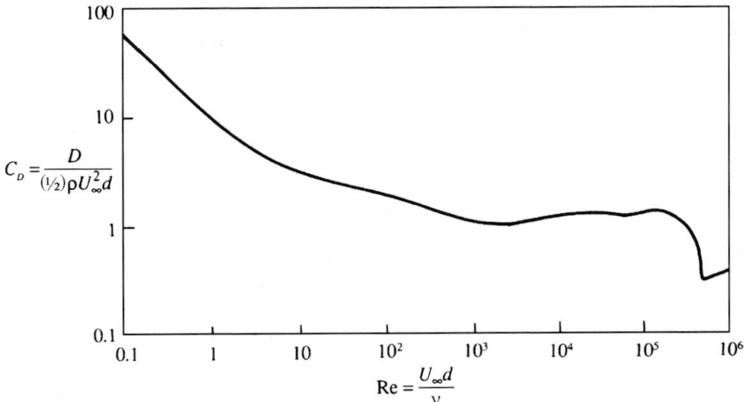

FIGURE 8.23 Measured drag coefficient, C_D, of a smooth circular cylinder vs. $\mathrm{Re} = U_\infty d/\nu$. The sharp dip in C_D near Re_{cr} is due to the transition of the boundary layer to turbulence, and the consequent downstream movement of the point of separation and change in the cylinder's surface pressure distribution.

It should be noted that the critical Reynolds number at which the boundary layer undergoes transition is strongly affected by two factors, namely the intensity of fluctuations existing in the approaching stream and the roughness of the surface, an increase in either decreases $\mathrm{Re}_{\mathrm{cr}}$. The value of 3×10^5 is found to be valid for a smooth circular cylinder at low levels of fluctuation of the oncoming stream.

We close this section by noting that this flow illustrates three instances where the solution is counterintuitive. First, small causes can have large effects. If we solve for the flow of a fluid with zero viscosity around a circular cylinder, we obtain the results of Section 6.3. The inviscid flow has fore-aft symmetry and the cylinder experiences zero drag. The bottom two panels of Fig. 8.18 illustrate the flow for small viscosity. In the limit as viscosity tends to zero, the flow must look like the last panel in which there is substantial fore-aft asymmetry, a significant wake, and significant drag. This is because of the necessity of a boundary layer and the satisfaction of the no-slip boundary condition on the surface so long as viscosity is not exactly zero. When viscosity is exactly zero, there is no boundary layer and there is slip at the surface. Thus, the resolution of d'Alembert's paradox lies in the existence of, and an understanding of, the boundary layer.

The second instance of counterintuitivity is that symmetric problems can have non-symmetric solutions. This is evident in the intermediate Reynolds number middle panel of Fig. 8.18. Beyond a Reynolds number of ≈ 40, the symmetric wake becomes unstable and a pattern of alternating vortices called a von Karman vortex street is established. Yet the equations and boundary conditions are symmetric about a central plane in the flow. If one were to solve only a half problem, assuming symmetry, a solution would be obtained, but it would be unstable to infinitesimal disturbances and unlikely to be observed in a laboratory.

The third instance of counterintuitivity is that there is a range of Reynolds numbers where roughening the surface of the body can reduce its drag, the reason that golf balls have dimples. This is true for all blunt bodies. In this range of Reynolds numbers, the boundary layer on the surface of a blunt body is laminar, but sensitive to disturbances such as surface roughness, which would cause earlier transition of the boundary layer to turbulence than would occur on a smooth body. Although the skin friction of a turbulent boundary layer is much larger than that of a laminar boundary layer, most of the drag on a bluff body is caused by incomplete pressure recovery on its downstream side as shown in Fig. 8.22, rather than by skin friction In fact, it is because the skin friction of a turbulent boundary layer is much larger – as a result of a larger velocity gradient at the surface – that a turbulent boundary layer can remain attached farther on the downstream side of a blunt body, leading to a narrower wake, more complete pressure recovery, and reduced drag. The drag reduction attributed to the turbulent boundary layer is shown in Fig. 8.23 for a circular cylinder and Fig. 8.24 for a sphere.

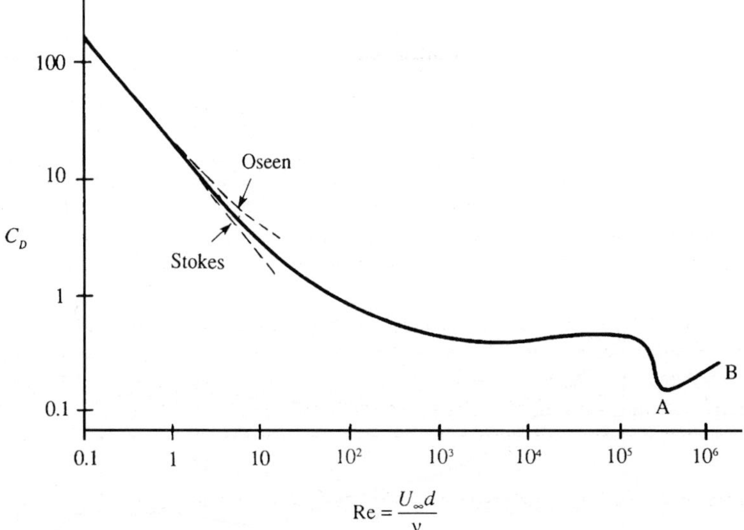

FIGURE 8.24 Measured drag coefficient, C_D, of a smooth sphere vs. $\mathrm{Re} = U_\infty d/\nu$. The Stokes solution is $C_D = 24/\mathrm{Re}$, and the Oseen solution is $C_D = (24/\mathrm{Re})(1 + 3\mathrm{Re}/16)$; these two solutions are discussed at the end of Chapter 7. The increase of drag coefficient in the range A–B has relevance in explaining why the flight paths of sports balls bend in the air.

Example 8.9

The phenomenon of a near-constant Strouhal number for Reynolds number greater than a few hundred has been exploited to produce volume flow rate meters with no moving parts. Typical designs involve a strut with cross-section dimension d that spans the inner diameter D of the pipe that conveys a volume flow rate Q of fluid having density ρ and viscosity μ (see Fig. 8.25). The frequency Ω of flow oscillations downstream of the strut are then sensed with one or more transducers, typically flush-mounted to the pipe's inner wall. What are the fluid mechanical design considerations for such a device?

Solution Fluid mechanics sets at least four primary performance features of vortex flow meters: dynamic range, calibration, signal amplitude, and static pressure losses. Before addressing these individually, determine how Ω must depend on the other five parameters. Using $U_{ave} = Q/(\pi D^2/4)$ in place of Q, dimensional analysis (see Section 1.11) produces:

$$\mathrm{St} = \frac{\Omega d}{U_{ave}} = \Psi\left(\frac{\rho U_{ave} d}{\mu}, \frac{D}{d}\right) = \Psi\left(\mathrm{Re}_d, \frac{D}{d}\right),$$

where Ψ is an undetermined function, and St and Re_d are dimensionless groups with topical significance. Thus, Ω will be directly proportional to U_{ave} (and Q) if Ψ has weak or no dependence on Re_d.

Based on boundary-layer fluid mechanics, Ψ should be Reynolds number independent over a wide range when two conditions are met. First, Re_d must be high enough so that thin boundary layers form on the strut and separate from it to produce regular oscillations in the strut's wake. In practice this means that Re_d must be at least a few hundred. And second, the boundary-layer separation points on the strut should be fixed so that the kinematics of the flow field do not change with increasing Reynolds number. Thus, robust calibration of the device should occur when the strut's cross section has sharp corners where the strut's boundary layers must separate. When both conditions are met, the device's calibration should follow:

$$Q = \frac{\pi}{4} D^2 d \frac{\Omega}{\mathrm{St}},$$

where a best-fit value of St must be measured for each strut shape and D/d considered. (In addition, the pipe-flow Reynolds number, Re_D, must be several thousand or higher, but this requirement does not originate from boundary-layer considerations.)

To generate a strong signal at the pipe sidewall the velocity perturbation from passing vortices should be large there. These velocity perturbations will be proportional to the vortex circulation Γ divided by D, since the transverse dimension from a vortex center to the pipe sidewall is proportional to D. Furthermore, since the presence of the strut leads to the vortices, Γ must be proportional to the product of the average flow-speed and the strut's transverse size, $U_{ave}d$. Thus, sidewall flow-speed fluctuations will be proportional to $U_{ave}d/D$, and this suggests the strut should be as large as possible to produce a readily transduced signal.

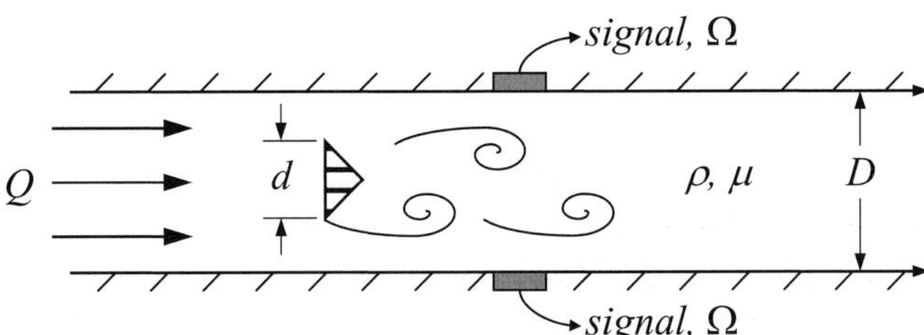

FIGURE 8.25 Schematic drawing of a generic vortex flow meter. The strut with cross section dimension d obstructs the flow and its unsteady vortex wake contains vortices that produce pressure fluctuations that can be measured by wall-mounted sensors.

However, the fluid-dynamic drag force on the strut, F_D, increases as its transverse dimension increases, a flow blockage effect. A simple control volume calculation shows that F_D causes a strut-induced static pressure drop, $\Delta p \cong F_D/(\pi D^2/4)$, that must be overcome with additional pumping power from an external source. Here F_D will be proportional to U_{ave}^2, so a overly-wide strut might produce unacceptable pressure losses at high flow rates.

8.9 Flow past a sphere

The behavior of the boundary layer around a smooth sphere is similar to that around a smooth circular cylinder. In particular it undergoes transition to turbulence at a critical Reynolds number of $\mathrm{Re}_{cr} \sim 5 \times 10^5$, which corresponds to a sudden dip of the drag coefficient (Fig. 8.24). As in the case of a circular cylinder, the separation point slowly moves upstream for postcritical Reynolds numbers, accompanied by a rise in the drag coefficient. A notable difference for flows over a three-dimensional body is the absence of a regular vortex structure in the wake. For flow around a sphere at low Reynolds numbers, there is an attached eddy in the form of a doughnut-shaped ring; in fact, an axial section of the flow looks similar to that shown in Fig. 8.18 for the range $4 < \mathrm{Re} < 40$. For $\mathrm{Re} > 130$ the ring-eddy oscillates, and some of it breaks off periodically in the form of distorted vortex loops.

Asymmetry in the pressure distribution over a sphere can give rise to lift forces that act perpendicular to its direction of motion (Fig. 8.26). A sphere placed in shear flow with translational velocity parallel to the streamlines will experience such a lift force. Another example is that of a spinning sphere moving through a fluid, referred to as the Magnus effect. This is often exploited in various sports, such as volleyball, baseball, soccer, tennis, and table tennis, to add curve to the ball's trajectory. The Magnus effect has also been explored as an alternative means of propulsion for ships (so-called rotor ships) and to reduce separation over airfoils. Drag and lift on a sphere, combined with the Basset history force discussed at the end of Chapter 7 and added mass discussed in Chapter 6, provide a starting point for studying multiphase flows involving bubbles, droplets, and solid particles. Although it is an important and fascinating sub-field of fluid mechanics, it is largely beyond the scope of this text, and its fundamentals are covered elsewhere (see Clift et al., 2005). In this section, shear-induced lift and the Magnus effect are briefly discussed.

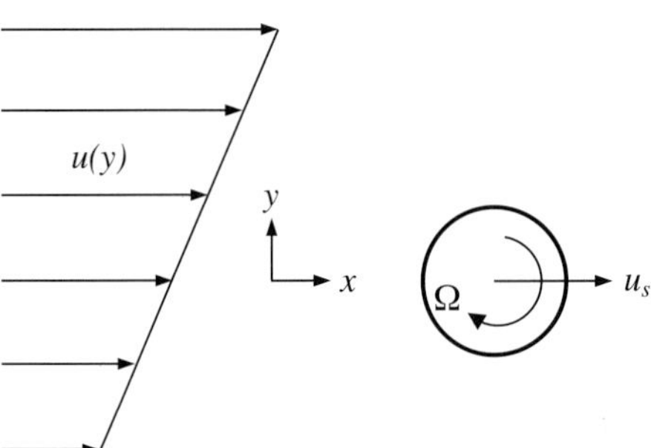

FIGURE 8.26 A sphere moving at speed u_s in planar shear flow $u(y)$. When $u > u_s$ and $\Omega = 0$, the drag force points to the right and the Saffman lift is upward.

Saffman lift

The Saffman lift force, named after British mathematician Philip G. Saffman who first formulated its expression (Saffman, 1965), is a consequence of the pressure distribution developed on a sphere in a velocity gradient. Consider a sphere in an unbounded linear shear flow with diameter d and velocity u_s parallel to the flow direction, as depicted in Fig. 8.26. The fluid velocity is defined as $\mathbf{u} = (u(y), 0)$, where $u = \dot{\gamma} y$ and $\dot{\gamma}$ is the (constant) shear rate. The high velocity above the sphere and low velocity below it results in a pressure difference that generates a vertical lift force. However, the pressure cannot be directly inferred from the Bernoulli equation due to the background flow being rotational and to boundary layer effects near the surface of the sphere.

Two Reynolds numbers can be defined in this case. The first is based on the sphere diameter and *relative* velocity between the fluid and sphere; $\mathrm{Re}_r = |\mathbf{u} - \mathbf{u}_s| d/\nu$, where $\mathbf{u}$ is evaluated at the center of the sphere. The second is a shear Reynolds number, $\mathrm{Re}_G = \dot{\gamma} d^2/\nu$, which can be thought of as the velocity difference between the top and bottom of the sphere. Because there is more than one Reynolds number, the small Reynolds number limit can be thought of as being due to large viscosity. When $\mathrm{Re}_r \ll \mathrm{Re}_G$ and both Reynolds numbers are much less than unity, Saffman (1965) found the magnitude of the lift force resulting from inertia effects to be

$$F_{\mathrm{Saffman}} = 1.61 \mu d (u - u_s) \sqrt{\mathrm{Re}_G}. \tag{8.53}$$

It can be seen that when the relative velocity is positive (i.e., the fluid is moving faster than the sphere), the lift force acts towards the higher velocity side. If the relative velocity is negative (sphere moving faster than the fluid), the lift force acts in the opposite direction. If the sphere is rotating with angular velocity $\boldsymbol{\Omega} = (0, 0, \Omega)$ as shown in Fig. 8.26, the resulting lift force is in the direction $\boldsymbol{\Omega} \times (\mathbf{u} - \mathbf{u}_s) = \Omega(u - u_s)$.

McLaughlin (1991) extended Saffman's analysis and showed that the magnitude of the lift force rapidly decreases when $\mathrm{Re}_r > \sqrt{\mathrm{Re}_G}$. For increasing Reynolds numbers above unity, analytic expressions for the force are inaccessible owing to boundary layer separation and the formation of an unsteady wake. Instead, extensions to higher Reynolds numbers are typically based on empirical fits to numerical data (Clift et al., 2005).

Magnus effect

The Magnus effect refers to the lift force that acts on a rotating object (commonly a sphere or cylinder) moving relative to a fluid as depicted in Fig. 8.27 with the sphere fixed and the fluid moving from left to right. Viscous flow around cylinders and spheres at high Reynolds numbers based on the relative velocity exhibit significant flow separation and drag (Fig. 8.18). Without sphere rotation, the flow is symmetric and the pressure difference between the top and bottom of the sphere is on average zero, so there is no lift force. For positive Magnus effect on spheres (and cylinders) rotating at high Re_r, the sphere's rotation pushes the separation point further downstream on the side of the sphere moving in the same direction as the flow and upstream on the side moving opposite to the flow direction (see Fig. 8.27b). Thus, rotating spheres and cylinders experience significant lift even in the absence of fluid shear.

The resulting lift force has significant effect on the path of sports balls. A ball thrown or hit with topspin curves downward, whereas a ball thrown or hit with underspin (backspin) travels along a much flatter trajectory (or even rises) relative to a ball without spin. The direction of the lateral force is therefore in the same sense as that of the Magnus effect experienced by a circular cylinder in potential flow with circulation (see Section 6.3). The mechanics, however, are different. The potential flow argument (involving the Bernoulli equation) offered to account for the lateral force around a circular cylinder cannot explain why a *negative* Magnus effect is universally observed at lower Reynolds numbers. (By a negative Magnus effect we mean a lateral force opposite to that experienced by a cylinder with a circulation of the same sense as the rotation of the sphere.) The correct argument seems to be the asymmetric boundary-layer separation caused by the spin. In fact, the phenomenon was not properly explained until boundary-layer concepts were understood in the twentieth century. Some pioneering experimental work on the bending paths of spinning spheres was conducted by British mathematician and military engineer Robins (1742) over two hundred years ago in studying the trajectories of bullets fired from muskets. Because of this, the deflection of rotating spheres is sometimes called the *Robins effect*.

With this background, we are now in a position to understand how a spinning sphere generates a negative Magnus effect at $\mathrm{Re} < \mathrm{Re}_{\mathrm{cr}}$ and a positive Magnus effect at $\mathrm{Re} > \mathrm{Re}_{cr}$. For clockwise rotation of a sphere, the fluid velocity *relative to the surface* is larger on the lower side (Fig. 8.27). For the lower Reynolds number case (Fig. 8.27a), this causes transition of the boundary layer on the sphere's lower side, while the boundary layer on the upper side remains laminar. The result is delayed separation and lower pressure on the bottom surface, and a consequent downward force on the sphere, opposite to that of the positive Magnus effect.

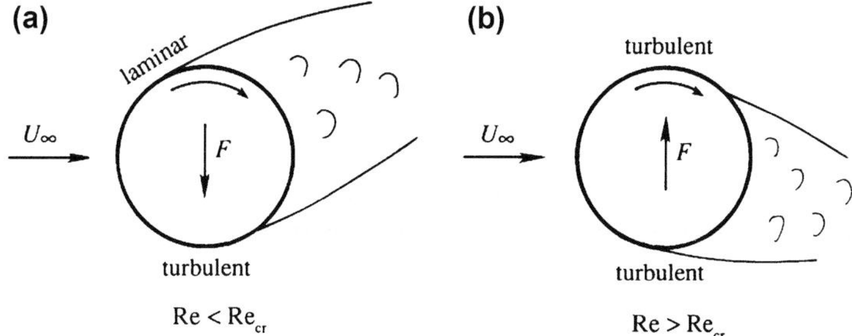

FIGURE 8.27 Rotation-induced forces on stationary rotating spheres when the flow moves from left to right. Here F indicates the force exerted on the sphere by the fluid: (a) negative Magnus effect; and (b) positive Magnus effect.

Roughness on the surface of a sphere (such as that of a golf or tennis ball) lowers the critical Reynolds number, so that the boundary layers on both sides of the sphere have already undergone transition. Due to the higher relative velocity, the flow near the bottom has a higher Reynolds number, and is therefore farther along the Re-axis of Fig. 8.24, in the range AB in which the separation point moves upstream with an increase of the Reynolds number. The separation therefore occurs *earlier* on the bottom side, resulting in a higher pressure there than on the top. This causes an upward lift force and a positive Magnus effect.

Assuming a sphere is rotating with angular velocity $\mathbf{\Omega}$ in a quiescent fluid far from any walls at low Reynolds number, Rubinow and Keller (1961) used Stokes and Oseen expansions (see Chapter 7) to derive an expression for the lift force, given by

$$\mathbf{F}_{\mathrm{Magnus}} = \frac{\pi}{8} d^3 \rho \mathbf{\Omega} \times (\mathbf{u}_s - \mathbf{u}) \left[1 + \mathcal{O}(\mathrm{Re})\right]. \tag{8.54}$$

A surprising result is that the expression is independent of viscosity and is similar to the lift force for two-dimensional potential flow. Eq. (8.54) can be rewritten in terms of a lift coefficient due to rotation, C_{LR}, as $\mathbf{F}_{\mathrm{Magnus}} = \rho C_{LR} A |\mathbf{u}_s - \mathbf{u}|(\mathbf{u}_s - \mathbf{u})/2$, where A is the projected area of the object. From Eq. (8.54) it can be seen that the lift coefficient acting in the y-direction of a sphere translating in the x-direction rotating with angular speed Ω in the z-plane at low Reynolds numbers is $C_{LR} = \Omega d/|u_s - u| = 2S$, where $S = \Omega d/(2|u_s - u|)$ is the so-called spin parameter. The spin parameter can be thought of as the maximum surface velocity normalized by the relative translational velocity. Fig. 8.28 shows measurements of C_{LR}

at moderate and high Reynolds numbers. Well below the critical Reynolds number ($\mathrm{Re} = 2 \times 10^3$) C_{LR} increases relatively linearly with S. At sufficiently high Reynolds numbers, a negative Magnus effect is observed when $S < 0.4$ due to an earlier transition to turbulence.

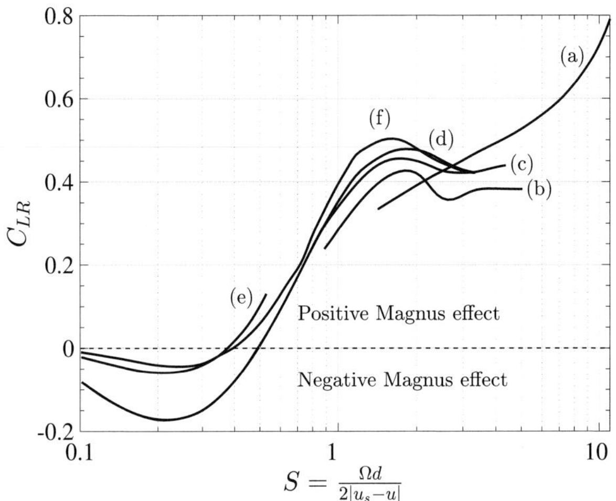

FIGURE 8.28 Experimental measurements of rotational lift coefficient as a function of spin parameter for rotating smooth spheres. (a) $\mathrm{Re}_r = 2 \times 10^3$, (b) $\mathrm{Re}_r = 4.6 \times 10^4$, (c) $\mathrm{Re}_r = 6.2 \times 10^4$, (d) $\mathrm{Re}_r = 7.7 \times 10^4$, (e) $\mathrm{Re}_r = 9 \times 10^4$, (f) $\mathrm{Re}_r = 1.1 \times 10^5$. Adapted from Clift et al. (2005) with permission.

Example 8.10

A tennis player hits a tennis ball horizontally with a modest swing such that the ball's initial velocity is $u_s = 30$ m/s in the x-direction. Accounting for lift and drag, numerically integrate the ball's trajectory when the ball is (a) hit with topspin Ω, (b) hit with backspin $-\Omega$, and (c) hit without spin. How much backspin is required to double the distance the ball travels before it hits the ground compared to the case without spin? How much shorter is the ball's trajectory if the same angular velocity is applied in the opposite direction?

Solution The equation of motion for a sphere with mass m_s and diameter d is

$$\frac{d\mathbf{x}_s}{dt} = \mathbf{u}_s \quad \text{and} \quad m_s \frac{d\mathbf{u}_s}{dt} = \mathbf{F}_D + \mathbf{F}_L + m_s \mathbf{g},$$

where $\mathbf{g} = (0, -g)$ is acceleration due to gravity. Here, Basset history and added mass have been neglected since the tennis ball is significantly denser than the air and the flow is assumed to be quiescent (see Exercise 7.57). The first term on the right-hand represents the drag force, which can be expressed in terms of a drag coefficient C_D according to $\mathbf{F}_D = \rho C_D A |\mathbf{u} - \mathbf{u}_s|(\mathbf{u} - \mathbf{u}_s)/2$. A tennis ball traveling at these speeds have a Reynolds number $\mathrm{Re} = \mathcal{O}(10^5)$, and thus from Fig. 8.24, it is reasonable to assume a constant drag coefficient of $C_D = 0.5$. The fluid velocity is taken to be $\mathbf{u} = 0$ and the lift force $\mathbf{F}_L$ is given by (8.54).

The International Tennis Federation defines the official diameter of a tennis ball to be $d = 6.54 - 6.86$ cm and mass $56.0 - 59.4$ g. Choosing a diameter $d = 6.7$ cm and mass $m = 57$ g, its corresponding density is $\rho_s \approx 362$ kg/m^3. A MATLAB® code for solving the problem is given below, where S was determined via trial-and-error.

```
global g rho mu d rhos CD Omega

% Set parameters
g=9.81; % Gravity (m/s^2)
rho=1.2; % Fluid density (kg/m^3)
mu=1.8e-5; % Dynamic viscosity (kg m/s)
d=6.7e-2; % Sphere diameter (m)
rhos=362; % Sphere density (kg/m^3)
CD=0.5; % Drag coefficient (-)
t=0:.001:2; % Time interval (s)

% Initial conditions
x0=0; y0=1;
```

```
u0=30; v0=0;

% Solver the ODEs
S=0.15;
Omega=2*S/d*u0
[t1,X1] = ode45(@sphere_calc,t,[x0; y0;u0; 0]);

%% Function to calculate the trajectory of a rotating sphere
function Xout = sphere_calc(t,Xin)
global g rho mu d rhos CD Omega

A=pi*d^2/3; % Projected area
m=pi*d^3/6*rhos; % Mass
Umag=sqrt(Xin(3)^2+Xin(4)^2); % Velocity magnitude
Xout=[Xin(3);...
      Xin(4);...
     -1/2*rho*CD*A*Umag*Xin(3)/m+pi/8*d^3*rho*Omega*Xin(4)/m;...
     -1/2*rho*CD*A*Umag*Xin(4)/m-pi/8*d^3*rho*Omega*Xin(3)/m-g];
return
end
```

The solution to the coupled set of ODEs is shown in Fig. 8.29. It can be seen that backspin with a spin parameter of $S = 0.15$, corresponding to an angular speed of $\Omega \approx 8000$ rpm is required to cause the ball to travel twice the distance it would have traveled without spin. The same value of S applied to backspin causes the ball to travel approximately 75% of its distance. While this angular speed might sound unrealistically high, slow-motion video has revealed some professional players to generate topspin at more than 4000 rpm!

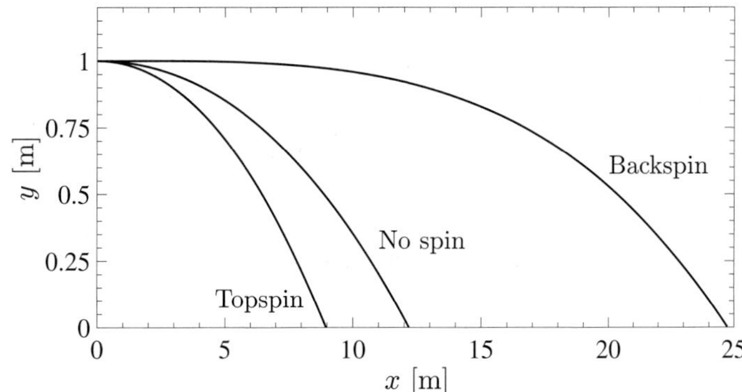

FIGURE 8.29 Solution to Example 8.10. A tennis ball initially traveling 30 m/s horizontally will require backspin with a spin parameter of $S = 0.15$ (≈ 8000 rpm) to travel twice the distance it would without spin.

8.10 Two-dimensional jets

The previous nine sections have considered boundary layers over solid surfaces. The concept of a boundary layer, however, is more general, and the approximations involved are applicable whenever the vorticity in the flow is confined in thin layers, even in the absence of a solid surface. Such a layer can be in the form of a jet of fluid ejected from an orifice, a wake (where the velocity is lower than the upstream velocity) behind a solid object, or a thin shear layer (vortex sheet) between two uniform streams of different speeds. As an illustration of the method of analysis of these *free shear flows*, we shall consider the case of a laminar two-dimensional jet, which is an efflux of fluid from a long and narrow orifice that issues into a large quiescent reservoir of the same fluid. Downstream from the orifice, some of the ambient fluid is carried along with the moving jet fluid through viscous vorticity diffusion at the outer edge of the jet (Fig. 8.30). The process of drawing reservoir fluid into the jet is called *entrainment*.

The velocity distribution near the opening of the jet depends on the conditions upstream of the orifice exit. However, because of the absence of an externally imposed length scale in the downstream direction, the velocity profile in the jet approaches a self-similar shape not far from where it emerges into the reservoir, regardless of the velocity distribution at the orifice.

For large Reynolds numbers, the jet is narrow and the boundary-layer approximation can be applied. Consider a control volume with sides cutting across the jet axis at two sections (Fig. 8.30); the other two sides of the control volume are taken at large distances from the jet axis. No external pressure gradient is maintained in the surrounding fluid so dp/dx is zero. According to the boundary-layer approximation, the same zero pressure gradient is also impressed upon the jet. There is, therefore, no net force acting on the surfaces of the control volume, and this requires the x-momentum flux at the two sections across the jet to be the same.

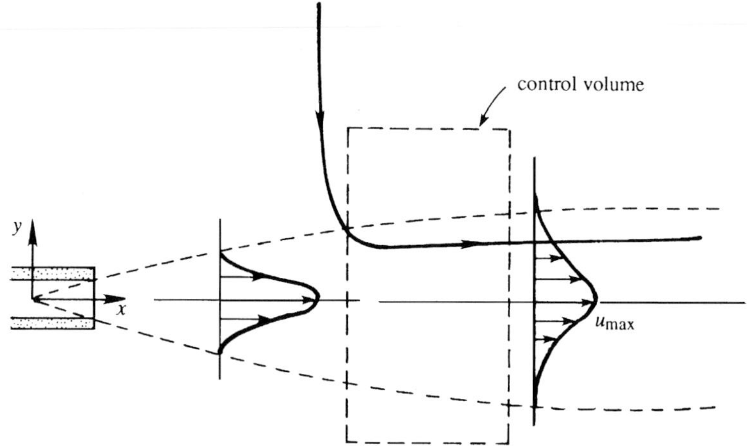

FIGURE 8.30 Simple laminar two-dimensional free jet. A narrow slot injects fluid horizontally with an initial momentum flux J (per unit span) into a nominally quiescent reservoir of the same fluid. The region of horizontally moving fluid slows and expands as x increases. A typical streamline showing entrainment of surrounding fluid is indicated.

Let $u_0(x)$ be the stream-wise velocity on the x-axis and assume $\mathrm{Re} = u_0 x/\nu$ is sufficiently large for the boundary-layer equations to be valid. The flow is steady, two-dimensional (x, y), without body forces, and with constant properties (ρ, μ). Then $\partial/\partial y \gg \partial/\partial x$, $v \ll u$, $\partial p/\partial y = 0$, so the fluid equations of motion are the same as for the Blasius boundary layer: (6.2) and (8.18). However, the boundary conditions are different here:

$$u = 0 \quad \text{for} \quad y \to \pm\infty \quad \text{and} \quad x > 0, \tag{8.55}$$

$$v = 0 \quad \text{on} \quad y = 0 \quad \text{for} \quad x > 0, \text{ and} \tag{8.56}$$

$$u = \tilde{u}(y) \quad \text{on} \quad x = x_0, \tag{8.57}$$

where $\tilde{u}$ is a known flow profile. Now partially follow the derivation of the von Karman boundary-layer integral equation. Multiply (6.2) by u and add it to the left side of (8.18) but this time integrate over all y to find:

$$\int_{-\infty}^{+\infty} 2u\frac{\partial u}{\partial x}dy + \int_{-\infty}^{+\infty}\left[u\frac{\partial v}{\partial y} + v\frac{\partial u}{\partial y}\right]dy = \int_{-\infty}^{+\infty}\frac{\partial \tau}{\partial y}dy, \quad \text{or} \quad \frac{d}{dx}\int_{-\infty}^{+\infty} u^2 dy + [uv]_{-\infty}^{+\infty} = [\tau]_{-\infty}^{+\infty}. \tag{8.58}$$

Since $u(y = \pm\infty) = 0$, all derivatives of u with respect to y must also be zero at $y = \pm\infty$. Thus, since $\tau = \mu(\partial u/\partial y)$, the second and third terms in the second equation of (8.58) are both zero. Hence, (8.58) reduces to:

$$\frac{d}{dx}\int_{-\infty}^{+\infty} u^2 dy = 0, \tag{8.59}$$

a statement that the stream-wise momentum flux is conserved. Thus, when integrated, (8.59) becomes:

$$\int_{-\infty}^{+\infty} u^2 dy = const. = \int_{-\infty}^{+\infty} \tilde{u}^2(y)dy = J/\rho, \tag{8.60}$$

where the second equality follows from (8.57). Here, the constant is the momentum flux in the jet per unit span, J, divided by the fluid density, ρ.

A similarity solution is obtained far enough downstream so that the boundary-layer equations are valid and $\tilde{u}(y)$ has been forgotten. Thus, we can seek a solution in the form of (7.31) or (8.19):

$$\psi = u_0(x)\delta(x)f(\eta), \quad \text{where} \quad \eta = y/\delta(x), \quad \delta(x) = [\nu x/u_0(x)]^{1/2}, \tag{8.61}$$

and $u_0(x)$ is the stream-wise velocity on $y = 0$. The stream-wise velocity throughout the field is obtained from differentiation:

$$u = \partial\psi/\partial y = [\nu x u_0(x)]^{1/2}(df/d\eta)[\nu x/u_0(x)]^{-1/2} = u_0(x)(df/d\eta). \tag{8.62}$$

The final equality here implies that $f' = 1$ on $\eta = 0$. When (8.62) is substituted into (8.60), the dependence of $u_0(x)$ on x is determined:

$$\frac{J}{\rho} = \int_{-\infty}^{+\infty} u^2 dy = u_0^2(x)\int_{-\infty}^{+\infty} f'^2(\eta)dy = u_0^2(x)\delta(x)\int_{-\infty}^{+\infty} f'^2(\eta)d\eta. \tag{8.63}$$

Since the integral is a dimensionless constant ($= C$), we must have:

$$Cu_o^2(x)\delta(x) = Cu_o^{3/2}(x)\cdot(\nu x)^{1/2} = J/\rho,$$

so:

$$u_0(x) = [J^2/C^2\rho^2\nu x]^{1/3}, \quad \text{and} \quad \delta(x) = [C\rho\nu^2x^2/J]^{1/3}. \tag{8.64, 8.65}$$

Thus, (8.61) becomes:

$$\psi = [J\nu x/C\rho]^{1/3} f(\eta) \quad \text{where} \quad \eta = y/[C\rho\nu^2x^2/J]^{1/3}. \tag{8.66}$$

In terms of the stream function, (8.18) becomes:

$$\frac{\partial\psi}{\partial y}\frac{\partial}{\partial x}\left(\frac{\partial\psi}{\partial y}\right) - \frac{\partial\psi}{\partial x}\frac{\partial}{\partial y}\left(\frac{\partial\psi}{\partial y}\right) = \nu\frac{\partial^2}{\partial y^2}\left(\frac{\partial\psi}{\partial y}\right). \tag{8.67}$$

Evaluating the derivatives using (8.66) and simplifying produces a differential equation for f:

$$3f''' + f''f + f'^2 = 0.$$

The boundary conditions for $x > 0$ are:

$$f' = 0 \text{ for } \eta \to \pm\infty, \qquad f' = 1 \text{ on } \eta = 0, \qquad \text{and} \quad f = 0 \text{ on } \eta = 0. \tag{8.68, 8.69, 8.70}$$

Integrating the differential equation for f once produces:

$$3f'' + f'f = C_1.$$

Evaluating at $\eta = \pm\infty$ implies $C_1 = 0$ from (8.68) since $f' = 0$ implies $f'' = 0$ too. Integrating again yields:

$$3f' + f^2/2 = C_2. \tag{8.71}$$

Evaluating on $\eta = 0$ implies $C_2 = 3$ from (8.69) and (8.70). The independent and dependent variables in (8.71) can be separated and integrated:

$$3\frac{df}{d\eta} = 3 - \frac{f^2}{2} \quad \text{or} \quad \int\frac{df}{1 - f^2/6} = \int d\eta.$$

The integral on the left in the second equality can be evaluated via the variable substitution $f = \sqrt{6}\tanh\beta$, and leads to:

$$\tanh^{-1}\left(\frac{f}{\sqrt{6}}\right) = \frac{\eta}{\sqrt{6}} + C_3, \quad \text{or} \quad f = \sqrt{6}\tanh\left(\frac{\eta}{\sqrt{6}} + C_3\right). \tag{8.72}$$

Evaluating the final expression on $\eta = 0$ implies $C_3 = 0$ from (8.70). Thus, using (8.62), (8.64), and (8.72), the stream-wise velocity field is:

$$u(x, y) = u_0(x)f'(\eta) = \left(\frac{J^2}{C^2\rho^2\nu x}\right)^{1/3} \operatorname{sech}^2\left(\frac{y}{\sqrt{6}}\left[\frac{J}{C\rho\nu^2x^2}\right]^{1/3}\right), \tag{8.73}$$

and the dimensionless constant, C, is determined from:

$$C = \int_{-\infty}^{+\infty} f'^2(\eta)d\eta = \int_{-\infty}^{+\infty} \operatorname{sech}^4\left(\frac{\eta}{\sqrt{6}}\right) d\eta = \frac{4\sqrt{6}}{3}. \tag{8.74}$$

The mass flux of the jet per unit span is:

$$\dot{m} = \int_{-\infty}^{+\infty} \rho u_0(x) f'(\eta) dy = \rho u_0(x)\delta(x) \int_{-\infty}^{+\infty} f'(\eta)d\eta = \rho u_0(x)\delta(x)[f]_{-\infty}^{+\infty} = \rho u_0(x)\delta(x)\cdot 2\sqrt{6}.$$

Using (8.64), (8.65), and (8.74), this simplifies to:

$$\dot{m} = (36J\rho^2 \nu x)^{1/3}, \tag{8.75}$$

which shows that the jet's mass flux increases with increasing downstream distance as the jet entrains ambient reservoir fluid via the action of viscosity. The jet's entrainment induces flow toward the jet within the reservoir. The vertical velocity is:

$$v = -\frac{\partial \psi}{\partial x} = -\frac{1}{3}\left(\frac{J\nu}{C\rho x^2}\right)^{1/3}[f - 2\eta f'], \quad \text{or} \quad \frac{v}{u_0(x)} = -\frac{[f - 2\eta f']}{3\sqrt{\mathrm{Re}_x}} \quad \text{where} \quad \mathrm{Re}_x = \frac{x u_0(x)}{\nu}. \tag{8.76}$$

Here, $f(\eta) \to \pm\sqrt{6}$ and $2\eta f'(\eta) = 2\eta \operatorname{sech}^2(\eta/\sqrt{6}) \to 0$ as $\eta \to \pm\infty$, so:

$$\frac{v}{u_0(x)} \to \mp \frac{\sqrt{6}}{3\sqrt{\mathrm{Re}_x}} \quad \text{as} \quad \eta \to \pm\infty. \tag{8.77}$$

Thus, the jet's entrainment field is a flow of reservoir fluid toward the jet from above and below.

The jet spreads as it travels downstream, and this can be deduced from (8.73). Following the definition of δ_{99} in Section 8.2, the 99% half width of the jet, h_{99}, may be defined as the y-location where the horizontal velocity falls to 1% of its value at $y = 0$. Thus, from (8.73) we can determine:

$$\operatorname{sech}^2\left(\frac{h_{99}}{\sqrt{6}}\left[\frac{J}{C\rho\nu^2 x^2}\right]^{1/3}\right) = 0.01 \to \frac{h_{99}}{\sqrt{6}}\left[\frac{J}{C\rho\nu^2 x^2}\right]^{1/3} \cong 2.2924 \to h_{99} \cong 5.6152\left[\frac{C\rho\nu^2 x^2}{J}\right]^{1/3}, \tag{8.78}$$

which shows the jet width grows with increasing downstream distance like $x^{2/3}$. Viscosity increases the jet's thickness but higher momentum jets are thinner. The Reynolds numbers based on the stream-wise (x) and cross-stream (h_{99}) dimensions of the jet are:

$$\mathrm{Re}_x = \frac{x u_0(x)}{\nu} = \left(\frac{3Jx}{4\sqrt{6}\rho\nu^2}\right)^{2/3} \quad \text{and} \quad \mathrm{Re}_{h_{99}} \frac{h_{99} u_0(x)}{\nu} = 5.6152\left(\frac{3Jx}{4\sqrt{6}\rho\nu^2}\right)^{1/3}.$$

Unfortunately, this steady-flow, two-dimensional laminar jet solution is not readily observable because the flow is unstable when $\mathrm{Re} \gg 1$. The low critical Reynolds number for instability of a jet or wake is associated with the existence of one or more inflection points in the stream-wise velocity profile, as discussed in Chapter 9. Nevertheless, the laminar solution has revealed at least two significant phenomena – constancy of jet momentum flux and increase of jet mass flux through entrainment – that also apply to round jets and turbulent jets. However, the cross-stream spreading rate of a turbulent jet is found to be independent of Reynolds number and is faster than the laminar jet, being more like $h_{99} \propto x$ rather than $h_{99} \propto x^{2/3}$ (see Chapter 10).

A second example of a two-dimensional jet that also shares some boundary-layer characteristics is the *wall jet*. The solution here is due to Glauert (1956). We consider fluid exiting a narrow slot with its lower boundary being a planar wall taken along the x-axis (see Fig. 8.31). Near the wall ($y = 0$) the flow behaves like a boundary layer, but far from the wall it behaves like a free jet. For large Re_x the jet is thin ($\delta/x \ll 1$) so $\partial p/\partial y \approx 0$ across it. The pressure is constant in the nearly stagnant outer fluid so $p \approx$ const. throughout the flow. Here again the fluid mechanical equations of motion are (6.2) and (8.18) This time the boundary conditions are:

$$u = v = 0 \quad \text{on} \quad y = 0 \quad \text{for} \quad x > 0, \quad \text{and} \quad u(x, y) \to 0 \quad \text{as} \quad y \to \infty. \tag{8.79, 8.80}$$

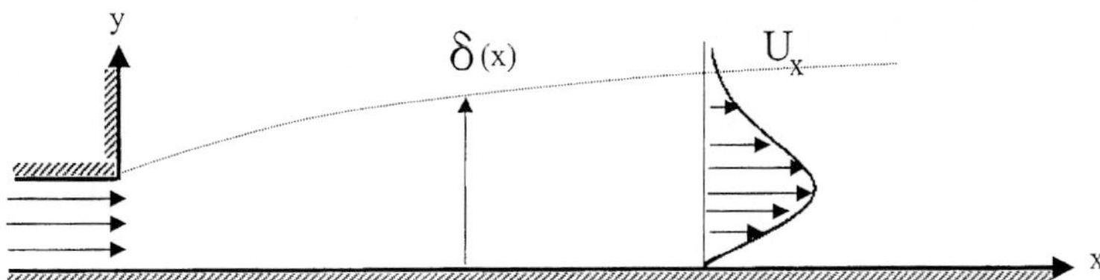

FIGURE 8.31 The laminar two-dimensional wall jet. A narrow slot injects fluid horizontally along a smooth flat wall. As for the free jet, the thickness of the region of horizontally moving fluid slows and expands as x increases but with different dependencies.

Here again, a similarity solution valid for $\mathrm{Re}_x \to \infty$ can be found under the assumption that the initial velocity distribution is forgotten by the flow. However, unlike the free jet, the momentum flux of the wall jet is not constant; it diminishes with increasing downstream distance because of the wall shear stress. To obtain the conserved property in the wall jet, integrate (8.18) in the wall normal direction from y to ∞:

$$\int_y^\infty u\frac{\partial u}{\partial x}dy + \int_y^\infty v\frac{\partial u}{\partial y}dy = \nu\int_y^\infty \frac{\partial^2 u}{\partial y^2}dy = \nu\left[\frac{\partial u}{\partial y}\right]_y^\infty = -\nu\frac{\partial u}{\partial y},$$

multiply this by u, and integrate from 0 to ∞ in the wall normal direction:

$$\int_0^\infty \left(u\frac{\partial}{\partial x}\int_0^\infty \frac{u^2}{2}dy\right)dy + \int_0^\infty \left(u\int_y^\infty v\frac{\partial u}{\partial y}dy\right)dy = -\frac{\nu}{2}\int_0^\infty \frac{\partial u^2}{\partial y}dy = -\frac{\nu}{2}\left[u^2\right]_0^\infty = 0.$$

The final equality follows from the boundary conditions (8.79) and (8.80). Integrating the interior integral of the second term on the left by parts and using (6.2) yields a term equal to the first term and one that lacks any differentiation:

$$\int_0^\infty \left(u\frac{\partial}{\partial x}\int_y^\infty u^2 dy\right)dy - \int_0^\infty u^2 v dy = 0. \tag{8.81}$$

Now consider:

$$\frac{d}{dx}\int_0^\infty \left(u\int_y^\infty u^2 dy\right)dy = \int_0^\infty \left(\frac{\partial u}{\partial x}\int_0^\infty u^2 dy\right)dy + \int_0^\infty \left(u\frac{\partial}{\partial x}\int_y^\infty u^2 dy\right)dy,$$

use (6.2) in the first term on the right side, integrate by parts, and combine this with (8.81) to obtain:

$$\frac{d}{dx}\int_0^\infty \left(u\int_y^\infty u^2 dy\right)dy = 0. \tag{8.82}$$

This says that the flux of exterior momentum flux remains constant with increasing downstream distance and is the necessary condition for obtaining similarity exponents.

As for the steady free laminar jet, the field equation is (8.67) and the solution is presumed to be in the similarity form specified by (8.61). Here $u_0(x)$ is to be determined and this similarity solution should be valid when $x \ll x_o$, where x_o is the location where the initial condition is specified, which we take to be the upstream extent of the validity of the boundary-layer momentum equation (8.18) or (8.67). Substituting $u = \partial\psi/\partial y = u_0(x) f'(\eta)$ from (8.61) into (8.82) produces:

$$\frac{d}{dx}\left[u_0^3(x)\cdot\frac{\nu x}{u_0(x)}\int_0^\infty \left(f'\int_\eta^\infty f'^2 d\eta\right)d\eta\right] = 0. \tag{8.83}$$

If the double integration is independent of x, then the factor outside the integral must be constant.

Therefore, set $xu_0^2(x) = C^2$, which implies $u_0(x) = Cx^{-1/2}$ so (8.61) becomes:

$$\psi(x, y) = [\nu C x^{1/2}]^{1/2} f(\eta) \quad \text{where} \quad \eta = y/\delta(x), \quad \delta(x) = [\nu x^{3/2}/C]^{1/2}. \tag{8.84}$$

After appropriately differentiating (8.84), substituting into (8.67), and canceling common factors, (8.67) reduces to:

$$f''' + ff'' + 2f'^2 = 0,$$

subject to the boundary conditions (8.79) and (8.80): $f(0)=0$; $f'(0)=0$; $f'(\infty)=0$. This third-order equation can be integrated once after multiplying by the integrating factor f, to yield $4ff''-2f'^2+f^2f'=0$, where the constant of integration has been evaluated at $\eta=0$. Dividing by the integrating factor $4f^{3/2}$ allows another integration. The result is:

$$f^{-1/2}f'+f^{3/2}/6=C_1\equiv f_\infty^{3/2}/6, \quad \text{where} \quad f_\infty=f(\infty).$$

The final integration can be performed by separating variables and defining $g^2(\eta)=f/f_\infty$:

$$\int \frac{df}{f_\infty^{3/2}f-f^2}=\frac{1}{6}\int d\eta, \quad \text{or} \quad \int \frac{dg}{1-g^3}=\frac{f_\infty}{12}\int d\eta.$$

The integration on the left may be performed via a partial fraction expansion using $1-g^3=(1-g)\cdot(1+g+g^2)$ with the final result left in implicit form (8.85):

$$-\ln(1-g)+\sqrt{3}\tan^{-1}\left(\frac{2g+1}{\sqrt{3}}\right)+\ln(1+g+g^2)^{1/2}=\frac{f_\infty}{4}\eta+\sqrt{3}\tan^{-1}\left(\frac{1}{\sqrt{3}}\right), \tag{8.85}$$

where the boundary condition $g(0)=0$ was used to evaluate the constant of integration. The profiles of f and f' are plotted vs. η in Fig. 8.32. We can verify easily that $f'\to 0$ exponentially fast in η from this solution for $g(\eta)$. As $\eta\to\infty$, $g\to 1$, so for large η the solution for g reduces to $-\ln(1-g)+\sqrt{3}\tan^{-1}\sqrt{3}+(1/2)\ln 3\cong f_\infty\eta/4+\sqrt{3}\tan^{-1}(1/\sqrt{3})$. The first term on each side of this equation dominates, leaving $1-g\approx e^{-(f_\infty/4)\eta}$. Thus, for $\eta\to\infty$, we must have: $f'=2f_\infty gg'\approx\frac{1}{2}f_\infty^2\exp[-f_\infty\eta/4]$. The mass flow rate per unit span in the steady laminar wall jet is:

$$\dot{m}=\int_0^\infty \rho u dy=\rho u_0(x)\delta(x)\int_0^\infty f'(\eta)d\eta=\rho\sqrt{\nu C}f_\infty x^{1/4}, \tag{8.86}$$

indicating that entrainment increases the mass flow rate in the jet with $x^{1/4}$. The two constants, C and f_∞, can be determined from the integrated form of (8.83) in terms of Ψ, the flux of the exterior momentum flux (a constant):

$$u_0^2(x)\nu x\int_0^\infty\left(f'\int_\eta^\infty f'^2d\eta\right)d\eta=C^2\nu\int_0^\infty\left(f'\int_\eta^\infty f'^2d\eta\right)d\eta=\Psi, \tag{8.87}$$

and knowledge of $\dot{m}$ at one downstream location.

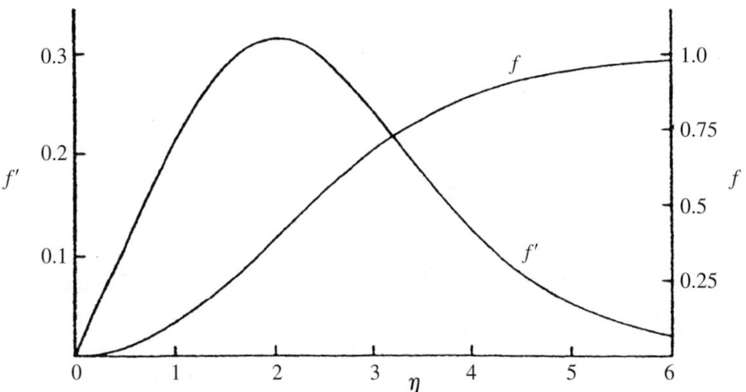

FIGURE 8.32 Variation of normalized mass flux (f) and normalized stream-wise velocity profile (f') with similarly variable η for the laminar two-dimensional wall jet. Reprinted with the permission of Cambridge University Press.

The entrainment into the steady laminar wall jet is evident from the form of $v=-\partial\psi/\partial x=-\sqrt{\nu C}(f-3\eta f')/4x^{3/4}$, which simplifies to $v\approx-\sqrt{\nu C}f_\infty/4x^{3/4}$ as $\eta\to\infty$, so, far above the jet, the flow is downward toward the jet.

8.11 Secondary flows

Large Reynolds number flows with curved streamlines tend to generate additional velocity components because of the properties of boundary layers. These additional components are commonly called secondary flows. An example of such a flow is made dramatically visible by randomly dispersing finely crushed tea leaves into a cup of water, and then stirring vigorously in a circular motion. When the motion has ceased, all of the particles have collected in a mound at the center of the bottom of the cup (see Fig. 8.33). An explanation of this phenomenon is given in terms of thin boundary layers. The stirring motion imparts a primary velocity, $u_\varphi(R)$ (see Appendix B.6 for coordinates), large enough for the Reynolds number to be large enough for the boundary layers on the cup's sidewalls and bottom to be thin. The two largest terms in the R-momentum equation are:

$$\partial p/\partial R = \rho u_\varphi^2/R.$$

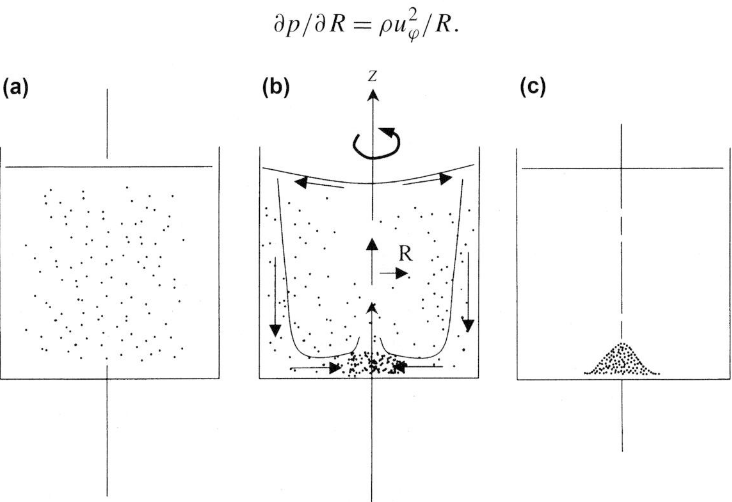

FIGURE 8.33 Secondary flow in a teacup. Tea leaf fragments are slightly denser than water. (a) Tea leaf fragments randomly dispersed—initial state; (b) stirred vigorously—transient motion; and (c) final state where all the tea leaf fragments are piled near the axis of rotation on the bottom of the cup.

Away from the walls, the flow is inviscid. As the boundary layer on the bottom is thin, boundary-layer theory yields $\partial p/\partial z = 0$ from the axial momentum equation. Thus, the pressure in the bottom boundary layer is the same as for the inviscid flow just outside the boundary layer. However, within the boundary layer, u_φ is less than the inviscid value at the edge. Thus $p(R)$ is everywhere larger in the boundary layer than that required for circular streamlines inside the boundary layer, and this pressure difference pushes the streamlines inward toward the center of the cup. That is, the pressure gradient within the boundary layer generates an inwardly directed u_R. This motion induces a downwardly directed flow in the sidewall boundary layer and an outwardly directed flow on the top surface. This secondary flow is closed by an upward flow along the cup's centerline. The visualization is accomplished by crushed tea leaves which are slightly denser than water. They descend by gravity or are driven outward by centrifugal acceleration. If they enter the sidewall boundary layer, they are transported downward and thence to the center by the secondary flow. If the tea particles enter the bottom boundary layer from above, they are quickly swept to the center and dropped as the flow turns upward. All the particles collect at the center of the bottom of the teacup. A practical application of this effect, illustrated in Exercise 8.33, relates to sand and silt transport by the Mississippi River.

Exercises

8.1 A thin flat plate 2 m long and 1 m wide is placed at zero angle of attack in a low-speed wind tunnel in the two positions sketched below.

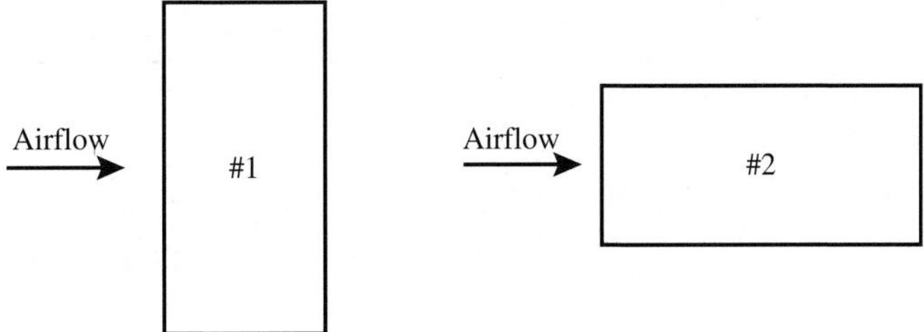

a) For steady airflow, what is the ratio: $\frac{\text{drag on the plate in position \#1}}{\text{drag on the plate in position \#2}}$?

b) For steady airflow at 10 m/sec, what is the total drag on the plate in position #1?

c) If the air flow is impulsively raised from zero to 10 m/sec at $t = 0$, will the initial drag on the plate in position #1 be greater or less than the steady-state drag value calculated for part b)?

d) Estimate how long it will take for drag on the plate in position #1 in the impulsively started flow to reach the steady-state drag value calculated for part b)?

8.2 Solve the Blasius equations (8.27) through (8.29) with a computer, using the Runge–Kutta scheme of numerical integration, and plot the results. What value of f'' at $\eta = 0$ leads to a successful profile?

8.3 A flat plate 4 m wide and 1 m long (in the direction of flow) is immersed in kerosene at $20\,°\text{C}(\upsilon = 2.29 \times 10^{-6}\ \text{m}^2/\text{s}$, $\rho = 800\ \text{kg/m}^3)$, flowing with an undisturbed velocity of 0.5 m/s. Verify that the Reynolds number is less than critical everywhere, so that the flow is laminar. Show that the thickness of the boundary layer and the shear stress at the center of the plate are $\delta = 0.74$ cm and $\tau_w = 0.2\ \text{N/m}^2$, and those at the trailing edge are $\delta = 1.05$ cm and $\tau_w = 0.14\ \text{N/m}^2$. Show also that the total frictional drag on one side of the plate is 1.14 N. Assume that the similarity solution holds for the entire plate.

8.4 Starting with the continuity and boundary-layer-momentum equations, (6.2) and (8.18), for a zero-pressure-gradient flat-plate boundary layer show that the conservation equation for the primary component of the boundary-layer's kinetic energy, $\rho u^2/2$, is:

$$\frac{\partial}{\partial x}\left(\frac{1}{2}\rho u^3\right) + \frac{\partial}{\partial y}\left(\frac{1}{2}\rho u^2 v\right) = u\mu\frac{\partial^2 u}{\partial y^2}.$$

From this equation, derive the integral equation for the deficit of $\rho u^2/2$:

$$\frac{d}{dx}\left(\frac{1}{2}\rho U^3\delta_3\right) = \int_0^\delta \mu\left(\frac{\partial u}{\partial y}\right)^2 dy,$$

where U is the flow speed above the boundary layer, δ is the full thickness of the boundary layer, and $\delta_3 = \int_0^\infty \frac{u}{U}\left(1 - \frac{u^2}{U^2}\right)dy$ is the kinetic energy deficit thickness. What does the second equation imply about kinetic energy in the boundary layer?

8.5 A fluid with constant density and viscosity flows with a constant horizontal speed U_∞ over an infinite flat porous plate placed at $y = 0$ through which fluid is drawn with a constant velocity V_s. For this flow the steady two-dimensional zero-pressure-gradient boundary-layer equations are (6.2) and (8.18) and the boundary conditions are $u(y = 0) = 0$, $v(y = 0) = -V_s$, and $u = U_\infty$ for $y \to \infty$.

a) Assuming u depends only on y, determine $u(y)$ in terms of ν, V_s, U_∞, and y.

b) What is the wall shear stress τ_w? How does it depend on μ?

c) What parametric change(s) decrease the boundary-layer thickness?

8.6 A square-duct wind tunnel test section of length $L = 1$ m is being designed to operate at room temperature and atmospheric conditions. A uniform air flow at $U = 1$ m/s enters through an opening of $D = 20$ cm. Due to the viscosity of air, it is necessary to design a variable cross-sectional area if a constant velocity is to be maintained in the middle part of the cross-section throughout the wind tunnel.

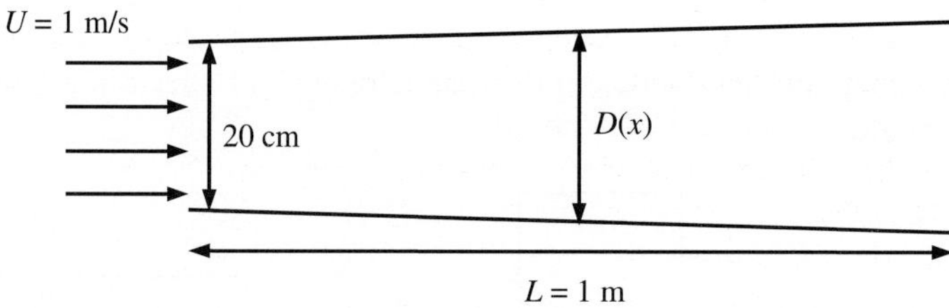

a) Determine the duct size, $D(x)$, as a function of x.

b) How will the result be affected if $U = 20$ m/s? At a given value of x, will $D(x)$ be larger or smaller than (or the same as) the value obtained in a)? Explain.

c) How will the result be affected if the wind tunnel is to be operated at 10 atm (and $U = 1$ m/s)? At a given value of x, will $D(x)$ be larger or smaller than (or the same as) the value obtained in a)? Explain. [*Hint*: the dynamic viscosity of air ($\mu[\text{N}\cdot\text{s/m}^2]$) is largely unaffected by pressure.]

d) Does the airflow apply a net force to the wind tunnel test section? If so, indicate the direction of the force.

8.7 Use the control volume shown to derive the definition of the momentum thickness, θ, for flow over a flat plate:

$$\rho U^2\theta = \rho U^2 \int_0^h \frac{u}{U}\left(1-\frac{u}{U}\right)dy = \frac{\text{Drag force on the plate from zero to } x}{\text{unit depth into the page}} = \int_0^x \tau_w dx$$

The words in the figure describe the upper and lower control volume boundaries.

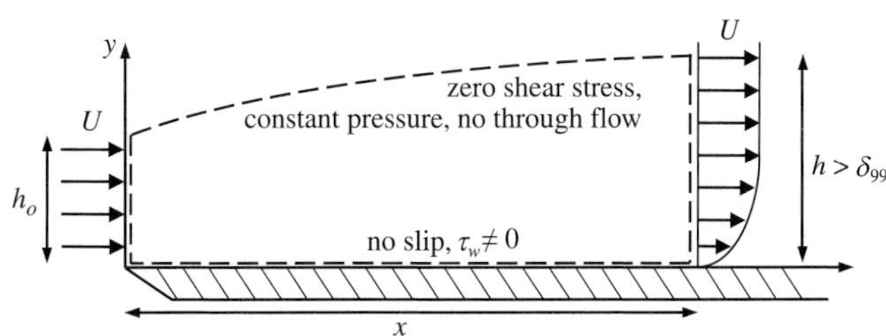

8.8 Estimate the 99% boundary-layer thickness on:

a) A paper airplane wing (length = 0.25 m, $U = 1$ m/sec)
b) The underside of a super tanker (length = 300 m, $U = 5$ m/sec)
c) An airport runway on a blustery day (length = 3 miles, $U = 10$ m/sec)
d) Will these estimates be accurate in each case? Explain.

8.9 Air at 20 °C and 100 kPa ($\rho = 1.167$ kg/m^3, $\nu = 1.5 \times 10^{-5}$ m^2/s) flows over a thin plate with a free-stream velocity of 6 m/s. At a point 15 cm from the leading edge, determine the value of y at which $u/U = 0.456$. Also calculate v and $\partial u/\partial y$ at this point. [*Answer:* $y = 0.857$ mm, $v = 0.384$ cm/s, $\partial u/\partial y = 3012$ s^{-1}.]

8.10 An incompressible fluid (density ρ, viscosity μ) flows steadily from a large reservoir into a long pipe with diameter D. Assume the pipe wall boundary-layer thickness is zero at $x = 0$. The Reynolds number based on D, Re_D, is greater than 10^4.

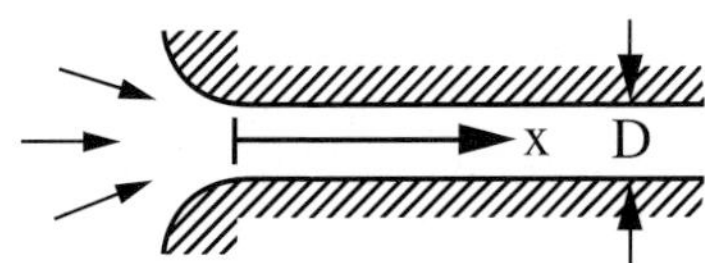

a) Estimate the necessary pipe length for establishing a parabolic velocity profile in the pipe.
b) Will the pressure drop in this entry length be larger or smaller than an equivalent pipe length in which the flow has a parabolic profile? Why?

8.11 [1]A variety of different dimensionless groups have been used to characterize the importance of a pressure gradient in boundary-layer flows. Develop an expression for each of the following parameters for the Falkner-Skan boundary-layer solutions in terms of the exponent n in $U_e(x) = ax^n$, $\text{Re}_x = U_e x/\nu$, integrals involving the profile function f', and $f''(0)$, the profile slope at $y = 0$. Here $u(x, y) = U_e(x) f'(y/\delta(x)) = U_e f'(\eta)$ and the wall shear stress $\tau_w = \mu(\partial u/\partial y)_{y=0} = (\mu U_e/\delta(x)) f''(0)$. What value does each parameter take in a Blasius boundary layer? What value does each parameter achieve at the separation condition?

a) $(\nu/U_e^2)(dU_e/dx)$, an inverse Reynolds number
b) $(\theta^2/\nu)(dU_e/dx)$, the Holstein and Bohlen correlation parameter
c) $(\mu/\sqrt{\rho\tau_w^3})(dp/dx)$, Patel's parameter
d) $(\delta^*/\tau_w)(dp/dx)$, Clauser's parameter

8.12 Consider the boundary layer that develops as a constant density viscous fluid is drawn to a point sink at $x = 0$ on an infinite flat plate in two dimensions (x, y). Here $U_e(x) = -U_o L_o/x$, so set $\eta = y/\sqrt{\nu x/|U_e|}$ and $\psi = \sqrt{\nu x|U_e|} f(\eta)$ and redo the steps leading to (8.36) to find $f''' - f'^2 + 1 = 0$. Solve this equation and utilize appropriate boundary conditions to find $f' = 3\left[\dfrac{1-\alpha e^{-\sqrt{2}\eta}}{1+\alpha e^{-\sqrt{2}\eta}}\right]^2 - 2$ where $\alpha = \dfrac{\sqrt{3}-\sqrt{2}}{\sqrt{3}+\sqrt{2}}$.

8.13 Start from the boundary-layer equations, (6.2), (8.9), and (8.10), and $\psi = U_e(x)\delta(x) f(\eta)$, where $\eta = y/\delta(x)$, with $\delta(x)$ unspecified, to complete the following items.

1. Inspired by problem 4.10 on page 330 of White (2006).

a) Show that the boundary-layer profile equation can be written:

$$f''' + \alpha ff'' + \beta(1 - f'^2) = 0, \quad \text{where} \quad \alpha = (\delta/\nu)d(U_e\delta)/dx, \quad \text{and} \quad \beta = (\delta^2/\nu)dU_e/dx.$$

b) The part a) equation will yield similarity solutions when α and β do not depend on x. Therefore, assume α and β are constants, set $U_e = ax^n$, and show that $n = \beta/(2\alpha - \beta)$.

c) Deduce the values of α and β that allow the profile equation to simplify to the Falkner-Skan profile equation (8.36).

8.14 Solve the Falkner-Skan profile equation (8.36) numerically for $n = -0.0904, -0.654, 0, 1/9, 1/3$, and 1 using boundary conditions (8.28) and (8.29) and the Runge–Kutta scheme of numerical integration. Plot the results and compare to Fig. 8.8. What values of f'' at $\eta = 0$ lead to successful profiles at these six values of n?

8.15 By completing the steps below, show that it is possible to derive von Karman's boundary-layer integral equation without integrating to infinity in the surface-normal direction using the three boundary-layer thicknesses commonly defined for laminar and turbulent boundary layers: 1) δ (or δ_{99}) = the full boundary-layer thickness that encompasses all (or 99%) of the region of viscous influence, 2) δ^* = the displacement thickness of the boundary layer, and 3) θ = momentum thickness of the boundary layer. Here, the definitions of the latter two involve the first: $\delta^*(x) = \int_{y=0}^{y=\delta}\left(1 - \frac{u(x,y)}{U_e(x)}\right)dy$ and $\theta(x) = \int_{y=0}^{y=\delta}\frac{u(x,y)}{U_e(x)}\left(1 - \frac{u(x,y)}{U_e(x)}\right)dy$, where $U_e(x)$ is the flow speed parallel to the wall outside the boundary layer, and δ is presumed to depend on x too.

a) Integrate the two-dimensional continuity equation from $y = 0$ to δ to show that the vertical velocity at the edge of the boundary layer is: $v(x, y = \delta) = \frac{d}{dx}(U_e(x)\delta^*(x)) - \delta\frac{dU_e}{dx}$.

b) Integrate the steady two-dimensional x-direction boundary-layer momentum equation from $y = 0$ to δ to show that: $\frac{\tau_w}{\rho} = \frac{d}{dx}(U_e^2(x)\theta(x)) + \frac{\delta^*(x)}{2}\frac{dU_e^2(x)}{dx}$.

[*Hint*: Use Leibniz's rule $\frac{d}{dx}\int_{a(x)}^{b(x)} f(x,y)dy = \left[f(x,b)\frac{db}{dx}\right] - \left[f(x,a)\frac{da}{dx}\right] + \int_{a(x)}^{b(x)}\frac{\partial f(x,y)}{\partial x}dy$ to handle the fact that $\delta = \delta(x)$]

8.16 Derive the von Karman boundary-layer integral equation by conserving mass and momentum in a control volume (C.V.) of width dx and height h that moves at the exterior flow speed $U_e(x)$ as shown. Here h is a constant distance that is comfortably greater than the overall boundary-layer thickness δ.

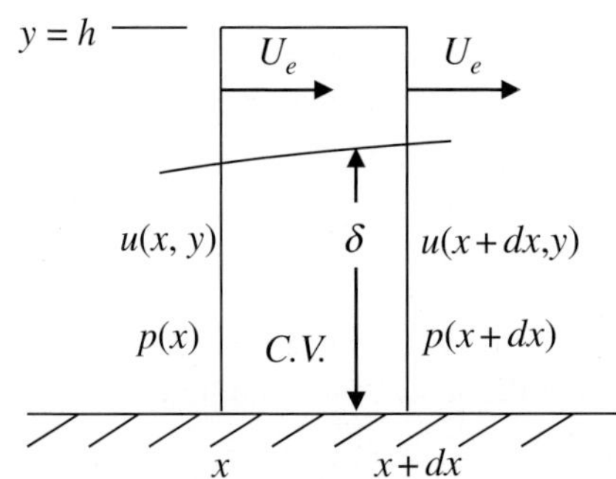

8.17 For the following approximate flat-plate boundary-layer profile:

$$\frac{u}{U} = \left\{\begin{array}{lr} \sin(\pi y/2\delta) & \text{for} \quad 0 \le y \le \delta \\ 1 & \text{for} \quad y > \delta \end{array}\right\},$$

where δ is the generic boundary-layer thickness, determine:

a) The displacement thickness δ^*, the momentum thicknesses θ, and the shape factor $H = \delta^*/\theta$.

b) Use the zero-pressure-gradient boundary-layer integral equation to find: $(\delta/x)\mathrm{Re}_x^{1/2}$, $(\delta^*/x)\mathrm{Re}_x^{1/2}$, $(\theta/x)\mathrm{Re}_x^{1/2}$, $c_f\mathrm{Re}_x^{1/2}$, and $C_D\mathrm{Re}_L^{1/2}$ for the approximate profile.

c) Compare these results to their equivalent Blasius boundary-layer values.

8.18 An incompressible viscous fluid with kinematic viscosity ν flows steadily in a long two-dimensional horn with cross-sectional area $A(x) = A_o \exp\{\beta x\}$. At $x = 0$, the fluid velocity in the horn is uniform and equal to U_o. The boundary-layer momentum thickness is zero at $x = 0$.

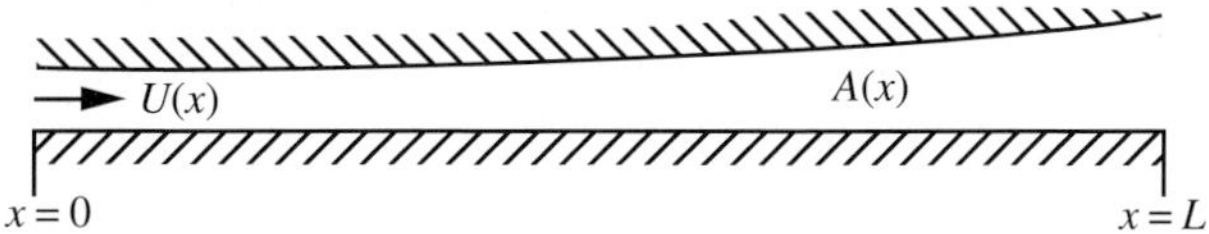

a) Assuming no separation, determine the boundary-layer momentum thickness, $\theta(x)$, on the lower horn boundary using Thwaites' method.

b) Determine the condition on β that makes the no-separation assumption valid for $0 < x < L$.

c) If $\theta(x = 0)$ was non-zero and positive, would the flow in the horn be more or less likely to separate than the $\theta(x = 0) = 0$ case with the same horn geometry?

8.19 The steady two-dimensional velocity potential for a source of strength q located a distance b above a large flat surface located at $y = 0$ is:

$$\phi(x, y) = \frac{q}{2\pi}\left(\ln\sqrt{x^2 + (y - b)^2} + \ln\sqrt{x^2 + (y + b)^2}\right)$$

a) Determine $U(x)$, the horizontal fluid velocity on $y = 0$.

b) Use this $U(x)$ and Thwaites' method to estimate the momentum thickness, $\theta(x)$, of the laminar boundary layer that develops on the flat surface when the initial momentum thickness θ_o is zero.

$$\left[\text{Potentially useful information: } \int_0^x \frac{\xi^5 d\xi}{(\xi^2 + b^2)^5} = \frac{x^6(x^2 + 4b^2)}{24b^4(x^2 + b^2)^4}\right]$$

c) Will boundary-layer separation occur in this flow? If so, at what value of x/b does Thwaites' method predict zero wall shear stress?

d) Using solid lines, sketch the streamlines for the ideal flow specified by the velocity potential given above. For comparison, on the same sketch, indicate with dashed lines the streamlines you expect for the flow of a real fluid in the same geometry at the same flow rate.

8.20 A fluid-mediated particle-deposition process requires a laminar boundary-layer flow with a *constant* shear stress, τ_w, on a smooth flat surface. The fluid has viscosity μ and density ρ (both constant). The flow is steady, incompressible, and two dimensional, and the flat surface extends from $0 < x < L$. The flow speed above the boundary layer is $U(x)$. Ignore body forces.

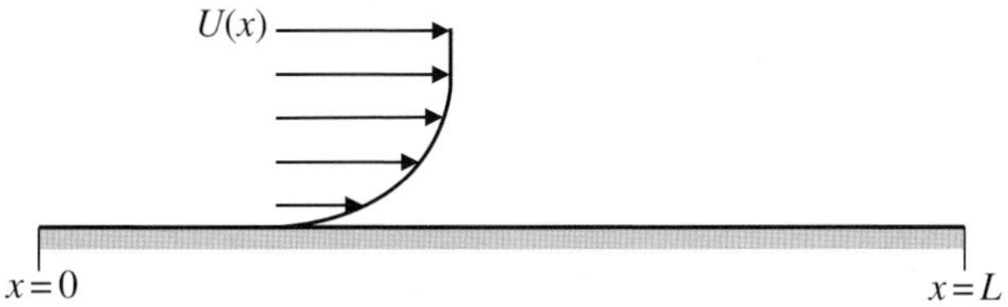

a) Assume the boundary-layer thickness is zero at $x = 0$, and use Thwaites' formulation for the shear stress, $\tau_w = (\mu U/\theta)l(\lambda)$ with $\lambda = (\theta^2/\nu)(dU/dx)$, to determine $\theta(x)$ and $U(x)$ in terms of λ, $\nu = \mu/\rho$, x, and $\tau_w/\mu =$ constant. [*Hint*: assume that $U/\theta = A$ and $l(\lambda)$ are both constants so that $\tau_w/\mu = Al(\lambda)$.]

b) Using the Thwaites integral (8.50) and the results of part a), determine λ.

c) Is boundary-layer separation a concern in this flow? Explain with words or equations.

8.21 The steady two-dimensional potential for incompressible flow at nominal horizontal speed U over a stationary but mildly wavy wall is: $\phi(x, y) = Ux - U\varepsilon \exp(-ky)\cos(kx)$, where $k\varepsilon \ll 1$. Here, ε is the amplitude of the waviness and $k = 2\pi/\Lambda$, where $\Lambda =$ wavelength of the waviness.

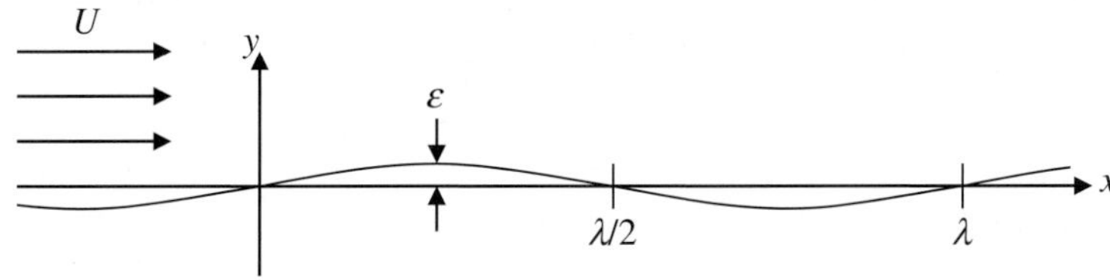

a) Use the potential to determine the horizontal velocity $u(x, y)$ on $y = 0$.
b) Assume that $u(x, 0)$ from part a) is the exterior velocity on the wavy wall and use Thwaites' method to approximately determine the momentum thickness, θ, of the laminar boundary layer that develops on the wavy wall when the fluid viscosity is μ, and $\theta = 0$ at $x = 0$. Keep only the linear terms in $k\varepsilon$ and ε/x to simplify your work.
c) Is the average wall shear stress higher for $\Lambda/2 \le x \le 3\Lambda/4$, or for $3\Lambda/4 \le x \le \Lambda$.
d) Does the boundary layer ever separate when $k\varepsilon = 0.01$?
e) In $0 \le x \le \Lambda$, determine where the wall pressure is the highest and the lowest.
f) If the wavy surface were actually an air-water interface, would a steady wind tend to increase or decrease water wave amplitudes? Explain.

8.22 Consider the boundary layer that develops in stagnation point flow: $U_e(x) = U_o x/L$.
a) With $\theta = 0$ at $x = 0$, use Thwaites' method to determine $\delta^*(x)$, $\theta(x)$, and $c_f(x)$.
b) This flow also has an exact similarity solution of the full Navier-Stokes equations. Numerical evaluation of the final nonlinear ordinary differential equation produces: $C_f\sqrt{\mathrm{Re}_x} = 2.4652$, where $\mathrm{Re}_x = U_e x/\nu = U_o x^2/L\nu$. Assess the accuracy of the predictions for $C_f(x)$ from the Thwaites' method for this flow.

8.23 A laminar boundary layer develops on a large smooth flat surface under the influence of an exterior flow velocity $U(x)$ that varies with downstream distance, x.

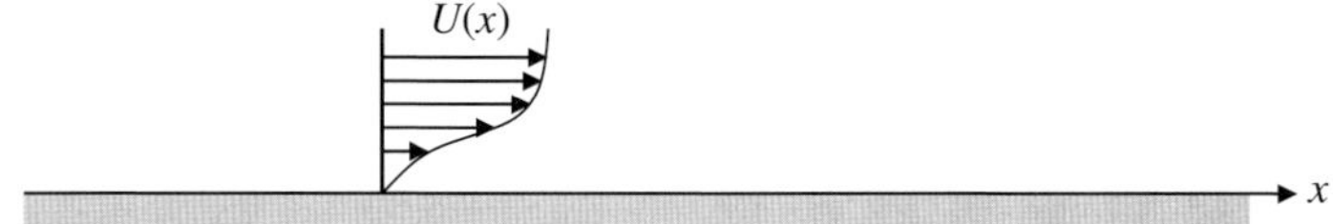

a) Using Thwaites' method, find a single integral-differential equation for $U(x)$ if the boundary layer is to remain perpetually right on the verge of separation so that the wall shear stress, τ_w, is zero. Assume that the boundary layer has zero thickness at $x = 0$.
b) Assume $U(x) = U_o(x/L)^\gamma$ and use the result of part a) to find γ.
c) Compute the boundary-layer momentum thickness $\theta(x)$ for this situation.
d) Determine the extent to which the results of parts b) and c) satisfy the von Karman boundary-layer integral equation, (8.43), when $\tau_w = 0$ by computing the residual of this equation. Interpret the meaning of your answer; is von Karman's equation well satisfied, or is the residual of sufficient size to be problematic?
e) Can the $U(x)$ determined for part b) be produced in a duct with cross-sectional area $A(x) = A_o(x/L)^{-\gamma}$? Explain your reasoning.

8.24 Consider the boundary layer that develops on a cylinder of radius a in a cross flow.

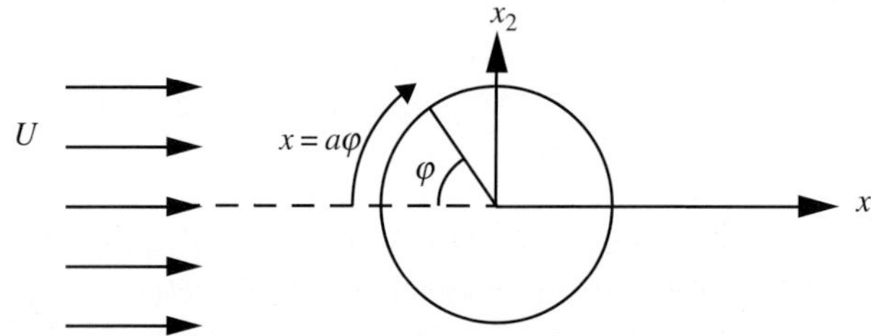

a) Using Thwaites' method, determine the momentum thickness as a function of φ, the angle from the upstream stagnation point (see drawing).
b) Make a sketch of c_f versus φ.
c) At what angle does the Thwaites method predict vanishing wall shear stress?

8.25 An ideal flow model predicts the following surface velocity for the suction (i.e., the upper) side of a thin airfoil with chord c placed in a uniform horizontal air stream of speed U_o:

$$U_e(x) = 2U_o[x/c]^{1/5}\exp(-x/c).$$

a) Assuming that x is the coordinate along the foil's suction surface, use Thwaites method to estimate the momentum thickness $\theta(x)$ of the laminar boundary layer that develops on this surface.

b) Using the results of part a) show that the correlation parameter λ is given by:

$$\lambda = \frac{0.45}{125(x/c)^2}\left[e^{5x/c} - 1 - \frac{5x}{c}\right]\left[1 - \frac{5x}{c}\right]$$

c) Does Thwaites' method predict boundary-layer separation in the range $1/5 < x/c < 1$?

d) If a laminar boundary layer is predicted to separate from the surface of this airfoil, suggest at least two changes that could be made to the foil that would tend to prevent separation.

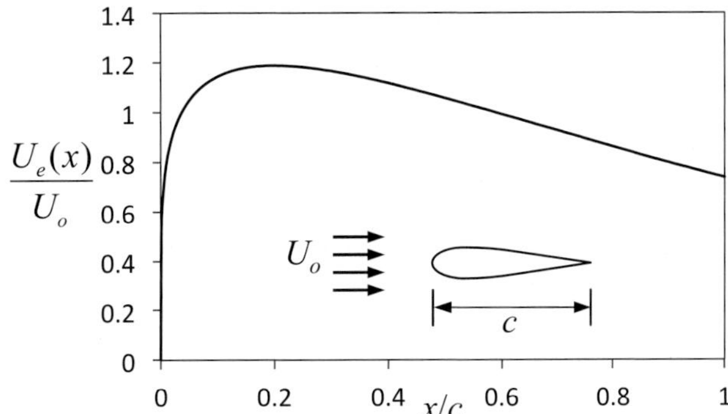

8.26 An incompressible viscous fluid flows steadily in a large duct with constant cross-sectional area A_o and interior perimeter b. A laminar boundary layer develops on the duct's sidewalls. At $x = 0$, the fluid velocity in the duct is uniform and equal to U_o, and the boundary-layer thickness is zero. Assume the thickness of the duct-wall boundary layer is small compared to A_o/b.

a) Calculate the duct-wall boundary-layer momentum and displacement thicknesses, $\theta(x)$ and $\delta^*(x)$ respectively, using Thwaites' method when $U(x) = U_o$.

b) Using the $\delta^*(x)$ found for part a), compute a more accurate version of $U(x)$ that includes boundary-layer displacement effects.

c) Using the $U(x)$ found for part b), recompute $\theta(x)$ and compare to the results of part a). To simplify your work, linearize all the power-law expressions, i.e., $(1 - b\delta^*/A_o)^n \cong 1 - nb\delta^*/A_o$.

d) If the duct area expanded mildly as the flow moved downstream, would the presence of the sidewall boundary layers be more likely to move boundary-layer separation upstream or downstream? Explain.

8.27 Water flows over a flat plate 30 m long and 17 m wide with a free-stream velocity of 1 m/s. Verify that the Reynolds number at the end of the plate is larger than the critical value for transition to turbulence. Using the drag coefficient in Fig. 8.12, estimate the drag on the plate.

8.28 A common means of assessing boundary-layer separation is to observe the surface streaks left by oil or paint drops that were smeared across a surface by the flow. Such investigations can be carried out in an elementary manner for cross-flow past a cylinder using a blow dryer, a cylinder 0.5 to 1 cm in diameter that is ~10 cm long (a common ball-point pen), and a suitable viscous liquid. Here, creamy salad dressing, shampoo, dish washing liquid, or molasses should work. And, for the best observations, the liquid should not be clear and the cylinder and liquid should be different colors. Dip your finger into the viscous liquid and wipe it over two thirds of the surface of the cylinder. The liquid layer should be thick enough so that you can easily tell where it is thick or thin. Use the remaining dry third of the cylinder to hold the cylinder horizontal. Now, turn on the blow dryer, leaving the heat off and direct its outflow across the wetted portion of the horizontal cylinder to mimic the flow situation in the drawing for Exercise 8.24.

a) Hold the cylinder stationary, and observe how the viscous fluid moves on the surface of the cylinder and try to determine the angle φ_s at which boundary-layer separation occurs. To get consistent results you may have to experiment with different liquids, different initial liquid thicknesses, different blow-dryer fan settings, and different distances between cylinder and blow dryer. Estimate the cylinder-diameter-based Reynolds number of the flow you've studied.

b) If you have completed Exercise 8.24, do your boundary-layer separation observations match the calculations? Explain any discrepancies between your experiments and the calculations.

8.29 Find the diameter of a parachute required to provide a fall velocity no larger than that caused by jumping from a 2.5 m height, if the total load is 80 kg. Assume that the properties of air are $\rho = 1.167$ kg/m^3, $\nu = 1.5 \times 10^{-5} m^2/s$, and treat the parachute as a hemispherical shell with $C_D = 2.3$. [*Answer:* 3.9 m]

8.30 The boundary-layer approximation is sometimes applied to flows that do not have a bounding surface. Here the approximation is based on two conditions: downstream fluid motion dominates over the cross-stream flow, and any moving layer thickness defined in the transverse direction evolves slowly in the downstream direction. Consider a laminar jet of momentum flux J that emerges from a small orifice into a large pool of stationary viscous fluid at $z=0$. Assume the jet is directed along the positive z-axis in a cylindrical coordinate system. In this case, the steady, incompressible, axisymmetric boundary-layer equations are:

$$\frac{1}{R}\frac{\partial(Ru_R)}{\partial r}+\frac{\partial w}{\partial z}=0,\quad \text{and}\quad w\frac{\partial w}{\partial z}+u_R\frac{\partial w}{\partial R}=-\frac{1}{\rho}\frac{\partial p}{\partial z}+\frac{\nu}{R}\frac{\partial}{\partial R}\left(R\frac{\partial w}{\partial R}\right),$$

where w is the (axial) z-direction velocity component, and R is the radial coordinate. Let $r(z)$ denote the generic radius of the cone of jet flow.

a) Let $w(R,z)=(\nu/z)f(\eta)$ where $\eta=R/z$, and derive the following equation for f: $\eta f'+f\int^{\eta}\eta f\,d\eta=0$.

b) Solve this equation by defining a new function $F=\int^{\eta}\eta f\,d\eta$. Determine constants from the boundary condition: $w\to 0$ as $\eta\to\infty$, and the requirement: $J=2\pi\rho\int_{R=0}^{R=r(z)}w^2(R,z)R\,dR=\text{const}$.

c) At fixed z, does $r(z)$ increase or decrease with increasing J?

[*Hints*: 1) The fact that the jet emerges into a pool of quiescent fluid should provide information about $\partial p/\partial z$, and 2) $f(\eta)\propto(1+const\cdot\eta^2)^{-2}$, but try to obtain this result without using it.]

8.31 A simple realization of a temporal boundary layer involves the spinning fluid in a cylindrical container. Consider a viscous incompressible fluid (density $=\rho$, viscosity $=\mu$) in solid body rotation (rotational speed $=\Omega$) in a cylindrical container of diameter d. The mean depth of the fluid is h. An external stirring mechanism forces the fluid to maintain solid body rotation. At $t=0$, the external stirring ceases. Denote the time for the fluid to spin-down (i.e., to stop rotating) by τ.

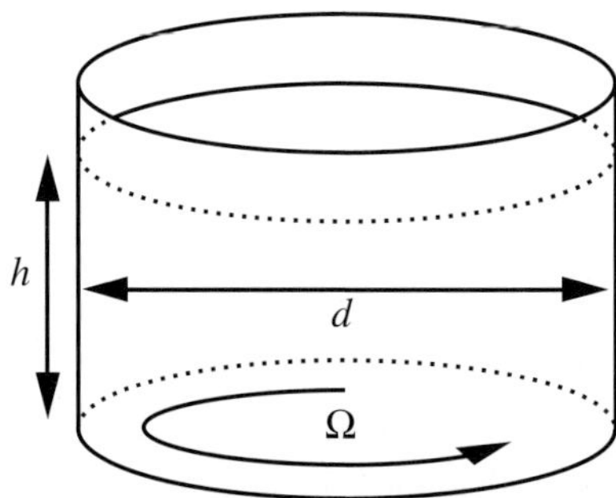

a) For $h\gg d$, write a simple laminar-flow scaling law for τ assuming that the velocity perturbation produced by the no-slip condition on the container's sidewall must travel inward a distance $d/2$ via diffusion.

b) For $h\ll d$, write a simple laminar-flow scaling law for τ assuming that the velocity perturbation produced by the no-slip condition on the container's bottom must travel upward a distance h via diffusion.

c) Using partially filled cylindrical containers of several different sizes (drinking glasses and pots and pans are suggested) with different amounts of water, test the validity of the above diffusion estimates. Use a spoon or a whirling motion of the container to bring the water into something approaching solid body rotation. You'll know when the fluid motion is close to solid body rotation because the fluid surface will be a paraboloid of revolution. Once you have this initial flow condition set up, cease the stirring or whirling and note how long it takes for the fluid to stop moving. Perform at least one test when d and h are several inches or more. Cookie or bread crumbs sprinkled on the water surface will help visualize surface motion. The judicious addition of a few drops of milk after the fluid starts slowing down may prove interesting.

d) Compute numbers from your scaling laws for parts a) and b) using the viscosity of water, the dimensions of the containers, and the experimental water depths. Are the scaling laws from parts a) and b) useful for predicting the experimental results? If not, explain why.

(The phenomena investigated here have some important practical consequences in atmospheric and oceanic flows and in IC engines where swirl and tumble are exploited to mix the fuel charge and increase combustion speeds.)

8.32 Consider a boundary layer formed over a bed of spherical particles as shown. The shear stress at the surface is $\tau=\mu(du/dy)_{y=h}$, where μ is the fluid viscosity, ρ is the density, and $(du/dy)_{y=0}$ is the velocity gradient at the surface of the bed. A friction velocity can be defined as $u_* = \sqrt{\tau/\rho}$ (this will be revisited in more detail in Chapter 10). The threshold friction velocity u_{*t} dictates the minimum value when erosion is initiated and represents a critical parameter in geophysical fluid mechanics for predicting sediment transport in rivers and the formation and migration of sand dunes in deserts. For this problem, assume each particle has diameter d and density $\rho_s \gg \rho$. Neglect added mass, Basset history, and inter-particle cohesion.

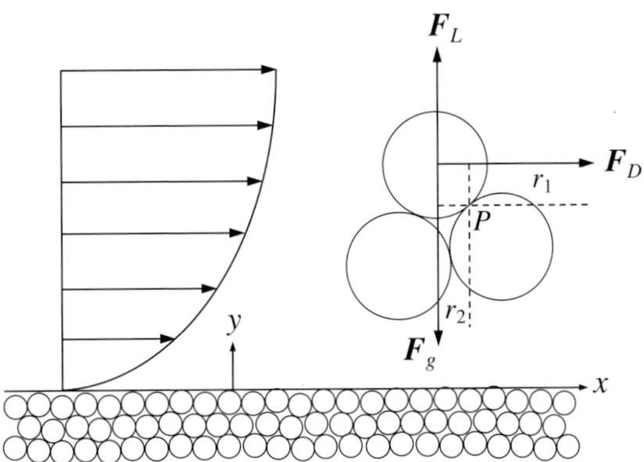

a) Using dimensional analysis, determine how u_* must depend on the other five parameters: ρ, ρ_s, d, g, μ.

b) The critical fluid velocity to initiate particle entrainment can be obtained by a moment balance accounting for drag $\mathbf{F}_D$, lift $\mathbf{F}_L$, and gravity $\mathbf{F}_g$. Consider an individual particle that begins to pivot in the downstream direction about a point P on a supporting particle as shown in the figure. Assume the moment arm lengths associated with the (horizontal) drag force and (vertical) lift and gravitational forces, r_1 and r_2, are equal and proportional to d. Write an expression that relates the minimum stream wise fluid velocity u to initiate motion in terms of d, ρ, ρ_s, g, C_D, and C_L.

c) Assuming a Blasius boundary layer profile, rewrite the solution to b) in terms of the friction velocity u_{*t}.

8.33 Mississippi River boatmen know that when rounding a bend in the river, they must stay close to the outer bank or else they will run aground. Explain in fluid mechanical terms the reason for the cross-sectional shape of the river at the bend.

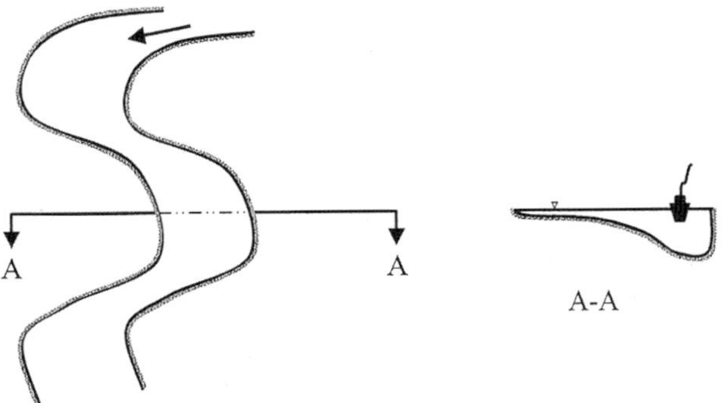

References

Batchelor, G.K., 1967. An Introduction to Fluid Dynamics. Cambridge University Press, London.

Bender, C.M., Orszag, S.A., 1978. Advanced Mathematical Methods for Scientists and Engineers. McGraw-Hill, New York.

Clift, R., Grace, J.R., Weber, M.E., 2005. Bubbles, Drops, and Particles. Dover Publications, Inc., Mineola, New York.

Curle, N., 1962. The Laminar Boundary Layer Equations. Oxford: Clarendon Press, London.

Falkner, V.W., Skan, S.W., 1931. Solutions of the boundary layer equations. Philosophical Magazine 12 (7), 865–896.

Gallo, W.F., Marvin, J.G., Gnos, A.V., 1970. Nonsimilar nature of the laminar boundary layer. AIAA Journal 8, 75–81.

Glauert, M.B., 1956. The wall jet. Journal of Fluid Mechanics 1, 625–643.

Goldstein, S. (Ed.), 1938. Modern Developments in Fluid Dynamics. Oxford University Press. Reprinted by Dover, New York, 1965.

Holstein, H., Bohlen, T., 1940. Ein einfaches Verfahren zur Berechnung laminarer Reibungsschichten die dem Näherungsverfahren von K. Pohlhausen genügen. Lilienthal-Bericht. S 10, 5–16.

Kevorkian, J., Cole, J.D., 1981. Perturbation Methods in Applied Mathematics. Springer-Verlag, New York.

McLaughlin, J.B., 1991. Inertial migration of a small sphere in linear shear flows. Journal of Fluid Mechanics 224, 261–274.

Nayfeh, A.H., 1981. Introduction to Perturbation Techniques. Wiley, New York.

Peletier, L.A., 1972. On the asymptotic behavior of velocity profiles in laminar boundary layers. Archive for Rational Mechanics and Analysis 45, 110–119.

Pohlhausen, K., 1921. Zur näherungsweisen Integration der Differentialgleichung der laminaren Grenzschicht. Zeitschrift für Angewandte Mathematik und Mechanik 1, 252–268.

Robins, B., 1742. New Principles of Gunnery: Containing the Determinations of the Force of Gun-Powder and Investigations of the Difference in the Resisting Power of the Air to Swift and Slow Motions. J. Nourse, London.

Rubinow, S.I., Keller, J.B., 1961. The transverse force on a spinning sphere moving in a viscous fluid. Journal of Fluid Mechanics 11 (3), 447–459.

Saffman, P.G., 1965. The lift on a small sphere in a slow shear flow. Journal of Fluid Mechanics 22, 385–400.

Schlichting, H., 1979. Boundary Layer Theory, 7th ed. McGraw-Hill, New York.

Serrin, J., 1967. Asymptotic behaviour of velocity profiles in the Prandtl boundary layer theory. Proceedings of the Royal Society A 299, 491–507.

Taneda, S., 1965. Experimental investigation of vortex streets. Journal of the Physical Society of Japan 20, 1714–1721.

Thwaites, B., 1949. Approximate calculation of the laminar boundary layer. Aeronautical Quarterly 1, 245–280.

van Dyke, M., 1975. Perturbation Methods in Fluid Mechanics. The Parabolic Press, Stanford, CA.

von Karman, T., 1921. Über laminare und turbulente Reibung. Zeitschrift für Angewandte Mathematik und Mechanik 1, 233–252.

Wen, C.-Y., Lin, C.-Y., 2001. Two-dimensional vortex shedding of a circular cylinder. Physics of Fluids 13, 557–560.

Further reading

Friedrichs, K.O., 1955. Asymptotic phenomena in mathematical physics. Bulletin of the American Mathematical Society 61, 485–504.

Lagerstrom, P.A., Casten, R.G., 1972. Basic concepts underlying singular perturbation techniques. SIAM Review 14, 63–120.

Meksyn, D., 1961. New Methods in Laminar Boundary Layer Theory. Pergamon Press, New York.

Panton, R.L., 2005. Incompressible Flow, 3rd ed. Wiley, New York.

Sherman, F.S., 1990. Viscous Flow. McGraw-Hill, New York.

White, F.M., 2006. Viscous Fluid Flow. McGraw-Hill, New York.

Yih, C.S., 1977. Fluid Mechanics: A Concise Introduction to the Theory. West River Press, Ann Arbor, MI.

Chapter 9

Instability⍟

Contents

9.1 Introduction 319
9.2 Method of normal modes 321
9.3 Kelvin-Helmholtz instability 322
9.4 Rayleigh-Plateau instability 326
9.5 Thermal instability: the Bénard problem 329
9.6 Double-diffusive instability 336
9.7 Centrifugal instability: Taylor problem 339
9.8 Instability of continuously stratified parallel flows 344
9.9 Squire's theorem and the Orr-Sommerfeld equation 349
9.10 Inviscid stability of parallel flows 351
9.11 Results for parallel and nearly parallel viscous flows 354
Two-stream shear layer 354
Plane Poiseuille flow 355
Plane Couette flow 355
Pipe flow 355
Boundary layers with pressure gradients 356
9.12 Experimental verification of boundary-layer instability 357
9.13 Comments on nonlinear effects 359
9.14 Transition 359
9.15 Deterministic chaos 360
Final remarks 364
Exercises 365
References 372
Further reading 372

Chapter objectives

- To present the mathematical theory of temporal flow instability
- To illustrate how this theory may be applied in a variety of confined and unconfined flows
- To present classic theoretical results for parallel flows
- To describe results for viscous flows
- To discuss nonlinearity and the possible role of chaos in the transition to turbulence

9.1 Introduction

Up to this point, fluid flows satisfying the conservation laws have been considered without taking into account the small disturbances that are invariably present in any real system. In many situations, flows may spontaneously develop fluctuations or transition to a new state when subjected to such disturbances, resulting in vastly different behavior. Consider the stability of two simple mechanical systems in a vertical gravitation field. A sharpened pencil may, in theory, be balanced on its point on a horizontal surface, but any small surface vibration or air pressure disturbance will knock it over. Thus, sharpened pencils on horizontal surfaces are commonly observed lying horizontally. Similarly, the position of a smooth ball resting on the inside surface of a hemispherical bowl is stable provided the bowl is concave upwards. However, the ball's position is unstable to small displacements if placed on the outer side of a hemispherical bowl when the bowl is concave downwards (Fig. 9.1). In fluid flows, smooth laminar flows are stable to small disturbances only when certain conditions are satisfied. For example, in the flow of a homogeneous viscous fluid in a channel, the Reynolds number must be less than some critical value for the flow to be stable, and in a stratified shear flow, the Richardson number must be larger than a critical value for stability. When these conditions are not satisfied, infinitesimal disturbances may grow spontaneously and completely change the character of the original flow. Sometimes the disturbances can grow to finite amplitude and reach a new steady-state equilibrium. The new state may then become unstable to other types of disturbances, and may evolve to yet another steady state, and so on. As a limit of this situation, the flow becomes a superposition of a variety of interacting nonlinear disturbances with nearly random phases, a state of chaotic or nearly chaotic fluctuations that is commonly described as *turbulence*. In fact, two primary motivations for studying fluid-flow stability are: 1) to understand the process of laminar to turbulent transition, and 2) to predict the onset of this transition.

⍟. This book has a companion website hosting complementary materials. Visit this URL to access it: https://www.elsevier.com/books-and-journals/book-companion/9780128198070.

Fluid Mechanics. https://doi.org/10.1016/B978-0-12-819807-0.00018-1

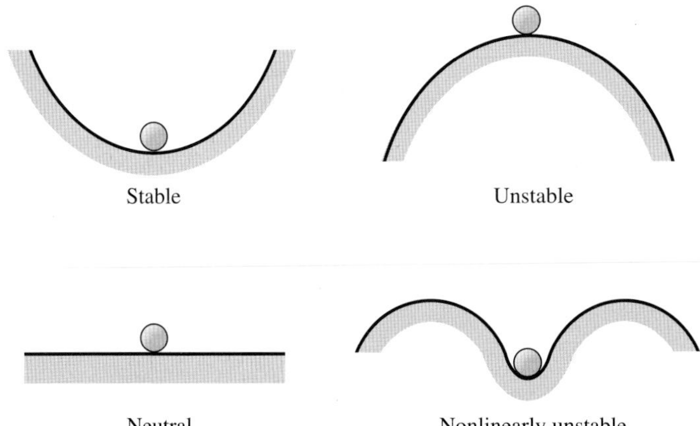

FIGURE 9.1 Stable and unstable mechanical systems. Here, gravity is presumed to act downward. In the upper-left and lower-right panels, a small displacement of the round object away from equilibrium will be opposed by the action of gravity and the object will move back toward its equilibrium location. These are linearly stable situations. In the upper right panel, a small displacement of the object will be enhanced by the action of gravity and the object will move away from its equilibrium location, an unstable situation. In the lower left, a small displacement of the object does not produce a new force, thus the situation is neutrally stable. In the lower right panel, a sufficiently large displacement of the object may place it beyond its region of stability; thus this situation is nonlinearly unstable.

The primary objective of this chapter is the examination of the stability of certain fluid flows with respect to infinitesimal disturbances. Finite-amplitude effects, including the development of chaotic solutions, are examined briefly later in this chapter. We shall introduce perturbations on a particular flow, and determine whether the equations of motion predict that the perturbations grow or decay. In this analysis, the perturbations are commonly assumed to be small enough so that linearization is possible through neglect of quadratic and higher order terms in the perturbation variables and their derivatives. While such linearization fruitfully allows the production of analytical results, it inherently limits the applicability of such results to the *initial* behavior of infinitesimal disturbances. The loss of stability does not in itself constitute a transition to turbulence since the linear theory can, at best, describe only the very beginning of the transition process. In addition, a real flow may be stable to infinitesimal disturbances (linearly stable), but may be unstable to sufficiently large disturbances (nonlinearly unstable); this is schematically represented in Fig. 9.1.

In spite of these limitations, linear stability theory enjoys considerable success. There is excellent agreement between experimental results and the theoretical prediction of the onset of thermal convection in a layer of fluid, and of the onset of Tollmien-Schlichting waves in a viscous boundary layer. Taylor's experimental verification of his own theoretical prediction of the onset of secondary flow in circular Couette flow is so striking that it has led people to suggest that Taylor's work is the first rigorous confirmation of the Navier-Stokes equations on which the calculations are based.

This chapter describes the temporal instability of confined and unconfined flows where spatially extended perturbations decay, persist, or grow in time. The complimentary situation where spatially confined disturbances decay, persist, or grow while traveling in space is more complicated and is described elsewhere (see Huerre and Monkewitz, 1990). The primary analysis technique used here, the method of normal modes, is described in the next section. Sections 9.3–9.12 utilize this technique to illustrate basic flow physics and to present results for a variety of flows important in engineering applications and geophysical situations. The final few sections describe transition and the onset of turbulence. None of the flow situations discussed in this chapter contain Coriolis effects. *Baroclinic instability*, which does contain the Coriolis frequency, is discussed in Chapter 12. The books by Chandrasekhar (1961, 1981), and Drazin and Reid (1981) provide further information on flow instability. The review article by Bayly et al. (1988) is recommended as well.

Example 9.1

Consider the linear differential equation for the position, $x(t)$, of a mass m constrained by a linear damper with coefficient γ, and a spring of stiffness k_s: $m(d^2x/dt^2) + \gamma(dx/dt) + k_s x = 0$. If the mass is given a slight displacement ε away from $x = 0$ and released from rest at $t = 0$, when is the subsequent $x(t)$ stable, neutrally stable, or unstable when $m > 0$, $k_s \geq 0$, but $\gamma \neq 0$ may have either sign?

Solution For the given initial conditions, the solution for $x(t)$ is:

$$x(t) = \frac{\varepsilon}{\beta_+ - \beta_-}(-\beta_- \exp\{\beta_+ t\} + \beta_+ \exp\{\beta_- t\}), \quad \text{where} \quad \beta_\pm = -\frac{\gamma}{2m} \pm \sqrt{\frac{\gamma^2}{4m^2} - \frac{k_s}{m}}.$$

There are three distinct cases. (1) If $\gamma > 0$ and $k_s > 0$, then $x(t)$ is stable because both β_+ and β_- will have negative real parts and the initial displacement decays exponentially with increasing time. (2) If $k_s = 0$, then β_+ will be zero for any value of γ, and the solution for $x(t)$ reduces to $x = \varepsilon$. This represents neutral stability since a new steady solution emerges after the initial displacement. (3) If $\gamma < 0$ and $k_s > 0$, then $x(t)$ is unstable since both β_+ and β_- have a positive real parts and the initial displacement grows exponentially with increasing time. Interestingly, real fluid flows can mimic this negative-damping destabilizing effect.

9.2 Method of normal modes

Basic linear stability analysis consists of presuming the existence of sinusoidal disturbances to a *basic state* (also called a background, initial, or equilibrium state), which is the flow whose stability is being investigated. For example, the velocity field of a basic state involving a flow parallel to the x-axis and varying along the z-axis is $\mathbf{U} = U(z)\mathbf{e}_x$. On this background flow we superpose a spatially extended disturbance of the form:

$$u(x, y, z, t) = \widehat{u}(z) \exp\{ikx + imy + \sigma t\} = \widehat{u}(z) \exp\{i|\mathbf{K}|(\mathbf{e}_K \cdot \mathbf{x} - ct)\}, \tag{9.1}$$

where $\widehat{u}(z)$ is a complex amplitude, $i = \sqrt{-1}$ is the imaginary root, $\mathbf{K} = (k, m, 0)$ is the disturbance wave number, $\mathbf{e}_K = \mathbf{K}/|\mathbf{K}|$, $\mathbf{x} = (x, y, z)$, σ is the temporal growth rate, c is the complex phase speed of the disturbance, and the real part of (9.1) is taken to obtain physical quantities. The complex notation used here is explained in Section 11.7. The two forms of (9.1) are useful when the disturbance is stationary, and when it takes the form of a traveling wave, respectively. The reason solutions exponential in (x, y, t) are allowed in (9.1) is that the coefficients of the differential equation governing the perturbation in this flow are independent of (x, y, t). The flow field is assumed to be unbounded in the x and y directions, hence the wave number components k and m can only be real (and $|\mathbf{K}|$ positive real) in order that the dependent variables remain bounded as x, $y \to \pm\infty$; however, $\sigma = \sigma_r + i\sigma_i$ and $c = c_r + ic_i$ are regarded as complex.

The behavior of the system for *all* possible disturbance wave numbers, $\mathbf{K}$, is examined in the analysis. If σ_r or c_i are positive for *any* value of the wave number, the system is unstable to disturbances of this wave number. If no such unstable state can be found, the system is stable. We say that:

$$\begin{aligned} \sigma_r < 0 \quad &\text{or} \quad c_i < 0 \quad \text{implies a } \textit{stable} \text{ flow,} \\ \sigma_r = 0 \quad &\text{or} \quad c_i = 0 \quad \text{implies a } \textit{neutrally stable} \text{ flow, and} \\ \sigma_r > 0 \quad &\text{or} \quad c_i > 0 \quad \text{implies an } \textit{unstable} \text{ flow.} \end{aligned}$$

The method of analysis involving the examination of Fourier components such as (9.1) is called the *normal mode method*. An arbitrary disturbance can be decomposed into a complete set of normal modes. In this method the stability of each of the modes is examined separately, as the linearity of the problem implies that the various modes do not interact. The method leads to an eigenvalue problem.

The boundary between stability and instability is called the *marginal state*, for which $\sigma_r = c_i = 0$. There can be two types of marginal states, depending on whether σ_i or c_r is also zero or nonzero in this state. If $\sigma_i = c_r = 0$ in the marginal state, then (9.1) shows that the marginal state is characterized by a *stationary pattern* of motion; we shall see later that the instability here appears in the form of *cellular convection* or *secondary flow* (see Fig. 9.21 later). If, on the other hand, $\sigma_i \neq 0$ or $c_r \neq 0$ in the marginal state, then the instability sets in as *traveling oscillations* of growing amplitude. Such a mode of instability is frequently called *overstability* because the restoring forces are so strong that the system overshoots its corresponding position on the other side of equilibrium. We prefer to avoid this term and instead call it the *oscillatory mode* of instability.

The difference between the *neutrally stable state* and the *marginal state* should be noted as both have $\sigma_r = c_i = 0$. However, the marginal state has the additional constraint that it lies at the *borderline* between stable and unstable solutions. That is, a slight change of parameters (such as the Reynolds number) from the marginal state can take the system into an unstable regime where $\sigma_r > 0$. In many cases we shall find the stability criterion by simply setting $\sigma_r = 0$ or $c_i = 0$, without formally demonstrating that these conditions define the borderline between unstable and stable states.

9.3 Kelvin-Helmholtz instability

Instability at the interface between two horizontal parallel fluid streams with different velocities and densities is called the *Kelvin-Helmholtz instability*. This is an inertial instability and it can be readily analyzed assuming ideal flow in each stream. The name is also commonly used to describe the instability of the more general case where the variations of velocity and density are continuous and occur over a finite thickness (see Section 9.8).

The Kelvin-Helmholtz instability leads to enhanced momentum, heat, and moisture transport in the atmosphere, plus it is routinely exploited in a variety of geometries for mixing two or more fluid streams in engineering applications. The simplest version is analyzed here in two dimensions (x, z), where x is the stream-wise coordinate and z is the vertical coordinate. Consider two fluid layers of infinite depth that meet at a zero-thickness interface located at $z = \zeta(x, t)$. Let U_1 and ρ_1 be the horizontal velocity and density of the basic state in the upper half-space, and U_2 and ρ_2 be those of the basic state in the lower half-space (Fig. 9.2). From Kelvin's circulation theorem, the perturbed flow must be irrotational in each half-space because the motion develops from an irrotational basic state, uniform velocity in each half-space. Thus, the infinitesimally perturbed flow above (subscript 1) and below (subscript 2) the interface can be described by the velocity potentials:

$$\tilde{\phi}_1 = U_1 x + \phi_1, \quad \text{and} \quad \tilde{\phi}_2 = U_2 x + \phi_2, \tag{9.2}$$

where the U_1 and U_2 terms represent the basic state, and tildes ($\sim$) denote the total flow potentials (background plus perturbations), a notation used throughout this chapter. Here $\tilde{\phi}_1$ and $\tilde{\phi}_2$ must satisfy the Laplace equation, so the perturbation potentials, ϕ_1 and ϕ_2, must also satisfy Laplace equations:

$$\nabla^2\phi_1 = 0 \quad \text{and} \quad \nabla^2\phi_2 = 0. \tag{9.3}$$

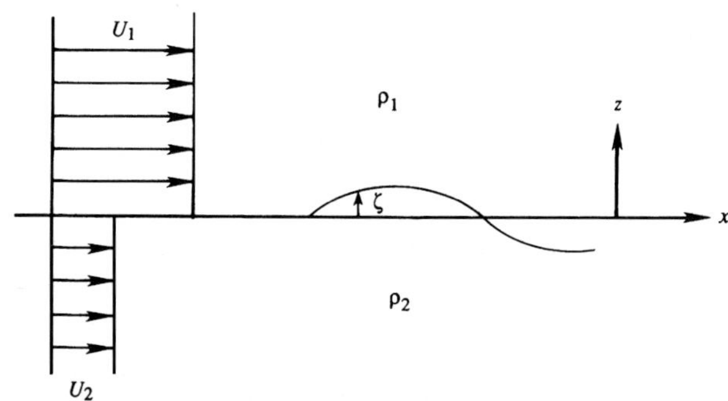

FIGURE 9.2 Basic flow configuration leading to the Kelvin-Helmholtz instability. Here the velocity and density profiles are discontinuous across an interface nominally located at $z = 0$. If the small vertical perturbation $\zeta(x, t)$ to this interface grows, then the flow is unstable.

There are a total of four boundary conditions:

$$\phi_1 \to 0 \quad \text{as} \quad z \to +\infty, \qquad \phi_2 \to 0 \quad \text{as} \quad z \to -\infty, \tag{9.4, 9.5}$$

$$\mathbf{n} \cdot \nabla\tilde{\phi}_1 = \mathbf{n} \cdot \mathbf{u}_s = \mathbf{n} \cdot \nabla\tilde{\phi}_2 \quad \text{on} \quad z = \zeta, \quad \text{and} \tag{9.6}$$

$$p_1 = p_2 \quad \text{on} \quad z = \zeta, \tag{9.7}$$

where $\mathbf{n}$ is the local normal to the interface, $\mathbf{u}_s$ is the velocity of the interface, and p_1 and p_2 are the pressures above and below the interface. Here, the kinematic and dynamic boundary conditions, (9.6) and (9.7) respectively, are conceptually similar to (11.13) and (11.19). The kinematic condition, (9.6), can be rewritten:

$$\mathbf{n} \cdot \left\{ \frac{\partial\tilde{\phi}_1}{\partial x}\mathbf{e}_x + \frac{\partial\tilde{\phi}_1}{\partial z}\mathbf{e}_z \right\} = \mathbf{n} \cdot \left\{ \frac{\partial\zeta}{\partial t}\mathbf{e}_z \right\} = \mathbf{n} \cdot \left\{ \frac{\partial\tilde{\phi}_2}{\partial x}\mathbf{e}_x + \frac{\partial\tilde{\phi}_2}{\partial z}\mathbf{e}_z \right\} \quad \text{on} \quad z = \zeta, \tag{9.8}$$

where $\mathbf{n} = \nabla f/|\nabla f| = [-(\partial\zeta/\partial x)\mathbf{e}_x + \mathbf{e}_z]/\sqrt{1 + (\partial\zeta/\partial x)^2}$ when $f(x, z, t) = z - \zeta(x, t) = 0$ defines the interface, and $\mathbf{u}_s = (\partial\zeta/\partial t)\mathbf{e}_z$ can be considered purely vertical. When the dot products are performed, the common square-root factor removed, and the derivatives of the potentials evaluated using (9.2), (9.8) reduces to:

$$-\left(U_1 + \frac{\partial\phi_1}{\partial x}\right)\frac{\partial\zeta}{\partial x} + \frac{\partial\phi_1}{\partial z} = \frac{\partial\zeta}{\partial t} = -\left(U_2 + \frac{\partial\phi_2}{\partial x}\right)\frac{\partial\zeta}{\partial x} + \frac{\partial\phi_2}{\partial z} \quad \text{on} \quad z = \zeta.$$

This condition can be linearized by applying it at $z=0$ instead of at $z=\zeta$ and by neglecting quadratic terms. Thus, the simplified version of (9.6) is:

$$-U_1\frac{\partial\zeta}{\partial x}+\frac{\partial\phi_1}{\partial z}=\frac{\partial\zeta}{\partial t}=-U_2\frac{\partial\zeta}{\partial x}+\frac{\partial\phi_2}{\partial z}\quad\text{on}\quad z=0. \tag{9.9}$$

The dynamic boundary condition at the interface requires the pressure to be continuous across the interface (when surface tension is neglected). The unsteady Bernoulli equations above and below the layer are:

$$\frac{\partial\widetilde{\phi}_1}{\partial t}+\frac{1}{2}|\nabla\widetilde{\phi}_1|^2+\frac{p_1}{\rho_1}+gz=C_1,\quad\text{and}\quad\frac{\partial\widetilde{\phi}_2}{\partial t}+\frac{1}{2}|\nabla\widetilde{\phi}_2|^2+\frac{p_2}{\rho_2}+gz=C_2. \tag{9.10}$$

So pressure matching requires:

$$p_1=\rho_1\left(C_1-\frac{\partial\widetilde{\phi}_1}{\partial t}-\frac{1}{2}|\nabla\widetilde{\phi}_1|^2-gz\right)=\rho_2\left(C_2-\frac{\partial\widetilde{\phi}_2}{\partial t}-\frac{1}{2}|\nabla\widetilde{\phi}_2|^2-gz\right)=p_2\quad\text{on}\quad z=\zeta. \tag{9.11}$$

In the undisturbed state ($\phi_1=\phi_2=0$, and $\zeta=0$), (9.11) implies:

$$(p_1)_{undisturbed}=\rho_1\left(C_1-\frac{1}{2}U_1^2\right)=\rho_2\left(C_2-\frac{1}{2}U_2^2\right)=(p_2)_{undisturbed}. \tag{9.12}$$

Subtracting (9.11) from (9.12), and inserting (9.2) leads to:

$$\rho_1\left(\frac{\partial\phi_1}{\partial t}+U_1\frac{\partial\phi_1}{\partial x}+\frac{1}{2}|\nabla\phi_1|^2+gz\right)=\rho_2\left(\frac{\partial\phi_2}{\partial t}+U_2\frac{\partial\phi_2}{\partial x}+\frac{1}{2}|\nabla\phi_2|^2+gz\right)\quad\text{on}\quad z=\zeta,$$

and this condition can be linearized by dropping quadratic terms and evaluating derivatives on $z=0$ to find:

$$\rho_1\left(\frac{\partial\phi_1}{\partial t}+U_1\frac{\partial\phi_1}{\partial x}+g\zeta\right)=\rho_2\left(\frac{\partial\phi_2}{\partial t}+U_2\frac{\partial\phi_2}{\partial x}+g\zeta\right)\quad\text{on}\quad z=0. \tag{9.13}$$

Thus, field equations (9.3) and the conditions (9.4), (9.5), (9.9), and (9.13) specify the linear stability of a velocity discontinuity between uniform flows of different speeds and densities.

Now apply the method of normal modes to look for oscillatory solutions for ϕ_1 and ϕ_2 in the second exponential form of (9.1) with $\mathbf{K}=(k,0,0)$:

$$\phi_1(x,z,t)=A_1(z)\exp\{ik(x-ct)\},\quad\text{and}\quad\phi_2(x,z,t)=A_2(z)\exp\{ik(x-ct)\}. \tag{9.14}$$

Insertion of (9.14) into (9.3) produces:

$$-k^2A_1+\frac{d^2A_1}{dz^2}=0,\quad\text{and}\quad-k^2A_2+\frac{d^2A_2}{dz^2}=0,$$

after common factors are divided out. These equations have exponential solutions: $A_\pm\exp(\pm kz)$. The boundary conditions (9.4) and (9.5) require the minus sign for $z>0$, and the positive sign for $z<0$, so (9.14) reduces to:

$$\phi_1=A_-\exp\{ik(x-ct)-kz\},\quad\text{and}\quad\phi_2=A_+\exp\{ik(x-ct)+kz\}. \tag{9.15}$$

Inserting these two equations and a matching form for the interface shape, $\zeta=\zeta_o\exp\{ik(x-ct)\}$, into (9.9) and (9.13) leads to:

$$\begin{aligned}&-iU_1k\zeta_o-kA_-=-ikc\zeta_o=-iU_2k\zeta_o+kA_+,\quad\text{and}\\&\rho_1(-ikcA_-+ikU_1A_-+g\zeta_o)=\rho_2(-ikcA_++ikU_2A_++g\zeta_o).\end{aligned} \tag{9.16, 9.17}$$

The remnant of the kinematic boundary condition (9.16) is sufficient to find $A_\pm$ in terms of ζ_o:

$$kA_-=-(ikU_1-ikc)\zeta_o,\quad\text{and}\quad kA_+=(ikU_2-ikc)\zeta_o.$$

Substituting these into the remnant of the dynamic boundary condition (9.17) leads to a quadratic equation for c:

$$\rho_1(-(-ikc+ikU_1)^2+gk)=\rho_2((-ikc+ikU_2)^2+gk)$$

after the common factor of ζ_o has been divided out. The two solutions for c are:

$$c=\frac{\rho_2 U_2+\rho_1 U_1}{\rho_2+\rho_1}\pm\left[\left(\frac{\rho_2-\rho_1}{\rho_2+\rho_1}\right)\frac{g}{k}-\frac{\rho_2\rho_1}{(\rho_2+\rho_1)^2}(U_2-U_1)^2\right]^{1/2}. \tag{9.18}$$

Both possible values for c imply neutral stability ($c_i=0$) as long as the second term within the square root is smaller than the first. However, one of these solutions will lead to exponential growth ($c_i>0$) when:

$$\left(\frac{\rho_2-\rho_1}{\rho_2+\rho_1}\right)\frac{g}{k}<\frac{\rho_2\rho_1}{(\rho_2+\rho_1)^2}(U_2-U_1)^2 \quad \text{or} \quad g(\rho_2^2-\rho_1^2)<k\rho_1\rho_2(U_2-U_1)^2,$$

which occurs when the free-stream velocity difference is high enough, the density difference is small enough, or the wave number k (presumed positive real) is large enough. In addition, for each growing solution there is a corresponding decaying solution. This happens because the coefficients of the differential equation and the boundary conditions are all real (see Section 9.8).

Although it is somewhat complicated, (9.18) includes several limiting cases with simple interpretations. First of all, setting $U_1=U_2=0$ simplifies (9.18) to:

$$c=\pm\left[\left(\frac{\rho_2-\rho_1}{\rho_2+\rho_1}\right)\frac{g}{k}\right]^{1/2}, \tag{9.19}$$

which indicates a neutrally stable situation as long as $\rho_2>\rho_1$. In this case, (9.19) is the dispersion relation for interface waves in an initially static medium, see (11.95). The situation when $\rho_2<\rho_1$ (heavier fluid accelerates into lighter fluid) is the *Rayleigh-Taylor instability*. When $U_1\neq U_2$, one can always find a value of k large enough to satisfy the requirement for instability. Because all wavelengths must be allowed in an instability analysis, we can say that the *flow is always unstable to short waves when* $U_1\neq U_2$. When $\rho_1=\rho_2$, the interface becomes a vortex sheet (see Section 5.7) with strength $\gamma=U_2-U_1$, and (9.18) reduces to:

$$c=\left(\frac{U_2+U_1}{2}\right)\pm i\left(\frac{U_2-U_1}{2}\right). \tag{9.20}$$

Here, there is always a positive imaginary value of c for every k, so a vortex sheet is unstable to disturbances of any wavelength. It is also seen that the unstable wave moves with a phase velocity, c_r, equal to the average velocity of the basic flow. This must be true from symmetry considerations. In a frame of reference moving with the average velocity, the basic flow is symmetric and the wave therefore should have no preference between the positive and negative x-directions (Fig. 9.3).

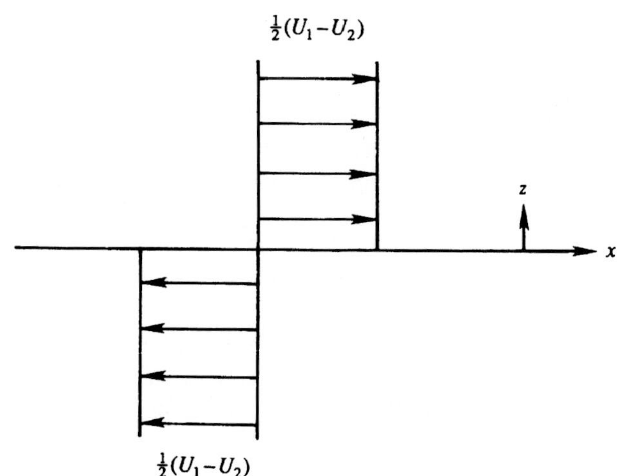

FIGURE 9.3 Background velocity field for the Kelvin-Helmholtz instability as seen by an observer traveling at the average velocity $(U_1+U_2)/2$ of the two layers. When the densities of the two layers are equal, a disturbance to the interface will be stationary in this frame of reference.

The Kelvin-Helmholtz instability is caused by the destabilizing effect of shear, which overcomes the stabilizing effect of stratification. This kind of instability can be generated in a laboratory by filling a horizontal glass tube (of rectangular cross section) containing two liquids of slightly different densities (one colored) and gently tilting it. This starts a current in the lower layer down the plane and a current in the upper layer up the plane. An example of instability generated in this manner is shown in Fig. 9.4.

FIGURE 9.4 Kelvin-Helmholtz instability generated by tilting a horizontal channel containing two liquids of different densities. The lower layer is dyed and 18 wavelengths of the developing interfacial disturbance are shown. The mean flow in the lower layer is down the plane (to the left) and that in the upper layer is up the plane (to the right). *Thorpe,* Journal of Fluid Mechanics, 46, *299–319, 1971; reprinted with the permission of Cambridge University Press.*

Shear instability of stratified fluids is ubiquitous in the atmosphere and the ocean and is believed to be a major source of internal waves. Fig. 9.5 is an acoustic backscatter image of Kelvin-Helmholtz vortical structures observed in a river estuary at ebb tide when nominally fresh water is flowing over saltier and denser ocean water. Similar images of injected dye have been recorded in oceanic thermoclines (Woods, 1969), and in billow cloud patterns (Drazin and Reid, 1981, p. 21).

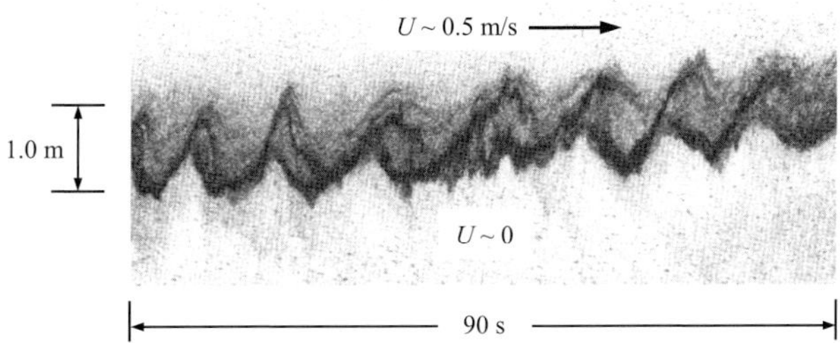

FIGURE 9.5 Natural overturning motions generated by the Kelvin-Helmholtz instability in a salt-stratified estuary at early ebb tide (see Lavery et al., 2013). The similarity of these structures to those shown in Figs. 9.4 and 9.6 is striking. The image was created from a downward looking backscatter sonar system operating in the 375 kHz to 600 kHz band on a research vessel moving upstream into the river mouth. The upper fresh water layer from the Connecticut River is ~2 m deep. The lower saltwater layer is ~6 m deep. The Reynolds number of the flow based on the vertical extent of the overturning structures is ~500,000. The gradient Richardson number here is ~1/2 so the structures are not growing at this point (see Section 9.8, and Example 9.5). This image provided by, and used with the permission of, *Jonathan Fincke & Andone C. Lavery* of *Woods Hole Oceanographic Institution.*

Figs. 9.4 and 9.5 show the advanced nonlinear stage of the instability in which the interface is a rolled-up layer of vorticity. Such evolution of the interface is in agreement with results of numerical calculations in which the nonlinear terms are retained (Fig. 9.6).

The energy source for the Kelvin-Helmholtz instability is the kinetic energy of the two streams. The disturbances evolve to smear out the gradients until they cannot grow further. Fig. 9.7 shows a typical behavior, in which the unstable waves at the interface have transformed the sharp density profile ACDF to ABEF and the sharp velocity profile MOPR to MNQR. The high-density fluid in the depth range DE has been raised upward (and mixed with the lower-density fluid in the depth range BC), which means that the potential energy of the system has increased after the action of the instability. The required energy has been drawn from the kinetic energy of the basic field, and the kinetic energy of the initial profile MOPR is larger than that of the final profile MNQR. To see this, assume that the initial velocity of the lower layer is zero and that of the upper layer is U_1. Then the linear velocity profiles after mixing are given by:

$$U(z) = \frac{U_1}{2}\left(1+\frac{z}{h}\right) \quad \text{and} \quad \overline{\rho}(z) = \rho_2 - \frac{(\rho_2-\rho_1)}{2}\left(1+\frac{z}{h}\right) \quad \text{for} \quad -h \le z \le h.$$

Consider the change in kinetic energy only in the depth range $-h < z < h$, as the energy outside this range does not change. Then the initial and final kinetic energies per unit width are:

$$E_{initial} = \frac{1}{2}\rho_1 U_1^2 h \quad \text{and} \quad E_{final} = \frac{1}{2}\int_{-h}^{+h} \overline{\rho}(z) U^2(z) dz = \frac{1}{3}\rho_1 U_1^2 h + \frac{1}{12}(\rho_2 - \rho_1) U_1^2 h.$$

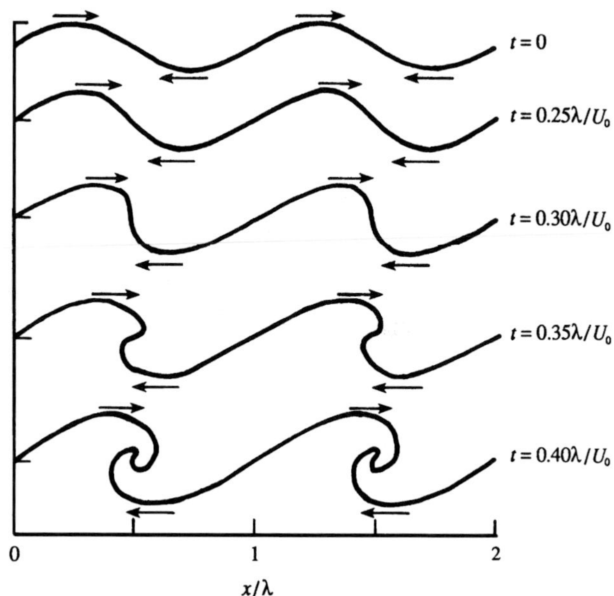

FIGURE 9.6 Nonlinear numerical calculation of the evolution of a vortex sheet that has been given a small transverse sinusoidal displacement with wavelength λ. The density difference across the interface is zero, and U_0 is the velocity difference across the vortex sheet. Here again, the similarity of the interface shape at the last time with the results shown in Figs. 9.4 and 9.5 is striking. The smaller vertical displacements shown Figs. 9.4 and 9.5 are consistent with the effects of stratification that are absent from the calculations shown in this figure. *Turner,* Buoyancy Effects in Fluids, *1973; reprinted with the permission of Cambridge University Press.*

When $(\rho_2 - \rho_1) \ll \rho_1$, the kinetic energy of the flow is clearly decreased, although the total momentum ($= \int U dz$) is nearly unchanged. This is a general result: If the integral of $U(z)$ does not change, then the integral of $U^2(z)$ decreases if the gradients decrease.

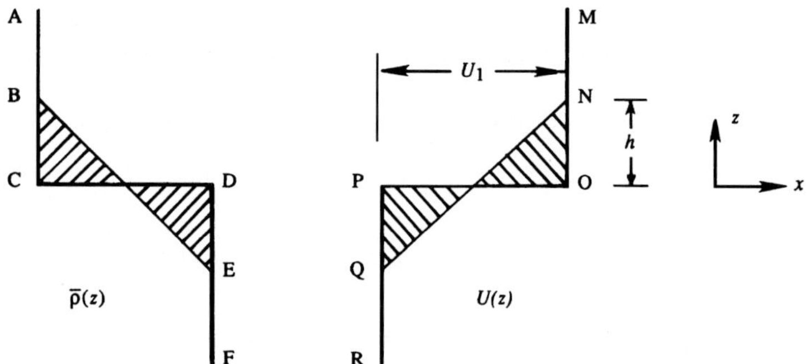

FIGURE 9.7 Smearing out of sharp density and velocity profiles, resulting in an increase of potential energy and a decrease of kinetic energy. When turbulent, the overturning eddies or structure shown in Figs. 9.4 and 9.5 lead to vertical (cross stream) momentum transport and fluid mixing. The discontinuous profiles ACDF and MOPR evolve toward ABEF and MNQR as the instability develops.

In this section, the case of a discontinuous variation across an infinitely thin interface is considered and the flow is always unstable. The case of continuous variation is considered in Section 9.8, and we shall see that one or more additional conditions must be satisfied in order for the flow to be unstable.

9.4 Rayleigh-Plateau instability

The early stage of the breakdown of a cylindrical liquid jet is referred to as the *Rayleigh-Plateau instability*. Unlike the Kelvin-Helmholtz instability, this instability is entirely driven by surface tension effects. Plateau (1873) showed experimentally that a vertically falling liquid jet is unstable to axisymmetric modes for sufficiently large wavelengths compared to its radius, eventually causing the jet to pinch-off into droplets. Rayleigh (1879) later described the dynamics of the instability, providing an expression for the growth rate of the disturbance in the inviscid case. In this section, the conditions necessary for the breakup of a liquid jet are presented.[1]

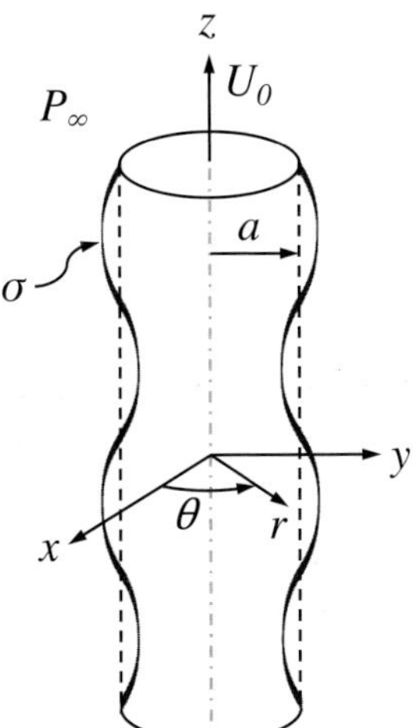

FIGURE 9.8 Basic flow configuration leading to the Rayleigh-Plateau instability. Surface tension σ acts at the interface $r = a + \zeta(\theta, z, t)$. Note: gravity is not part of this flow analysis.

Consider a cylindrical liquid column of density ρ_l subject to an initial sinusoidal perturbation as depicted in Fig. 9.8. The jet is moving with velocity U_0 in a stationary gas of density ρ_g and uniform pressure P_∞. The pressure inside the liquid can be deduced from the Laplace pressure alone:

$$p_i - p_o = \sigma\left(\frac{1}{R_1} + \frac{1}{R_2}\right), \tag{1.11}$$

where σ is the interfacial surface tension coefficient and R_1 and R_2 are the principle radii of curvature. Here, R_1 is the radius of the cylinder and R_2 is the radius of curvature of the perturbation along the z-axis. For brevity, assume both liquid and gas phases are inviscid. Thus, the unperturbed cylindrical jet will not change shape as it moves, and the liquid phase of the basic state is taken to be an infinite cylinder of radius a. From (1.11) with $R_1 = a$ and $R_2 = \infty$, the liquid pressure of the basic state is $P = P_\infty + \sigma/a$. The velocity of the basic flow (in the frame of reference of the liquid) is $\mathbf{U} = 0$ for $r < a$ and $\mathbf{U} = -U_0\mathbf{e}_z$ for $r > a$.

Use cylindrical coordinates, decompose the total flow field into its basic state plus perturbations:

$$\widetilde{\mathbf{u}} = 0 + \mathbf{u}(r, \theta, z, t) \quad \text{and} \quad \widetilde{p} = P_\infty + \frac{\sigma}{a} + p(r, \theta, z, t) \quad \text{in the liquid,} \tag{9.21}$$

$$\widetilde{\mathbf{u}} = -U_0\mathbf{e}_z + \mathbf{u}(r, \theta, z, t) \quad \text{and} \quad \widetilde{p} = P_\infty + p(r, \theta, z, t) \quad \text{in the gas.} \tag{9.22}$$

The disturbance velocity field $\mathbf{u}$ and the disturbance pressure p obey the linearized Euler equations:

$$\frac{\partial \mathbf{u}}{\partial t} + \frac{1}{\rho_l}\nabla p = 0 \quad \text{in the liquid,} \tag{9.23}$$

$$\frac{\partial \mathbf{u}}{\partial t} - U_0\frac{\partial \mathbf{u}}{\partial z} + \frac{1}{\rho_g}\nabla p = 0 \quad \text{in the gas.} \tag{9.24}$$

In addition, the velocity field in both phases satisfy $\nabla \cdot \mathbf{u} = 0$. Taking the divergence of (9.23) and (9.24) results in a Laplace equation for pressure in both phases:

$$\nabla^2 p = 0. \tag{9.25}$$

The perturbed liquid jet interface is defined at $r = a + \zeta(\theta, z, t)$. The kinematic boundary condition of the material surface is

$$\frac{D\zeta}{Dt} = u_r|_{r=a+\zeta}. \tag{9.26}$$

1. The derivation presented here is adapted from lecture notes by Prof. Olivier Desjardins.

Rewriting this condition at $r = a$ and neglecting nonlinear terms yields:

$$\frac{\partial \zeta}{\partial t} = u_r|_{r=a-0} \tag{9.27}$$

and

$$\frac{\partial \zeta}{\partial t} - U_0 \frac{\partial \zeta}{\partial z} = u_r|_{r=a+0}. \tag{9.28}$$

The dynamic boundary condition for pressure is

$$P_- + p_- = P_+ + p_+ + \sigma\kappa, \tag{9.29}$$

where $P_+ = P(a+0) = P_\infty$, $P_- = P(a-0) = P_\infty + \sigma/a$, and $\kappa = \nabla \cdot \mathbf{n}$ is the curvature of the interface. As before, the local normal to the interface can be expressed as $\mathbf{n} = [-\nabla\zeta + \mathbf{e}_r]/\sqrt{1+|\nabla\zeta|^2}$. Linearize the curvature for small departure from the initial cylinder by writing

$$\kappa \approx \frac{1}{a} - \left(\frac{\zeta}{a^2} + \frac{1}{a^2}\frac{\partial^2\zeta}{\partial\theta^2} + \frac{\partial^2\zeta}{\partial z^2}\right). \tag{9.30}$$

From (9.23) and (9.24), the pressure disturbances scale like

$$p_- \sim \rho_l \frac{l^2}{\tau^2} \quad \text{and} \quad p_+ \sim \rho_g \frac{l^2}{\tau^2}$$

for some length scale l and timescale τ. Since $\rho_l \gg \rho_g$, $p_- \gg p_+$, and thus from (9.29) and (9.30) the linearized dynamic condition is

$$p(r=a, z, t) = -\sigma\left(\frac{\zeta}{a^2} + \frac{1}{a^2}\frac{\partial^2\zeta}{\partial\theta^2} + \frac{\partial^2\zeta}{\partial z^2}\right). \tag{9.31}$$

At this point, kinematic boundary conditions at the gas-liquid interface are defined. Use the method of normal modes (see Section 9.2) to assume a solution of the form

$$\zeta(\theta, z, t) = \zeta_0 \exp\{ikz + im\theta + \gamma t\} \tag{9.32}$$

and

$$p(r, \theta, z, t) = \widehat{p}(r) \exp\{ikz + im\theta + \gamma t\}, \tag{9.33}$$

where k and m are the axial and orthoradial wave numbers, respectively, and γ is used here to denote the temporal growth rate in place of σ to avoid confusion with the surface tension coefficient. Plugging (9.33) into (9.25) leads to:

$$\frac{d^2\widehat{p}}{dr^2} + \frac{1}{r}\frac{d\widehat{p}}{dr} - \left(\frac{m^2}{r^2} + k^2\right)\widehat{p} = 0. \tag{9.34}$$

This is a form of the modified Bessel equation of order m with the radial coordinate r scaled by the axial wave number k. The solution is:

$$\widehat{p}(r) = AI_m(kr) + BK_m(kr), \tag{9.35}$$

where A and B are constants, and I_m and K_m are modified Bessel functions of the first and second kind, respectively. Here $\widehat{p}$ must be finite, so B must be zero since K_m grows without bound as its argument goes to zero. Differentiating the kinematic boundary condition (9.27) in time and applying the radial component of the momentum equation (9.23) yields

$$\frac{\partial^2\zeta}{\partial t^2} = \left.\frac{\partial u_r}{\partial t}\right|_{r=a-0} = -\frac{1}{\rho_l}\left.\frac{\partial p}{\partial r}\right|_{r=a-0}, \tag{9.36}$$

from which it follows that $\gamma^2\zeta_0 = AkI'_m(ak)/\rho_l$. Use the dynamic boundary condition (9.31) to write $AI_m(ak) = -\sigma(1 - m^2 - k^2a^2)\zeta_0/a^2$, which allows for A and ζ_0 to be determined. The growth (or decay) rate of the disturbance is therefore:

$$\gamma^2 \frac{\rho_l a^3}{\sigma} = ak(1 - m^2 - a^2k^2)\frac{I'_m(ak)}{I_m(ak)},$$

because $I_m \geq 0$ and $I'_m \geq 0$, $\gamma^2 < 0$ when $m^2 \geq 1$. Thus, no instability is possible for $m^2 \geq 1$. Setting $m = 0$ and noting $I'_0 = I_1$ (i.e., the modified Bessel function of order one) yields

$$\gamma^2 \frac{\rho_l a^3}{\sigma} = ak(1 - a^2k^2)\frac{I_1(ak)}{I_0(ak)}. \tag{9.37}$$

It can be seen that γ can only take on real values and therefore results in unbounded growth of the perturbations and eventual rupture of the jet if $ak < 1$. The corresponding non-dimensional growth rate as a function of axial wave number is shown in Fig. 9.9. It can be seen that the maximum growth rate occurs at $ak \approx 0.697$. At wave numbers $ak > 1$, *surface tension stabilizes the deflected interface*.

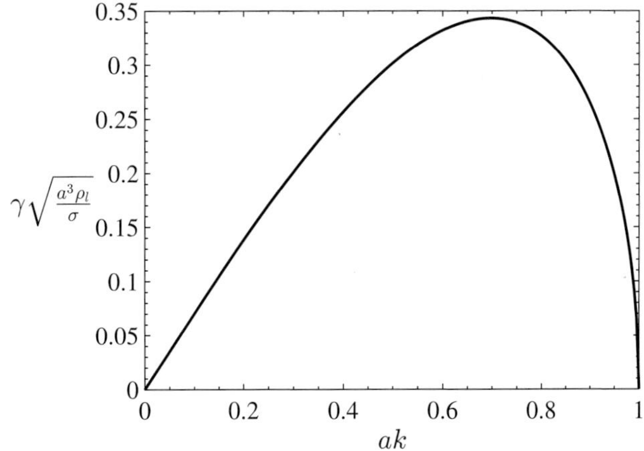

FIGURE 9.9 Growth rate γ as a function of axial wave number k for the Rayleigh-Plateau instability.

Example 9.2

A liquid column of coffee with radius a exits the spout of an espresso machine (at $z = 0$) with velocity U_0 and density ρ as depicted in Fig. 9.10. The surface tension coefficient at the coffee-air interface is σ. As seen in the figure, small disturbances along the liquid column grow downstream until the jet reaches a critical length λ_{cr} where it pinches off into a droplet. Predict λ_{cr} along with the characteristic breakup time τ_{cr}. Assume the liquid column has an initial radius $a \approx 8$ mm. Neglect gravity and viscous effects.

Solution From Fig. 9.9, the perturbation is most amplified when the wave number is $k_{cr} \approx 0.697/a$. The critical wavelength is thus

$$\lambda_{cr} = 2\pi/k_{cr} \approx 9.01a.$$

Using $a = 8$ mm, this implies $\lambda_{cr} \approx 72.12$ mm, which appears to be reasonable based on Fig. 9.10. The characteristic breakup time is $\tau_{cr} = 1/\max(\gamma)$. From Fig. 9.9, the maximum growth rate is $\max(\gamma) \approx 0.34\sqrt{\sigma/(a^3\rho_l)}$. It is interesting to note that higher surface tension causes the jet to breakup faster. Using properties for water at 90 °C ($\rho = 964$ kg/m^3 and $\sigma = 0.06$ N/m) the characteristic breakup time is $\tau_{cr} \approx 0.264$ s.

9.5 Thermal instability: the Bénard problem

In natural flows and engineering flows, heat addition to a nominally quiescent fluid from below can lead to a situation where cool dense fluid overlies warmer less-dense fluid. Eq. (9.19) indicates that such stratification will be unstable and lead to instability-driven motion when the fluid is ideal. However, when viscosity and thermal conduction are active, they may delay the onset of unstable convective motion, and only for large enough temperature gradients is the situation unstable. In this section, the conditions necessary for the onset of thermal instability in a layer of fluid are presented.

The first intensive experiments on instability caused by heating a layer of fluid from below were conducted by Bénard in 1900 (see Bénard, 1900). Bénard experimented on only very thin layers (a millimeter or less) that had a free surface,

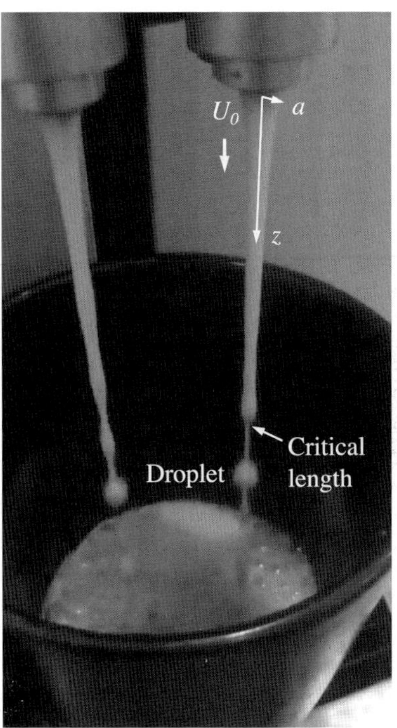

FIGURE 9.10 Rayleigh-Plateau instability resulting in the formation of droplets during a pour of an espresso shot.

and observed beautiful hexagonal cells when convection developed. Stimulated by these experiments, Rayleigh in 1916 (Rayleigh, 1916) derived the theoretical requirement for the development of convective motion in a layer of fluid with two free surfaces. He showed that the instability would occur when the adverse temperature gradient was large enough to make the ratio:

$$\mathrm{Ra} = g\alpha\Gamma d^4/\kappa\nu \tag{9.38}$$

exceed a certain critical value. Here, g is the acceleration due to gravity, α is the fluid's coefficient of thermal expansion, $\Gamma = -d\overline{T}/dz$ is the vertical temperature gradient of the background state, d is the depth of the layer, κ is the fluid's thermal diffusivity, and ν is the fluid's kinematic viscosity. The parameter Ra is called the *Rayleigh number*, and it represents a ratio of the destabilizing effect of buoyancy to the stabilizing effect of viscosity. Since Bénard's original experiments, it has been recognized that most of the motions he observed were *instabilities driven by the variation of surface tension with temperature and not the thermal instability due to a top-heavy density gradient* (Drazin and Reid, 1981, p. 34). The importance of instabilities driven by surface tension decreases as the layer becomes thicker. Later experiments on thermal convection in thicker layers (with or without a free surface) have obtained convective cells of many forms, not just hexagonal. Nevertheless, the phenomenon of thermal convection in a layer of fluid is still commonly called *Bénard convection*. Rayleigh's solution of the thermal convection problem is considered a major triumph of linear stability theory. The concept of a critical Rayleigh number finds application in such geophysical problems as solar convection, cloud formation in the atmosphere, and the motion of the earth's core.

The formulation of the problem starts with a fluid layer of thickness d confined between two isothermal walls where the lower wall is maintained at a higher temperature, T_0, than the upper wall, $T_0 - \Delta T$, where $\Delta T > 0$ (see Fig. 9.11). Use Cartesian coordinates centered in the middle of the fluid layer with the z-axis vertical, start from the Boussinesq set of equations:

$$\nabla\cdot\tilde{\mathbf{u}} = 0,\quad \frac{\partial\tilde{\mathbf{u}}}{\partial t} + (\tilde{\mathbf{u}}\cdot\nabla)\tilde{\mathbf{u}} = -\frac{1}{\rho_0}\nabla\tilde{p} - g[1-\alpha(\tilde{T}-T_0)]\mathbf{e}_z + \nu\nabla^2\tilde{\mathbf{u}},\quad \frac{\partial\tilde{T}}{\partial t} + (\tilde{\mathbf{u}}\cdot\nabla)\tilde{T} = \kappa\nabla^2\tilde{T}, \tag{4.10, 4.87, 4.90}$$

and the simplified equation for the density in terms of the temperature: $\rho = \rho_o[1-\alpha(\tilde{T}-T_0)]$, where ρ_0 and T_0 are the reference density and temperature. Here again, the total flow variables (background plus perturbation) carry a tilde ($\sim$). As before, decompose the total flow field into a motionless background plus perturbations:

$$\tilde{\mathbf{u}} = 0 + \mathbf{u}(\mathbf{x}, t), \quad \tilde{T} = \overline{T}(z) + T'(\mathbf{x}, t), \quad \text{and} \quad \tilde{p} = P(z) + p(\mathbf{x}, t). \tag{9.39}$$

The basic state is represented by a quiescent fluid with temperature and pressure distributions $\overline{T}(z)$ and $P(z)$ that satisfy the equations:

$$0 = -\frac{1}{\rho_0}\nabla P - g[1 - \alpha(\overline{T} - T_0)]\mathbf{e}_z \quad \text{and} \quad 0 = \kappa\frac{\partial^2 \overline{T}}{\partial z^2}. \tag{9.40}$$

The preceding thermal equation gives the linear vertical temperature distribution:

$$\overline{T}(z) = T_0 - \frac{1}{2}\Delta T - \Gamma z, \tag{9.41}$$

where $\Gamma \equiv \Delta T/d$ is the magnitude of the vertical temperature gradient. Substituting (9.39) into the Boussinesq equation set, and subtracting (9.40) produces:

$$\nabla\cdot\mathbf{u} = 0, \quad \frac{\partial \mathbf{u}}{\partial t} + (\mathbf{u}\cdot\nabla)\mathbf{u} = -\frac{1}{\rho_0}\nabla p + g\alpha T'\mathbf{e}_z + \nu\nabla^2\mathbf{u}, \quad \text{and} \quad \frac{\partial T'}{\partial t} - w\Gamma + (\mathbf{u}\cdot\nabla)T' = \kappa\nabla^2 T',$$

where w is the vertical component of the fluid velocity, and the $-w\Gamma$ term in the final equation comes from evaluating $(\mathbf{u}\cdot\nabla)\overline{T}$ using (9.41). For small perturbations, it is appropriate to linearize the second two equations by dropping quadratic and higher order terms:

$$\nabla\cdot\mathbf{u} = 0, \quad \frac{\partial \mathbf{u}}{\partial t} = -\frac{1}{\rho_0}\nabla p + g\alpha T'\mathbf{e}_z + \nu\nabla^2\mathbf{u}, \quad \text{and} \quad \frac{\partial T'}{\partial t} - w\Gamma = \kappa\nabla^2 T'. \tag{9.42, 9.43, 9.44}$$

These equations govern the behavior of perturbations to the basic state. A simple scaling analysis based on these equations leads to the Rayleigh number when $T' \sim \Delta T$, and $\nabla \sim 1/d$. From (9.44), the vertical velocity scale is found by equating the advective and diffusion terms:

$$w\Gamma \sim \kappa\nabla^2 T' \sim \kappa\frac{1}{d^2}\Delta T = \kappa\frac{1}{d}\frac{\Delta T}{d} = \kappa\frac{1}{d}\Gamma, \quad \text{so} \quad w \sim \kappa/d.$$

Forming a ratio of the last two terms in (9.43) leads to:

$$\frac{\textit{Buoyant force}}{\textit{viscous force}} \sim \frac{g\alpha T'}{\nu(1/d^2)w} \sim \frac{g\alpha(\Delta T/d)d}{\nu(1/d^2)(\kappa/d)} = \frac{g\alpha\Gamma d^4}{\nu\kappa} = \text{Ra}.$$

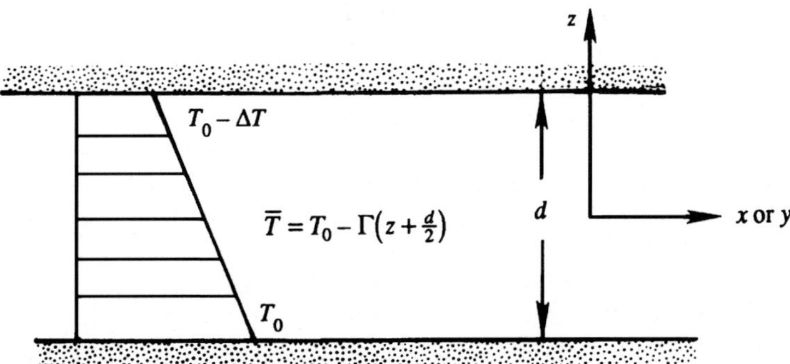

FIGURE 9.11 Flow geometry for the thermal convection between horizontal surfaces separated by a distance d. The lower surface is maintained at a higher temperature than the upper surface, and the coordinates are centered between them. For a given fluid and a fixed geometry, when the temperature difference ΔT is small, the fluid remains motionless and heat is transferred between the plates by thermal conduction. However, a sufficiently high ΔT will cause a cellular flow pattern to appear and thermal convection of heat to occur.

The perturbation equations can be written in terms of w and T' by taking the Laplacian of the z-component of (9.43):

$$\frac{\partial}{\partial t}\nabla^2 w = -\frac{1}{\rho_0}\nabla^2\frac{\partial p}{\partial z} + g\alpha\nabla^2 T' + \nu\nabla^4 w. \tag{9.45}$$

The pressure term in (9.45) can be eliminated by taking the divergence of (9.43) using (9.42):

$$\frac{\partial}{\partial t}\nabla\cdot\mathbf{u} = -\frac{1}{\rho_0}\nabla^2 p + g\alpha\frac{\partial}{\partial z}T' + \nu\nabla^2\nabla\cdot\mathbf{u}, \quad \text{or} \quad 0 = -\frac{1}{\rho_0}\nabla^2 p + g\alpha\frac{\partial}{\partial z}T',$$

and then differentiating with respect to z to obtain:

$$0 = -\frac{1}{\rho_0}\nabla^2\frac{\partial p}{\partial z} + g\alpha\frac{\partial^2 T'}{\partial z^2},$$

which can be subtracted from (9.45) to find:

$$\frac{\partial}{\partial t}\nabla^2 w = +g\alpha\nabla_H^2 T' + \nu\nabla^4 w, \tag{9.46}$$

where $\nabla_H^2 = \partial^2/\partial x^2 + \partial^2/\partial y^2$ is the horizontal Laplacian operator.

Eqs. (9.44) and (9.46) govern the development of perturbations on the system. The boundary conditions on the upper and lower rigid surfaces are that the no-slip condition is satisfied and that the walls are maintained at constant temperatures. These conditions require $u = v = w = T' = 0$ at $z = \pm d/2$. Because the conditions on u and v hold for all x and y, it follows from the continuity equation that $\partial w/\partial z = 0$ at the walls. The boundary conditions therefore can be written as:

$$w = \partial w/\partial z = T' = 0 \quad \text{on} \quad z = \pm d/2. \tag{9.47}$$

Dimensionless independent variables (denoted with an asterisk) are used in the rest of the analysis via the transformation:

$$t^* = (\kappa/d^2)t \quad \text{and} \quad (x^*, y^*, z^*) = (x/d, y/d, z/d).$$

Eqs. (9.44), (9.46), and (9.47) then become:

$$\left(\frac{\partial}{\partial t^*} - \nabla^{*2}\right)T' = \frac{\Gamma d^2}{\kappa}w, \quad \left(\frac{1}{\Pr}\frac{\partial}{\partial t^*} - \nabla^{*2}\right)\nabla^{*2}w = \frac{g\alpha d^2}{\nu}\nabla_H^{*2}T',$$
$$\text{and} \quad w = \frac{\partial w}{\partial z^*} = T' = 0 \quad \text{on} \quad z^* = \pm\frac{1}{2}, \tag{9.48, 9.49, 9.50}$$

where $\Pr \equiv \nu/\kappa$ is the Prandtl number.

The method of normal modes is now introduced. Because the coefficients in (9.48) and (9.49) are independent of x, y, and t, solutions exponential in these variables are allowed. We therefore assume normal modes given by the first version of (9.1) with $\mathbf{K} = (k, l, 0)$:

$$w = \widehat{w}(z^*)\exp\left\{ikx^* + ily^* + \sigma t^*\right\}, \quad \text{and} \quad T' = \widehat{T}(z^*)\exp\left\{ikx^* + ily^* + \sigma t^*\right\}.$$

The requirement that solutions remain bounded as $x^*, y^* \to \infty$ implies that the wave numbers k and l must be real. In other words, the normal modes must be oscillatory in the directions of unboundedness. The temporal growth rate $\sigma = \sigma_r + i\sigma_i$ is allowed to be complex. With this dependence, the differential operators in (9.48) and (9.49) primarily transform to algebraic multipliers via:

$$\partial/\partial t^* \to \sigma, \quad \nabla_H^{*2} \to -k^2 - l^2 \equiv -K^2, \quad \text{and} \quad \nabla^{*2} \to -K^2 + d^2/dz^{*2},$$

where $K = |\mathbf{K}|$ is the magnitude of the (dimensionless) horizontal wave number. Eqs. (9.48) and (9.49) then become:

$$\left(\sigma + K^2 - \frac{d^2}{dz^{*2}}\right)\widehat{T} = \frac{\Gamma d^2}{\kappa}\widehat{w} \quad \text{and} \quad \left(\frac{\sigma}{\Pr} + K^2 - \frac{d^2}{dz^{*2}}\right)\left(\frac{d^2}{dz^{*2}} - K^2\right)\widehat{w} = -\frac{g\alpha d^2 K^2}{\nu}\widehat{T}. \tag{9.51, 9.52}$$

Making the substitution $W \equiv (\Gamma d^2/\kappa)\widehat{w}$, allows (9.51) and (9.52) to be reduced to:

$$\left(\sigma + K^2 - \frac{d^2}{dz^{*2}}\right)\widehat{T} = W \quad \text{and} \quad \left(\frac{\sigma}{\Pr} + K^2 - \frac{d^2}{dz^{*2}}\right)\left(\frac{d^2}{dz^{*2}} - K^2\right)W = -RaK^2\widehat{T}. \tag{9.53, 9.54}$$

The boundary conditions (9.50) become:

$$W = \partial W/\partial z^* = \widehat{T} = 0 \quad \text{on} \quad z = \pm 1/2. \tag{9.55}$$

Here we note that σ is real for Ra > 0 (see Exercise 9.7). The Bénard problem is one of two well-known problems in which σ is real. (The other one is Taylor-Couette flow between rotating cylinders, discussed in the following section.) In most other problems σ is complex, and the marginal state ($\sigma_r = 0$) contains propagating waves (as is true for the Kelvin-Helmholtz instability). In the Bénard and Taylor problems, however, the marginal state corresponds to $\sigma = 0$, and is therefore *stationary* and does not contain propagating waves. In these flows, the onset of instability is marked by a transition from the background state to another *steady* state. In such a case we commonly say that the *principle of exchange of stabilities* is valid, and the instability sets in as a *cellular convection*, which will be explained shortly.

Two solutions for Rayleigh-Bénard flow are presented in the remainder of this section. First, the solution is presented for the case that is easiest to realize in a laboratory experiment, namely, a layer of fluid confined between two rigid plates on which no-slip conditions are satisfied. The solution to this problem was first given by Jeffreys in 1928. The second solution for a layer of fluid with two stress-free surfaces is presented after the first.

For the marginal state $\sigma = 0$, the equation pair (9.53) and (9.54) become:

$$\left(\frac{d^2}{dz^{*2}} - K^2\right)\widehat{T} = -W \quad \text{and} \quad \left(\frac{d^2}{dz^{*2}} - K^2\right)^2 W = \text{Ra}K^2\widehat{T}. \tag{9.56}$$

Eliminating $\widehat{T}$ leads to:

$$\left(\frac{d^2}{dz^{*2}} - K^2\right)^3 W = -\text{Ra}K^2 W, \tag{9.57}$$

and the boundary condition (9.55) becomes:

$$W = \partial W/\partial z^* = (d^2/dz^{*2} - K^2)^2 W = 0 \quad \text{on} \quad z^* = \pm 1/2. \tag{9.58}$$

We have a sixth-order homogeneous differential equation with six homogeneous boundary conditions. Non-zero solutions for such a system can only exist for a particular value of Ra (for a given K). It is therefore an eigenvalue problem. Note that the Prandtl number has dropped out of the marginal state.

The point to observe is that the problem is symmetric with respect to the two boundaries, thus the eigenfunctions fall into two distinct classes – those with the vertical velocity symmetric about the midplane $z = 0$, and those with the vertical velocity antisymmetric about the midplane (Fig. 9.12). The gravest even mode therefore has one row of cells, and the gravest odd mode has two rows of cells. It can be shown that the smallest critical Rayleigh number is obtained by assuming disturbances in the form of the gravest even mode, which also agrees with experimental findings of a single row of cells.

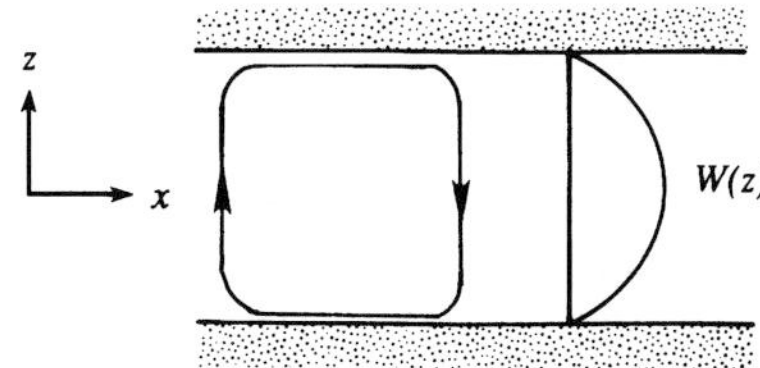

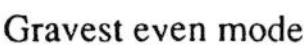

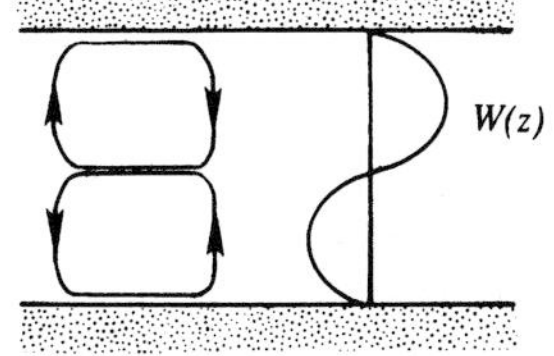

FIGURE 9.12 Flow pattern and eigenfunction structure of the gravest even mode and the gravest odd mode in the Bénard problem. The even mode is observed first as the temperature difference between the surfaces is increased.

Because the coefficients of the governing equation (9.57) are independent of z, the general solution can be expressed as a superposition of solutions of the form: $W \propto \exp(qz^*)$, where the six roots of q are found from:

$$(q^2 - K^2)^3 = -\text{Ra}K^2.$$

The three roots of this equation for q^2 are:

$$q^2 = -K^2\left[\left(\frac{\text{Ra}}{K^4}\right)^{1/3} - 1\right] \quad \text{and} \quad q^2 = K^2\left[1 + \frac{1}{2}\left(\frac{\text{Ra}}{K^4}\right)^{1/3}\left(1 \pm i\sqrt{3}\right)\right]. \tag{9.59}$$

Taking square roots, the six roots for q are $\pm iq_0$, $\pm q$, and $\pm q^*$, where:

$$q_0 = K\left[\left(\frac{\text{Ra}}{K^4}\right)^{1/3} - 1\right]^{1/2}$$

and q and its complex conjugate q^* are given by the two roots of the second part of (9.59).

The even solution of (9.57) is therefore:

$$W = A\cos q_o z^* + B\cosh q z^* + C\cosh q^* z^*,$$

where A, B, and C are constants. To apply the boundary conditions on this solution, we find the following derivatives:

$$dW/dz^* = -Aq_0\sin q_0 z^* + Bq\sinh q z^* + Cq^*\sinh q^* z^*, \quad \text{and}$$
$$(d^2/dz^{*2} - K^2)^2 W = A(q_0^2 + K^2)^2\cos q_0 z^* + B(q^2 - K^2)^2\cosh q z^* + B(q^2 - K^2)^2\cosh q^* z^*.$$

The boundary conditions (9.58) then require:

$$\begin{bmatrix} \cos\frac{q_0}{2} & \cosh\frac{q}{2} & \cosh\frac{q^*}{2} \\ -q_0\sin\frac{q_0}{2} & q\sinh\frac{q}{2} & q^*\sinh\frac{q^*}{2} \\ (q_0^2 + K^2)^2\cos\frac{q_0}{2} & (q^2 - K^2)^2\cosh\frac{q}{2} & (q^{*2} - K^2)^2\cosh\frac{q^*}{2} \end{bmatrix} \begin{bmatrix} A \\ B \\ C \end{bmatrix} = 0.$$

Here, A, B, and C cannot all be zero if we want to have a non-zero solution, which requires that the determinant of the matrix must vanish. This gives a relation between Ra and the corresponding eigenvalue K (Fig. 9.13). Points on the curve $K(\text{Ra})$ represent marginally stable states, which separate regions of stability and instability. The lowest value of Ra for marginal stability is found to be:

$$\text{Ra}_{\text{cr}} = 1708$$

attained at $K_{\text{cr}} = 3.12$. As *all* values of K are allowed by the system, the flow first becomes unstable when the Rayleigh number reaches Ra_{cr}. The wavelength at the onset of instability is: $\lambda_{\text{cr}} = 2\pi d/K_{\text{cr}} \cong 2d$. Laboratory experiments agree remarkably well with these predictions, and the solution of the Bénard problem is considered one of the major successes of the linear stability theory.

The solution for a fluid layer with stress-free surfaces is somewhat simpler and was first given by Rayleigh. This case can be approximately realized in a laboratory experiment if the layer of liquid heated from below is floating on top of a somewhat heavier liquid. Here the boundary conditions are $w = T' = \mu(\partial u/\partial z^* + \partial w/\partial x^*) = \mu(\partial v/\partial z^* + \partial w/\partial y^*) = 0$ at the surfaces, the latter two conditions resulting from zero stress. Because w vanishes (for all x and y) on the boundaries, it follows that the vanishing stress conditions require $\partial u/\partial z^* = \partial v/\partial z^* = 0$ at the boundaries. On differentiating the continuity equation with respect to z, it follows that $\partial^2 w/\partial z^{*2} = 0$ on the free surfaces. In terms of the complex amplitudes, the eigenvalue problem is therefore defined by (9.56) and with boundary conditions:

$$W = (d^2/dz^{*2} - K^2)^2 W = d^2W/dz^{*2} = 0 \quad \text{on} \quad z^* = \pm 1/2.$$

By expanding and simplifying the products of operators, the boundary conditions can be rewritten as:

$$W = d^2W/dz^{*2} = d^4W/dz^{*4} = 0 \quad \text{on} \quad z^* = \pm 1/2, \tag{9.60}$$

which should be compared with the conditions (9.58) for rigid boundaries.

Successive differentiation of (9.57) shows that *all* even derivatives of W vanish on the boundaries. The eigenfunctions must therefore be:

$$W = A\sin(n\pi z),$$

where A is any constant and n is an integer. Substitution into (9.57) leads to the eigenvalue relation:

$$\text{Ra} = (n^2\pi^2 + K^2)^3/K^2, \tag{9.61}$$

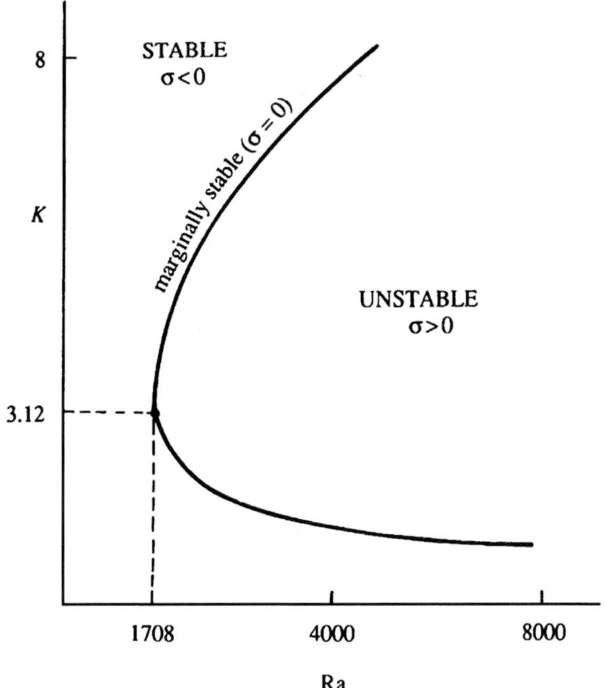

FIGURE 9.13 Stable and unstable regions for Bénard convection in a plot of the dimensionless wave number K vs. Ra, the Rayleigh number (9.38). The lowest possible Ra value for which the flow may be unstable is 1708, and the wave number of the first mode of instability is $3.12/d$, where d is the separation between the horizontal surfaces.

which gives the Rayleigh number in the marginal state. For a given K^2, the lowest value of Ra occurs when $n = 1$, which is the gravest mode. The critical Rayleigh number is obtained by finding the minimum value of Ra as K^2 is varied, that is, by setting $d\text{Ra}/dK^2 = 0$:

$$\frac{d\text{Ra}}{dK^2} = \frac{3(\pi^2 + K^2)^2}{K^2} - \frac{3(\pi^2 + K^2)^3}{K^4} = 0,$$

which requires $K_{cr}^2 = \pi^2/2$. The corresponding value of Ra is:

$$\text{Ra}_{cr} = (27/4)\pi^4 = 657.5.$$

For a layer with a free upper surface (where the stress is zero) and a rigid bottom wall, the solution of the eigenvalue problem gives $\text{Ra}_{cr} = 1101$ and $K_{cr} = 2.68$. This case is of interest in laboratory experiments having the most visual effects, as originally conducted by Bénard.

The linear theory specifies the horizontal wavelength at the onset of instability, but not the horizontal pattern of the convective cells. This is because a given wave number vector **K** can be decomposed into two orthogonal components in an infinite number of ways. If we assume that the experimental conditions are horizontally isotropic, with no preferred directions, then regular polygons in the form of equilateral triangles, squares, and regular hexagons are all possible structures. Bénard's original experiments showed only hexagonal patterns, but we now know that he was observing a different phenomenon. The observations summarized in Drazin and Reid (1981) indicate that hexagons frequently predominate initially. As Ra is increased, the cells tend to merge and form rolls, on the walls of which the fluid rises or sinks (Fig. 9.14). The cell structure becomes more chaotic as Ra is increased further, and the flow becomes turbulent when $\text{Ra} > 5 \times 10^4$.

The magnitude or direction of flow in the cells cannot be predicted by linear theory. After a short time of exponential growth, the flow becomes fast enough for the nonlinear terms to be important and it reaches a nonlinear equilibrium stage. The flow pattern for a hexagonal cell is sketched in Fig. 9.15. Particles in the middle of the cell usually rise in a liquid and fall in a gas. This has been attributed to the property that the viscosity of a liquid decreases with temperature, whereas that of a gas increases with temperature. The rising fluid loses heat by thermal conduction at the top wall, travels horizontally, and then sinks. For a steady cellular pattern, the continuous generation of kinetic energy is balanced by viscous dissipation.

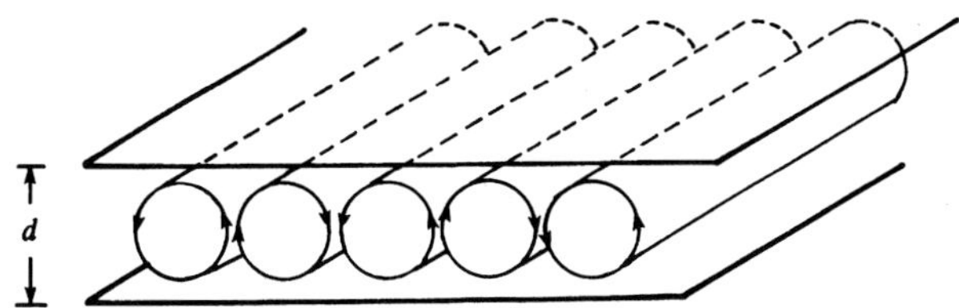

FIGURE 9.14 Two-dimensional convection rolls in Bénard convection. Fluid alternately ascends and descends between the rolls. The horizontal spacing between roll centers is nearly the same as the spacing between the horizontal surfaces.

The generation of kinetic energy is maintained by continuous release of potential energy due to heating at the bottom and cooling at the top.

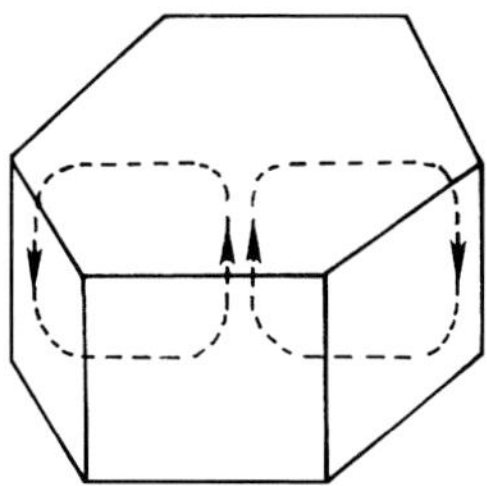

FIGURE 9.15 Above the critical Rayleigh number, complicated flow patterns may exist because a range of wave numbers is unstable for the first mode. A commonly observed Bénard-convection flow pattern involves hexagonal cells. One such cell is shown here.

Example 9.3

A horizontal sky-light is made from two panes of glass 2.0 cm apart that trap a layer of air at 1 atm pressure. In the winter when the air outside is cold and the building interior is warm, Bénard convection is possible between the panes. When it occurs, it increases the heat transfer rate out of the building interior. What is the onset temperature difference for Bénard convection? What changes to the skylight design would increase this temperature difference?

Solution The critical Rayleigh number for Bénard convection is 1708. Thus, set:

$$\mathrm{Ra} = 1708 = \frac{g\alpha\Gamma d^4}{\kappa\nu} = \frac{g\alpha d^4}{\kappa\nu}\left|\frac{\Delta T}{d}\right|.$$

Using the room-temperature properties of air (see Appendix A.4) for the given geometry, this implies:

$$|\Delta T| = 1708\frac{\kappa\nu}{g\alpha d^3} = 1708\frac{(2.08\times10^{-5}\ \mathrm{m^2s^{-1}})(1.50\times10^{-5}\ \mathrm{m^2s^{-1}})}{(9.81\ \mathrm{ms^{-2}})(3.41\times10^{-4}\ \mathrm{K^{-1}})(0.02\ \mathrm{m})^3} \approx 20\ \mathrm{K}.$$

Here, g and α cannot be readily altered. However, κ and ν may be increased by placing a different (lighter) gas between the panes or by lowering the pressure between the panes. However, the most effective means for increasing the onset value of ΔT is to decrease the gap d. This could be done by moving the two panes closer together, or by adding one (or more) additional panes between the two existing ones.

9.6 Double-diffusive instability

An interesting instability results when the density of the fluid depends on two opposing gradients. The possibility of this phenomenon was first suggested by Stommel et al. (1956), but the dynamics of the process was first explained by Stern (1960). Turner (1973), and review articles by Huppert and Turner (1981), and Turner (1985) discuss the dynamics of this phenomenon and its applications to various fields such as astrophysics, engineering, and geology. Historically, the phenomenon was first suggested with oceanic application in mind, and this is how we shall present it. For sea water, the density depends on the temperature $\widetilde{T}$ and salt content $\widetilde{s}$ (kilograms of salt per kilograms of water), so that the density is given by:

$$\widetilde{\rho} = \rho_0[1 - \alpha(\widetilde{T} - T_0) + \beta(\widetilde{s} - S_0)],$$

where the value of α determines how fast the density decreases with temperature, and the value of β determines how fast the density increases with salinity. As defined here, both α and β are positive. The key factor in this instability is that the diffusivity κ_s of salt in water is only 1% of the thermal diffusivity κ. *Such a system can be unstable even when the density decreases upwards.* By means of the instability, the flow releases the potential energy of the *component* that is "heavy at the top." Therefore, the effect of diffusion in such a system can be to *destabilize* a stable density gradient. This is in contrast to a medium containing a single diffusing component, for which the analysis of the preceding section shows that the effect of diffusion is to *stabilize* the system even when it is heavy at the top.

Consider the two situations of Fig. 9.16, both of which can be unstable although each is stably stratified in density ($d\overline{\rho}/dz < 0$). Consider first the case of hot and salty water lying over cold and fresh water (Fig. 9.16a), that is, when the system is top heavy in salt. In this case both $d\overline{T}/dz$ and dS/dz are positive, and we can arrange the composition of water such that the density decreases upward. Because $\kappa_s \ll \kappa$, a displaced particle would be near thermal equilibrium with the surroundings, but would exchange negligible salt. A rising particle therefore would be constantly lighter than the surroundings because of its salinity deficit, and would continue to rise. A parcel displaced downward would similarly continue to plunge downward because of its excess salinity. The basic state shown in Fig. 9.16a is therefore unstable. Laboratory observations show that the instability in this case appears in the form of a forest of long narrow convective cells, called *salt fingers* (Fig. 9.17). Shadowgraph images in the deep ocean have confirmed their existence in nature.

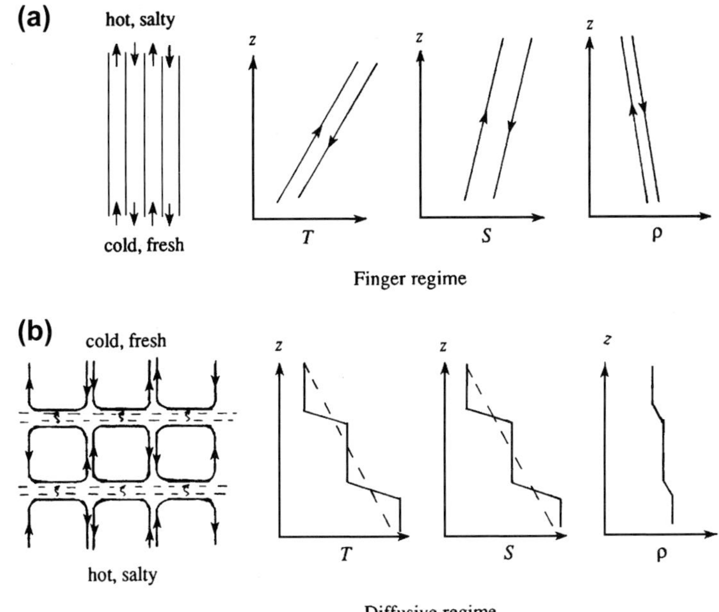

FIGURE 9.16 Two kinds of double-diffusive instabilities. (a) Finger instability, showing up- and down-going salt fingers and their temperature, salinity, and density. Arrows indicate the direction of fluid motion. (b) Oscillating instability, finally resulting in a series of convecting layers separated by "diffusive" interfaces. Across these interfaces T and S vary sharply, but heat is transported much faster than salt.

We can derive a criterion for instability by generalizing the Bénard convection analysis so as to include salt diffusion. Assume a layer of depth d confined between stress-free boundaries maintained at constant temperature and constant salinity. If we repeat the derivation of the perturbation equations for the normal modes of the system, the equations that replace (9.56) are found to be:

$$\left(\frac{d^2}{dz^{*2}} - K^2\right)\widehat{T} = -W, \quad \frac{\kappa_s}{\kappa}\left(\frac{d^2}{dz^{*2}} - K^2\right)\widehat{s} = -W, \quad \text{and}$$
$$\left(\frac{d^2}{dz^{*2}} - K^2\right)^2 W = -\mathrm{Ra}K^2\widehat{T} + Rs'K^2\widehat{s}, \tag{9.62}$$

where $\widehat{s}(z)$ is the complex amplitude of the salinity perturbation, and we have defined:

$$\mathrm{Ra} \equiv \frac{g\alpha d^4(d\overline{T}/dz)}{\nu\kappa} \quad \text{and} \quad \mathrm{Rs}' \equiv \frac{g\beta d^4(dS/dz)}{\nu\kappa}.$$

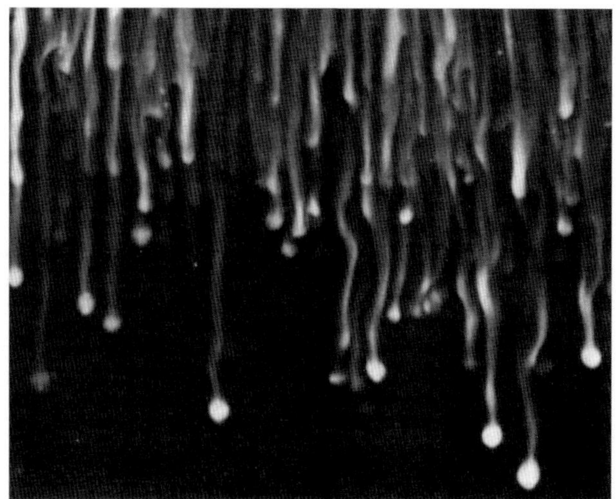

FIGURE 9.17 Salt fingers, produced by pouring a salt solution on top of a stable temperature gradient. Flow visualization by fluorescent dye and a horizontal beam of light. *J. Turner*, Naturwissenschaften, 72, *70–75, 1985; reprinted with the permission of Springer-Verlag GmbH & Co.*

Note that κ (and not κ_s) appears in the definition of Rs′. In contrast to (9.62), a positive sign appeared in (9.56) in front of Ra because in the preceding section Ra was defined to be positive for a top-heavy situation.

It is seen from the first two equations of (9.62) that the equations for $\widehat{T}$ and $\widehat{s}\kappa_s/\kappa$ are the same. The boundary conditions are also the same for these variables:

$$\widehat{T} = \widehat{s}\kappa_s/\kappa = 0 \quad \text{at} \quad z^* = \pm 1/2.$$

It follows that we must have $\widehat{T} = \widehat{s}\kappa_s/\kappa$ everywhere. Eqs. (9.62) therefore become:

$$(d^2/dz^{*2} - K^2)\widehat{T} = -W \quad \text{and} \quad (d^2/dz^{*2} - K^2)^2 W = (\mathrm{Rs} - \mathrm{Ra})K^2\widehat{T},$$

where:

$$\mathrm{Rs} \equiv \frac{\kappa}{\kappa_s}\mathrm{Rs}' = \frac{g\beta d^4(dS/dz)}{\nu\kappa_s}.$$

The preceding set is now identical to the set (9.56) for the Bénard convection, with (Rs – Ra) replacing Ra. For stress-free boundaries, the preceding section shows that the critical value is:

$$\mathrm{Rs} - \mathrm{Ra} = \frac{27}{4}\pi^4 = 657,$$

which can be written as:

$$\frac{gd^4}{\nu}\left[\frac{\beta}{\kappa_s}\frac{dS}{dz} - \frac{\alpha}{\kappa}\frac{d\overline{T}}{dz}\right] = 657. \tag{9.63}$$

Even if $\alpha(d\overline{T}/dz) - \beta(dS/dz) > 0$ (i.e., $\overline{\rho}$ decreases upward), the condition (9.63) can be quite easily satisfied because κ_s is much smaller than κ. The flow can therefore be made unstable simply by making d large enough and ensuring that the factor within [,]-brackets is positive.

The analysis predicts that the lateral width of the cell is of the order of d, but such wide cells are not observed at supercritical stages when (Rs – Ra) far exceeds 657. Instead, long thin salt fingers are observed, as shown in Fig. 9.17. If the salinity gradient is large, then experiments as well as calculations show that a deep layer of salt fingers becomes unstable and breaks down into a series of convective layers, with fingers confined to the interfaces. Oceanographic observations frequently show a series of staircase-shaped vertical distributions of salinity and temperature; with a positive overall dS/dz and $d\overline{T}/dz$ such distributions can indicate salt finger activity.

The double-diffusive instability may also occur when cold fresh water overlays hot salty water (Fig. 9.16b). In this case both $d\overline{T}/dz$ and dS/dz are negative, and we can choose their values such that the density decreases upward. Again the system is unstable, but the dynamics are different. A particle displaced upward loses heat but no salt. Thus it becomes heavier than the surroundings and buoyancy forces it back toward its initial position, resulting in an oscillation. However, a stability calculation shows that less than perfect heat conduction results in a growing oscillation, although some energy is dissipated. In this case the growth rate σ is complex, in contrast to the situation of Fig. 9.16a where it is real.

Laboratory experiments show that the initial oscillatory instability does not last long, and eventually results in the formation of a number of horizontal *convecting layers*, as sketched in Fig. 9.16b. Consider the situation when a stable salinity gradient in an isothermal fluid is heated from below (Fig. 9.18). The initial instability starts as a growing oscillation near the bottom. As the heating is continued beyond the initial appearance of the instability, a well-mixed layer develops, capped by a salinity step, a temperature step, and no density step. The heat flux through this step forms a thermal boundary layer, as shown in Fig. 9.18. As the well-mixed layer grows, the temperature step across the thermal boundary layer becomes larger. Eventually, the Rayleigh number across the thermal boundary layer becomes critical, and a second convecting layer forms on top of the first. The second layer is maintained by heat flux (and negligible salt flux) across a sharp laminar interface on top of the first layer. This process continues until a stack of horizontal layers forms one upon another. From comparison with the Bénard convection, it is clear that inclusion of a stable salinity gradient has prevented a complete overturning from top to bottom.

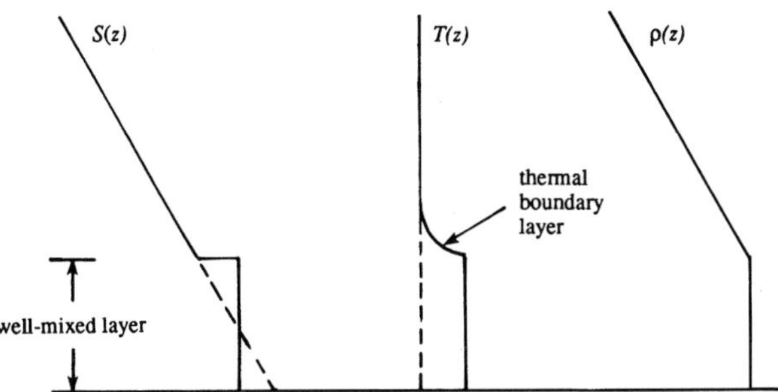

FIGURE 9.18 Distributions of salinity, temperature, and density generated by heating a linear salinity gradient from below. As heating continues the mixed layer depth will increase until a second mixed layer forms. Eventually, the flow pattern sketched and described in Fig. 9.16b forms. Top to bottom overturning motion is not possible because of the overall stratification.

The two examples in this section show that in a double-component system in which the diffusivities for the two components are different, the effect of diffusion can be destabilizing, even if the system is judged hydrostatically stable. In contrast, diffusion is stabilizing in a single-component system, such as the Bénard system. The two requirements for the double-diffusive instability are that the diffusivities of the components be different, and that the components make opposite contributions to the vertical density gradient.

9.7 Centrifugal instability: Taylor problem

In this section we shall consider the instability of Couette flow between concentric rotating cylinders, a problem first solved by G.I. Taylor in 1923. In many ways the problem is similar to the Bénard problem, in which there is a potentially unstable arrangement of temperature. In the Couette-flow problem the source of the instability is the unstable arrangement of angular momentum. Whereas convection in a heated layer is brought about by buoyant forces becoming large enough to overcome the viscous resistance, the convection in a Couette flow is generated by the centrifugal forces being able to overcome the viscous forces. We shall first present Rayleigh's discovery of an inviscid stability criterion for the problem and then outline Taylor's solution of the viscous case. Experiments indicate that the instability initially appears in the form of axisymmetric disturbances, for which $\partial/\partial\theta = 0$. Accordingly, we shall limit ourselves only to the axisymmetric case.

The problem was first considered by Rayleigh in 1888. Neglecting viscous effects, he discovered the source of instability for this problem and demonstrated a necessary and sufficient condition for instability. Let $U_\theta(r)$ be the angular-directed velocity in the r-θ plane at any radial distance from the origin. For inviscid flows $U_\theta(r)$ can be any function, but only certain distributions can be stable. Imagine that two fluid rings with equal mass at radial distances r_1 and $r_2(> r_1)$ are interchanged. As the motion is inviscid, Kelvin's theorem requires that the circulation $\Gamma = 2\pi r U_\theta$ (proportional to the angular momentum rU_θ) should remain constant during the interchange. That is, after the interchange, the fluid at r_2 will have the circulation (namely, Γ_1) that it had at r_1 before the interchange. Similarly, the fluid at r_1 will have the circulation (namely, Γ_2) that it had at r_2 before the interchange. Conservation of circulation requires that the kinetic energy E must change during the interchange. Because $E = U_\theta^2/2 = \Gamma^2/8\pi^2 r^2$, we have:

$$E_{\text{initial}} = \frac{1}{8\pi^2}\left[\frac{\Gamma_1^2}{r_1^2} + \frac{\Gamma_2^2}{r_2^2}\right] \quad \text{and} \quad E_{\text{final}} = \frac{1}{8\pi^2}\left[\frac{\Gamma_2^2}{r_1^2} + \frac{\Gamma_1^2}{r_2^2}\right],$$

so that the kinetic energy change per unit mass is:

$$\Delta E = E_{\text{final}} - E_{\text{initial}} = \frac{1}{8\pi^2}(\Gamma_2^2 - \Gamma_1^2)\left(\frac{1}{r_1^2} - \frac{1}{r_2^2}\right).$$

Because $r_2 > r_1$, a velocity distribution for which $\Gamma_2^2 > \Gamma_1^2$ would make ΔE positive, and this implies that an external source of energy would be necessary to perform the interchange of the fluid rings. Under this condition a *spontaneous* interchange of the rings is not possible, and the flow is stable. On the other hand, if Γ^2 decreases with r, then an interchange of rings will result in a release of energy; such a flow is unstable. It can be shown that in this situation the centrifugal force in the new location of an outwardly displaced ring is larger than the prevailing (radially inward) pressure gradient force.

Rayleigh's criterion can therefore be stated as follows: *An inviscid Couette flow is unstable if:*

$$d\Gamma^2/dr < 0 \quad \text{(unstable)}.$$

The criterion is analogous to the inviscid requirement for static instability in a density-stratified fluid:

$$d\overline{\rho}/dz > 0 \quad \text{(unstable)}.$$

Therefore, the stratification of angular momentum in a Couette flow is unstable if it decreases radially outwards. When the outer cylinder is held stationary and the inner cylinder is rotated, $d\Gamma^2/dr < 0$ and Rayleigh's criterion implies that the flow is inviscidly unstable. As in the Bénard problem, however, merely having a potentially unstable arrangement does not cause instability in a viscous medium.

This inviscid Rayleigh criterion is modified in a viscous version of the problem. Taylor's solution of the viscous problem is outlined in what follows. Using cylindrical polar coordinates (R, ϕ, z) and assuming axial symmetry, the equations of motion are:

$$\frac{D\tilde{u}_R}{Dt} - \frac{\tilde{u}_\phi^2}{R} = -\frac{1}{\rho}\frac{\partial\tilde{p}}{\partial R} + \nu\left(\nabla^2\tilde{u}_R - \frac{\tilde{u}_R}{R^2}\right), \quad \frac{D\tilde{u}_\phi}{Dt} + \frac{\tilde{u}_R\tilde{u}_\phi}{R} = \nu\left(\nabla^2\tilde{u}_\phi - \frac{\tilde{u}_\phi}{R^2}\right),$$
$$\frac{D\tilde{u}_z}{Dt} = -\frac{1}{\rho}\frac{\partial\tilde{p}}{\partial z} + \nu\nabla^2\tilde{u}_z, \quad \text{and} \quad \frac{1}{R}\frac{\partial}{\partial R}(R\tilde{u}_R) + \frac{\partial\tilde{u}_z}{\partial z} = 0, \tag{9.64}$$

where:

$$\frac{D}{Dt} = \frac{\partial}{\partial t} + \tilde{u}_R\frac{\partial}{\partial R} + \tilde{u}_z\frac{\partial}{\partial z} \quad \text{and} \quad \nabla^2 = \frac{\partial^2}{\partial R^2} + \frac{1}{R}\frac{\partial}{\partial R} + \frac{\partial^2}{\partial z^2}.$$

Decompose the motion into a background state plus perturbation:

$$\tilde{\mathbf{u}} = \mathbf{U} + \mathbf{u} \quad \text{and} \quad \tilde{p} = P + p. \tag{9.65}$$

The background state is given by (see Section 9.2):

$$U_R = U_z = 0, \quad U_\phi = AR + B/R \quad \text{and} \quad \frac{1}{\rho}\frac{dP}{dR} = \frac{U_\phi^2}{R}, \tag{9.66}$$

where:

$$A = \frac{\Omega_2 R_2^2 - \Omega_1 R_1^2}{R_2^2 - R_1^2}, \quad \text{and} \quad B = -\frac{(\Omega_2 - \Omega_1)R_1^2 R_2^2}{R_2^2 - R_1^2}.$$

Here, Ω_1 and Ω_2 are the angular speeds of the inner and outer cylinders, respectively, and R_1 and R_2 are their radii (Fig. 9.19).

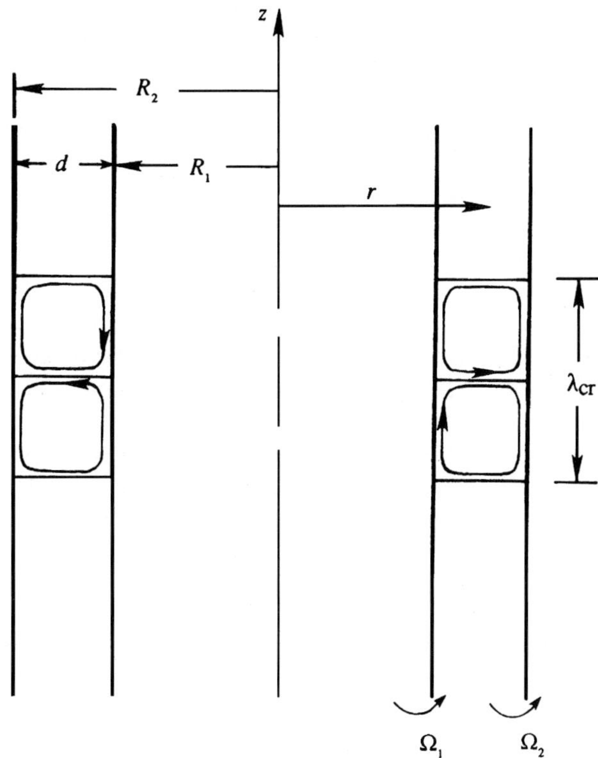

FIGURE 9.19 Geometry of the flow and the instability in rotating Couette flow. The fluid resides between rotating cylinders with radii R_1 and R_2. As for Bénard convection, the resulting instability forms as counter-rotating rolls with a wavelength that is approximately twice the gap between the cylinders.

Substituting (9.65) into (9.64), neglecting nonlinear terms, and subtracting the background state (9.66), we obtain the perturbation equations:

$$\frac{\partial u_R}{\partial t} - \frac{2U_\phi u_\phi}{R} = -\frac{1}{\rho}\frac{\partial p}{\partial R} + \nu\left(\nabla^2 u_R - \frac{u_R}{R^2}\right), \quad \frac{\partial u_\phi}{\partial t} + \left(\frac{dU_\phi}{dR} + \frac{U_\phi}{R}\right)u_R = \nu\left(\nabla^2 u_\phi - \frac{u_\phi}{R^2}\right),$$
$$\frac{\partial u_z}{\partial t} = -\frac{1}{\rho}\frac{\partial p}{\partial z} + \nu\nabla^2 u_z, \quad \text{and} \quad \frac{1}{R}\frac{\partial}{\partial R}(Ru_R) + \frac{\partial u_z}{\partial z} = 0. \tag{9.67}$$

As the coefficients in these equations depend only on R, the equations admit solutions that depend on z and t exponentially. We therefore consider normal mode solutions of the form:

$$(u_R, u_\phi, u_z, p) = (\widehat{u}_R(R), \widehat{u}_\phi(R), \widehat{u}_z(R), \widehat{p}(R))\exp\{ikz + \sigma t\}.$$

The requirement that the solutions remain bounded as $z \to \pm\infty$ implies that the axial wave number k must be real. After substituting the normal modes into (9.67) and eliminating $\widehat{u}_z$ and $\widehat{p}$, we get a coupled system of equations in $\widehat{u}_R$ and $\widehat{u}_\phi$. Under the *narrow-gap approximation*, for which $d = R_2 - R_1$ is much smaller than $(R_1 + R_2)/2$, these equations finally become (see Chandrasekhar, 1961 for details):

$$(d^2/dR^2 - k^2 - \sigma)(d^2/dR^2 - k^2)\widehat{u}_R = (1 + \alpha x)\widehat{u}_\phi, \quad \text{and} \quad (d^2/dR^2 - k^2 - \sigma)\widehat{u}_\phi = -\mathrm{Ta}k^2\widehat{u}_R, \tag{9.68}$$

where:

$$\alpha \equiv (\Omega_2/\Omega_1) - 1, \quad x \equiv (R - R_1)/d, \quad d \equiv R_2 - R_1,$$

and Ta is the Taylor number:

$$\mathrm{Ta} \equiv 4\left(\frac{\Omega_1 R_1^2 - \Omega_2 R_2^2}{R_2^2 - R_1^2}\right)\frac{\Omega_1 d^4}{\nu^2}. \tag{9.69}$$

It is the ratio of the centrifugal force to viscous force, and equals $2(\Omega_1 R_1 d/\nu)^2(d/R_1)$ when only the inner cylinder is rotating and the gap is narrow.

The boundary conditions are:

$$\widehat{u}_R = d\widehat{u}_R/dR = \widehat{u}_\phi = 0 \quad \text{at} \quad x = 0 \quad \text{and} \quad x = 1. \tag{9.70}$$

The eigenvalues k at the marginal state are found by setting the real part of σ to zero. On the basis of experimental evidence, Taylor assumed that the marginal states are given by $\sigma = 0$. This was later proven to be true for cylinders rotating in the same directions, but a general demonstration for all conditions is still lacking.

A solution of the eigenvalue problem (9.68), subject to (9.70), was obtained by Taylor. Fig. 9.20 shows the results of his calculations and his own experimental verification of the analysis. The vertical axis represents the angular velocity of the inner cylinder (taken positive), and the horizontal axis represents the angular velocity of the outer cylinder. Cylinders rotating in opposite directions are represented by a negative Ω_2. Taylor's solution of the marginal state is indicated, with the region above the curve corresponding to instability. Rayleigh's inviscid criterion is also indicated by the straight dashed line. Taylor's viscous solution indicates that the flow remains stable until a critical Taylor number of:

$$\mathrm{Ta}_{\mathrm{cr}} = \frac{1708}{(1/2)(1 + \Omega_2/\Omega_1)} \tag{9.71}$$

is attained. The non-dimensional axial wave number at the onset of instability is found to be $k_{\mathrm{cr}} = 3.12$, which implies that the wavelength at onset is $\lambda_{\mathrm{cr}} = 2\pi d/k_{\mathrm{cr}} \approx 2d$. The height of one cell is therefore nearly equal to d, so that the cross-section of a cell is nearly a square. In the limit $\Omega_2/\Omega_1 \to 1$, the critical Taylor number is identical to the critical Rayleigh number for thermal convection discussed in Section 9.5, for which the solution was given by Jeffreys five years later. The agreement is expected, because in this limit $\alpha = 0$, and the eigenvalue problem (9.68) reduces to that of the Bénard problem (9.56). For cylinders rotating in opposite directions the Rayleigh criterion predicts instability, but the viscous solution can be stable.

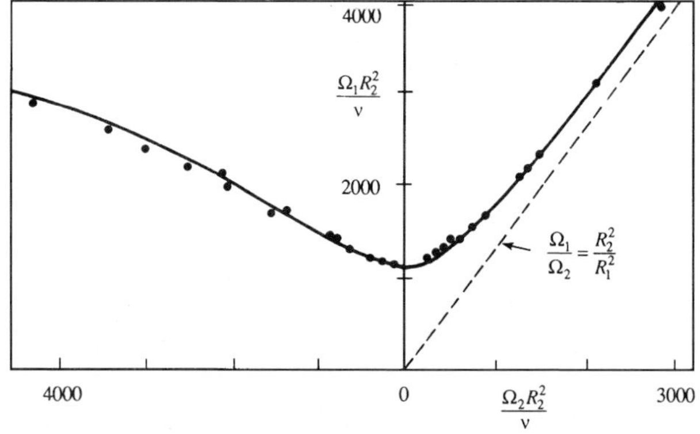

FIGURE 9.20 Taylor's observation and narrow-gap calculation of marginal stability in rotating Couette flow of water. The ratio of radii is $R_2/R_1 = 1.14$. The region above the curve is unstable. The dashed line represents Rayleigh's inviscid criterion, with the region to the left of the line representing instability. The experimental and theoretical results agree well and suggest that viscosity acts to stabilize the flow.

Taylor's analysis of the problem was enormously satisfying, both experimentally and theoretically. He measured the wavelength at the onset of instability by injecting dye and obtained an almost exact agreement with his calculations. The observed onset of instability in the Ω_1, Ω_2-plane (Fig. 9.20) was also in remarkable agreement. This has prompted remarks such as, "the closeness of the agreement between his theoretical and experimental results was without precedent in the history of fluid mechanics" (Drazin and Reid, 1981, p. 105). It even led some people to suggest happily that the agreement can be regarded as a verification of the underlying Navier-Stokes equations, which make a host of assumptions including a linearity between stress and strain rate.

The instability appears in the form of counter-rotating toroidal (or doughnut-shaped) vortices (Fig. 9.21a) called *Taylor vortices*. The streamlines are in the form of helixes, with axes wrapping around the annulus, somewhat like the stripes on a barber's pole. These vortices themselves become unstable at higher values of Ta, when they give rise to wavy vortices for which $\partial/\partial\varphi \neq 0$ (Fig. 9.21b). In effect, the flow has now attained the next higher mode. The number of waves around the annulus depends on the Taylor number, and the wave pattern travels around the annulus. More complicated patterns of vortices result at a higher rates of rotation, finally resulting in the occasional appearance of turbulent patches (Fig. 9.21d), and then fully turbulent flow.

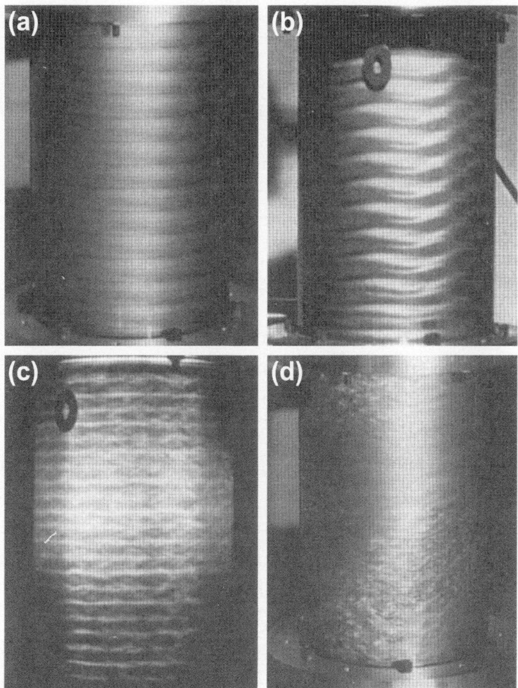

FIGURE 9.21 Instability of rotating Couette flow. Panels (a), (b), (c), and (d) correspond to increasing Taylor number. At first the instability appears as periodic rolls that do not vary with the azimuthal angle. Next, the rolls develop azimuthal waves with wavelengths that depend on the Taylor number. Eventually, the flow becomes turbulent. *Coles,* Journal of Fluid Mechanics, 21, *385–425, 1965; reprinted with the permission of Cambridge University Press.*

Phenomena analogous to the Taylor vortices are called *secondary flows* because they are superposed on a primary flow (such as the Couette flow in the present case). There are two other situations where a combination of curved streamlines (which give rise to centrifugal forces) and viscosity result in instability and steady secondary flows in the form of vortices. One is the flow through a curved channel, driven by a pressure gradient. The other is the appearance of *Görtler vortices* in a boundary-layer flow along a concave wall (Fig. 9.22). The possibility of secondary flows signifies that the *solutions of the Navier-Stokes equations are non-unique* in the sense that more than one steady solution is allowed under the same boundary conditions. We can derive the form of the primary flow only if we exclude the secondary flow by appropriate assumptions. For example, we can derive the expression (9.66) for Couette flow by *assuming* $U_r = 0$ and $U_z = 0$ and thereby rule out the secondary flow.

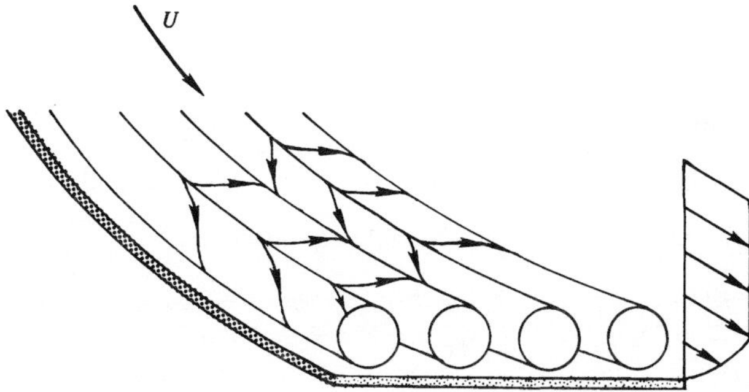

FIGURE 9.22 Görtler vortices in a boundary layer along a concave wall. The instability phenomenon here is essentially the same as that in Taylor-Couette flow, the only difference being the lack of the inner curved surface.

Example 9.4

A simple journal bearing is composed of a rotating shaft (or journal) held by a stationary housing with a thin layer or oil separating the two. What is the Taylor number for a water-lubricated journal bearing that holds a shaft with radius $R_1 = 25$ mm that rotates at 10,000 rpm with a clearance of $d = 25$ μm? Will the water flow in this bearing be unstable?

Solution Here, $\Omega_2 = 0$ and the kinematic viscosity of water is $1.0 \times 10^{-6}\ \mathrm{m^2s^{-1}}$, so Ta is:

$$\mathrm{Ta} = 2\left(\frac{\Omega_1 R_1 d}{\nu}\right)^2 \frac{d}{R_1} = 2\left(\frac{10^4\ \mathrm{rpm} \times (2\pi/60)(2.5 \times 10^{-2}\ \mathrm{m})(2.5 \times 10^{-5}\ \mathrm{m})}{1.0 \times 10^{-6}\ \mathrm{m^2s^{-1}}}\right)^2 \frac{2.5 \times 10^{-5} m}{2.5 \times 10^{-2} m} \cong 857.$$

For $\Omega_2 = 0$, the critical Taylor number is 3416, so this flow is stable. Given that lubricating oils are ten or more times more viscous than water, this centrifugal instability is not typically active in lubrication flows.

9.8 Instability of continuously stratified parallel flows

An instability of great geophysical importance is that of an inviscid stratified fluid in horizontal parallel flow. If the density and velocity vary discontinuously across an interface, the analysis in Section 9.3 shows that the flow is unconditionally unstable. Although only the discontinuous case was studied by Kelvin and Helmholtz, the more general case of continuous distribution is also commonly called the *Kelvin-Helmholtz instability.*

The problem has a long history. In 1915, Taylor, on the basis of his calculations with assumed distributions of velocity and density, *conjectured* that a gradient Richardson number (to be defined shortly) must be less than 1/4 for instability. Other values of the critical Richardson number (ranging from 2 to 1/4) were suggested by Prandtl, Goldstein, Richardson, Synge, and Chandrasekhar. Finally, Miles (1961) was able to prove Taylor's conjecture, and Howard (1961) immediately and elegantly generalized Miles' proof. A short record of the history is given in Miles (1986). In this section we shall prove the Richardson number criterion in the manner given by Howard.

Consider a horizontal parallel flow $U(z)$ directed along the x-axis. The z-axis is taken vertically upward. The basic flow is in equilibrium with the undisturbed density field $\overline{\rho}(z)$ and the basic pressure field $P(z)$. We shall only consider two-dimensional disturbances on this basic state, assuming that they are more unstable than three-dimensional disturbances; this is called *Squires' theorem* and is demonstrated in Section 9.9 in another context. The disturbed state has velocity, pressure, and density fields of:

$$\tilde{\mathbf{u}} = U\mathbf{e}_x + \mathbf{u} = (U + u, 0, w), \quad \tilde{p} = P + p, \quad \text{and} \quad \tilde{\rho} = \overline{\rho} + \rho,$$

where, as before, the tilde indicates a total flow variable. The continuity equation reduces to $\partial u/\partial x + \partial w/\partial z = 0$, and the disturbed velocity field is assumed to satisfy the inviscid Boussinesq momentum equation:

$$\frac{\partial \tilde{\mathbf{u}}}{\partial t} + (\tilde{\mathbf{u}} \cdot \nabla)\tilde{\mathbf{u}} = -\frac{1}{\rho_0}\nabla \tilde{p} - g\frac{(\overline{\rho} + \rho)}{\rho_0}\mathbf{e}_z,$$

where the density variations are neglected except in the vertical component. Here, ρ_0 is a reference density. The basic flow satisfies:

$$0 = -\frac{1}{\rho_0}\frac{\partial P}{\partial z} - g\frac{\overline{\rho}}{\rho_0}.$$

Subtracting the last two equations and dropping nonlinear terms, we obtain the perturbation equation of motion:

$$\frac{\partial \mathbf{u}}{\partial t} + (\mathbf{u} \cdot \nabla)(U\mathbf{e}_x) + U(\mathbf{e}_x \cdot \nabla)\mathbf{u} = -\frac{1}{\rho_0}\nabla p - g\frac{\rho}{\rho_0}\mathbf{e}_z.$$

The horizontal (x) and vertical (z) components of the preceding equation are:

$$\frac{\partial u}{\partial t} + w\frac{\partial U}{\partial z} + U\frac{\partial u}{\partial x} = -\frac{1}{\rho_0}\frac{\partial p}{\partial x} \quad \text{and} \quad \frac{\partial w}{\partial t} + U\frac{\partial w}{\partial x} = -\frac{1}{\rho_0}\frac{\partial p}{\partial x} - g\frac{\rho}{\rho_0}. \tag{9.72}$$

In the absence of diffusion the density of fluid particles does not change: $D\tilde{\rho}/Dt = 0$, or:

$$\frac{\partial}{\partial t}(\overline{\rho} + \rho) + (U + u)\frac{\partial}{\partial x}(\overline{\rho} + \rho) + w\frac{\partial}{\partial z}(\overline{\rho} + \rho) = 0.$$

Keeping only the linear terms, and using the fact that $\overline{\rho}$ is a function of z only, we obtain:

$$\frac{\partial \rho}{\partial t} + U\frac{\partial \rho}{\partial x} + w\frac{\partial \overline{\rho}}{\partial z} = 0,$$

which can be written as:

$$\frac{\partial \rho}{\partial t} + U\frac{\partial \rho}{\partial x} - \frac{\rho_o N^2 w}{g} = 0, \tag{9.73}$$

where N is the buoyancy frequency in an incompressible flow:

$$N^2 \equiv -\frac{g}{\rho_0}\frac{d\bar{\rho}}{dz}. \tag{8.126}$$

The last term in (9.73) represents the density change at a point due to the vertical advection of the basic density field across the point.

The continuity equation can be satisfied with a stream function $u = \partial\psi/\partial z$ and $w = -\partial\psi/\partial x$. Eqs. (9.72) and (9.73) then become:

$$\frac{\partial^2\psi}{\partial t\partial z} - \frac{\partial\psi}{\partial x}\frac{dU}{dz} + U\frac{\partial^2\psi}{\partial x\partial z} = -\frac{1}{\rho_0}\frac{\partial p}{\partial x}, \; -\frac{\partial^2\psi}{\partial t\partial x} - U\frac{\partial^2\psi}{\partial x^2} = -\frac{g\rho}{\rho_0} - \frac{1}{\rho_0}\frac{\partial p}{\partial z},$$
$$\frac{\partial \rho}{\partial t} + U\frac{\partial \rho}{\partial x} + \frac{\rho_o N^2}{g}\frac{\partial\psi}{\partial x} = 0. \tag{9.74}$$

Since the coefficients of derivatives in (9.74) are independent of x and t, exponential variations in these variables are allowed. Consequently, we assume traveling-wave normal mode solutions of the form:

$$[\rho, p, \psi] = [\hat{\rho}(z), \hat{p}(z), \hat{\psi}(z)]\exp\{ik(x-ct)\},$$

where quantities denoted by $\hat{()}$ are complex amplitudes. Because the flow is unbounded in x, the wave number k must be real. The eigenvalue $c = c_r + ic_i$ can be complex, and the solution is unstable if there exists a $c_i > 0$, similar to the development in Section 9.3. Substituting in the normal modes, (9.74) becomes:

$$(U-c)\frac{\partial\hat{\psi}}{\partial z} - \frac{\partial U}{\partial z}\hat{\psi} = -\frac{1}{\rho_0}\hat{p}, \quad k^2(U-c)\hat{\psi} = -g\frac{\hat{\rho}}{\rho_0} - \frac{1}{\rho_0}\frac{\partial\hat{p}}{\partial z}, \quad (U-c)\hat{\rho} + \frac{\rho_0 N^2}{g}\hat{\psi} = 0. \quad (9.75, 9.76, 9.77)$$

To reach a single equation for $\hat{\psi}$, the pressure can be eliminated by taking the z-derivative of (9.75) and subtracting (9.76). The density can be eliminated via substitution from (9.77) to produce:

$$(U-c)\left(\frac{d^2}{dz^2} - k^2\right)\hat{\psi} - \frac{\partial^2 U}{\partial z^2}\hat{\psi} + \frac{N^2}{U-c}\hat{\psi} = 0. \tag{9.78}$$

This is the *Taylor-Goldstein equation*, which governs the behavior of perturbations in a stratified parallel flow. Note that the complex conjugate of (9.78) is also a valid equation because we can take the imaginary part of the equation, change the sign, and add it to the real part of the equation. Now because the Taylor-Goldstein equation does not involve any i, a complex conjugate of the equation shows that if $\hat{\psi}$ is an eigenfunction with eigenvalue c for some k, then $\hat{\psi}^*$ is a possible eigenfunction with eigenvalue c^* for the same k. Therefore, to each eigenvalue with a positive c_i there is a corresponding eigenvalue with a negative c_i. In other words, *to each growing mode there is a corresponding decaying mode.* A non-zero c_i therefore ensures instability.

The boundary conditions are that $w = 0$ on rigid boundaries, presuming these are located at $z = 0$ and d. This requires $\partial\psi/\partial x = ik\hat{\psi}\exp\{ik(x-ct)\} = 0$ at the walls, which is possible only if:

$$\hat{\psi}(0) = \hat{\psi}(d) = 0. \tag{9.79}$$

A necessary condition involving the Richardson number for linear instability of inviscid stratified parallel flows can be derived by defining a new field variable ϕ (not the velocity potential) by:

$$\phi \equiv \hat{\psi}/(U-c)^{1/2} \quad \text{or} \quad \hat{\psi} = (U-c)^{1/2}\phi. \tag{9.80}$$

Then we obtain the derivatives:

$$\frac{\partial \widehat{\psi}}{\partial z} = (U-c)^{1/2}\frac{\partial \phi}{\partial z} + \frac{\phi}{2(U-c)^{1/2}}\frac{dU}{dz}, \quad \text{and}$$

$$\frac{\partial^2 \widehat{\psi}}{\partial z^2} = (U-c)^{1/2}\frac{\partial^2 \phi}{\partial z^2} + \frac{1}{(U-c)^{1/2}}\left(\frac{d\phi}{dz}\frac{dU}{dz} + \frac{1}{2}\phi\frac{d^2U}{dz^2}\right) - \frac{\phi}{4(U-c)^{3/2}}\left(\frac{dU}{dz}\right)^2.$$

The Taylor-Goldstein equation then becomes, after some rearrangement:

$$\frac{d}{dz}\left[(U-c)\frac{d\phi}{dz}\right] - \left\{k^2(U-c) + \frac{1}{2}\frac{d^2U}{dz^2} + \frac{(1/4)(dU/dz)^2 - N^2}{U-c}\right\}\phi = 0. \tag{9.81}$$

Now multiply (9.81) by ϕ^* (the complex conjugate of ϕ), integrate from $z=0$ to $z=d$, and use the boundary conditions $\phi(0)=\phi(d)=0$. The first term gives:

$$\int_0^d \frac{d}{dz}\left\{(U-c)\frac{d\phi}{dz}\right\}\phi^* dz = \int_0^d \left[\frac{d}{dz}\left\{(U-c)\frac{d\phi}{dz}\phi^*\right\} - (U-c)\frac{d\phi}{dz}\frac{d\phi^*}{dz}\right]dz = -\int_0^d (U-c)\left|\frac{d\phi}{dz}\right|^2 dz.$$

Integrals of the other terms in (9.81) are also simple to manipulate. We finally obtain:

$$\int_0^d \left\{\frac{N^2 - (1/4)(dU/dz)^2}{U-c}\right\}|\phi|^2 dz = \int_0^d (U-c)\left\{\left|\frac{d\phi}{dz}\right|^2 + k^2|\phi|^2\right\}dz + \int_0^d \frac{1}{2}\frac{d^2U}{dz^2}|\phi|^2 dz. \tag{9.82}$$

The last term in the preceding equation is real. The imaginary part of the first term can be found by noting that:

$$\frac{1}{U-c} = \frac{U-c^*}{|U-c|^2} = \frac{U-c_r+ic_i}{|U-c|^2}.$$

Taking the imaginary part of (9.82) leads to:

$$c_i\int_0^d \left\{\frac{N^2-(1/4)(dU/dz)^2}{|U-c|^2}\right\}|\phi|^2 dz = -c_i\int_0^d \left\{\left|\frac{d\phi}{dz}\right|^2 + k^2|\phi|^2\right\}dz.$$

The integral on the right side is positive. If the flow is such that $N^2 > (1/4)(dU/dz)^2$ everywhere, then the preceding equation states that c_i times a positive quantity equals c_i times a negative quantity; this is impossible and requires that $c_i = 0$ for such a case. Thus, defining the *gradient Richardson number*:

$$\mathrm{Ri}(z) \equiv N^2/(dU/dz)^2, \tag{9.83}$$

we can say that linear stability is guaranteed if the inequality:

$$\mathrm{Ri} > 1/4 \quad \text{(stable)} \tag{9.84}$$

is satisfied everywhere in the flow.

Note that the criterion does not state that the flow is necessarily unstable if $\mathrm{Ri} < 1/4$ somewhere, or even everywhere, in the flow. Thus $\mathrm{Ri} < 1/4$ is a *necessary* but not sufficient condition for instability. For example, in a jet-like velocity profile $u \propto \mathrm{sech}^2 z$ and an exponential density profile, the flow does not become unstable until the Richardson number falls below 0.214. A critical Richardson number lower than 1/4 is also found in the presence of boundaries, which stabilize the flow. In fact, there is no unique critical Richardson number that applies to all distributions of $U(z)$ and $N(z)$. However, several calculations show that in many shear layers (having linear, tanh, or error function profiles for velocity and density), the flow does become unstable to disturbances of certain wavelengths if the minimum value of Ri in the flow (which is generally at the center of the shear layer where $|dU/dz|$ is greatest) is less than 1/4. The *most unstable* wave, defined as the first to become unstable as Ri is reduced below 1/4, is found to have a wavelength $\lambda \approx 7h$, where h is the thickness of the shear layer. Laboratory (Scotti and Corcos, 1972) as well as geophysical observations (Eriksen, 1978) show that the requirement:

$$\mathrm{Ri}_{\min} < 1/4$$

is a useful guide for the prediction of instability of a stratified shear layer.

Similar to the previous analysis, another useful result concerning the behavior of the complex phase speed c in an inviscid parallel shear flow can be determined by considering an alternative version of (9.80):

$$F \equiv \hat{\psi}/(U-c), \tag{9.85}$$

which leads to derivatives:

$$\frac{\partial \hat{\psi}}{\partial z} = (U-c)\frac{\partial F}{\partial z} + \frac{dU}{dz}F, \quad \text{and} \quad \frac{\partial^2 \hat{\psi}}{\partial z^2} = (U-c)\frac{\partial^2 F}{\partial z^2} + 2\frac{dU}{dz}\frac{dF}{dz} + \frac{d^2U}{dz^2}F.$$

When (9.85) is substituted into the Taylor-Goldstein equation (9.78), the result is:

$$(U-c)\left[(U-c)\frac{d^2F}{dz^2} + 2\frac{dU}{dz}\frac{dF}{dz} - k^2(U-c)F\right] + N^2F = 0,$$

and the terms involving d^2U/dz^2 have canceled out. This can be rearranged into the form:

$$\frac{d}{dz}\left[(U-c)^2\frac{dF}{dz}\right] - k^2(U-c)F + N^2F = 0.$$

Multiplying by F^*, integrating (by parts when necessary) over the depth of the flow, and using the boundary conditions, we obtain:

$$-\int (U-c)^2\left|\frac{dF}{dz}\right|^2 dz - k^2\int (U-c)^2|F|^2dz + \int N^2|F|^2dz = 0,$$

which can be written as:

$$\int (U-c)^2Q\,dz = \int N^2|F|^2dz \quad \text{where} \quad Q \equiv |dF/dz|^2 + k^2|F|^2$$

is positive. Equating real and imaginary parts, we obtain:

$$\int\left[(U-c_r)^2 - c_i^2\right]Q\,dz = \int N^2|F|^2dz \quad \text{and} \quad c_i\int (U-c_r)Q\,dz = 0. \tag{9.86, 9.87}$$

For instability $c_i \neq 0$, for which (9.87) shows that $(U-c_r)$ must change sign somewhere in the flow:

$$U_{\min} < c_r < U_{\max}, \tag{9.88}$$

which states that c_r lies in the range of U. Recall that we have assumed solutions of the form:

$$\exp\{ik(x-ct)\} = \exp\{ik(x-c_rt)\}\exp\{+kc_it\},$$

which means that c_r is the phase velocity in the positive x direction, and kc_i is the growth rate. Eq. (9.88) shows that c_r is positive if U is everywhere positive, and is negative if U is everywhere negative. In these cases we can say that unstable waves propagate in the direction of the background flow.

Limits on the maximum growth rate can also be predicted. Eq. (9.86) gives:

$$\int\left[U^2 - 2Uc_r + c_r^2 - c_i^2\right]Q\,dz > 0,$$

which, on using (9.87), becomes:

$$\int [U^2 - c_r^2 - c_i^2]Q\,dz > 0. \tag{9.89}$$

Now because $(U_{\min} - U) < 0$ and $U_{\max} - U > 0$, it is always true that:

$$\int [U_{\min} - U][U_{\max} - U]\,dz \leq 0,$$

which can be recast as:

$$\int [U_{\max}U_{\min} + U^2 - U(U_{\max} + U_{\min})]Qdz \le 0.$$

Using (9.89), this gives:

$$\int [U_{\max}U_{\min} + c_r^2 + c_i^2 - U(U_{\max} + U_{\min})]Qdz \le 0,$$

and after using (9.87), this becomes:

$$\int [U_{\max}U_{\min} + c_r^2 + c_i^2 - c_r(U_{\max} + U_{\min})]Qdz \le 0.$$

Because the quantity within [,]-brackets is independent of z, and $\int Qdz > 0$, we must have $[\ ] \le 0$. With some rearrangement, this condition can be written as:

$$\left[c_r - \frac{1}{2}(U_{\max} + U_{\min})\right]^2 + c_i^2 \le \left[\frac{1}{2}(U_{\max} - U_{\min})\right]^2.$$

This shows that the complex wave velocity, c, of any unstable mode of a disturbance in parallel flows of an inviscid fluid must lie inside the semicircle in the upper half of the c-plane, which has the range of U as the diameter (Fig. 9.23). This result was first derived by Howard (1961) and is valid for flows with and without stratification. It is called the Howard semicircle theorem and states that the maximum growth rate is limited by:

$$kc_i < (k/2)(U_{\max} - U_{\min}).$$

The theorem is useful in searching for eigenvalues $c(k)$ in numerical solution of instability problems.

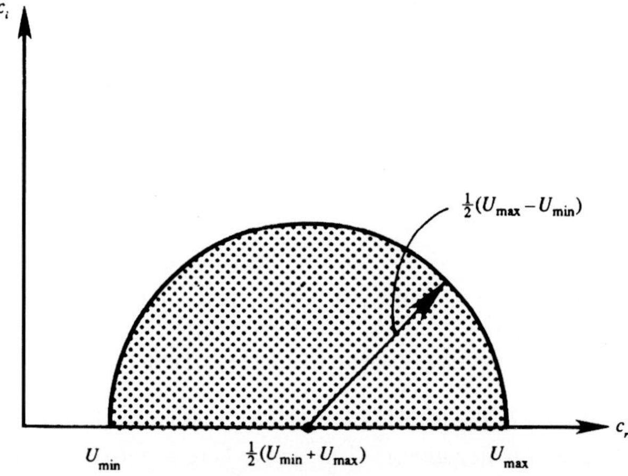

FIGURE 9.23 Depiction of the Howard semicircle theorem. In inviscid parallel flows the complex eigenvalue c must lie within the semicircle shown. This theorem limits both the real part (the phase speed) and the imaginary part (the growth rate divided by k) of the eigenvalue.

Example 9.5

A stream of nominally fresh water with density ρ_f travels horizontally at speed U above stationary salt water with density ρ_s. If the transition layer between the fresh and salt water is initially thin and the flow persists for a long time, estimate the final thickness h of the region of mixed-salinity water.

Solution If the shear layer between the fresh and salt water is initially thin, it will be unstable and overturning fluid motions at the interface, such as those shown in Figs. 9.4 through 9.6, will mix moving fresh water and stationary salt water together. This mixing will continue and the mixed layer will thicken until the flow is no longer unstable. The criterion for reaching stability when the background flow depends only on depth, see (9.83) and (9.84), is given in terms of the gradient Richardson number:

$$\mathrm{Ri}(z) \equiv \frac{N^2}{(dU/dz)^2} = -\frac{g}{\rho_0}\frac{d\bar{\rho}}{dz} \bigg/ \left(\frac{dU}{dz}\right)^2 > \frac{1}{4}.$$

To estimate a layer thickness, assume the vertical extent h of the velocity and density mixing regions are the same. The instability ceases when Ri(z) reaches 1/4, so the mixing region's final thickness can be estimated from:

$$-\frac{g}{\rho_0}\left(\frac{\Delta\rho}{h}\right) \bigg/ \left(\frac{U}{h}\right)^2 \approx \frac{1}{4}, \quad \text{which implies :} \quad h \approx \frac{1}{4}\frac{\rho_o U^2}{g|\Delta\rho|}.$$

The parameters for the estuary flow shown in Fig. 9.5 are $U_c \approx 0.5\ \mathrm{ms}^{-1}$, $\rho_f \approx 1008\ \mathrm{kgm}^{-3}$, and $\rho_s \approx 1021\ \mathrm{kgm}^{-3}$ (and $g = 9.81\ \mathrm{ms}^{-2}$). Thus, the layer thickness estimate is:

$$h \approx \frac{1}{4}\frac{(1014\ \mathrm{kgm}^{-3})(0.5\ \mathrm{ms}^{-1})^2}{(9.81\ \mathrm{ms}^{-2})(13\ \mathrm{kgm}^{-3})} = 0.50\ \mathrm{m},$$

which agrees well with the *average* observed layer thickness in Fig. 9.5.

Interestingly, the total height of the structures shown in Fig. 9.5 can be estimated from a simple mechanical energy balance. The largest possible vortical structure that can rotate in a stratified shear flow will convert upper- (or lower-) edge kinetic energy to lower- (or upper-) edge potential energy during a half rotation. Equating the kinetic ($\rho_o U^2/2$) and potential ($|\Delta\rho|gh$) energy differences between the top and bottom of one such vortical structure, leads to $h \sim (1/2)\rho_o U^2/g|\Delta\rho|$. This is twice the thickness estimate provided above and agrees well with the observed structure height shown in Fig. 9.5.

For both estimates, the final mixed region thickness is proportional to the square of the flow speed difference. Thus, better final-state mixing occurs when the initial velocity difference is higher, a fact that is often exploited in forced air heating and cooling to provide uniform air temperature in indoor spaces.

9.9 Squire's theorem and the Orr-Sommerfeld equation

The Bénard and Taylor problems are two flows in which viscosity has a stabilizing effect. Curiously, viscous effects can also be *destabilizing*, as indicated by several calculations of wall-bounded parallel flows. In this section we shall derive the equation governing the stability of parallel flows of a homogeneous (i.e., constant density ρ) viscous fluid. Let the primary flow be directed along the x-direction and vary in the y direction so that $\mathbf{U} = (U(y), 0, 0)$. We decompose the total flow as the sum of the basic flow plus the perturbation:

$$\tilde{\mathbf{u}} = (U + u, v, w), \quad \text{and} \quad \tilde{p} = P + p.$$

Together the basic and perturbation flows satisfy the horizontal Navier-Stokes momentum equation:

$$\frac{\partial u}{\partial t} + (U+u)\frac{\partial}{\partial x}(U+u) + v\frac{\partial}{\partial y}(U+u) + w\frac{\partial}{\partial z}(U+u) = -\frac{\partial}{\partial x}(P+p) + \nu\nabla^2(U+u) \tag{9.90}$$

and the background flow satisfies:

$$0 = -\frac{\partial P}{\partial x} + \nu\nabla^2 U.$$

Subtracting this from (9.90) and neglecting terms nonlinear in the perturbations produces the x-momentum equation for the perturbations:

$$\frac{\partial u}{\partial t} + U\frac{\partial u}{\partial x} + v\frac{\partial U}{\partial y} = -\frac{\partial p}{\partial x} + \nu\nabla^2 u. \tag{9.91}$$

Similarly the y-momentum, z-momentum, and continuity equations for the perturbations are:

$$\frac{\partial v}{\partial t} + U\frac{\partial v}{\partial x} = -\frac{\partial p}{\partial y} + \nu\nabla^2 v, \quad \frac{\partial w}{\partial t} + U\frac{\partial w}{\partial x} = -\frac{\partial p}{\partial z} + \nu\nabla^2 w, \quad \text{and} \quad \frac{\partial u}{\partial x} + \frac{\partial v}{\partial y} + \frac{\partial w}{\partial z} = 0. \tag{9.92}$$

The coefficients in (9.91) and (9.92) depend only on y, so that the equations admit solutions exponential in x, z, and t. Accordingly, we assume normal modes of the form:

$$[\mathbf{u}, p] = [\widehat{\mathbf{u}}(y), \widehat{p}(y)] \exp\{i(kx + mz - kct)\}. \tag{9.93}$$

As the flow is unbounded in x and z, the wave number components k and m must be real. However, the wave speed $c = c_r + ic_i$ may be complex. Without loss of generality, we can consider only positive values for k and m; the sense of propagation is then left open by keeping the sign of c_r unspecified. The normal modes represent waves that travel obliquely to the basic flow with a wave number magnitude $\sqrt{k^2 + m^2}$ and have an amplitude that varies as $\exp(kc_i t)$. Solutions are therefore stable if $c_i < 0$ and unstable if $c_i > 0$.

Substitution of (9.93) into the perturbation equations (9.91) and (9.92), and replacement of ν by U_0L/Re produces:

$$\begin{gathered} ik(U - c)\widehat{u} + \widehat{v}(dU/dy) = -ikp + (U_0L/\mathrm{Re})[d^2\widehat{u}/dy^2 - (k^2 + m^2)\widehat{u}], \\ ik(U - c)\widehat{v} = -d\widehat{p}/dy + (U_0L/\mathrm{Re})[d^2\widehat{v}/dy^2 - (k^2 + m^2)\widehat{v}], \\ ik(U - c)\widehat{w} = -im\widehat{p} + (U_0L/\mathrm{Re})[d^2\widehat{w}/dy^2 - (k^2 + m^2)\widehat{w}], \quad \text{and} \quad ik\widehat{u} + d\widehat{v}/dy + im\widehat{w} = 0, \end{gathered} \tag{9.94}$$

where U_0 is a characteristic velocity scale (such as the maximum velocity of the basic flow), L is a characteristic length scale (such as the cross-stream dimension of the basic flow), and the Reynolds number of the basic flow is $\mathrm{Re} = U_0L/\nu$. These are the normal mode equations for three-dimensional disturbances of the basic flow $\mathbf{U} = U(y)\mathbf{e}_x$.

Before proceeding further, we shall first prove Squire's Theorem (1933), which states that *to each unstable three-dimensional disturbance there corresponds a more unstable two-dimensional one*. To prove this theorem, consider the *Squire transformation*:

$$\overline{k} = \sqrt{k^2 + m^2}, \quad \overline{c} = c, \quad \overline{k}\overline{u} = k\widehat{u} + m\widehat{w}, \quad \overline{v} = \widehat{v}, \quad \overline{p}/\overline{k} = \widehat{p}/k, \quad \text{and} \quad \overline{k}\,\overline{\mathrm{Re}} = k\mathrm{Re}. \tag{9.95}$$

The finally equality of (9.95) sets $\overline{\mathrm{Re}} = (k/\sqrt{k^2 + m^2})\mathrm{Re}$, so $\overline{\mathrm{Re}} < \mathrm{Re}$ when $m \neq 0$. After substituting (9.95) into (9.94), and adding the first and third equations, the result is:

$$\begin{gathered} i\overline{k}(U - c)\overline{u} + \overline{v}(dU/dy) = -i\overline{k}\overline{p} + (U_0L/\overline{\mathrm{Re}})[d^2\overline{u}/dy^2 - \overline{k}^2\overline{u}], \\ i\overline{k}(U - c)\overline{v} = -d\widehat{p}/dy + (U_0L/\overline{\mathrm{Re}})[d^2\overline{v}/dy^2 - \overline{k}^2\overline{v}], \quad \text{and} \quad i\overline{k}\overline{u} + d\overline{v}/dy = 0. \end{gathered}$$

These equations are the same as (9.94) when $m = \widehat{w} = 0$ and $\overline{\mathrm{Re}}$ replaces Re. Thus, to each three-dimensional problem there corresponds an equivalent two-dimensional one at a *lower* Reynolds number. Therefore, the critical Reynolds number at which the instability starts is reached first by two-dimensional disturbances as Re increases, so we only need to consider a two-dimensional disturbance to determine the minimum Reynolds number for the onset of instability.

This contention about two-dimensional disturbances can also be understood as follows. The three-dimensional disturbance (9.93) is a wave propagating obliquely to the basic flow. If the coordinate system is rotated so that the new x-axis lies in this direction, the equations of motion are such that only the component of the basic flow in the new x-direction affects the disturbance. Thus, the effective Reynolds number for this oblique disturbance is reduced below that for a flow-aligned two-dimensional disturbance having $m = \widehat{w} = 0$ in the original coordinate system.

Interestingly, Squire's theorem also holds for several other problems that do not involve the Reynolds number. The transformation (9.95) leads to a growth rate for a two-dimensional disturbance of $\exp(\overline{k}\overline{c}_i t)$, whereas (9.93) shows that the growth rate of a three-dimensional disturbance is $\exp(kc_i t)$. The two-dimensional growth rate is therefore larger because Squire's transformation requires $\overline{k} > k$ and $c = \overline{c}$. Thus, two-dimensional disturbances are more unstable.

Because of Squire's theorem, we only need consider the equation set (9.94) with $m = \widehat{w} = 0$ to determine the stability of viscous parallel flow. The two-dimensionality allows the use of a stream function $\psi(x, y, t)$ for the perturbation field via the usual relationships: $u = \partial\psi/\partial y$ and $v = -\partial\psi/\partial x$. Again, use normal modes:

$$[u, v, \psi] = [\widehat{u}(y), \widehat{v}(y), \phi(y)] \exp\{ik(x - ct)\}.$$

(To be consistent, the dimensionless complex amplitude of ψ should be $\widehat{\psi}$; however, ϕ [not the potential] is used here to follow the standard notation for this variable in the literature.) Then, it follows that $\widehat{u} = \partial\phi/\partial y$ and $\widehat{v} = -ik\phi$, and a single equation in terms of ϕ can now be found by eliminating the pressure from (9.94). This effort yields a fourth-order ordinary differential equation:

$$(U-c)\left(\frac{d^2\phi}{dy^2}-k^2\phi\right)-\frac{d^2U}{dy^2}\phi=\frac{\nu}{ik}\left(\frac{d^4\phi}{dy^4}-2k^2\frac{d^2\phi}{dy^2}+k^4\phi\right). \tag{9.96}$$

For confined basic flows, the disturbance boundary conditions are no-slip on the confining walls at y_1 and y_2:

$$\phi \quad \text{and} \quad d\phi/dy=0 \quad \text{on} \quad y=y_1 \quad \text{and} \quad y=y_2. \tag{9.97a}$$

For unconfined basic flows having a confined region of non-zero shear near $y=0$, the disturbance must decay to zero away from the region of basic-flow shear, so the boundary conditions are:

$$\phi \quad \text{and} \quad d\phi/dy\to 0 \quad \text{as} \quad |y|\to\infty. \tag{9.97b}$$

Eq. (9.96) is the well-known *Orr-Sommerfeld equation*, which governs the stability of nearly parallel viscous flows such as those in a straight channel or in a boundary layer. It is essentially a vorticity equation because the pressure has been eliminated. Analytical solutions of the Orr-Sommerfeld equations are difficult to obtain, and only the results of some simple flows will be discussed in the later sections. However, we shall first present certain results obtained by ignoring the viscous terms on the right side of this equation.

9.10 Inviscid stability of parallel flows

Insight into the viscous stability of parallel flows can be obtained by first assuming that the disturbances obey inviscid dynamics. The governing equation can be found by setting $\nu=0$ in the Orr-Sommerfeld equation, (9.96), giving:

$$(U-c)\left(\frac{d^2\phi}{dy^2}-k^2\phi\right)-\frac{d^2U}{dy^2}\phi=0, \tag{9.98}$$

which is called the *Rayleigh equation*. If the flow is bounded by walls at y_1 and y_2 where $v=0$, then the boundary conditions are:

$$\phi=0 \quad \text{at} \quad y=y_1 \quad \text{and} \quad y_2. \tag{9.99a}$$

If the region of shear in the basic flow is localized near $y=0$, then the disturbance must decay away from this region so its boundary conditions are:

$$\phi\to 0 \quad \text{as} \quad |y|\to\infty. \tag{9.99b}$$

The set (9.98) and (9.99) defines an eigenvalue problem, with $c(k)$ as the eigenvalue and ϕ as the eigenfunction. As these equations do not involve the imaginary root, i, taking the complex conjugate shows that if ϕ is an eigenfunction with eigenvalue c for some k, then ϕ^* is also an eigenfunction with eigenvalue c^* for the same k. Therefore, to each eigenvalue with a positive c_i there is a corresponding eigenvalue with a negative c_i. In other words, *to each growing mode there is a corresponding decaying mode*. Stable solutions therefore can have only a real c. Note that this is true of inviscid flows only. The viscous term in the full Orr-Sommerfeld equation (9.96) involves an i, and the foregoing conclusion is no longer valid.

Starting from (9.98) it is possible to show that certain velocity distributions $U(y)$ are potentially unstable. In this discussion it should be noted that only the *disturbances* are assumed to obey inviscid dynamics; the background flow profile $U(y)$ may be that of a steady laminar viscous flow.

The first deduction that can be made from (9.98) is Rayleigh's inflection point criterion: a necessary (but not sufficient) condition for instability of an inviscid parallel flow is that the basic velocity profile $U(y)$ has a point of inflection. To prove the theorem, rewrite the Rayleigh equation (9.98) in the form:

$$\frac{d^2\phi}{dy^2}-k^2\phi-\frac{1}{U-c}\frac{d^2U}{dy^2}\phi=0,$$

and consider the unstable mode for which $c_i>0$, and therefore $U-c\neq 0$. Multiply this equation by ϕ^*, integrate from the lower to the upper boundary of the flow, by parts where necessary, and apply the boundary condition (9.99). The first term transforms as follows:

$$\int \phi^*(d^2\phi/dy^2)dy = \left[\phi^*(d\phi/dy)\right]_{y_1 \text{ or} -\infty}^{y_2 \text{ or} +\infty} - \int (d\phi^*/dy)(d\phi/dy)dy = -\int |d\phi/dy|^2 dy,$$

where the limits on the integrals have not been explicitly written. The Rayleigh equation then gives:

$$\int \left(|d\phi/dy|^2 + k^2|\phi|^2\right)dy + \int \frac{1}{U-c}\frac{d^2U}{dy^2}|\phi|^2 dy = 0. \tag{9.100}$$

The first term is real. The second term in (9.100) is complex, and its imaginary part can be found by multiplying the numerator and denominator by $(U - c^*)$. Thus, the imaginary part of (9.100) implies:

$$c_i \int \frac{1}{|U-c|^2}\frac{d^2U}{dy^2}|\phi|^2 dy = 0. \tag{9.101}$$

For the unstable case, for which $c_i \neq 0$, (9.101) can be satisfied only if d^2U/dy^2 changes sign at least once in the interval $y_1 < y < y_2$, or $-\infty < y < +\infty$. In other words, for instability the background velocity distribution must have at least one point of inflection (where $d^2U/dy^2 = 0$) within the flow. Clearly, the existence of a point of inflection does not guarantee a non-zero c_i. The inflection point is therefore a necessary but not sufficient condition for inviscid instability.

Some seventy years after Rayleigh's discovery, the Swedish meteorologist Fjortoft in 1950 discovered a stronger necessary condition for the instability of inviscid parallel flows. He showed that *a necessary condition for instability of inviscid parallel flows is that* $(U - U_1)(d^2U/dy^2) < 0$ *somewhere in the flow*, where U_1 is the value of U at the point of inflection. To prove the theorem, take the real part of (9.100):

$$\int \frac{U-c_r}{|U-c|^2}\frac{d^2U}{dy^2}|\phi|^2 dy = -\int \left(|d\phi/dy|^2 + k^2|\phi|^2\right) dy < 0. \tag{9.102}$$

Suppose that the flow is unstable, so that $c_i \neq 0$, and a point of inflection does exist according to the Rayleigh criterion. Then it follows from (9.101) that:

$$(c_r - U_1)\int \frac{1}{|U-c|^2}\frac{d^2U}{dy^2}|\phi|^2 dy = 0. \tag{9.103}$$

Adding Eqs. (9.102) and (9.103), we obtain:

$$\int \frac{U-U_1}{|U-c|^2}\frac{d^2U}{dy^2}|\phi|^2 dy = 0,$$

so that $(U - U_1)(d^2U/dy^2)$ must be negative somewhere in the flow.

Some common velocity profiles are shown in Fig. 9.24. Only the two flows shown in the bottom row can possibly be unstable, for only they satisfy Fjortoft's theorem. Flows (a), (b), and (c) do not have an inflection point: flow (d) does satisfy Rayleigh's condition but not Fjortoft's because $(U - U_1)(d^2U/dy^2)$ is positive. Note that an alternate way of stating Fjortoft's theorem is that *the basic flow's vorticity magnitude must have a maximum within the region of flow*, not at the boundary. In flow (d), the maximum magnitude of vorticity occurs at the walls.

The criteria of Rayleigh and Fjortoft indicate the importance of having a point of inflection in the velocity profile. They show that flows in jets, wakes, shear layers, and boundary layers with adverse pressure gradients, all of which have a point of inflection and satisfy Fjortoft's theorem, are potentially unstable. On the other hand, plane Couette flow, Poiseuille flow, and a boundary-layer flow with zero or favorable pressure gradient have no point of inflection in the velocity profile and are stable in the inviscid limit.

However, neither of the two conditions is sufficient for instability. An example is the sinusoidal profile $U = \sin(y)$, with boundaries at $y = \pm b$. It has been shown that the flow is stable if the width is restricted to $2b < \pi$, although the profile has an inflection point at $y = 0$.

Inviscid parallel flows satisfy Howard's semicircle theorem, which was proved in Section 9.8 for the more general case of a stratified shear flow. The theorem states that the phase speed c_r of an unstable mode with wave number k has a value that lies between the minimum and the maximum values of $U(y)$ in the flow field. Growing and decaying modes are characterized by a non-zero c_i, whereas neutral modes can have only a real $c = c_r$. Thus, it follows that neutral modes must have $U = c$ somewhere in the flow field. The neighborhood y around y_c at which $U = c = c_r$ is called a *critical layer*.

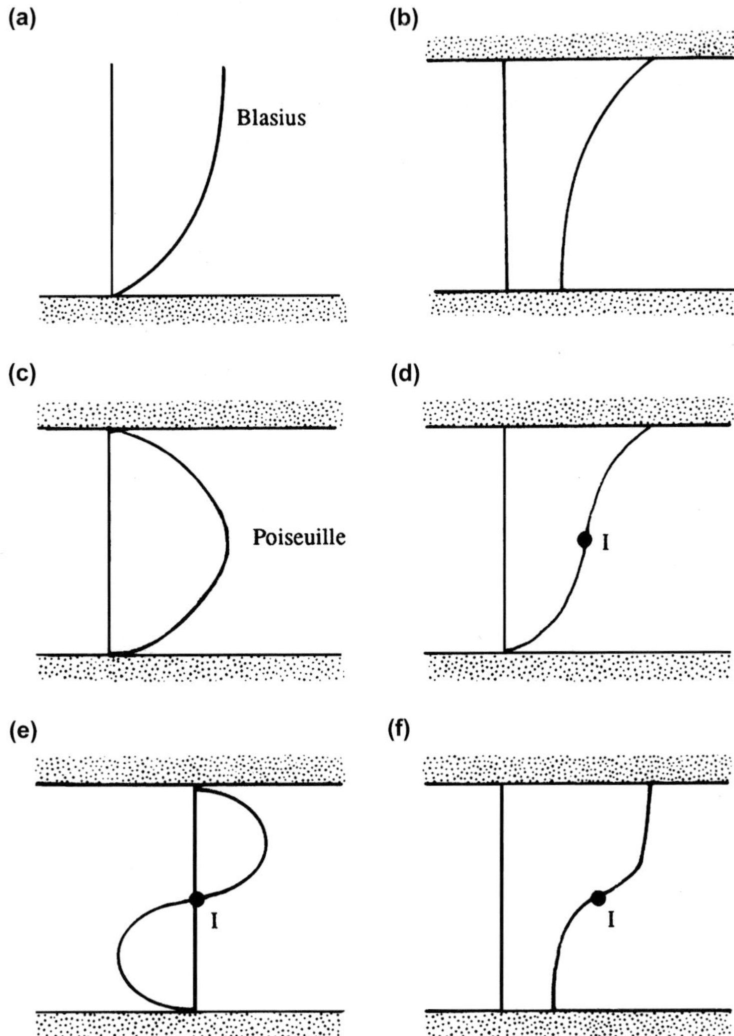

FIGURE 9.24 Examples of parallel flows. Points of inflection are denoted by I. Profiles (a), (b), and (c) are inviscidly stable. Profiles (d), (e), and (f) may be inviscidly unstable by Rayleigh's inflection point criterion. Only profiles (e) and (f) satisfy Fjortoft's criterion of inviscid instability.

The location y_c is a critical point of the inviscid governing equation (9.98), because the highest derivative drops out at this value of y, and the eigenfunction for this k and c may be discontinuous across this layer. The full Orr-Sommerfeld equation (9.96) has no such critical layer because the highest-order derivative does not drop out when $U = c$. It is apparent that in a real flow a viscous boundary layer must form at the location where $U = c$, and that the layer becomes thinner as $\mathrm{Re} \to \infty$.

The streamline pattern in the neighborhood of the critical layer where $U = c$ was given by Kelvin in 1888, and indicates the nature of the nearby unstable modes having the same k but small positive c_i. The discussion provided here is adapted from Drazin and Reid (1981). Consider a flow viewed by an observer moving with the phase velocity $c = c_r$. Then the basic velocity field seen by this observer is $(U - c)$, so that the stream function due to the basic flow is:

$$\Psi = \int (U - c)dy.$$

The total stream function is obtained by adding the perturbation:

$$\widehat{\psi} = \int (U - c)dy + A\phi(y)\exp\{ikx\}, \tag{9.104}$$

where A is an arbitrary constant, and the time factor in the second term is omitted because the disturbance is neutrally stable. Near the critical layer $y = y_c$, a Taylor series expansion of the real part of (9.104) is approximately:

$$\widehat{\psi} \cong \frac{(y-y_c)^2}{2}\left[\frac{dU}{dy}\right]_{y=y_c} + A\phi(y_c)\cos(kx),$$

where $\phi(y_c)$ is assumed to be real. The streamline pattern corresponding to this equation is sketched in Fig. 9.25, showing the so-called *Kelvin cat's eye pattern* that is visually similar to the illustrations of the Kelvin-Helmholtz instability given in Figs. 9.4–9.6.

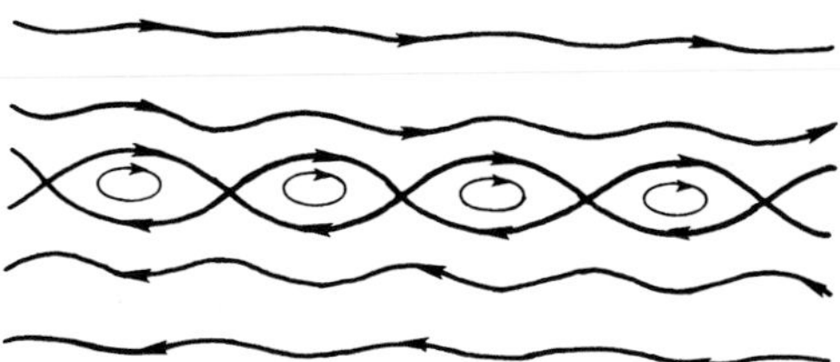

FIGURE 9.25 The Kelvin cat's eye pattern near a critical layer, showing streamlines as seen by an observer moving with a neutrally stable wave having $c = c_r$. This flow pattern is reminiscent of those shown in Figs. 9.4–9.6.

9.11 Results for parallel and nearly parallel viscous flows

The dominant intuitive expectation is that viscous effects are stabilizing. The stability of thermal and centrifugal convections discussed in Sections 9.5 and 9.7 confirm this expectation. However, the actual situation is more complicated. Consider the Poiseuille-flow and Blasius boundary-layer velocity profiles in Fig. 9.24. Neither has an inflection point so both are inviscidly stable. Yet, in experiments, these flows are known to undergo transition to turbulence at some Reynolds number, and this suggests that viscous effects are destabilizing in these flows. Thus, fluid viscosity may be stabilizing as well as destabilizing, a duality confirmed by stability calculations of parallel viscous flows.

Analytical solution of the Orr-Sommerfeld equation is notoriously complicated and will not be presented here. The viscous term in (9.96) contains the highest-order derivative, and therefore the eigenfunction may contain regions of rapid variation in which the viscous effects become important. Sophisticated asymptotic techniques are therefore needed to treat these boundary layers. Alternatively, solutions can be obtained numerically. For our purposes, we shall discuss only certain features of these calculations for the two-stream shear layer, plane Poiseuille flow, plane Couette flow, pipe flow, and boundary layers with pressure gradients. This section concludes with an explanation of how viscosity can act to destabilize a flow. Additional information can be found in Drazin and Reid (1981), and in the review article by Bayly et al. (1988).

Two-stream shear layer

Consider a shear layer with the velocity profile $U(y) = U_0 \tanh(y/L)$, so that $U(y) \to \pm U_0$ as $y/L \to \pm\infty$. This profile has its peak vorticity at its inflection point and is of the type shown in Fig. 9.24f. A stability diagram for solution of the Orr-Sommerfeld equation for this velocity distribution is sketched in Fig. 9.26. At all Reynolds numbers the flow is unstable to waves having low wave numbers in the range $0 < k < k_u$, where the upper limit k_u depends on the Reynolds number $\mathrm{Re} = U_0 L/\nu$. For high values of Re, the range of unstable wave numbers increases to $0 < k < 1/L$, which corresponds to a wavelength range of $\infty > \lambda > 2\pi L$. It is therefore essentially a long-wavelength instability. In the limit $kL \to 0$, these results simplify to those given in Section 9.3 for a vortex sheet.

Fig. 9.26 implies that the critical Reynolds number for the onset of instability in a shear layer is zero. In fact, viscous calculations for all flows with *inflectional profiles* show a small critical Reynolds number; for example, for a jet of the form $u = U\mathrm{sech}^2(y/L)$, it is $\mathrm{Re}_{\mathrm{cr}} = 4$. These wall-free shear flows therefore become unstable quickly, and the inviscid prediction that these flows are always unstable is a fairly good description. The reason the inviscid analysis works well in describing the stability characteristics of free shear flows can be explained as follows. For flows with inflection points the eigenfunction of the inviscid solution is smooth. On this zero-order approximation, the viscous term acts as a *regular* perturbation, and the resulting correction to the eigenfunction and eigenvalues can be computed as a perturbation expansion in powers of the small parameter 1/Re. This is true even though the viscous term in the Orr-Sommerfeld equation contains the highest-order derivative.

The instability in flows with inflection points is observed to form rolled-up regions of vorticity, much like in the calculations of Fig. 9.6 or in the pictures in Figs. 9.4 and 9.5. This behavior is robust and insensitive to the detailed experimental conditions. They are therefore easily observed. In contrast, the unstable waves in a wall-bounded shear flow are extremely difficult to observe, as discussed in the next section.

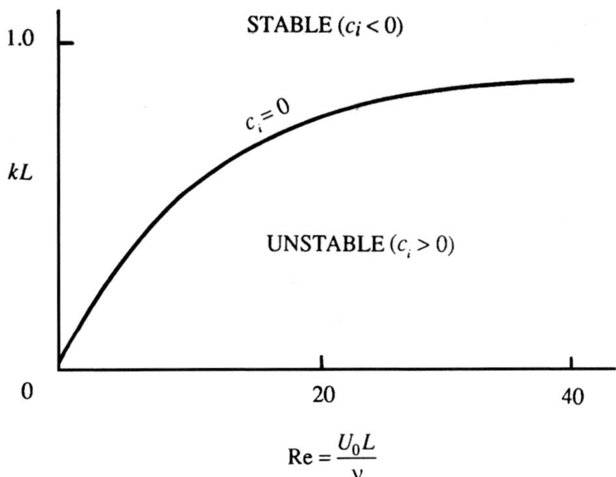

FIGURE 9.26 Marginal stability curve for a shear layer with a velocity profile of $U_0 \tanh(y/L)$ in terms of the Reynolds number $U_o L/\nu$ and the dimensionless wave number kL of the disturbance. This flow is only unstable to low wave number disturbances.

Plane Poiseuille flow

The flow in a channel with a parabolic velocity distribution has no point of inflection and is inviscidly stable. However, linear viscous calculations show that the flow becomes unstable at a critical Reynolds number of 5780. Nonlinear calculations, which consider the distortion of the basic profile by the finite amplitude of the perturbations, give a critical Reynolds number of 2510, which agrees better with the observations of transition. In any case, the interesting point is that viscosity is *destabilizing* for this flow. The solution of the Orr-Sommerfeld equation for Poiseuille flow and other parallel flows with rigid boundaries, which do not have an inflection point, is complicated. In contrast to flows with inflection points, the viscosity here acts as a *singular* perturbation, and the eigenfunction has viscous boundary layers on the channel walls and around critical layers where $U = c_r$. The disturbances that cause instability in these flows are called *Tollmien-Schlichting* waves, and their experimental detection is discussed in the next section. In his 1979 text, Yih gives a thorough discussion of the solution of the Orr-Sommerfeld equation using asymptotic expansions in the limit sequence $\mathrm{Re} \to \infty$, then $k \to 0$ (but $k\mathrm{Re} \gg 1$) (Yih, 1979). He follows closely the analysis of Heisenberg (1924). Yih presents Lin's (1955) improvements on Heisenberg's analysis with Shen's (1954) calculations of the stability curves.

Plane Couette flow

This is the flow confined between two parallel plates; it is driven by the motion of one of the plates parallel to itself. The basic velocity profile is linear, with $U \propto y$. Contrary to the experimentally observed fact that the flow does become turbulent at high Reynolds numbers, all linear analyses have shown that the flow is stable to small disturbances. It is believed that the observed instability is caused by disturbances of finite magnitude.

Pipe flow

The absence of an inflection point in the velocity profile signifies that the flow is inviscidly stable. All linear stability calculations of the *viscous* problem have also shown that the flow is stable to small disturbances. In contrast, most experiments show that the transition to turbulence takes place at a Reynolds number of about $\mathrm{Re} = U_{\max} d/\nu \sim 3000$. However, careful experiments, some of them performed by Reynolds in his classic investigation of the onset of turbulence, have been able to maintain laminar flow up to $\mathrm{Re} = 50{,}000$. Beyond this the observed flow is invariably turbulent. The observed transition has been attributed to one of the following effects: 1) It could be a finite amplitude effect; 2) the turbulence may be initiated at the entrance of the tube by boundary-layer instability (Fig. 7.2); and 3) the instability could be caused by a slow rotation of the inlet flow which, when added to the Poiseuille distribution, has been shown to result in instability. This is still under investigation. New insights into the instability and transition of pipe flow were described by Eckhardt et al. (2007) by analysis via dynamical systems theory and comparison with recent carefully crafted experiments by them and others. They characterized the turbulent state as a *chaotic saddle in state space*. The boundary between laminar and turbulent flow was found to be exquisitely sensitive to initial conditions. Because pipe flow is linearly stable, finite amplitude disturbances are necessary to cause transition, but as the Reynolds number increases, the amplitude of the critical disturbance diminishes.

The boundary between laminar and turbulent states appears to be characterized by a pair of vortices closer to the walls that give the strongest amplification of the initial disturbance.

Boundary layers with pressure gradients

Recall from Section 8.7 that when pressure decreases in the direction of flow, the pressure gradient is said to be *favorable*, and when pressure increases in the direction of flow the pressure gradient is said to be *adverse*. It was shown there that boundary layers developing in an adverse pressure gradient have a point of inflection in the velocity profile. This has a dramatic effect on stability characteristics. A schematic plot of the marginal stability curve for a boundary layer with favorable and adverse gradients of pressure is shown in Fig. 9.27. The ordinate in the plot represents the longitudinal wave number, and the abscissa represents the Reynolds number based on the free-stream velocity and the displacement thickness δ^* of the boundary layer. The marginal stability curve divides stable and unstable regions, with the region within the loop representing instability. Because the boundary layer thickness grows along the direction of flow, Re_δ increases with x, and points at various downstream distances are represented by larger values of Re_δ.

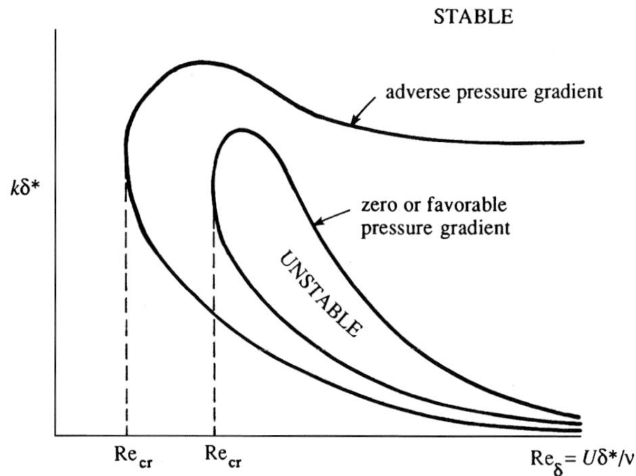

FIGURE 9.27 Sketch of marginal stability curves for a laminar boundary layers with favorable and adverse pressure gradients in terms of the displacement-thickness Reynolds number $U_o\delta^*/\nu$ and the dimensionless wave number $k\delta^*$ of the disturbance. The addition of the inflection point in the adverse-pressure gradient case increases the parametric realm of instability.

The following features can be noted in the figure. Boundary-layer flows are stable for low Reynolds numbers, but may become unstable as the Reynolds number increases. The effect of increasing viscosity is therefore stabilizing in this range. For boundary layers with a zero pressure gradient (Blasius flow) or a favorable pressure gradient, the instability loop shrinks to zero as $\text{Re}_\delta \to \infty$. This is consistent with the fact that these flows do not have a point of inflection in the velocity profile and are therefore inviscidly stable. In contrast, for boundary layers with an adverse pressure gradient, the instability loop does not shrink to zero; the upper branch of the marginal stability curve now becomes flat with a limiting value of k_∞ as $\text{Re}_\delta \to \infty$. The flow is then unstable to disturbances with wave numbers in the range $0 < k < k_\infty$. This is consistent with the existence of a point of inflection in the velocity profile, and the results of the shear layer calculations (Fig. 9.26). Note also that the critical Reynolds number is lower for flows with adverse pressure gradients.

Table 9.1 summarizes the results of the linear stability analyses of some common parallel viscous flows. The first two flows in the table have points of inflection in the velocity profile and are inviscidly unstable; the viscous solution shows either a zero or a small critical Reynolds number. The remaining flows are stable in the inviscid limit. Of these, the Blasius boundary layer and the plane Poiseuille flow are unstable in the presence of viscosity, but have high critical Reynolds numbers. Although the idealized tanh profile for a shear layer, assuming straight and parallel streamlines, is immediately unstable, more recent work by Bhattacharya et al. (2006), which allowed for the basic flow to be two dimensional, has yielded a finite critical Reynolds number.

While the results presented in the preceding paragraphs document flows where viscous effects are destabilizing, the mechanism of this destabilization has not been identified. One means of describing the destabilization mechanism relies on use of the equation for integrated kinetic energy of the disturbance:

$$\frac{d}{dt}\int \frac{1}{2}u_i^2 dV = -\int u_i u_j \frac{\partial U_i}{\partial x_j} dV - \Lambda, \tag{9.105}$$

TABLE 9.1 Linear stability results of common viscous parallel flows.

Flow	$U(y)/U_0$	Re_{cr}	Remarks
Jet	$\mathrm{sech}^2(y/L)$	4	
Shear layer	$\tanh(y/L)$	0	Always unstable
Blasius	(see Fig. 10.5)	520	Re based on δ^*
Plane Poiseuille	$1-(y/L)^2$	5780	$L=$ half − width
Pipe flow	$1-(r/R)^2$	∞	Always stable
Plane Couette	y/L	∞	Always stable

where V is a stationary volume having stream-wise control surfaces chosen to coincide with the walls where no-slip conditions are satisfied or where $u_i \to 0$, and having a length (in the stream-wise direction) that is an integer number of disturbance wavelengths (see Fig. 9.28). In (9.105), $\Lambda = \nu \int (\partial u_i/\partial x_i)^2 dV$ is the total viscous dissipation rate of kinetic energy in V. This disturbance kinetic energy equation can be derived from the incompressible Navier-Stokes momentum equation for the flow (see Exercise 9.16).

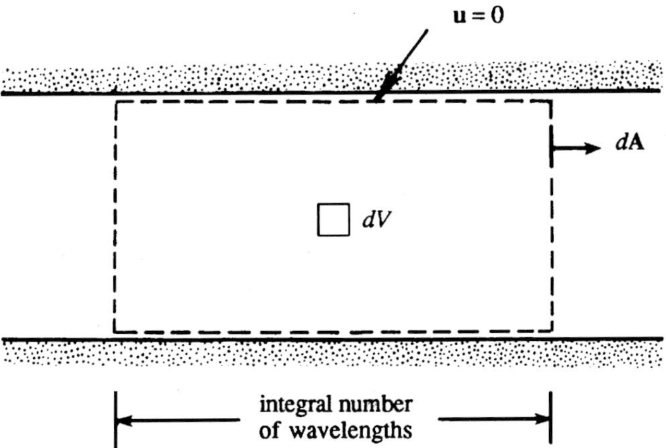

FIGURE 9.28 A control volume for deriving (9.105). Here, there is zero net flux across boundaries. This control volume can be extended to boundary-layer flow stability, when the boundary layer forms on the lower wall, by placing the upper control surface far enough from the lower wall so that the disturbance velocity $u_i \to 0$ on this control surface, even if this control surface may not abut the upper wall.

For two-dimensional disturbances in a shear flow defined by $\mathbf{U} = [U(y), 0, 0]$, the disturbance energy equation becomes:

$$\frac{d}{dt}\int \frac{1}{2}(u^2+v^2)dV = -\int uv\frac{\partial U}{\partial y}dV - \Lambda,$$

and has a simple interpretation. The first term is the rate of change of kinetic energy of the two-dimensional disturbance, and the second term is the rate of production of disturbance energy by the interaction of the product uv (also known as the Reynolds shear stress) and the mean shear $\partial U/\partial y$. (The concept of Reynolds stresses is explained in Chapter 10.) The point to note here is that the value of the product uv averaged over a period is zero if the velocity components u and v are out of phase; for example, the mean value of uv is zero if $u = \sin(t)$ and $v = \cos(t)$. In inviscid parallel flows without a point of inflection in the velocity profile, the u and v components are such that the disturbance field cannot extract energy from the basic shear flow, thus resulting in stability. The presence of viscosity, however, changes the phase relationship between u and v, which causes the spatial integral of $-uv(\partial U/\partial y)$ to be positive and larger than the viscous dissipation rate. This is how viscous effects can cause instability.

9.12 Experimental verification of boundary-layer instability

This section presents the results of stability calculations of the Blasius boundary-layer profile and compares them with experimental results. Because of the nearly parallel nature of the Blasius flow, most stability calculations are based on an analysis of the Orr-Sommerfeld equation, which assumes a parallel flow. The first calculations were performed by Tollmien in 1929 and Schlichting in 1933. Instead of assuming exactly the Blasius profile (which can be specified only numerically), they used the profile:

$$\frac{U}{U_\infty} = \left\{ \begin{array}{ll} 1.7(y/\delta) & 0 \le y/\delta \le 0.1724 \\ 1 - 1.03[1 - (y/\delta)^2] & 0.1724 \le y/\delta \le 1 \\ 1 & y/\delta \ge 1 \end{array} \right\},$$

which, like the Blasius profile, has a zero curvature at the wall. The calculations of Tollmien and Schlichting showed that unstable waves appear when the Reynolds number is high enough; the unstable waves in a viscous boundary layer are called *Tollmien-Schlichting waves*. Until 1947 these waves remained undetected, and the experimentalists of the period believed that the transition in a real boundary layer was probably a finite-amplitude effect. The speculation was that large disturbances cause locally adverse pressure gradients, which resulted in a local separation and consequent transition. The theoretical view, in contrast, was that small disturbances of the right frequency or wavelength can amplify if the Reynolds number is large enough.

Verification of the theory was finally provided by some clever experiments conducted by Schubauer and Skramstad (1947). The experiments were conducted in a wind tunnel specially designed to suppress fluctuations in the free-stream flow. The experimental technique used was novel. Instead of depending on natural disturbances, they introduced periodic disturbances of known frequency by means of a vibrating metallic ribbon stretched across the flow close to the wall. The ribbon was vibrated by passing an alternating current through it in the field of a magnet. The subsequent development of the disturbance was measured downstream via hot-wire anemometry. Such techniques later became standard.

The experimental data are shown in Fig. 9.29, which also shows the calculations of Schlichting and the more accurate calculations of Shen (1954). Instead of the wave number, the ordinate represents the frequency of the disturbance, which is easier to measure. It is apparent that the agreement between Shen's calculations and the experimental data is very good.

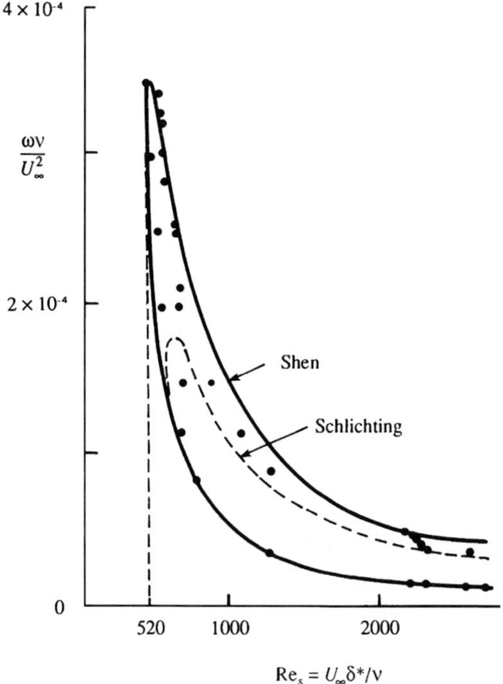

FIGURE 9.29 Marginal stability curve for a Blasius boundary layer. Theoretical solutions of Shen and Schlichting are compared with experimental data of Schubauer and Skramstad.

The detection of the Tollmien-Schlichting waves is regarded as a major accomplishment of linear stability theory. The ideal conditions for their existence are two dimensionality and negligible fluctuations in the free stream. These waves have been found to be sensitive to small deviations from the ideal conditions, and that is why they can be observed only under carefully controlled experimental conditions with artificial excitation. Today, Tollmien-Schlichting waves are often studied using numerical simulations. People who care about historical fairness have suggested that the waves should only be referred to as TS waves, to honor Tollmien, Schlichting, Schubauer, and Skramstad. TS waves have also been observed in natural flow (Bayly et al., 1988).

Fransson and Alfredsson (2003) have shown that the asymptotic suction profile (solved in Exercise 8.5) significantly delays transition stimulated by free-stream turbulence or by Tollmien-Schlichting waves. Specifically, the value of $\mathrm{Re}_{cr} = 520$ based on δ^* in Table 9.1 is increased for suction velocity ratio $v_0/U_\infty = -0.00288$ to more than 54,000. The large stabilizing effect is a result of the change in the shape of the stream-wise velocity profile from the Blasius profile to an exponential.

9.13 Comments on nonlinear effects

To this point we have discussed only linear stability theory, which considers infinitesimal perturbations and predicts exponential growth when the relevant parameter exceeds a critical value. The effect of the perturbations on the basic flow is neglected in the linear theory. An examination of (9.105) shows that the perturbation field must be such that the average uv (the average taken over a wavelength) must be non-zero for the perturbations to extract energy from the basic shear; similarly, the heat flux, the average of uT', must be non-zero in a thermal convection problem. These rectified fluxes of momentum and heat change the *basic* velocity and temperature fields. Linear instability theory neglects these changes of the basic state. A consequence of the constancy of the basic state is that the growth rate of the perturbations is also constant, leading to predictions of exponential growth. However, after some time, the perturbations eventually become so large that the rectified fluxes of momentum and heat significantly change the basic state, which in turn alters the growth of the perturbations.

A frequent effect of nonlinearity is to change the basic state in such a way as to arrest the growth of the disturbances after they have reached significant amplitude via their initial exponential growth. (Note, however, that the effect of nonlinearity can sometimes be destabilizing; for example, the instability in a pipe flow may be a finite-amplitude effect because the flow is stable to infinitesimal disturbances.) Consider thermal convection in the annular space between two vertical cylinders rotating at the same speed. The outer wall of the annulus is heated and the inner wall is cooled. For small heating rates the flow is steady. For large heating rates a system of regularly spaced waves develop and progress azimuthally at a uniform speed without changing their shape. (This is the equilibrated form of baroclinic instability, discussed in Section 12.15) At still larger heating rates an irregular, aperiodic, or chaotic flow develops. The chaotic response to constant forcing (in this case the heating rate) is an interesting nonlinear effect and is discussed further in Section 9.15. Meanwhile, a brief description of the transition from laminar to turbulent flow is given in the next section.

9.14 Transition

The process by which a laminar flow changes to a turbulent one is called *transition*. Instability of a laminar flow does not immediately lead to turbulence, which is a severely nonlinear and chaotic flow state. After the initial breakdown of laminar flow because of amplification of small disturbances, the flow goes through a complex sequence of changes, finally resulting in the chaotic state we call turbulence. The process of transition is greatly affected by such experimental conditions as the intensity of fluctuations of the free stream and the roughness of any walls. The sequence of events that lead to turbulence is also greatly dependent on flow geometry. For example, the scenario of transition in a wall-bounded shear flows is different from that in free shear flows such as jets and wakes.

Early stages of transition consist of a succession of instabilities on increasingly complex basic flows, an idea first suggested by Landau in 1944 (see Landau and Lifshitz, 1959). The basic state of wall-bounded parallel shear flows becomes unstable to two-dimensional TS waves, which grow and eventually reach equilibrium at some finite amplitude. This steady state can be considered a new background state, and calculations show that it is generally unstable to *three-dimensional* waves of short wavelength, which vary in the cross-stream or *span-wise* direction. (If x denotes the stream-wise flow direction and y denotes the wall-normal direction, then the z-axis lies in the *span-wise* direction.) We shall call this the *secondary instability*. Interestingly, the secondary instability does not reach equilibrium at finite amplitude but directly evolves to a fully turbulent flow. Recent calculations of the secondary instability have been quite successful in reproducing critical Reynolds numbers for various wall-bounded flows, as well as predicting three-dimensional structures observed in experiments.

A key experiment on the three-dimensional nature of the transition process in a boundary layer was performed by Klebanoff et al. (1962). They conducted a series of controlled experiments by which they introduced three-dimensional disturbances on a field of TS waves in a boundary layer. The TS waves were as usual artificially generated by an electromagnetically vibrated ribbon, and the three dimensionality of a particular span-wise wavelength was introduced by placing spacers (small pieces of transparent tape) at equal intervals underneath the vibrating ribbon (Fig. 9.30). When the amplitude of the TS waves became roughly 1% of the free-stream velocity, the three-dimensional perturbations grew rapidly and resulted in a span-wise irregularity of the stream-wise velocity displaying peaks and valleys in the amplitude of u. The three-dimensional disturbances continued to grow until the boundary layer became fully turbulent. The chaotic flow seems to result from the nonlinear evolution of the secondary instability, and numerical calculations have accurately reproduced several characteristic features of real flows (see Figs. 7 and 8 in Bayly et al., 1988).

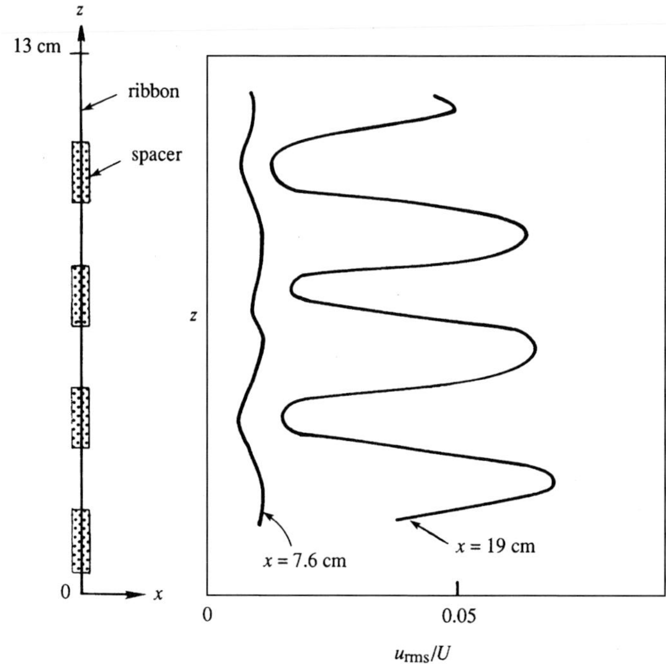

FIGURE 9.30 Three-dimensional unstable waves initiated by a vibrating ribbon. Measured distributions of intensity of the u-fluctuation at two distances from the ribbon are shown. Clearly, the span-wise variation enhances the signature of the instability. *P.S. Klebanoff et al.*, Journal of Fluid Mechanics, 12, *1–34, 1962; reprinted with the permission of Cambridge University Press.*

It is interesting to compare the chaos observed in turbulent shear flows with that in controlled low-order dynamical systems such as the Bénard convection or Taylor-Couette flow. In these low-order flows only a small number of modes participate in the dynamics because of the strong constraint of the boundary conditions. All but a few low modes are identically zero, and the chaos develops in an orderly way. As the constraints are relaxed (we can think of this as increasing the number of allowed Fourier modes), the evolution toward apparent chaos becomes less orderly.

Transition in a free shear layer, such as a jet or a wake, occurs in a different manner. Because of the inflectional velocity profiles involved, these flows are unstable at a very low Reynolds numbers, that is, of order 10 compared to about 10^3 for wall-bounded flows. The breakdown of the laminar flow therefore occurs quite readily and close to the origin of such a flow. Transition in a free shear layer is characterized by the appearance of a rolled-up row of vortices, whose wavelength corresponds to the one with the largest growth rate. Frequently, pairs of such vortices regroup themselves to produce a dominant wavelength twice the original wavelength. Small-scale fluctuations develop in the strain fields between and within these larger scale vortices, finally leading to turbulence.

9.15 Deterministic chaos

The discussion in the previous section has shown that dissipative nonlinear systems such as fluid flows reach a random or chaotic state when the parameter measuring nonlinearity (say, the Reynolds number or the Rayleigh number) is large. The evolution from laminar flow to the chaotic state generally takes place through a sequence of transitions, with the exact route depending on the flow geometry and other characteristics. It has been realized that chaotic behavior not only occurs in continuous systems having an infinite number of degrees of freedom, but also in discrete nonlinear systems having only a small number of degrees of freedom, governed by ordinary nonlinear differential equations. In this context, a *chaotic system* is defined as one in which the solution is *extremely sensitive to initial conditions*. That is, solutions with arbitrarily close initial conditions evolve into quite different states. Other symptoms of a chaotic system are that the solutions are *aperiodic*, and that the spectrum of fluctuations is broadband instead of being composed of a few discrete frequencies or wave numbers.

Numerical integrations (to be shown later in this section) have demonstrated that nonlinear systems governed by a finite set of deterministic ordinary differential equations allow chaotic solutions in response to a steady forcing. This fact is interesting because in a dissipative *linear* system a constant forcing ultimately (after the decay of the transients) leads to a constant response, a periodic forcing leads to a periodic response, and a random forcing leads to a random response. In the presence of nonlinearity, however, a constant forcing can lead to a variable response, both periodic and aperiodic. One example is the periodic oscillation in the flow behind a blunt body at Re ~ 40 (associated with the initial appearance of the von Karman vortex street) and the breakdown of the oscillation into turbulent flow at larger values of the Reynolds number.

It has been found that transition to chaos in the solution of ordinary nonlinear differential equations displays a certain *universal* behavior and proceeds in one of a few different ways. Transition to turbulence in fluid flows may be related to the development of chaos in the solutions of these simple systems. In this section, some of the elementary ideas involved are presented, starting with the definitions for phase space and attractors, moving on to the Lorenz model of thermal convection and scenarios for transition to chaos, and then concluding with a description of the implications of such phenomena. An introduction to the subject of chaos is given by Berge et al. (1984); a useful review is given in Lanford (1982). The subject has far-reaching cosmic consequences in physics and evolutionary biology, as discussed by Davies (1988).

Very few nonlinear equations have analytical solutions. For nonlinear systems, a typical procedure is to find a numerical solution and display its properties in a space whose axes are the *dependent* variables. Consider the equation governing the motion of a simple pendulum of length l:

$$\ddot{X} + (g/l)\sin X = 0,$$

where X is the *angular* displacement and $\ddot{X}$ is the angular acceleration. The equation is nonlinear because of the sin$X$ term. This second-order equation can be split into two coupled first-order equations:

$$\dot{X} = Y \quad \text{and} \quad \dot{Y} = -(g/l)\sin X. \tag{9.106}$$

Starting with some initial conditions on X and Y, one can integrate (9.106) forward in time. The behavior of the system can be studied by describing how the variables $Y(=\dot{X})$ and X vary as functions of time. For the pendulum problem, the space whose axes are $\dot{X}$ and X is called a *phase space,* and the evolution of the system is described by a *trajectory* in this space. The dimension of the phase space is called the *degree of freedom* of the system; it equals the number of independent initial conditions necessary to specify the system. For example, there are two degrees of freedom for the set (9.106).

Dissipative systems are characterized by the existence of *attractors,* which are structures in the phase space toward which neighboring trajectories approach as $t \to \infty$. An attractor can be a *fixed point* representing a stable steady flow or a closed curve (called a *limit cycle*) representing a stable oscillation (Fig. 9.31a, b). The nature of the attractor depends on the value of the nonlinearity parameter, which will be denoted by R in this section. As R is increased, the fixed point representing a steady solution may change from being an attractor to a repeller with spirally outgoing trajectories, signifying that the steady flow has become unstable to infinitesimal perturbations. Frequently, the trajectories are then attracted by a limit cycle, which means that the unstable steady solution gives way to a steady oscillation (Fig. 9.31b). For example, the steady flow behind a blunt body becomes oscillatory as Re is increased, resulting in the periodic von Karman vortex street (Fig. 8.19).

The branching of a solution at a critical value R_{cr} of the nonlinearity parameter is called a *bifurcation*. Thus, we say that the stable steady solution of Fig. 9.31a bifurcates to a stable limit cycle as R increases through R_{cr}. This can be represented on the graph of a dependent variable (say, X) versus R (Fig. 9.31c). At $R = R_{cr}$, the solution curve branches into two paths; the two values of X on these branches (say, X_1 and X_2) correspond to the maximum and minimum values of X in Fig. 9.31b. It is seen that the size of the limit cycle grows larger as $(R - R_{cr})$ becomes larger. Limit cycles, representing oscillatory response with amplitude independent of initial conditions, are characteristic features of nonlinear systems. Linear stability theory predicts an exponential growth of the perturbations if $R > R_{cr}$, but a nonlinear theory frequently shows that the perturbations eventually equilibrate to a steady oscillation whose amplitude increases with $(R - R_{cr})$.

A famous fluid-flow example involving these concepts comes from thermal convection in a layer heated from below (the Bénard problem). Lorenz (1963) demonstrated that the development of chaos is associated with the flow's attractor acquiring certain strange properties. He considered a layer with stress-free boundaries. Assuming nonlinear disturbances in the form of rolls invariant in the y direction, and defining a disturbance stream function in the $x - z$ plane by $u = -\partial\psi/\partial z$ and $w = \partial\psi/\partial x$, he substituted solutions of the form

$$\psi \propto X(t)\cos(\pi z)\sin(kx) \quad \text{and} \quad T' \propto Y(t)\cos(\pi z)\cos(kx) + Z(t)\sin(2\pi z) \tag{9.107}$$

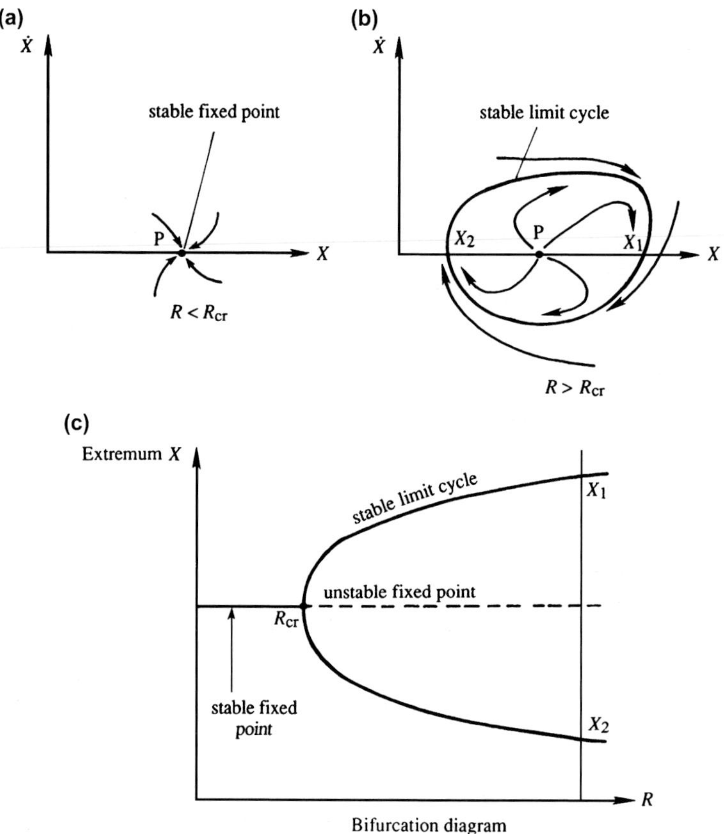

FIGURE 9.31 Attractors in a phase plane of X and $\dot{X}$. In (a), point P is an attractor. For a larger value of R, the nonlinearity parameter, panel (b) shows that P becomes an unstable fixed point (a *repeller*), and the trajectories are attracted to an orbit or limit cycle that encircles P. Panel (c) is the bifurcation diagram corresponding to this situation.

into the equations of motion. Here, T' is the departure of temperature from the state of no convection, k is the wave number of the perturbation, and the boundaries are at $z = \pm 1/2$. It is clear that X is proportional to the speed of convective motion, Y is proportional to the temperature difference between the ascending and descending currents, and Z is proportional to the distortion of the average vertical profile of temperature from linearity. (Note in (9.107) that the x-average of the term multiplied by $Y(t)$ is zero, so that this term does not cause distortion of the basic temperature profile.) As discussed in Section 9.5, Rayleigh's linear analysis showed that solutions of the form (9.107), with X and Y constants and $Z = 0$, would develop if Ra slightly exceeds the critical value $\mathrm{Ra_{cr}} = 27\pi^4/4$. Eqs. (9.107) are expected to give realistic results when Ra is slightly supercritical but not when strong convection occurs because only the lowest wave number terms are retained.

On substitution of (9.107) into the equations of motion, Lorenz finally obtained the system:

$$\dot{X} = \mathrm{Pr}(Y - X), \quad \dot{Y} = -XZ + rX - Y, \quad \text{and} \quad \dot{Z} = XY - bZ, \tag{9.108}$$

where Pr is the Prandtl number, $r = \mathrm{Ra}/\mathrm{Ra_{cr}}$, and $b = 4\pi^2/(\pi^2 + k^2)$. Eqs. (9.108) are a set of nonlinear equations with three degrees of freedom, which means that the phase space is three dimensional.

Eqs. (9.108) allow the steady solution $X = Y = Z = 0$, representing the state of no convection. For $r > 1$ the system possesses two additional steady-state solutions, which we shall denote by $\overline{X} = \overline{Y} = \pm\sqrt{b(r-1)}$, and $\overline{Z} = r - 1$; the two signs correspond to the two possible senses of rotation of the rolls. (The fact that these steady solutions satisfy (9.108) can easily be checked by substitution after setting $\dot{X} = \dot{Y} = \dot{Z} = 0$.) Lorenz showed that the steady-state convection becomes unstable if r is large. Choosing $\mathrm{Pr} = 10$, $b = 8/3$, and $r = 28$, he numerically integrated the set and found that the solution never repeats itself; it is aperiodic and wanders about in a chaotic manner. Fig. 9.32 shows the variation of $X(t)$, starting with some initial conditions. (The variables $Y(t)$ and $Z(t)$ also behave in a similar way.) It is seen that the amplitude of the convecting motion initially oscillates around one of the steady values $\overline{X} = \pm\sqrt{b(r-1)}$, with the oscillations growing in magnitude. When it is large enough, the amplitude suddenly goes through zero to start oscillations of opposite sign about the other value of $\overline{X}$. That is, the motion switches in a chaotic manner between two oscillatory limit cycles, with the number of oscillations between transitions seemingly random. Calculations show that the variables X, Y, and Z have continuous spectra and that the solution is extremely sensitive to initial conditions.

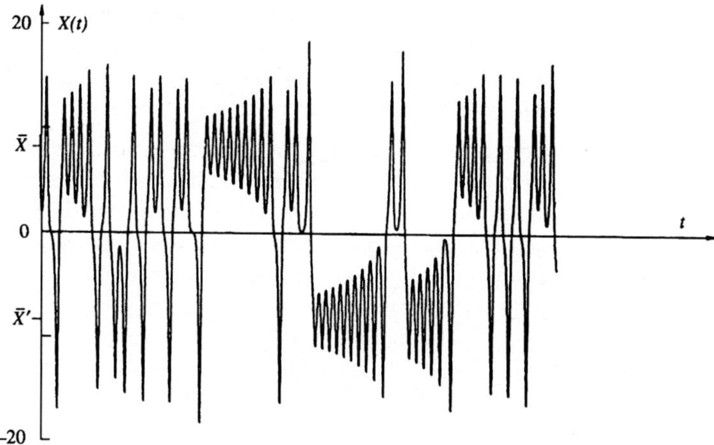

FIGURE 9.32 Variation of $X(t)$ in the Lorenz model. Note that the solution oscillates erratically around the two steady values $\overline{X}$ and $\overline{X}'$ and does not have a reliable period. *P. Bergé, Y. Pomeau, and C. Vidal,* Order Within Chaos, *1984; reprinting permitted by Heinemann Educational, a division of Reed Educational & Professional Publishing Ltd.*

The trajectories in the phase space of the Lorenz model of thermal convection are shown in Fig. 9.33. The centers of the two loops represent the two steady convections $\overline{X} = \overline{Y} = \pm\sqrt{b(r-1)}$ and $\overline{Z} = r - 1$. The structure resembles two rather flat loops of ribbon, one lying slightly in front of the other along a central band with the two joined together at the bottom of that band. The trajectories go clockwise around the left loop and counterclockwise around the right loop; two trajectories never intersect. The structure shown in Fig. 9.33 is an attractor because orbits starting with initial conditions *outside of the attractor* merge on it and then follow it. The attraction is a result of dissipation in the system. The aperiodic attractor, however, is unlike the normal attractor in the form of a fixed point (representing steady motion) or a closed curve (representing a limit cycle). This is because two trajectories *on the aperiodic attractor*, with infinitesimally different initial conditions, follow each other closely only for a while, eventually diverging to very different final states. This is the basic reason for sensitivity to initial conditions.

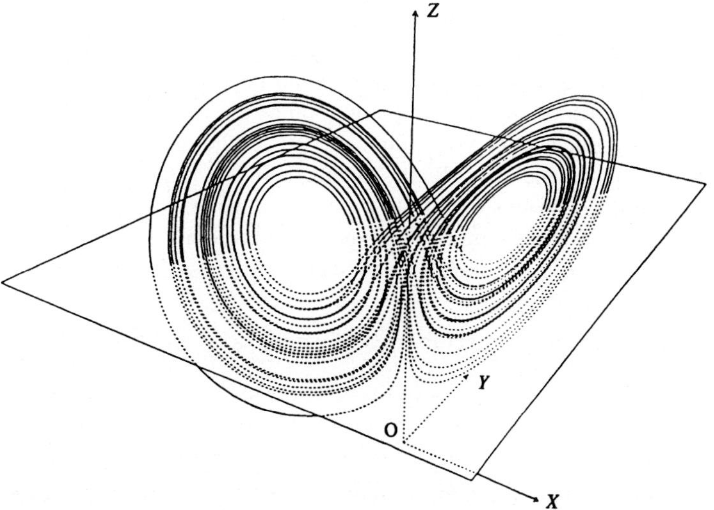

FIGURE 9.33 The Lorenz attractor. All nearby initial conditions are attracted to this double loop structure, but any two such trajectories will eventually diverge, even if they begin very close together. The centers of the two loops represent the two steady solutions $(\overline{X}, \overline{Y}, \overline{Z})$.

For these reasons the aperiodic attractor is called a *strange attractor*. The idea of a strange attractor is not intuitive because it has the dual property of attraction and divergence. Trajectories starting from the neighboring regions in phase space are drawn toward it, but once on the attractor the trajectories eventually diverge and result in chaos. An ordinary attractor in phase space allows the trajectories from slightly different initial conditions to merge, so that the *memory* of initial conditions is lost. However, the strange attractor ultimately accentuates small initial condition differences. The idea of the strange attractor was first conceived by Lorenz, and since then attractors of other chaotic systems have also been studied. They all have the common property of aperiodicity, continuous spectra, and sensitivity to initial conditions.

Thus far we have described discrete dynamical systems having only a small number of degrees of freedom and seen that aperiodic or chaotic solutions result when the nonlinearity parameter is large. Several routes or scenarios of transition to chaos in such systems have been identified. Two of these are described briefly here.

(1) *Transition through subharmonic cascade*: As R is increased, a typical nonlinear system develops a limit cycle of a certain frequency ω. With further increase of R, several systems are found to generate additional frequencies $\omega/2$, $\omega/4$, $\omega/8, \ldots$. The addition of frequencies in the form of *subharmonics* does not change the periodic nature of the solution, but the period doubles each time a lower harmonic is added. The period doubling takes place more and more rapidly as R is increased, until an *accumulation point* (Fig. 9.34) is reached, beyond which the solution wanders about in a chaotic manner. At this point the peaks disappear from the temporal-frequency spectrum, which becomes broadband. Many systems approach chaotic behavior through period doubling. Feigenbaum (1978) proved the important result that this kind of transition develops in a *universal* way, independent of the particular nonlinear systems studied. If R_n represents the value for development of a new subharmonic, then R_n converges in a geometric series with:

$$\frac{R_n - R_{n-1}}{R_{n+1} - R_n} \to 4.6692 \quad \text{as} \quad n \to \infty$$

That is, the horizontal gap between two bifurcation points is about a fifth of the previous gap. The vertical gap between the branches of the bifurcation diagram also decreases, with each gap about two-fifths of the previous gap. In other words, the bifurcation diagram (Fig. 9.34) becomes "self-similar" as the accumulation point is approached. (Note that Fig. 9.34 has not been drawn to scale, for illustrative purposes.) Experiments in low Prandtl number fluids (such as liquid metals) indicate that Bénard convection in the form of rolls develops oscillatory motion of a certain frequency ω at $\text{Ra} = 2\text{Ra}_{\text{cr}}$. As Ra is further increased, additional frequencies $\omega/2$, $\omega/4$, $\omega/8$, $\omega/16$, and $\omega/32$ have been observed. The convergence ratio has been measured to be 4.4, close to the value of 4.669 predicted by Feigenbaum's theory. The experimental evidence is discussed further in Berge et al. (1984).

(2) *Transition through quasi-periodic regime*: Ruelle and Takens (1971) have mathematically proven that certain systems need only a *small number* of bifurcations to produce chaotic solutions. As the nonlinearity parameter is increased, the steady solution loses stability and bifurcates to an oscillatory limit cycle with frequency ω_1. As R is increased, two more frequencies (ω_2 and ω_3) appear through additional bifurcations. In this scenario the ratios of the three frequencies (such as ω_1/ω_2) are *irrational* numbers, so that the motion consisting of the three frequencies is not exactly periodic. (When the ratios are rational numbers, the motion is exactly periodic. To see this, think of the Fourier series of a periodic function in which the various terms represent sinusoids of the fundamental frequency ω and its harmonics $2\omega, 3\omega, \ldots$. Some of the Fourier coefficients could be zero.) The spectrum for these systems suddenly develops broadband characteristics of chaotic motion as soon as the third frequency ω_3 appears. The exact point at which chaos sets in is not easy to detect in a measurement; in fact the third frequency may not be identifiable in the spectrum before it becomes broadband. The Ruelle-Takens theory is fundamentally different from that of Landau, who conjectured that turbulence develops due to an *infinite* number of bifurcations, each generating a new higher frequency, so that the spectrum becomes saturated with peaks and resembles a continuous one. According to Berge et al. (1984), the Bénard convection experiments in *water* seem to suggest that turbulence in this case probably sets in according to the Ruelle-Takens scenario.

The development of chaos in the Lorenz attractor is more complicated and does not follow either of the two routes mentioned in the preceding discussion.

Final remarks

Perhaps the most intriguing characteristic of a chaotic system is the extreme *sensitivity to initial conditions*. That is, solutions with arbitrarily close initial conditions evolve into two quite different states. Most nonlinear systems are susceptible to chaotic behavior. The extreme sensitivity to initial conditions implies that nonlinear phenomena (including the weather, in which Lorenz was primarily interested when he studied the convection problem) are essentially unpredictable, no matter how well we know the governing equations or the initial conditions. Although the subject of chaos has become a scientific revolution recently, Henri Poincaré conceived the central idea in 1908. He did not, of course, have the computing facilities to demonstrate it through numerical integration.

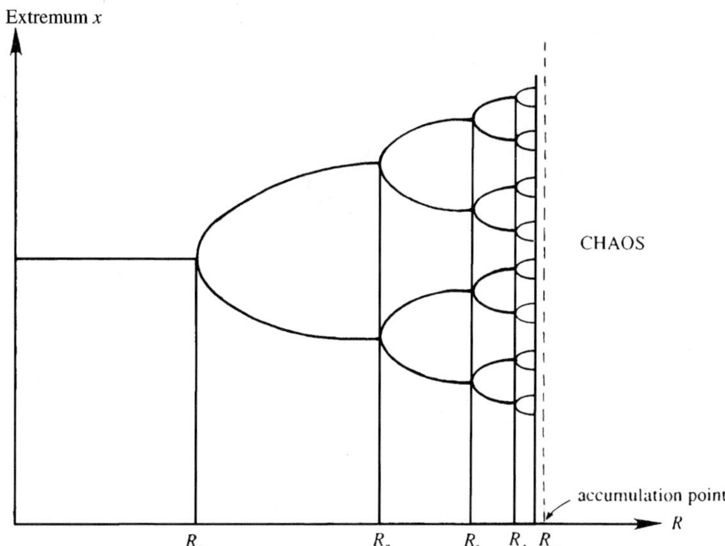

FIGURE 9.34 Bifurcation diagram during period doubling. The period doubles at each value R_n of the nonlinearity parameter. For large n the "bifurcation tree" becomes self-similar. Chaos sets in beyond the accumulation point R_∞. This process may mimic the transition from laminar to turbulent flow for some circumstances.

It is important to realize that the behavior of chaotic systems is not *intrinsically* non-deterministic; as such the implication of deterministic chaos is different from that of the uncertainty principle of quantum mechanics. In any case, the extreme sensitivity to initial conditions implies that the *future is essentially unknowable* because it is never possible to know the initial conditions *exactly*. As discussed by Davies (1988), this fact has interesting philosophical implications regarding the evolution of the universe, including that of living species.

We have examined certain elementary ideas about how chaotic behavior may result in simple nonlinear systems having only a small number of degrees of freedom. Turbulence in a continuous fluid medium is capable of displaying an infinite number of degrees of freedom, and it is unclear whether the study of chaos can throw a great deal of light on more complicated transitions such as those in pipe or boundary-layer flow. However, the fact that nonlinear systems can have chaotic solutions for a large value of the nonlinearity parameter (see Fig. 9.32) is an important result by itself.

Exercises

9.1 A perturbed vortex sheet nominally located at $y = 0$ separates inviscid flows of differing density in the presence of gravity with downward acceleration g. The upper stream is semi-infinite and has density ρ_1 and horizontal velocity U_1. The lower stream has thickness h density ρ_2, and horizontal velocity U_2. A smooth flat impenetrable surface located at $y = -h$ lies below the second layer. The interfacial tension between the two fluids is σ. Assume a disturbance occurs on the vortex sheet with wave number $k = 2\pi/\lambda$, and complex wave speed c, i.e., $[y]_{sheet} = f(x,t) = f_o \mathrm{Re}\left\{e^{ik(x-ct)}\right\}$. The four boundary conditions are:

1) $u_1, v_1 \to 0$ as $y \to +\infty$

2) $v_2 = 0$ on $y = -h$.

3) $\mathbf{u}_1 \cdot \mathbf{n} = \mathbf{u}_2 \cdot \mathbf{n} =$ normal velocity of the vortex sheet on both sides of the vortex sheet:

4) $p_1 - p_2 = \sigma \frac{\partial^2 f}{\partial x^2}$ on the vortex sheet ($\sigma =$ interfacial surface tension)

a) Following the development in Section 9.3, show that:

$$c = \frac{\rho_1 U_1 + \rho_2 U_2 \coth(kh)}{\rho_1 + \rho_2 \coth(kh)} \pm \left[\frac{(g/k)(\rho_2 - \rho_1) + \sigma k}{\rho_1 + \rho_2 \coth(kh)} - \frac{\rho_1 \rho_2 (U_1 - U_2)^2 \coth(kh)}{(\rho_1 + \rho_2 \coth(kh))^2}\right]^{1/2}.$$

b) Use the result of part a) to show that the vortex sheet is *unstable* when:

$$\left(\tanh(kh) + \frac{\rho_2}{\rho_1}\right)\left(\frac{g}{k}\frac{(\rho_2 - \rho_1)}{\rho_2} + \frac{\sigma k}{\rho_2}\right) < (U_1 - U_2)^2.$$

c) Will the sheet be stable or unstable to long wavelength disturbances ($k \to 0$) when $\rho_2 > \rho_1$ for a fixed velocity difference?

d) Will the sheet be stable or unstable to short wavelength disturbances ($k \to \infty$) for a fixed velocity difference?

e) Will the sheet ever be unstable when $U_1 = U_2$?

f) Under what conditions will the thickness h matter?

9.2 Consider a fluid layer of depth h and density ρ_2 lying under a lighter, infinitely deep fluid of density $\rho_1 < \rho_2$. By setting $U_1 = U_2 = 0$, in the results of Exercise 9.1, the following formula for the phase speed is found:

$$c = \pm\left[\frac{(g/k)(\rho_2 - \rho_1) + \sigma k}{\rho_1 + \rho_2 \coth(kh)}\right]^{1/2}.$$

Now invert the sign of gravity and consider why drops form when a liquid is splashed on the underside of a flat surface. Are long or short waves most unstable? Does a professional painter want interior ceiling paint with high or low surface tension? For a smooth finish should the painter apply thin or thick coats of paint? Assuming the liquid has the properties of water (surface tension ≈ 0.072 N/m, density $\approx 10^3$ kg/m) and that the lighter fluid is air, what is the longest stable wavelength on the underside of a horizontal surface? [This is the *Rayleigh-Taylor instability* and it occurs when density and pressure gradients point in opposite directions. It may be readily observed by accelerating rapidly downward an upward-open cup of water.]

9.3 Suppression of the Kelvin-Helmholtz instability is often attempted through the addition of a thin porous or perforated barrier between the two streams. Consider an inviscid constant-density idealization of this situation in an unbounded environment where the initial flow speeds in the half-spaces above and below the barrier are U_1, and U_2, respectively. Here the thin barrier coincides with the x-axis ($y = 0$) and only influences vertical motions between the two fluid streams. This idealized barrier changes the dynamic boundary condition to:

$$p_2' - p_1' = \rho\beta\,(\partial\zeta/\partial t) \quad \text{on} \quad y = 0,$$

where p_1' and p_2' are pressure fluctuations above and below the barrier, ρ is the fluid density, β is a positive constant (with units length/time) that characterizes the barrier, and $\zeta(x,t) = \zeta_0 \exp\{i(\omega t - kx) + \alpha t\}$ is the surface that separates the upper and lower fluid streams. This separating surface may pass through the porous barrier when there are vertical velocity fluctuations at $y = 0$. Here, ζ_0, ω, k, and α are all real constants.

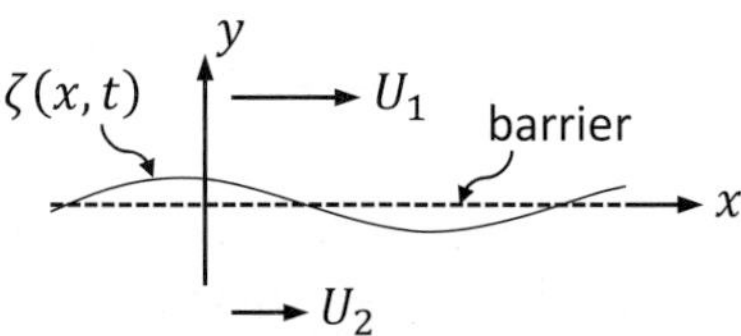

For small fluctuations, the velocity potentials above and below the separating surface are:

$$\phi_1(x,y,t) = A(y)\exp\{i(\omega t - kx) + \alpha t\} + U_1 x,$$
$$\phi_2(x,y,t) = B(y)\exp\{i(\omega t - kx) + \alpha t\} + U_2 x.$$

a) Use the given information, the linearized kinematic boundary condition, and the linearized dynamic boundary condition provided above (both applied on $y = 0$) to show that a potential flow analysis leads to the following relationship:

$$\frac{\omega - i\alpha}{k} = \frac{U_1 + U_2}{2} + \frac{i\beta}{4} \pm \frac{i}{2}\left[(U_1 - U_2)^2 - i\beta(U_1 + U_2) + \frac{\beta^2}{4}\right]^{1/2}.$$

b) Consider the part a) result with $U_2 = -U_1$, and answer the following questions.

i. How fast does the surface disturbance travel?

ii. What is the growth or decay rate of the surface disturbance?

iii. When $\beta = 0$, what flow condition(s) lead to neutrally stability?

iv. Should β be positive or negative to better control the Kelvin-Helmholtz instability?

v. Should the magnitude of β be increased or decreased to better control the Kelvin- Helmholtz instability when $U_2 - U_1 \neq 0$?

c) If $U_2 = -U_1 \neq 0$, can the Kelvin-Helmholtz instability be fully suppressed for a finite value of β?

d) In a real application where a perforated plate is placed between two fluid streams, there will be additional fluid-flow phenomena that are not present in the simplified analysis considered here. Mention and describe at least two of these.

9.4 Inviscid horizontal flow in the half space $y > 0$ moves at speed U over a porous surface located at $y = 0$. Here the fluid density ρ is constant and gravity plays no roll. A weak vertical velocity fluctuation occurs at the porous surface: $[v]_{surface} = v_o \mathrm{Re}\left\{e^{ik(x-ct)}\right\}$, where $v_o \ll U$.

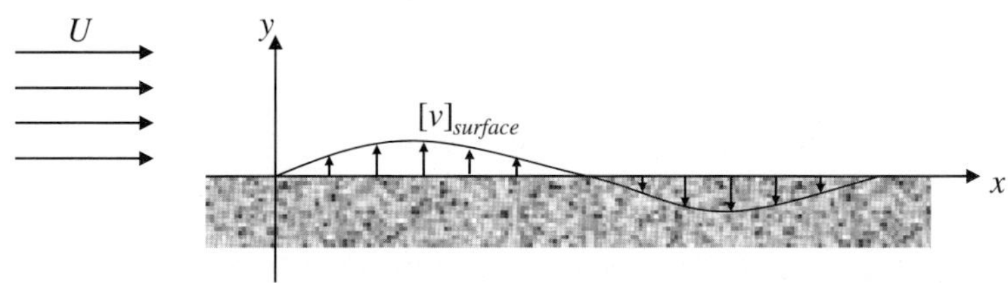

a) The velocity potential for the flow may be written $\widetilde{\phi} = Ux + \phi$, where ϕ leads to $[v]_{surface}$ at $y = 0$ and ϕ vanishes as $y \to +\infty$. Determine the perturbation potential ϕ in terms of v_o, U, ρ, k, c, and the independent variables (x, y, t).

b) The porous surface responds to pressure fluctuations in the fluid via: $[p - p_s]_{y=0} = -\gamma[v]_{surface}$, where p is the pressure in the fluid, p_s is the steady static pressure that is felt on the surface when the vertical velocity fluctuations are absent, and γ is a real material parameter that defines the porous surface's flow resistance. Determine a formula for c in terms of U, γ, ρ, and k.

c) What is the propagation velocity, $\mathrm{Re}\{c\}$, of the surface velocity fluctuation?

d) What sign should γ have for the flow to be stable? Interpret your answer.

9.5 Repeat Exercise 9.4 for a compliant surface nominally lying at $y = 0$ that is perturbed from equilibrium by a small surface wave: $[y]_{surface} = \zeta(x, t) = \zeta_o \mathrm{Re}\left\{e^{ik(x-ct)}\right\}$.

a) Determine the perturbation potential ϕ in terms of U, ρ, k, and c by assuming that ϕ vanishes as $y \to +\infty$, and that there is no flow through the compliant surface. Ignore gravity.

b) The compliant surface responds to pressure fluctuations in the fluid via: $[p - p_s]_{y=0} = -\gamma\zeta(x, t)$, where p is the pressure in the fluid, p_s is the steady pressure that is felt on the surface when the surface wave is absent, and γ is a real material parameter that defines the surface's compliance. Determine a formula for c in terms of U, γ, ρ, and k.

c) What is the propagation velocity, $\mathrm{Re}\{c\}$, of the surface waves?

d) If γ is positive, is the flow stable? Interpret your answer.

9.6 As a simplified version of flag waving, consider the stability of a simple membrane in a uniform flow. Here, the undisturbed membrane lies in the $x - z$ plane at $y = 0$, the flow is parallel to the x-axis at speed U, and the fluid has density ρ. The membrane has mass per unit area $= \rho_m$ and uniform tension per unit length $= T$. The membrane satisfies a dynamic equation based on pressure forces and internal tension combined with its local surface curvature:

$$\rho_m \frac{\partial^2 \zeta}{\partial t^2} = p_2 - p_1 + T\left(\frac{\partial^2 \zeta}{\partial x^2} + \frac{\partial^2 \zeta}{\partial z^2}\right).$$

Here, the vertical membrane displacement is given by $y = \zeta(x, z, t)$, and p_1 and p_2 are the pressures acting on the membrane from above and below, respectively. The velocity potentials for the undisturbed flow above (1) and below (2) the membrane are $\phi_1 = \phi_2 = Ux$. For the following items, assume a small amplitude wave is present on the membrane $\zeta(x, t) = \zeta_o \mathrm{Re}\left\{e^{ik(x-ct)}\right\}$ with k a real parameter, and assume that all deflections and other fluctuations are uniform in the z-direction and small enough for the usual linear simplifications. In addition, assume the static pressures above and below the membrane, in the absence of membrane motion, are matched.

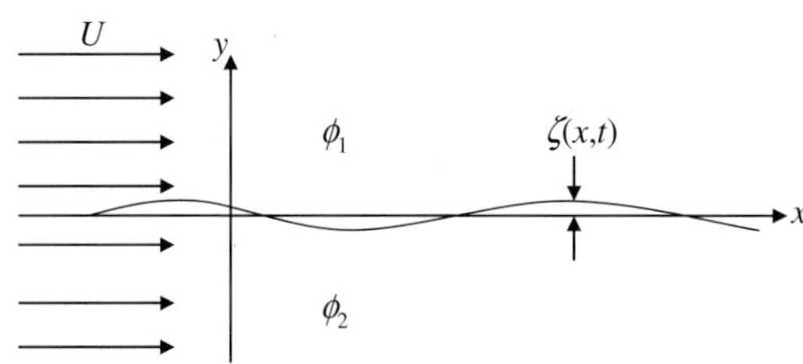

a) Using the membrane equation, determine the propagation speed of the membrane waves, Re{c}, in the absence of fluid loading (i.e., when $p_2 - p_1 = \rho = 0$).

b) Assuming inviscid flow above and below the membrane, determine a formula for c in terms of T, ρ_m, ρ, U, and k.

c) Is the membrane more or less unstable if U, T, ρ, and ρ_m are individually increased with the others held constant?

d) What is the propagation speed of the membrane waves when $U = 0$? Compare this to your answer for part a) and explain any differences.

9.7 Prove that $\sigma_r > 0$ for the thermal instability discussed in Section 9.5 via the following steps that include integration by parts and use of the boundary conditions (9.55).

a) Multiply (9.53) by $\widehat{T}^*$ and integrate the result from $z^* = -1/2$ to $z^* = +1/2$, where z^* is the dimensionless vertical coordinate, to find: $\sigma I_1 + I_2 = \int \widehat{T}^* W dz^*$, where $I_1 \equiv \int |\widehat{T}|^2 dz^*$, $I_2 \equiv \int [|d\widehat{T}/dz^*|^2 + K^2|\widehat{T}|^2] dz^*$, and the limits of the integrations have been suppressed for clarity.

b) Multiply (9.54) by $W*$ and integrate from $z^* = -1/2$ to $z^* = +1/2$ to find: $\frac{\sigma}{Pr} J_1 + J_2 = \mathrm{Ra} K^2 \int W^* \widehat{T} dz^*$ where $J_1 \equiv \int \left[|dW/dz^*|^2 + K^2|W|^2\right] dz^*$, $J_2 \equiv \int \left[|d^2W/dz^{*2}|^2 + 2K^2|dW/dz^*|^2 + K^4|W|^2\right] dz^*$, and again the limits of the integrations have been suppressed.

c) Combine the results of a) and b) to eliminate the mixed integral of W and $\widehat{T}$, and use the result of this combination to show that $\sigma_i = 0$ for Ra > 0. (*Note*: the integrals I_1, I_2, J_1, and J_2 are all positive definite.)

9.8 Consider the thermal instability of a fluid confined between two rigid plates, as discussed in Section 9.5. It was stated there without proof that the minimum critical Rayleigh number of $\mathrm{Ra}_{\mathrm{cr}} = 1708$ is obtained for the gravest *even* mode. To verify this, consider the gravest *odd* mode for which:

$$W = A\sin q_o z^* + B\sinh q z^* + C\sinh q^* z^*,$$

(Compare this with the gravest even mode structure: $W = A\cos q_o z^* + B\cosh q z^* + C\cosh q^* z^*$.) Following Chandrasekhar (1961, p. 39), show that the minimum Rayleigh number is now 17,610, reached at the wave number $K_{\mathrm{cr}} = 5.365$.

9.9 Consider the centrifugal instability problem of Section 9.7. Making the narrow-gap approximation, work out the algebra of going from (9.67) to (9.68).

9.10 Consider the centrifugal instability problem of Section 9.7. From (9.68) and (9.70), the eigenvalue problem for determining the marginal state ($\sigma = 0$) is:

$$(d^2/dR^2 - k^2)^2 \widehat{u}_R = (1 + \alpha x)\widehat{u}_\phi, \qquad (d^2/dR^2 - k^2)^2 \widehat{u}_\phi = -\mathrm{Ta}k^2 \widehat{u}_R, \tag{9.109, 9.110}$$

with $\widehat{u}_R = d\widehat{u}_R/dR = \widehat{u}_\phi = 0$ at $x = 0$ and 1. Conditions on $\widehat{u}_\phi$ are satisfied by assuming solutions of the form:

$$\widehat{u}_\phi = \sum_{m=1}^{\infty} C_m \sin(m\pi x). \tag{9.111}$$

Inserting this into (9.109), obtain an equation for $\widehat{u}_R$, and arrange so that the solution satisfies the four remaining conditions on $\widehat{u}_R$. With $\widehat{u}_R$ determined in this manner and $\widehat{u}_\phi$ given by (9.111), (9.110) leads to an eigenvalue problem for Ta(k). Following Chandrasekhar (1961, p. 300), show that the minimum Taylor number is given by (9.71) and is reached at $K_{\mathrm{cr}} = 3.12$.

9.11 For a Kelvin-Helmholtz instability in a continuously stratified ocean, obtain a globally integrated energy equation in the form:

$$\frac{1}{2}\frac{d}{dt}\int (u^2 + w^2 + g^2\rho^2/\rho_0^2 N^2) dV = -\int uw \frac{\partial U}{\partial z} dV.$$

(As in Fig. 9.28, the integration in x takes place over an integer number of wavelengths.) Discuss the physical meaning of each term and the mechanism of instability.

9.12 In two-dimensional (x, y)-Cartesian coordinates, consider the inviscid stability of horizontal parallel shear flow defined by two linear velocity gradients: $U(y) = \left\{ \begin{array}{c} S^{+}y \text{ for } y > 0 \\ S^{-}y \text{ for } y \leq 0 \end{array} \right\}$, where S^{+} and S^{-} are real constants. Assume an infinitesimal velocity perturbation with vertical component $v = f(y)\exp\{ik(x-ct)\}$, where k is positive real but ω may be complex.

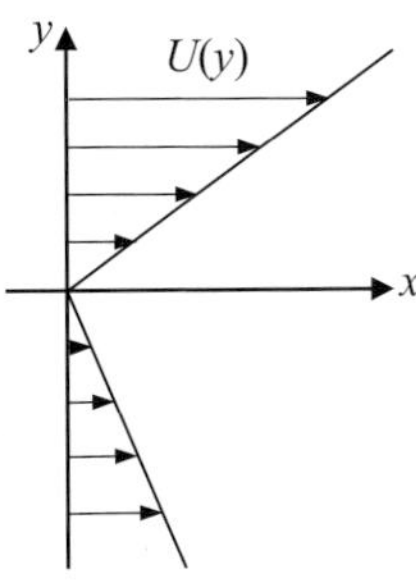

a) Use the Rayleigh equation $f'' - k^2 f - \frac{fU''}{U-c} = 0$ with $f(y) \to 0$ as $|y| \to \infty$ to find $f(y)$.

b) Require the pressure perturbation associated with v to be continuous across $y = 0$, and determine a single equation for the disturbance phase speed c in terms of the other parameters.

c) For what values of S^{+}, S^{-}, and k, is this flow stable, unstable, or neutrally stable?

d) What is special about the case $S^{+} = S^{-}$?

9.13 [2]Consider the inviscid stability of a constant vorticity layer of thickness h between uniform streams with flow speeds U_1 and U_3. Region 1 lies above the layer, $y > h/2$ with $U(y) = U_1$. Region 2 lies within the layer, $|y| \leq h/2$, $U(y) = \frac{1}{2}(U_1 + U_3) + (U_1 - U_3)(y/h)$. Region 3 lies below the layer, $y < -h/2$ with $U(y) = U_3$.

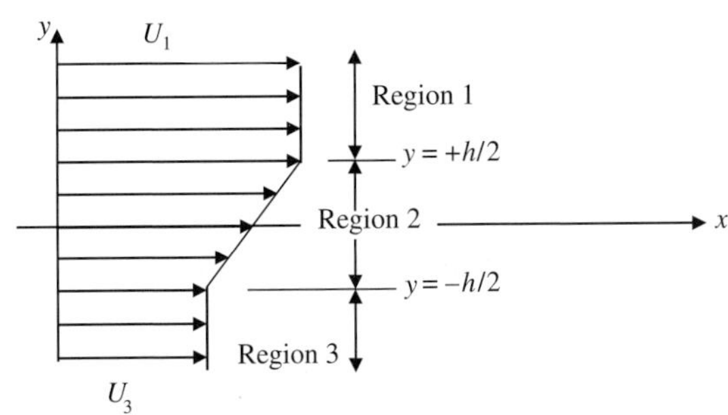

a) Solve the Rayleigh equation, $f'' - k^2 f - \frac{fU''}{U-c} = 0$, in each region, then use appropriate boundary and matching conditions to obtain:

$$f_1(y) = (A\cosh(kh/2) + B\sinh(kh/2))e^{-k(y-h/2)} \quad \text{for} \quad y > +h/2,$$

$$f_2(y) = A\cosh(ky) + B\sinh(ky) \quad \text{for} \quad |y| \leq h/2,$$

$$f_3(y) = (A\cosh(kh/2) - B\sinh(kh/2))e^{+k(y+h/2)} \quad \text{for} \quad y < -h/2,$$

where f defines the spatial extent of the disturbance: $v' = f(y)e^{ik(x-ct)}$ and $u' = -(f'/ik)e^{ik(x-ct)}$, and A and B are undetermined constants.

b) The linearized horizontal momentum equation is: $\frac{\partial u'}{\partial t} + U\frac{\partial u'}{\partial x} + v'\frac{\partial U}{\partial y} = -\frac{1}{\rho}\frac{\partial p'}{\partial x}$. Integrate this equation with respect to x, require the pressure to be continuous at $y = \pm h/2$, and simplify your results to find two additional constraint equations:

$$(c - U_1)f_1'(+h/2) = (c - U_1)f_2'(+h/2) + \frac{U_1 - U_3}{h}f_2(+h/2), \quad \text{and}$$

$$(c - U_3)f_3'(-h/2) = (c - U_3)f_2'(-h/2) + \frac{U_1 - U_3}{h}f_2(-h/2).$$

2. Developed from Sherman, F.S. (1990). *Viscous Fluid Flow*, McGraw-Hill, New York, pp. 466–467.

c) Define $c_o = c - \frac{1}{2}(U_1 + U_3)$ (this is the phase speed of the disturbance waves in a frame of reference moving at the average speed), and use the results of parts a) and b) to determine a single equation for c_o:

$$c_o^2 = \left(\frac{U_1 - U_3}{2kh}\right)^2 \left\{(kh-1)^2 - e^{-2kh}\right\}.$$

[This part of this problem requires patience and algebraic skill.]

d) From the result of part c), c_o will be real for $kh \gg 1$ (short wave disturbances), so the flow is stable or neutrally stable. However, for $kh \ll 1$ (long wave disturbances), use the result of part c) to show that:

$$c_o \cong \pm i\left(\frac{U_1 - U_3}{2}\right)\sqrt{1 - \frac{4}{3}kh + \ldots}$$

e) Determine the largest value of kh at which the flow is unstable.

9.14 A perturbed vortex sheet nominally located at $y = 0$ separates inviscid flows of uniform density as depicted in Fig. 9.3. The velocity difference between the upper and lower stream is $U_0 = U_1 - U_2$. At time $t = 0$ the interface of the vortex sheet is

$$y = y_0 \sin\left(\frac{2\pi x}{\lambda}\right),$$

where λ is the wavelength and y_0 is the initial amplitude.

a) Using (9.1) and (9.20), show that the maximum displacement of the wave y_{max} grows like $y_{max} = y_0 \exp(\sigma t)$, where the growth rate is $\sigma = U_0 k/2$ and $k = 2\pi/\lambda$ is the wave number.

b) Using the vortex method outlined in Example 5.8, solve the Biot–Savart vortex induction law numerically by discretizing the vortex sheet into $N = 100$ point vortices in a periodic domain of length $L = 1$. Initialize the vortices uniformly according to

$$x_i = \left(i - \frac{1}{2}\right)\frac{L}{N},$$

for $i = 1, N$. Take $\lambda = L/2$ and $y_0 = 0.1\lambda$. Run the simulation out to $t = 2\lambda/U_0$ and compare the solution to Fig. 9.6. Comment on the reason for any discrepancies that might be observed.

c) Estimate the temporal growth rate σ. Vary the vortex width parameter ϵ and comment on its role in the growth of the Kelvin-Helmholtz instability. [*Hint*: Relate ϵ to h from Exercise 9.13.]

9.15 Consider the inviscid instability of parallel flows given by the Rayleigh equation for the y-component of the perturbation velocity, $v = \hat{v}(y)\exp\{ik(x - ct)\}$:

$$(U - c)\left(\frac{d^2\hat{v}}{dy^2} - k^2\hat{v}\right) - \frac{d^2U}{dy^2}\hat{v} = 0, \tag{9.112}$$

a) Note that this equation is identical to the Rayleigh equation (9.98) for the stream function amplitude ϕ, as it must because $\hat{v}(y) = -ik\phi$. For a flow bounded by walls at y_1 and y_2, note that the boundary conditions are identical in terms of ϕ and $\hat{v}$.

b) Show that if c is an eigenvalue of (9.112), then so is its conjugate $c^* = c_r - ic_i$. What aspect of (9.112) allows this result to be valid?

c) Let $U(y)$ be *antisymmetric*, so that $U(y) = -U(-y)$. Demonstrate that if $c(k)$ is an eigenvalue, then $-c(k)$ is also an eigenvalue. Explain the result physically in terms of the possible directions of propagation of perturbations in such an antisymmetric flow.

d) Let $U(y)$ be *symmetric* so that $U(y) = U(-y)$. Show that in this case $\hat{v}$ is either symmetric or antisymmetric about $y = 0$.
[*Hint*: Letting $y \to -y$, show that the solution $\hat{v}(-y)$ satisfies (9.112) with the same eigenvalue c. Form a symmetric solution, $S(y) = \hat{v}(y) + \hat{v}(-y) = S(-y)$, and an antisymmetric solution, $A(y) = \hat{v}(y) - \hat{v}(-y) = -A(-y)$. Then write $A[S\text{-}eqn] - S[A\text{-}eqn] = 0$ where S-*eqn* indicates the differential equation (9.112) in terms of S. Canceling terms this reduces to $(SA' - AS')' = 0$, where the prime ($'$) indicates a y-derivative. Integration gives $SA' - AS' = 0$, where the constant of integration is zero because of the boundary conditions. Another integration gives $S = bA$, where b is a constant of integration. Because the symmetric and antisymmetric functions cannot be proportional, it follows that one of them must be zero.]

Comments: If v is symmetric, then the cross-stream velocity has the same sign across the entire flow, although the sign alternates every half wavelength along the flow. This mode is consequently called *sinuous*. On the other hand, if v is antisymmetric, then the shape of the jet expands and contracts along the length. This mode is now generally called the *sausage* instability because it resembles a line of linked sausages.

9.16 Derive (9.105) starting from the incompressible Navier-Stokes momentum equation for the disturbed flow:

$$\frac{\partial}{\partial t}(U_i + u_i) + (U_j + u_j)\frac{\partial}{\partial x_j}(U_i + u_i) = -\frac{1}{\rho}\frac{\partial}{\partial x_i}(P + p) + \nu\frac{\partial^2}{\partial x_j \partial x_j}(U_i + u_i), \tag{9.113}$$

where U_i and u_i represent the basic flow and the disturbance, respectively. Subtract the equation of motion for the basic state from (9.113), multiply by u_i, and integrate the result within a stationary volume having stream-wise control surfaces chosen to coincide with the walls where no-slip conditions are satisfied or where $u_i \to 0$, and having a length (in the stream-wise direction) that is an integer number of disturbance wavelengths.

9.17 [3]The process of transition from laminar to turbulent flow may be driven both by exterior flow fluctuations and nonlinearity. Both of these effects can be simulated with the simple nonlinear logistic map $x_{n+1} = Ax_n(1 - x_n)$ and a computer spreadsheet program. Here, x_n can be considered to be the flow speed at the point of interest with A playing the role of the nonlinearity parameter (Reynolds number), x_0 (the initial condition) playing the role of an external disturbance, and iteration of the equation playing the role of increasing time. The essential feature illustrated by this problem is that increasing the nonlinearity parameter or changing the initial condition in the presence of nonlinearity may fully alter the character of the resulting sequence of x_n values. Plotting x_n vs. n should aid understanding for parts b) through e).

a) Determine the background solution of the logistic map that occurs when $x_{n+1} = x_n$ in terms of A.
Now, using a spreadsheet program, set up a column that computes x_{n+1} for $n = 1$ to 100 for user selectable values of x_0 and A for $0 < x_0 < 1$, and $0 < A < 4$.

b) For $A = 1.0, 1.5, 2.0, 2.9$, choose a few different values of x_0 and numerically determine if the background solution is reached by $n = 100$. Is the flow *stable* for these values of A, i.e., does it converge toward the background solution?

c) For the slightly larger value, $A = 3.2$, choose $x_0 = 0.6875, 0.6874$, and 0.6876. Is the flow stable or oscillatory in these three cases? If it is oscillatory, how many iterations are needed for it to repeat?

d) For $A = 3.5$, is the flow stable or oscillatory? If it is oscillatory, how many iterations are needed for it to repeat? Does any value of x_0 lead to a stable solution?

e) For $A = 3.9$, is the flow stable, oscillatory, or chaotic? Does any value of x_0 lead to a stable solution?

9.18 The three-dimensional velocity field $\mathbf{u} = (u, v, w)$ of a spherical drop with radius $R = 0.5$ immersed in a linear flow in the limit of small Reynolds number takes the form (Stone et al., 1991)

$$\begin{aligned}
u &= \frac{1}{2}\left[(5|\mathbf{x}|^2 - 3)\frac{x}{1+\alpha} - 2x\left(\frac{x^2}{1+\alpha} + \frac{\alpha y^2}{1+\alpha} - z^2\right) + \omega_y z - \omega_z y\right], \\
v &= \frac{1}{2}\left[(5|\mathbf{x}|^2 - 3)\frac{\alpha y}{1+\alpha} - 2y\left(\frac{x^2}{1+\alpha} + \frac{\alpha y^2}{1+\alpha} - z^2\right) + \omega_z x - \omega_x z\right], \\
w &= \frac{1}{2}\left[(-5|\mathbf{x}|^2 - 3)z - 2z\left(\frac{x^2}{1+\alpha} + \frac{\alpha y^2}{1+\alpha} - z^2\right) + \omega_x y - \omega_y x\right],
\end{aligned}$$

where $\boldsymbol{\omega} = (\omega_x, \omega_y, \omega_z)$ is the angular velocity of the drop and $\alpha = S_{22}/S_{11}$. The path line of a fluid particle inside the drop is integrable for $\omega_x = \omega_y = 0$, and chaotic otherwise. Consider an axisymmetric flow ($\alpha = 1$) and solve the path line numerically out to $t = 100$ for a particle passing through a point near the origin ($x = y = z = 0.02R$) at $t = 0$ and another for a particle placed very close to the first ($x = y = z = 0.02R + \epsilon$) with $\epsilon = 0.0001R$. Plot the displacement between the two particles as a function of time. First consider $\boldsymbol{\omega} = (0, 0, 1.5)$ (integrable), then repeat this for $\boldsymbol{\omega} = (0.9742, 0, 1.1406)$ (chaotic).

3. Provided to the third author by Professor Werner Dahm.

References

Bayly, B.J., Orszag, S.A., Herbert, T., 1988. Instability mechanisms in shear-flow transition. Annual Review of Fluid Mechanics 20, 359–391.

Bénard, H., 1900. Les tourbillons cellulaires dans une nappe liquide (cellular vortices in a liquid sheet). Revue Générale des Sciences Pures et Appliquées 11, 1261–1271.

Berge, P., Pomeau, Y., Vidal, C., 1984. Order Within Chaos. Wiley, New York.

Bhattacharya, P., Manoharan, M.P., Govindarajan, R., Narasimha, R., 2006. The critical Reynolds number of a laminar incompressible mixing layer from minimal composite theory. Journal of Fluid Mechanics 565, 105–114.

Chandrasekhar, S., 1961. Hydrodynamic and Hydromagnetic Stability. Oxford University Press, London. Reprinted by Dover, New York, 1981.

Coles, D., 1965. Transition in circular Couette flow. Journal of Fluid Mechanics 21, 385–425.

Davies, P., 1988. Cosmic Blueprint. Simon and Schuster, New York.

Drazin, P.G., Reid, W.H., 1981. Hydrodynamic Stability. Cambridge University Press, London.

Eckhardt, B., Schneider, T.M., Hof, B., Westerweel, J., 2007. Turbulence transition in pipe flow. Annual Review of Fluid Mechanics 39, 447–468.

Eriksen, C.C., 1978. Measurements and models of fine structure, internal gravity waves, and wave breaking in the deep ocean. Journal of Geophysical Research 83, 2989–3009.

Feigenbaum, M.J., 1978. Quantitative universality for a class of nonlinear transformations. Journal of Statistical Physics 19, 25–52.

Fjortoft, R., 1950. Application of integral theorems in deriving criteria of instability for laminar flows and for the baroclinic circular vortex. Geofysiske Publikasjoner Oslo 17 (6), 1–52.

Fransson, J.H.M., Alfredsson, P.H., 2003. On the disturbance growth in an asymptotic suction boundary layer. Journal of Fluid Mechanics 482, 51–90.

Heisenberg, W., 1924. Über Stabilität und Turbulenz von Flüssigkeitsströmen. Annalen der Physik (Leipzig) 4 (74), 577–627.

Howard, L.N., 1961. Note on a paper of John W. Miles. Journal of Fluid Mechanics 13, 158–160.

Huerre, P., Monkewitz, P.A., 1990. Local and global instabilities in spatially developing flows. Annual Review of Fluid Mechanics 22, 473–537.

Huppert, H.E., Turner, J.S., 1981. Double-diffusive convection. Journal of Fluid Mechanics 106, 299–329.

Klebanoff, P.S., Tidstrom, K.D., Sargent, L.H., 1962. The three-dimensional nature of boundary layer instability. Journal of Fluid Mechanics 12, 1–34.

Landau, L.D., Lifshitz, E.M., 1959. Fluid Mechanics. Pergamon Press, Oxford, England, pp. 103–107.

Lanford, O.E., 1982. The strange attractor theory of turbulence. Annual Review of Fluid Mechanics 14, 347–364.

Lavery, A.C., Geyer, W.R., Scully, M.E., 2013. Broadband acoustic quantification of stratified turbulence. The Journal of the Acoustical Society of America 134, 40–54.

Lin, C.C., 1955. The Theory of Hydrodynamic Stability. Cambridge University Press, London.

Lorenz, E., 1963. Deterministic nonperiodic flows. Journal of the Atmospheric Sciences 20, 130–141.

Miles, J.W., 1961. On the stability of heterogeneous shear flows. Journal of Fluid Mechanics 10, 496–508.

Miles, J.W., 1986. Richardson's criterion for the stability of stratified flow. Physics of Fluids 29, 3470–3471.

Plateau, J.A.F., 1873. Statique expérimentale et théorique des liquides soumis aux seules forces moléculaires: Tome premier (Vol. 2). Gauthier-Villars.

Rayleigh, L., 1879. On the capillary phenomena of jets. Proceedings of the Royal Society of London 29, 71–97.

Rayleigh, L., 1916. On the convective currents in a horizontal layer of fluid when the higher temperature is on the under side. London, Edinburgh and Dublin Philosophical Magazine and Journal of Science 32 (192), 529–546.

Ruelle, D., Takens, F., 1971. On the nature of turbulence. Communications in Mathematical Physics 20, 167–192.

Schubauer, G.B., Skramstad, H.K., 1947. Laminar boundary-layer oscillations and transition on a flat plate. Journal of Research of the National Bureau of Standards 38, 251–292.

Scotti, R.S., Corcos, G.M., 1972. An experiment on the stability of small disturbances in a stratified free shear layer. Journal of Fluid Mechanics 52, 499–528.

Shen, S.F., 1954. Calculated amplified oscillations in plane Poiseuille and Blasius flows. Journal of the Aeronautical Sciences 21, 62–64.

Squire, H.B., 1933. On the stability of three dimensional disturbances of viscous flows between parallel walls. Proceedings of the Royal Society London A 142, 621–628.

Stern, M.E., 1960. The salt fountain and thermohaline convection. Tellus 12, 172–175.

Stommel, H., Arons, A.B., Blanchard, D., 1956. An oceanographic curiosity: the perpetual salt fountain. Deep-Sea Research 3, 152–153.

Stone, H.A., Nadim, A., Strogatz, S.H., 1991. Chaotic streamlines inside drops immersed in steady Stokes flows. Journal of Fluid Mechanics 232, 629–646.

Thorpe, S.A., 1971. Experiments on the instability of stratified shear flows: miscible fluids. Journal of Fluid Mechanics 46, 299–319.

Turner, J.S., 1973. Buoyancy Effects in Fluids. Cambridge University Press, London.

Turner, J.S., 1985. Convection in multicomponent systems. Naturwissenschaften 72, 70–75.

Woods, J.D., 1969. On Richardson's number as a criterion for turbulent—laminar transition in the atmosphere and ocean. Radio Science 4, 1289–1298.

Yih, C.S., 1979. Fluid Mechanics: A Concise Introduction to the Theory. West River Press, Ann Arbor, MI.

Further reading

Jefferies, H., 1928. Some cases of instability in fluid motion. Proceedings of the Royal Society London A 118, 195–208.

Nayfeh, A.H., Saric, W.S., 1975. Nonparallel stability of boundary layer flows. Physics of Fluids 18, 945–950.

Chapter 10

Turbulence⍟

Contents

10.1 **Introduction** 373
10.2 **Historical notes** 375
10.3 **Nomenclature and statistics for turbulent flow** 376
10.4 **Correlations and spectra** 379
10.5 **Averaged equations of motion** 383
10.6 **Homogeneous isotropic turbulence** 389
10.7 **Turbulent energy cascade and spectrum** 393
10.8 **Free turbulent shear flows** 399
10.9 **Wall-bounded turbulent shear flows** 405
Inner layer: law of the wall 407
Outer layer: velocity defect law 408
Overlap layer: logarithmic law 408
Turbulent flow in ducts 412
Rough surfaces 413
10.10 **Turbulence modeling** 417
A mixing length model 418
One-equation models 420
Two-equation models 420
Reynolds stress models 421
10.11 **Turbulence in a stratified medium** 422
The Richardson numbers 422
Monin-Obukhov length 423
Spectrum of temperature fluctuations 424
10.12 **Taylor's theory of turbulent dispersion** 426
Rate of dispersion of a single particle 426
Behavior for small t 428
Behavior for large t 428
Random walk 428
Behavior of a smoke plume in the wind 429
Turbulent diffusivity 429
Exercises 431
References 442
Further reading 443

Chapter objectives

- To introduce and describe turbulent flow
- To define the statistics and functions commonly used to quantify turbulent flow phenomena
- To derive the Reynolds-averaged equations
- To present assumptions and approximations leading to the classical scaling laws for turbulent flow
- To provide useful summaries of mean-flow results for free and wall-bounded turbulent shear flows
- To introduce the basic elements of turbulence modeling
- To describe basic turbulence phenomena relevant in atmospheric turbulence

10.1 Introduction

Nearly all macroscopic flows encountered in the natural world and in engineering practice are turbulent. Winds and currents in the atmosphere and ocean; flows through residential, commercial, and municipal water (and air) delivery systems; flows past transportation devices (cars, trains, aircraft, ships, etc.); and flows through turbines, engines, and reactors used for power generation and conversion are all turbulent. Turbulence is an enigmatic state of fluid flow that may be simultaneously beneficial and problematic. For example, in air-breathing combustion systems, it is exploited for mixing reactants but, within the same devices, it also leads to noise, and efficiency losses. Within the earth's ocean and atmosphere, turbulence sets the mass, momentum, and heat transfer rates involved in pollutant dispersion and climate regulation.

Turbulence involves fluctuations that are unpredictable in detail, and it has not been conquered by deterministic or statistical analysis. However, useful predictions about it are still possible and these may arise from physical intuition, dimensional arguments, direct numerical simulations, or empirical models and computational schemes. In spite of our everyday experience with it, turbulence is not easy to define precisely and there is a tendency to confuse turbulence with randomness. A turbulent fluid velocity field conserves mass, momentum, and energy while a purely random time-dependent vector field need not.

⍟. This book has a companion website hosting complementary materials. Visit this URL to access it: https://www.elsevier.com/books-and-journals/book-companion/9780128198070.

Fluid Mechanics. https://doi.org/10.1016/B978-0-12-819807-0.00019-3

This chapter presents basic features of turbulence beginning with this listing of generic characteristics.

(1) *Fluctuations*: Turbulent flows contain fluctuations in the dependent-field quantities (velocity, pressure, temperature, etc.) even when the flow's boundary conditions are steady. Turbulent fluctuations appear to be irregular, chaotic, and unpredictable.

(2) *Nonlinearity*: The momentum conservation equation contains the nonlinear advective-acceleration term, and even in ideal flows this nonlinearity causes pressure to depend on the square of the velocity. Turbulence represents an even further assertion of this nonlinearity, and occurs when the relevant nonlinearity parameter – the Reynolds number Re, the Rayleigh number Ra, or the inverse Richardson number Ri^{-1} – comfortably exceeds a critical value. The enhanced nonlinearity of turbulence is evident in vortex stretching, a key mechanism that produces three-dimensional fluctuations from energy input(s) commonly having lower dimensionality. Turbulence's nonlinearity also invalidates the superposition principle exploited in ideal flow (Chapter 6), and causes the time and length scales of the flow's initial and boundary conditions to be smeared into fluctuations having continuous spectra involving a range of frequencies and wave numbers.

(3) *Vorticity*: Turbulence is characterized by fluctuating vorticity. A cross-section view of a turbulent flow typically appears as a diverse collection of streaks, strain regions, and swirls of various sizes that deform, coalesce, divide, and spin. Identifiable structures in a turbulent flow, particularly those that spin, are called *eddies*. Turbulence always involves a range of eddy sizes and the size range increases with $\text{Re}^{3/4}$. The characteristic size of the largest eddies is the width of the turbulent region; in a turbulent boundary layer this is the thickness of the layer (Fig. 10.1). Such layer-spanning eddies commonly contain most of the fluctuation energy in a turbulent flow and may be orders of magnitude larger than the smallest eddies.

(4) *Dissipation*: On average, the vortex stretching mechanism transfers fluctuation energy and vorticity to smaller and smaller eddies via nonlinear interactions, until velocity gradients become so large that fluctuation energy is converted into heat (i.e., dissipated) by the action of viscosity and the motions of the smallest eddies. Persistent turbulence therefore requires a continuous supply of energy from an imposed velocity or pressure difference to make up for this energy loss.

(5) *Diffusivity*: Within a turbulent flow, the prevalence of fluctuations and vortical overturning motions leads to mixing and diffusion of chemical species, momentum, and heat that is orders of magnitude faster than molecular transport in equivalent laminar flows that lack such fluctuations and vortical motions.

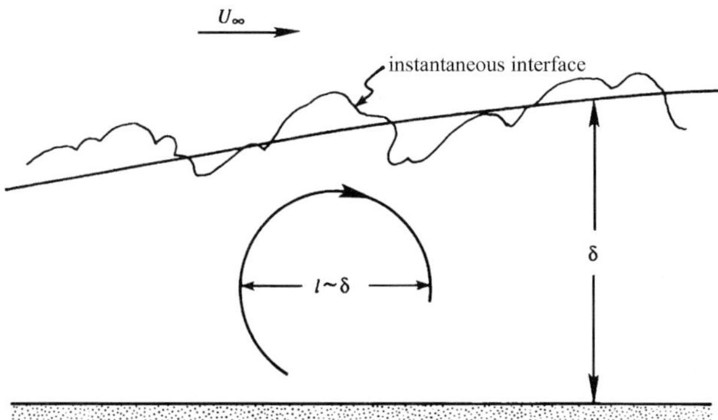

FIGURE 10.1 Turbulent boundary-layer flow showing a typical large eddy of size l, the average layer thickness δ, and the instantaneous interface between turbulent and non-turbulent (typically irrotational) fluid. Here, as in most turbulent flows, the size of the largest eddies is comparable to the overall layer thickness.

These features of turbulence suggest that many flows that seem random such as wind-driven ocean-surface waves, failing rain, or internal waves in the ocean or the atmosphere, are not turbulent because they are not simultaneously dissipative, vortical, and nonlinear.

Although imperfect, a simple definition of turbulence as *a dissipative flow state characterized by nonlinear fluctuating three-dimensional vorticity* is offered. Incompressible turbulent mean flows in systems not large enough to be influenced by the Coriolis force are emphasized in this chapter. The fluctuations in such flows are three dimensional. In large-scale geophysical systems, on the other hand, the existence of stratification and the Coriolis acceleration severely restricts vertical motion and leads to chaotic flow that may be nearly two dimensional or *geostrophic*. *Geostrophic turbulence* is briefly mentioned in Chapter 12. More extensive treatments of turbulence are provided in Monin and Yaglom (1971, 1975), Tennekes and Lumley (1972), Hinze (1975), and Pope (2000).

Example 10.1

The largest eddies in turbulent flows mix their interior contents in approximately one turnover time. Estimate the time necessary to mix the contents of a 10-cm-diameter engine cylinder if the average piston speed is 10 m/s, and estimate the time for molecular diffusive processes in air to accomplish the same mixing task.

Solution The flow past the intake valves into the engine cylinder will lead to a region of confined turbulence. The size of the largest possible eddies will be set by the cylinder diameter D, and their overturning velocity will be set by the piston speed U_p. Thus, the large-eddy turnover time in this case must be proportional to:

$$D/U_p = (0.10 \text{ m})/(10 \text{ m/s}) = 0.01 \text{ s}.$$

The molecular diffusive time scale for the same mixing task can be estimated using the diffusion scaling established in Section 7.4:

$$\textit{diffusion time} \sim D^2/\nu = (0.10 \text{ m})^2/(1.5 \times 10^{-5} \text{ m}^2/\text{s}) = 667 \text{ s} \sim 11 \text{ minutes},$$

where the kinematic viscosity, ν, is representative of gas diffusivities in air. The two time estimates differ by almost five orders of magnitude and together show that turbulence is essential for the proper functioning of internal combustion engines.

10.2 Historical notes

Turbulence is a leading topic in modern fluid dynamics research, and some of the best-known physicists have worked in this area during the last century. Among them are G.I. Taylor, Kolmogorov, Reynolds, Prandtl, von Karman, Heisenberg, Landau, Millikan, and Onsagar. A brief historical outline is given in what follows; further interesting details can be found in Monin and Yaglom (1971, 1975).

The first systematic work on turbulence was carried out by Osborne Reynolds in 1883. His experiments in pipe flows showed that the flow becomes turbulent or irregular when the dimensionless ratio $\text{Re} = UL/\nu$, later named the Reynolds number by Sommerfeld, exceeds a certain critical value (see Fig. 7.1). (Here, U is the velocity scale, L is the length scale, and ν is the kinematic viscosity.) This dimensionless number subsequently proved to be the parameter that determines the dynamic similarity of viscous flows. Reynolds also separated turbulent flow-dependent variables into mean and fluctuating components, and arrived at the concept of turbulent stress. The meaning of the Reynolds number and the existence of turbulent stresses are foundational elements in our present understanding of turbulence.

In 1921, the British physicist G.I. Taylor, in a simple and elegant study of turbulent diffusion, introduced the idea of a correlation function. He showed that the root-mean-square distance of a particle from its source point in a turbulent flow initially increases with time as t, and subsequently as $t^{1/2}$, as in a random walk. Taylor continued his outstanding work in a series of papers during 1935–1936 in which he laid down the foundation of the statistical theory of turbulence. Among the concepts he introduced were those of homogeneous and isotropic turbulence and of a turbulence spectrum. Although real turbulent flows are not isotropic (turbulent shear stresses, in fact, vanish for isotropic flows), the mathematical techniques involved have proved valuable for describing the *small scales* of turbulence, which are isotropic or nearly so. In 1915, Taylor also introduced the mixing length concept, although it is generally credited to Prandtl for making full use of the idea.

During the 1920s Prandtl and his student von Karman, working in Göttingen, Germany, developed semi-empirical theories of turbulence. The most successful of these was the mixing length theory, which is based on an analogy with the concept of mean free path in the kinetic theory of gases. By guessing at the correct form for the mixing length, Prandtl was able to deduce that the average turbulent velocity profile near a solid wall is logarithmic, one of the most reliable results for turbulent flows. However, it has subsequently become clear that the mixing length theory is not generally predictive since there is really no rational way of independently determining the mixing length. In fact, the near-wall logarithmic law can be justified from dimensional considerations alone.

Seminal work was done by the British meteorologist Lewis Richardson. In 1922 he wrote the first book on numerical weather prediction. In this book he proposed that turbulent kinetic energy is continually transferred from larger to smaller eddies, until it is destroyed by viscous dissipation. This idea of a spectral energy cascade is at the heart of our present understanding of turbulence. However, Richardson's work was largely ignored at the time, and it was not until some 20 years later that the idea of a spectral cascade took quantitative shape in the hands of Kolmogorov and Obukhov in Russia. Richardson also did another important piece of work that displayed his amazing physical intuition. On the basis of experimental data for the movement of balloons in the atmosphere, he proposed that the effective diffusion coefficient of a patch of turbulence is proportional to $l^{4/3}$, where l is the scale of the patch. This is called Richardson's four-thirds law, and has been subsequently found to be in agreement with Kolmogorov's famous five-thirds law for the energy spectrum.

The Russian mathematician Kolmogorov, generally regarded as the greatest probabilist of the twentieth century, followed up on Richardson's idea of a spectral energy cascade. He hypothesized that the statistics of the small scales are isotropic and depend on only two parameters, namely ν, the kinematic viscosity, and $\overline{\varepsilon}$, the average rate of kinetic energy dissipation per unit mass of fluid. On dimensional grounds, he derived that the smallest scales must be of size $\eta = (\nu^3/\overline{\varepsilon})^{1/4}$. His second hypothesis was that, at scales much smaller than l (see Fig. 10.1) and much larger than η, there must exist an inertial sub-range of turbulent eddy sizes for which ν plays no role; in this range the statistics depend only on a single parameter $\overline{\varepsilon}$. Using this idea, in 1941 Kolmogorov and Obukhov independently derived that the spectrum in the inertial sub-range must be proportional to $\overline{\varepsilon}^{2/3}k^{-5/3}$, where k is the wave number. The five-thirds law is one of the most important results of turbulence theory and is in agreement with high Reynolds number observations.

Recent decades have seen much progress in theory, simulations, and measurements. Among these may be mentioned the work on coherent structures, direct numerical and large-eddy simulations, and multidimensional diagnostics. Observations in the ocean and the atmosphere (which von Karman called "a giant laboratory for turbulence research"), in which the Reynolds numbers are very large, have shed new light on the structure of stratified turbulence.

Turbulence remains an area of classical fluid mechanics that is the subject of continuing research. A modern topical overview is provided in Davidson et al. (2013). Due to the large volume of data produced from experiments and numerical simulations of turbulence, researchers are now turning to scientific machine learning to augment and improve existing theories and models (see Duraisamy et al., 2019).

Example 10.2

What are the units of $\overline{\varepsilon}$? What combination of the large-eddy velocity-scale U and length scale L has the same units as $\overline{\varepsilon}$? If $\overline{\varepsilon}$ is proportional to this combination of U and L, how does the ratio of the Kolmogorov scale to the large scale, $\eta/L = (\nu^3/\overline{\varepsilon})^{1/4}/L$, depend on the large-eddy Reynolds number $\mathrm{Re}_L = UL/\nu$?

Solution The units of $\overline{\varepsilon}$ are energy loss rate (power) per unit mass: (length)^2/(time)^3. The combination of U and L that has the same units is U^3/L. When $\overline{\varepsilon}$ is proportional to U^3/L, then the length-scale ratio is:

$$\frac{(\nu^3/\overline{\varepsilon})^{1/4}}{L} \sim \frac{(\nu^3 L/U^3)^{1/4}}{L} \sim (\nu^3/U^3L^3)^{1/4} = Re_L^{-3/4}.$$

Thus, at high Reynolds numbers, the Kolmogorov-scale eddies may be many orders of magnitude smaller than the largest eddies. Interestingly, this simple result applies to high-Reynolds number turbulence, and indicates that high-spatial-resolution experimental techniques and computational meshes are required for accurately measuring or simulating all eddies in a high Reynolds number turbulent flow.

10.3 Nomenclature and statistics for turbulent flow

The dependent-field variables in a turbulent flow (velocity components, pressure, temperature, etc.) are commonly analyzed and described using definitions and nomenclature borrowed from the theory of *stochastic processes* and *random variables* even though fluid-dynamic turbulence is not entirely random. Thus, the characteristics of turbulent-flow field variables are commonly specified in terms of their *statistics* or *moments*. In particular, a turbulent field quantity, ϑ, is commonly separated into its first moment, $\overline{\vartheta}$, and its fluctuations, $\vartheta' \equiv \vartheta - \overline{\vartheta}$, which have zero mean. This separation is known as the *Reynolds decomposition* and is further described and utilized in Section 10.5 with different notation. In this chapter, the first moment of the velocity component u_i is also denoted by U_i.

To define moments precisely, specific terminology is needed. A *collection* of independent realizations of a random variable, obtained under identical conditions, is called an *ensemble*. The ordinary arithmetic average over the collection is called an *ensemble average* and is denoted herein by an over bar. When the number N of realizations in the ensemble is large, $N \to \infty$, the ensemble average is called an *expected value* and is denoted with angle brackets. With this terminology and notation, the mth-moment, $\overline{u^m}$, of the random variable u at location $\mathbf{x}$ and time t is defined as the ensemble average of u^m:

$$\langle u^m(\mathbf{x},t)\rangle = \lim_{N\to\infty} \overline{u^m(\mathbf{x},t)} \equiv \lim_{N\to\infty} \frac{1}{N}\sum_{n=1}^{N}(u(\mathbf{x},t:n))^m, \tag{10.1}$$

where $u(\mathbf{x}, t{:}n)$ is the nth realization in the ensemble. The limit $N \to \infty$ can only be taken formally in theoretical analysis, so when dealing with measurements, $\overline{u^m}$ is commonly used in place of $\langle u^m \rangle$ and good experimental design ensures that N is large enough for reliable determination of the first few moments of u. Thus, the over-bar notation for ensemble average is favored in the remainder of this chapter. Collectively, the moments for integer values of m are known as the *statistics* of $u(\mathbf{x}, t)$.

Under certain circumstances, ensemble averaging is not necessary for moment estimation. When u is *stationary in time*, its statistics do not depend on time, and $\overline{u^m}$ at $\mathbf{x}$ can be reliably estimated from time averaging:

$$\overline{u^m(\mathbf{x})} = \frac{1}{\Delta t}\int_{t-\Delta t/2}^{t+\Delta t/2} u^m(\mathbf{x}, t)dt, \tag{10.2}$$

when Δt is large enough. Time averages are relevant for turbulent flows that persist with the same boundary conditions for long periods of time, an example being the turbulent boundary-layer flow on the hull of a long-range ship that traverses a calm sea at constant speed. Example time histories of temporally stationary and non-stationary processes are shown in Fig. 10.2. When u is *homogeneous* or *stationary in space*, its statistics do not depend on location, and $\overline{u^m}$ at time t can be reliably estimated from spatial averaging in a volume V:

$$\overline{u^m(t)} = \frac{1}{V}\int_V u^m(\mathbf{x}, t)dV, \tag{10.3}$$

when V is large enough and defined appropriately. This type of average is often relevant in confined turbulent flows subject to externally imposed temporal variations, an example being the in-cylinder swirling and tumbling gas flow driven by piston motion and valves flows in an internal combustion piston engine.

Throughout this chapter, all moments denoted by over bars are ensemble averages determined from (10.1), unless otherwise specified. Eqs. (10.2) and (10.3) are provided here because they are commonly used to convert turbulent flow measurements into moment values. In particular, (10.2) or (10.3) are used in atmospheric and oceanic field measurements because ongoing natural phenomena like weather or the slow meandering of ocean currents make it practically impossible to precisely repeat field observations under identical circumstances. For such measurements, a judicious selection of Δt or V is necessary; they should be long or large enough for reliable moment estimation but small enough so that the resulting statistics are only weakly influenced by ongoing natural variations not of fluid mechanical origin.

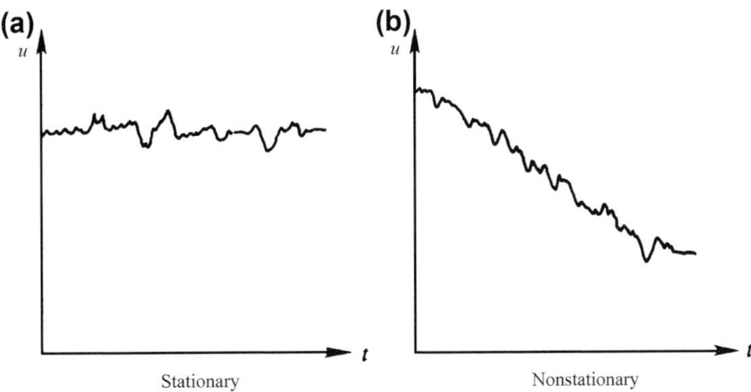

FIGURE 10.2 Sample time series indicating temporally stationary (a) and non-stationary processes (b). The time series in (b) shows that the average value of u decreases with time compared to the time series in (a).

Before defining and describing specific moments, several important properties of the process of ensemble averaging defined by (10.1) must be mentioned. First, ensemble averaging *commutes* with differentiation, that is, the application order of these two operators can be interchanged:

$$\overline{\frac{\partial u^m}{\partial t}} = \frac{1}{N}\sum_{n=1}^{N}\frac{\partial}{\partial t}(u(\mathbf{x}, t{:}n))^m = \frac{\partial}{\partial t}\left(\frac{1}{N}\sum_{n=1}^{N}(u(\mathbf{x}, t{:}n))^m\right) = \frac{\partial}{\partial t}\overline{u^m}.$$

Similarly, ensemble averaging commutes with addition, multiplication by a constant, time integration, spatial differentiation, and spatial integration. Thus the following are all true:

$$\overline{u^m + v^m} = \overline{u^m} + \overline{v^m}, \qquad \overline{Au^m} = A\overline{u^m}, \qquad \overline{\frac{\partial u^m}{\partial t}} = \frac{\partial}{\partial t}\overline{u^m}, \tag{10.4, 10.5, 10.6}$$

$$\overline{\int_a^b u^m dt} = \int_a^b \overline{u^m} dt, \qquad \overline{\frac{\partial u^m}{\partial x_j}} = \frac{\partial}{\partial x_j}\overline{u^m}, \qquad \overline{\int u^m d\mathbf{x}} = \int \overline{u^m} d\mathbf{x}, \tag{10.7, 10.8, 10.9}$$

where v is another random variable; a, b, m, and A are all constants; and $d\mathbf{x}$ represents a general spatial increment. In particular, (10.5) with $m = 0$ implies $\overline{A} = A$, so if $A = \overline{u}$ then $\overline{\overline{u}} = \overline{u}$; the ensemble average of an average is just the average. However, the ensemble average of a product of random variables is not necessarily the product of the ensemble averages. In general,

$$\overline{u^m} \neq \overline{u}^m \quad \text{and} \quad \overline{uv} \neq \overline{u}\,\overline{v},$$

when $m \neq 1$, and u and v are different random variables.

The simplest statistic of a random variable u is its *first moment*, *mean*, or *average*, $\overline{u}$. From (10.1) with $m = 1$, $\overline{u}$ is:

$$\overline{u(\mathbf{x}, t)} \equiv \frac{1}{N}\sum_{n=1}^{N} u(\mathbf{x}, t{:}n). \tag{10.10}$$

In general, $\overline{u}$ may depend on both space and time, and is obtained by summing the N separate realizations of the ensemble, $u(\mathbf{x}, t{:}n)$ for $1 \le n \le N$, at time t and location $\mathbf{x}$, and then dividing the sum by N. A graphical depiction of ensemble averaging, as specified by (10.10), is shown in Fig. 10.3 for time-series measurements recorded at the same point $\mathbf{x}$ in space. The left panel of the figure shows four members, $u(\mathbf{x}, t{:}n)$ for $1 \le n \le 4$, of the ensemble. Here the average value of u decreases with increasing time. Time records such as these might represent atmospheric temperature measurements during the first few hours after sunset on different days, or they might represent a component of the flow velocity from the cylinder of a compressor in the first 10 or 20 milliseconds after an exhaust valve opens. The right panel of Fig. 10.3 shows the ensemble average $\overline{u(\mathbf{x}, t)}$ obtained from the first two, four, and eight members of the ensemble. The solid smooth curve in the lower right panel of Fig. 10.3 is the expected value that would be obtained from ensemble averaging in the limit $N \to \infty$. The dashed curve is a time average computed from only the fourth member of the ensemble using (10.2) with $m = 1$ and Δt equal to one-tenth of the total time displayed for each time history. Fig. 10.3 shows the primary effect of averaging is to suppress fluctuations since they become less prominent as N increases and are absent from the expected value. In addition, it shows that differences between an ensemble average of many realizations and a finite-duration temporal average of a single realization may be small, even when the flow is not stationary in time.

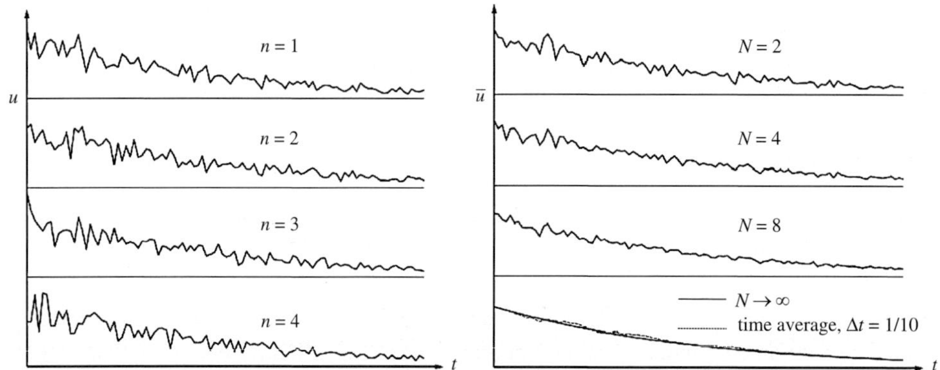

FIGURE 10.3 Illustration of ensemble and temporal averaging. The left panel shows four members of an ensemble of time series for the decaying random variable u. In all four cases, the fluctuations are different but the decreasing trend with increasing t is apparent in each. The right panel shows averages of two, four, and eight members of the ensemble in the upper three plots. As the sample number N increases, fluctuations in the ensemble average decreases. The lowest plot on the right shows the $N \to \infty$ curve – this is the expected value of $u(t)$ – and a simple sliding time average of the $n = 4$ curve where the duration of the time average is one-tenth of the time period shown. In this case, time and ensemble averaging produce nearly the same curve.

Although useful and important in many situations, the average or *first moment* alone does not directly provide information about turbulent fluctuations. Such information is commonly reported in terms of one or more *higher-order central moments* defined by:

$$\overline{(u-\langle u\rangle)^m} \equiv \frac{1}{N}\sum_{n=1}^{N}(u(\mathbf{x},t{:}n)-\langle u(\mathbf{x},t)\rangle)^m, \tag{10.11}$$

where in practice $\overline{u(\mathbf{x},t)}$ often replaces $\langle u(\mathbf{x},t)\rangle$. The central moments primarily carry information about the fluctuations since (10.11) explicitly shows that the mean is removed from each ensemble member. The first central moment is zero by definition. The next three have special names: $\overline{(u-\overline{u})^2}$ is the *variance* of u, $\overline{(u-\overline{u})^3}$ is the *skewness* of u, and $\overline{(u-\overline{u})^4}$ is the *kurtosis* of u. In addition, the square root of the variance is known as the *standard deviation* and is frequently denoted by the subscript *rms* for *root-mean-square of the fluctuations*: $\sqrt{\overline{(u-\overline{u})^2}} = u_{rms}$. In the study of turbulence, a field variable's first moment and variance are most important.

Example 10.3

Compute the time average of the function $u(t) = Ae^{-t/\tau} + B\cos(\omega t)$ using (10.2). Presuming this function is meant to represent a turbulent field variable with zero-mean fluctuations, $B\cos(\omega t)$, superimposed on a decaying time-dependent average, $Ae^{-t/\tau}$, what condition on Δt leads to an accurate recovery of the decaying average? And, what condition on Δt leads to suppression of the fluctuations?

Solution Start by directly substituting the given function into (10.2):

$$\overline{u(t)} = \frac{1}{\Delta t}\int_{t-\Delta t/2}^{t+\Delta t/2}(Ae^{-t/\tau} + B\cos(\omega t))dt,$$

and evaluating the integral:

$$\overline{u(t)} = \frac{1}{\Delta t}\left(-A\tau\exp\left(-\frac{t+\Delta t/2}{\tau}\right) + A\tau\exp\left(-\frac{t-\Delta t/2}{\tau}\right) + \frac{B}{\omega}\sin\left[\omega(t+\Delta t/2)\right] - \frac{B}{\omega}\sin\left[\omega(t-\Delta t/2)\right]\right).$$

This can be simplified to find:

$$\overline{u(t)} = \left[\frac{\sinh(\Delta t/2\tau)}{\Delta t/2\tau}\right]Ae^{-t/\tau} + \left[\frac{\sin(\omega\Delta t/2)}{\omega\Delta t/2}\right]B\cos(\omega t).$$

In the limit $\Delta t \to 0$, both factors in [,]-braces go to unity and the original function is recovered. Thus, the condition for properly determining the decaying average is $\Delta t \ll \tau$; the averaging interval Δt must be short compared to the time scale for decay, τ. However, to suppress the contribution of the fluctuations represented by the second term, its coefficient must be small. This occurs when $\omega\Delta t \gg 1$ which implies the averaging interval must be many fluctuation time periods long. Therefore, a proper averaging interval should satisfy: $1 \ll \omega\Delta t \ll \omega\tau$, but such a choice for Δt is not possible unless $\omega\tau \gg 1$.

10.4 Correlations and spectra

While moments of a random variable are important and interesting, they do not convey information about the temporal duration or spatial extent of fluctuations, nor do they indicate anything about relationships between one or more dependent-field variables at different places and times. In the study of turbulence, correlations and spectra are commonly used to further characterize fluctuations and are described in this section. Furthermore, since we seek to describe fluctuations, all the random variables in this section are assumed to have zero mean, an assumption that is consistent with the Reynolds decomposition described in the next section. The material presented here starts with general definitions that are simplified for a temporally stationary random variable sampled at the same point in space, or a spatially stationary random variable sampled at different points at the same time. Other approaches to specifying the temporal and spatial character of fluctuations, such as structure functions, fractal dimensions, multi-fractal spectra, and multiplier distributions, etc., are beyond the scope of this discussion.

In three spatial dimensions, the *correlation function* of the random variable u_i at location $\mathbf{x}_1$ and time t_1 with the random variable u_j at location $\mathbf{x}_2$ and time t_2 is defined as

$$R_{ij}(\mathbf{x}_1, t_1, \mathbf{x}_2, t_2) \equiv \overline{u_i(\mathbf{x}_1, t_1)u_j(\mathbf{x}_2, t_2)}, \tag{10.12}$$

where we will soon interpret u_i and u_j as turbulent-flow, velocity-component fluctuations. Note that this R_{ij} is not the rotation tensor defined in Chapter 3. The correlation function R_{ij} can be computed via (10.1) when each realization of the ensemble contains time history pairs: $u_i(\mathbf{x}, t{:}n)$ and $u_j(\mathbf{x}, t; n)$. First, the N pairs $u_i(\mathbf{x}_1, t_1{:}n)$ and $u_j(\mathbf{x}_2, t_2{:}n)$ are selected from the realizations and multiplied together. Then the N pair-products are summed and divided by N to complete the calculation of R_{ij}.

The correlation R_{ij} specifies how similar $u_i(\mathbf{x}_1, t_1)$ and $u_j(\mathbf{x}_2, t_2)$ are to each other. The magnitude of R_{ij} is zero when positive values of $u_i(\mathbf{x}_1, t_1{:}n)$ are associated with equal likelihood with both positive and negative values of $u_j(\mathbf{x}_2, t_2{:}n)$. In this case, $u_i(\mathbf{x}_1, t_1)$ and $u_j(\mathbf{x}_2, t_2)$ are said to be *uncorrelated* when $R_{ij} = 0$, or *weakly correlated* when R_{ij} is small and positive. If, a positive value of $u_i(\mathbf{x}_1, t_1{:}n)$ is mostly associated with a positive value of $u_j(\mathbf{x}_2, t_2{:}n)$, and a negative value of $u_i(\mathbf{x}_1, t_1{:}n)$ is mostly associated with a negative value of $u_j(\mathbf{x}_2, t_2{:}n)$, then the magnitude of R_{ij} is large and positive. In this case, $u_i(\mathbf{x}_1, t_1)$ and $u_j(\mathbf{x}_2, t_2)$ are said to be *strongly correlated*. It is also possible for $u_i(\mathbf{x}_1, t_1{:}n)$ to be mostly associated with values of $u_j(\mathbf{x}_2, t_2{:}n)$ having the opposite sign so that R_{ij} is negative. In this case, $u_i(\mathbf{x}_1, t_1)$ and $u_j(\mathbf{x}_2, t_2)$ are said to be *anticorrelated*.

When $i \neq j$ in (10.12) the resulting function is called a *cross-correlation* function. When $i = j$ in (10.12) and $u_j(\mathbf{x}_2, t_2)$ is replaced by $u_i(\mathbf{x}_2, t_2)$, the resulting function is called an *autocorrelation* function; for example, $i = j = 1$ implies:

$$R_{11}(\mathbf{x}_1, t_1, \mathbf{x}_2, t_2) \equiv \overline{u_1(\mathbf{x}_1, t_1)u_1(\mathbf{x}_2, t_2)}. \tag{10.13}$$

The two definitions, (10.12) and (10.13) may be normalized to define *correlation coefficients*, r_{ij}, and r_{11}. For example when $i = 1$ and $j = 2$, and $i = 1$ and $j = 1$, then:

$$r_{12}(\mathbf{x}_1, t_1, \mathbf{x}_2, t_2) \equiv \frac{R_{12}(\mathbf{x}_1, t_1, \mathbf{x}_2, t_2)}{\sqrt{R_{11}(\mathbf{x}_1, t_1, \mathbf{x}_1, t_1)R_{22}(\mathbf{x}_2, t_2, \mathbf{x}_2, t_2)}} = \frac{\overline{u_1(\mathbf{x}_1, t_1)u_2(\mathbf{x}_2, t_2)}}{\sqrt{\overline{u_1^2}(\mathbf{x}_1, t_1)}\sqrt{\overline{u_2^2}(\mathbf{x}_2, t_2)}} \quad \text{and} \tag{10.14}$$

$$r_{11}(\mathbf{x}_1, t_1, \mathbf{x}_2, t_2) \equiv \frac{R_{11}(\mathbf{x}_1, t_1, \mathbf{x}_2, t_2)}{\sqrt{R_{11}(\mathbf{x}_1, t_1, \mathbf{x}_1, t_1)R_{11}(\mathbf{x}_2, t_2, \mathbf{x}_2, t_2)}} = \frac{\overline{u_1(\mathbf{x}_1, t_1)u_1(\mathbf{x}_2, t_2)}}{\sqrt{\overline{u_1^2}(\mathbf{x}_1, t_1)}\sqrt{\overline{u_1^2}(\mathbf{x}_2, t_2)}}. \tag{10.15}$$

These correlation coefficients are restricted to lie between −1 (perfect anticorrelation) and +1 (perfect correlation). For any two functions u and v, it can be proved that

$$\overline{u(\mathbf{x}_1, t_1)v(\mathbf{x}_2, t_2)} \leq \sqrt{\overline{u^2(\mathbf{x}_1, t_1)}}\sqrt{\overline{v^2(\mathbf{x}_2, t_2)}}, \tag{10.16}$$

which is called the *Schwartz inequality*. It is analogous to the rule that the inner product of two vectors cannot be larger than the product of their magnitudes. From (10.15), $r_{11}(\mathbf{x}_1, t_1, \mathbf{x}_1, t_1)$ is unity.

For temporally stationary processes that are sampled at the same point in space, $\mathbf{x} = \mathbf{x}_1 = \mathbf{x}_2$, the above formulas simplify, and the listing of $\mathbf{x}$ as an argument may be dropped to streamline the notation. The statistics of temporally stationary random processes are independent of the time origin, so we can shift the time origin to t_1 when computing a correlation so that $\overline{u_i(t_1)u_j(t_2)} = \overline{u_i(0)u_j(t_2 - t_1)} = \overline{u_i(0)u_j(\tau)}$, where $\tau = t_2 - t_1$ is the *time lag*, without changing the correlation. Or, we can change t_1 in (10.13) into t, $\overline{u_i(t)u_j(t_2)} = \overline{u_i(t)u_j(t + \tau)}$, without changing the correlation. Thus, the correlation and autocorrelation functions can be written:

$$R_{ij}(\tau) = \overline{u_i(t)u_j(t + \tau)} \quad \text{and} \quad R_{11}(\tau) = \overline{u_1(t)u_1(t + \tau)}, \tag{10.17}$$

where the over bar can be regarded as either an ensemble or time average in this case. Furthermore, under these conditions, the autocorrelation is symmetric:

$$R_{11}(\tau) = \overline{u_1(t)u_1(t + \tau)} = \overline{u_1(t - \tau)u_1(t)} = \overline{u_1(t)u_1(t - \tau)} = R_{11}(-\tau).$$

However, this is not the case for cross-correlations, $R_{ij}(\tau) \neq R_{ij}(-\tau)$ when $i \neq j$. The value of a cross-correlation function at $\tau = 0$, $\overline{u_i(t)u_j(t)}$, is simply written as $\overline{u_iu_j}$ and called the *correlation* of u_i and u_j.

Fig. 10.4 illustrates several of these concepts for two temporally stationary random variables $u(t)$ and $v(t)$. The left panel shows $u(t)$, $u(t+t_o)$, $v(t)$, and the time shift t_o is indicated near the top. The right panel shows the autocorrelation of u, the cross correlation of u and u with an imposed time shift of t_o, and the cross-correlation of u and v. The tic-mark spacing represents the same amount of time in both panels. The time shift necessary for $\overline{u(t)u(t+\tau)}$ to reach zero is comparable to the width of peaks or valleys of $u(t)$. As expected, the autocorrelation is maximum when the two time arguments of u under the ensemble average are equal, and this correlation peak is symmetric about this time shift. Correlation is a mathematical shape-comparison indicator that is sensitive to time alignment. Consider $u(t)$, $u(t+t_o)$, and $\overline{u(t+t_o)u(t+\tau)}$. When $\tau = 0$ the peaks and valleys of $u(t)$ and $u(t+t_o)$ – which are of course are identical – are not temporally aligned so $\overline{u(t+t_o)u(t+\tau)}$ in Fig. 10.4 is nearly zero when $\tau = 0$. However, as τ increases – this corresponds to moving the time history of $u(t)$ to the left – the peaks and valleys of $u(t)$ and $u(t+t_o)$ come closer into temporal alignment. Perfect alignment is reached when $\tau = t_o$ and this produces the correlation maximum in $\overline{u(t+t_o)u(t+\tau)}$ at $\tau = t_o$. The cross-correlation function results in Fig. 10.4 can be understood in a similar manner by looking for temporal alignment in u and v as v slides to left with increasing τ. As shown in the left panel, the largest peak of u is temporally aligned with the largest peak of v when $\tau = t_+$ and this leads to the positive correlation maximum in $\overline{u(t)v(t+\tau)}$ at $\tau = t_+$. However, as τ increases further, the largest peak of u becomes temporally aligned with the deepest valley of v, and this leads to the cross-correlation minimum at $\tau = t_-$. Thus, the zeros and extrema of correlation functions indicate the time shifts necessary to temporally misalign, align, or anti-align field-variable fluctuations. Such timing information cannot be obtained from moments alone.

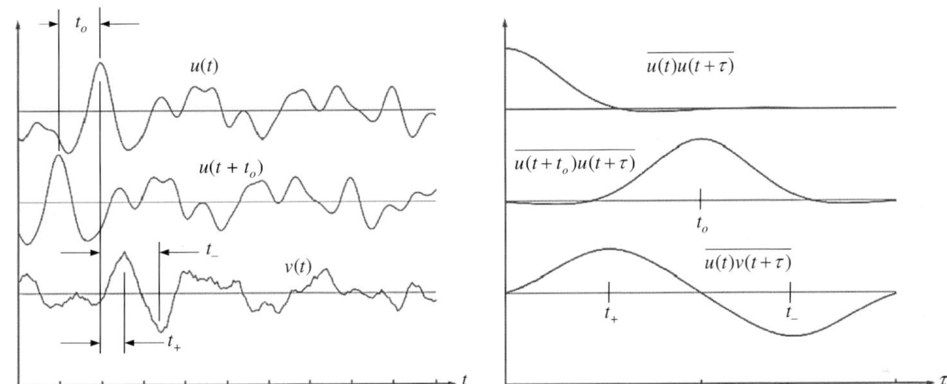

FIGURE 10.4 Sample results for auto- and cross-correlation functions of $u(t)$, $u(t+t_o)$, and $v(t)$. These three time series are shown on the left. The upper curve on the right is the autocorrelation function, $\overline{u(t)u(t+\tau)}$, of the upper time series on the left. The tic marks on the axes represent the same time interval so the width of a peak of $u(t)$ is about equal to the correlation time determined from $\overline{u(t)u(t+\tau)}$. The correlation of $u(t+t_o)$ and $u(t)$ is shown as the middle curve on the right, and it is just a shifted replica of $\overline{u(t)u(t+\tau)}$. The cross-correlation of $u(t)$ and $v(t)$ is the lower curve on the right. Here the maximum cross-correlation occurs when $\tau = t_+$ and the peaks of u and v coincide. Similarly, u and v are most anti-correlated when peaks in u align with valleys in v at $\tau = t_-$.

Several time scales can be determined from the autocorrelation function. For turbulence, the most important of these is the *integral time scale* Λ_t. Under normal conditions R_{11} goes to 0 as $\tau \to \infty$ because the turbulent fluctuation u_1 becomes uncorrelated with itself after a long time. The integral time scale is found by equating the area under the autocorrelation coefficient curve to a rectangle of unity height and duration Λ_t:

$$\Lambda_t \equiv \int_0^\infty r_{11}(\tau)d\tau = (1/R_{11}(0))\int_0^\infty R_{11}(\tau)d\tau, \tag{10.18}$$

where $r_{11}(\tau) = R_{11}(\tau)/R_{11}(0)$ is the autocorrelation coefficient for the stream-wise velocity fluctuation u_1. Of course, (10.18) can be written in terms of r_{22} or r_{33}, but for the purposes at hand this is not necessary. The calculation in (10.18) is shown graphically in Fig. 10.5. The integral time scale is a generic specification of the time over which a turbulent fluctuation is correlated with itself. In other words, Λ_t is a measure of the *memory* of the turbulence. The correlation time t_c is also shown on Fig. 10.5 as the time when $r_{11}(\tau)$ first reaches zero. When temporally averaging a single random-variable time history of length Δt to mimic an ensemble average, the equivalent number of ensemble members can be estimated from $N \approx \Delta t/t_c$. A third time scale, the Taylor microscale λ_t, can also be extracted from $r_{11}(\tau)$. It is obtained from the curvature of the autocorrelation peak at $\tau = 0$ and is given by:

$$\lambda_t^2 \equiv -2/\left[d^2r_{11}/d\tau^2\right]_{\tau=0} \tag{10.19}$$

(see Exercise 10.9). The Taylor microscale λ_t is much less than Λ_t in high Reynolds number turbulence, and it indicates where a turbulent fluctuation, $u_1(t)$ in (10.19), is well correlated with itself.

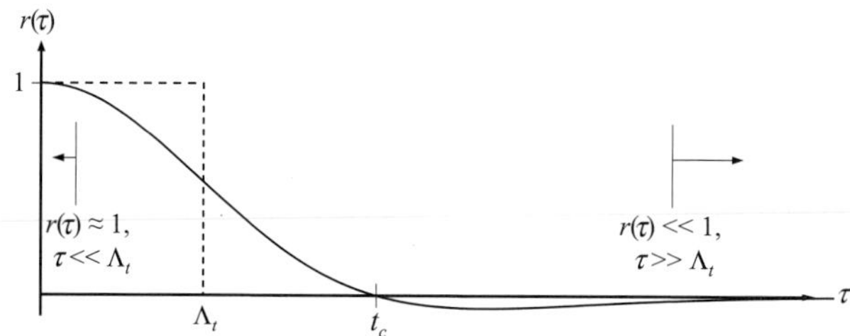

FIGURE 10.5 Sample plot of an autocorrelation coefficient showing the integral time scale Λ_t, and the correlation time t_c. The normalization requires $r(0) = 1$. In the limit $\tau \to \infty$, $r(\tau) \to 0$ and thereby indicates that the random process used to construct r becomes uncorrelated with itself when the time shift τ is large enough.

A second (and equivalent) means of describing the characteristics of turbulent fluctuations, which also complements the information provided by moments, is the *energy spectrum* $S_e(\omega)$ defined as the Fourier transform of the autocorrelation function $R_{11}(\tau)$:

$$S_e(\omega) \equiv \frac{1}{2\pi}\int_{-\infty}^{+\infty} R_{11}(\tau)\exp\{-i\omega\tau\}\,d\tau. \tag{10.20}$$

Thus, $S_e(\omega)$ and $R_{11}(\tau)$ are a Fourier transform pair:

$$R_{11}(\tau) \equiv \int_{-\infty}^{+\infty} S_e(\omega)\exp\{+i\omega\tau\}\,d\omega. \tag{10.21}$$

The relationships (10.20) and (10.21) are not special for $S_e(\omega)$ and $R_{11}(\tau)$ alone, but hold for many function pairs for which a Fourier transform exists. In general terms, a Fourier transform can be defined if the function decays to zero fast enough as its argument goes to infinity. Since $R_{11}(\tau)$ is real and symmetric, then $S_e(\omega)$ is real and symmetric (see Exercise 10.6). Substitution of $\tau = 0$ in (10.21) gives

$$\overline{u_1^2} \equiv \int_{-\infty}^{+\infty} S_e(\omega)d\omega. \tag{10.22}$$

This shows that the integrand increment $S_e(\omega)d\omega$ is the contribution to the variance (or fluctuation energy) of u_1 from the frequency band $d\omega$ centered at ω. Therefore, the function $S_e(\omega)$ represents the way fluctuation energy is distributed across frequency ω. Hence, $S_e(\omega)$ and $R_{11}(\tau)$ are further constrained since $S_e(\omega)$ must be non-negative (see Papoulis, 1965). From (10.20) it also follows that

$$S_e(0) = \frac{1}{2\pi}\int_{-\infty}^{\infty} R_{11}(\tau)d\tau = \frac{\overline{u_1^2}}{\pi}\int_{0}^{\infty} r_{11}(\tau)d\tau = \frac{\overline{u_1^2}}{\pi}\Lambda_t,$$

which shows that the spectral value at zero frequency is proportional to the variance of u_1 and the integral time scale.

From (10.16) to this point, u_i and u_j have been considered stationary functions of time measured at the same point in space. In a similar manner, we now consider u_i and u_j to be stationary functions in space measured at the same instant in time t. For simplicity we drop the listing of t as an independent variable. In this case, the correlation tensor only depends on the vector separation between $\mathbf{x}_1$ and $\mathbf{x}_2$, $\mathbf{r} = \mathbf{x}_2 - \mathbf{x}_1$,

$$R_{ij}(\mathbf{r}) \equiv \overline{u_i(\mathbf{x})u_j(\mathbf{x}+\mathbf{r})}. \tag{10.23}$$

An instantaneous field measurement of $u_i(\mathbf{x})$ is needed to calculate the spatial autocorrelation $R_{ij}(\mathbf{x})$. This is a difficult task in three dimensions, but planar particle imaging velocimetry (PIV) makes it possible in two. However, single-point measurements of a time series $u_1(t)$ in turbulent flows are still common and spatial results may be obtained approximately by rapidly moving a probe in a desired direction. If the speed U_0 of the probe is high enough, we can assume that the field of turbulence is *frozen* and does not change while the probe moves through it during the measurement. Although the probe actually records a time series $u_1(t)$, it can be approximately transformed into a spatial series $u_1(x)$ by replacing t by x/U_0.

The assumption that the turbulent fluctuations at a point are caused by the advection of a frozen field past the point is called *Taylor's hypothesis*, and the accuracy of this approximation increases as the ratio u_{rms}/U_0 decreases (see Exercise 10.11).

Example 10.4

Consider the autocorrelation function $R_{11}(\tau) = \overline{u_1^2}/(1+\omega_o^2\tau^2)$ for the random variable $u_1(t)$. What are the integral (Λ_t) and Taylor (λ_t) length scales, and the spectrum (S_e) of $u_1(t)$?

Solution The integral scale can be found directly by substituting the given autocorrelation function into (10.18) and evaluating the integral using an integration variable $\beta = \omega_o\tau$:

$$\Lambda_t = \frac{1}{R_{11}(0)}\int_0^{+\infty} R_{11}(\tau)d\tau = \frac{1}{\overline{u_1^2}}\int_0^{+\infty}\frac{\overline{u_1^2}}{1+\omega_o^2\tau^2}d\tau = \frac{1}{\omega_o}\int_0^{+\infty}\frac{d\beta}{1+\beta^2} = \frac{1}{\omega_o}\left[\tan^{-1}\beta\right]_0^{+\infty} = \frac{\pi}{2\omega_o}.$$

The Taylor scale can be found from (10.19). Here, $r_{11}(\tau) = 1/(1+\omega_o^2\tau^2)$, which can be expanded around $\tau = 0$ to find: $r_{11}(\tau) = 1 - \omega_o^2\tau^2 + \cdots$. Thus, $[d^2 r_{11}/d\tau^2]_{\tau=0} = -2\omega_o^2$, so (10.19) implies

$$\lambda_t \equiv \left(-2/[d^2 r_{11}/d\tau^2]_{\tau=0}\right)^{1/2} = \frac{1}{\omega_o}.$$

As expected, the integral length scale is larger than the Taylor length scale. In high-Reynolds number turbulence, the ratio Λ_t/λ_t can be much greater than that found here. The spectrum $S_e(\omega)$ is obtained from (10.20):

$$S_e(\omega) \equiv \frac{\overline{u_1^2}}{2\pi}\int_{-\infty}^{+\infty}\frac{\exp\{-i\omega\tau\}}{1+\omega_o^2\tau^2}d\tau = \frac{\overline{u_1^2}}{2\omega_o}\exp\left\{-\frac{|\omega|}{\omega_o}\right\},$$

where the integral is readily evaluated using complex contour integration techniques.

10.5 Averaged equations of motion

In this section, the equations of motion for the mean state in a turbulent flow are derived. The contribution of turbulent fluctuations appears in these equations as a correlation of velocity-component fluctuations. A turbulent flow instantaneously satisfies the Navier-Stokes equations. However, it is virtually impossible to predict the flow in detail at high Reynolds numbers, as there is an enormous range of length scales to be resolved (see Example 10.2) at each instant in time. Perhaps more importantly, knowledge of all these details is typically not necessary. If a commercial aircraft must fly from Los Angeles, California to Sydney, Australia, and turbulent skin-friction fluctuations occur in a frequency range from a few Hz to more than 10^4 Hz, the economically important parameter is the average skin friction because the time of the flight (many hours) is much longer than the even the longest fluctuation time scale. Here, the integrated effect of the fluctuations approaches zero when compared to the integral of the average. This situation where the overall duration of the flow far exceeds turbulent-fluctuation time scales is very common in engineering and geophysical science.

The following development of the mean-flow equations is for incompressible turbulent flow with constant viscosity where density fluctuations are caused by temperature fluctuations alone. The first step is to separate the dependent-field quantities into components representing the mean (capital letters and those with over bars) and those representing the deviation from the mean (lower case letters and those with primes):

$$\tilde{u}_i = U_i + u_i, \quad \tilde{p} = P + p, \quad \tilde{\rho} = \overline{\rho} + \rho', \quad \text{and} \quad \tilde{T} = \overline{T} + T', \tag{10.24}$$

where – as in the preceding chapter – the complete field quantities are denoted by a tilde (~). As mentioned in Section 10.3, this separation into mean and fluctuating components is called the *Reynolds decomposition*. Although it doubles the number of dependent field variables, this decomposition remains useful and relevant more than a century after it was first proposed. However, it leads to a closure problem in the resulting equation set that has still not been resolved without empiricism and modeling. The mean quantities in (10.24) are regarded as expected values:

$$\overline{\tilde{u}_i} = U_i, \quad \overline{\tilde{p}} = P, \quad \overline{\tilde{\rho}} = \overline{\rho}, \quad \text{and} \quad \overline{\tilde{T}} = \overline{T}, \tag{10.25}$$

and the fluctuations have zero mean:

$$\overline{u_i} = 0, \quad \overline{p} = 0, \quad \overline{\rho'} = 0, \quad \text{and} \quad \overline{T'} = 0. \tag{10.26}$$

The equations satisfied by the mean flow are obtained by substituting (10.24) into the governing equations and averaging. Here, the starting point is the Boussinesq set:

$$\frac{\partial \tilde{u}_i}{\partial x_i} = 0, \quad \frac{\partial \tilde{u}_i}{\partial t} + \tilde{u}_j \frac{\partial \tilde{u}_i}{\partial x_j} = \frac{\partial \tilde{u}_i}{\partial t} + \frac{\partial}{\partial x_j}(\tilde{u}_j \tilde{u}_i) = -\frac{1}{\rho_0}\frac{\partial \tilde{p}}{\partial x_i} - g\left[1 - \alpha(\tilde{T} - T_0)\right]\delta_{i3} + \nu \frac{\partial^2 \tilde{u}_i}{\partial x_j^2}, \tag{4.10, 4.87}$$

$$\text{and} \quad \frac{\partial \tilde{T}}{\partial t} + \tilde{u}_j \frac{\partial \tilde{T}}{\partial x_j} = \frac{\partial \tilde{T}}{\partial t} + \frac{\partial}{\partial x_j}(\tilde{u}_j \tilde{T}) = \kappa \frac{\partial^2 \tilde{T}}{\partial x_j^2}, \tag{4.90}$$

where the first equality in (4.87) and (4.90) follows from adding $\tilde{u}_i(\partial \tilde{u}_j/\partial x_j) = 0$ and $\tilde{T}(\partial \tilde{u}_j/\partial x_j) = 0$, respectively, to the left-most sides of these equations. Simplifications for constant-density flow are readily obtained at the end of this equation-generation effort.

The continuity equation for the mean flow is obtained by putting the velocity decomposition of (10.24) into (4.10) and averaging:

$$\overline{\frac{\partial \tilde{u}_i}{\partial x_i}} = \overline{\frac{\partial}{\partial x_i}(U_i + u_i)} = \frac{\partial}{\partial x_i}\overline{(U_i + u_i)} = \frac{\partial}{\partial x_i}(U_i + \overline{u}_i) = \frac{\partial U_i}{\partial x_i} = 0, \tag{10.27}$$

where (10.8), $\overline{U}_i = U_i$, and $\overline{u}_i = 0$ have been used. Subtracting (10.27) from (4.10) produces:

$$\partial u_i / \partial x_i = 0. \tag{10.28}$$

Thus, the mean and fluctuating velocity fields are each divergence free.

The procedure for generating the mean momentum equation is similar but involves more terms. Substituting (10.24) into (4.87) produces:

$$\frac{\partial (U_i + u_i)}{\partial t} + \frac{\partial}{\partial x_j}((U_j + u_j)(U_i + u_i)) = -\frac{1}{\rho_0}\frac{\partial (P + p)}{\partial x_i} - g[1 - \alpha(\overline{T} + T' - T_0)]\delta_{i3} + \nu \frac{\partial^2 (U_i + u_i)}{\partial x_j^2}. \tag{10.29}$$

The averages of each term in this equation can be determined by using (10.26) and the properties of an ensemble average: (10.4) through (10.6) and (10.8). The term-by-term results are:

$$\overline{\frac{\partial (U_i + u_i)}{\partial t}} = \frac{\partial \overline{(U_i + u_i)}}{\partial t} = \frac{\partial (U_i + \overline{u}_i)}{\partial t} = \frac{\partial U_i}{\partial t},$$

$$\overline{\frac{\partial}{\partial x_j}((U_j + u_j)(U_i + u_i))} = \frac{\partial}{\partial x_j}\overline{(U_i U_j + U_i u_j + u_i U_j + u_i u_j)}$$

$$= \frac{\partial}{\partial x_j}(U_i U_j + \overline{u}_i U_j + U_i \overline{u}_j + \overline{u_i u_j}) = \frac{\partial}{\partial x_j}(U_i U_j + \overline{u_i u_j}),$$

$$\overline{\frac{1}{\rho_0}\frac{\partial (P + p)}{\partial x_i}} = \frac{1}{\rho_0}\frac{\partial \overline{(P + p)}}{\partial x_i} = \frac{1}{\rho_0}\frac{\partial (P + \overline{p})}{\partial x_i} = \frac{1}{\rho_0}\frac{\partial P}{\partial x_i},$$

$$\overline{g\left[1 - \alpha(\overline{T} + T' - T_0)\right]\delta_{i3}} = g\left[1 - \alpha(\overline{T} + \overline{T}' - T_0)\right]\delta_{i3} = g[1 - \alpha(\overline{T} - T_0)]\delta_{i3}, \quad \text{and}$$

$$\overline{\nu \frac{\partial^2 (U_i + u_i)}{\partial x_j^2}} = \nu \frac{\partial^2 \overline{(U_i + u_i)}}{\partial x_j^2} = \nu \frac{\partial^2 (U_i + \overline{u}_i)}{\partial x_j^2} = \nu \frac{\partial^2 U}{\partial x_j^2}.$$

Collecting terms, the ensemble average of the momentum equation is:

$$\frac{\partial U_i}{\partial t} + \frac{\partial}{\partial x_j}(U_i U_j) + \frac{\partial}{\partial x_j}(\overline{u_i u_j}) = -\frac{1}{\rho_0}\frac{\partial P}{\partial x_i} - g\left[1 - \alpha(\overline{T} - T_0)\right]\delta_{i3} + \nu \frac{\partial^2 U_i}{\partial x_j^2}.$$

This equation can be mildly rearranged using the final result of (10.27) and combining the gradient terms together to form the mean stress tensor $\overline{\tau}_{ij}$:

$$\begin{aligned}\frac{\partial U_i}{\partial t}+U_j\frac{\partial U_i}{\partial x_j}&=-g\left[1-\alpha(\overline{T}-T_0)\right]\delta_{i3}+\frac{1}{\rho_0}\frac{\partial \overline{\tau}_{ij}}{\partial x_j}\\&=-g\left[1-\alpha(\overline{T}-T_0)\right]\delta_{i3}+\frac{1}{\rho_0}\frac{\partial}{\partial x_j}\left(-P\delta_{ij}+2\mu\overline{S}_{ij}-\rho_0\overline{u_iu_j}\right),\end{aligned} \tag{10.30}$$

where $\overline{S}_{ij}=\frac{1}{2}\left(\partial U_i/\partial x_j+\partial U_j/\partial x_i\right)$ is the mean strain-rate tensor, and (4.41) has been used to put the mean viscous stress in the form shown in (10.30). The correlation tensor $\overline{u_iu_j}$ in (10.30) is generally non-zero even though $\overline{u}_i=0$. Its presence in (10.30) is important because it has no counterpart in the instantaneous momentum equation (4.87) and it links the character of the fluctuations to the mean flow. Unfortunately, the process of reaching (10.30) does not provide any new equations for this correlation tensor. Thus, the final equality of (10.27), and (10.30) do not comprise a closed system of equations, even when the flow is isothermal.

The new tensor in (10.30), $-\rho_0\overline{u_iu_j}$, plays the role of a stress and is called the *Reynolds stress tensor*. When present, Reynolds stresses are often much larger than viscous stresses, $\mu(\partial U_i/\partial x_j+\partial U_j/\partial x_i)$, except very close to a solid surface where the fluctuations go to zero and mean-flow gradients are large. The Reynolds stress tensor is symmetric since $\overline{u_iu_j}=\overline{u_ju_i}$, so it has six independent Cartesian components. Its diagonal components $\overline{u_1^2}$, $\overline{u_2^2}$, and $\overline{u_3^2}$ are normal stresses that augment the mean pressure, while its off-diagonal components $\overline{u_1u_2}$, $\overline{u_1u_3}$, and $\overline{u_2u_3}$ are shear stresses.

An explanation why the average product of the velocity fluctuations in a turbulent flow is not expected to be zero follows. Consider a shear flow where the mean shear dU/dy is positive (Fig. 10.6). Assume that a fluid particle at level y travels upward because of a fluctuation ($v>0$). On average this particle retains its original horizontal velocity during the migration, so when it arrives at level $y+dy$ it finds itself in a region where a larger horizontal velocity prevails. Thus the particle is on average slower ($u<0$) than neighboring fluid particles after it has reached the level $y+dy$. Conversely, fluid particles that travel downward ($v<0$) tend to cause a positive u at their new level $y-dy$. Taken together, a positive v is associated with a negative u, and a negative v is associated with a positive u. Therefore, the correlation $\overline{uv}$ is negative for the velocity field shown in Fig. 10.6, where $dU/dy>0$. This makes sense, since in this case the x-momentum should tend to flow in the negative y-direction as the turbulence tends to diffuse the gradients and decrease dU/dy.

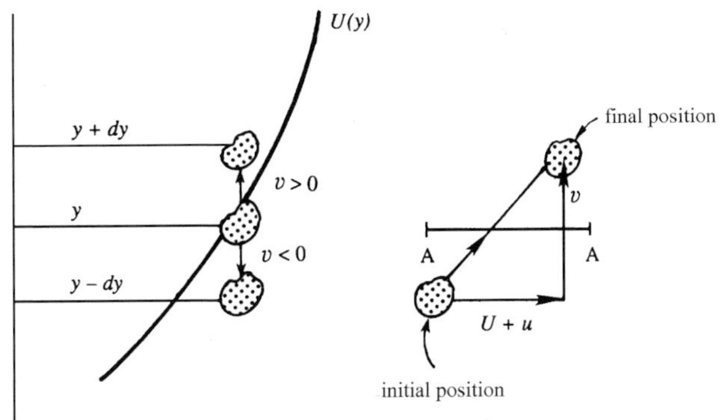

FIGURE 10.6 A schematic illustration of the development of non-zero Reynolds shear stress in a simple shear flow. A fluid particle that starts at y and is displaced upward to $y+dy$ by a positive vertical velocity fluctuation v brings an average horizontal fluid velocity of $U(y)$ that is lower than $U(y+dy)$. Thus, a positive vertical velocity fluctuation v is correlated with negative horizontal velocity fluctuation u, so $\overline{uv}<0$. Similarly, a negative v displaces the fluid particle to $y-dy$ where it arrives on average with positive u, so again $\overline{uv}<0$. Thus, turbulent fluctuations in shear flow are likely to produce negative non-zero Reynolds shear stress.

Reynolds stresses arise from the nonlinear advection term $u_j\left(\partial u_i/\partial x_j\right)$ of the momentum equation, and are the average stress exerted by turbulent fluctuations on the mean flow. Another way to interpret the Reynolds stress is that it is the rate of mean momentum transfer by turbulent fluctuations. Consider again the shear flow $U(y)$ shown in Fig. 10.6, where the instantaneous velocity is $(U+u,\ v,\ w)$. The fluctuating velocity components constantly transport fluid particles, and associated momentum, across a plane AA normal to the y-direction. The instantaneous rate of mass transfer across a unit area is $\rho_0 v$, and consequently the instantaneous rate of x-momentum transfer is $\rho_0(U+u)v$. Per unit area, the average rate of flow of x-momentum in the y-direction is therefore

$$\rho_0\overline{(U+u)v}=\rho_0U\overline{u}+\rho_0\overline{uv}=\rho_0\overline{uv}.$$

Generalizing, $\rho_0\overline{u_i u_j}$ is the average flux of i-momentum along the j-direction, which also equals the average flux of j-momentum along the i-direction.

The sign convention for the Reynolds stress is the same as that explained in Section 2.4. On a surface whose outward normal points in the positive i-direction, a positive τ_{ij} points along the j-direction. According to this convention, the Reynolds shear stresses $-\rho_0\overline{u_i u_j}(i \neq j)$ on a rectangular element are directed as in Fig. 10.7, if they are positive. Such a Reynolds stress causes mean transport of x-momentum along the negative y-direction.

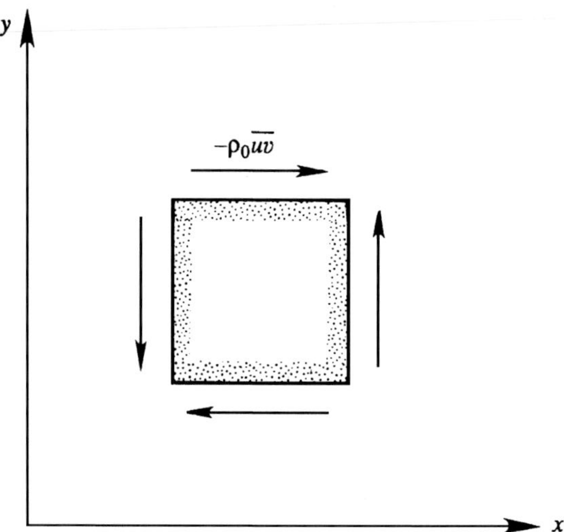

FIGURE 10.7 Positive directions of Reynolds stresses on a square element. These stress components are consistent with those drawn in Fig. 2.4.

The mean-flow thermal energy equation comes from substituting the velocity and temperature decompositions of (10.24) into (4.90) and averaging. The substitution step produces:

$$\frac{\partial}{\partial t}(\overline{T}+T')+\frac{\partial}{\partial x_j}((U_j+u_j)(\overline{T}+T'))=\kappa\frac{\partial^2}{\partial x_j^2}(\overline{T}+T').$$

The averages of each term in this equation are:

$$\overline{\frac{\partial}{\partial t}(\overline{T}+T')}=\frac{\partial}{\partial t}(\overline{T}+\overline{T'})=\frac{\partial\overline{T}}{\partial t},$$

$$\overline{\frac{\partial}{\partial x_j}((U_j+u_j)(\overline{T}+T'))}=\frac{\partial}{\partial x_j}(U_j\overline{T}+\overline{u}_j\overline{T}+U_j\overline{T'}+\overline{u_jT'})=U_j\frac{\partial\overline{T}}{\partial x_j}+\frac{\partial}{\partial x_j}(\overline{u_jT'}),\quad\text{and}$$

$$\overline{\kappa\frac{\partial^2}{\partial x_j^2}(\overline{T}+T')}=\kappa\frac{\partial^2}{\partial x_j^2}(\overline{T}+\overline{T'})=\kappa\frac{\partial^2\overline{T}}{\partial x_j^2},$$

where the final equality of (10.27), (10.4) through (10.6), and (10.8) have been used. Collecting terms, the mean temperature equation takes the form:

$$\frac{\partial\overline{T}}{\partial t}+U_j\frac{\partial\overline{T}}{\partial x_j}+\frac{\partial}{\partial x_j}(\overline{u_jT'})=\kappa\frac{\partial^2\overline{T}}{\partial x_j^2}. \tag{10.31}$$

When multiplied by $\rho_o c_p$ and rearranged, (10.31) becomes the heat transfer equivalent of (10.30) and can be stated in terms of the mean heat flux Q_j:

$$\rho_0 c_p\left(\frac{\partial\overline{T}}{\partial t}+U_j\frac{\partial\overline{T}}{\partial x_j}\right)=-\frac{\partial Q_j}{\partial x_j}=-\frac{\partial}{\partial x_j}\left(-k\frac{\partial\overline{T}}{\partial x_j}+\rho_0 c_p\overline{u_jT'}\right), \tag{10.32}$$

where $k=\rho_o c_p\kappa$ is the thermal conductivity. Eq. (10.32) shows that the fluctuations cause an additional mean *turbulent heat flux of* $\rho_0 c_p\overline{u_jT'}$ that has no equivalent in (4.90). The turbulent heat flux is the thermal equivalent of the Reynolds stress $-\rho_0\overline{u_i u_j}$ found in (10.30). Unfortunately, the process of reaching (10.31) and (10.32) has not provided any new equations for the turbulent heat flux. However, some understanding of the turbulent heat flux can be gained by considering diurnal heating of the earth's surface. During daylight hours, the sun may heat the surface of the earth, resulting in a mean temperature that decreases with height and in the potential for turbulent convective air motion. When such motions occur, an upward velocity fluctuation is mostly associated with a positive temperature fluctuation, giving rise to an upward heat flux $\rho_0 c_p \overline{u_3 T'} > 0$.

The final mean-flow equation commonly considered for turbulent flows is that for transport of a dye or a non-reacting molecular species that is merely carried by the turbulent flow without altering the flow. Such passive contaminants are commonly called *passive scalars* or *conserved scalars* and the rate at which they are mixed with non-turbulent fluid is often of significant technological interest for pollutant dispersion and premixed combustion. Consider a simple binary mixture composed of a primary fluid with density ρ and a contaminant fluid (the passive scalar) with density ρ_s. The density ρ_m that results from mixing these two fluids is $\rho_m = \tilde{\upsilon}\rho_s + (1-\tilde{\upsilon})\rho$, where $\tilde{\upsilon}$ is the volume fraction of the passive scalar. The relevant conservation equation for the passive scalar is:

$$\frac{\partial}{\partial t}(\rho_m \tilde{Y}) + \frac{\partial}{\partial x_j}\left(\rho_m \tilde{Y}\tilde{u}_j\right) = \frac{\partial}{\partial x_j}\left(\rho_m \kappa_m \frac{\partial}{\partial x_j}\tilde{Y}\right) \tag{10.33}$$

(see Kuo, 1986), where $\tilde{u}_j$ is the instantaneous mass-averaged velocity of the mixture, $\tilde{Y}$ is the mass fraction of the passive scalar, and κ_m is the mass-based molecular diffusivity of the passive scalar (see (1.1)). If the mean and fluctuating mass fraction of the conserved scalar are $\overline{Y}$ and Y', and the mixture density is constant, then the mean-flow passive scalar conservation equation is (see Exercise 10.12):

$$\frac{\partial \overline{Y}}{\partial t} + U_j \frac{\partial \overline{Y}}{\partial x_j} = \frac{\partial}{\partial x_j}\left(\kappa_m \frac{\partial \overline{Y}}{\partial x_j} - \overline{u_j Y'}\right), \tag{10.34}$$

where $\overline{u_j Y'}$ is the turbulent flux of the passive scalar. This equation is valid when the mixture density is constant, and this occurs when $\rho = \rho_s =$ constant and when the contaminant is dilute so that $\rho_m \approx \rho =$ constant. If the amount of a passive scalar is characterized by a concentration (mass per unit volume), molecular number density, or mole fraction – instead of a mass fraction – the forms of (10.33) and (10.34) are unchanged but the diffusivity may need to be adjusted and molecular number or mass density factors may appear (see Bird et al., 1960; Kuo, 1986). Eq. (10.34) is of the same form as (10.32), and temperature may be considered a passive scalar in turbulent flows when it does not induce buoyancy, cause chemical reactions, or lead to significant density changes.

To summarize, (10.27), (10.30), (10.32), and (10.34) are the mean-flow equations for incompressible turbulent flow (in the Boussinesq approximation). The process of reaching these equations is known as *Reynolds averaging*, and it may be applied to the full compressible-flow equations of fluid motion as well. The equations that result from **R**eynolds **a**veraging of any form of the **N**avier-**S**tokes equations are commonly known as *RANS equations*. The constant-density mean-flow RANS equations commonly used in hydrodynamics are obtained from the results provided in this section by dropping the gravity term and the "0" from ρ_0 in (10.30), and reinterpreting the mean pressure as the deviation from hydrostatic (as explained in Section 4.9).

The primary problem with RANS equations is that there are more unknowns than equations. The system of equations for the first moments depends on correlations involving pairs of variables (second moments). And, RANS equations developed for these pair correlations involve triple correlations. For example, the conservation equation for the Reynolds stress correlation, $\overline{u_i u_j}$, is:

$$\begin{aligned}\frac{\partial \overline{u_i u_j}}{\partial t} + U_k \frac{\partial \overline{u_i u_j}}{\partial x_k} + \frac{\partial \overline{u_i u_j u_k}}{\partial x_k} &= -\overline{u_i u_k}\frac{\partial U_j}{\partial x_k} - \overline{u_j u_k}\frac{\partial U_i}{\partial x_k} - \frac{1}{\rho}\left(\overline{u_i \frac{\partial p}{\partial x_j}} + \overline{u_j \frac{\partial p}{\partial x_i}}\right) \\ &\quad - 2\nu \overline{\frac{\partial u_i}{\partial x_k}\frac{\partial u_j}{\partial x_k}} + \nu \frac{\partial^2}{\partial x_k^2}\overline{u_i u_j} + g\alpha\left(\overline{u_j T'}\delta_{i3} + \overline{u_i T'}\delta_{j3}\right)\end{aligned} \tag{10.35}$$

(see Exercise 10.16), and triple correlations appear in the third term on the left. Similar conservation equations for the triple correlations involve quadruple correlations, and the equations for the quadruple correlations depend on fifth-order correlations, and so on. This problem persists at all correlation levels and is known as the *closure problem* in turbulence. At the present time there are three approaches to the closure problem. The first, known as RANS closure modeling (see

Section 10.10), involves terminating the equation hierarchy at a given level and closing the resulting system of equations with model equations developed from dimensional analysis, intuition, symmetry requirements, and experimental results. The second, known as direct numerical simulations (DNS) involves numerically solving the time-dependent equations of motion and then Reynolds averaging the computational output to determine mean-flow quantities. The third, known as large-eddy simulation (LES), combines elements of the other two and involves some modeling and some numerical simulation of large-scale turbulent fluctuations (see Exercise 10.19).

A secondary problem associated with the RANS equations is that the presence of the Reynolds stresses in (10.30) excludes the possibility of converting (10.30) into a Bernoulli equation, even when the density is constant and the terms containing $\partial/\partial t$ and ν are zero.

Example 10.5

Assume steady constant-density two-dimensional mean flow, and use (10.27) and (10.30) to redo the control volume analysis of Example 4.1 shown in Fig. 4.2 to reach an integral formula for the average drag force $\overline{F}_D$ on a two-dimensional body with span l in terms of the mean stream-wise velocity U and the normal Reynolds stress components in body's turbulent wake.

Solution Here the analysis is similar to that in Example 4.1 with slightly different equations and integrands. To take advantage of the averaging in (10.27) and (10.30), their control volume form must be regenerated. For steady constant-density mean flow, (10.27) and (10.30) reduce to:

$$\frac{\partial U_j}{\partial x_j} = 0 \quad \text{and} \quad U_j \frac{\partial U_i}{\partial x_j} = +\frac{1}{\rho}\frac{\partial}{\partial x_j}(\overline{\tau}_{ij}),$$

where $\overline{\tau}_{ij}$ is defined by the second equality of (10.30). Multiply the first of these with U_i and combine the result with the left side of the second equation to reach:

$$\frac{\partial}{\partial x_j}(U_i U_j) = +\frac{1}{\rho}\frac{\partial}{\partial x_j}(\overline{\tau}_{ij}), \quad \text{or for } \rho = \text{constant}: \quad \frac{\partial}{\partial x_j}\left(\rho U_i U_j - \overline{\tau}_{ij}\right) = 0.$$

Now integrate the final equation within the control volume for Example 4.1 and use Gauss' divergence theorem to reduce to volume integral to an integral over its surface. Here, as in Example 4.1, the two vertical surfaces parallel to the flow upstream of the object each contribute half of the average drag force, $\overline{F}_D$. For conservation of horizontal momentum these steps lead to:

$$\left(\int_{inlet} + \int_{top} + \int_{bottom} + \int_{outlet}\right)\left(\rho U_1 U_j - \overline{\tau}_{1j}\right) n_j dA = -\overline{F}_D \mathbf{e}_1,$$

where "1" indicates the horizontal stream-wise direction. The equivalent conservation of mass statement is obtained from (10.27) by again using Gauss' divergence theorem and noting that the average mass flux is zero on the two vertical surfaces parallel to the flow upstream of the object.

$$\left(\int_{inlet} + \int_{top} + \int_{bottom} + \int_{outlet}\right) U_j n_j dA = 0.$$

To reach the desired formula, choose "2" as the vertical direction, denote $U_1 = U$ and $U_2 = V$, and subtract $\int P_\infty \mathbf{n} dA = 0$ from the momentum equation, where P_∞ (= constant) is the pressure outside the turbulent wake far from the body. For two-dimensional mean flow, $U_3 = 0$ so conservation of mass and horizontal momentum for the control volume of Example 4.1 imply:

$$-\int_{inlet} U_\infty l dy + \int_{top} V l dx - \int_{bottom} V l dx + \int_{outlet} U l dy = 0, \quad \text{and}$$

$$-\int_{inlet} \rho U_\infty^2 l dy + \int_{top} \rho U_\infty V l dx - \int_{bottom} \rho U_\infty V l dx + \int_{outlet} \rho\left(U^2(y) + (P - P_\infty) - 2\nu S_{11} + \overline{u^2}\right) l dy = -\overline{F}_D,$$

where l is the width of the control volume transverse to the flow, U_∞ is the steady horizontal flow speed upstream of the object, and $\overline{u^2}$ is the first normal Reynolds stress. Here the viscous and Reynolds stresses are only non-zero on the control volume's outlet surface, and the pressure on the inlet, top, and bottom of the control volume is P_∞. To eliminate the integrals over the inlet, top, and bottom surfaces, multiply the conservation-of-mass equation by $-\rho U_\infty$ and add it to the horizontal momentum equation to find:

$$\overline{F}_D = \int_{-H/2}^{+H/2} \rho\left(U_\infty U - U^2 + (P_\infty - P) + 2\nu S_{11} - \overline{u^2}\right) l dy.$$

Here, the remaining integral is over the outlet surface and H is the vertical height of the control volume.

Compared to the result of Example 4.1, this relationship for $\overline{F}_D$ contains three extra terms. The final extra term is the turbulent normal stress from stream-wise velocity fluctuations and must be retained. The second extra term is the viscous normal stress and may be neglected compared to the final term when the flow is turbulent; the viscous stress was ignored in Example 4.1 as well. The first extra term involves the pressure difference, and represents the contribution of the second normal Reynolds stress when the mean flow is nearly parallel and the boundary-layer approximation applies. Consider the vertical component of (10.30) for constant density flow written in terms of (x, y)-coordinates:

$$U\frac{\partial V}{\partial x} + V\frac{\partial V}{\partial y} = -\frac{1}{\rho}\frac{\partial P}{\partial y} - \frac{\partial}{\partial x}\overline{uv} - \frac{\partial}{\partial y}\overline{v^2},$$

where again the viscous stresses have been ignored and v is the vertical velocity fluctuation. For nearly parallel flow in the horizontal direction ($V \ll U$ and $\partial/\partial x \ll \partial/\partial y$), the boundary-layer approximation of this equation is:

$$0 + 0 \cong -\frac{1}{\rho}\frac{\partial P}{\partial y} - 0 - \frac{\partial}{\partial y}\overline{v^2} \quad \text{or} \quad 0 \cong \frac{\partial}{\partial y}\left(P + \rho\overline{v^2}\right),$$

which can be integrated and evaluated far from the turbulent wake where $v = 0$ and $P = P_\infty$ to find: $P_\infty \cong P + \rho\overline{v^2}$. Substituting this into the integral relationship for the drag force produces:

$$\frac{\overline{F}_D}{\rho l} \cong \int_{-H/2}^{+H/2}\left(U(U_\infty - U) + \overline{v^2} - \overline{u^2}\right)dy,$$

which implies that the result of Example 4.1 is acceptable when the first and second normal Reynolds stresses are small or equal.

10.6 Homogeneous isotropic turbulence

From (10.27), (10.30), (10.32), and (10.34), even with suitable boundary conditions the RANS equations for the mean flow are not directly solvable (even numerically) because of the closure problem. However, the idealization of turbulence as being homogeneous (or spatially stationary) and isotropic allows some significant simplifications. Turbulence behind a grid towed through a nominally quiescent fluid bath is approximately homogeneous and isotropic, and turbulence in the interior of a real inhomogeneous turbulent flow is commonly assumed to be locally homogeneous and isotropic (see Fig. 10.8). Compared to the overview provided here, the topic of homogeneous isotropic turbulence is covered in greater detail in Batchelor (1953) and Hinze (1975).

If turbulent fluctuations are completely isotropic, that is, if they do not have any directional preference, then the off-diagonal components of $\overline{u_i u_j}$ vanish, and the normal stresses are equal. This is illustrated in Fig. 10.9, which shows a cloud of data points (sometimes called a scatter plot) on a uv-plane. The dots represent the instantaneous values of the (u, v)-velocity component pair at different times in a turbulent flow. In the isotropic case there is no directional preference, and the dots form a symmetric pattern. In this case positive u is equally likely to be associated with both positive and negative v. Consequently, *the average value of the product uv is zero when the turbulence is isotropic*. In contrast, the scatter plot in an anisotropic turbulent field has an orientation. The figure shows a case where a positive u is mostly associated with a negative v, giving $\overline{uv} < 0$.

If, in addition, the turbulence is homogeneous, then there are no spatial variations in the flow's statistics and all directions are equivalent:

$$\frac{\partial}{\partial x_i}\overline{u_j^n} = 0, \quad \overline{u_1^2} = \overline{u_2^2} = \overline{u_3^2}, \quad \text{and} \quad \overline{\left(\frac{\partial u_1}{\partial x_1}\right)^n} = \overline{\left(\frac{\partial u_2}{\partial x_2}\right)^n} = \overline{\left(\frac{\partial u_3}{\partial x_3}\right)^n}, \tag{10.36}$$

but relative directions must be respected:

$$\overline{\left(\frac{\partial u_1}{\partial x_2}\right)^n} = \overline{\left(\frac{\partial u_1}{\partial x_3}\right)^n} = \overline{\left(\frac{\partial u_2}{\partial x_1}\right)^n} = \overline{\left(\frac{\partial u_2}{\partial x_3}\right)^n} = \overline{\left(\frac{\partial u_3}{\partial x_1}\right)^n} = \overline{\left(\frac{\partial u_3}{\partial x_2}\right)^n}. \tag{10.37}$$

Note that the continuity equation requires derivative moments in the third set of equalities of (10.36) to be zero when $n = 1$.

The spatial structure of the flow may be ascertained by considering the two-point correlation tensor, defined by (10.23), which reduces to the Reynolds stress correlation when $\mathbf{r} = 0$. In homogeneous flow, R_{ij} does not depend on $\mathbf{x}$, and can only depend on $\mathbf{r}$. If the turbulence is also isotropic, the direction of $\mathbf{r}$ cannot matter. In this special situation, only two different

FIGURE 10.8 Turbulent plume from the Mount Saint Helens eruption on May 18, 1980 (via U.S. Geological Survey). The inset shows the magnitude of velocity fluctuations in a direct numerical simulation of homogeneous isotropic turbulence with $R_\lambda = 87$. Despite large velocity gradients at the edge of the plume, Kolmogorov's hypothesis of local isotropy states that at sufficiently high Reynolds number, small-scale eddies ($l \ll L$) are statistically isotropic.

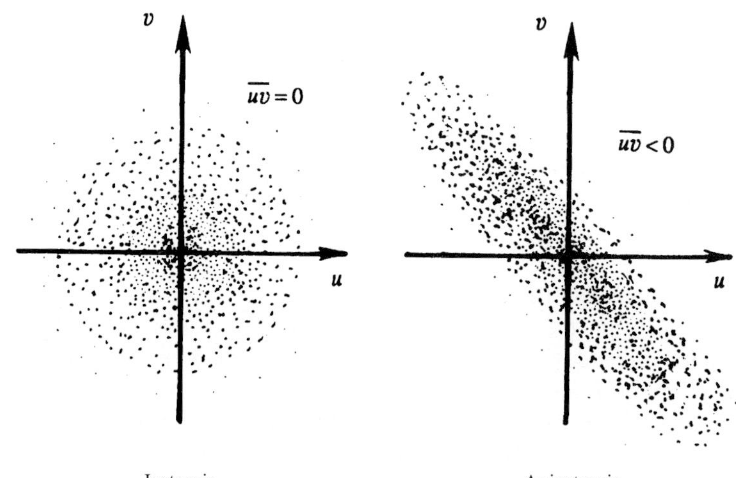

FIGURE 10.9 Scatter plots of velocity fluctuation samples in isotropic and anisotropic turbulent fields. Each dot represents a (u, v)-pair at a sample time and many sample times are represented in each panel. The isotropic case produces a symmetric cloud of points and indicates $\overline{uv} = 0$. The anisotropic case shows the data clustering around the line $v = -u$ and this indicates a negative correlation of u and v; $\overline{uv} < 0$.

types of velocity-field correlations survive. These are described by the longitudinal (f) and transverse (g) correlation coefficients defined by:

$$f(r) \equiv \overline{u_{||}(\mathbf{x}+\mathbf{r})u_{||}(\mathbf{x})}/\overline{u_{||}^2} \quad \text{and} \quad g(r) \equiv \overline{u_{\perp}(\mathbf{x}+\mathbf{r})u_{\perp}(\mathbf{x})}/\overline{u_{\perp}^2}, \tag{10.38}$$

where $u_{||}$ is parallel to $\mathbf{r}$, and $u_{\perp}$ is perpendicular to $\mathbf{r}$, $\overline{u_{||}^2} = \overline{u_{\perp}^2} = \overline{u^2}$, and, $f(0) = g(0) = 1$. The geometries for these two correlation functions are shown in Fig. 10.10, where solid vectors indicate velocities and the dashed vector represents $\mathbf{r}$. Longitudinal and transverse integral scales and Taylor microscales are defined by:

$$\Lambda_f \equiv \int_0^\infty f(r)dr, \quad \Lambda_g \equiv \int_0^\infty g(r)dr, \quad \lambda_f^2 \equiv -2/[d^2 f/dr^2]_{r=0}, \quad \text{and} \quad \lambda_g^2 \equiv -2/[d^2 g/dr^2]_{r=0}, \tag{10.39}$$

similar to the temporal integral scale and temporal Taylor microscale defined in (10.18) and (10.19), respectively. For homogeneous isotropic turbulence, the most general possible form of $R_{ij}(r)$ that satisfies all the necessary symmetries is:

$$R_{ij} = F(r)r_i r_j + G(r)\delta_{ij}, \tag{10.40}$$

where the components of $\mathbf{r}$ are r_i, $|\mathbf{r}| = r$, and the functions $F(r) = \overline{u^2}(f(r) - g(r))r^{-2}$ and $G(r) = \overline{u^2}g(r)$ can be found by equating the diagonal components of R_{ij} (Exercise 10.20). For incompressible flow, $g(r)$ can be eliminated from (10.40) to find:

$$R_{ij} = \overline{u^2}\left\{ f(r)\delta_{ij} + \frac{r}{2}\frac{df}{dr}\left(\delta_{ij} - \frac{r_i r_j}{r^2}\right)\right\} \tag{10.41}$$

(Exercise 10.21), and $\Lambda_g = \Lambda_f/2$ and $\lambda_g = \lambda_f/\sqrt{2}$.

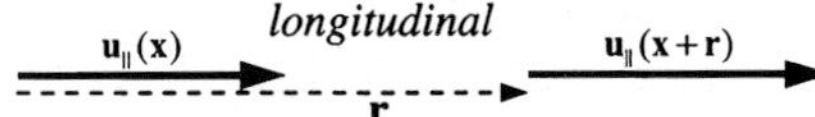

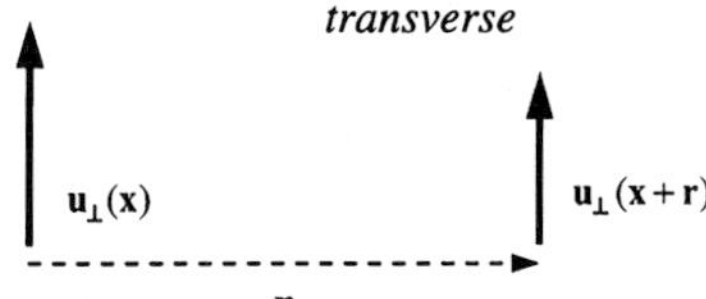

FIGURE 10.10 Longitudinal and transverse correlation geometries. In the longitudinal case, $\mathbf{u}_{||}$ is parallel to the displacement $\mathbf{r}$. In the transverse case, $\mathbf{u}_\perp$ is perpendicular to $\mathbf{r}$. Here, $\mathbf{r}$ is shown horizontal but it may point in any direction.

Admittedly, the preceding formulae do not readily produce insights; however, the trace of R_{ij} evaluated at $r = 0$ is twice the average kinetic energy $\overline{e}$ (per unit mass) of the turbulent fluctuations:

$$R_{ii}(0) = \overline{u_i u_i} = 2\cdot\frac{1}{2}\left(\overline{u_1^2} + \overline{u_2^2} + \overline{u_3^2}\right) = 2\overline{e},$$

and $\overline{e}$ is an important element in understanding and modeling turbulence. We know from Section 4.8 that the kinetic energy of a flowing fluid may be converted into heat (dissipated) by the action of viscosity. Thus, the average kinetic energy dissipation rate $\overline{\varepsilon}$ (per unit mass) in an incompressible turbulent flow comprised entirely of fluctuations is the average of (4.59):

$$\overline{\varepsilon} = \frac{\nu}{2}\overline{(\partial u_i/\partial x_j + \partial u_j/\partial x_i)^2}. \tag{10.42}$$

When the flow is isotropic, the various directional symmetries, (10.36), and (10.37) imply:

$$\overline{\varepsilon} = 6\nu\left\{\overline{\left(\frac{\partial u_1}{\partial x_1}\right)^2} + \overline{\left(\frac{\partial u_1}{\partial x_2}\right)^2} + \overline{\left(\frac{\partial u_1}{\partial x_2}\right)\left(\frac{\partial u_2}{\partial x_1}\right)}\right\} = -15\nu\overline{u^2}\left[\frac{d^2 f}{dr^2}\right]_{r=0} = 30\nu\frac{\overline{u^2}}{\lambda_f^2} = 15\nu\frac{\overline{u^2}}{\lambda_g^2}, \tag{10.43}$$

where everything inside the {,}-braces has been put in terms of the first and second directions, and the second equality follows from the results of Exercise 10.22. Until the development of modern multidimensional measurement techniques, (10.43) was the primary means available for estimating $\overline{\varepsilon}$ from measurements in turbulent flows. Even today, fully resolved three-dimensional turbulent flow measurements are seldom possible, so reduced dimensionality relationships like (10.43), based on some assumed homogeneity, symmetry, or isotropy, commonly appear in the literature. In addition, Taylor-scale Reynolds numbers:

$$R_\lambda \equiv \lambda_{(g \text{ or } f)}\sqrt{\overline{u^2}}/\nu, \tag{10.44}$$

are occasionally quoted with $R_\lambda > 10^2$ being a nominal condition for fully turbulent flow (Dimotakis, 2000).

These concepts from homogeneous isotropic turbulence also allow the energy spectrum $S_{11}(k_1)$ of stream-wise velocity fluctuations along a stream-wise line through the turbulent field to be defined in terms of the autocorrelation function (10.23) when $i = j = 1$:

$$S_{11}(k_1) = \frac{1}{2\pi}\int_{-\infty}^{+\infty} R_{11}(r_1)\exp\{-ik_1r_1\}dr_1 = \frac{\overline{u_1^2}}{2\pi}\int_{-\infty}^{+\infty} f(r_1)\exp\{-ik_1r_1\}dr_1, \quad (10.45)$$

where "1" implies the stream-wise flow direction. Measured spectra reported in the turbulence literature are commonly produced using (10.45) or its alternative involving finite-window Fourier transformations (see Exercise 10.8). The basic procedure is to collect time-series measurements of u_1, convert them to spatial measurements using Taylor's frozen turbulence hypothesis, compute R_{11} from the spatial series, and then use (10.45) to determine $S_{11}(k_1)$. As described in the next section, the functional dependence of a portion of S_{11} on k_1 and $\overline{\varepsilon}$ can be anticipated from dimensional analysis and insights derived from the progression or cascade of fluctuation kinetic energy through a turbulent flow. Additional relationships for R_{ij} and its associated spectrum tensor are also available (Hinze, 1975).

Example 10.6

Consider the structured periodic two-dimensional Taylor-Green vortex velocity field, $(u, v) = (A\sin(kx)\cos(ky), -A\cos(kx)\sin(ky))$, where A is function of time and k is a constant, as an idealized case of turbulent velocity fluctuations. Streamlines for one period in the x- and y-directions are shown in Fig. 10.11. If the spatial average centered on (x', y') is defined by $\overline{()} = (1/\ell)^2 \int_{x'-\ell/2}^{x'+\ell/2}\int_{y'-\ell/2}^{y'+\ell/2} ()dxdy$, for what values of ℓ is this flow field isotropic?

Solution First, compute the three independent Reynolds stresses ($\overline{u^2}$, $\overline{v^2}$, and $\overline{uv}$) for the given two-dimensional flow field. The integrals may be evaluated and simplified to reach:

$$\overline{u^2} = \frac{1}{\ell^2}\int_{x'-\ell/2}^{x'+\ell/2}\int_{y'-\ell/2}^{y'+\ell/2} A^2\sin^2(kx)\cos^2(ky)dxdy = \frac{A^2}{4}\left(1 - \cos(2kx')\frac{\sin(k\ell)}{k\ell}\right)\left(1 + \cos(2ky')\frac{\sin(k\ell)}{k\ell}\right),$$

$$\overline{v^2} = \frac{1}{\ell^2}\int_{x'-\ell/2}^{x'+\ell/2}\int_{y'-\ell/2}^{y'+\ell/2} A^2\sin^2(kx)\sin^2(ky)dxdy = \frac{A^2}{4}\left(1 + \cos(2kx')\frac{\sin(k\ell)}{k\ell}\right)\left(1 - \cos(2ky')\frac{\sin(k\ell)}{k\ell}\right),$$

and

$$\overline{uv} = \frac{1}{\ell^2}\int_{x'-\ell/2}^{x'+\ell/2}\int_{y'-\ell/2}^{y'+\ell/2} -A^2\sin(kx)\cos(kx)\cos(ky)\sin(kx)dxdy = -\frac{A^2}{4}\sin(2kx')\sin(2ky')\left(\frac{\sin(k\ell)}{k\ell}\right)^2.$$

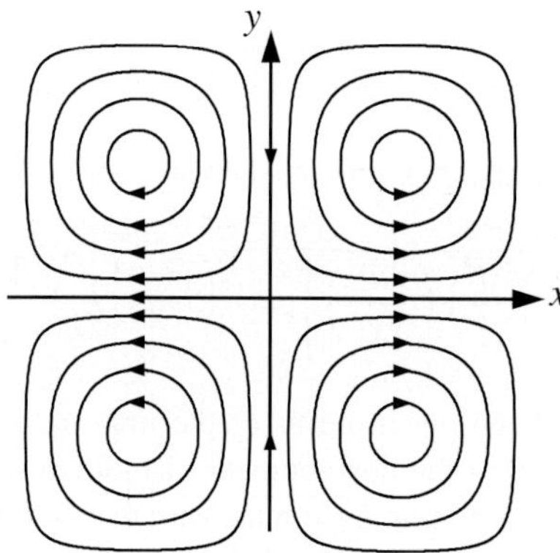

FIGURE 10.11 Streamlines for the two-dimensional Taylor-Green vortex flow. The flow is an infinite periodic array of vortices having counter-rotating neighbors. One full period in both the x- and y-directions is shown.

These results show that the Reynolds stresses may depend on where they are measured as well as the interval over which the averaging is performed. In a two-dimensional isotropic flow, the two normal Reynolds stresses will be equal and the Reynolds shear stress will be zero. For the results provided above, $\overline{u^2} = \overline{v^2} = A^2/4$ and $\overline{uv} = 0$ when $k\ell \to \infty$ or when $k\ell = m\pi$, where m is a positive integer. In practice, turbulent flow fields are irregular (not periodic) so the first possibility, $k\ell \to \infty$, is the most relevant, the implication being that turbulence statistics should be obtained from measurements or simulations that include many fundamental eddies. The second possibility for finding isotropy, $k\ell = m\pi$, arises from the periodic structure of the Taylor-Green vortex flow field.

10.7 Turbulent energy cascade and spectrum

As mentioned in the introductory section of this chapter, turbulence rapidly dissipates kinetic energy, and an understanding of how this happens is possible via a term-by-term inspection of the equations that govern the kinetic energy in the mean flow and the average kinetic energy of the fluctuations.

An equation for the mean-flow's kinetic energy per unit mass, $\overline{E} = \frac{1}{2}U_i^2$, can be obtained by multiplying (10.30) by U_i, and averaging (Exercise 10.15). With $\overline{S_{ij}} = \frac{1}{2}\left(\partial U_i/\partial x_j + \partial U_j/\partial x_i\right)$ defining the mean strain-rate tensor, the resulting *energy-balance* or *energy-budget* equation for $\overline{E}$ is:

$$\underbrace{\frac{\partial \overline{E}}{\partial t} + U_j\frac{\partial \overline{E}}{\partial x_j}}_{\substack{\text{Time rate of change}\\ \text{of } \overline{E} \text{ following}\\ \text{the mean flow}}} = \underbrace{\frac{\partial}{\partial x_j}\left(-\frac{U_j P}{\rho_0} + 2\nu U_i \overline{S_{ij}} - \overline{u_i u_j} U_i\right)}_{\text{Transport}} \underbrace{- 2\nu \overline{S_{ij}}\,\overline{S_{ij}}}_{\substack{\text{Viscous}\\ \text{dissipation}}} + \underbrace{\overline{u_i u_j}\frac{\partial U_i}{\partial x_j}}_{\substack{\text{Loss to}\\ \text{turbulence}}} - \underbrace{\frac{g}{\rho_0}\overline{\rho} U_3}_{\substack{\text{Loss to}\\ \text{potential}\\ \text{energy}}}. \tag{10.46}$$

The left side is merely the total time derivative of $\overline{E}$ following a mean-flow fluid particle, while the right side represents the various mechanisms that bring about changes in $\overline{E}$.

The first three divergence terms on the right side of (10.46) represent *transport* of mean kinetic energy by pressure, viscous stresses, and Reynolds stresses. If (10.46) is integrated over the volume occupied by the flow to obtain the rate of change of the total (or global) mean-flow kinetic energy, then these transport terms can be transformed into a surface integral by Gauss' theorem. Thus, these terms do not contribute to the total rate of change of $\overline{E}$ if $U_i = 0$ on the boundaries of the flow. Therefore, these three terms only *transport* or redistribute mean-flow kinetic energy from one region to another; they do not generate it or dissipate it.

The fourth term is the product of the mean flow's viscous stress (per unit mass) $2\nu\overline{S}_{ij}$ and the mean strain rate $\overline{S}_{ij}$. It represents the *direct viscous dissipation* of mean kinetic energy via its conversion into heat.

The fifth term is analogous to the fourth term. It can be written as $\overline{u_i u_j}(\partial U_i/\partial x_j) = \overline{u_i u_j}\overline{S}_{ij}$ so that it is a product of the turbulent stress (per unit mass) and the mean strain rate. Here, the doubly contracted product of a symmetric tensor $\overline{u_i u_j}$ and the tensor $\partial U_i/\partial x_j$ is equal to the product of $\overline{u_i u_j}$ and the *symmetric* part of $\partial U_i/\partial x_j$, namely $\overline{S}_{ij}$, as proved in Section 2.10. If the mean flow is given by $U(y)$ alone, then $\overline{u_i u_j}(\partial U_i/\partial x_j) = \overline{uv}(dU/dy)$. From the preceding section, $\overline{uv}$ is likely to be negative if dU/dy is positive. Thus, the fifth term is likely to be negative in shear flows. So, by analogy with the fourth term, it must represent a mean-flow kinetic energy loss to the fluctuating velocity field. Indeed, this term appears on the right side of the equation for the rate of change of the turbulent kinetic energy, but *with the sign reversed*. Therefore, this term generally results in a loss of mean kinetic energy and a gain of turbulent kinetic energy. It is commonly known as the *shear production* term.

The sixth term represents the work done by gravity on the mean vertical motion. For example, an upward mean motion results in a loss of mean kinetic energy, which is accompanied by an increase in the potential energy of the mean field.

The two viscous terms in (10.46), namely, the viscous transport $2\nu\partial(U_i\overline{S}_{ij})/\partial x_j$ and the mean-flow viscous dissipation $-2\nu\overline{S}_{ij}\overline{S}_{ij}$, are small compared to the equivalent turbulence terms in a fully turbulent flow at high Reynolds numbers. To see this, compare the mean-flow viscous dissipation and the shear-production terms:

$$\frac{2\nu\overline{S}_{ij}\overline{S}_{ij}}{\overline{u_i u_j}(\partial U_i/\partial x_j)} \sim \frac{\nu(U/L)^2}{u_{rms}^2(U/L)} \sim \frac{\nu}{UL} = \frac{1}{\text{Re}} \ll 1,$$

where U is the velocity scale for the mean flow, L is a length scale for the mean flow (e.g., the overall thickness of a boundary layer), and u_{rms} is presumed to be of the same order of magnitude as U, a presumption commonly supported by experimental evidence. The direct influence of viscous terms is therefore negligible on the *mean* kinetic energy budget. However, this is *not* true for the *turbulent* kinetic energy budget, in which viscous terms play a major role. What happens is as follows: The mean flow loses kinetic energy to the turbulent field by means of the shear production term and the *turbulent* kinetic energy so generated is then dissipated by viscosity.

An equation for the mean kinetic energy $\overline{e} = \frac{1}{2}\overline{u_i^2}$ of the turbulent velocity fluctuations can be obtained by setting $i = j$ in (10.35) and dividing by two. With $S'_{ij} = \frac{1}{2}(\partial u_i/\partial x_j + \partial u_j/\partial x_i)$ defining the fluctuation strain-rate tensor, the resulting energy-budget equation for $\overline{e}$ is:

$$\underbrace{\frac{\partial \overline{e}}{\partial t} + U_j \frac{\partial \overline{e}}{\partial x_j}}_{\substack{\text{Time rate of change}\\ \text{of } \overline{e} \text{ following}\\ \text{the mean flow}}} = \underbrace{\frac{\partial}{\partial x_j}\left(-\frac{1}{\rho_0}\overline{pu_j} + 2\nu\overline{u_i S'_{ij}} - \frac{1}{2}\overline{u_i^2 u_j}\right)}_{\text{Transport}} - \underbrace{2\nu\overline{S'_{ij}S'_{ij}}}_{\substack{\text{Viscous}\\ \text{dissipation}}} - \underbrace{\overline{u_i u_j}\frac{\partial U_i}{\partial x_j}}_{\substack{\text{Gain from}\\ \text{mean flow}}} - \underbrace{g\alpha\overline{u_3 T'}}_{\substack{\text{Buoyant}\\ \text{production}}} . \tag{10.47}$$

The first three terms on the right side are in divergence form and consequently represent the spatial transport of turbulent kinetic energy via turbulent pressure fluctuations, viscous diffusion, and turbulent stresses.

The fourth term $\overline{\varepsilon} = 2\nu\overline{S'_{ij}S'_{ij}}$ is the *viscous dissipation of turbulent kinetic energy*, and it is *not* negligible in the turbulent kinetic energy budget (10.47), although the analogous term $2\nu\overline{S}_{ij}\overline{S}_{ij}$ is negligible in the mean-flow kinetic energy budget (10.46). In fact, the viscous dissipation $\overline{\varepsilon}$ is always positive and its magnitude is typically similar to that of the turbulence-production terms in most locations.

The fifth term $\overline{u_i u_j}(\partial U_i/\partial x_j)$ is the shear-production term and it represents the rate at which kinetic energy is lost by the mean flow and gained by the turbulent fluctuations. It appears in the mean-flow kinetic energy budget (10.46) with the other sign.

The sixth term $g\alpha\overline{u_3 T'}$ can have either sign, depending on the nature of the background temperature distribution $\overline{T}(x_3)$. In a stable situation in which the background temperature increases upward (as found, e.g., in the atmospheric boundary layer at night), rising fluid elements are likely to be associated with a negative temperature fluctuation, resulting in $\overline{u_3 T'} < 0$, which means a downward turbulent heat flux. In such a stable situation $g\alpha\overline{u_3 T'}$ represents the rate of turbulent energy loss via work against the stable background density gradient. In the opposite case, when the background density profile is unstable, the turbulent heat flux correlation $\overline{u_3 T'}$ is positive upward, and convective motions cause an increase of turbulent kinetic energy (Fig. 10.12). Thus, $g\alpha\overline{u_3 T'}$ is the *buoyant production* of turbulent kinetic energy; it can also be a buoyant *destruction* when the turbulent heat flux is downward. In isotropic turbulence, the upward thermal flux correlation $\overline{u_3 T'}$ is zero because there is no preference between the upward and downward directions.

The buoyant generation of turbulent kinetic energy lowers the potential energy of the mean field. This can be understood from Fig. 10.12, where it is seen that the heavier fluid has moved downward in the final state as a result of the heat flux. This can also be demonstrated by deriving an equation for the mean potential energy, in which the term $g\alpha\overline{u_3 T'}$ appears with a *negative* sign on the right-hand side. Therefore, the *buoyant generation* of turbulent kinetic energy by the upward heat flux occurs at the expense of the mean *potential* energy. This is in contrast to the *shear production* of turbulent kinetic energy, which occurs at the expense of the mean *kinetic* energy.

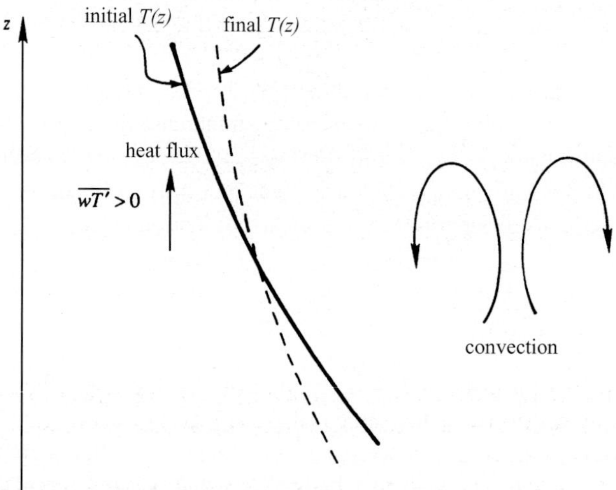

FIGURE 10.12 Heat flux in an unstable environment. Here warm air from below may rise and cool air may sink thereby generating turbulent kinetic energy by lowering the mean potential energy. In the final state, the upper air is warmer and less dense, and the lower air is cooler and denser.

The kinetic energy budgets for constant density flow are recovered from (10.46) and (10.47) by dropping the terms with gravity and re-interpreting the mean pressure as the deviation from hydrostatic (see Section 4.9).

The shear-production term represents an essential link between the mean and fluctuating fields. For it to be active (or non-zero), the flow must have mean shear and the turbulence must be anisotropic. When the turbulence is isotropic, the off-diagonal components of the Reynolds stress $\overline{u_i u_j}$ are zero (see Section 10.6) and the on-diagonal ones are equal (10.36). Thus, the double sum implied by $\overline{u_i u_j}(\partial U_i/\partial x_j)$ reduces to:

$$\overline{u_1^2}(\partial U_1/\partial x_1) + \overline{u_2^2}(\partial U_2/\partial x_2) + \overline{u_3^2}(\partial U_3/\partial x_3) = \overline{u_1^2}(\partial U_i/\partial x_i) = 0,$$

where the final equality holds from (10.27). Experimental observations suggest the largest eddies in a turbulent shear flow generally span the cross-stream distance L between those locations in a turbulent flow giving the maximum average velocity difference ΔU (Fig. 10.13). In a layer with only one sign for the mean shear, L spans the layer as in Fig. 10.13a, but for consistency when the shear has both signs, such as in turbulent pipe flow, L is the pipe radius as in Fig. 10.13b. These largest eddies feel the mean shear – which must be of order $\Delta U/L$ – and are distorted or made anisotropic by it. Energy is provided to these largest eddies by the mean flow as it forces them to deform and turn over. In this situation, turbulent velocity fluctuations are also of order ΔU, so the energy input rate $\dot{W}$ to a region of turbulence by the mean flow (per unit mass of fluid) is:

$$\dot{W} \sim \overline{u_i u_j}(\partial U_i/\partial x_j) \sim (\Delta U)^2[\Delta U/L] = (\Delta U)^3/L, \tag{10.48}$$

where L and ΔU are commonly called the *outer* length scale and velocity difference. Of course the details of $\dot{W}$ will vary with flow geometry but its parametric dependence is set by (10.48). In reaching (10.48), it was implicitly assumed that the outer scale Reynolds number $\mathrm{Re}_L = \Delta U L/\nu$ is so large that viscosity plays no role in the interaction between the mean flow and the largest eddies of the turbulent shear flow.

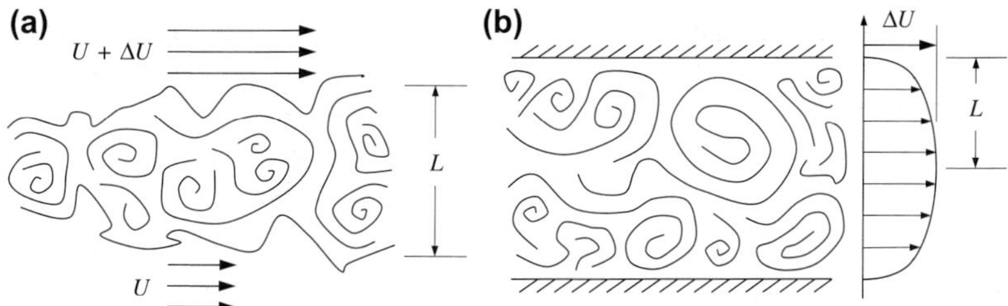

FIGURE 10.13 Schematic drawings of a turbulent flow without boundaries (a), and one with boundaries (b). Here the outer scale of the turbulence L spans the cross-stream distance over which the outer scale velocity difference ΔU occurs. Here, L may be the half width of the flow when the flow is symmetric as in (b). This choice of L and ΔU ensures that mean-flow velocity gradients will be of order $\Delta U/L$.

In temporally stationary turbulence, the turbulent kinetic energy $\overline{e}$ cannot build up (or shrink to zero) so the work input at the largest scales from the mean flow must be balanced by the kinetic energy dissipation rate:

$$\dot{W} = \overline{\varepsilon}, \quad \text{so} \quad \overline{\varepsilon} \sim (\Delta U)^3/L. \tag{10.49}$$

Thus, $\overline{\varepsilon}$ does not depend on ν – in spite of its definition (see (10.42)) – but is determined instead by the *inviscid* properties of the largest eddies, which extract energy from the mean flow. Second-tier eddies that are somewhat smaller than the largest ones are distorted and forced to roll over by the strain field of the largest eddies, and these thereby extract energy from the largest eddies by the same mechanism that the largest eddies extract energy from the mean flow. Thus the average turbulent-kinetic-energy cascade pattern is set, and third-tier eddies extract energy from second-tier eddies, fourth-tier eddies extract energy from third-tier eddies, and so on. So, *turbulent kinetic energy is on average cascaded down from large to small eddies by interactions between eddies of neighboring size.* Small eddies are essentially advected in the velocity field of large eddies, since the scales of the strain-rate field of the large eddies are much larger than the size of a small eddy. Therefore, small eddies do not interact directly with the large eddies or the mean field, and are therefore nearly isotropic. The turbulent kinetic-energy cascade process is essentially inviscid with decreasing eddy scale size l' and eddy-velocity u' as long as the eddy Reynolds number $u'l'/\nu$ is much greater than unity. The cascade terminates when the eddy Reynolds number becomes of order unity and viscous effects are important. This average cascade process was first discussed by Richardson (1922), and is a foundational element in the understanding of turbulence.

In 1941, Kolmogorov suggested that the dissipating eddies are essentially homogeneous and isotropic, and that their size depends on those parameters that are relevant to the smallest eddies. These parameters are $\overline{\varepsilon}$, the rate at which kinetic energy is supplied to the smallest eddies, and ν, the kinematic viscosity that smears out the velocity gradients of the smallest eddies. Since the units of $\overline{\varepsilon}$ are $\mathrm{m}^2/\mathrm{s}^3$, dimensional analysis only allows one way to construct a length scale η and a velocity scale u_K from $\overline{\varepsilon}$ and ν:

$$\eta = (\nu^3/\overline{\varepsilon})^{1/4} \quad \text{and} \quad u_K = (\nu\overline{\varepsilon})^{1/4}. \tag{10.50}$$

These are called the *Kolmogorov microscale* and *velocity scale*, and the Reynolds number determined from them is:

$$\eta u_K/\nu = (\nu^3/\overline{\varepsilon})^{1/4}(\nu\overline{\varepsilon})^{1/4}/\nu = 1,$$

which appropriately suggests a balance of inertial and viscous effects for Kolmogorov-scale eddies. The relationship (10.50) and the recognition that ν does not influence $\overline{\varepsilon}$ suggests that a *decrease of* ν *merely decreases the eddy size at which viscous dissipation takes place*. In particular, the size of η relative to L can be determined by eliminating $\overline{\varepsilon}$ from (10.49) and the first equation of (10.50) to find:

$$\eta/L \sim \mathrm{Re}_L^{-3/4}, \quad \text{where} \quad \mathrm{Re}_L = \Delta UL/\nu, \tag{10.51}$$

which is the result in Example 10.2. Therefore, the sizes of the largest and smallest eddies in high Reynolds number turbulence potentially differ by many orders of magnitude. For flow in a fixed-size device, the length scale L is fixed, so increasing the input velocity that leads to shear (or decreasing ν) leads to an increase in Re_L and a decrease in the size of the Kolmogorov eddies. In the ocean and the atmosphere, the Kolmogorov microscale η is commonly of order millimeters. However, in engineering flows η may be much smaller because of the larger power densities and dissipation rates. Landahl and Mollo-Christensen (1986) give a nice illustration of this. Suppose a 100-W kitchen mixer is used to churn 1 kg of water in a cube 0.1 m (= L) on a side. Since all the power is used to generate turbulence, the rate of energy dissipation is $\overline{\varepsilon} = 100\ \mathrm{W/kg} = 100\ \mathrm{m}^2/\mathrm{s}^3$. Using $\nu = 10^{-6}\ \mathrm{m}^2/\mathrm{s}$ for water, we obtain $\eta = 10^{-5}$ m from (10.50).

Interestingly, the path that leads to (10.51) can also be used for either of the Taylor microscales (generically labeled λ_T here). Eliminating $\overline{\varepsilon}$ from (10.43) and (10.49) produces:

$$\frac{(\Delta U)^3}{L} \propto \frac{\nu\overline{u^2}}{\lambda_T^2} \rightarrow \frac{\lambda_T^2}{L^2} \propto \frac{\nu\overline{u^2}}{(\Delta U)^3 L} = \frac{\overline{u^2}}{(\Delta U)^2}\left(\frac{\nu}{\Delta UL}\right) \propto \frac{1}{\mathrm{Re}_L}, \quad \text{or} \quad \frac{\lambda_T}{L} \propto \mathrm{Re}_L^{-1/2}, \tag{10.52}$$

where the final two proportionalities are valid when the fluctuation velocity is proportional to the ΔU. The negative half-power of the outer-scale Reynolds number matches that for laminar boundary-layer thicknesses (see (8.30)). Thus, the Taylor microscale can be interpreted as an internal boundary-layer thickness that develops at the edge of a large eddy during a single rotational movement having a path length L. However, it not a distinguished length scale in the partition of turbulent kinetic energy even though (10.43) associates λ_T with $\overline{\varepsilon}$. The reason for this anonymity is that the velocity fluctuation appearing in (10.43) is not appropriate for eddies that dissipate turbulent kinetic energy. The appropriate dissipation-scale velocity is given by the second equality of (10.50). Thus, in high Reynolds number turbulence, λ_T is larger than η, as is clear from a comparison of (10.51) and (10.52) with $\mathrm{Re} \rightarrow \infty$. In addition, (10.52) implies $\mathrm{Re}_\lambda \sim (\mathrm{Re}_L)^{1/2}$, so a nominal condition for fully turbulent flow is $\mathrm{Re}_L > 10^4$ (Dimotakis, 2000). Above such a Reynolds number, the following ordering of length scales should occur: $\eta < \lambda_T < \Lambda_{(f \text{ or } g)} < L$.

Richardson's cascade, Kolmogorov's insights, the simplicity of homogeneous isotropic turbulence, and dimensional analysis lead to perhaps the most famous and prominent feature of high Reynolds number turbulence: the universal power law form of the energy spectrum in the inertial sub-range. Consider the one-dimensional energy spectrum $S_{11}(k_1)$ – it is the one most readily determined from experimental measurements – and associate eddy size l with the inverse of the wave number: $l \sim 2\pi/k_1$. For large-eddy sizes (small wave numbers), the energy spectrum will not be universal because these eddies are directly influenced by the geometry-dependent mean flow. However, smaller eddies a few tiers down in the cascade may approach isotropy. In this case the mean shear no longer matters, so their spectrum of fluctuations $S_{11}(k_1)$ can only depend on the kinetic energy cascade rate $\overline{\varepsilon}$, the fluid's kinematic viscosity ν, and the wave number k_1. From (10.45), the units of S_{11} are found to be $\mathrm{m}^3/\mathrm{s}^2$; therefore dimensional analysis using S_{11}, $\overline{\varepsilon}$, ν, and k_1 requires:

$$\frac{S_{11}(k_1)}{\nu^{5/4}\overline{\varepsilon}^{1/4}} = \Phi\left(\frac{k_1\nu^{3/4}}{\overline{\varepsilon}^{1/4}}\right), \quad \text{or} \quad \frac{S_{11}(k_1)}{u_K^2\eta} = \Phi(k_1\eta) \quad \text{for} \quad k_1 \gg 2\pi/L, \tag{10.53}$$

where Φ is an undetermined function, and both parts of (10.50) have been used to reach the second form of (10.53). Furthermore, for eddy sizes somewhat less than L, but also somewhat greater than η, $2\pi/L \ll k_1 \ll 2\pi/\eta$, the spectrum must be independent of both the mean shear and the kinematic viscosity. This wave number range is known as the *inertial sub-range*. Turbulent kinetic energy is transferred through this range of length scales without much loss to viscosity. Thus, the form of the spectrum in the inertial sub-range is obtained from dimensional analysis using only S_{11}, $\overline{\varepsilon}$, and k_1:

$$S_{11}(k_1) = const \cdot \overline{\varepsilon}^{2/3} \cdot k_1^{5/3} \quad \text{for} \quad 2\pi/L \ll k_1 \ll 2\pi/\eta. \tag{10.54}$$

The *constant* has been found to be universal for all turbulent flows and is approximately 0.25 for $S_{11}(k_1)$ subject to a *double-sided* normalization:

$$\int_{-\infty}^{+\infty} S_{11}(k_1)dk_1 = \overline{u_1^2} = \int_0^{+\infty} 2S_{11}(k_1)dk_1. \tag{10.55}$$

Eq. (10.54) is usually called *Kolmogorov's $k^{-5/3}$ law* and it is one of the most important results of turbulence theory. When the spectral form (10.54) is subject to the normalization:

$$\overline{e} = \int_0^{+\infty} S(K)dK,$$

where K is the magnitude of the three-dimensional wave-number vector, the constant in the three-dimensional form of (10.54) is approximately 1.5 (see Pope, 2000). If the Reynolds number of the flow is large, then the dissipating eddies are much smaller than the energy-containing eddies, and the inertial sub-range is broad.

Fig. 10.14 shows a plot of experimental spectral measurements of $2S_{11}$ from several different types of turbulent flows (Chapman, 1979). The normalizations of the axes follow (10.53), $\overline{\varepsilon}$ is calculated from (10.43), η is calculated from (10.50), and the Taylor-Reynolds numbers (labeled R_λ in the figure) come from the longitudinal autocorrelation $f(r)$. The collapse of the data at high wave numbers to a single curve indicates the universal character of (10.53) at high wave numbers. The spectral form of Pao (1965) adequately fits the data and indicates how the spectral amplitude decreases faster than $k^{-5/3}$ as $k_1\eta$ approaches and then exceeds unity. The scaled wave number at which the data are approximately a factor of two below the $-5/3$ power law is $k_1\eta \approx 0.2$ (dashed vertical line), so the actual eddy size where viscous dissipation is clearly felt is $l_D \approx 30\eta$. The various spectra shown on Fig. 10.14 turn horizontal with decreasing $k_1\eta$ where $k_1 L$ is of order unity.

Because very large Reynolds numbers are difficult to generate in an ordinary laboratory, the Kolmogorov spectral law (10.54) was not verified for many years. In fact, doubts were raised about its theoretical validity. The first confirmation of the Kolmogorov law came from the oceanic observations of Grant et al. (1962), who obtained a velocity spectrum in a tidal flow through a narrow passage between two islands just off the west coast of Canada. The velocity fluctuations were measured by hanging a hot film anemometer from the bottom of a ship. Based on the water depth and the average flow velocity, the outer-scale Reynolds number was of order 10^8. Such large Reynolds numbers are typical of geophysical flows, since the length scales are very large. Thus, the tidal channel data and results from other geophysical flows prominently display the $k^{-5/3}$ spectral form in Fig. 10.14.

For the purpose of formulating predictions, the universality of the high wave number portion of the energy spectrum of turbulent fluctuations suggests that a single-closure model might adequately represent the effects of inertial sub-range and smaller eddies on the non-universal large-scale eddies. This possibility has inspired the development of a wide variety of RANS-equation closure models, and it provides justification for the central idea behind large-eddy simulations (LES) of turbulent flow (see Exercise 10.19). Such models are described in Pope (2000).

Example 10.7

Consider again the periodic two-dimensional Taylor-Green vortex velocity field, $(u, v) = (A(t)\sin(kx)\cos(ky), -A(t)\cos(kx)\sin(ky))$ as an idealized case of turbulent velocity fluctuations. If the averaging area is chosen so that the flow's statistics are homogeneous and isotropic, evaluate each of the terms in (10.47) and solve the resulting differential equation to determine $A(t)$.

Solution First, simplify (10.47) for flow statistics that are homogeneous and isotropic. In this case, the various averages and moments appearing in (10.47) ($\overline{e}$, U_j, $\overline{pu_j}$, $\overline{u_i S'_{ij}}$, $\overline{u_i^2 u_j}$, $\overline{S'_{ij} S'_{ij}}$, $\overline{u_i u_j}$, $\overline{u_3 T'}$) are either uniform in space or zero. Thus, all the terms in (10.47) that include spatial differentiation of these averages and moments are zero, leaving:

$$\frac{d\overline{e}}{dt} = -2\nu\overline{S'_{ij}S'_{ij}} = -\overline{\varepsilon}.$$

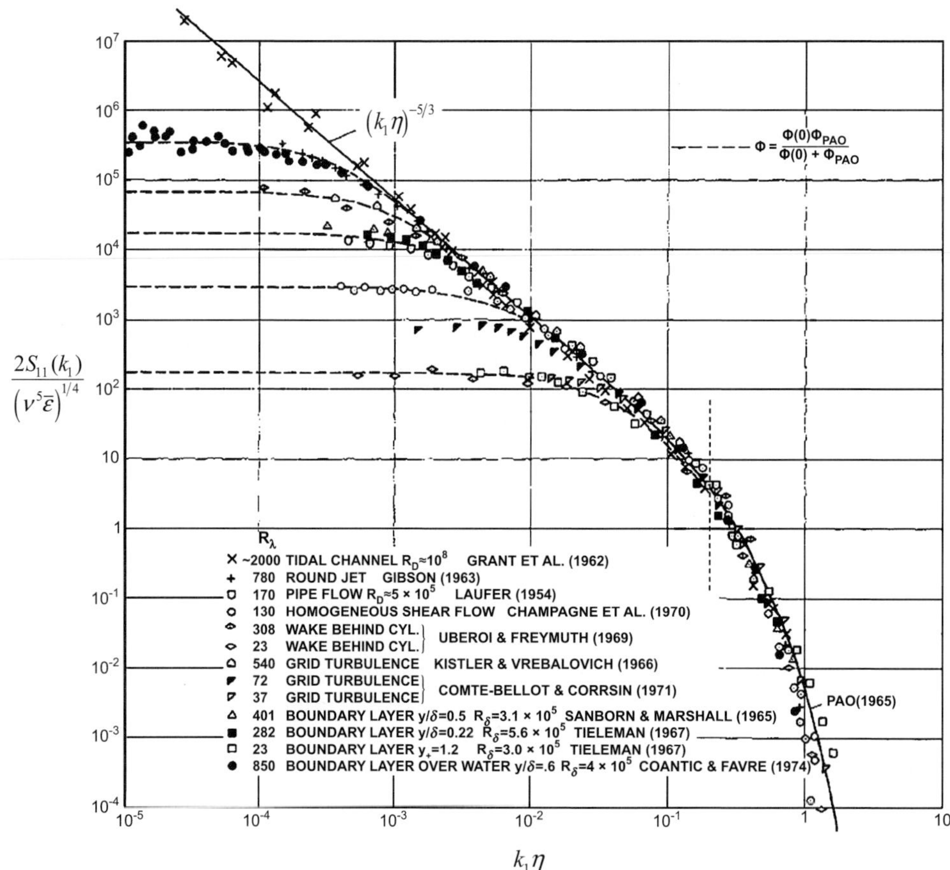

FIGURE 10.14 One-dimensional energy spectra $S_{11}(k_1)$ from a variety of turbulent flows plotted in Kolmogorov normalized form, reproduced from Chapman (1979). Here k_1 is the stream-wise wave number, η is the Kolmogorov scale defined by (10.50), and $\bar{\varepsilon}$ is the average kinetic energy dissipation rate determined from (10.43). Kolmogorov's -5/3 power law is indicated by the sloping line. The collapse of the various spectra to this line and to each other as $k_1\eta$ approaches and then exceeds unity strongly suggests that high-wave-number turbulent velocity fluctuations are universal when the Reynolds number is high enough. The dashed vertical line indicates the location where the spectral data are a factor of two below the −5/3 line established at lower wave numbers.

Using the results from Example 10.6 with $k\ell = \pi$, $\bar{e}$ in two dimensions is simply:

$$\bar{e} = \frac{1}{2}\left(\overline{u^2} + \overline{v^2}\right) = \frac{1}{2}\left(\frac{A^2(t)}{4} + \frac{A^2(t)}{4}\right) = \frac{A^2(t)}{4}.$$

Evaluation of $\bar{\varepsilon}$ from (10.42) or (10.47) using the given velocity field is tedious but straightforward:

$$\bar{\varepsilon} = 2\nu\left[\overline{S_{11}'^2} + \overline{S_{12}'^2} + \overline{S_{21}'^2} + \overline{S_{22}'^2}\right] = 2\nu\left[\overline{\left(\frac{\partial u}{\partial x}\right)^2} + \frac{1}{2}\overline{\left(\frac{\partial u}{\partial y}\right)^2} + \overline{\left(\frac{\partial u}{\partial y}\right)\left(\frac{\partial v}{\partial x}\right)} + \frac{1}{2}\overline{\left(\frac{\partial v}{\partial x}\right)^2} + \overline{\left(\frac{\partial v}{\partial y}\right)^2}\right]$$
$$= 2\nu\left[1 + \frac{1}{2} - 1 + \frac{1}{2} + 1\right]\left(\frac{k^2A^2}{4}\right) = \nu k^2 A^2.$$

Thus, the remnant of (10.47) becomes a simple first-order differential equation with an exponentially decaying solution:

$$\frac{1}{4}\frac{d}{dt}A^2(t) = -\nu k^2 A^2, \quad \text{or} \quad A(t) = A(0)\exp\left\{-2\nu k^2 t\right\},$$

where $A(0)$ is the initial amplitude of the velocity fluctuations. This final result is qualitatively consistent with actual turbulence since it shows that small eddies (with large k) dissipate rapidly compared to large eddies (with small k).

10.8 Free turbulent shear flows

Persistent turbulence is maintained by the presence of mean-flow shear. This shear may exist because of a mismatch of fluid velocity within a flow, or because of the presence of one or more solid boundaries near the moving fluid. Turbulent flows in the former category are called *free* turbulent shear flows, and those in the latter are called *wall-bounded* turbulent shear flows. This section covers free turbulent flows that develop away from solid boundaries. Such flows include jets, wakes, shear layers, and plumes; the first three are depicted in Fig. 10.15. A plume is a buoyancy-driven jet that develops vertically so its appearance is similar to that shown in Fig. 10.15a when the flow direction is rotated to be vertical. Jets, wakes, and plumes may exist in planar and axisymmetric geometries. Although idealized, such free shear flows are important for mixing reactants and in remote sensing, and are scientifically interesting because their development can sometimes be described by a single length scale and one boundary condition or origin parameter. Such a description commonly results from a similarity analysis in which the mean flow is assumed to be *self-preserving*. This section presents one such similarity analysis for a single free turbulent shear flow (the planar jet), and then summarizes the similarity characteristics of a variety of planar and axisymmetric free turbulent shear flows. In most circumstances, free turbulent shear flows are simpler than wall-bounded turbulent shear flows. However, the outer portion of a turbulent boundary layer (from $y \sim 0.2\delta$ to its unconstrained edge) is similar to a free shear flow.

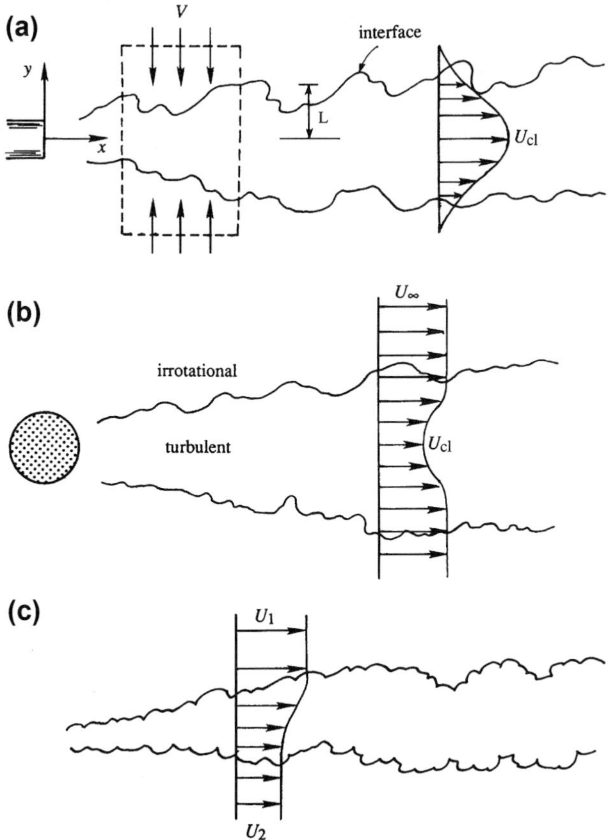

FIGURE 10.15 Three generic free turbulent shear flows: (a) jet, (b) wake, and (c) shear layer. In each case, the region of turbulence coincides with the region of shear in the mean-velocity profile, and entrainment causes the cross-stream dimension L or δ of each flow to increase with increasing downstream distance. The fluid outside the region of turbulence is assumed to be irrotational.

In snapshots and laser-pulse images, free shear flows usually appear with an erratic boundary that divides nominally turbulent from irrotational (or non-turbulent) fluid. Locally, the motion of this boundary is determined by the velocity induced by the turbulent vortices inside the region of turbulence. Typically, these vortices induce the surrounding non-turbulent fluid to flow toward the region of turbulence, and this induced flow, commonly called *entrainment*, causes the cross-stream size (L or δ) of the turbulent region to increase with increasing downstream distance. Because of entrainment, a passive scalar in the body of the turbulent flow is diluted with increasing downstream distance. The actual mechanism of entrainment involves both large- and small-eddy motions, and it may be altered within limits in some free shear flows by introducing velocity, pressure, or geometrical perturbations.

When a time-lapse image or an ensemble average of measurements from a free shear flow is examined, the edge of the region of turbulence is diffuse and the average velocity field and average passive scalar fields are found to be smooth functions. Significantly, the shapes of these mean profiles from different downstream locations within the same flow are commonly found to be self-similar when scaled appropriately. When this happens, the flow is in a state of *moving equilibrium*, in which both the mean and the turbulent fields are determined solely by the *local* length and velocity scales, a situation called *self-preservation*.

Some characteristics of the self-preserving state may be determined from a similarity analysis of the mean momentum equation (10.30) for a variety of free turbulent shear flows. The details of such an analysis are provided here for the plane turbulent jet. The similarity scalings for other free turbulent shear flows are listed in Table 10.1, and are covered in this chapter's exercises. A plane turbulent jet is formed by fast-moving fluid that emerges into a quiescent reservoir from a long slot of height d, as shown in Fig. 10.15a. Here, the long dimension of the slot is perpendicular to the page so the mean-velocity field has only U and V components. Using the x-y coordinates shown in Fig. 10.15a, the self-preserving form for the jet's mean stream-wise velocity and Reynolds shear-stress correlation is:

$$U(x,y)=U_{CL}(x)F(y/\delta(x)), \quad \text{and} \quad -\overline{uv}=\Psi(x)G(y/\delta(x)), \tag{10.56, 10.57}$$

where $U_{CL}(x)$ is the mean stream-wise velocity on the centerline of the flow ($y=0$), $\Psi(x)$ is a function that sets the amplitude of the turbulent shear stress with increasing downstream distance, F and G are undetermined profile functions, and $\delta(x)$ is a characteristic cross-stream length scale. The profile functions must confine the region of turbulence, so F, $G \to 0$ as $y/\delta \to \pm\infty$, and they must allow the jet to spread equally upward and downward, so F must be even and G must be odd; thus $F(0)=1$ and $G(0)=0$. When the self-preserving forms (10.56) and (10.57) are successful, the turbulence is said to have one characteristic length scale. These two equations are the similarity-solution forms (see (11.32)) for the steady mean-flow RANS equations when x and y are the independent variables.

For two-dimensional, constant-density flow with steady boundary conditions, the mean-flow equations are:

$$\frac{\partial U}{\partial x}+\frac{\partial V}{\partial y}=0, \qquad U\frac{\partial U}{\partial x}+V\frac{\partial U}{\partial y}=-\frac{1}{\rho}\frac{\partial P}{\partial x}+\nu\left(\frac{\partial^2 U}{\partial x^2}+\frac{\partial^2 U}{\partial y^2}\right)-\frac{\partial\overline{u^2}}{\partial x}-\frac{\partial\overline{uv}}{\partial y}, \quad \text{and} \tag{10.58, 10.59}$$

$$U\frac{\partial V}{\partial x}+V\frac{\partial V}{\partial y}=-\frac{1}{\rho}\frac{\partial P}{\partial y}+\nu\left(\frac{\partial^2 V}{\partial x^2}+\frac{\partial^2 V}{\partial y^2}\right)-\frac{\partial\overline{uv}}{\partial x}-\frac{\partial\overline{v^2}}{\partial y}. \tag{10.60}$$

For this analysis, the simplest possible form of these equations is adequate. Thus, the jet flow is assumed to be thin, so the boundary-layer approximations are made: $U \gg V$ and $\partial/\partial y \gg \partial/\partial x$. In addition, pressure gradients are presumed small within the nominally quiescent reservoir fluid so that $\partial P/\partial x \approx 0$, and the jet flow's Reynolds number is assumed to be high enough so that viscous stresses can be ignored compared to Reynolds stresses. With these simplifications, the two momentum equations (10.59) and (10.60) become:

$$U\frac{\partial U}{\partial x}+V\frac{\partial U}{\partial y}\cong-\frac{\partial\overline{uv}}{\partial y}, \quad \text{and} \quad 0\cong-\frac{1}{\rho}\frac{\partial}{\partial y}\left(P+\rho\overline{v^2}\right). \tag{10.61}$$

Because the viscous terms have been dropped, these equations are independent of the Reynolds number and should be valid for all Re that are high enough to justify this approximation. Multiplying (10.58) by U and adding it to the first part of (10.61) produces:

$$\frac{\partial}{\partial x}(U^2)+\frac{\partial}{\partial y}(VU+\overline{uv})\cong 0,$$

which can be integrated in the cross-stream direction between infinite limits to obtain:

$$\frac{\partial}{\partial x}\int_{-\infty}^{+\infty}U^2 dy+[VU+\overline{uv}]_{y=-\infty}^{y=+\infty}\cong 0.$$

When evaluated, the terms in [,]-brackets are zero because U, and $\overline{uv}$ are all presumed to go to zero as $y \to \pm\infty$. This equation can be integrated in the stream-wise direction from 0 to x to find:

$$J_s \equiv \rho_s\int_{-\infty}^{+\infty}\left[U^2\right]_{x=0}dy \cong \rho\int_{-\infty}^{+\infty}U^2 dy = const. \tag{10.62}$$

TABLE 10.1 Self-similar far-field results for some free turbulent shear flows.

Flow	Mean Fields	ξ	Profile Widths
Planar Jet	$\frac{U(x,y)}{U_0}=\frac{U_{CL}(x)}{U_0}F(\xi)=2.4\left(\frac{\rho_s}{\rho}\right)^{1/2}\left(\frac{x}{d}\right)^{-1/2}F(\xi)$ $\frac{\overline{Y}(x,y)}{Y_0}=\frac{Y_{CL}(x)}{Y_0}H(\xi)=2.0\left(\frac{\rho_s}{\rho}\right)^{1/2}\left(\frac{x}{d}\right)^{-1/2}H(\xi)$	y/x	$(\xi_{1/2})_U=0.11$ $(\xi_{1/2})_Y=0.14$
Planar Plume	$U(x,y)=U_{CL}(x)F(\xi)=1.9\left(\frac{g(\rho-\rho_s)U_0 d}{\rho}\right)^{1/3}F(\xi)$ $\frac{\overline{Y}(x,y)}{Y_0}=\frac{Y_{CL}(x)}{Y_0}H(\xi)=2.4\left(\frac{\rho U_0^2}{g(\rho-\rho_s)d}\right)^{1/3}\left(\frac{x}{d}\right)^{-1}H(\xi)$	y/x	$(\xi_{1/2})_U=0.12$ $(\xi_{1/2})_Y=0.13$
Round Jet	$\frac{U(R,z)}{U_0}=\frac{U_{CL}(z)}{U_0}F(\xi)=6.0\left(\frac{\rho_s}{\rho}\right)^{1/2}\left(\frac{z}{d}\right)^{-1}F(\xi)$ $\frac{\overline{Y}(R,z)}{Y_0}=\frac{Y_{CL}(z)}{Y_0}H(\xi)=5.0\left(\frac{\rho_s}{\rho}\right)^{1/2}\left(\frac{z}{d}\right)^{-1}H(\xi)$	R/z	$(\xi_{1/2})_U=0.090$ $(\xi_{1/2})_Y=0.11$
Round Plume	$U(R,z)=U_{CL}(z)F(\xi)=3.5\left(\frac{g(\rho-\rho_s)U_0 d}{\rho}\right)^{1/3}\left(\frac{z}{d}\right)^{-1/3}F(\xi)$ $\frac{\overline{Y}(R,z)}{Y_0}=\frac{Y_{CL}(z)}{Y_0}H(\xi)=9.4\left(\frac{\rho U_0^2}{g(\rho-\rho_s)d}\right)^{1/3}\left(\frac{z}{d}\right)^{-5/3}H(\xi)$	R/z	$(\xi_{1/2})_U=0.11$ $(\xi_{1/2})_Y=0.10$
Shear Layer	$U(x,y)=U_2+(U_1-U_2)\frac{\int_{-\infty}^{\xi}F(\xi')d\xi'}{\int_{-\infty}^{+\infty}F(\xi')d\xi'}$	$\frac{y-y_{CL}(x)}{x}$	$(\Delta\xi_{80})_U=\frac{0.085(U_1-U_2)}{\frac{1}{2}(U_1+U_2)}$
Planar Wake	$U(x,y)=U_\infty-\Delta U_{CL}(x)F(\xi)$ $\frac{\Delta U_{CL}(x)}{U_\infty}=1.8\left(\frac{x}{\theta_p}\right)^{-1/2};\ \theta_p=\frac{\text{drag force}}{\rho U_\infty^2\cdot\text{span}}$	$y/\sqrt{\theta_p x}$	$(\xi_{1/2})_U=0.31$
Round Wake	$U(R,z)=U_\infty-\Delta U_{CL}(z)F(\xi)$ $\frac{\Delta U_{CL}(z)}{U_\infty}=(0.4\ to\ 2.0)\left(\frac{z}{\theta_r}\right)^{-2/3};\ \theta_r^2=\frac{\text{drag force}}{\rho U_\infty^2}$	$R/(\theta_r^2 z)^{1/3}$	$(\xi_{1/2})_U=0.4$ to 0.9

Nomenclature
d = slot- or nozzle-exit width or diameter
U_0 = slot- or nozzle-exit fluid velocity
ρ_s = slot- or nozzle-exit fluid density
ρ = nominally quiescent reservoir fluid density
Y_0 = slot- or nozzle-exit passive scalar mass fraction
x = stream-wise centerplane coordinate (planar mean flow)
y = distance from the flow's centerplane
z = stream-wise centerline coordinate (axisymmetric mean flow)
R = radial distance from the flow's centerline
$y_{CL}(x)$ = location of the point in the shear layer where $U=(U_1+U_2)/2$
$(\Delta\xi_{80})_U$ = difference in ξ that spans the central 80% of the velocity difference U_1-U_2
U_∞ = uniform velocity outside the wake
ξ = profile similarity variable
$F(\xi)$, $H(\xi)$ = velocity and mass-fraction profiles, approximately $=\exp\left\{-\ln(2)\xi^2/\xi_{1/2}^2\right\}$

In (10.62), J_s is the momentum injected into the flow per unit span of the slot and the two integrals in (10.62) come from evaluating J_s at $d=0$ and at a location well downstream in the jet. Here, ρ_s is the density of the fluid that emerges from the nozzle, and any difference between ρ and ρ_s is presumed to be insignificant downstream in the jet because the fluid that comes from the slot is mixed with and diluted by the nominally quiescent fluid entrained into the jet. The basis for this presumption is provided further on in this section. Overall, (10.62) can be regarded as a constraint that requires the turbulent flow to contain the same amount of stream-wise momentum at all locations downstream of the nozzle.

To determine the form of the similarity solution for the plane jet, first eliminate V from (10.61) using an integrated form of (10.58), $V=-\int_0^y(\partial U/\partial x)dy$, to find

$$U\frac{\partial U}{\partial x}-\left[\int_0^y\left(\frac{\partial U}{\partial x}\right)dy\right]\frac{\partial U}{\partial y}\cong\frac{\partial\overline{uv}}{\partial y},$$

where $V(0)=0$ by symmetry and y is an integration variable. Then, substitute (10.56) and (10.57) into this equation to reach a single equation involving the two amplitude functions, U_{CL} and Ψ, and the two profile functions, F and G:

$$U_{CL}F\frac{\partial}{\partial x}(U_{CL}F)-\left[\int_0^y\frac{\partial}{\partial x}(U_{CL}F)dy\right]\frac{\partial}{\partial y}(U_{CL}F)\cong\frac{\partial}{\partial y}(\Psi G).$$

Although somewhat tedious, the terms of this equation can be expanded and simplified to find:

$$\left\{\frac{\delta U'_{CL}}{U_{CL}}\right\} F^2 - \left\{\frac{\delta U'_{CL}}{U_{CL}} + \delta'\right\} F' \int_0^{\xi} F d\xi = \left\{\frac{\Psi}{U_{CL}^2}\right\} G', \tag{10.63}$$

where a prime indicates differentiation of a function with respect to its is argument, $\xi = y/\delta$, and ξ is an integration variable. For a simple similarity solution to exist, the coefficients inside {,}-braces in (10.63) should not be functions of x. Setting each equal to a constant produces two ordinary differential equations and an algebraic one:

$$\frac{\delta U'_{CL}}{U_{CL}} = C_1, \quad \frac{\delta U'_{CL}}{U_{CL}} + \delta' = C_2, \quad \text{and} \quad \frac{\Psi}{U_{CL}^2} = C_3. \tag{10.64}$$

The first two of these imply $\delta' = C_2 - C_1$, which is readily integrated to find: $\delta(x) = (C_2 - C_1)(x - x_o)$, where x_o is a constant and is known as the virtual origin of the flow. It is traditional, to choose $C_2 - C_1 = 1$, and to presume that x_o is small so that $\delta = x$. In experiments, x_o is typically found to be of order d, the height of the slot. With $\delta = x$, the first equation of (10.64) may be integrated to determine: $U_{CL} = C_4 x \gamma$, where C_4 and γ are constants. Substituting this into (10.62) leads to:

$$J_s = \rho \int_{-\infty}^{+\infty} U^2 dy = \rho U_{CL}^2 \int_{-\infty}^{+\infty} F^2(\xi) dy = \rho U_{CL}^2 \delta \int_{-\infty}^{+\infty} F^2(\xi) d\xi = \rho C_4^2 x^{2\gamma+1} \int_{-\infty}^{+\infty} F^2(\xi) d\xi. \tag{10.65}$$

Here the final definite integral is just a dimensionless number, so the final form of (10.65) can only be independent of x when $2\gamma + 1 = 0$, or $\gamma = -1/2$. Thus, the results of (10.64) imply that (10.56) and (10.57) may be rewritten:

$$U(x, y) = C_5 (J_s/\rho)^{1/2} x^{-1/2} F(y/x) \quad \text{and} \quad -\overline{uv} = C_3 U_{CL}^2 G(y/x) = C_3 C_5^2 (J_s/\rho) x^{-1} G(y/x). \tag{10.66, 10.67}$$

where the constants C_3 and C_5, and the profile functions F and G must be determined from experimental measurements, direct numerical simulations, or an alternate theory. They cannot be determined from this type of simple similarity analysis because (10.63) is one equation for two unknown profile functions, a situation that is a direct legacy of the closure problem. However, the parametric dependencies shown in (10.66) and (10.67) are those found in experiments, and this is the primary reason for seeking self-preserving forms via a similarity analysis.

The result (10.66) may be used to determine the volume flux (per unit span) $\dot{V}$ in the jet via a simple integration,

$$\dot{V}(x) = \int_{-\infty}^{+\infty} U(x, y) dy = C_5 (J_s/\rho)^{1/2} x^{+1/2} \int_{-\infty}^{+\infty} F(\xi) d\xi, \tag{10.68}$$

where again the definite integral is just a dimensionless number. Therefore, the volume flux in the jet increases with increasing downstream distance like $x^{1/2}$, so the dilution assumption made about J_s in (10.62) should be valid sufficiently far from the jet nozzle. At such distances, commonly known as the *far field* of the jet, the mean mass fraction $\overline{Y}(x, y)$ of slot fluid (or any other suitably defined passive scalar like a dye concentration) will also follow a similarity form:

$$\overline{Y}(x, y) = Y_{CL}(x) H(y/x), \tag{10.69}$$

where Y_{CL} is the centerline nozzle-fluid mass fraction, and H is another profile function defined so that $H(0) = 1$ and $H \to 0$ as $y/\delta \to \pm\infty$. Conservation of slot fluid requires:

$$\dot{M}_s = \rho_s \int_{-\infty}^{+\infty} [U]_{y=0} dy \cong \rho \int_{-\infty}^{+\infty} \overline{Y}(x, y) U(x, y) dy = \rho Y_{CL} C_5 (J_s/\rho)^{1/2} x^{+1/2} \int_{-\infty}^{+\infty} H(\xi) F(\xi) d\xi, \tag{10.70}$$

where $\dot{M}_s$ is the slot-fluid mass injection rate per unit span. In (10.70), the stream-wise turbulent scalar transport term $\overline{uY'}$ has been neglected because it tends to be much smaller than the stream-wise mean scalar transport term $\overline{Y}U$. Reducing (10.70) to a single relationship for Y_{CL}, and substituting this into (10.69) produces:

$$\overline{Y}(x, y) = C_6 (\dot{M}_s/\sqrt{\rho J_s}) x^{-1/2} H(y/x), \tag{10.71}$$

where C_6 is another constant. Eqs. (10.66), (10.67), and (10.71) represent the outcomes from this similarity analysis and can be compared with the $u \propto x^{-1/3}$, $\delta \propto x^{2/3}$ behavior of a planar laminar jet derived in Section 8.10.

Over the years some success has been achieved in determining the various profile shapes and the constants. For example, when the slot exit velocity is uniform and equal to U_0, then $J_s = \rho_s U_0^2 d$ and $\dot{M}_s = \rho_s U_0 d$ so (10.66) and (10.71) reduce to:

$$U(x, y) = C_5 U_0 (\rho_s/\rho)^{1/2} (x/d)^{-1/2} F(y/x), \quad \text{and} \quad \overline{Y}(x, y) = C_7 Y_0 (\rho_s/\rho)^{1/2} (x/d)^{-1/2} H(y/x), \qquad (10.72, 10.73)$$

where Y_0 is the mass fraction of a passive scalar in the slot fluid. Here $Y_0 = 1$ if the slot fluid is the passive scalar, but Y_0 may be much less than one if it represents a trace contaminant or a dye concentration. In addition, the profile functions F and H are smooth, bell-shaped curves commonly specified by their one-sided half-widths $(\xi_{1/2})_U$ and $(\xi_{1/2})_Y$, the values of y/δ that produce F and $H = 1/2$, respectively. For example when a Gaussian function is fit to mean profiles of U, the function F becomes:

$$F(y/x) = \exp\left\{-\ln(2)(y/x)^2/\left(\xi_{1/2}\right)_U^2\right\}.$$

Approximate empirical values for C_5, C_6, $(\xi_{1/2})_U$, and $(\xi_{1/2})_Y$ from Chen and Rodi (1980) and Pope (2000) are provided in Table 10.1 for the plane turbulent jet along with results for other free turbulent shear flows. The similarity forms shown in this table for the planar and round wakes should also be followed in the far-field of jets in coflowing streams. Unfortunately, variations in the empirical constants between experiments may be ±20% (or even more; see the round-wake results) and these variations are thought to be caused by unintentional experimental artifacts, such as unmeasured vibrations, geometrical imperfections, or fluctuations in one of the input flows.

Interestingly, as pointed out in George (1989), such variation in similarity constants is consistent with the type of similarity analysis presented in this section. The three equations (10.64) determined from the coefficients of the similarity momentum equation specify the simplest possibility leading to self-similarity of the mean flow. A more general version of (10.64) that also leads to self-similarity is:

$$\frac{\delta U'_{CL}}{U_{CL}} = C_8\left(\frac{\delta U'_{CL}}{U_{CL}} + \delta'\right) = C_9\frac{\Psi}{U_{CL}^2}, \qquad (10.74)$$

which specifies that the x-dependence of the three coefficients must be equal. Here, $\delta \sim x^m$, $U_{CL} \sim x^n$, and $\Psi \sim x^{2n+m1}$ satisfy (10.74) as do $\delta \sim \exp\{-ax\}$, $U_{CL} \sim \exp\{-ax\}$, and $\Psi \sim \exp\{-ax\}$; thus, multiple possibilities are allowed by (10.63) for the plane jet's similarity solution. While the second law of thermodynamics and the constraint (10.62) rule out some of these possibilities, (10.74) or its equivalent for other free shear flows, and conditions at the flow's origin ($d = 0$) apparently allow the expected self-similar states for a particular shear flow to vary somewhat from experiment to experiment.

Example 10.8

Pure methane gas issues from a round nozzle with diameter $d = 1$ cm at a speed of $U_0 = 20$ m/s into a combustion chamber nominally filled with quiescent air at room temperature and pressure. Assuming the volume fraction of oxygen in air is 0.21, use the round turbulent jet similarity law to estimate the centerline distance from the nozzle exit to the location where the stoichiometric condition is reached, and the centerline speed and nominal width of the jet flow at that location.

Solution Using subscripts "A" for air and "M" for methane, and molecular weights of 28.96 and 16.04 for air and methane, respectively, the Reynolds number of the flow is:

$$\sqrt{J_s/\rho_A}/\nu \sim (\rho_M/\rho_A)^{1/2} U_0 d/\nu_A = (16.04/28.96)^{1/2}(20\ m/s)(0.01\ m)/(1.5 \times 10^{-5}\ m^2 s^{-1}) \sim 10^4,$$

which is high enough to form a turbulent jet. Since the relevant chemical reaction is $CH_4 + 2O_2 \rightarrow CO_2 + 2H_2O$, the mole fraction of methane is half that of oxygen at the stoichiometric condition. Thus, at the location of interest x, there are two equations relating mean volume fractions: $\overline{\upsilon}_A + \overline{\upsilon}_M = 1$ and $\overline{\upsilon}_M = 0.5(0.21\overline{\upsilon}_A)$, that are readily solved to find: $\overline{\upsilon}_A = 1/1.105 = 0.905$, and $\overline{\upsilon}_M = 1 - (1/1.105) = 0.095$. With this composition, the mixture density is within 4% or so of the density of air. The requisite mass fraction of methane is:

$$Y_M = (0.095)(16.04)/[(0.905)(28.96) + (0.095)(16.04)] = 0.0549.$$

The axial location z is found from the entry in Table 10.1 for the mass-fraction field of a round turbulent jet. This means setting $Y_M = 5.0 Y_0 (\rho_s/\rho_A)^{1/2}(z/d)^{-1}$ and solving for z to find:

$$z = 5.0(Y_0/Y_M)(\rho_s/\rho_A)^{1/2} d = 5.0(1.0/0.0549)(16.04/28.96)^{1/2}(0.01\ m) \cong 0.68\ m.$$

Using Table 10.1 entry for the velocity field of a round turbulent jet, the jet centerline velocity at this location is:

$$U_{CL}(z) = 6.0U_o(\rho_s/\rho_A)^{1/2}(z/d)^{-1} = 6.0(20\ m/s)(16.04/28.96)^{1/2}(68)^{-1} \cong 1.3\ m/s.$$

The nominal width of the flow will be approximately four times larger than the jet's mean concentration profile half radius $R_{1/2}$. From the round jet profile width entry in Table 10.1, we have:

$$(\xi_{1/2})_Y = 0.11 = (R_{1/2}/z)_Y \quad \text{so} \quad 4R_{1/2} = 4(0.11)(0.68\ m) \cong 0.30\ m.$$

Thus, the width of a jet's cone of turbulence is a little less than half the downstream distance.

When the mean-velocity and mass-fraction fields of a free turbulent shear flow are self-similar, their corresponding fluctuations are commonly self-similar with the same dependence on the downstream coordinate as that found for the mean fields. However, the profile functions for the various Reynolds stress components or the passive scalar variance are typically not bell-shaped curves. Sample free shear flow measurements for the Reynolds stress components for the plane turbulent jet are shown in Fig. 10.16. Results such as these indicate how fluctuation energy varies within a turbulent flow and may be used to develop and test closure models for RANS equations. Here $\overline{u^2}$, $\overline{v^2}$, and $\overline{w^2}$ are the velocity component variances (commonly called *turbulent intensities*) in the stream-wise (x), slot-normal (y), and slot-parallel (z) directions, respectively. The Reynolds stresses $\overline{uw}$ and $\overline{vw}$ are zero throughout the planar jet since the flow is homogeneous in the z-direction and there is no reason for w to be mostly of one sign if u or v is either positive or negative. Similarly, the Reynolds stress $\overline{uv}$ is zero on the jet centerline by symmetry. In Fig. 10.16, the Reynolds stress reaches a maximum magnitude roughly where $\partial U/\partial y$ is maximum. This is also close to the region where the turbulent kinetic energy $\overline{e}$ reaches a maximum. Such correspondences are commonly exploited in the development of turbulence models.

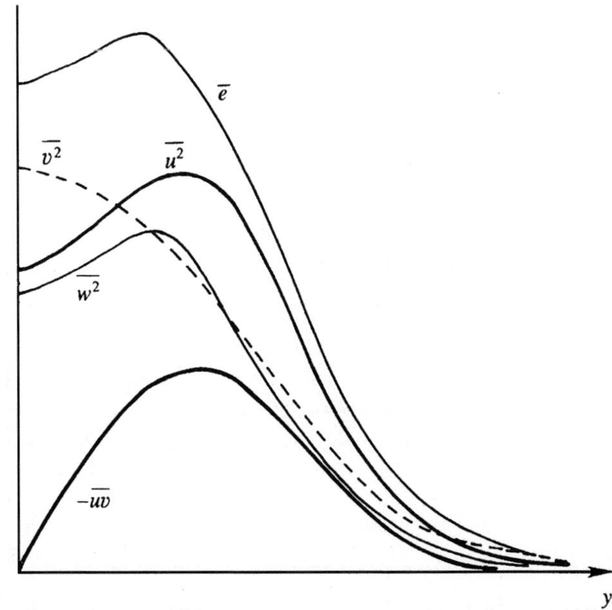

FIGURE 10.16 Sketch of the observed variation of the turbulent kinetic energy $\overline{e}$ and the non-zero Reynolds stress components across a planar jet. Here, $\overline{uv} = 0$ on the jet's centerline ($y = 0$), and $\overline{uw} = \overline{vw} = 0$ throughout the flow because the mean flow is symmetric about $y = 0$, and because the flow is homogeneous in the z-direction.

The terms in the turbulent kinetic energy budget for a two-dimensional jet are shown in Fig. 10.17. Under the boundary-layer assumption for derivatives, $\partial/\partial y \gg \partial/\partial x$, the budget equation (10.47) becomes:

$$0 = -U\frac{\partial \overline{e}}{\partial x} - V\frac{\partial \overline{e}}{\partial y} - \overline{uv}\frac{\partial U}{\partial y} - \frac{\partial}{\partial y}\left(\frac{1}{\rho_0}\overline{pv} + \overline{ev}\right) - \overline{\varepsilon}, \tag{10.75}$$

where the left side represents $\partial \overline{e}/\partial t$. Here, the viscous transport and the term $(\overline{v^2} - \overline{u^2})(\partial U/\partial x)$ arising out of the shear production have been neglected because they are small. The balance of terms shown in this figure is analyzed in Townsend

(1976). Here, T denotes turbulent transport represented by the fourth term on the right side of (10.75). The shear production is zero on the jet centerline where both $\partial U/\partial y$ and $\overline{uv}$ are zero, and reaches a maximum close to the position of the maximum Reynolds stress. Near the center of the jet, the dissipation is primarily balanced by the downstream advection $-U(\partial\overline{e}/\partial x)$, which is positive because the turbulent kinetic energy $\overline{e}$ decays downstream. Away from the jet's center, but not too close to the jet's outer edge, the production and dissipation terms balance. In the outer parts of the jet, the transport term balances the cross-stream advection. In this region V is negative (i.e., toward the center) due to entrainment of the surrounding fluid, and also $\overline{e}$ decreases with increasing y. Therefore, the cross-stream advection $-V(\partial\overline{e}/\partial y)$ is negative, signifying that the entrainment velocity V tends to decrease the turbulent kinetic energy at the outer edge of the jet. A temporally stationary state is therefore maintained by the transport term T carrying $\overline{e}$ away from the jet's center (where $T < 0$) into the outer parts of the jet (where $T > 0$).

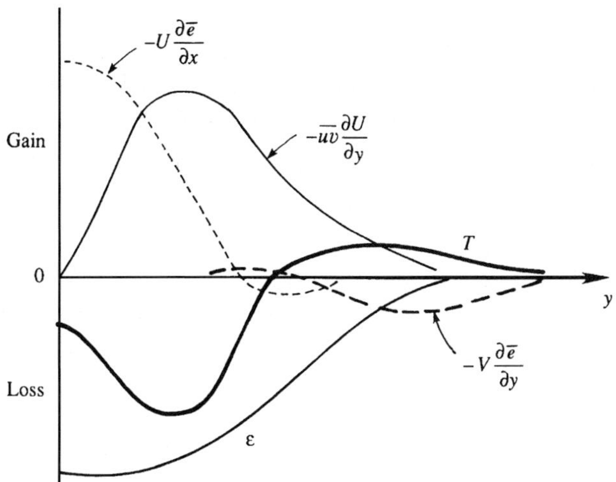

FIGURE 10.17 Sketch of measurements of the terms in the kinetic energy budget of a planar turbulent jet. Here the turbulent transport terms are lumped together and indicated by T. Information of this type is used to build, adjust, and validate closure models for RANS equations.

10.9 Wall-bounded turbulent shear flows

At sufficiently high Reynolds number, the characteristics of free turbulent shear flows discussed in the preceding section are independent of Reynolds number and may be self-similar based on a single length scale. However, neither of these simplifications occurs when the flow is bounded by one or more solid surfaces. The effects of viscosity are always felt near the wall where turbulent fluctuations go to zero, and this gives rise to a second fundamental length scale l_ν that complements the turbulent layer thickness δ. In addition, the persistent effects of viscosity are reflected in the fact that the skin-friction coefficient for a smooth flat plate or smooth round pipe depends on Re, even when Re $\rightarrow \infty$, as seen in Fig. 8.12. Therefore, Re independence of the flow as Re $\rightarrow \infty$ does not occur in wall-bounded turbulent shear flows when the wall(s) is(are) smooth.

The importance of wall-bounded turbulence in engineering applications and geophysical situations is hard to overstate since it sets fundamental limits for the efficiency of transportation systems and on the exchange of mass, momentum, and heat at the earth's surface. Thus, the literature on wall-bounded turbulent flows is large and the material provided here merely covers the fundamentals of the mean flow. A more extensive presentation that includes turbulence intensities is provided in Chapter 7 of Pope (2000). Vortical structures in wall-bounded turbulence are discussed in Kline et al. (1967), Cantwell (1981), and Adrian (2007). The review articles by George (2006), Marusic et al. (2010), and Smits et al. (2011) are also recommended.

Three generic wall-bounded turbulent shear flows are described in this section: pressure-driven channel flow between stationary parallel plates, pressure-driven flow through a round pipe, and the turbulent boundary-layer flow that develops from nominally uniform flow over a flat plate. The first two are fully confined while the boundary layer has one free edge. The main differences between turbulent and laminar wall-bounded flows are illustrated on Fig. 10.18. In general, mean turbulent-flow profiles (solid curves) are blunter, and turbulent-flow wall-shear stresses are higher than those of equivalent steady laminar flows (dashed curves). In addition, a turbulent boundary-layer, mean-velocity profile approaches the free-stream speed very gradually with increasing y so the full thickness of the profile shown in the right panel of Fig. 10.18 lies beyond the extent of the figure. Throughout this section, the density of the flow is taken to be constant.

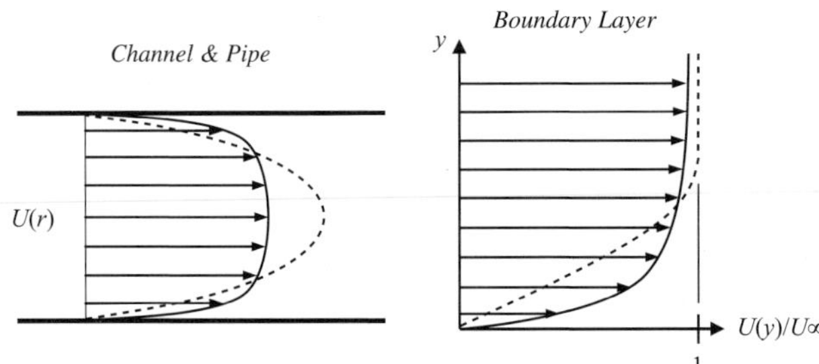

FIGURE 10.18 Sample profiles for wall-bounded turbulent flows (solid curves) compared to equivalent laminar profiles (dashed curves). In general turbulent profiles are blunter with higher skin friction; that is, $\mu(dU/dy)$ evaluated at the wall is greater in turbulent flows than in equivalent laminar ones. In channel and pipe flows, the steady laminar profile is parabolic while a mean turbulent flow profile is more uniform across the central 80% of the channel or pipe. For boundary layers having the same displacement thickness, the steady laminar profile remains linear farther above the wall and converges to the free-stream speed more rapidly than the mean turbulent profile.

Fully developed channel flow is perhaps the simplest wall-bounded turbulent flow. Here, the modifier *fully developed* implies that the statistics of the flow are independent of the downstream direction. The analysis provided here is readily extended to pipe flow, after a suitable redefinition of coordinates. Further extension of channel flow results to boundary-layer flows is not as direct, but can be made when the boundary-layer approximation replaces the fully developed flow assumption. If we align the x-axis with the flow direction, and chose the y-axis in the cross-stream direction perpendicular to the plates so that $y=0$ and $y=h$ define the plate surfaces, then fully developed channel flow must have $\partial U/\partial x=0$. Hence, U can only depend on y, and it is the only mean-velocity component because the remainder of (10.27) implies $\partial V/\partial y=0$, and the boundary conditions $V=0$ on $y=0$ and h, then require $V=0$ throughout the channel. Under these circumstances, the mean-flow momentum equations are:

$$0=-\frac{\partial P}{\partial x}+\frac{\partial \overline{\tau}}{\partial y} \quad \text{and} \quad 0=-\frac{\partial}{\partial y}\left(P+\rho\overline{v^2}\right), \tag{10.76}$$

where $\overline{\tau}=\mu(\partial U/\partial y)-\rho\overline{uv}$ is the total average stress and it cannot depend on x. Integrating the second of these equations from the lower wall up to y produces:

$$P(x,y)-P(x,0)=-\rho\overline{v^2}+\rho\left[\overline{v^2}\right]_{y=0}=-\rho\overline{v^2},$$

where the final equality follows because the variance of the vertical velocity fluctuation is zero at the wall ($y=0$). Differentiating this with respect to x produces:

$$\frac{\partial}{\partial x}P(x,y)-\frac{d}{dx}P(x,0)=-\rho\frac{\partial}{\partial x}\overline{v^2}=0, \tag{10.77}$$

where $P(x,0)$ is the ensemble-average pressure on $y=0$ and the final equality follows from the fully developed flow assumption. Thus, the stream-wise pressure gradient is only a function of x; $\partial P(x,y)/\partial x=dP(x,0)/dx$. Therefore, the only way for the first equation of (10.76) to be valid is for $\partial P/\partial x$ and $\partial\overline{\tau}/\partial y$ to each be constant, so the total average stress distribution $\overline{\tau}(y)$ in turbulent channel flow is linear as shown in Fig. 10.19a. Away from the wall, $\overline{\tau}$ is due mostly to the Reynolds stress, close to the wall the viscous contribution dominates, and at the wall the stress is entirely viscous.

For a boundary layer on a flat plate, the stream-wise mean-flow momentum equation is:

$$U\frac{\partial U}{\partial x}+V\frac{\partial U}{\partial y}=-\frac{1}{\rho}\frac{\partial P}{\partial x}+\frac{1}{\rho}\frac{\partial \overline{\tau}}{\partial y}, \tag{10.78}$$

where $\overline{\tau}$ is a function of x and y. The variation of the stress across a boundary layer is sketched in Fig. 10.19b for the zero-pressure-gradient (ZPG) condition. Here, a constant stress layer, $\partial\overline{\tau}/\partial y\approx 0$, occurs near the wall since both U and $V\to 0$ as $y\to 0$. When the pressure gradient is not zero, the stress profile approaches the wall with a constant slope. Although it is not shown in the figure, the structure of the near-wall region of the turbulent boundary layer is similar to that depicted for the channel flow in Fig. 10.19a with viscous stresses dominating at and near the wall.

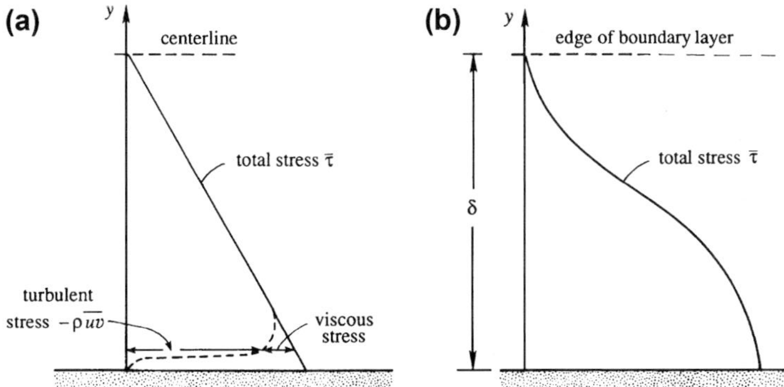

FIGURE 10.19 Variation of total shear stress across a turbulent channel flow (a) and through a zero-pressure-gradient turbulent boundary layer (b). In both cases, the Reynolds shear stress dominates away from the wall but the viscous shear stress takes over close to the wall. The shape of the two stress curves is set by momentum transport between the fast-moving part of the flow and the wall where $U = 0$.

The partitioning of the stress based on its viscous and turbulent origins leads to the identification of two different scaling laws for wall-bounded turbulent flows. The first is known as the *law of the wall* and it applies throughout the region of the boundary layer where viscosity matters and the largest relevant length scale is y, the distance from the wall. This region of the flow is typically called the *inner layer*. The second scaling law is known as the *velocity defect law*, and it applies where the flow is largely independent of viscosity and the largest relevant length scale is the overall thickness of the turbulent layer δ. This region of the flow is typically called the *outer layer*. Fortunately, the inner and outer layers of wall-bounded turbulent flow overlap, and in this overlap region the form of the mean stream-wise velocity profile may be deduced from dimensional analysis.

Inner layer: law of the wall

Consider the flow near the wall of a channel, pipe, or boundary layer. Let U_∞ be the centerline velocity in the channel or pipe, or the free-stream velocity outside the boundary layer. Let δ be the thickness of the flow between the wall and the location where $U = U_\infty$. Thus, δ may be the channel half width, the radius of the pipe, or the boundary-layer thickness. Assume that the wall is smooth, so that any surface roughness is too small to affect the flow. Physical considerations suggest that the near-wall velocity profile should depend only on the near-wall parameters and not on U_∞ or the thickness of the flow δ. Thus, very near the smooth surface, we expect:

$$U = U(\rho, \tau_w, \nu, y), \tag{10.79}$$

where τ_w is the shear stress on the smooth surface. This equation may be recast in dimensionless form as:

$$U^+ \equiv \frac{U}{u_*} = f\left(\frac{yu_*}{\nu}\right) = f\left(\frac{y}{l_\nu}\right) = f(y^+) \quad \text{where} \quad u_*^2 \equiv \frac{\tau_w}{\rho}, \tag{10.80, 10.81}$$

f is an undetermined function, u_* is the *friction velocity* or *shear velocity*, and $l_\nu = \nu/u_*$ is the *viscous wall unit*. Eq. (10.80) is the *law of the wall* and it states that U/u_* should be a universal function of yu_*/ν near a smooth wall. The superscript plus signs are standard notation for indicating a dimensionless law-of-the-wall variable.

The inner part of the wall layer, right next to the wall, is dominated by viscous effects and is called the *viscous sub-layer*. In spite of the fact that it contains fluctuations, the Reynolds stresses are small here because the presence of the wall quells wall-normal velocity fluctuations. At high Reynolds numbers, the viscous sub-layer is thin enough so that the stress is uniform within the layer and equal to the wall shear stress τ_w. Therefore the mean-velocity gradient in the viscous sub-layer is given by:

$$\mu(dU/dy) = \tau_w \rightarrow U = \tau_w y/\mu \quad \text{or} \quad U^+ = y^+, \tag{10.82}$$

where the second two equalities follow from integrating the first. Eq. (10.82) shows that the velocity distribution is linear in the viscous sub-layer, and experiments confirm that this linearity holds up to $yu_*/\nu \approx 5$, which may be taken to be the limit of the viscous sub-layer.

Outer layer: velocity defect law

Now consider the velocity distribution in the outer part of a turbulent boundary layer. The gross characteristics of the turbulence in the outer region are inviscid and resemble those of a free shear flow. The existence of Reynolds stresses in the outer region results in a drag on the flow and generates a *velocity defect* $\Delta U = U_\infty - U$, just like the planar wake. Therefore, in the outer layer we expect,

$$U = U(\rho, \tau_w, \delta, y), \tag{10.83}$$

and by dimensional analysis can write:

$$\frac{U_\infty - U}{u_*} = F\left(\frac{y}{\delta}\right) = F(\xi) \tag{10.84}$$

so that the defect velocity, $U_\infty - U$, is proportional to the friction velocity u_* and a profile function. This is called the *velocity defect law*, and this is its traditional form. In the last two decades, it has been the topic of considerable discussion in the research community, and alternative velocity and length scales have been proposed for use in (10.84), especially for turbulent boundary-layer flows.

Overlap layer: logarithmic law

From the preceding discussion, the mean-velocity profiles in the inner and outer layers of a wall-bounded turbulent flow are governed by different laws, (10.80) and (10.84), in which the independent coordinate y is scaled differently. Distances in the outer part are scaled by δ, whereas those in the inner part are scaled by the much smaller viscous wall unit $l_\nu = \nu/u_*$. Thus, wall-bounded turbulent flows involve at least two turbulent length scales, and this prevents them from reaching the same type of self-similar form with increasing Reynolds number as that found for simple free turbulent shear flows.

Interestingly, a region of overlap in the two profile forms can be found by taking the limits $y^+ \to \infty$ and $\xi \to 0$ simultaneously. Instead of matching the mean velocity directly, in this case it is more convenient to match mean-velocity gradients. (The following short derivation closely follows that in Tennekes and Lumley, 1972.) From (10.80) and (10.84), dU/dy in the inner and outer regions is given by:

$$\frac{dU}{dy} = \frac{u_*^2}{\nu}\frac{df}{dy^+} \quad \text{and} \quad -\frac{dU}{dy} = \frac{u_*}{\delta}\frac{dF}{d\xi}, \tag{10.85, 10.86}$$

respectively. Equating these and multiplying by y/u_* produces:

$$-\xi\frac{dF}{d\xi} = y^+\frac{df}{dy^+}, \tag{10.87}$$

an equation that should be valid for large y_+ and small ξ. As the left-hand side can only be a function of ξ and the right-hand side can only be a function of y^+, both sides must be equal to the same universal constant, say $1/\kappa$, where κ is the *von Karman constant* (not the thermal diffusivity). Experiments show that $\kappa \approx 0.4$ with some dependence on flow type and pressure gradient, as is discussed further on in this section. Setting each side of (10.87) equal to $1/\kappa$, integrating, and using (10.80) gives:

$$U^+ \equiv \frac{U}{u_*} = f(y^+) = \frac{1}{\kappa}\ln(y^+) + B \quad \text{and} \quad F(\xi) = -\frac{1}{\kappa}\ln(\xi) + A, \tag{10.88, 10.89}$$

where B and A are constants with values around 4 or 5, and 1, respectively, again with some dependence on flow type and pressure gradient. Eq. (10.88) or (10.89) is the mean-velocity profile in the *overlap layer* or the *logarithmic layer.* In addition, the constants in (10.88), κ and B, are known as the logarithmic-law (or log-law) constants. As the derivation shows, (10.88) and (10.89) are only valid for large y^+ and small y/δ, respectively. The foregoing method of justifying the logarithmic velocity distribution near a wall was first given by Clark B. Millikan in 1938. The logarithmic law, however, was known from experiments conducted by the German researchers, and several derivations based on semi-empirical theories were proposed by Prandtl and von Karman. One such derivation using the so-called mixing length theory is presented in the following section.

The logarithmic law (10.88) may be the best-known and most important result for wall-bounded turbulent flows. Experimental confirmation of this law is shown in Figs. 10.20 and 10.21 in law-of-the-wall and velocity-defect coordinates, respectively, for the turbulent boundary-layer data reported in Oweis et al. (2010) and Winkel et al. (2012). Nominal specifications for the extent of the viscous sub-layer, the buffer layer, the logarithmic layer, and the wake region are shown there as well. On Fig. 10.20, the linear viscous sub-layer profile appears as a curve for $y^+ < 5$. However, a logarithmic velocity profile appears as a straight line on a log-linear plot, and such a linear region is evident for approximately two decades in y^+ starting near $y^+ \sim 10^2$. The extent of this logarithmic region increases in these coordinates with increasing Reynolds number. The region $5 < y^+ < 30$, where the velocity distribution is neither linear nor logarithmic, is called the *buffer layer*. Neither the viscous stress nor the Reynolds stresses are negligible here, and this layer is dynamically important because turbulence production reaches a maximum here. For the data shown, the measured profiles collapse well to a single curve below $y^+ \sim 10^4$ (or $y/\delta \sim 0.2$) in conformance with the law of the wall. For larger values of y^+, this profile collapse ends where the overlap region ends and the boundary layer's wake flow begins. These velocity profiles do not collapse in the wake region when plotted with law-of-the-wall normalizations because the wake-flow similarity variable is y/δ (not y/l_ν) and the ratio $\delta^+ = \delta/l_\nu$ (commonly known as Re_τ) is different at the three different Reynolds numbers. The fitted curves shown in Fig. 10.20 are mildly adjusted versions of those recommended in Monkewitz et al. (2007) for smooth-flat-plate ZPG turbulent boundary layers.

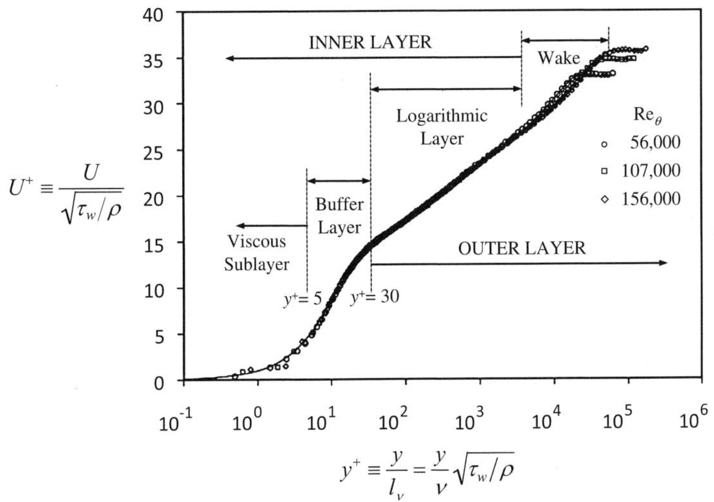

FIGURE 10.20 Mean-velocity profile of a smooth-flat-plate turbulent boundary layer plotted in log-linear coordinates with law-of-the-wall normalizations. The data are replotted from Oweis et al. (2010) and represent three Reynolds numbers. The extent of the various layers within a wall-bounded turbulent flow is indicated by vertical dashed lines. The log-layer-to-wake-region boundary is usually assumed to begin at $y/\delta \approx 0.10$ to 0.20 in turbulent boundary layers. Overall the data collapse well for the inner layer region, as expected, and the logarithmic layer extends for approximately two decades. The wake region shows differences between the Reynolds numbers because its similarity variable is y/δ, and δ/l_ν differs between the various Reynolds numbers.

Although the wake region appears to be smaller than the log-region on Fig. 10.20, this is an artifact of the logarithmic horizontal axis. A turbulent boundary layer's wake region typically occupies the outer 80% of the flow's full thickness and this is shown more clearly in Fig. 10.21 where measured mean-flow profiles are plotted in the velocity deficit form of (10.84). Here, again, the collapse of the various profiles to a single curve is excellent, and the measured profiles diverge from the log-law near $y/\delta \sim 0.2$.

For fully developed channel and pipe flows, the mean stream-wise velocity profile does not evolve with increasing downstream distance. However, turbulent boundary layers do thicken. The following parameter results are developed from the systematic fitting and expansion efforts for ZPG turbulent boundary layers described in Monkewitz et al. (2007), and are intended for use when $\mathrm{Re}_x > 10^6$:

$$\text{Momentum thickness} = \theta/x \approx 0.016\,\mathrm{Re}_x^{-0.15}, \tag{10.90}$$

$$\text{Displacement thickness} = \delta^* \approx \theta \exp\left\{\frac{7.11\kappa}{\ln(\mathrm{Re}_\theta)}\right\}, \tag{10.91}$$

$$99\%\ \text{thickness} = \delta_{99} = 0.2\delta^*[\kappa^{-1}\ln(\mathrm{Re}_{\delta*}) + 3.30],\ \text{and} \tag{10.92}$$

$$\text{Skin friction coefficient} = C_f = \frac{\tau_w}{\frac{1}{2}\rho U_\infty^2} \cong \frac{2.0}{[\kappa^{-1}\ln(\mathrm{Re}_{\delta^*}) + 3.30]^2}, \tag{10.93a}$$

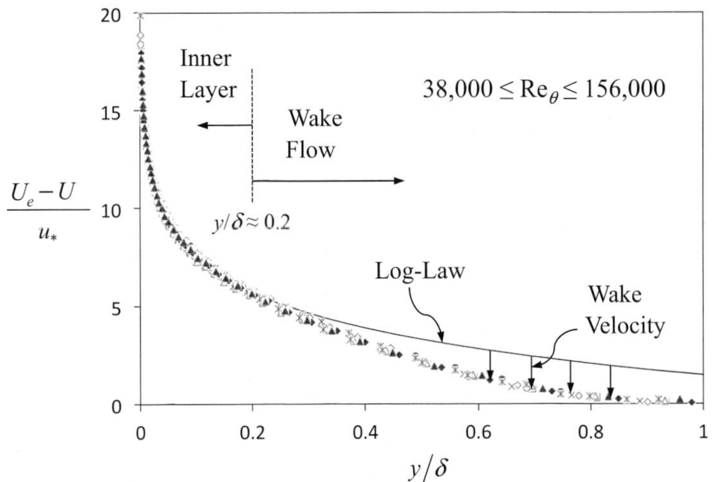

FIGURE 10.21 Mean-velocity profile of a smooth-flat-plate turbulent boundary layer plotted using the velocity defect coordinates of (10.84). The plotted data represent twelve different velocity profiles from the experiments reported in Oweis et al. (2010) and Winkel et al. (2012) covering the Reynolds number range $38{,}000 \le Re_\theta \le 156{,}000$. Here, the log-law diverges from the measured profiles at $y/\delta \sim 0.20$. The measurement-log law difference represents the wake component of the mean-velocity profile. In these experiments, there was a slight favorable pressure gradient so the strength of the wake flow is about ~25% lower than that in a zero-pressure-gradient boundary layer.

where x is the downstream distance, $\mathrm{Re}_x = U_\infty x/\nu$, $\mathrm{Re}_\theta = U_\infty \theta/\nu$, and $\mathrm{Re}_{\delta^*} = U_\infty \delta^*/\nu$. Other common ZPG turbulent boundary-layer skin-friction correlations are those by Schultz-Grunow (1941) and White (2006):

$$C_f \cong 0.370(\log_{10}\mathrm{Re}_x)^{-2.584} \quad \text{and} \quad C_f \cong \frac{0.455}{[\ln(0.06\mathrm{Re}_x)]^2}, \qquad (10.93\text{b}, 10.93\text{c})$$

respectively. These formulae should be used cautiously because the influence of a boundary layer's virtual origin has not been explicitly included and it may be substantial (Chauhan et al., 2009; see also Marusic et al., 2010).

For the purpose of completeness, the following approximate mean-velocity profile functions are offered for wall-bounded turbulent flows:

$$\text{inner profile}: y^+ = U^+_{inner} + e^{-\kappa B}\left[\exp(\kappa U^+_{inner}) - 1 - \kappa U^+_{inner} - \frac{1}{2}(\kappa U^+_{inner})^2 - \frac{1}{6}(\kappa U^+_{inner})^3\right],$$
$$\text{and} \quad \text{outer profile}: U^+_{outer} = \frac{1}{\kappa}\ln(y^+) + B + \frac{2\Pi}{\kappa}W(y/\delta), \qquad (10.94, 10.95)$$

where the inner profile from Spalding (1961) is specified in implicit form, κ and B are the log-law constants from (10.88), and Π and W are the wake strength parameter and a wake function, respectively, both introduced by Coles (1956). The wake function W and the length scale δ in its argument are empirical and are typically determined by fitting curves to experimental profile data.

Example 10.9

Estimate the boundary-layer thicknesses on the underside of the wing of a large commercial airliner on its landing approach. Use the flat-plate results provided above, a chord-length distance of $x = 8$ m, a flow speed of 100 m/s, and a nominal value of $\kappa \approx 0.4$.

Solution First, compute the downstream-distance Reynolds number Re_x using the nominal kinematic viscosity of air: $\mathrm{Re}_x = (100\text{ m/s})(8\text{ m})/(1.5 \times 10^{-5}\text{ m}^2/\text{s}) = 53 \times 10^6$. This Reynolds number is clearly high enough for turbulent flow, so the estimates are:

$$\theta \approx 0.016 \cdot \mathrm{Re}_x^{-0.15} \cdot x = 0.016(53 \times 10^6)^{-0.15}(8\text{ m}) = 0.0089\text{ m},$$

$$\delta^* \approx \theta \exp\left\{\frac{7.11\kappa}{\ln(\mathrm{Re}_\theta)}\right\} = (0.0089\text{ m})\exp\left\{\frac{7.11(0.4)}{\ln((0.0089\text{ m})(100\text{ m/s})/1.5 \times 10^{-5}\text{ m}^2/\text{s})}\right\} \cong 0.0115\text{ m}, \quad \text{and}$$

$$\delta_{99} = 0.2\delta^*[\kappa^{-1}\ln(\mathrm{Re}_{\delta^*}) + 3.30] = 0.2(0.0115\text{ m})\left[0.4^{-1}\ln\left(\frac{(0.0115\text{ m})(100\text{ m/s})}{1.5 \times 10^{-5}\text{ m}^2/\text{s}}\right) + 3.30\right] \cong 0.072\text{ m}.$$

Here, we note that θ and δ^* are almost an order of magnitude smaller than δ_{99}, and that all three boundary-layer thicknesses are miniscule compared to the wing's chord length of 8 m. The latter finding is a primary reason why boundary-layer thicknesses are commonly ignored in aerodynamic analyses.

Of the three generic wall-bounded turbulent flows, the boundary layer's wake is typically the most prominent. For ZPG boundary layers the wake strength is $\Pi = 0.44$ (Chauhan et al., 2009). When the pressure gradient is favorable, Π is lower, and when the pressure gradient is adverse, Π is higher. The wake function is typically chosen to go smoothly from zero to unity as y goes from zero to δ. Among the simplest possibilities for $W(\xi)$ are $3\xi^2 - 2\xi^3$ and $\sin^2(\pi\xi/2)$, however more sophisticated fits are currently in use (see Monkewitz et al., 2007; Chauhan et al., 2009). In the outer profile form given above, δ is interpreted as the 100% boundary-layer thickness where U first equals the local free-stream velocity as y increases. In practice, this requirement cannot be evaluated perfectly with finite-precision experimental data so δ is often approximated as being the 99% or the 99.5% thickness, δ_{99} or $\delta_{99.5}$, respectively. Of course, for channel or pipe flows, δ is half the channel height or the pipe radius, respectively.

As of this writing, new and important concepts and results for wall-bounded turbulence continue to emerge. These include the possibility that the overlap layer might instead be of power law form (Barenblatt, 1993; George and Castillo, 1997) and a reinterpretation of the layer structure in terms of stress gradients (Wei et al., 2005; Fife et al., 2005). The comparisons in Monkewitz et al. (2008) suggest that the logarithmic law should be favored over a power law, while the implications of the stress gradient balance approach are still under consideration. These and other topics in the current wall-bounded turbulent flow literature are discussed in Marusic et al. (2010) and Smits et al. (2011).

Perhaps the most fundamental unanswered question concerns the universality of wall-bounded turbulent flow profiles; are all wall-bounded turbulent flows universal (statistically the same) when scaled appropriately? To answer this question, consider the inner, outer, and overlap layers separately. First of all, the viscous sub-layer profile $U^+ = y^+$ (10.82) is universal using law-of-the-wall normalizations. However, geometrical differences suggest that the wake flow region is not universal. Consider the zone of maximum average fluid velocity at the outer edge of the wake portion of a wall-bounded flow. This maximum velocity zone occurs on the centerline of a channel flow (a plane), on the centerline of a pipe flow (a line), and at the edge of a boundary layer (a slightly tilted, nearly planar surface). Thus, the ratio of the maximum-velocity area to the bounding-wall surface area is one-half for channel flow, vanishingly small for pipe flow, and slightly greater than unity for boundary-layer flow. On this basis, the three wake flows are distinguished. Additionally, the boundary layer differs from the other two flows because it is bounded on one side only. The boundary layer's wake-flow region entrains irrotational fluid at its free edge and does not collide or interact with turbulence arising from an opposing wall, as is the case for channel and pipe flows. Thus, the wake-flow regions of these wall-bounded turbulent flows should all be different and not universal.

Now consider the overlap layer in which the mean-velocity profile takes a logarithmic form. Logarithmic profiles have been observed in all three generic wall-bounded turbulent flows, and Coles and Hirst (1968) recommended values of $\kappa = 0.41$ and $B = 5.0$ for the log-law constants. However, in each circumstance, the log layer inherits properties from the universal viscous sub-layer and from a non-universal wake flow. Thus, the log-law (10.88) may imperfectly approach universality, and this situation is found in experiments. In particular, an assessment of published literature (Nagib and Chauhan, 2008) supports the following values for the logarithmic-law constants at high Reynolds numbers:

Channel flow:	$\kappa = 0.37$	$B = 3.7$
Pipe flow:	$\kappa = 0.41$	$B = 5.0$
ZPG boundary layer:	$\kappa = 0.384$	$B = 4.17$

Yet, the situation remains unsettled. Recent channel-flow experiments at $\mathrm{Re}_h = U_{av}h/\nu$ up to 300,000 (where h is the full channel height, and U_{av} is the time- and height-averaged flow speed) again find $\kappa = 0.41$ and $B = 5.0$ (Flack et al., 2012). Plus, an assessment of the highest Reynolds number data available suggest that turbulent pipe and boundary-layer flows may share log-law constants ($\kappa = 0.39$ and $B = 4.3$) in the range $3\sqrt{u_*\delta/\nu} < y^+$ and $y < 0.15\delta$, where δ is the full boundary-layer thickness (Marusic et al., 2013).

The observed flow-to-flow variation in log-law constants is not anticipated by the analysis presented earlier in this section because geometric differences in the wake-flow regions were not accounted for in (10.83). However, the previous analysis remains valid for each outer-layer flow geometry. Thus, the log-law (10.88) does describe the overlap layer of wall-bounded turbulent flows when the log-law constants are appropriate for that flow's geometry.

Interestingly, there is another issue at play here for turbulent boundary layers. From a flow-parameter perspective, a turbulent boundary layer differs from fully developed channel and pipe flows because the pressure gradient that may exist in a boundary-layer flow is not directly linked to the wall shear stress τ_w. In fully developed channel and pipe flow, a stationary control volume calculation (see Exercise 10.39) requires:

$$dP/dx = -2\tau_w/h \quad \text{or} \quad dP/dx = -4\tau_w/d, \tag{10.96, 10.97}$$

respectively, where h is the channel height and d is the pipe diameter. Thus, the starting points for the dimensional analysis of the inner and outer layers of the mean-velocity profile, (10.79) and (10.83), need not include dP/dx for pipe and channel flows because τ_w is already included. Yet, there is no equivalent to (10.96) or (10.97) for turbulent boundary layers. More general forms of (10.79) and (10.83) that would be applicable to all turbulent boundary layers need to include $\partial P/\partial x$, especially since $\partial P/\partial x$ does not drop from the mean stream-wise momentum equation, (10.78), for any value of y when $\partial P/\partial x$ is non-zero. The apparent outcome of this situation is that the log-law constants in turbulent boundary layers depend on the pressure gradient. Surprisingly, the following empirical correlation, offered by Nagib and Chauhan (2008):

$$\kappa B = 1.6[\exp(0.1663B) - 1], \tag{10.98}$$

collapses measured values of κ and B from all three types of wall-bounded shear flows for $0.15 < \kappa < 0.80$, and $-4 < B < 12$. Here, the most extreme values of κ and B arise from turbulent boundary layers in adverse (low values of κ and B) and favorable (high values of κ and B) pressure gradients.

Turbulent flow in ducts

Fully enclosed turbulent flows through tubes, pipes, ducts, and other conduits have historical, scientific, and practical importance. The study of wall-bounded turbulence originates in the pipe flow studies of Hagen, Poiseuille, Darcy, and Reynolds (see Mullin, 2011). Modern laboratory pipe flow experiments (Zagarola & Smits 1998, McKeon et al., 2004, Hultmark et al., 2012) have reached higher Reynolds numbers than equivalent studies of channel and boundary-layer flows, and thereby have been pivotal in developing a more nuanced understanding of wall-bounded turbulence. Plus, water, air, gasoline, natural gas, crude oil, and variety of other liquids and gases are commonly conveyed from place to place in the developed world through pipelines and other fully enclosed conduits for economic, and health- and safety-related reasons.

A general understanding of duct flows can be developed by considering nominally steady fully developed constant-density flow in a straight fully enclosed duct with smooth walls and constant cross-sectional area A. Consider a simple momentum balance for a duct segment of length L shown in Fig. 10.22. Time-averaging the horizontal component of (4.17) with this duct segment as the control volume and with $\mathbf{b} = 0$ when gravity is vertical leads to:

$$-\int_A \rho\left(U^2 + \overline{u^2}\right)_u dA + \int_A \rho\left(U^2 + \overline{u^2}\right)_d dA = (P_u - P_d)A - L\int_0^l \tau_w\, ds, \tag{10.99}$$

where the u- and d-subscripts denote the upstream and downstream duct cross-sections, respectively, and the interior perimeter of the duct cross section has length l. For fully developed flow, the momentum flux terms on the left of (10.99) cancel, so the equation reduces to:

$$\frac{P_u - P_d}{L} = \frac{l}{A}\left(\frac{1}{l}\int_0^l \tau_w\, ds\right) = \frac{l}{A}\overline{\tau}_w, \tag{10.100}$$

where $\overline{\tau}_w$ is the perimeter- and time-averaged wall shear stress. Thus, the pressure difference necessary to maintain the flow is directly proportional to $\overline{\tau}_w$. Eq. (10.100) is commonly recast in terms of the Darcy friction factor, $\overline{f}$, by scaling the pressure difference with the duct length L and $\frac{1}{2}\rho U_{av}^2$:

$$P_u - P_d = \frac{1}{2}\rho U_{av}^2 \cdot \frac{L}{d_h} \cdot \overline{f} \quad \text{or} \quad \overline{f} = 4\overline{C}_f \tag{10.101, 10.102}$$

where $U_{av} = (1/A)\int_{Area} U dA$ is the time and cross-sectional-area averaged flow speed in the duct, $\overline{C}_f = \overline{\tau}_w/\frac{1}{2}\rho U_{av}^2$ is the average skin friction coefficient (or Fanning friction factor), and d_h is the *hydraulic diameter* of the duct:

$$d_h = 4A/l, \tag{10.103}$$

defined so that it reduces to pipe-interior diameter when the duct is round.

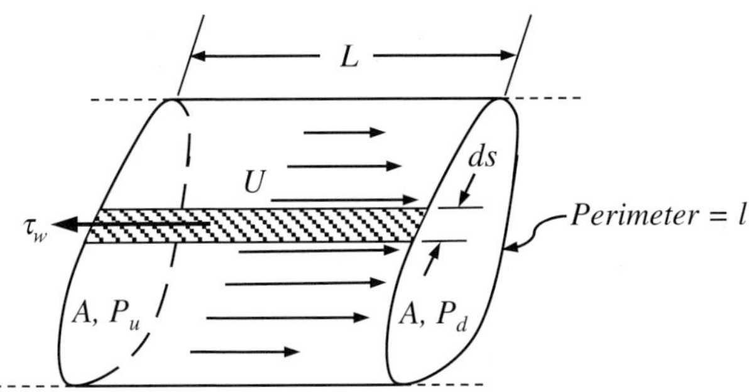

FIGURE 10.22 A segment L of a straight duct of non-circular but constant cross-sectional area A. The average momentum flux into and out of the duct will be equal so conservation of momentum within the duct reduces to a balance of the pressure force $(P_u - P_d)A$, and the integrated wall-shear stress $L\int_0^l \tau_w ds$ acting on the fluid in the duct segment.

In this smooth-wall formulation, the friction factor, $\overline{f}$, may depend on Reynolds number and duct geometry. Consider first the implications of the outer flow profile (10.91) and the simplified momentum balance (10.101) for round pipes where $d_h = d =$ the geometrical pipe diameter. For turbulent flow, the viscous sub-layer is thin and the wake component in (10.101) is weak in round pipes. Thus, (10.88) may be integrated throughout the cross-section of the pipe to find:

$$U_{av} \cong u_*[(1/\kappa)\ln(du_*/2\nu) + B - 3/2\kappa], \tag{10.104}$$

and this can be converted into an implicit relationship for the Darcy friction factor using the generic log-law constants $\kappa = 0.41$ and $B = 5.0$:

$$\overline{f}^{-1/2} \cong 2.0\log_{10}\left(\mathrm{Re}_d \overline{f}^{1/2}\right) - 1.0, \tag{10.105}$$

(see Exercise 10.42) to reach a formula first derived by Prandtl in 1935. To compensate for neglecting the viscous sub-layer and the wake contribution near the pipe's centerline, he modified the second constant:

$$\overline{f}^{-1/2} \cong 2.0\log_{10}\left(\mathrm{Re}_d \overline{f}^{1/2}\right) - 0.8, \tag{10.106}$$

to better match the available experimental data. This empirical formula is valid for $\mathrm{Re}_d \geq 4000$ (White, 2006) and yields $\overline{f}$-values substantially larger than the laminar pipe flow result $\overline{f} = 64/\mathrm{Re}_d$. Although some adjustments to the constants have been recommended (Zagarola & Smits 1998; McKeon et al., 2004), (10.106) still provides a worthwhile quantitative means for determining the Reynolds number dependence of $\overline{f}$.

The dependence of $\overline{f}$ on duct geometry is commonly managed for engineering purposes by using a non-circular conduit's hydraulic diameter in (10.106). The implicit assumption here is that $\overline{\tau}_w$ has the same relationship to U_{av} in non-circular ducts as it does in circular ones. However, this approach looses accuracy when the duct cross-section has sharp corners or when it has a high width-to-height aspect ratio because the conduit's Reynolds number based on hydraulic diameter, $U_{av}d_h/\nu$, is too large. Such difficulties can be partially corrected in rectangular and annual ducts by adjusting the Reynolds number in (10.106) downward using laminar flow results:

$$\overline{f}_{duct,turb}^{-1/2} \cong 2.0\log_{10}\left(\left(\frac{\overline{f}_{pipe}\cdot \mathrm{Re}_d}{\overline{f}_{duct}\cdot \mathrm{Re}_{d_h}}\right)_{laminar}\cdot \mathrm{Re}_{d_h}\cdot \overline{f}_{duct,turb}^{-1/2}\right) - 0.8 \tag{10.107}$$

(see Jones, 1976; Jones and Leung, 1981). The added laminar-flow factor in (10.107) may be obtained from tabulations of laminar flow results (see White, 2006) and is 2/3 when the duct is a high-aspect-ratio channel. A review of flow friction in non-circular ducts and an alternative correction scheme for (10.106) are provided in Obot (1988).

Rough surfaces

In deriving the logarithmic law (10.88), we assumed that the flow closest to the wall is determined by viscosity. This is true only for *hydrodynamically smooth* surfaces, for which the average height of the surface roughness elements is less than the thickness of the viscous sub-layer. For a hydrodynamically rough surface, on the other hand, the roughness elements are taller than the viscous sub-layer (if it exists), and may prevent its formation. An extreme example is the atmospheric boundary layer, where vegetation, buildings, etc., act as roughness elements. In such fully rough situations, the boundary-layer flow impinges directly on the roughness elements leading to wake formation behind each element. Here, shear stress is transmitted to the wall by the resulting drag on the roughness elements, and it nearly always exceeds equivalent smooth wall values. For such fully rough conditions, viscosity is irrelevant for determining either the velocity distribution or the overall friction drag on the surface. This is why the friction coefficients for rough-wall pipes become constant as $\mathrm{Re} \to \infty$.

Although turbulent wall bounded flows over rough surfaces have been of interest for more than a century (Jimenez, 2004), a full understanding of such flows remains elusive and empirical correlations are commonly used for predicting the character of such flows (Flack and Schultz, 2010). The phenomenology of turbulent flow near a rough wall is depicted in Figs. 10.23 and 10.24, which show mean stream-wise velocity profiles in physical and law-of-the-wall coordinates, respectively. When the surface is fully rough, the viscous sub-layer is lost and the velocity distribution above the roughness elements is logarithmic, but the log-law intercept constant is lower than the equivalent smooth-wall value. This downward shift in the log-law, ΔU^+, is known as the roughness function (see Fig. 10.24). Thus, rough-wall turbulent velocity profiles are adequately described by:

$$U^+ = \frac{U(y)}{u_*} = \frac{U(y)}{U_e}\sqrt{\frac{2}{C_f}} = \frac{1}{\kappa}\ln\left(\frac{yu_*}{\nu}\right) + B + \frac{2\Pi}{\kappa}W\left(\frac{y}{\delta}\right) - \Delta U^+, \tag{10.108}$$

where ΔU^+ provides the roughness correction to the smooth wall result (10.91) For pipes and channels, U_e is typically chosen equal to U_{av}, while for boundary layers it is the average wall-parallel velocity at the location $y = \delta$. Similarly, C_f is the skin friction coefficient based on U_{av} or U_e, and δ is understood to be the channel half-height, pipe radius, or overall boundary-layer thicknesses, as appropriate for the flow's geometry. In geophysical flows, the first, second, and last terms on the right-most side of (10.108) are commonly combined and written in terms of a roughness height k_0 that is defined as the value of y at which the logarithmic velocity profile gives $U = 0$ (Fig. 10.23b):

$$U^+ = \frac{1}{\kappa}\ln\left(\frac{y}{k_0}\right) + \frac{2\Pi}{\kappa}W\left(\frac{y}{\delta}\right). \tag{10.109}$$

Here, k_0 carries the roughness correction, and the two formulations, (10.108) and (10.109), are equivalent. The wake portion of wall-bounded turbulent flows is consistently found to be unaltered by the presence of wall roughness (Flack and Schultz, 2010) even though roughness does tend to increase δ for boundary-layer flows.

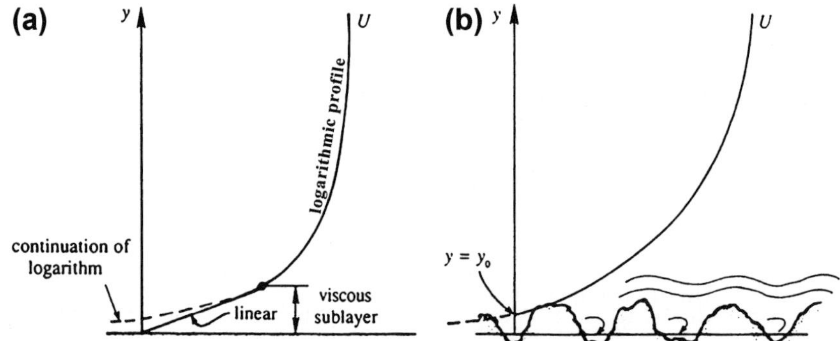

FIGURE 10.23 Logarithmic velocity distributions near smooth (a) and rough (b) surfaces. The presence of surface roughness may eliminate the viscous sub-layer when the roughness elements protrude higher than several l_ν. In this case the log-law may be extended to a virtual wall location $y = k_0$ where U appears to go to zero.

This phenomenology introduces at least two conundrums. The first is the location of $y = 0$. In experimental work, this location is typically chosen to lie somewhere between the peaks and valleys of the roughness elements to maximize the quality of a logarithmic fit to the measured velocity profile. The second, and more important, is the quantitative connection between the actual spatial profile of the rough surface and the resulting surface friction for a given flow speed and flow geometry (channel, pipe, boundary layer, etc.). A rough surface may have structured (patterned) or random roughness, and, in general, must be characterized by multiple length scales such as average or root-mean-square roughness height (k or k_{rms}), and surface correlation lengths in the stream-wise and cross-stream directions. The first quantitative work on this topic was conducted by Niuradse (1933) who studied the impact of uniform-size sand-grain roughness on pipe flow friction using average-sand-grain diameter to characterize the roughness height of the surface. His work has remained important and

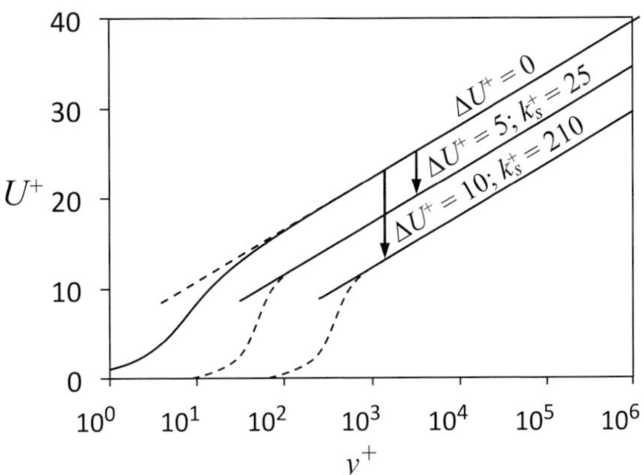

FIGURE 10.24 Average stream-wise velocity profiles, $U^+ = U(y)/u_*$, near smooth ($\Delta U^+ = 0$) and rough ($\Delta U^+ > 0$) surfaces in law-of-the-wall coordinates where y is the wall normal distance, and $y^+ = yu_*/\nu$ [see (10.80) and (10.81)]. When the wall is rough, the skin friction is higher and this causes the log-law portion of the velocity profile to shift downward by an amount ΔU^+, known as the roughness function.

compelling so that essentially all subsequent rough-wall friction measurements have been reported in terms of equivalent sand-grain roughness height, k_s.

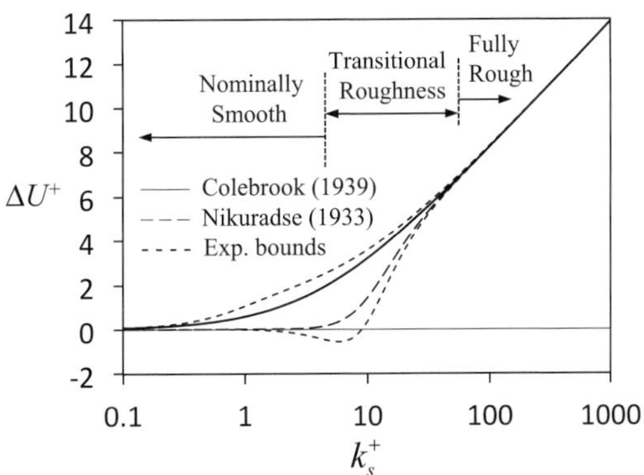

FIGURE 10.25 Roughness function, ΔU^+, as a function of law-of-the-wall-scaled equivalent sand-grain roughness height, $k_s^+ = k_s u_*/\nu$. The solid curve is the correlation of Colebrook (1939) for surfaces typical of commercial pipes. The long-dash curve follows the sand-grain roughness results of Niuradse (1933). The short-dash curves provide approximate upper and lower bounds for experimental results from a variety of rough surfaces. Although the chosen normalizations produce consistent results below k_s^+ of unity and above k_s^+ of ~50, this figure shows that k_s alone is insufficient to describe the effects of wall roughness in between these nominal limiting values.

Fig. 10.25 shows how the roughness function ΔU^+ depends on the law-of-the-wall-scaled equivalent sand-grain roughness height $k_s^+ = k_s u_*/\nu$. The sand-grain results are indicated with long dashes, and are commonly used to define three regimes:

$$\text{nominally smooth: } k_s^+ < 4,$$
$$\text{transitionally rough: } 4 < k_s^+ < 70, \text{ and}$$
$$\text{fully rough: } 70 < k_s^+.$$

In the nominally smooth (also known as hydraulically smooth) regime, sand-grain roughness has no effect, but other types of wall roughness may still cause a roughness effect (Colebrook, 1939). More recently, the measured onset of roughness effects has been found to occur at $k_t^+ \sim 9$ (Flack et al., 2012), where k_t is the peak-to-trough roughness height. In the transitional roughness regime, the roughness function cannot be characterized by k_s^+ alone. Here, surfaces typical of commercial piping produce the higher values of ΔU^+, while triangular riblets may even produce small negative values of ΔU^+ corresponding to skin friction reduction (see Fig. 3 and discussion in Jimenez [2004]). In the fully rough regime, the effects of roughness are independent of the Reynolds number, and ΔU^+ depends logarithmically on k_s^+.

In 1939, Colebrook devised an interpolation formula for the Darcy friction factor for surface roughness typical of commercial piping that spans the three regimes:

$$\overline{f}^{-1/2} \cong -2.0\log_{10}\left(\frac{k_s/d}{3.7} + \frac{2.51}{\mathrm{Re}_d \overline{f}^{1/2}}\right) \tag{10.110}$$

This interpolation formula reduces to (10.106) when $k_s = 0$, and it results in the well-known Moody diagram (Moody, 1944) when $\overline{f}$ is plotted vs. Re_d for different values of k_s/d. An alternative form of (10.110), provided in Jimenez (2004):

$$\Delta U^+ = \frac{1}{\kappa}\ln(1 + 0.26k_s^+), \tag{10.111}$$

appears on Fig. 10.25 as the solid curve. In (10.111) the value of κ is presumed to be 0.40, so that when (10.110) is substituted into (10.108), Nikuradse's fully-rough velocity profile:

$$U^+ = \frac{1}{\kappa}\ln\left(\frac{y}{k_s}\right) + 8.5 + \frac{2\Pi}{\kappa}W\left(\frac{y}{\delta}\right), \tag{10.112}$$

is recovered. As expected, this profile is independent of ν.

With Colebrook's interpolation formula, the flow friction associated with a rough surface can be estimated when that surface's equivalent sand-grain roughness is known. Traditionally, determination of equivalent sand grain roughness required experimental tests. However, Flack and Schultz (2010) have recently suggested that k_s can be determined from the root-mean-square roughness height k_{rms}, and the standardized skewness, $s_k = (\text{skewness})/k_{rms}^3$, of the surface elevation probability density function:

$$k_s = 4.43k_{rms}(1 + s_k)^{1.37}. \tag{10.113}$$

Example 10.10

A zero-pressure-gradient (ZPG) turbulent boundary layer (TBL) develops from $d = 0$ as water flows over a flat plate. For $U_e = 10$ m/s and $x = 10$ m, estimate the skin friction coefficient when the surface is smooth and when the plate surface has been roughened so that $k_{rms} = 70$ μm and $s_k = 0.50$.

Solution The purpose of this example is to show how to combine the various empirical results to estimate the impact of surface roughness. First, evaluate (10.108) at the rough boundary-layer edge, $y = \delta_R$. At this vertical location, $U = U_e$ and the wake function is unity, so:

$$\frac{U_e}{u_{*R}} = \frac{1}{\kappa}\ln\left(\frac{\delta_R u_{*R}}{\nu}\right) + B + \frac{2\Pi}{\kappa} - \Delta U^+,$$

where the R subscript denotes rough surface conditions.

This equation can be simplified and revised to obtain as single implicit equation for the skin friction coefficient, C_{fR}. First, the definition of u_*, implies $U_e/u_{*R} = \sqrt{2/C_{fR}}$. Second, the outer-flow velocity defect law (10.84):

$$\left(\frac{U_e - U(y)}{u_*}\right)_R = F\left(\frac{y}{\delta_R}\right),$$

can be vertically integrated from 0 to ∞ to find:

$$\int_0^\infty \left(\frac{U_e - U(y)}{u_*}\right)_R dy = \int_0^\infty F\left(\frac{y}{\delta_R}\right)dy \to \frac{U_e}{(u_*)_R}\delta_R^* = \delta_R\int_0^\infty F(\eta)d\eta.$$

where δ_R^* is the displacement thickness of the rough-wall boundary layer. Thus, the argument of the natural logarithm is:

$$\frac{\delta_R u_{*R}}{\nu} = \frac{U_e\delta_R^*/\nu}{\int_0^\infty F(\eta)d\eta} \cong \frac{U_e\delta_R^*/\nu}{3.5},$$

where the integral has been approximately evaluated using the velocity deficit profile shown in Fig. 10.21. And third, (10.111) can be used for ΔU^+ for the purposes of estimation. Substituting these relationships into the mean-velocity-profile-edge equation produces:

$$\sqrt{\frac{2}{C_{fR}}} = \frac{1}{\kappa}\ln\left(\frac{U_e\delta_R^*/\nu}{3.5}\right) + B + \frac{2\Pi}{\kappa} - \frac{1}{\kappa}\ln\left(1 + 0.26\frac{k_s U_e}{\nu}\sqrt{\frac{C_{fR}}{2}}\right).$$

For a ZPG boundary layer, the von Karman boundary-layer integral equation is simply $C_f = 2d\theta/dx$. When integrated and multiplied by the boundary-layer shape factor H, this equation provides an estimate of the boundary-layer displacement thickness in terms of the skin friction:

$$\delta_R^*(x) = H_R\int_0^x (C_{fR}/2)dx \approx H_R(C_{fR}/2)x,$$

where the approximate equality is valid when H_R and C_{fR} vary little with increasing x. Although this approximation is not strictly accurate, the inaccuracy it introduces is suppressed by its appearance within the argument of a logarithmic function. Thus, the mean-velocity-profile-edge equation becomes:

$$\sqrt{\frac{2}{C_{fR}}} \cong \frac{1}{\kappa}\ln\left(\frac{(H_R/3.5)\mathrm{Re}_x C_{fR}/2}{1 + 0.26\mathrm{Re}_{ks}\sqrt{C_{fR}/2}}\right) + B + \frac{2\Pi}{\kappa}, \tag{10.114}$$

where $\mathrm{Re}_x = U_e x/\nu$ and $\mathrm{Re}_{ks} = U_e k_s/\nu$.

This equation can be solved implicitly for C_{fR} when the other parameters are known. With $\nu = 10^{-6}\ \mathrm{m^2/s}$, the given information leads to $\mathrm{Re}_x = 10^8$ and (10.113) leads to $k_s = 540\ \mu\mathrm{m}$, so $\mathrm{Re}_{ks} = 5400$. Using $\kappa = 0.4$, $B = 5.0$, $\Pi = 0.44$, and $H_R = 1.3$, a generic high-Reynolds number value for the shape factor, (10.114) leads to $C_{fR} \approx 0.0033$.

Interestingly, as shown on Fig. 10.26, C_{fR} from (10.114) calculated with $\mathrm{Re}_{ks} = 0$ (and the other constants as specified in this example) falls within 2% of the more modern smooth-wall skin friction results from (10.93) evaluated with the log-law constants recommended by Marusic et al. (2013), $\kappa = 0.39$ and $B = 4.3$. Plus, (10.114) provides C_{fR} values that are within +3 and −6% of the classical rough-wall data correlations found in Schlichtling (1979) for the fully rough regime when $10^2 \le x/k_s \le 10^6$ (see Exercise 10.46).

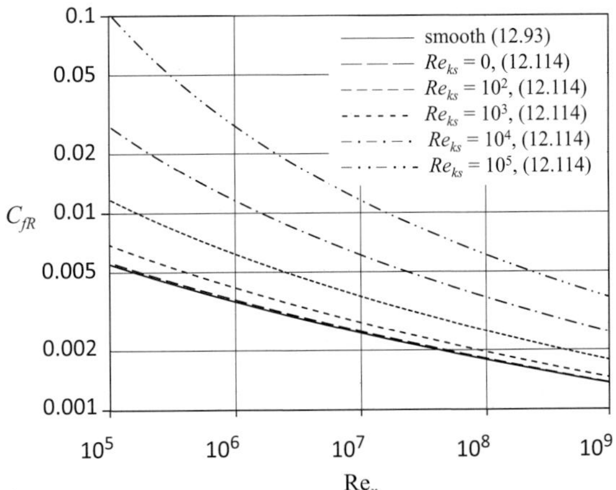

FIGURE 10.26 Rough surface skin friction coefficient, C_{fR}, for a zero-pressure-gradient flat-plate turbulent boundary layer vs. Re_x, the Reynolds number based on downstream distance. The solid curve corresponds to (10.93) evaluated using log-law constants $\kappa = 0.39$ and $B = 4.3$ (as recommended by Marusic et al., 2013). The dashed and dash-dot curves come from implicit evaluation of (10.114) for equivalent-sand-grain roughness-height Reynolds numbers of $\mathrm{Re}_{ks} = 0, 10^2, 10^3, 10^4$, and 10^5. The C_{fR} values produced by (10.114) agree within engineering accuracy (± 5% or so) with prior rough-plate results.

10.10 Turbulence modeling

The closure problem arising from Reynolds-averaging of the equations of fluid motion has led to the development of approximate models to close systems of RANS equations. Because of the practical importance of such models for weather forecasting and performance prediction for engineered devices, RANS-closure modeling efforts have existed for more than a century and continue to this day. This section presents a terse development of the so-called k-ε closure model for the RANS mean-flow momentum equation (10.30) and the foundational elements of second-order closures. The many details associated with these and other RANS closure schemes, and large-eddy simulations are described in Pope (2000) and Wilcox (2006). The earlier review article by Speziale (1991) is also recommended.

The primary purpose of a turbulent-mean-flow closure model is to relate the Reynolds stress correlations, $\overline{u_i u_j}$ to the mean-velocity field U_i. Prandtl and von Karman developed certain semi-empirical theories that attempted to provide this relationship. These theories are based on drawing an analogy between molecular-motion-based laminar momentum and scalar transport, and eddy-motion-based turbulent momentum and scalar transport. The outcome of such modeling efforts is typically an *eddy viscosity* ν_T (first introduced by Boussinesq in 1877) and *eddy diffusivities* κ_T and κ_{mT} for the closure-model equations:

$$\overline{u_i u_j} = \frac{2}{3}\overline{e}\delta_{ij} - \nu_T\left(\frac{\partial U_i}{\partial x_j} + \frac{\partial U_j}{\partial x_i}\right), \qquad \overline{u_i T'} = -\kappa_T \frac{\partial \overline{T}}{\partial x_i}, \quad \text{and} \quad \overline{u_i Y'} = -\kappa_{mT}\frac{\partial \overline{Y}}{\partial x_i}. \tag{10.115, 10.116, 10.117}$$

Eq. (10.115) is mathematically analogous to the stress-rate-of-strain relationship for a Newtonian fluid (4.37) with the term that includes the turbulent kinetic energy $\overline{e}$ playing the role of a turbulent pressure. It represents the *turbulent viscosity hypothesis*. Similarly, (10.116) and (10.117) are mathematically analogous to Fourier's law and Fick's law for molecular diffusion of heat and species, respectively, and these equations represent the *gradient diffusion hypothesis* for turbulent transport of heat and a passive scalar.

To illustrate the implications of such hypotheses, substitute (10.115) into (10.30) to find:

$$\frac{\partial U_i}{\partial t} + U_j\frac{\partial U_i}{\partial x_j} = -\frac{1}{\rho}\frac{\partial P}{\partial x_j} + \frac{\partial}{\partial x_j}\left([\nu + \nu_T]\left(\frac{\partial U_i}{\partial x_j} + \frac{\partial U_j}{\partial x_i}\right) - \frac{2}{3}\overline{e}\delta_{ij}\right) \tag{10.118}$$

for constant-density flow. The factor in [,]-brackets is commonly known as the *effective viscosity*, and the correspondence between this mean-flow equation and its unaveraged counterpart, (4.86) simplified for constant density, is clear and compelling. Mean-flow equations for $\overline{T}$ and $\overline{Y}$ similar to (10.118) are readily obtained by substituting (10.116) and (10.117) into (10.32) and (10.34), respectively. Unfortunately, the molecular-dynamics-to-eddy-dynamics analogy is imperfect. Molecular sizes are typically much less than fluid-flow gradient length scales while turbulent eddy sizes are typically comparable to fluid-flow gradient length scales. For ordinary molecule sizes, averages taken over small volumes include many molecules and these averages converge adequately for macroscopic transport predictions. Equivalent averages over eddies may be unsuccessful because turbulent eddies are so much larger than molecules. Thus, ν_T, κ_T, and κ_{mT} are *not* properties of the fluid or fluid mixture, as ν, κ, and κ_m are. Instead, ν_T, κ_T, and κ_{mT} are properties of the flow, and this transport-flow relationship must be modeled. Hence, (10.118) and its counterparts for $\overline{T}$ and $\overline{Y}$ must be regarded as approximate because (10.115) through (10.117) have inherent limitations. Nevertheless, RANS closure models involving (10.115) through (10.117) are sufficiently accurate for many tasks involving computational fluid dynamics.

From dimensional considerations, ν_T, κ_T, and κ_{mT} should all be proportional to the product of a characteristic turbulent length scale l_T and a characteristic turbulent velocity u_T:

$$\nu_T, k_T, \text{ or } \kappa_{mT} \sim l_T u_T. \tag{10.119}$$

For simplicity, consider fully-developed, temporally-stationary unidirectional shear flow $U(y)$ where y is a cross-stream wall-normal coordinate (Fig. 10.27). The mean-flow momentum equation in this case is:

$$0 = -\frac{1}{\rho}\frac{dP}{dx} + \frac{\partial}{\partial y}\left(\nu\frac{\partial U}{\partial y} - \overline{uv}\right) = -\frac{1}{\rho}\frac{dP}{dx} + \frac{\partial}{\partial y}\left([\nu + \nu_T]\left(\frac{\partial U}{\partial y}\right)\right). \tag{10.120}$$

A mixing length model

An eddy viscosity for this equation can be constructed by interpreting l_T as a *mixing length*, defined as the cross-stream distance traveled by a fluid particle before it gives up its momentum and loses its identity. In this situation, an eddy of size l_T driven by a local shear rate of dU/dy produces a velocity fluctuation of $u_T \sim l_T(dU/dy)$ as it turns over, so

$$-\overline{uv} = \nu_T\frac{dU}{dy} \sim l_T u_T \frac{dU}{dy} \sim l_T\left(l_T\frac{dU}{dy}\right)\frac{dU}{dy} = l_T^2\left(\frac{dU}{dy}\right)^2.$$

The mixing-length concept was first introduced by Taylor (1915), but the approach was fully developed by Prandtl and his coworkers. For a wall-bounded flow it makes sense to assume that l_T is proportional to y when $y = 0$ defines the wall. Thus, setting $l_T = \kappa y$, where κ is presumed to be constant, completes a simple mixing-length turbulence model, and (10.120) becomes:

$$0 = -\frac{1}{\rho}\frac{dP}{dx} + \frac{\partial}{\partial y}\left(\nu\frac{dU}{dy} + \kappa^2 y^2\left(\frac{dU}{dy}\right)^2\right). \tag{10.121}$$

When the pressure gradient is zero or small enough to be ignored, (10.121) can be integrated once to find:

$$\nu\frac{dU}{dy} + \kappa^2 y^2\left(\frac{dU}{dy}\right)^2 = const. = \frac{\tau_w}{\rho},$$

where the final equality comes from evaluating the expression on the left at $y = 0$. For points outside the viscous sub-layer, where the turbulence term dominates, the last equation reduces to a simple ordinary differential equation that is readily integrated to reach:

$$\frac{dU}{dy} \cong \sqrt{\frac{\tau_w}{\rho}}\frac{1}{\kappa y}, \quad \text{or} \quad \frac{U}{u_*} \cong \frac{1}{\kappa}\ln y + const., \tag{10.122}$$

which replicates the log-law (10.88) This simplest-level turbulence model is known as an *algebraic* or *zero-equation* model. Such mixing length models can be generalized to a certain extent by using a contracted form of the mean strain-rate tensor or the mean rotation-rate tensor in place of $(dU/dy)^2$. However, there is no rational approach for relating l_T to the mean-flow field in general.

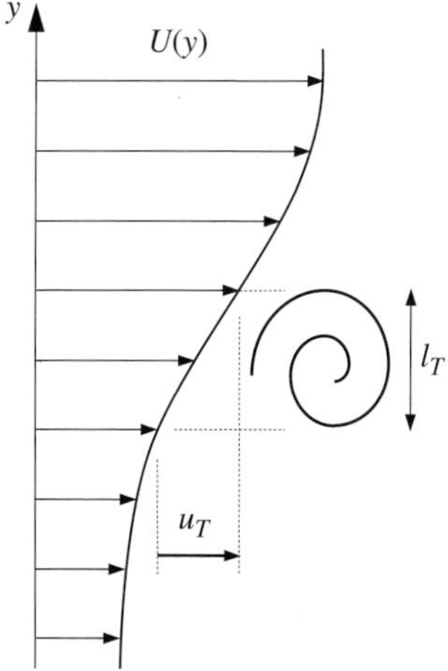

FIGURE 10.27 Schematic drawing of an eddy of size l_T in a shear flow with mean-velocity profile $U(y)$. A velocity fluctuation, u or v, that might be produced by this eddy must be of order $l_T(dU/dy)$. Therefore, we expect that the Reynolds shear stress will scale like $\overline{uv} \sim l_T^2(dU/dy)^2$.

Since the development of modern computational techniques for solving partial differential equations, the need for simple intuitive approaches like the mixing length theory has essentially vanished, and Prandtl's derivation of the empirically known logarithmic velocity distribution has only historical value. However, the relationship (10.119) remains useful for estimating the order of magnitude of the eddy diffusivity in turbulent flows, and for development of more sophisticated RANS closure models (see below). Consider the estimation task first via the specific example of thermal convection between two horizontal plates in air. If the plates are separated by a distance $L = 3$ m, and the lower plate is warmer by $\Delta T = 1\,°\text{C}$. The equation for the vertical velocity fluctuation gives the vertical acceleration as:

$$Dw/Dt \sim g\alpha T' \sim g\Delta T/T, \tag{10.123}$$

since T' is expected to be of order ΔT and $\alpha = 1/T$ for a perfect gas. The time t_r to rise through a height L will be proportional to L/w, so (10.123) gives a characteristic vertical velocity acceleration of:

$$w/t_r = w^2/L \sim Dw/Dt \sim g\Delta T/T \to w \sim \sqrt{gL\Delta T/T} \cong 0.3 \text{ m/s}.$$

The largest eddies will scale with the plate separation L, so the thermal eddy diffusivity, κ_T, is:

$$\kappa_T \sim wL \sim 0.9\ \text{m}^2/\text{s},$$

which is significantly larger than the molecular thermal diffusivity, $2 \times 10^{-5}\ \text{m}^2/\text{s}$.

One-equation models

Independently, Kolmogorov and Prandtl suggested that the velocity scale in (10.98) should be determined from the turbulent kinetic energy:

$$u_T = c\sqrt{\overline{e}},$$

where c is a model constant. The turbulent viscosity is then obtained from an algebraic specification of the turbulent length scale l_T, and the solution of a transport equation for $\overline{e}$ that is based on its exact transport equation (10.47). In this case, the dissipation rate $\overline{\varepsilon}$ and the transport terms must be modeled. For high Reynolds number turbulence, the scaling relationship (10.48) and the gradient diffusion hypothesis lead to the following model equations for the dissipation and the transport of turbulent kinetic energy:

$$\overline{\varepsilon} = C_\varepsilon \overline{e}^{3/2}/l_T \quad \text{and} \quad -\frac{1}{\rho_0}\overline{pu_j} + 2\nu\overline{u_j S'_{ij}} - \frac{1}{2}\overline{u_i^2 u_j} = \frac{\nu_T}{\sigma_e}\frac{\partial \overline{e}}{\partial x_j},$$

where C_ε, and σ_e are model constants. So, for constant density, the turbulent kinetic energy model equation is:

$$\frac{\partial \overline{e}}{\partial t} + U_j\frac{\partial \overline{e}}{\partial x_j} = \frac{\partial}{\partial x_j}\left(\frac{\nu_T}{\sigma_e}\frac{\partial \overline{e}}{\partial x_j}\right) - \overline{\varepsilon} - \overline{u_i u_j}\frac{\partial U_i}{\partial x_j}, \tag{10.124}$$

and this represents *one* additional nonlinear second-order partial-differential equation that must be solved, hence the name *one-equation model*. As mentioned in Pope, one-equation models provide a modest accuracy improvement over the simpler algebraic models.

Two-equation models

These models eliminate the need for a specified turbulent length scale by generating l_T from the solutions of transport equations for $\overline{e}$ and $\overline{\varepsilon}$. The popular k-ε closure model of Jones and Launder (1972) is described here. A k-ω closure model also exists. (Throughout much of the turbulence modeling literature "k" used for the turbulent kinetic energy, so the model name "k-ε" is merely a specification of the dependent-field variables in the two extra partial differential equations.) The k-ε model is based on the turbulent viscosity hypothesis (10.115) with ν_T specified by (10.119), $l_T = \overline{e}^{3/2}/\overline{\varepsilon}$, and $u_T = \overline{e}^{1/2}$:

$$\nu_T = C_\mu\left[\overline{e}^{3/2}/\overline{\varepsilon}\right]\sqrt{\overline{e}} = C_\mu \overline{e}^2/\overline{\varepsilon}, \tag{10.125}$$

where C_μ is one of five model constants. The first additional partial-differential equation is (10.124) for $\overline{e}$. The second additional partial-differential equation is an empirical construction for the dissipation:

$$\frac{\partial \overline{\varepsilon}}{\partial t} + U_j\frac{\partial \overline{\varepsilon}}{\partial x_j} = \frac{\partial}{\partial x_j}\left(\frac{\nu_T}{\sigma_\varepsilon}\frac{\partial \overline{\varepsilon}}{\partial x_j}\right) - C_{\varepsilon 1}\left(\overline{u_i u_j}\frac{\partial U_i}{\partial x_j}\right)\frac{\overline{\varepsilon}}{\overline{e}} - C_{\varepsilon 2}\frac{\overline{\varepsilon}^2}{\overline{e}}. \tag{10.126}$$

The standard model constants are from Launder and Sharma (1974):

$$C_\mu = 0.09, \quad C_{\varepsilon 1} = 1.44, \quad C_{\varepsilon 2} = 1.92, \quad \sigma_e = 1.0, \quad \text{and} \quad \sigma_\varepsilon = 1.3,$$

and these have been set so the model's predictions reasonably conform to experimentally determined mean-velocity profiles, fluctuation profiles, and energy budgets of the type shown in Figs. 10.16 and 10.17 for a variety of simple turbulent flows. More recently renormalization group theory has been used to justify (10.126) with slightly modified constants (Yakhot and Orszag, 1986; Lam, 1992; see also Smith and Reynolds, 1992).

When the density is constant, (10.27), (10.30), (10.115), and (10.124) through (10.126) represent a closed set of equations. Ideally, the usual viscous boundary conditions would be applied to U_i. However, steep near-wall gradients of the dependent field variables pose a significant computational challenge. Thus, boundary conditions on solid surfaces are commonly applied slightly above the surface using empirical *wall functions* intended to mimic the inner layer of a wall-bounded turbulent flow. Wall functions allow the mean-flow momentum equation (10.30) and the turbulence model equations, (10.124) and (10.126), to be efficiently, but approximately, evaluated near a solid surface. Unfortunately, wall functions that perform well with attached turbulent boundary layers are of questionable validity for separating, impinging, and adverse-pressure-gradient flows. Furthermore, the use of wall functions introduces an additional model parameter, the distance above the wall where boundary conditions are applied.

Overall, the k-ε turbulence model is complete and versatile. It is commonly used to rank the performance of fluid dynamic system designs before experimental tests are undertaken. Limitations on its accuracy arise from the turbulent viscosity hypothesis, the $\overline{\varepsilon}$ equation, and wall functions when they are used. In addition, variations in inlet boundary conditions for $\overline{e}$ and $\overline{\varepsilon}$, which may not be known precisely, can produce changes in predicted results. In recent decades, two equation turbulence models based on the eddy viscosity hypothesis have begun to be eclipsed by *Reynolds stress models* or *second-order closures* that directly compute the Reynolds stress tensor from a modeled version of its exact transport equation (10.35).

Reynolds stress models

Reynolds stress closures for RANS equations are attractive because they eliminate the need for an eddy viscosity, but the resulting equations are considerably more complicated than those of two-equation closures. Reynolds stress models are generally superior to two-equation closures because they incorporate flow-history effects since $\overline{u_i u_j}$ is not directly linked to $\partial U_i/\partial x_j$ as it is in (10.115); they include streamline curvature and rotation effects through their direct use of $D\overline{u_i u_j}/Dt$; and (3) they do not require the normal stresses to be equal when $\partial U_i/\partial x_j$ vanishes, as is the case for (10.115). For Reynolds stress models, every dependent field variable in the Reynolds-averaged momentum equation (10.30) is computed, so the pallet of scalars, vectors, and tensors for the creation of closure models includes: P, U_i, and $\overline{u_i u_j}$. Additionally, a model equation like (10.126) for the kinetic energy dissipation rate is typically solved as well, so $\overline{\varepsilon}$ is also used in Reynolds stress closure models. The summary provided here is drawn largely from Speziale (1991), Pope (2000), and Wilcox (2006).

To illustrate the form of common Reynolds stress models, consider the following equivalent form of (10.35) for constant density:

$$\frac{\partial \overline{u_i u_j}}{\partial t} + U_k \frac{\partial \overline{u_i u_j}}{\partial x_k} = -\overline{u_i u_k}\frac{\partial U_j}{\partial x_k} - \overline{u_j u_k}\frac{\partial U_i}{\partial x_k} - \varepsilon_{ij} + M_{ij} + N_{ij} \tag{10.127}$$

(see Exercise 10.50). The first two terms on the right side are the Reynolds-stress production terms and do not need to be modeled. The third term on the right side (10.127) is the Reynolds-stress dissipation rate tensor:,

$$\varepsilon_{ij} \equiv 2\nu \overline{\frac{\partial u_i}{\partial x_k}\frac{\partial u_j}{\partial x_k}} \cong \frac{2}{3}\overline{\varepsilon}\delta_{ij}. \tag{10.128}$$

It is typically modeled as being isotropic via the approximate equality in (10.128). The fourth term on the right side of (10.127) is the Reynolds-stress transport tensor:

$$M_{ij} \equiv \frac{\partial}{\partial x_k}\left(\nu \frac{\partial}{\partial x_k}\overline{u_i u_j} - \overline{u_i u_j u_k} - \frac{\overline{u_i p}}{\rho}\delta_{jk} - \frac{\overline{u_j p}}{\rho}\delta_{ik}\right), \tag{10.129}$$

which includes viscous-, turbulent-, and pressure-transport contributions. The viscous contribution need not be modeled, and is typically negligible except near walls. The other contributions are commonly modeled by an equation representing gradient diffusion that embodies appropriate symmetries, such as:

$$-\overline{u_i u_j u_k} - \frac{\overline{u_i p}}{\rho}\delta_{jk} - \frac{\overline{u_j p}}{\rho}\delta_{ik} \cong \frac{2}{3}C_s \frac{\overline{e}^2}{\overline{\varepsilon}}\left[\frac{\partial \overline{u_i u_j}}{\partial x_k} + \frac{\partial \overline{u_j u_k}}{\partial x_i} + \frac{\partial \overline{u_k u_i}}{\partial x_j}\right], \tag{10.130}$$

where $C_s \approx 0.11$ (Mellor and Herring, 1973; Launder et al., 1975). The final term on the right of (10.127) is the pressure-rate-of-strain tensor:

$$N_{ij} \equiv \overline{\frac{p}{\rho}\left(\frac{\partial u_i}{\partial x_j} + \frac{\partial u_j}{\partial x_i}\right)}. \tag{10.131}$$

It is important for the solution of (10.127) and difficult to model because simultaneous measurements of velocity-fluctuation gradients and pressure fluctuations within a turbulent flow are essentially non-existent.

Insight into the character of the pressure-rate-of-strain tensor can be obtained from the Poisson equation for the pressure fluctuation p:

$$\frac{1}{\rho}\frac{\partial^2 p}{\partial x_k^2} = -2\frac{\partial U_i}{\partial x_j}\frac{\partial u_j}{\partial x_i} - \frac{\partial^2}{\partial x_i \partial x_j}\left(u_i u_j - \overline{u_i u_j}\right), \tag{10.132}$$

(see Exercise 10.51). On the basis of this equation, pressure fluctuations are presumed to fall into three categories: (1) *rapid* pressure fluctuations that respond immediately to changes in mean-flow gradients via the first term on the right of (10.132), (2) *slow* pressure fluctuations that occur in response to changes the second term on the right of (10.132), and (3) *harmonic* pressure fluctuations for which $\partial^2 p/\partial x_k^2 = 0$ that arise to satisfy boundary conditions on p. Overall, (10.132) indicates that pressure fluctuations may be influenced by the entire flow field; thus, an accurate local closure scheme for (10.127) is impossible. However, an assumption of locally homogeneous turbulence allows a Green's function solution of (10.132) to be combined with (10.131) to reach:

$$N_{ij} = \frac{1}{4\pi}\iiint_V \overline{\left(\frac{\partial u_i}{\partial x_j} + \frac{\partial u_j}{\partial x_i}\right)\frac{\partial^2 (u_k u_l)}{\partial y_k \partial y_l}}\frac{d^3 y}{|\mathbf{x}-\mathbf{y}|} + \frac{1}{2\pi}\frac{\partial U_k}{\partial x_l}\iiint_V \overline{\left(\frac{\partial u_i}{\partial x_j} + \frac{\partial u_j}{\partial x_i}\right)\frac{\partial u_l}{\partial y_k}}\frac{d^3 y}{|\mathbf{x}-\mathbf{y}|}, \tag{10.133}$$

where $\mathbf{x}$ is the position vector, $\mathbf{y}$ is a position-vector integration variable, and V is a volume that encompasses the entire flow field (see Exercise 10.52). Most Reynolds stress models use a form of (10.133) with modeled terms replacing the volume integrals (see Wilcox, 2006).

Thus, a complete Reynolds stress closure model for constant-density turbulent flow consists of (10.27), (10.30), (10.126) or its equivalent, (10.127)–(10.129), (10.130) or its equivalent, and (10.133) or a model equation that replaces it. The performance of Reynolds-stress models is described in Pope (2000) and Wilcox (2006).

10.11 Turbulence in a stratified medium

The effects of stratification are often important in geophysical flows and may be important in laboratory flows as well. Some effects of stratification on turbulent flows are described in this section. Further discussion can be found in Tennekes and Lumley (1972), Phillips (1977), and Panofsky and Dutton (1984).

As is customary in the geophysical literature, the z-direction points upward opposing gravity so the mean velocity of a horizontal shear flow will be denoted by $U(z)$. For simplicity, U is assumed to be independent of x and y. Turbulence in a stratified medium depends critically on the stability of the vertical density profile. In the neutrally stable state of a compressible environment the density decreases upward, because of the decrease of pressure, at a rate $d\rho_a/dz$ called the *adiabatic density gradient*, as discussed in Section 1.10. A medium is statically stable if the density decreases faster than the adiabatic decrease. The effective density gradient that determines the stability of the environment is then determined by the sign of $d(\rho - \rho_a)/dz$, where $\rho - \rho_a$ is called the *potential density*. In the following discussion, it is assumed that the adiabatic variations of density have been subtracted out, so that "density" or "temperature" really mean potential density or potential temperature.

The Richardson numbers

First, examine the equation for turbulent kinetic energy (10.47). Omitting the viscous transport and assuming that the flow is independent of x and y, it reduces to:

$$\frac{\partial \overline{e}}{\partial t} + U\frac{\partial \overline{e}}{\partial x} = -\frac{\partial}{\partial z}\left(\frac{1}{\rho_0}\overline{pw} + \overline{ew}\right) - \overline{uw}\frac{\partial U}{\partial z} + g\alpha\overline{wT'} - \overline{\varepsilon}, \tag{10.134}$$

where x increases in the downstream direction. The first term on the right side is the transport of turbulent kinetic energy by vertical velocity fluctuations. The second term is the production of turbulent energy by the interaction of Reynolds stress and the mean shear; this term is almost always positive. The third term is the production of turbulent kinetic energy by the vertical heat flux; it is called the *buoyant production*, and was discussed in Section 10.7. In an unstable environment, in which the mean temperature $\overline{T}$ decreases upward, the heat-flux correlation $\overline{wT'}$ is positive (upward), signifying that the turbulence is generated convectively by upward heat fluxes. In the opposite case of a stable environment, the turbulence is suppressed by stratification. The ratio of the buoyant destruction of turbulent kinetic energy to the shear production is called the *flux Richardson number*:

$$\mathrm{Rf} = \frac{-g\alpha\overline{wT'}}{-\overline{uw}(dU/dz)} = \frac{\text{buoyant destruction}}{\text{shear production}}. \tag{10.135}$$

As the shear production is positive with the minus sign displayed, the sign of Rf depends on the sign of $\overline{wT'}$. For an unstable environment in which the heat flux is upward Rf is negative and for a stable environment it is positive. For Rf > 1, buoyant destruction removes turbulence at a rate larger than the rate at which it is produced by shear production. However, the critical value of Rf at which the turbulence ceases to be self-supporting is less than unity, as dissipation is necessarily a large fraction of the shear production. Observations indicate that the critical value is $\mathrm{Rf}_{\mathrm{cr}} \approx 0.25$ (Panofsky and Dutton, 1984, p. 94). If measurements indicate the presence of turbulent fluctuations, but at the same time a value of Rf much larger than 0.25, then a fair conclusion is that the turbulence is decaying. When Rf is negative, a large −Rf means strong convection and weak mechanical turbulence.

Instead of Rf, it is easier to measure the *gradient Richardson number*, defined as:

$$\mathrm{Ri} \equiv \frac{N^2}{(dU/dz)^2} = \frac{\alpha g(d\overline{T}/dz)}{(dU/dz)^2}, \tag{10.136}$$

where N is the buoyancy frequency and the second equality follows for stratification by thermal variations. If we make the turbulent viscosity and gradient diffusion assumptions (10.115) and (10.116), then the two Richardson numbers are related by:

$$\mathrm{Ri} = (\upsilon_T/\kappa_T)\mathrm{Rf}. \tag{10.137}$$

The ratio υ_T/κ_T is the *turbulent Prandtl number*, which determines the relative efficiency of the vertical turbulent exchanges of momentum and heat. Stable stratification damps vertical transport of both heat and momentum; however, the momentum flux is reduced less because the internal waves in a stable environment can transfer momentum (by moving vertically from one region to another) but not heat. Therefore, $\upsilon_T/\kappa_T > 1$ for a stable environment. Eq. (10.137) then shows that turbulence can persist even when Ri > 1, if the critical value of 0.25 applies on the *flux* Richardson number (Turner, 1981; Bradshaw and Woods, 1978). In an unstable environment, on the other hand, υ_T/κ_T becomes small. In a neutral environment it is usually found that $\upsilon_T \approx \kappa_T$; the idea of equating the eddy coefficients of heat and momentum is called the *Reynolds analogy*.

Monin-Obukhov length

The Richardson numbers are ratios that compare the relative importance of mechanical and thermal-convective turbulence. Another parameter used for the same purpose is not a ratio, but a length, the *Monin-Obukhov length*, defined as:

$$L_M \equiv -u_*^3/\kappa\alpha g\overline{wT'}, \tag{10.138}$$

where u_* is the friction velocity, $\overline{wT'}$ is the heat flux correlation, α is the coefficient of thermal expansion, and κ is the von Karman constant introduced for convenience. Although $\overline{wT'}$ is a function of z, the parameter L_M is effectively a constant for the flow, as it is used only in the logarithmic region of the earth's atmospheric boundary layer in which both $\overline{uw}$ and $\overline{wT'}$ are nearly constant. The Monin-Obukhov length then becomes a parameter determined from the boundary conditions of friction and the heat flux at the surface. Like Rf, it is positive for stable conditions and negative for unstable conditions.

The significance of L_M within the atmospheric boundary layer becomes clearer if we write Rf in terms of L_M, using the logarithmic velocity distribution (10.88), from which $dU/dz = u_*/\kappa z$. (Note that z is the distance perpendicular to the surface.) Using $\overline{uw} = u_*^2$ because of the near uniformity of stress in the logarithmic layer, (10.135) becomes:

$$\mathrm{Rf} = z/L_M. \tag{10.139}$$

As Rf is the ratio of buoyant destruction to shear production of turbulence, (10.111) shows that L_M is the height at which these two effects are of the same order. For both stable and unstable conditions, the effects of stratification are slight if $z \ll |L_M|$. At these small heights, then, the velocity profile is logarithmic, as in a neutral environment. This is called a *forced convection* region, because the turbulence is mechanically forced. For $z \gg |L_M|$, the effects of stratification dominate. In an unstable environment, it follows that the turbulence is generated mainly by buoyancy at heights $z \gg -L_M$, and the shear production is negligible. The region beyond the forced convecting layer is therefore called a zone of *free convection* (Fig. 10.28), containing thermal plumes (columns of hot rising gases) characteristic of free convection from a heated horizontal plate in the absence of shear flow.

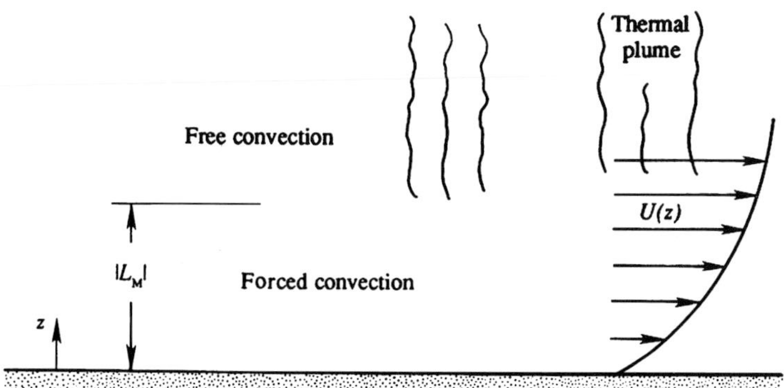

FIGURE 10.28 Forced and free convection zones in an unstable atmosphere. In strongly sheared regions, the turbulence will not include buoyant effects (forced convection). However, where shear is weak, buoyant convection will set the turbulent scales (free convection).

Observations as well as analysis show that the effect of stratification on the velocity distribution in the surface layer is given by the log-linear profile (Turner, 1973):

$$U = \frac{u_*}{\kappa}\left[\ln\frac{z}{z_o} + 5\frac{z}{L_M}\right]. \tag{10.140}$$

The form of this profile is sketched in Fig. 10.29 for stable and unstable conditions. It shows that the velocity is more uniform than $\ln(z)$ in the unstable case because of the enhanced vertical mixing due to buoyant convection.

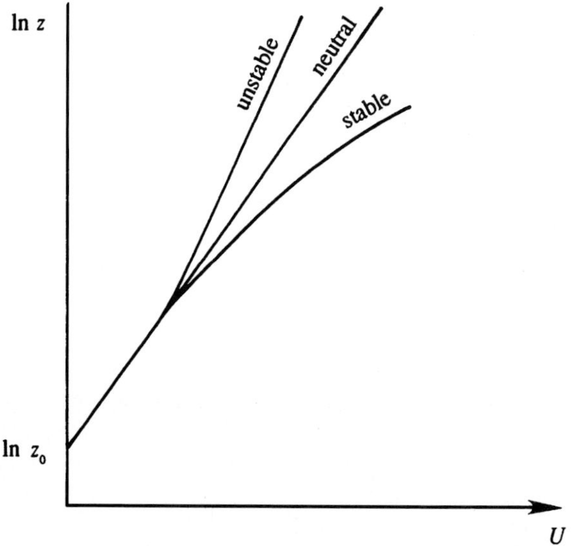

FIGURE 10.29 Effect of stability on velocity profiles in the surface layer. When the atmospheric boundary layer is neutrally stable, the mean-velocity profile is logarithmic. When it is stable, vertical turbulent motions are suppressed so higher shear may exist in the mean flow; this is shown as the lower curve labeled stable. When the atmospheric boundary layer is unstable, vertical turbulent motions are enhanced, mean-flow shear is reduced, and $U(z)$ becomes more nearly uniform; this is shown as the upper curve labeled unstable.

Spectrum of temperature fluctuations

An equation for the intensity of temperature fluctuations $\overline{T'^2}$ can be obtained in a manner identical to that used for obtaining the turbulent kinetic energy. The procedure is therefore to obtain an equation for DT'/Dt by subtracting those for $D\tilde{T}/Dt$

and $D\overline{T}/Dt$, multiplying the resulting equation for DT'/Dt by T', and then averaging this equation. The result is:

$$\frac{\partial}{\partial t}\left(\frac{1}{2}\overline{T'^2}\right)+U\frac{\partial}{\partial x}\left(\frac{1}{2}\overline{T'^2}\right)=-\overline{wT'}\frac{d\overline{T}}{dz}-\frac{\partial}{\partial z}\left(\frac{1}{2}\overline{T'^2w}-\kappa\frac{\partial\overline{T'^2}}{\partial z}\right)-\overline{\varepsilon}_T, \tag{10.141}$$

where κ = thermal diffusivity (not the von Karman constant), and $\overline{\varepsilon}_T=\kappa\overline{(\partial T'/\partial x_j)^2}$ is the *dissipation rate of temperature fluctuations*, analogous to the dissipation of turbulent kinetic energy $\overline{\varepsilon}$ defined within (10.47). The first term on the right side is the generation of $\overline{T'^2}$ by the mean temperature gradient, $\overline{wT'}$ being positive if $d\overline{T}/dz$ is negative. The second term on the right side is the turbulent transport of $\overline{T'^2}$.

A wave number spectrum of temperature fluctuations can be defined such that:

$$\overline{T'}\equiv\int_0^\infty S_T(K)dK.$$

As in the case of the kinetic energy spectrum, an inertial range of wave numbers exists in which neither the production by large-scale eddies nor the dissipation by conductive and viscous effects are important. As the temperature fluctuations are intimately associated with velocity fluctuations, $S_T(K)$ in this range must depend not only on ε_T but also on the variables that determine the velocity spectrum, namely ε and K. Therefore:

$$S_T=S_T(K,\overline{\varepsilon},\overline{\varepsilon}_T)\quad\text{for}\quad 2\pi/L\ll K\ll 2\pi/\eta,$$

where L is the size of the largest eddies. The units of S_T are °C^2 m, and the units of $\overline{\varepsilon}_T$ are °C^2/s, so dimensional analysis requires:

$$S_T\propto\overline{\varepsilon}_T\overline{\varepsilon}^{-1/3}K^{-5/3}\quad\text{for}\quad 2\pi/L\ll K\ll 2\pi/\eta, \tag{10.142}$$

which was first derived by Obukhov (1949). Comparing with (10.54), it is apparent that the spectra of both velocity and temperature fluctuations in the inertial sub-range have the same $K^{-5/3}$ form.

The spectrum beyond the inertial sub-range depends on whether the Prandtl number ν/κ of the fluid is smaller or larger than one. We shall only consider the case of $\nu/\kappa\gg 1$, which applies (at least approximately) to water for which the Prandtl number is 7.1. Let η_T be the scale responsible for smearing out the temperature gradients, the thermal equivalent of η (the Kolmogorov microscale) at which the velocity gradients are smeared out. For $\nu/\kappa\gg 1$ we expect that $\eta_T\ll\eta$, because then the conductive effects are important at scales smaller than the smallest viscous scale. In fact, Batchelor (1959) showed that $\eta_T=\eta(\kappa/\nu)^{1/2}\ll\eta$. In such a case, there exists a range of wave numbers $2\pi/\eta\ll K\ll 2\pi/\eta_T$, in which the scales are small enough for viscosity to suppress velocity fluctuations but not small enough for the thermal diffusivity to suppress temperature fluctuations. Therefore, $S_T(K)$ continues up to wave numbers of order $2\pi/\eta_T$, although the kinetic energy spectrum has dropped off sharply. This is called the *viscous convective sub-range*, because the spectrum is dominated by viscosity but is still actively convective. Batchelor (1959) showed that the spectrum in the viscous convective sub-range is:

$$S_T\propto K^{-1}\quad\text{for}\quad 2\pi/\eta\ll K\ll 2\pi/\eta_T, \tag{10.143}$$

Fig. 10.30 shows a comparison of velocity and temperature spectra, observed in a tidal flow through a narrow channel. The temperature spectrum shows that the spectral slope increases from −5/3 in the inertial sub-range to −1 in the viscous convective sub-range.

Example 10.11

In the late 1990s very-high-Reynolds-number atmospheric boundary-layers measurements were made on the nearly-smooth salt flats of the Great Salt Lake Dessert in western Utah (Metzger et al., 2001; Metzger and Klewicki, 2001). The measurements were made when the atmosphere was neutrally stable (or nearly so) and the mean flow agreed with the log-law (10.88) using $\kappa=0.41$ and $B=5.0$ for vertical distances from several meters to approximately 100 m. Using $U_e=4.8$ m/s, $u_*=0.13$ m/s, and $\nu=1.7\times10^{-5}$ m^2/s to account for the 1300 m elevation at the experimental site, what are the smallest values of the Monin-Obukhov length that would lead to 5% change in the log-law at $z=10$ m and 100 m?

Solution Rewrite the log-linear profile (10.140) to replace z_o with B.

$$\frac{U}{u_*}=\frac{1}{\kappa}\ln\frac{z}{z_o}+5\frac{z}{\kappa L_M}=\frac{1}{\kappa}\ln\frac{zu_*}{\nu}-\frac{1}{\kappa}\ln\frac{z_ou_*}{\nu}+5\frac{z}{\kappa L_M}=\frac{1}{\kappa}\ln\frac{zu_*}{\nu}+B+5\frac{z}{\kappa L_M}.$$

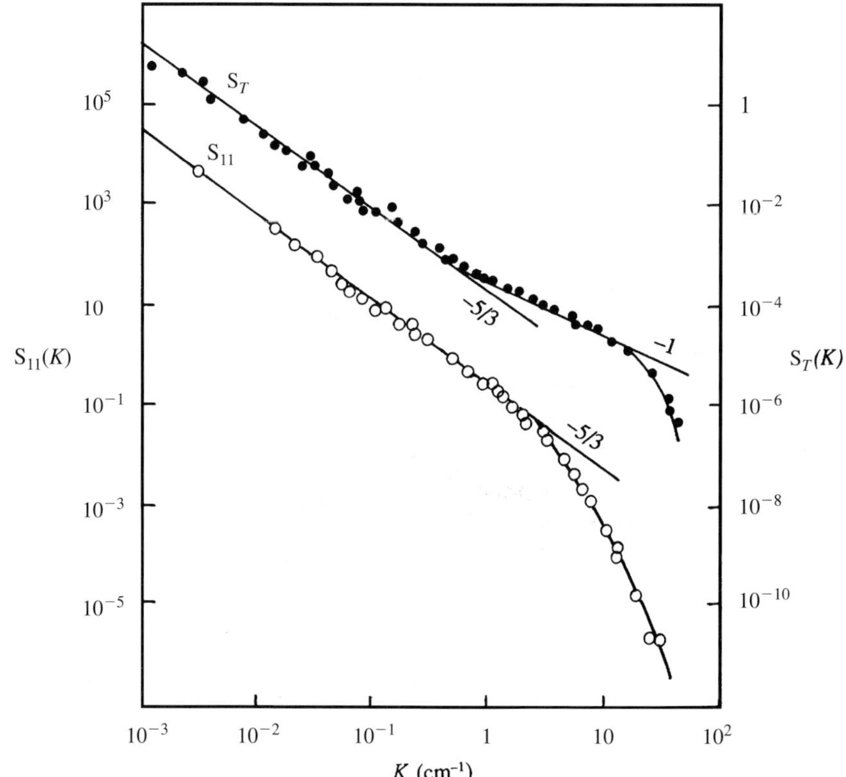

FIGURE 10.30 Temperature and velocity spectra measured by Grant et al. (1968). The measurements were made at a depth of 23 m in a tidal passage through islands near the coast of British Columbia, Canada. The wave number K is in cm^{-1}. Solid points represent S_T in $(°C)^2/cm^{-1}$, and open points represent S_{11} in $(cm/s)^2/cm^{-1}$. Powers of K that fit the observation are indicated by straight lines. *Phillips, O.M. (1997).* The Dynamics of the Upper Ocean, *reprinted with the permission of Cambridge University Press.*

At $z = 10$ m, the log-law portion of the mean-velocity profile is $(1/\kappa)\ln(z^+) + B = (1/0.41)\ln[10(0.13)/1.7 \times 10^{-5}] + 5.0 = 32.4$, and 5% of this is 1.62. Thus, the L_M that leads to 5% of the log-law contribution at this height is given by $5(10\text{ m})/[0.41|L_M|] = 1.62$, or $|L_M| = 5(10\text{ m})/[0.41(1.62)] = 75$ m. Similarly, at $z = 100$ m, the log-law portion of the mean-velocity profile is $(1/\kappa)\ln(z^+) + B = (1/0.41)\ln[100(0.13)/1.7 \times 10^{-5}] + 5.0 = 38.0$, and 5% of this is 1.90. Thus, the L_M that leads to 5% of the log-law contribution at this height is given by $5(100\ m)/[0.41|L_M|] = 1.90$ or, $|L_M| = 5(100\text{ m})/[0.41(1.90)] = 642$ m. Thus, achieving a log-law profile further from the earth's surface requires a larger Monin-Obukhov length, which is possible when u_* is high and $\overline{wT'}$ is small.

10.12 Taylor's theory of turbulent dispersion

The large mixing rate in a turbulent flow is due to the fact that the fluid particles wander away from their initial location. Taylor (1921) studied this problem and calculated the rate at which a particle disperses (i.e., moves away) from its initial location. The presentation here is directly adapted from his classic paper. He considered a point source emitting particles, say a chimney emitting smoke. The particles are emitted into a stationary and homogeneous turbulent medium in which the mean velocity is zero. Taylor used Lagrangian coordinates $\mathbf{X}(\mathbf{a}, t)$, which is the present location at time t of a particle that was at location $\boldsymbol{a}$ at time $t = 0$ (see Section 3.2). We shall take the point source to be the origin of coordinates and consider an ensemble of experiments in which we evaluate the location $\mathbf{X}(\mathbf{0}, t)$ at time t of all the particles that started from the origin (Fig. 10.31). For notational simplicity the first argument in $\mathbf{X}(\mathbf{0}, t)$ will be dropped from here on so that $\mathbf{X}(\mathbf{0}, t) = \mathbf{X}(t)$.

Rate of dispersion of a single particle

Consider the behavior of a single component of $\mathbf{X}$, say $X_\alpha(\alpha = 1, 2, \text{ or } 3)$. (Recall that a Greek subscript means that the summation convention is *not* followed.) The average rate at which the *magnitude* of X_α increases with time can be found

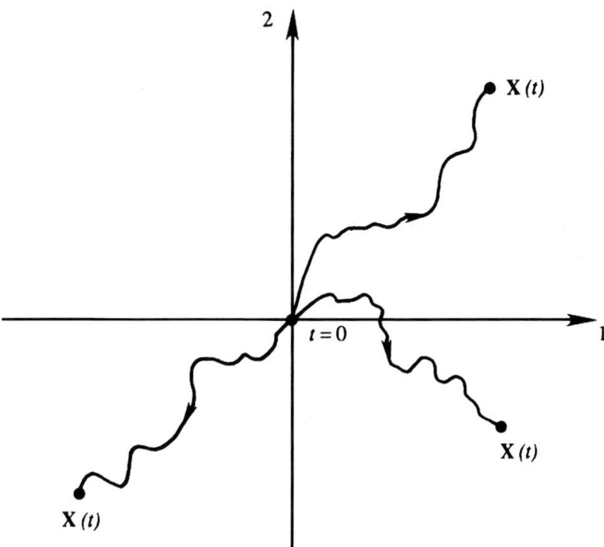

FIGURE 10.31 Three members of an ensemble of particle trajectories, $\mathbf{X}(t)$, at time t for particles released at the origin of coordinates at $t = 0$ in a turbulent flow with zero mean velocity. The distance traveled by the particles indicates how tracer particles disperse in a turbulent flow.

by finding $\overline{d(X_\alpha^2)/dt}$, where the over bar denotes an ensemble average and not a time average. We can write:

$$\frac{d}{dt}\left(\overline{X_\alpha^2}\right) = 2\overline{X_\alpha \frac{dX_\alpha}{dt}}, \tag{10.144}$$

where we have used the commutation rule (10.6). Defining $u_\alpha = dX_\alpha/dt$ as the *Lagrangian* velocity component of a fluid particle at time t, (10.144) becomes:

$$\frac{d}{dt}\left(\overline{X_\alpha^2}\right) = 2\overline{X_\alpha u_\alpha} = 2\overline{\left[\int_0^t u_\alpha(t')dt'\right] u_\alpha} = 2\int_0^t \overline{u_\alpha(t')u_\alpha(t)}dt', \tag{10.145}$$

where (10.7) has been used for averaging and integration, and the integration occurs along the particle's trajectory:

$$X_\alpha(t) = \int_0^t u_\alpha(t')dt',$$

which is valid when X_α and u_α are associated with the same particle. Because the flow is assumed to be stationary, $\overline{u_\alpha^2}$ is independent of time, and the autocorrelation of $u_\alpha(t)$ and $u_\alpha(t')$ is only a function of the time difference $t - t'$. Defining:

$$r_\alpha(\tau) = \overline{u_\alpha(t)u_\alpha(t+\tau)}/\overline{u_\alpha^2},$$

to be the autocorrelation coefficient of the Lagrangian velocity components of a particle, (10.145) becomes:

$$\frac{d}{dt}\left(\overline{X_\alpha^2}\right) = 2\overline{u_\alpha^2}\int_0^t r_\alpha(t'-t)dt' = 2\overline{u_\alpha^2}\int_0^t r_\alpha(\tau)d\tau, \tag{10.146}$$

via change in integration variable from t' to $\tau = t - t'$. Integrating (10.146) in time leads to:

$$\overline{X_\alpha^2}(t) = 2\overline{u_\alpha^2}\int_0^t \left(\int_0^{t'} r_\alpha(\tau)d\tau\right)dt', \tag{10.147}$$

which shows how the variance of the particle position changes with time.

Another useful form of Eq. (10.147) is obtained by integrating it by parts:

$$\int_0^t \left[\int_0^{t'} r_\alpha(\tau)d\tau\right]dt' = \left[t'\int_0^{t'} r_\alpha(\tau)d\tau\right]_{t'=0}^{t} - \int_0^t t' r_\alpha(t')dt'$$
$$= t\int_0^t r_\alpha(\tau)d\tau - \int_0^t t' r_\alpha(t')dt' = t\int_0^t \left(1-\frac{\tau}{t}\right) r_\alpha(\tau)d\tau,$$

which implies:

$$\overline{X_\alpha^2}(t) = 2\overline{u_\alpha^2}t\int_0^t \left(1-\frac{\tau}{t}\right) r_\alpha(\tau)d\tau. \tag{10.148}$$

Two limiting cases are examined in what follows.

Behavior for small t

If t is small compared to Λ_t = the integral time scale determined from the Lagrangian particle velocity correlation $r_\alpha(\tau)$, then $r_\alpha(\tau) \approx 1$ throughout the integral in (10.148) (see Fig. 10.5 for $\tau \ll \Lambda_t$). This circumstance leads to:

$$\overline{X_\alpha^2}(t) \cong \overline{u_\alpha^2}t^2. \tag{10.149}$$

Taking the square root of both sides, we obtain:

$$(X_\alpha)_{rms} = (u_\alpha)_{rms}t \quad \text{for} \quad t \ll \Lambda_t, \tag{10.150}$$

which shows that the *rms* displacement increases linearly with time and is proportional to the standard deviation of the turbulent fluctuations in the medium.

Behavior for large t

If t is large compared with Λ_t, then τ/t in (10.148) is negligible, and $\int_0^t r_\alpha(\tau)d\tau \approx \int_0^\infty r_\alpha(\tau)d\tau$ (see Fig. 10.5 for $\tau \gg \Lambda_t$). This a problem of turbulent dispersion of a patch of particles issuing from a point is given by:

$$\overline{X_\alpha^2}(t) = 2\overline{u_\alpha^2}\Lambda_t t, \quad \text{where} \quad \Lambda_t = \int_0^\infty r_\alpha(\tau)d\tau. \tag{10.151, 10.18}$$

Taking the square root of (10.151) gives:

$$(X_\alpha)_{rms} = (u_\alpha)_{rms}\sqrt{2\Lambda_t t} \quad \text{for} \quad t \gg \Lambda_t. \tag{10.152}$$

The $t^{1/2}$ behavior of (10.152) at large times is similar to the behavior in a *random walk*, in which the average distance traveled in a series of random (i.e., uncorrelated) steps increases as $t^{1/2}$. This similarity is due to the fact that for large t the fluid particles have *forgotten* their initial behavior at $t = 0$. In contrast, the small time behavior described by (10.121) is due to complete correlation, with *each realization* giving $X_\alpha \cong u_\alpha t$. The random walk concept is discussed in what follows.

Random walk

The description provided here is adapted from Feynman et al. (1963, pp. 6–5, 41–48). Imagine a person who starts walking in a random manner from the origin of coordinates, so that there is no correlation between the directions of two consecutive steps. Let the vector $\mathbf{R}_n$ represent the distance from the origin after n steps, and the vector $\mathbf{L}$ represent the nth step (Fig. 10.32). We assume that each step has the same magnitude L. Then:

$$\mathbf{R}_n = \mathbf{R}_{n-1} + \mathbf{L},$$

which gives:

$$R_n^2 = \mathbf{R}_n \cdot \mathbf{R}_n = (\mathbf{R}_{n-1} + \mathbf{L}) \cdot (\mathbf{R}_{n-1} + \mathbf{L}) = R_{n-1}^2 + L^2 + 2\mathbf{R}_{n-1} \cdot \mathbf{L}.$$

Averaging this equation leads to:

$$\overline{R_n^2} = \overline{R_{n-1}^2} + L^2 + 2\overline{\mathbf{R}_{n-1} \cdot \mathbf{L}}. \tag{10.153}$$

The last term is zero because there is no correlation between the direction of the nth step and the location reached after $n-1$ steps. Using rule (10.153) successively, we get:

$$\overline{R_n^2} = \overline{R_{n-1}^2} + L^2 = \overline{R_{n-2}^2} + 2L^2 = \overline{R_1^2} + (n-1)L^2 = nL^2.$$

The *rms* distance from the origin after n uncorrelated steps, each of length L, is therefore:

$$(R_n)_{rms} = L\sqrt{n}, \tag{10.154}$$

which is called a random walk.

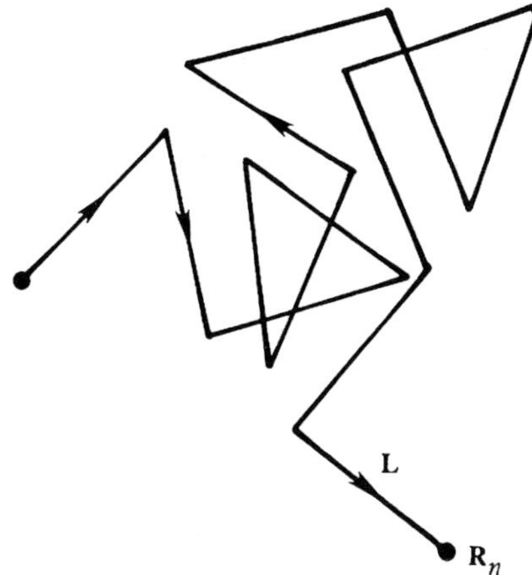

FIGURE 10.32 A sample realization of a random walk where the step length L is a uniform distance, but the step direction is random. After n steps, the vector distance from the starting point is $\mathbf{R}_n$. However, the root-mean-square distance from the starting point is only $L\sqrt{n}$ (not Ln) because many of the n steps lie in nearly opposite directions.

Behavior of a smoke plume in the wind

Taylor's analysis can be easily adapted to account for the presence of a constant mean velocity. Consider the dispersion of smoke into a wind blowing in the x-direction (Fig. 10.33). A photograph of the smoke plume, in which the film is exposed for a long time, would outline the average width Z_{rms}. As the x-direction in this problem is similar to time in Taylor's problem, the limiting behavior in (10.150) and (10.152) shows that the smoke plume is parabolic with a *pointed* vertex.

Turbulent diffusivity

An equivalent eddy diffusivity can be estimated from Taylor's analysis. The equivalence is based on considering the spreading of a concentrated line source in a fluid of *constant* diffusivity. What should the diffusivity be in order that the spreading rate equals that predicted by (10.146)? The problem of the sudden introduction of a line vortex of strength Γ (Exercise 7.39) is such a problem of diffusion from a concentrated line source. The tangential velocity in this flow is given by:

$$u_\theta = (\Gamma/2\pi r)\exp(-r^2/4\nu t).$$

The solution is therefore proportional to $\exp(-r^2/4\nu t)$, which has a Gaussian shape in the radial direction r, with a characteristic width of $\sigma = \sqrt{2\nu t}$. It follows that the momentum diffusivity ν in this problem is related to the variance σ^2 as:

$$\nu = (1/2)(d\sigma^2/dt), \tag{10.155}$$

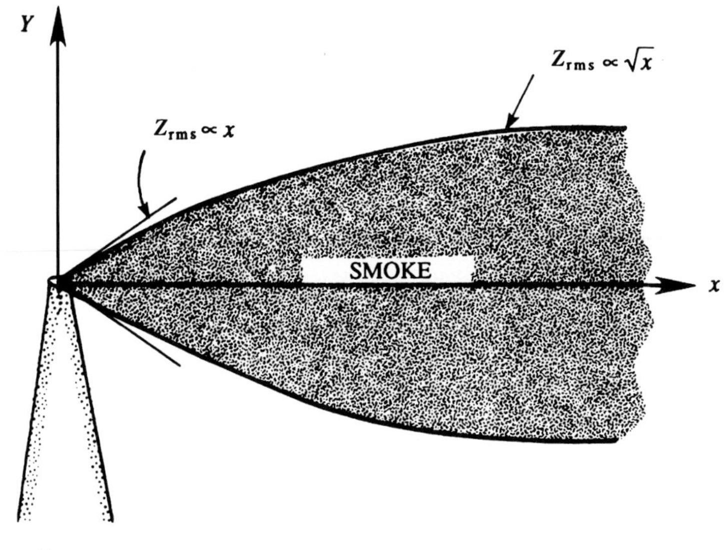

FIGURE 10.33 Average cross-sectional shape of a smoke plume in a turbulent wind blowing uniformly along the x-axis. Close to the chimney outlet, the *rms* width Z_{rms} of the smoke plume is proportional to x. Far from the chimney, Z_{rms} is proportional to $x^{1/2}$. *Taylor, G.I. (1921).* Proc. London Mathematical Society, *20, 196–211.*

which can be calculated if $\sigma^2(t)$ is known. Generalizing (10.155), the effective diffusivity D_T in a problem of turbulent dispersion of a patch of particles issuing from a point is given by:

$$D_T \equiv \frac{1}{2}\frac{d}{dt}\left(\overline{X_\alpha^2}\right) = \overline{u_\alpha^2}\int_0^t r_\alpha(\tau)d\tau, \tag{10.156}$$

where we have used (10.146). From (10.149) and (10.151), the two limiting cases of (10.156) are:

$$D_T \cong \overline{u_\alpha^2}t \quad \text{for} \quad t \ll \Lambda_t, \quad \text{and} \quad D_T \cong \overline{u_\alpha^2}\Lambda_t \quad \text{for} \quad t \gg \Lambda_t. \tag{10.157, 10.158}$$

Eq. (10.157) shows the interesting fact that the eddy diffusivity initially increases with time, a behavior different from that in molecular diffusion with constant diffusivity. This can be understood as follows. The dispersion (or separation) of particles in a patch is caused by eddies with scales less than or equal to the scale of the patch, since the larger eddies simply advect the patch and do not cause any separation of the particles. As the patch size becomes larger, an *increasing* range of eddy sizes is able to cause dispersion, giving $D_T \propto t$. This behavior shows that *it is frequently impossible to represent turbulent diffusion by means of a large but constant eddy diffusivity*. Turbulent diffusion does not behave like molecular diffusion. For large times, on the other hand, the patch size becomes larger than the largest eddies present, in which case the diffusive behavior becomes similar to that of molecular diffusion with a constant diffusivity given by (10.158).

Example 10.12

For the Lagrangian particle velocity correlation $r_\alpha(\tau) = \exp\{-|\tau|/T_u\}$, where T_u is a correlation time scale for a particle's velocity, how is T_u related to Λ_t and what is $\overline{X_\alpha^2}(t)$?

Solution Use the definition (10.18) to find:

$$\Lambda_t = \int_0^\infty r_\alpha(\tau)d\tau = \int_0^\infty \exp\{-\tau/T_u\}d\tau = T_u.$$

Thus, the velocity correlation time scale is equal to the integral time scale in this case, $T_u = \Lambda_t$. The variance of particle location can be determined from (10.148):

$$\overline{X_\alpha^2}(t) = 2\overline{u_\alpha^2}t\int_0^t\left(1-\frac{\tau}{t}\right)r_\alpha(\tau)d\tau = 2\overline{u_\alpha^2}\int_0^t(t-\tau)\exp\{-\tau/T_u\}d\tau,$$

for $t > 0$. Evaluating the integral leads to:

$$\overline{X_\alpha^2}(t) = 2\overline{u_\alpha^2}T_u t\left(1 + \frac{\exp\{-t/T_u\}}{t/T_u} - \frac{1}{t/T_u}\right).$$

For small times, $t/T_u \ll 1$, this reduces to $\overline{X_\alpha^2}(t) = \overline{u_\alpha^2}t^2$, and for large times $t/T_u \gg 1$ it simplifies to $\overline{X_\alpha^2}(t) = 2\overline{u_\alpha^2}T_u t$, as expected from (10.150) and (10.152).

Exercises

10.1 Determine general relationships for the second, third, and fourth central moments (variance $= \sigma^2$, skewness $= S$, and kurtosis $= K$) of the random variable u in terms of its first four ordinary moments: $\overline{u}$, $\overline{u^2}$, $\overline{u^3}$, and $\overline{u^4}$.

10.2 Calculate the mean, mean square, variance, and *rms* value (or standard deviation) of the periodic time series $u(t) = \overline{U} + U_0\cos(\omega t)$, where $\overline{U}$, U_0, and ω are positive real constants.

10.3 Show that the autocorrelation function $\overline{u(t)u(t+\tau)}$ of a periodic series $u = U\cos(\omega t)$ is itself periodic.

10.4 Calculate the zero-lag cross-correlation $\overline{u(t)v(t)}$ between two periodic series $u(t) = \cos(\omega t)$ and $v(t) = \cos(\omega t + \varphi)$ by performing at time average over one period $= 2\pi/\omega$. For values of $\phi = 0, \pi/4$, and $\pi/2$, plot the scatter diagrams of u vs v at different times, as in Fig. 10.9. Note that the plot is a straight line if $\phi = 0$, an ellipse if $\phi = \pi/4$, and a circle if $\phi = \pi/2$; the straight line, as well as the axes of the ellipse, are inclined at 45° to the uv-axes. Argue that the straight line signifies a perfect correlation, the ellipse a partial correlation, and the circle a zero correlation.

10.5 If $u(t)$ is a stationary random signal, show that $u(t)$ and $du(t)/dt$ are uncorrelated.

10.6 Let $R(\tau)$ and $S(\omega)$ be a Fourier transform pair. Show that $S(\omega)$ is real and symmetric if $R(\tau)$ is real and symmetric.

10.7 Compute the power spectrum, integral time scale, and Taylor time scale when $R_{11}(\tau) = \overline{u_1^2}\exp(-\alpha\tau^2)\cos(\omega_o\tau)$, assuming that α and ω_o are real positive constants.

10.8 Two formulae for the energy spectrum $S_e(\omega)$ of the stationary zero-mean signal $u(t)$ are:

$$S_e(\omega) = \frac{1}{2\pi}\int_{-\infty}^{+\infty} R_{11}(\tau)\exp\{-i\omega\tau\}d\tau \quad \text{and} \quad S_e(\omega) = \lim_{T\to\infty}\frac{1}{2\pi T}\left|\int_{-T/2}^{+T/2} u(t)\exp\{-i\omega t\}dt\right|^2.$$

Prove that these two are identical *without* requiring the existence of the Fourier transform of $u(t)$.

10.9 Derive the formula for the temporal Taylor microscale λ_t by expanding the definition of the temporal correlation function (10.17) into a two-term Taylor series and determining the time shift, $\tau = \lambda_t$, where this two-term expansion equals zero.

10.10 When x, r, and k_1 all lie in the stream-wise direction, the wave number spectrum $S_{11}(k_1)$ of the stream-wise velocity fluctuation $u_1(x)$ defined by (10.45) can be interpreted as a distribution function for energy across stream-wise wave number k_1. Show that the energy-weighted mean-square value of the stream-wise wave number is:

$$\overline{k_1^2} \equiv \frac{1}{\overline{u^2}}\int_{-\infty}^{+\infty} k_1^2 S_{11}(k_1)dk_1 = -\frac{1}{\overline{u^2}}\left[\frac{d^2}{dr^2}R_{11}(r)\right]_{r=0}, \quad \text{and that } \lambda_f = \sqrt{2\Big/\overline{k_1^2}}.$$

10.11 In many situations, measurements are only possible of one velocity component at one point in a turbulent flow, but consider a flow that has a nonzero mean velocity and moves past the measurement point. Thus, the experimenter obtains a time history of $u_1(t)$ at fixed point. In order to estimate spatial velocity gradients, Taylor's frozen-turbulence hypothesis can be invoked to estimate a spatial gradient from a time derivative: $\frac{\partial u_1}{\partial x_1} \approx -\frac{1}{U_1}\frac{\partial u_1}{\partial t}$ where the "1"-axis must be aligned with the direction of the average flow, i.e., $U_i = (U_1, 0, 0)$. Show that this approximate relationship is true when $\sqrt{\overline{u_i u_i}}/U_1 \ll 1$, $p \sim \rho u_1^2$, and Re is high enough to neglect the influence of viscosity.

10.12 **a)** Starting from (10.33), derive (10.34) via an appropriate process of Reynolds decomposition and ensemble averaging.
b) Determine an equation for the scalar fluctuation energy $= \frac{1}{2}\overline{Y'^2}$, one-half the scalar variance.
c) When the scalar variance goes to zero, the fluid is well mixed. Identify the term in the equation from part b) that dissipates scalar fluctuation energy.

10.13 Measurements in an atmosphere at 20 °C show an *rms* vertical velocity of $w_{rms} = 1$ m/s and an *rms* temperature fluctuation of $T_{rms} = 0.1$ °C. If the correlation coefficient is 0.5, calculate the heat flux $\rho c_p\overline{wT'}$.

10.14 **a)** Compute the divergence of the constant-density Navier-Stokes momentum equation $\frac{\partial u_i}{\partial t} + u_j \frac{\partial u_i}{\partial x_j} = -\frac{1}{\rho}\frac{\partial p}{\partial x_i} + \nu \frac{\partial^2 u_i}{\partial x_j \partial x_j}$ to determine a Poisson equation for the pressure.

b) If the equation $\frac{\partial^2 G}{\partial x_j \partial x_j} = \delta\left(x_j - \tilde{x}_j\right)$ has solution: $G(x_j, \tilde{x}_j) = -\left(4\pi\sqrt{(x_j - \tilde{x}_j)^2}\right)^{-1}$, then use this and the result from part a) to show that the equation for the average pressure $P(x_j)$ in turbulent flow is: $P(x_j) = \frac{\rho}{4\pi}\int_{\tilde{x}} \frac{1}{\sqrt{(x_j - \tilde{x}_j)^2}} \frac{\partial^2}{\partial \tilde{x}_j \partial \tilde{x}_i}(U_i U_j + \overline{u_i u_j}) d^3\tilde{x}.$

10.15 Starting with the RANS momentum equation (10.30), derive the equation for the kinetic energy of the average flow field (10.46).

10.16 Derive the RANS transport equation for the Reynolds stress correlation (10.35) via the following steps.

a) By subtracting (10.30) from (4.86), show that the instantaneous momentum equation for the fluctuating turbulent velocity u_i is:

$$\frac{\partial u_i}{\partial t} + u_k \frac{\partial U_i}{\partial x_k} + U_k \frac{\partial u_i}{\partial x_k} + u_k \frac{\partial u_i}{\partial x_k} = -\frac{1}{\rho_0}\frac{\partial p}{\partial x_i} + \nu \frac{\partial^2 u_i}{\partial x_k^2} + g\alpha T' \delta_{i3} + \frac{\partial}{\partial x_k}\overline{u_i u_k}.$$

b) Show that: $\overline{u_i \frac{Du_j}{Dt} + u_j \frac{Du_i}{Dt}} = \frac{\partial \overline{u_i u_j}}{\partial t} + U_k \frac{\partial \overline{u_i u_j}}{\partial x_k} + \frac{\partial \overline{u_i u_j u_k}}{\partial x_k} + \overline{u_k u_j}\frac{\partial U_i}{\partial x_k} + \overline{u_i u_k}\frac{\partial U_j}{\partial x_k}.$

c) Combine and simplify the results of parts a) and b) to reach (10.35).

10.17 In two dimensions, the RANS equations for constant-viscosity constant-density turbulent boundary-layer flow are:

$$\frac{\partial U}{\partial x} + \frac{\partial V}{\partial y} = 0, \quad U\frac{\partial U}{\partial x} + V\frac{\partial U}{\partial y} \cong -\frac{1}{\rho}\frac{\partial}{\partial x}\left(P + \rho\overline{u^2}\right) + \frac{\partial}{\partial y}\left(\nu\frac{\partial U}{\partial y} - \overline{uv}\right),$$

$$\text{and} \quad 0 \cong -\frac{\partial}{\partial y}\left(P + \rho\overline{v^2}\right),$$

where x & y are the stream-wise and wall-normal coordinates, U & V are the average stream-wise and wall-normal velocity components, u & v are the stream-wise and wall normal velocity fluctuations, P is the average pressure, and an overbar denotes a time average.

a) Assume that the fluid velocity $U_e(x)$ above the turbulent boundary layer is steady and not turbulent so that the average pressure, P_e, at the upper edge of the boundary layer can be determined from the simple Bernoulli equation: $P_e + \frac{1}{2}\rho U_e^2 = const$. Use this assumption, the given Bernoulli equation, and the wall-normal momentum equation to show that:

$$-\frac{1}{\rho}\frac{\partial P}{\partial x} = U_e \frac{dU_e}{dx} + \frac{\partial \overline{v^2}}{\partial x}.$$

b) Use the part a) result, the continuity equation, and the stream-wise momentum equation to derive the turbulent-flow von Karman boundary-layer momentum-integral equation:

$$\frac{\tau_w}{\rho} = \frac{d}{dx}(U_e^2 \overline{\theta}) + U_e \overline{\delta^*}\frac{dU_e}{dx} + \frac{d}{dx}\int_0^\infty \left(\overline{v^2} - \overline{u^2}\right) dy,$$

where: $\overline{\delta^*} = \int_0^\infty \left(1 - \frac{U}{U_e}\right) dy$, and $\overline{\theta} = \int_0^\infty \frac{U}{U_e}\left(1 - \frac{U}{U_e}\right) dy$. In practice, the final term is typically small enough to ignore, but the efforts here should include it.

10.18 Consider turbulent flow of a fluid with constant density ρ inside a sealed stationary container with constant volume V and surface area A. The location (X_j) of the center of mass of the fluid in the container is:

$$X_j = \frac{1}{\rho V}\int_V r_j dm = \frac{1}{V}\int_V r_j dV,$$

where $dm = \rho dV$ is a fluid mass element, and r_j is the position vector from the origin to dm. For this problem, use the following Reynolds decomposition for the fluid velocity $u_j = U_j + u'_j$, where:

$$U_j(t) = \frac{1}{V}\int_V u_j dV$$

is the volume-averaged fluid velocity, and u'_j is the fluid velocity fluctuation.

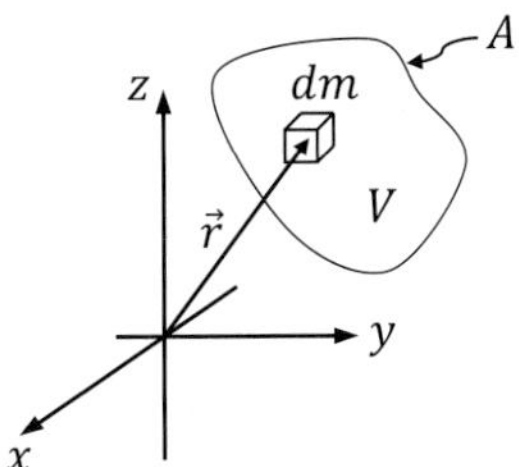

a) Show that $U_j = 0$ [This means $u_j = u'_j$ so 'primes' are dropped below in parts b)-e)].

b) Use the part a) result and the following version of the Cauchy momentum equation,

$$\rho\frac{\partial u_j}{\partial t} + \rho\frac{\partial}{\partial x_i}\left(u_i u_j\right) = \frac{\partial}{\partial x_i}T_{ij},$$

where T_{ij} is the stress tensor, to show that $\int_A T_{ij} n_j dA = 0$, i.e. there is no net surface force on the fluid in the volume.

c) For an incompressible Newtonian fluid with constant viscosity μ, the stress tensor is:

$$T_{ij} = -p\delta_{ij} + 2\mu S_{ij},$$

where p is the pressure and S_{ij} is the strain-rate tensor. Substitute this T_{ij} into the Cauchy momentum equation above, multiply the result by u_j, and modify terms to reach the following equation for the turbulent kinetic energy per unit mass, $e = (u_j u_j)/2$:

$$\rho\frac{\partial e}{\partial t} + \rho\frac{\partial}{\partial x_i}(eu_i) = -\frac{\partial}{\partial x_j}(u_j p) + 2\mu u_j\frac{\partial S_{ij}}{\partial x_i}.$$

d) Volume average the result of part c), and use the viscous-fluid velocity boundary conditions on the surface A to show that:

$$\frac{d\bar{e}}{dt} = -\int_V S_{ij} S_{ij} dV,$$

where $\bar{e}$ is the volume average of e.

e) If the turbulence in the container has energy $\bar{e}_0$ at $t = 0$, will $\bar{e}$ increase, decrease, or stay the same as t increases?

10.19 In large-eddy simulation (LES), large-scale motions are resolved and smaller scale motions are modeled. The computational cost thus lies between RANS and DNS. Similar to Reynolds decomposition, the velocity field is decomposed into $\tilde{u}_i = \widehat{u}_i + u'_i$, where the filtered component is obtained by the convolution

$$\widehat{u}_i(\mathbf{x}, t) \equiv \int G(\mathbf{r})\tilde{u}_i(\mathbf{x} - \mathbf{r}, t)\, d\mathbf{r}.$$

Here, $G(\mathbf{r}) > 0$ is a homogeneous low-pass filter that satisfies $\int G(\mathbf{r})\, d\mathbf{r} = 1$.

a) Show that the filtered continuity and momentum equations for a constant density flow are

$$\frac{\partial\widehat{u}_i}{\partial x_i} = 0 \quad \text{and} \quad \frac{\partial\widehat{u}_i}{\partial t} + \widehat{u}_j\frac{\partial\widehat{u}_i}{\partial x_j} = -\frac{1}{\rho}\frac{\partial\widehat{p}}{\partial x_i} + \nu\frac{\partial^2\widehat{u}_i}{\partial x_j^2} - \frac{\partial\tau_{ij}^R}{\partial x_j},$$

where $\widehat{p}$ is the filtered pressure and $\tau_{ij}^R = \widehat{\tilde{u}_i\tilde{u}_j} - \widehat{u}_i\widehat{u}_j$ is the *residual stress tensor*.

b) Although the solution from part a) closely resembles the Reynolds-averaged momentum equation (10.30), an important difference is that the filtered velocity $\widehat{u}_i$ is a random field so that $\widehat{\widehat{u}}_i \neq \widehat{u}_i$. Show that in general $\widehat{u'_i} \neq 0$.

c) Show that the residual stress tensor can be decomposed into

$$\tau_{ij}^R \equiv L_{ij} + C_{ij} + R_{ij},$$

where $L_{ij} \equiv \widehat{\widehat{u}_i\widehat{u}_j} - \widehat{u}_i\widehat{u}_j$ are the *Leonard stresses*, $C_{ij} \equiv \widehat{\widehat{u}_i u'_j} + \widehat{u'_i\widehat{u}_j}$ are the *cross stresses*, and $R_{ij} \equiv \widehat{u'_i u'_j}$ are the sub-filtered *Reynolds stresses*.

10.20 Starting from (10.38) and (10.40), set $\mathbf{r} = r\mathbf{e}_1$ and use $R_{11} = \overline{u^2} f(r)$, and $R_{22} = \overline{u^2} g(r)$, to show that $F(r) = \overline{u^2}(f(r) - g(r))r^{-2}$ and $G(r) = \overline{u^2} g(r)$.

10.21 **a)** Starting with R_{ij} from (10.39), compute $\partial R_{ij}/\partial r_j$ for incompressible flow.

b) For homogeneous-isotropic turbulence use the part a) result to show that the longitudinal, $f(r)$, and transverse, $g(r)$, correlation functions are related by $g(r) = f(r) + (r/2)(df(r)/dr)$.

c) Use part b), and the integral length scale and Taylor microscale definitions to find $2\Lambda_g = \Lambda_f$ and $\sqrt{2}\lambda_g = \lambda_f$.

10.22 In homogeneous turbulence: $R_{ij}(\mathbf{r}_b - \mathbf{r}_a) = \overline{u_i(\mathbf{x}+\mathbf{r}_a)u_j(\mathbf{x}+\mathbf{r}_b)} = R_{ij}(\mathbf{r})$, where $\mathbf{r} = \mathbf{r}_b - \mathbf{r}_a$.

a) Show that $\overline{(\partial u_i(\mathbf{x})/\partial x_k)(\partial u_j(\mathbf{x})/\partial x_l)} = -(\partial^2 R_{ij}/\partial r_k \partial r_l)_{r=0}$.

b) If the flow is incompressible and isotropic, show that

$$-\overline{(\partial u_1(\mathbf{x})/\partial x_1)^2} = -\frac{1}{2}\overline{(\partial u_1(\mathbf{x})/\partial x_2)^2} = +2\overline{(\partial u_1(\mathbf{x})/\partial x_2)(\partial u_2(\mathbf{x})/\partial x_1)} = \overline{u'^2}(d^2 f/dr^2)_{r=0}.$$

[*Hint*: expand $f(r)$ about $r = 0$ *before* taking any derivatives.]

10.23 The turbulent kinetic energy equation contains a pressure-velocity correlation, $K_j = \overline{p(\mathbf{x})u_j(\mathbf{x}+\mathbf{r})}$. In homogeneous isotropic turbulent flow, the most general form of this correlation is: $K_j = K(r)r_j$. If the flow is also incompressible, show that $K(r)$ must be zero.

10.24 The velocity potential for two-dimensional water waves of small amplitude ξ_o on a deep pool can be written: $\phi(x_1, x_2, t) = (\omega\xi_o/k)e^{+kx_2}\cos(\omega t - kx_1)$, where x_1 and x_2 are the horizontal and vertical coordinates with $x_2 = 0$ defining the average free surface. Here, ω is the temporal radian frequency of the waves and k is their wave number.

a) Compute the two-dimensional velocity field: $\mathbf{u} = (u_1, u_2) = (\partial\phi/\partial x_1, \partial\phi/\partial x_2)$.

b) Show that this velocity field is a solution of the two-dimensional continuity and Navier-Stokes equations for incompressible fluid flow.

c) Compute the strain-rate tensor $S_{ij} = \frac{1}{2}(\partial u_i/\partial x_j + \partial u_j/\partial x_i)$.

d) Although this flow is not turbulent, it must still satisfy the turbulent kinetic energy equation that contains an energy dissipation term. Denote the kinematic viscosity by ν, and compute the kinetic energy dissipation rate in this flow: $\varepsilon = 2\nu\overline{S_{ij}S_{ij}}$, where the over bar implies a time average over one wave period ($= 2\pi/\omega$.) Only time averages of even powers of the trig-functions are non-zero, for example: $\overline{\cos^2(\omega t - kx)} = \overline{\sin^2(\omega t - kx)} = 1/2$ while $\overline{\cos(\omega t - kx)} = \overline{\sin(\omega t - kx)} = 0$.

e) The original potential does not include any viscous effects. Explain how this situation can occur when the kinetic-energy dissipation rate is not zero.

10.25 Consider homogeneous turbulence containing a large number of solid particles. Unlike in single-phase turbulence, velocity fluctuations may originate at small scales from particle wakes, and is thus termed *pseudo-turbulence*. The goal of this problem is to rederive the energy-budget (10.47) for turbulent dispersed two-phase flows.

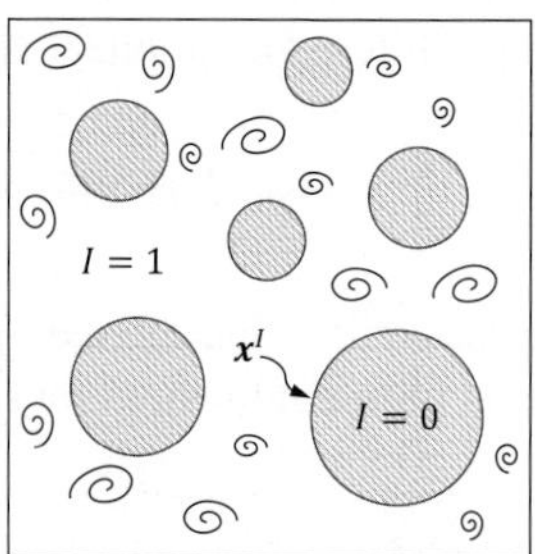

a) Define an indicator function $I(\mathbf{x}, t)$ such that $I = 1$ in the fluid and $I = 0$ inside a particle. Show that the expectation $\langle I(\mathbf{x}, t)\rangle = \epsilon$, where ϵ is the mean fluid-phase volume fraction.

b) Starting from the momentum equation for an incompressible, constant-density fluid:

$$\rho\left(\frac{\partial \tilde{u}_i}{\partial t}+\tilde{u}_j\frac{\partial \tilde{u}_i}{\partial x_j}\right)=-\frac{\partial \tilde{p}}{\partial x_i}+\mu\frac{\partial^2 \tilde{u}_i}{\partial x_j^2},$$

decompose the velocity into a mean and fluctuation: $\tilde{u}_i=\overline{u}_i+u_i$. Here, the over bar denotes a *phase average*: $\overline{()}=\langle I()\rangle/\langle I\rangle$. Then multiply each term in the equation by Iu_i. Show that the for homogeneous flow the left-hand side reduces to $\rho\epsilon\partial\overline{e}_f/\partial t$, where $\overline{e}_f=\frac{1}{2}\langle Iu_i^2\rangle/\langle I\rangle$ is the phase-averaged kinetic energy of the turbulent velocity fluctuations (within the fluid).

c) Repeat part b) for the right-hand side of the momentum equation and show that the energy budget reduces to:

$$\rho\epsilon\frac{d\overline{e}_f}{dt}=\underbrace{\left\langle u_iT'_{ij}n_j\delta\left(\mathbf{x}-\mathbf{x}^I\right)\right\rangle}_{\text{Gain from interphase momentum transfer}}-\underbrace{2\mu\epsilon\overline{S'_{ij}S'_{ij}}}_{\text{Viscous dissipation}}$$

where $T'_{ij}=-p\delta_{ij}+2\mu S'_{ij}$ is the fluctuation stress tensor, $S'_{ij}=\frac{1}{2}(\partial u_i/\partial x_j+\partial u_j/\partial x_i)$ is the fluctuation strain-rate tensor, n_i is the unit normal vector pointing outward from the particle, and $\delta(\mathbf{x}-\mathbf{x}^I)$ is the Dirac delta function that is non-zero at the fluid-particle interface $\mathbf{x}^I$. [*Hint*: Use the divergence theorem to find $\partial I/\partial x_i=-n_i\delta(\mathbf{x}-\mathbf{x}^I)$.]

10.26 A mass of 10 kg of water is stirred by a mixer. After one hour of stirring, the temperature of the water rises by 1.0 °C. What is the power output of the mixer in watts? What is the size η of the dissipating eddies?

10.27 A direct numerical simulation (DNS) resolves all of the scales of motion present in a turbulent flow. Consider a turbulent flow simulated in a cubic domain of length $\mathcal{L}\approx L$ discretized using N grid points in each direction.

a) Using (10.49) and (10.50), show that the total number of grid points required for DNS scales like

$$N^3\sim \mathrm{Re}_L^{9/4}.$$

b) In order to advance the simulation in time accurately, the time step size Δt must be limited by the *Courant condition*: $\Delta t<\Delta x/\max|\mathbf{u}|$. In other words, the time step size much be small enough to prevent a fluid particle from traveling more than the length of the grid cell over that time. Show that the total computational cost to simulate one large-eddy turnover time scales like

$$MN^3\sim \mathrm{Re}_L^3,$$

where M is the total number of time steps.

10.28 In addition to correlations and spectra, the spatial structure of velocity fluctuations are often characterized by the velocity structure function. The second-order velocity structure function: $D_{ij}(\mathbf{x},\mathbf{r},t)=\langle(u_i(\mathbf{x}+\mathbf{r},t)-u_j(\mathbf{x},t))^2\rangle$, represents the correlation between the difference in velocity at two points separated by a distance $\mathbf{r}$ at time t. Assuming sufficiently large Reynolds numbers, use the Kolmogorov hypotheses to show that for homogeneous isotropic turbulence in the inertial sub-range ($\eta\ll r\ll L$), the longitudinal structure function, $D_{11}(\mathbf{e}_1r,t)$, scales like

$$D_{11}(\mathbf{e}_1r,t)=C\left(\overline{\varepsilon}r\right)^{2/3},$$

where C is a constant.

10.29 In locally isotropic turbulence, Kolmogorov determined that the wave number spectrum can be represented by $S_{11}(k)/(\nu^5\overline{\varepsilon})^{1/4}=\Phi(k\nu^{3/4}/\overline{\varepsilon}^{1/4})$ in the inertial sub-range and dissipation range of turbulent scales, where Φ is an undetermined function.

a) Determine the equivalent form for the temporal spectrum $S_e(\omega)$ in terms of the average kinetic energy dissipation rate $\overline{\varepsilon}$, the fluid's kinematic viscosity ν, and the temporal frequency ω.

b) Simplify the results of part a) for the inertial range of scales where ν is dropped from the dimensional analysis.

c) To obtain the results for parts a) and b), an implicit assumption has been made that leads to the neglect of an important parameter. Add the missing parameter and redo the dimensional analysis of part a).

d) Use the missing parameter and ω to develop an equivalent wave number. Insist that your result for S_e only depends on this equivalent wave number and $\overline{\varepsilon}$ to recover the minus-five-thirds law.

10.30 [1]Estimates for the importance of anisotropy in a turbulent flow can be developed by assuming that fluid velocities and spatial derivatives of the average-flow (or RANS) equation are scaled by the average velocity difference ΔU that drives the largest eddies in the flow having a size L, and that the fluctuating velocities and spatial derivatives in the turbulent kinetic energy (TKE) equation are scaled by the kinematic viscosity ν and the Kolmogorov scales η and u_K (see (10.50)). Thus, the scaling for a mean-velocity gradient is: $\partial U_i/\partial x_j \sim \Delta U/L$, while the mean-square turbulent velocity gradient scales as: $\overline{(\partial u_i/\partial x_j)^2} \sim (u_K/\eta)^2 = \nu^2/\eta^4$, where the "$\sim$" sign means "scales as." Use these scaling ideas in parts a) and d):

a) The total energy dissipation rate in a turbulent flow is $2\nu\overline{S}_{ij}\overline{S}_{ij} + 2\nu\overline{S'_{ij}S'_{ij}}$, where $\overline{S}_{ij} = \frac{1}{2}\left(\frac{\partial U_i}{\partial x_j} + \frac{\partial U_j}{\partial x_i}\right)$ and $S'_{ij} = \frac{1}{2}\left(\frac{\partial u_i}{\partial x_j} + \frac{\partial u_j}{\partial x_i}\right)$. Determine how the ratio $\frac{\overline{S'_{ij}S'_{ij}}}{\overline{S}_{ij}\overline{S}_{ij}}$ depends on the outer-scale Reynolds number: $\mathrm{Re}_L = \Delta U \cdot L/\nu$.

b) Is average-flow or fluctuating-flow energy dissipation rate more important?

c) Show that the turbulent kinetic energy dissipation rate, $\overline{\varepsilon} = 2\nu\overline{S'_{ij}S'_{ij}}$ can be written:

$$\overline{\varepsilon} = \nu\left[\overline{\frac{\partial u_i}{\partial x_j}\frac{\partial u_i}{\partial x_j}} + \frac{\partial^2}{\partial x_i \partial x_j}\overline{u_i u_j}\right].$$

d) For homogeneous isotropic turbulence, the second term in the result of part c) is zero but it is non-zero in a turbulent shear flow. Therefore, estimate how $\frac{\partial^2}{\partial x_i \partial x_j}\overline{u_i u_j} \Big/ \overline{\frac{\partial u_i}{\partial x_j}\frac{\partial u_i}{\partial x_j}}$ depends on Re_L in turbulent shear flow as a means of assessing how much impact anisotropy has on the turbulent kinetic energy dissipation rate.

e) Is an isotropic model for the turbulent dissipation appropriate at high Re_L in a turbulent shear flow?

10.31 Determine the self-preserving form of the average stream-wise velocity $U_z(z, R)$ of a round turbulent jet using cylindrical coordinates where z increases along the jet axis and R is the radial coordinate. Ignore gravity in your work. Denote the density of the nominally quiescent reservoir fluid by ρ.

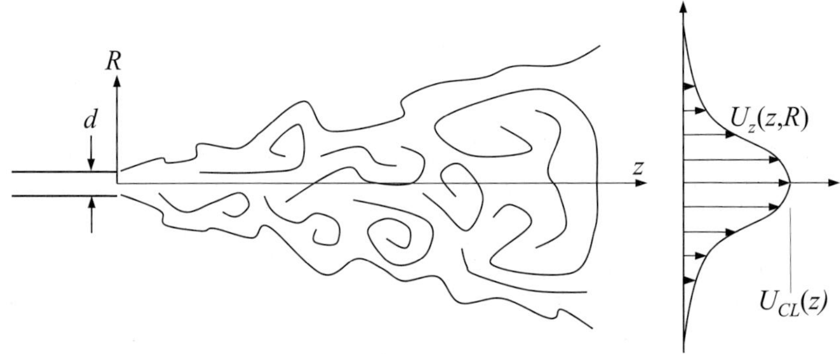

a) Place a stationary cylindrical control volume around the jet's cone of turbulence so that circular control surfaces slice all the way through the jet flow at its origin and at a distance z downstream where the fluid density is ρ. Assuming that the fluid outside the jet is nearly stationary so that pressure does not vary in the axial direction and that the fluid entrained into the volume has negligible z-direction momentum, show

$$J_0 \equiv \int_0^{d/2} \rho_0 U_0^2 2\pi R dR = \int_0^{D/2} \rho_0 U_z^2(z, R) 2\pi R dR,$$

where J_0 is the jet's momentum flux, ρ_0 is the density of the jet fluid, U_0 is the jet exit velocity, and D is the diameter of the jet's cone of turbulence.

b) Simplify the exact mean-flow equations

$$\frac{\partial U_z}{\partial z} + \frac{1}{R}\frac{\partial}{\partial R}(RU_R) = 0,$$

1. Obtained from Prof. Werner Dahm.

$$U_z \frac{\partial U_z}{\partial z} + U_R \frac{\partial U_z}{\partial R} = -\frac{1}{\rho}\frac{\partial P}{\partial z} + \frac{\upsilon}{R}\frac{\partial}{\partial R}\left(R\frac{\partial U_z}{\partial R}\right) - \frac{1}{R}\frac{\partial}{\partial R}\left(R\overline{u_z u_R}\right) - \frac{\partial}{\partial z}\left(R\overline{u_z^2}\right),$$

when $\partial P/\partial z \approx 0$, the jet is slender enough for the boundary-layer approximation $\partial/\partial R \gg \partial/\partial z$ to be valid, and the flow is at high Reynolds number so that the viscous terms are negligible.

c) Eliminate the average radial velocity from the simplified equations to find:

$$U_z \frac{\partial U_z}{\partial z} - \left\{\frac{1}{R}\int_0^R \mathcal{R}\frac{\partial U_z}{\partial z} d\mathcal{R}\right\}\frac{\partial U_z}{\partial R} = -\frac{1}{R}\frac{\partial}{\partial R}\left(R\overline{u_z u_R}\right),$$

where $\mathcal{R}$ is just an integration variable.

d) Assume a similarity form: $U_z(z, R) = U_{CL}(z) f(\xi)$, $-\overline{u_z u_R} = \Psi(z) g(\xi)$, where $\xi = R/\delta(z)$ and f and g are undetermined functions, use the results of parts a) and c), and choose constant values appropriately to find $U_z(z, R) = const.(J_0/\rho)^{1/2} z^{-1} f(R/z)$.

e) Determine a formula for the volume flux in the jet. Will the jet fluid from the nozzle be diluted with increasing z?

10.32 Consider the turbulent wake far from a two-dimensional body placed perpendicular to the direction of a uniform flow.

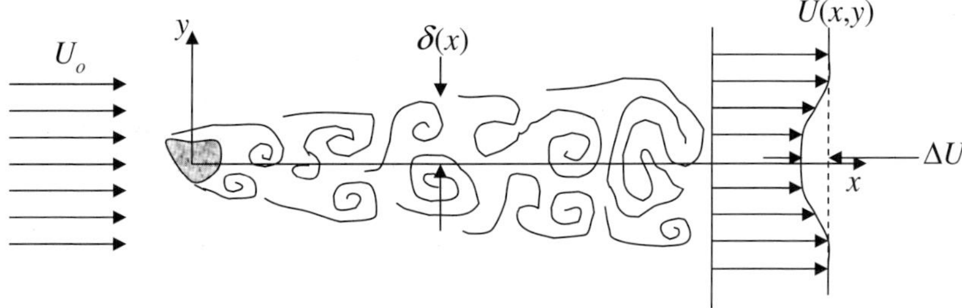

Using the notation defined in the figure, the result of Example 10.5 may be written:

$$\frac{\overline{F}_D/l}{\rho U_o^2} = \theta = \int_{-\infty}^{+\infty}\left[\frac{U(x, y)}{U_o}\left(1 - \frac{U(x, y)}{U_o}\right) + \frac{\overline{v^2} - \overline{u^2}}{U_o^2}\right] dy,$$

where θ is the momentum thickness of the wake flow (a constant), and $U(x, y)$ is the average horizontal velocity profile a distance x downstream of the body.

a) When $\Delta U \ll U_o$, find the conditions necessary for a self-similar form for the wake's velocity deficit, $U(x, y) = U_o - \Delta U(x) f(\xi)$, to be valid based on the equation above and the steady two-dimensional continuity and boundary-layer RANS equations. Here, $\xi = y/\delta(x)$ and δ is the transverse length scale of the wake.

b) Determine how ΔU and δ must depend on x in the self-similar region. State your results in appropriate dimensionless form using θ and U_o as appropriate.

10.33 Consider the two-dimensional shear layer that forms between two steady streams with flow speed U_2 above and U_1 below $y = 0$, that meet at $d = 0$, as shown. Assume a self-similar form for the average horizontal velocity:

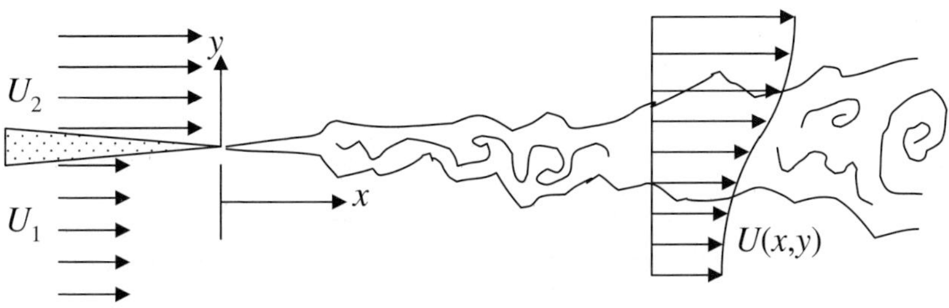

$$U(x, y) = U_1 + (U_2 - U_1) f(\xi) \quad \text{with} \quad \xi = y/\delta(x). \tag{10.159}$$

a) What are the boundary conditions on $f(\xi)$ as $y \to \pm\infty$?

b) If the flow is laminar, use $\frac{\partial U}{\partial x}+\frac{\partial V}{\partial y}=0$ and $U\frac{\partial U}{\partial x}+V\frac{\partial U}{\partial y}=\nu\frac{\partial^2 U}{\partial y^2}$ with $\delta(x)=\sqrt{\nu x/U_1}$ to obtain a single equation for $f(\xi)$. There is no need to solve this equation.

c) If the flow is turbulent, use: $\frac{\partial U}{\partial x}+\frac{\partial V}{\partial y}=0$ and $U\frac{\partial U}{\partial x}+V\frac{\partial U}{\partial y}=-\frac{\partial}{\partial y}(\overline{uv})$ with $-\overline{uv}=(U_2-U_1)^2 g(\xi)$ to obtain a single equation involving f and g. Determine how δ must depend on x for the flow to be self-similar.

d) Does the laminar or the turbulent mixing layer grow more quickly as x increases?

10.34 Consider an orifice of diameter d that emits an incompressible fluid of density ρ_o at speed U_o into an infinite half space of fluid with density ρ_∞. With gravity acting and $\rho_\infty > \rho_o$, the orifice fluid rises, mixes with the ambient fluid, and forms a buoyant plume with a diameter $D(z)$ that grows with increasing height above the orifice. Assuming that the plume is turbulent and self-similar in the far-field ($z \gg d$), determine how the plume diameter D, the mean centerline velocity U_{CL}, and the mean centerline mass fraction of orifice fluid Y_{CL} depend on the vertical coordinate z via the steps suggested below. Ignore the initial momentum of the orifice fluid. Use both dimensional and control-volume analysis as necessary. Ignore stream-wise turbulent fluxes to simplify your work. Assume uniform flow from the nozzle.

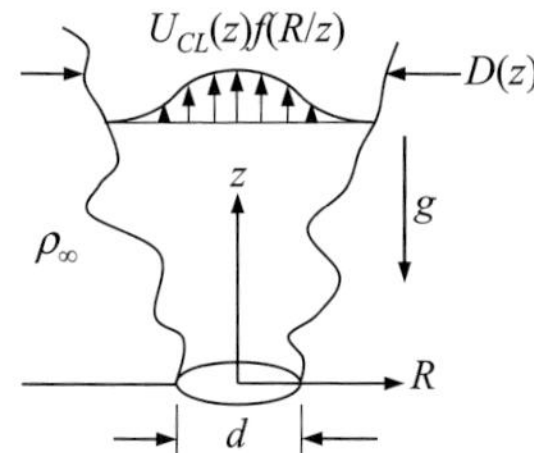

a) Place a stationary cylindrical control volume around the plume with circular control surfaces that slice through the plume at its origin and at height z. Use similarity forms for the average vertical velocity $U_z(z,R)=U_{CL}(z)f(R/z)$ and nozzle fluid mass fraction $\overline{Y}(z,R)=(\rho_\infty-\overline{\rho})/(\rho_\infty-\rho_o)=Y_{CL}(z)h(R/z)$ to conserve the flux of nozzle fluid in the plume, and find: $\dot{m}_o=\int_{source}\rho_o U_o dA=\int_0^{D/2}\rho_o\overline{Y}(z,R)U_z(z,R)2\pi R dR$.

b) Conserve vertical momentum using the same control volume assuming that all entrained fluid enters with negligible vertical momentum, to determine:

$$-\int_{source}\rho_o U_o^2 dA+\int_0^{D/2}\overline{\rho}(z,R)U_z^2(z,R)2\pi R dR=\int_{volume} g[\rho_\infty-\overline{\rho}(z,R)]dV,$$

where $\overline{\rho}=\overline{Y}\rho_o+(1-\overline{Y})\rho_\infty$.

c) Ignore the source momentum flux, assume z is large enough so that $Y_{CL}\ll 1$, and use the results of parts a) and b) to find: $U_{CL}(z)=C_1\cdot\sqrt[3]{B/\rho_\infty z}$ and $((\rho_\infty-\rho_o)/\rho_\infty)Y_{CL}(z)=C_2\sqrt[3]{B^2/g^3\rho_\infty^2 z^5}$, where C_1 and C_2 are dimensionless constants, and $B=\int_{source}(\rho_\infty-\rho_o)gU_o dA$.

10.35 Laminar and turbulent boundary-layer skin friction are very different. Consider skin-friction correlations from zero-pressure-gradient (ZPG) boundary-layer flow over a flat plate placed parallel to the flow.

$$\text{Laminar boundary layer: } C_f=\frac{\tau_w}{\frac{1}{2}\rho U^2}=\frac{0.664}{\sqrt{Re_x}}\ \text{(Blasius boundary layer)}$$

Turbulent boundary layer: see correlations in Section 10.9.
Create a table of computed results at $Re_x=Ux/\nu=10^4$, 10^5, 10^6, 10^7, 10^8, and 10^9 for the laminar and turbulent skin-friction coefficients, and the friction force acting on 1.0 m^2 plate surface in sea-level air at 100 m/s and in water at 20 m/s assuming laminar and turbulent flow.

10.36 Derive the following logarithmic velocity profile for a smooth wall: $U^+=(1/\kappa)\ln y^+ + 5.0$ by starting from $U=(u_*/\kappa)\ln y^+ + const.$ and matching the profile to the edge of the viscous sub-layer assuming the viscous sub-layer ends at $y=10.7\upsilon/u_*$.

10.37 [2]Derive the log-law for the mean-flow profile in a zero-pressure gradient (ZPG) flat-plate turbulent boundary layer (TBL) through the following mathematical and dimensional arguments.

2. Inspired by exercise 7.20 in Pope (2000) p. 311.

a) Start with the law of the wall, $U/u_* = f(yu_*/\nu)$ or $U^+ f(y^+)$, for the near-wall region of the boundary layer, and the defect law for the outer region, $\frac{U_e - U}{u_*} = F\left(\frac{y}{\delta}\right)$. These formulae must overlap when $y^+ \to +\infty$ and $y/\delta \to 0$. In this matching or overlap region, set U and $\partial U/\partial y$ from both formulas equal to get two equations involving f and F.

b) In the limit as $y^+ \to +\infty$, the kinematic viscosity must drop out of the equation that includes df/dy^+. Use this fact, to show that $U/u_* = A_I \ln(yu_*/\nu) + B_I$ as $y^+ \to +\infty$ where A_I and B_I are constants for the near-wall or *inner* boundary-layer scaling.

c) Use the result of part b) to determine $F(\xi) = -A_I \ln(\xi) - B_O$ where $\xi = y/\delta$, and A_I and B_O are constants for the wake flow or *outer* boundary-layer scaling.

d) It is traditional to set $A_I = 1/\kappa$, and to keep B_I but to drop its subscript. Using these new requirements determine the two functions, f_I and F_O, in the matching region. Which function explicitly depends on the Reynolds number of the flow?

10.38 For zero pressure gradient, the Von Karman boundary-layer integral equation simplifies to $C_f = 2d\theta/dx$. Use this fact, and (12.90) to determine C_f and numerically compare this result to C_f obtained from (12.93). Do the results match well for $10^6 < \text{Re}_x < 10^9$? What difference does the choice of log-law constants make? Consider (κ, B) pairs representative of the nominal modern values for pipes (0.41, 5.2) and boundary layers: (0.38, 4.2).

10.39 Prove (10.96) and (10.97) by considering a stationary control volume that resides inside the channel or pipe and has stream-normal control surfaces separated by a distance dx and stream-parallel surfaces that coincide with the wall or walls that confine the flow.

10.40 The log-law occurs in turbulent channel, pipe, or boundary-layer flows and should be absent in laminar flows in the same geometries. The extent of the log-law is governed by $\text{Re}_\tau = \delta^+ = \delta/l_\nu = \delta u_*/\nu$, where δ is the channel half-height $(h/2)$, pipe radius $(d/2)$, or full boundary-layer thickness, as appropriate for each flow geometry.

a) For laminar channel flow, show that $\text{Re}_\tau = \sqrt{(3/2)\text{Re}_h}$, and compute Re_τ at an approximate transition Reynolds number of $\text{Re}_h \sim 3000$.

b) For laminar pipe flow, show that $\text{Re}_\tau = \sqrt{2\text{Re}_d}$, and compute Re_τ at an approximate transition Reynolds number of $\text{Re}_d \sim 4000$.

c) For the Blasius boundary layer, show that $\text{Re}_\tau \cong 2.9\text{Re}_x^{1/4}$, and compute $\text{Re}\tau$ at a transition Reynolds number of $\text{Re}_x \sim 10^6$.

d) If mean profile measurements are made in a wall-bounded turbulent flow at $\text{Re}_\tau \sim 10^2$, do you expect the profiles to display the log-law? Why or why not?

e) Repeat part d) when $\text{Re}_\tau > 10^3$.

f) The log-law constants (κ and B) are determined from fitting (10.88) to experimental data. At which Re_τ are κ and B most likely to be accurately determined: 10^2, 10^3, or 10^4?

10.41 A horizontal smooth pipe 20 cm in diameter carries water at a temperature of 20 °C. The drop of pressure is $dp/dx = -8\ \text{N/m}^2$ per meter. Assuming turbulent flow, verify that the thickness of the viscous sub-layer is ≈ 0.25 mm. [*Hint*: Use dp/dx as given by (10.97) to find $\tau_w = 0.4\ \text{N/m}^2$, and therefore $u_* = 0.02$ m/s.]

10.42 The cross-section averaged flow speed U_{av} in a round pipe of radius a may be written:

$$U_{av} \equiv \frac{volume\ flux}{area} = \frac{1}{\pi a^2}\int_0^a U(y)2\pi r\,dr = \frac{2}{a^2}\int_0^a U(y)(a-y)dy,$$

where r is the radial distance from the pipe's centerline, and $y = a - r$ is the distance inward from the pipe's wall. Turbulent pipe flow has very little wake, and the viscous sub-layer is very thin at high Reynolds number; therefore assume the log-law profile, $U(y) = (u_*/\kappa)\ln(yu_*/\nu) + B$, holds throughout the pipe to find

$$U_{av} \cong u_*[(1/\kappa)\ln(au_*/\nu) + B - 3/2\kappa].$$

Now use the definitions: $C_f = \tau_w/\frac{1}{2}\rho U_{av}^2$, $Re_d = 2U_{av}a/\nu$, $\overline{f} = 4C_f =$ the Darcy friction factor, $\kappa = 0.41$, $B = 5.0$, and switch to base-10 logarithms to reach (10.105).

10.43 The cross-section averaged flow speed U_{av} in a wide channel of full height b may be written:

$$U_{av} \equiv \frac{2}{b}\int_0^{b/2} U(y)dy,$$

where y is the vertical distance from the channel's lower wall. Turbulent channel flow has very little wake, and the viscous sub-layer is very thin at high Reynolds number; therefore assume the log-law profile, $U(y) = (u_*/\kappa)\ln(yu_*/\nu) + B$, holds throughout the channel to find

$$U_{av} \cong u_*[(1/\kappa)\ln(bu_*/2\nu) + B - 1/\kappa].$$

Now use the definitions: $C_f = \tau_w/\frac{1}{2}\rho U_{av}^2$, $\mathrm{Re}_b = U_{av}b/\nu$, $\overline{f} = 4C_f =$ the Darcy friction factor, $\kappa = 0.41$, $B = 5.0$, and switch to base-10 logarithms to reach: $\overline{f}^{-1/2} = 2.0\log_{10}(\mathrm{Re}_b\overline{f}^{1/2}) - 0.59$.

10.44 For laminar flow, the hydraulic diameter concept is successful when the ratio $(\overline{f}\cdot U_{av}d_h/\nu)_{duct}/(\overline{f}\cdot U_{av}d/\nu)_{round\ pipe}$ is near unity. Show that this ratio is 1.5 when the duct is a wide channel.

10.45 **a)** Rewrite the final friction factor equation in Exercise 10.43 in terms of the channel's hydraulic diameter instead of its height b.

b) Using the friction factor-Reynolds number ratio given in Exercise 10.44, evaluate (10.107) for a wide channel.

c) Are the results of parts a) and b) in good agreement?

10.46 **a)** Simplify (10.114) when the roughness Reynolds number is large $\mathrm{Re}_{ks} \gg 1$ to show that C_{fR} is independent of ν in the fully rough regime.

b) Reconcile the finding of part a) with the results in Fig. 10.26 which appear to show that C_{fR} depends on Re_x for all values of Re_{ks}.

c) For this fully rough regime, compare C_{fR} computed from (10.114) with the empirical formula provided in Schlichtling (1979): $C_{fR} = (2.87 + 1.58\cdot\log_{10}(x/k_s))^{-2.5}$.

10.47 Perhaps the simplest way to model turbulent flow is to develop an eddy viscosity from dimensional analysis and physical reasoning. Consider turbulent Couette flow with wall spacing h. Assume that eddies of size l produce velocity fluctuations of size $l(\partial U/\partial y)$ so that the turbulent shear stress correlation can be modeled as: $-\overline{uv} \propto l^2(\partial U/\partial y)^2$. Unfortunately, l cannot be a constant because it must disappear near the walls. Thus, more educated guessing is needed, so for this problem assume $\partial U/\partial y$ will have some symmetry about the channel centerline (as shown) and try: $l = Cy$ for $0 \le y \le h/2$ where C is a positive dimensionless constant and y is the vertical distance measured from the lower wall. With this turbulence model, the horizontal RANS momentum equation for $0 \le y \le h/2$ becomes:

$$U\frac{\partial U}{\partial x} + V\frac{\partial U}{\partial y} = -\frac{1}{\rho}\frac{d\overline{p}}{dx} + \frac{1}{\rho}\frac{\partial\tau_{xy}}{\partial y} \quad \text{where} \quad \tau_{xy} = \mu\frac{\partial U}{\partial y} + \rho C^2y^2\left(\frac{\partial U}{\partial y}\right)^2.$$

Determine an analytic form for $U(y)$ after making appropriate simplifications of the RANS equation for fully developed flow assuming the pressure gradient is zero. Check to see that your final answer recovers the appropriate forms as $y \to 0$ and $C \to 0$. Use the fact that $U(y = h/2) = U_o/2$ in your work if necessary.

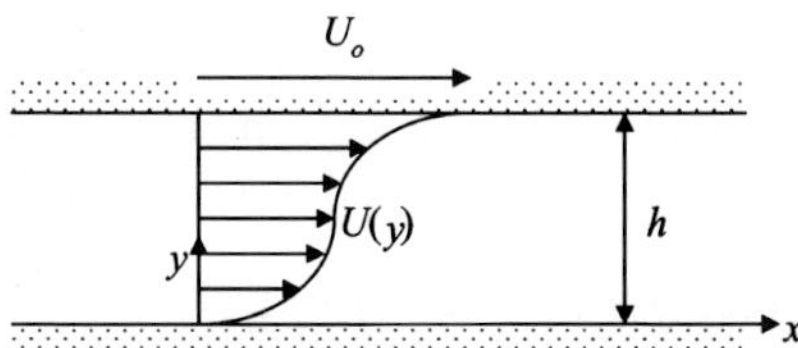

10.48 Incompressible, constant-density-and-viscosity, fully developed, pressure-gradient-driven, turbulent channel flow is often used to test turbulence models for wall-bounded flows. Thus, for this flow, investigate the following simplified mixing-length model for the Reynolds shear stress: $-\overline{u'v'} = \beta y\sqrt{\tau_w/\rho}(\partial U/\partial y)$ for $0 \le y \le h/2$ where y is measured from the lower wall of the channel, β is a positive dimensionless constant, $\tau_w =$ wall shear stress (a constant), and $\rho =$ fluid density.

a) Use this turbulence model, the fully-developed flow assumption $\mathbf{U} = U(y)\mathbf{e}_x$, the assumption of a constant downstream pressure gradient, and the x-direction RANS momentum equation,

$$U\frac{\partial U}{\partial x} + V\frac{\partial U}{\partial y} = -\frac{1}{\rho}\frac{\partial P}{\partial x} + \nu\left(\frac{\partial^2 U}{\partial x^2} + \frac{\partial^2 U}{\partial y^2}\right) - \frac{\partial}{\partial x}(\overline{u'^2}) - \frac{\partial}{\partial y}(\overline{u'v'})$$

to find: $U(y) = \dfrac{u_*}{\beta}\left[\left(1 + \dfrac{2\nu}{\beta u_* h}\right)\ln\left(1 + \dfrac{\beta u_* y}{\nu}\right) - \dfrac{2y}{h}\right]$ for $0 \le y \le h/2$ where $u_* = \sqrt{\tau_w/\rho}$.

b) Does this velocity profile have the proper gradient at $y = 0$ and $y = h/2$?
c) Show that this velocity profile returns to a parabolic flow profile as $\beta \to 0$.
d) How should the constant β be determined?

10.49 The model equations for the two-equation "k-ε" turbulence model, (10.124) and (10.126), include 5 empirical constants. One of these, $C\varepsilon_2$, can be estimated independently of the others by fitting a solution of the model equations to experimental results for the decay of the turbulent kinetic energy, $\overline{e}$, downstream of a random grid placed at the inlet of a wind-tunnel test section.

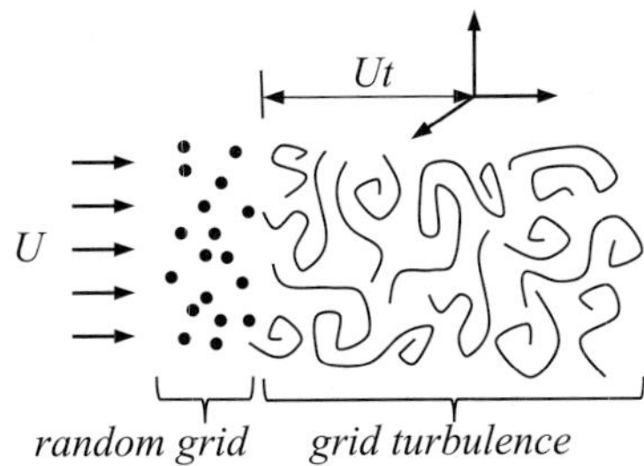

The development of this estimate is further simplified by use of a coordinate system that translates with the average flow velocity in the wind-tunnel. In these translating coordinates $U_i = 0$, and $\overline{e}$ and $\overline{\varepsilon}$ are both functions of time t alone.
a) Simplify (10.124) and (10.126) for random grid turbulence when $U_i = 0$.
b) Assume $\overline{e}(t)$ follows a power-law solution, $\overline{e} = e_o t^{-n}$, where e_o and n are positive constants, and determine a formula for the model constant $C\varepsilon_2$ in terms of n.
c) The experimental value of n is approximately 1.3, so the part b) formula then predicts $C\varepsilon_2 = 1.77$, which is below the standard value ($C\varepsilon_2 = 1.92$ from Launder and Sharma, 1974). Provide at least two reasons that justify this discrepancy.

10.50 Derive (10.127) from (10.35) using the definition equalities in (10.128), (10.129), and (10.131).

10.51 Derive (10.132) by taking the divergence of the constant-density Navier-Stokes momentum equation, computing its average, using the continuity equation, and then subtracting the averaged equation from the instantaneous equation.

10.52 Using the Green's function given in Exercise 10.14 and the properties of homogeneous turbulence, formally solve (10.132) and then use (10.131) to reach (10.133).

10.53 Turbulence largely governs the mixing and transport of water vapor (and other gases) in the atmosphere. Such processes can sometimes be assessed by considering the conservation law (10.34) for a passive scalar.

a) Appropriately simplify (10.34) for turbulence at high Reynolds number that is characterized by: an outer length scale of L, a large-eddy turnover time of T, and a mass-fraction magnitude of Y_o. In addition, assume that the molecular diffusivity D_Y is at most as large as $\nu = \mu/\rho =$ the fluid's kinematic viscosity.
b) Now consider a simple model of how a dry turbulent wind collects moisture as it blows over a nominally flat water surface ($x_1 > 0$) from a dry surface ($x_1 < 0$). Assume the mean velocity is steady and has a single component with a linear gradient, $U_j = (Sx_2, 0, 0)$, and use a simple gradient diffusion model: $-\overline{u_j Y'} = \Delta U L(0, \partial\overline{Y}/\partial x_2, 0)$, where ΔU and L are (constant) velocity and length scales that characterize the turbulent diffusion in this case. This turbulence model allows the turbulent mean flow to be treated like a laminar flow with a large diffusivity $= \Delta UL$ (a turbulent diffusivity). For the simple boundary conditions: $\overline{Y}(x_j) = 0$ for $x_1 < 0$, $\overline{Y}(x_j) = 1$ at $x_2 = 0$ for $x_1 > 0$, and $\overline{Y}(x_j) \to 0$ as $x_2 \to \infty$, show that

$$\overline{Y}(x_1, x_2, x_3) = \int_{\xi}^{\infty} \exp\left(-\frac{1}{9}\zeta^3\right) d\zeta \Big/ \int_0^{\infty} \exp\left(-\frac{1}{9}\zeta^3\right) d\zeta \quad \text{where} \quad \xi = x_2\left(\frac{S}{\Delta U L x_1}\right)^{1/3}$$

for $x_1, x_2 > 0$.

10.54 Estimate the Monin-Obukhov length in the atmospheric boundary layer if the surface stress is 0.1 N/m^2 and the upward heat flux is 200 W/m^2.

10.55 Consider one-dimensional turbulent diffusion of particles issuing from a point source. Assume a Gaussian-Lagrangian correlation function of particle velocity, $r(\tau) = \exp\{-\tau^2/t_c^2\}$, where t_c is a constant. By integrating the correlation function from $\tau = 0$ to ∞, find the integral time scale Λ_t in terms of t_c. Using the Taylor theory, estimate the eddy diffusivity at large times $t/\Lambda_t \gg 1$, given that the *rms* fluctuating velocity is 1 m/s and $t_c = 1$ s.

References

Adrian, R.J., 2007. Hairpin vortex organization in wall turbulence. Physics of Fluids 19, 041301.

Barenblatt, G.I., 1993. Scaling laws for fully developed shear flows. Part I. Basic hypotheses and analysis. Journal of Fluid Mechanics 248, 513–520.

Batchelor, G.K., 1953. The Theory of Homogeneous Turbulence. Cambridge University Press, New York.

Batchelor, G.K., 1959. Small scale variation of convected quantities like temperature in turbulent fluid. Part I: general discussion and the case of small conductivity. Journal of Fluid Mechanics 5, 113–133.

Bird, R.B., Stewart, W.E., Lightfoot, E.N., 1960. Transport Phenomena. John Wiley and Sons, New York.

Bradshaw, P., Woods, J.D., 1978. Geophysical turbulence and buoyant flows. In: Bradshaw, P. (Ed.), Turbulence. Springer-Verlag, New York.

Cantwell, B.J., 1981. Organized motion in turbulent flow. Annual Review of Fluid Mechanics 13, 457–515.

Chapman, D.R., 1979. Computational aerodynamics development and outlook. AIAA Journal 17, 1293–1313.

Chauhan, K.A., Monkewitz, P.A., Nagib, H.M., 2009. Criteria for assessing experiments in zero pressure gradient boundary layers. Fluid Dynamics Research 41, 021404.

Chen, C.J., Rodi, W., 1980. Vertical Turbulent Buoyant Jets – A Review of Experimental Data. Pergamon Press, Oxford.

Colebrook, C.F., 1939. Turbulent flow in pipes, with particular reference to the transitional regime between smooth and rough pipe laws. Journal of the Institution of Civil Engineers 11, 133–156.

Coles, D.E., 1956. The law of the wake in the turbulent boundary layer. Journal of Fluid Mechanics 1, 191–226.

Coles, D.E., Hirst, E.A., 1968. Computation of turbulent boundary layers – 1968 AFOSR-IFP Stanford Conference. In: Proceedings 1968 Conference, vol. 2. Stanford Univ., Stanford, California.

Davidson, P.A., Kaneda, Y., Sreenivasan, K.R. (Eds.), 2013. Ten Chapters in Turbulence. Cambridge University Press, Cambridge.

Dimotakis, P.E., 2000. The mixing transition in turbulent flows. Journal of Fluid Mechanics 409, 69–98.

Duraisamy, K., Iaccarino, G., Xiao, H., 2019. Turbulence modeling in the age of data. Annual Review of Fluid Mechanics 51, 357–377.

Feynman, R.P., Leighton, R.B., Sands, M., 1963. The Feynman Lectures on Physics. Addison-Wesley, New York.

Fife, P., Wei, T., Klewicki, J., McMurtry, P., 2005. Stress gradient balance layers and scale hierarchies in wall-bounded turbulent flows. Journal of Fluid Mechanics 532, 165–189.

Flack, K.A., Schultz, M.P., 2010. Review of hydraulic roughness scales in the fully rough regime. Journal of Fluids Engineering 132, 041203.

Flack, K.A., Schultz, M.P., Rose, W.B., 2012. The onset of roughness effects in the transitionally rough regime. International Journal of Heat and Fluid Flow 35, 160–167.

George, W.K., 1989. The self-preservation of turbulent flows and its relation to initial conditions and coherent structures. In: George, W.K., Arndt, R. (Eds.), Advances in Turbulence. Hemisphere Publishing Corp., New York, pp. 39–73.

George, W.K., 2006. Recent advancements toward the understanding of turbulent boundary layers. AIAA Journal 44, 2435–2449.

George, W.K., Castillo, L., 1997. Zero-pressure-gradient turbulent boundary layer. Applied Mechanics Reviews 50, 689–729.

Grant, H.L., Hughes, B.A., Vogel, W.M., Moilliet, A., 1968. The spectrum of temperature fluctuation in turbulent flow. Journal of Fluid Mechanics 34, 423–442.

Grant, H.L., Stewart, R.W., Moilliet, A., 1962. The spectrum of a cross-stream component of turbulence in a tidal stream. Journal of Fluid Mechanics 13, 237–240.

Hinze, J.O., 1975. Turbulence, 2nd ed. McGraw-Hill, New York.

Jimenez, J., 2004. Turbulent flows over rough walls. Annual Review of Fluid Mechanics 36, 173–196.

Jones, O.C., 1976. An improvement in the calculation of turbulent friction in rectangular ducts. Journal of Fluid Engineering 98, 173–181.

Jones, O.C., Leung, J.C.M., 1981. An improvement in the calculation of turbulent friction in smooth concentric annuli. Journal of Fluid Engineering 103, 615–623.

Jones, W.P., Launder, B.E., 1972. The prediction of laminarization with a two-equation model of turbulence. International Journal of Heat and Mass Transfer 15, 301–314.

Kline, S.J., Reynolds, W.C., Schraub, F.A., Runstadler, P.W., 1967. The structure of turbulent boundary layers. Journal of Fluid Mechanics 30, 741–773.

Kuo, K.K., 1986. Principles of Combustion. John Wiley and Sons, New York.

Lam, S.H., 1992. On the RNG theory of turbulence. Physics of Fluids A 4, 1007–1017.

Landahl, M.T., Mollo-Christensen, E., 1986. Turbulence and Random Processes in Fluid Mechanics. Cambridge University Press, London.

Launder, B.E., Reece, G.J., Rodi, W., 1975. Progress in the development of a Reynolds-stress turbulence closure. Journal of Fluid Mechanics 68, 537–566.

Launder, B.E., Sharma, B.I., 1974. Application of the energy dissipation model of turbulence to the calculation of flow near a spinning disk. Letters in Heat and Mass Transfer 1, 131–137.

Marusic, I., Mckeon, B.J., Monkewitz, P.A., Nagib, H.M., Smits, A.J., Sreenivasan, K.R., 2010. Wall bounded turbulent flows at high Reynolds numbers: recent advances and key issues. Physics of Fluids 22, 065103.

Marusic, I., Monty, J.P., Hultmark, M., Smits, A.J., 2013. On the logarithmic region in wall turbulence. Journal of Fluid Mechanics 716, R3-1–R3-11.
Mellor, G.L., Herring, H.J., 1973. A survey of mean turbulent field closure models. AIAA Journal 11, 590–599.
Metzger, M.M., Klewicki, J.C., 2001. A comparative study of near-wall turbulence in high and low Reynolds number boundary layers. Physics of Fluids 13, 692–701.
Metzger, M.M., Klewicki, J.C., Bradshaw, K.L., Sadr, R., 2001. Scaling the near wall axial turbulent stress in the zero pressure gradient boundary layers. Physics of Fluids 13, 1819–1821.
Monin, A.S., Yaglom, A.M., 1971, 1975. In: Lumley, J.L. (Ed.), Statistical Fluid Mechanics, vols. I and II. MIT Press, Cambridge, MA.
Monkewitz, P.A., Chauhan, K.A., Nagib, H.M., 2007. Self-consistent high-Reynolds-number asymptotics for zero-pressure-gradient turbulent boundary layers. Physics of Fluids 19, 115101.
Monkewitz, P.A., Chauhan, K.A., Nagib, H.M., 2008. Comparison of mean flow similarity laws in zero pressure gradient turbulent boundary layers. Physics of Fluids 20, 105102.
Moody, L.F., 1944. Friction factors for pipe flow. Transactions of the American Society of Mechanical Engineers 66, 671–684.
Mullin, T., 2011. Experimental studies of transition to turbulence in a pipe. Annual Review of Fluid Mechanics 43, 1–24.
Nagib, H.M., Chauhan, K.A., 2008. Variations of von Kármán coefficient in canonical flows. Physics of Fluids 20, 101518.
Niuradse, J., 1933. Laws of flow in rough pipes. NACA Technical Memorandum 1292.
Obot, N.T., 1988. Determination of incompressible flow friction in smooth circular and noncircular passages: a generalized approach including validation of the nearly century old hydraulic diameter concept. Journal of Fluids Engineering 110, 431–440.
Obukhov, A.M., 1949. Structure of the temperature field in turbulent flow. Izvestiya Akademii Nauk SSSR, Georgr. and Geophys. Series 13 (1), 58–69.
Oweis, G.F., Winkel, E.S., Cutbirth, J.M., Ceccio, S.L., Perlin, M., Dowling, D.R., 2010. The mean velocity profile of a smooth-flat-plate turbulent boundary layer at high Reynolds number. Journal of Fluid Mechanics 665, 357–381.
Panofsky, H.A., Dutton, J.A., 1984. Atmospheric Turbulence. Wiley, New York.
Pao, Y.H., 1965. Structure of turbulent velocity and scalar fields at large wave numbers. Physics of Fluids 8, 1063–1075.
Papoulis, A., 1965. Probability, Random Variables, and Stochastic Processes. McGraw-Hill, New York.
Phillips, O.M., 1977. The Dynamics of the Upper Ocean. Cambridge University Press, London.
Pope, S.B., 2000. Turbulent Flows. Cambridge University Press, London.
Richardson, L.F., 1922. Weather Prediction by Numerical Process. Cambridge University Press, Cambridge.
Schlichtling, H., 1979. Boundary Layer Theory, 7th ed. McGraw-Hill, New York, pp. 652–654.
Schultz-Grunow, F., 1941. New frictional resistance law for smooth plates. NACA Technical Memorandum 17 (8), 1–24.
Smith, L.M., Reynolds, W.C., 1992. On the Yakhot-Orszag renormalization group method for deriving turbulence statistics and models. Physics of Fluids A 4, 364–390.
Smits, A.J., Mckeon, B.J., Marusic, I., 2011. High-Reynolds number wall turbulence. Annual Review of Fluid Mechanics 43, 353–375.
Spalding, D.B., 1961. A single formula for the law of the wall. Journal of Applied Mechanics 28, 455–457.
Speziale, C.G., 1991. Analytical methods for the development of Reynolds-stress closures in turbulence. Annual Review of Fluid Mechanics 23, 107–157.
Taylor, G.I., 1915. Eddy motion in the atmosphere. Philosophical Transactions of the Royal Society of London A 215, 1–26.
Taylor, G.I., 1921. Diffusion by continuous movements. Proceedings of the London Mathematical Society 20, 196–211.
Tennekes, H., Lumley, J.L., 1972. A First Course in Turbulence. MIT Press, Cambridge, MA.
Townsend, A.A., 1976. The Structure of Turbulent Shear Flow. Cambridge University Press, London.
Turner, J.S., 1973. Buoyancy Effects in Fluids. Cambridge University Press, London.
Turner, J.S., 1981. Small-scale mixing processes. In: Warren, B.A., Wunch, C. (Eds.), Evolution of Physical Oceanography. MIT Press, Cambridge, MA.
Wei, T., Fife, P., Klewicki, J., McMurtry, P., 2005. Properties of the mean momentum balance in turbulent boundary layer, pipe and channel flows. Journal of Fluid Mechanics 522, 303–327.
White, F.M., 2006. Viscous Fluid Flow, 3rd ed. McGraw-Hill, Boston. Chs. 3, 6.
Wilcox, D.C., 2006. Turbulence Modeling for CFD, 3rd ed. DCW Industries, La Cañada, CA.
Winkel, E.S., Cutbirth, J.M., Ceccio, S.L., Perlin, M., Dowling, D.R., 2012. Turbulence profiles from a smooth flat-plate turbulent boundary layer at high Reynolds number. Experimental Thermal and Fluid Science 40, 140–149.
Yakhot, V., Orszag, S.A., 1986. Renormalization group analysis of turbulence. I. Basic theory. Journal of Scientific Computing 1, 3–51.

Further reading

Hunt, J.C.R., Sandham, N.D., Vassilicos, J.C., Launder, B.E., Monkewitz, P.A., Hewitt, G.F., 2001. Developments in turbulence research: are view based on the 1999 Programme of the Isaac Newton Institute, Cambridge. Published in Journal of Fluid Mechanics 436, 353–391.
Proceedings of the Boeing Symposium on Turbulence. Published in Journal of Fluid Mechanics 41 (1970). Parts 1 (March) and 2 (April).
George, W.K., Arndt, R. (Eds.), 1989. Advances in Turbulence. Hemisphere Publishing Corp., New York.
Kolmogorov, A.N., 1941a. Dissipation of energy in locally isotropic turbulence. Doklady Akademii Nauk SSSR 32, 19–21.
Kolmogorov, A.N., 1941b. The local structure of turbulence in incompressible viscous fluid for very large Reynolds numbers. Doklady Akademii Nauk SSSR 30, 299–303.
Lesieur, M., 1987. Turbulence in Fluids. Martinus Nijhoff Publishers, Dordrecht, Netherlands.
Symposium on Fluid Mechanics of Stirring and Mixing, IUTAM. Published in Physics of Fluids A 3 (5) (1991). May, Part 2.
The Turbulent Years: John Lumley at 70, A Symposium in Honor of John L. Lumley on his 70th Birthday. Published in Physics of Fluids 14 (2002) 2424–2557.

Chapter 11

Gravity waves⍟

Contents

11.1 **Introduction** 445
11.2 **Linear liquid-surface gravity waves** 447
Approximations for deep and shallow water 454
11.3 **Influence of surface tension** 457
11.4 **Standing waves** 459
11.5 **Group velocity, energy flux, and dispersion** 461
11.6 **Nonlinear waves in shallow and deep water** 466
11.7 **Waves on a density interface** 472
11.8 **Internal waves in a continuously stratified fluid** 477
Internal waves in a stratified fluid 479
Dispersion of internal waves in a stratified fluid 481
Energy considerations for internal waves in a stratified fluid 482
Exercises 485
References 488

Chapter objectives

- To develop the equations and boundary conditions for surface, interface, and internal waves
- To derive relationships for linear capillary-gravity wave propagation speed(s), pressure fluctuations, dispersion, particle motion, and energy flux for surface waves on a liquid layer of arbitrary but constant depth
- To describe and highlight wave refraction and nonlinear gravity wave results in shallow and deep water
- To determine linear density-interface wave characteristics with and without an additional free surface
- To present the characteristics of gravity waves on a density gradient with constant buoyancy frequency

11.1 Introduction

There are three types of waves commonly considered in the study of fluid mechanics: interface waves, internal waves, and compression and expansion waves. In all cases, the waves are traveling fluid oscillations, impulses, or pressure changes sustained by the interplay of fluid inertia and a restoring force or a pressure imbalance. For interface waves the restoring forces are gravity and surface tension. For internal waves, the restoring force is gravity. For expansion and compression waves, the restoring force comes directly from the compressibility of the fluid. The basic elements of linear and nonlinear compression and expansion waves are presented in Chapter 14, which covers compressible fluid dynamics. This chapter covers interface and internal waves with an emphasis on gravity as the restoring force. The approach and results from Chapter 6 will be exploited here since the wave physics and wave phenomena presented in this chapter primarily involve irrotational flow.

Perhaps the simplest and most readily observed fluid waves are those that form and travel on the density discontinuity provided by an air-water interface. Such *surface capillary-gravity waves*, sometimes simply called *water waves*, involve fluid particle motions parallel and perpendicular to the direction of wave propagation. Thus, the waves are neither longitudinal nor transverse. When generalized to internal waves that propagate in a fluid medium having a continuous density gradient, the situation may be more complicated. This chapter presents some basic features of wave motion illustrated using water waves. Throughout this chapter, the wave frequency will be assumed much higher than the Coriolis frequency so the wave motion is unaffected by the Earth's rotation. Waves affected by planetary rotation are considered in Chapter 12. And, unless specified otherwise, wave amplitudes are assumed small enough so that the governing equations and boundary conditions are linear.

For such linear waves, Fourier superposition of sinusoidal waves allows arbitrary wave-forms to be constructed. Such waveforms arise naturally from the linearized equations (see Exercise 11.4). Consequently, a simple sinusoidal traveling

⍟. This book has a companion website hosting complementary materials. Visit this URL to access it: https://www.elsevier.com/books-and-journals/book-companion/9780128198070.

Fluid Mechanics. https://doi.org/10.1016/B978-0-12-819807-0.00020-X

wave of the form:

$$\eta(x,t) = a\cos\left[\frac{2\pi}{\lambda}(x - ct)\right] \tag{11.1}$$

is a foundational element for what follows. In Cartesian coordinates with x horizontal and z vertical, $z = \eta(x,t)$ specifies the *waveform* or surface shape where a is the wave *amplitude*, λ is the *wavelength*, c is the *phase speed*, and $2\pi(x - ct)/\lambda$ is the *phase*. In addition, the spatial frequency $k \equiv 2\pi/\lambda$, with units of rad/m, is known as the *wave number*. If (11.1) describes the vertical deflection of an air-water interface, then the height of wave crests is $+a$ and the depth of the wave troughs is $-a$ compared to the undisturbed water-surface location $z = 0$. At any instant in time, the distance between successive wave crests is λ. At any fixed x-location, the time between passage of successive wave crests is the *period*, $T = 2\pi/kc = \lambda/c$. Thus, the wave's *cyclic frequency* is $\nu = 1/T$ with units of Hz, and its *radian frequency* is $\omega = 2\pi\nu$ with units of rad/s. In terms of k and ω, (11.1) can be written:

$$\eta(x,t) = a\cos[kx - \omega t]. \tag{11.2}$$

The wave propagation speed is readily deduced from (11.1) or (11.2) by determining the travel speed of wave crests. This can be done by setting the phase in (11.1) or (11.2) so that the cosine function is unity and $\eta = +a$. This occurs when the phase is $2n\pi$ where n is an integer:

$$\frac{2\pi}{\lambda}(x_{crest} - ct) = 2n\pi = kx_{crest} - \omega t, \tag{11.3}$$

and x_{crest} is the time-dependent location where $\eta = +a$. Solving for the crest location produces:

$$x_{crest} = (\omega/k)t + 2n\pi/k.$$

Therefore, in a time increment Δt, a wave crest moves a distance $\Delta x_{crest} = (\omega/k)\Delta t$, so:

$$c = \omega/k = \lambda\nu \tag{11.4}$$

is known as the phase speed because it specifies the travel speed of constant-phase wave features, like wave crests or troughs.

Although instructive, (11.1) and (11.2) are limited to propagation in the positive-x direction only. In general, waves may propagate in any direction. A useful three-dimensional generalization of (11.2) is:

$$\eta = a\cos(kx + ly + mz - \omega t) = a\cos(\mathbf{K}\cdot\mathbf{x} - \omega t), \tag{11.5}$$

where $\mathbf{K} = (k, l, m)$ is a vector, called the *wave number vector*, whose magnitude K is given by:

$$K^2 = k^2 + l^2 + m^2. \tag{11.6}$$

The wavelength derived from (11.5) is:

$$\lambda = 2\pi/K, \tag{11.7}$$

which is illustrated in Fig. 11.1 in two dimensions. The magnitude of the phase velocity is $c = \omega/K$, and the direction of propagation is parallel to $\mathbf{K}$, so the phase velocity vector is:

$$\mathbf{c} = (\omega/K)\mathbf{e}_K, \tag{11.8}$$

where $\mathbf{e}_K = \mathbf{K}/K$.

From Fig. 11.1, $c_x = \omega/k$, $c_y = \omega/l$, and $c_z = \omega/m$ are each larger than the resultant $c = \omega/K$, because k, l, and m are individually smaller than K when all three are non-zero, as required by (11.6). Thus, c_x, c_y, and c_z are not vector components of the phase velocity in the usual sense, but they do reflect the fact that constant-phase surfaces appear to travel faster along directions not coinciding with the direction of propagation, the x and y directions in Fig. 11.1 for example. Any of the three axis-specific phase speeds, is sometimes called the *trace velocity* along its associated axis.

If sinusoidal waves exist in a fluid moving with uniform speed $\mathbf{U}$, then the observed phase speed is $\mathbf{c}_0 = \mathbf{c} + \mathbf{U}$. Forming a dot product of $\mathbf{c}_0$ with $\mathbf{K}$, and using (11.8), produces:

$$\omega_0 = \omega + \mathbf{U}\cdot\mathbf{K}, \tag{11.9}$$

where ω_0 is the *observed frequency* at a fixed point, and ω is the *intrinsic frequency* measured by an observer moving with the flow. The frequency of a wave is *Doppler shifted* by an amount $\mathbf{U} \cdot \mathbf{K}$ in non-zero flow. Eq. (11.9) may be understood by considering a situation in which the intrinsic frequency ω is zero, but the flow pattern has a periodicity in the x direction of wavelength $2\pi/k$. If this sinusoidal pattern is translated in the x direction at speed U, then the observed frequency at a fixed point is $\omega_0 = Uk$. The effects of uniform flow on frequency will not be considered further, and all frequencies in the remainder of this chapter should be interpreted as intrinsic frequencies.

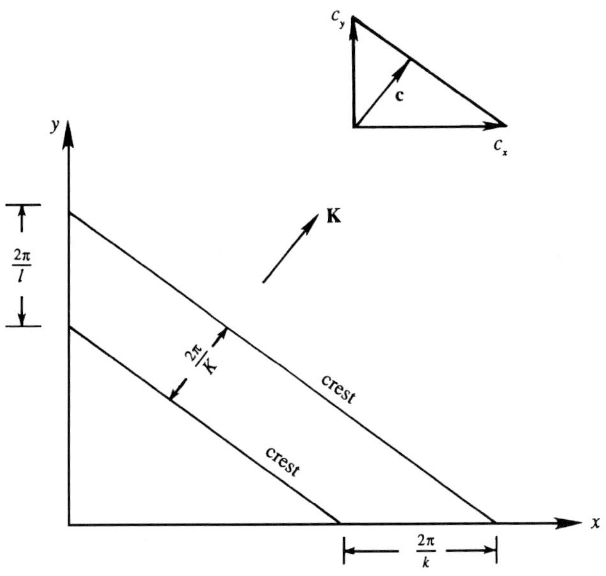

FIGURE 11.1 Wave crests propagating in the x-y plane. The crest spacing along the coordinate axes is larger than the wavelength $\lambda = 2\pi/K$. The inset shows how the trace velocities c_x and c_y are combined to give the phase velocity vector $\mathbf{c}$.

Example 11.1

If a surface wave described by (11.2) has small surface slope, $|\partial\eta/\partial x| \ll 1$, what does this imply about the vertical velocity of the surface, $\partial\eta/\partial t$?

Solution Differentiate (11.2) with respect to x and t to find:

$$\text{surface slope} = \frac{\partial\eta}{\partial x} = -ka\sin[kx - \omega t] \quad \text{and} \quad \text{surface velocity} = \frac{\partial\eta}{\partial t} = +\omega a\sin[kx - \omega t].$$

Thus, the surface slope has amplitude ka. These relationships and $k = \omega/c$ imply:

$$\frac{1}{c}\frac{\partial\eta}{\partial t} = -\frac{\partial\eta}{\partial x}.$$

So when $ka = 2\pi a/\lambda$ (the amplitude of $\partial\eta/\partial x$) is much less than unity, the vertical velocity of the surface ($\partial\eta/\partial t$) is much less than the wave speed, c.

11.2 Linear liquid-surface gravity waves

Starting from the equations for ideal flow, this section develops the properties of small-slope, small-amplitude gravity waves on the free surface of a constant-density liquid layer of uniform depth H, which may be large or small compared to the wavelength λ. The limitation to waves with small slopes and amplitudes implies $a/\lambda \ll 1$ and $a/H \ll 1$, respectively. These two conditions allow the problem to be linearized. In this first assessment of wave motion, surface tension is neglected for simplicity; in water its effect is limited to wavelengths less than 5 to 10 centimeters, as discussed in Section 11.3. In addition, the air above the liquid is ignored, and the liquid's motion is presumed to be irrotational and entirely caused by the surface waves.

To get started, choose the x-axis in the direction of wave propagation with the z-axis vertical so that the motion is two dimensional in the $x - z$ plane (Fig. 11.2). Let $\eta(x, t)$ denote the vertical liquid-surface displacement from its undisturbed location $z = 0$. Because the liquid's motion is irrotational, a velocity potential $\phi(x, z, t)$ can be defined such that:

$$u = \partial\phi/\partial x, \quad \text{and} \quad w = \partial\phi/\partial z, \tag{11.10}$$

so the incompressible continuity equation $\partial u/\partial x + \partial w/\partial z = 0$ implies:

$$\partial^2\phi/\partial x^2 + \partial^2\phi/\partial z^2 = 0. \tag{11.11}$$

There are three boundary conditions. The condition at the bottom of the liquid layer is zero normal velocity, that is:

$$w = \partial\phi/\partial z = 0 \quad \text{on} \quad z = -H. \tag{11.12}$$

At the free surface, a *kinematic boundary condition* is applied that requires the fluid-particle velocity normal to the surface, $\mathbf{u} \cdot \mathbf{n}$, and on the surface be the same as the velocity of the surface $\mathbf{u}_s$ normal to itself:

$$(\mathbf{n} \cdot \mathbf{u})_{z=\eta} = \mathbf{n} \cdot \mathbf{u}_s, \tag{11.13}$$

where $\mathbf{n}$ is the surface normal. This is a simplified version of (4.91), and it ensures that the liquid elements that define the interface do not become separated from the interface while still allowing these interface elements to move along the interface.

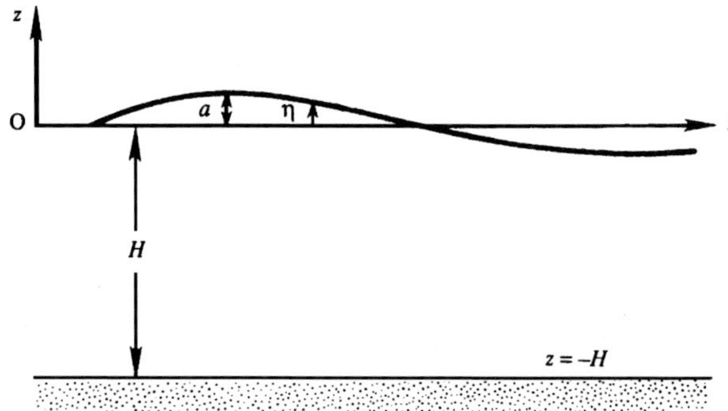

FIGURE 11.2 Geometry for determining the properties of linear gravity waves on the surface of a liquid layer of depth H. Gravity points downward along the z-axis. The undisturbed liquid surface location is $z = 0$ so the bottom is located at $z = -H$. The surface's vertical deflection or waveform is $\eta(x, t)$. When η is sinusoidal, its peak deflection from $z = 0$ is the sinusoid's amplitude a.

For the current situation, the equation for the surface may be written $f(x, z, t) = z - \eta(x, t) = 0$, so the surface normal $\mathbf{n}$, which points upward out of the liquid, will be:

$$\mathbf{n} = \nabla f/|\nabla f| = (-(\partial\eta/\partial x)\mathbf{e}_x + \mathbf{e}_z)/\sqrt{(\partial\eta/\partial x)^2 + 1}. \tag{11.14}$$

The velocity of the surface $\mathbf{u}_s$ at any location x can be considered purely vertical:

$$\mathbf{u}_s = (\partial\eta/\partial t)\mathbf{e}_z. \tag{11.15}$$

Thus, (11.13) multiplied by $|\nabla f|$ implies $(\nabla f \cdot \mathbf{u})_{z=\eta} = \nabla f \cdot \mathbf{u}_s$, which can be evaluated using (11.15) and $\mathbf{u} = u\mathbf{e}_x + w\mathbf{e}_z$ to find:

$$\left(-u\frac{\partial\eta}{\partial x} + w\right)_{z=\eta} = \frac{\partial\eta}{\partial t}, \quad \text{or} \quad \left(\frac{\partial\phi}{\partial z}\right)_{z=\eta} = \frac{\partial\eta}{\partial t} + \frac{\partial\eta}{\partial x}\left(\frac{\partial\phi}{\partial x}\right)_{z=\eta}, \tag{11.16}$$

where (11.10) has been used for the fluid velocity components to achieve the second form of (11.16). For small slope waves, the final term in (11.16) is small compared to the other two, so the kinematic boundary condition can be approximated:

$$\left(\frac{\partial\phi}{\partial z}\right)_{z=\eta} \cong \frac{\partial\eta}{\partial t}. \tag{11.17}$$

For consistency, the left side of (11.17) must also be approximated for small wave slopes, and this is readily accomplished via a Taylor series expansion around $z = 0$:

$$\left(\frac{\partial \phi}{\partial z}\right)_{z=\eta} = \left(\frac{\partial \phi}{\partial z}\right)_{z=0} + \eta\left(\frac{\partial^2 \phi}{\partial z^2}\right)_{z=0} + \cong \frac{\partial \eta}{\partial t}.$$

Thus, when a/λ is small enough, the most simplified version of (11.13) is:

$$\left(\frac{\partial \phi}{\partial z}\right)_{z=0} \cong \frac{\partial \eta}{\partial t}. \tag{11.18}$$

These simplifications of the kinematic boundary are justified when $ka = 2\pi a/\lambda \ll 1$ (see Exercise 11.3).

In addition to the kinematic condition at the surface, there is a *dynamic condition* that the pressure just below the liquid surface be equal to the ambient pressure, with surface tension neglected. Taking the ambient air pressure above the liquid to be a constant atmospheric pressure, the dynamic surface condition can be stated:

$$(p)_{z=\eta} = 0, \tag{11.19}$$

where p in (11.19) is the gauge pressure. Eq. (11.19) follows from the boundary condition (4.94) when the flow is inviscid and surface tension is neglected. Eq. (11.19) and the neglect of any shear stresses on $z = \eta$ define a stress-free boundary. Thus, the water surface in this ideal case is commonly called a *free surface*. For consistency, this condition should also be simplified for small-slope waves by dropping the nonlinear term $|\nabla\phi|^2$ in the relevant Bernoulli equation (4.84):

$$\frac{\partial \phi}{\partial t} + \frac{p}{\rho} + gz \cong 0, \tag{11.20}$$

where the Bernoulli constant has been evaluated on the undisturbed liquid surface far from the surface wave. Evaluating (11.20) on $z = \eta$ and applying (11.19) produces:

$$\left(\frac{\partial \phi}{\partial t} + \frac{p}{\rho} + gz\right)_{z=\eta} \cong \left(\frac{\partial \phi}{\partial t}\right)_{z=0} + g\eta \cong 0, \quad \text{or} \quad \left(\frac{\partial \phi}{\partial t}\right)_{z=0} \cong -g\eta. \tag{11.21}$$

The first approximate equality follows because $(\partial\varphi/\partial t)_{z=0}$ is the first term in a Taylor series expansion of $(\partial\varphi/\partial t)_{z=\eta}$ in powers of η about $\eta = 0$. This approximation is consistent with (11.18).

Interestingly, even with the specification of the field equation (11.11) and the three boundary conditions, (11.12), (11.18), and (11.21), the overall linear surface-wave problem is not fully defined without initial conditions for the surface shape (Exercise 11.4). For simplicity, choose $\eta(x, t = 0) = a\cos(kx)$, since it matches the simple sinusoidal wave (11.2), which now becomes a foundational part of the solution. To produce a cosine dependence for η on the phase $(kx - \omega t)$ in (11.2), (11.18), and (11.21) require ϕ to be a sine function of $(kx - \omega t)$. Consequently, a solution is sought for ϕ in the form:

$$\phi(x, z, t) = f(z)\sin(kx - \omega(k)t), \tag{11.22}$$

where the functions $f(z)$ and $\omega(k)$ are to be determined. Substitution of (11.22) into the Laplace equation (11.11) gives:

$$d^2 f/dz^2 - k^2 f = 0,$$

which has the general solution $f(z) = Ae^{kz} + Be^{-kz}$, where A and B are constants. Thus, (11.22) implies:

$$\phi(x, z, t) = (Ae^{+kz} + Be^{-kz})\sin(kx - \omega t). \tag{11.23}$$

The constants A and B can be determined by substituting (11.23) into (11.12):

$$k(Ae^{-kH} - Be^{+kH})\sin(kx - \omega t) = 0 \quad \text{or} \quad B = Ae^{-2kH}, \tag{11.24}$$

and by substituting (11.2) and (11.23) into (11.18):

$$k(A - B)\sin(kx - \omega t) = \omega a\sin(kx - \omega t) \quad \text{or} \quad k(A - B) = \omega a. \tag{11.25}$$

Solving (11.24) and (11.25) for A and B produces:

$$A = \frac{a\omega}{k(1 - e^{-2kH})}, \quad \text{and} \quad B = \frac{a\omega e^{-2kH}}{k(1 - e^{-2kH})}.$$

The velocity potential (11.23) then becomes:

$$\phi = \frac{a\omega}{k} \frac{\cosh(k(z+H))}{\sinh(kH)} \sin(kx - \omega t), \tag{11.26}$$

from which the fluid velocity components are found as:

$$u = a\omega \frac{\cosh(k(z+H))}{\sinh(kH)} \cos(kx - \omega t), \quad \text{and} \quad w = a\omega \frac{\sinh(k(z+H))}{\sinh(kH)} \sin(kx - \omega t). \tag{11.27}$$

This solution of the Laplace equation has been found using kinematic boundary conditions alone, and this is typical of irrotational constant-density flows where fluid pressure is determined through a Bernoulli equation after the velocity field has been found. Here, the dynamic surface boundary condition (11.21) enforces $p = 0$ on the liquid surface, and substitution of (11.2) and (11.26) into (11.21) produces:

$$\left(\frac{\partial \phi}{\partial t}\right)_{z=0} = -\frac{a\omega^2}{k} \frac{\cosh(kH)}{\sinh(kH)} \cos(kx - \omega t) \cong -g\eta = -ag\cos(kx - \omega t),$$

which simplifies to a relation between ω and k (or equivalently, between the wave period T and the wave length λ):

$$\omega = \sqrt{gk\tanh(kH)} \quad \text{or} \quad T = \sqrt{\frac{2\pi\lambda}{g}\coth\left(\frac{2\pi H}{\lambda}\right)}. \tag{11.28}$$

The first part of (11.28) specifies how temporal and spatial frequencies of the surface waves are related, and it is known as a *dispersion relation*. The phase speed c of these surface waves is given by:

$$c = \frac{\omega}{k} = \sqrt{\frac{g}{k}\tanh(kH)} = \sqrt{\frac{g\lambda}{2\pi}\tanh\left(\frac{2\pi H}{\lambda}\right)}. \tag{11.29}$$

This result is of fundamental importance for water waves. It shows that surface waves are *dispersive* because their propagation speed depends on wave number, with lower k (longer wavelength) waves traveling faster. (*Dispersion* is a term borrowed from optics, where it signifies separation of different colors due to the speed of light in a medium depending on the wavelength.) Thus, a concentrated wave packet made up of many different wavelengths (or frequencies) will not maintain a constant waveform or shape. Instead, it will disperse or spread out as it travels. The longer wavelength components will travel faster than the shorter wavelength ones so that an initial impulse evolves into a wide wave train. This is precisely what happens when an object is dropped onto the surface of a quiescent pool, pond, or lake. The radial extent of the circular waves increases with time, and the longest wavelengths appear furthest from the point of impact while the shortest wavelengths are seen closest to the point of impact.

The rest of this section covers some implications of the linear surface-wave solution (11.26) and the dispersion relation (11.28). Given the ease with which it can be measured, the pressure below the liquid surface is considered first. In particular, the time-dependent perturbation pressure:

$$p' \equiv p + \rho g z, \tag{11.30}$$

produced by surface waves is of interest. Using this and (11.26) in the linearized Bernoulli equation (11.20) leads to:

$$p' = -\rho\frac{\partial \phi}{\partial t} = \rho\frac{a\omega^2}{k}\frac{\cosh(k(z+H))}{\sinh(kH)}\cos(kx - \omega t) = \rho g a \frac{\cosh(k(z+H))}{\cosh(kH)}\cos(kx - \omega t), \tag{11.31}$$

where the second equality follows when (11.28) is used to eliminate ω^2. The perturbation pressure therefore decreases with increasing depth, and the extent of this decrease depends on the wavelength through k.

Another interesting feature of linear surface waves is the fact that they travel and cause fluid elements to move, but they do not cause fluid elements to travel. To ascertain what happens when a linear surface wave passes, consider the fluid element that follows a path $\mathbf{x}_p(t) = x_p(t)\mathbf{e}_x + z_p(t)\mathbf{e}_z$. The path-line equations (3.8) for this fluid element are:

$$\frac{dx_p(t)}{dt} = u(x_p, z_p, t), \quad \text{and} \quad \frac{dz_p(t)}{dt} = w(x_p, z_p, t), \tag{11.32}$$

which imply:

$$\frac{dx_p}{dt} = a\omega\frac{\cosh(k(z_p + H))}{\sinh(kH)}\cos(kx_p - \omega t), \quad \text{and} \quad \frac{dz_p}{dt} = a\omega\frac{\sinh(k(z_p + H))}{\sinh(kH)}\sin(kx_p - \omega t), \tag{11.33}$$

when combined with (11.27). To be consistent with the small amplitude approximation, these equations can be linearized by setting $x_p(t) = x_0 + \xi(t)$ and $z_p(t) = z_0 + \zeta(t)$, where (x_0, z_0) is the average fluid element location and the element excursion vector (ξ, ζ) (see Fig. 11.3) is assumed to be small compared to the wavelength. Thus, the linearized versions of (11.33) are obtained by evaluating the right side of each equation at (x_0, z_0):

$$\frac{d\xi}{dt} \cong a\omega\frac{\cosh(k(z_0 + H))}{\sinh(kH)}\cos(kx_0 - \omega t), \quad \text{and} \quad \frac{d\zeta}{dt} \cong a\omega\frac{\sinh(k(z_0 + H))}{\sinh(kH)}\sin(kx_0 - \omega t), \tag{11.34a, 11.34b}$$

where x_0 and z_0 have been assumed independent of time. This linearization is valid when the velocity of the fluid element along its path is nearly equal to the fluid velocity at (x_0, z_0) at that instant. It is accurate when $a \ll \lambda$. Eqs. (11.34a, 11.34b) are reminiscent of those in Example 3.3, and are readily time-integrated:

$$\xi \cong -a\frac{\cosh(k(z_0 + H))}{\sinh(kH)}\sin(kx_0 - \omega t), \quad \text{and} \quad \zeta \cong a\frac{\sinh(k(z_0 + H))}{\sinh(kH)}\cos(kx_0 - \omega t). \tag{11.35a, 11.35b}$$

Here, $\xi(t)$ and $\zeta(t)$ are entirely oscillatory. Neither contains a term that increases with time so the assumption that x_0 and z_0 are time independent is self consistent when $a \ll \lambda$. Elimination of the phase $(kx_0 - \omega t)$ from (11.35a, 11.35b) gives:

$$\xi^2 \Big/ \left[a\frac{\cosh(k(z_0 + H))}{\sinh(kH)}\right]^2 + \zeta^2 \Big/ \left[a\frac{\sinh(k(z_0 + H))}{\sinh(kH)}\right]^2 = 1, \tag{11.36}$$

which represents an ellipse. Both the semi-major axis, $a\cosh[k(z_0 + H)]/\sinh(kH)$ and the semi-minor axis, $a\sinh[k(z_0 + H)]/\sinh(kH)$ decrease with depth, the minor axis vanishing at $z_0 = -H$ (Fig. 11.4b). The distance between foci remains constant with depth. Eqs. (11.35a, 11.35b) show that the phase of the motion is independent of z_0, so fluid elements in any vertical column move in phase. That is, if one of them is at the top of its orbit, then all elements at the same x_0 are at the top of their orbits.

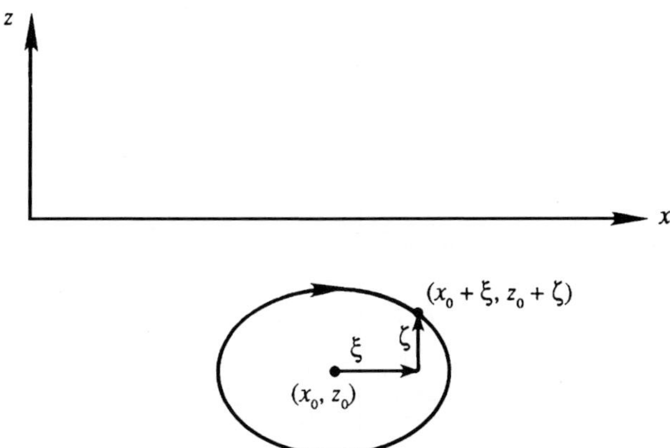

FIGURE 11.3 Orbit of a fluid particle below a linear surface wave. The average position is of the particle is (x_0, z_0), and $\xi(t)$ and $\zeta(t)$ are small time-dependent displacements in the horizontal and vertical directions, respectively. When the surface wave is sinusoidal, travels to the right, and has small amplitude, fluid particles below the surface traverse closed elliptical orbits in the clockwise direction.

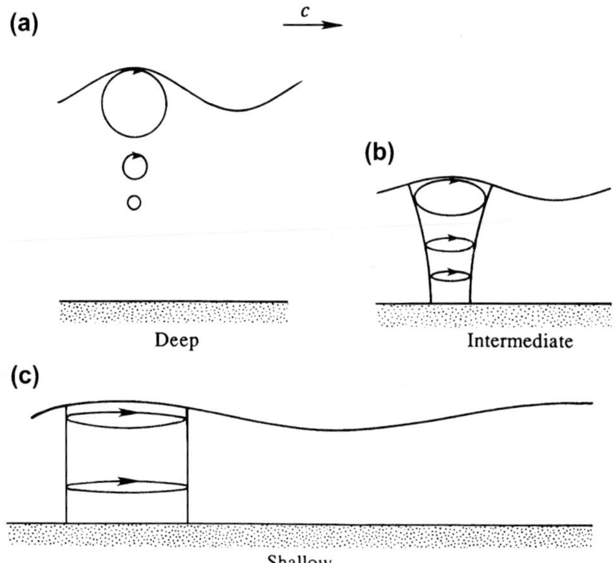

FIGURE 11.4 Fluid particle orbits caused by a linear sinusoidal surface wave traveling to the right for three liquid depths. (a) When the liquid is deep and $\tanh(kH) \approx 1$, then particle obits are circular and decrease in size with increasing depth. (b) At intermediate depths, the particle orbits are broad ellipses that narrow and contract with increasing depth. (c) When the water is shallow and $\tanh(kH) \approx \sinh(kH) \approx kH$, the orbits are thin ellipses that become thinner with increasing depth.

Streamlines may be found from the stream function ψ, which can be determined by integrating the velocity component equations $\partial\psi/\partial z = u$ and $-\partial\psi/\partial x = w$ when u and w are given by (11.27):

$$\psi = \frac{a\omega}{k}\frac{\sinh(k(z+H))}{\sinh(kH)}\cos(kx - \omega t) \tag{11.37}$$

(Exercise 11.5). To understand the streamline structure, consider a particular time, $t = 0$, when:

$$\psi \propto \sinh[k(z+H)]\cos kx.$$

It can be seen that $\psi = 0$ at $z = -H$, so that the bottom wall is a part of the $\psi = 0$ streamline. However, ψ is also zero at $kx = \pm\pi/2, \pm 3\pi/2, \ldots$ for any z. At $t = 0$ and at these values of kx, η from (11.2) vanishes. The resulting streamline pattern is shown in Fig. 11.5. Here, the *velocity is in the direction of propagation* (*and horizontal*) *at all depths below the crests, and opposite to the direction of propagation at all depths below troughs.*

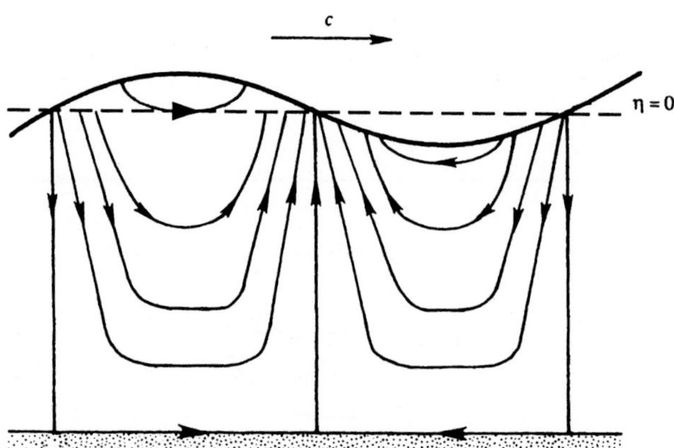

FIGURE 11.5 Instantaneous streamline pattern for a sinusoidal surface wave propagating to the right. Here, the $\psi = 0$ streamline follows the bottom and jumps up to contact the surface where $\eta = 0$. The remaining streamlines start and end on the liquid surface with purely horizontal motion found in the $+x$ direction below a wave crest and in $-x$ direction below a wave trough.

Surface gravity waves possess kinetic energy in the motion of the fluid and potential energy in the vertical deformation of the free surface. The *kinetic energy* per unit horizontal area, E_k, of the wave system is found by integrating over the depth and averaging over a wavelength:

$$E_k = \frac{\rho}{2\lambda}\int_0^\lambda \int_{-H}^0 (u^2 + w^2)dzdx. \tag{11.38}$$

Here, the z-integral is taken from $z = -H$ to $z = 0$, consistent with the linearization performed to reach (11.27); integrating from $z = -H$ to $z = \eta$ merely introduces a higher-order term. Substitution of the velocity components from (11.27), use of the dispersion relationship (11.28), and evaluation of the integrals gives:

$$E_k = \frac{1}{2}\rho g\overline{\eta^2}, \tag{11.39}$$

where $\overline{\eta^2}$ is the mean-square vertical surface displacement (see Exercise 11.9).

The *potential energy* per unit horizontal area, E_p, of the wave system is defined as the work done per unit area to deform a horizontal free surface into the disturbed state. It is therefore equal to the *difference* of potential energies of the system in the disturbed and undisturbed states. As the potential energy of an element in the fluid (per unit length in y) is $\rho gz\,dx\,dz$ (Fig. 11.6), E_p can be calculated as:

$$E_p = \frac{\rho g}{\lambda}\int_0^\lambda \int_{-H}^\eta z\,dzdx - \frac{\rho g}{\lambda}\int_0^\lambda \int_{-H}^0 z\,dzdx = \frac{\rho g}{\lambda}\int_0^\lambda \int_0^\eta z\,dzdx = \frac{\rho g}{2\lambda}\int_0^\lambda \eta^2\,dx, \tag{11.40}$$

and this can also be written in terms of the mean square displacement as:

$$E_p = \frac{1}{2}\rho g\overline{\eta^2}. \tag{11.41}$$

Thus, the average kinetic and potential energies are equal. This is called the *principle of equipartition of energy* and is valid in conservative dynamical systems undergoing small oscillations that are unaffected by planetary rotation. However, it is not valid when the Coriolis acceleration is included, as described in Chapter 12. The total wave energy in the water column per unit horizontal area is:

$$E = E_k + E_p = \rho g\overline{\eta^2} = \frac{1}{2}\rho g a^2, \tag{11.42}$$

where the last form in terms of the amplitude a is valid if η is assumed sinusoidal, since the average over a wavelength of the square of a sinusoid is 1/2.

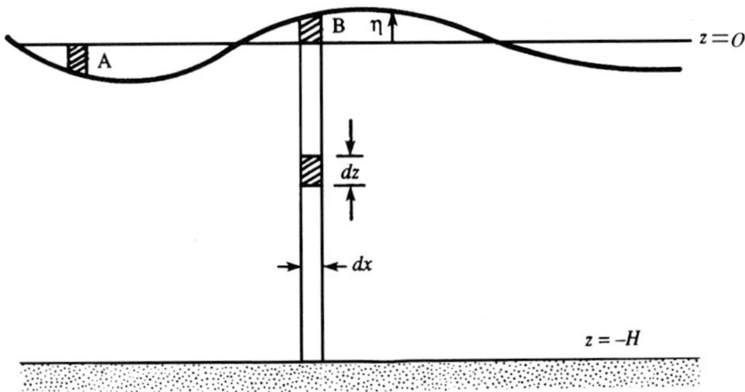

FIGURE 11.6 Calculation of potential energy of a fluid column. Here, work must be done to push the liquid surface down below $z = 0$ (A), and lift the liquid surface up above $z = 0$ (B).

Next, consider the rate of transmission of energy due to a single sinusoidal component of wave number k. The *energy flux* across the vertical plane $x = 0$ is the pressure work done by the fluid in the region $x < 0$ on the fluid in the region $x > 0$. The time average energy flux EF per unit length of crest is (writing p as the sum of a perturbation p' and a background pressure $= -\rho g z$):

$$EF = \frac{\omega}{2\pi}\int_0^{2\pi/\omega}\int_{-H}^{0} pu\,dz\,dt = \frac{\omega}{2\pi}\int_0^{2\pi/\omega}\int_{-H}^{0}(p' - \rho g z)u\,dz\,dt$$
$$= \frac{\omega}{2\pi}\int_0^{2\pi/\omega}\int_{-H}^{0} p'u\,dz\,dt + \frac{\omega}{2\pi}\frac{\rho g H^2}{2}\int_0^{2\pi/\omega} u\,dt, \tag{11.43}$$

where the wave period is $2\pi/\omega$. The final integral in (11.43) is zero because the time average of u over one wave period is zero. Substituting for p' from (11.31) and u from (11.27), (11.43) becomes:

$$EF = \frac{\omega}{2\pi}\int_0^{2\pi/\omega}\cos^2(kx - \omega t)dx \cdot \frac{\rho a^2\omega^3}{k\sinh^2 kH}\int_{-H}^{0}\cosh^2[k(z+h)]dz = \frac{\rho g a^2}{2}\left[\frac{c}{2}\left(1 + \frac{2kH}{\sinh(2kH)}\right)\right]. \tag{11.44}$$

The first factor of the final form for the energy flux is the wave energy per unit area given in (11.42). Therefore, the second factor must be the speed of propagation of the wave energy of component k. This energy propagation speed is called the *group speed*, and is further discussed in Section 11.5.

Approximations for deep and shallow water

The preceding analysis is applicable for any value of H/λ. However, interesting simplifications are provided in the next few paragraphs for deep water, $H/\lambda \gg 1$, and shallow water, $H/\lambda \ll 1$.

Consider deep water first. The general expression for the phase speed is (11.29), but $\tanh(x) \to 1$ for $x \to \infty$. However, x need not be very large for this approximation to be valid. In fact, $\tanh(x) = 0.96403$ for $x = 2.0$, so it follows that (11.29) can be approximated with 2% accuracy by:

$$c = \sqrt{g/k} = \sqrt{g\lambda/2\pi}, \tag{11.45}$$

for $H > 0.32\lambda$ (corresponding to $kH > 2.0$). Surface waves are therefore classified as *deep-water waves* if the depth is more than one-third of the wavelength. Deep-water waves are dispersive since their phase speed depends on wavelength.

A common period of wind-generated surface gravity waves in the ocean is ~10 s, which, via the dispersion relation (11.29), corresponds to a wavelength of 150 m. The water depth on a typical continental shelf is ~100 m, and in the open ocean it is ~4 km. Thus, the dominant wind waves in the ocean, even over the continental shelf, act as deep-water waves and do not feel the effects of the ocean bottom until they arrive near a coastline. This is not true of the very long wavelength gravity waves or tsunamis generated by tidal forces or seismic activity. Such waves may have wavelengths of hundreds of kilometers.

In deep water, the semi-major and semi-minor axes of particle orbits produced by small-amplitude gravity waves are nearly equal to ae^{kz} since

$$\frac{\cosh(k(z+H))}{\sinh(kH)} \approx \frac{\sinh(k(z+H))}{\sinh(kH)} \approx e^{kz}$$

for $kH > 2.0$, so the deep water wave-induced fluid particle motions are:

$$\xi \cong -ae^{kz_0}\sin(kx_0 - \omega t), \quad \text{and} \quad \zeta \cong ae^{kz_0}\cos(kx_0 - \omega t). \tag{11.46}$$

These particle orbits are circles (Fig. 11.4a). At the surface, their radius is a, the amplitude of the wave.

The fluid velocity components for deep-water waves are

$$u = a\omega e^{kz}\cos(kx - \omega t), \quad \text{and} \quad w = a\omega e^{kz}\sin(kx - \omega t). \tag{11.47}$$

At a fixed spatial location, the velocity vector rotates clockwise (for a wave traveling in the positive x direction) at frequency ω, while its magnitude remains constant at $a\omega e^{kz}$.

For deep-water waves, the perturbation pressure from (11.31) simplifies to:

$$p' = \rho g a e^{kz}\cos(kx - \omega t), \tag{11.48}$$

which shows the wave-induced pressure change decays exponentially with depth, reaching $e^{-\pi} \approx 4\%$ of its surface magnitude at a depth of $\lambda/2$. Thus, a bottom-mounted sensor used to record wave-induced pressure fluctuations will respond as a low-pass filter. Its signal will favor long waves while rejecting short ones.

The shallow water limit is also important and interesting. It is obtained by noting $\tanh(x) \approx x$ as $x \to 0$, so for $H/\lambda \ll 1$:

$$\tanh(2\pi H/\lambda) \approx 2\pi H/\lambda,$$

in which case the phase speed from (11.29) simplifies to:

$$c = \sqrt{gH}, \tag{11.49}$$

and this matches the control volume result from Example 4.6. The approximation gives better than 3% accuracy if $H < 0.07\lambda$. Therefore, surface waves are regarded as *shallow-water waves* only if they are 14 times longer than the water depth. For such long waves, (11.49) shows that the wave speed increases with water depth, and that it is independent of wavelength, so shallow-water waves are *non-dispersive*.

To determine the approximate form of particle orbits for shallow-water waves, substitute the following hyperbolic-function approximations into (11.35):

$$\cosh(k(z+H)) \cong 1, \quad \sinh(k(z+H)) \cong k(z+H), \quad \text{and} \quad \sinh(kH) \cong kH.$$

The particle excursions then become:

$$\xi \cong -\frac{a}{kH}\sin(kx_0 - \omega t), \quad \text{and} \quad \zeta \cong a\left(1 + \frac{z}{H}\right)\cos(kx_0 - \omega t). \tag{11.50}$$

These represent thin ellipses (Fig. 11.4c), with a depth-independent semi-major axis a/kH and a semi-minor axis $a(1 + z/H)$ that linearly decreases to zero at the bottom wall.

From (11.27), the velocity field is:

$$u = \frac{a\omega}{kH}\cos(kx - \omega t), \quad \text{and} \quad w = a\omega\left(1 + \frac{z}{H}\right)\sin(kx - \omega t), \tag{11.51}$$

which shows that the vertical component is much smaller than the horizontal component.

The pressure change from the undisturbed state is found from (11.31) to be:

$$p' = \rho g a \cos(kx - \omega t) = \rho g \eta, \tag{11.52}$$

where (11.2) has been used to express the pressure change in terms of η. This shows that the pressure change at any point is independent of depth, and equals the hydrostatic increase of pressure due to the surface elevation change η. *The pressure field is therefore completely hydrostatic in shallow-water waves.* Vertical accelerations are negligible because of the small w-field. For this reason, shallow water waves are also called *hydrostatic waves*. Any worthwhile pressure sensor mounted on the bottom will sense these waves.

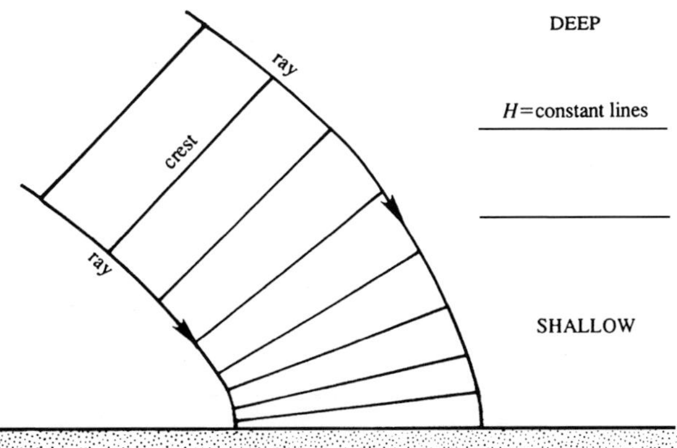

FIGURE 11.7 Refraction of a surface gravity wave approaching a sloping beach caused by changes in depth. In deep water, wave crests are commonly misaligned with isobaths. However, as a wave approaches the shore from any angle, the portion of the wave in shallower water will be slowed compared to that in deeper water. Thus, the wave crests will rotate and tend to become parallel to the shore as they approach it.

The depth-dependent wave speed (11.49) in shallow water leads to the phenomenon of shallow-water wave *refraction* observed at coastlines around the world. Consider a sloping beach, with depth contours parallel to the coastline (Fig. 11.7).

Assume that waves are propagating toward the coast from the deep ocean, with their crests at an angle to the coastline. Sufficiently near the coastline they begin to feel the effect of the bottom and finally become shallow-water waves. Their frequency does not change along the path, but their speed of propagation $c = (gH)^{1/2}$ and their wavelength λ become smaller. Consequently, the crest lines, which are perpendicular to the local direction of c, tend to become parallel to the coast. This is why the waves coming toward a gradually sloping beach always seem to have their crests parallel to the coastline.

An interesting example of wave refraction occurs when a deep-water wave train with straight crests approaches an island (Fig. 11.8). Assume that the water gradually becomes shallower as the island is approached, and that the constant depth contours are circles concentric with the island. Fig. 11.8 shows that the waves always come in *toward* the island, even on the shadowed-side marked A.

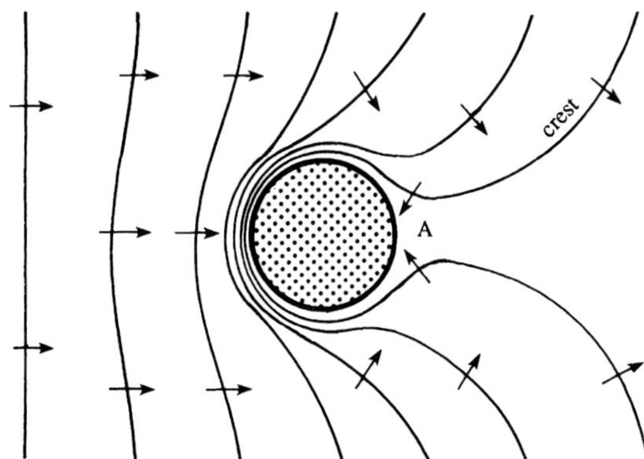

FIGURE 11.8 Refraction of surface gravity waves approaching a circular island with a gradually sloping beach. Crest lines are shown and are observed to travel toward the island, even on its shadow side A. *Reprinted with the permission of Mrs. Dorothy Kinsman Brown: Kinsman,* Wind Waves, *Prentice-Hall, Englewood Cliffs, NJ, 1965.*

The bending of wave paths in an inhomogeneous medium is called *wave refraction*. In this case the source of inhomogeneity is the spatial dependence of H. The analogous phenomenon in optics is the bending of light due to refractive index changes along its path.

Example 11.2

Surfers, sailors, and even dolphins know that steep surface waves can provide propulsion (in the direction of wave travel) to objects located near the free surface. Determine the direction of the near-surface pressure gradient of the linear sinusoidal water wave given by (11.2) to explain the origins of this wave propulsion.

Solution For a surface wave traveling in the positive x-direction, pressure in the water is a sum of the hydrostatic pressure and the pressure perturbation p' produced by the water wave (11.32):

$$p = -\rho g z + p' = -\rho g z + \rho g a \frac{\cosh(k(z+H))}{\cosh(kH)} \cos(kx - \omega t).$$

Thus, the pressure gradient is:

$$\nabla p = -\rho g \mathbf{e}_z - \rho g k a \frac{\cosh(k(z+H))}{\cosh(kH)} \sin(kx - \omega t)\mathbf{e}_x + \rho g k a \frac{\sinh(k(z+H))}{\cosh(kH)} \cos(kx - \omega t)\mathbf{e}_z,$$

which can be rearranged and evaluated on $z = 0$ (the approximate location of the surface) to find:

$$-\frac{1}{\rho}[\nabla p]_{z=0} = gka \sin(kx - \omega t)\mathbf{e}_x + g(1 - ka \tanh(kH)\cos(kx - \omega t))\mathbf{e}_z \cong gka \sin(kx - \omega t)\mathbf{e}_x + g\mathbf{e}_z,$$

where the final approximate equality follows when $ka \ll 1$. The term on the left is the pressure force (per unit mass) on fluid elements near the free surface, and this equation shows that it has both horizontal and vertical components. These can be better understood by noting that for $ka \ll 1$, the surface normal, $\mathbf{n}$, can be approximated by:

$$\mathbf{n} \cong \nabla(z - \eta(x,t)) = \mathbf{e}_z - \mathbf{e}_x(\partial\eta/\partial x) = \mathbf{e}_z + \mathbf{e}_x ka \sin(kx - \omega t) \quad \text{so that} \quad -\frac{1}{\rho}[\nabla p]_{z=0} \cong g\mathbf{n}.$$

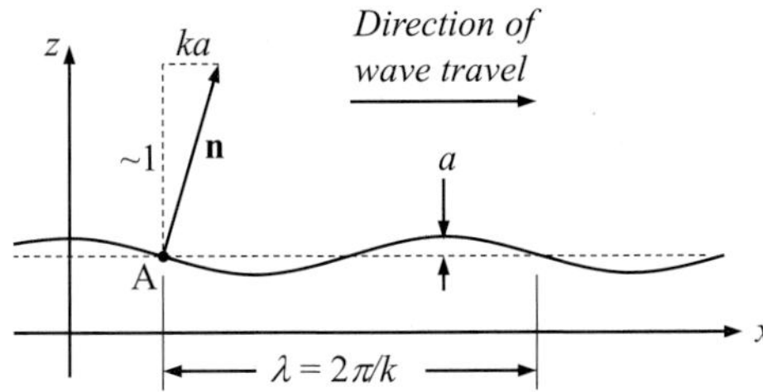

FIGURE 11.9 Sinusoidal wave profile showing the surface normal **n**. The near surface pressure force per unit mass is approximately $g\mathbf{n}$. Thus, an object located in the water near point A is pushed to the right by the pressure gradient in the water. The object may even be transported by the wave when this propulsive force exceeds the object's drag when moving at the wave's speed near the water surface.

Thus, as depicted in Fig. 11.9, the pressure gradient provides a propulsive force in the direction of wave travel that is proportional to ka, the wave's slope (or steepness). Thus, steep waves, such as those near the bow of a moving ship or those that have been slowed by a shallow and upward sloping ocean bottom are most likely to provide propulsion to objects near the free surface; a spectacular example of such propulsion from an ocean wave that abruptly encountered shallow water is shown on the cover of the 6th-edition of this textbook.

11.3 Influence of surface tension

As described in Section 1.6, the interface between two immiscible fluids is in a state of tension. The tension acts as another restoring force on surface deformation, enabling the interface to support waves in a manner analogous to waves on a stretched membrane or string. Waves that occur and propagate because of surface tension are called *capillary waves*. Although gravity is not needed to support these waves, the existence of surface tension alone without gravity is uncommon in terrestrial environments. Thus, the preceding results for pure gravity waves are modified to include surface tension in this section.

As shown in Section 4.10, there is a pressure difference $\Delta p = \sigma(1/R_1 + 1/R_2)$ across a curved interface with non-zero surface tension σ when the surface's principal radii of curvature are R_1 and R_2. The pressure is greater on the side of the surface with the centers of curvature of the interface, and this pressure difference modifies the free-surface boundary condition (11.19).

For straight-crested surface waves that produce fluid motion in the x-z plane, there is no variation in the y-direction, so one of the radii of curvature is infinite, and the other, denoted R, lies in the x-z plane. Thus, if the pressure above the liquid is atmospheric, p_a, then pressure p in the liquid at the surface $z = \eta$ can be found from (1.11):

$$p_a - (p)_{z=\eta} = \sigma\frac{1}{R} = \sigma\frac{\partial^2\eta/\partial x^2}{\left[1 + (\partial\eta/\partial x)^2\right]^{3/2}} \cong \sigma\frac{\partial^2\eta}{\partial x^2}, \tag{11.53}$$

where the second equality follows from the definition of the curvature $1/R$ and the final approximate equality holds when the liquid surface slope $\partial\eta/\partial x$ is small. As before we can choose p to be a gauge pressure and this means setting $p_a = 0$ in (11.53), which leaves:

$$(p)_{z=\eta} = -\sigma\frac{\partial^2\eta}{\partial x^2} \tag{11.54}$$

as the pressure-matching boundary condition at the liquid surface for small slope surface waves. As before, this can be combined with the linearized unsteady Bernoulli equation (11.20) and evaluated on $z = 0$ for small slope surface waves:

$$\left(\frac{\partial\phi}{\partial t}\right)_{z=0} = \frac{\sigma}{\rho}\frac{\partial^2\eta}{\partial x^2} - g\eta. \tag{11.55}$$

The linear capillary-gravity, surface-wave solution now proceeds in an identical manner to that for pure gravity waves, except that the dynamic boundary condition (11.21) is replaced by (11.55). This modification only influences the dispersion relation $\omega(k)$, which is found by substitution of (11.2) and (11.26) into (11.55), to give:

$$\omega = \sqrt{k\left(g + \frac{\sigma k^2}{\rho}\right)\tanh(kH)}, \tag{11.56}$$

so the phase velocity is:

$$c=\sqrt{\left(\frac{g}{k}+\frac{\sigma k}{\rho}\right)\tanh(kH)}=\sqrt{\left(\frac{g\lambda}{2\pi}+\frac{2\pi\sigma}{\lambda\rho}\right)\tanh\left(\frac{2\pi H}{\lambda}\right)}. \tag{11.57}$$

A plot of (11.57) is shown in Fig. 11.10. The primary effect of surface tension is to increase c above its value for pure gravity waves at all wavelengths. This increase occurs because there are two restoring forces that act together on the surface, instead of just one. However, the effect of surface tension is only appreciable for small wavelengths. The nominal size of these wavelengths is obtained by noting that there is a minimum phase speed at $\lambda=\lambda_m$, and surface tension dominates for $\lambda<\lambda_m$ (Fig. 11.10). Setting $dc/d\lambda=0$ in (11.57), and assuming deep water, $H>0.32\lambda$ so $\tanh(2\pi H/\lambda)\approx 1$, produces:

$$c_{\min}=[4g\sigma/\rho]^{1/4} \quad \text{at} \quad \lambda_m=2\pi\sqrt{\sigma/\rho g}. \tag{11.58}$$

For an air-water interface at 20 °C, the surface tension is $\sigma=0.073$ N/m, giving:

$$c_{\min}=23.1 \text{ cm/s} \quad \text{at} \quad \lambda_m=1.71 \text{ cm}. \tag{11.59}$$

Therefore, only short-wavelength waves ($\lambda <\sim 7$ cm for an air-water interface), called *ripples*, are affected by surface tension. The waves specified by (11.59) are readily observed as the wave rings closest to the point of impact after an object is dropped onto the surface of a quiescent pool, pond, or lake of clean water. Surfactants and surface contaminants may lower σ or even introduce additional surface properties like surface viscosity or elasticity. Water-surface wavelengths below 4 mm are dominated by surface tension and are essentially unaffected by gravity. From (11.57), the phase speed of *pure capillary waves* is:

$$c=\sqrt{2\pi\sigma/\rho\lambda}, \tag{11.60}$$

where again $\tanh(2\pi H/\lambda)\approx 1$ has been assumed.

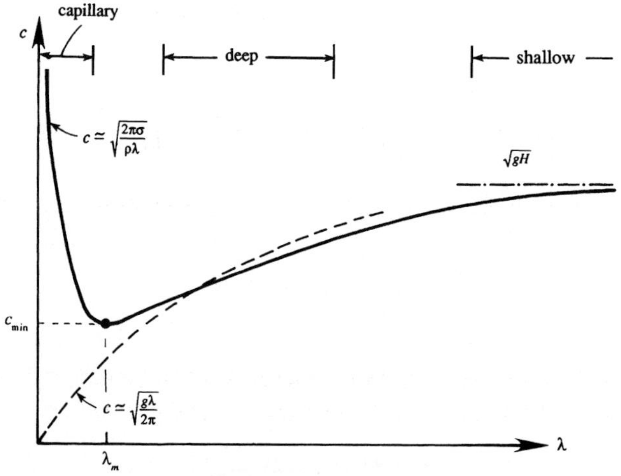

FIGURE 11.10 Generic sketch of the phase velocity c vs. wavelength λ for waves on the surface of liquid layer of depth H. The phase speed of the shortest waves is set by the liquid's surface tension σ and density ρ. The phase speed of the longest waves is set by gravity g and depth H. In between these limits, the phase speed has a minimum that typically occurs when the effects of surface tension and gravity are both important.

Example 11.3

For linear sinusoidal capillary-gravity waves on the surface of a liquid with density ρ and surface tension σ, what water depth H most nearly produces non-dispersive waves when kH is small?

Solution Wave propagation is non-dispersive when the phase speed c is independent of the wavelength λ (or wave number $k=2\pi/\lambda$). For capillary-gravity waves with $kH \ll 1$, $c=[gH]^{1/2}$ is independent of k. However, as k increases for fixed H, c decreases for pure gravity waves but c increases for pure capillary waves. The answer to the question is the liquid depth that causes these competing effects to best cancel as kH increases.

With this background information, start from c^2 from (11.57) and expand the tanh-function for small argument:

$$c^2 = \left(\frac{g}{k} + \frac{\sigma k}{\rho}\right)\tanh(kH) = \left(\frac{g}{k} + \frac{\sigma k}{\rho}\right)\left(kH - \frac{1}{3}(kH)^3 + \frac{2}{15}(kH)^5 + \ldots\right)$$
$$= \frac{g}{k}kH + \frac{\sigma k}{\rho}kH - \frac{g}{3k}(kH)^3 - \frac{\sigma k}{3\rho}(kH)^3 + \frac{2g}{15k}(kH)^5 + \ldots$$

Rearrange the terms to form a power series in kH:

$$c^2 = gH + \left(\frac{\sigma}{\rho H} - \frac{gH}{3}\right)(kH)^2 + \left(-\frac{\sigma}{3\rho H} + \frac{2gH}{15}\right)(kH)^4 + \ldots$$

The first term on the right does not depend on k and represents non-dispersive shallow-water wave propagation. The third term on the right is proportional to $(kH)^4$ and will likely be very small when kH is small. The second term on the right is the one that first causes wave dispersion unless the coefficient of $(kH)^2$ is zero. This occurs when $H = [3\sigma/\rho g]^{1/2}$, the liquid depth that balances the effects of surface tension and gravity, and leads to the least dispersion for capillary-gravity waves. When evaluated for water at 20 °C, this least-dispersive depth is 4.7 mm.

11.4 Standing waves

The wave motion results presented so far are for one propagation direction ($+x$) as specified by (11.2). However, a small-amplitude sinusoidal wave with phase $(kx + \omega t)$ can be an equally valid solution of (11.11). Such a waveform:

$$\eta(x, t) = a\cos[kx + \omega t], \tag{11.61}$$

only differs from (11.2) in its direction of propagation. Its wave crests move in the $-x$-direction with increasing time. Interestingly, non-propagating waves can be generated by superposing two waves with the same amplitude and wavelength that travel in opposite directions. The resulting surface displacement is

$$\eta = a\cos[kx - \omega t] + a\cos[kx + \omega t] = 2a\cos(kx)\cos(\omega t).$$

Here, it follows that $\eta = 0$ at $kx = \pm\pi/2, \pm 3\pi/2$, etc., for all time. Such locations of zero surface displacement are called *nodes*. In this case, deflections of the liquid surface do not travel. The surface simply oscillates up and down at frequency ω with spatially varying amplitude, keeping the nodal points fixed. Such waves are called *standing waves*. The corresponding stream function, a direct extension of (11.37), includes both the $\cos(kx - \omega t)$ and $\cos(kx + \omega t)$ components:

$$\psi = \frac{a\omega}{k}\frac{\sinh(k(z+H))}{\sinh(kH)}[\cos(kx - \omega t) - \cos(kx + \omega t)] = \frac{2a\omega}{k}\frac{\sinh(k(z+H))}{\sinh(kH)}\sin(kx)\sin(\omega t) \tag{11.62}$$

The instantaneous streamline pattern shown in Fig. 11.11 should be compared with the streamline pattern for a propagating wave (Fig. 11.5).

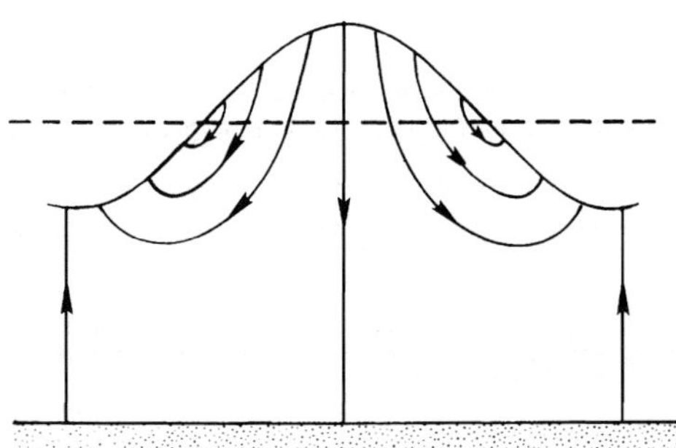

FIGURE 11.11 Instantaneous streamline pattern in a standing surface gravity wave. Here, the $\psi = 0$ streamline follows the bottom and jumps up to contact the surface at wave crests and troughs where the horizontal velocity is zero. If this standing wave represents the $n = 1$ mode of a reservoir of length L with vertical walls, then $L = \lambda/2$ is the distance between a crest and a trough. If it represents the $n = 2$ mode, then $L = \lambda$ is the distance between successive crests or successive troughs.

Standing waves may form in a limited body of water such as a tank, pool, or lake when traveling waves reflect from its walls, sides, or shores. A standing-wave oscillation in a lake is called a *seiche* (pronounced "saysh"), in which only certain wavelengths and frequencies ω (eigenvalues) are allowed by the system. Consider a rectangular tank (an ideal lake) of length L with uniform depth H and vertical walls (shores), and assume that the waves are invariant along y. The possible wavelengths are found by setting $u = 0$ at the two walls. Here, $u = \partial\psi/\partial z$, so (11.62) gives:

$$u = 2a\omega\frac{\cosh(k(z+H))}{\sinh(kH)}\sin(kx)\sin(\omega t). \tag{11.63}$$

Taking the walls at $x = 0$ and L, the condition of no flow through the vertical sidewalls requires $u(x = 0) = u(x = L) = 0$. For non-trivial wave motion, this means $\sin(kL) = 0$, which requires:

$$kL = n\pi, \quad \text{for} \quad n = 1, 2, 3, ...,$$

so the allowable wavelengths are:

$$\lambda = 2L/n. \tag{11.64}$$

The largest possible wavelength ($n = 1$) is $2L$ and the next largest ($n = 2$) is L (Fig. 11.12). The allowed frequencies can be found from the dispersion relation (11.28):

$$\omega = \sqrt{\frac{n\pi g}{L}\tanh\left(\frac{n\pi H}{L}\right)}, \tag{11.65}$$

and these are the natural frequencies of the tank.

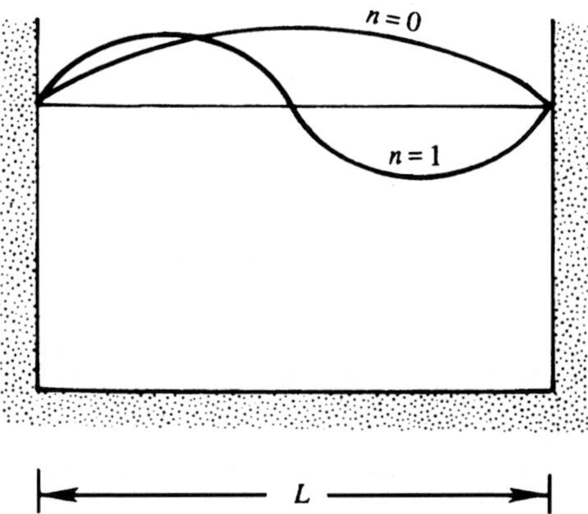

FIGURE 11.12 Distributions of horizontal velocity u for the first two normal modes in a lake or reservoir with vertical sides. Here, the boundary conditions require $u = 0$ on the vertical sides. These distributions are consistent with the streamline pattern of Fig. 11.11.

Example 11.4

What is the average horizontal energy flux of a standing wave?

Solution From (11.43), the time average energy flux EF per unit length of crest can be expressed as a double integral of the product $p'u$. In this case, $u = \partial\phi/\partial x$ is given by (11.63), and $\phi = \int u dx$. For linear surface waves, $p' = -\rho\partial\phi/\partial t$, which means that $p' = -\rho(\partial/\partial t)\int u dx$. Putting all this together produces:

$$EF = \frac{\omega}{2\pi}\int_0^{2\pi/\omega}\int_{-H}^{0} p'u\,dz\,dt$$

$$= \frac{\omega}{2\pi}\int_0^{2\pi/\omega}\int_{-H}^{0}\left(\frac{2\rho a\omega^2}{k}\frac{\cosh(k(z+H))}{\sinh(kH)}\cos(kx)\cos(\omega t)\right)\left(2a\omega\frac{\cosh(k(z+H))}{\sinh(kH)}\sin(kx)\sin(\omega t)\right)dz\,dt.$$

This double integral can be rewritten as a product of integrals:

$$EF = \frac{2}{\pi}\frac{\rho a^2\omega^4}{k}\cos(kx)\sin(kx)\int_0^{2\pi/\omega}\cos(\omega t)\sin(\omega t)dt\int_{-H}^{0}\left(\frac{\cosh(k(z+H))}{\sinh(kH)}\right)^2 dz,$$

but the time integral is zero, so $EF = 0$. Thus, standing waves do not convey wave energy; instead, they represent trapped wave energy.

11.5 Group velocity, energy flux, and dispersion

A variety of interesting phenomena take place when waves are dispersive and their phase speed depends on wavelength. Such wavelength-dependent propagation is common for waves that travel on interfaces between different materials (Graff, 1975). Examples are Rayleigh waves (vacuum and a solid), Stonely waves (a solid and another material), or interface waves (two different immiscible liquids). Here we consider only air-water interface waves and emphasize deep water gravity waves for which c is proportional to $\sqrt{\lambda}$.

In a dispersive system, the energy of a wave component does not propagate at the phase velocity $c = \omega/k$, but at the *group velocity* defined as $c_g = d\omega/dk$. To understand this, consider the superposition of two sinusoidal wave components of equal amplitude but slightly different wave number (and consequently slightly different frequency because $\omega = \omega(k)$). The waveform of the combination is:

$$\eta = a\cos(k_1x - \omega_1 t) + a\cos(k_2x - \omega_2 t) = 2a\cos\left(\frac{1}{2}\Delta kx - \frac{1}{2}\Delta\omega x\right)\cos(kx - \omega t), \tag{11.66}$$

where the trigonometric identity for the sum of cosines of different arguments has been used, $\Delta k = k_2 - k_1$ and $\Delta\omega = \omega_2 - \omega_1$, $k = (k_1 + k_2)/2$, and $\omega = (\omega_1 + \omega_2)/2$. Here, $\cos(kx - \omega t)$ is a progressive wave with a phase speed of $c = \omega/k$. However, its amplitude $2a$ is modulated by a *slowly varying* function $\cos(\Delta kx/2 - \Delta\omega t/2)$, which has a large wavelength $4\pi/\Delta k$, a long period $4\pi/\Delta\omega$, and propagates at a speed (wavelength/period) of:

$$c_g = \Delta\omega/\Delta k \cong d\omega/dk, \tag{11.67}$$

where the approximate equality becomes full in the limit as Δk and $\Delta\omega \to 0$. Multiplication of a rapidly varying sinusoid and a slowly varying sinusoid, as in (11.66), generates repeating wave groups (Fig. 11.13). The individual wave crests (and troughs) propagate with the speed $c = \omega/k$, but the envelope of the wave groups travels with the speed c_g, which is therefore called the *group velocity*. If $c_g < c$, then individual wave crests appear spontaneously at a nodal point, proceed forward through the wave group, and disappear at the next nodal point. If, on the other hand, $c_g > c$, then individual wave crests emerge from a forward nodal point and vanish at a backward nodal point.

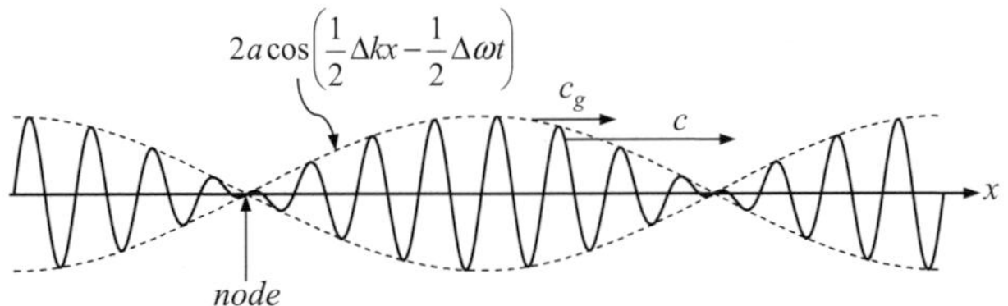

FIGURE 11.13 Linear combination of two equal amplitude sinusoids of nearly the same frequency that form a modulated wave train. Individual wave crests or troughs travel at the phase speed c. However, the nodal locations which partition the wave train into groups, travel at the group speed $c_g = c/2$.

Eq. (11.67) shows that the group speed of waves of a certain wave number k is given by the slope of the *tangent* to the dispersion curve $\omega(k)$. In contrast, the phase velocity is given by the slope of the radius or distance vector on the same plot (Fig. 11.14).

A particularly illuminating example of the idea of group velocity is provided by the concept of a *wave packet*, formed by combining all wave numbers in a certain narrow band δk around a central value k. In physical space, the wave appears nearly sinusoidal with wavelength $2\pi/k$, but the amplitude *dies away* over a distance proportional to $1/\delta k$ (Fig. 11.15). If the spectral width δk is narrow, then decay of the wave amplitude in physical space is slow. The concept of such a wave packet is more realistic than the one in Fig. 11.13, which is rather unphysical because the wave groups repeat themselves.

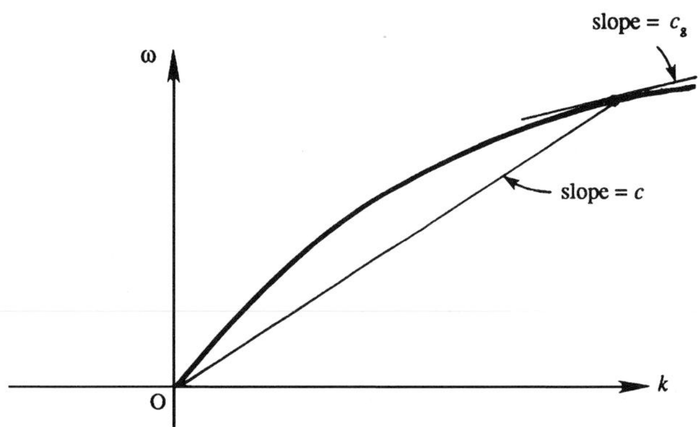

FIGURE 11.14 Graphical depiction of the phase speed, c, and group speed, c_g, on a generic plot of a gravity wave dispersion relation, $\omega(k)$ vs. k. If a sinusoidal wave has frequency ω and wave number k, then the phase speed c is the slope of the straight line through the points (0,0) and (k, ω). While the group speed c_g is the tangent to the dispersion relation at the point (k, ω). For the dispersion relation depicted here, c_g is less than c.

Suppose that, at some initial time, the wave group is represented by $\eta = a(x)\cos(kx)$. It can be shown (see, for example, Phillips, 1977, p. 25) that for small times, the subsequent evolution of the wave profile is approximately described by

$$\eta = a(x - c_g t)\cos(kx - \omega t), \tag{11.68}$$

where $c_g = d\omega/dk$. This shows that the *amplitude of a wave packet travels with the group speed.* It follows that c_g must equal the speed of propagation of *energy* of a certain wavelength. The fact that c_g is the speed of energy propagation is also evident in Fig. 11.13 because the nodal points travel at c_g and no energy crosses nodal points because $p' = 0$ there.

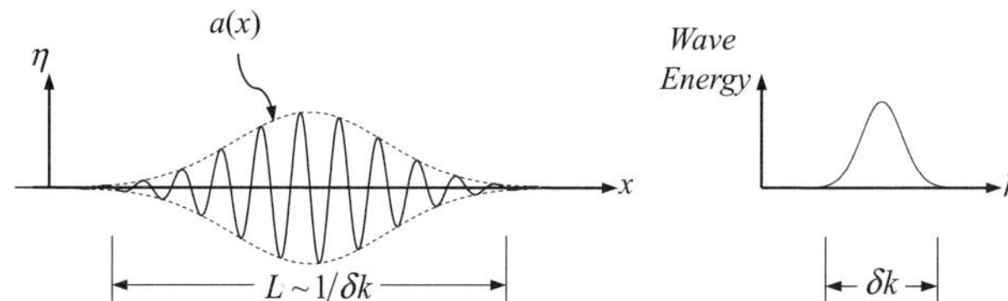

FIGURE 11.15 A wave packet composed of wave numbers lying in a confined bandwidth δk. The length of the wave packet in physical space is proportional to $1/\delta k$. Thus, narrowband packets are longer than broadband packets.

For surface gravity waves having the dispersion relation (11.28), the group velocity is:

$$c_g = \frac{c}{2}\left[1 + \frac{2kH}{\sinh(2kH)}\right], \tag{11.69}$$

which has two limiting cases:

$$c_g = c/2 \text{ (deep water)}, \quad \text{and} \quad c_g = c \text{ (shallow water)}. \tag{11.70}$$

The group velocity of deep-water gravity waves is half the deep-water phase speed while shallow-water waves are non-dispersive with $c = c_g$. For a linear non-dispersive system, any waveform preserves its shape as it travels because all the wavelengths that make up the waveform travel at the same speed. For a pure capillary wave, the group velocity is $c_g = 3c/2$ (Exercise 11.10).

The energy flux for gravity waves is given by (11.44), namely:

$$EF = E\frac{c}{2}\left[1 + \frac{2kH}{\sinh(2kH)}\right] = Ec_g, \tag{11.71}$$

where $E = \rho g a^2/2$ is the average energy in the water column per unit horizontal area. This signifies that the *rate of transmission of energy of a sinusoidal wave component is the wave energy times the group velocity*, and reinforces the interpretation of the group velocity as the speed of propagation of wave energy.

In three dimensions, the dispersion relation $\omega = \omega(k, l, m)$ may depend on all three components of the wave number vector $\mathbf{K} = (k, l, m)$. Here, using index notation, the group velocity vector is given by:

$$c_{gi} = \partial\omega/\partial K_i,$$

so the group velocity vector is the gradient of ω in the wave number space.

As mentioned in connection with (11.28) and (11.59), deep-water wave dispersion readily explains the evolution of the surface disturbance generated by dropping a stone into a quiescent pool, pond, or lake. Here, the initial disturbance can be thought of as being composed of a great many wavelengths, but the longer ones travel faster. A short time after impact, at $t = t_1$, the water surface may have the rather irregular profile shown in Fig. 11.16. The appearance of the surface at a later time t_2, however, is more regular, with the longer components (which travel faster) out in front. The waves in front are the longest waves produced by the initial disturbance. Their length, $\lambda_{\max}$, is typically a few times larger than the dropped object. The leading edge of the wave system therefore propagates at the group speed of these wavelengths:

$$c_{g\max} = \frac{1}{2}\sqrt{g\lambda_{\max}/2\pi}.$$

Pure capillary waves can propagate faster than this speed, but they may have small amplitudes and are dissipated quickly. Interestingly, the region of the impact becomes calm because there is a minimum group velocity for water waves due to the influence of surface tension, namely $c_{g\min} = 17.8$ cm/s (Exercise 11.11), and the trailing edge of the wave system travels at this speed. With $c_{g\min} > 17.8$ cm/s for ordinary hand-size stones, the length of the disturbed region gets larger, as shown in Fig. 11.16. The wave heights become correspondingly smaller because there is a fixed amount of energy in the wave system. (Wave dispersion, therefore, makes the linearity assumptions more accurate.) The smoothing of the waveform and the spreading of the region of disturbance continue until the amplitudes become imperceptible or the waves are damped by viscous dissipation (Exercise 11.13). The *initial superposition of various wavelengths, running for some time, will sort themselves from slowest to fastest traveling components* since the different sinusoidal components, differing widely in their wave numbers, become spatially *separated*, with the slow ones closer to the point of impact and the fast ones further away. This is a basic feature of the behavior of dispersive wave propagation.

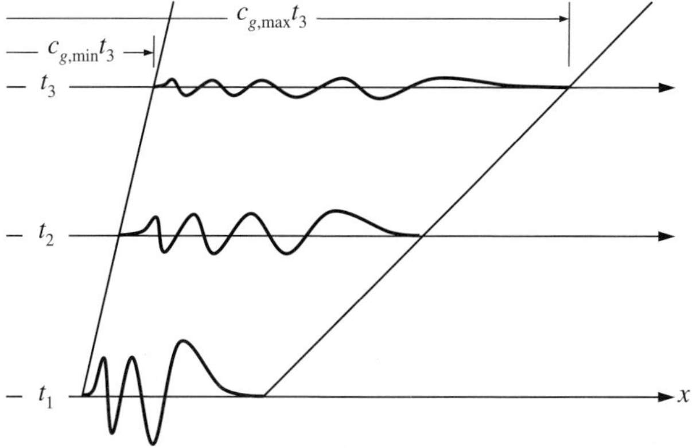

FIGURE 11.16 Generic surface profiles at three successive times of the wave train produced by dropping a stone into a deep quiescent pool. As time increases, the initial disturbance's long-wave (low-frequency) components travel faster than its short-wave (high-frequency) components. Thus, the wave train lengthens, the number of crests and troughs increases, and amplitudes fall (to conserve energy).

In the case of deep-water surface waves described here, the wave group as a whole travels slower than individual crests. Therefore, if we try to follow the last crest at the rear of the wave train, quite soon it is the second one from the rear; a new crest has appeared behind it. In fact, new crests are constantly appearing at the rear of the train, propagating through the wave train, and finally disappearing at the front of the wave train. This is because, by following a particular crest, we are traveling at roughly twice the speed at which the wave energy is traveling. Consequently, *we do not see a wave of fixed wavelength if we follow a particular crest.* In fact, an individual wave oscillation constantly becomes longer as it propagates through the train. When its length becomes equal to the longest wave generated initially, it cannot evolve further and disappears. The waves at the front of the train are the longest Fourier components present in the initial disturbance. In addition, the temporal frequencies of the highest and lowest speed wave components of the wave group are typically different enough so that the number of wave crests in the train increases with time.

Another way to understand the group velocity is to consider the k or λ determined by an observer traveling at speed c_g with a slowly varying wave train described by:

$$\eta = a(x,t)\cos[\theta(x,t)], \tag{11.72}$$

in an otherwise quiescent pool of water with constant depth H. Here, $a(x,t)$ is a slowly varying amplitude and $\theta(x,t)$ is the local phase. For a specific wave number k and frequency ω, the phase is $\theta = kx - \omega t$. For a slowly varying wave train, define the *local* wave number $k(x,t)$ and the *local* frequency $\omega(x,t)$ as the rate of change of phase in space and time, respectively:

$$k(x,t) \equiv (\partial/\partial x)\theta(x,t) \quad \text{and} \quad \omega(x,t) \equiv -(\partial/\partial t)\theta(x,t). \tag{11.73}$$

Cross-differentiation leads to:

$$\partial k/\partial t + \partial\omega/\partial x = 0, \tag{11.74}$$

but when there is a dispersion relationship $\omega = \omega(k)$, the spatial derivative of ω can be rewritten using the chain rule, $\partial\omega/\partial x = (d\omega/dk)\partial k/\partial x = c_g \partial k/\partial x$, so that (11.74) becomes:

$$\partial k/\partial t + c_g \partial k/\partial x = 0. \tag{11.75}$$

The left-hand side of (11.75) is similar to the material derivative and gives the rate of change of k as seen by an observer traveling at speed c_g, which in this case is zero. Therefore, such an observer will always see the same wavelength. The *group velocity is therefore the speed at which wave numbers travel.* This is shown in the xt-diagram of Fig. 11.17, where wave crests follow lines with $dx/dt = c$ and wavelengths are preserved along the lines $dx/dt = c_g$. Note that the width of the disturbed region, bounded by the first and last thick lines in Fig. 11.17, increases with time, and that the crests constantly appear at the back of the group and vanish at the front.

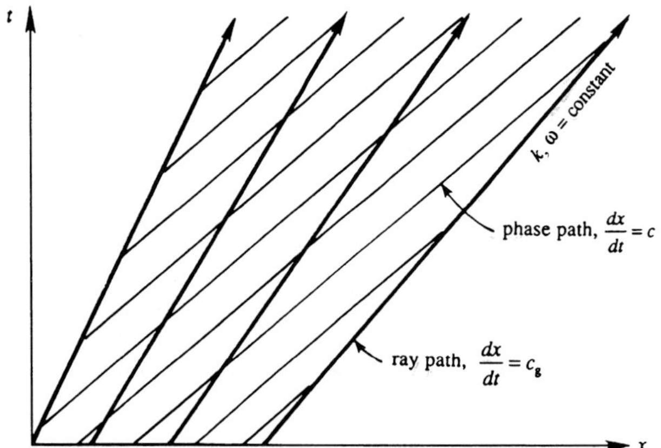

FIGURE 11.17 Propagation of a wave group in a homogeneous medium, represented on an x-t plot. Thin lines indicate paths taken by wave crests, and thick lines represent paths along which k and ω are constant. *M.J. Lighthill,* Waves in Fluids, *1978, reprinted with the permission of Cambridge University Press, London.*

Now consider the same traveling observer, but allow there to be smooth variations in the water depth $H(x)$. Such depth variation creates an inhomogeneous medium when the waves are long enough to feel the presence of the bottom. Here, the dispersion relationship will be:

$$\omega = \sqrt{gk\tanh[kH(x)]},$$

which is of the form:

$$\omega = \omega(k,x). \tag{11.76}$$

Thus, a local value of the group velocity can be defined:

$$\partial\omega(k,x)/\partial k = c_g, \tag{11.77}$$

which on multiplication by $\partial k/\partial t$ gives:

$$c_g(\partial k/\partial t) = (\partial\omega/\partial k)(\partial k/\partial t) = \partial\omega/\partial t. \qquad (11.78)$$

Multiplying (11.74) by c_g and using (11.78) we obtain

$$\partial\omega/\partial t + c_g \partial\omega/\partial x = 0. \qquad (11.79)$$

In three dimensions, this implies:

$$\partial\omega/\partial t + \mathbf{c}_g \cdot \nabla\omega = 0,$$

which shows that ω, the frequency of the wave, remains constant to an observer traveling with the group velocity in an inhomogeneous medium.

Summarizing, an observer traveling at c_g in a homogeneous medium sees constant values of k, $\omega(k)$, c, and $c_g(k)$. Consequently, ray paths describing the group velocity in the x-t plane are straight lines (Fig. 11.17). In an inhomogeneous medium ω remains constant along the lines $dx/dt = c_g$, but k, c, and c_g can change. Consequently, ray paths are not straight in this case (Fig. 11.18).

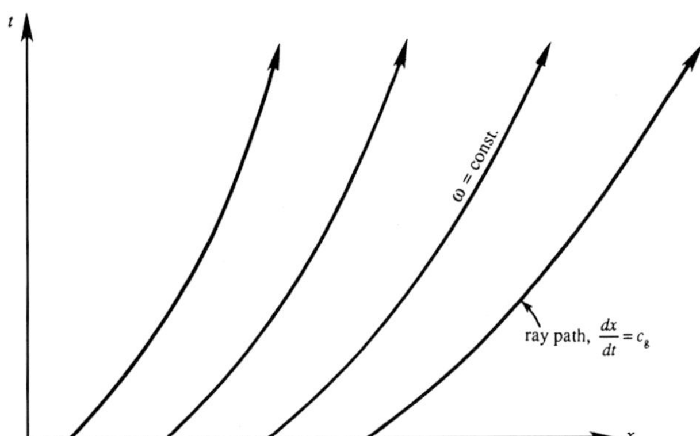

FIGURE 11.18 Propagation of a wave group in an inhomogeneous medium represented on an x-t plot. Only ray paths along which ω is constant are shown. *M.J. Lighthill,* Waves in Fluids, *1978, reprinted with the permission of Cambridge University Press, London.*

Example 11.5

In deep water, the trailing wave field produced by a ship moving at a constant velocity is found to be confined within a wedge of half angle slightly less than 20° (the Kelvin wedge) that is nearly independent of the ship's velocity. Explain this phenomenon by considering a straight ship trajectory at constant speed U, and the deep-water gravity-wave relationship for group and phase speeds: $c_g = c/2$.

Solution The important facts to consider here are that an individual wave crest travels at speed c, while the disturbance energy from the moving ship travels at speed $c_g = c/2$. An overhead diagram of the wave-front geometry on the starboard side of the ship is shown on the left in Fig. 11.19, where the ship is represented by a point disturbance moving to the left at constant speed U. Such a spatially concentrated disturbance generates waves that may propagate in all directions. Thus, if the ship generates waves at location A, all possible wave propagation angles β must be considered at subsequent times to determine the width of the ship's trailing wave field.

The kinematics of the ship and its trailing wave field may be developed as follows. If the ship moves from A to B in time t, the distance between these two points is Ut. The wave fronts generated by the ship at point A that travel at the phase speed c in the direction indicated by the angle β will have the orientation given by the segment BC. However, these wave fronts will not appear at point C since the wave energy originating at A and traveling at angle β will only have reached point D in time t since $c_g = c/2$. Thus, looking aft over the ship's stern, the azimuthal half angle, φ, that indicates where these waves appear is given by:

$$\tan\varphi = \frac{DE}{BE} = \frac{(ct/2)\sin\beta}{Ut - (ct/2)\cos\beta} = \frac{\sin\beta}{(2U/c) - \cos\beta} = \frac{\sin\beta}{(2/\cos\beta) - \cos\beta} = \frac{\sin(2\beta)}{3 - \cos(2\beta)},$$

where the second to last equality is possible because $\cos\beta = CA/BA = c/U$, and the final equality follows after a little algebraic rearrangement and use of double-angle trigonometric identities.

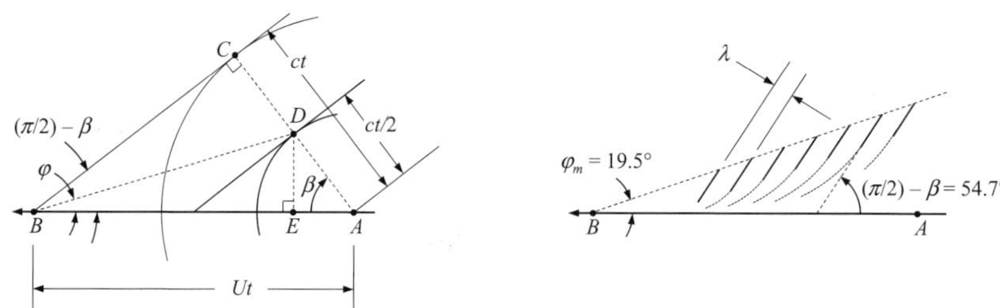

FIGURE 11.19 Overhead view of the wave kinematics for a point disturbance (a ship) moving steadily at speed U in deep water where $c_g = c/2$. The wave field is symmetrical about the disturbance trajectory, so only the upper (starboard) half of the wave field is shown. In time t, the disturbance moves from A to B. The diagram on the left shows that waves generated at point A that travel in the direction indicated by angle β reach point D in time t. When viewed from the location of the disturbance, the azimuthal angle of these waves is φ. When all possible values of β are considered, the disturbance's trailing wave field will lie within the maximum-possible azimuthal angle $\varphi_m = 19.5°$. Features of the resulting wave field are illustrated in the diagram on the right. Wave fronts at the edge of the wave field are shown as short solid upward-sloping segments and have an orientation of 54.7°. The wavelength λ is proportional to U^2/g.

Since the ship sends waves in all directions as it travels, the observed angular width φ_m of the ship's trailing wave field will be the maximum φ that occurs as β is varied. To find this maximum, compute $d(\tan\varphi)/d\beta$, set it equal to zero, solve for β, and use this value of β to determine φ_m. These steps are readily completed using the above formula for $\tan\varphi$:

$$\frac{d\tan\varphi}{d\beta} = \frac{2\cos(2\beta)}{3-\cos(2\beta)} - \frac{\sin(2\beta)}{(3-\cos(2\beta))^2}2\sin(2\beta) = 0.$$

After combining terms, this equation becomes $6\cos(2\beta) - 2[\cos^2(2\beta) + \sin^2(2\beta)] = 0$ which reduces to $\cos(2\beta) = 1/3$, so $\sin(2\beta) = \sqrt{8/9} = 2\sqrt{2}/3$. Thus, when $c_g = c/2$, the wave front orientation at the edge of the ship's trailing wave field is set by $\beta = (1/2)\cos^{-1}(1/3) = 35.3°$, and the angular width of the trailing wave field of a steadily moving ship is:

$$\varphi_m = \tan^{-1}\frac{2\sqrt{2}/3}{3-(1/3)} = \tan^{-1}\left(\frac{1}{2\sqrt{2}}\right) \cong 19.5°.$$

This is the desired result, and these wave-field features are illustrated in the diagram on the right in Fig. 11.19. The observed wavelength at the wave field's edge is determined by the ship's speed U, the wave-field geometry ($U\cos\beta = c$), and (11.45):

$$\lambda = 2\pi\frac{U^2}{g}\cos^2\beta = \frac{\pi U^2}{3g}.$$

At a speed of 12 knots (6.17 m/s), this leads to a wavelength of 4.06 m. Further description of a ship's trailing wave field may be found in Whitham (1974) or Lighthill (1978).

11.6 Nonlinear waves in shallow and deep water

In the first five sections of this chapter, the wave slope has been assumed to be small enough so that neglect of higher-order terms in the Bernoulli equation and application of the boundary conditions at $z = 0$ instead of at the free surface $z = \eta$ are acceptable approximations. One consequence of such linear analysis has been that shallow-water waves of arbitrary shape propagate unchanged in form. The unchanging form results from the fact that all wavelengths composing the initial waveform propagate at the same speed, $c = (gH)^{1/2}$, provided all the sinusoidal components satisfy the shallow-water and linear wave approximations $kH \ll 1$, and $ka \ll 1$, respectively. Such waveform invariance no longer occurs if *finite amplitude* effects are considered. This and several other nonlinear effects are discussed in this section.

Finite amplitude effects in gas dynamics can be formally treated by the *method of characteristics*; this briefly introduced in Section 14.7. Instead, a qualitative approach is initially adopted here. Consider a finite amplitude surface displacement consisting of a wave crest and trough, propagating in shallow-water of undisturbed depth H (Fig. 11.20). Let a little wavelet be superposed on the crest at point x', at which the water depth is H' and the fluid velocity due to the wave motion is $u(x')$. Relative to an observer moving with the fluid velocity u, the wavelet propagates at the local shallow-water speed $c' = \sqrt{gH'}$. The speed of the wavelet relative to a frame of reference fixed in the undisturbed fluid is therefore $c = c' + u$. The local wave speed c is no longer constant because $c'(x)$ and $u(x)$ are variables. This is in contrast to the linearized theory in which u is negligible and c' is constant because $H' \approx H$.

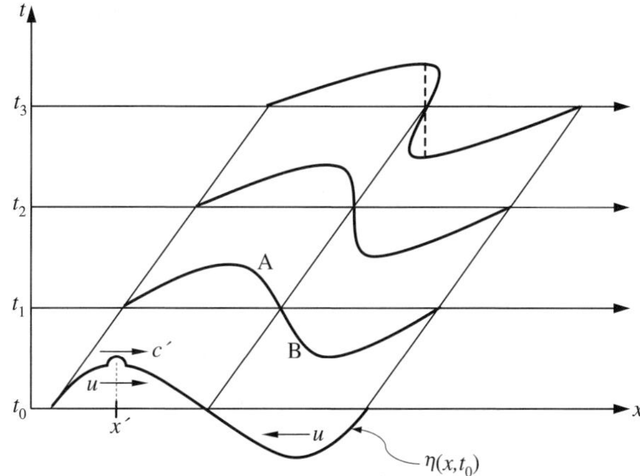

FIGURE 11.20 Finite-amplitude surface wave profiles at four successive times. When the wave amplitude is large enough, the fluid velocity below a crest or trough may be an appreciable fraction of the phase speed. This will cause wave crests to overtake wave troughs and will steepen the compressive portion of the wave (section A-B at time t_1). As this steepening continues, the wave-compression surface slope may become very large (t_2), or the wave may overturn and become a plunging breaker (t_3). Depending on the dynamics of the actual wave, the conditions shown at t_2 and t_3 may or may not occur since additional nonlinear processes (not described here) may contribute to the wave's evolution after t_1. When the waves are longitudinal (as in one-dimensional gas dynamics), the waveform at t_2 would represent a nascent shockwave, while the wave waveform at t_3 would represent a fully formed shockwave and would follow the dashed line to produce a single-valued profile.

Let us now examine the effect of variable phase speed on the wave profile. The value of c' is larger for points near the wave crest than for points in the wave trough. From Fig. 11.5 we also know that the fluid velocity u is positive (i.e., in the direction of wave propagation) under a wave crest and negative under a trough. It follows that wave speed c is larger for points on the crest than for points on the trough, so that the waveform deforms as it propagates, the crest region tending to overtake the trough region (Fig. 11.20).

We shall call the front face AB a *compression region* because the surface here is rising with time and this implies an increase in pressure at any depth within the liquid. Fig. 11.20 shows that the net effect of nonlinearity is a steepening of the compression region. For finite amplitude waves in a non-dispersive medium like shallow water, therefore, there is an important distinction between compression and expansion regions. A compression region tends to steepen with time, while an expansion region tends to flatten out. This eventually would lead to the wave shape shown at the top of Fig. 11.20, where there are three values of surface elevation at a point. While this situation is certainly possible for time-evolving waves and is readily observed as plunging breakers develop in the surf zone along ocean coastlines, the actual wave dynamics of such a situation lie beyond the scope of this discussion. However, even before the formation of a plunging breaker, the wave slope becomes infinite (profile at t_2 in Fig. 11.20), so that additional physical processes including wave breaking, air entrainment, and foaming become important, and the current ideal flow analysis becomes inapplicable. Once the wave has broken, it takes the form of a front that propagates into still fluid at a constant speed that lies between $\sqrt{gH_1}$ and $\sqrt{gH_2}$ where H_1 and H_2 are the water depths on the two sides of the front (Fig. 11.21). Such a wave is called a *hydraulic jump*, and it is similar to a *shockwave* in a compressible flow. Here it should be noted that the t_3 wave profile shown in Fig. 11.20 is not possible for longitudinal gas-dynamic compression waves. Such a profile instead leads to a shockwave with a front shown by the dashed line.

To analyze a hydraulic jump, consider the flow in a shallow canal of depth H. If the flow speed is u, we may define a dimensionless speed via the Froude number, Fr:

$$\mathrm{Fr} \equiv u \Big/ \sqrt{gH} = u/c. \tag{4.105}$$

The Froude number is analogous to the *Mach number* in compressible flow. The flow is called *supercritical* if Fr > 1, and *subcritical* if Fr < 1. For the situation shown in Fig. 11.21b, where the jump is stationary, the upstream flow is supercritical while the downstream flow is subcritical, just as a compressible flow changes from supersonic to subsonic by going through a shockwave (see Chapter 14). The depth of flow is greater downstream of a hydraulic jump, just as the gas pressure is greater downstream of a shockwave. However, dissipative processes act within shockwaves and hydraulic jumps so that mechanical energy is converted into thermal energy in both cases. An example of a stationary hydraulic jump is found at the foot of a dam, where the flow almost always reaches a supercritical state because of the freefall (Fig. 11.21a). A tidal bore propagating into a river mouth is an example of a propagating hydraulic jump. A circular hydraulic jump can be made by directing a vertically falling water stream onto a flat horizontal surface (Exercise 4.29).

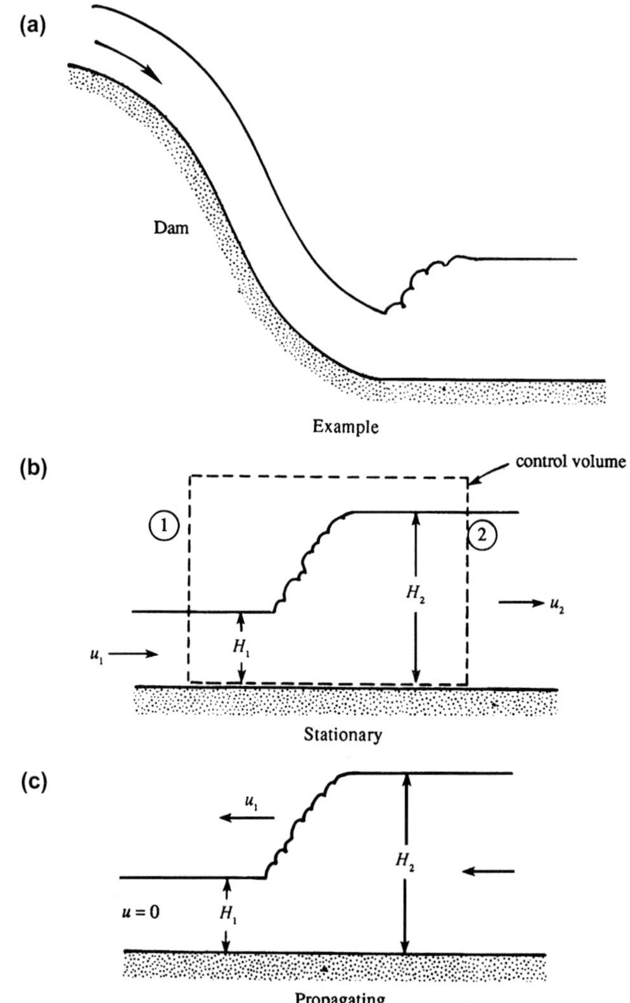

FIGURE 11.21 Schematic cross-section drawings of hydraulic jumps. (a) A stationary hydraulic jump formed at the bottom of a damn's spillway. (b) A stationary hydraulic jump and a stationary rectangular control volume with vertical inlet surface (1) and vertical outlet surface (2). (c) A hydraulic jump moving into a quiescent fluid layer of depth H_1. The flow speed behind the jump is non-zero.

The planar hydraulic jump shown in cross-section in Fig. 11.21b can be analyzed by using the dashed control volume shown, the goal being to determine how the depth ratio depends on the upstream Froude number. As shown, the depth rises from H_1 to H_2 and the velocity falls from u_1 to u_2. If the velocities are uniform through the depth and Q is the volume flow rate per unit width normal to the plane of the paper, then mass conservation requires:

$$Q = u_1 H_1 = u_2 H_2.$$

Conserving momentum with the same control volume via (4.17) with $d/dt = 0$ and $\mathbf{b} = 0$, produces:

$$\rho Q(u_2 - u_1) = \frac{1}{2}\rho g(H_1^2 - H_2^2),$$

where the left-hand terms come from the outlet and inlet momentum fluxes, and the right-hand terms are the hydrostatic pressure forces. Substituting $u_1 = Q/H_1$ and $u_2 = Q/H_2$ on the right side yields:

$$Q^2\left(\frac{1}{H_2} - \frac{1}{H_1}\right) = \frac{1}{2}g(H_1^2 - H_2^2). \tag{11.80}$$

After canceling out a common factor of $H_1 - H_2$, this can be rearranged to find:

$$\left(\frac{H_2}{H_1}\right)^2 + \frac{H_2}{H_1} - 2\mathrm{Fr}_1^2 = 0,$$

where $\mathrm{Fr}_1^2 = Q^2/gH_1^3 = u_1^2/gH_1$. The physically meaningful solution is:

$$H_2/H_1 = \frac{1}{2}\left(-1 + \sqrt{1 + 8\mathrm{Fr}_1^2}\right). \tag{11.81}$$

For supercritical flows $\mathrm{Fr}_1 > 1$, for which (11.81) requires that $H_2 > H_1$, and this verifies that water depth increases through a hydraulic jump.

Although a solution with $H_2 < H_1$ for $\mathrm{Fr}_1 < 1$ is mathematically allowed, such a solution violates the second law of thermodynamics, because it implies an increase of mechanical energy through the jump. To see this, consider the mechanical energy of a fluid particle at the surface, $E = u^2/2 + gH = Q^2/2H^2 + gH$. Using this definition of E and eliminating Q by using (11.80) leads to:

$$E_2 - E_1 = -(H_2 - H_1)\frac{g(H_2 - H_1)^2}{4H_1H_2}.$$

This shows that $H_2 < H_1$ implies $E_2 > E_1$, which violates the second law of thermodynamics. The mechanical energy, in fact, *decreases* in a hydraulic jump because of the action of viscosity.

Hydraulic jumps are not limited to air-water interfaces and may also appear at density interfaces in a stratified fluid, in the laboratory as well as in the atmosphere and the ocean. (For example, see Turner, 1973, Fig. 3.11, for a photograph of an internal hydraulic jump on the lee side of a mountain.)

In a non-dispersive medium, nonlinear effects may continually accumulate until they become large changes. Such an accumulation is prevented in a dispersive medium because the different Fourier components propagate at different speeds and tend to separate from each other. In a dispersive system, then, nonlinear steepening could cancel out the dispersive spreading, resulting in finite amplitude waves of constant form. This is indeed the case. A brief description of the phenomenon is given here; further discussion can be found in Whitham (1974), Lighthill (1978), and LeBlond and Mysak (1978).

In 1847, Stokes showed that periodic waves of finite amplitude are possible in deep water. In terms of a power series in the amplitude a, he showed that the surface deflection of irrotational waves in deep water is given by:

$$\eta = a\cos[k(x - ct)] + \frac{1}{2}ka^2\cos[2k(x - ct)] + \frac{3}{8}k^2a^3\cos[3k(x - ct)] + ..., \tag{11.82}$$

where the speed of propagation is:

$$c = \sqrt{\frac{g}{k}(1 + k^2a^2 + ...)}. \tag{11.83}$$

Eq. (11.82) shows the first three terms in a Fourier series for the waveform η. The addition of Fourier components of different wavelengths in (11.82) shows that the wave profile η is no longer exactly sinusoidal. The arguments in the cosine terms show that all the Fourier components propagate at the same speed c, so that the wave profile propagates unchanged in time. It has now been established that the existence of periodic wave trains of unchanging form is a typical feature of nonlinear dispersive systems. Another important result, generally valid for nonlinear systems, is that the wave speed depends on the amplitude, as in (11.83).

Periodic finite-amplitude irrotational waves in deep water are frequently called *Stokes waves*. They have flattened troughs and peaked crests (Fig. 11.22). The maximum possible amplitude is $a_{max} = 0.07\lambda$, at which point the crest becomes a sharp 120° angle. Attempts at generating waves of larger amplitude result in the appearance of foam (*white caps*) at these sharp crests.

FIGURE 11.22 The waveform of a Stokes wave. Stokes waves are finite-amplitude, periodic irrotational waves in deep water with crests that are more pointed and troughs that are broader than sinusoidal waves.

When finite amplitude waves are present, fluid particles no longer trace closed orbits, but undergo a slow drift in the direction of wave propagation. This is called *Stokes drift*. It is a second-order or finite-amplitude effect that causes fluid particle orbits to no longer close and instead take a shape like that shown in Fig. 11.23. The mean velocity of a fluid particle is therefore not zero, although the mean velocity at a fixed point in space must be zero if the wave motion is periodic. The drift occurs because the particle moves forward faster when at the top of its trajectory than it does backward when at the bottom of its trajectory.

To find an expression for the Stokes drift, start from the path-line equations (11.32) for the fluid particle trajectory $\mathbf{x}_p(t) = x_p(t)\mathbf{e}_x + z_p(t)\mathbf{e}_z$, but this time include first-order variations in the u and w fluid velocities via a first-order Taylor series in $\xi = x_p - x_0$, and $\zeta = z_p - z_0$:

$$\frac{dx_p(t)}{dt} = u(x_p, z_p, t) = u(x_0, z_0, t) + \xi\left(\frac{\partial u}{\partial x}\right)_{x_0,z_0} + \zeta\left(\frac{\partial u}{\partial z}\right)_{x_0,z_0} + \ldots, \text{ and} \tag{11.84a}$$

$$\frac{dz_p(t)}{dt} = w(x_p, z_p, t) = w(x_0, z_0, t) + \xi\left(\frac{\partial w}{\partial x}\right)_{x_0,z_0} + \zeta\left(\frac{\partial w}{\partial z}\right)_{x_0,z_0} + \ldots, \tag{11.84b}$$

where (x_0, z_0) is the fluid element location in the absence of wave motion. The Stokes drift is the time average of (11.84a). However, the time average of $u(x_0, z_0, t)$ is zero; thus, the Stokes drift is given by the time average of the next two terms of (11.84a). These terms were neglected in the fluid particle trajectory analysis in Section 11.2, and the result was closed orbits.

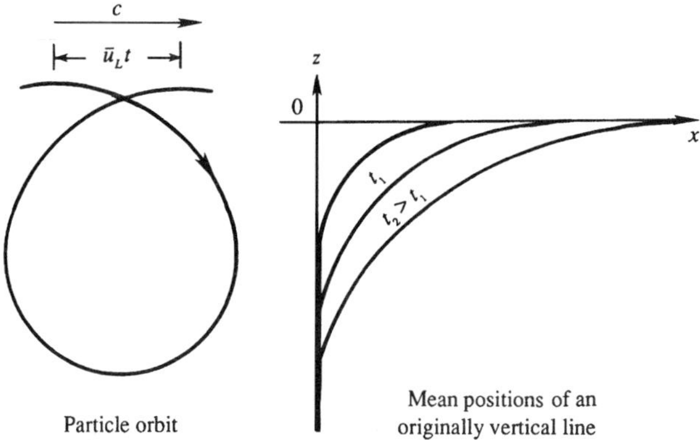

FIGURE 11.23 The Stokes drift. The drift velocity $\overline{u}_L$ is a finite-amplitude effect and occurs because near-surface fluid particle paths are no longer closed orbits. The mean position of an initially vertical line of fluid particles extending downward from the liquid surface will increasingly bend in the direction of wave propagation with increasing time.

For deep-water gravity waves, the Stokes drift speed $\overline{u}_L$ can be estimated by evaluating the time average of (11.84a) using (11.47) to produce:

$$\overline{u}_L = a^2 \omega k e^{2kz_0}, \tag{11.85}$$

which is the Stokes drift speed in deep water. Its surface value is $a^2\omega k$, and the vertical decay rate is twice that for the fluid velocity components. It is therefore confined very close to the sea surface. For arbitrary water depth, (11.85) may be generalized to:

$$\overline{u}_L = a^2 \omega k \frac{\cosh(2k(z_0 + H))}{2\sinh^2(kH)} \tag{11.86}$$

(Exercise 11.16). As might be expected, the vertical component of the Stokes drift is zero.

The Stokes drift causes mass transport in the fluid so it is also called the *mass transport velocity*. A vertical column of fluid elements marked by some dye gradually bend near the surface (Fig. 11.23). In spite of this mass transport, the mean fluid velocity at any point that resides within the liquid for the entire wave period is exactly zero (to any order of accuracy), if the flow is irrotational. This follows from the condition of irrotationality $\partial u/\partial z = \partial w/\partial x$, a vertical integral of which gives:

$$u = [u]_{z=-H} + \int_{-H}^{z} (\partial w/\partial x)dz,$$

showing that the mean of u is proportional to the mean of $\partial w/\partial x$ over a wavelength, which is zero for periodic flows.

There also exits a variety of wave analyses for specialized circumstances that involve dispersion, nonlinearity, and viscosity to varying degrees. So, before moving on to internal waves, one of the classical examples of this type of specialization is presented here for nonlinear waves that are slightly dispersive. In 1895, Korteweg and de Vries showed that waves with λ/H in the range between 10 and 20 satisfy:

$$\frac{\partial \eta}{\partial t}+c_0\frac{\partial \eta}{\partial x}+\frac{3}{8}c_0\frac{\eta}{H}\frac{\partial \eta}{\partial x}+\frac{1}{6}c_0H^2\frac{\partial^3 \eta}{\partial x^3}=0, \tag{11.87}$$

where $c_0=\sqrt{gH}$. This is the *Korteweg–de Vries equation*. The first two terms are linear and non-dispersive. The third term is nonlinear and represents finite amplitude effects. The fourth term is linear and results from weak dispersion due to the water depth not being shallow enough. If the nonlinear term in (11.87) is neglected, then setting $\eta=a\cos(kx-\omega t)$ leads to the dispersion relation $c=c_0(1-(1/6)k^2H^2)$. This agrees with the first two terms in the Taylor series expansion of $c^2=(g/k)\tanh(kH)$ for small kH, verifying that weak dispersive effects are indeed properly accounted for by the last term in (11.87).

The ratio of nonlinear and dispersion terms in (11.87) is:

$$\frac{\eta}{H}\frac{\partial \eta}{\partial x}\Big/ H^2\frac{\partial^3 \eta}{\partial x^3}\sim\frac{a\lambda^2}{H^3}.$$

When $a\lambda^2/H^3$ is larger than ~16, nonlinear effects sharpen the forward face of the wave, leading to a hydraulic jump, as discussed earlier in this section. For lower values of $a\lambda^2/H^3$, a balance can be achieved between nonlinear steepening and dispersive spreading, and waves of unchanging form become possible.

Analysis of the Korteweg–de Vries equation shows that two types of solutions are then possible – a periodic solution and a solitary wave solution. The periodic solution is called a *cnoidal wave*, because it is expressed in terms of elliptic functions denoted by $cn(x)$. Its waveform is shown in Fig. 11.24. The other possible solution of the Korteweg–de Vries equation involves only a single wave crest and is called a *solitary wave* or *soliton*. Its profile is given by:

$$\eta=a\,\mathrm{sech}^2\left[\left(\frac{3a}{4H^3}\right)^{1/2}(x-ct)\right],\quad \text{where } c=c_0\left(1+\frac{a}{2H}\right), \tag{11.88}$$

showing that the propagation velocity increases with amplitude. The validity of (11.88) can be checked by substitution into (11.87) (Exercise 11.17). The waveform of the solitary wave is shown in Fig. 11.24.

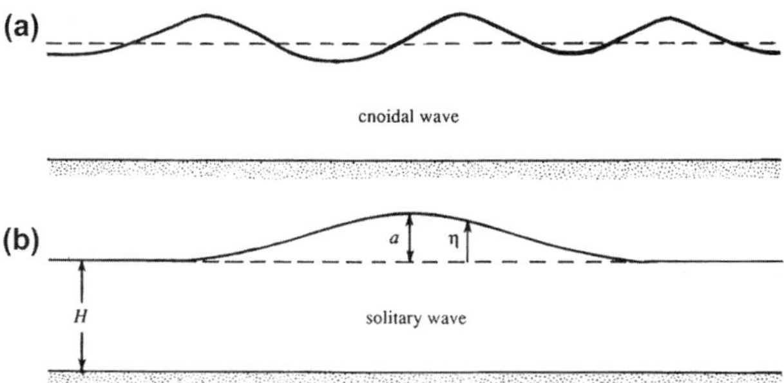

FIGURE 11.24 Finite-amplitude waves of unchanging form: (a) cnoidal waves and (b) a solitary wave. In both cases, the processes of nonlinear steepening and dispersive spreading balance so that the waveform is unchanged.

An isolated single-hump water wave propagating at constant speed with unchanging form and in fairly shallow water was first observed experimentally by S. Russell in 1844. Solitons have been observed to exist not only as surface waves, but also as internal waves in stratified fluids, in the laboratory as well as in the ocean (see Turner, 1973, Fig. 3.3).

Example 11.6

Tsunamis are unusual ocean waves caused by seismic activity or other processes that displace vast volumes of ocean water. Their wavelengths are so long that they travel through the open ocean (nominal depth of 4 to 5 km) as linear shallow-water gravity waves where they pose no threat to watercraft or sea life. However, as they approach the shore where water depth decreases their propagation speed slows, their wavelength decreases, their amplitude increases, and their potential menace rises to catastrophic levels. If a tsunami wave train has wavelength $\lambda = 100$ km, and amplitude $a = 1.0$ m when $H = 4$ km, what will its wavelength and amplitude be in water 10 m deep if the energy flux and period of the wave train are the same in both cases?

Solution The presumption of constant energy flux is approximate, but sufficient for the present purposes. First, determine the wave speed and ka value when $H = 4$ km:

$$c = \sqrt{gH} = \sqrt{(9.81\ \mathrm{ms}^{-2})(4000\ \mathrm{m})} = 198\ \mathrm{ms}^{-1},\ ka = 2\pi a/\lambda = 2\pi(1.0\ \mathrm{m})/(100\ \mathrm{km}) = 6.3 \times 10^{-5}.$$

Here $a/H = 2.5 \times 10^{-4}$ and ka are both small enough to use linear wave results. So, from (11.44), the energy flux, EF, for these shallow water waves is:

$$EF = \frac{\rho g a^2}{2} c = \frac{(10^3\ \mathrm{kgm}^{-3})(9.81\ \mathrm{ms}^{-2})(1.0\ \mathrm{m})^2}{2}(198\ \mathrm{ms}^{-1}) = 0.97\ \mathrm{MW/m}.$$

The wave period here is $\lambda/c = (100\ \mathrm{km})/(198\ \mathrm{ms}^{-1}) = 505$ s, and this will be the same (or nearly so) when the wave train reaches shallower water where its speed will be:

$$c = \sqrt{gH} = \sqrt{(9.81\ \mathrm{ms}^{-2})(10\ \mathrm{m})} = 9.90\ \mathrm{ms}^{-1}.$$

The wavelength in shallow water will be $(9.90\ \mathrm{ms}^{-1})(505\ \mathrm{s}) = 5.0$ km. For constant energy flux, the wave amplitude at the shallower depth will be:

$$a = \left(\frac{2EF}{\rho g c}\right)^{1/2} = \left(\frac{2 \cdot 0.97 \times 10^6\ \mathrm{Wm}^{-1}}{(10^3\ \mathrm{kgm}^{-3})(9.81\ \mathrm{ms}^{-2})(9.90\ \mathrm{ms}^{-1})}\right)^{1/2} = 4.47\ \mathrm{m},$$

and this is substantially larger than the wave's open-ocean amplitude. Here, the amplitude-depth ratio a/H is $(4.47\ \mathrm{m})/(10\ \mathrm{m}) = 0.447$, and this is too large for the linear theory; a nonlinear theory is needed to better determine the wave shape and its amplitude. An alternative approach to this amplitude-increase calculation is provided as Exercise 11.18.

11.7 Waves on a density interface

To this point, waves at the surface of a liquid have been considered without regard to the gas (or liquid) above the surface. Yet, gravity and capillary waves can also exist at the interface between two immiscible liquids of different densities. A sharp-density gradient can be readily generated in the laboratory (at least temporarily) between gases with different densities, and between oil and water. In the ocean sharp density gradients may be generated by solar heating of the upper layer, or in an estuary (that is, a river mouth) or fjord into which fresh river water flows over salty oceanic water, which is heavier. The basic situation can be idealized by considering a lighter fluid of density ρ_1 lying over a heavier fluid of density ρ_2 (Fig. 11.25).

For simplicity ignore interfacial (surface) tension, and assume that only small-slope linear waves exist on the interface and that both fluids are infinitely deep, so that only those solutions that decay exponentially from the interface are allowed. In this section and in the rest of this chapter, *complex notation* will be used to ease the algebraic and trigonometric effort. This means that (11.2) will be replaced by:

$$\zeta(x,t) = \mathrm{Re}\{a \exp[i(kx - \omega t)]\},$$

where Re{} is the operator that extracts the real part of the complex function in { }-braces, and $i = \sqrt{-1}$ (it is not an index here). When using complex numbers and variables in linear mathematical analyses it is customary to drop Re{ } and simply write:

$$\zeta(x,t) = a \exp[i(kx - \omega t)] \tag{11.89}$$

until reporting the final results when Re{} commonly reappears. Any analysis done with (11.89) includes an imaginary part, sometimes denoted Im{ }, that winds up being of no consequence in the final results.

To determine wave properties in this situation, the Laplace equation for the velocity potential must be solved in both fluids subject to the continuity of p and w at the interface. The equations are:

$$\frac{\partial^2 \phi_1}{\partial x^2} + \frac{\partial^2 \phi_1}{\partial z^2} = 0 \quad \text{and} \quad \frac{\partial^2 \phi_2}{\partial x^2} + \frac{\partial^2 \phi_2}{\partial z^2} = 0, \tag{11.90}$$

subject to:

$$\phi_1 \to 0 \quad \text{as} \quad z \to \infty, \qquad \phi_2 \to 0 \quad \text{as} \quad z \to -\infty, \tag{11.91, 11.92}$$

$$\partial\phi_1/\partial z = \partial\phi_1/\partial z = \partial\zeta/\partial t \quad \text{at} \quad z = 0, \quad \text{and} \tag{11.93}$$

$$\rho_1(\partial\phi_1/\partial t) + \rho_1 g\zeta = \rho_2(\partial\phi_2/\partial t) + \rho_2 g\zeta \quad \text{at} \quad z = 0. \tag{11.94}$$

Eq. (11.93) follows from equating the vertical velocity of the fluid on both sides of the interface to the rate of rise of the interface. Eq. (11.94) follows from the continuity of pressure across the interface in the absence of interfacial (surface) tension, $\sigma = 0$. As in the case of surface waves, the boundary conditions are linearized and applied at $z = 0$ instead of at $z = \zeta$. Conditions (11.91) and (11.92) require that the solutions of (11.90) must be of the form:

$$\phi_1 = Ae^{-kz}e^{i(kx-\omega t)} \quad \text{and} \quad \phi_2 = Be^{+kz}e^{i(kx-\omega t)},$$

because a solution proportional to e^{kz} is not allowed in the upper fluid, and a solution proportional to e^{-kz} is not allowed in the lower fluid. Here the amplitudes A and B can be complex. As in Section 11.2, the constants are determined from the kinematic boundary conditions (11.93), giving:

$$A = -B = i\omega a/k.$$

The dynamic boundary condition (11.94) then leads to the dispersion relation:

$$\omega = \sqrt{gk\left(\frac{\rho_2 - \rho_1}{\rho_2 + \rho_1}\right)} = \varepsilon\sqrt{gk}, \tag{11.95}$$

where $\varepsilon^2 \equiv (\rho_2 - \rho_1)/(\rho_2 + \rho_1)$ is a small number if the density difference between the two liquids is small. The case of small density difference is relevant in geophysical situations; for example, a 10 °C temperature change causes the density of an upper layer of the ocean to decrease by 0.3%. Eq. (11.95) shows that waves at the interface between two liquids of infinite thickness travel like deep-water surface waves, with ω proportional to $\sqrt{gk}$, but at a frequency that is lower by the factor ε. In general, *internal waves have a lower frequency and slower phase speed, than surface waves with the same wave number*. As expected, (11.95) divided by k recovers (11.45) as $\varepsilon \to 1$ when $\rho_1/\rho_2 \to 0$.

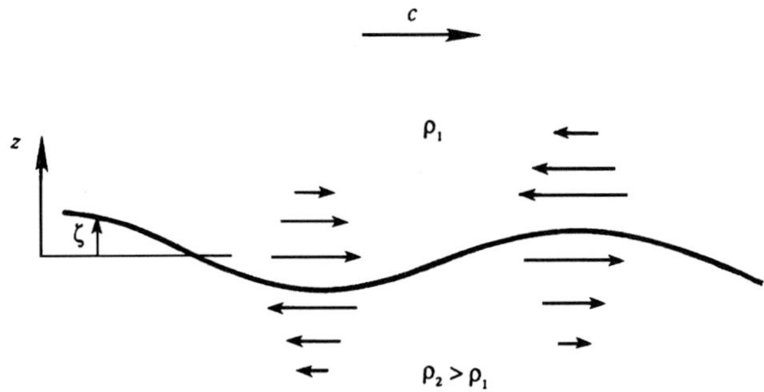

FIGURE 11.25 Internal wave at a density interface between two infinitely deep fluids. Here the horizontal velocity is equal and opposite above and below the interface, so there is a time-dependent vortex sheet at the interface.

The kinetic energy E_k per unit area of interface of the field can be found by integrating $\rho(u^2 + w^2)/2$ over the range $z = \pm\infty$ (Exercise 11.21):

$$E_k = \frac{1}{4}(\rho_2 - \rho_1)ga^2.$$

The potential energy can be calculated from the displacement of fluid particles. With $z = 0$ at the undisturbed interface, the potential energy of a displaced fluid element (per unit length into the page) is $(\rho_2 - \rho_1)gz\,dx\,dz$ (see Fig. 11.6 with ζ replacing η, ρ_1 above the interface, and ρ_2 below it). Thus, potential energy E_p per unit horizontal area is:

$$E_p = \frac{(\rho_2 - \rho_1)g}{\lambda}\int_0^\lambda\int_0^\zeta z\,dz\,dx = \frac{(\rho_2 - \rho_1)g}{2\lambda}\int_0^\lambda \zeta^2\,dx = \frac{1}{4}(\rho_2 - \rho_1)ga^2.$$

The total wave energy per unit horizontal area is:

$$E = E_k + E_p = \frac{1}{2}(\rho_2 - \rho_1)ga^2. \tag{11.96}$$

In a comparison with (11.42), it follows that the amplitude of ocean internal waves is usually much larger than those of surface waves for the same amount of energy per unit interface area when $(\rho_2 - \rho_1) \ll \rho_2$.

The horizontal velocity components in the two layers are:

$$u_1 = \frac{\partial \phi_1}{\partial x} = -\omega a e^{-kz} e^{i(kx-\omega t)} \quad \text{and} \quad u_2 = \frac{\partial \phi_2}{\partial x} = \omega a e^{+kz} e^{i(kx-\omega t)},$$

and are oppositely directed (Fig. 11.25). The interface is therefore a time-dependent *vortex sheet* and the tangential velocity is discontinuous across it. It can be expected that a continuously stratified medium, in which the density varies continuously as a function of z, will support internal waves whose vorticity is distributed throughout the flow. Consequently, *internal waves in a continuously stratified fluid are not irrotational and do not satisfy the Laplace equation.*

The existence of internal waves at a density discontinuity has explained an interesting phenomenon observed in Norwegian fjords (Gill, 1982). It was known for a long time that ships experienced unusually high drags on entering these fjords. The phenomenon was a mystery (and was attributed to "dead water") until Bjerknes, a Norwegian oceanographer, explained it as due to the internal waves at the interface generated by the motion of the ship (Fig. 11.26). (Note that the product of the drag times the speed of the ship gives the rate of generation of wave energy, with other sources of resistance neglected.)

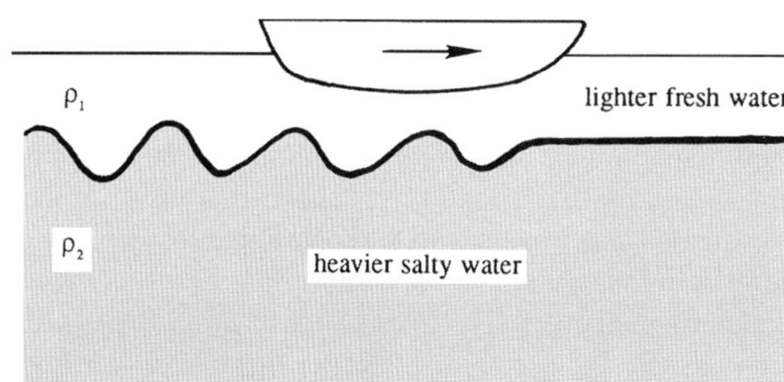

FIGURE 11.26 Schematic explanation for the phenomenon of *dead water* in Norwegian fjords. The ship on the ocean surface may produce waves on the ocean surface and on an interface between lighter fresher water and denser saltier water. Wave production leads to drag on the ship and both types of waves are generated under certain conditions.

As a second example of an internal wave at a density discontinuity, consider the case in which the upper layer is not infinitely thick but has a finite thickness; the lower layer is initially assumed to be infinitely thick. The case of two infinitely deep liquids, treated on the preceding pages, is then a special case of the present situation. Whereas only waves at the interface were allowed with two semi-infinite fluid layers, the presence of a free surface now allows surface waves to enter the problem. The present configuration will allow two modes of oscillation where the free-surface and interface waves are in or out of phase.

To analyze this situation, let H be the thickness of the upper layer, and let the origin be placed at the mean position of the free surface (Fig. 11.27). The field equations are (11.90) and the boundary conditions are (11.92),

$$\partial \phi_1/\partial z = \partial \eta/\partial t \quad \text{and} \quad \partial \phi_1/\partial t + g\eta = 0 \quad \text{at} \quad z = 0, \tag{11.97, 11.98}$$

$$\partial \phi_1/\partial z = \partial \phi_2/\partial z = \partial \zeta/\partial t \quad \text{and} \quad \rho_1(\partial \phi_1/\partial t) + \rho_1 g\zeta = \rho_2(\partial \phi_2/\partial t) + \rho_2 g\zeta \quad \text{at} \quad z = -H. \tag{11.99, 11.100}$$

In addition, assume free-surface and interface displacements of the form:

$$\eta = ae^{i(kx-\omega t)} \quad \text{and} \quad \zeta = be^{i(kx-\omega t)}, \tag{11.101, 11.102}$$

respectively. Without losing generality, a can be considered real, which means that the surface wave matches (11.2). The interface-wave amplitude b should be left complex since η and ζ may not be in phase, and the solution of the problem should determine this phase difference.

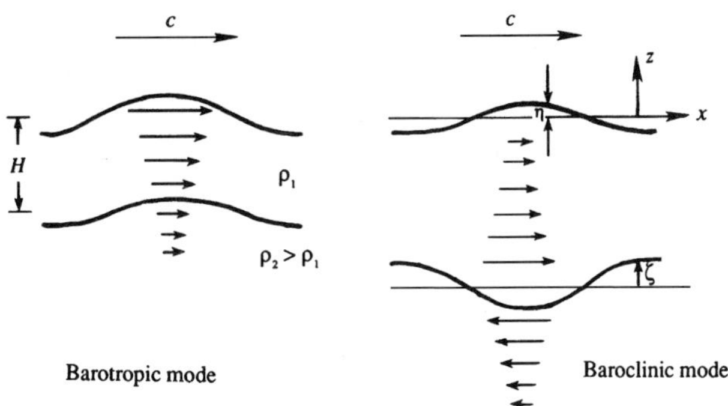

FIGURE 11.27 The two modes of motion of a layer of fluid overlying an infinitely deep fluid. The barotropic mode is an extension of the surface wave motion discussed in the first six sections of this chapter. The baroclinc mode is the extension of the interface wave motion described earlier in this section and shown in Fig. 11.25. The baroclinic mode includes vorticity at the density interface; the barotropic mode does not.

The velocity potentials in the layers must be of the forms:

$$\phi_1 = (Ae^{+kz} + Be^{-kz})e^{i(kx-\omega t)}, \quad \text{and} \quad \phi_2 = Ce^{+kz}e^{i(kx-\omega t)}. \tag{11.103, 11.104}$$

The form (11.104) satisfies (11.92) Conditions (11.97) through (11.99) allow a solution for the constants in terms of a, ω, k, g, and H:

$$A = -\frac{ia}{2}\left(\frac{\omega}{k} + \frac{g}{\omega}\right), \quad B = \frac{ia}{2}\left(\frac{\omega}{k} - \frac{g}{\omega}\right), \quad C = -\frac{ia}{2}\left(\frac{\omega}{k} + \frac{g}{\omega}\right) - \frac{ia}{2}\left(\frac{\omega}{k} - \frac{g}{\omega}\right)e^{2kH}, \tag{11.105, 11.106, 11.107}$$

$$\text{and} \quad b = \frac{a}{2}\left(1 + \frac{gk}{\omega^2}\right)e^{-kH} + \frac{a}{2}\left(1 - \frac{gk}{\omega^2}\right)e^{kH}. \tag{11.108}$$

Substitution into (11.100) leads to the dispersion relation $\omega(k)$. After some algebraic manipulations, the result can be written as (see Exercise 11.22):

$$\left(\frac{\omega^2}{gk} - 1\right)\left\{\frac{\omega^2}{gk}[\rho_1 \sinh(kH) + \rho_2 \cosh(kH)] - (\rho_2 - \rho_1)\sinh(kH)\right\} = 0. \tag{11.109}$$

One possible root of (11.109) is:

$$\omega^2 = gk, \tag{11.110}$$

which is the same as that for a deep-water gravity wave. Substituting (11.110) into (11.108) leads to:

$$b = ae^{-kH}, \tag{11.111}$$

which implies that the interface waves are in phase with the surface waves but are reduced in amplitude by the factor e^{-kH}. This mode is similar to a gravity wave propagating on the free surface of the upper liquid, in which the motion decays as e^{-kz} from the free surface. It is called the *barotropic mode* because the surfaces of constant pressure and density coincide.

The other root of (11.109) is:

$$\omega^2 = \frac{gk(\rho_2 - \rho_1)\sinh(kH)}{\rho_1 \sinh(kH) + \rho_2 \cosh(kH)}, \tag{11.112}$$

which reduces to (11.95) when $kH \to \infty$. Substituting (11.112) into (11.108) leads to:

$$\eta = -\zeta\left(\frac{\rho_2 - \rho_1}{\rho_1}\right)e^{-kH}, \tag{11.113}$$

which demonstrates that η and ζ have opposite signs and that the interface displacement (ζ) is much larger than the surface displacement (η) if the density difference is small. This is the *baroclinic* or *internal mode* because the surfaces of constant pressure and density do not coincide. Here the horizontal velocity u changes sign across the interface. The existence of a density difference has therefore generated a motion that is quite different from the barotropic mode, (11.110) and (11.111). The case described at the beginning of this section, where the fluids have infinite depth and no free surface, has only a baroclinic mode and no barotropic mode.

A very common simplification, frequently made in geophysical situations, involves assuming that the wavelengths are large compared to the upper layer depth. For example, the depth of the oceanic upper layer, below which there is a sharp-density gradient, could be ≈50 m thick, but interfacial waves much longer than this may be of interest. The relevant approximation in this case, $kH \ll 1$, is called the *shallow-water* or *long-wave approximation* and is implemented via:

$$\sinh(kH) \approx kH \quad \text{and} \quad \cosh(kH) \approx 1,$$

so the dispersion relation (11.112) corresponding to the baroclinic mode reduces to:

$$\omega^2 = gk\left(\frac{\rho_2 - \rho_1}{\rho_2}\right)kH \tag{11.114}$$

to lowest order in the small parameter kH. The phase velocity of waves at the interface is:

$$c = [g'H]^{1/2}, \quad \text{where} \quad g' = g(\rho_2 - \rho_1)/\rho_2, \tag{11.115, 11.116}$$

is the reduced gravity. Eq. (11.115) is similar to the corresponding expression for *surface* waves in a shallow homogeneous layer of thickness H, except that the phase speed is reduced by the factor $\sqrt{(\rho_2 - \rho_1)/\rho_2}$. This agrees with the previous conclusion that internal waves propagate slower than surface waves. Under the shallow-water approximation, (11.113) reduces to:

$$\eta = -\zeta(\rho_2 - \rho_1)/\rho_1. \tag{11.117}$$

In Section 11.2, the shallow-water approximation for surface waves is found equivalent to a hydrostatic approximation and results in a depth-independent horizontal velocity. This conclusion also holds for interfacial waves. The fact that u_1 is independent of z follows from (11.103) on noting that $e^{kz} \approx e^{-kz} \approx 1$. To see that pressure is hydrostatic, the perturbation pressure p' in the upper layer determined from (11.103) is:

$$p' = -\rho_1(\partial\phi_1/\partial t) = i\rho_1\omega(A + B)e^{i(kx-\omega t)} = \rho_1 g\eta, \tag{11.118}$$

where the constants given in (11.105) and (11.106) have been used. This shows that p' is independent of z and equals the hydrostatic pressure change due to the free-surface displacement.

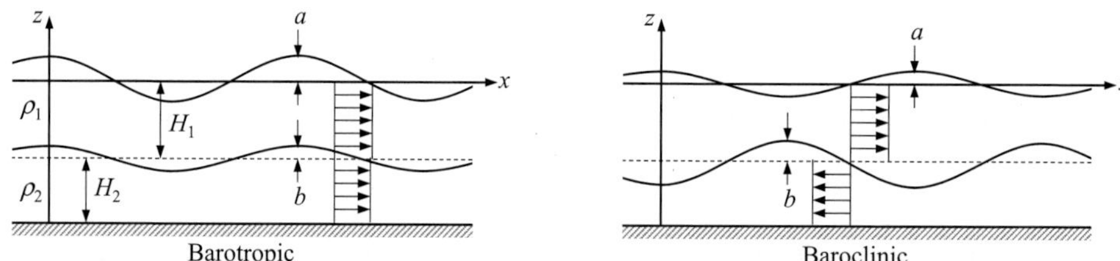

FIGURE 11.28 The two modes of motion in a shallow-water, two-layer system in the Boussinesq limit. These profiles are the limiting case of those in Fig. 11.27 when the lower fluid layer depth is shallow. As before, the surface and interface deflections are in-phase for the baroclinic mode, and out-of-phase for the baroclinic.

So far, the lower fluid has been assumed to be infinitely deep, resulting in an exponential decay of the flow field from the interface into the lower layer, with a decay scale of the order of the wavelength. If the lower layer is now considered thin compared to the wavelength, then the horizontal velocity will be depth independent, and the flow hydrostatic, in the lower layer. If *both* layers are considered thin compared to the wavelength, then the flow is hydrostatic (and the horizontal velocity field is depth independent) in *both* layers. This is the *shallow-water* or *long-wave approximation* for a two-layer fluid. In such a case the horizontal velocity field in the barotropic mode has a discontinuity at the interface, which vanishes in the Boussinesq limit $(\rho_2 - \rho_1)/\rho_1 \ll 1$. Under these conditions the two modes of a two-layer system have a simple structure (Fig. 11.28): a barotropic mode in which the horizontal velocity is depth independent across the entire water column; and a baroclinic mode in which the horizontal velocity is directed in opposite directions in the two layers, but is depth independent in each layer (see Exercise 11.24).

Example 11.7

Is it possible for the barotropic and baroclinic waves to cancel at the ocean surface so that only internal waves on the interface exist?

Solution This is an interesting possibility since it would imply that waves on an interface between fluids of different densities could propagate without any surface signature. The only way this would possible is if the surface deflections of the barotropic wave and the baroclinic wave were of equal but opposite amplitude, and the waves had the same frequency and wave number. The later two requirements can be used to answer the question. They imply that:

$$\omega^2 = gk = \frac{gk(\rho_2 - \rho_1)\sinh(kH)}{\rho_1 \sinh(kH) + \rho_2 \cosh(kH)},$$

where the first equality follows from (11.110) and the second from (11.112). Here, in the second equality, gk is a common factor so it implies:

$$\rho_1 \sinh(kH) + \rho_2 \cosh(kH) = (\rho_2 - \rho_1)\sinh(kH),$$

and this equation can be solved for k to find:

$$k = \frac{1}{H}\tanh^{-1}\left(\frac{\rho_2}{\rho_2 - 2\rho_1}\right).$$

When both densities are positive, the factor in big parentheses is either greater than one ($\rho_2 > 2\rho_1$) or negative ($\rho_2 < 2\rho_1$). When the factor is greater than one, there is no real solution for k, because the range of the hyperbolic tangent is -1 to $+1$. When the factor is negative, then k will be negative, so there is no real solution for ω since (11.110) would require $\omega^2 < 0$. Therefore, internal waves on the interface between fluids of different density must have a surface signature.

11.8 Internal waves in a continuously stratified fluid

Waves may also exist in the interior of a pool, reservoir, lake, or ocean when the fluid's density in a quiescent state is a continuous function of the vertical coordinate z. The equations of motion for internal waves in such a stratified medium presented here are simplifications of the Boussinesq set specified at the end of Section 4.9. The Boussinesq approximation treats the density as constant, except in the vertical momentum equation. For simplicity, we shall also assume that: 1) the wave motion is effectively inviscid because the velocity gradients are small and the Reynolds number is large, 2) the wave amplitudes are small enough so that the nonlinear advection terms can be neglected, and 3) the frequency of wave motion is much larger than the Coriolis frequency so it does not affect the wave motion. Effects of the earth's rotation are considered in Chapter 12. The Boussinesq set then simplifies to:

$$\frac{D\rho}{Dt} = 0, \quad \frac{\partial u}{\partial x} + \frac{\partial v}{\partial y} + \frac{\partial w}{\partial z} = 0, \tag{4.9, 4.10}$$

$$\frac{\partial u}{\partial t} = -\frac{1}{\rho_0}\frac{\partial p}{\partial x}, \quad \frac{\partial v}{\partial t} = -\frac{1}{\rho_0}\frac{\partial p}{\partial y}, \quad \text{and} \quad \frac{\partial w}{\partial t} = -\frac{1}{\rho_0}\frac{\partial p}{\partial z} - \frac{\rho g}{\rho_0}, \tag{11.119, 11.120, 11.121}$$

where ρ_0 is a constant reference density. Here, (4.9) expresses constancy of fluid-particle density while (4.10) is the condition for incompressible flow. If temperature is the only cause of density change, then $D\rho/Dt = 0$ follows from the heat equation in the non-diffusive form $DT/Dt = 0$ and a temperature-only equation of state, in the form $\delta\rho/\rho = -\alpha\delta T$, where α is the coefficient of thermal expansion. If the density changes are due to changes in the concentration S of a constituent (e.g., salinity in the ocean or water vapor in the atmosphere), then $D\rho/Dt = 0$ follows from $DS/Dt = 0$ (the non-diffusive form of the constituent conservation equation) and a concentration-only equation of state, $\rho = \rho(S)$, in the form of $\delta\rho/\rho = \beta\delta S$, where β is the coefficient describing how the density changes due to concentration of the constituent. In both cases, the principle underlying $D\rho/Dt = 0$ is an equation of state that does not include pressure. In terms of common usage, this equation is frequently called the *density equation*, as opposed to the *continuity equation* (4.10).

The five equations (4.9), (4.10), and (11.119) through (11.121) contain five unknowns (u, v, w, p, ρ). Before considering wave motions, first define the quiescent density $\overline{\rho}(z)$ and pressure $\overline{p}(z)$ profiles in the medium as those that satisfy a hydrostatic balance:

$$0 = -\frac{1}{\rho_0}\frac{\partial \overline{p}}{\partial z} - \frac{\overline{\rho} g}{\rho_0}. \tag{11.122}$$

When the motion develops, the pressure and density will change relative to their quiescent values:

$$p=\overline{p}(z)+p', \quad \text{and} \quad \rho=\overline{\rho}(z)+\rho'. \tag{11.123}$$

The density equation (4.9) then becomes:

$$\frac{\partial}{\partial t}(\overline{\rho}+\rho')+u\frac{\partial}{\partial x}(\overline{\rho}+\rho')+v\frac{\partial}{\partial y}(\overline{\rho}+\rho')+w\frac{\partial}{\partial z}(\overline{\rho}+\rho')=0. \tag{11.124}$$

Here, $\partial\overline{\rho}/\partial t=\partial\overline{\rho}/\partial x=\partial\overline{\rho}/\partial y=0$, and the nonlinear terms (namely, $u\partial\rho'/\partial x$, $v\partial\rho'/\partial y$, and $w\partial\rho'/\partial z$) are also negligible for small-amplitude motions. The *linear* part of the fourth term, $wd\overline{\rho}/dz$, must be retained, so the linearized version of (4.9) is:

$$\frac{\partial\rho'}{\partial t}+w\frac{\partial\overline{\rho}}{\partial z}=0, \tag{11.125}$$

which states that the density perturbation at a point is generated only by the vertical advection of the *background* density distribution. We now introduce the *Brunt–Väisälä frequency*, or *buoyancy frequency*:

$$N^2=-\frac{g}{\rho_0}\frac{\partial\overline{\rho}}{\partial z}. \tag{11.126}$$

This is (1.36) when the adiabatic density gradient is zero. As described in Section 1.10, $N(z)$ has units of rad/s and is the oscillation frequency of a vertically displaced fluid particle released from rest in the absence of fluid friction. Using (11.122) and (11.126) in (11.119) through (11.121) and (11.125) produces:

$$\frac{\partial u}{\partial t}=-\frac{1}{\rho_0}\frac{\partial p'}{\partial x}, \quad \frac{\partial v}{\partial t}=-\frac{1}{\rho_0}\frac{\partial p'}{\partial y}, \quad \text{and} \quad \frac{\partial w}{\partial t}=-\frac{1}{\rho_0}\frac{\partial p'}{\partial z}-\frac{\rho' g}{\rho_0}, \tag{11.127, 11.128, 11.129}$$

$$\text{and} \quad \frac{\partial\rho'}{\partial t}-\frac{N^2\rho_0}{g}w=0. \tag{11.130}$$

Comparing (11.119) through (11.121) and (11.127) through (11.129), we see that the only difference is the replacement of the total density ρ and pressure p with the perturbation density ρ' and pressure p'.

The full set of equations for linear wave motion in a stratified fluid are (4.10) and (11.127) through (11.130), where ρ may be a function of temperature T and concentration S of a constituent, but not of pressure. At first this does not seem to be a good assumption. The compressibility effects in the atmosphere are certainly not negligible; even in the ocean the density changes due to the huge changes in the background pressure are as much as 4%, which is $\approx$10 times the density changes due to the variations of the salinity and temperature. The effects of compressibility, however, can be handled within the Boussinesq approximation if we regard $\overline{\rho}$ in the definition of N as the background *potential density*, that is, the density distribution from which the adiabatic changes of density, due to the changes of pressure, have been subtracted out. The concept of potential density is explained in Chapter 1. Oceanographers account for compressibility effects by converting all their density measurements to the standard atmospheric pressure; thus, when they report variations in density (what they call "sigma tee") they are generally reporting variations due only to changes in temperature and salinity.

A useful condensation of the above equations involving only w can be obtained by taking the time derivative of (4.10) and using the horizontal momentum equations (11.127) and (11.128) to eliminate u and v. The result is:

$$\frac{1}{\rho_0}\nabla_H^2 p'=\frac{\partial^2 w}{\partial z\partial t}, \tag{11.131}$$

where $\nabla_H^2=\partial^2/\partial x^2+\partial^2/\partial y^2$ is the *horizontal* Laplacian operator. Elimination of ρ' from (11.129) and (11.130) gives:

$$\frac{1}{\rho_0}\frac{\partial^2 p'}{\partial t\partial z}=-\frac{\partial^2 w}{\partial t^2}-N^2 w. \tag{11.132}$$

Finally, p' can be eliminated by taking ∇_H^2 of (11.132), and inserting the result in (11.131) to find:

$$\frac{\partial^2}{\partial t\partial z}\left(\frac{\partial^2 w}{\partial t\partial z}\right)=-\nabla_H^2\left(\frac{\partial^2 w}{\partial t^2}+N^2 w\right), \quad \text{or} \quad \frac{\partial^2}{\partial t^2}(\nabla^2 w)+N^2\nabla_H^2 w=0, \tag{11.133}$$

where $\nabla^2 = \nabla_H^2 + \partial^2/\partial z^2$ is the three-dimensional Laplacian operator. This equation for the vertical velocity w can be used to derive the dispersion relation for internal gravity waves.

Internal waves in a stratified fluid

The situation embodied in (11.133) is fundamentally different from that of interface waves because there is no obvious direction of propagation. For interface waves constrained to follow a horizontal surface with the x-axis chosen along the direction of wave propagation, a dispersion relation $\omega(k)$ was obtained that is independent of the wave direction. Furthermore, wave crests and wave groups propagate in the same direction, although at different speeds. However, in the current situation, the fluid is *continuously* stratified and internal waves might propagate in any direction and at any angle to the vertical. In such a case the *direction* of the wave number vector $\mathbf{K} = (k, l, m)$ becomes important and the dispersion relationship is *anisotropic* and depends on the wave number components:

$$\omega = \omega(k, l, m) = \omega(\mathbf{K}). \tag{11.134}$$

Consequently, the wave number, phase velocity, and group velocity are no longer scalars and the prototype sinusoidal wave (11.2) must be replaced with its three-dimensional extension (11.5). However, (11.134) must still be isotropic in k and l, the wave number components in the two horizontal directions.

The propagation of internal waves is a baroclinic process, in which the surfaces of constant pressure do not coincide with the surfaces of constant density. It was shown in Section 5.2, in connection with Kelvin's circulation theorem, that baroclinic processes generate vorticity. Internal waves in a continuously stratified fluid are therefore rotational. Waves at a density interface constitute a limiting case in which all the vorticity is concentrated in the form of a velocity discontinuity at the interface. The Laplace equation can therefore be used to describe the flow field within each layer. However, internal waves in a continuously stratified fluid cannot be described by the Laplace equation.

To reveal the structure of the situation described by (11.133) and (11.134), consider the complex version of (11.5) for the vertical velocity:

$$w = w_0 e^{i(kx+ly+mz-\omega t)} = w_0 e^{i(\mathbf{K}\cdot\mathbf{x}-\omega t)} \tag{11.135}$$

in a fluid medium having a constant buoyancy frequency. Substituting (11.135) into (11.133) with constant N leads to the dispersion relation:

$$\omega^2 = \frac{k^2 + l^2}{k^2 + l^2 + m^2} N^2. \tag{11.136}$$

For simplicity choose the x-z plane so it contains $\mathbf{K}$ and $l = 0$. No generality is lost through this choice because the medium is horizontally isotropic, but k now represents the entire horizontal wave number and (11.136) can be written:

$$\omega = \frac{kN}{\sqrt{k^2 + m^2}} = \frac{kN}{K}. \tag{11.137}$$

This is the dispersion relation for internal gravity waves and can also be written as:

$$\omega = N \cos\theta, \tag{11.138}$$

where $\theta = \tan^{-1}(m/k)$ is the angle between the phase velocity vector $\mathbf{c}$ (and therefore $\mathbf{K}$) and the horizontal direction (Fig. 11.29). Interestingly, (11.138) states that the frequency of an internal wave in a stratified fluid depends only on the *direction* of the wave number vector and not on its magnitude. This is in sharp contrast with surface and interfacial gravity waves, for which frequency depends only on the magnitude. In addition, the wave frequency lies in the range $0 < \omega < N$, and this indicates that *N is the maximum possible frequency of internal waves in a stratified fluid.*

Before further investigation of the dispersion relation, consider particle motion in an incompressible internal wave. For consistency with (11.135), the horizontal fluid velocity is written as:

$$u = u_0 e^{i(kx+ly+mz-\omega t)}, \tag{11.139}$$

plus two similar expressions for v and w. Differentiating produces:

$$\partial u/\partial x = iku_0 e^{i(kx+ly+mz-\omega t)} = iku.$$

Thus, (4.10) then requires that $ku + lv + mw = 0$, that is:

$$\mathbf{K} \cdot \mathbf{u} = 0, \tag{11.140}$$

showing that *particle motion is perpendicular to the wave number vector* (Fig. 11.29). Note that only two conditions have been used to derive this result, namely the incompressible continuity equation and trigonometric behavior in *all* spatial directions. As such, the result is valid for many other wave systems that meet these two conditions. These waves are called *shear waves* (or transverse waves) because the fluid moves parallel to the constant phase lines. Surface or interfacial gravity waves do not have this property because the field varies *exponentially* in the vertical.

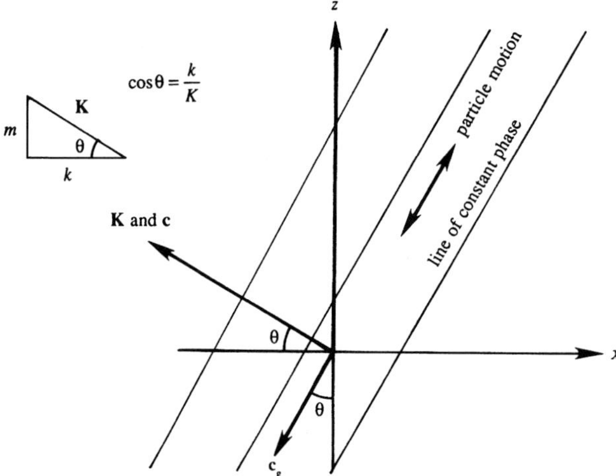

FIGURE 11.29 Geometric parameters for internal waves. Here, z is vertical and x is horizontal. Note that $\mathbf{c}$ and $\mathbf{c}_g$ are at right angles and have opposite vertical components while $\mathbf{u}$ is parallel to the group velocity. Thus, internal wave packets slide along their crests.

We can now interpret θ in the dispersion relation (11.138) as the angle between the particle motion and the *vertical* direction (Fig. 11.29). The maximum frequency $\omega = N$ occurs when $\theta = 0$, that is, when the particles move up and down vertically. This case corresponds to $m = 0$ (see (11.137)), showing that the motion is independent of the z-coordinate. The resulting motion consists of a series of vertical columns, all oscillating at the buoyancy frequency N, with the flow field varying in the horizontal direction only.

At the opposite extreme we have $\omega = 0$ when $\theta = \pi/2$, that is, when the particle motion is completely horizontal. In this limit our internal wave solution (11.137) would seem to require $k = 0$, that is, horizontal independence of the motion. However, such a conclusion is not valid; pure horizontal motion is not a limiting case of internal waves, and it is necessary to examine the basic equations to draw any conclusion for this case. An examination of the governing set, (4.10) and (11.127) through (11.130), shows that a possible steady solution is $w = p' = \rho' = 0$, with u and v and *any* functions of x and y satisfying the two-dimensional incompressible continuity equation:

$$\partial u/\partial x + \partial v/\partial y = 0. \tag{6.2}$$

The z-dependence of u and v is arbitrary. The motion is therefore two dimensional in the horizontal plane, with the motion in the various horizontal planes decoupled from each other. This is why clouds in the upper atmosphere seem to move in flat horizontal sheets, as often observed in airplane flights (Gill, 1982). For a similar reason a cloud pattern pierced by a mountain peak sometimes shows *Karman vortex streets*, a two-dimensional flow feature; see the striking photograph in Fig. 8.21. A restriction of strong stratification is necessary for such almost horizontal flows, because (11.130) suggests that the vertical motion is small if N is large.

The foregoing discussion leads to the interesting phenomenon of *blocking* in a strongly stratified fluid. Consider a two-dimensional body placed in such a fluid, with its axis horizontal (Fig. 11.30). The two dimensionality of the body requires $\partial v/\partial y = 0$, so that the continuity equation (6.2) reduces to $\partial u/\partial x = 0$. A horizontal layer of fluid ahead of the body, bounded by tangents above and below it, is therefore blocked and held motionless. (For photographic evidence, see Fig. 3.18 in the book by Turner [1973].) This happens because the strong stratification suppresses the w field and prevents the fluid from going below or over the body.

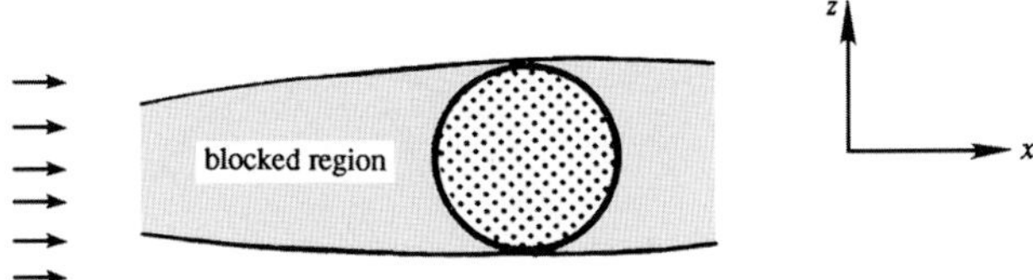

FIGURE 11.30 Blocking in strongly stratified flow. The circular region represents a two-dimensional body with its axis along the y direction (perpendicular to the page). Horizontal flow in the shaded region is blocked by the body when the stratification is strong enough to prevent fluid in the blocked layer from going over or under the body.

Dispersion of internal waves in a stratified fluid

The dispersion relationship (11.137) for linear internal waves with constant buoyancy frequency contains a few genuine surprises that challenge our imaginations and violate the intuition acquired by observing surface or interface waves. One of these surprises involves the phase, $\mathbf{c}$, and group, $\mathbf{c}_g$, velocity vectors. In multiple dimensions, these are defined by:

$$\mathbf{c} = (\omega/K)\mathbf{e}_K \quad \text{and} \quad \mathbf{c}_g = \mathbf{e}_x(\partial\omega/\partial k) + \mathbf{e}_y(\partial\omega/\partial l) + \mathbf{e}_z(\partial\omega/\partial m), \tag{11.8, 11.141}$$

where $\mathbf{e}_K = \mathbf{K}/K$. For interface waves $\mathbf{c}$ and $\mathbf{c}_g$ are in the same direction, although their magnitudes can be different. For internal waves, (11.137), (11.8), and (11.141) can be used to determine:

$$\mathbf{c} = \frac{\omega}{K^2}(k\mathbf{e}_x + m\mathbf{e}_z), \quad \text{and} \quad \mathbf{c}_g = \frac{Nm}{K^3}(m\mathbf{e}_x - k\mathbf{e}_z). \tag{11.142, 11.143}$$

Forming the dot product of these two equations produces:

$$\mathbf{c}_g \cdot \mathbf{c} = 0! \tag{11.144}$$

Thus, the *phase and group velocity vectors are perpendicular* as shown on Fig. 11.29. Eqs. (11.142) and (11.143) do place the horizontal components of $\mathbf{c}$ and $\mathbf{c}_g$ in the same direction, but their vertical components are equal and opposite. In fact, $\mathbf{c}$ and $\mathbf{c}_g$ form two sides of a right triangle whose hypotenuse is horizontal (Fig. 11.31). Consequently, the phase velocity has an upward component when the group velocity has a downward component, and vice versa. Eqs. (11.140) and (11.144) are consistent because $\mathbf{c}$ and $\mathbf{K}$ are parallel and $\mathbf{c}_g$ and $\mathbf{u}$ are parallel. The fact that $\mathbf{c}$ and $\mathbf{c}_g$ are perpendicular, and have opposite vertical components, is illustrated in Fig. 11.32. It shows that the phase lines are propagating toward the left and upward, whereas the wave group is propagating to the left and downward. Wave crests are constantly appearing at one edge of the group, propagating through the group, and vanishing at the other edge.

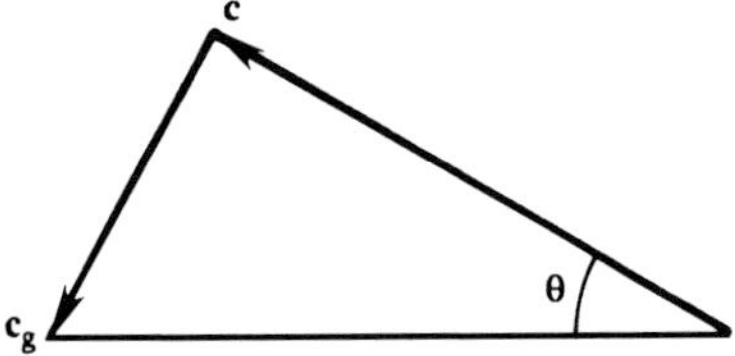

FIGURE 11.31 Orientation of phase and group velocity for internal waves. The vertical components of the phase and group velocities are equal and opposite.

The group velocity here has the usual significance of being the velocity of propagation of energy of a certain sinusoidal component. Suppose a source is oscillating at frequency ω. Then its energy will only be transmitted outward along four beams oriented at an angle θ with the vertical, where $\cos\theta = \omega/N$. This has been verified in a laboratory experiment (Fig. 11.33). The source in this case was a vertically oscillating cylinder with its axis perpendicular to the plane of paper. The frequency was $\omega < N$. The light and dark lines in the photograph are lines of constant density, made visible by an optical technique. The experiment showed that the energy radiated along four beams that became more vertical as the frequency was increased, which agrees with $\cos\theta = \omega/N$.

These results were obtained by assuming that N is depth independent, an assumption that may seem unrealistic at first. Fig. 12.2 shows N vs. depth for the deep ocean, and $N < 0.01$ everywhere, but N is largest between ~200 m and ~2 km. These results can be considered *locally* valid if N varies slowly over the vertical wavelength $2\pi/m$ of the motion. The so-called *WKB approximation* for internal waves, in which such a slow variation of $N(z)$ is not neglected, is discussed in Chapter 12.

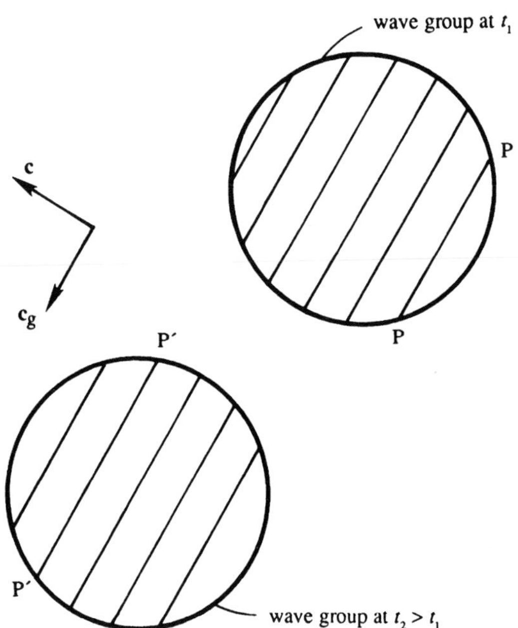

FIGURE 11.32 Illustration of phase and group propagation in a circular internal-wave packet. Positions of the wave packet at two times are shown. The constant-phase line PP (a crest perhaps) at time t_1 propagates to P′P′ at t_2.

Energy considerations for internal waves in a stratified fluid

The energy carried by an internal wave travels in the direction and at the speed of the group velocity. To show this is the case, construct a mechanical energy equation from (11.127) through (11.129) by multiplying the first equation by $\rho_0 u$, the second by $\rho_0 v$, the third by $\rho_0 w$, and summing the results to find:

$$\frac{\partial}{\partial t}\left[\frac{\rho_0}{2}(u^2+v^2+w^2)\right]+g\rho' w+\nabla\cdot(p'\mathbf{u})=0. \tag{11.145}$$

Here, the continuity equation has been used to write $u(\partial p'/\partial x)+v(\partial p'/\partial y)+w(\partial p'/\partial z)=\nabla\cdot(p'\mathbf{u})$, which represents the net work done by pressure forces. Another interpretation is that $\nabla\cdot(p'u)$ is the divergence of the *energy flux* $p'\mathbf{u}$, which must change the wave energy at a point. As the first term in (11.145) is the rate of change of kinetic energy, we can anticipate that the second term $g\rho' w$ must be the rate of change of potential energy, E_p. This is consistent with the energy principle derived in Chapter 4 (see (4.57)), except that ρ' and p' replace ρ and p because we have subtracted the mean quiescent state here. Using the density equation (11.130), the rate of change of potential energy can be written as:

$$\frac{\partial E_p}{\partial t}=g\rho' w=\frac{\partial}{\partial t}\left[\frac{g^2\rho'^2}{2\rho_0 N^2}\right], \tag{11.146}$$

which shows that the potential energy per unit volume must be the positive quantity $E_p=g^2\rho'^2/2\rho_0 N^2$. The potential energy can also be expressed in terms of the displacement ζ of a fluid particle, given by $w=\partial\zeta/\partial t$. Using (11.130), we can write:

$$\frac{\partial\rho'}{\partial t}=\frac{N^2\rho_0}{g}\frac{\partial\zeta}{\partial t},\ \text{which requires that}\ \rho'=\frac{N^2\rho_0\zeta}{g}. \tag{11.147}$$

The potential energy *per unit volume* is therefore:

$$E_p=\frac{g^2\rho'^2}{2\rho_0 N^2}=\frac{1}{2}N^2\rho_0\zeta^2. \tag{11.148}$$

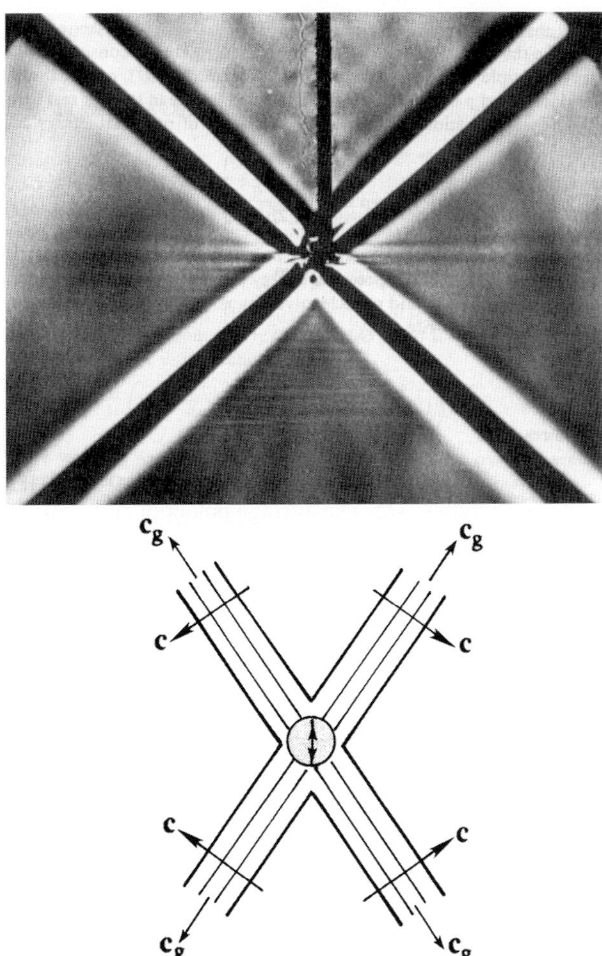

FIGURE 11.33 Waves generated in a stratified fluid of uniform buoyancy frequency $N = 1$ rad/s. The forcing agency is a horizontal cylinder, with its axis perpendicular to the plane of the paper, oscillating vertically at frequency $\omega = 0.71$ rad/s. With $\omega/N = 0.71 = \cos\theta$, this agrees with the observed angle of $\theta = 45°$ made by the beams with the horizontal. The vertical dark line in the upper half of the photograph is the cylinder support and should be ignored. The light and dark radial lines represent contours of constant ρ' and are therefore constant phase lines. The schematic diagram below the photograph shows the directions of $\mathbf{c}$ and $\mathbf{c}_g$ for the four beams. *Reprinted with the permission of Dr. T. Neil Stevenson, University of Manchester.*

This expression is consistent with our previous result from (11.96) for two infinitely deep fluids, for which the average potential energy of the entire water column *per unit horizontal area* was shown to be:

$$\frac{1}{4}(\rho_2 - \rho_1)ga^2, \tag{11.149}$$

where the interface displacement is of the form $\zeta = a\cos(kx - \omega t)$ and $(\rho_2 - \rho_1)$ is the density discontinuity. To see the consistency, we shall symbolically represent the buoyancy frequency of a density discontinuity at $z = 0$ as:

$$N^2 = -\frac{g}{\rho_0}\frac{d\overline{\rho}}{dz} = \frac{g}{\rho_0}(\rho_2 - \rho_1)\delta(z), \tag{11.150}$$

where $\delta(z)$ is the Dirac delta function (see Appendix B.4). (As with other relations involving the delta function, (11.150) is valid in the *integral* sense, that is, the integral (across $z = 0$) of the last two terms is equal because $\int \delta(z)dz = 1$.) Using (11.150), a vertical integral of (11.148), coupled with horizontal averaging over a wavelength, gives the expression (11.149). Note that for surface or interfacial waves, E_k and E_p represent kinetic and potential energies of the entire water column, per unit horizontal area. In a continuously stratified fluid, they represent energies per unit volume.

We shall now demonstrate that the average kinetic and potential energies are equal for internal wave motion. Assume periodic solutions:

$$[u, w, p', \rho'] = [\widehat{u}, \widehat{w}, \widehat{p}, \widehat{\rho}]e^{i(kx+ly+mz-\omega t)}. \tag{11.151}$$

Then all variables can be expressed in terms of w:

$$p' = -\frac{\omega m \rho_0}{k^2}\widehat{w}e^{i(kx+mz-\omega t)}, \quad \rho' = i\frac{N^2\rho_0}{\omega g}\widehat{w}e^{i(kx+mz-\omega t)}, \quad \text{and} \quad u = -\frac{m}{k}\widehat{w}e^{i(kx+mz-\omega t)}, \tag{11.152}$$

where p' is derived from (11.131), ρ' from (11.130), and u from (11.127). The average kinetic energy per unit volume is therefore:

$$E_k = \frac{1}{2}\rho_0\overline{(u^2 + w^2)} = \frac{1}{4}\rho_0\left(\frac{m^2}{k^2} + 1\right)\widehat{w}^2, \tag{11.153}$$

where we have taken real parts of the various expressions in (11.151) before computing quadratic quantities and used the fact that the average of $\cos^2()$ over a wavelength is 1/2. The average potential energy per unit volume is:

$$E_p = \frac{g^2\overline{p'^2}}{2\rho_0 N^2} = \frac{N^2\rho_0}{4\omega^2}\widehat{w}^2, \tag{11.154}$$

where we have used $\overline{\rho'2} = \widehat{w}^2N^4\rho_0^2/2\omega^2g^2$, found from (11.152) after taking its real part. Use of the dispersion relation $\omega^2 = k^2N^2/(k^2 + m^2)$ shows that:

$$E_k = E_p, \tag{11.155}$$

which is a general result for small oscillations of a conservative system without Coriolis forces. The total wave energy is:

$$E = E_k + E_p = \frac{1}{2}\rho_0\left(\frac{m^2}{k^2} + 1\right)\widehat{w}^2. \tag{11.156}$$

Last, we shall show that $\mathbf{c}_g$ times the wave energy equals the energy flux. The average energy flux $\mathbf{F}$ across a unit area can be found from (11.151):

$$\mathbf{F} = \overline{p'\mathbf{u}} = \mathbf{e}_x\overline{p'u} + \mathbf{e}_z\overline{p'w} = \frac{\rho_0\omega m\widehat{w}^2}{2k^2}\left(\mathbf{e}_x\frac{m}{k} - \mathbf{e}_z\right). \tag{11.157}$$

Using Eqs. (11.143) and (11.156), group velocity times wave energy is:

$$\mathbf{c}_g E = \frac{Nm}{K^3}(m\mathbf{e}_x - k\mathbf{e}_z)\left[\frac{1}{2}\rho_0\left(\frac{m^2}{k^2} + 1\right)\widehat{w}^2\right],$$

which reduces to (11.157) on using the dispersion relation (11.137), so it follows that:

$$\mathbf{F} = \mathbf{c}_g E. \tag{11.158}$$

This result also holds for surface or interfacial gravity waves. However, in that case $\mathbf{F}$ represents the flux per unit width perpendicular to the propagation direction (integrated over the entire depth), and E represents the energy per unit horizontal area. In (11.157), on the other hand, $\mathbf{F}$ is the flux per unit area, and E is the energy per unit volume.

Example 11.8

If the wave's frequency is $\omega \le N$ and the phase speed vector makes an angle θ with the horizontal direction as shown in Fig. 11.29, what are the fluid particle trajectories in the x-z plane as an internal wave with vertical velocity amplitude w_o passes?

Solution In this case, the horizontal and vertical velocities are:

$$u = -\frac{m}{k}w_o\cos(kx + mz - \omega t) \quad \text{and} \quad w = w_o\cos(kx + mz - \omega t),$$

as follows from (11.151) and (11.152) after real parts are taken. Fluid particle paths are given by x and z components of (3.8):

$$\frac{dx_p}{dt} = -\frac{m}{k}w_o \cos(kx_p + mz_p - \omega t) \quad \text{and} \quad \frac{dz_p}{dt} = w_o \cos(kx_p + mz_p - \omega t),$$

where the subscript p indicates the fluid particle's coordinates. The ratio of these two equations leads to: $dz_p/dx_p = -k/m$, or $kx_p + mz_p = constant = kx_o + mz_o$, if the fluid particle of interest is at x_o and z_o at $t = 0$, then:

$$k(x_p - x_o) = -m(z_p - z_o) = \frac{mw_o}{\omega}(\cos(kx_o + mz_o - \omega t) - \cos(kx_o + mz_o)).$$

Thus, this fluid particle oscillates along a line segment centered on (x_o, z_o) with horizontal amplitude $mw_o/k\omega = (w_o/\omega)\tan\theta$, and vertical amplitude w_o/ω. The orientation of this line segment is perpendicular to the wave number vector $\mathbf{K} = (k, 0, m)$. These fluid particle trajectories are different from the circles or ellipses found in the water column below ordinary water-surface waves; see (11.35).

Exercises

11.1 The kinematic boundary condition on a moving interface can be derived geometrically by considering the interface shape, $z = \eta(x, t)$, at two closely spaced instants in time.

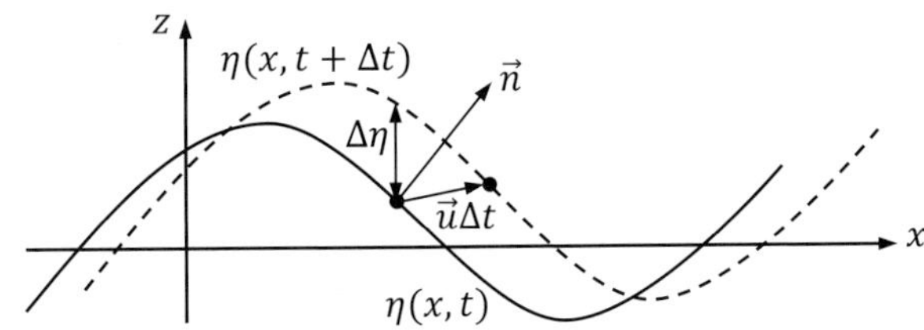

Use two-dimensional (x, z)-coordinates and let the interface be defined by $f(x, z, t) = \eta(x, t) - z = 0$. The point marked by a black dot is a material element on the surface. Use (11.13) for the surface normal $\mathbf{n}$, and equate the normal components of the interface displacement $\Delta\eta\mathbf{e}_z$ and the fluid particle displacement $\mathbf{u}\Delta t$. Then, let $\Delta t \to 0$ to reach (11.16):

$$[-u\partial\eta/\partial x + w]_{on\ the\ surface} = \partial\eta/\partial t.$$

11.2 Starting from (11.5) and working in (x, y, z) Cartesian coordinates, determine an equation that specifies the locus of points that defines a wave crest. Verify that the travel speed of the crests in the direction of $\mathbf{K} = (k, l, m)$ is $c = \omega/|\mathbf{K}|$. Can anything be determined about the wave crest travel speed in other directions?

11.3 For $ka \ll 1$, use the potential for linear deep water waves, $\phi(z, x, t) = a(\omega/k)e^{kz}\sin(kx - \omega t)$ and the waveform $\eta(x, t) = a\cos(kx - \omega t) + \alpha ka^2\cos[2(kx - \omega t)]$ to show that:

a) With an appropriate choice of the constant α, the kinematic boundary condition (11.16) can be satisfied for terms proportional to $(ka)^0$ and $(ka)^1$ once the common factor of $a\omega$ has been divided out.

b) With an appropriate choice of the constant γ, the dynamic boundary condition (11.19) can be satisfied for terms proportional to $(ka)^0$, $(ka)^1$, and $(ka)^2$ when $\omega^2 = gk(1 + \gamma k^2a^2)$ once the common factor of ag has been divided out.

11.4 The field equation for surface waves on a deep fluid layer in two dimensions (x, z) is: $\partial^2\phi/\partial x^2 + \partial^2\phi/\partial z^2 = 0$, where ϕ is the velocity potential, $\nabla\phi = (u, w)$. The linearized free-surface boundary conditions and the bottom boundary condition are:

$$(\partial\phi/\partial z)_{z=0} \cong \partial\eta/\partial t, \quad (\partial\phi/\partial t)_{z=0} + g\eta \cong 0, \quad \text{and} \quad (\partial\phi/\partial z)_{z\to-\infty} = 0,$$

where $z = \eta(x, t)$ defines the free surface, gravity g points downward along the z-axis, the undisturbed free surface lies at $z = 0$. The goal of this problem is to develop the general solution for these equations without assuming a sinusoidal form for the free surface as was done in Section 11.1 and 11.2.

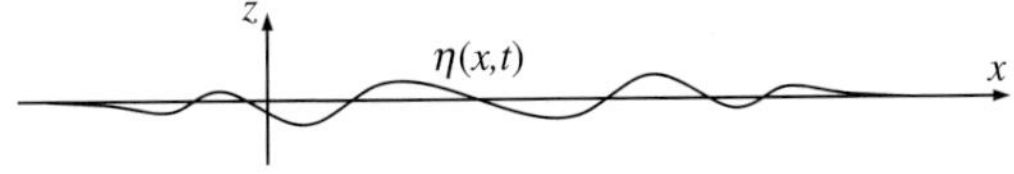

a) Assume $\phi(x,z,t) = \Lambda(x,t)Z(z)$, and use the field equation and bottom boundary condition to show that $\phi(x,z,t) = \Lambda(x,t)\exp(+kz)$, where k is a positive real constant.

b) Use the results of part a) and the remaining boundary conditions to show:

$$\frac{\partial^2 \Lambda}{\partial t^2} + gk\Lambda = 0 \quad \text{and} \quad \frac{\partial^2 \Lambda}{\partial x^2} + k^2\Lambda = 0.$$

c) For a fixed value of k, find $\Lambda(x,t)$ in terms of four unknown amplitudes A, B, C, and D.

d) For the initial conditions: $\eta = h(x)$ and $\partial\eta/\partial t = \dot{h}(x)$ at $t = 0$, determine the general form of $\phi(x,z,t)$.

11.5 Derive (11.37) from (11.27).

11.6 Consider stationary surface gravity waves in a rectangular container of length L and breadth b, containing water of undisturbed depth H. Show that the velocity potential $\phi = A\cos(m\pi x/L)\cos(n\pi y/b)\cosh[k(z+H)]e^{-i\omega t}$, satisfies the Laplace equation and the wall boundary conditions, if $(m\pi/L)^2 + (n\pi/b)^2 = k^2$. Here m and n are integers. To satisfy the linearized free-surface boundary condition, show that the allowable frequencies must be $\omega^2 = gk\tanh(kH)$. [*Hint*: combine the two boundary conditions (11.18) and (11.21) into a single equation $\partial^2\phi/\partial t^2 = -g\partial\phi/\partial z$ at $z = 0$.]

11.7 A lake has the following dimensions: $L = 30$ km, $b = 2$ km, and $H = 100$ m. If the wind sets up the mode $m = 1$ and $n = 0$, show that the period of the oscillation is 32 min.

11.8 Fill a square or rectangular cake pan half-way with water. Do the same for a round frying pan of about the same size. Agitate the water by carrying the two pans while walking briskly (outside) at a consistent pace on a horizontal surface.

a) Which shape lends itself better to spilling?

b) At what portion of the perimeter of the rectangular pan does spilling occur most readily?

c) Explain your observations in terms of standing wave modes.

11.9 Use (11.27), (11.28), and (11.38), to prove (11.39).

11.10 Show that the group velocity of pure capillary waves in deep water, for which the gravitational effects are negligible, is $c_g = (3/2)c$.

11.11 Assuming deep water, plot the group velocity of surface gravity waves, including surface tension σ, as a function of λ for water at 20 °C($\rho = 1000$ kg/m^3 and $\sigma = 0.074$ N/m),

a) Show that the group velocity is $c_g = \dfrac{1}{2}\sqrt{\dfrac{g}{k}}\dfrac{1 + 3\sigma k^2/\rho g}{\sqrt{1+\sigma k^2/\rho g}}$.

b) Show that this becomes minimum at a wave number given by $\dfrac{\sigma k^2}{\rho g} = \dfrac{2}{\sqrt{3}} - 1$.

c) Verify that $c_{g\,min} = 17.8$ cm/s.

11.12 The energy propagation characteristics of sinusoidal deep-water capillary-gravity waves follow those of pure gravity waves when corrections are made for the influence of surface tension. Assume a waveform shape of $\eta(x,t) = a\cos(kx - \omega t)$ for the following items.

a) Show that the sum of the wave's kinetic and potential energies (per unit surface area) can be written: $E = E_k + E_p = \frac{1}{2}\rho a^2\tilde{g}$ where $\tilde{g} = g + k^2\sigma/\rho$.

b) Determine the wave energy flux EF (per unit length) from

$$EF = \frac{\omega}{2\pi}\int_0^{2\pi/\omega}\left(\int_{-\infty}^{0} p'u dz - \sigma\mathbf{t}\cdot[\mathbf{u}]_{z=0}\right)dt,$$

where $\mathbf{t}$ is the surface tangent vector, and the extra term involving σ represents energy transfer via surface tension.

c) Use the results of parts a) and b) to show that $EF = Ec_g$, where c_g is the group velocity determined in Exercise 11.11 part a).

11.13 The effect of viscosity on the energy of linear deep-water surface waves can be determined from the wave motion's velocity components and the viscous dissipation (4.59).

a) For incompressible flow, the viscous dissipation of energy per unit mass of fluid is $\varepsilon = 2(\mu/\rho)S_{ij}^2$, where S_{ij} is the strain-rate tensor and μ is the fluid's viscosity. Determine ε using (11.47).

b) The total wave energy per unit surface area, E, for a linear sinusoidal water wave with amplitude a is given by (11.42). Assume that a is function of time, set $dE/dt = -\varepsilon$, and show that $a(t) = a_0\exp[-2(\mu/\rho)k^2t]$, where a_0 is the wave amplitude at $t = 0$.

c) Using a nominal value of $\mu/\rho = 10^{-6}$ m^2/s for water, determine the time necessary for an amplitude reduction of 50% for water-surface waves having $\lambda = 1$ mm, 1 cm, 10 cm, 1 m, 10 m, and 100 m.

d) Convert the times calculated in c) to travel distances by multiplication with an appropriate group speed. Remember to include surface tension. Can a typhoon located near New Zealand produce increased surf on the west coast of North America? [The circumference of the earth is approximately 40,000 km].

11.14 Consider a deep-water wave train with a Gaussian envelope that resides near $x = 0$ at $t = 0$ and travels in the positive-x direction. The surface shape at any time is a Fourier superposition of waves with all possible wave numbers:

$$\eta(x,t) = \int_{-\infty}^{+\infty} \tilde{\eta}(k) \exp\left[i(kx - (g|k|)^{1/2}t)\right] dk, \qquad (\dagger)$$

where $\tilde{\eta}(k)$ is the amplitude of the wave component with wave number k, and the dispersion relation is $\omega = (gk)^{1/2}$. For the following items assume the surface shape at $t = 0$ is:

$$\eta(x,0) = \frac{a}{\sqrt{2\pi}\alpha} \exp\left\{-\frac{x^2}{2\alpha^2} + ik_d x\right\}.$$

Here, $k_d > 0$ is the dominant wave number, and α sets the initial horizontal extent of the wave train, with larger α producing a longer wave train.

a) Plot Re$\{\eta(x,0)\}$ for $|x| \le 40$ m when $\alpha = 10$ m and $k_d = 2\pi/\lambda_d = 2\pi/10$ m^{-1}.

b) Use the inverse Fourier transform at $t = 0$, $\tilde{\eta}(k) = (1/2\pi)\int_{-\infty}^{+\infty} \eta(x,0)\exp[-ikx]dk$, to find the wave amplitude distribution: $\tilde{\eta}(k) = (1/2\pi)\exp\left\{-\frac{1}{2}(k - k_d)^2\alpha^2\right\}$, and plot this function for $0 < k < 2k_d$ using the numerical values from part a). Does the dominant contribution to the wave activity come from wave numbers near k_d for the part a) values?

c) For large x and t, the integrand of ($\dagger$) will be highly oscillatory unless the phase $\Phi \equiv kx - (g|k|)^{1/2}t$ happens to be constant. Thus, for any x and t, the primary contribution to η will come from the region where the phase in ($\dagger$) does not depend on k. Thus, set $d\Phi/dk = 0$, and solve for k_s (= the wave number where the phase is independent of k) in terms of x, t, and g.

d) Using the result of part b), set $k_s = k_d$ to find the x-location where the dominant portion of the wave activity occurs. At this location, the ratio x/t is the propagation speed of the dominant portion of the wave activity. Is this propagation speed the phase speed, the group speed, or another speed altogether?

11.15 Show that the vertical component of the Stokes drift is zero.

11.16 Extend the deep water Stokes drift result (11.85) to arbitrary depth to derive (11.86).

11.17 Explicitly show through substitution and differentiation that (11.88) is a solution of (11.87).

11.18 A sinusoidal long-wavelength shallow-water wave with amplitude A and wave number $k_1 = 2\pi/\lambda_1$ travels to the right in water of depth H_1 until it encounters a mild depth transition at $x = 0$ to a slightly shallower depth H_2. A portion of the incident wave continues to the right with amplitude B and a portion is reflected and propagates to the left.

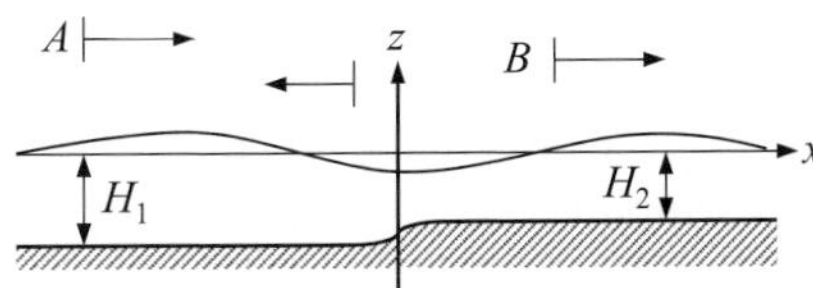

a) By requiring the water surface deflection and the horizontal volume flux in the water column to be continuous at $x = 0$, show that $B/A = 2/(1 + \sqrt{H_2/H_1})$.

b) If a tsunami wave starts with $A = 2.0$ m in water 5 km deep, estimate its amplitude when it reaches water 10 m deep if the ocean depth change can be modeled as a large number of discrete depth changes. [Recommendation: consider multiple steps of $H_2/H_1 = 0.90$, since $(0.90)^{59} \approx 0.002 = (10 \text{ m})/(5 \text{ km})$].

c) Redo part b) when the wave's energy flux remains constant at both depths when λ_1 is 100 km (as in Example 11.6).

d) Compare the results of parts b) and c). Which amplitude in shallow water is lower and why is it lower?

11.19 A *thermocline* is a thin layer in the upper ocean across which water temperature and, consequently, water density change rapidly. Suppose the thermocline in a very deep ocean is at a depth of 100 m from the ocean surface, and that the temperature drops across it from 30 °C to 20 °C. Show that the reduced gravity is $g' = 0.025\ \text{m/s}^2$. Neglecting Coriolis effects, show that the speed of propagation of long gravity waves on such a thermocline is 1.58 m/s.

11.20 Consider internal waves in a continuously stratified fluid of buoyancy frequency $N = 0.02\ \text{s}^{-1}$ and average density 800 kg/m^3. What is the direction of ray paths if the frequency of oscillation is $\omega = 0.01\ \text{s}^{-1}$? Find the energy flux per unit area if the amplitude of vertical velocity is $\widehat{w} = 1$ cm/s and the horizontal wavelength is π meters.

11.21 Consider internal waves at a density interface between two infinitely deep fluids, and show that the average kinetic energy per unit horizontal area is $E_k = (\rho_2 - \rho_1)ga^2/4$.

11.22 Consider waves in a finite layer overlying an infinitely deep fluid. Using the constants given in (11.105) through (11.108), prove the dispersion relation (11.109).

11.23 A simple model of oceanic internal waves involves two ideal incompressible fluids ($\rho_2 > \rho_1$) trapped between two horizontal surfaces at $z = h_1$ and $z = -h_2$, and having an average interface location of $z = 0$. For traveling waves on the interface, assume that the interface deflection from $z = 0$ is $\xi = \xi_o \text{Re}\{\exp(i(\omega t - kx))\}$. The phase speed of the waves is $c = \omega/k$.

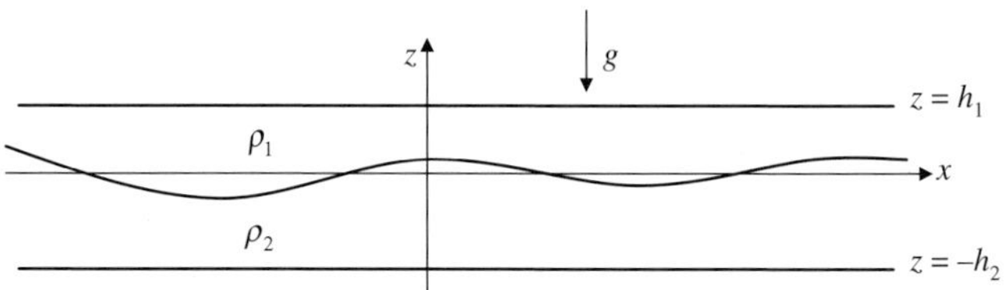

a) Show that the dispersion relationship is $\omega^2 = \dfrac{gk(\rho_2 - \rho_1)}{\rho_2 \coth(kh_2) + \rho_1 \coth(kh_1)}$, where g is the acceleration of gravity.

b) Determine the limiting form of c for short (i.e., unconfined) waves, kh_1 and $kh_2 \to \infty$.

c) Determine the limiting form of c for long (i.e., confined) waves, kh_1 and $kh_2 \to 0$.

d) At fixed wavelength λ (or fixed $k = 2\pi/\lambda$), do confined waves go faster or slower than unconfined waves?

e) At a fixed frequency, what happens to the wavelength and phase speed as $\rho_2 - \rho_1 \to 0$?

f) What happens if $\rho_2 < \rho_1$?

11.24 Consider the long-wavelength limit of surface and interface waves with amplitudes a and b, respectively, that occur when two ideal fluids with densities ρ_1 and $\rho_2 (> \rho_1)$ are layered as shown in Fig. 11.28. Here, the velocity potentials – accurate through second order in kH_1 and KH_2 – in the two fluids are:

$$\phi_1 \cong A_1\left(1 + \frac{1}{2}k^2(z - z_1)^2\right)e^{i(kx-\omega t)} \quad \text{and} \quad \phi_2 \cong A_2\left(1 + \frac{1}{2}k^2(z - z_2)^2\right)e^{i(kx-\omega t)}$$
$$\text{for} \quad kH_1,\ KH_2 \ll 1,$$

where b, A_1, z_1, A_2, and z_2 are constants to be found in terms of a, g, k, ω, H_1 and H_2. From analysis similar to that in Section 11.7, find the dispersion relationship $\omega = \omega(k)$. When $H_1 = H_2 = H/2$, show that the surface and interface waves are in phase with $a > b$ for the barotropic mode, and out-of-phase with $|b| > |a|$ for the baroclinic mode. For this case, what are the phase speeds of the two modes? Which mode travels faster? What happens to the baroclinic mode's phase speed and amplitude as $\rho_2 - \rho_1 \to 0$?

References

Gill, A., 1982. Atmosphere–Ocean Dynamics. Academic Press, New York.
Graff, K.A., 1975. Wave Motion in Elastic Solids. Oxford University Press, Oxford.
Kinsman, B., 1965. Wind Waves. Prentice-Hall, Englewood Cliffs, New Jersey.
LeBlond, P.H., Mysak, L.A., 1978. Waves in the Ocean. Elsevier Scientific Publishing, Amsterdam.
Lighthill, M.J., 1978. Waves in Fluids. Cambridge University Press, London.
Phillips, O.M., 1977. The Dynamics of the Upper Ocean. Cambridge University Press, London.
Turner, J.S., 1973. Buoyancy Effects in Fluids. Cambridge University Press, London.
Whitham, G.B., 1974. Linear and Nonlinear Waves. Wiley, New York.

Chapter 12

Geophysical fluid dynamics ⊛

Contents

12.1 **Introduction** 489
12.2 **Vertical variation of density in the atmosphere and ocean** 491
12.3 **Equations of motion for geophysical flows** 493
f-Plane model 495
β-Plane model 495
12.4 **Geostrophic flow** 496
Thermal wind 497
Taylor-Proudman theorem 497
12.5 **Ekman layers** 499
Ekman layer at a free surface 500
Explanation in terms of vortex tilting 502
Ekman layer on a rigid surface 503
12.6 **Shallow-water equations** 505
12.7 **Normal modes in a continuously stratified layer** 507
Boundary conditions on ψ_n 508
Vertical mode solution for uniform N 508
Summary 510
12.8 **High- and low-frequency regimes in shallow-water equations** 510
12.9 **Gravity waves with rotation** 512
Particle orbit 513
Inertial motion 514
12.10 **Kelvin wave** 514
12.11 **Potential vorticity conservation in shallow-water theory** 517
12.12 **Internal waves** 520
WKB solution 521
Particle orbit 523
Discussion of the dispersion relation 523
Lee wave 526
12.13 **Rossby wave** 528
Quasi-geostrophic vorticity equation 528
Dispersion relation 529
12.14 **Barotropic instability** 531
12.15 **Baroclinic instability** 532
Perturbation vorticity equation 533
Wave solution 535
Instability criterion 535
Energetics 536
12.16 **Geostrophic turbulence** 537
Exercises 540
References 541
Further reading 542

Chapter objectives

- To introduce the approximations and phenomena that are common in geophysical fluid dynamics
- To describe flows near air-water interfaces and solid surfaces in steadily rotating coordinate systems
- To specify the effects of planetary rotation on waves in stratified fluids
- To describe the instabilities of very long waves that span a significant range of latitudes
- To provide an introduction to geostrophic turbulence and the reverse energy cascade

12.1 Introduction

The subject of geophysical fluid dynamics deals with the dynamics of the atmosphere and the ocean. Motions within these fluid masses are intimately connected through continual exchanges of momentum, heat, and moisture, and cannot be considered separately on a global scale. The field has been largely developed by meteorologists and oceanographers, but non-specialists have also been interested in the subject. Taylor was not a geophysical fluid dynamicist, but he held the position of a meteorologist for some time, and through this involvement he developed a special interest in the problems of turbulence and instability. Although Prandtl was mainly interested in the engineering aspects of fluid mechanics, his well-known textbook (Prandtl, 1952) contains several sections dealing with meteorological aspects of fluid mechanics. Notwithstanding the pressure for technical specialization, it is worthwhile to learn something of this fascinating field even if one's primary interest is in another area of fluid mechanics.

⊛. This book has a companion website hosting complementary materials. Visit this URL to access it: https://www.elsevier.com/books-and-journals/book-companion/9780128198070.

Fluid Mechanics. https://doi.org/10.1016/B978-0-12-819807-0.00021-1

Together the atmosphere and ocean have a large and consequential impact on humanity. The combined dynamics of the atmosphere and ocean are leading contributors to global climate. We all live within the atmosphere and are almost helplessly affected by the weather and its rather chaotic behavior that modulates agricultural success. Ocean currents affect navigation, fisheries, and pollution disposal. Populations that occupy coastlines can do little to prevent hurricanes, typhoons, or tsunamis. Thus, understanding and reliably predicting geophysical fluid dynamic events and trends are scientific, economic, humanitarian, and even political priorities. This chapter provides the basic elements necessary for developing an understanding of geophysical fluid dynamics.

The two features that distinguish geophysical fluid dynamics from other areas of fluid dynamics are the rotation of the earth and vertical density stratification of the media. These two effects dominate the dynamics to such an extent that entirely new classes of phenomena arise, which have no counterpart in the laboratory-scale flows emphasized in the preceding chapters. (For example, the dominant mode of flow in the atmosphere and the ocean is *along* the lines of constant pressure, not from high to low pressures.) The motion of the atmosphere and the ocean is naturally studied in a coordinate frame rotating with the earth. This gives rise to the Coriolis acceleration (see Section 4.7). The density stratification gives rise to buoyancy forces (Section 4.11 and Chapter 11). In addition, important relevant material includes vorticity, boundary layers, instability, and turbulence (Chapters 5, 8, 9, and 10). The reader should be familiar with these topics before proceeding further with the present chapter.

Because the Coriolis acceleration and fluid stratification play dominating roles in both the atmosphere and the ocean, there is a great deal of similarity between the dynamics of these two media; this makes it possible to study them together. There are also significant differences, however. For example the effects of lateral boundaries, due to the presence of continents, are important in the ocean but less so in the atmosphere. The intense currents (like the Gulf Stream and the Kuroshio) along the western ocean boundaries have no atmospheric analog. On the other hand, phenomena like cloud formation and latent heat release due to moisture condensation are solely atmospheric phenomena. Plus, processes are generally slower in the ocean, in which a typical horizontal velocity is 0.1 m/s, although velocities of the order of 1–2 m/s are found within the intense western boundary currents. In contrast, typical velocities in the atmosphere are 10–20 m/s. The nomenclature can also be different in the two fields. Meteorologists refer to a flow directed to the west as an "easterly wind" (i.e., *from* the east), while oceanographers refer to such a flow as a "westward current." Atmospheric scientists refer to vertical positions by *heights* measured upward from the earth's surface, while oceanographers refer to *depths* measured downward from the sea surface. In this chapter, the vertical coordinate z increases upward, following the atmospheric science convention.

The rotational effects arising from the Coriolis acceleration have opposite signs in the two hemispheres. Note that *all figures and descriptions given here are valid for the northern hemisphere.* In some cases the sense of the rotational effect for the southern hemisphere has been explicitly mentioned. When the sense of the rotational effect is left unspecified for the southern hemisphere, it should be assumed as opposite to that in the northern hemisphere.

Example 12.1

The fluids (air & water) and velocities (a few cm/s to tens of m/s) involved in geophysical fluid mechanics are the same as those of many laboratory flows. However, the length scales of geophysical flows are considerably larger. Calculate the Reynolds numbers associated with two relatively *small-scale* geophysical phenomena: (*i*) a single-cell thunderstorm 10 km in diameter with a vertical velocity of 5 m/s, and (*ii*) a 100-m-wavelength deep-ocean water wave having an amplitude of 4 m. Are both flows turbulent? Should they be?

Solution For the thunderstorm, the Reynolds number should be computed using the radius of the rising column of air and a mid-troposphere ($z \approx 5\,\text{km}$) value of the kinematic viscosity since such storms may span the troposphere:

$$\mathrm{Re} = \frac{U(D/2)}{\nu} \cong \frac{(5\,\text{m/s})(5\times 10^3\,\text{m})}{2.2\times 10^{-5}\,\text{m}^2/\text{s}} \sim 10^9.$$

This is a free shear flow driven by buoyancy and it has some of the character of a buoyant plume, so at this high Reynolds number it is most-definitely turbulent.

For the water wave, the characteristic fluid velocity within the wave, ωA, is set by the wave's frequency from (11.28), $\omega^2 = 2\pi g/\lambda$, and the wave's amplitude A. This leads to:

$$\mathrm{Re} = \frac{\omega A\lambda}{\nu} = \frac{[2\pi g/\lambda]^{1/2}A\lambda}{\nu} = \frac{[2\pi g\lambda]^{1/2}A}{\nu} = \frac{[2\pi(9.81\,\text{m/s}^2)100\,\text{m}]^{1/2}(4\,\text{m})}{1.\times 10^{-6}\,\text{m}^2/\text{s}} \sim 3\times 10^8.$$

In spite of this high Reynolds number, the water motion associated with such a wave may be well described by potential flow. Thus, the high Reynolds number in this case indicates that such waves see only miniscule viscous lossless and therefore may travel enormous distances without much amplitude decay (see Exercise 11.13).

12.2 Vertical variation of density in the atmosphere and ocean

An important characteristic of geophysical fluid dynamics is density stratification. As described in Section 1.10, the static stability of a fluid medium is determined by the sign of the potential density gradient:

$$\frac{d\rho_\theta}{dz} = \frac{d\rho}{dz} + \frac{g\rho}{c^2}, \tag{1.42}$$

where c is the speed of sound. A medium is statically stable if the potential density decreases with height. The first term on the right side of (1.42) corresponds to the *in situ* density change due to all sources such as pressure, temperature, and concentration of a constituent such as the salinity in the sea or the water vapor in the atmosphere. The second term on the right side is the density gradient due to the pressure decrease with height in an adiabatic environment and is called the *adiabatic density gradient*. The corresponding temperature gradient is called the *adiabatic temperature gradient*. For incompressible fluids $c = \infty$ and the adiabatic density gradient is zero.

The temperature of a dry adiabatic atmosphere decreases upward at the rate of approximately 10 °C/km, and that of a moist atmosphere decreases at the rate of ≈ 5–6 °C/km. In the ocean, the adiabatic density gradient is $g\rho/c^2 \sim 4.4 \times 10^{-3}\,\text{kg/m}^4$, for a typical sound speed of $c = 1520\,\text{m/s}$. The potential density in the ocean increases with depth at a much smaller rate of $0.6 \times 10^{-3}\,\text{kg/m}^4$, so it follows that most of the *in situ* density increase with depth in the ocean is due to the compressibility effects and not to changes in temperature or salinity. As potential density is the variable that determines static stability, oceanographers take into account the compressibility effects by referring all their density measurements to the sea-level pressure. Unless specified otherwise, throughout the present chapter potential density will simply be referred to as "density," omitting the qualifier "potential."

The mean vertical distribution of the *in situ* temperature in the lower 50 km of the atmosphere is shown in Fig. 12.1. The lowest 10 to 15 km is called the *troposphere*, in which the temperature decreases with height at the rate of 6.5 °C/km. This is close to the moist adiabatic lapse rate, which means that the troposphere is close to being neutrally stable. Neutral stability is expected because turbulent mixing due to frictional and convective effects in the lower atmosphere keeps it well stirred and therefore close to neutral stratification. Practically all the clouds, weather changes, and water vapor of the atmosphere are found in the troposphere. The layer is capped by the *tropopause*, at an average height of approximately 10 km, above which temperature increases with increasing height. The altitude of the tropopause varies with latitude from 8 to 9 km at the poles to more than 15 km at the equator. The layer above the tropopause is called the *stratosphere*, because it is highly stably stratified. The increase of temperature with height in this layer is caused by the absorption of the sun's ultraviolet rays by ozone. The stability of the layer inhibits mixing and consequently acts as a lid on the turbulence and convective motion of the troposphere. The positive temperature gradient stops at the *stratopause* at a height of nearly 50 km. The altitude of the stratopause varies little with latitude.

The vertical structure of density in the ocean is sketched in Fig. 12.2, showing typical profiles of potential density and temperature. Most of the temperature increase with height is due to the absorption of solar radiation within the upper layer of the ocean. The density distribution in the ocean is also affected by the salinity. However, there is no characteristic variation of salinity with depth, and a decrease with depth is found to be as common as an increase with depth. In most cases, however, the vertical structure of density in the ocean is determined mainly by that of temperature, the salinity effects being secondary. The upper 50–200 m of ocean is relatively well mixed, due to the turbulence generated by the wind, waves, current shear, and the convective overturning caused by surface cooling. Temperature gradients decrease with depth, becoming quite small below a depth of 1500 m. There is usually a large temperature gradient in the depth range of 100–500 m. This layer of high stability is called the *thermocline*. Fig. 12.2 also shows the profile of *buoyancy frequency* N, defined by:

$$N^2 \equiv -\frac{g}{\rho_0}\frac{d\rho}{dz}, \tag{12.1}$$

where ρ is the potential density and ρ_0 is a constant reference density (*cf.* (1.36) and (11.126)). The buoyancy frequency reaches a typical maximum value of $N_{\max} \sim 0.01$ rad/s (period ~ 10 min) in the thermocline and decreases both upward and downward.

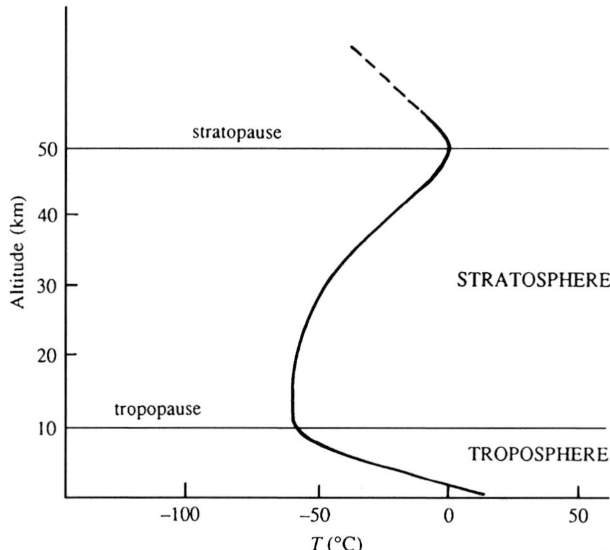

FIGURE 12.1 Sketch of the vertical distribution of temperature in the lower 50 km of the atmosphere. In the lowest layer, the troposphere, the temperature decreases with height and this is where nearly all weather occurs. The next layer is the stratosphere where temperature increases with height. The troposphere is separated from the stratosphere by the tropopause, and the stratosphere ends at the stratopause.

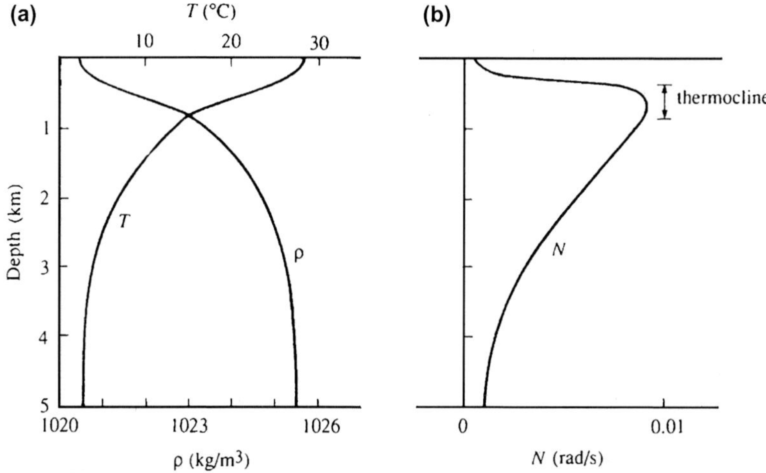

FIGURE 12.2 Typical vertical distributions of: (a) temperature and density, and (b) buoyancy frequency in the ocean. Temperature falls while density increases with increasing depth. The buoyancy frequency peaks in the region of the thermocline where temperature changes most rapidly with depth.

Example 12.2

For a neutrally-stable ocean having a surface-water density of $\rho_0 = 1030\,\mathrm{kg/m^3}$ what is the water density at a depth of 5 km assuming a constant sound speed of $c = 1520\,\mathrm{m/s}$?

Solution For a neutrally stable ocean, $\partial\rho_\theta/\partial z = 0$, so (1.42) implies $\partial\rho/\partial z = -g\rho/c^2$ and this is readily integrated to find $\rho(z) = \rho_0 \exp(-gz/c^2)$. Therefore, at $z = -5\,\mathrm{km}$:

$$\rho = (1030\,\mathrm{kg/m^3})\exp(-(9.81\,\mathrm{m/s^2})(-5\times10^3\,\mathrm{m})/(1520\,\mathrm{m/s})^2) = 1052\,\mathrm{kg/m^3},$$

which is just a little more than a 2% increase over the surface water density, even though the pressure at $z = -5$ km is more than 500 times higher than at the surface.

12.3 Equations of motion for geophysical flows

From Section 4.7, the equations of motion for a stratified fluid, observed in a system of coordinates rotating at a constant angular velocity Ω with respect to the "fixed stars" are:

$$\nabla \cdot \mathbf{u} = 0, \qquad \frac{D\mathbf{u}}{Dt} + 2\boldsymbol{\Omega} \times \mathbf{u} = -\frac{1}{\rho_0}(\nabla p + \rho g \mathbf{e}_z) + \mathbf{F}, \quad \text{and} \quad \frac{D\rho}{Dt} = 0, \tag{4.10, 12.2, 4.9}$$

where $\mathbf{F}$ is the friction force per unit mass of fluid, the centrifugal acceleration is combined into the body force acceleration $g\mathbf{e}_z$, and diffusive effects in the density equation are omitted. This equation set embodies the *Boussinesq approximation*, discussed in Section 4.9, in which the density variations are neglected everywhere except in the gravity term and the vertical scale of the motion is assumed less than the "scale height" of the medium c^2/g, where c is the speed of sound. This assumption is very good in the ocean, in which $c^2/g \sim 200$ km. In the atmosphere it is less applicable, because $c^2/g \sim 10$ km. Under the Boussinesq approximation, the principle of mass conservation is expressed by $\nabla \cdot \mathbf{u} = 0$ (4.10), and the density equation $D\rho/Dt = 0$ (4.9) follows from the non-diffusive heat or species equation $DT/Dt = 0$ or $DS/Dt = 0$ and an incompressible equation of state of the form $\delta\rho/\rho_0 = -\alpha\delta T$ or $\delta\rho/\rho_0 = \beta\delta S$, where S the concentration of a constituent such as water vapor in the atmosphere or the salinity in the ocean. Fortunately, (4.9) and (4.10) are consistent with each other, as described in Section 4.2, even if they occur together here for a different reason.

For a closed set of equations, the friction force per unit mass, $\mathbf{F}$ in (12.2), must be appropriately related to the (average) velocity field. Geophysical flows are commonly turbulent and anisotropic with vertical velocities that are typically much smaller than horizontal ones. From Section 4.5, the friction force is given by $F_i = \partial\tau_{ij}/\partial x_j$, where τ_{ij} is the viscous stress tensor. In large-scale geophysical flows, however, the frictional forces are typically provided by turbulent momentum exchange and viscous effects are negligible. Yet, the complexity of turbulence makes it impossible to relate the stress to the (average) velocity field in a simple way. Thus, to proceed in a rudimentary manner that includes anisotropy, the eddy viscosity hypothesis (10.115) is adopted but the turbulent viscosity is presumed to have directional dependence. In particular, geophysical fluid media are commonly in the form of stratified layers that inhibit vertical transport of horizontal momentum. This means that the exchange of momentum upward or downward across a horizontal surface is much weaker than that in either horizontal direction across a vertical surface. To reflect this phenomenology in $\mathbf{F}$, the vertical eddy viscosity ν_V is assumed to be much smaller than the horizontal eddy viscosity ν_H, and the turbulent stress components are assumed to be related to the fluid velocity $\mathbf{u} = (u, v, w)$ by:

$$\tau_{xz} = \tau_{zx} = \rho\nu_V \frac{\partial u}{\partial z} + \rho\nu_H \frac{\partial w}{\partial x}, \quad \tau_{yz} = \tau_{zy} = \rho\nu_V \frac{\partial v}{\partial z} + \rho\nu_H \frac{\partial w}{\partial y}, \quad \tau_{xy} = \tau_{yx} = \rho\nu_H \left(\frac{\partial u}{\partial y} + \frac{\partial v}{\partial x}\right),$$
$$\tau_{xx} = 2\rho\nu_H \frac{\partial u}{\partial x}, \quad \tau_{yy} = 2\rho\nu_H \frac{\partial v}{\partial y}, \quad \text{and} \quad \tau_{zz} = 2\rho\nu_V \frac{\partial w}{\partial z}. \tag{12.3}$$

There are two difficulties with the set (12.3). First, the expressions for τ_{xz} and τ_{yz} depend on the *rotation* of fluid elements in a vertical plane and not just their deformation. As stated in Section 4.5, a requirement for a constitutive equation for a fluid is that the stresses should be independent of fluid element rotation and should depend only on element deformation. Therefore, τ_{xz} should depend only on the combination $(\partial u/\partial z + \partial w/\partial x)$, whereas the expression in (12.3) depends on both deformation and rotation. A tensorially correct geophysical treatment of the frictional terms is discussed, for example, in Kamenkovich (1967). Second, the eddy viscosity assumption for modeling momentum transport in turbulent flow is of questionable validity (Pedlosky (1971) describes it as a "rather disreputable and desperate attempt"). However, (12.3) provides a simple approximate formulation for viscous effects suitable for the current level of inquiry. So, using the set (12.3) and further assuming ν_V and ν_H are constants, the components of the frictional force $F_i = \partial\tau_{ij}/\partial x_j$ become:

$$F_x = \frac{\partial\tau_{xx}}{\partial x} + \frac{\partial\tau_{xy}}{\partial y} + \frac{\partial\tau_{xz}}{\partial z} = \nu_H\left(\frac{\partial^2 u}{\partial x^2} + \frac{\partial^2 u}{\partial y^2}\right) + \nu_V\left(\frac{\partial^2 u}{\partial z^2}\right),$$
$$F_y = \frac{\partial\tau_{yx}}{\partial x} + \frac{\partial\tau_{yy}}{\partial y} + \frac{\partial\tau_{yz}}{\partial z} = \nu_H\left(\frac{\partial^2 v}{\partial x^2} + \frac{\partial^2 v}{\partial y^2}\right) + \nu_V\left(\frac{\partial^2 v}{\partial z^2}\right), \quad \text{and} \tag{12.4}$$
$$F_z = \frac{\partial\tau_{zx}}{\partial x} + \frac{\partial\tau_{zy}}{\partial y} + \frac{\partial\tau_{zz}}{\partial z} = \nu_H\left(\frac{\partial^2 w}{\partial x^2} + \frac{\partial^2 w}{\partial y^2}\right) + \nu_V\left(\frac{\partial^2 w}{\partial z^2}\right).$$

Estimates of the eddy coefficients vary greatly. Typical suggested values are $\nu_V \sim 10\,\text{m}^2/\text{s}$ and $\nu_H \sim 10^5\,\text{m}^2/\text{s}$ for the lower atmosphere, and $\nu_V \sim 0.01\,\text{m}^2/\text{s}$ and $\nu_H \sim 100\,\text{m}^2/\text{s}$ for the upper ocean. In comparison, the molecular values are $\nu = 1.5 \times 10^{-5}\,\text{m}^2/\text{s}$ for air and $\nu = 10^{-6}\,\text{m}^2/\text{s}$ for water at atmospheric pressure and 20 °C.

When (12.4) is used in (12.2), the set (4.9), (4.10) and (12.2) provides five equations for the five field variables u, v, w, p, and ρ. However, for most geophysical flows, these five equations are commonly solved after several additional simplifications and approximations.

In general, geophysical flow problems should be solved using spherical polar coordinates attached to earth. However, the vertical scales of the ocean and the troposphere are of order 5 to 15 km while their horizontal scales are of order of hundreds, or even thousands, of kilometers. Thus, the trajectories of fluid elements in atmospheric and oceanic flows are nearly horizontal, $|u|$, $|v| \gg |w|$, and most geophysical flows can be considered to occur in thin layers. In fact, (4.10) suggests that:

$$|u|/L \sim |w|/H,$$

where H is the vertical scale and L is the horizontal length scale. Stratification and Coriolis effects usually constrain the vertical velocity to be even smaller than $|u|H/L$. If, in addition, the horizontal length scales of interest are much smaller than the radius of the earth (= 6371 km), then the curvature of the earth can be ignored, and the motion can be studied by adopting a *local* Cartesian system on a *tangent plane*. Fig. 12.3 shows this tangent-plane xyz coordinate system: with x increasing eastward (into the page), y northward, and z upward. The corresponding velocity components are u (eastward), v (northward), and w (upward).

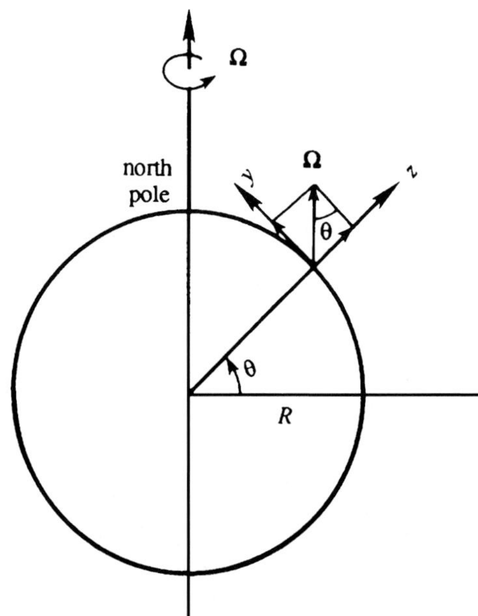

FIGURE 12.3 Tangent-plane Cartesian coordinates. The x-axis points into the plane of the paper. The y-axis is tangent to the earth's surface and points toward the north pole. The z-axis is vertical, opposing gravity. The earth's angular rotation vector has positive y and z components in the northern hemisphere. The angle θ is the geographic latitude and is defined with respect to the local surface normal; thus, it is not quite the same as the geocentric latitude indicated near the center of the figure.

The tangent-plane approximation allows the components of the Coriolis acceleration in (12.2) to be evaluated. The earth rotates at a rate:

$$\Omega \equiv |\mathbf{\Omega}| = 2\pi\ (\text{radians/day}) = 7.27 \times 10^{-5}\ \text{rad/s}$$

around the polar axis, in a counterclockwise sense looking from above the north pole. From Fig. 12.3, the components of angular velocity of the earth in the local tangent-plane Cartesian system are $\mathbf{\Omega} = (\Omega_x, \Omega_y, \Omega_z) = (0, \Omega\cos\theta, \Omega\sin\theta)$, where θ is the geographic latitude. If the earth were perfectly spherical, θ would be the geocentric latitude as well. With this geometry, the Coriolis acceleration appearing in (12.2) is:

$$2\mathbf{\Omega}\times\mathbf{u}=\begin{vmatrix}\mathbf{e}_x & \mathbf{e}_y & \mathbf{e}_z\\ 0 & 2\Omega\cos\theta & 2\Omega\sin\theta\\ u & v & w\end{vmatrix}=2\Omega\left[\mathbf{e}_x(w\cos\theta-v\sin\theta)+\mathbf{e}_y u\sin\theta-\mathbf{e}_z u\cos\theta\right].$$

The thin-fluid layer approximation, $|w|\ll|v|$, allows $w\cos\theta$ to be ignored compared to $v\sin\theta$ in the term multiplied by $\mathbf{e}_x$ away from the equator ($\theta=0$). Thus, the three components of the Coriolis acceleration are:

$$2\mathbf{\Omega}\times\mathbf{u}\cong(-2\Omega v\sin\theta,\ 2\Omega u\sin\theta,\ -2\Omega u\cos\theta)=(-fv,\ fu,\ -2\Omega u\cos\theta),\quad\text{where}\quad f=2\Omega\sin\theta \tag{12.5, 12.6}$$

is twice the local *vertical* component of $\mathbf{\Omega}$. Since vorticity is twice the angular velocity, f is referred to as the *planetary vorticity*, or, more commonly, as the *Coriolis parameter* or the *Coriolis frequency*. It is positive in the northern hemisphere and negative in the southern hemisphere, varying from $\pm1.45\times10^{-4}\,\text{s}^{-1}$ at the poles to zero at the equator. This makes sense, since an object at the north pole spins around a vertical axis at a counterclockwise rate Ω, whereas a object at the equator does not spin around a vertical axis. The quantity, $T_i=2\pi/f$, is called the *inertial period*, for reasons that will be clear in Section 12.9; it does not represent the components of a vector.

The pressure and gravity terms in (12.2) can also be simplified by writing them in terms of the pressure and density *perturbations* from a state of rest:

$$\rho(\mathbf{x},t)=\overline{\rho}(z)+\rho'(\mathbf{x},t)\quad\text{and}\quad p(\mathbf{x},t)=\overline{p}(z)+p'(\mathbf{x},t), \tag{12.7}$$

where the static distribution of density, $\overline{\rho}(z)$, and pressure, $\overline{p}(z)$, follow the hydrostatic law (1.14). When (12.7) is substituted into (12.2), the first two terms inside parentheses on the right side become:

$$\nabla p+g\rho\mathbf{e}_z=\nabla(\overline{p}+p')+g(\overline{\rho}+\rho')\mathbf{e}_z=\left[\frac{d\overline{p}}{dz}+g\overline{\rho}\right]\mathbf{e}_z+\nabla p'+g\rho'\mathbf{e}_z=\nabla p'+g\rho'\mathbf{e}_z \tag{12.8}$$

because the terms in [,]-braces sum to zero from (1.14).

In addition, the vertical component of the Coriolis acceleration, namely $-2\Omega u\cos\theta$, is generally negligible compared to the dominant terms in the vertical equation of motion, namely $g\rho'/\rho_0$ and $\rho_0^{-1}(\partial p'/\partial z)$. Thus, combining (12.2), (12.4), (12.5), (12.6), and (12.8), while ignoring $-2\Omega u\cos\theta$ in the vertical momentum equation, leads to the simplified equation set:

$$\begin{aligned}
\frac{Du}{Dt}-fv&=-\frac{1}{\rho_0}\frac{\partial p'}{\partial x}+\nu_\text{H}\left(\frac{\partial^2u}{\partial x^2}+\frac{\partial^2u}{\partial y^2}\right)+\nu_\text{V}\left(\frac{\partial^2u}{\partial z^2}\right),\\
\frac{Dv}{Dt}+fu&=-\frac{1}{\rho_0}\frac{\partial p'}{\partial y}+\nu_\text{H}\left(\frac{\partial^2v}{\partial x^2}+\frac{\partial^2v}{\partial y^2}\right)+\nu_\text{V}\left(\frac{\partial^2v}{\partial z^2}\right),\quad\text{and}\\
\frac{Dw}{Dt}&=-\frac{1}{\rho_0}\frac{\partial p'}{\partial z}-\frac{g\rho'}{\rho_0}+\nu_\text{H}\left(\frac{\partial^2w}{\partial x^2}+\frac{\partial^2w}{\partial y^2}\right)+\nu_\text{V}\left(\frac{\partial^2w}{\partial z^2}\right).
\end{aligned} \tag{12.9}$$

These are the equations for primarily horizontal fluid motion within a thin layer on a locally-flat rotating earth. Note that only the *vertical* component of the earth's angular velocity appears as a consequence of the flatness of the fluid trajectories.

f-Plane model

Although the Coriolis parameter $f=2\Omega\sin\theta$ varies with latitude θ, this variation is important only for phenomena having relatively long time scales (several weeks) or relatively long length scales (thousands of kilometers). For many purposes, f can be well approximated as constant, say $f_0=2\Omega\sin\theta_0$, where θ_0 is the central latitude of the region under study. A model using a constant Coriolis parameter is called an *f-plane model*.

β-Plane model

The variation of f with latitude can be approximately represented by expanding (12.6) in a Taylor series about the central latitude θ_0:

$$f\cong f_0+y\left(\frac{\partial f}{\partial y}\right)_{\theta_0}=f_0+y\left(\frac{df}{d\theta}\frac{d\theta}{dy}\right)_{\theta_0}=2\Omega\sin\theta_0+\frac{2\Omega\cos\theta_0}{R}y\equiv f_0+\beta y. \tag{12.10}$$

Here, $df/d\theta = 2\Omega\cos\theta$ and $d\theta/dy = 1/R$, where $R = 6371$ km is the radius of the earth. A model that takes into account the variation of the Coriolis parameter in the simplified form given by the final quality of (12.10) with β as constant is called a *β-plane model.*

Example 12.3

Compute the magnitude of the vertical Coriolis acceleration for a 185 km/hr (100 knot) horizontal air speed at latitude 45°. What temperature fluctuation in an air mass at 300 K would produce an equivalent buoyant acceleration? Is neglect of the vertical Coriolis acceleration likely to be justified in most situations? Explain.

Solution For the given speed and latitude, the magnitude of the vertical Coriolis acceleration is:

$$2\Omega u\cos\theta = 2(7.27\times10^{-5}\,\mathrm{s}^{-1})(185\times10^{3}\,\mathrm{m}/3600\,\mathrm{s})\cos 45^\circ = 5.28\times10^{-3}\,\mathrm{ms}^{-2}.$$

The vertical (or buoyant) acceleration term in the final equation of the set (12.9) is $g\rho'/\rho_0$. For constant air pressure, temperature and density fluctuations are related by $\delta T/T_0 = -\delta\rho/\rho_0$, so an equivalent buoyant acceleration is will be produced by a temperature fluctuation T' that satisfies:

$$5.28\times10^{-3}\,\mathrm{ms}^{-2} = gT'/T_0$$

In this case, with $T_0 = 300$ K, T' is just 0.16 K. Thus, as long as naturally occurring temperature fluctuations are several degrees Kelvin or more, the vertical Coriolis acceleration can be neglected in a simplified analysis involving this wind speed. However, a 100-knot wind represents an extreme situation; near the ground it corresponds to category-three hurricane force winds while aloft it corresponds to a strong jet stream. Thus, neglect of the vertical Coriolis acceleration is likely to be well justified in more ordinary atmospheric situations.

12.4 Geostrophic flow

Consider quasi-steady, large-scale motions in the atmosphere or the ocean, away from boundaries. For these flows an excellent approximation for the horizontal equilibrium is a *geostrophic balance* where the Coriolis acceleration matches the horizontal pressure-gradient acceleration:

$$-fv = -\frac{1}{\rho_0}\frac{\partial p'}{\partial x} = -\frac{1}{\rho_0}\frac{\partial p}{\partial x}, \quad \text{and} \quad fu = -\frac{1}{\rho_0}\frac{\partial p'}{\partial y} = -\frac{1}{\rho_0}\frac{\partial p}{\partial y}. \qquad (12.11,\ 12.12)$$

The second equality in each case follows from (12.7). These are the first two equations of the set (12.9) when the friction terms, and the unsteady and nonlinear acceleration terms are neglected. If U is the horizontal velocity scale, and L is the horizontal length scale, then ratio of the nonlinear term to the Coriolis term, called the *Rossby number*, is:

$$\text{Rossby Number} = \frac{\text{Nonlinear acceleration}}{\text{Coriolis acceleration}} \sim \frac{U^2/L}{fU} = \frac{U}{fL} = \mathrm{Ro}. \qquad (12.13)$$

For a typical atmospheric value of $U \sim 10\,\mathrm{m/s}$ with $f \sim 10^{-4}\,\mathrm{s}^{-1}$, and $L \sim 1000\,\mathrm{km}$, Ro is 0.1, and it is even smaller for many flows in the ocean. Thus, neglect of the nonlinear terms is justified for many geophysical flows. Geostrophic equilibrium is lost near the equator (within a latitude belt of ±3°), where f becomes small, and it also breaks down if frictional effects or unsteadiness become important.

For steady flow, (12.11) and (12.12) can be used to understand some of the unique phenomena associated with the Coriolis acceleration. For example, when these equations apply, the velocity distribution can be determined from a measured distribution of the pressure field. In particular, these equations imply that velocities in a geostrophic flow are perpendicular to the horizontal pressure gradient. Forming $\mathbf{u}\cdot\nabla p$ using (12.11) and (12.12) produces:

$$(u\mathbf{e}_x + v\mathbf{e}_y)\cdot\nabla p = \frac{1}{\rho_0 f}\left(-\mathbf{e}_x\frac{\partial p}{\partial y} + \mathbf{e}_y\frac{\partial p}{\partial x}\right)\cdot\left(\mathbf{e}_x\frac{\partial p}{\partial x} + \mathbf{e}_y\frac{\partial p}{\partial y}\right) = 0.$$

Thus, the horizontal velocity is *along*, and not across, the lines of constant pressure. If f is regarded as constant, then the geostrophic balance, (12.11) and (12.12), shows that $p/(f\rho_0)$ can be regarded as a stream function. Therefore, the isobars on a weather map are nearly the streamlines of the flow.

Fig. 12.4 shows the geostrophic flow around low- and high-pressure centers in the northern hemisphere. Here the Coriolis acceleration acts to the right of the velocity vector. This requires the flow to be counterclockwise (viewed from above) around a low-pressure region and clockwise around a high-pressure region. The sense of circulation is opposite in the southern hemisphere, where the Coriolis acceleration acts to the left of the velocity vector. Frictional forces become important at lower levels in the atmosphere and result in a flow partially *across* the isobars. For example, frictional effects on an otherwise geostrophic flow cause the flow around a low-pressure center to spiral *inward* (see Section 12.5).

Thermal wind

In the presence of a *horizontal* gradient of density, the geostrophic velocity develops a *vertical* shear. This can be shown from the geostrophic and hydrostatic balances by differentiating both parts of (12.11) with respect to z and then using (1.14), $\partial p/\partial z = -\rho g$, to eliminate p from both equations. The result is:

$$\frac{\partial v}{\partial z} = -\frac{g}{\rho_0 f}\frac{\partial \rho}{\partial x}, \quad \text{and} \quad \frac{\partial u}{\partial z} = \frac{g}{\rho_0 f}\frac{\partial \rho}{\partial y}. \qquad (12.14, 12.15)$$

Meteorologists call these the *thermal wind* equations because they give the vertical variation of wind from measurements of horizontal temperature (and pressure) gradients. The thermal wind is a baroclinic phenomenon because the surfaces of constant p and ρ do not coincide.

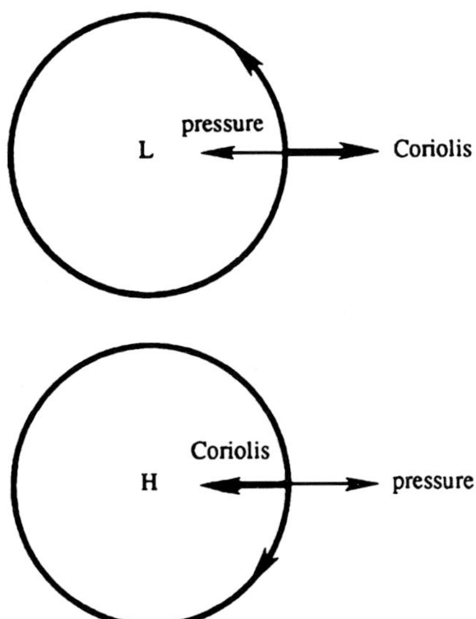

FIGURE 12.4 Circular geostrophic flow around ideal low- and high-pressure centers in the northern hemisphere. The pressure-gradient acceleration, $-(1/\rho_0)\nabla p$, is indicated by a thin arrow, and the Coriolis acceleration is indicated by a thick arrow.

Taylor-Proudman theorem

A striking phenomenon occurs in the geostrophic flow of a *homogeneous* fluid. It can only be observed in a laboratory because stratification effects cannot be avoided in natural flows. Consider then a laboratory experiment in which a tank of fluid is steadily rotated at a high angular speed Ω and a solid body is moved slowly along the bottom of the tank. The purpose of making Ω large and the movement of the solid body slow is to make the Coriolis acceleration much larger than the advective acceleration terms, which must be made negligible for geostrophic equilibrium. Away from the frictional effects of boundaries, the balance is geostrophic in the horizontal and hydrostatic in the vertical. Setting $f = 2\Omega$ in (12.11) and (12.12) produces:

$$-2\Omega v = -\frac{1}{\rho}\frac{\partial p}{\partial x} \quad \text{and} \quad 2\Omega u = -\frac{1}{\rho}\frac{\partial p}{\partial y}. \qquad (12.16, 12.17)$$

It is useful to define an Ekman number, E, as the ratio of viscous to Coriolis accelerations:

$$\text{Ekman number} = \frac{\text{viscous force per unit mass}}{\text{Coriolis acceleration}} \sim \frac{\rho\nu U/L^2}{\rho f U} = \frac{\nu}{fL^2} = \text{E}. \tag{12.18}$$

Under the circumstances already described here, both Ro and E are small.

Elimination of p from (12.16) and (12.17) by cross-differentiation leads to:

$$2\Omega(\partial v/\partial y + \partial u/\partial x) = 0,$$

which can be combined with the continuity equation (4.10) to reach:

$$\partial w/\partial z = 0. \tag{12.19}$$

Also, differentiating (12.16) and (12.17) with respect to z, and using (1.14) with $\rho =$ constant, leads to:

$$\partial u/\partial z = \partial v/\partial z = 0. \tag{12.20}$$

Taken together, (12.19) and (12.20) imply:

$$\partial \mathbf{u}/\partial z = 0. \tag{12.21}$$

Thus, the fluid velocity cannot vary in the direction of Ω. In other words, steady slow motions in a rotating, homogeneous, inviscid fluid are two-dimensional. This is the *Taylor-Proudman theorem*, first derived by Proudman in 1916 and demonstrated experimentally by Taylor soon afterward.

In Taylor's experiment, a tank was made to rotate as a solid body, and a small cylinder was slowly dragged along the bottom of the tank (Fig. 12.5). Dye was introduced from point A above the cylinder and directly ahead of it. In a non-rotating fluid, the water would pass over the top of the moving cylinder. In the rotating experiment, however, the dye divides at a point S, as if it had been blocked by a vertical extension of the cylinder, and flows around this *imaginary* cylinder, called the *Taylor column*. Dye released from a point B within the Taylor column remained there and moved with the cylinder. The conclusion was that the flow outside the upward extension of the cylinder is the same as if the cylinder extended across the entire water depth and that a column of water directly above the cylinder moves with it. The motion is two dimensional, although the solid body does not extend across the entire water depth. Taylor did a second experiment, in which he dragged a solid body *parallel* to the axis of rotation. In accordance with $\partial w/\partial z = 0$, he observed that a column of fluid is pushed ahead. The lateral velocity components u and v were zero. In both of these experiments, there are shear layers at the edge of the Taylor column.

In summary, Taylor's experiment established the following striking fact for steady inviscid motion of a homogeneous fluid in a strongly rotating system: bodies moving either parallel or perpendicular to the axis of rotation carry along with their motion a so-called Taylor column of fluid, oriented parallel to the axis of rotation. The phenomenon is analogous to the horizontal *blocking* caused by a solid body (say a mountain) in a strongly stratified system, shown in Fig. 11.30.

Example 12.4

Continent-scale weather maps commonly report the 500 mb height, Z, the vertical distance from sea level to the point in the atmosphere where the pressure $p = 50$ kPa (mb = milli-bar; 1 mb = 100 Pa). Typical values for Z lie between 4.5 and 6 km where the geostrophic balance dominates wind patterns in the mid-latitudes, especially in winter. In this height range, how are the horizontal wind components u and v related to Z?

Solution The 500 mb height function, $z = Z(x, y)$, is defined by $p(x, y, Z) = 500\,\text{mb} = \text{const}$. Partial differentiation of this equation with respect to x produces:

$$0 = \frac{\partial p}{\partial x} + \frac{\partial p}{\partial z}\frac{\partial Z}{\partial x} = \frac{\partial p}{\partial x} - \rho_0 g\frac{\partial Z}{\partial x}, \quad \text{or} \quad \frac{\partial p}{\partial x} = \rho_0 g\frac{\partial Z}{\partial x},$$

where (1.14) with $\rho = \rho_0$ has been used to eliminate $\partial p/\partial z$. Similarly, partial differentiation with respect to y leads to:

$$\partial p/\partial y = \rho_0 g \partial Z/\partial y.$$

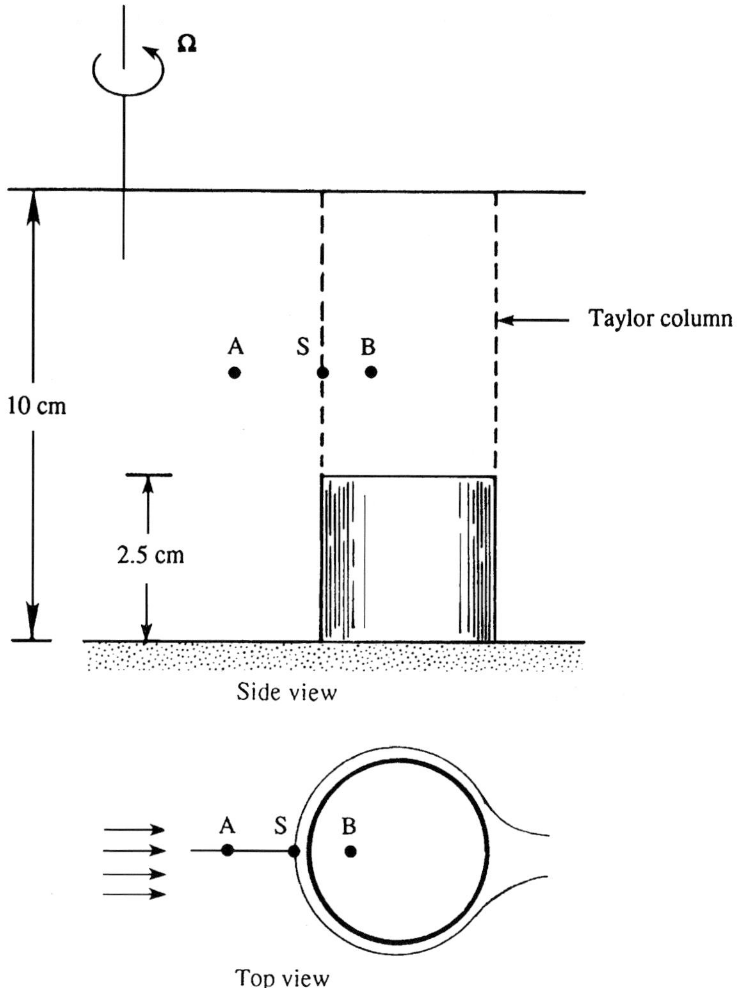

FIGURE 12.5 Taylor's experiment in a strongly rotating flow of a homogeneous fluid. When the short cylinder is moved toward the axis of rotation, an extension of the cylinder forms in the fluid above it. Dye released above the cylinder at point A flows around the extension of cylinder as if it were a solid object. Dye released above the cylinder at point B follows the motion of the short cylinder.

Substituting these results for $\partial p/\partial x$ and $\partial p/\partial y$ into (12.11) and (12.12) produces:

$$-fv = -\frac{1}{\rho_0}\left(\rho_0 g\frac{\partial Z}{\partial x}\right) = -g\frac{\partial Z}{\partial x} \quad \text{and} \quad fu = -\frac{1}{\rho_0}\left(\rho_0 g\frac{\partial Z}{\partial y}\right) = -g\frac{\partial Z}{\partial y},$$

$$\text{or}: \quad (u, v) = \left(-\frac{g}{f}\frac{\partial Z}{\partial y}, \frac{g}{f}\frac{\partial Z}{\partial x}\right).$$

Thus, $-gZ/f$ can also serve as a stream function for geostrophic flow. Weather maps of Z (such as Fig. 12.28) simultaneously show regions of high and low pressure (Z increases with increasing surface pressure) and the mid-troposphere horizontal winds that steer weather systems.

12.5 Ekman layers

In the preceding section, geostrophic balance was found to occur at low Rossby number when the flow is steady and frictionless. This section extends this analysis to fluid motion within the frictional layers that develop on horizontal surfaces. In viscous flows unaffected by Coriolis accelerations and pressure gradients, the only terms in the equation of motion that can balance viscous friction are either the unsteady acceleration $\partial\mathbf{u}/\partial t$, or the advective acceleration $(\mathbf{u}\cdot\nabla)\mathbf{u}$. The balance of $\partial\mathbf{u}/\partial t$ and viscous friction gives rise to a viscous layer having characteristics that evolve with time, as in the case of a moving flat plate (see Sections 7.4 and 7.5). Alternatively, the balance of $(\mathbf{u}\cdot\nabla)\mathbf{u}$ and viscous friction for flow past a flat surface gives rise to a boundary layer having a thickness that may vary in the direction of flow (see Sections 8.3 and 8.4). In a steadily rotating coordinate system, a third possibility arises, a balance between the Coriolis acceleration and friction. Here, the viscous layer, known as an Ekman layer, can be invariant in time and space, and two examples of such layers are provided in this section.

Ekman layer at a free surface

Consider first the frictional layer near the free surface of the ocean that is formed by a steady wind stress τ on the ocean surface in the x-direction. For simplicity, only the steady solution is examined for horizontally homogeneous flow without horizontal pressure gradients. Under these conditions, the first two equations of (12.9) reduce to:

$$-fv = \nu_V \frac{d^2u}{dz^2} \quad \text{and} \quad fu = \nu_V \frac{d^2v}{dz^2}. \tag{12.22, 12.23}$$

Defining $z = 0$ on surface of the ocean, the boundary conditions are:

$$\rho\nu_V(du/dz) = \tau \quad \text{at} \quad z = 0, \qquad dv/dz = 0 \quad \text{at} \quad z = 0, \qquad \text{and} \quad u, v \to 0 \quad \text{as} \quad z \to -\infty. \tag{12.24, 12.25, 12.26}$$

Eqs. (12.22) and (12.23) are linear and can be solved via a complex superposition. Multiply (12.23) by the imaginary root, $i = \sqrt{-1}$, and add (12.22) to reach:

$$\frac{d^2V}{dz^2} = \frac{if}{\nu_V}V, \tag{12.27}$$

where $V \equiv u + iv$ is the *complex velocity*. The solution of (12.27) is:

$$V = A\exp\{(1+i)z/\delta\} + B\exp\{-(1+i)z/\delta\}$$

where A and B are constants, and:

$$\delta = \sqrt{2\nu_V/f} \tag{12.28}$$

sets the thickness of the *Ekman layer*. To satisfy (12.26), the constant B must be zero. The surface boundary conditions (12.24) and (12.25) can be combined as $\rho\nu_V(dV/dz) = \tau$ at $z = 0$, from which (12.28) with $B = 0$ gives:

$$A = (1-i)\tau\delta/2\rho\nu_V.$$

Combining this with (12.28) and the definition $V = u + iv$ leads to velocity components:

$$u = \frac{\tau}{\rho\sqrt{f\nu_V}}\exp\left(\frac{z}{\delta}\right)\cos\left(-\frac{z}{\delta}+\frac{\pi}{4}\right) \quad \text{and} \quad v = -\frac{\tau}{\rho\sqrt{f\nu_V}}\exp\left(\frac{z}{\delta}\right)\sin\left(-\frac{z}{\delta}+\frac{\pi}{4}\right). \tag{12.29}$$

The Swedish oceanographer Ekman worked out this solution in 1905. The solution is shown in Fig. 12.6 for the case of the northern hemisphere, in which f is positive. The velocities at various depths within the ocean are plotted in Fig. 12.6a where each arrow represents the velocity vector at a certain depth. Such a plot of v versus u is sometimes called a *hodograph*. The vertical distributions of u and v are shown in Fig. 12.6b. The hodograph shows that the surface velocity is deflected 45° to the right of the applied wind stress. (In the southern hemisphere, the deflection is to the left of the surface stress.) The velocity vector rotates clockwise (looking down) with depth, and the magnitude exponentially decays with an e-folding length of δ, the Ekman layer thickness. The tips of the velocity vector at various depths form a spiral, called the *Ekman spiral*.

The components of the volume transport in the Ekman layer are:

$$\int_{-\infty}^{0} u\,dz = 0, \quad \text{and} \quad \int_{-\infty}^{0} v\,dz = -\tau/\rho f. \tag{12.30}$$

This shows that the *net transport is to the right of the applied stress and is independent of* ν_V. In fact, the second part of (12.30) follows directly from a vertical integration of (12.22) in the form $-\rho fv = d\tau/dz$ so that the result does not depend on the eddy viscosity assumption. The fact that the transport is to the right of the applied stress makes sense because then

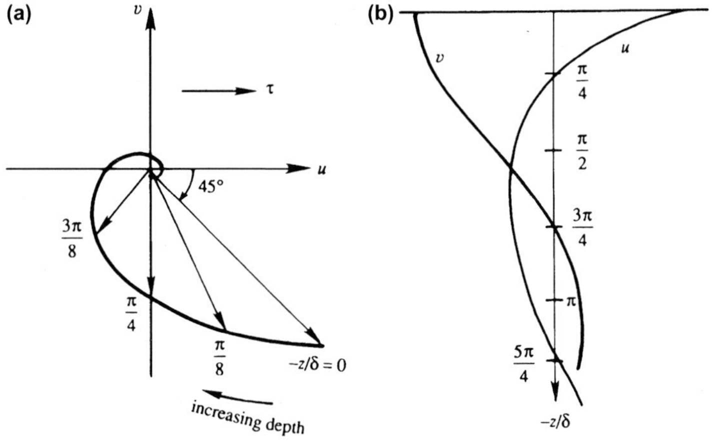

FIGURE 12.6 Ekman layer below a water surface on which a shear stress τ is applied in the x-direction. The left panel (a) shows the horizontal fluid velocity components (u, v) at various depths; values of $-z/\delta$ are indicated along the curve traced out by the tip of the velocity vector. The flow speed is highest near the surface. The right panel (b) shows vertical distributions of u and v. Here, the Coriolis acceleration produces significant depth dependence in the fluid velocity even though τ is constant and unidirectional.

the net (depth-integrated) effect of the Coriolis acceleration, which is directed to the right of the depth-integrated transport, balances the wind stress.

The horizontal uniformity assumed in the solution is not a serious limitation since Ekman layers near the ocean surface have a thickness (~50 m) that is much smaller than the scale of horizontal variation ($L > 100$ km). The assumed absence of a horizontal pressure gradient can also be reconsidered. Because of the thinness of the layer, any imposed horizontal pressure gradient remains constant across the layer. The presence of a horizontal pressure gradient merely adds a depth-independent geostrophic velocity to the Ekman solution. Suppose the sea surface slopes down to the north, so that there is a pressure force acting northward throughout the Ekman layer and below (Fig. 12.7). This means that at the bottom of the Ekman layer ($z/\delta \to -\infty$) there is a geostrophic velocity U to the right of the pressure force. The surface Ekman spiral forced by the wind stress joins smoothly to this geostrophic velocity as $z/\delta \to -\infty$.

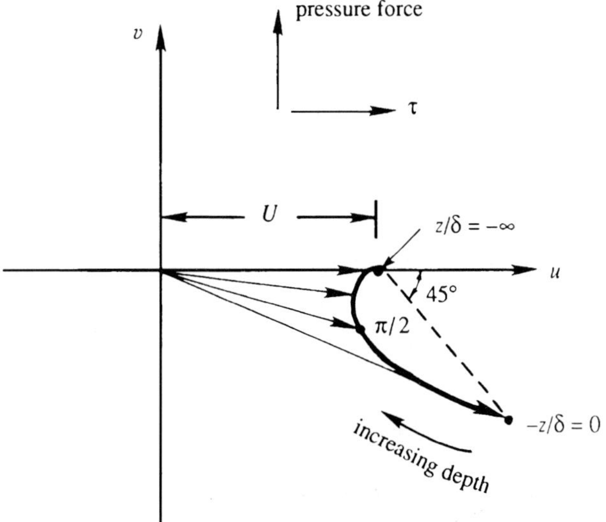

FIGURE 12.7 Ekman layer at a free surface in the presence of a pressure gradient. The geostrophic velocity forced by the pressure gradient is U. The flow profile in this case is the sum of U and the profile shown in Fig. 12.6.

Pure Ekman spirals are not observed in the surface layer of the ocean, mainly because the assumptions of constant eddy viscosity and steadiness are particularly restrictive. When the flow is averaged over a few days, however, several instances have been found in which the current does look like a spiral. One such example is shown in Fig. 12.8.

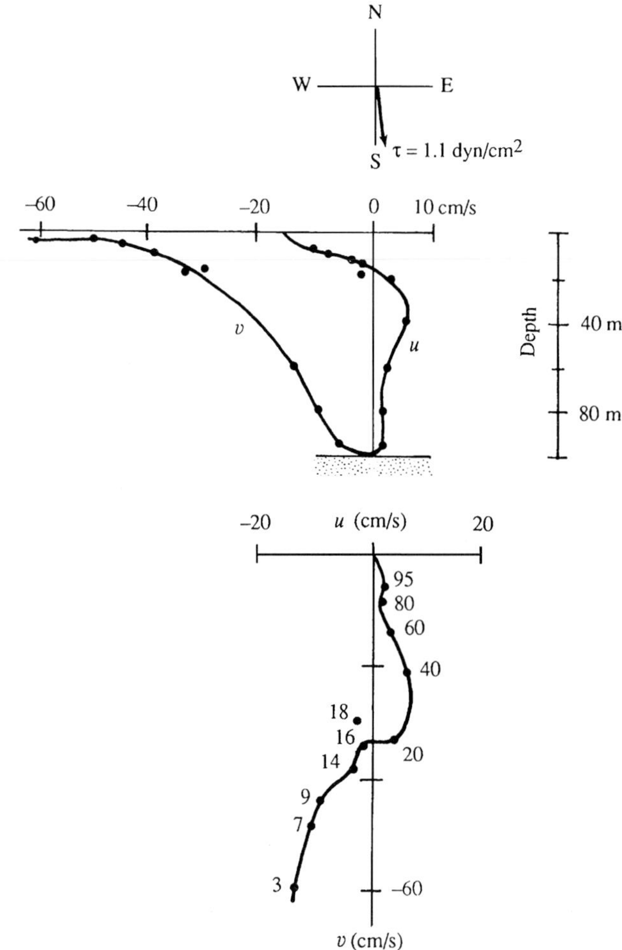

FIGURE 12.8 An observed velocity distribution near the coast of Oregon. Velocity is averaged over 7 days. Wind stress had a magnitude of 1.1 dyn/cm^2 and was directed nearly southward, as indicated at the top of the figure. The upper panel shows vertical distributions of u and v, and the lower panel shows the hodograph in which depths are indicated in meters. The hodograph is similar to that of a surface Ekman layer (of depth 16 m) lying over the bottom Ekman layer (extending from a depth of 16 m to the ocean bottom). *Kundu (1977); reprinted with the permission of Jacques C.J. Nihoul.*

Explanation in terms of vortex tilting

In flows without rotation, the thickness of a viscous layer usually grows in time or in downstream distance. The Ekman solution, in contrast, results in a viscous layer that does not grow either in time or space. This can be explained by examining the vorticity equation (Pedlosky, 1987). The vorticity components in the x- and y-directions are:

$$\omega_x = \frac{\partial w}{\partial y} - \frac{\partial v}{\partial z} = -\frac{dv}{dz} \quad \text{and} \quad \omega_y = \frac{\partial u}{\partial z} - \frac{\partial w}{\partial x} = \frac{du}{dz},$$

when $w = 0$. Using these, the z-derivative of (12.22) and (12.23) are:

$$-f\frac{dv}{dz} = \nu_V \frac{d^2\omega_y}{dz^2} \quad \text{and} \quad -f\frac{du}{dz} = \nu_V \frac{d^2\omega_x}{dz^2}. \tag{12.31}$$

The right sides of these equations represent diffusion of vorticity. Without Coriolis effects this diffusion would cause a thickening of the viscous layer. The presence of planetary rotation, however, means that vertical fluid lines coincide with the planetary vortex lines. The tilting of vertical fluid lines, represented by terms on the left sides of Eqs. (12.31), then causes a rate of change of the horizontal component of vorticity that just cancels the diffusion term.

Ekman layer on a rigid surface

As a second case of a viscous layer that is invariant in time and space, consider a steady viscous layer on a solid surface in a rotating flow that is independent of the horizontal coordinates x and y. This can be the atmospheric boundary layer over the ground that develops from winds aloft, or the boundary layer on the ocean bottom that develops below a uniform current in the water column. As for the first Ekman layer, assume the fluid velocity is in the x-direction with magnitude U at large distances from the surface. Viscous forces are negligible far from the surface, so that the Coriolis acceleration can be balanced only by a pressure gradient and (12.12) applies with $u = U$:

$$fU = -\frac{1}{\rho}\frac{\partial p}{\partial y}. \tag{12.32}$$

This simply states that the flow outside the viscous layer is in geostrophic balance, U being the geostrophic velocity. For positive U and f, dp/dy must be negative, so that the pressure falls with increasing y—that is, the pressure force is directed along the positive y direction, resulting in a geostrophic flow U to the right of the pressure force in the northern hemisphere. The horizontal pressure gradient remains constant within the thin boundary layer.

Near the solid surface friction forces are important, so that the balance within the boundary layer is:

$$-fv = \nu_V\frac{d^2u}{dz^2} \quad \text{and} \quad fu = \nu_V\frac{d^2v}{dz^2} + fU, \tag{12.33, 12.34}$$

where $-\rho^{-1}(dp/dy)$ has been replaced by fU in accordance with (12.32). The boundary conditions are:

$$u = U \quad \text{and} \quad v = 0 \quad \text{as} \quad z \to \infty, \quad \text{and} \quad u = v = 0 \quad \text{at} \quad z = 0, \tag{12.35, 12.36}$$

where $z = 0$ on the solid surface and is positive upward. As for the free-surface Ekman layer, multiply (12.34) by i and add (12.33), to find the equivalent of (12.27):

$$\frac{d^2V}{dz^2} = \frac{if}{\nu_V}(V - U), \tag{12.37}$$

where $V = u + iv$. The boundary conditions (12.35) and (12.36) in terms of V are:

$$V = U \quad \text{as} \quad z \to \infty, \quad \text{and} \quad V = 0 \quad \text{at} \quad z = 0. \tag{12.38, 12.39}$$

The particular solution of the linear differential equation (12.37) is $V = U$, so the total solution is:

$$V = U + A\exp\{(1+i)z/\delta\} + B\exp\{-(1+i)z/\delta\}, \tag{12.40}$$

where δ is given by (12.28). To satisfy (12.38), A must be zero, so (12.39) then requires $B = -U$. Thus, the velocity components are:

$$u = U[1 - \exp(-z/\delta)\cos(z/\delta)] \quad \text{and} \quad v = U\exp(-z/\delta)\sin(z/\delta). \tag{12.41}$$

In this case, the tip of the velocity vector again describes a spiral for various values of z (Fig. 12.9a). As with the free-surface Ekman layer, the frictional effects are confined within a layer of thickness δ, which increases with ν_V and decreases with the rotation rate f. Interestingly, the layer thickness is independent of the magnitude of the free-stream velocity U; this behavior is quite different from that of a steady non-rotating boundary layer on a semi-infinite plate (see Section 8.3) in which the thickness is proportional to $U^{-1/2}$. And, the velocity fields for both Ekman layers, (12.29) and (12.41), are in the form $\mathbf{u} = (u(z), v(z), 0)$ so that all the fluid acceleration terms, $D\mathbf{u}/Dt = \partial\mathbf{u}/\partial + (\mathbf{u}\cdot\nabla)\mathbf{u}$, in (12.9) are zero; thus, (12.29) and (12.41) are exact solutions of (12.9).

Fig. 12.9b shows the vertical distribution of the velocity components. Far from the wall the velocity is entirely in the x-direction, and the Coriolis acceleration balances the pressure gradient. As the wall is approached, frictional effects decrease u and the associated Coriolis acceleration, so that the pressure gradient (which is independent of z) produces a component v in the direction of the pressure force. Using (12.41), the net transport in the Ekman layer normal to the uniform stream outside the layer is:

$$\int_0^\infty v\,dz = U(\nu_V/2f)^{1/2} = \frac{1}{2}U\delta,$$

which is directed to the *left* of the free-stream velocity, in the direction of the pressure force.

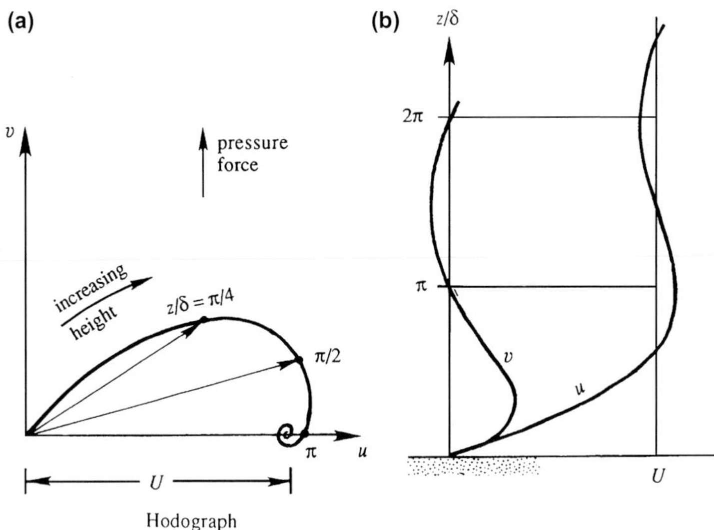

FIGURE 12.9 Ekman layer above a rigid surface for a steady outer-flow velocity of U (parallel to the x-axis). The left panel shows velocity vectors at various heights; values of z/δ are indicated along the curve traced out by the tip of the velocity vectors. The right panel shows vertical distributions of u and v.

If the atmosphere were in laminar motion, ν_V would be equal to its molecular value for air, and the Ekman layer thickness at a latitude of 45° (where $f \simeq 10^{-4}\,\mathrm{s}^{-1}$) would be $\approx \delta \sim 0.4$ m. The observed thickness of the atmospheric boundary layer is of order 1 km, which implies an eddy viscosity of order $\nu_V \sim 50\,\mathrm{m}^2/\mathrm{s}$. In fact, Taylor (1915) tried to estimate the eddy viscosity by matching the predicted velocity distributions (12.41) with the observed wind at various heights.

The Ekman layer solution on a solid surface demonstrates that the three-way balance among the Coriolis, the pressure gradient, and friction terms within the boundary layer results in a component of flow directed toward the lower pressure. This balance of forces within the boundary layer is illustrated in Fig. 12.10. The net frictional force on an element is oriented approximately opposite to the velocity vector **u**. It is clear that a balance of forces is possible only if the velocity vector has a component from high to low pressure, as shown. Frictional forces therefore cause the flow around a low-pressure center to spiral *inward*. Mass conservation requires that the inward converging flow rise within a low-pressure system, resulting in cloud formation and rainfall. This is what happens in a *cyclone*, a low-pressure system. In contrast, within a high-pressure system the air sinks as it spirals outward due to frictional effects. The arrival of high-pressure systems therefore brings in clear skies and fair weather, because the sinking air suppresses cloud formation.

Frictional effects, in particular the Ekman transport by surface winds, play a fundamental role in the theory of wind-driven ocean circulation. Possibly the most important result of such theories was given by Henry Stommel in 1948 (Stommel, 1948). He showed that the northward increase of the Coriolis parameter f is responsible for making the currents along western ocean boundaries (e.g., the Gulf Stream in the Atlantic and the Kuroshio in the Pacific) much stronger than the currents on the eastern side. These are discussed in books on physical oceanography and are not presented here.

Example 12.5

For the Ekman layer on a rigid surface, determine the maximum flow speed in terms of U, the flow speed above the layer, the altitude where this maximum occurs in terms of δ, and the direction of the flow at this altitude.

Solution From (12.41), the altitude of the maximum flow speed can be obtained from:

$$\frac{d}{d\eta}\left\{u^2 + v^2\right\} = \frac{d}{d\eta}\left\{U^2[1 - \exp(-\eta)\cos\eta]^2 + U^2\exp(-2\eta)\sin^2\eta\right\} = 0,$$

where $\eta = z/\delta$. By simplifying inside the {,}-braces and differentiating, this equation can be reduced to the simple transcendental form:

$$\cos\eta + \sin\eta - \exp(-\eta) = 0,$$

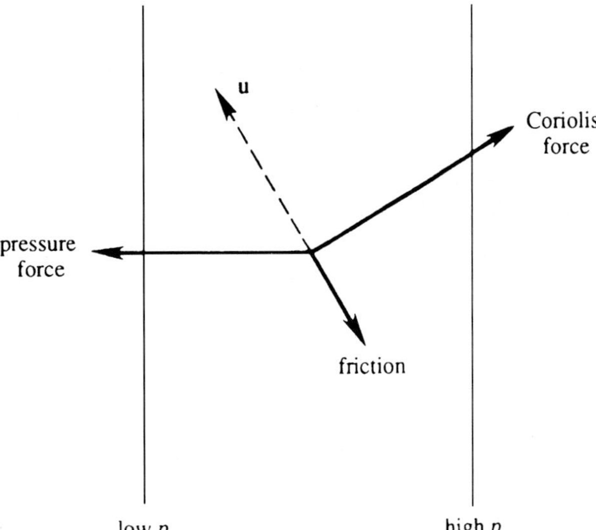

FIGURE 12.10 Balance of forces within an Ekman layer. For steady flow without friction, pressure gradient and Coriolis terms would balance. When friction is added, the pressure gradient and Coriolis terms must counteract it. Since friction acts opposite the direction of flow, the velocity **u** must have a component toward lower pressure when friction is present.

which has a root at $z = 2.2841\delta$. For this value of η, $u = 1.0667U$ and $v = 0.07703U$, so the maximum flow speed is $1.0694U$ and it occurs 4.13° to the left (or northward) from the x-direction (due east). At a latitude of 45° with $\upsilon_V \sim 50\,\text{m}^2/\text{s}$, δ is approximately 1 km, so the altitude of maximum speed is ~2 km.

12.6 Shallow-water equations

The characteristics of surface and internal gravity waves are discussed in Chapter 11 when the effect of planetary rotation is ignored because the wave frequency ω is much larger than the Coriolis parameter f. Here, inviscid wave motion is considered when ω is low enough to be comparable to f, and the effect of planetary rotation must be included. The emphasis here is on long-wavelength (λ) gravity waves in a shallow layer of homogeneous fluid whose mean depth is H. For $\lambda \gg H$, vertical velocities are much smaller than the horizontal velocities. The pressure distribution is hydrostatic in this circumstance (see Section 11.2), and fluid particles execute a horizontal rectilinear motion that is independent of z. When the effects of planetary rotation are included, the horizontal velocity is still depth independent, although the particle orbits are no longer rectilinear but elliptic in the horizontal plane, as described in the following section.

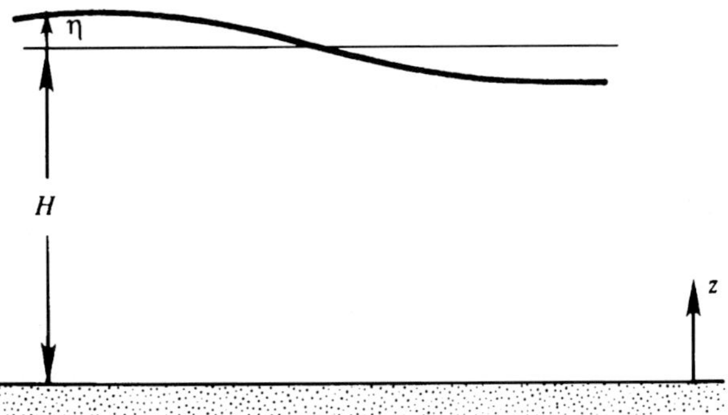

FIGURE 12.11 Geometry for a fluid layer of average thickness H above a flat stationary bottom coincident with $z = 0$. At any horizontal location the liquid's surface height is $H + \eta$.

Consider a homogeneous layer of fluid of average depth H lying over a flat horizontal bottom (Fig. 12.11). Set $z = 0$ on the bottom surface, and let $\eta(x, y, t)$ be the displacement of the free surface. When the pressure on the fluid's surface is set to zero, the pressure at height z from the bottom, which is hydrostatic, is given by $p = \rho g(H + \eta(x, y, t) - z)$, so the pressure gradient is:

$$\nabla p = \mathbf{e}_x \rho g(\partial\eta/\partial x) + \mathbf{e}_y \rho g(\partial\eta/\partial y) - \mathbf{e}_z \rho g. \tag{12.42}$$

Since these are independent of z, the resulting horizontal motion is also depth independent so $\partial u/\partial x$ and $\partial v/\partial y$ are independent of z. Therefore, the continuity equation, $\partial u/\partial x + \partial v/\partial y + \partial w/\partial z = 0$, requires that w vary linearly with z, from zero at the bottom to the maximum value at the free surface. Integrating the continuity equation vertically across the water column from $z = 0$ to $z = H + \eta$, and noting that u and v are depth independent, leads to:

$$(H+\eta)\frac{\partial u}{\partial x} + (H+\eta)\frac{\partial v}{\partial y} + w(\eta) - w(0) = 0, \tag{12.43}$$

where $w(\eta)$ is the vertical velocity at the surface and $w(0) = 0$ is the vertical velocity at the bottom. The vertical surface velocity is given by:

$$w(\eta) = \frac{D\eta}{Dt} = \frac{\partial\eta}{\partial t} + u\frac{\partial\eta}{\partial x} + v\frac{\partial\eta}{\partial y},$$

which is the exact kinematic boundary condition on a free surface with two independent horizontal dimensions (*cf.* (11.16)). The continuity equation (12.43) then becomes:

$$\frac{\partial\eta}{\partial t} + \frac{\partial}{\partial x}[u(H+\eta)] + \frac{\partial}{\partial y}[u(H+\eta)] = 0, \tag{12.44}$$

when $w(0) = 0$, and x- and y-derivative terms are combined. This equation requires the divergence of the horizontal fluid transport to depresses the free surface. For small amplitude waves ($\eta \ll H$), the quadratic nonlinear terms can be neglected in comparison to the linear terms, so that the divergence terms in (12.44) reduce to $H(\partial u/\partial x + \partial v/\partial y)$.

The linearized continuity and momentum equations are then:

$$\frac{\partial\eta}{\partial t} + H\left(\frac{\partial u}{\partial x} + \frac{\partial v}{\partial y}\right) = 0, \quad \frac{\partial u}{\partial t} - fv = -g\frac{\partial\eta}{\partial x}, \quad \text{and} \quad \frac{\partial v}{\partial t} + fu = -g\frac{\partial\eta}{\partial y}. \tag{12.45, 12.46, 12.47}$$

In the two momentum equations, (12.46) and (12.47), the pressure gradient terms are from (12.42) and the nonlinear advective terms in (12.9) have been neglected under the small amplitude assumption. Eqs. (12.45)–(12.47), called the *shallow-water equations*, govern the motion of a layer of fluid in which the horizontal scale is much larger than the depth of the layer. These equations are used in the following sections for studying various types of gravity waves.

Although the preceding analysis has been formulated for a layer of *homogeneous* fluid, (12.45) to (12.47) are applicable to internal waves in a stratified medium when H is replaced by an *equivalent depth* H_e, defined by $c^2 = gH_e$, where c is the speed of long non-rotating *internal* gravity waves. This correspondence is further developed in the following section.

Example 12.6

Use (12.46) and (12.47) to determine u and v for the steady surface deflection $\eta = \eta_o \exp(-(x^2 + y^2)/2\sigma^2)$. Is there anything unusual about this flow field?

Solution Here, the surface deflection is steady, so (12.46) and (12.47) imply:

$$v = \frac{g}{f}\frac{\partial\eta}{\partial x} = -\frac{gx}{f\sigma^2}\eta_o \exp\left\{\frac{-x^2 - y^2}{2\sigma^2}\right\}, \quad \text{and} \quad u = -\frac{g}{f}\frac{\partial\eta}{\partial y} = \frac{gy}{f\sigma^2}\eta_o \exp\left\{\frac{-x^2 - y^2}{2\sigma^2}\right\}.$$

This is a swirling flow with circular streamlines, as can be determined from the first equality of (3.7) and these velocity field results:

$$dy/dx = v/u = -x/y \quad \text{which integrates to: } x^2 + y^2 = \text{constant}.$$

Here, when η_o is negative (a depression in the free surface) the flow is counterclockwise when viewed from above, and when η_o is positive (an elevation of the free surface) the flow is clockwise when viewed from above. As expected, these results are consistent with wind directions around depressions and elevations of the 500 mb isobar (Z) in the atmosphere (see Example 12.4). However, elevation of the free surface at the center of a swirling flow is beyond our everyday experience since the liquid surface in an ordinary cup of water, coffee, tea, etc. is always depressed in the middle when the liquid is swirled with a spoon, regardless of the direction of swirling.

12.7 Normal modes in a continuously stratified layer

In the preceding section the governing equations were derived for waves of wavelength larger than the depth of the fluid layer. Now consider a continuously stratified medium and assume that the horizontal scale of motion is much larger than the vertical scale. The pressure distribution is therefore hydrostatic, and the linearized equations of motion are the incompressible flow continuity equation $\nabla \cdot \mathbf{u} = 0$ (4.10); the linearized inviscid versions of (12.9):

$$\frac{\partial u}{\partial t} - fv = -\frac{1}{\rho_0}\frac{\partial p'}{\partial x}, \qquad \frac{\partial v}{\partial t} + fu = -\frac{1}{\rho_0}\frac{\partial p'}{\partial y}, \qquad 0 = -\frac{\partial p'}{\partial z} - g\rho'; \qquad (12.48, 12.49, 12.50)$$

and the linearized density equation (4.9):

$$\frac{\partial \rho'}{\partial t} - \rho_0 \frac{N^2}{g} w = 0, \qquad (12.51)$$

where p' and ρ' represent *perturbations* of pressure and density from the state of rest (see (12.7)). The advective term in the density equation (12.51) is written in the linearized form $w(d\overline{\rho}/dz) = -\rho_0 N^2 w/g$, where $N(z)$ is the buoyancy frequency. In this form, the rate of change of density at a point is due only to the vertical advection of the background density distribution $\overline{\rho}(z)$, as discussed in Section 11.8.

In a continuously stratified medium, it is convenient to use a separation-of-variables expansion to write $q = \sum q_n(x, y, t)\psi_n(z)$ for a dependent-field variable q in terms of the vertical *normal modes* $\psi_n(z)$, which are orthogonal to each other. The vertical structure of a mode is described by ψ_n while q_n describes the horizontal propagation of the mode. Although each mode propagates only horizontally, the *sum* of a number of modes can also propagate vertically if the various q_n are out of phase.

Assume variables-separable solutions of the form:

$$[u, v, p'/\rho_0] = \sum_{n=0}^{\infty} [u_n, v_n, p_n]\, \psi_n(z), \qquad w(z) = \sum_{n=0}^{\infty} w_n \int_{-\infty}^{z} \psi_n(z)dz, \quad \text{and} \quad \rho' = \sum_{n=0}^{\infty} \rho_n (d\psi_n/dz),$$
(12.52, 12.53, 12.54)

where the amplitudes u_n, v_n, p_n, w_n, and ρ_n are functions of (x, y, t). The z-axis is measured from the upper free surface of the fluid layer, and $z = -H$ represents a flat bottom wall. The reasons for assuming the various forms of z-dependence in (12.52)–(12.54) are as follows. The variables u, v, and p have the same vertical structure in order to be consistent with (12.48) and (12.49) The continuity equation (4.10) requires that the vertical structure of w should be the integral of $\psi_n(z)$. And, (12.50) requires that the vertical structure of ρ' must be the z-derivative of the vertical structure of p'.

The formal solution for the ψ_n requires several steps. Substitution of (12.53) and (12.54) into (12.51) gives:

$$\sum_{n=0}^{\infty}\left[\frac{\partial \rho_n}{\partial t}\frac{\partial \psi_n}{\partial z} - \frac{\rho_0 N^2}{g} w_n \int_{-H}^{z} \psi_n(z)dz\right] = 0.$$

This is valid for all values of z, and the modes are linearly independent, so the quantity within brackets must vanish for each mode, which implies:

$$\frac{\partial \psi_n/\partial z}{N^2 \int_{-H}^{z} \psi_n(z)dz} = \frac{\rho_0}{g(\partial \rho_n/\partial t)} w_n \equiv -\frac{1}{c_n^2}. \qquad (12.55)$$

As the first term is a function of z alone and the second term is a function of (x, y, t) alone, for consistency both terms must be equal to a constant that we take to be $-1/c_n^2$. The vertical structure of the normal modes is then given by:

$$\frac{1}{N^2}\frac{d\psi_n}{dz} = -\frac{1}{c_n^2}\int_{-H}^{z} \psi_n(z)dz, \quad \text{or} \quad \frac{d}{dz}\left(\frac{1}{N^2}\frac{d\psi_n}{dz}\right) + \frac{1}{c_n^2}\psi_n = 0, \qquad (12.56)$$

and this equation has the so-called Sturm-Liouville form, for which the various solutions ψ_n are orthogonal. Continuing the development, (12.55) also gives:

$$w_n \equiv -\frac{g}{\rho_0 c_n^2}\frac{\partial \rho_n}{\partial t}.$$

Substitution of (12.52) through (12.54) into (4.10) and (12.48) through (12.51) finally gives the normal mode equations:

$$\frac{\partial u_n}{\partial x}+\frac{\partial v_n}{\partial y}+\frac{1}{c_n^2}\frac{\partial p_n}{\partial t}=0,\quad \frac{\partial u_n}{\partial t}-fv_n=-\frac{\partial p_n}{\partial x},\quad \frac{\partial v_n}{\partial t}+fu_n=-\frac{\partial p_n}{\partial y},\quad p_n=-\frac{g}{\rho_0}\rho_n,\quad \text{and}\quad w_n=\frac{1}{c_n^2}\frac{\partial p_n}{\partial t}.$$

(12.57, 12.58, 12.59, 12.60, 12.61)

Once (12.57) through (12.59) have been solved for u_n, v_n, and p_n, the amplitudes ρ_n and w_n can be obtained from (12.60) and (12.61) The set (12.57) through (12.59) is identical to the set (12.45) to (12.47) governing the motion of a *homogeneous* layer, provided p_n is identified with $g\eta$ and c_n^2 is identified with gH. In a stratified flow, each mode (having a fixed vertical structure) behaves, in the horizontal dimensions and in time, just like a homogeneous layer, with an *equivalent depth* H_e defined by:

$$c_n^2 \equiv gH_e. \tag{12.62}$$

Boundary conditions on ψ_n

At the bottom of the fluid layer, the boundary condition is $w=0$ at $z=-H$. To write this condition in terms of ψ_n, combine the hydrostatic equation (12.50) and the density equation (12.51) to give w in terms of p':

$$w=\frac{(\partial\rho'/\partial t)}{\rho_0 N^2}=-\frac{1}{\rho_0 N^2}\frac{\partial^2 p'}{\partial z\partial t}=-\frac{1}{N^2}\sum_{n=0}^{\infty}\frac{\partial p_n}{\partial t}\frac{\partial\psi_n}{\partial z}. \tag{12.63}$$

The requirement $w=0$ then yields the bottom boundary condition:

$$d\psi_n/dz=0\quad\text{at}\quad z=-H. \tag{12.64}$$

Following the development in Section 11.2, the linearized surface boundary conditions are $w=\partial\eta/\partial t$ and $p'=\rho_0 g\eta$ at $z=0$, where η is the free-surface displacement. These conditions can be combined into $\partial p'/\partial t=\rho_0 g w$ at $z=0$, and using (12.63) this surface boundary condition becomes:

$$\frac{g}{N^2}\frac{\partial^2 p'}{\partial z\partial t}+\frac{\partial p'}{\partial t}=0\quad\text{at}\quad z=0.$$

Substituting in the normal mode decomposition (12.52) gives:

$$\frac{d\psi_n}{dz}+\frac{N^2}{g}\psi_n=0\quad\text{at}\quad z=0. \tag{12.65}$$

Thus, the boundary conditions on ψ_n are (12.64) and (12.65).

Vertical mode solution for uniform N

The character of the normal mode solutions of these equations can be illustrated by determining ψ_n, c_n, and H_e for the first few modes when N is constant. In this case, (12.56) simplifies to:

$$\frac{d^2\psi_n}{dz^2}+\frac{N^2}{c_n^2}\psi_n=0, \tag{12.66}$$

and the boundary conditions are (12.64) and (12.65). The set (12.64) through (12.66) defines an eigenvalue problem, with ψ_n as the eigenfunction and c_n as the eigenvalue. The solution of (12.66) is:

$$\psi_n=A_n\cos(Nz/c_n)+B_n\sin(Nz/c_n). \tag{12.67}$$

The surface and bottom boundary conditions, (12.65) and (12.64), give:

$$B_n=-(c_nN/g)A_n\quad\text{and}\quad\tan(NH/c_n)=c_nN/g=(c_n/NH)(N^2H/g). \tag{12.68, 12.69}$$

The solutions of (12.69) define the eigenvalues of the problem and are indicated graphically in Fig. 12.12. The first root ($n=0$) occurs for $NH/c_n<\pi/2$. In most geophysical flows, $N^2H/g\ll 1$, so both sides of (12.69) are much less than 1.

Thus, the tangent can be expanded for small argument, $\tan(NH/c_n) \approx NH/c_n$, so (12.69) gives (indicating this root by $n = 0$):

$$c_0 = \sqrt{gH}. \tag{12.70}$$

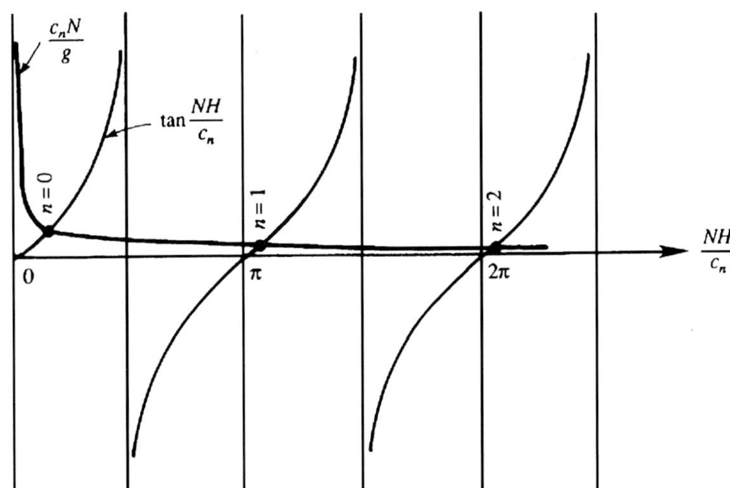

FIGURE 12.12 Calculation of eigenvalues c_n of vertical normal modes in a fluid layer of depth H and uniform stratification N. The eigenvalues occur where the curves defined by $c_n N/g$ and $\tan(NH/c_n)$ cross. As drawn, these crossing points lie slightly above $n\pi$ for $n = 0$, 1, and 2.

The vertical modal structure is found from (12.67). Because the magnitude of an eigenfunction is arbitrary, we can set $A_0 = 1$, obtaining:

$$\psi_0 = \cos(Nz/c_0) - (c_0 N/g)\sin(Nz/c_0) \cong 1 - (N^2 z/g) \cong 1,$$

where we have used $N|z|/c_0 \ll 1$ (with $NH/c_0 \ll 1$), and $N^2 z/g \ll 1$ (with $N^2 H/g = (NH/c_0)(c_0 N/g) \ll 1$). For this mode, the vertical structure of u, v, and p' is nearly depth independent. The corresponding structure for w, given by $\int \psi_0\, dz$, as indicated in (12.53), is linear in z, with w equal to zero at the bottom and a maximum at the upper free surface. A stratified medium therefore has a mode of motion that behaves like that in an unstratified medium; this mode does not feel the stratification. The $n = 0$ mode is called the *barotropic mode*.

The remaining modes $n \geq 1$ are *baroclinic*. For these modes $c_n N/g \ll 1$ but NH/c_n is not small, as can be seen in Fig. 12.12, so that the baroclinic roots of (12.69) are nearly given by:

$$\tan(NH/c_n) = 0, \quad \text{which implies } c_n = NH/n\pi \quad \text{for} \quad n = 1, 2, 3, \ldots \tag{12.71}$$

Taking a nominal depth-averaged oceanic value of $N \sim 2 \times 10^{-3}\,\text{s}^{-1}$ and $H \sim 5\,\text{km}$, the eigenvalue for the first baroclinic mode is $c_1 \sim 3.2\,\text{m/s}$. The corresponding equivalent depth is $H_e = c_1^2/g \sim 1.1\,\text{m}$. For the second baroclinic mode, $c_2 \sim 1.6\,\text{m/s}$. The corresponding equivalent depth is $H_e = c_2^2/g \sim 0.26\,\text{m}$.

An examination of the algebraic steps leading to (12.69) shows that neglecting the right side is equivalent to replacing the upper boundary condition (12.65) by $w = 0$ at $z = 0$. This is called the rigid lid approximation. The *baroclinic modes are negligibly distorted by the rigid lid approximation*. In contrast, the rigid lid approximation applied to the *barotropic* mode would yield $c_0 = \infty$, as (12.71) shows for $n = 0$. Note that the rigid lid approximation does *not* imply that the free-surface displacement corresponding to the baroclinic modes is negligible in the ocean. In fact, excluding wind waves and tides, much of the free-surface displacements in the ocean are due to baroclinic motions. The rigid lid approximation merely implies that, for baroclinic motions, the vertical displacements at the surface are much smaller than those within the fluid column. A valid baroclinic solution can therefore be obtained by setting $w = 0$ at $z = 0$. Further, the rigid lid approximation does not imply that the pressure is constant at the level surface $z = 0$; if a rigid lid were actually imposed at $z = 0$, then the pressure on the lid would vary due to the baroclinic motions.

The vertical mode shapes under the rigid lid approximation are given by:

$$\psi_n = \cos(n\pi z/H) \quad \text{for} \quad n = 1, 2, 3, \ldots$$

because it satisfies $d\psi_n/dz = 0$ at $z = 0, -H$. The nth mode ψ_n has n zero crossings within the layer (Fig. 12.13).

A decomposition into normal modes is only possible in the absence of topographic variations and mean currents with shear. It is valid with or without Coriolis effects and with or without the β-effect. However, the hydrostatic approximation here means that the frequencies are much smaller than N. Under this condition the eigenfunctions are independent of the frequency ω, as (12.56) shows. Without the hydrostatic approximation the eigenfunctions ψ_n become dependent on the frequency ω. This is discussed, for example, in LeBlond and Mysak (1978).

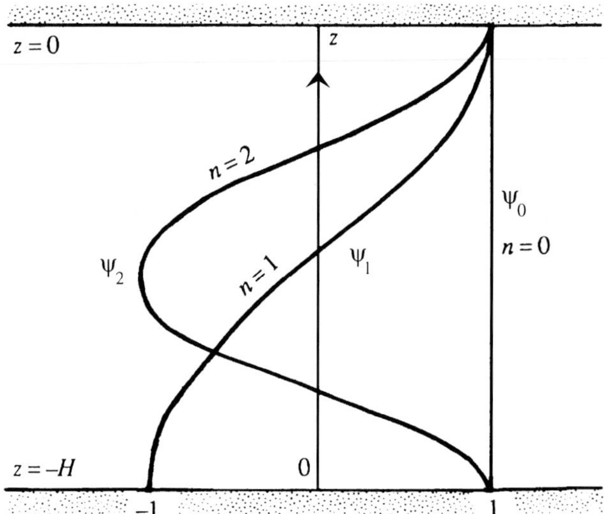

FIGURE 12.13 Vertical distributions of the first three normal modes in a stratified medium of uniform buoyancy frequency for a fluid layer of depth H. The first mode ($n = 0$) is nearly uniform through the depth. The second mode ($n = 1$) shows one-half wavelength in $-H < z < 0$. The third mode ($n = 2$) shows one full wavelength in $-H < z < 0$. Note that all modes must have $d\psi_n/dz = 0$ on $z = -H$, while $d\psi_n/dz$ is only approximately zero a $z = 0$.

Summary

Small amplitude motion in a frictionless continuously stratified ocean can be decomposed into non-interacting vertical normal modes. The vertical structure of each mode is defined by an eigenfunction $\psi_n(z)$. If the horizontal scale of the waves is much larger than the vertical scale, then the equations governing the horizontal propagation of each mode are identical to those of a shallow *homogeneous* layer, with the layer depth H replaced by an equivalent depth H_e defined by (12.62). For a medium of constant N, the baroclinic ($n \geq 1$) eigenvalues are given by $c_n = NH/\pi n$, while the barotropic eigenvalue is $c_0 = \sqrt{gH}$. The rigid lid approximation is quite good for the baroclinic modes.

Example 12.7

Using $N = 2 \times 10^{-3}\,\text{s}^{-1}$ and $H = 5\,\text{km}$, find the first three roots of (12.69) and determine c_0, c_1, and c_2 without linearizing the tangent function or employing the rigid lid approximation. Compare these values to those determined from (12.70) and (12.71).

Solution The transcendental equation (12.69) for the c_n's may be rewritten:

$$\tan(NH/c_n) = c_n N/g \rightarrow \xi \tan(\xi) = N^2 H/g = 2.03874 \times 10^{-3},$$

where $\xi = NH/c_n$ and the numerical value on the right comes from evaluating N^2H/g using $g = 9.81\,\text{m/s}^2$ and the given values of N and H. The first three roots of this equation are $\xi_0 = 0.0451371$, $\xi_1 = 3.14807$, and $\xi_2 = 6.28643$. Thus, with $NH = 10.0\,\text{m/s}$, the first three $c_n = NH/\xi_n$ may be found: $c_0 = 221.547\,\text{m/s}$, $c_1 = 3.17655\,\text{m/s}$, and $c_2 = 1.59073\,\text{m/s}$. The results from (12.70) and (12.71) are: $c_0 = 221.472\,\text{m/s}$, $c_1 = 3.18310\,\text{m/s}$, and $c_2 = 1.59155\,\text{m/s}$, which are just fractions of a percent different. Thus, (12.70) and (12.71) are quite accurate in this case.

12.8 High- and low-frequency regimes in shallow-water equations

Having established that the shallow-water equations (12.45) to (12.47) apply to both uniform-density and stratified fluid layers, consider next which terms are negligible in the shallow-water equations for the various frequency ranges. The analysis presented here primarily describes results for a single homogeneous fluid layer, but it is readily extended to stratified fluid layers when H is interpreted as the equivalent depth H_e defined by (12.62) and c is interpreted as c_n, the speed of long non-rotating *internal* gravity waves. When N is uniform, c_n can be found from (12.69) The β-effect (see (12.10)) is considered here. As f varies only northward, horizontal isotropy is lost whenever the β-effect is included, and it becomes necessary to distinguish between the different horizontal directions. Here again, the usual geophysical convention is followed: the x-axis is directed eastward and the y-axis is directed northward, with u and v the corresponding velocity components.

The starting point for the analysis is the construction of a single equation for v from (12.45) to (12.47) First, time differentiate the momentum equations (12.46) and (12.47), then use (12.45) to eliminate $\partial\eta/\partial t$. These steps produce:

$$\frac{\partial^2 u}{\partial t^2} - f\frac{\partial v}{\partial t} = gH\frac{\partial}{\partial x}\left(\frac{\partial u}{\partial x} + \frac{\partial v}{\partial y}\right), \quad \text{and} \quad \frac{\partial^2 v}{\partial t^2} + f\frac{\partial u}{\partial t} = gH\frac{\partial}{\partial y}\left(\frac{\partial u}{\partial x} + \frac{\partial v}{\partial y}\right). \tag{12.72, 12.73}$$

Now apply $\partial/\partial t$ to (12.73) and use (12.72) to reach:

$$\frac{\partial^3 v}{\partial t^3} + f\left[f\frac{\partial v}{\partial t} + gH\frac{\partial}{\partial x}\left(\frac{\partial u}{\partial x} + \frac{\partial v}{\partial y}\right)\right] = gH\frac{\partial^2}{\partial y \partial t}\left(\frac{\partial u}{\partial x} + \frac{\partial v}{\partial y}\right). \tag{12.74}$$

To eliminate u, first develop a vorticity equation by cross-differentiating and subtracting (12.46) and (12.47):

$$\frac{\partial}{\partial t}\left(\frac{\partial u}{\partial y} - \frac{\partial v}{\partial x}\right) - f_0\left(\frac{\partial u}{\partial x} + \frac{\partial v}{\partial y}\right) - \beta v = 0.$$

Here, the customary β-plane approximation has been made. It is valid if the y-scale is small enough so that $\Delta f/f \ll 1$. Accordingly, f is treated as constant (and replaced it by an average value f_0) *except* when df/dy appears; thus, f_0 appears in the second term of the last equation. Taking the x-derivative this equation, multiplying by gH, and adding the result to (12.74), produces a vorticity equation in terms of v only:

$$\frac{\partial^3 v}{\partial t^3} - gH\frac{\partial}{\partial t}\nabla_H^2 v + f_0^2\frac{\partial v}{\partial t} - gH\beta\frac{\partial v}{\partial x} = 0, \tag{12.75}$$

where $\nabla_H^2 \equiv \partial^2/\partial x^2 + \partial^2/\partial y^2$ is the horizontal Laplacian operator.

Eq. (12.75) is linear, hydrostatic, and based on the Boussinesq approximation, but is otherwise quite general in the sense that it is applicable to waves of any frequency. Thus, consider traveling wave solutions of the form $v = \widehat{v}\exp\{i(kx + ly - \omega t)\}$, where k is the eastward wave number and l is the northward wave number. Use of this exponential form for v, reduces (12.75) to an algebraic equation:

$$\omega^3 - c^2\omega K^2 - f_0^2\omega - c^2\beta k = 0, \tag{12.76}$$

where $K^2 = k^2 + l^2$ and $c^2 = gH$. Interestingly, all the roots of (12.76) are real; two are superinertial ($\omega \gg f$) and the third is subinertial ($\omega \ll f$). Eq. (12.76) is the complete dispersion relation for the linear shallow-water equations. In various parametric ranges it takes simpler forms, representing simpler waves.

First, consider high-frequency waves $\omega \gg f$. Here, the third term of (12.76) is negligible compared to the first term. Moreover, the fourth term is also negligible in this range. For example, the ratio of fourth and second terms is:

$$\frac{c^2\beta k}{c^2\omega K^2} \sim \frac{\beta}{\omega K} \sim 10^{-3},$$

where the numerical value is comes from typical values of $\beta = 2 \times 10^{-11}\,\text{m}^{-1}\,\text{s}^{-1}$, $\omega = 3f \sim 3 \times 10^{-4}\,\text{s}^{-1}$, and $2\pi/K \sim 100\,\text{km}$. For $\omega \gg f$, therefore, the balance is between the first and second terms in (12.76), and the roots are $\omega = \pm K[gH]^{1/2}$, which correspond to a propagation speed of $\omega/K = [gH]^{1/2}$. The effects of both f and β are therefore negligible for high-frequency waves, as is expected since they are too fast to be affected by the Coriolis acceleration.

Next, consider $\omega > f$, but $\omega \sim f$. Then the third term in Eq. (12.76) is not negligible, but the β-effect is. These are gravity waves influenced by the Coriolis acceleration and are discussed in the next section. However, the time scales are still too short for the motion to be influenced by the β-effect.

Last, consider slow waves for which $\omega \ll f$. Then the β-effect becomes important, and the first term in (12.76) becomes negligible. For this frequency range, the ratio of the first and the last terms is:

$$\omega^3/c^2\beta k \ll 1.$$

Typical values for the ocean are $c \sim 200\,\text{m/s}$ for the barotropic mode, $c \sim 2\,\text{m/s}$ for the baroclinic mode, $\beta = 2 \times 10^{-11}\,\text{m}^{-1}\,\text{s}^{-1}$, $2\pi/k \sim 100$ km, and $\omega \sim 10^{-5}\,\text{s}^{-1}$. This makes the aforementioned ratio about 0.2×10^{-4} for the barotropic mode and 0.2 for the baroclinic mode. The first term in (12.76) is therefore negligible for $\omega \ll f$.

Eq. (12.75) governs the dynamics of a variety of wave motions in the ocean and the atmosphere, and the discussion in this section shows what terms can be dropped under various limiting conditions. An understanding of these limiting conditions will be useful in the following sections.

Example 12.8

Consider waves with $k = 0$ having crests lying along latitude lines. What are the roots of (12.76) in this case? What is the effect of rotation on these waves?

Solution Eq. (12.76) evaluated with $k = 0$ and $K^2 = l^2$ no longer includes β, and can be factored:

$$\omega^3 - c^2\omega l^2 - f_0^2\omega = 0, \quad \text{or} \quad \omega\left(\omega - \sqrt{c^2l^2 + f_0^2}\right)\left(\omega + \sqrt{c^2l^2 + f_0^2}\right) = 0.$$

The first root, $\omega = 0$, corresponds to steady geostrophic flow. Use of $v = \widehat{v}\exp\{ily\}$, (12.46), and (12.47) leads to $\eta = x(\widehat{v}f/g)\exp\{ily\}$ and $u = -ilx\widehat{v}\exp\{ily\}$ when $\eta = 0$ at $x = 0$. Although possible, this flow field requires a divergent ocean surface deflection and east-west velocity, so it cannot persist over a large area.

The second two roots, $\omega = \pm\sqrt{c^2l^2 + f_0^2}$, correspond to northward and southward traveling waves that propagate at a speed, $\omega/l = \pm\sqrt{c^2 + f_0^2/l^2}$, that is augmented by the Coriolis effect.

12.9 Gravity waves with rotation

The focus of this section is shallow-water gravity waves with frequencies in the range $\omega > f$, for which the β-effect is negligible so the Coriolis frequency f is regarded as constant. As in the prior section, consider progressive waves of the form:

$$(u, v, \eta) = (\widehat{u}, \widehat{v}, \widehat{\eta})\exp\{i(kx + ly - \omega t)\}, \tag{12.77}$$

where $\widehat{u}$, $\widehat{v}$, and $\widehat{\eta}$ are the complex amplitudes, and the real part of the right side is physically meaningful. For such waves, (12.45)–(12.47) give:

$$-i\omega\widehat{\eta} + iH(k\widehat{u} + l\widehat{v}) = 0, \qquad -i\omega\widehat{u} - f\widehat{v} = -ikg\widehat{\eta}, \quad \text{and} \quad -i\omega\widehat{v} + f\widehat{u} = -ilg\widehat{\eta}. \tag{12.78, 12.79, 12.80}$$

Solving for $\widehat{u}$ and $\widehat{v}$ in terms of $\widehat{\eta}$ between (12.79) and (12.80) leads to:

$$\widehat{u} = \frac{g\widehat{\eta}}{\omega^2 - f^2}(\omega k + ifl) \quad \text{and} \quad \widehat{v} = \frac{g\widehat{\eta}}{\omega^2 - f^2}(-ifk + \omega l). \tag{12.81}$$

Substituting these into (12.78) produces:

$$\omega^2 - f^2 = gH(k^2 + l^2) \quad \text{or} \quad \omega^2 = f^2 + gHK^2, \tag{12.82}$$

where $K = [k^2 + l^2]^{1/2}$ is the magnitude of the horizontal wave number.

This is the dispersion relation of gravity waves in the presence of Coriolis effects. (The relation can be most simply derived by setting the determinant of the set of linear homogeneous equations (12.77) through (12.79) to zero.) This dispersion relation shows that the waves can propagate in any horizontal direction and must have $\omega > f$. Gravity waves affected by the Coriolis acceleration are called *Poincaré waves*, *Sverdrup waves*, or simply *rotational gravity waves*. (Sometimes the name "Poincaré wave" is used to describe those rotational gravity waves that satisfy the boundary conditions in a channel.) In spite of their name, the solution was first worked out by Kelvin (Gill, 1982, p. 197). A plot of (12.82) is shown in

Fig. 12.14. The waves are dispersive except for $\omega \gg f$ where Eq. (12.82) gives $\omega^2 \approx gHK^2$, so that the propagation speed is $\omega/K = [gH]^{1/2}$. This high-frequency limit agrees with the previous discussion of surface gravity waves unaffected by the Coriolis acceleration.

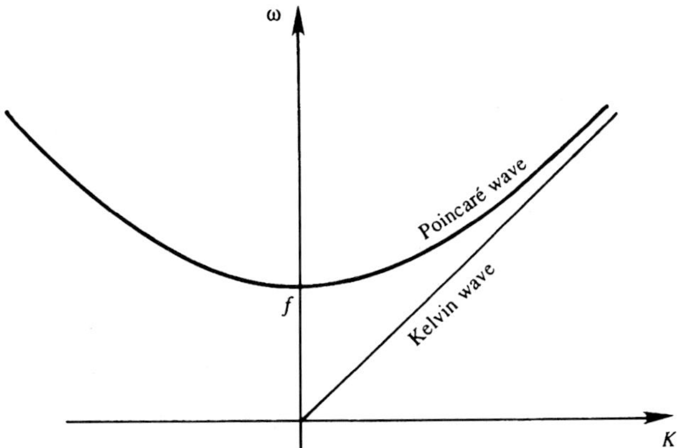

FIGURE 12.14 Dispersion relations for Poincaré and Kelvin waves. Here, ω is the wave frequency, K is the magnitude of the wave number, and f is the local inertial frequency. At frequencies $\omega \gg f$, the ordinary shallow-water wave dispersion relationship $\omega^2 = gHK^2$ is recovered.

Particle orbit

The symmetry of the dispersion relation (12.82) with respect to k and l means that the x- and y-directions are not felt differently by the wave field. This horizontal isotropy is a result of treating f as constant. (Rossby waves, which depend on the β-effect, are not horizontally isotropic; see Section 12.13.) Thus, choose the wave number vector along the x-axis and set $l = 0$, so that the wave field is invariant along the y-axis. To find the particle orbits, work with real quantities based on $\hat{\eta}$ being real and positive, so that $\eta = \hat{\eta}\cos(kx - \omega t)$. In this case, the assumed exponential form (12.77) and the relations (12.81) imply:

$$u = \frac{\omega\hat{\eta}}{kH}\cos(kx - \omega t), \quad \text{and} \quad v = \frac{f\hat{\eta}}{kH}\sin(kx - \omega t). \tag{12.83}$$

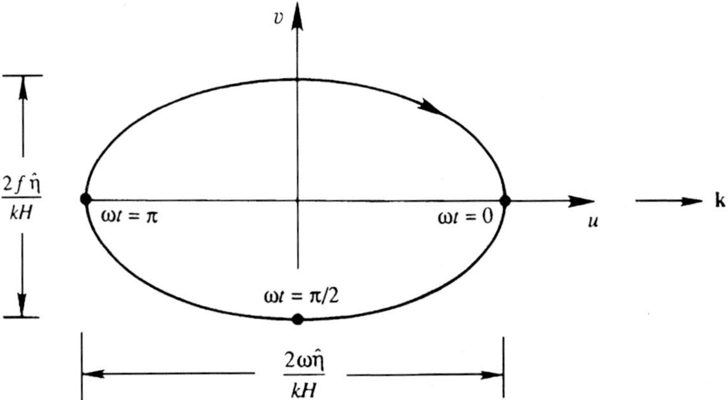

FIGURE 12.15 Particle orbit in a gravity wave traveling in the positive x-direction. Looking down on the surface, the orbit is an ellipse having major and minor axes proportional to the wave frequency ω and the inertial frequency f. Velocity components corresponding to $\omega t = 0, \pi/2$, and π are indicated.

To find the particle paths, take $x = 0$ and consider three values of time corresponding to $\omega t = 0, \pi/2$, and π. The corresponding values of u and v from (12.83) show that the velocity vector rotates clockwise (in the northern hemisphere) in elliptic paths (Fig. 12.15). The ellipticity is expected, since the presence of the Coriolis acceleration means that fu must generate $\partial v/\partial t$ according to (12.47), since $\partial\eta/\partial y = 0$ when the wave number is oriented along the x-axis and $l = 0$. Fluid particles are therefore constantly deflected to the right by the Coriolis acceleration, resulting in elliptic orbits. *The ellipses have an axis ratio of ω/f and the major axis is oriented in the direction of wave propagation.* The ellipses become narrower as ω/f increases, approaching the rectilinear orbit of gravity waves unaffected by planetary rotation. However, the sea surface deflection of a rotational gravity wave is no different from that for ordinary gravity waves, namely oscillatory in the direction of propagation and invariant in the perpendicular direction.

Inertial motion

In the limit $\omega \to f$, the waves develop an entirely different character compared to faster-oscillating gravity waves. Here, fluid particle paths are circular and the dispersion relation (12.82) requires that $K \to 0$, implying horizontal uniformity of the flow field. Furthermore, (12.78) shows that $\widehat{\eta}$ must tend to zero in this limit because k and l must approach zero, so that there are no horizontal pressure gradients. For horizontally uniform flow, $\partial u/\partial x = \partial v/\partial y = 0$ so the continuity equation and the bottom boundary condition require then $w = 0$. Thus, fluid particles must move in horizontal planes, each layer decoupled from the one above and below it. The balance of forces from (12.46) and (12.47) is:

$$\partial u/\partial t - fv = 0 \quad \text{and} \quad \partial v/\partial t + fu = 0.$$

The solution of this set is of the form:

$$u = q\cos(ft) \quad \text{and} \quad v = -q\sin(ft),$$

where the speed $q = [u^2 + v^2]^{1/2}$ is constant along the path. The radius r of the orbit can be found by adopting a Lagrangian point of view, and noting that the equilibrium of forces is between the Coriolis acceleration fq and the centrifugal acceleration $r\omega^2 = rf^2$, giving $r = q/f$. The limiting case of motion in circular orbits at a frequency f is called *inertial motion*, because in the absence of pressure gradients a particle moves by virtue of its inertia alone. The corresponding period $2\pi/f$ is called the *inertial period*. In the absence of planetary rotation such motion would be along straight lines; in the presence of Coriolis effects the motion is along circular paths, called *inertial circles*. Near-inertial motion is frequently generated in the surface layer of the ocean by sudden changes of the wind field, essentially because the equations of motion (12.45) to (12.47) have a natural frequency f. Taking a typical current magnitude of $q \sim 0.1$ m/s, the radius of the orbit is $r \sim 1$ km.

Example 12.9

Determine the wavelength λ and frequency ω of a shallow-water gravity wave in a rotating system with $f \sim 10^{-4}$ rad/s so that its propagation speed is 1% higher than $[gH]^{1/2}$ when $H = 4$ km, and $H_e = 1.0$ m.

Solution The dispersion relationship for gravity waves with rotation is (12.82). After taking a square root, it can be divided by K to yield the phase speed, c:

$$c \equiv \omega/K = \sqrt{gH + f^2/K^2} = \sqrt{gH}\sqrt{1 + f^2/gHK^2} = \sqrt{gH}\sqrt{1 + f^2\lambda^2/4\pi^2 gH}.$$

A 1% augmentation of the phase speed implies:

$$1.01 = \sqrt{1 + f^2\lambda^2/4\pi^2 gH}, \quad \text{or} \quad \lambda = (2\pi/f)\sqrt{\left[(1.01)^2 - 1\right]gH}.$$

For $f \sim 10^{-4}$ rad/s, the final equation produces $\lambda \approx 1800$ km when $H = 4$ km, and $\lambda \approx 28$ km when $H = 1.0$ m. Interestingly, the wave frequency is the same in both cases, $\omega \approx 7.1 \times 10^{-4}$ rad/s (or 0.41 cycles per hour) as determined from (12.82). At shorter wavelengths and higher frequencies, the speed correction is less than 1%.

12.10 Kelvin wave

The characteristics of shallow-water gravity waves propagating in a horizontally *unbounded* ocean were presented in the preceding section. The crests are horizontal and oriented in a direction perpendicular to the direction of propagation. The *absence* of a transverse pressure gradient proportional to $\partial\eta/\partial y$ resulted in oscillatory transverse flow and elliptic fluid-particle orbits. In this section, we consider a gravity wave propagating parallel to a wall, whose presence allows non-zero $\partial\eta/\partial y$ that decays away from the wall. This situation permits a gravity wave in which fu is geostrophically balanced by $-g(\partial\eta/\partial y)$ with $v = 0$. Consequently fluid particle orbits are not elliptic but rectilinear.

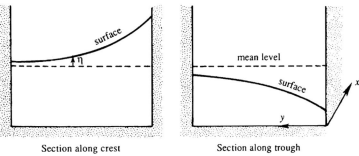

Section along crest Section along trough

FIGURE 12.16 Free-surface distribution in a Kelvin gravity wave propagating into the plane of the paper (the x-direction) within a channel aligned with x-direction. The wave crests and troughs are enhanced on the right side of the channel.

Consider first a gravity wave propagating in the x-direction in a channel aligned with the x-direction. From Fig. 11.5 (and its related discussion), the fluid velocity under a crest is in the direction of wave propagation, and that under a trough is opposite the direction of propagation. Fig. 12.16 shows two transverse sections of the wave, one through a crest (left panel) and the other through a trough (right panel). The wave is propagating into the plane of the paper so that the fluid velocity under the crest is into the plane of the paper and that under the trough is out of the plane of the paper. The constraints of the sidewalls require that $v = 0$ at the walls, and we are exploring the possibility of a wave motion in which v is zero everywhere. Then, the linearized momentum equation along the y-direction (12.47) requires that fu can only be geostrophically balanced by a transverse slope of the sea surface across the channel:

$$fu = -g\partial\eta/\partial y.$$

In the northern hemisphere, the surface must slope as indicated in the figure, that is, downward to the left under the crest and upward to the left under the trough, so that the pressure force has the current directed to its right. The result is that the amplitude of the wave is larger on the right-hand side of the channel, looking in the direction of propagation, as indicated in Fig. 12.16. The current amplitude, like the surface displacement, also decays to the left.

If the left wall in Fig. 12.16 is moved away to infinity, what remains is a gravity wave trapped to the coast (Fig. 12.17). Such coastally trapped long gravity waves, in which the transverse velocity $v = 0$ everywhere, are called *Kelvin waves*. It is clear that such waves can propagate only in a direction such that the coast is to the right (looking in the direction of propagation) in the northern hemisphere and to the left in the southern hemisphere. The opposite direction of propagation would result in a sea surface displacement increasing exponentially away from the coast, which is not possible.

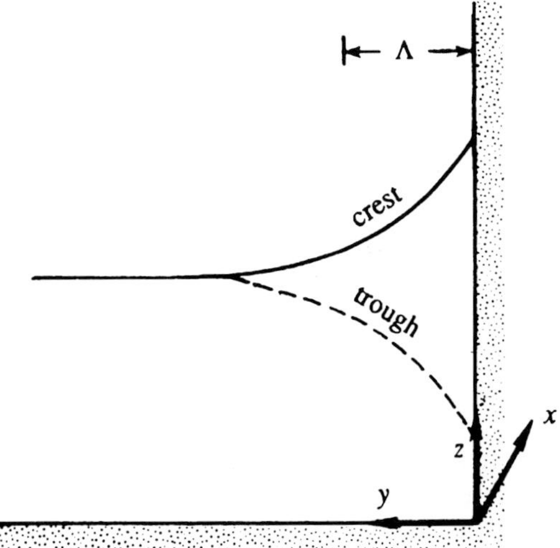

FIGURE 12.17 Coastal Kelvin wave propagating along the x-axis. The sea surface shape across a section through a crest is indicated by the continuous line, and that along a trough is indicated by the dashed line.

An examination of (12.47), $\partial v/\partial t + fu = -g\partial\eta/\partial y$, reveals fundamental differences between Poincaré waves away from boundaries and Kelvin waves. For a Poincaré wave, the crests are horizontal, and the absence of a transverse pressure gradient requires a $\partial v/\partial t$ to balance the Coriolis acceleration, resulting in elliptic orbits. In a Kelvin wave, a transverse velocity is prevented by a geostrophic balance of fu and $-g(\partial\eta/\partial y)$.

From the shallow-water set (12.45)–(12.47), the equations of motion for a Kelvin wave propagating along a coast aligned with the x-axis (Fig. 12.17) are:

$$\frac{\partial\eta}{\partial t}+H\frac{\partial u}{\partial x}=0,\quad \frac{\partial u}{\partial t}=-g\frac{\partial\eta}{\partial x},\quad \text{and}\quad fu=-g\frac{\partial\eta}{\partial y}. \tag{12.84}$$

Assume a solution of the form: $(u,\eta)=(\widehat{u}(y),\widehat{\eta}(y))\exp\{i(kx-\omega t)\}$, to reduce the set (12.84) to three algebraic equations:

$$-i\omega\widehat{\eta}+ikH\widehat{u}=0,\quad -i\omega\widehat{u}=-ikg\widehat{\eta},\quad \text{and}\quad f\widehat{u}=-g\partial\widehat{\eta}/\partial y. \tag{12.85}$$

The dispersion relation arises from the first two of these equations; the third equation then determines the transverse dependence. Eliminate $\widehat{u}$ between the first two equations to obtain:

$$\widehat{\eta}(\omega^2-gHk^2)=0.$$

A non-trivial solution is therefore possible only if $\omega=\pm k[gH]^{1/2}$, so that the wave propagates with a non-dispersive speed:

$$c=\sqrt{gH}. \tag{12.86}$$

The propagation speed of a Kelvin wave is therefore identical to that of non-rotating gravity waves. Its dispersion equation is a straight line and is shown in Fig. 12.14. All frequencies are possible.

To determine the transverse dependence, eliminate $\widehat{u}$ between the first and third equation of (12.85), giving:

$$\partial\widehat{\eta}/\partial y\pm(f/c)\widehat{\eta}=0.$$

The solution that decays away from the coast is $\widehat{\eta}=\eta_0e^{-fy/c}$, where η_0 is the amplitude at the coast located along $y=0$. Therefore, the sea surface slope and the velocity field for a Kelvin wave have the form:

$$\eta=\eta_0e^{-fy/c}\cos(k(x-ct)),\quad \text{and}\quad u=\eta_0\sqrt{g/H}e^{-fy/c}\cos(k(x-ct)), \tag{12.87}$$

where real parts have been taken, and (12.85) has been used in obtaining the u-field.

Eqs. (12.87) show that the transverse decay scale of the Kelvin wave is:

$$\Lambda\equiv c/f=\sqrt{gH}/f,$$

which is called the (external) *Rossby radius of deformation*. For an ocean depth of $H=5$ km, and a mid-latitude value of $f=10^{-4}\,\mathrm{s}^{-1}$, $c\approx 220\,\mathrm{m/s}$ so $\Lambda=c/f=2200$ km. Tides are frequently in the form of coastal Kelvin waves of semi-diurnal frequency. The tides are forced by the periodic changes in the gravitational attraction of the moon and the sun. These waves propagate along the boundaries of an ocean basin and cause sea level fluctuations at coastal stations.

Analogous to the surface or "external" Kelvin waves discussed in the preceding paragraphs, *internal Kelvin waves* at the interface between two fluids of different densities can also exist (Fig. 12.18). If the lower layer is very deep, then the speed of propagation is given by (11.115) and (11.116), $c=[g'H]^{1/2}$, where H is the thickness of the upper layer and $g'=g(\rho_2-\rho_1)/\rho_2$ is the reduced gravity. For a continuously stratified medium of depth H and buoyancy frequency N internal Kelvin waves can propagate at any of the normal mode speeds given by (12.71). The decay scale for *internal* Kelvin waves is again $\Lambda=c/f$, but it is called the *internal Rossby radius of deformation*. The value of Λ for internal Kelvin waves is much smaller than the external Rossby radius of deformation. For $n=1$, a typical value in the ocean is $\Lambda=NH/\pi f\sim 50$ km; a typical atmospheric value is much larger, being of order $\Lambda\sim 1000$ km.

Internal Kelvin waves in the ocean are frequently forced by wind changes near coastal areas. For example, a southward wind along the west coast of a continent in the northern hemisphere (say, California) generates an Ekman layer at the ocean surface, in which the mass flow is *away* from the coast (to the right of the applied wind stress). The mass flux in the near-surface layer is compensated by the movement of deeper water toward the coast, which raises the thermocline. An upward movement of the thermocline, as indicated by the dashed line in Fig. 12.18, is called *upwelling*. The vertical movement of the thermocline in the wind-forced region then propagates poleward along the coast as an internal Kelvin wave.

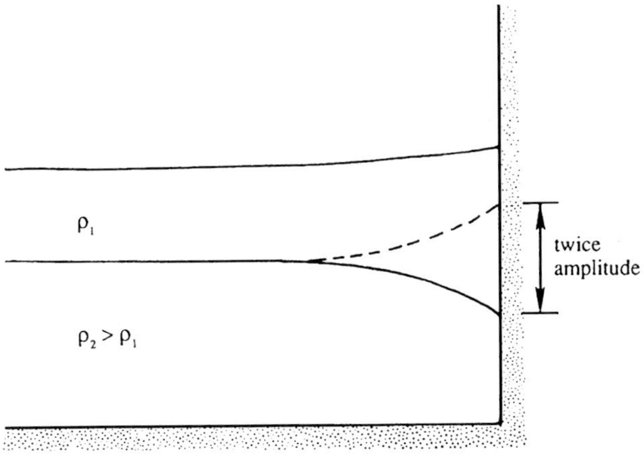

FIGURE 12.18 Internal Kelvin wave at an interface. The dashed line indicates the position of the interface when it is at its maximum height. The displacement of the free surface is much smaller than that of the interface and is oppositely directed.

Example 12.10

Redo the analysis of the surface Kelvin wave assuming it travels along a north-south coastline so that $u = 0$. In what direction do Kelvin waves travel if they abut the west coast of Chile in South America?

Solution When $u = 0$, the equation set (12.45)–(12.47) simplifies to:

$$\frac{\partial \eta}{\partial t} + H\frac{\partial v}{\partial y} = 0, \quad -fv = -g\frac{\partial \eta}{\partial x}, \quad \text{and} \quad \frac{\partial v}{\partial t} = -g\frac{\partial \eta}{\partial y}.$$

Here assume a solution of the form: $(v, \eta) = (\hat{v}(x), \hat{\eta}(x))\exp\{i(ly - \omega t)\}$; these waves travel northward when l is positive. Inserting the assumed solution form into the simplified equations produces:

$$-i\omega\hat{\eta} + ilH\hat{v} = 0, \quad -f\hat{v} = -g\partial\hat{\eta}/\partial x, \quad \text{and} \quad -i\omega\hat{v} = -ilg\hat{\eta}.$$

The first and third equations require $\omega^2 - gHl^2 = 0$ or $\omega = \pm l[gH]^{1/2}$ when $\hat{\eta} \neq 0$. Using this result, solve the first or third equation for $\hat{v} = \pm\sqrt{g/H}\hat{\eta}$ and substitute this into the second equation to obtain:

$$\pm f\sqrt{g/H}\hat{\eta} = g\partial\hat{\eta}/\partial x, \quad \text{or} \quad \eta = \eta_o \exp\left\{\pm fx/\sqrt{gH}\right\}.$$

When f is positive, as would be the case along the west coast of North America, the '+' sign must be selected since x becomes increasingly negative moving westward away from the coast and η must go to zero as $x \to -\infty$. Using this sign choice and taking the real part of the assumed solution form leads to the Kelvin wave surface profile:

$$\eta = \eta_o \exp\left\{-f|x|/\sqrt{gH}\right\}\cos\left\{l\left(y - t\sqrt{gH}\right)\right\},$$

and this wave travels northward. When f is negative, as would be the case along the west coast of Chile, the '−' sign must be selected, and the Kelvin wave surface profile becomes:

$$\eta = \eta_o \exp\left\{-|fx|/\sqrt{gH}\right\}\cos\left\{l\left(y + t\sqrt{gH}\right)\right\},$$

and this wave that travels southward.

12.11 Potential vorticity conservation in shallow-water theory

In this section, a useful conservation law is derived for the vorticity in a variable-depth shallow layer of homogeneous fluid that flows without friction. As in Section 12.8, the constant density of the layer and the hydrostatic pressure distribution cause the vertical velocity to be linear in z, and cause the horizontal pressure gradient to be depth independent, so that only a depth-independent current is generated. The equations of motion are the two horizontal momentum equations of

(12.9) simplified for negligible vertical velocity with $\nu_H = \nu_V = 0$ and (12.42) used for the pressure gradient terms, and the continuity equation (12.44) with $h(x, y, t) = H + \eta$ as the overall depth of the flow:

$$\frac{\partial u}{\partial t} + u\frac{\partial u}{\partial x} + v\frac{\partial u}{\partial y} - fv = -g\frac{\partial \eta}{\partial x}, \quad \frac{\partial v}{\partial t} + u\frac{\partial v}{\partial x} + v\frac{\partial v}{\partial y} + fu = -g\frac{\partial \eta}{\partial y}, \quad \frac{\partial h}{\partial t} + \frac{\partial (uh)}{\partial x} + \frac{\partial (vh)}{\partial y} = 0. \tag{12.88, 12.89, 12.90}$$

Here, all the nonlinear terms have been retained; $\eta(x, y, t)$ is the height of the sea surface measured from a convenient horizontal reference plane (Fig. 12.19); the x-axis is taken eastward; the y-axis is taken northward; u and v are the corresponding velocity components; and the Coriolis frequency $f = f_0 + \beta y$ is regarded as latitude dependent.

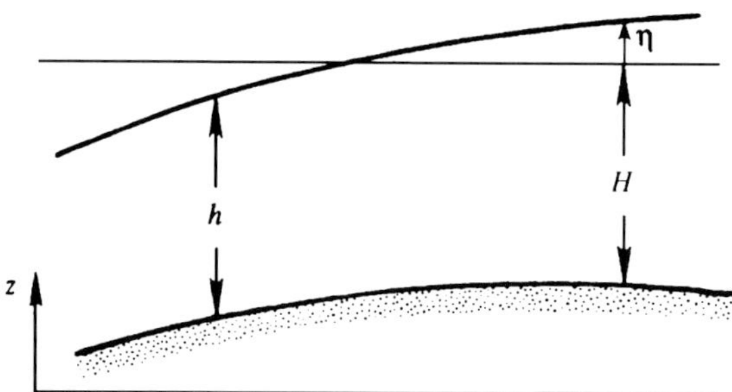

FIGURE 12.19 Shallow layer of instantaneous depth $h(x, y, t)$ when the ocean bottom is not flat. Here η is the sea surface deflection measured from a conveniently chosen horizontal plane.

A vorticity equation can be derived by differentiating (12.88) with respect to y, (12.89) with respect to x, and subtracting. As expected, these steps eliminate η and result in:

$$\frac{\partial}{\partial t}\left(\frac{\partial v}{\partial x} - \frac{\partial u}{\partial y}\right) + \frac{\partial}{\partial x}\left(u\frac{\partial v}{\partial x} + v\frac{\partial v}{\partial y}\right) - \frac{\partial}{\partial y}\left(u\frac{\partial u}{\partial x} + v\frac{\partial u}{\partial y}\right) + f_0\left(\frac{\partial u}{\partial x} + \frac{\partial v}{\partial y}\right) + \beta v = 0. \tag{12.91}$$

Following the customary β-plane approximation, f has been treated as constant (and replaced by an average value f_0) *except* when df/dy appears. Now introduce:

$$\zeta = \frac{\partial v}{\partial x} - \frac{\partial u}{\partial y},$$

as the vertical component of *relative vorticity*, that is, the vorticity measured relative to the rotating earth. Using ζ, the nonlinear terms in (12.91) can be rearranged in the form:

$$u\frac{\partial \zeta}{\partial x} + v\frac{\partial \zeta}{\partial y} + \left(\frac{\partial u}{\partial x} + \frac{\partial v}{\partial y}\right)\zeta,$$

so that (12.91) becomes:

$$\frac{\partial \zeta}{\partial t} + u\frac{\partial \zeta}{\partial x} + v\frac{\partial \zeta}{\partial y} + \left(\frac{\partial u}{\partial x} + \frac{\partial v}{\partial y}\right)(\zeta + f_0) + \beta v = 0, \quad \text{or} \quad \frac{D\zeta}{Dt} + (\zeta + f_0)\left(\frac{\partial u}{\partial x} + \frac{\partial v}{\partial y}\right) + \beta v = 0, \tag{12.92}$$

where (for this section) D/Dt is the derivative following only the horizontal motion of the layer:

$$\frac{D}{Dt} = \frac{\partial}{\partial t} + u\frac{\partial}{\partial x} + v\frac{\partial}{\partial y}.$$

The horizontal divergence $(\partial u/\partial x + \partial v/\partial y)$ in (12.92) can be eliminated by using the continuity equation (12.90), which can be written:

$$\frac{Dh}{Dt} + h\left(\frac{\partial u}{\partial x} + \frac{\partial v}{\partial y}\right) = 0.$$

Eq. (12.92) then becomes:

$$\frac{D\zeta}{Dt} = \left(\frac{\zeta + f_0}{h}\right)\frac{Dh}{Dt} - \beta v \quad \text{or} \quad \frac{D(\zeta + f)}{Dt} = \left(\frac{\zeta + f_0}{h}\right)\frac{Dh}{Dt}, \tag{12.93}$$

where the second equation involves moving the $-\beta v$ term inside the D/Dt-differentiation on the right since:

$$\frac{Df}{Dt} = \frac{\partial f}{\partial t} + u\frac{\partial f}{\partial x} + v\frac{\partial f}{\partial y} = 0 + 0 + \beta v.$$

Because of the absence of vertical shear, the vorticity in a shallow-water model is purely vertical and independent of depth. The relative vorticity measured with respect to the rotating earth is ζ, while f is the planetary vorticity, so that the *absolute vorticity* is $\zeta + f$. Eq. (12.93) shows that the rate of change of absolute vorticity is proportional to the absolute vorticity times the vertical stretching Dh/Dt of the water column. It is apparent that $D\zeta/Dt$ can be non-zero even if $\zeta = 0$ initially. This is different from a non-rotating flow in which stretching a fluid line changes its vorticity only if the line has an *initial* vorticity. (This is why the process was called the *vortex* stretching; see Section 5.3.) The difference arises because vertical lines in a geophysical flow contain the Earth's planetary vorticity even when $\zeta = 0$. The vortex *tilting* term, discussed in Section 5.3, is absent in shallow-water theory because the water moves in the form of vertical columns without ever tilting.

Interestingly, (12.93) can be written in the approximate compact form:

$$\frac{D}{Dt}\left(\frac{\zeta + f}{h}\right) = 0, \tag{12.94}$$

where $f = f_0 + \beta y$, and $\beta y \ll f_0$ has been assumed. The ratio $(\zeta + f)/h$ is called the *potential vorticity* in shallow-water theory, and (12.94) shows that the *potential vorticity is conserved along the horizontal trajectory of a fluid particle*, an important principle in geophysical fluid dynamics. In the ocean, outside regions of strong current vorticity such as coastal boundaries, the magnitude of ζ is much smaller than that of f. In such a case, $\zeta + f$ has the sign of f. The principle of conservation of potential vorticity means that an increase in h must make $\zeta + f$ more positive in the northern hemisphere and more negative in the southern hemisphere.

To illustrate the implications of (12.94), consider eastward flow at uniform speed U over a step change in depth (at $x = 0$) running north–south, across which the layer thickness changes discontinuously from h_0 to h_1 (Fig. 12.20). The flow upstream of the step has no relative vorticity. To conserve the ratio $(\zeta + f)/h$, the flow must suddenly acquire negative (clockwise) relative vorticity due to the sudden decrease in layer thickness. The relative vorticity of a fluid element just after passing the step can be found from:

$$f/h_0 = (\zeta + f)/h_1 \quad \text{which implies} \quad \zeta = f(h_1 - h_0)/h_0 < 0,$$

where f is evaluated at the upstream latitude of the streamline. Because of the clockwise vorticity, the fluid starts to move south at $x = 0$. The southward movement decreases f, so that ζ must correspondingly increase to keep $f + \zeta$ constant. This means that the clockwise curvature of the stream reduces, and eventually becomes a counterclockwise curvature. In this manner an eastward flow over a step generates stationary undulatory flow on its downstream side. In Section 12.13, this stationary oscillation is identified as a Rossby wave, with wavelength $2\pi[U/\beta]^{1/2}$, generated at the step having a westward phase velocity that counteracts the eastward current with speed U.

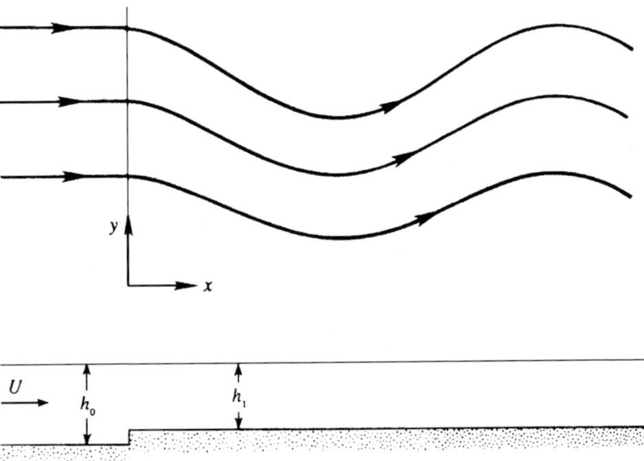

FIGURE 12.20 Eastward flow over a step change in depth. Looking down from above, the step causes southward deflection of the streamlines that is eventually countered by latitude change and results in stationary spatial oscillations of wavelength $2\pi[U/\beta]^{1/2}$.

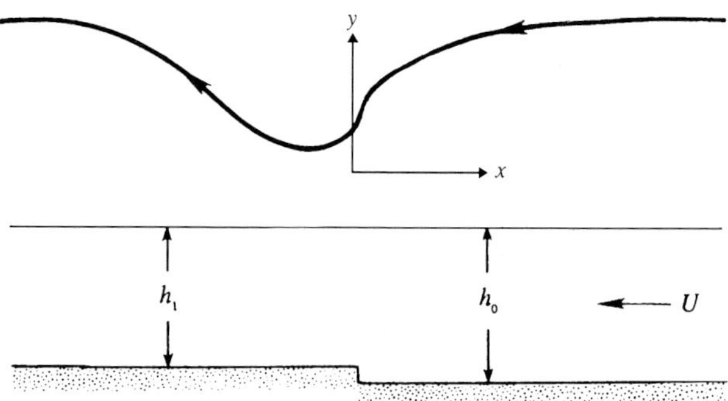

FIGURE 12.21 Westward flow over a step change in depth. Unlike the eastward flow depicted in Fig. 12.20, the westward flow is not oscillatory and feels the upstream influence of the step. Looking down from above, the step causes one southward deflection that starts before the step and recovers after it.

The situation is fundamentally different for a *westward* flow over a step. In this case, a fluid particle would suddenly acquire clockwise vorticity as the depth of the flow decreases at $x = 0$, which would require the particle to move north. It would then come into a region of larger f, which would require ζ to decrease further leading to an exponential divergence, suggesting that the given line of reasoning is flawed. Unlike an eastward flow, a westward one feels the *upstream* influence of the step so that it acquires a counterclockwise curvature *before* it encounters the step (Fig. 12.21). The positive vorticity is balanced by a reduction in f, which is consistent with conservation of potential vorticity. At the location of the step the vorticity decreases suddenly. Finally, far downstream of the step the fluid particle is again moving westward at its original latitude. The westward flow over a topographic step is *not* oscillatory.

Example 12.11

If an Atlantic hurricane at latitude $\theta = 30°$N draws equally from nominally quiescent air masses at latitudes 5° to the north and the south, is the storm likely to weaken or intensify?

Solution In the northern hemisphere, hurricanes are cyclonic storms, with significant positive relative vorticity ($\zeta > 0$), that typically originate over the ocean from tropical depressions (regions of low atmospheric pressure) a few degrees north or south of the equator. Hurricanes commonly fill the vertical extent of the troposphere and grow to 100's of kilometers in both horizontal directions. These storms may achieve sustained wind speeds of 200 km/hr or more, and travel thousands of kilometers during a lifetime of several weeks. Although atmospheric thermodynamics plays a critical role in storm intensification and the actual airflow patterns near to and within a hurricane are fully three dimensional, an examination of the potential vorticity determined from the horizontal wind components u and v alone is sufficient to indicate a likely answer. Here, the tropopause height (in km) is assumed to be $h(\theta) \approx 13 + 4\cos(2\theta)$, and this height is assumed to be the correct vertical length scale for computing the potential vorticity.

The goal of this effort is to determine if the combined flux of relative vorticity from north and south of the storm is positive or negative. Eqs. (12.8) and (12.94) imply the following for air at latitude θ with $\zeta = 0$ that is pulled into the storm at $\theta = 30°$:

$$\left(\frac{0+f}{h}\right)_\theta = \left(\frac{\zeta+f}{h}\right)_{30°}, \quad \text{or} \quad \zeta(30°) = 2\Omega\left(\frac{h(30°)}{h(\theta)}\sin\theta - \sin 30°\right),$$

where Ω is the rotation rate of the earth. Evaluating at $\theta = 35°$ produces: $\zeta(30°) = +0.198\Omega$. Thus, quiescent air entrained from the north brings positive relative vorticity to the storm. A similar evaluation using $\theta = 25°$ produces: $\zeta(30°) = -0.186\Omega$, so quiescent air from the south brings negative relative vorticity to the storm. However, if the storm entrains equally from both air masses, the northern air with higher ζ will dominate, so the hurricane is likely to intensify because of the net influx of positive vorticity.

12.12 Internal waves

Section 11.8 describes internal gravity waves unaffected by the Coriolis acceleration. Such waves are not isotropic; in fact the direction of propagation with respect to the vertical determines their frequency. We also saw that their frequency satisfies the inequality $\omega \le N$, where N is the buoyancy frequency. Their phase-velocity vector $\mathbf{c}$ and the group-velocity vector $\mathbf{c}_g$ are perpendicular and have oppositely directed vertical components (Fig. 11.29 and Fig. 11.31). That is, phases propagate upward if the groups propagate downward, and vice versa. In this section, the local impact of the earth's rotation on internal waves is presented, assuming that the Coriolis parameter f is independent of latitude.

Internal waves are ubiquitous in the atmosphere and the ocean. In the lower atmosphere turbulent motions dominate, so that internal wave activity represents a minor component of the motion. In contrast, the stratosphere contains a great deal of internal wave activity, and very little convective motion, because of its stable density distribution. Internal waves generally propagate upward from the lower atmosphere, where they are generated. In the ocean, internal waves are as common as the waves on the surface, and measurements show that they can cause the isotherms to go up and down by tens of meters. Sometimes the internal waves break and generate smaller-scale turbulence in a somewhat similar manner to bubble and foam generation by breaking surface waves.

The equations of motion for linear internal waves including Coriolis effects are the continuity equation for incompressible flow (4.10); the linearized and frictionless momentum equations (12.48), (12.49), and:

$$\frac{\partial w}{\partial t} = -\frac{1}{\rho_0}\frac{\partial p'}{\partial z} - \frac{g\rho'}{\rho_0}, \tag{12.95}$$

which is obtained from (12.9); and the linearized density equation (12.51). Here the hydrostatic assumption is not made because the horizontal and vertical extent of the wave field may be comparable. And, to be somewhat more general than in Section 11.8, let the buoyancy frequency, $N(z)$, be depth dependent because internal wave activity is more intense near the thermocline where N varies appreciably (Fig. 12.2).

An equation for the vertical velocity w can be derived from the set (4.10), (12.48), (12.49), (12.51), (12.95) by eliminating all other dependent variables (see Exercise 12.8). The derivation is similar to that provided in Section 11.8 but here includes Coriolis terms, and produces:

$$\frac{\partial^2}{\partial t^2}\nabla^2 w + N^2\nabla_{\mathrm{H}}^2 w + f^2\frac{\partial^2 w}{\partial z^2} = 0, \tag{12.96}$$

where $\nabla_{\mathrm{H}}^2 \equiv \partial^2/\partial x^2 + \partial^2/\partial y^2 = \nabla^2 - \partial^2/\partial z^2$. Because the coefficients in (12.96) are independent of the horizontal directions, (12.96) can have solutions that are trigonometric in x and y. Therefore, assume a complex exponential traveling-wave solution of the form:

$$(u, v, w) = (\widehat{u}(z), \widehat{v}(z), \widehat{w}(z))\exp\{i(kx + ly - \omega t)\}. \tag{12.97}$$

Substitution of (12.97) into (12.96) leads to an ordinary differential equation:

$$(-i\omega)^2\left[(ik)^2 + (il)^2 + \frac{d^2}{dz^2}\right]\widehat{w} + N^2\left[(ik)^2 + (il)^2\right]\widehat{w} + f^2\frac{d^2\widehat{w}}{dz^2} = 0, \tag{12.98}$$

that can be simplified to:

$$\frac{d^2\widehat{w}}{dz^2} + m^2(z)\widehat{w} = 0, \quad \text{where} \quad m^2(z) \equiv \frac{(k^2 + l^2)(N^2(z) - \omega^2)}{\omega^2 - f^2}. \tag{12.99, 12.100}$$

For $m^2 < 0$, the solutions of (12.99) must be exponentially decaying (evanescent) with increasing depth signifying that the resulting wave motion is surface-trapped and corresponds to a surface wave propagating horizontally. For $m^2 > 0$, the solutions of (12.99) are trigonometric in z and correspond to internal waves propagating vertically as well as horizontally. From (12.100), therefore, internal waves are only possible in the frequency range $f < \omega < N$, provided that $N > f$, as is true for much of the atmosphere and the ocean.

WKB solution

Even though exact analytical solutions of (12.99) are only known for specific $m^2(z)$, approximate solutions are possible when $N(z)$ changes mildly over a vertical wavelength. Thus, consider only those internal waves whose vertical wavelength is short compared to the scale of variation of N. If H is a characteristic vertical distance over which N varies appreciably, then this restriction means $Hm \gg 1$. For such slowly varying $N(z)$, $m(z)$ given by (12.100) should also be slowly varying, that is, $m(z)$ changes by a small fraction in a distance $1/m$. Under this assumption the internal waves *locally* behave like plane waves, as if m is constant. This is the so-called *WKB approximation* (after Wentzel-Kramers-Brillouin), which applies when the properties of the medium (in this case N) are slowly varying.

To derive the approximate WKB solution of (12.99), look for a solution in the form:

$$\widehat{w}(z) = A(z)\exp\{i\varphi(z)\},$$

where the phase ϕ and the (slowly varying) amplitude A are real functions. No generality is lost by assuming A to be real since its complex phase may be included in ϕ. Substitution into (12.99) gives:

$$\frac{d^2A}{dz^2}+A\left[m^2-\left(\frac{d\phi}{dz}\right)^2\right]+2i\frac{dA}{dz}\frac{d\phi}{dz}+iA\frac{d^2\phi}{dz^2}=0.$$

Equating the real and imaginary parts of this equation leads to:

$$\frac{d^2A}{dz^2}+A\left[m^2-\left(\frac{d\phi}{dz}\right)^2\right]=0, \quad \text{and} \quad 2\frac{dA}{dz}\frac{d\phi}{dz}+A\frac{d^2\phi}{dz^2}=0. \tag{12.101, 12.102}$$

In (12.101), the term d^2A/dz^2 is negligible because it is small compared to the second term:

$$\frac{d^2A/dz^2}{Am^2}\sim\frac{1}{H^2m^2}\ll 1.$$

Eq. (12.101) then becomes approximately:

$$d\phi/dz=\pm m, \tag{12.103}$$

which has solution:

$$\phi=\pm\int^z m(z')\,dz',$$

the lower limit of the integral being arbitrary.

The amplitude is determined by writing (12.102) in the form:

$$\frac{dA}{A}=-\frac{(d^2\phi/dz^2)}{2(d\phi/dz)}dz=-\frac{(dm/dz)}{2m}dz=-\frac{1}{2}\frac{dm}{m},$$

where (12.103) has been used. Integrating and exponentiating leads to:

$$\ln A=-\frac{1}{2}\ln m+const., \quad \text{and then to} \quad A=\frac{A_0}{\sqrt{m}},$$

where A_0 is a constant. The WKB solution of (12.99) is therefore:

$$\widehat{w}(z)=\frac{A_0}{\sqrt{m(z)}}\exp\left\{\pm i\int^z m(z')dz'\right\}. \tag{12.104}$$

Because of neglect of the β-effect, the waves must behave similarly in x and y, as indicated by the symmetry of the dispersion relation (12.100) in k and l. Therefore, no generality is lost by orienting the x-axis in the direction of propagation and taking $k>0$, $l=0$, and $\omega>0$. In this case, u (and v) can be found from w by using the continuity equation $\partial u/\partial x+\partial w/\partial z=0$, noting that the y-derivatives are zero because of the choice $l=0$. Substituting the assumed solution (12.97) with $l=0$ into the continuity equation gives:

$$ik\widehat{u}+d\widehat{w}/dz=0. \tag{12.105}$$

The z-derivative of $\widehat{w}$ in (12.104) can be obtained by treating the denominator $\sqrt{m}$ as approximately constant because the variation of $\widehat{w}$ is dominated by the oscillatory behavior of the local plane wave solution. This gives:

$$\frac{d\widehat{w}}{dz}=\left(\frac{A_0}{2}\frac{(dm/dz)}{(m)^{3/2}}+\frac{A_0(\pm im)}{\sqrt{m}}\right)\exp\left\{\pm i\int^z m\,dz'\right\}\cong\pm iA_0\sqrt{m}\exp\left\{\pm i\int^z m\,dz'\right\},$$

so that (12.105) becomes:

$$\widehat{u}=\mp\frac{A_0\sqrt{m}}{k}\exp\left\{\pm i\int^z m\,dz'\right\}. \tag{12.106}$$

An expression for $\widehat{v}$ can now be obtained from the horizontal equations of motion (12.48) and (12.49) by cross-differentiating and adding, to obtain a linearized vorticity equation:

$$\frac{\partial}{\partial t}\left(\frac{\partial v}{\partial x}-\frac{\partial u}{\partial y}\right)+f\left(\frac{\partial u}{\partial x}+\frac{\partial v}{\partial y}\right)=0. \tag{12.107}$$

Using the wave solution (12.97) with $l=0$, this reduces to $\widehat{u}/\widehat{v}=i\omega/f$ so (12.106) requires:

$$\widehat{v}=\pm\frac{if}{\omega}\frac{A_0\sqrt{m}}{k}\exp\left\{\pm i\int^z m dz'\right\}. \tag{12.108}$$

The velocity field is then found from the real parts of (12.104), (12.106), and (12.108):

$$u=\mp\frac{A_0\sqrt{m}}{k}\cos\mathbf{\Phi},\quad v=\mp\frac{A_0 f\sqrt{m}}{\omega k}\sin\mathbf{\Phi},\quad\text{and}\quad w=\frac{A_0}{\sqrt{m}}\cos\mathbf{\Phi},$$
$$\text{where } \mathbf{\Phi}=kx\pm\int^z m(z')dz'-\omega t, \tag{12.109}$$

and the dispersion relation is (12.100) with $l=0$. The meaning of $m(z)=\partial\mathbf{\Phi}/\partial z$ is now clear, it is the vertical wave number at depth z. In addition, for k, m, $\omega>0$, the *upper signs* in (12.109) represent waves with *upward* phase propagation, and the *lower signs* represent waves with *downward* phase propagation.

Particle orbit

To find the shape of the hodograph in the horizontal plane, evaluate (12.109) at $x=z=0$:

$$u=\mp\frac{A_0\sqrt{m}}{k}\cos(\omega t),\quad\text{and}\quad v=\pm\frac{A_0\sqrt{m}}{k}\frac{f}{\omega}\sin(\omega t). \tag{12.110}$$

Taking the upper signs in (12.110), the values of u and v are indicated in Fig. 12.22a for three values of time corresponding to $\omega t=0$, $\pi/2$, and π. It is clear that the horizontal hodographs are clockwise ellipses, with the major axis in the direction of phase-front propagation x, and the axis ratio is f/ω. The same conclusion applies for the lower signs in (12.110). The particle orbits in the horizontal plane are therefore identical to those of Poincaré waves (Fig. 12.15).

However, the plane of the motion is no longer horizontal. From the velocity component equations (12.109), we note that:

$$u/w=\mp m/k=\mp\tan\theta, \tag{12.111}$$

where $\theta=\tan^{-1}(m/k)$ is the angle made by the wave number vector $\mathbf{K}$ with the horizontal (Fig. 12.23). For upward phase propagation, (12.111) gives $u/w=-\tan\theta$, so that w is negative if u is positive, as indicated in Fig. 12.23. A three-dimensional sketch of the particle orbit is shown in Fig. 12.22b. It can be shown (Exercise 12.9) that the phase velocity vector $\mathbf{c}$ is in the direction of $\mathbf{K}$, that $\mathbf{c}$ and $\mathbf{c}_g$ are perpendicular, and that the fluid motion $\mathbf{u}$ is parallel to $\mathbf{c}_g$; these facts are discussed in Chapter 11 for internal waves unaffected by the Coriolis acceleration.

The velocity vector at any location rotates clockwise with time. Because of the vertical propagation of phase, the tips of the *instantaneous* vectors also turn with *depth*. Consider the turning of the velocity vectors with depth when the phase velocity is upward, so that the deeper currents have a phase lead over the shallower currents (Fig. 12.24). Because the currents at all depths rotate clockwise in *time* (whether the vertical component of $\mathbf{c}$ is upward or downward), it follows that the tips of the instantaneous velocity vectors should fall on a helical spiral that turns clockwise with *depth*. Only such a turning in depth, coupled with a clockwise rotation of the velocity vectors with time, can result in a phase lead of the deeper currents. In the opposite case of a *downward* phase propagation, the helix turns *counterclockwise* with depth. The direction of turning of the velocity vectors can also be found from (12.109), by considering $x=t=0$ and finding u and v at various values of z.

Discussion of the dispersion relation

The dispersion relation (12.100) with $l=0$ can be written:

$$\omega^2-f^2=\frac{k^2}{m^2}(N^2-\omega^2). \tag{12.112}$$

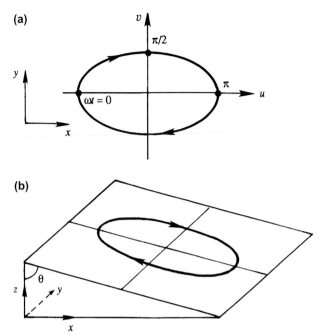

FIGURE 12.22 Particle orbit in an internal wave having x-direction wave number $k \neq 0$, and y-direction wave number $l = 0$. The upper panel (a) shows a projection on a horizontal plane; points corresponding to $\omega t = 0$, $\pi/2$, and π are indicated. The sense of rotation is the same as that of the surface-gravity-wave particle orbit shown in Fig. 12.15 and is valid for the northern hemisphere. The lower panel (b) shows a three-dimensional view of the orbit.

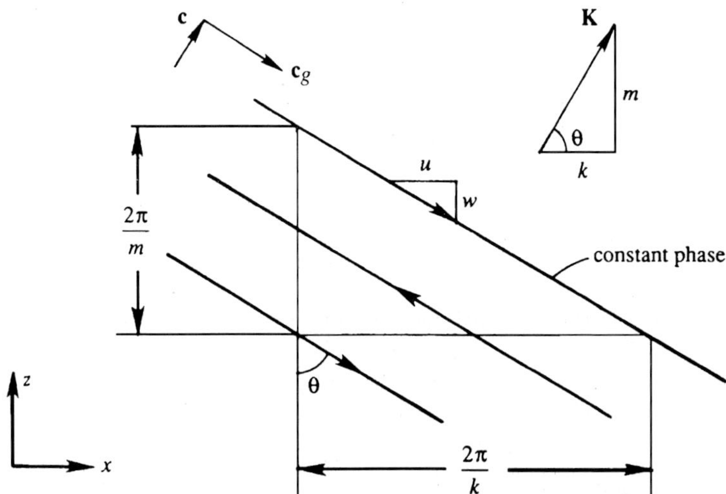

FIGURE 12.23 Vertical section of an internal wave. The three parallel lines are constant phase lines corresponding to one full wavelength, with the arrows indicating fluid motion along the lines. The phase velocity is perpendicular to these lines. The group velocity is parallel to these lines. The angle θ of the wave number with respect to the horizontal depends on the wave frequency ω, the buoyancy frequency N, and the local inertial frequency f.

Introducing $\tan\theta = m/k$, (12.112) becomes:

$$\omega^2 = f^2 \sin^2\theta + N^2 \cos^2\theta,$$

which shows that ω is a function of the angle made by the wave number with the horizontal and is not a function of the magnitude of $\mathbf{K}$. For $f = 0$ the aforementioned expression reduces to $\omega = N\cos\theta$, derived in Section 11.8 without the Coriolis acceleration.

A plot of the dispersion relation (12.112) is presented in Fig. 12.25, showing ω as a function of k for various values of m. All curves pass through the point $\omega = f$, which represents inertial oscillations. Typically, $N \gg f$ in most of the atmosphere

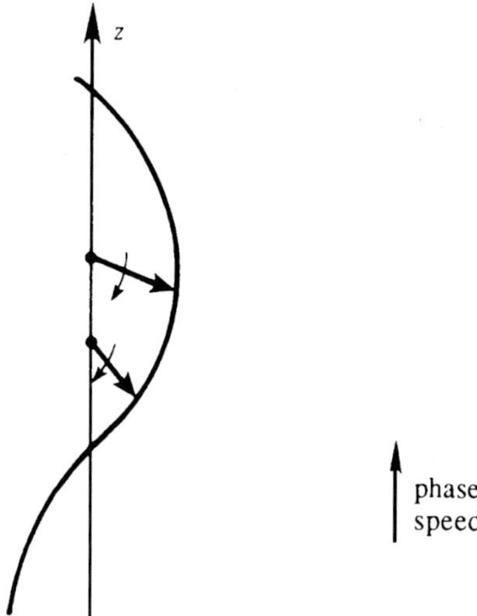

FIGURE 12.24 Helical-spiral traced out by the tips of instantaneous velocity vectors in an internal wave with upward phase speed. Heavy arrows show the velocity vectors at two depths, and light arrows indicate that they are rotating clockwise with increasing time. Note that the instantaneous vectors turn clockwise with increasing depth.

and the ocean. Because of the wide separation of the upper and lower limits of the internal wave range $f \leq \omega \leq N$, various limiting cases are possible, as indicated in Fig. 12.25. They are:

(1) *High-frequency regime* ($\omega \sim N$, *but* $\omega \leq N$): In this range f^2 is negligible in comparison with ω^2 in the denominator of the dispersion relation (12.109), which reduces to:

$$m^2 \cong \frac{k^2}{\omega^2}(N^2 - \omega^2) \quad \text{or} \quad \omega^2 \cong \frac{N^2 k^2}{m^2 + k^2}.$$

Using $\tan\theta = m/k$, this gives $\omega = N\cos\theta$. Thus, high-frequency internal waves are the same as the non-rotating internal waves discussed in Chapter 11.

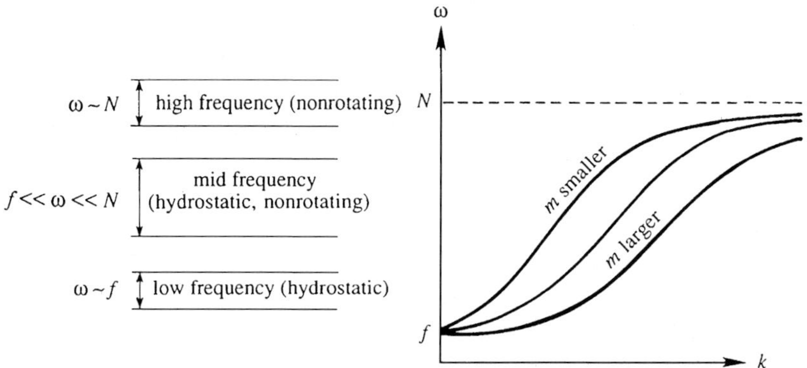

FIGURE 12.25 Dispersion relation for internal waves. The different regimes are indicated on the left-hand side of the figure. The wave frequency ω increases monotonically with increasing horizontal wave number k. The buoyancy frequency N and the local inertial frequency f set the upper and lower limits for ω.

(2) *Low-frequency regime* ($\omega \sim f$, but $\omega \geq f$): In this range, ω^2 can be neglected in comparison to N^2 in the dispersion relation (12.109), which becomes:

$$m^2 \cong \frac{k^2 N^2}{\omega^2 - f^2} \quad \text{or} \quad \omega^2 \cong f^2 + \frac{k^2 N^2}{m^2}.$$

The low-frequency limit is obtained by making the hydrostatic assumption, that is, neglecting $\partial w/\partial t$ in the vertical equation of motion.

(3) *Mid-frequency regime* ($f \ll \omega \ll N$): In this range the dispersion relation (12.100) with $l = 0$ simplifies to:

$$m^2 \cong k^2 N^2/\omega^2,$$

so that *both* the hydrostatic and the non-rotating assumptions are applicable.

Lee wave

Internal waves in the atmosphere are frequently found in the *lee* (that is, the downstream side) of mountains. In stably stratified conditions, the flow of air over a mountain causes a vertical displacement of fluid particles, which sets up internal waves as the air moves downstream of the mountain. If the amplitude is large and the air is moist, the upward motion causes condensation and cloud formation.

Due to the effect of a mean flow, such lee waves are stationary with respect to the ground. This is shown in Fig. 12.26, where the westward phase speed is canceled by the eastward mean flow. We shall determine what wave parameters make this cancellation possible. The frequency of lee waves is much larger than f, so that rotational effects are negligible. The dispersion relation is therefore:

$$\omega^2 = \frac{N^2 k^2}{m^2 + k^2}. \tag{12.113}$$

However, we now have to introduce the effects of the mean flow. The dispersion relation (12.113) is still valid if ω is interpreted as the *intrinsic frequency*, that is, the frequency measured in a frame of reference moving with the mean flow. In a medium moving with a velocity $\mathbf{U}$, the *observed frequency* of waves at a fixed point is Doppler shifted to:

$$\omega_0 = \omega + \mathbf{K} \cdot \mathbf{U},$$

where ω is the intrinsic frequency; this is discussed further in Section 11.1. For a stationary wave $\omega_0 = 0$, which requires that the intrinsic frequency is $\omega = -\mathbf{K} \cdot \mathbf{U} = kU$. (Here $-\mathbf{K} \cdot \mathbf{U}$ is positive because $\mathbf{K}$ is westward and $\mathbf{U}$ is eastward.) The dispersion relation (12.113) then gives:

$$U = \frac{N}{\sqrt{k^2 + m^2}}.$$

If the flow speed U is given, and the mountain introduces a typical horizontal wave number k, then the preceding equation determines the vertical wave number m that generates stationary waves. Waves that do not satisfy this condition would radiate away.

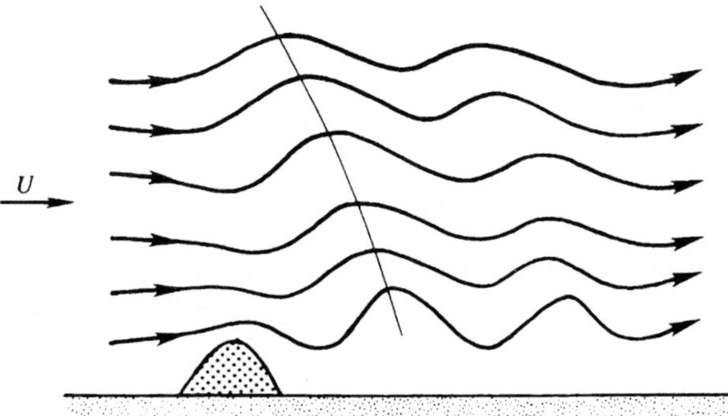

FIGURE 12.26 Schematic streamlines in a lee wave downstream of mountain. The thin line drawn through crests shows that the phase propagates downward and westward when the eastward velocity U is accounted for.

The energy source of lee waves is at the surface. The energy therefore must propagate upward, and consequently the phases propagate downward. The intrinsic phase speed is therefore westward and downward as shown in Fig. 12.26. With this information, we can determine which way the constant phase lines should tilt in a stationary lee wave. Note that the wave pattern in Fig. 12.26 would propagate to the left in the absence of a mean velocity, and only with the constant phase lines tilting backward with height would the flow at larger height lead the flow at a lower height.

Further discussion of internal waves can be found in Phillips (1977) and Munk (1981); lee waves are discussed in Holton (1979).

Example 12.12

The buoyancy frequency in the ocean is typically highest near but not at the surface (see Fig. 12.2) and then decreases with depth. A simple profile the embodies this decrease is:

$$N^2(z) = \left\{ \begin{array}{ccc} N_0^2(1 + z/z_0) & \text{for} & -z_0 < z \le 0 \\ 0 & \text{for} & z < -z_0 \end{array} \right\}.$$

Use this profile of $N^2(z)$ to determine the equation of the constant phase curve $\Phi(x, z)$ that intersects the origin for an internal wave with frequency ω.

Solution The final equation of (12.109) defines the phase of an internal wave within the WKB approximation. For consistency with the constant-phase lines shown in Fig. 12.23, chose the '+' sign in this definition. Here, the lower limit of the integral of $m(z)$ and the time t must be chosen so that the constant phase curve intersects the origin. Thus, the requisite curve is given by:

$$C = kx + \int_z m(z')\,dz' = kx + \frac{k}{\sqrt{\omega^2 - f^2}} \int \sqrt{N_0^2 - \omega^2 + (N_0^2/z_0)z}\,dz,$$

where C is an integration constant. Evaluation of the integral leads to:

$$C = kx + 2k\frac{\left[N_0^2 - \omega^2 + (N_0^2/z_0)z\right]^{3/2}}{3\sqrt{\omega^2 - f^2}(N_0^2/z_0)}.$$

Setting $x = z = 0$ in this equation allows C to be determined:

$$C = \frac{2kz_0\left[1 - \omega^2/N_0^2\right]^{3/2}}{3\sqrt{\omega^2/N_0^2 - f^2/N_0^2}}.$$

Eliminating C from the last two equations, along with some algebraic rearrangement, produces:

$$\frac{z}{z_0} = \left\{ \frac{3}{2}\left(\frac{2\left[1 - \omega^2/N_0^2\right]^{3/2}}{3\sqrt{\omega^2/N_0^2 - f^2/N_0^2}} - \frac{x}{z_o} \right)\sqrt{\omega^2/N_0^2 - f^2/N_0^2} \right\}^{2/3} - (1 - \omega^2/N_0^2),$$

a result that does not depend on the horizontal wave number k.

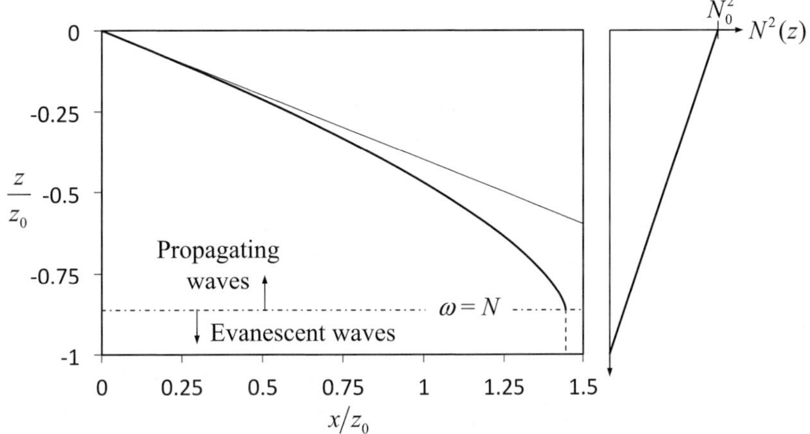

FIGURE 12.27 Internal-wave constant phase curve $\Phi(x, z)$ from (12.109) in normalized range, x/z_0, and depth, z/z_0, coordinates. The heavier solid curve is the constant phase curve for the depth-dependent buoyancy frequency $N(z)$ shown at the right when $f/N_0 = 0.01$ and $\omega/N_0 = 0.37$. The thinner solid line is the constant phase curve when $N(z) = N_0 =$ constant, as in Fig. 12.23. The depth where $\omega = N$ is indicated. Above this depth, internal waves with frequency ω can propagate; below this depth they are evanescent (exponentially decaying). The vertical dashed line indicates the extension of the constant phase curve into the evanescent region.

A sample constant phase curve is shown in Fig. 12.27 for $f/N_0 = 0.01$ and $\omega/N_0 = 0.37$. When N depends on depth, the constant phase curve is no longer linear, and, given that $\mathbf{K} = (k, m)$ is perpendicular to this curve, the phase velocity $\mathbf{c}$ of internal waves rotates toward the horizontal as the depth where $\omega = N$ is approached from above. Recalling that $\mathbf{c}_g$ is perpendicular to $\mathbf{c}$, $\mathbf{c}_g$ must be vertically upwards at the depth where $\omega = N$ since internal waves cannot propagate downward below this depth. Thus, internal waves are reflected from the depth where $\omega = N$.

12.13 Rossby wave

To this point, wave motions with a constant Coriolis frequency f have been considered and these waves all have frequencies larger than f. However, there are wave motions at lower frequencies that owe their existence to the variation of f with latitude. These waves are known as *Rossby waves*. Their spatial scales are so large in the atmosphere that they usually have only a few wavelengths around the entire globe (Fig. 12.28). This is why Rossby waves are also called *planetary waves*. In the ocean, however, their wavelengths are only about 100 km. Rossby-wave frequencies obey the inequality $\omega \ll f$. Because of this slowness, the time derivative terms are an order of magnitude smaller than the Coriolis acceleration and the pressure gradients in the horizontal equations of motion. Such *nearly* geostrophic flows are called *quasi-geostrophic motions*.

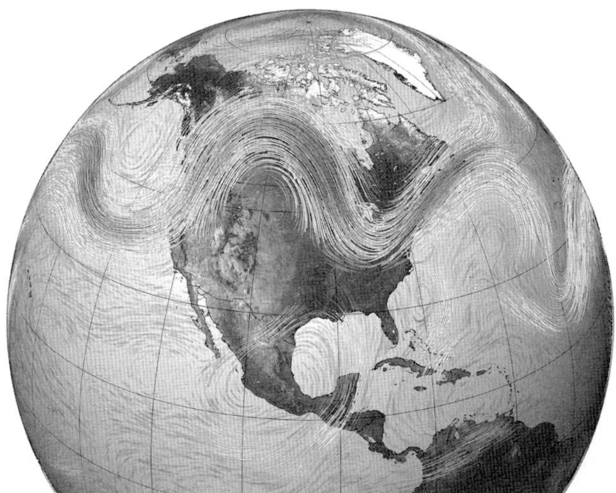

FIGURE 12.28 A model of the upper atmospheric circulation over the Northern Hemisphere using weather and climate observations from NASA's MERRA dataset, showing the northern jet stream. Contour lines represent wind speed (increasing speed from light to dark). The undulations are due to Rossby waves. Image credit: NASA/Goddard Space Flight Center Scientific Visualization Studio.

Quasi-geostrophic vorticity equation

The first step is to derive the governing equation for quasi-geostrophic motions using the customary β-plane approximation valid for $\beta y \ll f_0$, keeping in mind that the approximation is not an especially good one for atmospheric Rossby waves, which have planetary scales. Although Rossby waves are frequently superposed on a mean flow, the equations are derived here without a mean flow. Instead, a uniform mean flow is added at the end, assuming that the perturbations are small and that a linear superposition is valid. The first step is to simplify the vorticity equation for quasi-geostrophic motions, assuming that the *velocity is geostrophic to the lowest order*. The small departures from geostrophy, however, are important because they determine the *evolution* of the flow with time.

Start with the shallow-water potential vorticity equation (12.94), and rewrite it as:

$$h\frac{D}{Dt}(\zeta + f) - (\zeta + f)\frac{Dh}{Dt} = 0.$$

Expand the material derivatives and substitute $h = H + \eta$, where H is the uniform undisturbed depth of the layer, and η is the surface displacement. This gives:

$$(H+\eta)\left(\frac{\partial \zeta}{\partial t} + u\frac{\partial \zeta}{\partial x} + v\frac{\partial \zeta}{\partial y} + \beta v\right) - (\zeta + f_0)\left(\frac{\partial \eta}{\partial t} + u\frac{\partial \eta}{\partial x} + v\frac{\partial \eta}{\partial y}\right) = 0, \tag{12.114}$$

where $Df/Dt = v(df/dy) = \beta v$ has been used, and f has been replaced by f_0 in the second term because the usual β-plane approximation neglects the variation of f *except* for terms involving df/dy. For small perturbations, neglect the quadratic nonlinear terms in (12.114) to obtain:

$$H\frac{\partial \zeta}{\partial t} + H\beta v - f_0\frac{\partial \eta}{\partial t} = 0. \tag{12.115}$$

This is the linearized form of the potential vorticity equation. Its quasi-geostrophic version is obtained by inserting the approximate geostrophic expressions for velocity components:

$$u \cong -\frac{g}{f_0}\frac{\partial \eta}{\partial y}, \quad \text{and} \quad v \cong \frac{g}{f_0}\frac{\partial \eta}{\partial x}. \tag{12.116}$$

From these the vertical vorticity is found as:

$$\zeta = \frac{\partial v}{\partial x} - \frac{\partial u}{\partial y} = \frac{g}{f_0}\left(\frac{\partial^2 \eta}{\partial x^2} + \frac{\partial^2 \eta}{\partial y^2}\right),$$

so that the linearized potential vorticity equation (12.115) becomes:

$$\frac{gH}{f_0}\frac{\partial}{\partial t}\left(\frac{\partial^2 \eta}{\partial x^2} + \frac{\partial^2 \eta}{\partial y^2}\right) + \frac{gH\beta}{f_0}\frac{\partial \eta}{\partial x} - f_0\frac{\partial \eta}{\partial t} = 0. \tag{12.117}$$

Denoting $c^2 = gH$, this equation becomes:

$$\frac{\partial}{\partial t}\left(\frac{\partial^2 \eta}{\partial x^2} + \frac{\partial^2 \eta}{\partial y^2} - \frac{f_0^2}{c^2}\eta\right) + \beta\frac{\partial \eta}{\partial x} = 0, \tag{12.118}$$

which is the quasi-geostrophic form of the linearized vorticity equation for flow fields that span a significant range of latitude. The ratio c/f_0 is recognized as the Rossby radius. Note that $\partial\eta/\partial t$ was not set to zero in (12.115) to reach (12.118), although a strict validity of the geostrophic relations (12.116) would require that the horizontal divergence, and hence $\partial\eta/\partial t$, be zero. This is because the *departure* from strict geostrophy determines the evolution of the flows described by (12.118). The geostrophic relations for the velocity can be used everywhere except in the horizontal divergence term in the vorticity equation.

Dispersion relation

As for the prior wave motions considered in this chapter, assume the solutions of (12.118) will be in the form $\eta = \widehat{\eta}\exp\{i(kx + ly - \omega t)\}$, and regard ω as positive; the signs of k and l then determine the direction of phase propagation. Substituting this assumed solution form into (12.118) gives:

$$\omega = -\frac{\beta k}{k^2 + l^2 + f_0^2/c^2}. \tag{12.119}$$

This is the dispersion relation for *Rossby waves*. The asymmetry of the dispersion relation with respect to k and l signifies that the wave motion is not isotropic in the horizontal, as is expected because of the β-effect. Although (12.119) was derived for a single homogeneous layer, it is equally applicable to stratified flows if c is replaced by the corresponding *internal* value, which is $c^2 = g'H$ for the reduced-gravity model (see Section 11.7) and $c = NH/n\pi$ for the nth mode of a continuously stratified model. For the barotropic mode c is very large, so f_0^2/c^2 is usually negligible compared to other terms in the denominator of (12.119).

Using (12.119), $\omega(k, l)$ and can be displayed as a surface, taking k and l along Cartesian axes and plotting contours of constant ω. The section of this surface along $l = 0$ is indicated in the upper panel of Fig. 12.29, and contours of the surface for three values of ω are indicated in the bottom panel. These contours are circles because (12.119) can be written as:

$$\left(k + \frac{\beta}{2\omega}\right)^2 + l^2 = \left(\frac{\beta}{2\omega}\right)^2 - \frac{f_0^2}{c^2}.$$

In the lower panel of Fig. 12.29, the arrows perpendicular to the constant-ω contours indicate directions of group velocity vector $\mathbf{c}_g$:

$$\mathbf{c}_g = \mathbf{e}_x c_{gx} + \mathbf{e}_y c_{gy} = \mathbf{e}_x \frac{\partial \omega}{\partial k} + \mathbf{e}_y \frac{\partial \omega}{\partial l},$$

which is the gradient of ω in the wave number space. The direction of $\mathbf{c}_g$ is therefore perpendicular to the ω contours. For $l = 0$, the maximum frequency and zero group speed are attained at $kc/f_0 = -1$, corresponding to $\omega_{\max} f_0/\beta c = 0.5$. The maximum frequency is much smaller than the Coriolis frequency. For example, in the ocean the ratio $\omega_{\max}/f_0 = 0.5\beta c/f_0^2$ is of order 0.1 for the barotropic mode, and of order 0.001 for a baroclinic mode, taking a typical mid-latitude value of $f_0 \sim 10^{-4}\,\mathrm{s}^{-1}$, a barotropic gravity wave speed of $c \sim 200\,\mathrm{m/s}$, and a baroclinic gravity wave speed of $c \sim 2\,\mathrm{m/s}$. The shortest period of mid-latitude baroclinic Rossby waves in the ocean can therefore be more than a year.

The eastward phase speed is:

$$c_x = \frac{\omega}{k} = -\frac{\beta}{k^2 + l^2 + f_0^2/c^2}. \tag{12.120}$$

The negative sign shows that the *phase propagation is always westward.* The phase speed reaches a maximum when $k^2 + l^2 \to 0$, corresponding to very large wavelengths represented by the region near the origin of Fig. 12.29. In this region the waves are nearly non-dispersive and have an eastward phase speed $c_x \cong -\beta c^2/f_0^2$. With $\beta = 2 \times 10^{-11}\,\mathrm{m}^{-1}\,\mathrm{s}^{-1}$, a typical baroclinic value of $c \sim 2\,\mathrm{m/s}$, and a mid-latitude value of $f_0 \sim 10^{-4}\,\mathrm{s}^{-1}$, this gives $c_x \sim 10^{-2}\,\mathrm{m/s}$. At these slow speeds the Rossby waves would take years to cross the width of the ocean at mid-latitudes. Rossby waves in the ocean are therefore more important at lower latitudes, where they propagate faster. However, the dispersion relation (12.119), is not valid within a latitude band of 3° from the equator for which the assumption of a near geostrophic balance breaks down. A different analysis is needed in the tropics. A discussion of the wave dynamics of the tropics is given in Gill (1982) and in the review paper by McCreary (1985). In the atmosphere, c is much larger, and consequently the Rossby waves propagate faster. A typical large atmospheric disturbance can propagate as a Rossby wave at a speed of several meters per second.

Frequently, Rossby waves are superposed on a strong eastward mean current, such as the atmospheric jet stream. If U is the speed of this eastward current, then the observed eastward phase speed is:

$$c_x = U - \frac{\beta}{k^2 + l^2 + f_0^2/c^2}. \tag{12.121}$$

Stationary Rossby waves can therefore form when the eastward current cancels the westward phase speed, giving $c_x = 0$. This is how stationary waves are formed downstream of the topographic step in Fig. 12.20. A simple expression for the wavelength results if we assume $l = 0$ and the flow is barotropic, so that f_0^2/c^2 is negligible in (12.121). This gives $U = \beta/k^2$ for stationary Rossby waves, so that the wavelength is $2\pi[U/\beta]^{1/2}$.

Finally, the derivation of the quasi-geostrophic vorticity equation provided in this section has not been rigorously justified in the sense that approximations have been made without a formal ordering of the scales. Gill (1982) provides a more rigorous derivation, expanding in terms of a small parameter. Another way to justify the dispersion relation (12.119) is to obtain it from the general dispersion relation (12.76). For $\omega \ll f$, the first term in (12.76) is negligible compared to the third, and when this term is dropped (12.76) reduces to (12.119).

Example 12.13

If the west coast of North America experiences relatively clear and dry winter weather (a ridge in the 500 mb isobar height) while the center of the continent 3000 km to the east experiences relatively cool and cloudy conditions (a depression in the 500 mb isobar height), estimate the average eastward convection speed U of the atmosphere assuming these weather phenomena result from a stationary barotropic Rossby wave.

Solution For an estimate, use the simplified version of (12.121), $U = \beta/k^2 = \beta\lambda^2/(2\pi)^2$, and recognize that 3000 km represents the peak-to-valley distance or half of the Rossby wavelength. The numerical values then imply:

$$U = \frac{\beta\lambda^2}{(2\pi)^2} \cong \frac{(2 \times 10^{-11}\,\mathrm{m}^{-1}\mathrm{s}^{-1})(2 \cdot 3 \times 10^6\,\mathrm{m})^2}{4\pi^2} = 18\,\mathrm{ms}^{-1},$$

where a mid-latitude value of $\beta = 2 \times 10^{-11}\,\mathrm{m}^{-1}\,\mathrm{s}^{-1}$, has been used. Although this speed is lower than values typically associated with the polar jet stream (30 to 50 m/s) in the northern hemisphere, it is in the right range given that it should represent a latitude- and altitude-averaged atmospheric convection speed.

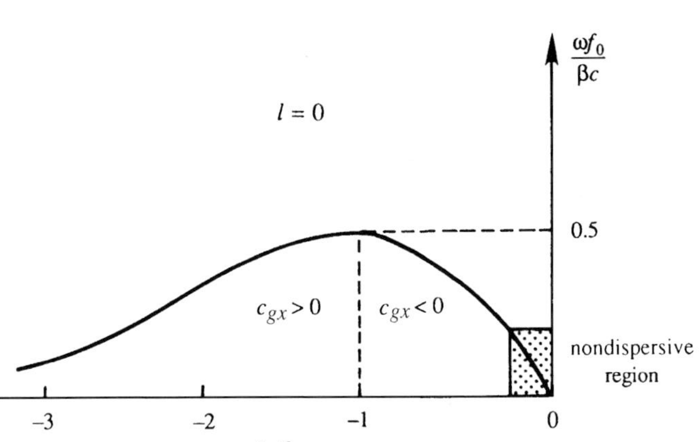

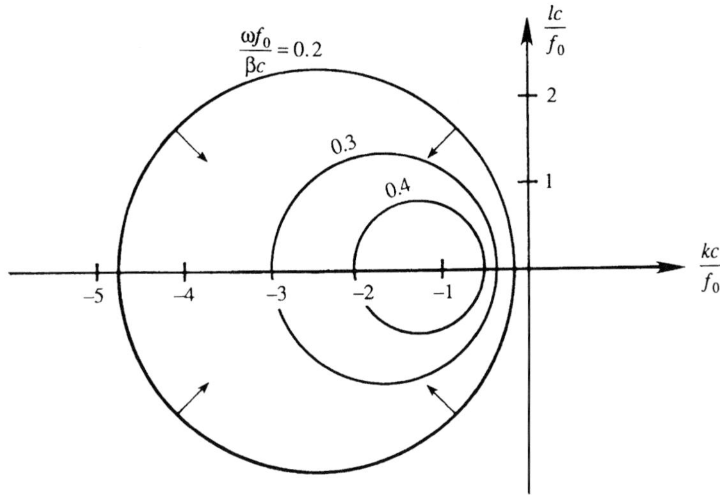

FIGURE 12.29 Dispersion relation $\omega(k, l)$ for a Rossby wave. The upper panel shows ω versus k for $l = 0$. Regions of positive and negative group velocity c_{gx} are indicated. The lower panel shows a plane view of the surface $\omega(k, l)$, showing contours of constant ω on a kl-plane. The values of $\omega f_0/\beta c$ for the three circles are 0.2, 0.3, and 0.4. Arrows perpendicular to the largest circular constant-ω contour indicate directions of the group velocity vector $\mathbf{c}_g$. *A.E. Gill,* Atmosphere–Ocean Dynamics, *1982; reprinted with the permission of Academic Press and Mrs. Helen Saunders-Gill.*

12.14 Barotropic instability

In Section 9.10, the inviscid stability of a shear flow $U(y)$ in a non-rotating frame of reference was analyzed and it was found that a necessary condition for instability is that d^2U/dy^2 must change sign somewhere in the flow. This condition is called *Rayleigh's inflection point criterion*. In terms of mean flow vorticity, $\overline{\zeta} = -dU/dy$, the criterion states that $d\overline{\zeta}/dy$ must change sign somewhere in the flow. That analysis is extended here to a rotating earth to find that the criterion requires that $d(\overline{\zeta} + f)/dy$ must change sign somewhere within the flow.

Consider a horizontal wind profile or current $U(y)$ in a medium of uniform density. In the absence of horizontal density gradients only the barotropic mode is allowed, and $U(y)$ does not vary with depth. The vorticity equation is:

$$\left(\frac{\partial}{\partial t} + \mathbf{u} \cdot \nabla\right)(\zeta + f) = 0, \tag{12.122}$$

which is (12.94), $D/Dt[(\zeta + f)/h] = 0$, with the added simplification that the layer depth h is constant because $w = 0$. Let the total flow be decomposed into a background flow plus a disturbance:

$$u = U(y) + u', \quad \text{and} \quad v = v'.$$

The total vorticity is then:

$$\zeta = \bar{\zeta} + \zeta' = -\frac{dU}{dy} + \left(\frac{\partial v'}{\partial x} - \frac{\partial u'}{\partial y}\right) = -\frac{dU}{dy} - \nabla^2\psi, \tag{12.123}$$

where ψ is the stream function for the disturbance, $u' = \partial\psi/\partial y$ and $v' = -\partial\psi/\partial x$, defined to be consistent with (6.3). Substituting these relationships into (12.122) and linearizing, leads to the perturbation vorticity equation:

$$\frac{\partial}{\partial t}(\nabla^2\psi) + U\frac{\partial}{\partial x}(\nabla^2\psi) + \left(\beta - \frac{d^2U}{dy^2}\right)\frac{\partial\psi}{\partial x} = 0. \tag{12.124}$$

Because the coefficients of (12.124) are independent of x and t, it's solutions can be of the form $\psi = \widehat{\psi}(y)\exp\{ik(x - ct)\}$. Here, the phase speed, $c = c_r + ic_i$, may be complex and solutions are unstable when $c_i > 0$. The perturbation vorticity equation (12.124) then becomes:

$$(U - c)\left[\frac{d^2}{dy^2} - k^2\right]\widehat{\psi} + \left[\beta - \frac{d^2U}{dy^2}\right]\widehat{\psi} = 0.$$

Comparing this with (9.98) derived without the Coriolis acceleration, the effect of planetary rotation is the replacement of $-d^2U/dy^2$ by $(\beta - d^2U/dy^2)$. The analysis of Section 9.10 therefore carries over to the present case, resulting in the following criterion: *A necessary condition for the inviscid instability of a barotropic current $U(y)$ is that the gradient of the absolute vorticity:*

$$\frac{d}{dy}(\bar{\zeta} + f) = -\frac{d^2U}{dy^2} + \beta, \tag{12.125}$$

must change sign somewhere in the flow, a result first derived by Kuo (1949).

Barotropic instability quite possibly plays an important role in the instability of currents in the atmosphere and in the ocean. The instability has no preference for any latitude, because the criterion involves β and not f. However, the mechanism presumably dominates in the tropics because mid-latitude disturbances prefer the *baroclinic instability* mechanism discussed in the following section. An unstable distribution of westward tropical wind is shown in Fig. 12.30.

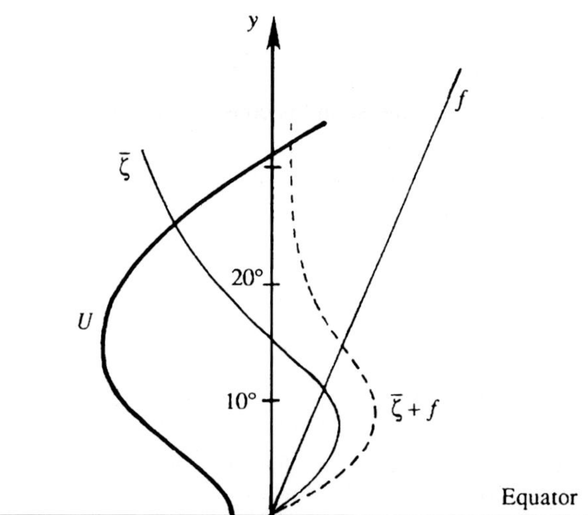

FIGURE 12.30 Profiles of velocity $U(y)$, vorticity $\bar{\zeta}$, and Coriolis parameter f in a westward tropical wind as a function of latitude. The velocity distribution is barotropically unstable as $d(\bar{\zeta} + f)/dy$ changes sign within the flow near 8°. *Houghton,* The Physics of the Atmosphere, *1986; reprinted with the permission of Cambridge University Press.*

12.15 Baroclinic instability

Weather maps at mid-latitudes invariably show the presence of wavelike horizontal excursions of temperature and pressure contours, superposed on eastward mean flows such as the jet stream. Similar undulations are also found in the ocean on eastward currents such as the Gulf Stream in the north Atlantic. A typical wavelength of these disturbances is observed to be of the order of the internal Rossby radius, that is, about 4000 km in the atmosphere and 100 km in the ocean. They seem to be propagating as Rossby waves, but their erratic and unexpected appearance suggests that they are not forced by any external agency, but are due to an inherent *instability* of mid-latitude eastward flows. In other words, the eastward flows have a spontaneous tendency to develop wavelike disturbances. In this section we shall investigate the instability mechanism that is responsible for the spontaneous meandering of large-scale eastward flows.

The poleward decrease of solar irradiation results in a poleward decrease of air temperature and a consequent increase of air density. An idealized distribution of the atmospheric density in the northern hemisphere is shown in Fig. 12.31. The density increases northward due to the lower temperatures near the poles and decreases upward because of static stability. According to the thermal wind relation (12.15), an eastward flow (such as the jet stream in the atmosphere or the Gulf Stream in the Atlantic) in equilibrium with such a density structure must have a velocity that increases with height. A system with inclined density surfaces, such as the one in Fig. 12.31, has more potential energy than a system with horizontal density surfaces, just as a system with an inclined free surface has more potential energy than a system with a horizontal free surface. Thus, this arrangement of atmospheric mass is possibly unstable because it can release the stored potential energy by means of an instability that would cause the density surfaces to flatten out. In the process, vertical shear of the mean flow $U(z)$ would decrease, and perturbations would gain kinetic energy.

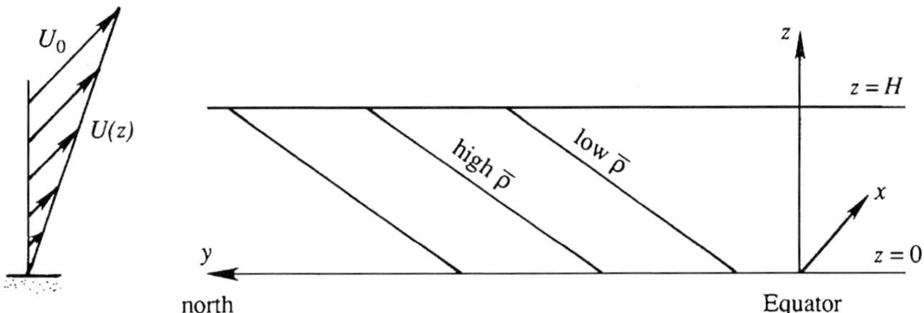

FIGURE 12.31 Lines of constant density in the northern hemispheric atmosphere. The lines are nearly horizontal and the slopes are greatly exaggerated in the figure. The velocity $U(z)$ shown at the left is into the plane of paper.

Instability of baroclinic flows that releases potential energy by flattening out constant density surfaces is called the *baroclinic instability*. The analysis provided here shows that the preferred scale of such unstable waves is indeed of the order of the Rossby radius, as observed for the mid-latitude weather disturbances. The theory of baroclinic instability was developed in the 1940s by Vilhem Bjerknes and others, and is considered one of the major triumphs of geophysical fluid mechanics. The presentation provided here is based on the review article by Pedlosky (1971).

Consider a simple basic state in which the density increases northward at a *constant* rate $\partial\bar{\rho}/\partial y$ and is stably stratified in the vertical with a *uniform* buoyancy frequency N. According to the thermal wind relation, the constancy of $\partial\bar{\rho}/\partial y$ requires that the vertical shear of the basic eastward flow $U(z)$ also be constant. The β-effect is neglected here since it is not essential for the instability. (The β-effect does modify the instability, however.) This is borne out by the spontaneous appearance of undulations in laboratory experiments in a rotating annulus, in which the inner wall is maintained at a higher temperature than the outer wall. The β-effect is absent in such an experiment.

Perturbation vorticity equation

The equations for the total flow are the continuity equation (4.10), the horizontal momentum equations of (12.9) simplified for frictionless flow with negligible vertical velocity, vertical hydrostatic equilibrium (1.14), and the density equation (4.9). The total flow is assumed to be composed of an eastward wind $U(z)$ in geostrophic equilibrium with the basic density structure $\bar{\rho}(y, z)$ shown in Fig. 12.31, plus perturbations:

$$\begin{aligned} u &= U(z) + u'(\mathbf{x}, t), \quad v = v'(\mathbf{x}, t), \quad w = w'(\mathbf{x}, t), \\ \rho &= \bar{\rho}(y, z) + \rho'(\mathbf{x}, t), \quad \text{and} \quad p = \bar{p}(y, z) + p'(\mathbf{x}, t). \end{aligned} \tag{12.126}$$

The basic flow is in geostrophic and hydrostatic balance:

$$fU = -\frac{1}{\rho_0}\frac{\partial \bar{p}}{\partial y}, \quad \text{and} \quad 0 = -\frac{\partial \bar{p}}{\partial z} - \bar{\rho}g. \qquad (12.127, 12.128)$$

Eliminating the pressure, we obtain the thermal wind relation:

$$\frac{dU}{dz} = \frac{g}{\rho_0 f}\frac{\partial \overline{\rho}}{\partial y}, \tag{12.129}$$

which requires the eastward flow to increase with height because $\partial\overline{\rho}/\partial y > 0$. Here, for simplicity, assume that $\partial\overline{\rho}/\partial y$ is constant, and that $U = 0$ at the surface $z = 0$. Thus, the background flow is:

$$U(z) = U_0 z/H,$$

where U_0 is the velocity at the top of the layer of interest, $z = H$.

Next form the vorticity equation by cross-differentiating and adding the frictionless horizontal momentum equations to eliminate the pressure. Then, use (4.10) to replace $\partial u/\partial x + \partial v/\partial y$ with $-\partial w/\partial z$. The result is:

$$\frac{\partial \zeta}{\partial t} + u\frac{\partial \zeta}{\partial x} + v\frac{\partial \zeta}{\partial y} - (\zeta + f)\frac{\partial w}{\partial z} = 0. \tag{12.130}$$

The development follows that leading to (12.92), except the β-effect is excluded here. Substitute the decompositions (12.126) into (12.130), drop nonlinear terms, and note that $\zeta = \zeta'$ because the basic flow $U = U_0 z/H$ has no vertical component of vorticity. After these steps, (12.130) becomes:

$$\frac{\partial \zeta'}{\partial t} + U\frac{\partial \zeta'}{\partial x} + - f\frac{\partial w'}{\partial z} = 0. \tag{12.131}$$

This is the perturbation vorticity equation, and it can be written in terms of p'.

Assume that the perturbations are large-scale and slow, so that the velocity is nearly geostrophic:

$$u' \cong -\frac{1}{\rho_0 f}\frac{\partial p'}{\partial y}, \quad \text{and} \quad v' \cong \frac{1}{\rho_0 f}\frac{\partial p'}{\partial x}, \tag{12.132}$$

from which the perturbation vorticity is found as:

$$\zeta' = \frac{1}{\rho_0 f}\left(\frac{\partial^2}{\partial x^2} + \frac{\partial^2}{\partial y^2}\right)p' = \frac{1}{\rho_0 f}\nabla_{\mathrm{H}}^2 p'. \tag{12.133}$$

Next, develop an expression for w' in terms of p' using the density equation (4.9):

$$\frac{\partial}{\partial t}(\overline{\rho} + \rho') + (U + u')\frac{\partial}{\partial x}(\overline{\rho} + \rho') + v'\frac{\partial}{\partial y}(\overline{\rho} + \rho') + w'\frac{\partial}{\partial z}(\overline{\rho} + \rho') = 0.$$

Evaluate derivatives and linearize, to obtain:

$$\frac{\partial \rho'}{\partial t} + U\frac{\partial \rho'}{\partial x} + v'\frac{\partial \overline{\rho}}{\partial y} - \frac{\rho_0 N^2 w'}{g} = 0, \tag{12.134}$$

where N^2 is given by (12.1). The pressure is presumed to be hydrostatic, so the perturbation density ρ' can be written in terms of p' by using (1.14) and subtracting the background state (12.128) to reach:

$$0 = -\partial p'/\partial z - \rho' g. \tag{12.135}$$

Substituting this into (12.134) leads to:

$$w' = -\frac{1}{\rho_0 N^2}\left[\left(\frac{\partial}{\partial t} + U\frac{\partial}{\partial x}\right)\frac{\partial p'}{\partial z} - \frac{dU}{dz}\frac{\partial p'}{\partial x}\right], \tag{12.136}$$

where (12.129) has been used to write $\partial\overline{\rho}/\partial y$ in terms of the thermal wind dU/dz. Using (12.133) and (12.136), the perturbation vorticity equation (12.131) becomes:

$$\left(\frac{\partial}{\partial t} + U\frac{\partial}{\partial x}\right)\left[\nabla_{\mathrm{H}}^2 p' + \frac{f^2}{N^2}\frac{\partial^2 p'}{\partial z^2}\right] = 0. \tag{12.137}$$

This is the equation that governs quasi-geostrophic perturbations on an eastward flow $U(z)$.

Wave solution

Assume that (12.137) has traveling wave solutions,

$$p' = \widehat{p}(z)\exp\{i(kx + ly - \omega t)\}, \tag{12.138}$$

confined between horizontal planes at $z = 0$ and $z = H$ that are unbounded in x and y. Real flows are likely to be bounded in the y direction, especially in a laboratory situation of flow in an annular channel, where the walls set boundary conditions parallel to the flow. Boundedness in y, however, simply sets up normal modes in the form $\sin(n\pi y/L)$, where L is the width of the channel. Each of these modes can be replaced by a periodicity in y.

Inserting (12.138) into (12.137), reduces (12.137) to an ordinary differential equation for $\widehat{p}$:

$$\frac{d^2\widehat{p}}{dz^2} + \alpha^2\widehat{p} = 0, \quad \text{where} \quad \alpha^2 \equiv \frac{N^2}{f^2}(k^2 + l^2). \tag{12.139, 12.140}$$

The solution of (12.139) can be written as:

$$\widehat{p} = A\cosh[\alpha(z - H/2)] + B\sinh[\alpha(z - H/2)], \tag{12.141}$$

and is completely specified when the boundary conditions $w' = 0$ at $z = 0$ & H are satisfied. The boundary conditions on p' corresponding to those on w' are found from (12.136) and $U(z) = U_0 z/H$:

$$\left(\frac{\partial}{\partial t} + \frac{U_0 z}{H}\frac{\partial}{\partial x}\right)\frac{\partial p'}{\partial z} - \frac{U_0}{H}\frac{\partial p'}{\partial x} = 0 \quad \text{at} \quad z = 0 \quad \text{and} \quad z = H.$$

In particular, these two boundary conditions are:

$$\frac{\partial^2 p'}{\partial t \partial z} - \frac{U_0}{H}\frac{\partial p'}{\partial x} = 0 \quad \text{at} \quad z = 0, \quad \text{and} \quad \frac{\partial^2 p'}{\partial t \partial z} + U_o\frac{\partial^2 p'}{\partial x \partial z} - \frac{U_0}{H}\frac{\partial p'}{\partial x} = 0 \quad \text{at} \quad z = H.$$

Instability criterion

Using (12.138) and (12.141), the two boundary conditions require:

$$A\left[\alpha c\sinh\frac{\alpha H}{2} - \frac{U_0}{H}\cosh\frac{\alpha H}{2}\right] + B\left[-\alpha c\cosh\frac{\alpha H}{2} + \frac{U_0}{H}\sinh\frac{\alpha H}{2}\right] = 0, \quad \text{and}$$

$$A\left[\alpha(U_0 - c)\sinh\frac{\alpha H}{2} - \frac{U_0}{H}\cosh\frac{\alpha H}{2}\right] + B\left[\alpha(U_0 - c)\cosh\frac{\alpha H}{2} - \frac{U_0}{H}\sinh\frac{\alpha H}{2}\right] = 0,$$

where $c = \omega/k$ is the eastward phase velocity.

This is a pair of homogeneous equations for the constants A and B. For non-trivial solutions to exist, the determinant of the coefficients must vanish. This gives, after some algebra, the phase velocity:

$$c = \frac{U_0}{2} \pm \frac{U_0}{\alpha H}\sqrt{\left(\frac{\alpha H}{2} - \tanh\frac{\alpha H}{2}\right)\left(\frac{\alpha H}{2} - \coth\frac{\alpha H}{2}\right)}. \tag{12.142}$$

Whether the solution grows with time depends on the sign of the radicand. The behavior of the functions under the radical sign is sketched in Fig. 12.32. It is apparent that the first factor in the radicand is positive because $\alpha H/2 > \tanh(\alpha H/2) for$ all values of αH. However, the second factor is negative for small values of αH for which $\alpha H/2 < \coth(\alpha H/2)$. In this range the roots of c are complex conjugates, with $c = U_0/2 \pm ic_i$. Because we have assumed that the perturbations are of the form $\exp(-ikct)$, the existence of a non-zero c_i implies the possibility of a perturbation that grows as $\exp(kc_i t)$, and the solution is unstable. The marginal stability is given by the critical value of α satisfying:

$$\alpha_c H/2 = \coth(\alpha_c H/2),$$

whose solution is $\alpha_c H = 2.4$, so the flow is unstable if $\alpha H < 2.4$. Using the definition of α in (12.140), it follows that the flow is unstable if:

$$HN/f < 2.4\Big/\sqrt{k^2 + l^2}\,.$$

Since all values of k and l are allowed, a value of $k^2 + l^2$ low enough to satisfy this inequality can always be found. *The flow is therefore always unstable (to low wave number disturbances).* For a north-south wave number $l = 0$, instability is ensured if the east-west wave number k is small enough such that:

$$HN/f < 2.4/k. \tag{12.143}$$

In a continuously stratified ocean, the speed of a long internal wave for the $n = 1$ baroclinic mode is $c = NH/\pi$, so that the corresponding internal Rossby radius is $c/f = NH/\pi f$. It is usual to omit the factor π and define the Rossby radius Λ in a continuously stratified fluid as:

$$\Lambda \equiv HN/f.$$

The condition (12.143) for baroclinic instability is therefore that the east-west wavelength be large enough so that $\lambda > 2.6\Lambda$.

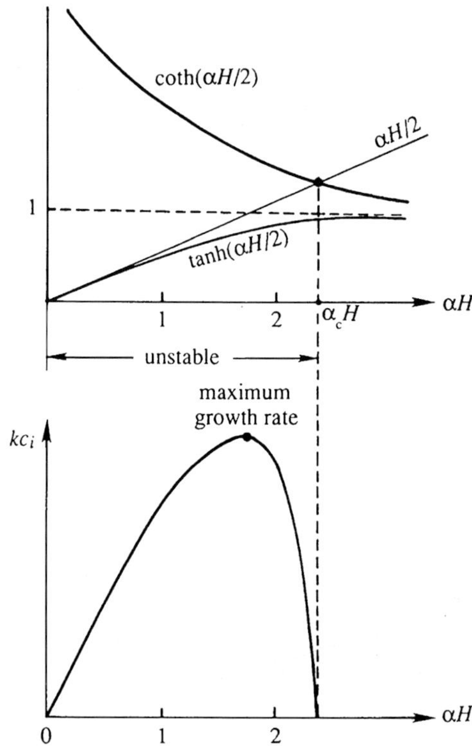

FIGURE 12.32 Baroclinic instability. The upper panel shows behavior of the functions in (12.142) and the lower panel shows growth rates of unstable waves.

However, the wavelength $\lambda = 2.6\Lambda$ does not grow at the fastest rate. It can be shown from (12.142) that the wavelength with the largest growth rate is:

$$\lambda_{\max} = 3.9\Lambda.$$

This is therefore the wavelength that is observed when the instability develops. Typical values for f, N, and H suggest that $\lambda_{\max} \sim 4000$ km in the atmosphere and 200 km in the ocean, which agree with observations. Waves much smaller than the Rossby radius do not grow, and the ones much larger than the Rossby radius grow very slowly.

Energetics

The foregoing analysis suggests that the existence of planet-encircling weather waves is due to the fact that small perturbations can grow spontaneously when superposed on an eastward current maintained by the sloping density surfaces (Fig. 12.31). Although the basic current does have a vertical shear, the perturbations do not grow by extracting energy from the vertical shear field. Instead, they extract their energy from the *potential energy* stored in the system of sloping density surfaces. The energetics of the baroclinic instability are therefore quite different than those of the Kelvin-Helmholtz instability where the perturbation Reynolds stress $\overline{u'w'}$ extracts energy from the mean-flow's vertical shear. The baroclinic instability is *not* a shear-flow instability; the Reynolds stresses are too small because of the small w' in quasi-geostrophic large-scale flows.

The energetics of the baroclinic instability can be understood by examining the equation for the perturbation kinetic energy. Such an equation can be derived by multiplying the equations for $\partial u'/\partial t$ and $\partial v'/\partial t$ by u' and v', respectively, adding the two, and integrating over the volume of the flow. Because of the assumed periodicity in x and y, the extent of the volume integration is appropriately confined to one wavelength in either direction. To complete this integration, the boundary conditions of zero normal flow on the upper and lower surfaces and periodicity in x and y are used repeatedly. The procedure is similar to that for the derivation of (9.105) and is not repeated here. The result is:

$$\frac{d}{dt}\left(\frac{\rho_0}{2}\int (u'^2 v'^2) dx\,dy\,dz\right) = \frac{dKE}{dt} = -g\int w'\rho' dx\,dy\,dz,$$

where KE is the global perturbation kinetic energy. In unstable flows, dKE/dt must be greater than zero, which requires the volume integral of $w'\rho'$ to be negative. Denote the volume average of $w'\rho'$ by $\overline{w'\rho'}$. A negative $\overline{w'\rho'}$ means that on average the lighter fluid rises and the heavier fluid sinks. By such an interchange the center of gravity of the system, and therefore its potential energy, is lowered. The interesting point is that this cannot happen in a stably stratified system with *horizontal* density surfaces; in that case an exchange of fluid particles *raises* the potential energy. Moreover, a basic state with inclined density surfaces (Fig. 12.31) cannot have $\overline{w'\rho'} < 0$ if the particle excursions are only vertical. If, however, the particle excursions include northward and southward displacements, and fall within the wedge formed by the constant density lines and the horizontal (Fig. 12.33), then an exchange of fluid particles takes lighter particles upward (and northward) and denser particles downward (and southward). Such an interchange would tend to make the density surfaces more horizontal, releasing potential energy from the mean density field with a consequent growth of the perturbation energy. This type of convection is called *sloping convection*. According to Fig. 12.33 the exchange of fluid particles within this *wedge of instability* results in a net poleward transport of heat from the tropics, which serves to redistribute the larger solar heat received by the tropics.

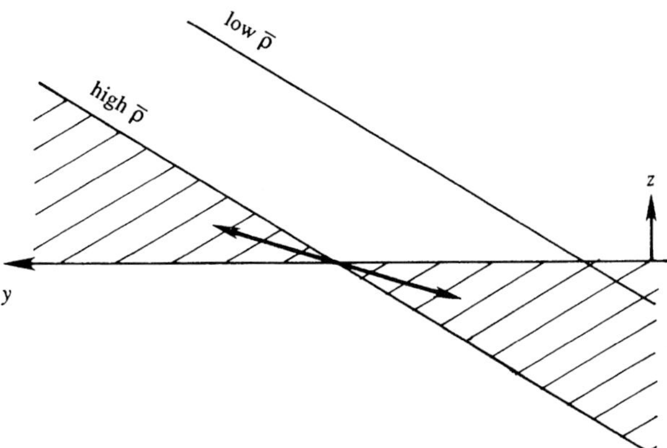

FIGURE 12.33 Wedge of instability (shaded) in a baroclinic instability. The wedge is bounded by constant density lines and the horizontal. Unstable waves have particle trajectories that fall within the wedge and cause lighter fluid particles to move upward and northward, and heaver fluid particles to move downward and southward.

In summary, baroclinic instability draws energy from the potential energy of the mean density field. The resulting eddy motion has particle trajectories that are oriented at a small angle with the horizontal, so that the resulting heat transfer has a poleward component. The preferred scale of the disturbance is the Rossby radius.

12.16 Geostrophic turbulence

Two common modes of instability of a large-scale wind or current system were presented in the preceding sections. When the flow is strong enough, such instabilities can cause a flow to become chaotic or turbulent. A peculiarity of large-scale turbulence in the atmosphere or the ocean is that it is essentially two dimensional in nature. The existence of the Coriolis acceleration, stratification, and the relatively small thickness of geophysical media severely restricts the vertical velocity in large-scale flows, which tend to be quasi-geostrophic, with the Coriolis acceleration balancing the horizontal pressure gradient to the lowest order. Because vortex stretching, a key mechanism by which ordinary three-dimensional turbulent flows transfer energy from large to small scales, is absent in two-dimensional flow, one expects that the dynamics of geostrophic turbulence are likely to be fundamentally different from that of three-dimensional, laboratory-scale turbulence discussed in Chapter 10. However, such motion can still be considered *turbulent* because it is unpredictable and diffusive.

A key result on the subject was discovered by the meteorologist Fjortoft (1953), and since then Kraichnan, Leith, Batchelor, and others have contributed to various aspects of the problem. A good discussion is given in Pedlosky (1987), to which the reader is referred for a fuller treatment. The present discussion merely highlights a few important results.

In two-dimensional turbulence, the vorticity, ζ, normal to the plane of fluid motion is of special interest and its mean square value, $\overline{\zeta^2}$, is known as *enstrophy*. In an isotropic turbulent field we can define an energy spectrum $S(K)$ so that:

$$\overline{u^2} = \int_0^\infty S(K)\,dK,$$

where K is the magnitude of the wave number. It can be shown that the enstrophy spectrum is $K^2S(K)$, so that:

$$\overline{\zeta^2} = \int_0^\infty K^2S(K)\,dK,$$

which makes sense because vorticity involves the spatial derivatives of velocity.

Consider a freely evolving turbulent field in which the shape of the velocity spectrum changes with time. The large scales are essentially inviscid, so that both energy and enstrophy are conserved (or nearly so):

$$\frac{d}{dt}\int_0^\infty S(K)dK \cong 0, \quad \text{and} \quad \frac{d}{dt}\int_0^\infty K^2S(K)dK \cong 0, \tag{12.144, 12.145}$$

where terms proportional to the molecular viscosity υ have been neglected on the right-hand sides of these equations. Enstrophy conservation is unique to two-dimensional turbulence because of the absence of vortex stretching.

Suppose that the energy spectrum initially contains all its energy at wave number K_0. Nonlinear interactions transfer this energy to other wave numbers, so that the sharp spectral peak smears out. For the sake of argument, suppose that all of the initial energy goes to two neighboring wave numbers K_1 and K_2, with $K_1 < K_0 < K_2$. Conservation of energy and enstrophy requires that:

$$S(K_0) = S(K_1) + S(K_2) \quad \text{and} \quad K_0^2S(K_0) = K_1^2S(K_1) + K_2^2S(K_2).$$

From this we can find the ratios of energy and enstrophy spectra after the transfer:

$$\frac{S(K_1)}{S(K_2)} = \frac{K_2 - K_0}{K_0 - K_1}\frac{K_2 + K_0}{K_1 + K_0}, \quad \text{and} \quad \frac{K_1^2S(K_1)}{K_2^2S(K_2)} = \frac{K_1^2}{K_2^2}\frac{K_2^2 - K_0^2}{K_0^2 - K_1^2}. \tag{12.146}$$

As an example, suppose that nonlinear smearing transfers energy to wave numbers $K_1 = K_0/2$ and $K_2 = 2K_0$. Then (12.146) shows that $S(K_1)/S(K_2) = 4$ and $K_1^2S(K_1)/K_2^2S(K_2) = 1/4$, so that more energy goes to lower wave numbers (large scales), whereas more enstrophy goes to higher wave numbers (smaller scales). This important result for two-dimensional turbulence was derived by Fjortoft (1953). Clearly, the constraint of enstrophy conservation in two-dimensional turbulence has prevented a symmetric spreading of the initial energy peak at K_0.

The unique character of two-dimensional turbulence is evident here. In three-dimensional turbulence, the primary topic of Chapter 10, the energy goes to smaller and smaller scales until it is dissipated by viscosity. In geostrophic turbulence, on the other hand, the energy goes to larger scales, where it is less susceptible to viscous dissipation. Numerical calculations are indeed in agreement with this behavior and show that energy-containing eddies grow in size by coalescing. On the other hand, the vorticity becomes increasingly confined to thin shear layers on the eddy boundaries; these shear layers contain very little energy. The backward (or inverse) energy cascade and forward enstrophy cascade are represented schematically in Fig. 12.34. It is clear that there are two *inertial* regions in the spectrum of a two-dimensional turbulent flow, namely, the energy cascade region and the enstrophy cascade region. If energy is injected into the system at a rate ε, then the energy spectrum in the energy cascade region has the form $S(K) \propto \varepsilon^{2/3}K^{-5/3}$; the argument is essentially the same as in the case of the Kolmogorov spectrum in three-dimensional turbulence (Section 10.7), except that the transfer is (backward) to lower wave numbers. A dimensional argument also shows that the energy spectrum in the enstrophy cascade region is of the form

$S(K) \propto \alpha^{2/3} K^{-3}$, where α is the forward enstrophy flux to higher wave numbers. There is negligible energy flux in the enstrophy cascade region.

As the eddies grow in size, they become increasingly immune to viscous dissipation, and the inviscid assumption implied in (12.144) becomes increasingly applicable. (This would not be the case in three-dimensional turbulence in which the eddies continue to decrease in size until viscous effects drain energy out of the flow.) In contrast, the corresponding assumption in the enstrophy conservation equation (12.145) becomes less and less valid as enstrophy goes to smaller scales, where viscous dissipation drains enstrophy out of the system. At later stages in the evolution, then, (12.145) may not be a good assumption. However, it can be shown (see Pedlosky, 1987) that the dissipation of enstrophy actually *intensifies* the process of energy transfer to larger scales, so that the *red* cascade (that is, transfer to larger scales) of energy is a general result of two-dimensional turbulence.

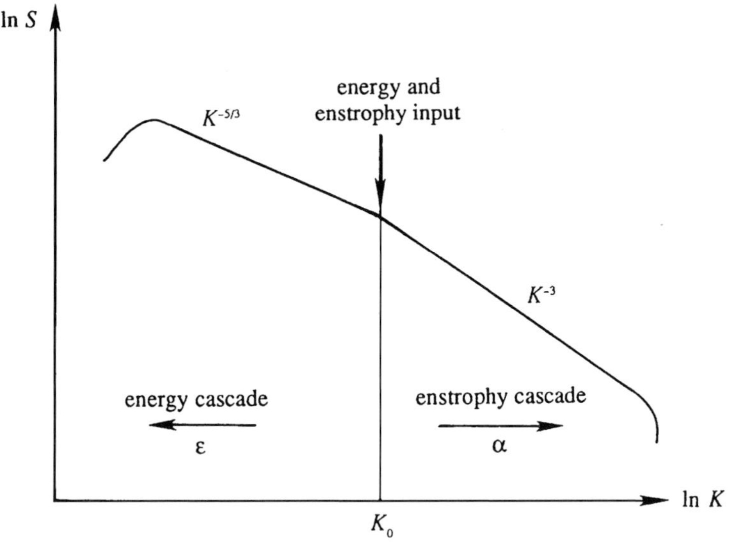

FIGURE 12.34 Energy and enstrophy cascade in two-dimensional turbulence. Here the two-dimensional character of the turbulence causes turbulent kinetic energy to cascade to larger scales, while enstrophy cascades to smaller scales.

The eddies, however, do not grow in size indefinitely. They become increasingly slower as their length scale l increases, while their velocity scale u remains constant. The slower dynamics makes them increasingly wavelike, and the eddies transform into Rossby-wave packets as their length scale becomes of order (Rhines, 1975):

$$l \sim \sqrt{u/\beta} \quad \text{(Rhines length)},$$

where $\beta = df/dy$ and u is the rms fluctuating speed. The Rossby-wave propagation results in an anisotropic elongation of the eddies in the east–west ("zonal") direction, while the eddy size in the north–south direction stops growing at $\sqrt{u/\beta}$. Finally, the velocity field consists of zonally directed jets whose north–south extent is of order $\sqrt{u/\beta}$. This has been suggested as an explanation for the existence of zonal jets in the atmosphere of the planet Jupiter (Williams, 1979). The inverse energy cascade regime may not occur in the earth's atmosphere and the ocean at mid-latitudes because the Rhines length (about 1000 km in the atmosphere and 100 km in the ocean) is of the order of the internal Rossby radius, where the energy is injected by baroclinic instability. (For the inverse cascade to occur, $\sqrt{u/\beta}$ needs to be larger than the scale at which energy is injected.)

Eventually, however, the kinetic energy has to be dissipated by molecular effects at the Kolmogorov microscale η, which is of the order of a few millimeters in the ocean and the atmosphere. A fair hypothesis is that processes such as internal waves drain energy out of the mesoscale eddies, and breaking internal waves generate three-dimensional turbulence that finally cascades energy to molecular scales.

Example 12.14

Eqs. (12.146) describe the reverse energy cascade of geostrophic turbulence in spectral terms. Redo this analysis by considering the merger of two large-scale cyclones idealized as identical disks of air with radius r_o rotating as solid bodies with rate Ω_o. Estimate the radius r_f and rotation rate Ω_f of the final cyclone if it also rotates as a solid body. How are these answers changed if half of the enstrophy is lost during the merger?

Solution For an atmosphere of height H, the kinetic energy of one disk of air with radius r_o undergoing solid body rotation with rate Ω_o is:

$$\frac{1}{2}\int_{volume}\rho|\mathbf{u}|^2 dV = \frac{1}{2}\int_0^H \rho \int_0^{r_0} (\Omega_o r)^2 2\pi r\, dr\, dz = \frac{\pi}{4}\overline{\rho}_o H \Omega_o^2 r_o^4,$$

where $\overline{\rho}_o$ is the altitude-averaged density. Thus, conservation of energy for two such disks merging into one implies:

$$2\cdot\frac{\pi}{4}\overline{\rho}_o H\Omega_o^2 r_o^4 = \frac{\pi}{4}\overline{\rho}_o H\Omega_f^2 r_f^4.$$

The vorticity inside each of the two initial disks is $2\Omega_o$, so conservation of enstrophy requires:

$$\frac{2\cdot\pi r_o^2}{A}(2\Omega_o)^2 = \frac{\pi r_f^2}{A}(2\Omega_f)^2,$$

where A is the relevant horizontal area for averaging the square of the vorticity. Simultaneous solution of these two equations leads to:

$$r_f = r_o \quad \text{and} \quad \Omega_f = \sqrt{2}\Omega_o.$$

Thus, the single merged cyclone is the same size as the original two but it rotates more quickly. From a spectral point of view, this merger represents a reduction in the wave number – even though the merged cyclone is the same size – because the final flow field contains one cyclonic event in a nominal horizontal distance of $A^{1/2}$ while the initial field contained two cyclonic events in the same distance.

When only half the enstrophy survives the merging process, the prior steps may be redone to find that the final cyclonic disk is larger and rotates slower than either of the original cyclonic disks.

$$r_f = \sqrt{2}r_o \quad \text{and} \quad \Omega_f = \Omega_o/\sqrt{2}.$$

Exercises

12.1 The Gulf Stream flows northward along the east coast of the United States with a surface current of average magnitude 2 m/s. If the flow is assumed to be in geostrophic balance, find the average slope of the sea surface across the current at a latitude of 45°N. [*Answer*: 2.1 cm per km]

12.2 An American football player kicks a 40 yard (36.58 meters) field goal in a north-south facing stadium located at latitude 40°. If the football is traveling at 20 m/s, how far will it deflect due to earth's rotation? Assume the speed of the football remains constant throughout its trajectory.

12.3 A plate with water ($\nu = 10^{-6}\,\text{m}^2/\text{s}$) above it rotates at a rate of 10 revolutions per minute. Find the depth of the Ekman layer, assuming that the flow is laminar.

12.4 Assume that the atmospheric Ekman layer over the earth's surface at a latitude of 45°N can be approximated by an eddy viscosity of $\nu_V = 10\,\text{m}^2/\text{s}$. If the geostrophic velocity above the Ekman layer is 10 m/s, what is the Ekman transport across isobars? [*Answer*: 2203 m^2/s]

12.5 **a)** From the set (12.45)–(12.47), develop the following equation for the water surface elevation $\eta(x, y, t)$:

$$\frac{\partial}{\partial t}\left\{\frac{\partial^2}{\partial t^2} + f^2 - gH\left(\frac{\partial^2}{\partial x^2} + \frac{\partial^2}{\partial y^2}\right)\right\}\eta(x, y, t) = 0$$

b) Using $\eta(x, y, t) = \hat{\eta}\exp\{i(kx + ly - \omega t)\}$ show that the dispersion relationship reduces to $\omega = 0$ or (12.82).

c) What type of flows have $\omega = 0$?

12.6 Find the axis ratio of a hodograph plot for a semi-diurnal tide in the middle of the ocean at a latitude of 45°N. Assume that the mid-ocean tides are rotational surface gravity waves of long wavelength and are unaffected by the proximity of coastal boundaries. If the depth of the ocean is 4 km, find the wavelength, the phase velocity, and the group velocity. Note, however, that the wavelength is comparable to the width of the ocean, so that the neglect of coastal boundaries is not very realistic.

12.7 An internal Kelvin wave on the thermocline of the ocean propagates along the west coast of Australia. The thermocline has a depth of 50 m and has a nearly discontinuous density change of 2 kg/m^3 across it. The layer below the thermocline is deep. At a latitude of 30°S, find the direction and magnitude of the propagation speed and the decay scale perpendicular to the coast.

12.8 Derive (12.96) for the vertical velocity w from (4.10), (12.48), (12.49), (12.51), (12.95) by eliminating all other dependent variables.

12.9 Using the dispersion relation $m^2 = k^2(N^2 - \omega^2)/(\omega^2 - f^2)$ for internal waves, show that the group velocity vector is given by $[c_{gx}, c_{gz}] = \dfrac{(N^2 - f^2)km}{(m^2 + k^2)^{3/2}(m^2 f^2 + k^2 N^2)^{1/2}}[m, -k]$. [*Hint*: Differentiate the dispersion relation partially with respect to k and m.] Show that $\mathbf{c}_g$ and $\mathbf{c}$ are perpendicular and have oppositely directed vertical components. Verify that $\mathbf{c}_g$ is parallel to $\mathbf{u}$.]

12.10 Suppose the atmosphere at a latitude of 45°N is idealized by a uniformly stratified layer of height 10 km, across which the potential temperature increases by 50 °C.

a) What is the value of the buoyancy frequency N?

b) Find the speed of a long gravity wave corresponding to the $n = 1$ baroclinic mode.

c) For the $n = 1$ mode, find the westward speed of non-dispersive (i.e., very large wavelength) Rossby waves. [*Answer*: $N = 0.01279\,\text{s}^{-1}$; $c_1 = 40.71\,\text{m/s}$; $c_x = -3.12\,\text{m/s}$]

12.11 Consider a steady flow rotating between plane parallel boundaries a distance L apart. The angular velocity is Ω and a small rectilinear velocity U is superposed. There is a protuberance of height $h \ll L$ in the flow. The Ekman and Rossby numbers are both small: $Ro \ll 1$, $E \ll 1$. Obtain an integral of the relevant equations of motion that relates the modified pressure and the streamfunction for the motion, and show that the modified pressure is constant on streamlines.

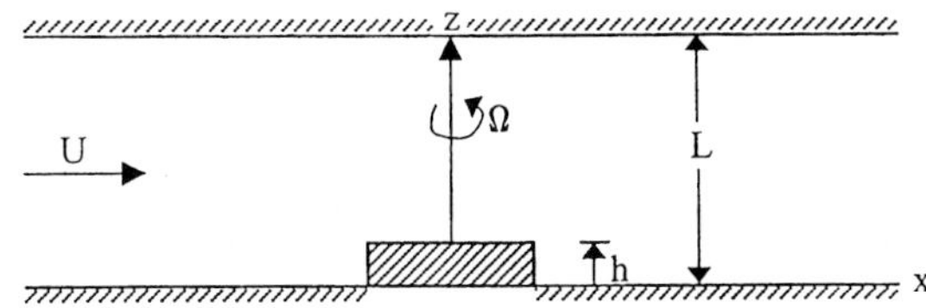

12.12 Consider an atmosphere of height H that initially contains quiescent air and N different cyclonic disks of height H and radius R_i inside which the air rotates at rate Ω_i. After some time, the various cyclonic disks merge into one because of the reverse energy cascade of geostrophic turbulence. Show that the radius R_f and rotation rate Ω_f of the single final disk is:

$$R_f^2 = \sum_{i=1}^{N} \Omega_i^2 R_i^4 \Big/ \sum_{i=1}^{N} \Omega_i^2 R_i^2 \quad \text{and} \quad \Omega_f^2 = \left(\sum_{i=1}^{N} \Omega_i^2 R_i^2\right)^2 \Big/ \sum_{i=1}^{N} \Omega_i^2 R_i^4$$

by conserving energy and enstrophy. How are these answers different if all of the energy but only a fraction $\varepsilon(0 < \varepsilon < 1)$ of the enstrophy is retained after the merging process? Assume the relevant horizontal area is the same at the start and end of the disk-merging process.

References

Fjortoft, R., 1953. On the changes in the spectral distributions of kinetic energy for two-dimensional non-divergent flow. Tellus 5, 225–230.

Gill, A.E., 1982. Atmosphere–Ocean Dynamics. Academic Press, New York.

Holton, J.R., 1979. An Introduction to Dynamic Meteorology. Academic Press, New York.

Houghton, J.T., 1986. The Physics of the Atmosphere. Cambridge University Press, London.

Kamenkovich, V.M., 1967. On the coefficients of eddy diffusion and eddy viscosity in large-scale oceanic and atmospheric motions. Izvestiya. Atmospheric and Oceanic Physics 3, 1326–1333.

Kundu, P.K., 1977. On the importance of friction in two typical continental waters: off Oregon and Spanish Sahara. In: Nihoul, J.C.J. (Ed.), Bottom Turbulence. Elsevier, Amsterdam.

Kuo, H.L., 1949. Dynamic instability of two-dimensional nondivergent flow in a barotropic atmosphere. Journal of Meteorology 6, 105–122.

LeBlond, P.H., Mysak, L.A., 1978. Waves in the Ocean. Elsevier, Amsterdam.

McCreary, J.P., 1985. Modeling equatorial ocean circulation. Annual Review of Fluid Mechanics 17, 359–409.

Munk, W., 1981. Internal waves and small-scale processes. In: Warren, B.A., Wunch, C. (Eds.), Evolution of Physical Oceanography. MIT Press, Cambridge, MA.

Pedlosky, J., 1971. Geophysical fluid dynamics. In: Reid, W.H. (Ed.), Mathematical Problems in the Geophysical Sciences. American Mathematical Society, Providence, Rhode Island.

Pedlosky, J., 1987. Geophysical Fluid Dynamics. Springer-Verlag, New York.

Phillips, O.M., 1977. The Dynamics of the Upper Ocean. Cambridge University Press, London.

Prandtl, L., 1952. Essentials of Fluid Dynamics. Hafner Publ. Co., New York.

Rhines, P.B., 1975. Waves and turbulence on a β-plane. Journal of Fluid Mechanics 69, 417–443.

Stommel, H., 1948. The westward intensification of wind-driven ocean currents. Transactions - American Geophysical Union 29 (2), 202–206.

Taylor, G.I., 1915. Eddy motion in the atmosphere. Philosophical Transactions of the Royal Society of London A 215, 1–26.

Williams, G.P., 1979. Planetary circulations: 2. The Jovian quasi-geostrophic regime. Journal of the Atmospheric Sciences 36, 932–968.

Further reading

Chan, J.C.L., 2005. The physics of tropical cyclone motion. Annual Review of Fluid Mechanics 37, 99–128.

Haynes, P., 2005. Stratospheric dynamics. Annual Review of Fluid Mechanics 37, 263–293.

Wiggins, S., 2005. The dynamical systems approach to Lagrangian transport in oceanic flows. Annual Review of Fluid Mechanics 37, 295–328.

Chapter 13

Aerodynamics⊛

Contents

13.1 Introduction 543
13.2 Aircraft terminology 544
Control surfaces 544
13.3 Characteristics of airfoil sections 547
Historical notes 551
13.4 Conformal transformation for generating airfoil shapes 552
13.5 Lift of a Zhukhovsky airfoil 555
13.6 Elementary lifting line theory for wings of finite span 558
Lanchester vs. Prandtl 564
13.7 Lift and drag characteristics of airfoils 565
13.8 Propulsive mechanisms of fishes and birds 567
13.9 Sailing against the wind 569
Exercises 570
References 574
Further reading 574

Chapter objectives

- To introduce the fundamental concepts and vocabulary associated with aircraft and aerodynamics
- To quantify the ideal-flow performance of simple two-dimensional airfoil sections
- To present the lifting line theory of Prandtl and Lanchester for a finite-span wing
- To describe the means by which fish, birds, insects, and sails exploit lift forces for flight and/or propulsion

13.1 Introduction

Aerodynamics is the branch of fluid mechanics that deals with the determination of the fluid mechanical forces and moments on bodies of interest moving through a gas (typically air). The subject is called *incompressible aerodynamics* if the flow speeds are low enough (Mach number < 0.3) for the compressibility effects to be negligible. At larger Mach numbers where gas-compressibility effects are important the subject is normally called *gas dynamics*. Aerodynamic parametric ranges of interest are usually consistent with: 1) neglecting buoyancy forces and fluid stratification, 2) assuming uniform constant-density flow upstream of the body, and 3) presuming viscous effects are confined to thin boundary layers adjacent to the body surface (Fig. 8.1). Airfoil stall is a notable exception to this last presumption.

This chapter emphasizes the elementary aspects of incompressible aerodynamics. Thus, with the simplifications just stated, the flows considered here are primarily ideal flows, and a significant portion of the material in Chapter 6 is relevant here. The aerodynamic force F on a moving body can be resolved into a *drag force* D parallel to the oncoming stream, and a *lift force* L perpendicular to the oncoming stream. The primary means for quantifying aerodynamic performance are the coefficients of drag and lift:

$$C_D \equiv \frac{D}{(1/2)\rho U^2 A}, \quad \text{and} \quad C_L \equiv \frac{L}{(1/2)\rho U^2 A}, \qquad (4.108, 4.109)$$

where A is a reference area that may be chosen differently for each coefficient. In addition, much of the material in this chapter also applies to ship propellers and to turbomachines (e.g., fans, turbines, compressors, and pumps) since the blades of these devices may all have similar cross-sections.

Example 13.1

A sphere with radius a moves along the x-axis on a trajectory given by $\mathbf{x}_p(t) = x_p(t)\mathbf{e}_x$ in a fluid moving with uniform velocity parallel to the y-axis: $\mathbf{u} = V\mathbf{e}_y$. Determine a formula for the mechanical power, W, necessary to overcome the aerodynamic drag force on the sphere in terms of a, V, x_p, $\rho =$ the density of the air, and $C_D =$ the drag coefficient of the sphere.

⊛. This book has a companion website hosting complementary materials. Visit this URL to access it: https://www.elsevier.com/books-and-journals/book-companion/9780128198070.

Fluid Mechanics. https://doi.org/10.1016/B978-0-12-819807-0.00022-3

Solution The drag force on the sphere acts in the same direction as the fluid velocity when observed from the sphere. For the situation described, the velocity of the fluid with respect to the sphere is: $-d\mathbf{x}_p/dt + V\mathbf{e}_y$. Thus, the drag force is:

$$\mathbf{D} = \frac{\pi a^2}{2}\rho C_D\left[\left(\frac{dx_p}{dt}\right)^2 + V^2\right]^{1/2}\left[-\frac{dx_p}{dt}\mathbf{e}_x + V\mathbf{e}_y\right],$$

where the area $A = \pi a^2$.

The dot product of $\mathbf{D}$ with $d\mathbf{x}_p/dt$ is the power delivered to the sphere by the fluid, so the power necessary for the sphere to overcome the drag force is:

$$W = -\mathbf{D}\cdot\left(\frac{d\mathbf{x}_p}{dt}\right) = \frac{\pi a^2}{2}\rho C_D\left[\left(\frac{dx_p}{dt}\right)^2 + V^2\right]^{1/2}\left(\frac{dx_p}{dt}\right)^2.$$

This result has at least two practical consequences. First of all, for steady rectilinear motion of the sphere, the presence of a pure cross-wind increases the power necessary to move the sphere. Thus, cross-winds can be anticipated to reduce the fuel economy of ground vehicles, even when their AC_D-product does not depend strongly on direction. Second, W depends only on $(dx_p/dt)^2$ and V^2, so a cross-wind increases the power necessary to keep a pendulum swinging. Hence, the pendulums of grandfather clocks are commonly enclosed, ensuring $V \approx 0$, so that the clockwork can keep the clock running for the longest possible time.

13.2 Aircraft terminology

Modern commercial aircrafts embody nearly all the principles of aerodynamics presented in this chapter. Thus, a review of aircraft terminology and control strategies is provided in this section. Fig. 13.1 shows three views of a commercial airliner. The body of the aircraft, which houses the passengers, crew, and cargo, is called the *fuselage*. The engines (jet or propeller) are often attached to the wings but they may be mounted on the fuselage or tail, too. Fig. 13.2 shows an overhead (or planform) view of an airliner wing. The location where a wing attaches to the fuselage is called the *wing root*. The outer end of a wing furthest from the fuselage is called the *wing tip*, and the distance between the wing tips is called the *wingspan*, s. The distance between the leading and trailing edges of the wing is called the *chord length*, c, and it varies in the span-wise direction. The area of the wing when viewed from above is called the *planform area*, A. The slenderness of the wing planform is measured by its *aspect ratio*:

$$\Lambda \equiv s^2/A = s/\bar{c}, \quad \text{where} \quad \bar{c} = A/s \qquad (13.1, 13.2)$$

is the average chord length.

The various possible rotational motions of an aircraft can be referred to three aircraft-fixed axes, called the *pitch axis*, the *roll axis*, and the *yaw axis* (Fig. 13.3). A positive aerodynamic drag force points opposite to the direction of flight. Negative drag is called *thrust* and it must be produced by the aircraft's engines for full execution of the aircraft's flight envelope (takeoff, cruise, landing, etc.). Lift is the aerodynamic force that points perpendicular to the direction of flight. It must be generated by the wings to counter the weight of the aircraft in flight. Movable surfaces on the wings and tail fin, known as control surfaces, can alter the distribution of lift and drag forces on the aircraft and provide the primary means for controlling the direction of flight. Variation of engine thrust can also be used to steer the aircraft.

Control surfaces

The aircraft is controlled by the pilot seated in the *cockpit*, who – with hydraulic assistance – sets the engine thrust and moves the control surfaces described in the following paragraphs. For the most part, these control surfaces act to change the local *camber* or curvature of the wings or fins to alter the lift force generated in the vicinity of the control surface.

***Aileron*:** These are flaps near each wing tip (Fig. 13.1), joined to the main wing by a hinged connection, as shown in Fig. 13.4. They move differentially in the sense that one moves up while the other moves down. A depressed aileron increases the lift, and a raised aileron decreases the lift, so that a rolling moment results when they are differentially actuated. Ailerons are located near each wing tip to generate a large rolling moment with minimal angular deflection.

***Elevator*:** The elevators are hinged to the trailing edge of the horizontal stabilizers (tail fins). Unlike ailerons, they move together, and their movement generates a pitching motion of the aircraft.

***Rudder*:** The yawing motion of the aircraft is governed by the hinged rear portion of the vertical tail fin, called the rudder.

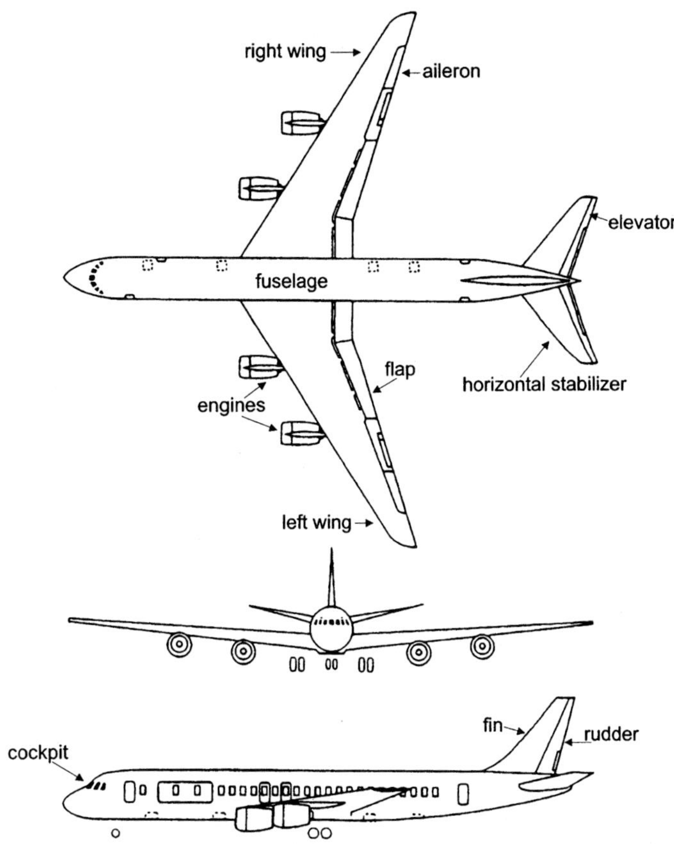

FIGURE 13.1 Three views of a commercial airliner and its control surfaces (NASA). The top view shows the wing planform. The wing is both backward swept and tapered. The various control surfaces shown modify the trailing edge geometry of the wing and tail fins. Landing gear details have been omitted.

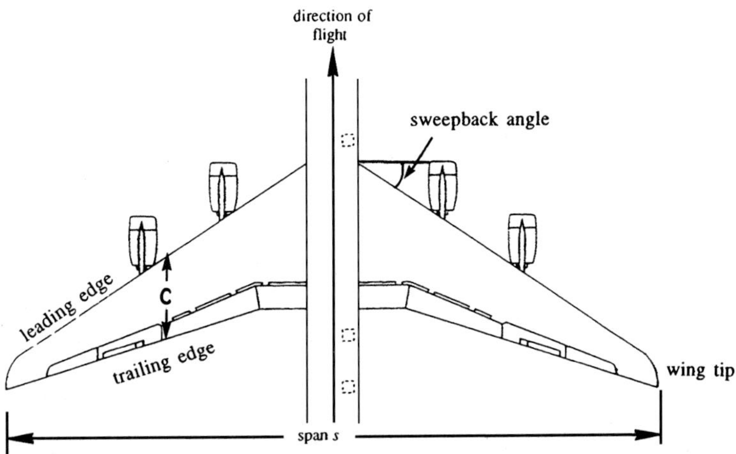

FIGURE 13.2 Wing planform geometry. The span, s, is the straight-line distance between wingtips and is shown at the bottom of the figure. The sweepback angle is shown near the starboard wing root. The chord c depends on location along the span.

Flap: During take off, the speed of the aircraft may be too small for a single-piece wing to generate enough lift to support the weight of the aircraft. To overcome this, a section of the rear of the wing is split, so that it can be rotated downward and moved aft to increase the lift (Fig. 13.5). A further function of the flap is to increase both lift and drag during landing.

Modern airliners also have *spoilers* on the top surface of each wing. When raised slightly, they cause early boundary-layer separation on part of the top of the wing and this decreases or *spoils* the wing's lift. They can be deployed together or individually. Reducing the lift on one wing will bank the aircraft so that it would turn in the direction of the lower-lift

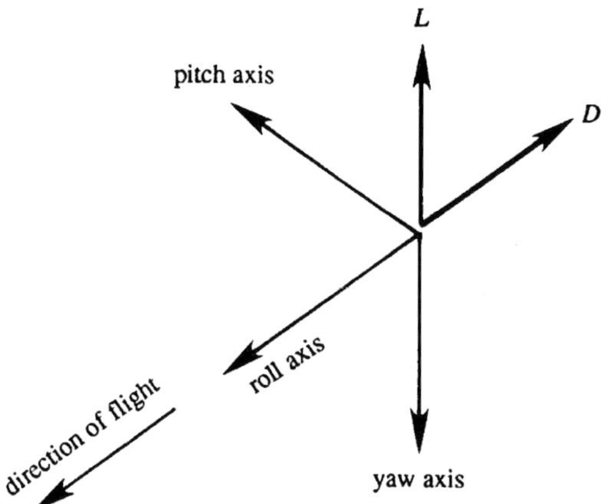

FIGURE 13.3 Aircraft axes. These are defined by the names of aircraft rotations about these axes. Positive pitch raises the aircraft's nose. Positive roll banks the aircraft for a right turn. Positive yaw moves the aircraft's nose to the right from the point of view of the pilot.

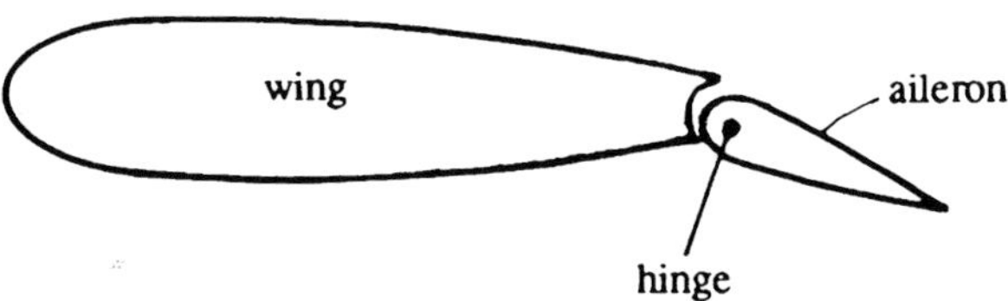

FIGURE 13.4 The aileron. As shown, this aileron deflection would increase lift by increasing the camber of the effective foil shape.

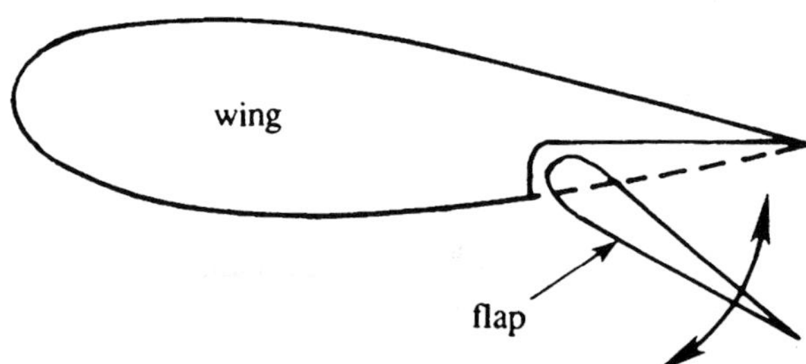

FIGURE 13.5 The flap. As shown, this flat deflection would increase lift by increasing the camber of the effective foil shape. The design of flaps often exploits the flow in the slot formed between the main wing and the rotating flap element to increase lift and delay stall.

wing. When deployed together, overall lift is decreased and the aircraft descends. Spoilers have another function as well. During landing immediately after touchdown, they are deployed fully to eliminate a significant fraction the aircraft's wing lift and thereby ensure that the aircraft stays on the ground and does not become unintentionally airborne again, even in gusty winds. In addition, the spoilers increase drag and slow the aircraft to shorten the length of its landing roll.

An aircraft is said to be in trimmed flight when there are no moments about its center of gravity and the drag force is minimal. Trim tabs are small adjustable surfaces within or adjacent to the major control surfaces – ailerons, elevators, and rudder. Deflections of these surfaces may be set and held to adjust for a change in the aircraft's center of gravity in flight due to consumption of fuel or a change in the direction of the prevailing wind with respect to the flight path. These are set for steady-level flight on a straight path with minimum deflection of the major control surfaces.

Example 13.2

The mass, wing span, and wing aspect ratio of a Boeing 747-400 are approximately $M = 400{,}000$ kg, $s = 64$ m, and $\Lambda = 7.4$, respectively. What lift coefficients are needed for this aircraft to cruise at 900 km/hr at an altitude of 11 km, and land at 300 km/hr at sea level?

Solution For both flight conditions, the aircraft's lift $= L$ must balance its weight $= Mg$:

$$L = \frac{1}{2}\rho U^2 A C_L = Mg, \quad \text{or} \quad C_L = \frac{2Mg}{\rho U^2 A} = \frac{2Mg\Lambda}{\rho U^2 s^2},$$

where the final equality follows from (13.1). Using Appendix A.5, standard-atmospheric air density is 0.362 and 1.225 kg/m^3 at altitudes of 11 km and sea level, respectively. Thus, the requisite lift coefficients are:

$$(C_L)_{11\,\text{km}} = \frac{2(4\times 10^5\,\text{kg})(9.81\,\text{ms}^{-2})7.4}{(0.362\,\text{kgm}^{-3})(9\times 10^5\,\text{m}/3600\,\text{s})^2(64\,\text{m})^2} \cong 0.627, \quad \text{and}$$

$$(C_L)_{sea\ level} = \frac{2(4\times 10^5\,\text{kg})(9.81\,\text{ms}^{-2})7.4}{(1.225\,\text{kgm}^{-3})(3\times 10^5\,\text{m}/3600\,\text{s})^2(64\,\text{m})^2} \cong 1.67.$$

Modern articulated wings that can change geometry in flight produce such variation in lift coefficient with only minor changes in the aircraft fuselage's orientation.

13.3 Characteristics of airfoil sections

Fig. 13.6 shows the shape of the cross-section of a wing, called an *airfoil* section (spelled *aerofoil* in the British literature). The leading edge of the profile is generally rounded, whereas the trailing edge is sharp. The straight line joining the centers of curvature of the leading and trailing edges is called the *chord*. The meridian line of the section passing midway between the upper and lower surfaces is called the *camber line*. The maximum height of the camber line above the chord line is called the *camber* of the section. Normally the camber varies from nearly zero for high-speed supersonic wings, to $\approx 5\%$ of chord length for low-speed wings. The angle α between the chord line and the direction of flight (i.e., the direction of the undisturbed stream) is called the *angle of attack* or *angle of incidence*.

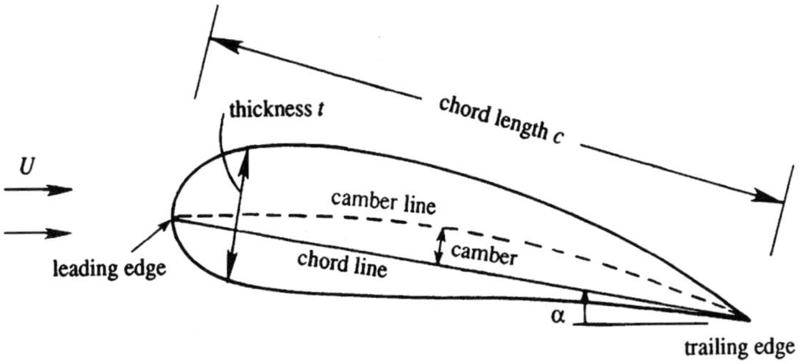

FIGURE 13.6 Airfoil geometry. A rounded leading edge and a sharp trailing edge are essential geometrical features of airfoils. For the material discussed in this chapter, the most important parameters are: the angle of attack α, the chord length c, and the maximum camber. An airfoil's thickness distribution is often modified to minimize drag and/or prevent stall.

The forces on airfoils are usually studied in a foil-fixed frame of reference with a uniform flow approaching the foil along the x-axis with the y-axis pointing vertically upward. Fig. 13.7 shows this geometrical arrangement, and the net aerodynamic force F on an airfoil, which is composed of the drag force D and the lift force L. In steady-level flight the overall lift equals the weight of the aircraft while its drag is balanced by engine thrust. Measurements or specifications of C_D and C_L (see (4.108) and (4.109)) are the primary means for stating airfoil performance. Drag forces result from viscous stresses and pressure distributions on the foil's surface. These are called the *friction drag* and the *pressure drag* (or *form drag*), respectively. The lift is almost entirely due to the pressure distribution. Fig. 13.8 shows the distribution of the pressure coefficient $C_p = (p - p_\infty)/[(1/2)\rho U^2]$ on an airfoil at a moderate angle of attack. The outward arrows correspond to a negative C_p, while a positive C_p is represented by inward arrows. It is seen that the pressure coefficient is negative over most of the surface, except over small regions near the nose and the tail. However, the pressures over most of the upper surface are smaller than those over the bottom surface, which results in a net lift force. The top and bottom surfaces of an airfoil are popularly referred to as the *suction side* and the *compression* (or *pressure*) *side*, respectively.

In steady ideal flow, the *Kutta-Zhukhovsky lift theorem* (see Section 6.5) requires the lift (per unit span) on a two dimensional airfoil to be:

$$L = \rho U \Gamma, \tag{6.40}$$

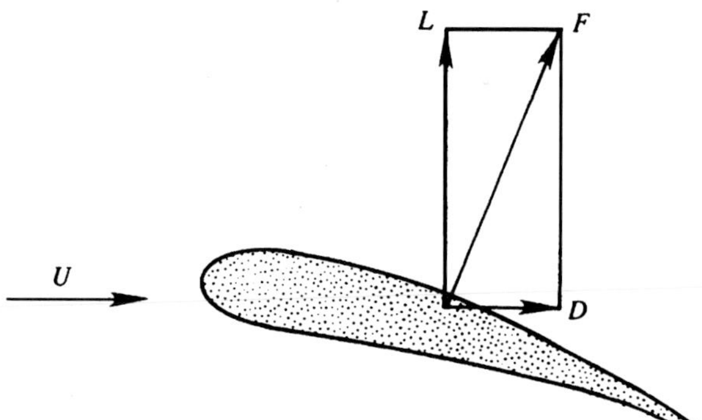

FIGURE 13.7 Forces on an airfoil. Lift L acts perpendicular to the oncoming stream and may be positive or negative. Drag D acts parallel to the oncoming stream and is positive for passive objects.

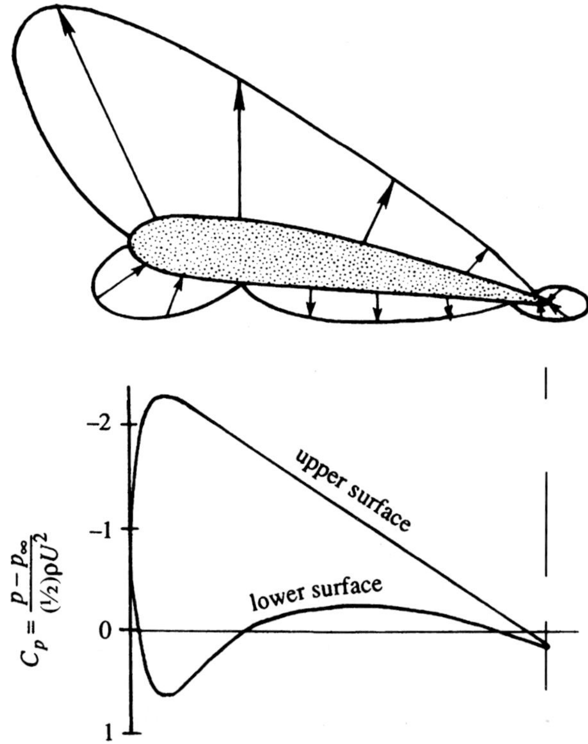

FIGURE 13.8 Distribution of the pressure coefficient C_p over an airfoil. The upper panel shows C_p plotted normal to the surface and the lower panel shows C_p plotted normal to the chord line. Note that negative values appear on the upper half of the vertical axis in the lower panel. And, on the upper or suction foil surface, a pressure minimum occurs near the foil's leading edge. Thus, the suction-side boundary layer faces an adverse pressure gradient over most of the upper surface of the foil.

where U is the free-stream velocity and Γ is the clockwise circulation around the body. Thus, lift development on an airfoil is synonymous with circulation development. As was seen in Section 6.3 for $0 < \Gamma < 4\pi aU$, the amount of circulation held by a cylinder determines the location of stagnation points where the oncoming stream attaches to and separates from the cylinder's surface. This is also true for an airfoil with circulation; foil-surface flow attachment and separation locations are set by the foil's circulation strength. In subsonic aerodynamics, airfoil circulation is determined by the net vorticity trapped in the foil's viscous boundary layers. Thus, asymmetrical foil shapes intended for positive lift generation are designed to place more vorticity in the suction-side boundary layer than in the pressure-side boundary layer. For fixed chord length and free-stream flow speed, three common strategies are followed for robust lift generation and control. The first allows the other two to be effective.

For reliable subsonic lift generation, a foil should have a sharp tailing edge. At low to moderate angles of attack, $|\alpha|$ up to approximately 15° to 20°, a sharp trailing edge causes the suction and pressure side boundary layers to leave the foil surface together at the foil's trailing edge. Thus, a sharp trailing edge becomes the downstream flow separation point, so its location thereby determines the foil's circulation for a given foil shape and free-stream speed. The actual fluid-dynamic interaction leading to this situation involves the foil's viscous boundary layers and is described later. However, this possibility for controlling circulation was experimentally observed before the development of boundary-layer theory. In 1902, the German aerodynamicist Wilhelm Kutta proposed the following rule: *in flow over a two-dimensional body with a sharp trailing edge, there develops a circulation of magnitude just sufficient to move the rear stagnation point to the trailing edge*. This statement is called the *Kutta condition*, sometimes also called the *Zhukhovsky hypothesis*. It is applied in ideal-flow aerodynamics as a simple means of capturing the viscous flow effects of a sharp-trailing-edged foil's attached boundary layers.

A second strategy for controlling a foil's lift is to change its angle of attack α. For $|\alpha| < 15°$ to $20°$, increasing α increases the lift, and nominal extreme C_L values of ± 2 can be obtained from well-designed, single-piece airfoils at high Reynolds number.

The final strategy for controlling a foil's lift is to change its camber. For a fixed angle of attack, increasing camber increases the lift. This is the primary reason for moveable control surfaces at the trailing edges of the wings and tail fins of an aircraft. Angular rotation of such control surfaces locally changes a foil's camber line and thereby changes the lift force generated by the portion of the wing or fin spanned by the control surface.

Two additional considerations are worth mentioning here. First, most foils have a rounded leading edge to keep the foil's suction-side boundary layer attached, and this increases lift and decreases drag. A properly designed leading edge recovers nearly all of the ideal-flow leading edge suction that occurs on foils of negligible thickness (see Exercise 13.8). And second, when a foil is pitched upward to a sufficiently high angle of attack, the Kutta condition will fail and the foil's suction-side boundary layer will separate upstream of the foil's trailing edge. This situation is called *stall* and its onset depends on: the foil's shape, the Reynolds number of the flow, the foil's surface roughness, and other three-dimensional effects. Stall occurs when the suction-side boundary layer cannot overcome the adverse pressure gradient aft of the pressure minimum on the foil's suction side. For small violations of the Kutta condition where the suction-side boundary layer separates at ~80% or 90% of the chord length, a typical foil's lift is not strongly affected but its drag increases. For more severe violations of the Kutta condition, where the suction-side boundary layer separates upstream of the mid-chord location, the foil's lift is noticeably reduced and its drag is greatly increased. In nearly all cases, stall leads to such undesirable foil performance that its onset places important limitations on an aircraft's operating envelope.

The physical reason for the Kutta condition is illustrated in Fig. 13.9 where the same simple airfoil and nearby streamlines are shown at three different times. Here, the foil is held fixed and flow is impulsively accelerated to speed U at $t = 0$. Fig. 13.9a shows streamlines near the foil immediately after the fluid has started moving but before boundary layers have developed on either its suction or pressure sides. The fluid velocity at this stage has a near discontinuity adjacent to the foil's surface. And, the fluid goes around the foil's trailing edge with a high velocity and overcomes a steep deceleration and pressure rise from the trailing edge to the rear flow-separation (and stagnation) point at B. The flow is able to turn the sharp trailing-edge corner because the vorticity on the foil's surface near the trailing edge at this instant is nearly singular at the trailing edge and it induces counterclockwise fluid motion (shown in Fig. 13.9a by a dashed arrow). Overall at this time, the flow is irrotational away from the foil's surface, the foil's net circulation is zero, it generates no lift, the forward flow-attachment stagnation point at A is close to the nose of the foil, and the rear stagnation point at B resides on the foil's suction surface.

Fig. 13.9b shows the flow a short time later in a hypothetical situation where the foil's pressure-side boundary layer has developed first. In this case, the points A and B have not moved much. However, the pressure-side boundary layer now separates at the sharp trailing edge because the slowly moving boundary-layer fluid near the foil's surface does not have sufficient kinetic energy to negotiate the steep pressure rise near the stagnation point B nor can it turn the sharp trailing-edge corner. Furthermore, the separated pressure-side boundary-layer flow has carried the near singularity of vorticity, which initially resided on the foil's surface at its trailing edge, into the foil's near wake as a concentrated vortex. Two phenomena near the trailing edge now act to eliminate the zone of separated flow caused by pressure-side boundary-layer separation at the trailing edge. First, the Bernoulli equation ensures that the stagnation pressure at B is higher than the pressure in the moving fluid that is leaving the trailing edge from the pressure side of the foil. The resulting pressure gradient between B and the trailing edge pushes the stationary fluid near B toward the foil's trailing edge. Second, the induced velocities from the vorticity in the separated pressure-side boundary layer and from the near-wake concentrated vortex both induce the stationary fluid near B to move toward the foil's trailing edge. Together these two phenomena cause the rear stagnation point at B to move to the foil's trailing edge. Although an actual impulsively started flow involves simultaneous suction- and pressure-side boundary layer development, the outcome is the same; the rear stagnation point winds up at the trailing edge.

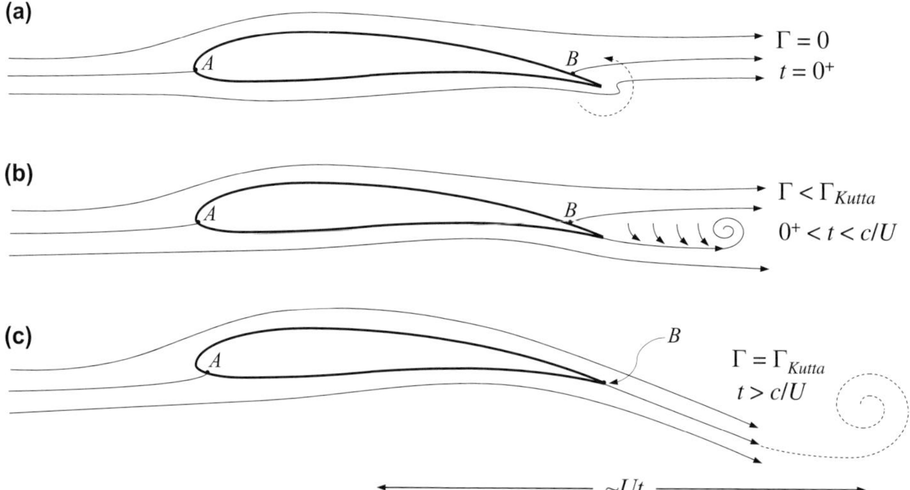

FIGURE 13.9 Flow patterns over a stationary airfoil at a low angle of attack in an impulsively started horizontal flow. (a) Streamlines immediately after the velocity jumps to a positive value. Here the boundary layers on the foil have not had a chance to develop and the rear stagnation (separation) point B occurs on the suction surface of the foil. The foil-surface vorticity at the trailing edge is nearly singular and induces a counterclockwise fluid velocity that draws fluid around the sharp trailing edge. (b) If the pressure-side boundary layer develops first, it will separate from the trailing edge as shown. However, the pressure distribution near the trailing edge and the induced velocities from the foil's near-wake vorticity both act to bring B to the trailing edge. (c) Steady-flow pattern established after the flow has moved a chord length or two. Here the leading edge stagnation point A has moved onto the pressure side of the foil and the net circulation trapped in the foil's boundary layers satisfies the Kutta condition. In this case the rear stagnation (separation) point lies at the foil's trailing edge. The net circulation of the whole flow field remains zero because the unsteady flow process leading to this flow pattern produces a counter-rotating starting vortex, shown in (c) as a dashed spiral.

Fig. 13.9c shows the final condition after the flow has traveled a chord length or two past the foil. The leading-edge stagnation point has traveled under the nose of the foil and onto the foil's pressure side, and the suction-surface separation point B has been drawn to the foil's trailing edge. (The ideal airfoil trailing edge is a perfect cusp with zero included angle that allows the pressure and suction side flows to meet and separate from the foil without changing direction and without a stagnation point. However, structural requirements cause real foils to have finite included-angle trailing edges, thus point B is a stagnation point even when the trailing edge's included angle is small; see Section 6.4 and Exercise 13.3). Once the flow shown in Fig. 13.9c is established, the foil now carries more vorticity in its suction-side boundary layer than it does in its pressure-side boundary layer. This difference causes the flow to sweep upward ahead of the foil and downward behind it. The foil's net circulation is that necessary to satisfy the Kutta condition, Γ_{Kutta}. If the foil's circulation is further increased beyond Γ_{Kutta}, the rear stagnation point moves under the foil and onto the *pressure* surface. Although it is an ideal-flow possibility, $\Gamma > \Gamma_{Kutta}$ is not observed in real airfoil flows.

The net circulation in the impulsively started flow described in this section and illustrated in Fig. 13.9 is maintained at zero by the presence of an opposite sign vortex, known as a *starting vortex*, in the fluid that was near the foil when the flow began moving. In the scenario described earlier, this vortex is the remnant of the vorticity shed by the pressure-side boundary layer before point B moved to the foil's trailing edge and the cast-off concentrated vorticity that initially caused the flow to fully turn the foil's sharp trailing-edge corner.

The equivalence of the final circulation magnitude bound to the foil and that in the starting vortex is illustrated in Fig. 13.10 where the sense of the foil's circulation is clockwise and that in the starting vortex is counterclockwise. For the flow shown in this figure, imagine that the fluid is stationary and the airfoil is moving to the left. Consider a material circuit ABCD large enough to enclose both the initial and final locations of the airfoil. Initially the trailing edge was within the region BCD, which now contains the starting vortex only. According to Kelvin's circulation theorem, the circulation around any material circuit remains constant, if the circuit remains in a region of inviscid flow (although viscous processes may go on *inside* the region enclosed by the circuit). The circulation around the large circuit ABCD therefore remains zero, since it was zero initially. Consequently the counterclockwise circulation of the starting vortex around DBC is balanced by an equal clockwise circulation around ADB. The wing is therefore left with a circulation Γ equal and opposite to the circulation of the starting vortex.

From the discussion and illustrations in Fig. 13.9, a value of circulation other than Γ_{Kutta} would result a readjustment of the flow. Thus, with every change in flow speed, angle of attack, or airfoil camber (via flap deflection) a new starting

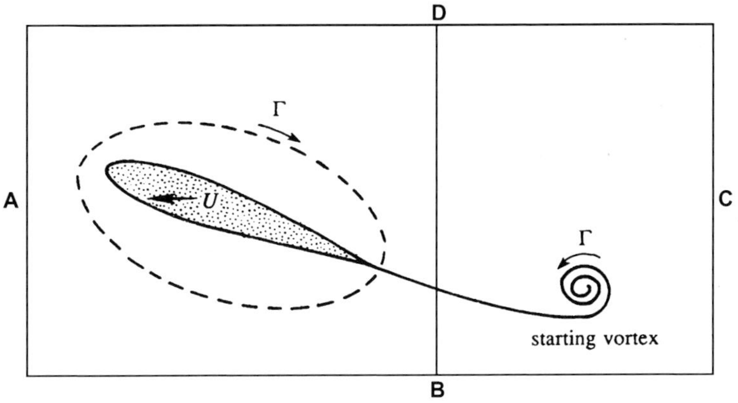

D
A
C
B
Γ
U
Γ
starting vortex

FIGURE 13.10 A material circuit in a stationary fluid that contains an impulsively started airfoil moving to the left. The entire outer part of the circuit was initially in stationary fluid. Thus, the circulation on ABCD must be zero. Therefore, if the sub-circuit ABD contains the airfoil with circulation Γ, then the other sub-circuit BCD must contain a starting vortex with circulation $-\Gamma$.

vortex is cast off and left behind the foil. A new value of circulation around the foil is established to once again place the rear stagnation point at the foil's trailing edge.

Interestingly, *fluid viscosity is not only responsible for the drag, but also for the development of circulation and lift.* In developing the circulation, the flow leads to a steady state where further boundary-layer separation is prevented. The establishment of circulation around an airfoil-shaped body in a real fluid is truly remarkable.

Historical notes

According to von Karman (1954), the connection between the lift of airplane wings and the circulation around them was recognized and developed by three persons. One of them was the Englishman Frederick Lanchester (1887–1946). He was a multisided and imaginative person, a practical engineer as well as an amateur mathematician. His trade was automobile building; in fact, he was the chief engineer and general manager of the Lanchester Motor Company. He once took von Karman for a ride around Cambridge in an automobile that he built himself, but von Karman "felt a little uneasy discussing aerodynamics at such rather frightening speed" (p. 34). The second person is the German mathematician Wilhelm Kutta (1867–1944), well known for the Runge-Kutta scheme used in the numerical integration of ordinary differential equations. He started out as a pure mathematician, but later became interested in aerodynamics. The third person is the Russian physicist Nikolai Zhukhovsky, who developed the mathematical foundations of the theory of lift for wings of infinite span, independently of Lanchester and Kutta. An excellent history of flight and the science of aerodynamics is provided by Anderson (1998).

Example 13.3

As a simplified means to explain how a flying aircraft's weight is transmitted to the ground, consider two-dimensional ideal flow with density ρ and speed U past an ideal vortex of strength Γ a distance H above an infinite flat surface (see Fig. 13.11). Integrate the pressure distribution on the surface to show that it carries a load of $\rho U\Gamma$ when $H \to \infty$.

Solution Choose the flat surface at $y = 0$, and presume the uniform inflow is horizontal along the x-axis so that the stream function (see Sections 6.2 and 6.3) may be written:

$$\psi(x, y) = Uy + \frac{\Gamma}{2\pi}\ln\left(\sqrt{x^2 + (y - H)^2}\right) - \frac{\Gamma}{2\pi}\ln\left(\sqrt{x^2 + (y + H)^2}\right),$$

where Γ is positive in the clockwise direction, and the method of images (see Section 6.3) has been used to represent the surface. In Fig. 13.11, the actual and image vortices appear as solid and open circles, respectively. With this stream function the velocity components are:

$$u(x, y) = \frac{\partial\psi}{\partial y} = U + \frac{\Gamma}{2\pi}\left(\frac{y - H}{x^2 + (y - H)^2} - \frac{y + H}{x^2 + (y + H)^2}\right), \quad \text{and}$$

$$v(x, y) = -\frac{\partial\psi}{\partial x} = -\frac{\Gamma}{2\pi}\left(\frac{x}{x^2 + (y - H)^2} - \frac{x}{x^2 + (y + H)^2}\right).$$

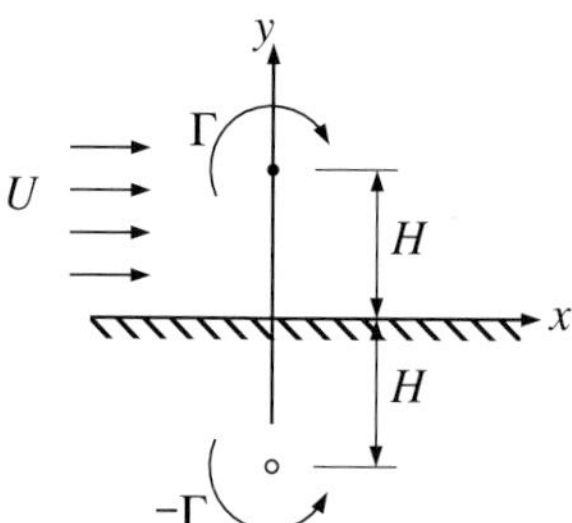

FIGURE 13.11 Two-dimensional ideal flow geometry for Example 13.3. A uniform horizontal stream with speed U passes a vortex (the solid circle) of strength Γ a distance H above a solid surface. The pressure distribution on the solid surface matches the lift load on the vortex. An opposite strength image vortex (the open circle) is located a distance H below the surface to satisfy the no-through-flow boundary condition on the surface.

For steady two-dimensional ideal flow in Cartesian coordinates, the simplest version of the Bernoulli equation applies along the surface:

$$p(x,y)+\frac{1}{2}\rho(u^2(x,y)+v^2(x,y))=p_\infty+\frac{1}{2}\rho U^2,\quad \text{or for}\quad y=0:\ p(x,0)-p_\infty=\frac{\rho}{2}(U^2-u^2(x,0)),$$

where the second form occurs because $v(x,0)=0$. From above, the horizontal velocity on the surface is:

$$u(x,0)=U-\frac{\Gamma}{\pi}\left(\frac{H}{x^2+H^2}\right),\quad \text{so}\quad U^2-u^2(x,0)=2U\frac{\Gamma}{\pi}\left(\frac{H}{x^2+H^2}\right)-\frac{\Gamma^2}{\pi^2}\left(\frac{H}{x^2+H^2}\right)^2.$$

Thus, pressure force (per unit length) is:

$$\int_{-\infty}^{+\infty}(p(x,0)-p_\infty)dx=\frac{\rho U\Gamma}{\pi}\int_{-\infty}^{+\infty}\frac{H}{x^2+H^2}dx-\frac{\rho\Gamma^2}{2\pi^2}\int_{-\infty}^{+\infty}\left(\frac{H}{x^2+H^2}\right)^2dx.$$

Both integrals can be evaluated using the variable substitution $x=H\tan\xi$, to find:

$$\int_{-\infty}^{+\infty}(p(x,0)-p_\infty)dx=\frac{\rho U\Gamma}{\pi}[\xi]_{-\pi/2}^{+\pi/2}-\frac{\rho\Gamma^2}{2\pi^2 H}\int_{-\pi/2}^{+\pi/2}\cos^2\xi\, d\xi=\rho U\Gamma-\rho\frac{\Gamma^2}{4\pi H}.$$

The first term of the final answer balances the lift force on the vortex and is independent of how far the vortex is from the surface. This lift force is transmitted to the surface through the combined effects of the vortex's induced velocity and the free-stream flow. The second term of the final answer is an interference term that is negligible as $H\to\infty$. For common aircraft geometries in three dimensions, this term is more complicated and includes contributions from multiple vortices. It leads to what pilots call *ground effect* when landing (see Exercise 13.18).

13.4 Conformal transformation for generating airfoil shapes

In the study of airfoils, one is interested in finding the flow pattern and the surface-pressure distribution. The *direct* solution of the Laplace equation for the prescribed boundary shape of the airfoil is straightforward using a computer, but analytically it is more difficult. In general, analytical solutions are possible only when the airfoil is assumed thin. This is called *thin airfoil theory*, in which the airfoil is replaced by a vortex sheet coinciding with the camber line. An integral equation is developed for the local vorticity distribution from the condition that the camber line be a streamline (velocity tangent to the camber line). The velocity at each point on the camber line is the superposition (i.e., integral) of velocities induced at that point due to the vorticity distribution at all other points on the camber line plus that from the oncoming stream. Since the maximum camber is small, evaluations are made on the x-axis of the $x-y$-plane. The Kutta condition is enforced by requiring the strength of the vortex sheet at the trailing edge to be zero. Thin airfoil theory is treated in detail in Kuethe and Chow (1998, Chapter 5) and Anderson (2007, Chapter 4). An *indirect* way to solve the problem involves the method of conformal transformation, in which a mapping function is determined such that the airfoil shape is transformed into a circle. Then a study of the flow around the circle determines the flow pattern around the airfoil. This is called *Theodorsen's method*, which is complicated and will not be discussed here.

Instead, we shall deal with the case in which a *given* transformation maps a circle into an airfoil-like shape and determines the properties of the airfoil generated thereby. This is the *Zhukhovsky transformation*:

$$z=\zeta+b^2/\zeta, \tag{13.3}$$

where b is a positive real constant. It maps regions of the ζ-plane into the z-plane, some examples of which are discussed in Section 6.6. Here, we shall consider circles in different configurations in the ζ-plane and examine their transformed shapes in the z-plane. It will be seen that one of them will result in an airfoil shape.

First consider the transformation of a circle into a straight line. Start from a circle, centered at the origin in the ζ-plane, whose radius b is the same as the constant in the Zhukhovsky transformation (Fig. 13.12). For a point $\zeta = be^{i\theta}$ on the circle, the corresponding point in the z-plane is:

$$z = be^{i\theta} + be^{-i\theta} = 2b\cos\theta.$$

As θ varies from 0 to π, z goes along the x-axis from $2b$ to $-2b$. As θ varies from π to 2π, z goes from $-2b$ to $2b$. The circle of radius b in the ζ-plane is thus transformed into a straight line of length $4b$ in the z-plane. Rhe region *outside* the circle in the ζ-plane is mapped into the *entire* z-plane. (It can be shown that the region inside the circle is also transformed into the entire z-plane. This, however, is of no concern to us, since we shall not consider the interior of the circle in the ζ-plane.)

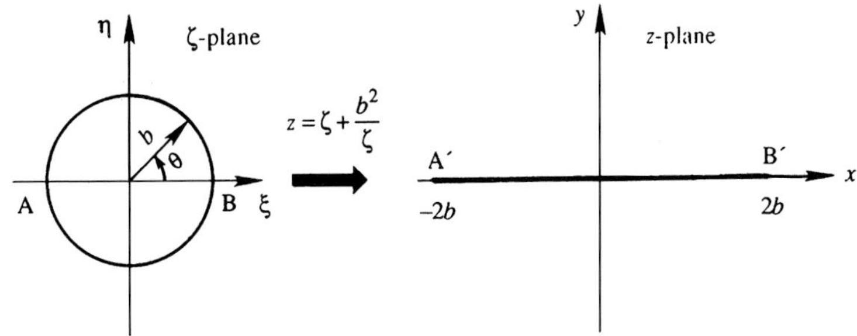

FIGURE 13.12 Transformation of a circle into a straight line. Here the ζ-plane contains the circle of radius b and the transformation $z = \zeta + b^2/\zeta$ converts it into a line segment of length $4b$ in the z-plane.

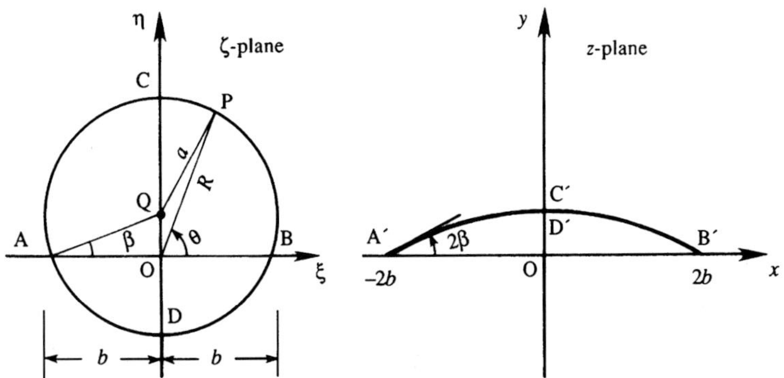

FIGURE 13.13 Transformation of a circle into a circular arc. This situation is similar to that shown in Fig. 13.12 except that here the circle is displaced upward and its radius is larger. The object created in the z-plane is a circular arc.

Next consider the transformation of a circle into a circular arc. Again start with a circle in the ζ-plane, but this time let its radius be a, let it be centered at point Q along the vertical the η-axis, and let it cut the horizontal ξ-axis at $(\pm b, 0)$, as shown in Fig. 13.13. If a point on the circle in the ζ-plane is represented by $\zeta = Re^{i\theta}$, then the corresponding point in the z-plane is:

$$z = \mathrm{R}e^{i\theta} + (b^2/R)\mathrm{e}^{-i\theta},$$

whose real and imaginary parts are:

$$x = (R + b^2/R)\cos\theta, \quad \text{and} \quad y = (R - b^2/R)\sin\theta. \tag{13.4}$$

Eliminate R to obtain:

$$x^2\sin^2\theta - y^2\cos^2\theta = 4b^2\sin^2\theta\cos^2\theta. \tag{13.5}$$

To understand the shape of the curve represented by (13.5), express θ in terms of x, y, and the known constants. From triangle OQP, we obtain:

$$\mathrm{QP}^2 = \mathrm{OP}^2 + \mathrm{OQ}^2 - 2(\mathrm{OQ})(\mathrm{OP})\cos(\angle \mathrm{QOP}).$$

Using $\mathrm{QP} = a = b/\cos\beta$ and $\mathrm{OQ} = b\tan\beta$, this becomes:

$$b^2/\cos^2\beta = R^2 + b^2\tan^2\beta - 2Rb\tan\beta\cos(90^\circ - \theta),$$

which simplifies to

$$2b\tan\beta\sin\theta = R - b^2/R = y/\sin\theta,$$

where β is known from $\cos\beta = b/a$. This is the equation of a circle in the z-plane, having the center at $(0, -2b\cot 2\beta)$ and a radius of $2b\csc 2\beta$. The Zhukhovsky transformation has thus mapped a complete circle into a circular arc.

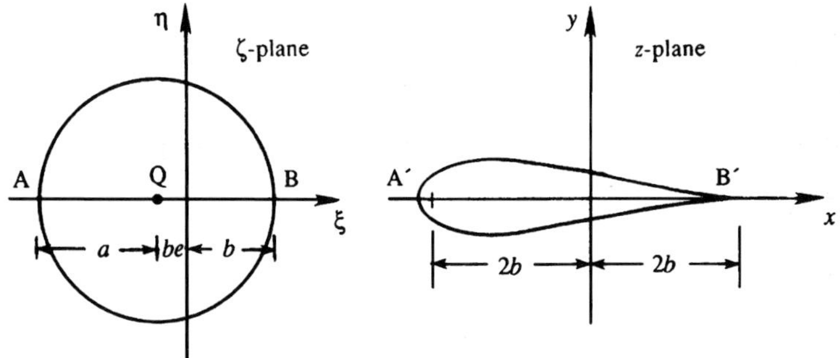

FIGURE 13.14 Transformation of a circle into a symmetric airfoil. This situation is similar to that shown in Fig. 13.12 except that here the circle is displaced to the left and its radius is larger. The object created in the z-plane has a symmetric (zero camber) airfoil shape.

Now consider what happens when the center of the circle in the ζ-plane is displaced to a point Q on the real axis (Fig. 13.14). The radius of the circle is again a, and we assume that a is slightly larger than b:

$$a \equiv b(1+e), \quad \text{with} \quad e \ll 1. \tag{13.6}$$

A numerical evaluation of the Zhukhovsky transformation (13.3), with assumed values for a and b, shows that the corresponding shape in the z-plane is a symmetrical airfoil shape, a streamlined body that is symmetrical about the x-axis. Note that the airfoil in Fig. 13.14 has a rounded nose and thickness, while the one in Fig. 13.13 has camber but no thickness.

Therefore, a potentially realistic airfoil shape with both thickness and camber can be generated by starting from a circle in the ζ-plane that is displaced in both η and ξ directions (Fig. 13.15). The following relations can be proved for $e \ll 1$:

$$c \cong 4b, \quad \text{camber} = \beta c/2, \quad \text{and} \quad t_{max}/c = 1.3e. \tag{13.7}$$

Here t_{max} is the maximum thickness, which is reached nearly at the quarter chord position $x = -b$, and *camber* as defined in Fig. 13.6 is indicated in Fig. 13.15.

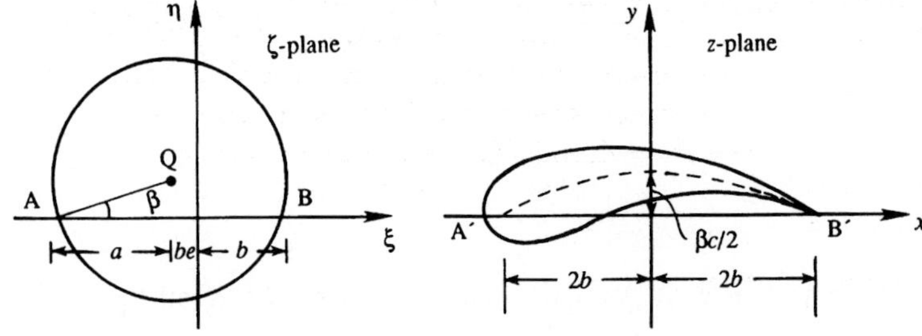

FIGURE 13.15 Transformation of a circle into a cambered airfoil. This situation combines the effects illustrated in Figs. 13.12–13.14. The circle is displaced upward and leftward, and its radius is larger. The resulting shape in the z-plane is that of an airfoil.

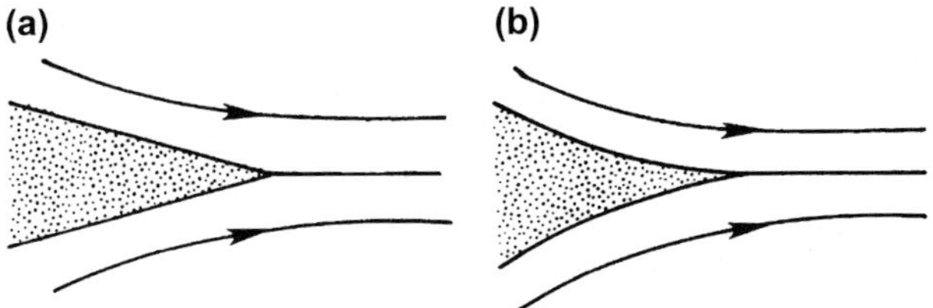

FIGURE 13.16 Shapes of the trailing edge: (a) trailing edge with finite angle; and (b) cusped trailing edge. Application of the Kutta condition to a trailing edge with a finite included angle results in a stagnation point at the trailing edge. A cusped trailing edge avoids the stagnation point.

Such airfoils generated from the Zhukhovsky transformation are called *Zhukhovsky airfoils*. They have the property that the trailing edge is a *cusp*, which means that the upper and lower surfaces are tangent to each other at the trailing edge. Without the Kutta condition, the trailing edge is a point of infinite velocity. If the trailing edge angle is non-zero (Fig. 13.16a), then a stagnation point occurs at the trailing edge because the suction and pressure side flows must change direction when they meet (Exercise 13.3). However, the cusped trailing edge of a Zhukhovsky airfoil (Fig. 13.16b) does not require any flow deflection so it is not a stagnation point. In that case, the tangents to the upper and lower surfaces coincide at the trailing edge, and the fluid leaves the trailing edge smoothly. The trailing edge for the Zhukhovsky airfoil is simply an ordinary point where the velocity is neither zero nor infinite.

Example 13.4

Into what shape in the z-plane is the circle defined by $\zeta = |b|e^{i\theta}$ in the ζ-plane mapped by the transformation $z = \zeta + b^2/\zeta$ when b is a complex number $b = |b|e^{-i\alpha}$ when α is a positive real constant?

Solution Set $\zeta = |b|e^{i\theta}$ in the given transformation and then separate z into real and imaginary parts using $z = x + iy$:

$$z = \zeta + \frac{b^2}{\zeta} = |b|e^{i\theta} + |b|^2 e^{-2i\alpha} \cdot \frac{1}{|b|e^{i\theta}} = |b|e^{-i\alpha}\left(e^{i(\theta+\alpha)} + e^{-i(\theta+\alpha)}\right) = 2|b|e^{-i\alpha}\cos(\theta+\alpha), \quad \text{or}$$

$$x = 2|b|\cos\alpha\cos(\theta+\alpha), \quad \text{and} \quad y = -2|b|\sin\alpha\cos(\theta+\alpha).$$

The ratio $y/x = -\tan\alpha$ is constant, and the extreme (x, y)-coordinates are: $(2|b|\cos\alpha, -2|b|\sin\alpha)$ and $(-2|b|\cos\alpha, 2|b|\sin\alpha)$, which occur when $\theta + \alpha = 0$ and π, respectively. Thus, the shape is a line segment of length $4|b|$ centered at the origin with a slope of $-\tan\alpha$. If there were a uniform flow along the x-axis in the z-plane, this transformation would map a circle in the ζ-plane into a flat plate at angle of attack α in the z-plane.

13.5 Lift of a Zhukhovsky airfoil

The preceding section has shown how a circle in the ζ-plane can be transformed into an airfoil in the z-plane with the help of the Zhukhovsky transformation. The performance of such an airfoil can be determined with the aid of the transformation. Start with flow around a circle with clockwise circulation Γ in the ζ-plane, in which the approach velocity is inclined at an angle α with the ξ-axis (Fig. 13.17). The corresponding pattern in the z-plane is the flow around an airfoil with circulation Γ and angle of attack α. It can be shown that the circulation does not change during a conformal transformation. If $w = \phi + i\psi$ is the complex potential, then the velocities in the two planes are related by:

$$\frac{dw}{dz} = \frac{dw}{d\zeta}\frac{d\zeta}{dz}.$$

Using the Zhukhovsky transformation (13.3), this becomes:

$$\frac{dw}{dz} = \frac{dw}{d\zeta}\frac{\zeta^2}{\zeta^2 - b^2}. \tag{13.8}$$

Here $dw/dz = u - iv$ is the complex velocity in the z-plane, and $dw/d\zeta$ is the complex velocity in the ζ-plane. Eq. (13.8) shows that the velocities in the two planes become equal as $\zeta \to \infty$, which means that the free-stream velocities are inclined at the same angle α in the two planes.

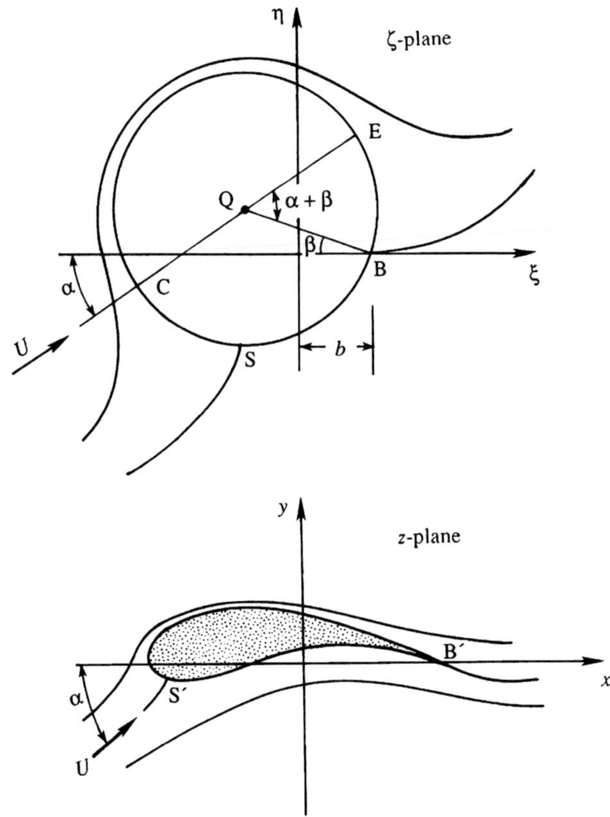

FIGURE 13.17 Transformation of flow around a circle with circulation in the ζ-plane into flow around a Zhukhovsky airfoil in the z-plane. The stagnation points S and B in the upper panel are mapped into the stagnation points $S\prime$ and B' in the lower panel. The angle of attack α is the same in both complex planes.

Point B with coordinates $(b, 0)$ in the ζ-plane is transformed into the trailing edge B′ of the airfoil. Because $\zeta^2 - b^2$ vanishes there, it follows from (13.8) that the velocity at the trailing edge will in general be infinite. If, however, we arrange that B is a stagnation point in the ζ-plane at which $dw/d\zeta = 0$, then dw/dz at the trailing edge will have a zero-over-zero form. Our discussion of Fig. 13.16b has shown that this will in fact result in a finite velocity *at* B′.

From (6.37), the tangential velocity at the surface of the circle in the ζ-plane is given by:

$$u_\theta = -2U \sin\theta - \Gamma/2\pi a, \tag{13.9}$$

where θ is measured from the free-stream-aligned diameter CQE. At point B, we have $u_\theta = 0$ and $\theta = -(\alpha + \beta)$. Therefore, (13.9) gives:

$$\Gamma = 4\pi U a \sin(\alpha + \beta), \tag{13.10}$$

which is the clockwise circulation required by the Kutta condition. It shows that the circulation around an airfoil depends on the speed U, the chord length c ($\approx 4a$), the angle of attack α, and the camber/chord ratio $\beta/2$. The coefficient of lift is:

$$C_L = \frac{L}{(1/2)\rho U^2 A} \cong 2\pi(\alpha + \beta), \tag{13.11}$$

where we have used $4a \approx c$, and $\sin(\alpha + \beta) \approx (\alpha + \beta)$ for small angles of attack. Eq. (13.11) shows that the lift can be increased by adding a certain amount of camber. The lift is zero at a negative angle of attack $\alpha = -\beta$, so that the angle $(\alpha + \beta)$ can be called the *absolute* angle of attack. The fact that the lift of an airfoil is proportional to the angle of attack allows the pilot to control the lift simply by adjusting the attitude (orientation) of the airfoil with respect to its flight direction.

A comparison of the theoretical lift equation (13.11) with typical experimental results for a Zhukhovsky airfoil is shown in Fig. 13.18. The small disagreement can be attributed to the finite thickness of the foil-surface boundary layers whose displacement thicknesses change the effective shape of the airfoil. The sudden drop of the lift at $\alpha + \beta \approx 20°$ is the signature of stall, and it is caused by early suction-side boundary-layer separation that worsens with increasing angle of attack. Stall is further discussed in Section 13.7.

Zhukhovsky airfoils are not practical for two basic reasons. First, they demand a cusped trailing edge, which cannot be practically constructed or maintained. Second, the camber line in a Zhukhovsky airfoil is nearly a circular arc, and therefore the maximum camber lies close to mid-chord. However, a maximum camber within the forward portion of the chord is usually preferred so as to obtain a desirable pressure distribution. To get around these difficulties, other families of airfoils have been generated from circles by means of more complicated transformations. Nevertheless, the results for a Zhukhovsky airfoil given here have considerable application as reference values, and the conformal mapping technique remains an efficient means for assessing airfoil designs.

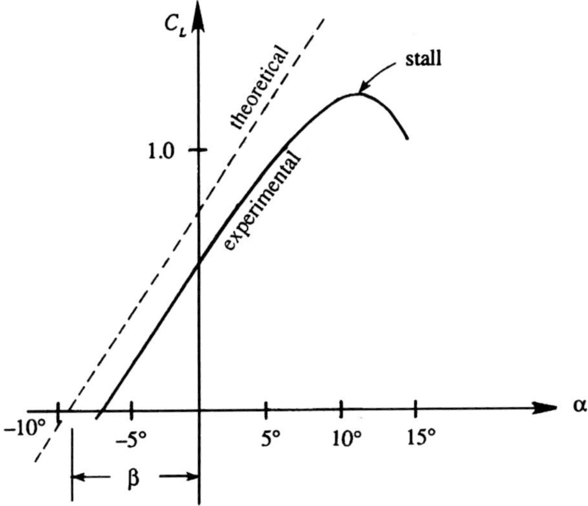

FIGURE 13.18 Comparison of theoretical and experimental lift coefficients for a cambered Zhukhovsky airfoil. The lift curve slopes match well and boundary-layer thicknesses may account for the offset between theoretical and measured curves. The most important difference is that the real airfoil stalls while the ideal one does not.

Example 13.5

If the complex potential in the ζ-plane represents uniform horizontal flow past a cylinder with radius a and clockwise circulation Γ (see Fig. 6.12a):

$$w(\zeta) = U(\zeta + a^2/\zeta) + \frac{i\Gamma}{2\pi}\ln(\zeta/a),$$

set $|b| = a$ and use the transformation from Example 13.4, $z = \zeta + (ae^{i\alpha})^2/\zeta$, and the Kutta condition to show that $C_L = 2\pi \sin\alpha$ for ideal flow past a flat plate at angle of attack α.

Solution The complex velocity in the ζ-plane is:

$$\frac{dw}{d\zeta} = U(1 - a^2/\zeta^2) + \frac{i\Gamma}{2\pi\zeta}.$$

To find the angle(s), θ_s, of the cylinder-surface stagnation points, set $dw/d\zeta = 0$ and substitute in $\zeta = a\exp(i\theta_s)$ to find:

$$0 = U(1 - a^2/a^2\exp(2i\theta_s)) + \frac{i\Gamma}{2\pi a\exp(i\theta_s)}, \text{ which implies } \sin\theta_s = -\Gamma/4\pi Ua.$$

To satisfy the Kutta condition in the z-plane, choose Γ so that the downstream cylinder-surface stagnation point (the S on the right in Fig. 6.12a) maps into the flat-plate's trailing edge. From Example 13.4 with $|b| = a$, the (x, y)-location of the plate's trailing edge is $(2a\cos\alpha, -2a\sin\alpha)$ and it is mapped from the cylinder-surface stagnation point defined by $\theta_s + \alpha = 0$, or $\theta_s = -\alpha$. Combine this result with that for the stagnation point location in the ζ-plane to find: $\sin\theta_s = -\sin\alpha = -\Gamma/4\pi Ua$ or $\Gamma = 4\pi Ua\sin\alpha$. Thus, the plate's coefficient of lift is:

$$C_L = \frac{L}{(1/2)\rho U^2(4a)} = \frac{\rho U\Gamma}{(1/2)\rho U^2(4a)} = \frac{\rho U \cdot 4\pi Ua\sin\alpha}{(1/2)\rho U^2(4a)} = 2\pi\sin\alpha,$$

where the chord length of the plate is $4a$.

While this result is readily anticipated from the formal results for Zhukhovsky airfoils, it does not require any approximations beyond those inherent to ideal flow, and is provided here to show that the body in the z-plane can be rotated by the transformation so that the free stream can be horizontal in both the ζ- and z-planes.

13.6 Elementary lifting line theory for wings of finite span

The foregoing two-dimensional results apply only to wings of infinite span. However, many of the concepts of two-dimensional aerodynamics can be extrapolated to three-dimensional flow and wings of finite span when the vorticity shed from a three-dimensional wing is accounted for. The lifting line theory of Prandtl and Lanchester is the simplest means for accomplishing this task and it provides useful insights into how lift and drag develop on finite span wings. Lifting line theory is based on several approximations to the three-dimensional flow field of a finite wing, so our starting point is a description of such a flow.

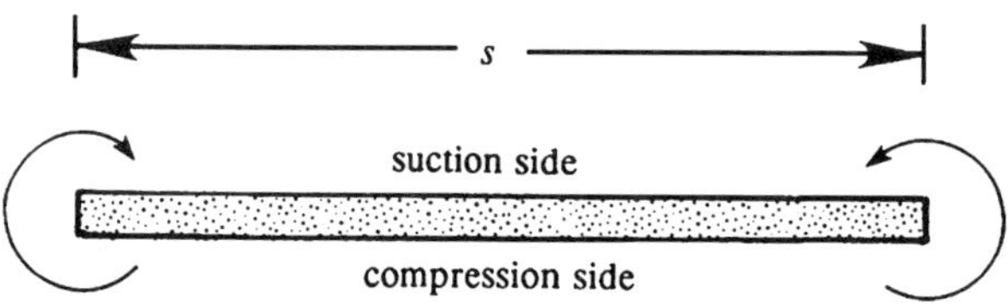

FIGURE 13.19 Flow around wing tips. Low suction-side pressures and high pressure-side pressures cause fluid to move toward the wing tips on the underside of a finite wing, and to move away from the wing tips on the topside of a finite wing. This three-dimensional flow eventually produces the tip vortices.

Fig. 13.19 shows a schematic view of a finite-span wing, looking downstream from the aircraft. As the pressure on the lower surface of the wing is greater than that on the upper surface, air flows around the wing tips from the lower into the upper side. Therefore, there is a span-wise component of velocity toward the wing tip on the underside of the wing and toward the wing root on the upper side, as shown by the streamlines in Fig. 13.20a. The span-wise momentum acquired as the fluid passes the wing continues into the wake downstream of the trailing edge. On the stream surface extending downstream from the wing, therefore, the lateral component of the flow is outward (toward the wing tips) on the underside and inward on the upper side. On this stream surface, then, there is vorticity oriented in the stream-wise direction. This stream-wise vorticity has opposite signs on the two sides of the wing-center axis OQ. The stream-wise vortex filaments downstream of the wing are called *trailing vortices*, which form a *vortex sheet* (Fig. 13.20b) in the near wake of the wing. As discussed in Section 5.7, a vortex sheet is composed of closely spaced vortex filaments that generate a discontinuity in tangential velocity.

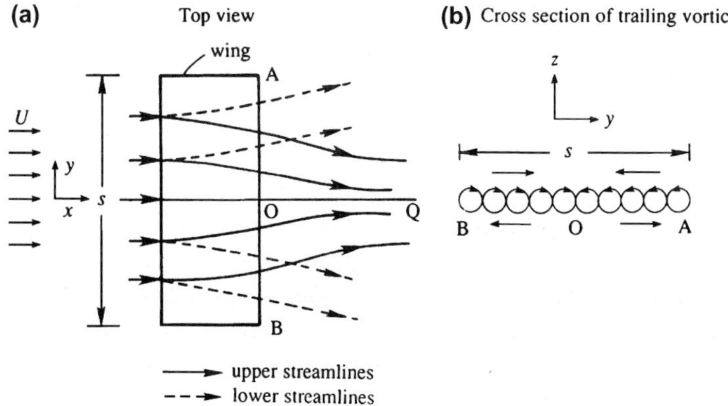

FIGURE 13.20 Flow over a wing of finite span: (a) top view of streamline patterns on the upper and lower surfaces of the wing; and (b) cross-section of trailing vortices behind the wing. The trailing vortices change sign at O, the center of the wing.

Downstream of the wing, each half of the vortex sheet rolls up on itself and forms two distinct counter-rotating vortices called *tip vortices* (Fig. 13.21). The circulation of each tip vortex is equal to Γ_0, the circulation at the center of the wing. Tip vortices may become visually evident when an aircraft flies in humid air. The decreased pressure (due to the high velocity) and temperature in the core of the tip vortices may cause atmospheric moisture to condense into droplets or ice crystals, which may be seen in the form of *vapor trails* extending for many kilometers behind an aircraft traversing a clear sky. As an aircraft proceeds after takeoff, the tip vortices get longer, which means that kinetic energy is being constantly supplied to generate them. Thus, an additional drag force must be experienced by a wing of finite span. This is called the *induced drag*, and it can be predicted with lifting line theory.

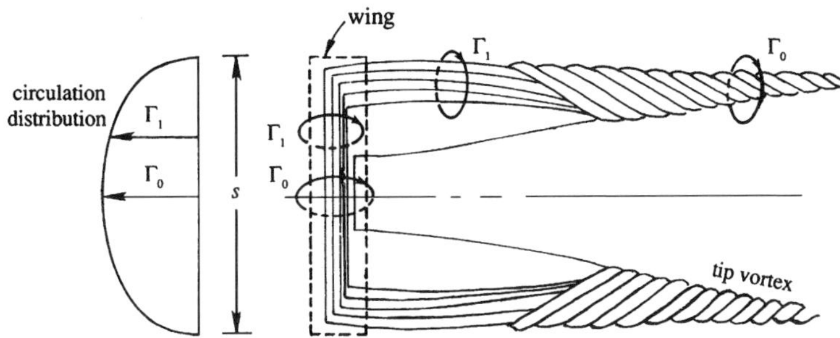

FIGURE 13.21 Rolling up of trailing vortices to form tip vortices. The mutual interaction of the trailing vortices eventually produces two counter-rotating wing-tip vortices having the same circulation as that bound to the center of the main wing.

One of Helmholtz's vortex theorems states that a vortex filament cannot end in the fluid, but must either end at a solid surface or form a closed vortex loop or ring. In the case of the finite wing, the tip vortices are the extension of the vorticity trapped in the wing's boundary layers. The tip vortices start at the wing and are joined together downstream of the aircraft by the various starting vortices of the wing. Starting vortices are left behind at the point where the aircraft took off and where the wing's lift was changed for aircraft maneuvers (ascent, descent, turns, etc.). In any case, the starting vortices are usually so far behind the wing that their effect on the wing's performance may be neglected and the tip vortices may be regarded as extending an infinite distance aft of the wing.

Three assumptions are needed for the simple version of lifting line theory presented here. The first is that the wing's aspect ratio, span/(average chord), is so large that the flow at any span-wise location may be treated as two dimensional. A second assumption is that the actual physical structure of the aircraft does not matter and that the aircraft's main wing may be replaced by a single (straight) vortex segment of variable strength. This vortex segment is called the *bound vortex*. It moves with the aircraft and lies along the aircraft's wings, nominally located at the center of lift at any span-wise location along the wing. The bound vortex forms the *lifting line* segment from which the theory draws its name. In general, the bound vortex is strongest near mid-span and weakest near the wing tips. As mentioned above, one of the Helmholtz theorems (Section 5.2) states that a vortex cannot begin or end in the fluid; it must end at a wall or form a closed loop. Therefore, as the bound vortex weakens from wing root to wingtip it releases vortex filaments that turn parallel to the stream-wise direction and are advected downstream, eventually coalescing to form the tip vortices. A third assumption made in lifting line theory is that the interaction of these trailing vortex filaments with each other can be ignored. Thus, each trailing vortex filament starts at the bound vortex and is assumed to lie along a straight semi-infinite horizontal line parallel to the upstream flow direction. Although a formal mathematical account of the theory was first published by Prandtl, many of the important underlying ideas were first conceived by Lanchester. The historical controversy regarding the credit for the theory is noted at the end of this section.

With these assumptions and the geometry shown in Fig. 13.22, a relation can be derived between the distribution of circulation along the wingspan and the strength of the trailing vortex filaments. Suppose that the clockwise circulation of the bound vortex changes from Γ to $\Gamma - d\Gamma$ at a certain point (Fig. 13.22a). Then another vortex AC of strength $d\Gamma$ must emerge from the location of the change. In fact, the strength and sign of the circulation around AC is such that, when AC is folded back onto AB, the circulation is uniform along the composite vortex tube. (Recall the vortex theorem of Helmholtz, which says that the strength of a vortex tube is constant along its length). Now consider the vortex strength or circulation distribution $\Gamma(y)$ that represents the main wing (Fig. 13.22b). The change in circulation in length dy is $d\Gamma$, which is a decrease if $dy > 0$. It follows that the magnitude of the trailing vortex filament of width dy is $-(d\Gamma/dy)dy$. For simple wings, the trailing vortices will be stronger near the wing tips where $d\Gamma/dy$ is the largest.

The critical contribution of lifting line theory is that it allows an approximate means of assessing the impact of the trailing vortex filaments on the performance of the bound vortex representing the aircraft's wing. The simplest means of

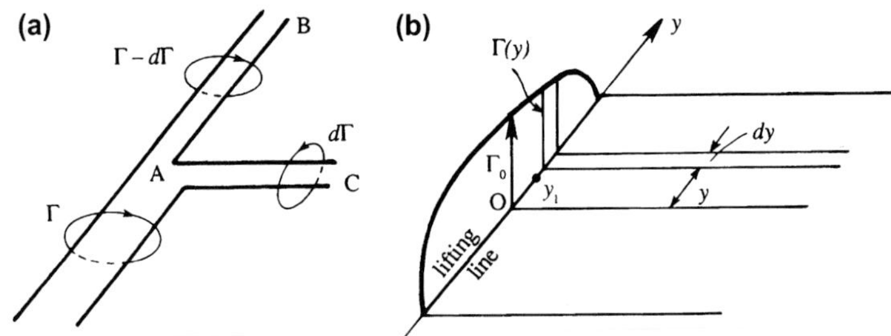

FIGURE 13.22 The mechanism leading to trailing vortices. (a) When the bound vortex having strength Γ weakens, it sheds a vortex filament AC of strength $d\Gamma$ into the wing's wake and continues along the wing as the vortex AB with strength $\Gamma - d\Gamma$. (b) The shed vortex filament that leaves the bound vortex at location y induces a downward velocity at location y_1 of the bound vortex when $y > y_1$. The induced velocity from all trailing vortex filaments is known as downwash.

assessing this impact is to determine the velocity induced at a point y_1 on the lifting line by the trailing vortex filament that leaves the wing at location y, and then integrating over the trailing filament contributions from all possible y values. Based on the Biot-Savart law (5.28), a straight semi-infinite trailing vortex filament that leaves the wing at y with strength $-(d\Gamma/dy)dy$ and remains horizontal induces a downward velocity of magnitude:

$$dw(y_1) = \frac{-(d\Gamma/dy)dy}{4\pi(y - y_1)}$$

at location $y(< y_1)$ along the lifting line (Example 5.4 with $\theta_1 = 0$ and $\theta_2 = 90°$). This velocity increment is *half* the velocity induced by an infinitely long vortex element. The bound vortex does not induce a velocity on itself, so for a wing of span s, the total downward velocity w at y_1 due to the entire trailing vortex sheet is therefore:

$$w(y_1) = \frac{1}{4\pi}\int_{-s/2}^{+s/2} \frac{d\Gamma}{dy}\frac{dy}{(y_1 - y)}, \tag{13.12}$$

which is called the *downwash* at y_1 on the lifting line. The vortex sheet also induces a smaller downward velocity in front of the airfoil and a larger one behind the airfoil (Fig. 13.23).

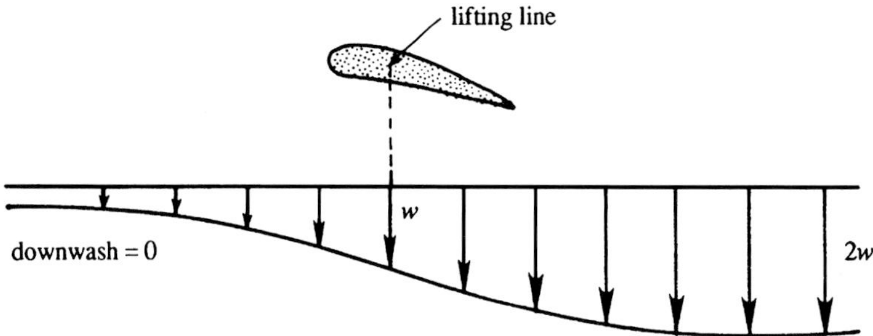

FIGURE 13.23 Variation of downwash ahead of and behind an airfoil. The downwash is weaker upstream of the wing and stronger downstream of it. The actual profile can be determined from the Biot-Savart law (5.28).

This downwash velocity adds to the free-stream velocity so that the incident flow at any location along the wing is the vector resultant of U and w (Fig. 13.24). The downwash therefore changes the local angle of attack of the airfoil, decreasing it by the angle:

$$\varepsilon = \tan(w/U) \approx w/U,$$

where the approximate equality follows when $w \ll U$, the most common situation in applications. Thus, the *effective angle of attack* at any span-wise location is:

$$\alpha_e = \alpha - \varepsilon = \alpha - w/U. \tag{13.13}$$

Because the aspect ratio is assumed large, ε is assumed to be small. Each element dy of the finite wing may then be assumed to act as though it is an isolated two-dimensional section set in a stream of uniform velocity U_e, at an angle of

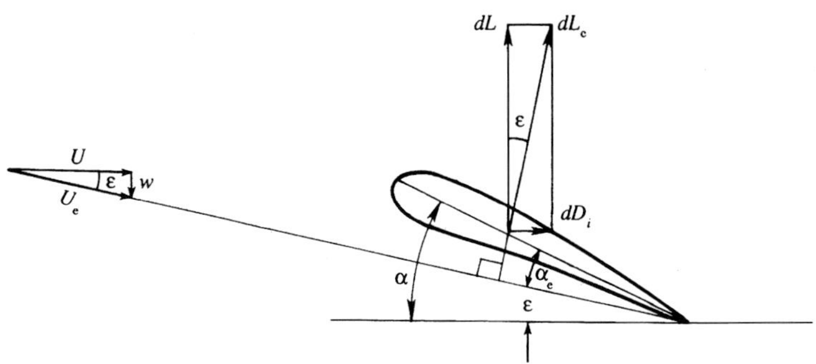

FIGURE 13.24 Lift and lift-induced drag on a wing element dy in the presence of a downwash velocity w. The downwash velocity locally lowers the angle of attack of the free stream and rotates the lift vector backward to produce the lift-induced drag.

attack α_e. According to the Kutta-Zhukhovsky lift theorem, a circulation Γ superimposed on the actual resultant velocity U_e generates an elemental aerodynamic force $dL_e = \rho U_e \Gamma dy$, which acts normal to U_e. This force may be resolved into two components, the conventional lift force dL normal to the direction of flight and a component dD_i parallel to the direction of flight (Fig. 13.24). Therefore:

$$dL = dL_e \cos\varepsilon = \rho U_e \Gamma dy \cos\varepsilon \approx \rho U \Gamma dy, \quad \text{and} \quad dD_i = dL_e \sin\varepsilon = \rho U_e \Gamma dy \sin\varepsilon \approx \rho w \Gamma dy.$$

In general w, Γ, U_e, ε, and α_e are all functions of y, so that for the entire wing:

$$L = \int_{-s/2}^{+s/2} \rho U \Gamma \, dy \quad \text{and} \quad D_i = \int_{-s/2}^{+s/2} \rho w \Gamma \, dy. \tag{13.14}$$

These expressions have a simple interpretation: whereas the interaction of U and Γ generates L, which acts normal to U, the interaction of w and Γ generates D_i, which acts normal to w.

The drag force D_i induced by the trailing vortices is called the *induced drag* and is zero for a wing of infinite span. It arises on a wing of finite span because finite-span wings continuously create trailing vortices and the rate of generation of trailing-vortex kinetic energy must equal the rate of work done against the induced drag, namely $D_i U$. For this reason, the induced drag is also known as the *vortex drag*. It is analogous to the *wave drag* experienced by a ship, which continuously radiates gravity waves while it travels. As we shall see, the induced drag is the largest part of the total drag experienced by an airfoil (away from stall).

A basic reason why there must be a downward velocity behind the wing is the following: The fluid exerts an upward lift force on the wing, and therefore the wing exerts a downward force on the fluid. The fluid must therefore constantly gain downward momentum as it goes past the wing.

For a given $\Gamma(y)$, $w(y)$ can be determined from (13.12) and D_i can then be determined from (13.14) However, $\Gamma(y)$ itself depends on the distribution of $w(y)$ because the effective angle of attack is changed due to $w(y)$. To see how $\Gamma(y)$ may be estimated, first note that the lift coefficient for a two-dimensional Zhukhovsky airfoil is nearly $C_L = 2\pi(\alpha + \beta)$. For a finite wing we may assume:

$$C_L = K\left[\alpha - \frac{w(y)}{U} + \beta(y)\right], \tag{13.15}$$

where $(\alpha - w/U)$ is the effective angle of attack, $-\beta(y)$ is the angle of attack for zero lift (found from experimental data such as Fig. 13.18), and K is the lift-curve slope, a constant whose value is nearly six for most airfoil sections ($K = 2\pi$ for Zhukhovsky and thin airfoils.) An expression for the circulation can be obtained by noting that the lift coefficient is related to the circulation as $C_L = L/[(1/2)\rho U^2 c] = \Gamma/[(1/2)Uc]$, so that $\Gamma = (1/2)UcC_L$. Eq. (13.15) is then equivalent to the assumption that the circulation for a wing of finite span is:

$$\Gamma(y) = \frac{K}{2} U c(y) \left[\alpha - \frac{w(y)}{U} + \beta(y)\right]. \tag{13.16}$$

For a given U, α, $c(y)$, and $\beta(y)$, (13.12) and (13.16) define an integral equation for determining $\Gamma(y)$.

An approximate solution to these two equations can be obtained by changing y and y_1 to angular variables γ and γ_1:

$$y = -(s/2)\cos\gamma \quad \text{and} \quad y_1 = -(s/2)\cos\gamma_1,$$

so that $\gamma = 0$ and $\gamma = \pi$ correspond to the left (port) and right (starboard) wing tips, respectively, and then assuming a Fourier series form for the circulation strength of the lifting line:

$$\Gamma = \sum_{n=1}^{\infty} \Gamma_n \sin(n\gamma), \tag{13.17}$$

where the Γ_n are undetermined coefficients. When (13.17) is substituted into (13.12), the resulting equation is:

$$w(y_1) = \frac{1}{2\pi s}\int_0^{\pi} \sum_{n=1}^{\infty} n\Gamma_n \frac{\cos(n\gamma)d\gamma}{\cos\gamma_1 - \cos\gamma} = \frac{1}{2\pi s}\sum_{n=1}^{\infty} n\Gamma_n \int_0^{\pi} \frac{\cos(n\gamma)d\gamma}{\cos\gamma_1 - \cos\gamma} = \frac{1}{2s}\sum_{n=1}^{\infty} n\Gamma_n \frac{\sin(n\gamma_1)}{\sin\gamma_1}, \tag{13.18}$$

where the final equality comes from evaluating the integral. Combining (13.16) through (13.18) and dropping the subscript '1' from γ, produces a single equation for the coefficients Γ_n:

$$\frac{K}{2}Uc(\alpha + \beta) = \sum_{n=1}^{\infty}\left(1 + \frac{nKc}{4s\sin\gamma}\right)\Gamma_n \sin(n\gamma), \tag{13.19}$$

where K, c, α, and β may all be functions of the transformed span coordinate γ. Thus, (13.19) is not a typical Fourier series solution because the coefficients of $\sin(n\gamma)$ inside the sum depend on γ. In practice, (13.19) can be solved approximately by truncating the sum after N terms, and then requiring its validity at N points along the wing to convert it into N algebraic equations for $\Gamma_1, \Gamma_2, \ldots \Gamma_N$. Fortunately in many circumstances, just few terms in the sum are needed to adequately represent $\Gamma(y)$.

With an approximate solution for $\Gamma(y)$ provided by several Γ_n computed algebraically from (13.19), the wing's lift and induced drag computed from (13.14) are:

$$L = \frac{\pi s}{4}\rho U\Gamma_1, \quad \text{and} \quad D_i = \frac{\pi}{8}\rho\sum_{n=1}^{N} n\Gamma_n^2. \tag{13.20, 13.21}$$

Thus, the wing's lift-to-drag ratio is maximized when $\Gamma_1 \neq 0$ and $\Gamma_n = 0$ for all $n > 1$. In this case, (13.17) reduces to:

$$\Gamma = \Gamma_1 \sin(\gamma) = \Gamma_1\sqrt{1 - (2y/s)^2}, \tag{13.22}$$

which is known as an *elliptical lift distribution*. For such a lift distribution, the three-dimensional wing's lift coefficient is:

$$C_{L,3D} = K\alpha/(1 + K/\pi\Lambda) \tag{13.23}$$

(see Exercise 13.15), where Λ is given by (13.1). The downwash for an elliptical lift distribution is constant across the wingspan:

$$w(y) = \Gamma_1/2s,$$

as can be found from (13.18) and (13.22) The induced drag for an elliptical lift distribution is:

$$D_i = \frac{\pi}{8}\rho\Gamma_1^2 = \frac{2L^2}{\pi\rho U^2 s^2}, \tag{13.24}$$

where (13.20) has been used to introduce L in the second equality. Thus, the induced drag coefficient for an elliptical lift distribution is:

$$C_{D_i}\frac{D_i}{(1/2)\rho U^2 A} = \frac{C_L^2}{\pi(s^2/A)} = \frac{C_L^2}{\pi\Lambda}, \tag{13.25}$$

where C_L and C_D are given by (4.108) and (4.109) in Section 4.11, A is the wing's planform area, and Λ is the wing's aspect ratio. Eq. (13.25) shows that $C_{D_i} \to 0$ when the flow is two dimensional, that is, in the limit $\Lambda \to \infty$. More importantly, it shows that the *induced drag coefficient increases as the square of the lift coefficient*. We shall see in the following section that the induced drag generally makes the largest contribution to the total drag of an airfoil.

Since an elliptic circulation distribution minimizes the induced drag, it is of interest to determine the circumstances under which such a circulation can be established. Consider an element dy of the wing (Fig. 13.24). The lift on the element is:

$$dL = \rho U \Gamma dy = C_L (1/2) \rho U^2 c dy, \tag{13.26}$$

where cdy is a wing area element. If the circulation distribution is elliptic, then the downwash is independent of y. In addition, if the wing profile is geometrically similar at every point along the span and has the same geometrical angle of attack α, then the effective angle of attack and hence the lift coefficient C_L will be independent of y. Eq. (13.26) shows that the chord length c is then simply proportional to Γ, and so $c(y)$ is also elliptically distributed. Thus, an untwisted wing with elliptic planform, or composed of two semi-ellipses (Fig. 13.25), will generate an elliptic circulation distribution. However, the same effect can also be achieved with non-elliptic planforms if the angle of attack varies along the span, that is, if the wing has twist (see Exercise 13.14).

The results of lifting line theory have had an enormous impact on the design and development of subsonic aircraft. However, the results presented here are approximate because of the geometrical assumptions made about the aircraft's wings, its trailing vortices, and the tip vortices. Thus, an elliptical lift distribution is only approximately optimal, and a more general theory would produce refinements. Yet, with suitable geometric modifications lifting line theory can be applied to multiple-wing aircraft and rotating propellers. Furthermore, its implications help explain near-ground effects for landing aircraft, and the Λ-pattern commonly formed by flocks of migrating birds.

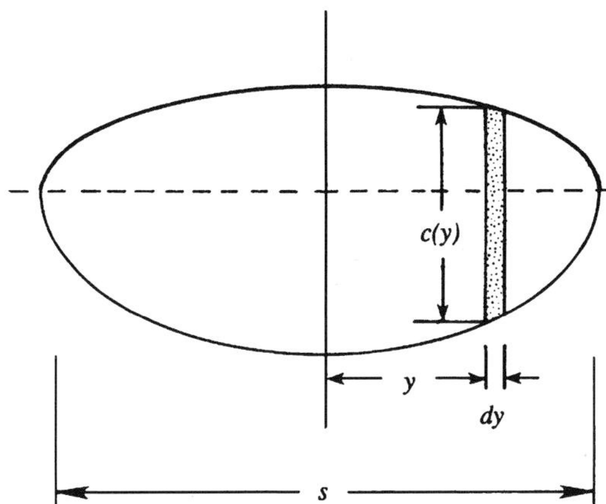

FIGURE 13.25 Wing with an elliptic planform. Here the variation in the chord over the span can produce an elliptical lift distribution. This planform is similar to that of the *British Spitfire*, a WWII combat aircraft.

Example 13.6

Consider an uncambered untwisted rectangular-planform wing with an aspect ratio = Λ = span/chord = $s/c = 6$, and a constant two-dimensional lift curve slope $K = 6$ flying at angle of attack α. For symmetrical loading, use three terms of (13.17) to solve (13.19) at $\gamma = \pi/6$, $\pi/3$, and $\pi/2$ to find Γ_1, Γ_3, Γ_5, C_L and C_{D_i}. Compare these coefficients to equivalent results from an elliptically-loaded wing.

Solution For an uncambered untwisted rectangular-planform wing, β will be zero, and α and c will be constants. Thus, (13.19) can be written:

$$1 = \sum_{n=1}^{\infty} \left(1 + \frac{nK}{4\Lambda \sin\gamma}\right) \Gamma'_n \sin(n\gamma) = \sum_{n=1}^{\infty} \left(1 + \frac{n}{4\sin\gamma}\right) \Gamma'_n \sin(n\gamma),$$

where $\Gamma'_n = 2\Gamma_n / KUc\alpha$, and the second equality follows in the given situation because $K = \Lambda = 6$. For symmetrical wing loading, only the odd-numbered Γ'_n will contribute, so the equation above reduces to:

$$1 \cong \left(1 + \frac{1}{4\sin\gamma}\right) \Gamma'_1 \sin(n\gamma) + \left(1 + \frac{3}{4\sin\gamma}\right) \Gamma'_3 \sin(3\gamma) + \left(1 + \frac{5}{4\sin\gamma}\right) \Gamma'_5 \sin(5\gamma),$$

when the first three odd numbered terms are kept. To find the Γ'_n, evaluate this equation at $\gamma = \pi/6$, $\pi/3$, and $\pi/2$ to generate a 3-by-3 system of linear algebraic equations:

$$\begin{Bmatrix} 1 \\ 1 \\ 1 \end{Bmatrix} \cong \begin{Bmatrix} 0.7500 & 2.5000 & 1.7500 \\ 1.1160 & 0.0 & -2.1160 \\ 1.2500 & -1.7500 & 2.2500 \end{Bmatrix} \begin{Bmatrix} \Gamma'_1 \\ \Gamma'_3 \\ \Gamma'_5 \end{Bmatrix}.$$

The solution is: $\Gamma'_1 = 0.9264,$; $\Gamma'_3 = 0.1109$, and $\Gamma'_5 = 0.0160$.

The rectangular wing's lift and drag coefficients can be calculated from these results. From (13.20), the wing's lift coefficient is:

$$C_L = \frac{(\pi/4)s\rho U \Gamma_1}{(1/2)\rho U^2 sc} = \frac{(\pi/4)s\rho U K \cdot (1/2) K U c \alpha \Gamma'_1}{(1/2)\rho U^2 sc} = \frac{\pi}{4} K \alpha \Gamma'_1 = 4.366\alpha.$$

From (13.23), the lift coefficient for an elliptically loaded wing with the same aspect ratio is:

$$C_L = K\alpha/(1 + K/\pi\Lambda) = 6\alpha/(1 + 6/6\pi) = 4.551\alpha,$$

which is approximately 4% higher. From (13.21), the rectangular wing's drag coefficient is:

$$C_D = \frac{(\pi/8)\rho(\Gamma_1^2 + 3\Gamma_3^2 + 5\Gamma_5^2)}{(1/2)\rho U^2 sc} = \frac{\pi K^2 \alpha^2}{16\Lambda}(\Gamma_1'^2 + 3\Gamma_3'^2 + 5\Gamma_5'^2) = 1.056\alpha^2.$$

From (13.25), the equivalent drag coefficient for an elliptically loaded wing with the same lift coefficient as the rectangular wing is:

$$C_D = \frac{C_L^2}{\pi\Lambda} = \frac{(4.366\alpha)^2}{6\pi} = 1.011\alpha^2,$$

which is also approximately 4% lower.

Although the differences between a rectangular wing and the more-efficient elliptically loaded wing may seem small, they can be critical for the commercial viability of long-haul aircraft intended for overseas routes where an early stop for refueling is impossible.

Lanchester vs. Prandtl

There is some controversy in the literature about who should get more credit for developing lifting line theory. Since Prandtl in 1918 first published the theory in a mathematical form, textbooks for a long time have called it the "Prandtl Lifting Line Theory." Lanchester was bitter about this, because he felt that his contributions were not adequately recognized. The controversy has been discussed by von Karman (1954, p. 50), who witnessed the development of the theory. He gives a lot of credit to Lanchester, but falls short of accusing his teacher Prandtl of being deliberately unfair. Here we shall note a few facts that von Karman brings up.

Lanchester was the first person to study a wing of finite span. He was also the first person to conceive that a wing can be replaced by a bound vortex, which bends backward to form the tip vortices. Last, Lanchester was the first to recognize that the minimum power necessary to fly is that required to generate the kinetic energy field of the downwash field. It seems, then, that Lanchester had conceived all of the basic ideas of the wing theory, which he published in 1907 in the form of a book called *Aerodynamics*. In fact, a figure from his book looks similar to the current Fig. 13.21.

Many of these ideas were explained by Lanchester in his talk at Göttingen, long before Prandtl published his theory. Prandtl, his graduate student von Karman, and Carl Runge were all present. Runge, well known for his numerical integration scheme of ordinary differential equations, served as an interpreter, because neither Lanchester nor Prandtl could speak the other's language. As von Karman said, "both Prandtl and Runge learned very much from these discussions."

However, Prandtl did not want to recognize Lanchester for priority of ideas, saying that he conceived of them before he saw Lanchester's book. Such controversies cannot be settled, and great intellects have been involved in controversies before.

In view of the fact that Lanchester's book was already in print when Prandtl published his theory, and the fact that Lanchester had all the ideas but not a formal mathematical theory, we have called it the "Lifting Line Theory of Prandtl and Lanchester" at the outset of this section.

13.7 Lift and drag characteristics of airfoils

Before an aircraft is built its wing design is tested in a wind tunnel, and the results are generally given as plots of C_L and C_D versus the angle of attack α. A typical plot for a simple rectangular-planform wing is shown in Fig. 13.26 where it is seen that, for $-9° < \alpha < 12°$, the variation of C_L with α is approximately linear, a typical value of $dC_L/d\alpha(= K)$ being ~ 0.1 per degree. The lift reaches a maximum value at $\alpha \approx 15°$. If the angle of attack is increased further, the steep adverse pressure gradient on the upper surface of the airfoil causes the flow to separate before reaching the wing's trailing edge, and a large wake is formed (Fig. 13.27). The drag coefficient increases and the lift coefficient drops. The wing is said to *stall* as the suction-side boundary-layer separation point moves toward the leading edge. Beyond the stalling incidence angle the lift coefficient levels off again and remains at $\approx 0.7-0.8$ up to α values of tens of degrees. The wing's C_L-curve slope, maximum lift coefficient, and stall characteristics could all be improved from that shown in Fig. 13.26 with a more sophisticated wing design.

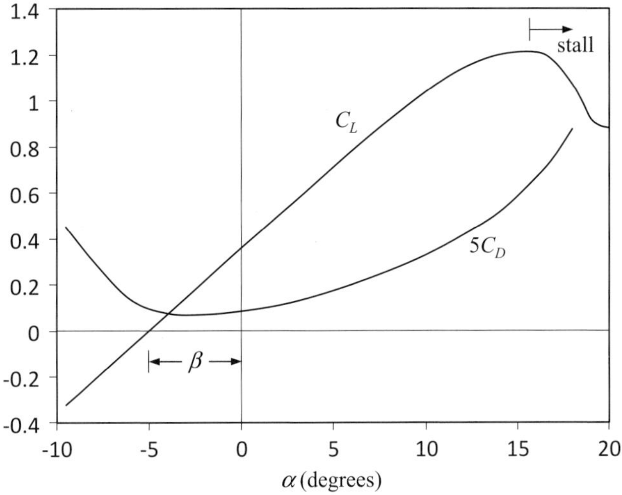

FIGURE 13.26 Generic lift and drag coefficients for a rectangular-planform finite-aspect-ratio wing vs. angle of attack. There is lift at $\alpha = 0$ so the foil shape has non-zero camber. The drag increase is almost quadratic with increasing angle of attack in accordance with (13.25).

For a fixed-shape wing, the maximum possible lift coefficient depends largely on the Reynolds number Re. For chord-based Reynolds numbers of Re $\sim 10^5 - 10^6$, the suction-side boundary layer may separate before it undergoes transition, and stall may begin before α reaches 10° leading to maximum lift coefficients < 0.9. At larger Reynolds numbers, say Re $> 10^7$, the suction-side boundary layer transitions to turbulence before it separates and is therefore able to stay attached up to α-values approaching or exceeding 20°. Maximum lift coefficients near or even slightly above two may be obtained at the highest Reynolds numbers.

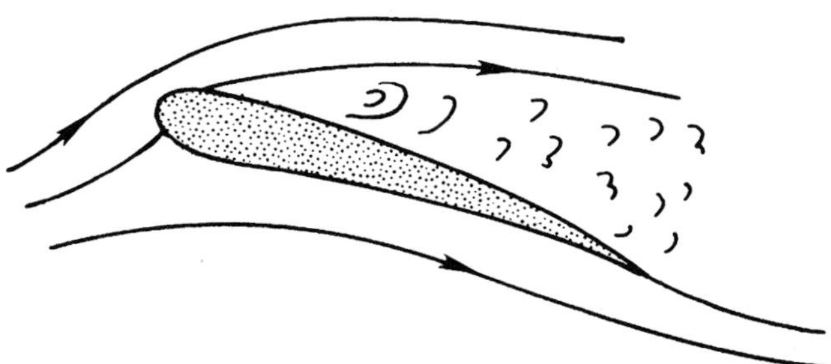

FIGURE 13.27 Stalling of an airfoil. Here, the Kutta condition is no longer satisfied, and the flow separates near the leading edge on the foil's suction side. In this situation, a foil's lift-to-drag ratio may fall by a factor of 4 or 5 from its design condition.

The angle of attack at zero lift, denoted by $-\beta$ here, is a f unction of the airfoil-section's camber. (For a Zhukhovsky airfoil, $\beta = 2$(camber)/chord.) The effect of increasing the airfoil camber is to raise the entire graph of C_L versus α, thus increasing the maximum values of C_L without stalling. A cambered profile delays stall because its leading edge points into the airstream while the rest of the airfoil is inclined to the stream. Rounding an airfoil's nose is also essential to prevent leading-edge separation in real fluids. (In ideal flow, a leading edge singularity may prevent separation; see Exercise 13.8.)

Trailing edge flaps act to increase the camber and thereby the lift coefficient when they are deployed, and this allows lower aircraft landing speeds.

Various terms are in common usage to describe the different components of the drag. The total drag of a body can be divided into a *friction drag* due to the tangential stresses on the surface and *pressure drag* due to the normal stresses. The pressure drag can be further subdivided into an *induced drag* and a *form drag*. The induced drag is the drag that results from the work done by the body to supply the kinetic energy of the down-wash field as the trailing vortices increase in length. The form drag is defined as the part of the total pressure drag that remains after the induced drag is subtracted out. (Sometimes the skin friction and form drags are grouped together and called the *profile drag*, which represents the drag due to the wing's geometrical profile alone and not due to the finiteness of the wing.) The form drag depends strongly on the shape and orientation of the airfoil and can be minimized by good design. In contrast, relatively little can be done about the induced drag if the wing's aspect ratio is fixed.

Normally the induced drag constitutes the major part of the total drag of a wing. As C_{D_i} is nearly proportional to C_L^2, and C_L is nearly proportional to α, it follows that $C_{D_i} \propto \alpha^2$. This is why the drag coefficient in Fig. 13.26 seems to increase quadratically with angle of attack.

For high-speed aircraft, the appearance of shock waves can adversely affect the behavior of the lift and drag characteristics. In such cases the maximum *flow* speeds can be close to or higher than the speed of sound even when the aircraft is flying at subsonic speeds. Shock waves can form when the local flow speed exceeds the local speed of sound. To reduce their effect, the wings are given a *sweepback angle*, as shown in Fig. 13.2. The maximum flow speeds depend primarily on the component of the oncoming stream perpendicular to the leading edge; this component is reduced as a result of the sweepback. Thus, increased flight speeds are achievable with highly swept wings. This is particularly true when the aircraft flies at supersonic speeds in which there is invariably a shock wave in front of the nose of the fuselage, extending downstream in the form of a cone. Highly swept wings are then used in order that the wing does not penetrate this shock wave. For flight speeds exceeding Mach number 2 or so, the wings have such large sweepback angles that they resemble the Greek letter Δ; these wings are sometimes called *delta wings*.

Example 13.7

The aerodynamic performance of wings is sometimes presented as C_L vs. C_D with the angle of attack a being the free parameter that varies along the curve. The relationship between C_L and C_D is known as a *drag polar* and such plots are called *polar plots*. Replot the aerodynamic results in Fig. 13.26 to create a polar plot, and compare with the drag polar for an elliptical lift distribution, (13.25), evaluated with an aspect ratio of $\Lambda = 6.5$.

Solution Using an angle of attack increment of 2°, the curves provided in Fig. 13.26 are approximately described by:

α (°)	C_D	C_L
−8.0	0.060	−0.22
−6.0	0.027	−0.07
−4.0	0.015	0.07
−2.0	0.014	0.22
0.0	0.017	0.36
2.0	0.022	0.50
4.0	0.030	0.64
6.0	0.040	0.78
8.0	0.052	0.92
10.0	0.066	1.04
12.0	0.083	1.14
14.0	0.103	1.20
16.0	0.133	1.21
18.0	0.175	1.07

The polar plot for this data is shown in Fig. 13.28, along with the curve predicted by (13.25) for an aspect ratio of 6.5. The horizontal shift between the plotted drag polar and the predicted curve is the *profile drag coefficient*, $C_{D,o}$, of the wing. Here, $C_{D,o} \approx 0.01$ and it is less than half of the total drag coefficient, C_D, when $C_L > 0.5$.

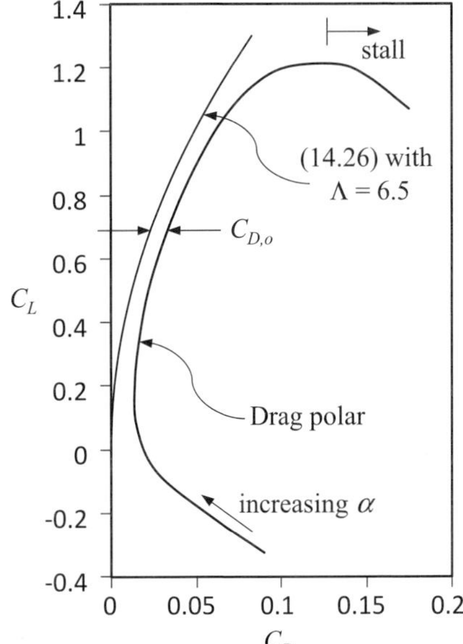

FIGURE 13.28 Polar plot of the lift and drag coefficients shown in Fig. 13.26 along with the prediction of (13.25) with $\Lambda = 6.5$. The horizontal shift between the two curves is the wing's profile drag coefficient, $C_{D,o}$. The foil is cambered so the minimum C_D does not occur at $C_L = 0$.

13.8 Propulsive mechanisms of fishes and birds

The propulsive mechanisms of many animals are based on lift generation by wing-like surfaces. Just the basic ideas of this interesting subject are presented here. More detail is provided by Lighthill (1986).

First consider swimming fish. They develop *forward* thrust by horizontally oscillating their tails from *side to side*. Fish tails like that shown in Fig. 13.29a have a cross-section resembling that of a symmetric airfoil. One-half of the oscillation is represented in Fig. 13.29b, which shows the top view of the tail. The sequence 1 to 5 represents the positions of the tail during the tail's motion to the left. A quick change of *orientation* occurs at one extreme position of the oscillation during 1 to 2; the tail then moves to the left during 2 to 4, and another quick change of orientation occurs at the other extreme during 4 to 5.

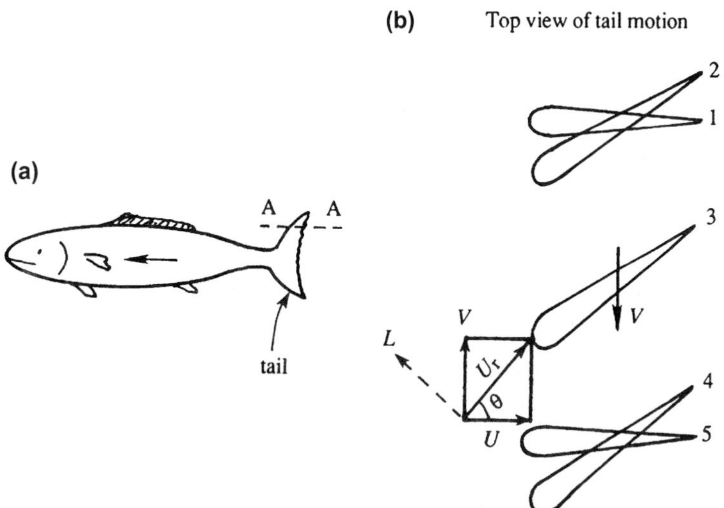

FIGURE 13.29 Propulsion of fish. (a) The cross-section of the tail along AA is that of a symmetric airfoil. Five positions of the tail during its motion to the left are shown in (b). The lift force L is normal to the resultant speed U_r of water with respect to the tail.

Suppose the tail is moving to the left at speed V, and the fish is moving forward at speed U. The fish controls these magnitudes so that the resultant fluid velocity U_r (relative to the tail) is inclined to the tail surface at a positive angle of attack. The resulting lift L is perpendicular to U_r and has a forward component $L\sin\theta$. (It is easy to verify that there is a similar forward propulsive force when the tail moves from left to right.) This thrust, working at the rate $UL\sin\theta$, propels the fish. To achieve this propulsion, the tail of the fish pushes sideways on the water against a force of $L\cos\theta$, which requires work at the rate $VL\cos\theta$. Since $V/U = \tan\theta$, the conversion of energy is ideally perfect – all of the oscillatory work done by the fish tail goes into the translation. In practice, however, this is not the case because of the presence of induced drag and other effects that generate a wake.

Most fish stay afloat by controlling the buoyancy of an internal swim bladder. In contrast, some large marine mammals such as whales and dolphins develop *both* a forward thrust and a vertical lift by moving their tails *vertically*. They are able to do this because their tail surface is *horizontal*, in contrast to the vertical tail shown in Fig. 13.29. A review by Fish and Lauder (2006) provided evidence that leading-edge tubercles as seen on humpback whale flippers increase lift and reduce drag at high angles of attack. This is because separation is delayed due to the creation of stream-wise vortices on the suction side. Cetacean flukes or flippers and fish tail fins as well as dorsal and pectoral fins are flexible and can vary their camber during a stroke. As a result they are efficient propulsive devices.

Now consider flying birds, who flap their wings to generate *both* the lift to support their body weight and the forward thrust to overcome drag. Fig. 13.30 shows a vertical section of the wing positions during the upstroke and downstroke of the wing. (Birds have cambered wings, but this is not shown in the figure.) The angle of inclination of the wing with the airstream changes suddenly at the end of each stroke, as shown. The important point is that the upstroke is inclined at a greater angle to the airstream than the downstroke. As the figure shows, the downstroke develops a lift force L perpendicular to the resultant velocity of the air relative to the wing. Both a forward thrust and an upward force result from the downstroke. In contrast, little aerodynamic force is developed during the upstroke, as the resultant velocity is then nearly parallel to the wing. Birds therefore do most of the work necessary for flight during the downstroke.

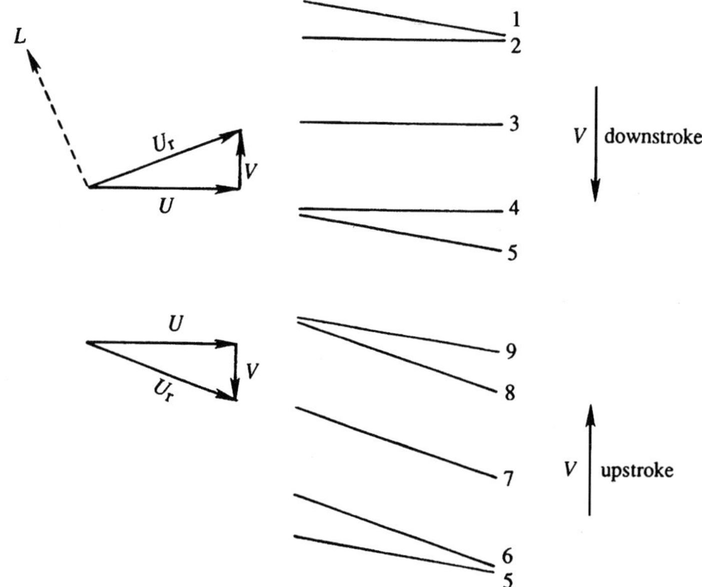

FIGURE 13.30 Propulsion of a bird. A cross-section of the wing is shown during upstroke and downstroke. During the downstroke, a lift force L acts normal to the resultant speed U_r of air with respect to the wing. During the upstroke, U_r is nearly parallel to the wing and very little aerodynamic force is generated.

Liu et al. (2006) provide the most complete description to date of wing planform, camber, airfoil section, and span-wise twist distribution of seagulls, mergansers, teals, and owls. Moreover, flapping as viewed by video images from free flight was digitized and modeled by a two-jointed wing at the quarter chord point. The data from this paper can be used to model the aerodynamics of bird flight.

Using previously measured kinematics and experiments on an approximately 100-times upscaled model, Ramamurti and Sandberg (2001) calculated the flow about a Drosophila (fruit fly) in flight. They matched Reynolds number (based on wing-tip speed and average chord) and found that viscosity had negligible effect on thrust and drag at a flight Reynolds number of 120. The wings were near elliptical plates with axis ratio 3:1.2 and thickness about 1/80 of the span. Averaged over a cycle, the mean thrust coefficient (thrust/[dynamic pressure × wing surface]) was 1.3 and the mean drag coefficient close to 1.5.

Example 13.8

Is there evidence that the natural realm has exploited the underlying physics of lifting line theory for aero- and hydrodynamic efficiency?

Solution The overwhelming answer is 'yes.' The natural world's soaring champion, the albatross, a sea bird, also has the highest aspect ratio wings (15+). Interestingly, terrestrial soaring birds (hawks, eagles, vultures, etc.) achieve nearly the same efficiency, as measured by lift-to-drag ratio, at approximately half that aspect ratio. Here, multiple non-interlocking wingtip feathers, having gaps in between, increase wingtip-shed vortex core size and thereby reduce vortex kinetic energy while maintaining the same vortex circulation. Wingtip modifications on modern commercial airliners serve the same purpose.

The story is similar in the ocean. The natural world's fastest swimmer, the sailfish, likely has the highest aspect ratio tailfin ($\sim$ 12). When viewed from the side, a sailfish's tailfin looks remarkably similar to the airliner wing planform shown in Figs. 13.1 and 13.2. Other fish noted for their speed and cruising efficiency (marlin, tuna, etc.) also have high-aspect ratio tailfins (7–10).

13.9 Sailing against the wind

People have sailed without the aid of an engine for thousands of years and have known how to reach an upwind destination. Actually, it is not possible to sail exactly against the wind, but it is possible to sail at $\approx 40-45°$ to the wind. Fig. 13.31 shows how this is made possible by the aerodynamic lift on the sail. The wind speed is U, and the sailing speed is V, so that the apparent wind speed relative to the boat is U_r. If the sail is properly oriented, this gives rise to a lift force perpendicular to U_r and a drag force parallel to U_r. The resultant force F can be resolved into a driving component (thrust) along the motion of the boat and a lateral component. The driving component performs work in moving the boat; most of this work goes into overcoming the frictional drag and in generating the gravity waves that radiate outward from the hull. The lateral component does not cause much sideways drift because of the shape of the hull. The thrust decreases as the angle θ decreases and normally vanishes when θ is $\approx 40-45°$. The energy for sailing comes from the wind field, which loses kinetic energy after passing the sail.

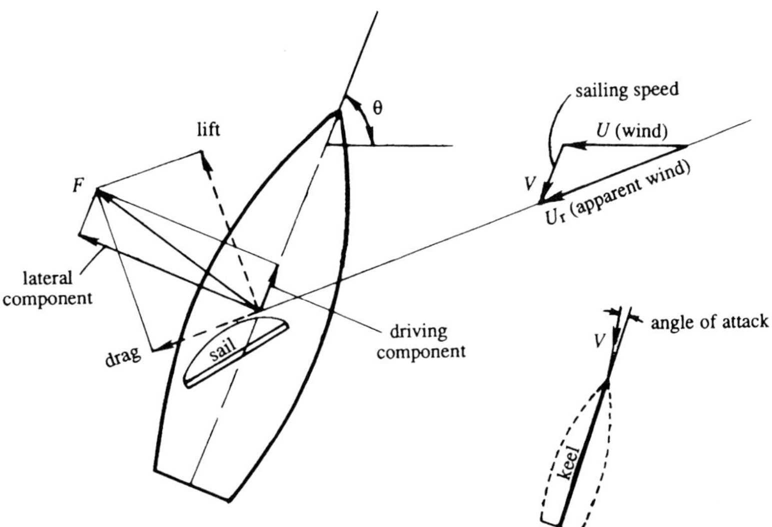

FIGURE 13.31 Principle of sailing against the wind. A small component of the sail's lift pushes the boat forward at an angle $\theta < 90°$ to the wind. Thus by traversing a zig-zag course at angles $\pm\theta$, a sailboat can reach an upwind destination. A sailboat's keel may make a contribution to its upwind progress, too.

In the foregoing discussion we have not considered the hydrodynamic forces exerted by the water on the hull. At constant sailing speed the net hydrodynamic force must be equal and opposite to the net aerodynamic force on the sail. The hydrodynamic force can be decomposed into drag (parallel to the direction of motion) and lift. Lift is provided by the sailboat's *keel*, which is a thin, but often heavy, vertical surface extending downward from the bottom of the hull. For the keel to act as a lifting surface, the longitudinal axis of the boat points at a small angle to the direction of motion of the boat, as indicated near the bottom right part of Fig. 13.31. This keel-angle of attack is generally $< 3°$ and is not noticeable.

The hydrodynamic lift developed by the keel opposes the aerodynamic lateral force on the sail. Without the keel the lateral aerodynamic force on the sail would topple the boat around its longitudinal axis.

To arrive at a destination directly up wind, one has to sail in a zig-zag path, always maintaining an angle of $\approx 45°$ to the wind. For example, if the wind is coming from the east, we can first proceed northeastward as shown, then change the orientation of the sail to proceed southeastward, and so on. In practice, a combination of a number of sails is used for effective maneuvering. The mechanics of sailing yachts is discussed in Herreshoff and Newman (1966).

Example 13.9

Two captains with identical sailing yachts and comparable crews decided to race downstream on a river. Unfortunately, the wind was calm as the horn sounded to start the race. On the first yacht, the captain decided to let the river's current do all the work so this captain ordered all sails fully lowered and instructed the helmsmen to steer a straight course to the finish line. The strategy on the second boat was entirely different; the second captain ordered the largest sails hoisted, and then kept the crew busy tacking back and forth, i.e., steering a zigzag path, across the river. Who won the race? Why?

Solution Near a river, a stationary shore station will be used to specify the wind speed. Therefore, on a calm day, a velocity difference exists between the stationary air and the river's moving water. An air-water speed difference is all that is needed for effective sailing, and as described in this section's text, sailors are able to make progress upwind by setting their boat to travel at acute angles on either side of the direction that is straight upwind. In the situation described here, straight upwind is directly down river when riding in either sailboat. Thus, the second yacht won the race by using the available air-water velocity difference to make upwind progress beyond that supplied by the river's current.

Exercises

13.1 As an extension of Example 13.1, consider a sphere with radius a that moves along the x-axis on a trajectory given by $\mathbf{x}_p(t) = x_p(t)\mathbf{e}_x$ in a fluid moving with uniform velocity: $\mathbf{u} = U\mathbf{e}_x + V\mathbf{e}_y$. Determine a formula for the mechanical power, W, necessary to overcome the aerodynamic drag force on the sphere in terms of a, U, V, x_p, $\rho =$ the density of the air, and $C_D =$ the drag coefficient of the sphere. If dx_p/dt is constant, under what conditions is W reduced by the presence of non-zero $\mathbf{u}$?

13.2 Consider the elementary aerodynamics of a projectile of mass m with $C_L = 0$ and $C_D =$ constant. In Cartesian coordinates with gravity g acting downward along the y-axis, a set of equations for such a projectile's motion are:

$$m\frac{dV_x}{dt} = -D\cos\theta,\ m\frac{dV_y}{dt} = -mg - D\sin\theta,$$

$$\tan\theta = V_y/V_x, \quad \text{and} \quad D = \frac{1}{2}\rho\left(V_x^2 + V_y^2\right)AC_D,$$

where V_x and V_y are the horizontal and vertical components of the projectile's velocity, θ is the angle of the projectile's trajectory with respect to the horizontal, D is the drag force on the projectile, ρ is the air density, and A is projectile's frontal area. Assuming a shallow trajectory, where $V_x^2 \gg V_y^2$ and $mg \gg D\sin\theta$, show that the distance traveled by the projectile over level ground is:

$$x \cong \frac{2m}{\rho AC_D}\ln\left(1 + \frac{\rho AC_D V_o^2\cos\theta_o\sin\theta_o}{mg}\right)$$

if it is launched from ground level with speed of V_o at an angle of θ_o with respect to the horizontal. Does this answer make sense as $C_D \to 0$?

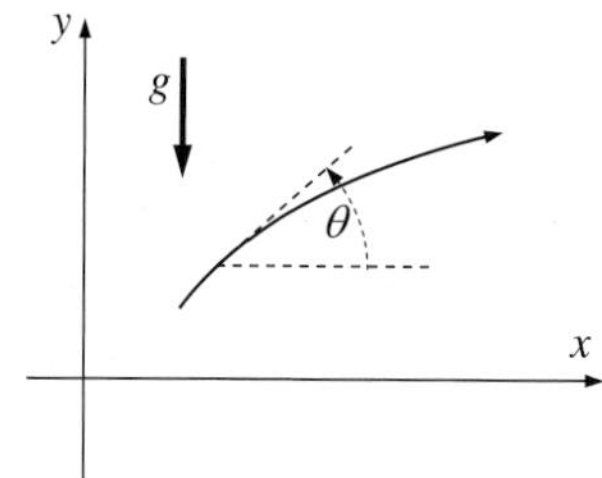

13.3 As a model of a two-dimensional airfoil's trailing edge flow, consider the potential $\phi(r,\theta) = (Ud/n)(r/d)^n \cos(n\theta)$ in the usual r-θ coordinates (Fig. 3.3a). Here U, d, and n are positive constants, the fluid has density ρ, and the foil's trailing edge lies at the origin of coordinates.

a) Sketch the flow for $n = 3/2, 5/4$, and 9/8 in the angle range $|\theta| < \pi/n$, and determine the full included angle of the foil's trailing edge in terms of n.

b) Determine the fluid velocity at $r = d$ and $\theta = 0$.

c) If p_0 is the pressure at the origin of coordinates and p_d is the pressure at $r = d$ and $\theta = 0$, determine the pressure coefficient: $C_p = (p_0 - p_d)/\left[(1/2)\rho U^2\right]$ as a function of n. In particular, what is C_p when $n = 1$ and when $n > 1$?

13.4 Consider an airfoil section in the xy-plane, the x-axis being aligned with the chord line. Examine the pressure forces on an element $d\mathbf{s} = (dx, dy)$ on the surface, and show that the net force (per unit span) in the y-direction is:

$$F_y = -\int_0^c p_u\, dx + \int_0^c p_l\, dx,$$

where p_u and p_l are the pressure on the upper and the lower surfaces, and c is the chord length. Show that this relation can be rearranged in the form:

$$C_y = \frac{F_y}{(1/2)\rho U^2 c} = \oint C_p d\left(\frac{x}{c}\right),$$

where $C_p = (p_0 - p_\infty)/[(1/2)\rho U^2]$, and the integral represents the area enclosed in a C_p versus x/c diagram, such as Fig. 13.8. Neglect shear stresses. [Note that C_y is not exactly the lift coefficient, since the airstream is inclined at a small angle α with respect to the x-axis.]

13.5 The measured pressure distribution over a section of a two-dimensional airfoil at 4° incidence has the following form:

Upper Surface: C_p is constant at −0.8 from the leading edge to a distance equal to 60% of chord and then increases linearly to 0.1 at the trailing edge.

Lower Surface: C_p is constant at −0.4 from the leading edge to a distance equal to 60% of chord and then increases linearly to 0.1 at the trailing edge.

Using the results of Exercise 13.4, show that the lift coefficient is nearly 0.32.

13.6 The Zhukhovsky transformation $z = \zeta + b^2/\zeta$ transforms a circle of radius b, centered at the origin of the ζ-plane, into a flat plate of length $4b$ in the z-plane. The circulation around the cylinder is such that the Kutta condition is satisfied at the trailing edge of the flat plate. If the plate is inclined at an angle α to a uniform stream U, show that:

(i) The complex potential in the ζ-plane is $w = U\left(\zeta e^{-i\alpha} + \frac{b^2}{\zeta} e^{+i\alpha}\right) + \frac{i\Gamma}{2\pi} \ln\left(\zeta e^{-i\alpha}\right)$, where $\Gamma = 4\pi U b \sin\alpha$.

Note that this represents flow over a circular cylinder with circulation in which the oncoming velocity is oriented at an angle α.

(ii) The velocity components at point P $(-2b, 0)$ in the ζ-plane are $\left[\frac{3}{4}U\cos\alpha, \frac{9}{4}U\sin\alpha\right]$.

(iii) The coordinates of the transformed point P′ in the xy-plane are $[-5b/2, 0]$.

(iv) The velocity components at $[-5b/2, 0]$ in the xy-plane are $[U\cos\alpha, 3U\sin\alpha]$.

13.7 In Fig. 13.13, the angle at A′ has been marked 2β. Prove this. [*Hint*: Locate the center of the circular arc in the z-plane.]

13.8 Ideal flow past a flat plate inclined at angle α with respect to a horizontal free stream produces lift but no drag when the Kutta condition is applied at the plate's trailing edge. However, pressure forces can only act in the plate-normal direction and this direction is *not* perpendicular to the flow. Therefore, to achieve zero drag, another force must act on the plate. This extra force is known as *leading-edge suction* and its existence can be assessed from the potential for flow around the tip of a flat plate that is coincident with the x-axis for $x > 0$. In two-dimensional polar coordinates, this velocity potential is $\phi = 2U_o\sqrt{ar}\cos(\theta/2)$ where U_o and a are velocity and length scales, respectively, that characterize the flow.

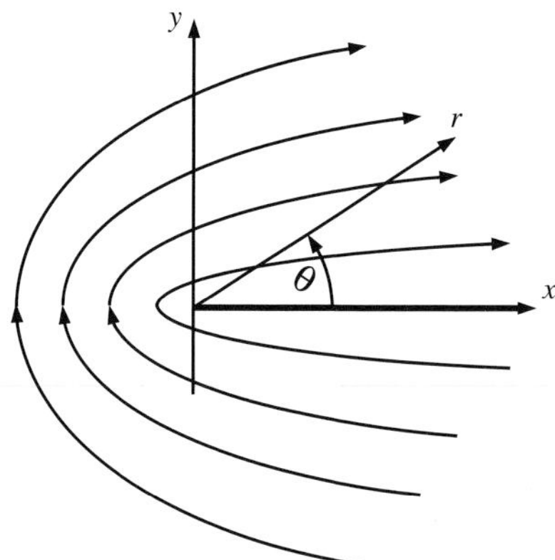

a) Determine u_r and u_θ, the radial and angular-directed velocity components, respectively.
b) If the pressure far from the origin is p_∞, determine the pressure p at any location (r, θ).
c) Use the given potential, a circular control volume of radius ε centered at the origin of coordinates, and the control volume version of the ideal flow momentum equation, $\int_C \rho \mathbf{u}(\mathbf{u} \cdot \mathbf{n}) d\xi = -\int_C p\mathbf{n} d\xi + \mathbf{F}$, to determine the force $\mathbf{F}$ (per unit depth into the page) that holds the plate stationary when $\varepsilon \to 0$. Here, $\mathbf{n}$ is the outward unit normal vector to the control volume surface, and $d\xi$ is the length increment of the circular control surface.
d) If the plate is released from rest, in what direction will it initially accelerate?

13.9 Consider a cambered Zhukhovsky airfoil determined by the following parameters: $a = 1.1, b = 1.0$, and $\beta = 0.1$. Using a computer, plot its contour by evaluating the Zhukhovsky transformation. Also plot a few streamlines, assuming an angle of attack of 5°.

13.10 A thin Zhukhovsky airfoil has a lift coefficient of 0.3 at zero incidence. What is the lift coefficient at 5° incidence?

13.11 [1]The simplest representation of a three-dimensional aircraft wing in flight is the rectangular horseshoe vortex.

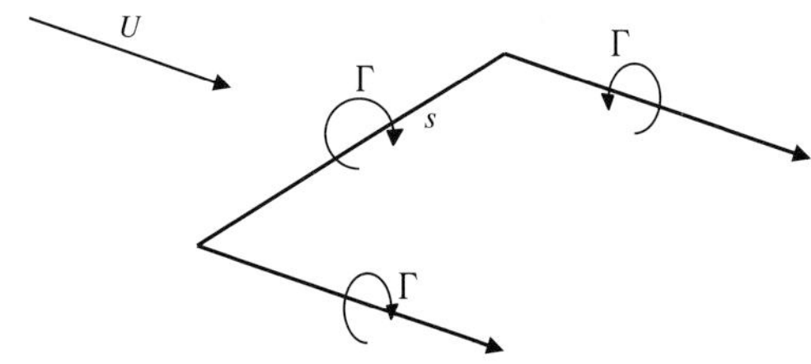

a) Calculate the induced downwash at the center of the wing.
b) Assuming the result of part a) applies along the entire wingspan, estimate C_{D_i}, the lift-induced coefficient of drag, in terms of the wing's aspect ratio: $\Lambda = s^2/A$, and the wing's coefficient of lift $C_L = L/\left[(1/2)\rho U^2 A\right]$, where A is the planform area of the wing.
c) Explain why the result of part b) appears to surpass the performance of the optimal elliptic lift distribution.

13.12 The circulation across the span of a wing follows the parabolic law $\Gamma = \Gamma_0(1 - (2y/s)^2)$. Calculate the induced velocity w at mid-span, and compare the value with that obtained when the distribution is elliptic.

13.13 An untwisted elliptic wing of 20-m span supports a weight of 80,000 N in a level flight at 300 km/hr. Assuming sea level conditions, find a) the induced drag and b) the circulation around sections halfway along each wing.

13.14 [1]A wing with a rectangular planform (span $= s$, chord $= c$) and uniform airfoil section without camber is twisted so that its geometrical angle, α_w, decreases from α_r at the root ($y = 0$) to zero at the wing tips ($y = \pm s/2$) according to the distribution: $\alpha_w(y) = \alpha_r\sqrt{1 - (2y/s)^2}$.

a) At what global angle of attack, α_t, should this wing be flown so that it has an elliptical lift distribution? The local angle of attack at any location along the span will be $\alpha_t + \alpha_w$. Assume the two-dimensional lift curve slope of the foil section is K.
b) Evaluate the lift and the lift-induced drag forces on the wing at the angle of attack determined in part a) when: $\alpha_r = 2°$, $K = 5.8$ rad^{-1}, $c = 1.5$ m, $s = 9$ m, the air density is 1.0 kg/m^3, and the airspeed is 150 m/s.

1. Obtained by the third author while a student in a course taught by Professor Fred Culick.

13.15 Consider the wing shown in Fig. 13.25. If the foil section is uniform along the span and the wing is not twisted, show that the three-dimensional lift coefficient, $C_{L,3D}$ is related to the two-dimensional lift curve slope and lift coefficient of the foil section, K and $C_{L,2D}$, respectively, by: $C_{L,3D} = K\alpha/(1+K/\pi\Lambda) \cong C_{L,2D}/(1+2/\Lambda)$, where $\Lambda = s^2/A$ is the aspect ratio of the wing.

13.16 The wing-tip vortices from large, heavy aircraft can cause a disruptive rolling torque on smaller, lighter ones. Lifting line theory allows the roll torque to be estimated when the small airplane's wing is modeled as a single linear vortex with strength $\Gamma(y)$ that resides at $x = 0$ between $y = -s/2$ and $y = +s/2$. Here, the small airplane's wing will be presumed rectangular (span s, chord c) with constant foil-shape, and the trailing vortex from the heavy airplane's wing will be assumed to lie along the x-axis and produce a vertical velocity distribution at $x = 0$ given by: $w(y) = \frac{\Gamma'}{2\pi y}[1 - \exp(-|y|/\ell)]$. To simplify your work for the following items, ignore the trailing vortices (shown as dashed lines) from the small airplane's wing and assume $U \gg w$.

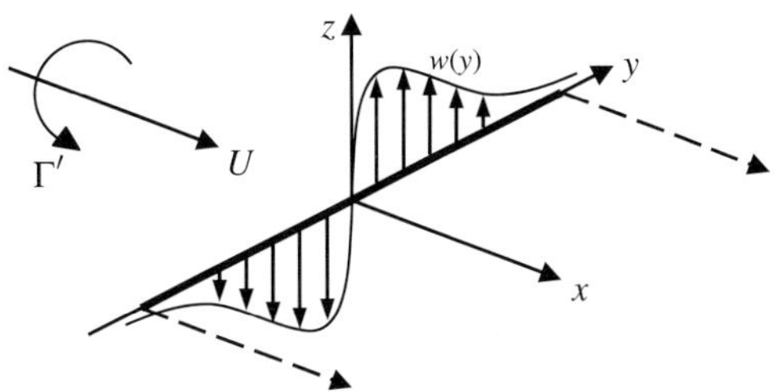

a) Determine a formula for the rolling moment, $M = \int_{-s/2}^{+s/2} \rho U y \Gamma(y) dy$, on the small aircraft's wing in terms of Γ', s, c, ℓ, the air density ρ, the flight speed of the small aircraft U, and the lift-curve slope of the small aircraft's wing section $K = dC_{L,2D}/d\alpha$, where α is the small-aircraft-wing angle of attack.

b) Calculate M when $\rho = 1.2$ kg/m^3, $U = 150$ m/s, $K = 6.0$/rad, $b = 9$ m, $c = 1.5$ m, $\Gamma' = 50$ m^2/s, and $s/(2\ell) = 1$. Comment on the magnitude of this torque.

13.17 Consider the ideal rectilinear horseshoe vortex of a simple wing having span s. Use the (x, y, z) coordinates shown for the following items.

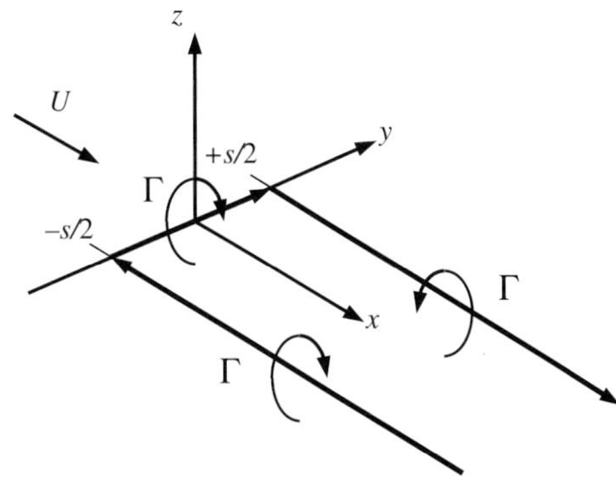

a) Determine a formula for the induced vertical velocity w at $(x, y, 0)$ for $x > 0$ and $y > 0$.

b) Using the results of part a), evaluate the induced vertical velocity at the following three locations $(s, 0, 0)$, $(0, s, 0)$, and $(s, s, 0)$.

c) Imagine that you are an efficiency-minded migrating bird and that the rectilinear horseshoe vortex shown is produced by another member of your flock. Describe where you would choose to center your own wings. List the coordinates of the part b) location that is closest to your chosen location.

13.18 As an airplane lands, the presence of the ground changes the plane's aerodynamic performance. To address the essential features of this situation, consider uniform flow past a horseshoe vortex (heavy solid lines below) with wingspan b located a distance h above a large, flat boundary defined by $z = 0$. From the method of images, the presence of the boundary can be accounted for by an image horseshoe vortex (heavy dashed lines below) of opposite strength located a distance h below the boundary.

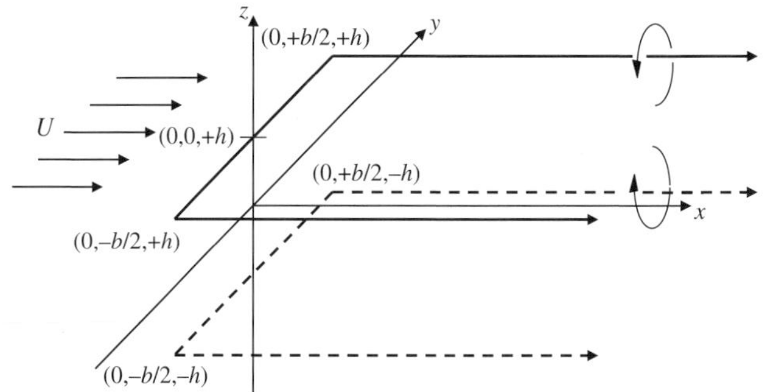

a) Determine the direction and the magnitude of the induced velocity at $\mathbf{x} = (0, 0, h)$, the center of the wing.

b) Assuming the result of part a) applies along the entire wingspan, estimate L and D_i, the lift and lift-induced drag, respectively, in terms of b, h, Γ, and $\rho =$ fluid density.

c) Compare the result of part b) to that obtained for the horseshoe vortex without a large, flat surface: $L = \rho U \Gamma b$ and $D_i = \rho \Gamma^2 / \pi$. Which configuration has more lift? Which one has less drag? Why?

13.19 Before modifications, an ordinary commercial airliner with wingspan $s = 30$ m generates two tip vortices of equal and opposite circulation having Rankine velocity profiles (see (3.28)) and a core size $\sigma_o = 0.5$ m for test-flight conditions. The addition of wingtip treatments (sometimes known as *winglets*) to both of the aircraft's wingtips doubles the tip vortex core size at the test condition. If the aircraft's weight is negligibly affected by the change, has the lift-induced drag of the aircraft been increased or decreased? Justify your answer. Estimate the percentage change in the induced drag.

13.20 Determine a formula for the range, R, of a long-haul jet-engine aircraft in steady level flight at speed U in terms of: $M_F =$ the initial mass of usable fuel; $M_A =$ the mass of the airframe, crew, passengers, cargo, and reserve fuel; $C_L/C_D =$ the aircraft's lift-to-drag ratio; $g =$ the acceleration of gravity; and $\eta =$ the aircraft's propulsion system *thrust-specific fuel consumption* (with units of time/length) defined by: $dM_F/dt = -\eta D$, where $D =$ the aircraft's aerodynamic drag. For simplicity, assume that U, the ratio C_L/C_D, and η are constants. [Hints. If $M(t)$ is the instantaneous mass of the flying aircraft, then $L = \text{Lift} = Mg$, $M = M_F + M_A$, and $dM/dt = dM_F/dt$. The final formula is known as the Breguet range equation.]

References

Anderson Jr., John D., 1998. A History of Aerodynamics. Cambridge University Press, London.

Anderson Jr., John D., 2007. Fundamentals of Aerodynamics. McGraw-Hill, New York.

Fish, F.E., Lauder, G.V., 2006. Passive and active control by swimming fishes and mammals. Annual Review of Fluid Mechanics 38, 193–224.

Herreshoff, H.C., Newman, J.N., 1966. The study of sailing yachts. Scientific American 215 (August issue), 61–68.

von Karman, T., 1954. Aerodynamics. McGraw-Hill, New York (A delightful little book, written for the nonspecialist, full of historical anecdotes and at the same time explaining aerodynamics in the easiest way).

Kuethe, A.M., Chow, C.Y., 1998. Foundations of Aerodynamics: Basis of Aerodynamic Design. Wiley, New York.

Lighthill, M.J., 1986. An Informal Introduction to Theoretical Fluid Mechanics. Clarendon Press, Oxford, England.

Liu, T., Kuykendoll, K., Rhew, R., Jones, S., 2006. Avian wing geometry and kinematics. AIAA Journal 44, 954–963.

Ramamurti, R., Sandberg, W.C., 2001. Computational Study of 3-D Flapping Foil Flows. AIAA Paper, 2001–0605.

Further reading

Ashley, H., Landahl, M., 1965. Aerodynamics of Wings and Bodies. Addison-Wesley, Reading, MA.

Batchelor, G.K., 1967. An Introduction to Fluid Dynamics. Cambridge University Press, London.

Karamcheti, K., 1980. Principles of Ideal-Fluid Aerodynamics. Krieger Publishing Co., Melbourne, FL.

Millikan, C.B., 1941. Aerodynamics of the Airplane. John Wiley & Sons, New York, NYL.

Prandtl, L., 1952. Essentials of Fluid Dynamics. Blackie & Sons Ltd, London (This is the English edition of the original German edition. It is easy to understand, and much of it is still relevant today. Printed in New York by Hafner Publishing Co. If this is unavailable, see the following reprints in paperback that contain much if not all of this material).

Prandtl, L., Tietjens, O.G., 1934a. Fundamentals of Hydro and Aero-Mechanics. Dover Publ. Co., New York [original publication date].

Prandtl, L., Tietjens, O.G., 1934b. Applied Hydro and Aeromechanics. Dover Publ. Co, New York [original publication date] (This contains many original flow photographs from Prandtl's laboratory).

Chapter 14

Compressible flow®

Contents

14.1 Introduction 575
Perfect gas thermodynamic relations 576
14.2 Acoustics 577
14.3 One-dimensional steady isentropic compressible flow in variable-area ducts 582
14.4 Normal shock waves 591
Stationary normal shock wave in a moving medium 591
Moving normal shock wave in a stationary medium 594
Normal-shock structure 595
14.5 Operation of nozzles at different back pressures 597
Convergent nozzle 597
Convergent–divergent nozzle 597
14.6 Effects of friction and heating in constant-area ducts 600
Effect of friction 600
Effect of heat transfer 601
14.7 One-dimensional unsteady compressible flow in constant-area ducts 603
14.8 Two-dimensional steady compressible flow 606
14.9 Thin-airfoil theory in supersonic flow 612
Exercises 614
References 619
Further reading 619

Chapter objectives

- To introduce the fundamental compressible flow interactions between velocity, pressure, density, and temperature
- To describe the features of isentropic flows in ducts with smoothly varying cross-sectional area
- To derive the jump conditions across normal shock waves from the conservation equations
- To describe the effects of friction and heat transfer in compressible flows through constant-area ducts
- To extend the prior steady flow analysis to unsteady one-dimensional gas dynamics
- To indicate how wall geometry produces pressure waves that cause fluid compression, expansion, and turning in steady supersonic flows near walls

14.1 Introduction

Up to this point, this text has primarily covered incompressible flows where changes in fluid momentum and pressure are closely related. Thermodynamics describes how variations in fluid density and pressure are related. This chapter presents some of the elementary aspects of *compressible flow* or *gas dynamics* where changes in fluid momentum produce important variations in fluid pressure and density, and the fluid's thermodynamic characteristics play a direct role in the flow's development. In compressible flows, the sound speed in the fluid becomes an important parameter and cannot be treated as infinite (the incompressible flow limit). This branch of fluid mechanics has wide application in high-speed flows around objects of engineering interest. These include *external flows* such as those around projectiles, rockets, re-entry vehicles, and airplanes; and *internal flows* in ducts and passages such as nozzles and diffusers used in jet engines, rocket motors, and compressed gas systems. Compressibility effects are also important in astrophysics. More complete treatments of gas dynamics are available in other texts: Shapiro (1953), Liepmann and Roshko (1957), and Thompson (1972); portions of the material presented here are drawn from these references.

Several startling and fascinating phenomena, which defy intuition and expectations developed from incompressible flows, arise in compressible flows and are described in this chapter. Near discontinuities (shock waves) may appear within the flow. An increase (or decrease) in flow area may accelerate (or decelerate) a uniform stream. Friction may increase a flow's speed. And, heat addition may lower a flow's temperature. These phenomena are therefore worthy of our attention because they either have no counterpart or act oppositely in low-speed flows. Except for the treatment of friction in constant-area ducts in Section 14.6, the material presented here is limited to that of frictionless flows outside boundary layers. In spite of this simplification, the results presented here have a great deal of practical value because boundary layers are especially thin in high-speed flows. Gravitational effects, which are of minor importance in compressible flows, are also neglected.

®. This book has a companion website hosting complementary materials. Visit this URL to access it: https://www.elsevier.com/books-and-journals/book-companion/9780128198070.

Fluid Mechanics. https://doi.org/10.1016/B978-0-12-819807-0.00023-5

As discussed in Section 4.11, the importance of compressibility for moving fluids can be assessed by considering the Mach number M, defined as:

$$M \equiv U/c, \tag{4.113}$$

where U is a representative flow speed, and c is the speed of sound, a thermodynamic quantity defined by:

$$c^2 \equiv (\partial p/\partial \rho)_s. \tag{1.25}$$

Here, the subscript s signifies that the partial derivative is taken at constant entropy. In particular, the dimensionless scaling (4.111) of the compressible-flow continuity equation for isentropic conditions leads to:

$$\nabla \cdot u = -M^2 \left(\frac{\rho_0}{\rho}\right) \frac{D}{Dt} \left(\frac{p - p_0}{\rho_0 U^2}\right), \tag{4.112}$$

where ρ_0 and p_0 are appropriately chosen reference values for density and pressure. In (4.112), the pressure is scaled by fluid inertia parameters as is appropriate for primarily frictionless high-speed flow. In engineering practice, the incompressible flow assumption is presumed valid if $M < 0.3$. Eq. (4.112) suggests that $M = 0.3$ corresponds to ~10% departure from perfectly incompressible flow behavior when the remainder of the right side of (4.112) is of order unity.

Although the significance of the ratio U/c was known for a long time, the Swiss aerodynamist Jacob Ackeret introduced the term *Mach number*, just as the term Reynolds number was introduced by Sommerfeld many years after Reynolds' experiments. The name of the Austrian physicist Ernst Mach (1836–1916) was chosen because of his pioneering studies on supersonic motion and his invention of the so-called *Schlieren method* for optical visualization of flows involving density changes; see von Karman (1954, p. 106). (Mach distinguished himself equally well in philosophy. Einstein acknowledged that his own thoughts on relativity were influenced by "Mach's principle," which states that properties of space had no independent existence but are determined by the mass distribution within it. Strangely, Mach never accepted either the theory of relativity or the atomic structure of matter.)

Using the Mach number, compressible flows can be nominally classified as follows:

(i) *Incompressible flow*: $M = 0$. Fluid density does not vary with pressure in the flow field. The flowing fluid may be a compressible gas but its density may be regarded as constant.

(ii) *Subsonic flow*: $0 < M < 1$. The Mach number does not exceed unity anywhere in the flow field. Shock waves do not appear in the flow. In engineering practice, subsonic flows for which $M < 0.3$ are often treated as being incompressible.

(iii) *Transonic flow*: The Mach number in the flow lies in the range 0.8–1.2. Shock waves may appear. Analysis of transonic flows is difficult because the governing equations are inherently nonlinear, and also because a separation of the inviscid and viscous aspects of the flow is often impossible. (The word "transonic" was invented by von Karman and Hugh Dryden, although the latter argued in favor of spelling it "transsonic." von Karman [1954] stated, "I first introduced the term in a report to the U.S. Air Force. I am not sure whether the general who read the word knew what it meant, but his answer contained the word, so it seemed to be officially accepted" [p. 116].)

(iv) *Supersonic flow*: $M > 1$. Shock waves are generally present. In many ways analysis of a flow that is supersonic everywhere is easier than analysis of a subsonic or incompressible flow as we shall see. This is because information propagates along certain trajectories, called characteristics, and a determination of these trajectories greatly facilitates the computation of the flow field.

(v) *Hypersonic flow*: $M > 5$.[1] Very high flow speeds combined with friction or shock waves may lead to sufficiently large increases in a fluid's temperature so that molecular dissociation and other chemical effects occur, and perfect-gas-based analytical formulae may no longer be accurate.

Perfect gas thermodynamic relations

As density changes are accompanied by temperature changes, thermodynamic principles are constantly used throughout this chapter. Most of the necessary concepts and relations have been summarized in Sections 1.8–1.9, which may be reviewed before proceeding further. The most frequently used relations, valid for a perfect gas with constant specific heats, are listed

1. Although $M > 5$ is often used to demarcate supersonic from hypersonic flow, the precise value is frequently disputed, and it may be more appropriate to classify hypersonic flows based on speeds that lead to the dissociation of gas molecules.

here for quick reference:

$$\textit{Internal energy}: e = c_v T, \quad \textit{Enthalpy}: h = c_p T, \quad \textit{Thermal equation of state}: p = \rho R T, \tag{14.1a – c}$$

$$\textit{Specific heats}: c_v = \frac{R}{\gamma - 1}, \quad c_p = \frac{\gamma R}{\gamma - 1}, \quad c_p - c_v = R, \quad \gamma = c_p/c_v, \tag{14.1d – g}$$

$$\textit{Speed of Sound}: c = \sqrt{\gamma R T} = \sqrt{\gamma p/\rho}, \quad \text{and}$$

$$\textit{Entropy change}: s_2 - s_1 = c_p \ln\left(\frac{T_2}{T_1}\right) - R \ln\left(\frac{p_2}{p_1}\right) = c_v \ln\left(\frac{T_2}{T_1}\right) - R \ln\left(\frac{\rho_2}{\rho_1}\right). \tag{14.1h, 14.1i}$$

Eq. (14.1h) implies that c is larger in monotonic and low-molecular weight gases (where γ and R are larger), and that it increases with increasing temperature. An isentropic process involving a perfect gas between states 1 and 2 obeys the following relations:

$$\frac{p_2}{p_1} = \left(\frac{\rho_2}{\rho_1}\right)^{\gamma}, \quad \text{and} \quad \frac{T_2}{T_1} = \left(\frac{\rho_2}{\rho_1}\right)^{\gamma-1} = \left(\frac{p_2}{p_1}\right)^{(\gamma-1)/\gamma}. \tag{14.2a, 14.2b}$$

Some important properties of air at ordinary temperatures and pressures are:

$$R = 287\,\mathrm{m^2/(s^2\,K)}, \quad c_v = 717\,\mathrm{m^2/(s^2K)}, \quad c_p = 1004\,\mathrm{m^2/(s^2K)}, \quad \text{and} \quad \gamma = 1.40; \tag{14.3a – d}$$

these values are useful for solution of the exercises at the end of this chapter.

Example 14.1

Which of the following are compressible flows? a) A weather balloon rises at 5 m/s from sea level to an altitude of more than 15 km. b) Water flows through the nozzle of a water jet cutter and the gage pressure drops from 100 MPa to zero. c) The piston of an internal combustion engine moves at 15 m/s and compresses air and gaseous fuel from 40 kPa to 1300 kPa. d) Liquid nitrogen at 100 kPa with density 807 kg/m^3 evaporates from a stationary dewar in a 1 m/s airflow to become nitrogen gas with density 1.16 kg/m^3. e) A 1.0 MPa compression wave travels at 1.0 km/s into air at 15 °C and 1.0 atm.

Solution The essential feature of compressible flow (as defined in this section) is coupling between fluid velocity, pressure, and density. For a), the pressure and density of the air outside the balloon and the helium inside it will vary by a factor of eight before the balloon reaches an altitude of 15 km. However, such changes primarily depend on altitude and are largely independent of the low-Mach number rise speed, so this is not a compressible flow. For b), the water experiences an enormous pressure drop and accelerates to more than 400 m/s. However, the density change is modest, as can be estimated from a discretized evaluation of (1.25) using $c_{water} \sim 1480\,\mathrm{m/s}$:

$$\Delta\rho/\rho \cong \Delta p/\rho c^2 \cong 100\,\mathrm{MPa}/\left(1000\,\mathrm{kgm^{-3}} \cdot (1480\,\mathrm{ms^{-1}})^2\right) = 0.046.$$

This flow might qualify as compressible depending on the level of accuracy that is sought. For c), the pressure and density of the combustion gases rise by factors of more than 30 and 10, respectively. However, like the weather balloon, such changes primarily depend on piston location and are largely independent of the low-Mach number piston speed, so this is not a compressible flow. For d), the nitrogen accelerates to 1 m/s and its density drops by a factor ~700, but there are no significant pressure variations, so this flow is not compressible. For e), the wave speed is supersonic and the increase in air pressure (a factor of ~10) and density (a factor of ~4) across the wave are significant. This is a compressible flow.

14.2 Acoustics

Perhaps the simplest and most common form of compressible flow is found when the variations in velocity, pressure, and density are small compared to steady reference values and the variations in pressure and density are isentropic. This branch of compressible flow is known as *acoustics* and is concerned with the study of sound waves. Acoustics is the small disturbance theory of compressible fluid dynamics and is a broad field with its own rich history (see Pierce, 1989). The primary concern in this section is to show how the speed of sound enters the equations for compressible flow, to deduce how pressure disturbances may arise in such flows, and to develop some insight into the behavior of pressure disturbances in compressible flow by considering solutions of the linearized equations of motion.

The development starts from the mass and momentum conservation equations for a moving single-component viscous fluid [see (4.8) and (4.38)] modified to include source terms:

$$\frac{1}{\rho}\frac{D\rho}{Dt}+\frac{\partial u_i}{\partial x_i}=q, \quad \text{and} \quad \frac{Du_j}{Dt}+\frac{1}{\rho}\frac{\partial p}{\partial x_j}=g_j+\frac{1}{\rho}\frac{\partial \tau_{ij}}{\partial x_i}+f_j, \tag{14.4, 14.5}$$

where τ_{ij} is the viscous stress tensor given by (4.36). The added source term in (14.4), $q(x_i,t)$, is an unsteady volume source distribution (per unit volume), and the added source term in (14.5), $f_j(x_i,t)$, is an unsteady body-force distribution (per unit mass of fluid). Both represent acoustic sources not produced directly by fluid motion. These terms may be non-zero because of naturally occurring or anthropogenic acoustic sources (voices, animal calls, loud speakers, transducers, unsteady combustion, unsteady heat transfer, vibrating structures, rotating fan or propeller blades, etc.). Alternatively, these source terms can often be specified through initial and boundary conditions.

Acoustic waves in fluids are lightly damped small-amplitude isentropic pressure fluctuations so pressure and density fluctuations are assumed to be isentropic following a fluid particle. Thus, τ_{ij} in (14.5) is commonly ignored unless the acoustic frequency is high or the propagation distance is long. The equation associated with the isentropic assumption can be obtained by D/Dt-differentiating the equation of state for the pressure $p=p(\rho,s)$, setting $Ds/Dt=0$, and using (1.25) for c^2:

$$\frac{Dp}{Dt}=\left(\frac{\partial p}{\partial \rho}\right)_s\frac{D\rho}{Dt}+\left(\frac{\partial p}{\partial s}\right)_\rho\frac{Ds}{Dt}=c^2\frac{D\rho}{Dt}. \tag{14.6}$$

The variations in p and ρ accommodated by (14.6) include acoustic waves and other isentropic density and pressure variations. Non-isentropic fluctuations, such as strong shock waves and compressible turbulence, are omitted. However, the acoustic effects of many non-isentropic processes such as thermal conduction, viscous dissipation, radiative energy transfer, chemical reactions, and phase change can be reintroduced to this formulation via the q and f_j source terms in (14.4) and (14.5).

A general wave equation for the pressure p can be assembled from (14.4), (14.5), and (14.6) by substitution and cross differentiation (see Exercise 14.1):

$$\frac{D}{Dt}\left(\frac{1}{\rho c^2}\frac{Dp}{Dt}\right)-\frac{\partial}{\partial x_j}\left(\frac{1}{\rho}\frac{\partial p}{\partial x_j}\right)=\frac{Dq}{Dt}-\frac{\partial f_j}{\partial x_j}+\frac{\partial u_i}{\partial x_j}\frac{\partial u_j}{\partial x_i}, \tag{14.7}$$

where τ_{ij} has been dropped to be consistent with (14.6), and the steady body force (gravity) has been presumed spatially uniform so that $\partial g_j/\partial x_j=0$. Although (14.7) incorporates several dependent variables (p, ρ, u_i), it serves to identify the acoustic source terms. The left side of (14.7) is a convected wave operator modified to account for varying fluid velocity, varying fluid density, and varying sound speed. The right side of (14.7) displays three source terms corresponding to unsteady expansion within the fluid domain (monopole sources), the divergence of spatially varying fluctuating body forces (dipole sources), and the interaction of the moving fluid with itself (quadrupole sources), respectively. For convenience, these may be combined:

$$\dot{q}\equiv\frac{Dq}{Dt}-\frac{\partial f_j}{\partial x_j}+\frac{\partial u_i}{\partial x_j}\frac{\partial u_j}{\partial x_i}, \tag{14.8}$$

since all three are scalars and since dipoles and quadrupoles can be constructed from weighted super positions of monopoles.

The final term in (14.8) represents sound generation by the flow itself, and is known as the aero-acoustic source term. Aerodynamic sound generation often occurs in circumstances where this source term is non-zero in a confined region that is embedded in a quiescent environment with uniform density ρ_0 and sound speed c_0 (see Fig. 14.1). In this case, with $q=0$ and $f_j=0$, a revised version of (14.7) can be obtained that is easier to interpret (see Lighthill, 1952, or Crighton, 1975). The starting point is again the continuity equation (4.7) and the inviscid momentum equation without the body force, (4.22) with $g=0$ and $T_{ij}=-p\delta_{ij}$,

$$\frac{\partial \rho}{\partial t}+\frac{\partial \rho u_i}{\partial x_i}=0, \quad \text{and} \quad \frac{\partial \rho u_i}{\partial t}+\frac{\partial \rho u_i u_j}{\partial x_j}=-\frac{\partial p}{\partial x_i}. \tag{14.9}$$

Add $c_0^2\partial\rho/\partial x_i$ to both sides of the second equation (14.9), and eliminate ρu_i between the two equations by cross-differentiation and subtraction. Combining terms and simplification then leads to:

$$\frac{\partial^2\rho}{\partial t^2} - c_0^2\frac{\partial^2\rho}{\partial x_i^2} = \frac{\partial^2}{\partial x_i\partial x_j}\left(\rho u_i u_j + p\delta_{ij} - c_0^2\rho\delta_{ij}\right), \tag{14.10}$$

which is an inhomogeneous wave equation for the density. The right-side term in parentheses is known as the *Lighthill stress tensor*, $T_{ij}^L = \rho u_i u_j + (p - c_0^2\rho)\delta_{ij}$. It represents the stress distribution in the confined region that produces the observed density fluctuations.

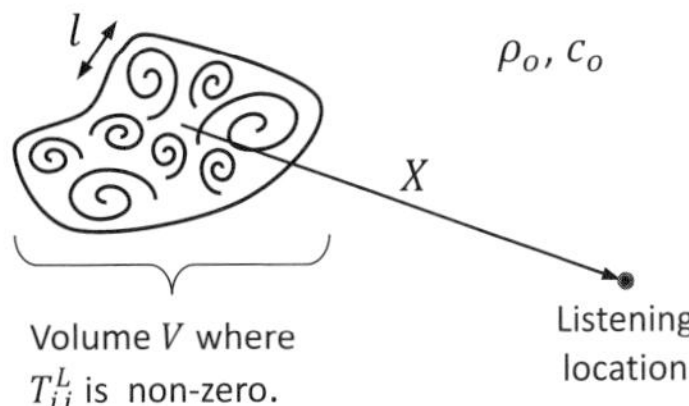

FIGURE 14.1 Turbulent aero-acoustic sound generation in a confined region V that produces sound heard by an observer at an average distance X.

In an unbounded space, the formal solution of (14.10) can be obtained in terms of the distance, $r = \sqrt{(x_i - y_i)^2} = |\mathbf{x} - \mathbf{y}|$, between the listening point, x_i, and the source location, y_i; and the retarded time, $t_r = t - r/c_0$:

$$\rho(x_i,t) - \rho_0 = \frac{1}{4\pi c_0^2}\int_V \left[\frac{\partial^2 T_{ij}^L}{\partial y_i \partial y_j}\right]_{t_r} \frac{d^3y}{r} \tag{14.11}$$

(see Exercise 14.2). Gauss's divergence theorem allows this to be rewritten:

$$\rho(x_i,t) - \rho_0 = \frac{1}{4\pi c_0^2}\frac{\partial^2}{\partial x_i\partial x_j}\int_V \frac{T_{ij}^L(y_i, t - r/c_0)}{r}d^3y \tag{14.12}$$

(see Exercise 14.3). This final equation embodies Lighthill's acoustic analogy, and is well known for its scaling predictions for aero-acoustic sound production from turbulent flows.

Consider the situation depicted in Fig. 14.1, where V is the volume of the region of turbulent aero-acoustic sound generation, l is the length scale of turbulence, u' is the velocity fluctuation of the turbulence, and the average distance from V to the listening location is X. In practice, V might be the exhaust region of a jet engine of a taxiing aircraft, while the listener might be some distance away at the same airport. With these parameters, a factor-by-factor scaling of (14.12) appears as:

$$\rho(x_i,t) - \rho_0 \sim \frac{1}{c_0^2}\frac{1}{(l/u')^2}\frac{1}{c_0^2}\rho_0 u'^2 \frac{1}{X}V = \rho_0\left(\frac{u'}{c_0}\right)^4\frac{V}{l^2X}, \tag{14.13}$$

where T_{ij}^L scales as $\rho_0 u'^2$, and the two spatial derivatives each scale as $(l/u')^{-1}(1/c_0)^{-1}$ because they act on the r within the time argument of T_{ij}^L in (14.12). The resulting scaling law (14.13) shows that the amplitude of aero-acoustic density fluctuations is proportional to the fluctuation Mach number to the fourth power. Thus, the aero-acoustic sound intensity is predicted to increase with fluctuation Mach number to the eighth power. Interestingly, the optical impact of these aero-acoustic density fluctuations is apparent in this textbook's cover photograph. Aero-acoustic sound waves radiated from the intense turbulence in the rocket's plume cause the boarders of the solar disc to be irregular (most prominently at and below the sun's equator) because the sun's light is refracted by the radiated sound waves' density fluctuations. In other situations, the radiated intensity's eighth-power dependence on u'/c_o may or may not be observed because of the presence of nearby solid boundaries (see Exercise 14.4). Furthermore, there are other acoustic analogies involving fixed and moving surfaces, and a variety of approximations for T_{ij}^L that allow (14.12) to be used with flow simulations for radiated noise predictions and other engineering tasks.

Up to this point, the small-fluctuation character of acoustic phenomena, beyond that inherent in (14.6), has not been invoked. However, to reduce (14.7) or (14.10) to a solvable equation for acoustic pressure fluctuations, several additional simplifications are commonly made. First, the various dependent field variables are separated into nominally steady and fluctuating acoustic values:

$$u_i = U_i + u'_i,\ p = p_0 + p',\ \rho = \rho_0 + \rho', \quad \text{and} \quad T = T_0 + T', \tag{14.14}$$

where U_i, p_0, ρ_0, and T_0 are constants applicable to the region of interest, and all the fluctuating quantities – denoted by primes in (14.14) – are considered to be small compared to these. In addition, if the flow is isentropic in the realm of interest, the pressure can be Taylor expanded about the reference thermodynamic state specified by p_0 and ρ_0:

$$p = p_0 + p' = p_0 + \left(\frac{\partial p}{\partial \rho}\right)_s (\rho - \rho_0) + \frac{1}{2}\left(\frac{\partial^2 p}{\partial \rho^2}\right)_s (\rho - \rho_0)^2 + \ldots = p_0 + c^2\rho' + \frac{1}{2}\left(\frac{\partial^2 p}{\partial \rho^2}\right)_s \rho'^2 + \ldots$$

For small isentropic variations, the second-order and higher terms can be neglected, and this leads to a simple relationship between acoustic pressure and density fluctuations that replaces (14.6):

$$p' = c^2\rho'. \tag{14.15}$$

This equation is valid when the fractional density change or *condensation* $= \rho'/\rho_0 = p'/\rho_0 c^2$ is small:

$$p'/\rho_0 c^2 \ll 1. \tag{14.16}$$

For ordinary sound levels in air, acoustic-pressure magnitudes are of order 1 Pa or less, so the ratio specified in (14.16) is typically less than 10^{-5} since $\rho_0 c^2 = \gamma p_0 \approx 1.4 \times 10^5$ Pa. Additionally, positive p' is called *compression* while negative p' is called *expansion* (or *rarefaction*). Acoustic pressure disturbances are commonly composed of equal amounts of compression and expansion.

The first approximate expression for c was found by Newton, who assumed that p'/p_0 was equal to ρ'/ρ_0 (Boyle's law) as would be true if the process undergone by a fluid particle was isothermal. In this manner Newton arrived at the expression $c = [RT]^{1/2}$. He attributed the discrepancy of this formula with experimental measurements as due to "unclean air." However, the science of thermodynamics was virtually non-existent at the time, so that the idea of an isentropic process was unknown to Newton. The correct expression for the sound speed was first given by Laplace.

The field equation for acoustic pressure disturbances can be obtained from (14.7) without source terms by inserting (14.14) and linearizing with U_i, p_0, ρ_0 and T_0 treated as time-invariant and spatially uniform. The resulting equation is the field equation for acoustic pressure disturbances in a uniform flow:

$$\frac{1}{c^2}\left(\frac{\partial}{\partial t} + U_i \frac{\partial}{\partial x_i}\right)^2 p' - \frac{\partial^2 p'}{\partial x_i \partial x_i} = 0 \tag{14.17}$$

(see Exercise 14.6).

To highlight the importance of the sound speed, consider a stationary fluid and one-dimensional pressure disturbances. For a stationary fluid ($U_i = 0$), (14.17) reduces to:

$$\frac{1}{c^2}\frac{\partial^2 p'}{\partial t^2} - \frac{\partial^2 p'}{\partial x_i \partial x_i} = 0, \tag{14.18}$$

and this is the classical wave equation for acoustic waves in a lossless uniform medium. Under these same circumstances, the simplified and linearized version of (14.5) provides the relationship between acoustic pressure and fluid velocity fluctuations:

$$\frac{\partial u'_j}{\partial t} + \frac{1}{\rho_0}\frac{\partial p'}{\partial x_j} = 0 \quad \text{or} \quad u'_j(x,t) = -\frac{1}{\rho_o}\int \frac{\partial p'}{\partial x_j} dt. \tag{14.19}$$

When the pressure disturbances are one-dimensional and only vary along the x_1-axis, then (14.18) reduces to the one-dimensional wave equation and its solutions are of the form:

$$p'(x,t) = f(x - ct) + g(x + ct), \tag{14.20}$$

where $x_1 = x$, and the functions f and g are determined by initial conditions (see Exercise 14.7). Eq. (14.20) is known as *d'Alembert's solution*, and $f(x - ct)$ and $g(x + ct)$ represent traveling pressure disturbances that propagate to the right and left, respectively, with increasing time. Consider a pressure pulse $p'(x,t)$ that propagates to the right and is centered at $x=0$ with shape $f(x)$ at $t=0$ as shown in Fig. 14.2. An arbitrary time t later, the wave is centered at $x=ct$ and its shape is described by $f(x-ct)$. Similarly, when $p'(x,t)=g(x+ct)$, the pressure disturbance propagates to the left and is located at $x=-ct$ at time t. Thus, the speed at which acoustic pressure disturbances travel is c, and this is independent of the shape of the pressure disturbance waveform.

However, the disturbance waveform does influence the fluid velocity, u', along the x-axis. It can be determined from (14.19) and (14.20) with $x_1=x$, and is given by:

$$u_1'(x,t)=\frac{1}{\rho_0 c}\left(f(x-ct)-g(x+ct)\right) \tag{14.21}$$

(see Exercise 14.8). Thus, the fluid velocity includes rightward- and leftward-propagating components that are matched to the pressure disturbance. Moreover, (14.21) shows that the compression portions of f and g lead to fluid velocity in the same direction as wave propagation; the fluid velocity from $f(x-ct)$ is to the right when $f>0$, and the fluid velocity from $g(x+ct)$ is to the left when $g>0$. Similarly, the expansion portions of f and g lead to fluid velocity in the direction opposite of wave propagation; the fluid velocity from $f(x-ct)$ is to the left when $f<0$, and the fluid velocity from $g(x+ct)$ is to the right when $g<0$. These fluid velocity directions are worth noting because they persist with the same signs when the wave amplitudes exceed those allowed by the approximation (14.16).

Now consider one-dimensional pressure waves $p'(x,t)$ when $U_i=(U,0,0)$ so that (14.17) becomes:

$$\frac{1}{c^2}\left(\frac{\partial}{\partial t}+U\frac{\partial}{\partial x}\right)^2 p'-\frac{\partial^2 p'}{\partial x^2}=0.$$

The general solution of this equation is:

$$p'(x,t)=f(x-(c+U)t)+g(x+(c-U)t). \tag{14.22}$$

When $U>0$, the travel speed of the downstream-propagating waves is enhanced and that of the upstream-propagating waves is reduced. However, when the flow is supersonic, $U>c$, both portions of (14.22) travel downstream, and this represents a major change in the character of the flow. In subsonic flow, both upstream and downstream pressure disturbances may influence the flow at the location of interest, while in supersonic flow only upstream disturbances may influence the flow. For aircraft moving through a nominally quiescent atmosphere, this means that a ground-based observer below the aircraft's flight path may hear a subsonic aircraft before it is overhead. However, a supersonic aircraft does not radiate sound forward in the direction of flight so the same ground-based observer will only hear a supersonic aircraft after it has passed overhead (see Section 14.8).

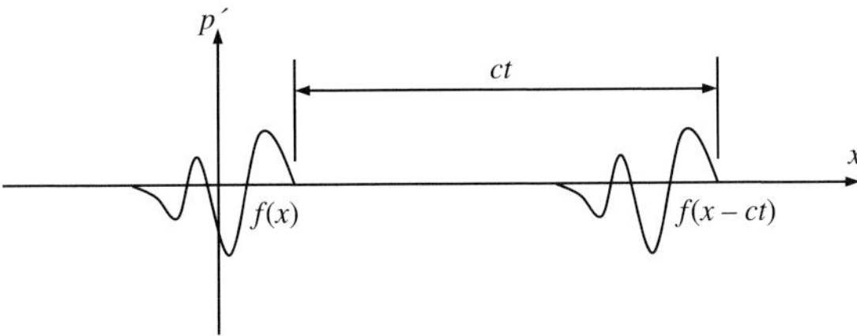

FIGURE 14.2 Propagation of an acoustic pressure disturbance p' that travels to the right with increasing time. At $t=0$ the disturbance is centered at $x=0$ and has waveform $f(x)$. At time t later, the disturbance has moved a distance ct but its waveform shape has not changed.

Linear acoustic theory is valuable and effective for weak pressure disturbances, and it also indicates how nonlinear phenomena arise as pressure-disturbance amplitudes increase. The speed of sound in gases depends on the local temperature, $c=[\gamma RT]^{1/2}$. For air at 15 °C, this gives $c=340\,\text{m/s}$. The nonlinear terms that were dropped in the linearization leading to (14.17) may change the waveform of a propagating nonlinear pressure disturbance depending on whether it is a compression or expansion. Because $\gamma>1$, the isentropic relations show that if $p'>0$ (compression), then $T'>0$ so the sound speed c increases within a compression disturbance. Therefore, pressure variations within a region of nonlinear compression travel faster than a zero-crossing of p' where $c=[\gamma RT_0]^{1/2}$ and therefore may catch up with the leading edge of the compressed region. Such compression-induced changes in c cause nonlinear compression waves to spontaneously steepen as they travel. The opposite is true for nonlinear expansion waves where $p'<0$ and $T'<0$, so c decreases. Here, any pressure variations within the region of expansion fall farther behind the leading edge of the expansion. This causes nonlinear expansion waves to spontaneously spread or flatten as they travel. When combined these effects cause a nonlinear sinusoidal pressure disturbance involving equal amounts of compression and expansion to evolve into a saw-tooth shape (see Chapter 11 in Pierce, 1989). Pressure disturbances that do not satisfy the approximation (14.16) are called *finite* amplitude waves.

The limiting form of a finite-amplitude compression wave is a discontinuous change of pressure, commonly known as a *shock wave*. In Section 14.6, it will be shown that the finite-amplitude compression waves are not isentropic and that they propagate through a still fluid *faster* than acoustic waves.

Example 14.2

For airborne sound, the decibel scale is defined by: $SPL = 20\log_{10}[p_{rms}/p_{ref}]$, where SPL is sound pressure level in dB (the abbreviation for decibels), p_{rms} is the root-mean-square pressure amplitude of the sound of interest, and $p_{ref} = 20\,\mu\text{Pa}$ (an international standard). Here, p_{ref} is chosen so that 0 dB corresponds to quiet sounds at the nominal threshold of human hearing. Using this information, assess the validity of (14.16) for sounds with $SPL = 30\,\text{dB}$ (soft whispering), 60 dB (normal conversation), 90 dB (noisy factory), 115 dB (good seat at a rock concert), 130 dB (near an aircraft engine), and 160 dB (inside of an automobile exhaust system) at atmospheric pressure.

Solution The definition of SPL can be inverted to find $p_{rms} = p_{ref}10^{(SPL/20)}$, and $\rho_0 c^2 = \gamma p_0 \approx 140$ kPa at atmospheric pressure. Thus, for $SPL = 30$ dB, (14.16) becomes:

$$p_{rms}/\rho_0 c^2 = (20\,\mu\text{Pa})10^{(30/20)}/(140\,\text{kPa}) = 4.5\times 10^{-9} \ll 1,$$

and the small amplitude requirement for linear acoustics is well satisfied. The remaining SPL values lead to the following numbers:

$$\begin{aligned}
60\,\text{dB}: &\quad p_{rms}/\rho_0 c^2 = 1.4\times 10^{-7},\\
90\,\text{dB}: &\quad p_{rms}/\rho_0 c^2 = 4.5\times 10^{-6},\\
115\,\text{dB}: &\quad p_{rms}/\rho_0 c^2 = 8.0\times 10^{-5},\\
130\,\text{dB}: &\quad p_{rms}/\rho_0 c^2 = 4.5\times 10^{-4},\\
160\,\text{dB}: &\quad p_{rms}/\rho_0 c^2 = 1.4\times 10^{-2}.
\end{aligned}$$

Thus, even at decibel levels that cause hearing damage (130 dB) or deafness (160 dB), the small-amplitude requirement for linear acoustics is met. Therefore, the linear theory of acoustics applies throughout (and somewhat beyond) the amplitude range of human hearing.

14.3 One-dimensional steady isentropic compressible flow in variable-area ducts

This section presents fundamental results for steady compressible flows that can be analyzed using one spatial dimension. The specific emphasis here is for flow through a duct having a sufficiently straight centerline with a cross-section that varies slowly enough so that all dependent flow-field variables (u, p, ρ, T) are well approximated at any location as being equal to their cross-section-averaged values. If the duct area $A(x)$ varies with the distance x along the duct, as shown in Fig. 14.3, the dependent flow-field variables are taken as $u(x)$, $p(x)$, $\rho(x)$, and $T(x)$. Unsteadiness (and much complexity) can be introduced by including t as an additional independent variable (see Section 14.7).

In this situation a control volume development of the basic equations is appropriate. Start with scalar equations representing conservation of mass and energy using the stationary control volume shown in Fig. 14.3. For steady flow within this control volume, the integral form of the continuity equation (4.5) requires:

$$\rho_1 u_1 A_1 = \rho_2 u_2 A_2, \quad \text{or} \quad \rho u A = \dot{m} = const., \tag{14.23}$$

where $\dot{m}$ is the mass flow rate in the duct, and the second form follows from the first because the locations 1 and 2 are arbitrary. Forming a general differential of the second form and dividing the result by $\dot{m}$, produces:

$$\frac{d\rho}{\rho} + \frac{du}{u} + \frac{dA}{A} = 0. \tag{14.24}$$

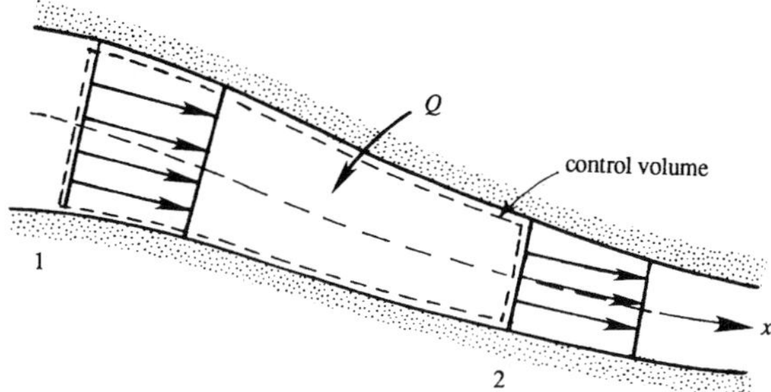

FIGURE 14.3 One-dimensional compressible flow in a duct with smoothly varying centerline direction and cross-sectional area. A stationary control volume in this duct is indicated by dotted lines. Conditions at the upstream and downstream control surfaces are denoted by "1" and "2" respectively. In some circumstances, heat Q may be added to the fluid in the volume. When the control surfaces normal to the flow are only a differential distance apart, then $x_2 = x_1 + dx$, $A_2 = A_1 + dA$, $u_2 = u_1 + du$, $p_2 = p_1 + dp$, $\rho_2 = \rho_1 + d\rho$, etc., where x is the duct's centerline coordinate, A is the duct's cross-sectional area, and u, p, and ρ are the cross-section averaged flow speed, pressure, and density.

For steady flow in a stationary control volume, the integral form of the energy equation (4.49) simplifies to:

$$\int_{A^*} \rho\left(e + \frac{1}{2}u_i^2\right) u_j n_j dA = \int_{A^*} u_i(-p\delta_{ij} + \tau_{ij})n_j dA - \int_{A^*} q_j n_j dA, \tag{14.25}$$

where e is the internal energy per unit mass, A^* is the control surface, n_j is the outward normal on the control surface, the body force has been neglected, τ_{ij} is the viscous stress tensor, and q_j is heat flux vector. The term on the left side represents the net flux of internal and kinetic energy out of the control volume. The first term on the right side represents the rate of work done on the control surface, and the second term on the right-hand side represents the heat *input* through the control surface. Here the minus sign in front of the final term occurs because $q_j n_j$ is positive when heat leaves the control volume. A term-by-term evaluation of (14.25) with the chosen control volume produces:

$$-\left(e + \frac{1}{2}u^2\right)_1 \dot{m} + \left(e + \frac{1}{2}u^2\right)_2 \dot{m} = (upA)_1 - (upA)_2 + \dot{m}Q, \tag{14.26}$$

where $\dot{m} = \rho_1 u_1 A_1 = \rho_2 u_2 A_2$ has been used, and Q is the heat added per unit mass of flowing fluid so that:

$$-\int_{A^*} q_j n_j dA = \dot{m}Q.$$

Here the wall shear stress does no work, because $u_i = 0$ in (14.25) at the wall. Thus the surface work done on the control volume comes from the pressure on the control surfaces lying perpendicular to the flow direction. Dividing (14.26) by $\dot{m}$ and noting that $upA/\dot{m} = p/\rho$ allows it to be simplified to:

$$\left(e + \frac{p}{\rho} + \frac{u^2}{2}\right)_2 - \left(e + \frac{p}{\rho} + \frac{u^2}{2}\right)_1 = Q, \quad \text{or} \quad h_2 + \frac{u_2^2}{2} - h_1 - \frac{u_1^2}{2} = Q, \tag{14.27}$$

where $h = e + p/\rho$, is the enthalpy per unit mass (1.19). This energy equation is valid even if friction or other non-isentropic processes (e.g., shock waves) occur between sections 1 and 2. It implies that the *sum of enthalpy and kinetic energy remains constant in an adiabatic flow*. Therefore, enthalpy plays the same role in a flowing system that internal energy plays in a non-flowing system. The difference between the two types of systems is the *flow work* required to push matter along the duct.

Now consider momentum conservation without the body force using the same control volume. The simplified version of (4.17) is:

$$\int_{A^*} \rho u_i u_j n_j dA = \int_{A^*} (-p\delta_{ij} + \tau_{ij})n_j dA. \tag{14.28}$$

The term on the left side represents the net flux of momentum out of the control volume and the term on the right side represents forces on the control surface. When applied to the control volume in Fig. 14.3 for the x-direction, (14.28) becomes:

$$-\dot{m}u_1 + \dot{m}u_2 = (pA)_1 - (pA)_2 + F, \tag{14.29}$$

where F is the x-component of the force exerted on the fluid in the control volume by the walls of the duct between locations 1 and 2. When the control volume has differential length, $x_2 = x_1 + dx$, then (14.29) can be written:

$$\dot{m}\frac{du}{dx} = -\frac{d}{dx}(pA) + p\frac{dA}{dx} - F_f = -A\frac{dp}{dx} - F_f, \tag{14.30}$$

where F_f is the perimeter friction force per unit length along the duct, and the second term in the middle portion of (14.30) is the pressure force on the control volume that occurs when the duct walls expand or contract. This term also appears in the derivation of (4.19), the inviscid steady-flow constant-density Bernoulli equation. For inviscid flow, F_f is zero and (14.30) simplifies to:

$$\rho u A\frac{du}{dx} = -A\frac{dp}{dx}, \quad \text{or} \quad u\,du + \frac{dp}{\rho} = 0, \tag{14.31}$$

where $\dot{m}$ in (14.30) has been replaced by $\rho u A$. The second equation of (14.31) is the Euler equation without a body force. A frictionless and adiabatic flow is isentropic, so the property relation (1.24) implies:

$$T\,dS = dh - dp/\rho = 0, \quad \text{so} \quad dh = dp/\rho.$$

Inserting the last relationship into the second equation of (14.31) and integrating produces:

$$h + u^2/2 = \text{const.}$$

This is the steady Bernoulli equation for isentropic compressible flow (4.79) without the body-force term. It is identical to (14.27) when $Q = 0$.

The equations for steady one-dimensional compressible flow in a nominally-straight duct with slowly varying area are (14.24), (14.27), and (14.30) The dependent flow variables are u, h, c, T, p, and ρ. A system of equations is closed by thermal and caloric equations of state (1.18), and the definitions of enthalpy (1.19) and sound speed (1.25). Thus, the next step in the solution of this system must be the specification of appropriate boundary conditions.

In incompressible flows, boundary conditions or known properties or profiles typically provide reference values for h, c, T, p, and ρ. In compressible flows, these thermodynamic variables depend on the flow's speed, u. Thus, reference values for thermodynamic variables in compressible flow must include a specification of the flow speed. The two most common reference conditions are the stagnation state ($u = 0$) and the sonic condition ($u = c$), and these are discussed in turn in the next few paragraphs.

If the properties of a compressible flow (h, ρ, u, etc.) are known at a certain point, the reference stagnation properties at that point are defined as those that would be obtained if the local flow were *imagined* to slow down to zero velocity *isentropically*. Stagnation properties are denoted by a subscript zero in gas dynamics. Thus, the *stagnation enthalpy* is defined as:

$$h_0 \equiv h + u^2/2.$$

For a perfect gas where $h = c_p T$, this implies:

$$c_p T_0 \equiv c_p T + u^2/2, \tag{14.32}$$

which defines the *stagnation temperature*. Ratios of local and stagnation variables are often sought, and these can be expressed in terms of the Mach number, M. For example, (14.32) can be rearranged:

$$\frac{T_0}{T} = 1 + \frac{u^2}{2c_p T} = 1 + \frac{\gamma - 1}{2}\frac{u^2}{\gamma RT} = 1 + \frac{\gamma - 1}{2}M^2, \tag{14.33}$$

where (14.1e) has been used for c_p. Thus the stagnation temperature T_0 can be found for a given T and M. The isentropic relations (14.2) can then be used to obtain the *stagnation pressure* and *stagnation density*:

$$\frac{p_0}{p} = \left(\frac{T_0}{T}\right)^{\gamma/(\gamma-1)} = \left[1 + \frac{\gamma - 1}{2}M^2\right]^{\gamma/(\gamma-1)}, \quad \text{and} \quad \frac{\rho_0}{\rho} = \left(\frac{T_0}{T}\right)^{1/(\gamma-1)} = \left[1 + \frac{\gamma - 1}{2}M^2\right]^{1/(\gamma-1)}. \tag{14.34, 14.35}$$

In a general flow, the stagnation properties can vary throughout the flow field. If, however, the flow is adiabatic (but not necessarily isentropic), then $h + u^2/2$ is constant throughout the flow as shown by (14.27) It follows that h_0, T_0, *and*

$c_0 (= [\gamma R T_0]^{1/2})$ *are constant throughout an adiabatic flow*, even in the presence of friction. In contrast, the stagnation pressure p_0 and density ρ_0 *decrease if there is friction*. To understand this, consider the entropy change in an adiabatic flow between sections 1 and 2 in a smoothly varying duct, with 2 being the downstream section. Let the flow at both sections hypothetically be brought to rest by isentropic processes, giving the local stagnation conditions p_{01}, p_{02}, T_{01}, and T_{02}. For this circumstance, the entropy change between the two sections can be expressed as:

$$s_2 - s_1 = s_{02} - s_{01} = -R \ln\left(\frac{p_{02}}{p_{01}}\right) + c_p \ln\left(\frac{T_{02}}{T_{01}}\right),$$

from (14.1i) The final term is zero for an adiabatic flow in which $T_{02} = T_{01}$. As the second law of thermodynamics requires that $s_2 > s_1$, it follows that:

$$p_{02} < p_{01}$$

which shows that the stagnation pressure falls due to friction. And, from $p_0 = \rho_0 R T_0$, ρ_0 must fall as well, for constant T_0.

It is apparent that all stagnation properties are constant along an isentropic flow. If such a flow happens to start from a large reservoir where the fluid is practically at rest, then the properties in the reservoir equal the stagnation properties everywhere in the flow (Fig. 14.4).

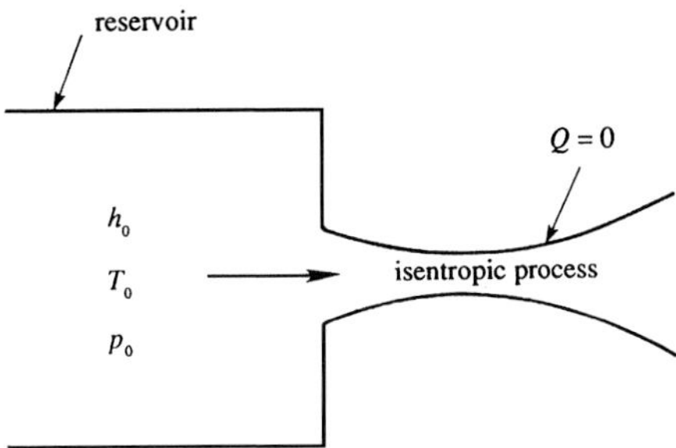

FIGURE 14.4 Schematic of an isentropic compressible-flow process starting from a reservoir. An isentropic process is both adiabatic (no heat exchange) and frictionless. Stagnation properties, indicated with a subscript 0, are uniform everywhere and are equal to the properties in the reservoir.

In addition to the stagnation properties, there is another useful set of reference quantities. These are called *sonic* or *critical* conditions and are commonly denoted by an asterisk. Thus, p^*, ρ^*, c^*, and T^* are properties attained if the local fluid is imagined to expand or compress isentropically until it reaches $M = 1$. The sonic area A^* is often the most useful or important because the stagnation area is infinite for any compressible duct flow. If M is known where the duct area is A, the passage area, A^*, at which the sonic conditions are attained can be determined to be:

$$\frac{A}{A^*} = \frac{1}{M}\left[\frac{2}{\gamma+1}\left(1 + \frac{\gamma-1}{2}M^2\right)\right]^{\frac{\gamma+1}{2(\gamma-1)}} \tag{14.36}$$

(see Exercise 14.12). Note, A^* in (14.36) is a separate concept from the control surface area in (14.25).

We shall see in the following section that sonic conditions can only be reached at the *throat* of a duct, where the area is minimum. However, a throat need not actually exist in the flow; the sonic variables are simply reference values that are reached *if* the flow were brought to the sonic state isentropically. From its definition, the value of A^* remains constant in isentropic flow. The presence of shock waves, friction, or heat transfer changes the value of A^* along the flow.

The values of T_0/T, p_0/p, ρ_0/ρ, and A/A^* at a point can be determined from (14.33) through (14.36) if the local Mach number is known. These ratios are tabulated in Table 14.1 for $\gamma = 1.4$. The reader should examine this table at this point. Examples 14.3 and 14.4 illustrate the use of this table.

Based on the foregoing development, consider some of the surprising consequences of compressibility that are found in isentropic flow through a duct of varying area. The natural application area for this topic is in the design of *nozzles* and *diffusers*. A nozzle is a device through which the flow expands from high to low pressure to generate a high-speed unidirectional jet of fluid. Examples of simple nozzles are the flow-exit fitting of a water jet cutter or a rocket motor. A diffuser's function is opposite that of a nozzle (and it has little or nothing to do with the diffusive transport of heat or species by molecular motion). In a diffuser a high-speed stream is decelerated and compressed. For example, air may enter the jet engine of an aircraft after passing through a diffuser, which raises the pressure and temperature of the air. In incompressible

TABLE 14.1 Isentropic flow of a perfect gas ($\gamma = 1.4$).

M	p/p_0	ρ/ρ_0	T/T_0	A/A^*	M	p/p_0	ρ/ρ_0	T/T_0	A/A^*
0	1	1	1	∞	1	0.5283	0.6339	0.8333	1
0.02	0.9997	0.9998	0.9999	28.9421	1.02	0.516	0.6234	0.8278	1.0003
0.04	0.9989	0.9992	0.9997	14.4815	1.04	0.5039	0.6129	0.8222	1.0013
0.06	0.9975	0.9982	0.9993	9.6659	1.06	0.4919	0.6024	0.8165	1.0029
0.08	0.9955	0.9968	0.9987	7.2616	1.08	0.48	0.592	0.8108	1.0051
0.1	0.993	0.995	0.998	5.8218	1.1	0.4684	0.5817	0.8052	1.0079
0.12	0.99	0.9928	0.9971	4.8643	1.12	0.4568	0.5714	0.7994	1.0113
0.14	0.9864	0.9903	0.9961	4.1824	1.14	0.4455	0.5612	0.7937	1.0153
0.16	0.9823	0.9873	0.9949	3.6727	1.16	0.4343	0.5511	0.7879	1.0198
0.18	0.9776	0.984	0.9936	3.2779	1.18	0.4232	0.5411	0.7822	1.0248
0.2	0.9725	0.9803	0.9921	2.9635	1.2	0.4124	0.5311	0.7764	1.0304
0.22	0.9668	0.9762	0.9904	2.7076	1.22	0.4017	0.5213	0.7706	1.0366
0.24	0.9607	0.9718	0.9886	2.4956	1.24	0.3912	0.5115	0.7648	1.0432
0.26	0.9541	0.967	0.9867	2.3173	1.26	0.3809	0.5019	0.759	1.0504
0.28	0.947	0.9619	0.9846	2.1656	1.28	0.3708	0.4923	0.7532	1.0581
0.3	0.9395	0.9564	0.9823	2.0351	1.3	0.3609	0.4829	0.7474	1.0663
0.32	0.9315	0.9506	0.9799	1.9219	1.32	0.3512	0.4736	0.7416	1.075
0.34	0.9231	0.9445	0.9774	1.8229	1.34	0.3417	0.4644	0.7358	1.0842
0.36	0.9143	0.938	0.9747	1.7358	1.36	0.3323	0.4553	0.73	1.094
0.38	0.9052	0.9313	0.9719	1.6587	1.38	0.3232	0.4463	0.7242	1.1042
0.4	0.8956	0.9243	0.969	1.5901	1.4	0.3142	0.4374	0.7184	1.1149
0.42	0.8857	0.917	0.9659	1.5289	1.42	0.3055	0.4287	0.7126	1.1262
0.44	0.8755	0.9094	0.9627	1.474	1.44	0.2969	0.4201	0.7069	1.1379
0.46	0.865	0.9016	0.9594	1.4246	1.46	0.2886	0.4116	0.7011	1.1501
0.48	0.8541	0.8935	0.9559	1.3801	1.48	0.2804	0.4032	0.6954	1.1629
0.5	0.843	0.8852	0.9524	1.3398	1.5	0.2724	0.395	0.6897	1.1762
0.52	0.8317	0.8766	0.9487	1.3034	1.52	0.2646	0.3869	0.684	1.1899
0.54	0.8201	0.8679	0.9449	1.2703	1.54	0.257	0.3789	0.6783	1.2042
0.56	0.8082	0.8589	0.941	1.2403	1.56	0.2496	0.371	0.6726	1.219
0.58	0.7962	0.8498	0.937	1.213	1.58	0.2423	0.3633	0.667	1.2344
0.6	0.784	0.8405	0.9328	1.1882	1.6	0.2353	0.3557	0.6614	1.2502
0.62	0.7716	0.831	0.9286	1.1656	1.62	0.2284	0.3483	0.6558	1.2666
0.64	0.7591	0.8213	0.9243	1.1451	1.64	0.2217	0.3409	0.6502	1.2836
0.66	0.7465	0.8115	0.9199	1.1265	1.66	0.2151	0.3337	0.6447	1.301
0.68	0.7338	0.8016	0.9153	1.1097	1.68	0.2088	0.3266	0.6392	1.319
0.7	0.7209	0.7916	0.9107	1.0944	1.7	0.2026	0.3197	0.6337	1.3376
0.72	0.708	0.7814	0.9061	1.0806	1.72	0.1966	0.3129	0.6283	1.3567
0.74	0.6951	0.7712	0.9013	1.0681	1.74	0.1907	0.3062	0.6229	1.3764
0.76	0.6821	0.7609	0.8964	1.057	1.76	0.185	0.2996	0.6175	1.3967
0.78	0.669	0.7505	0.8915	1.0471	1.78	0.1794	0.2931	0.6121	1.4175
0.8	0.656	0.74	0.8865	1.0382	1.8	0.174	0.2868	0.6068	1.439
0.82	0.643	0.7295	0.8815	1.0305	1.82	0.1688	0.2806	0.6015	1.461
0.84	0.63	0.7189	0.8763	1.0237	1.84	0.1637	0.2745	0.5963	1.4836
0.86	0.617	0.7083	0.8711	1.0179	1.86	0.1587	0.2686	0.591	1.5069
0.88	0.6041	0.6977	0.8659	1.0129	1.88	0.1539	0.2627	0.5859	1.5308
0.9	0.5913	0.687	0.8606	1.0089	1.9	0.1492	0.257	0.5807	1.5553
0.92	0.5785	0.6764	0.8552	1.0056	1.92	0.1447	0.2514	0.5756	1.5804
0.94	0.5658	0.6658	0.8498	1.0031	1.94	0.1403	0.2459	0.5705	1.6062
0.96	0.5532	0.6551	0.8444	1.0014	1.96	0.136	0.2405	0.5655	1.6326
0.98	0.5407	0.6445	0.8389	1.0003	1.98	0.1318	0.2352	0.5605	1.6597

continued on next page

TABLE 14.1 *(continued)*

M	p/p_0	ρ/ρ_0	T/T_0	A/A^*	M	p/p_0	ρ/ρ_0	T/T_0	A/A^*
2	0.1278	0.23	0.5556	1.6875	3	0.0272	0.0762	0.3571	4.2346
2.02	0.1239	0.225	0.5506	1.716	3.02	0.0264	0.0746	0.3541	4.316
2.04	0.1201	0.22	0.5458	1.7451	3.04	0.0256	0.073	0.3511	4.399
2.06	0.1164	0.2152	0.5409	1.775	3.06	0.0249	0.0715	0.3481	4.4835
2.08	0.1128	0.2104	0.5361	1.8056	3.08	0.0242	0.07	0.3452	4.5696
2.1	0.1094	0.2058	0.5313	1.8369	3.1	0.0234	0.0685	0.3422	4.6573
2.12	0.106	0.2013	0.5266	1.869	3.12	0.0228	0.0671	0.3393	4.7467
2.14	0.1027	0.1968	0.5219	1.9018	3.14	0.0221	0.0657	0.3365	4.8377
2.16	0.0996	0.1925	0.5173	1.9354	3.16	0.0215	0.0643	0.3337	4.9304
2.18	0.0965	0.1882	0.5127	1.9698	3.18	0.0208	0.063	0.3309	5.0248
2.2	0.0935	0.1841	0.5081	2.005	3.2	0.0202	0.0617	0.3281	5.121
2.22	0.0906	0.18	0.5036	2.0409	3.22	0.0196	0.0604	0.3253	5.2189
2.24	0.0878	0.176	0.4991	2.0777	3.24	0.0191	0.0591	0.3226	5.3186
2.26	0.0851	0.1721	0.4947	2.1153	3.26	0.0185	0.0579	0.3199	5.4201
2.28	0.0825	0.1683	0.4903	2.1538	3.28	0.018	0.0567	0.3173	5.5234
2.3	0.08	0.1646	0.4859	2.1931	3.3	0.0175	0.0555	0.3147	5.6286
2.32	0.0775	1.1609	0.4816	2.2333	3.32	0.017	0.0544	0.3121	5.7358
2.34	0.0751	0.1574	0.4773	2.2744	3.34	0.0165	0.0533	0.3095	5.8448
2.36	0.0728	0.1539	0.4731	2.3164	3.36	0.016	0.0522	0.3069	5.9558
2.38	0.0706	0.1505	0.4688	2.3593	3.38	0.0156	0.0511	0.3044	6.0687
2.4	0.0684	0.1472	0.4647	2.4031	3.4	0.0151	0.0501	0.3019	6.1837
2.42	0.0663	0.1439	0.4606	2.4479	3.42	0.0147	0.0491	0.2995	6.3007
2.44	0.0643	0.1408	0.4565	2.4936	3.44	0.0143	0.0481	0.297	6.4198
2.46	0.0623	0.1377	0.4524	2.5403	3.46	0.0139	0.0471	0.2946	6.5409
2.48	0.0604	0.1346	0.4484	2.588	3.48	0.0135	0.0462	0.2922	6.6642
2.5	0.0585	0.1317	0.4444	2.6367	3.5	0.0131	0.0452	0.2899	6.7896
2.52	0.0567	0.1288	0.4405	2.6865	3.52	0.0127	0.0443	0.2875	6.9172
2.54	0.055	0.126	0.4366	2.7372	3.54	0.0124	0.0434	0.2852	7.0471
2.56	0.0533	0.1232	0.4328	2.7891	3.56	0.012	0.0426	0.2829	7.1791
2.58	0.0517	0.1205	0.4289	2.842	3.58	0.0117	0.0417	0.2806	7.3135
2.6	0.0501	0.1179	0.4252	2.896	3.6	0.0114	0.0409	0.2784	7.4501
2.62	0.0486	0.1153	0.4214	2.9511	3.62	0.0111	0.0401	0.2762	7.5891
2.64	0.0471	0.1128	0.4177	3.0073	3.64	0.0108	0.0393	0.274	7.7305
2.66	0.0457	0.1103	0.4141	3.0647	3.66	0.0105	0.0385	0.2718	7.8742
2.68	0.0443	0.1079	0.4104	3.1233	3.68	0.0102	0.0378	0.2697	8.0204
2.7	0.043	0.1056	0.4068	3.183	3.7	0.0099	0.037	0.2675	8.1691
2.72	0.0417	0.1033	0.4033	3.244	3.72	0.0096	0.0363	0.2654	8.3202
2.74	0.0404	0.101	0.3998	3.3061	3.74	0.0094	0.0356	0.2633	8.4739
2.76	0.0392	0.0989	0.3963	3.3695	3.76	0.0091	0.0349	0.2613	8.6302
2.78	0.038	0.0967	0.3928	3.4342	3.78	0.0089	0.0342	0.2592	8.7891
2.8	0.0368	0.0946	0.3894	3.5001	3.8	0.0086	0.0335	0.2572	8.9506
2.82	0.0357	0.0926	0.386	3.5674	3.82	0.0084	0.0329	0.2552	9.1148
2.84	0.0347	0.0906	0.3827	3.6359	3.84	0.0082	0.0323	0.2532	0.2817
2.86	0.0336	0.0886	0.3794	3.7058	3.86	0.008	0.0316	0.2513	9.4513
2.88	0.0326	0.0867	0.3761	3.7771	3.88	0.0077	0.031	0.2493	9.6237
2.9	0.0317	0.0849	0.3729	3.8498	3.9	0.0075	0.0304	0.2474	9.799
2.92	0.0307	0.0831	0.3696	3.9238	3.92	0.0073	0.0299	0.2455	9.9771
2.94	0.0298	0.0813	0.3665	3.9993	3.94	0.0071	0.0293	0.2436	10.1581
2.96	0.0289	0.0796	0.3633	4.0763	3.96	0.0069	0.0287	0.2418	10.342
2.98	0.0281	0.0779	0.3602	4.1547	3.98	0.0068	0.0282	0.2399	10.5289

continued on next page

TABLE 14.1 *(continued)*

M	p/p_0	ρ/ρ_0	T/T_0	A/A^*	M	p/p_0	ρ/ρ_0	T/T_0	A/A^*
4	0.0066	0.0277	0.2381	10.7188	4.52	0.0034	0.0171	0.1966	16.8449
4.02	0.0064	0.0271	0.2363	10.9117	4.54	0.0033	0.0168	0.1952	17.1317
4.04	0.0062	0.0266	0.2345	11.1077	4.56	0.0032	0.0165	0.1938	17.4228
4.06	0.0061	0.0261	0.2327	11.3068	4.58	0.0031	0.0163	0.1925	17.7181
4.08	0.0059	0.0256	0.231	11.5091	4.6	0.0031	0.016	0.1911	18.0178
4.1	0.0058	0.0252	0.2293	11.7147	4.62	0.003	0.0157	0.1898	18.3218
4.12	0.0056	0.0247	0.2275	11.9234	4.64	0.0029	0.0154	0.1885	18.6303
4.14	0.0055	0.0242	0.2258	12.1354	4.66	0.0028	0.0152	0.1872	18.9433
4.16	0.0053	0.0238	0.2242	12.3508	4.68	0.0028	0.0149	0.1859	19.2608
4.18	0.0052	0.0234	0.2225	12.5695	4.7	0.0027	0.0146	0.1846	19.5828
4.2	0.0051	0.0229	0.2208	12.7916	4.72	0.0026	0.0144	0.1833	19.9095
4.22	0.0049	0.0225	0.2192	13.0172	4.74	0.0026	0.0141	0.182	20.2409
4.24	0.0048	0.0221	0.2176	13.2463	4.76	0.0025	0.0139	0.1808	20.577
4.26	0.0047	0.0217	0.216	13.4789	4.78	0.0025	0.0137	0.1795	20.9179
4.28	0.0046	0.0213	0.2144	13.7151	4.8	0.0024	0.0134	0.1783	21.2637
4.3	0.0044	0.0209	0.2129	13.9549	4.82	0.0023	0.0132	0.1771	21.6144
4.32	0.0043	0.0205	0.2113	14.1984	4.84	0.0023	0.013	0.1759	21.97
4.34	0.0042	0.0202	0.2098	14.4456	4.86	0.0022	0.0128	0.1747	22.3306
4.36	0.0041	0.0198	0.2083	14.6965	4.88	0.0022	0.0125	0.1735	22.6963
4.38	0.004	0.0194	0.2067	14.9513	4.9	0.0021	0.0123	0.1724	23.0671
4.4	0.0039	0.0191	0.2053	15.2099	4.92	0.0021	0.0121	0.1712	23.4431
4.42	0.0038	0.0187	0.2038	15.4724	4.94	0.002	0.0119	0.17	23.8243
4.44	0.0037	0.0184	0.2023	15.7388	4.96	0.002	0.0117	0.1689	24.2109
4.46	0.0036	0.0181	0.2009	16.0092	4.98	0.0019	0.0115	0.1678	24.6027
4.48	0.0035	0.0178	0.1994	16.2837	5	0.0019	0.0113	0.1667	25
4.5	0.0035	0.0174	0.198	16.5622					

flow, a nozzle profile converges in the direction of flow to increase the flow velocity, while a diffuser profile diverges. We shall see that such convergence and divergence must be reversed for supersonic flows in nozzles and diffusers.

Conservation of mass for compressible flow in a duct with smoothly varying area is specified by (14.24). For constant density flow, $d\rho/dx = 0$ and (14.24) implies $(1/A)dA/dx + (1/u)du/dx = 0$, so a decreasing area leads to an increase of velocity. When the flow is compressible, frictionless, and adiabatic then (14.31) implies:

$$udu = -dp/\rho = -c^2 d\rho/\rho, \tag{14.37}$$

because the flow is isentropic under these circumstances. Thus, the Euler equation requires that an increasing speed ($du > 0$) in the direction of flow must be accompanied by a fall of pressure ($dp < 0$). In terms of the Mach number, (14.37) becomes:

$$d\rho/\rho = -M^2 du/u. \tag{14.38}$$

This shows that for $M \ll 1$, the percentage change of density is much smaller than the percentage change of velocity. The density changes in the continuity equation (14.24) can therefore be neglected in low Mach number flows. Substituting (14.38) into (14.24), we obtain a velocity-area differential relationship that is valid in compressible flow:

$$\frac{du}{u} = -\frac{1}{1 - M^2}\frac{dA}{A}. \tag{14.39}$$

This relation leads to the following important conclusions about compressible flows:

(i) At subsonic speeds ($M < 1$) a decrease of area increases the speed of flow. A subsonic nozzle therefore must have a convergent profile, and a subsonic diffuser must have a divergent profile (upper row of Fig. 14.5). The behavior is qualitatively the same as in incompressible ($M = 0$) flows.

(ii) At supersonic speeds ($M > 1$) the denominator in (14.39) is negative, and we arrive at the conclusion that an *increase* in area leads to an increase of speed. The reason for such a behavior can be understood from (14.38), which shows that for $M > 1$ the density decreases faster than the velocity increases, thus the area must increase in an accelerating flow in order for $\rho u A$ to remain constant.

Therefore, the supersonic portion of a nozzle must have a divergent profile, and the supersonic part of a diffuser must have a convergent profile (bottom row of Fig. 14.5).

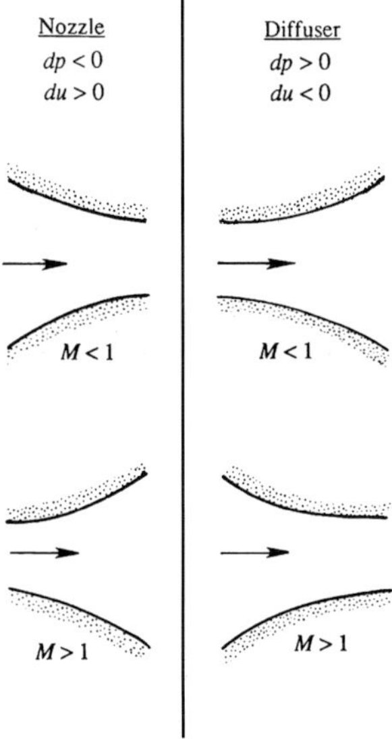

FIGURE 14.5 Shapes of nozzles and diffusers in subsonic and supersonic regimes. Nozzles are devices that accelerate the flow and are shown in the left column. Diffusers are devices that decelerate the flow and are shown in the right column. The area change with increasing downstream distance, dA/dx, switches sign for nozzles and diffusers and when the flow switches from subsonic to supersonic.

Suppose a nozzle is used to generate a supersonic stream, starting from a low-speed, high-pressure air stream at its inlet (Fig. 14.6). Then the Mach number must increase continuously from $M = 0$ near the inlet to $M > 1$ at the exit. The foregoing discussion shows that the nozzle must converge in the subsonic portion and diverge in the supersonic portion. Such a nozzle is called a *convergent–divergent nozzle*. From Fig. 14.6, it is clear that the Mach number must be unity at the *throat*, where the area is neither increasing nor decreasing ($dA \to 0$). This is consistent with (14.39), which shows that du can be non-zero at the throat only if $M = 1$. Hence, for steady isentropic compressible flow, *sonic velocity can be achieved only at the throat of a nozzle or a diffuser and nowhere else.*

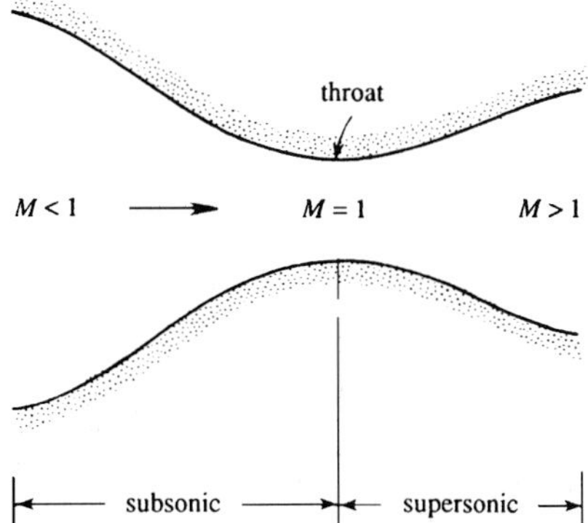

FIGURE 14.6 A convergent–divergent nozzle. When the pressure difference between the nozzle inlet and outlet is large enough, a compressible flow may be continuously accelerated from low speed to a supersonic Mach number through such a nozzle. When this happens the Mach number is unity at the minimum area, known as the nozzle's *throat*.

It does not, however, follow that M must necessarily be unity at the throat. According to (14.39), we may have a case where $M \neq 1$ at the throat if $du = 0$ there. As an example, the flow in a convergent–divergent tube may be subsonic everywhere, with M increasing in the convergent portion and decreasing in the divergent portion, with $M \neq 1$ at the throat (Fig. 14.7a). In this case the nozzle may also be known as a *venturi tube*. For entirely subsonic flow, the first half of the tube acts as a nozzle, whereas the second half acts as a diffuser. Alternatively, there may be supersonic flow everywhere in a convergent–divergent tube, with M decreasing in the convergent portion and increasing in the divergent portion, and again $M \neq 1$ at the throat (Fig. 14.7b).

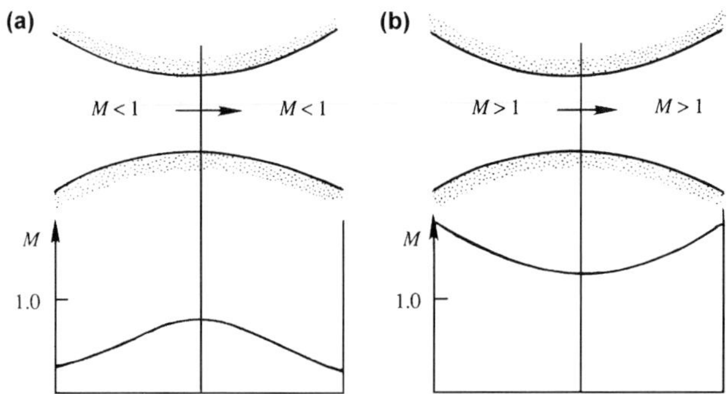

FIGURE 14.7 Convergent–divergent passages in which the condition at the throat is not sonic. This occurs when the flow is entirely subsonic as in (a), and when it is entirely supersonic as in (b).

Example 14.3

The nozzle of a rocket motor is designed to generate a thrust of 30,000 N when operating at an altitude of 20 km. The pressure and temperature inside the combustion chamber are 1000 kPa and 2500 K. The gas constant of the gas in the jet is $R = 280\,\text{m}^2/(\text{s}^2\text{K})$, and $\gamma = 1.4$. Assuming that the flow in the nozzle is isentropic, calculate the throat and exit areas. Use the isentropic table (Table 14.1).

Solution At an altitude of 20 km, the pressure of the standard atmosphere (Section A.4 in Appendix A) is 5467 Pa. If subscripts 0 and e refer to the stagnation and exit conditions, then a summary of the information given is as follows:

$$p_e = 5467\,\text{Pa},\ p_0 = 1000\,\text{kPa},\ T_0 = 2500\,\text{K},\quad \text{and}\quad \text{Thrust} = \rho_e u_e^2 A_e = 30\,\text{kN}.$$

Here, we have used the facts that the thrust equals mass flow rate times the exit velocity, and the pressure inside the combustion chamber is nearly equal to the stagnation pressure. The pressure ratio at the exit is:

$$\frac{p_e}{p_0} = \frac{5467\ \text{Pa}}{10^6\ \text{Pa}} = 5.467 \times 10^{-3}.$$

For this ratio of p_e/p_0, Table 14.1 gives:

$$M_e = 4.15,\ A_e/A^* = 12.2,\quad \text{and}\quad T_e/T_0 = 0.225.$$

The exit temperature and density are therefore:

$$T_e = 0.225(2500\,\text{K}) = 562\,\text{K},\quad \text{and}$$

$$\rho_e = p_e/RT_e = (5467\,\text{Pa})/(280\,\text{m}^2\text{s}^{-2}\text{K}^{-1})(562\,\text{K}) = 0.0347\,\text{kgm}^{-3}.$$

The exit velocity is:

$$u_e = M_e[\gamma RT_e]^{1/2} = 4.15[1.4(280\,\text{m}^2\text{s}^{-2}\text{K}^{-1})(562\,\text{K})]^{1/2} = 1948\,\text{ms}^{-1}.$$

The exit area is found from the expression for thrust:

$$A_e = \frac{\text{Thrust}}{\rho_e u_e^2} = \frac{30{,}000\,\text{N}}{(0.0347\,\text{kgm}^{-3})(1948\,\text{ms}^{-1})^2} = 0.0228\,\text{m}^2.$$

Because $A_e/A^* = 12.2$, the throat area is:

$$A^* = (0.0228\,\text{m}^2)/12.2 = 0.0187\,\text{m}^2.$$

14.4 Normal shock waves

A shock wave is similar to a step-change compression acoustic wave except that it has finite strength. The thickness of such waves is typically of the order of the molecular mean free path, so that fluid properties vary almost discontinuously across a shock wave. The high gradients of velocity and temperature result in entropy production within the wave so isentropic relations cannot be used across a shock. This section presents the relationships between properties of the flow upstream and downstream of a *normal shock*, where the shock is perpendicular to the direction of flow. Here, the shock wave is treated as a discontinuity and the actual process by which entropy is generated is not addressed. However, the entropy rise across the shock predicted by this analysis is correct. The internal structure of a shock, as predicted by the Navier-Stokes equations under certain simplifying assumptions, is given at the end of this section.

Stationary normal shock wave in a moving medium

To get started, consider the thin control volume shown in Fig. 14.8 that encloses a stationary shock wave. The control surface locations 1 and 2, shown as dashed lines in the figure, can be taken close to each other because of the discontinuous nature of the shock wave. In this case, the area change and the wall-surface friction between the upstream and the downstream control volume surfaces can be neglected. Furthermore, external heat addition is not of interest here so the basic equations are (14.23) and (14.29) with $F=0$, both simplified for constant area, and (14.27) with $Q=0$:

$$\rho_1 u_1 = \rho_2 u_2, \qquad p_1 - p_2 = -\rho_1 u_1^2 + \rho_2 u_2^2, \quad \text{and} \quad h_1 + \frac{1}{2}u_1^2 = h_2 + \frac{1}{2}u_2^2. \tag{14.40, 14.41, 14.42}$$

The Bernoulli equation cannot be used here because the process inside the shock wave is dissipative. Eqs. (14.40) through (14.42) contain four unknowns (h_2, u_2, p_2, ρ_2). The necessary additional relationship comes from the thermodynamics of a perfect gas (14.1):

$$h = c_p T = \frac{\gamma R}{\gamma - 1}\frac{p}{\rho R} = \frac{\gamma p}{(\gamma - 1)\rho},$$

so that (14.42) becomes:

$$\frac{\gamma}{(\gamma - 1)}\frac{p_1}{\rho_1} + \frac{u_1^2}{2} = \frac{\gamma}{(\gamma - 1)}\frac{p_2}{\rho_2} + \frac{u_2^2}{2}. \tag{14.43}$$

There are now three unknowns (u_2, p_2, ρ_2) and three equations: (14.40), (14.41), and (14.43), so the remainder of the effort to link the conditions upstream and downstream of a shock is primarily algebraic. Elimination of ρ_2 and u_2 from these equations leads to:

$$\frac{p_2}{p_1} = 1 + \frac{2\gamma}{\gamma + 1}\left[\frac{\rho_1 u_1^2}{\gamma p_1} - 1\right] = 1 + \frac{2\gamma}{\gamma + 1}[M_1^2 - 1], \tag{14.44}$$

where the second equality follows because $\rho u^2/\gamma p = u^2/\gamma RT = M^2$.

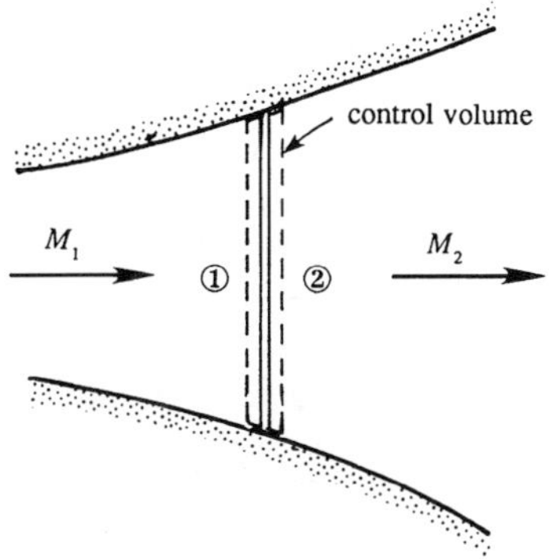

FIGURE 14.8 A normal shock wave trapped in a steady nozzle flow. Here, a control volume is shown that has control surfaces immediately upstream (1) and downstream (2) of the shock wave. Shock waves are very thin in most gases, so the area change and wall friction of the duct need not be considered as the flow traverses the shock wave.

With this relationship, an equation can be derived for M_2 in terms of M_1. Because $\rho u^2 = \rho c^2 M^2 = \rho(\gamma p/\rho)M^2 = \gamma p M^2$, so the momentum equation (14.41) can be written:

$$p_1 + \gamma p_1 M_1^2 = p_2 + \gamma p_2 M_2^2.$$

Using (14.44), this gives:

$$M_2^2 = \frac{(\gamma - 1)M_1^2 + 2}{2\gamma M_1^2 + 1 - \gamma}, \tag{14.45}$$

which is plotted in Fig. 14.9. Because $M_2 = M_1$ (state 2 = state 1) is a solution of (14.40), (14.41), and (14.43), there exists two possible solutions for M_2 for all $M_1 > [(\gamma - 1)/2\gamma]^{1/2}$. As is shown below, M_1 must be greater than unity to avoid violation of the second law of thermodynamics, so the two possibilities for the downstream state are: 1) no change from upstream, and 2) a sudden transition from supersonic to subsonic flow with consequent increases in pressure, density, and temperature. The density, velocity, and temperature ratios can be similarly obtained from the equations provided so far. They are:

$$\frac{\rho_2}{\rho_1} = \frac{u_1}{u_2} = \frac{(\gamma + 1)M_1^2}{(\gamma - 1)M_1^2 + 2}, \quad \text{and} \quad \frac{T_2}{T_1} = 1 + \frac{2(\gamma - 1)}{(\gamma + 1)^2}\frac{\gamma M_1^2 + 1}{M_1^2}(M_1^2 - 1). \tag{14.46, 14.47}$$

The normal shock relations (14.44) through (14.48) were worked out independently by the British engineer W.J.M. Rankine (1820–1872) and the French ballistician Pierre Henry Hugoniot (1851–1887). These equations are sometimes known as the *Rankine-Hugoniot relations*. The results of (14.43), (14.45), and (14.47) are tabulated for diatomic gases ($\gamma = 1.4$) in Table 14.2.

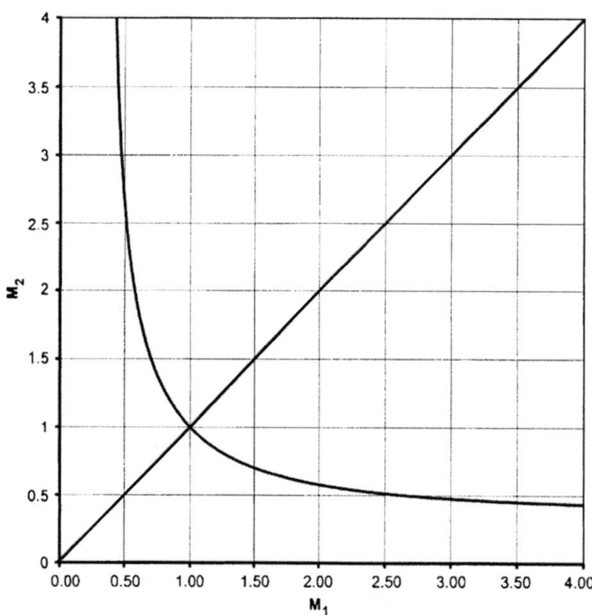

FIGURE 14.9 Normal shock wave solution for M_2 as function of M_1 for $\gamma = 1.4$. The trivial (no change) solution is also shown as the straight line with unity slope. Asymptotes are $[(\gamma - 1)/2\gamma]^{1/2} = 0.378$ for M_1 or $M_2 \to \infty$. The second law of thermodynamics limits valid shock wave solutions to those having $M_1 > 1$.

In terms of stagnation properties, T_0 and h_0 are constant across the shock because of the adiabatic nature of the process. In contrast, the stagnation properties p_0 and ρ_0 decrease across the shock due to the dissipative processes inside the shock zone that increase the entropy (see Exercise 14.19). Using (14.1) and (14.2), the entropy change is:

$$\frac{s_2 - s_1}{c_v} = \ln\left\{\frac{p_2}{p_1}\left(\frac{\rho_1}{\rho_2}\right)^\gamma\right\} = \ln\left\{\left[1 + \frac{2\gamma}{\gamma + 1}(M_1^2 - 1)\right]\left(\frac{(\gamma - 1)M_1^2 + 2}{(\gamma + 1)M_1^2}\right)^\gamma\right\}, \tag{14.48}$$

TABLE 14.2 One-dimensional normal-shock relations ($\gamma = 1.4$).

M_1	M_2	p_2/p_1	T_2/T_1	$(p_0)_2/(p_0)_1$	M_1	M_2	p_2/p_1	T_2/T_1	$(p_0)_2/(p_0)_1$
1	1	1	1	1	2.02	0.574	4.594	1.704	0.711
1.02	0.98	1.047	1.013	1	2.04	0.571	4.689	1.72	0.702
1.04	0.962	1.095	1.026	1	2.06	0.567	4.784	1.737	0.693
1.06	0.944	1.144	1.039	1	2.08	0.564	4.881	1.754	0.683
1.08	0.928	1.194	1.052	0.999	2.1	0.561	4.978	1.77	0.674
1.1	0.912	1.245	1.065	0.999	2.12	0.558	5.077	1.787	0.665
1.12	0.896	1.297	1.078	0.998	2.14	0.555	5.176	1.805	0.656
1.14	0.882	1.35	1.09	0.997	2.16	0.553	5.277	1.822	0.646
1.16	0.868	1.403	1.103	0.996	2.18	0.55	5.378	1.837	0.637
1.18	0.855	1.458	1.115	0.995	2.2	0.547	5.48	1.857	0.628
1.2	0.842	1.513	1.128	0.993	2.22	0.544	5.583	1.875	0.619
1.22	0.83	1.57	1.14	0.991	2.24	0.542	5.687	1.892	0.61
1.24	0.818	1.627	1.153	0.988	2.26	0.539	5.792	1.91	0.601
1.26	0.807	1.686	1.166	0.986	2.28	0.537	5.898	1.929	0.592
1.28	0.796	1.745	1.178	0.983	2.3	0.534	6.005	1.947	0.583
1.3	0.786	1.805	1.191	0.979	2.32	0.532	6.113	1.965	0.575
1.32	0.776	1.866	1.204	0.976	2.34	0.53	6.222	1.984	0.566
1.34	0.766	1.928	1.216	0.972	2.36	0.527	6.331	2.003	0.557
1.36	0.757	1.991	1.229	0.968	2.38	0.525	6.442	2.021	0.549
1.38	0.748	2.055	1.242	0.963	2.4	0.523	6.553	2.04	0.54
1.4	0.74	2.12	1.255	0.958	2.42	0.521	6.666	2.06	0.532
1.42	0.731	2.186	1.268	0.953	2.44	0.519	6.779	2.079	0.523
1.44	0.723	2.253	1.281	0.948	2.46	0.517	6.894	2.098	0.515
1.46	0.716	2.32	1.294	0.942	2.48	0.515	7.009	2.118	0.507
1.48	0.708	2.389	1.307	0.936	2.5	0.513	7.125	2.138	0.499
1.5	0.701	2.458	1.32	0.93	2.52	0.511	7.242	2.157	0.491
1.52	0.694	2.529	1.334	0.923	2.54	0.509	7.36	2.177	0.483
1.54	0.687	2.6	1.347	0.917	2.56	0.507	7.479	2.198	0.475
1.56	0.681	2.673	1.361	0.91	2.58	0.506	7.599	2.218	0.468
1.58	0.675	2.746	1.374	0.903	2.6	0.504	7.72	2.238	0.46
1.6	0.668	2.82	1.388	0.895	2.62	0.502	7.842	2.26	0.453
1.62	0.663	2.895	1.402	0.888	2.64	0.5	7.965	2.28	0.445
1.64	0.657	2.971	1.416	0.88	2.66	0.499	8.088	2.301	0.438
1.66	0.651	3.048	1.43	0.872	2.68	0.497	8.213	2.322	0.431
1.68	0.646	3.126	1.444	0.864	2.7	0.496	8.338	2.343	0.424
1.7	0.641	3.205	1.458	0.856	2.72	0.494	8.465	2.364	0.417
1.72	0.635	3.285	1.473	0.847	2.74	0.493	8.592	2.386	0.41
1.74	0.631	3.366	1.487	0.839	2.76	0.491	8.721	2.407	0.403
1.76	0.626	3.447	1.502	0.83	2.78	0.49	8.85	2.429	0.396
1.78	0.621	3.53	1.517	0.821	2.8	0.488	8.98	2.451	0.389
1.8	0.617	3.613	1.532	0.813	2.82	0.487	9.111	2.473	0.383
1.82	0.612	3.698	1.547	0.804	2.84	0.485	9.243	2.496	0.376
1.84	0.608	3.783	1.562	0.795	2.86	0.484	9.376	2.518	0.37
1.86	0.604	3.869	1.577	0.786	2.88	0.483	9.51	2.541	0.364
1.88	0.6	3.957	1.592	0.777	2.9	0.481	9.645	2.563	0.358
1.9	0.596	4.045	1.608	0.767	2.92	0.48	9.781	2.586	0.352
1.92	0.592	4.134	1.624	0.758	2.94	0.479	9.918	2.609	0.346
1.94	0.588	4.224	1.639	0.749	2.96	0.478	10.055	2.632	0.34
1.96	0.584	4.315	1.655	0.74	2.98	0.476	10.194	2.656	0.334
1.98	0.581	4.407	1.671	0.73	3	0.475	10.333	2.679	0.328
2	0.577	4.5	1.688	0.721					

which is plotted in Fig. 14.10. This figure shows that the entropy would decrease across an expansion shock in a perfect gas, which is impermissible. However, expansion shocks may be possible when the gas follows a different equation of state (Fergason et al., 2001). When the upstream Mach number is close to unity, Fig. 14.10 shows that the entropy change may be very small. The dependence of $s_2 - s_1$ on M_1 in the neighborhood of $M_1 = 1$ can be ascertained by treating $M_1^2 - 1$ as a small quantity and expanding (14.48) in terms of it (see Exercise 14.14) to find:

$$\frac{s_2 - s_1}{c_v} \cong \frac{2\gamma(\gamma - 1)}{3(\gamma + 1)^2}(M_1^2 - 1)^3. \tag{14.49a}$$

This equation explicitly shows that $s_2 - s_1$ will only be positive for a perfect gas when $M_1 > 1$. Thus, stationary shock waves do not occur when $M_1 < 1$ because of the second law of thermodynamics. However, when $M_1 > 1$, then (14.45) requires that $M_2 < 1$. Thus, the *Mach number changes from supersonic to subsonic values across a normal shock*, and this is the only possibility. A shock wave is therefore analogous to a hydraulic jump (see Section 11.6) in a gravity current, in which the Froude number jumps from supercritical to subcritical values; see Fig. 11.21. Eqs. (14.44), (14.46), and (14.47) then show that the jumps in p, ρ, and T are also from lower to higher values, so that a shock wave leads to compression and increased fluid temperature at the expense of stream-wise velocity.

Interestingly, terms involving $(M_1^2 - 1)$ and $(M_1^2 - 1)^2$ do not appear in (14.49a). Using the pressure ratio from (14.44), (14.49a) can be rewritten:

$$\frac{s_2 - s_1}{c_v} \cong \frac{\gamma^2 - 1}{12\gamma^2}\left(\frac{p_2 - p_1}{p_1}\right)^3. \tag{14.49b}$$

This shows that as the wave amplitude $\Delta p = p_2 - p_1$ decreases the entropy jump goes to zero like $(\Delta p)^3$. Thus, weak shock waves are nearly isentropic and this is the primary reason that loud acoustic disturbances are successfully treated as isentropic.

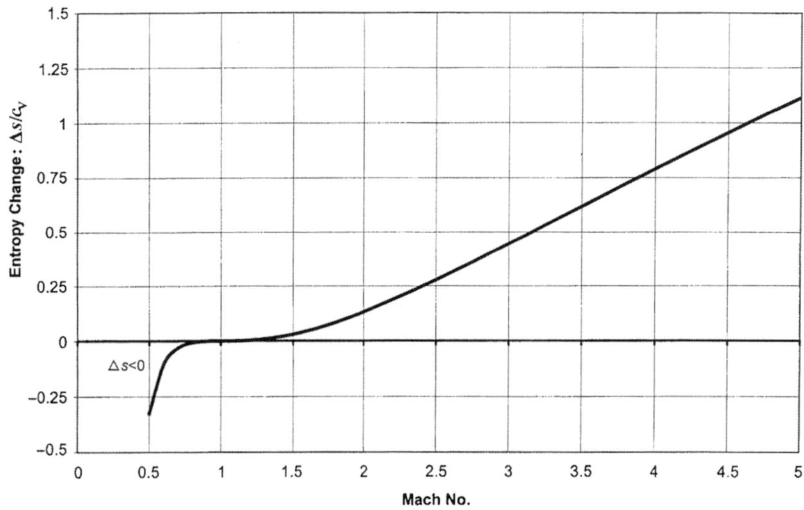

FIGURE 14.10 Entropy change $(s_2 - s_1)/c_v$ as a function of M_1 for $\gamma = 1.4$. Note higher-order contact at $M = 1$ to the horizontal line corresponding to zero entropy change as $M_1 \to 1$ from above. Negative entropy changes (a violation of the second law of thermodynamics) are predicted for $M_1 < 1$, so shock waves do not occur unless the upstream speed is supersonic, $M_1 > 1$.

Moving normal shock wave in a stationary medium

Frequently, one needs to calculate the properties of flow due to the propagation of a shock wave through a still medium, for example, that caused by an explosion. The Galilean transformation necessary to analyze this problem is indicated in Fig. 14.11. The left panel shows a stationary shock, with incoming and outgoing velocities u_1 and u_2, respectively. To this flow we add a velocity u_1 directed to the left, so that the fluid ahead of the shock is stationary, and the fluid behind the shock is moving to the *left* at a speed $u_1 - u_2$, as shown in the right panel of the figure. This is consistent with acoustic results in Section 14.2 where it was found that the fluid within a compression wave moves in the direction of the wave propagation. The shock speed is therefore u_1, with a supersonic Mach number $M_1 = u_1/c_1 > 1$. It follows that a *finite pressure disturbance propagates through a still fluid at supersonic speed*, in contrast to infinitesimal waves that propagate at the sonic speed. The expressions for all the thermodynamic properties of the flow, such as (14.44) through (14.49), are still applicable since their values are frame-independent.

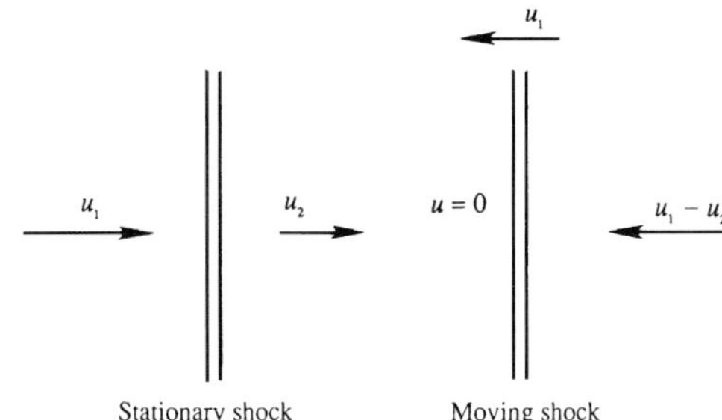

FIGURE 14.11 Stationary and moving shocks. The stationary shock shown on the left corresponds to a situation like that depicted in Fig. 14.8 where the incoming flow moves toward the shock. The moving shock situation shown on the right corresponds to blast wave that propagates away from an explosion into still air.

Normal-shock structure

We conclude this section on normal shock waves with a look into the structure of a shock wave. The viscous and heat conductive processes within the shock wave result in an entropy increase across the wave. However, the magnitudes of the viscosity μ and thermal conductivity k only determine the thickness of the shock wave and not the magnitude of the entropy increase. The entropy increase is determined solely by the upstream Mach number as shown by (14.48). We shall also see later that the *wave drag* experienced by a body due to the appearance of a shock wave is independent of viscosity or thermal conductivity. (The situation here is analogous to the viscous energy dissipation in fully turbulent flows, Section 10.7, in which the average kinetic-energy dissipation rate $\overline{\varepsilon}$ is determined by the velocity and length scales of a large-scale turbulence field (10.49) and not by the magnitude of the viscosity; a change in viscosity merely changes the length scale at which the dissipation takes place, namely, the Kolmogorov microscale.)

A shock wave can be considered a very thin boundary layer involving a large stream-wise velocity gradient du/dx, in contrast to the cross-stream (or wall-normal) velocity gradient involved in a viscous boundary layer near a solid surface. Analysis shows that the thickness δ of a shock wave is given by:

$$(u_1 - u_2)\delta/\nu \sim 1,$$

where the left side is a Reynolds number based on the velocity change across the shock, its thickness, and the average kinematic viscosity. Taking a typical value for air of $\nu \sim 10^{-5}\,\mathrm{m^2/s}$, and a velocity jump of $\Delta u \sim 100\,\mathrm{m/s}$, we obtain a shock thickness of 10^{-7} m. This is not much larger than the mean-free path (average distance traveled by a molecule between collisions), which suggests that the continuum hypothesis and the assumption of local thermodynamic equilibrium are both of questionable validity in analyzing shock structure.

With these limitations noted, some insight into the structure of shock waves may be gained by considering the one-dimensional steady Navier-Stokes equations, including heat conduction and Newtonian viscous stresses, in a shock-fixed coordinate system. The solution we obtain provides a smooth transition between upstream and downstream states, looks reasonable, and agrees with experiments and kinetic theory models for upstream Mach numbers less than about 2. The equations for conservation of mass, momentum, and energy, respectively, are the steady one-dimensional versions of (4.7), (4.38) without a body force, and (4.61) written in terms of enthalpy h:

$$\frac{d(\rho u)}{dx} = 0, \quad \rho u \frac{du}{dx} + \frac{dp}{dx} = \frac{d}{dx}\left(\left(\frac{4}{3}\mu + \mu_v\right)\frac{du}{dx}\right), \quad \text{and}$$

$$\rho u \frac{dh}{dx} - u\frac{dp}{dx} = \left(\frac{4}{3}\mu + \mu_v\right)\left(\frac{du}{dx}\right)^2 + \frac{d}{dx}\left(k\frac{dT}{dx}\right).$$

By adding the product of u and the momentum equation to the energy equation, these can be integrated once to find:

$$\rho u = m, \quad mu + p = \mu''\frac{du}{dx} + mV, \quad \text{and} \quad m\left(h + \frac{1}{2}u^2\right) = \mu'' u\frac{du}{dx} + k\frac{dT}{dx} + mI,$$

where m, V, and I are the constants of integration and $\mu'' = \frac{4}{3}\mu + \mu_v$. When these are evaluated upstream (state 1) and downstream (state 2) of the shock where gradients vanish, they yield the Rankine-Hugoniot relations derived earlier. We also need the equations of state for a perfect gas with constant specific heats to solve for the shock structure: $h = c_p T$, and $p = \rho RT$. Multiplying the energy equation by c_p/k we obtain the form:

$$m\frac{c_p}{k}\left(c_p T + \frac{1}{2}u^2\right) - \frac{\mu'' c_p}{2k}\frac{du^2}{dx} - c_p\frac{dT}{dx} = m\frac{c_p}{k}I.$$

This equation has an exact integral in the special case $\mathrm{Pr}'' \equiv \mu'' c_p/k = 1$ that was found by Becker in 1922. For most simple gases, Pr'' is likely to be near unity so it is reasonable to proceed assuming $\mathrm{Pr}'' = 1$. The Becker integral is $c_p T + u^2/2 = I$. Eliminating all variables but u from the momentum equation, using the equations of state, mass conservation, and the energy integral, we reach:

$$mu + (m/u)(R/c_p)(I - u^2/2) - \mu''(du/dx) = mV.$$

With $c_p/R = \gamma/(\gamma - 1)$, multiplying by u/m, leads to:

$$-\left[\frac{2\gamma}{\gamma+1}\right]\frac{\mu''}{m}\frac{udu}{dx} = -u^2 + \left[\frac{2\gamma}{\gamma+1}\right]uV - 2I\frac{\gamma-1}{\gamma+1} \equiv (U_1 - U)(U - U_2).$$

Divide by V^2 and let $u/V = U$. The equation for the structure becomes:

$$-U(U_1 - U)^{-1}(U - U_2)^{-1}dU = [(\gamma+1)/2\gamma](m/\mu'')dx,$$

where the roots of the quadratic are:

$$U_{1,2} = \left[\frac{\gamma}{\gamma+1}\right]\left\{1 \pm \left[1 - 2(\gamma^2 - 1)I/(\gamma^2 V^2)\right]^{1/2}\right\},$$

the dimensionless speeds far up- and downstream of the shock. The left-hand side of the equation for the structure is rewritten in terms of partial fractions and then integrated to obtain:

$$[U_1 \ln(U_1 - U) - U_2 \ln(U - U_2)]/(U_1 - U_2) = [(\gamma+1)/2\gamma]m\int dx/\mu'' \equiv [(\gamma+1)/2\gamma]\eta.$$

The resulting shock structure is shown in Fig. 14.12 in terms of the stretched coordinate $\eta = \int (m/\mu'')dx$ where μ'' is often a strong function of temperature and thus of x. A similar structure is obtained for all except quite small values of Pr''. In the limit $\mathrm{Pr}'' \to 0$, Hayes (1958) points out that there must be a "shock within a shock" because heat conduction alone cannot provide the entire structure. In fact, Becker (1922, footnote, p. 341) credits Prandtl for originating this idea. Cohen and Moraff (1971) provided the structure of both the outer (heat conducting) and inner (isothermal viscous) shocks. Here, the variable η is a dimensionless length scale measured very roughly in units of mean-free paths. We see that a measure of shock thickness is of the order of 5 mean-free paths from this analysis.

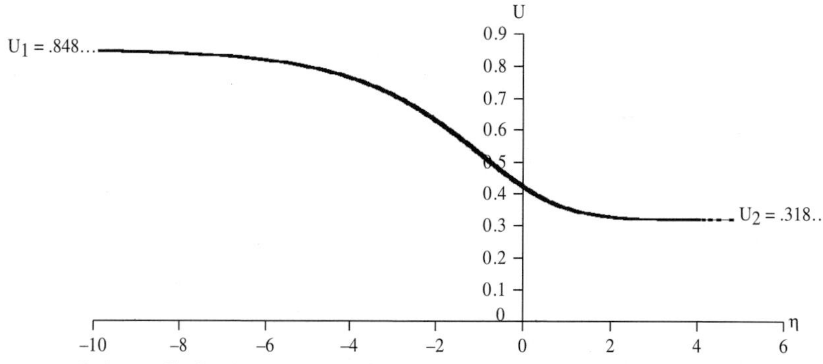

FIGURE 14.12 Shock structure velocity profile for the case $U_1 = 0.848485$, $U_2 = 0.31818$, corresponding to $M_1 = 2.187$. The units of the horizontal coordinate may be approximately interpreted as mean-free paths. Thus, a shock wave is typically a small countable number of mean-free-paths thick.

Example 14.4

A normal shock wave forms just ahead of a bullet as it travels at 750 m/s through still air at 100 kPa and 295 K. What are the pressure, temperature, and density of the air immediately behind the shock wave?

Solution Ahead of the bullet, the density is:

$$\rho_1 = p_1/RT_1 = (100\,\text{kPa})/[(287\,\text{m}^2\text{s}^{-2}\text{K}^{-1})(295\,\text{K})] = 1.181\,\text{kgm}^{-3},$$

and the sound speed is:

$$c_1 = \sqrt{\gamma RT_1} = \sqrt{1.4(287\,\text{m}^2\text{s}^{-2}\text{K}^{-1})295\,\text{K}} = 344\,\text{m/s}.$$

Thus, the shock Mach number is $M_1 = 750/344 = 2.18$. So, from Table 14.2, or Eqs. (14.44) to (14.47), $M_2 = 0.550$ and the ratios across the shock are: $p_2/p_1 = 5.378$, $T_2/T_1 = 1.837$, and $\rho_2/\rho_1 = 2.924$. Hence, $p_2 = 538\,\text{kPa}$, $T_2 = 542\,\text{K}$, and $\rho_2 = 3.453\,\text{kgm}^{-3}$.

14.5 Operation of nozzles at different back pressures

Nozzles are used to accelerate a fluid stream and are employed in such systems as wind tunnels, rocket motors, ejector pumps, water jet cutters, and steam turbines. A pressure drop is maintained across the nozzle to accelerate fluid through it. This section presents the behavior of the flow though a nozzle as the back pressure p_B on the nozzle is varied when the nozzle-supply pressure is maintained at a constant value p_0 (the stagnation pressure). Here the p_B is the pressure in the nominally quiescent environment into which the nozzle flow is directed. In the following discussion, the pressure p_{exit} at the exit plane of the nozzle equals the back pressure p_B if the flow at the exit plane is subsonic, but *not* if it is supersonic. This must be true because subsonic flow allows the downstream pressure p_B to be communicated up into the nozzle exit, and sharp pressure changes are only allowed in a supersonic flow.

Convergent nozzle

Consider first the case of a convergent nozzle shown in Fig. 14.13, which presents a sequence of states *a* through *c* during which the back pressure is gradually lowered. For curve *a*, the flow throughout the nozzle is subsonic. As p_B is lowered, the Mach number increases everywhere and the mass flux through the nozzle also increases. This continues until sonic conditions are reached at the exit, as represented by curve *b*. Further lowering of the back pressure has no effect on the flow inside the nozzle. This is because the fluid at the exit is now moving downstream at the velocity at which no pressure changes can propagate upstream. Changes in p_B therefore cannot propagate upstream after sonic conditions are reached at the nozzle exit. We say that the nozzle at this stage is *choked* because the mass flux cannot be increased by further lowering of back pressure. If p_B is lowered further (curve *c* in Fig. 14.13), supersonic flow is generated downstream of the nozzle, and the jet pressure adjusts to p_B by means of a series of oblique compression and expansion waves, as schematically indicated by the oscillating pressure distribution for curve *c*. Oblique compression and expansion waves are explained in Section 14.8. It is only necessary to note here that they are oriented at an angle to the direction of flow, and that the pressure increases through an oblique compression wave and decreases through an oblique expansion wave.

Convergent–divergent nozzle

Now consider the case of a convergent–divergent passage, also known as a Laval nozzle (Fig. 14.14). Completely subsonic flow applies to curve *a*. As p_B is lowered to p_b, the sonic condition is reached at the throat. On further reduction of the back pressure, the flow upstream of the throat does not respond, and the nozzle flow is choked in the sense that it has reached the maximum mass flow rate for the given values of p_0 and throat area. There is a range of back pressures, shown by curves *c* and *d*, in which the flow initially becomes supersonic in the divergent portion, but then adjusts to the back pressure by means of a normal-shock standing inside the nozzle. The flow downstream of the shock is, of course, subsonic. In this range the position of the shock moves downstream as p_B is decreased, and for curve *d* the normal-shock stands right at the exit plane. The flow in the entire divergent portion up to the exit plane is now supersonic and remains so on further reduction of p_B. When the back pressure is further reduced to p_e, there is no normal shock anywhere within the nozzle, and the jet pressure adjusts to p_B by means of oblique compression waves downstream of the nozzle's exit plane. These oblique waves vanish when $p_B = p_f$. On further reduction of the back pressure, the adjustment to p_B takes place outside the exit plane by means of oblique expansion waves.

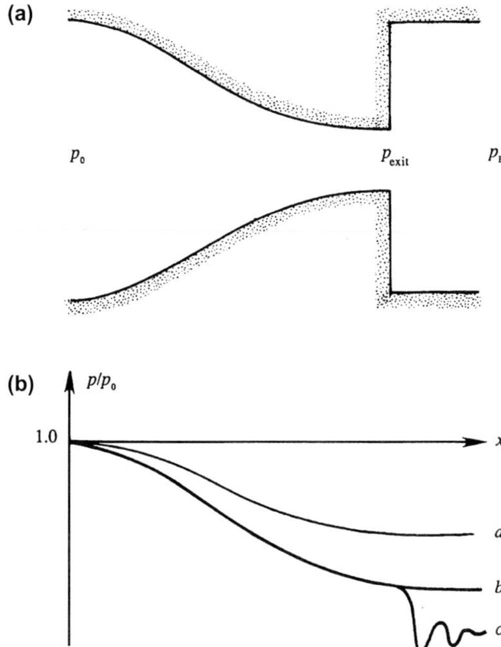

FIGURE 14.13 Pressure distribution along a convergent nozzle for different values of back pressure p_B: (a) diagram of the nozzle, and (b) pressure distributions as p_B is lowered. Here the highest possible flow speed at the nozzle exit is sonic. When p_B is lowered beyond the point of sonic flow at the nozzle exit, the flow continues to accelerate outside the nozzle via expansion waves that lead to non-uniform pressures (curve c).

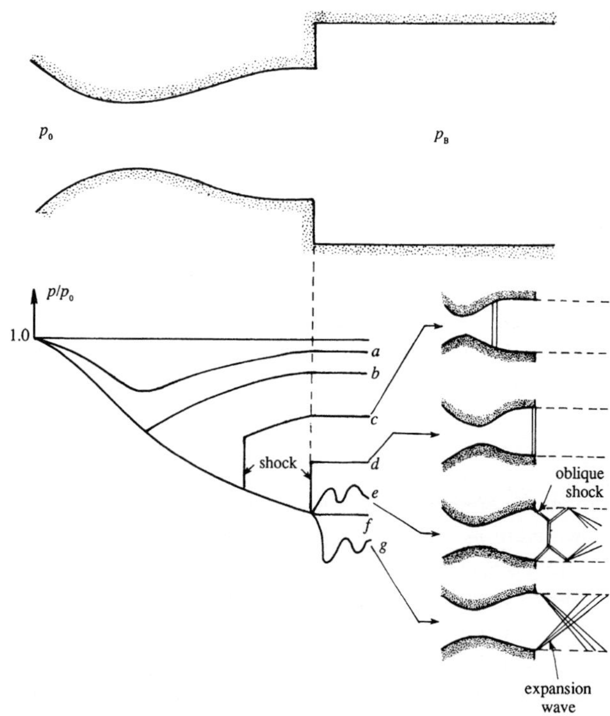

FIGURE 14.14 Pressure distribution along a convergent–divergent (aka Laval) nozzle for different values of the back pressure p_B. Flow patterns for cases c, d, e, and g are indicated schematically on the right. The condition f is the *pressure matched* case and usually corresponds to the nozzle's design condition. For this case, the flow looks like that of c or d without the shock wave. *H.W. Liepmann and A. Roshko,* Elements of Gas Dynamics, *Wiley, New York 1957; reprinted with the permission of Dr. Anatol Roshko.*

Example 14.5

A convergent–divergent nozzle is operating under off-design conditions, resulting in the presence of a shock wave in the diverging portion. A reservoir containing air at 400 kPa and 800 K supplies the nozzle, whose throat area is 0.2 m^2. The Mach number upstream of the shock is $M_1 = 2.44$. The area at the nozzle exit is 0.7 m^2. Find the area at the location of the shock and the exit temperature.

Solution Fig. 14.15 shows the profile of the nozzle, where sections 1 and 2 represent conditions across the shock. As a shock wave can exist only in a supersonic stream, we know that sonic conditions are reached at the throat, and the throat area equals the critical area A^*. The values given are therefore:

$$p_0 = 400\,\text{kPa},\ T_0 = 800\,\text{K}, A_{\text{throat}} = A_1^* = 0.2\,\text{m}^2,\ M_1 = 2.44, \quad \text{and} \quad A_3 = 0.7\,\text{m}^2.$$

Note that A^* is constant upstream of the shock because the flow isentropic there; this is why $A_{\text{throat}} = A_1{}^*$.

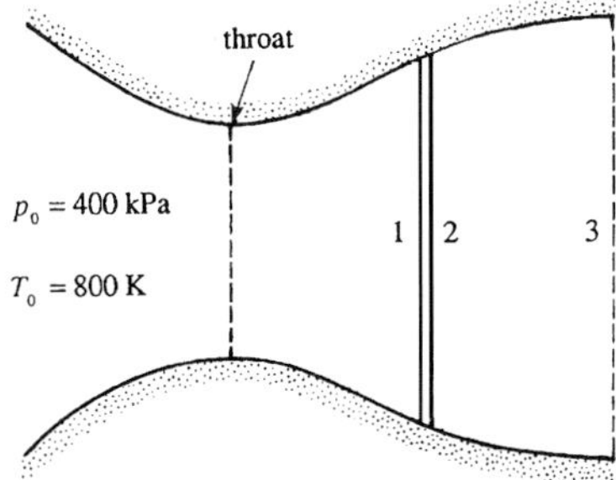

FIGURE 14.15 Drawing for Example 14.5 This is case c from Fig. 14.14 where a normal shock occurs in the nozzle.

The technique of solving this problem is to proceed downstream from the given stagnation conditions. For $M_1 = 2.44$, the isentropic table (Table 14.1) gives:

$$A_1/A_1^* = 2.5, \quad \text{so that} \quad A_1 = A_2 = (2.5)(0.2) = 0.5\,\text{m}^2.$$

This is the nozzle's cross-section area at the location of the shock. For $M_1 = 2.44$, the normal shock Table 14.2 gives:

$$M_2 = 0.519, \quad \text{and} \quad p_{02}/p_{01} = 0.523.$$

There is no loss of stagnation pressure up to section 1, so $p_{01} = p_0$, which implies:

$$p_{02} = 0.523\,p_0 = 0.523(400) = 209.2\,\text{kPa}.$$

The value of A^* changes across a shock wave. The ratio $A_2/A_2{}^*$ can be found from the *isentropic* table (Table 14.1) corresponding to a Mach number of $M_2 = 0.519$. (Note that $A_2{}^*$ simply denotes the area that would be reached if the flow from state 2 were accelerated isentropically to sonic conditions.) For $M_2 = 0.519$, Table 14.1 gives:

$$A_2/A_2^* = 1.3, \text{ which leads to } A_2^* = A_2/1.3 = 0.5/1.3 = 0.3846\,\text{m}^2.$$

The flow from section 2 to section 3 is isentropic, during which A^* remains constant, so:

$$A_3/A_3^* = A_3/A_2^* = 0.7/0.3846 = 1.82.$$

Now find the conditions at the nozzle exit from the isentropic table (Table 14.1). However, the value of $A/A^* = 1.82$ may be found either in the supersonic or the subsonic branch of the table. Since the flow downstream of a normal shock can only be subsonic, use the subsonic branch. For $A/A^* = 1.82$, Table 14.1 gives:

$$T_3/T_{03} = 0.977.$$

The stagnation temperature remains constant in an adiabatic process, so that $T_{03} = T_0$. Thus:

$$T_3 = 0.977(800) = 782\,\text{K}.$$

14.6 Effects of friction and heating in constant-area ducts

The results presented in the prior sections are valid for steady adiabatic compressible flows where discontinuous jumps in entropy are possible across shock waves. The subject of this section is steady non-isentropic compressible flow when the duct area is constant but friction and heat transfer may both influence the flow.

For steady one-dimensional compressible flow in a duct of constant cross-sectional area, the equations of mass, momentum, and energy conservation between an upstream location (1) and a downstream location (2) are:

$$\rho_1 u_1 = \rho_2 u_2,\ p_1 + \rho_1 u_1^2 = p_2 + \rho_2 u_2^2 + p_1 f, \quad \text{and} \quad h_1 + \frac{1}{2}u_1^2 + h_1 q = h_2 + \frac{1}{2}u_2^2, \tag{14.50}$$

where $f = F/(p_1 A)$ is a dimensionless friction parameter and $q = Q/h_1$ is a dimensionless heating parameter. In terms of Mach number, for a perfect gas with constant specific heats, the momentum and energy equations become, respectively:

$$p_1(1+\gamma M_1^2 - f) = p_2(1+\gamma M_2^2), \quad \text{and} \quad h_1\left(1+\frac{\gamma-1}{2}M_1^2 + q\right) = h_2\left(1+\frac{\gamma-1}{2}M_2^2\right).$$

Using mass conservation, the thermal equation of state $p = \rho RT$, and the definition of the Mach number, all thermodynamic variables can be implicitly eliminated from these equations, resulting in

$$\frac{M_2}{M_1} = \frac{1+\gamma M_2^2}{1+\gamma M_1^2 - f}\left[\frac{1+((\gamma-1)/2)M_1^2 + q}{1+((\gamma-1)/2)M_2^2}\right]^{1/2}.$$

Bringing the unknown M_2 to the left-hand side and assuming q and f are specified along with M_1 leads to:

$$\frac{M_2^2[1+((\gamma-1)/2)M_2^2]}{(1+\gamma M_2^2)^2} = \frac{M_1^2[1+((\gamma-1)/2)M_1^2 + q]}{(1+\gamma M_1^2 - f)^2} \equiv A,$$

where A is known. This is a biquadratic equation for M_2 with the solution:

$$M_2^2 = \frac{-(1-2A\gamma) \pm [1-2A(\gamma+1)]^{1/2}}{(\gamma-1) - 2A\gamma^2}. \tag{14.51}$$

Figs. 14.16 and 14.17 are plots of M_2 versus M_1 from (14.51), first with f as a parameter and $q = 0$ (Fig. 14.16), and then with q as a parameter and $f = 0$ (Fig. 14.17). Generally, flow properties are known at the inlet station (1) and the flow properties at the outlet station (2) are sought. Here, the dimensionless friction f and heat transfer q are presumed to be specified. Thus, once M_2 is calculated from (14.51), all of the other properties may be obtained from the conservation laws shown above. When q and $f = 0$, two solutions are possible: the trivial solution $M_1 = M_2$ and the normal-shock solution given in Section 14.4. The upper left branch of the solution $M_2 > 1$ when $M_1 < 1$ is inaccessible because it violates the second law of thermodynamics, that is, it results in a spontaneous decrease of entropy.

Effect of friction

Referring to Fig. 14.16 for M_1 and M_2 both subsonic, the solution indicates the surprising result that friction accelerates a subsonic flow leading to $M_2 > M_1$. This happens because friction causes the pressure, and therefore the density, to drop rapidly enough so that the fluid velocity must increase to maintain a constant mass flow. For this case of adiabatic flow with friction, the relevant equations for differential changes in pressure, velocity, and density in terms of the local Mach number $M = u/c$ are:

$$-\frac{dp}{p_1} = \frac{1+(\gamma-1)M^2}{1-M^2}df, \quad \text{and} \quad \frac{du}{u} = -\frac{d\rho}{\rho} = \frac{p_1}{p}\frac{df}{1-M^2}, \tag{14.52}$$

and these may be derived from (14.40) with $q = 0$ (Exercise 14.21). In particular since $df > 0$, (14.52) implies that dp/p_1 may have a large negative magnitude compared to df as M approaches unity from below. We will discuss in what follows what actually happens when there is no apparent solution for M_2. When M_1 is supersonic, two solutions are generally possible—one for which $1 < M_2 < M_1$ and the other where $M_2 < 1$. They are connected by a normal shock. Whether or not a shock occurs depends on the downstream pressure. There is also the possibility of M_1 insufficiently large or f

too large so that no solution is indicated. We will discuss that in a following paragraph but note that the two solutions coalesce when $M_2 = 1$ and the flow is choked. At this condition the maximum mass flow is passed by the duct. In the case $1 < M_2 < M_1$, the flow is decelerated and the pressure, density, and temperature all increase in the downstream direction. The stagnation pressure is always decreased by friction as the entropy is increased. In summary, friction's net effect is to drive a compressible duct flow toward $M_2 = 1$ for any value of M_1.

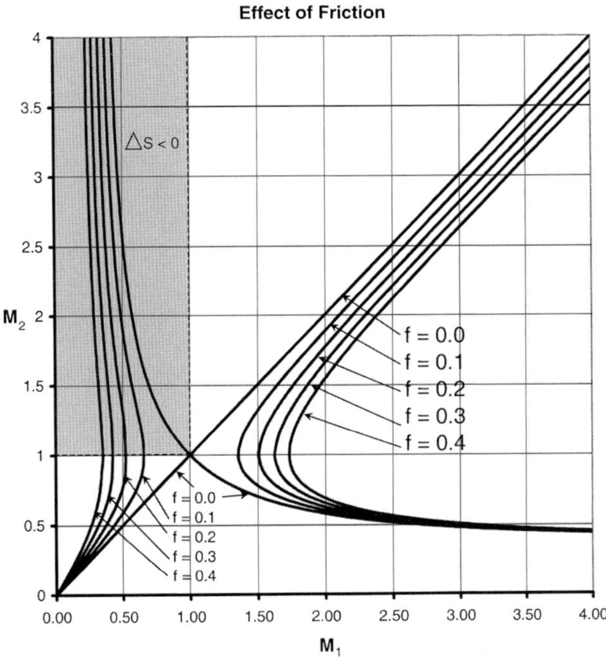

FIGURE 14.16 Flow in a constant-area duct with the dimensionless friction f as a parameter without heat exchange ($q = 0$) at $\gamma = 1.4$. The shaded region in the upper left is inaccessible because $\Delta s < 0$. For any duct inlet value of M_1 the curves indicate possible outlet states. Interestingly, for $M_1 < 1$, all possible M_2 values are at a higher Mach number. For $M_1 > 1$, the two possible final states are both at lower Mach numbers.

Effect of heat transfer

The range of solutions is twice as rich in this case as q may take both signs while f must be positive. Fig. 14.17 shows that for $q > 0$ solutions are similar in most respects to those with friction ($f > 0$). Heating accelerates a subsonic flow and lowers the pressure and density. However, heating generally increases the fluid temperature except in the limited range $1/\gamma < M_1^2 < 1$ in which the fluid temperature decreases with heat addition. The relevant equations for differential changes in temperature and flow speed in terms of the local Mach number $M = u/c$ are:

$$\frac{dT}{T_1} = \left(\frac{1 - \gamma M^2}{1 - M^2}\right) dq, \quad \text{and} \quad \frac{u du}{h_1} = \frac{(\gamma - 1)M^2}{1 - M^2} dq, \tag{14.53}$$

and these can be derived from (14.50) with $f = 0$ (Exercise 14.22). When $1/\gamma < M_1^2 < 1$, the energy from heat addition goes preferentially into increasing the velocity of the fluid. The supersonic branch $M_2 > 1$ when $M_1 < 1$ is inaccessible because those solutions violate the second law of thermodynamics. Again, as with f too large or M_1 too close to 1, there is a possibility of no indicated solution when q is too large; this is discussed in what follows. When $M_1 > 1$, two solutions for M_2 are generally possible and they are connected by a normal shock. The shock is absent if the downstream pressure is low and present if the downstream pressure is high. Although $q > 0$ (and $f > 0$) does not always indicate a solution (if the flow has been choked), there will always be a solution for $q < 0$. Cooling a supersonic flow accelerates it, thus decreasing its pressure, temperature, and density. If no shock occurs, $M_2 > M_1$. Conversely, cooling a subsonic flow decelerates it so that the pressure and density increase. The temperature decreases when heat is removed from the flow except in the limited range $1/\gamma < M_1^2 < 1$ in which the heat removal decelerates the flow so rapidly that the temperature increases.

For high molecular-weight gases, near critical conditions (high pressure, low temperature), the gas dynamic relationships may be completely different from those developed here for perfect gases. Cramer and Fry (1993) found that

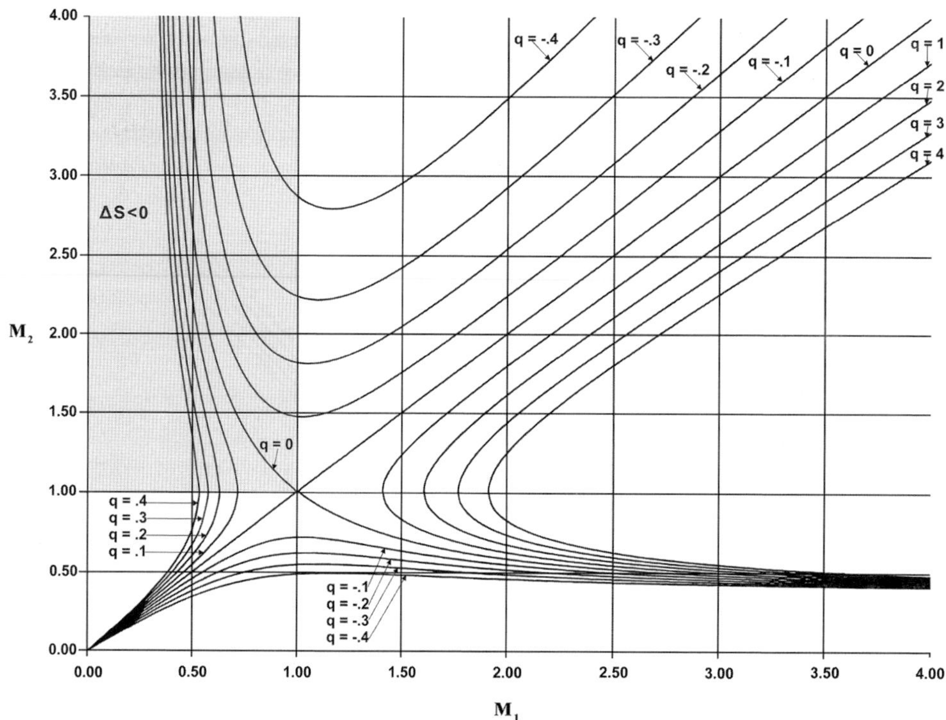

FIGURE 14.17 Flow in a constant-area duct with the dimensionless heat exchange q as a parameter without friction ($f = 0$) at $\gamma = 1.4$. The shaded region in the upper left is inaccessible because $\Delta s < 0$. Here, heat addition is seen to have much the same effect as friction.

non-perfect gases may support expansion shocks, accelerated flow through "antithroats," and generally behave in unfamiliar ways.

Figs. 14.16 and 14.17 show that friction or heat input in a constant-area duct both drive a compressible flow in the duct toward the sonic condition. For any given M_1, the maximum f or $q > 0$ that is permissible is the one for which $M = 1$ at the exit station. The flow is then said to be choked, and the mass flow rate through that duct cannot be increased without increasing p_1 or decreasing p_2. This is analogous to flow in a convergent duct. Imagine pouring liquid through a funnel from one container into another. There is a maximum volumetric flow rate that can be passed by the funnel, and beyond that flow rate, the funnel overflows. The same thing happens here. If f or q is too large, such that no (steady-state) solution is possible, external adjustment must be made that reduces the mass flow rate to that for which the exit speed is just sonic. For $M_1 < 1$ and $M_1 > 1$ the limiting curves for f and q indicating choked flow intersect $M_2 = 1$ at right angles. Qualitatively, the effect is the same as choking by area contraction.

Example 14.6

Consider adiabatic compressible flow of a perfect gas in a round duct with constant diameter D, and interior skin friction coefficient, C_f (the Fanning friction factor). Starting from (14.30), use the definition of the Mach number, $u = Mc$, and (14.33) to derive an equation for dM^2/dx, where x is the distance along the duct in the direction of flow.

Solution The perimeter friction per unit length in the duct is $F_f = \pi D(1/2)\rho u^2 C_f$, so (14.30) becomes:

$$\dot{m}\frac{du}{dx} = \rho u A\frac{du}{dx} = -A\frac{dp}{dx} - \frac{\pi D}{2}\rho u^2 C_f,$$

where the first equality follows from (14.23). Here, $A = \pi D^2/4$ is constant so the equation can be simplified:

$$\rho u\frac{du}{dx} = -\frac{dp}{dx} - \frac{2C_f}{D}\rho u^2, \text{ and rearranged}: \frac{1}{u^2}\frac{du^2}{dx} = -\frac{2}{\rho u^2}\frac{dp}{dx} - \frac{4C_f}{D}.$$

First, convert the u^2-derivative into one involving the Mach number. Start with (14.33), note that T_0 will be constant for adiabatic flow, multiply by c^2, and differentiate:

$$\frac{T_0}{T}=\frac{c_0^2}{c^2}=1+\frac{\gamma-1}{2}M^2,\text{ which implies}: c_0^2=c^2+\frac{\gamma-1}{2}u^2 \quad\text{or}\quad 0=\frac{dc^2}{dx}+\frac{\gamma-1}{2}\frac{du^2}{dx}.$$

Now differentiate $u^2=M^2c^2$, substitute in the differentiated result from (14.33), and isolate $(1/u^2)(du^2/dx)$:

$$\frac{du^2}{dx}=c^2\frac{dM^2}{dx}+M^2\frac{dc^2}{dx}=c^2\frac{dM^2}{dx}-M^2\frac{\gamma-1}{2}\frac{du^2}{dx},\quad\text{or}\quad \frac{1}{u^2}\frac{du^2}{dx}=\left(1+\frac{\gamma-1}{2}M^2\right)^{-1}\frac{1}{M^2}\frac{dM^2}{dx}.$$

To similarly convert the pressure gradient term in the rearranged version of (14.23) to one involving only u and M, use (14.1h) to eliminate p in favor of c and ρ:

$$\frac{2}{\rho u^2}\frac{dp}{dx}=\frac{2}{\rho u^2}\frac{d}{dx}\left(\frac{\rho c^2}{\gamma}\right)=\frac{2}{\gamma u^2}\frac{dc^2}{dx}+\frac{2}{\gamma M^2}\frac{1}{\rho}\frac{d\rho}{dx}.$$

Substitute for dc^2/dx from above, and note that $(1/\rho)(d\rho/dx)=-(1/u)(du/dx)$ for constant-area flow where $\rho u=$ constant. Thus, the pressure gradient term becomes:

$$-\frac{2}{\rho u^2}\frac{dp}{dx}=\frac{\gamma-1}{\gamma u^2}\frac{du^2}{dx}+\frac{1}{\gamma M^2}\frac{1}{u^2}\frac{du^2}{dx}=\left[\frac{(\gamma-1)M^2+1}{\gamma M^2}\right]\frac{1}{u^2}\frac{du^2}{dx}.$$

Using this result and that from (14.33), $(1/u^2)(du^2/dx)$ can be eliminated from the rearranged version of the momentum equation (14.23) so that – after some algebra – it takes the final form:

$$\frac{4C_f}{D}=\left[\frac{1-M^2}{\gamma M^4}\right]\left(1+\frac{\gamma-1}{2}M^2\right)^{-1}\frac{dM^2}{dx}.$$

The left-side term will always be positive. Thus, when $M<1$, dM^2/dx must be positive and the flow's Mach number must increase toward unity. And, when $M>1$, dM^2/dx must be negative and the flow's Mach number must decrease toward unity. Therefore, this equation supports the contention that the effect friction on a compressible flow is to drive the Mach number toward unity. Interestingly, this equation can be integrated, too (see Thompson, 1972, p. 299).

14.7 One-dimensional unsteady compressible flow in constant-area ducts

The results presented in the prior four sections are valid for steady compressible flow. In this section, a few of the fundamental features of unsteady compressible flow are presented for the simple situation of a long duct containing an inviscid perfect gas with uniform constant entropy (homentropic). The goal here is show how nonlinearity influences the velocity, pressure, and density within the duct when the flow's fluctuations exceed the acoustic limit and linearization of the equations of motion is not valid. Here, as in prior sections, body forces are ignored.

For this situation, the equations of motion are the one-dimensional continuity and inviscid momentum equations:

$$\frac{\partial\rho}{\partial t}+u\frac{\partial\rho}{\partial x}+\rho\frac{\partial u}{\partial x}=0,\quad\text{and}\quad\frac{\partial u}{\partial t}+u\frac{\partial u}{\partial x}+\frac{1}{\rho}\frac{\partial p}{\partial x}=0. \qquad (14.54, 14.55)$$

There are three dependent variables (u, p, ρ), so the constant-entropy relationship between p and ρ (14.2) closes the system of equations. Using $c=[\gamma RT]^{1/2}$, the thermodynamic equations can be rewritten in terms of the sound speed:

$$\rho=\rho_o(c/c_o)^{\frac{2}{\gamma-1}}\quad\text{and}\quad p=p_o(c/c_o)^{\frac{2\gamma}{\gamma-1}}, \qquad (14.56, 14.57)$$

and differentiated to find:

$$\frac{\partial\rho}{\partial t}=\frac{1}{\gamma-1}\frac{\rho}{c}\frac{\partial c}{\partial t},\ \frac{\partial\rho}{\partial x}=\frac{1}{\gamma-1}\frac{\rho}{c}\frac{\partial c}{\partial x}\quad\text{and}\quad\frac{1}{\rho}\frac{\partial p}{\partial x}=\frac{2\gamma}{\gamma-1}\frac{p}{\rho c}\frac{\partial c}{\partial x}=\frac{2c}{\gamma-1}\frac{\partial c}{\partial x},$$

where $c=[\gamma p/\rho]^{1/2}$ has been used as well. Substitute these into (14.54) and (14.55), cancel common factors, and multiply or divide by c to reach:

$$\frac{\partial}{\partial t}\left(\frac{2c}{\gamma-1}\right)+u\frac{\partial}{\partial x}\left(\frac{2c}{\gamma-1}\right)+c\frac{\partial u}{\partial x}=0,\quad\text{and}\quad\frac{\partial u}{\partial t}+u\frac{\partial u}{\partial x}+c\frac{\partial}{\partial x}\left(\frac{2c}{\gamma-1}\right)=0. \qquad (14.58, 14.59)$$

These are coupled nonlinear first-order differential equations for the flow velocity u, and the thermodynamic quantity $2c/(\gamma-1)$. When both are known at $t=0$ for all x, these equations can be solved using the *method of characteristics*.

The method of characteristics is a solution technique for first-order hyperbolic partial differential equations that is applicable when the equations can be combined so that they represent a total derivative along a *characteristic* curve that need not be straight or parallel to either of the independent coordinate axes. To apply this method to the present situation, construct a linear combination of (14.58) and (14.59):

$$\frac{\partial}{\partial t}\left(u+\alpha\frac{2c}{\gamma-1}\right)+u\frac{\partial}{\partial x}\left(u+\alpha\frac{2c}{\gamma-1}\right)+c\frac{\partial}{\partial x}\left(\alpha u+\frac{2c}{\gamma-1}\right)=0,$$

where α is a real constant. Here, the combination of dependent variables inside the (,)-parentheses is the same when $\alpha=\pm 1$, so the (14.58) and (14.59) can be rewritten:

$$\frac{\partial}{\partial t}\left(u\pm\frac{2c}{\gamma-1}\right)+(u\pm c)\frac{\partial}{\partial x}\left(u\pm\alpha\frac{2c}{\gamma-1}\right)=0,\quad \text{or}\quad \frac{\partial I_\pm}{\partial t}+(u\pm c)\frac{\partial I_\pm}{\partial x}=0, \tag{14.60}$$

where $I_\pm = u\pm 2c/(\gamma-1)$ are the two *Riemann invariants* of this problem named for Georg F.B. Riemann who developed this theory in the 1850s.

The solution of (14.60) is obtained by postulating the existence of curves along which the invariants remain constant, defined by $C=C(x,t)=C(x(s),t(s))$, where s is the arc-length along such a curve. The total derivative of $I_\pm$ with respect to s on such a curve is:

$$\left(\frac{dI_\pm}{ds}\right)_C=\frac{\partial I_\pm}{\partial t}\frac{dt}{ds}+\frac{\partial I_\pm}{\partial x}\frac{dx}{ds}. \tag{14.61}$$

A comparison of the coefficients of $\partial I_\pm/\partial t$ and $\partial I_\pm/\partial x$ in (14.60) and (14.61) produces the equations for $x(s)$ and $t(s)$ that ensure $(dI_\pm/ds)_C=0$. These equations are $dt/ds=1$ and $dx/ds=u\pm c$, which can be combined to find:

$$dx/dt=u\pm c. \tag{14.62}$$

This equation defines the characteristic curves $C_\pm$ in x-t domain. Here, the invariant $I_+=u+2c/(\gamma-1)$ is constant along the C_+-characteristic curves defined by $dx/dt=u+c$, and the invariant $I_-=u-2c/(\gamma-1)$ is constant along the C_- characteristic curves defined by $dx/dt=u-c$.

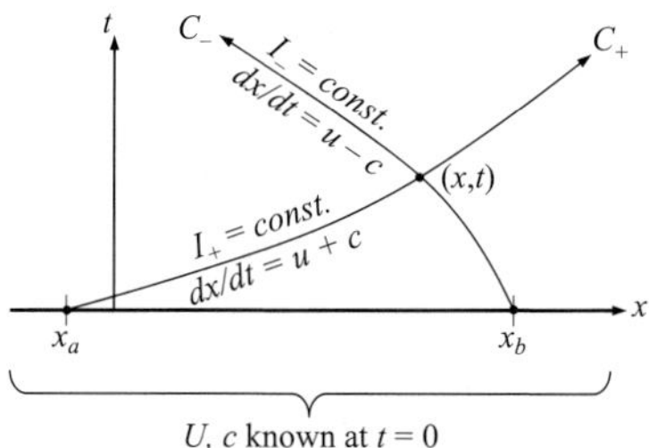

FIGURE 14.18 Method of characteristics construction to determine gas-dynamic flow properties at (x,t) when such properties are known for all x at $t=0$. The $C_\pm$ characteristics are curves on which $dx/dt=u\pm c$ and $I_\pm=u\pm 2c/(\gamma-1)$, respectively, are constants. If the C_+ characteristic emerging from $(x_a,0)$ and the C_- characteristic emerging from $(x_b,0)$ cross at (x,t), then $u(x,t)$ and $c(x,t)$ can be determined using the known values of $I_+(x_a,0)$ and $I_-(x_b,0)$ via (14.63) and (14.64).

Together, the two invariants and the two sets of characteristic curves in principle allow the solution of the initial value problem to be constructed as shown in the x-t diagram Fig. 14.18. The values of u and c at the point (x,t) must be consistent with the initial invariant values $I_+(x_a,0)$ and $I_-(x_b,0)$. For the situation shown in the figure:

$$u(x,t)=\frac{I_+(x_a,0)+I_-(x_b,0)}{2}\quad\text{and}\quad c(x,t)=\frac{\gamma-1}{4}\left(I_+(x_a,0)-I_-(x_b,0)\right). \tag{14.63, 14.64}$$

However, these relationships can only be applied after the characteristic curves through (x,t) have been identified by integrating (14.62). Thus, this solution is implicit because u and c may vary throughout the $x-t$ domain so (14.62), (14.63), and (14.64) must be solved simultaneously. The characteristic curves may be physically identified with the paths followed by small disturbances (sound waves), and this solution can be reduced to (14.22) when the fluctuations in the flow speed and the thermodynamic quantities are small.

Example 14.7

Consider a quiescent inviscid perfect gas to the right of an ideal piston in a long duct with constant cross-section. At $t = t_1$, the piston located at $x = c_o t_1$ accelerates to the left until it reaches $x = 0$, then its speed remains constant. The piston's trajectory, $x_p(t)$, is such that the C_+ characteristics for the time the piston is accelerating all pass through the origin of coordinates (see Fig. 14.19). Determine the gas velocity $u(x,t)$ and sound speed $c(x,t)$ in the gas to the right of the piston.

Solution The piston moves to the left, so the gas that feels the piston motion expands and moves to the left as well; thus, $u(x,t)$ will be negative (or zero) throughout the flow field. The method of characteristics can be used to determine $u(x,t)$ and $c(x,t)$ in all three regions of the flow field.

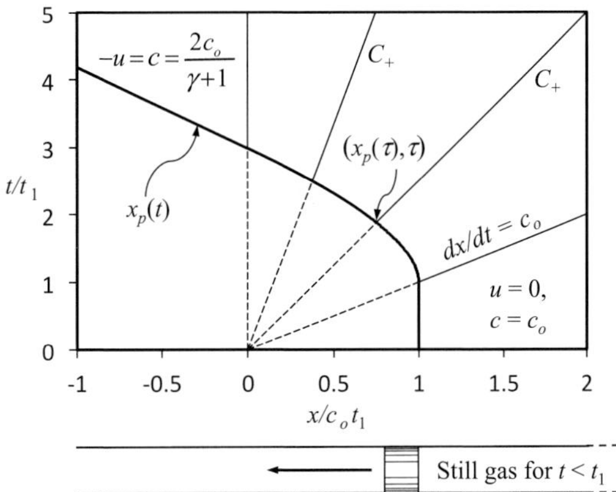

FIGURE 14.19 Distance-time diagram for unsteady expansion of a perfect gas with $\gamma = 1.4$ gas by piston motion. The initial piston location, $x = c_o t_1$, and direction of travel are shown below the $x - t$ diagram. The dark solid curve is the piston trajectory, and C_+ characteristics are shown as thin solid lines. The piston is stationary for $t < t_1$, after which it accelerates to the left. The piston's trajectory is specially chosen so that the C_+ characteristics that emerge from it for $0 \le x_p(t) < c_o t_1$ intersect the origin. For $x_p(t) \le 0$, the piston moves at constant speed. As the piston moves to the left, the gas is drawn in the same direction and expands, so its temperature falls. Thus, $|u|$ increases and c falls as t increases at any fixed location $x > 0$ (a vertical path), and as x decreases for any fixed time $t > t_1$ (a horizontal path).

The first region, $x > c_o t_1$ and $t < x/c_o$, lies at the lower right of Fig. 14.19 and is the simplest because it is not influenced by the piston's motion. Here, the field values are $u = 0$ and $c = c_o$, and this region is so labeled in Fig. 14.19.

The second region, $x \ge x_p(t)$, $x > 0$, and $t > x/c_o$, is more interesting. Here both u and c vary but the $I_\pm$ invariants are constant along the $C_\pm$ characteristics. For this region, all the $C-$ characteristics (not shown on Fig. 14.19) originate on x-axis where $u = 0$ and $c = c_o$, so the I_- invariant implies:

$$I_- = u(x,t) - \frac{2c(x,t)}{\gamma - 1} = u(x,0) - \frac{2c(x,0)}{\gamma - 1} = -\frac{2c_o}{\gamma - 1}.$$

In the second region, all the C_+ characteristics originate on the moving piston, so the I_+ invariant implies:

$$I_+ = u(x,t) + \frac{2c(x,t)}{\gamma - 1} = \dot{x}_p(\tau) + \frac{2c(x_p(\tau), \tau)}{\gamma - 1} \equiv I_+(\tau),$$

where τ is the time when the C_+ characteristic of interest touches the piston trajectory.

With these evaluations of $I_\pm$, (14.63) and (14.64) imply:

$$u_+(x,t) = \frac{1}{2}\left(I_+(\tau) - \frac{2c_o}{\gamma - 1}\right) \quad \text{and} \quad c_+(x,t) = \frac{\gamma - 1}{4}\left(I_+(\tau) + \frac{2c_o}{\gamma - 1}\right),$$

where the subscript "+" indicates a result that applies along a C_+ characteristics. Thus, $u(x,t)$ and $c(x,t)$ are both constant along each C_+ characteristics (but different for different C_+ characteristics). Evaluating the equation for u_+ on the piston implies:

$$u_+(x_p(\tau), \tau) = \dot{x}_p(\tau) = \frac{1}{2}\left(I_+(\tau) - \frac{2c_o}{\gamma - 1}\right) \quad \text{so} \quad I_+(\tau) = 2\dot{x}_p(\tau) + \frac{2c_o}{\gamma - 1}.$$

Using this value of $I_+(\tau)$, the sound speed on the C_+ characteristic is:

$$c_+ = c_o + \frac{\gamma - 1}{2}\dot{x}_p(\tau).$$

So, (14.62) can be integrated to find the equation for the C_+ characteristics:

$$x = \int \left(\frac{dx}{dt}\right) dt = (u_+ + c_+)t = \left(\dot{x}_p(\tau) + c_o + \frac{\gamma - 1}{2}\dot{x}_P(\tau)\right)t = \left(c_o + \frac{\gamma + 1}{2}\dot{x}_p(\tau)\right)t,$$

where the constant of integration is zero because the C_+ characteristic passes through the origin. Now, invert this last equation to solve for $\dot{x}_p(\tau)$ in term of x and t, and substitute this result into the prior results for u_+ and c_+ to find:

$$u(x,t) = \dot{x}_p(\tau) = \frac{2}{\gamma + 1}\left(\frac{x}{t} - c_o\right) \quad \text{and} \quad c(x,t) = c_o + \frac{\gamma - 1}{2}\dot{x}_p(\tau) = \frac{2}{\gamma + 1}c_o + \frac{\gamma - 1}{\gamma + 1}\frac{x}{t},$$

where the "+" subscripts have been dropped from u and c because the parameter τ that specifies a particular C_+ characteristic has been eliminated in favor of (x, t)-variables that apply to the whole region.

The third region $x < 0$, lies at the upper left of Fig. 14.19. Here the flow velocity and sound speed are constant, $-u = c = 2c_o/(\gamma + 1)$, because the piston speed is constant. These values for u and c can be obtained by evaluating the above equations at $x = 0$, and are indicated in Fig. 14.19, too. In this region the C_+ characteristics are vertical lines.

Interestingly, the solution given here is also valid for piston motion to the right, leading to compression of the gas, for a finite period of time. The situation is shown in Fig. 14.20. The piston is stationary for $t/t_1 < -1$, after which it accelerates to the right until it reaches speed c_o (at $t/t_1 = -0.3349$), and then continues at constant speed. The main difference between this gas-compression piston motion and the gas expansion piston motion shown on Fig. 14.19 is the formation of a shock wave beyond the location where the characteristics converge. Here, the homentropic assumption is violated, so a shock wave and entropy discontinuity must be fitted into the solution (see Exercise 14.27).

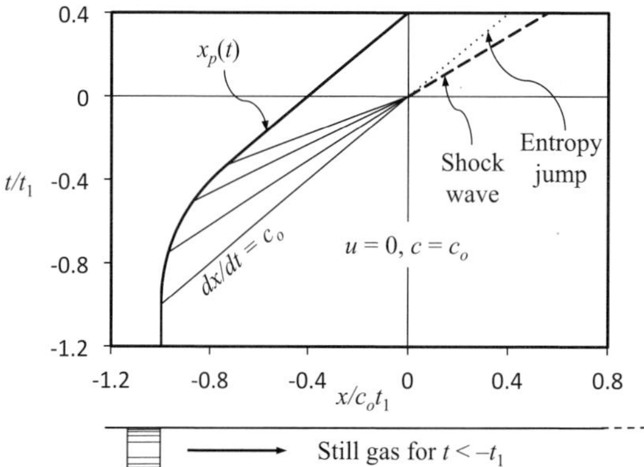

FIGURE 14.20 Distance-time diagram for unsteady compression of a perfect gas with $\gamma = 1.4$ by piston motion. The initial piston location, $x = -c_o t_1$, and direction of travel are shown below the x-t diagram. The dark solid curve is the piston trajectory, and C_+ characteristics are shown as thin solid lines that converge at the origin. The piston is stationary for $t/t_1 < -1$, after which it accelerates to the right until it reaches speed c_o at $t/t_1 = -0.3349$ (where the uppermost C_+ characteristic touches the piston trajectory). As in Fig. 14.19, the piston's trajectory is specially chosen so that the C_+ characteristics that emerge from it while the piston accelerates pass through the origin. For $t/t_1 > -0.3349$, the piston's speed is constant, $u = c_o$. As the piston moves to the right, the gas is pushed in the same direction and is compressed, so its temperature increases. Thus, u increases and c increases as t increases at any fixed location $x > -c_o t_1$ (a vertical path), and as x increases for any fixed time $t > -t_1$ (a horizontal path). Here, a shock wave and an entropy discontinuity are present for $x > 0$ and $t > 0$.

14.8 Two-dimensional steady compressible flow

To this point, the emphasis in this chapter has been on one-dimensional flows in which flow properties vary only in the direction of flow. This section presents steady compressible flow results for more than one spatial dimension. To get started, consider a point source emitting infinitesimal pressure (acoustic) disturbances in a quiescent compressible fluid in which the speed of sound is c. If the point source is stationary, then the pressure-disturbance wavefronts are concentric spheres. Fig. 14.21a shows the intersection of these wavefronts with a plane containing the source at times corresponding to integer multiples of Δt.

When the source propagates to the left at speed $U < c$, the wavefront diagram changes to look like Fig. 14.21b, which shows four locations of the source separated by equal time intervals Δt, with point 4 being the present location of the source. At the first point, the source emitted a wave that has spherically expanded to a radius of $3c\Delta t$ in the time interval $3\Delta t$. During this time the source has moved to the fourth location, a distance of $3U\Delta t$ from the first point of wavefront emission. The figure also shows the locations of the wavefronts emitted while the source was at the second and third points. Here, the wavefronts do not intersect because $U < c$. As in the case of the stationary source, the wavefronts propagate vertically upward and downward, and horizontally upstream and downstream from the source. Thus, *a body moving at a subsonic speed influences the entire flow field.*

Now consider the case depicted in Fig. 14.21c where the source moves supersonically, $U > c$. Here, the centers of the spherically expanding wavefronts are separated by more than $c\Delta t$, and no pressure disturbance propagates ahead of the source. Instead, the edges of the wavefronts form a conical tangent surface called the *Mach cone*. In planar two-dimensional flow, the tangent surface is in the form of a wedge, and the tangent lines are called *Mach lines*. An examination of the figure shows that the half-angle of the Mach cone (or wedge), called the *Mach angle* μ, is given by $\sin\mu = (c\Delta t)/(U\Delta t)$, so that:

$$\sin\mu = 1/M. \tag{14.65}$$

The Mach cone becomes wider as M decreases and becomes a plane front (i.e., $\mu = 90°$) when $M = 1$.

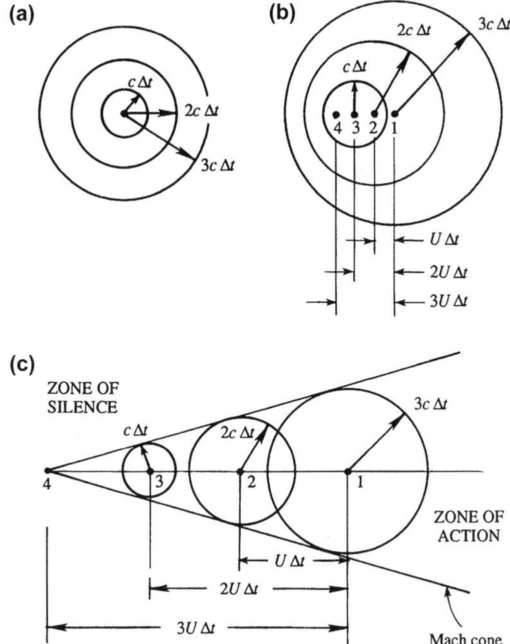

FIGURE 14.21 Wavefronts emitted by a point source in a still fluid when the source speed U is: (a) $U = 0$; (b) $U < c$; and (c) $U > c$. In each case the wavefronts are emitted at integer multiples of Δt. At subsonic source speeds, the wavefronts do not overlap and they spread ahead of the source. At supersonic source speeds, all the wavefronts lie behind the source within the Mach cone having a half angle $\sin^{-1}(1/M)$.

The situation depicted in Fig. 14.21 has at least two interpretations. The point source could be part of a solid body, which sends out pressure waves as it moves through the fluid. Or, after a Galilean transformation, Figs. 14.21b and c apply equally well to a stationary point source with a compressible fluid moving past it at speed U. From Fig. 14.21c we see that in a supersonic flow an observer outside the source's Mach cone would not detect or hear a pressure signal emitted by the source, hence this region is called the *zone of silence*. In contrast, the region inside the Mach cone is called the *zone of action*, within which the effects of the disturbance are felt. Thus, the sound of a supersonic aircraft passing overhead does not reach an observer on the ground until its Mach cone reaches the observer, and this arrival occurs *after* the aircraft has passed overhead.

At every point in a planar supersonic flow there are two Mach lines, oriented at $\pm\mu$ to the local direction of flow. Pressure disturbance information propagates along these lines, which are the *characteristics* of the governing differential equation. It can be shown that the nature of the governing differential equation is hyperbolic in a supersonic flow and elliptic in a subsonic flow. In addition, the method of characteristics may be applied to steady two-dimensional flow but this approach is not pursued here.

When pressure disturbances from the source are of finite amplitude, they may evolve into a shock wave that is not normal to the flow direction. Such *oblique* shock waves are commonly encountered in ballistics and supersonic flight, and differ from normal shock waves because they change the upstream flow velocity's magnitude *and* direction. A generic depiction of an oblique shock wave is provided in Fig. 14.22 in two coordinate systems. Fig. 14.22a shows the stream-aligned coordinate system where the shock wave resides at an angle σ from the horizontal. Here the velocity upstream of the shock is horizontal with magnitude V_1, while the velocity downstream of the shock is deflected from the horizontal by an angle δ and has magnitude V_2. Fig. 14.22b shows the same shock wave in a shock-aligned coordinate system where the shock wave is vertical, and the fluid velocities upstream and downstream of the shock are (u_1, v) and (u_2, v), respectively. Here v is parallel to the shock wave and is not influenced by it (see Exercise 14.30). Thus an oblique shock may be analyzed as a normal shock involving u_1 and u_2 to which a constant shock-parallel velocity v is added. Using the Cartesian coordinates in Fig. 14.22b where the shock coincides with the vertical axis, the relationships between the various components and angles are:

$$(u_1, v) = \sqrt{u_1^2 + v^2}(\sin\sigma, \cos\sigma) = V_1(\sin\sigma, \cos\sigma), \quad \text{and}$$
$$(u_2, v) = \sqrt{u_2^2 + v^2}(\sin(\sigma - \delta), \cos(\sigma - \delta)) = V_2(\sin(\sigma - \delta), \cos(\sigma - \delta)).$$

The angle σ is called the *shock angle* or *wave angle* and δ is called the *deflection angle*. The normal Mach numbers upstream (1) and downstream (2) of the shock are:

$$M_{n1} = u_1/c_1 = M_1 \sin\sigma > 1, \quad \text{and} \quad M_{n2} = u_2/c_2 = M_2 \sin(\sigma - \delta) < 1.$$

Because $u_2 < u_1$, there is a sudden change of direction of flow across the shock and the flow is turned *toward* the shock by angle δ.

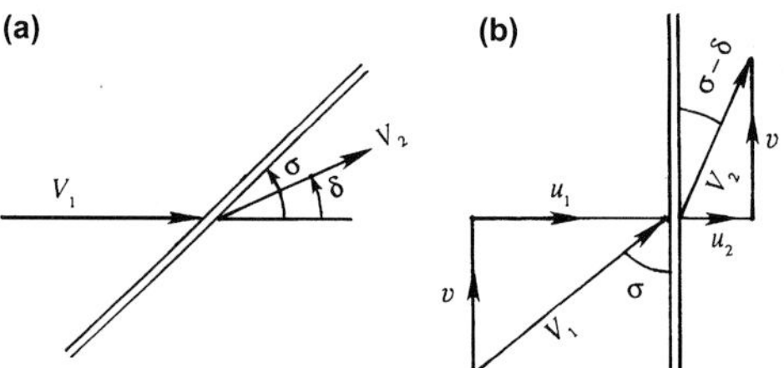

FIGURE 14.22 Two coordinate systems for an oblique shock wave. (a) Stream-aligned coordinates where the oblique shock wave lies at shock angle = σ and produces a flow-deflection of angle = δ. (b) Shock-normal coordinates which are preferred for analysis because an oblique shock wave is merely a normal shock with a superimposed shock-parallel velocity v.

Superposition of the tangential velocity v does not affect the *static* properties, which are therefore the same as those for a normal shock. The expressions for the ratios p_2/p_1, ρ_2/ρ_1, T_2/T_1, and $(s_2 - s_1)/c_v$ are therefore those given by (14.44) and (14.46) through (14.48), if M_1 is replaced by $M_{n1} = M_1 \sin\sigma$. For example:

$$\frac{p_2}{p_1} = 1 + \frac{2\gamma}{\gamma + 1}[M_1^2 \sin^2\sigma - 1], \quad \text{and} \quad \frac{\rho_2}{\rho_1} = \frac{u_1}{u_2} = \frac{(\gamma + 1)M_1^2 \sin^2\sigma}{(\gamma - 1)M_1^2 \sin^2\sigma + 2} = \frac{\tan\sigma}{\tan(\sigma - \delta)}. \tag{14.66, 14.67}$$

Thus, the normal-shock table, Table 14.2, is applicable to oblique shock waves when $M_1 \sin\sigma$ is used in place of M_1.

The relation between the upstream and downstream Mach numbers can be found from (14.45) by replacing M_1 by $M_1 \sin\sigma$ and M_2 by $M_2 \sin(\sigma - \delta)$. This gives:

$$M_2^2 \sin^2(\sigma - \delta) = \frac{(\gamma - 1)M_1^2 \sin^2\sigma + 2}{2\gamma M_1^2 \sin^2\sigma + 1 - \gamma}. \tag{14.68}$$

An important relation is that between the deflection angle δ and the shock angle σ for a given M_1, given in (15.62). Using the trigonometric identity for tan $(\sigma - \delta)$, this becomes:

$$\tan\delta = 2\cot\sigma \frac{M_1^2 \sin^2\sigma - 1}{M_1^2(\gamma + \cos 2\sigma) + 2}. \tag{14.69}$$

A plot of this relation is given in Fig. 14.23. The curves represent δ versus σ for constant M_1. The value of M_2 varies along the curves, and the locus of points corresponding to $M_2 = 1$ is indicated. It is apparent that there is a maximum deflection angle δ_{max} for oblique shock solutions to be possible; for example, $\delta_{max} = 23°$ for $M_1 = 2$. For a given M_1, δ becomes zero at $\sigma = \pi/2$ corresponding to a normal shock, and at $\sigma = \mu = \sin^{-1}(1/M_1)$ corresponding to the Mach angle. For a fixed M_1 and $\delta < \delta_{max}$, there are two possible solutions: a *weak shock* corresponding to a smaller σ and a *strong shock* corresponding to a larger σ. The flow downstream of a strong shock is always subsonic; in contrast, the flow downstream of a weak shock is generally supersonic, except in a small range in which δ is slightly smaller than δ_{max}.

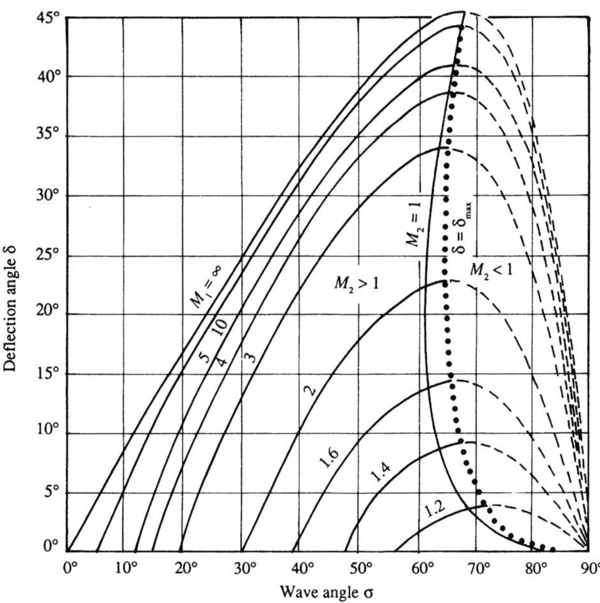

FIGURE 14.23 Plot of oblique shock solutions. The strong-shock branch is indicated by dashed lines on the right, and the heavy dotted line indicates the maximum deflection angle δ_{max}. *(From Ames Research Staff, 1953, NACA Report 1135.)*

Oblique shock waves are commonly generated when a supersonic flow is forced to change direction to go around a structure where the flow area cross-section is reduced. Two examples are shown in Fig. 14.24 that show supersonic flow past a wedge of half-angle δ, or the flow past a compression bend where the wall turns into the flow by an angle δ. If M_1 and δ are known, then σ can be obtained from Fig. 14.23, and M_{n2} (and therefore $M_2 = M_{n2}/\sin(\sigma - \delta)$) can be obtained from the shock table (Table 14.2). An attached shock wave, corresponding to the weak solution, forms at the nose of the wedge, such that the flow is parallel to the wedge after turning through an angle δ. The shock angle σ decreases to the Mach angle $\mu_1 = \sin^{-1}(1/M_1)$ as the deflection δ tends to zero. It is interesting that the corner velocity in a supersonic flow is finite. In contrast, the corner velocity in a subsonic (or incompressible) flow is either zero or infinite, depending on whether the wall shape is concave or convex. Moreover, the streamlines in Fig. 14.24 are straight, and computation of the field is easy. By contrast, the streamlines in a subsonic flow are curved, and the computation of the flow field is not as easy. The basic reason for this is that, in a supersonic flow, small pressure disturbances do not propagate upstream of Mach lines or shock waves, hence the flow field can be constructed step by step, *proceeding downstream*. In contrast, disturbances propagate both upstream and downstream in a subsonic flow so that all features in the entire flow field are related to each other.

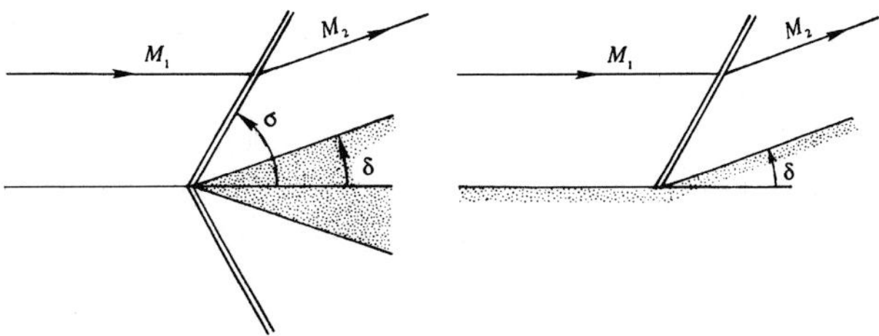

FIGURE 14.24 Two possible means for producing oblique shocks in a supersonic flow. In both cases a solid surface causes the flow to turn, and the flow area is reduced. The geometry shown in the right panel is sometimes called a compression corner.

As δ is increased beyond δ_{max}, attached oblique shocks are not possible, and a detached curved shock stands in front of the body (Fig. 14.25). The central streamline goes through a normal shock and generates a subsonic flow in front of the wedge. The *strong*-shock solution of Fig. 14.23 therefore holds near the nose of the body. Farther out, the shock angle decreases, and the weak-shock solution applies. If the wedge angle is not too large, then the curved detached shock in Fig. 14.25 becomes an oblique attached shock as the Mach number is increased. In the case of a blunt-nosed body, however, the shock at the leading edge is always detached, although it moves closer to the body as the Mach number is increased.

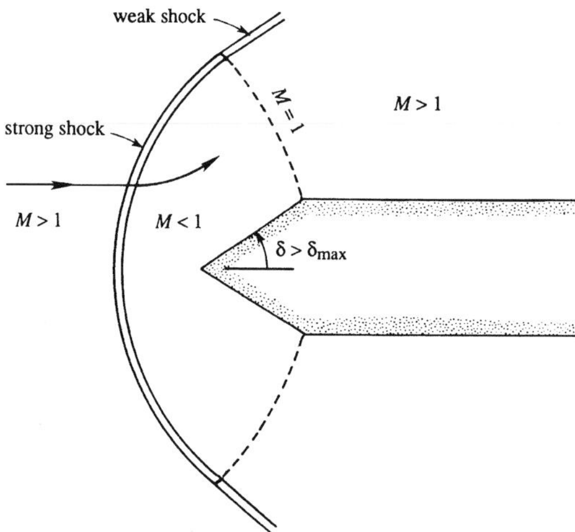

FIGURE 14.25 A detached shock wave. When angle of the wedge shown in the left panel of Fig. 14.24 is too great for an oblique shock, a curved shock wave will form that does not touch body. A portion of this detached shock wave will have the properties of a normal-shock wave.

We see that shock waves may exist in supersonic flows and their location and orientation adjust to satisfy boundary conditions. In external flows, such as those just described, the boundary condition is that streamlines at a solid surface must be tangent to that surface. In duct flows, the boundary condition locating the shock is usually the downstream pressure.

From the foregoing analysis, it is clear that large-angle supersonic flow deflections should be avoided when designing efficient devices that produce minimal total pressure losses. Efficient devices tend to be slender and thin, and their performance may be analyzed using a weak oblique shock approximation that can be obtained from the results above in the limit of small flow deflection angle, $\delta \ll 1$. To obtain this expression, simplify (14.69) by noting that as $\delta \to 0$, the shock angle σ tends to the Mach angle $\mu_1 = \sin^{-1}(1/M_1)$. And, from (14.66), we note that $(p_2 - p_1)/p_1 \to 0$ as $M_1^2 \sin^2\sigma - 1 \to 0$ (as $\sigma \to \mu$ and $\delta \to 0$). Then from (14.66) and (14.69):

$$\tan\delta = 2\cot\sigma \frac{\gamma+1}{2\gamma}\left(\frac{p_2 - p_1}{p_1}\right)\frac{1}{M_1^2(\gamma + 1 - 2\sin^2\sigma) + 2}. \tag{14.70}$$

As $\delta \to 0$, $\tan\delta \approx \delta$, $\cot\mu = [M_1^2 - 1]^{1/2}$, $\sin\sigma \approx 1/M_1$ and:

$$\frac{p_2 - p_1}{p_1} = \frac{\gamma M_1^2}{\sqrt{M_1^2 - 1}}\delta. \tag{14.71}$$

An interesting point is that the relation (14.71) is also applicable to weak *expansion* waves and not just weak compression waves. By this we mean that the pressure increase due to a small deflection of the wall toward the flow is the same as the pressure *decrease* due to a small deflection of the wall *away* from the flow. This extended range of validity of (14.71) occurs because the entropy change across a weak shock may be negligible even when the pressure change is appreciable (see (14.49b) and the related discussion). Thus, weak shock waves can be treated as isentropic or reversible. Relationships for a weak shock wave can therefore be applied to a weak expansion wave, except for some sign changes. In the final section of this chapter, (14.71) is used to estimate the lift and drag of a thin airfoil in supersonic flow.

When an initially horizontal supersonic flow follows a curving wall, the wall radiates compression and expansion waves into the flow that modulate the flow's direction and Mach number. When the wall is smoothly curved these compression and expansion waves follow Mach lines, inclined at an angle of $\mu = \sin^{-1}(1/M)$ to the *local* direction of flow (Fig. 14.26). In this simple circumstance where there is no upper wall that radiates compression or expansion waves downward into the region of interest, the flow's orientation and Mach number are constant on each Mach line. In the case of compression, the Mach number decreases along the flow, so that the Mach angle increases. The Mach lines may therefore coalesce and form an oblique shock as in Fig. 14.26a. In the case of a gradual expansion, the Mach number increases along the flow and the Mach lines diverge as in Fig. 14.26b.

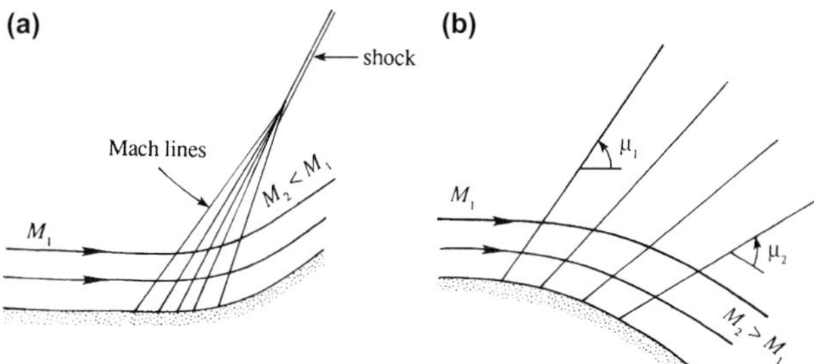

FIGURE 14.26 Gradual compression and expansion in supersonic flow. (a) A gradual compression corner like the one shown will eventually result in an oblique shock wave as the various Mach lines merge, each carrying a fraction of the overall compression. (b) A gradual expansion corner like the one shown produces Mach lines that diverge so the expansion spreads to become even more gradual farther from the wall.

If the wall has a sharp deflection (a corner) away from the approaching stream, then the pattern of Fig. 14.26b takes the form of Fig. 14.27 where all the Mach lines originate from the corner. In this case, this portion of the flow where it expands and turns, and is not parallel to the wall upstream or downstream of the corner, is known as a *Prandtl-Meyer expansion fan*. The Mach number increases through the fan, with $M_2 > M_1$. The first Mach line is inclined at an angle of μ_1 to the upstream wall direction, while the last Mach line is inclined at an angle of μ_2 to the downstream wall direction. The pressure falls gradually along any streamline through the fan. Along the wall, however, the pressure remains constant along the upstream wall, falls discontinuously at the corner, and then remains constant along the downstream wall. Fig. 14.27 should be compared with Fig. 14.26, in which the wall turns *inward* and generates an oblique shock wave. By contrast, the expansion in Fig. 14.27 is gradual and isentropic away from the wall.

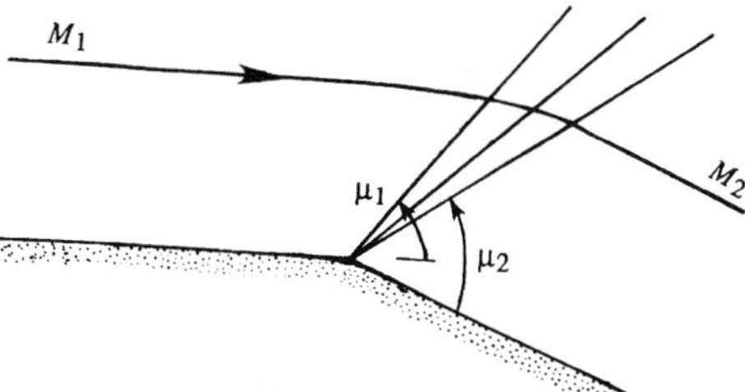

FIGURE 14.27 The Prandtl-Meyer expansion fan. This is the flow field developed by a sharp expansion corner. Here the flow area increases downstream of the corner so it accelerates a supersonic flow.

The flow through a Prandtl-Meyer expansion fan is calculated as follows. From Fig. 14.23b, conservation of momentum tangential to the shock shows that the tangential velocity is unchanged, or:

$$V_1 \cos\sigma = V_2 \cos(\sigma - \delta) = V_2(\cos\sigma \cos\delta + \sin\sigma \sin\delta).$$

We are concerned here with very small deflections, $\delta \to 0$ so $\sigma \to \mu$. Here, $\cos\delta \approx 1$, $\sin\delta \approx \delta$, $V_1 \approx V_2(1 + \delta \tan\sigma)$, so $(V_2 - V_1)/V_1 \approx -\delta \tan\sigma \approx -\delta/[M_1^2 - 1]^{1/2}$, where $\tan\sigma \approx 1/[M_1^2 - 1]^{1/2}$. Thus, the velocity change dV for an infinitesimal wall deflection $d\delta$ can be written as $d\delta = -(dV/V)[M_1^2 - 1]^{1/2}$ (first quadrant deflection). Because $V = Mc$, $dV/V = dM/M + dc/c$. With $c = \sqrt{\gamma RT}$ for a perfect gas, $dc/c = dT/2T$. Using (14.33) for adiabatic flow of a perfect gas, $dT/T = -(\gamma - 1)M\,dM/[1 + ((\gamma - 1)/2)M^2]$, then:

$$d\delta = -\frac{\sqrt{M^2 - 1}}{M} \frac{dM}{1 + \frac{1}{2}(\gamma - 1)M^2}.$$

Integrating δ from 0 (radians) and M from 1 gives $\delta + \nu(M) = \text{const.}$, where

$$\nu(M) = \int_1^M \frac{\sqrt{M^2 - 1}}{1 + \frac{1}{2}(\gamma - 1)M^2} \frac{dM}{M} = \frac{\sqrt{\gamma + 1}}{\gamma - 1} \tan^{-1} \sqrt{\frac{\gamma - 1}{\gamma + 1}(M^2 - 1)} - \tan^{-1} \sqrt{M^2 - 1} \qquad (14.72)$$

is called the Prandtl-Meyer function. The sign of $[M^2-1]^{1/2}$ originates from the identification of $\tan\sigma = \tan\mu = [M_1^2 - 1]^{-1/2}$ for a first quadrant deflection (upper half-plane). For a fourth quadrant deflection (lower half-plane), $\tan\mu = -[M_1^2 - 1]^{-1/2}$. For example, for Fig. 14.26a or b with δ_1, δ_2, and M_1 given, we would write:

$$\delta_1 + \nu(M_1) = \delta_2 + \nu(M_2), \quad \text{and then} \quad \nu(M_2) = \delta_1 - \delta_2 + \nu(M_1),$$

would determine M_2. In Fig. 14.26a, $\delta_1 - \delta_2 < 0$, so $\nu_2 < \nu_1$ and $M_2 < M_1$. In Fig. 14.26b, $\delta_1 - \delta_2 > 0$, so $\nu_2 > \nu_1$ and $M_2 > M_1$.

Example 14.8

A uniform flow at atmospheric pressure having $M_1 = 3.0$ is deflected by 20°. What are the Mach number and pressure in the flow after the deflection if it occurs through (a) an oblique shock wave from a compression corner (Fig. 14.24 right side panel), (b) an isentropic compression from a curved wall (Fig. 14.26a), and (c) an isentropic expansion (Fig. 14.26b).

Solution For (a) an oblique shock wave must be considered. Using Fig. 14.23, $M_1 = 3$ and $\delta = 20°$, leads to $\sigma = 37.5°$, so $M_1 \sin\sigma = 1.83$. Thus, from (14.66) and (14.68):

$$p_2 = \frac{p_2}{p_1}p_1 = \left(1 + \frac{2\gamma}{\gamma+1}\left[M_1^2\sin^2\sigma - 1\right]\right)(1.0\,\text{atm}) = 3.74\,\text{atm}, \quad \text{and}$$

$$M_2 = \frac{1}{\sin(\sigma-\delta)}\left[\frac{(\gamma-1)M_1^2\sin^2\sigma + 2}{2\gamma M_1^2\sin^2\sigma + 1 - \gamma}\right]^{1/2} = 2.03.$$

For (b), the Prandtl-Meyer function may be used. Here the initial flow angle is 0° and $\nu(M_1 = 3) = 49.76°$. Thus, $\nu(M_2) = 0 - 20° + 49.76 = 29.76°$, for which $M_2 = 2.125$. The downstream pressure can be recovered from Table 14.1:

$$p_2 = \frac{p_2}{p_0}\frac{p_0}{p_1}p_1 = 0.1051\frac{1}{0.0272}(1.0\text{ atm}) = 3.86\text{ atm}.$$

Here, both M_2 and p_2 are larger than those for (a) because this flow is isentropic while the oblique shock in (a) is not.

For (c), the Prandtl-Meyer function may again be used. Here again the initial flow angle is 0° and $\nu(M_1 = 3) = 49.76°$. Thus, $\nu(M_2) = 0 + 20° + 49.76 = 69.76°$, for which $M_2 = 4.31$. The downstream pressure can be recovered from Table 14.1:

$$p_2 = \frac{p_2}{p_0}\frac{p_0}{p_1}p_1 = 0.0044\frac{1}{0.0272}(1.0\text{ atm}) = 0.162\text{ atm}.$$

14.9 Thin-airfoil theory in supersonic flow

Simple expressions can be derived for the lift and drag coefficients of an airfoil in supersonic flow if the thickness and angle of attack are small. Under these circumstances the pressure disturbances caused by the airfoil are small, and the total flow can be built up by superposition of small disturbances emanating from points on the body. Such a linearized theory of lift and drag was developed by Ackeret. Because all flow inclinations are small, the relation (14.71) can be used to calculate the pressure changes due to a change in flow direction. For the current purposes this relation is rewritten as:

$$\frac{p - p_\infty}{p_\infty} = \frac{\gamma M_\infty^2 \delta}{\sqrt{M_\infty^2 - 1}}, \tag{14.73}$$

where p_∞ and M_∞ refer to the properties of the free stream, and p is the pressure at a point where the flow is inclined at an angle δ to the free-stream direction. The sign of δ in (14.73) determines the sign of $(p - p_\infty)$.

To see how the lift and drag of a thin body in a supersonic stream can be estimated, consider a flat plate inclined at a small angle α to a horizontal stream (Fig. 14.28). At the leading edge there is a weak expansion fan above the top surface and a weak oblique shock below the bottom surface. The streamlines ahead of these waves are straight. The streamlines above the plate turn through an angle α by expanding through an expansion fan, downstream of which they become parallel to the plate with a pressure $p_2 < p_\infty$. The upper streamlines then turn sharply across an oblique shock emanating from the trailing edge, becoming parallel to the free stream once again. Opposite features occur for the streamlines below the plate where the flow first undergoes compression across an oblique shock coming from the leading edge, which results in a pressure $p_3 > p_\infty$. It is, however, not important to distinguish between shock and expansion waves in Fig. 14.28, because the linearized theory treats them the same way, except for the sign of the pressure changes they produce.

The pressures above and below the plate can be found from (14.52), giving:

$$\frac{p_2 - p_\infty}{p_\infty} = -\frac{\gamma M_\infty^2 \alpha}{\sqrt{M_\infty^2 - 1}}, \quad \text{and} \quad \frac{p_3 - p_\infty}{p_\infty} = \frac{\gamma M_\infty^2 \alpha}{\sqrt{M_\infty^2 - 1}}.$$

The pressure difference across the plate is therefore:

$$\frac{p_3 - p_2}{p_\infty} = \frac{2\gamma M_\infty^2 \alpha}{\sqrt{M_\infty^2 - 1}}.$$

If b is the chord length, then the lift L and drag D forces per unit span are:

$$L = (p_3 - p_2)b\cos\alpha \cong \frac{2\alpha\gamma M_\infty^2 p_\infty b}{\sqrt{M_\infty^2 - 1}}, \quad \text{and} \quad D = (p_3 - p_2)b\sin\alpha \cong \frac{2\alpha^2\gamma M_\infty^2 p_\infty b}{\sqrt{M_\infty^2 - 1}}. \tag{14.74}$$

Using the relationship $\rho U^2 = \gamma p M^2$, the lift and drag coefficients are:

$$C_L = \frac{L}{(1/2)\rho_\infty U_\infty^2 b} \cong \frac{4\alpha}{\sqrt{M_\infty^2 - 1}}, \quad \text{and} \quad C_D = \frac{D}{(1/2)\rho_\infty U_\infty^2 b} \cong \frac{4\alpha^2}{\sqrt{M_\infty^2 - 1}}. \tag{14.75}$$

These expressions do not hold at transonic speeds $M_\infty \to 1$, when the process of linearization used here breaks down. The expression for the lift coefficient should be compared to the incompressible expression $C_L = 2\pi\alpha$ derived in the preceding chapter. Note that the flow in Fig. 14.28 does have a circulation because the velocities at the upper and lower surfaces are parallel but have different magnitudes. However, in a supersonic flow it is not necessary to invoke the Kutta condition (discussed in the preceding chapter) to determine the magnitude of the circulation. The flow in Fig. 14.28 does leave the trailing edge smoothly.

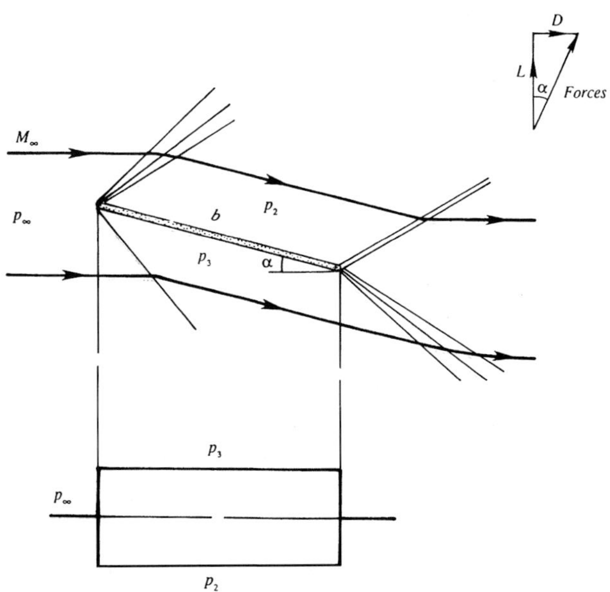

FIGURE 14.28 Inclined flat plate in a supersonic stream as a simple illustration of supersonic aerodynamics. The upper panel shows the flow pattern and the lower panel shows the pressure distribution on the suction and pressure sides of the simple foil. Here, an ideal compressible flow analysis does predict a drag component, unlike an equivalent ideal incompressible flow.

The drag in (14.75) is the *wave drag* experienced by a body in a supersonic stream, and exists even in an inviscid flow. The d'Alembert paradox therefore does not apply in a supersonic flow. The supersonic wave drag is analogous to the gravity wave drag experienced by a ship moving at a speed greater than the velocity of surface gravity waves, in which a system of bow waves is carried with the ship. The magnitude of the supersonic wave drag is independent of the value of the viscosity, although the energy spent in overcoming this drag is finally dissipated through viscous effects within the shock waves. In addition to the wave drag, additional drags due to viscous and finite-span effects, considered in the preceding chapter, act on a real wing.

In this connection, it is worth noting the difference between the airfoil shapes used in subsonic and supersonic airplanes. Low-speed airfoils have a streamlined shape, with a rounded nose and a sharp trailing edge. These features are not helpful in supersonic airfoils. The most effective way to reduce the drag of a supersonic airfoil is to reduce its thickness. Supersonic wings are characteristically thin and have sharp leading edges.

Example 14.9

Determine the lift and drag coefficients of an infinitely-thin but mildly-cambered airfoil at zero angle of attack in a horizontal supersonic flow at speed M_∞.

Solution Fig. 14.14 shows an infinitely thin cambered foil at zero angle of attack. Use the x-y coordinates shown there, let the camber line of the foil be $y_c(x)$, and assume the foil extends from $x = -2b$ to $x = +2b$. Here, zero angle of attack implies: $y_c(\pm 2b) = 0$. To compute the lift coefficient, C_L, first determine the pressure coefficient starting from is definition, then eliminate the pressure difference using (14.73):

$$C_p = \frac{p - p_\infty}{(1/2)\rho_\infty U_\infty^2} = \frac{\gamma M_\infty^2 \delta}{\sqrt{M_\infty^2 - 1}} \frac{p_\infty}{(1/2)\rho_\infty U_\infty^2} = \frac{\gamma M_\infty^2 \delta}{\sqrt{M_\infty^2 - 1}} \frac{2}{\gamma M_\infty^2} = \frac{2\delta}{\sqrt{M_\infty^2 - 1}}.$$

For small flow deflection angles: $\delta \approx \sin\delta \approx \tan\delta = dy_c/dx$ and $\cos\delta \approx 1$. Thus:

$$C_L = \frac{1}{4b}\int_{-2b}^{+2b} (C_{p,l} - C_{p,u})\cos\delta dx \cong \frac{1}{4b}\int_{-2b}^{+2b} \left(\frac{-2}{\sqrt{M_\infty^2 - 1}} - \frac{2}{\sqrt{M_\infty^2 - 1}} \right)\left(\frac{dy_c}{dx}\right) dx = \frac{-1}{b\sqrt{M_\infty^2 - 1}} [y_c]_{-2b}^{+2b} = 0,$$

where the extra 'l' and 'u' subscripts indicate the lower and upper foil surfaces, and δ is measured from the horizontal and is positive when it leads to flow compression. So, a cambered foil at zero angle of attack does not produce lift in a supersonic flow.

To determine the coefficient of drag, the local component of the pressure force in the direction of flow must be considered:

$$C_D = \frac{1}{4b}\int_{-2b}^{+2b} (C_{p,u} - C_{p,l})\sin\delta dx \cong \frac{1}{4b}\int_{-2b}^{+2b} \left(\frac{2}{\sqrt{M_\infty^2 - 1}} - \frac{-2}{\sqrt{M_\infty^2 - 1}} \right)\left(\frac{dy_c}{dx}\right)^2 dx = \frac{4}{\sqrt{M_\infty^2 - 1}} \overline{\left(\frac{dy_c}{dx}\right)^2},$$

where the overbar on the last term implies a spatial average over the chord length. This result suggests that airfoil camber only leads to drag in supersonic flow. Thus, the wings and control surfaces of supersonic aircraft and missiles are nearly flat.

Exercises

14.1 Use (14.4), (14.5), and (14.6) to derive (14.7) when the body force is spatially uniform and the effects of viscosity are negligible.

14.2 Show that the outward-propagating-wave solution to the inhomogeneous wave equation,

$$\frac{\partial^2 p'}{\partial t^2} - c_0^2 \nabla^2 p' = Q(x_i, t)$$

for acoustic pressure $p' = p - p_0$ in a uniform unbounded environment with ambient pressure p_0 and sound speed c_0 is given by

$$p'(x_i, t) = \frac{1}{4\pi c_0^2}\int_V Q\left(y_i, t - \frac{r}{c_0}\right)\frac{d^3y}{r},$$

where $r = \sqrt{(x_1 - y_1)^2 + (x_2 - y_2)^2 + (x_3 - y_3)^2}$ and $Q(x_i, t)$ is non-zero within the volume V that is centered on the origin of coordinates.

14.3 Use Gauss's divergence theorem to show that (14.11) may be rewritten as (14.12).

14.4 Consider the situation shown in Fig. 14.1 when the volume of non-zero T_{ij}^{L} is partially bounded by an impenetrable rigid surface, S. In this case an additional surface integral term,

$$\frac{1}{4\pi c_0^2}\frac{\partial}{\partial x_i}\int_{all\, \mathbf{y}\, on\, S} n_j[p\delta_{ij}]_{t-r/c_0}\frac{d^2y}{r},$$

must be added to the right side of (14.12). Here, p is the fluctuating pressure on the solid surface caused by the flow, and n_j is the normal to the solid surface. If the solid surface is flat, how do density fluctuations arising from this surface integral scale with ρ_0, M, l, V, and X far from the region of fluctuations? Will the volume-integral term shown in (14.12) or the surface integral term given here dominate sound radiation for subsonic flow fluctuations when $M \ll 1$?

14.5 Consider a two-phase mixture of fluid and particles (where particles might represent solid particles, liquid droplets, or bubbles). The continuity equation can be expressed as

$$\frac{\partial \epsilon\rho}{\partial t}+\frac{\partial}{\partial x_i}(\epsilon\rho u_i)=0,$$

where ρ is the density of the carrier-phase fluid, u_i is its velocity, and ϵ is the volume fraction of the carrier phase. The momentum equation for an inviscid two-phase flow in the absence of body forces is

$$\frac{\partial \epsilon\rho u_i}{\partial t}+\frac{\partial}{\partial x_i}\left(\epsilon\rho u_i u_j\right)=-\epsilon\frac{\partial p}{\partial x_j}\delta_{ij}+F_i,$$

where F_i is an interphase exchange force due to drag.

a) Show that the continuity equation can be rewritten as

$$\frac{\partial \rho}{\partial t}+\frac{\partial \rho u_i}{\partial x_i}=-\rho\frac{D(\ln\epsilon)}{Dt}$$

and provide a physical interpretation.

b) Recompute the dimensionless scaling of the compressible-flow continuity equation (4.112) taking into account the contribution from ϵ. Is the flow compressible when $M\to 0$?

c) Using the continuity and momentum equations given here, show how particles modify the inhomogeneous wave equation (14.10) and what additional terms are included in the Lighthill stress tensor.

14.6 Derive (14.17) through the following substitution and linearization steps. Set q and f_i to zero in (14.7) and insert the decompositions (14.14). Treat U_i, p_0, ρ_0 and T_0 as time-invariant and spatially uniform, and drop quadratic and higher order terms involving the fluctuations u_i', p', ρ', and T'.

14.7 The field equation for acoustic pressure fluctuations in a stationary ideal compressible fluid is (14.18). Consider one-dimensional solutions where $p=p(x,t)$ and $x=x_1$.

a) Drop the x_2 and x_3 dependence in (14.18), and change the independent variables x and t to $\xi=x-ct$ and $\zeta=x+ct$ to simplify (14.18) to $\partial^2 p'/\partial\xi\partial\zeta=0$.

b) Use the simplified equation in part a) to find the original field equation's general solution: $p'(x,t)=f(x-ct)+g(x+ct)$, where f and g are undetermined functions.

c) When the initial conditions are: $p'=F(x)$ and $\partial p'/\partial t=G(x)$ at $t=0$, show that:

$$f(x)=\frac{1}{2}\left[F(x)-\frac{1}{c}\int_0^x G(\ast)d\ast\right],\quad \text{and}\quad g(x)=\frac{1}{2}\left[F(x)-\frac{1}{c}\int_0^x G(\ast)d\ast\right],$$

where $\ast$ is just an integration variable.

14.8 Starting from (14.20) use (14.19) to prove (14.21)

14.9 At what value of the Mach number, $M=U/c$, does the dynamic pressure, $\rho U^2/2$, reach a maximum for an isentropic flow?

14.10 Consider two approaches to determining the upper Mach number limit for incompressible flow.

a) First consider pressure errors in the simplest possible steady-flow Bernoulli equation. Expand (14.34) for small Mach number to determine the next term in the expansion: $p_0=p+\frac{1}{2}\rho u^2+\ldots$ and determine the Mach number at which this next term is 5% of p when $\gamma=1.4$.

b) Second consider changes to the density. Expand (14.35) for small Mach number and determine the Mach number at which the density ratio ρ_0/ρ differs from unity by 5% when $\gamma = 1.4$.

c) Which criterion is correct? Explain why the criteria for incompressibility determined in a) and b) differ, and reconcile them if you can.

14.11 For steady frictionless compressible gas flow in a constant area duct with heat transfer, show that

$$\frac{dh}{ds} = T\left(1 - \frac{\alpha c^2}{c_p}\frac{M^2}{1-M^2}\right),$$

where $\alpha = -(1/\rho)(\partial\rho/\partial T)_p$ is the coefficient of thermal expansion of the gas, and $c_p = T(\partial s/\partial T)_p$ is the gas specific heat at constant pressure.

14.12 The critical area A^* of a duct flow was defined in Section 14.3. Show that the relation between A^* and the actual area A, at a section where the Mach number equals M, is that given by (14.36). This relation was not proved in the text. [*Hint*: Write:

$$\frac{A}{A^*} = \frac{\rho^* c^*}{\rho c} = \frac{\rho^*}{\rho_0}\frac{\rho_0}{\rho}\frac{c^*}{c}\frac{c}{u} = \frac{\rho^*}{\rho_0}\frac{\rho_0}{\rho}\sqrt{\frac{T^*}{T_0}\frac{T_0}{T}}\frac{1}{M};$$

then use the other relations given in Section 14.3.]

14.13 A perfect gas is stored in a large tank at the conditions specified by p_0, T_0. Calculate the maximum mass flow rate that can exhaust through a duct of cross-sectional area A. Assume that A is small enough that during the time of interest p_0 and T_0 do not change significantly and that the flow is isentropic.

14.14 The entropy change across a normal shock is given by (14.48) Show that this reduces to expressions (14.49) for weak shocks. [*Hint*: Let $M_1^2 - 1 \ll 1$. Write the terms within the two sets of brackets in Eq. (14.48) in the form $[1+\varepsilon_1][1+\varepsilon_2]^\gamma$, where ε_1 and ε_2 are small quantities. Then use the series expansion $\ln(1+\varepsilon) = \varepsilon - \varepsilon^2/2 + \varepsilon^3/3 + \ldots$. This gives Eq. (14.49) times a function of M_1 which can be evaluated at $M_1 = 1$.]

14.15 Show that the maximum velocity generated from a reservoir in which the stagnation temperature equals T_0 is $u_{max} = [2c_pT_0]^{1/2}$. What are the corresponding values of T and M?

14.16 In an adiabatic flow of air through a duct, the conditions at two points are $u_1 = 250\,\text{m/s}$, $T_1 = 300\,\text{K}$, $p_1 = 200\,\text{kPa}$, $u_2 = 300\,\text{m/s}$, $p_2 = 150\,\text{kPa}$. Show that the loss of stagnation pressure is nearly 34.2 kPa. What is the entropy increase?

14.17 A shock wave generated by an explosion propagates through a still atmosphere. If the pressure downstream of the shock wave is 700 kPa, estimate the shock speed and the flow velocity downstream of the shock.

14.18 Prove the following formulae for the jump the conditions across a stationary normal shock wave:

$$\frac{p_2 - p_1}{p_1} = \frac{2\gamma}{\gamma+1}(M_1^2 - 1),\quad \frac{u_2 - u_1}{c_1} = -\frac{2}{\gamma+1}\left(M_1 - \frac{1}{M_1}\right),$$

$$\text{and}\quad \frac{\upsilon_2 - \upsilon_1}{\upsilon_1} = -\frac{2}{\gamma+1}\left(1 - \frac{1}{M_1^2}\right),$$

where $\upsilon = 1/\rho$, and the subscripts '1' and '2' imply upstream and downstream conditions, respectively.

14.19 Using (14.1i), and (14.48), determine formula for p_{02}/p_{01} and ρ_{02}/ρ_{01} for a normal shock wave in terms of M_1 and γ. Is there anything notable about the results?

14.20 Using dimensional analysis, G.I. Taylor deduced that the radius $r(t)$ of the blast wave from a large explosion would be proportional to $(E/\rho_1)^{1/5}t^{2/5}$ where E is the explosive energy, ρ_1 is the quiescent air density ahead of the blast wave, and t is the time since the blast (see Example 1.10). The goal of this problem is to (approximately) determine the constant of proportionality assuming perfect-gas thermodynamics.

a) For the strong-shock limit where $M_1^2 \gg 1$, show:

$$\frac{\rho_2}{\rho_1} \cong \frac{\gamma+1}{\gamma-1},\quad \frac{T_2}{T_1} \cong \frac{\gamma-1}{\gamma+1}\frac{p_2}{p_1},\quad \text{and}\quad u_1 = M_1c_1 \cong \sqrt{\frac{\gamma+1}{2}\frac{p_2}{\rho_1}}.$$

b) For a perfect gas with internal energy per unit mass e, the internal energy per unit volume is ρe. For a hemispherical blast wave, the volume inside the blast wave will be $\frac{2}{3}\pi r^3$. Thus, set $\rho_2 e_2 = E/\frac{2}{3}\pi r^3$, determine p_2, set $u_1 = dr/dt$, and integrate the resulting first-order differential equation to show that $r(t) = K(E/\rho_1)^{1/5}t^{2/5}$ when $r(0) = 0$ and K is a constant that depends on γ.

c) Evaluate K for $\gamma = 1.4$. A full similarity solution of the non-linear gas-dynamic equations in spherical coordinates produces $K = 1.033$ for $\gamma = 1.4$ (see Thompson, 1972, p. 501). What is the percentage error in this exercise's approximate analysis?

14.21 Starting from the set (14.50) with $q = 0$, derive (14.52) by letting station (2) be a differential distance downstream of station (1).

14.22 Starting from the set (14.50) with $f = 0$, derive (14.53) by letting station (2) be a differential distance downstream of station (1).

14.23 For flow of a perfect gas entering a constant-area duct at Mach number M_1, calculate the maximum admissible values of f and q for the same mass flow rate. Case (a) $f = 0$; case (b) $q = 0$.

14.24 Show that the accelerating portion of the piston trajectory $(0 \le x_p(t) \le c_o t_1)$ shown in Fig. 14.19 is:

$$x_p(t) = \left(\frac{\gamma+1}{\gamma-1}\right) c_o t_1 \left(\frac{t}{t_1}\right)^{\frac{2}{\gamma+1}} - \frac{2c_o t}{\gamma-1} \quad \text{for} \quad 1 \le \frac{t}{t_1} \le \left(\frac{2}{\gamma+1}\right)^{\frac{1+\gamma}{1-\gamma}}.$$

14.25 For $t < 0$, an ideal (frictionless) piston with mass per unit area m is restrained at $x = 0$. This piston separates a quiescent perfect gas (to the left, $x < 0$) from perfect vacuum (to the right, $x > 0$) in a long tube with constant cross-sectional area. The initial sound speed and density in the perfect gas are c_o and ρ_o, respectively, and the gas has specific heat ratio γ. At $t = 0$, the piston restraint is removed, and the piston accelerates from rest to the right to achieve speed $U(t)$ at time $t > 0$. Assume the gas flow is isentropic for the following items.

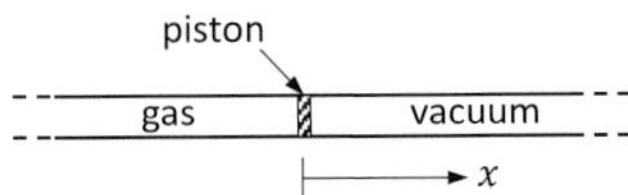

a) For $t > 0$, show that the gas pressure, $p(t)$, on the left side of the piston is:

$$p(t) = \frac{\rho_o c_o^2}{\gamma}\left[1 - \left(\frac{\gamma-1}{2}\right)\frac{U(t)}{c_o}\right]^{2\gamma/\gamma-1}.$$

b) Using the part a) result and Newton's second law determine $U(t)$.

c) As $m \to 0$ or $t \to \infty$, $U(t) \to 2c_o/(\gamma-1)$. How does this limiting speed compare to that from steady expansion of the same perfect gas into vacuum? Explain any difference you identify between unsteady and steady gas expansion with words or equations.

14.26 Consider the case of unsteady inviscid one-dimensional isothermal gas dynamics where (14.54) and (14.55) are the field equations, $p/p_o = \rho/\rho_o$, the speed-of-sound squared is $c_T^2 = p/\rho = RT_o = const.$, and T_o is the constant temperature.

a) Formulate a method of characteristics solution for these equations and show that:

$$I_\pm = u \pm c_T \ln(\rho/\rho_o)$$

are constant on characteristic curves $C_\pm$ defined by: $(dx/dt)_\pm = u \pm c_T$, where ρ_o is a constant reference density for the isothermal gas.

b) A long tube has a diaphragm at $x = 0$ that separates stationary gas with uniform density ρ_o (on the left), from vacuum (on the right). At $t = 0$, the diaphragm fully ruptures and the gas expands to the right. Use the results from part a) to find the following time-dependent velocity and density fields in the tube for $t > 0$:

$$u = 0, \qquad \rho = \rho_o \qquad \text{for} \quad x < -c_T t, \text{ and}$$

$$u = c_T(1 + x/c_T t), \qquad \rho = \rho_o \exp\{-(1 + x/c_T t)\} \qquad \text{for} \quad x > -c_T t.$$

c) Under the isothermal assumption, the leading edge of the expanding gas is predicted to travel with infinite speed. Explain with words how this is possible in this idealized situation. What phenomenon or phenomena prevent this infinite speed from occurring in reality?

14.27 For the flow conditions of Fig. 14.19, plot u/c_o and p/p_o as functions of $x/c_o t_1$ for $x_p(t) < x < c_o t$ at $t/t_1 = 2, 3$, and 4 for $\gamma = 1.4$, where c_o and p_o are the sound speed and pressure of the quiescent gas upstream of any disturbance from the moving piston. Does the progression of these waveforms indicate expansion wave steepening or spreading as t increases?

14.28 Consider the field properties in Fig. 14.20 before the formation of the shock wave.

a) Using the piston trajectory from Exercise 14.24, show that the time at which the piston reaches speed c_o is $-t_1((\gamma+1)/2)^{(1+\gamma)/(1-\gamma)} = -0.3349t_1$ for $\gamma = 1.4$.

b) Plot u/c_o and p/p_o as functions of $x/c_o t_1$ for $x_p(t) < x < c_o t_1$ at: $t/t_1 = -1/3, -1/6$ and $-1/25$ for $\gamma = 1.4$, where c_o and p_o are the sound speed and pressure of the quiescent gas upstream of any disturbance from the moving piston. Does the progression of these waveforms indicate compression wave steepening or spreading as $t \to 0$?

14.29 For the flow conditions of Fig. 14.20, assume the flow speed downstream of the shock wave is c_o and determine the shock Mach number, its x-t location, and the pressure, temperature and density ratios across the shock. Are these results well matched to the isentropic compression that occurred for $t < 0$? What additional adjustment is needed?

14.30 Write momentum conservation for the volume of the small rectangular control volume shown in Fig. 4.22 where the interface is a shock with flow from side 1 to side 2. Let the two end faces approach each other as the shock thickness $\to 0$ and assume viscous stresses may be neglected on these end faces (outside the structure). Show that the **n** component of momentum conservation yields (14.41) and the **t** component gives $\mathbf{u}\cdot\mathbf{t}$ is conserved or v is continuous across the shock.

14.31 Consider a normal shockwave traveling in a perfect gas with ratio of specific heats $= \gamma$. In the frame of reference where the shock wave is stationary, the ratio of the gas' upstream horizontal velocity, u_1, divided by the gas' downstream horizontal velocity, u_2, is given by (14.46)

$$\frac{u_1}{u_2} = \frac{(\gamma+1)M_1^2}{(\gamma-1)M_1^2+2},$$

when u_1 and u_2 are both positive, and $M_1 = (\textit{shock propagation speed})/c_1 \geq 1$.

a) This relationship can be useful in other (inertial) frames of reference, too. Show that the given equation can be revised to:

$$\frac{u_2-u_1}{c_1} = \pm\frac{2}{\gamma+1}\left(M_1 - \frac{1}{M_1}\right),$$

where '+' is for shock waves propagating to the right and '−' is for shock waves propagating to the left.

b) Now consider what happens when an incident normal shockwave propagating to the left into quiescent gas reflects from an impenetrable flat vertical surface. Determine the following relationship for the Mach number, M_r, of the reflected shock wave that propagates to the right (as shown).

$$M_r - \frac{1}{M_r} = \frac{c_1}{c_2}\left(M_i - \frac{1}{M_1}\right)$$

c) If $M_i = 1.20$ and $\gamma = 1.40$, determine the numerical value of the pressure ratio p_3/p_1.

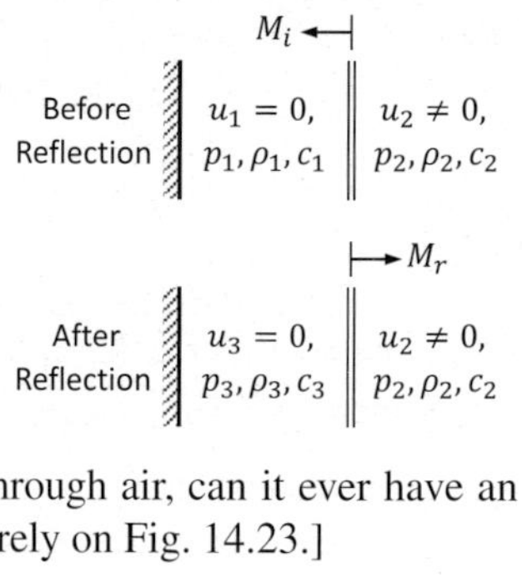

M_i

Before Reflection | $u_1 = 0$, p_1, ρ_1, c_1 | $u_2 \neq 0$, p_2, ρ_2, c_2

M_r

After Reflection | $u_3 = 0$, p_3, ρ_3, c_3 | $u_2 \neq 0$, p_2, ρ_2, c_2

14.32 A wedge has a half-angle of 50°. Moving through air, can it ever have an attached shock? What if the half-angle were 40°? [*Hint*: The argument is based entirely on Fig. 14.23.]

14.33 Air at standard atmospheric conditions is flowing over a surface at a Mach number of $M_1 = 2$. At a downstream location, the surface takes a sharp inward turn by an angle of 20°. Find the wave angle σ and the downstream Mach number. Repeat the calculation by using the weak-shock assumption and determine its accuracy by comparison with the first method.

14.34 A curved channel carrying a supersonic flow is designed so that the expansion turn caused by the lower wall is precisely canceled by the shape of the upper wall. The inlet Mach number is $M_1 = 1.5$. Find M_2, P_2/P_1, and h_2/h_1.

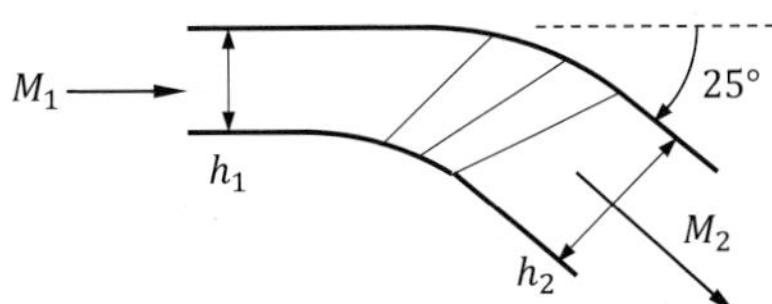

14.35 A flat plate is inclined at 10° to an airstream moving at $M_\infty = 2$. If the chord length is $b = 3$ m, find the lift and wave drag per unit span.

14.36 Using thin-airfoil theory calculate the lift and drag on the airfoil shape given by $y_u = t\sin(\pi x/c)$ for the upper surface and $y_l = 0$ for the lower surface. Assume a supersonic stream parallel to the x-axis, and that the thickness/chord $\ll 1$.

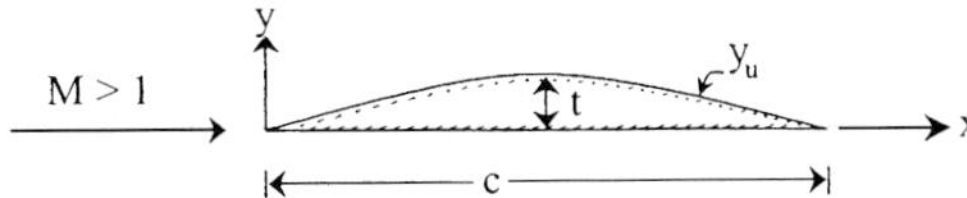

14.37 Consider a thin airfoil with chord length l at a small angle of attack in a horizontal supersonic flow at speed M_∞. The foil's upper and lower surface contours, $y_u(x)$ and $y_l(x)$, respectively, are defined by:

$$y_u(x) = t(x)/2 + y_c(x) - \alpha x, \quad \text{and} \quad y_l(x) = -t(x)/2 + y_c(x) - \alpha x,$$

where: $t(x)$ = the foil's thickness distribution, α = the foil's angle of attack, and $y_c(x)$ = the foil's camber line. Use these definitions to show that the foil's coefficients of lift and drag are:

$$C_L = \frac{4\alpha}{\sqrt{M_\infty^2 - 1}}, \quad \text{and} \quad C_D = \frac{4}{\sqrt{M_\infty^2 - 1}}\left[\frac{1}{4}\overline{\left(\frac{dt}{dx}\right)^2} + \overline{\left(\frac{dy_c}{dx}\right)^2} + \alpha^2\right].$$

References

Ames Research Staff, 1953. Equations, Tables, and Charts for Compressible Flow. NACA Report 1135.
Becker, R., 1922. Stosswelle und Detonation. Zealand Physik 8, 321–362.
Cohen, I.M., Moraff, C.A., 1971. Viscous inner structure of zero Prandtl number shocks. Physical Fluids 14, 1279–1280.
Cramer, M.S., Fry, R.N., 1993. Nozzle flows of dense gases. The Physics of Fluids A 5, 1246–1259.
Crighton, D.G., 1975. Basic principles of aerodynamic noise generation. Progress in Aerospace Sciences 16, 31–96.
Fergason, S.H., Ho, T.L., Argrow, B.M., Emanuel, G., 2001. Theory for producing a single-phase rarefaction shock wave in a shock tube. Journal of Fluid Mechanics 445, 37–54.
Hayes, W.D., 1958. The basic theory of gasdynamic discontinuities. In: Emmons, H.W. (Ed.), High Speed Aerodynamics and Jet Propulsion, vol. III, Sect. D of Fundamentals of Gasdynamics. Princeton University Press, Princeton, NJ.
Liepmann, H.W., Roshko, A., 1957. Elements of Gas Dynamics. Wiley, New York.
Lighthill, M.J., 1952. On sound generated aerodynamically I. General theory. Proceedings Royal Society A 211, 564–587.
Pierce, A.D., 1989. Acoustics. Acoustical, New York.
Shapiro, A.H., 1953. The Dynamics and Thermodynamics of Compressible Fluid Flow (vol. 2). Ronald Press, New York.
Thompson, P.A., 1972. Compressible Fluid Dynamics. McGraw-Hill, New York.
von Karman, T., 1954. Aerodynamics. McGraw-Hill, New York.

Further reading

Courant, R., Friedrichs, K.O., 1977. Supersonic Flow and Shock Waves. Springer-Verlag, New York.
Yahya, S.M., 1982. Fundamentals of Compressible Flow. Wiley Eastern, New Delhi.

Chapter 15

Computational fluid dynamics⊛

Grétar Tryggvason

Contents

15.1 Introduction 621
15.2 The advection-diffusion equation 623
Finite difference approximation 625
Accuracy 627
Finite volume approximation 629
Stability 629
Other schemes 632
Steady-state and boundary-value problems 633
Two-dimensional advection-diffusion equation 634
Multidimensional steady-state and boundary-value problems 636
15.3 Incompressible flows in rectangular domains 637
Vorticity-stream function methods 637
Numerical approximations 638
Boundary conditions 639
Simulations using ψ and ω 639
Velocity-pressure method for incompressible flows 641
Spatial discretization 643
The computational domain and boundary condition 647
Numerical code and results 648
15.4 Flow in complex domains 650
Immersed boundary methods 651
Mapped grid methods 652
The generation of body-fitted grids 655
Unstructured grids 655
15.5 Velocity-pressure method for compressible flow 656
One-dimensional flow 658
Two-dimensional flow 661
15.6 More to explore 662
Other methods 662
More complex physics 664
Direct numerical simulations 664
Multiscale modeling 665
Exercises 666
References 668
Further reading 668

Chapter objectives

- To introduce computational fluid dynamics (CFD) and how it has expanded the range of fluid dynamics problems that can be solved
- To review a few elementary numerical concepts needed to solve initial and boundary-value problems
- To show how to solve the Navier-Stokes equations in simple and complicated geometries
- To discuss how the elementary methods presented are expanded to solve much more complicated problems of practical interest

15.1 Introduction

The conservation principles for mass, momentum, and energy, along with constitutive assumptions for the viscous stresses and heat conduction, result in equations that should, at least in principle, make it possible to predict fluid flow in all circumstances. Unfortunately, as we see in other parts of the book, the equations are so complex that we can only find analytical solutions for a few relatively simple cases. Various approximations expand the range of solvable problems somewhat, but detailed solutions for complex flows were not possible until the development of computers, around the middle of the twentieth century, and the emergence of computational fluid dynamics (CFD).

The definition of CFD depends on whom you ask, but in this chapter we will take CFD to mean finding numerical solutions to the Navier-Stokes or the Euler equations in two or three dimensions, with the goals of obtaining physical insight into the dynamics and quantitative predictions. If we define CFD this way, identifying the people and the place where CFD was initiated becomes straightforward. The convergence of powerful computers, important problems, and creative people took place at the Los Alamos National Laboratory in the late 1950s. The extraordinary creativity of the group led to many ideas and methods that are still alive and well. Although the work of the group was quickly available in the open literature,

⊛. This book has a companion website hosting complementary materials. Visit this URL to access it: https://www.elsevier.com/books-and-journals/book-companion/9780128198070.

Fluid Mechanics. https://doi.org/10.1016/B978-0-12-819807-0.00024-7

it spread slowly in the beginning, and initial applications were focused on problems of immediate concern to the national laboratories in the USA. Nevertheless, about a decade later, the application of CFD to industrial problems started at the Imperial College in the UK. The 1970s and 1980s saw rapid developments of many methods and by the 1990s CFD had become a well-established field with computations routinely used for scientific inquiries and engineering predictions. CFD is now used for essentially all engineering designs, greatly reducing (although not eliminating!) the need for expensive prototyping and testing. Both Boeing and Airbus use CFD for the design of new airplanes and all automakers do the same. The makers of internal combustion engines and turbomachinery rely on CFD, as do designers of buildings and bridges. CFD is used to predict aerodynamic performance, structural loads, and noise, the performance of consumer goods and power plants, and increasingly the dynamics of living things. In short, CFD is used everywhere. While CFD has now become a well-established part of the scientific and engineering "toolbox," the increasing availability of computers of unprecedented power is redefining the field. Not only are we rapidly learning to model systems of enormous complexity, we now expect the results to be reliable and accurate and include an assessment of the uncertainty as well as the sensitivity to the various input parameters.

Computers work with discrete numbers so we must approximate the continuous flow field by discrete values and the partial differential equations must be replaced by algebraic equations that relate the discrete values to each other. There are several ways to do this but here we will focus on two approaches: finite differences and finite volumes. Although those arise from somewhat different considerations, they are complementary and often result in the same or similar discrete equations. Fig. 15.1 shows the general idea. The domain where we need to find the flow is divided by a grid that defines small control volumes or connected grid points. The principles of conservation of mass, momentum, and energy are then applied to each control volume or a grid point to find the point-wise or the average velocity. In the finite difference approach we start with the governing equations in differential form and approximate the continuum equations for the values at each point by approximate formula for the various derivatives, usually using a Taylor series expansion. In the finite volume approach we divide the flow domain into small control volumes and use the governing equations in integral form to derive discrete approximate equations for the average values in each control volume. In both cases the grid must be sufficiently fine so that the velocity, pressure, and other variables in each control volume, or in between the points, are changing slowly and are well described by simple functions determined by one or few parameters. The discrete equations are then used to determine these parameters.

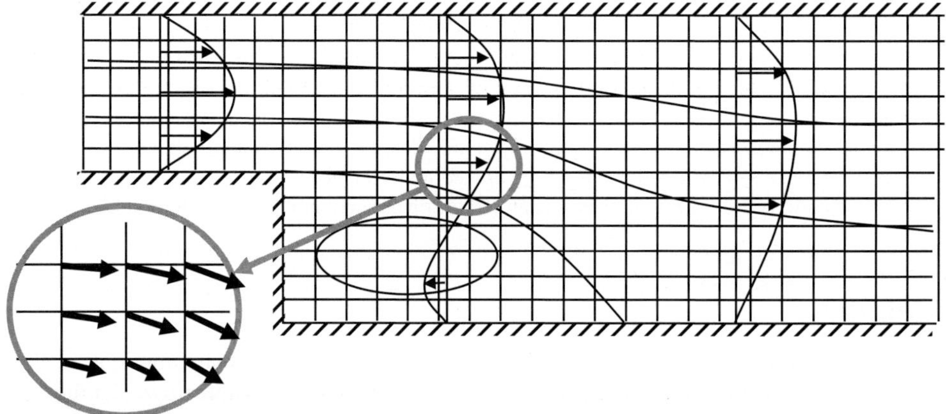

FIGURE 15.1 The approximation of a flow field on a discrete grid, where the grid points are either where we approximate the equations using finite differences or the centers of small control volumes. The horizontal component of the fluid velocity at three cross-sections and a few streamlines are shown.

To fully appreciate what is now possible and the impact it is having, we need to look at the size of the problems that we can deal with and the speed with which we can do them. On the computer, information are stored as bits (1 or 0) and 64 bits make eight bytes, corresponding to a double-precision floating point number (with about 16 significant digits), which is what we generally need for large simulations. For fully three-dimensional computations we need roughly ten numbers per grid point (for two copies of the three velocity components, the pressure, and a few auxiliary quantities). In 1990, a grid with about one million points (100^3) was a big computation and required roughly 100^3 grid points $\times 8$ bytes/number $\times 10$ numbers/grid points ≈ 0.1 gigabytes (GB) of memory. Today, a billion (1000^3) grid points constitute a large–but not impossibly large–simulation and to store everything at one time step takes $1000^3 \times 8 \times 10 \approx 100$ GB. Similarly, the growth in computer speed has been enormous. Measured in FLOPS (floating point operations per second), the maximum speed of the CRAY-1 (1976) was 133 megaflops, for ASCI White (2000) it was about 12 teraflops, Titan (2012) could achieve nearly 20 petaflops, and Frontier (2022) can exceed one exaflop, or 10^{10} times the speed of the CRAY-1.

We start this chapter by introducing finite difference and finite volume methods and use them for the simple advection-diffusion equation to explore accuracy, consistency, and stability of numerical methods. We then develop two methods to solve the Navier-Stokes equations for incompressible flows. The first is a finite difference method for the stream function/vorticity form of the equations and the second is a finite volume method for the equations in the more familiar velocity/pressure form. In both cases we use regular structured grids and focus on rectangular domains. The first method is then extended to more complex domains, using coordinate mapping. After that we move on to compressible flows, with a particular focus on how to capture shocks accurately. This is an area that has received significant attention and where significant progress has been made over the last few decades. We will not attempt to cover this progress in any detail but simply present one very simple (although not very accurate) method. At the end of the chapter we discuss briefly some of the topics that fall under the general area of CFD, but which are beyond the intent of the current chapter.

15.2 The advection-diffusion equation

Before attempting to solve the full Navier-Stokes equations, it is helpful to examine a simpler equation and here we will consider the unsteady advection-diffusion equation in one and two dimensions. This equation describes how a quantity f is carried with a flow as it also diffuses. To simplify the situation even more, we will assume that the velocity is constant. This results in a linear equation that has an analytical solution, providing us with a benchmark against which to compare our numerical results. Although simple, the linear advection-diffusion equation has considerable similarities with the fluid equations and much of what we learn here will carry directly over to the full equations.

Consider the situation sketched in Fig. 15.2. Here an arbitrary spatial distribution of a quantity f is carried downstream with a constant velocity U, as the profile also changes its shape due to diffusion. The question to be answered is: Given the distribution of f at time $t = t_1$, what is its distribution at a later time $t_2 = t_1 + \Delta t$? Or, given $f(t_1, x)$, what is $f(t_2, x)$? To predict that, we need to derive an equation for the evolution of f. If f is a conserved quantity, we can derive an equation for its evolution in the same way as we derived equations for the fluid flow earlier in the book. Although we have already derived the governing equations for more complex situations, it is useful to repeat the derivation here. We start by identifying a specific region of space as our *control volume*. A control volume has a specific shape, location, and well-defined boundaries, the *control surface*. In our case, the control volume is a region of length Δx, shown in Fig. 15.2, and the control surface consists of the two end points. The amount of f in the control volume at any given time is given by the integral of f, and the rate of change of the total amount of f in the control volume is given by the time derivative of the integral, or:

$$\frac{d}{dt}\int_{\Delta x} f\,dx. \tag{15.1}$$

If f is conserved, then the rate of change of the integral must be balanced by the difference in the flux, F, of f in and out of the control volume, or the difference between F_{in} and F_{out}. Here the flux F is taken to be positive in the x-direction so the equation governing the evolution of f is:

$$\frac{d}{dt}\int_{\Delta x} f\,dx = F_{in} - F_{out}. \tag{15.2}$$

By introducing the average value of f:

$$f_{av} = \frac{1}{\Delta x}\int_{\Delta x} f(x)dx, \tag{15.3}$$

we can write:

$$\frac{df_{av}}{dt} = \frac{1}{\Delta x}(F_{in} - F_{out}). \tag{15.4}$$

This is our equation in finite volume form and, once we have defined F, is the starting point for finite volume methods.

Although (15.4) is often the most useful starting point for a numerical approximation, it is sometimes better to work with the corresponding partial differential equation. We can rewrite the above equation using the fact that the integral of the

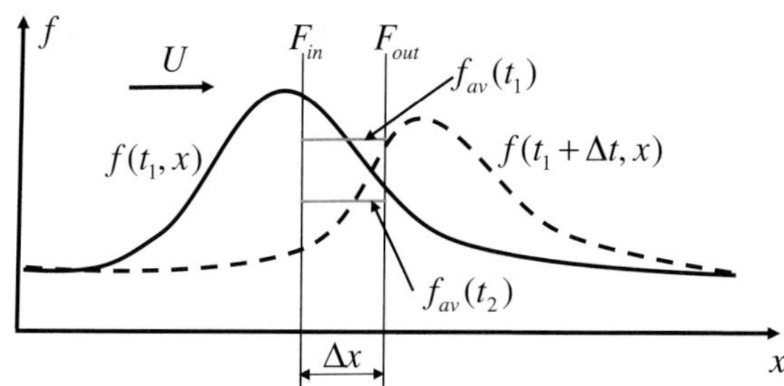

FIGURE 15.2 The advection-diffusion equation describes how a function $f(t, x)$ evolves in time and space. To derive the equation we focus our attention on a small control volume of size Δx and a small time increment Δt, and consider the limits $\Delta x \to 0$ and $\Delta t \to 0$. When approximating the equation numerically, Δx and Δt are small but finite.

derivative of F is equal to the difference in the values of F at the end points of the interval:

$$F_{out} - F_{in} = \int_{\Delta x} \frac{\partial F}{\partial x} dx. \tag{15.5}$$

Here, we use the partial derivative since F is both a function of time and space and the derivative is with respect to space only. Thus, (15.2) can be written as:

$$\frac{d}{dt} \int_{\Delta x} f\, dx = - \int_{\Delta x} \frac{\partial F}{\partial x} dx. \tag{15.6}$$

The integral is over a fixed region of space so we can move the time derivative under the integral. However, unlike the integral, f itself is a function of both space and time so we need to change the time derivative into a partial derivative. Rearranging the terms we have:

$$\int_{\Delta x} \left(\frac{\partial f}{\partial t} + \frac{\partial F}{\partial x} \right) dx = 0. \tag{15.7}$$

The integral must be zero for any choice of a control volume, and this can only be true if the integrand is identically zero. The partial differential equation stating that f is conserved is therefore:

$$\frac{\partial f}{\partial t} + \frac{\partial F}{\partial x} = 0. \tag{15.8}$$

Before attempting to solve this equation, we need to specify the fluxes F. In our case, we assume that F consists of advective fluxes, governing how f is carried with the flow velocity U, and diffusive fluxes. In the limit of a small time interval, $\Delta t \to 0$, we can take f crossing the control surface to be a constant during Δt, so the amount of f that crosses the surface during the time interval is $U f \Delta t$. Dividing by Δt gives the flow of f per unit time, or the flux. Thus:

$$F_{advec} = U f. \tag{15.9}$$

The diffusive fluxes are taken to be proportional to the gradient of f so we write:

$$F_{diff} = -D \frac{\partial f}{\partial x}, \tag{15.10}$$

where the negative sign indicates that the flow of f is from high values to low values and D is a positive material dependent diffusion coefficient.

Substituting the advective and diffusive fluxes into (15.8), taking D to be constant, and moving the diffusive fluxes to the right-hand side yields:

$$\frac{\partial f}{\partial t} + U \frac{\partial f}{\partial x} = D \frac{\partial^2 f}{\partial x^2}. \tag{15.11}$$

This equation can be solved analytically. A simple sine wave will, in particular, evolve according to:

$$f(t, x) = e^{-Dk^2 t} \sin(2\pi k(x - Ut)), \tag{15.12}$$

as can be verified by a direct substitution. Thus, a sine wave will move with a velocity U and decay with time at a rate determined by the diffusion coefficient D and the wave number squared. If the diffusivity is zero, it should be clear that the initial f profile simply moves with the constant velocity U. The evolution of an initial sine wave is shown in Fig. 15.3 at three different times.

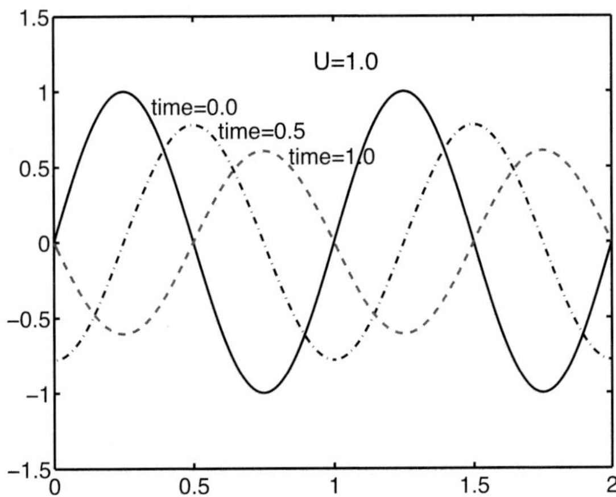

FIGURE 15.3 The evolution of sine wave by the advection diffusion equation. The initial amplitude is 1 and the domain is periodic with length 2, $k=1$, $U=1$, and $D=1.0$. The wave is shown at times $t_0=0$, $t_1=0.5$, and $t_2=1.0$.

Finite difference approximation

We will find a numerical approximation to the time-dependent solution of the advection-diffusion equation (15.11) in two ways. First, we solve it using a standard finite difference method and then we will do so using a finite volume method. Both approaches result in exactly the same discrete equation, as is often the case, but the "philosophy" differs. The finite difference approach is likely to be more familiar, but the finite volume point of view is often more intuitive.

We solve (15.11) numerically on the discrete grid shown in Fig. 15.4. Here the spatial dimension (horizontal axis) is discretized by grid points labeled $j=1...N$ that are assumed to be evenly spaced and separated by Δx. If the coordinate of point j is x_j, then the coordinate of point $j+1$ is $x_j+\Delta x$, and so on. Time is indicated by the vertical axis and f at the current time t is denoted by superscript n and the next one, at time $t+\Delta t$, is denoted by $n+1$.

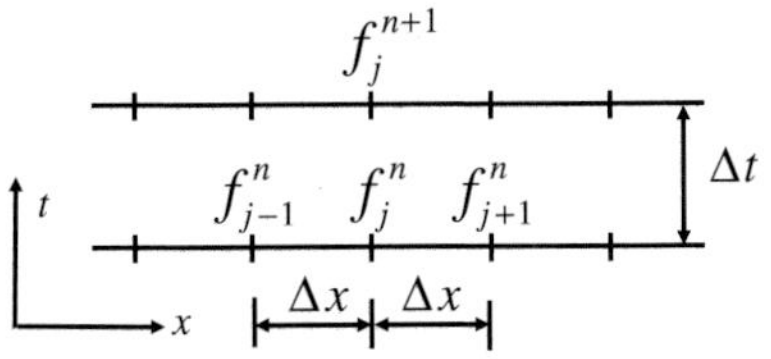

f_j^{n+1} $\quad f_{j-1}^n \quad f_j^n \quad f_{j+1}^n \quad \Delta t \quad t \quad x \quad \Delta x \quad \Delta x$

FIGURE 15.4 The grid layout used for finite difference approximations.

We start by writing (15.11) at grid point j at time n:

$$\left.\frac{\partial f}{\partial t}\right)_j^n + U\left.\frac{\partial f}{\partial x}\right)_j^n = D\left.\frac{\partial^2 f}{\partial x^2}\right)_j^n. \tag{15.13}$$

To approximate the derivatives based on the discrete values of f that are available we use a Taylor series. While we assume that the reader is familiar with the derivation of discrete approximations for derivatives in this way, it is very short so we include it here for completeness.

Consider the function $f(x)$, given at points x_{j-1}, x_j, x_{j+1}, and so on. Denote $f_{j-1}=f(x_{j-1})$, $f_j=f(x_j)$, and $f_{j+1}=f(x_{j+1})$. The values at different points can be related to each other by a Taylor series:

$$f_{j+1} = f_j + \frac{\partial f_j}{\partial x}\Delta x + \frac{\partial^2 f_j}{\partial x^2}\frac{\Delta x^2}{2} + \frac{\partial^3 f_j}{\partial x^3}\frac{\Delta x^3}{6} + \frac{\partial^4 f_j}{\partial x^4}\frac{\Delta x^4}{24} + \cdots, \tag{15.14}$$

where the derivatives are taken at x_j. This equation can be rearranged to give an expression for the first derivative:

$$\frac{\partial f_j}{\partial x} = \frac{f_{j+1} - f_j}{\Delta x} - \frac{\partial^2 f_j}{\partial x^2}\frac{\Delta x}{2} - \frac{\partial^3 f_j}{\partial x^3}\frac{\Delta x^2}{6} - \frac{\partial^4 f_j}{\partial x^4}\frac{\Delta x^3}{24} + \cdots. \tag{15.15}$$

As $\Delta x \to 0$, all the terms on the right-hand side go to zero, except for the first one. Since we have already decided that Δx must be small, it is therefore reasonable to approximate the first derivative of f at x_j simply as the difference between f_{j+1} and f_j divided by Δx, as we might have guessed. But (15.15) also shows us that how the error behaves as $\Delta x \to 0$. For small Δx the error is dominated by the Δx term and taking Δx half as small will cut the error in half. Since the error is directly related to the first power of Δx, (15.15) is a *first-order* approximation to the first derivative.

Instead of writing f_{j+1} as a Taylor series expansion of f_j to get an approximation for the derivative at x_j, we could just as well have written f_{j-1} as a Taylor series expansion of f_j, by replacing Δx by $-\Delta x$. This gives us:

$$f_{j-1} = f_j - \frac{\partial f_j}{\partial x}\Delta x + \frac{\partial^2 f_j}{\partial x^2}\frac{\Delta x^2}{2} - \frac{\partial^3 f_j}{\partial x^3}\frac{\Delta x^3}{6} + \frac{\partial^4 f_j}{\partial x^4}\frac{\Delta x^4}{24} + \cdots, \tag{15.16}$$

which can be rearranged to give us an expression for the derivative. However, we can also subtract it from (15.14), eliminating the second derivative term. Rearranging the results gives us another approximate equation for the first derivative:

$$\frac{\partial f_j}{\partial x} = \frac{f_{j+1} - f_{j-1}}{2\Delta x} - \frac{\partial^3 f_j}{\partial x^3}\frac{\Delta x^2}{6} + \cdots, \tag{15.17}$$

where we now approximate the first derivative by the difference between f_{j+1} and f_{j-1}, divided by $2\Delta x$. Here, the first error term contains Δx^2, instead of Δx, showing that if we divide Δx by two, the error will go down by a factor of four. Since the error is directly related to the second power of Δx, (15.17) is a *second-order* approximation to the first derivative. The second-order approximation can also be derived by taking the average of the first-order approximations using f_{i+1} and f_{i-1}, and it is presumably intuitive that the average would give us a better approximation.

By adding (15.14) and (15.16) we eliminate the first derivative and obtain a second-order approximation for the second derivative:

$$\frac{\partial^2 f_j}{\partial x^2} = \frac{f_{j+1} - 2f_j + f_{j-1}}{\Delta x^2} - \frac{\partial^4 f_j}{\partial x^4}\frac{\Delta x^2}{12} + \cdots. \tag{15.18}$$

By using a larger number of points we can derive approximations for higher derivatives and with higher-order error terms.

Here we have focused on the spatial derivative of f, but these considerations obviously also apply to the time derivative and, for reasons to be discussed later, we will start by using a one-sided first-order approximation for the time derivative (linking f_j^n and f_j^{n+1}) and second-order centered approximations for the spatial derivatives. Thus, we approximate:

$$\begin{aligned} \left.\frac{\partial f}{\partial t}\right)_j^n &\approx \frac{f_j^{n+1} - f_j^n}{\Delta t}; \\ \left.\frac{\partial f}{\partial x}\right)_j^n &\approx \frac{f_{j+1} - f_{j-1}}{2\Delta x}; \\ \left.\frac{\partial^2 f}{\partial x^2}\right)_j^n &\approx \frac{f_{j+1} - 2f_j + f_{j-1}}{\Delta x^2}, \end{aligned} \tag{15.19}$$

and the discrete approximation of the advection-diffusion equation, (15.11), becomes:

$$\frac{f_j^{n+1} - f_j^n}{\Delta t} + U\left(\frac{f_{j+1}^n - f_{j-1}^n}{2\Delta x}\right) = D\left(\frac{f_{j+1}^n - 2f_j^n + f_{j-1}^n}{\Delta x^2}\right). \tag{15.20}$$

Given the discrete values of f at one time $(\ldots, f_{j-1}^n, f_j^n, f_{j+1}^n, \ldots)$, this equation allows us to update all the f's by computing f_j^{n+1} as a linear combination of f_{j-1}^n, f_j^n, and f_{j+1}^n.

Now we have all the elements to construct a numerical code. The code is extremely simple: we need two loops, one over time and one over the spatial points. We also need an initial condition and boundary conditions. The simplest case is if the values of the endpoints are given, say $f(1:N-1) = 0$ and $f(N) = 1.0$, where N is the total number of grid points. Here, however, we will write a code to simulate the periodic sine wave in Fig. 15.3, so the first and the last point in the domain are the same. To include the periodic boundary we add an equation for the last point that substitutes $f(2)$ for $f(N+1)$ and then set $f(1) = f(N)$. Since we include both endpoints, the grid spacing is found by $\Delta x = L/(N-1)$, where L is the length of the domain.

CODE 1: SOLUTION OF THE UNSTEADY ONE-DIMENSIONAL LINEAR ADVECTION EQUATION USING AN EXPLICIT FORWARD IN TIME, CENTERED IN SPACE (FTCS) SCHEME

```
% one-dimensional advection-diffusion by the FTCS scheme
N=21; nstep=10; L=2.0; dt=0.05;U=1;D=0.05; k=1;
dx=L/(N-1); for j=1:N, x(j)=dx*(j-1);end
f=zeros(N,1); fo=zeros(N,1); time=0.0;
for j=1:N, f(j)=0.5*sin(2*pi*k*x(j)); end;                % initial cond.

for m=1:nstep, m, time                                     % time loop
  plot(x,f,'linewidth',2); axis([0 L -1.5, 1.5]); pause
  fo=f;                                                    % store sol.
  for j=2:N-1                                              % spatial loop
    f(j)=fo(j)-(0.5*U*dt/dx)*(fo(j+1)-fo(j-1))+...        % centered
           D*(dt/dx^2)*(fo(j+1)-2*fo(j)+fo(j-1));          % diff.
   end;
   f(N)=fo(N)-(0.5*U*dt/dx)*(fo(2)-fo(N-1))+...           % periodic
         D*(dt/dx^2)*(fo(2)-2*fo(N)+fo(N-1));              % boundary
  f(1)=f(N);                                               % condition
  time=time+dt;
end;
```

Fig. 15.5(a) shows the results of evolving a sine wave computed with this code, along with the analytical solution. Here we have used $N = 21$ and $\Delta t = 0.05$ and while it is clear that the numerical solution reproduces the analytical solution, there obviously are some differences. The numerical solution does, in particular, decay slower than it should. To see how the results change as we change the resolution (Δt and Δx), we have repeated the computation using $N = 201$ (so that $\Delta x = 0.01$) and $\Delta t = 0.0005$. The results are shown in Fig. 15.5(b), where we had to increase the thickness of the line showing the exact solution, since otherwise it would be completely covered by the numerical solution. Obviously, for small enough Δt and Δx, we reproduce the exact solution almost perfectly.

Accuracy

Fig. 15.5(a) showed that while the evolution of the wave was reproduced reasonably well when we used a moderate number of grid points and a large time step, the numerical solution did not match the analytical solution exactly. However, as Fig. 15.5(b) shows, when we increase the number of grid points and reduce the time step, the accuracy is improved. To understand the error and how it depends on the number of grid points and the size of the time step we first need to decide how we measure the error. In those cases where we have the analytical solution this is relatively straightforward, since the difference between the numerical solution and the analytical value is available at every grid point and time step. The total error can be defined in different ways, but here we will use the root mean square error, where we add the squares of the error and then take the square root of the sum. The integrated error at each grid point, or in each control volume, is the difference between the numerical and the exact solution, multiplied by Δx, thus allowing us to compare the error as we change the number of grid points:

$$E = \Delta x \sqrt{\sum_{j=1}^{N} \left(f_j - f_{exact}\right)^2}, \tag{15.21}$$

where we used the fact that Δx is a constant to put it outside the summation. We have repeated the computation shown in Fig. 15.5 using a different number of grid points and show the error at time 0.5 in Fig. 15.6, as a function of the size of Δx. Here we focus on the effect of the spatial resolution so we have done the simulation with a very small time step

($\Delta t = 0.0005$) to ensure that the error from the time step is negligible. To understand where the error comes from it is important to remember that when we replaced (15.11) by its discrete counterpart, (15.20), we left out some terms. For the approximations to the spatial derivatives the first neglected terms are proportional to Δx^2 and once Δx is small enough we expect all higher-order terms to be much smaller. Thus, it seems reasonable to expect the error to be proportional to Δx^2, or:

$$E = C\Delta x^2, \tag{15.22}$$

where C is some constant. Taking the logarithm of this equation gives us:

$$ln E = ln\left(C\Delta x^2\right) = lnC + 2ln\Delta x, \tag{15.23}$$

and plotting lnE versus $ln\Delta x$, as we have done in Fig. 15.6, we expect to get a line with slope 2. That is the slope of the solid line and it is clear that the error follows it closely. In this particular case, even the error for the coarsest grid aligns with the line. In practice, this is not always the case since the rest of the error terms have to be small before we expect the asymptotic behavior where the higher-order error terms are unimportant. For the finest grid, however, the error deviates from the line. This is also common and is due to the error that we are looking at becoming so small that other errors become important. In our case it is the error from the finite size of the time step but if we kept reducing the time step and increasing the number of grid points, eventually roundoff errors would become important and take us away from the line. Checking that the error behaves as expected is an important aspect of testing numerical codes since a deviation from the expected behavior is usually due to a programming error.

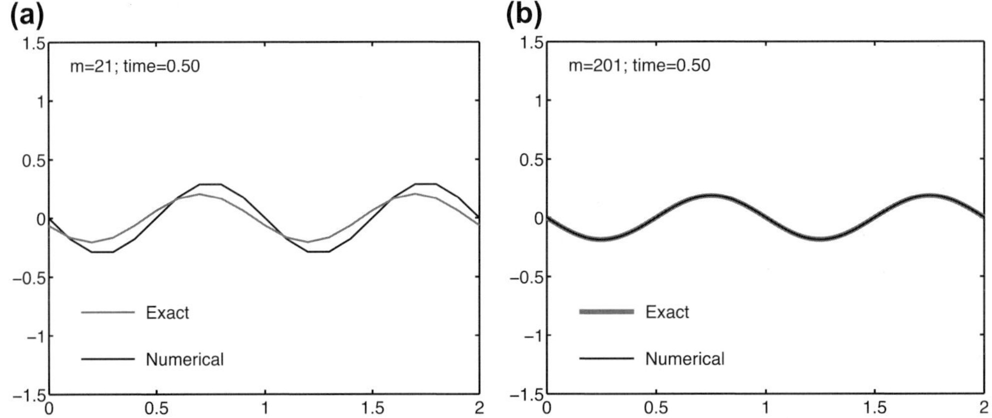

FIGURE 15.5 (a) A comparison of the numerical solution of the advection-diffusion equation with the analytical solution for $L = 2$, $U = 1$; $D = 0.05$; $k = 1$, at time 0.5, for relatively low resolution. (b) a comparison of the numerical solution of the advection-diffusion equation with the analytical solution for high resolution. (See the text for the numerical parameters used.)

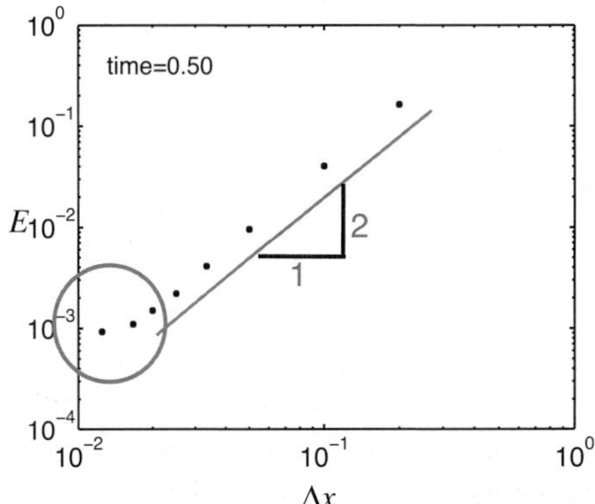

FIGURE 15.6 A log-log plot of the error versus Δx for numerical approximations of the advection-diffusion equation at time 0.5, as computed using a very small Δt. The solid line corresponds to a second-order method.

To gain added insight into how well the discrete equation approximates the original partial differential equation it is sometimes useful to derive the so-called modified equation. To do so we substitute the Taylor series expansion for each quantity in the discrete equation and manipulate it to isolate the original equation. For Eq. (15.20) the modified equation is:

$$\frac{\partial f}{\partial t}+U\frac{\partial f}{\partial x}-D\frac{\partial^2 f}{\partial x^2}=-\frac{\partial^2 f}{\partial t^2}\frac{\Delta t}{2}-U\frac{\partial^3 f}{\partial x^3}\frac{\Delta x^2}{6}+D\frac{\partial^4 f}{\partial x^4}\frac{\Delta x^2}{12}+\cdots. \tag{15.24}$$

The right-hand side includes terms with Δx and Δt that go to zero as $\Delta x \to 0$ and $\Delta t \to 0$, and we recover our original partial differential equation—as we hoped to do! The modified equation is sometimes also written as:

$$\frac{\partial f}{\partial t}+U\frac{\partial f}{\partial x}-D\frac{\partial^2 f}{\partial x^2}=O(\Delta t, \Delta x^2) \tag{15.25}$$

to emphasize that our particular approximation is first order in time and second order in space.

Numerical schemes where the left-hand side goes to zero – as we want them to do – are generally called *consistent*. Although almost all schemes used in practice are consistent, sometimes minor modifications lead to inconsistent schemes, so checking the consistency of new schemes is important.

Finite volume approximation

Instead of deriving the numerical approximation by replacing the derivatives in (15.11) with the finite difference approximations we start with (15.2). This is the *finite volume* approach, where we work directly with the conservation principle. In Fig. 15.7 we use dashed lines between the node points to divide the x-axis into *control volumes* centered around the node points. Thus, the control volumes are of width Δx and the boundaries (the dashed lines) are located half way between the nodes, at $x_{j\pm 1/2}$. If we approximate the integral in (15.2) by $\Delta x f_j$ and use the same one-sided approximation for the time derivative as we used for the finite difference approach, then the conservation law becomes:

$$\frac{f_j^{n+1}-f_j^n}{\Delta t}=\frac{1}{\Delta x}(F_{in}-F_{out}), \tag{15.26}$$

where we have divided through by Δx. Evaluating the fluxes at the current time step and identifying $F_{out}=F^n_{j+1/2}$ and $F_{in}=F^n_{j-1/2}$, we can rewrite (15.26) as:

$$\frac{f_j^{n+1}-f_j^n}{\Delta t}=\frac{F^n_{j+1/2}-F^n_{j-1/2}}{\Delta x}. \tag{15.27}$$

To compute the fluxes, we approximate f at the $j+1/2$ boundary as the average of the average value of f in the j and the $j+1$ control volume and the first derivative of f as the difference between f_{j+1} and f_j, divided by Δx. Thus:

$$F_{j+1/2}=\frac{U}{2}(f_{j+1}+f_j)-D\left(\frac{f_{j+1}-f_j}{\Delta x}\right). \tag{15.28}$$

The fluxes at the other boundary are obtained by the same equation, replacing j by $j-1$. Substituting the fluxes into (15.27) and rearranging the terms we end up with exactly the same equation as before (15.20). Although the discrete equation is the same, there are subtle differences in how we interpret the variables. In the finite volume approach, in particular, f_j is the average value of f in each control volume instead of the point value at x_j.

Stability

If accuracy is not a great concern and we only want to see the general behavior of the solution, or if we want to quickly evolve the solution to steady state, it would be natural to take large time steps. In Fig. 15.8 we have re-run the code used for Fig. 15.5 but with $\Delta t = 0.2$, and show the solution at times 0.0, 0.6, and 1.4. Obviously, something bad happened and the solution at later times does not look anything like in Fig. 15.3. The bad thing is *instability,* where times steps that are too large make the solution grow instead of decay, eventually leading to a number larger than the computer can handle (NaN or Not a Number).

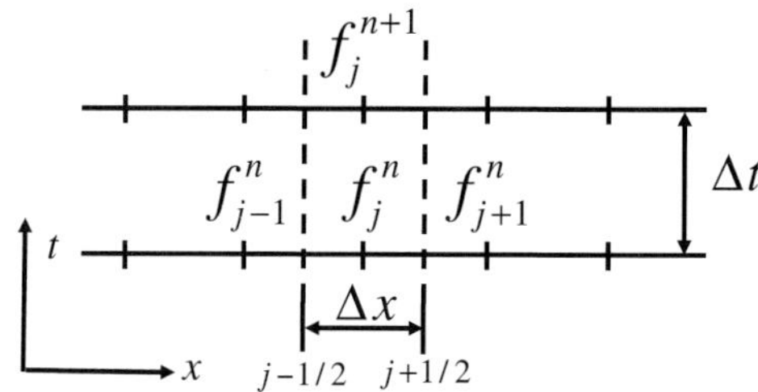

FIGURE 15.7 The grid layout for finite volume discretization.

We can get some insight into why the simulation goes unstable for a large time step by looking at a simple situation where f changes abruptly from 1 to 0. To simplify the situation further, we will assume that $U = 0$ so that our equation reduces to the diffusion equation. We take a finite volume point of view and examine what happens at the boundary where $f = 1$ in cell j and all cells to the left, and $f = 0$ in cell $j + 1$ and all cells to the right. Using the simple one-sided approximation for the time derivative, f_j^{n+1} is equal to the old value f_j^n plus Δt times the difference in the diffusive fluxes in and out of the cell. The amount of f that flows out of cell j and into cell $j + 1$ is $\Delta t F_{j+1/2} = (D/\Delta x)(f_{j+1} - f_j)$. The flux into cell j and the flux out of cell $j + 1$ are both zero. Thus, the only thing that happens for the initial conditions used here is that f flows out of cell j and into cell $j + 1$. The important thing is that the fluxes are found at the beginning of the time step and remain constant. Thus, if we wait long enough, the value of f in cell j goes negative and the value in cell $j + 1$ exceeds unity. The fluxes at the next step, found from the new values of f, will be even larger and the maximum and minimum of f will become even larger, until the computer cannot take it any more (and complains with an NaN). If we take the rather sensible point of view that we should limit our Δt in such a way that f in cells j and $j + 1$ are at most equal, then we stop when $f_j^{n+1} = f_{j+1}^{n+1}$, or:

$$f_j^n + \left(D\Delta t/\Delta x^2\right)\left(f_{j+1}^n + 1 - 2f_j^n + f_{j-1}^n\right) = f_{j+1}^n + \left(D\Delta t/\Delta x^2\right)\left(f_{j+2}^n + 1 - 2f_{j+1}^n + f_j^n\right).$$

Since $f_{j-1}^n = f_j^n = 1$ and $f_{j+1}^n = f_{j+2}^n = 0$, we find that the time step must be limited in such a way that:

$$\frac{D\Delta t}{\Delta x^2} \leq \frac{1}{2}.$$

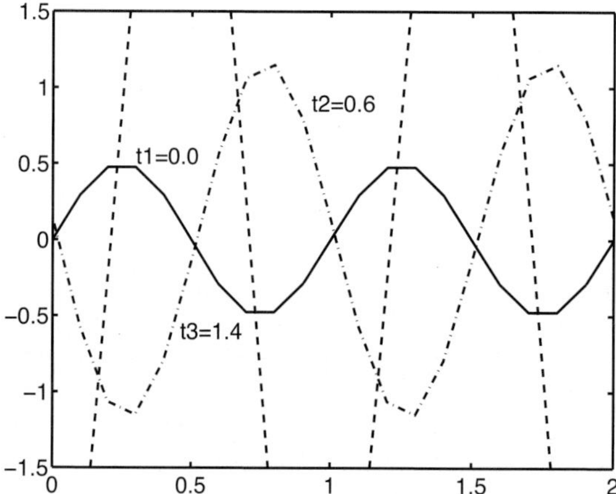

FIGURE 15.8 The numerical solution of the advection-diffusion equation for a large Δt. The amplitude of the solution keeps increasing, instead of decreasing as it does for small Δt, and as predicted by the analytical solution.

A general method to examine the stability of numerical approximations to linear equations was introduced by von Neumann (although he did not publish it) and is generally the second step in the analysis of a numerical scheme (after accuracy and consistency). Since the scheme is linear, any solution can be expressed as the sum of Fourier modes that evolve independently. An error can potentially contain any mode resolved by the grid and if any one mode grows without bounds the scheme is unstable. Thus, we examine the evolution of one mode:

$$\epsilon_j^n = \epsilon_k^n e^{ikx_j} \tag{15.29}$$

where k can be any wave number. Substituting into (15.20), we have:

$$\frac{\epsilon_k^{n+1} e^{ikx_j} - \epsilon_k^n e^{ikx_j}}{\Delta t} = -U\left(\frac{\epsilon_k^n e^{ikx_{j+1}} - \epsilon_k^n e^{ikx_{j-1}}}{2\Delta x}\right) + D\left(\frac{\epsilon_k^n e^{ikx_{j+1}} - 2\epsilon_k^n e^{ikx_j} + \epsilon_k^n e^{ikx_{j-1}}}{\Delta x^2}\right). \tag{15.30}$$

Using that:

$$\epsilon_{j+1}^n = \epsilon_k^n e^{ikx_{j+1}} = \epsilon_k^n e^{ik(x_j+\Delta x)} = \epsilon_k^n e^{ikx_j} e^{ik\Delta x}$$
$$\epsilon_{j-1}^n = \epsilon_k^n e^{ikx_{j-1}} = \epsilon_k^n e^{ik(x_j-\Delta x)} = \epsilon_k^n e^{ikxj} e^{-ik\Delta x}$$

and canceling a common factor of e^{ikx_j} results in:

$$\frac{\epsilon^{n+1} - \epsilon^n}{\Delta t} + U\frac{\epsilon^n}{2\Delta x}\left(e^{ik\Delta x} - e^{-ik\Delta x}\right) = D\frac{\epsilon^n}{\Delta x^2}\left(e^{ik\Delta x} + e^{-ik\Delta x} - 2\right), \tag{15.31}$$

where we have dropped the subscript k on the amplitude with the understanding that it applies to any wave number. Rearranging we have:

$$\begin{aligned}\frac{\epsilon^{n+1}}{n} &= 1 - \frac{U\Delta t}{2\Delta x}(e^{ik\Delta x} - e^{-ik\Delta x}) + \frac{D\Delta t}{\Delta x^2}(e^{ik\Delta x} + e^{-ik\Delta x} - 2)\\ &= 1 - \frac{U\Delta t}{2\Delta x}2i\sin k\Delta x + \frac{D\Delta t}{\Delta x^2}2(\cos k\Delta x - 1)\\ &= 1 - 4\frac{D\Delta t}{\Delta x^2}\sin^2\frac{k\Delta x}{2} - i\frac{U\Delta t}{2\Delta x}\sin k\Delta x.\end{aligned} \tag{15.32}$$

Stability requires that the absolute value of the error decreases in each time step, or:

$$\left|\frac{\epsilon^{n+1}}{\epsilon^n}\right| < 1. \tag{15.33}$$

Since the amplification factor is a complex number, and k, the wave number of the error can take any value, the determination of the stability limit is slightly involved. We will therefore only look at two special cases.

First, assume that $U = 0$, so the problem reduces to a pure diffusion. Since $\sin^2() \le 1$, the amplification factor is always less than 1, and we find that it is bigger than -1 if:

$$\frac{\Delta t D}{\Delta x^2} \le \frac{1}{2}. \tag{15.34}$$

Therefore, for a given diffusion coefficient D and spatial resolution Δx the method is stable if the time step Δt is small enough. Notice that Δt must decrease if Δx is decreased, and that the time step limitation is most severe for high diffusion coefficients. We also note that this criterion is the same as we found by our earlier heuristic argument.

Consider then the other limit where $D = 0$ and we have pure advection. Since the amplification factor has the form $1 + i()$, the absolute value of this complex number is always larger than unity and the method is unconditionally unstable for pure advection.

For the general case we must investigate (15.32) in more detail. We will not do so here, but simply quote the results: We must have:

$$\frac{\Delta t D}{\Delta x^2} \le \frac{1}{2}; \quad \frac{U^2\Delta t}{D} \le 2 \tag{15.35}$$

for the numerical approximation (15.20) to remain stable. Notice that high velocity and low viscosity lead to instability according to the second restriction.

Other schemes

The numerical scheme presented here, (15.20), is of course, not the only possible numerical scheme for the advection-diffusion equation. The first question is obviously why did we select a first-order scheme for the temporal derivative? Why not use the centered second-order scheme, resulting in:

$$\frac{f_j^{n+1}-f_j^{n-1}}{2\Delta t}+U\left(\frac{f_{j+1}^n-f_{j-1}^n}{2\Delta x}\right)=D\left(\frac{f_{j+1}^n-2f_j^n+f_{j-1}^n}{\Delta x^2}\right)? \tag{15.36}$$

Or, we could conversely ask, if the first-order scheme was good enough for the temporal derivative, why don't we use that one for the first-order spatial derivative? If we pick the "upwind" direction (the point to the left of j if $U>0$), such a scheme is given by:

$$\frac{f_j^{n+1}-f_j^{n}}{\Delta t}+U\left(\frac{f_{j}^n-f_{j-1}^n}{\Delta x}\right)=D\left(\frac{f_{j+1}^n-2f_j^n+f_{j-1}^n}{\Delta x^2}\right). \tag{15.37}$$

The first of those schemes, usually called the Leap-Frog scheme, works for the pure advection problem but is, unfortunately, unstable for the diffusion-only case. The second scheme, where we used one-sided approximation for the advection term, and picked the points from the direction that the fluid comes from, is called the upwind scheme for obvious reasons. The upwind scheme is sometimes used, since it is very robust, but unfortunately it is not very accurate.

The schemes that we have examined so far are all *explicit*. That is, the value of f_j at the new time level is given as a combination of the neighboring values at the old time. The problem with explicit schemes is that they are (at best) conditionally stable. Even when we do not care about the accuracy of the time integration—such as when we are only interested in the steady state—we are limited to small time steps by stability. This can be overcome by *implicit* schemes, where the spatial derivatives are found at the new time level:

$$\frac{f_j^{n+1}-f_j^{n}}{\Delta t}+U\left(\frac{f_{j+1}^{n+1}-f_{j-1}^{n+1}}{2\Delta x}\right)=D\left(\frac{f_{j+1}^{n+1}-2f_j^{n+1}+f_{j-1}^{n+1}}{\Delta x^2}\right). \tag{15.38}$$

The drawback is that now we need to solve a system of linear equations at every time step and in two- and three dimensions that is computationally expensive. Indeed, we may find that solving the linear system sometimes take just as much time as it would take to do many small time steps. However if, for example, diffusion is very small so that it limits the time step greatly, but we expect that the accuracy of the diffusion calculation is not very important, then an implicit scheme may be beneficial.

The implicit scheme presented above is only first order, but by taking the average of the first-order forward and the first-order backward scheme we get a second-order one. Computationally this scheme is very similar to the backward first-order scheme and therefore it is generally preferred. It is also the basis for an explicit second-order scheme based on first taking a first-order explicit time step and then repeating the step using the average of the spatial derivatives at the old time step and the first-order predicted values. It can be shown that the following two-step scheme is second order in both time and space:

$$\begin{aligned}
f_j^{*}&=f_j^n+\Delta t\left(-U\left(\frac{f_{j+1}^n-f_{j-1}^n}{2\Delta x}\right)+D\left(\frac{f_{j+1}^n-2f_j^n+f_{j-1}^n}{\Delta x^2}\right)\right)\\
f_j^{n+1}&=f_j^n+\frac{\Delta t}{2}\left(-U\left(\frac{f_{j+1}^n-f_{j-1}^n}{2\Delta x}\right)+D\left(\frac{f_{j+1}^n-2f_j^n+f_{j-1}^n}{\Delta x^2}\right)\right.\\
&\quad\left.-U\left(\frac{f_{j+1}^*-f_{j-1}^*}{2\Delta x}\right)+D\left(\frac{f_{j+1}^*-2f_j^*+f_{j-1}^*}{\Delta x^2}\right)\right).
\end{aligned} \tag{15.39}$$

The scheme can be rewritten in such a way that a first-order explicit method can be extended to second order in just a few lines of code (see Exercises). Given how easily a first-order method in time is made second-order, we will focus mostly on first-order in time methods in the rest of the chapter.

Steady-state and boundary-value problems

In many cases we are concerned with the solution of equations that do not have a time derivative and thus represent either a steady-state solution to a time-dependent problem or an equilibrium solution. For the advection-diffusion equation the steady state is given by:

$$\frac{\partial^2 f}{\partial x^2} - \frac{U}{D}\frac{\partial f}{\partial x} = 0. \tag{15.40}$$

The solution, which can be found analytically, clearly depends on U/D or the relative importance of advection versus diffusion.

The discrete version of the partial differential equation, using centered differences, results in:

$$\frac{U}{D}\left(\frac{f_{j+1} - f_{j-1}}{2\Delta x}\right) = \left(\frac{f_{j+1} - 2f_j + f_{j-1}}{\Delta x^2}\right). \tag{15.41}$$

Rearranging this slightly, we have:

$$(R_\Delta - 2)f_{j+1} + 4f_j - (R_\Delta + 2)f_{j-1} = 0, \tag{15.42}$$

where $R_\Delta = U\Delta x/D$. Thus, the discretization results in a system of algebraic equations, one equation for each grid point. The equations can be solved in many different ways, including using direct elimination. While direct elimination, or other direct methods, can be used for relatively small systems of linear equations, for large systems we usually use iterative methods. We will discuss elementary iterative methods shortly, but in our case it turns out that Eq. (15.42) has an analytical solution given by:

$$f_j = \frac{\left(\frac{2+R_\Delta}{2-R_\Delta}\right)^j - \left(\frac{2+R_\Delta}{2-R_\Delta}\right)}{\left(\frac{2+R_\Delta}{2-R_\Delta}\right)^N - \left(\frac{2+R_\Delta}{2-R_\Delta}\right)}, \tag{15.43}$$

as can be verified by direct substitution. In Fig. 15.9(a) we plot f_j for a domain of size 1, where $f(0) = f_1 = 0$ and $f(1) = f_N = 1$, found using $N = 11$ grid points, for a few values of U/D. When U/D is very small, diffusion dominates over advection and the profile is almost linear, as one would expect since a linear profile is the solution to the pure diffusion problem. As U/D increases, advection becomes more important and "pushes" the profile to the right, eventually resulting in a thin boundary layer near the right boundary. For the highest U/D the solution does, however, dip below zero near the right boundary. This is an artifact of the low resolution as can be seen in Fig. 15.9(b), where the solution is plotted for three different resolutions, or R_Δ. The finest resolution is essentially identical to the exact analytical solution. The unphysical behavior of the solution in the boundary layer due to lack of resolution is actually clear in the analytical solution. Once $R_\Delta > 2$, the denominator in the term raised to the jth power is negative and rising it to odd powers can result in negative values of f_j. The negative values can be avoided by using an "upwind" scheme for the advection term so that the discrete equation is:

$$\left(\frac{f_j - f_{j-1}}{\Delta x}\right) = \frac{D}{U}\left(\frac{f_{j+1} - 2f_j + f_{j-1}}{\Delta x^2}\right), \tag{15.44}$$

which can also be solved analytically, in the same way as (15.42):

$$f_j = \frac{(1+R_\Delta)^j - (1+R_\Delta)}{(1+R_\Delta)^N - (1+R_\Delta)}. \tag{15.45}$$

Here it is clear that $(1 + R_\Delta)^j$ is always positive. The solution is plotted in Fig. 15.9(c) for $U/D = 30$ and $N = 11$ (or $R_\Delta = 3$), along with a solution using the centered difference approximation and the exact solution. Although the solution no longer shows oscillatory behavior near the right-hand boundary, there is clearly a significant difference between it and the exact solution. Thus, one approach leads to oscillation and the other to large errors and in both cases a finer grid is needed.

The quantity R_Δ is usually called the cell Reynolds number if D is assumed to represent the kinematic viscosity, and the results suggest that when advection is balanced by a little bit of diffusion we should expect oscillations when we use centered differences for the advection terms and $R_\Delta > 2$. For unsteady flows, we only expect this behavior in thin boundary layers or internal strained diffusion layers, and while it is sometimes possible to tolerate some localized oscillations we can

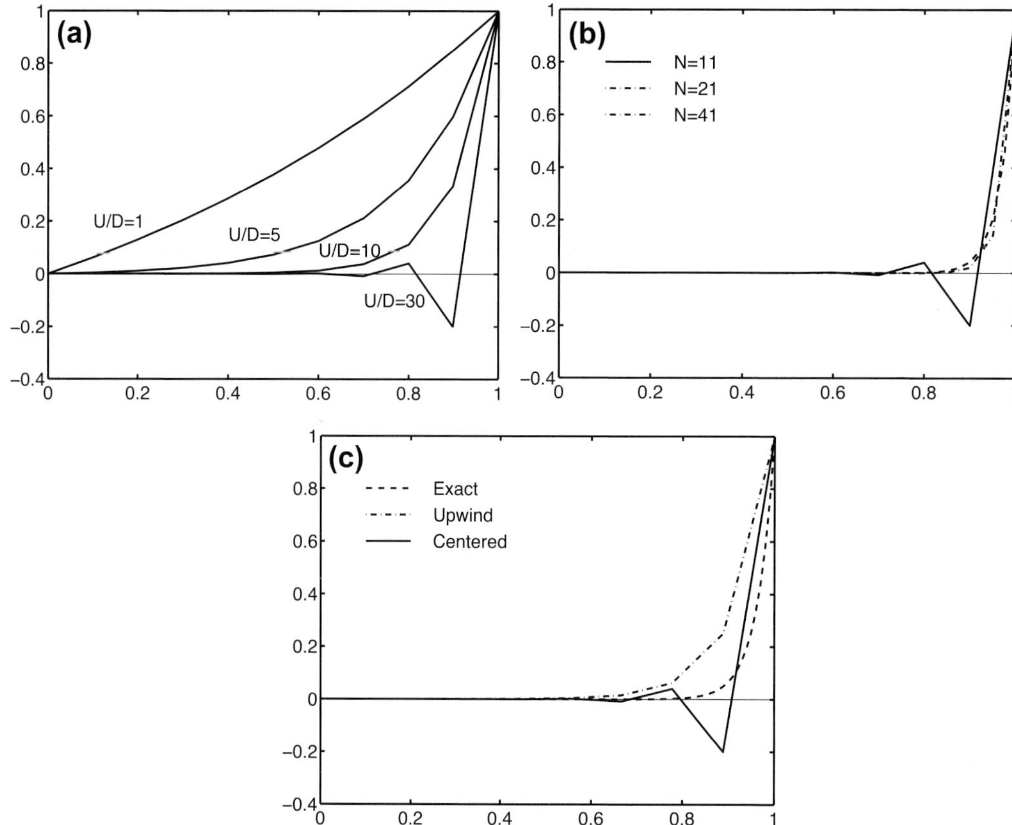

FIGURE 15.9 Numerical solutions to the steady-state advection-diffusion equation. (a) the solution using $N = 11$ for four different values of U/D, corresponding to $R_\Delta = 0.1, 0.5, 1$, and 3. (b) The solution for $U/D = 30$ and three different resolutions, corresponding to $R_\Delta = 3$, 1.5, and 0.75. (c) The solution for $U/D = 30$ and $R_\Delta = 3$, using a centered difference scheme and an upstream difference scheme compared with the analytical solution.

eliminate it by using a lower-order scheme, but at the cost of decreased accuracy. The best solution is, of course, to reduce Δx, if that is possible.

Two-dimensional advection-diffusion equation

For two-dimensional flow the advection-diffusion equation is given by:

$$\frac{\partial f}{\partial t} + \nabla \cdot \mathbf{F} = 0, \tag{15.46}$$

where the flux is now a vector $\mathbf{F} = (F_x, F_y)$. Substituting the vector and writing out the divergence we have:

$$\frac{\partial f}{\partial t} + \frac{\partial F_x}{\partial x} + \frac{\partial F_y}{\partial y} = 0. \tag{15.47}$$

Taking the fluxes to be a sum of advective and diffusive fluxes $\mathbf{F} = \mathbf{u} f - D\nabla f$, where $\mathbf{u} = (u, v)$, we have:

$$\frac{\partial f}{\partial t} + \mathbf{u} \cdot \nabla f = D\nabla^2 f. \tag{15.48}$$

Or, writing the velocity in terms of its components:

$$\frac{\partial f}{\partial t} + u\frac{\partial f}{\partial x} + v\frac{\partial f}{\partial y} = D\left(\frac{\partial^2 f}{\partial x^2} + \frac{\partial^2 f}{\partial y^2}\right). \tag{15.49}$$

The integral form can be obtained by integrating (15.46) over a volume V and using the divergence theorem to convert the volume integral into a surface integral:

$$\frac{d}{dt}\int_V f dv + \oint_S \mathbf{F}\cdot\mathbf{n}ds = 0. \tag{15.50}$$

Here, $\mathbf{n}$ is normal to the control surface S. Substituting for the fluxes gives:

$$\frac{d}{dt}\int_V f dv + \oint_S f\mathbf{u}\cdot\mathbf{n}ds = \oint_S D\nabla f\cdot\mathbf{n}ds. \tag{15.51}$$

We will solve (15.49) numerically on a discrete grid, where the grid points are labeled (i, j), with $i = 1...N_x$ changing in the x-direction and $j = 1...N_y$ changing in the y-direction. The grid points are separated by Δx in the x-direction and by Δy in the y-direction. The coordinates of point (i, j) are (x_i, y_j), so the coordinates of point $(i+1, j)$ are $(x_i + \Delta x, y_j)$, the coordinates of $(i, j+1)$ are $(x_i, y_j + \Delta y)$, and so on. The notation is shown in Fig. 15.10. The current time t is denoted by superscript n and the next one, after one time step at time $t + \Delta t$, is denoted by subscript $n+1$, as before.

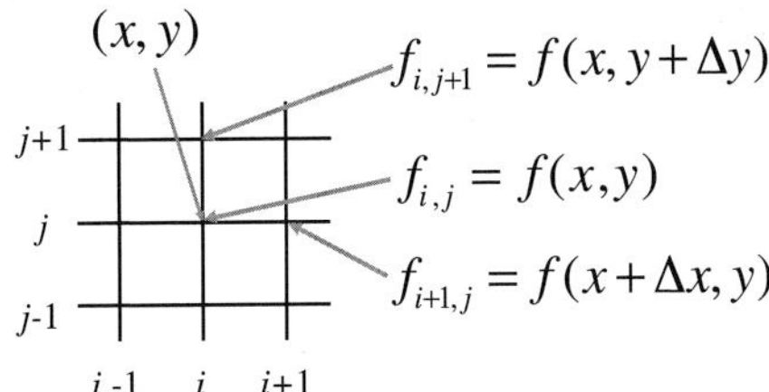

FIGURE 15.10 The notation used for the two-dimensional grid.

To derive a discrete approximation to (15.49) we approximate each term by the finite difference approximations given by (15.15), (15.17), and (15.18), replacing x by y where needed. The result is:

$$\begin{aligned}\frac{f_{i,j}^{n+1} - f_{i,j}^n}{\Delta t} &= -u\left(\frac{f_{i+1,j}^n - f_{i-1,j}^n}{2\Delta x}\right) - v\left(\frac{f_{i,j+1}^n - f_{i,j-1}^n}{2\Delta y}\right) \\ &+ D\left(\frac{f_{i+1,j}^n - 2f_i^n + f_{i-1,j}^n}{\Delta x^2} + \frac{f_{i,j+1}^n - 2f_{i,j}^n + f_{i,j-1}^n}{\Delta y^2}\right).\end{aligned} \tag{15.52}$$

Starting with the integral form and assuming that $f_{i,j}$ is the average value over a control volume whose area is $\Delta x\ \Delta y$, as well as a good approximation for the value at the center, results in exactly the same discrete equation.

The Matlab® code that follows is written for a rectangular domain of size $L_x \times L_y$, resolved by $N_x \times N_y$ grid points. At the left boundary we take f to be given, with $f = 0$, except in the middle where $f = 1$. At other boundaries we approximate a zero gradient by putting the value of f on the boundary equal to the value of the next interior point. The velocities are taken to be $u = 0$ and $v = -1$, and in the interior $f = 0$ everywhere at time zero. For two-dimensional problems the limitation on the time step size is more restrictive than in the one-dimensional cases and the first condition in (15.35) must be replaced by $\Delta t D/\Delta x^2 \le \frac{1}{4}$.

CODE 2: SOLUTION OF THE UNSTEADY TWO-DIMENSIONAL LINEAR ADVECTION EQUATION USING AN EXPLICIT FORWARD IN TIME, CENTERED IN SPACE SCHEME

```
% Two-dimensional unsteady diffusion by the FTCS scheme
Nx=32;Ny=32;nstep=70;D=0.025;Lx=2.0;Ly=2.0;
dx=Lx/(Nx-1); dy=Ly/(Ny-1); dt=0.02;
f=zeros(Nx,Ny);fo=zeros(Nx,Ny);time=0.0;
for i=1:Nx,for j=1:Ny,x(i,j)=dx*(i-1);y(i,j)=dy*(j-1);end,end;
u=-0.; v=-1.0; f(12:21,Nx)=1.0;
for l=1:nstep,l,time
```

```
hold off;mesh(x,y,f); axis([1 Nx 1 Ny 0 1.5]);pause(0.01);
    fo=f;
    for i=2:Nx-1, for j=2:Ny-1
        f(i,j)=fo(i,j)-(0.5*dt*u/dx)*(fo(i+1,j)-fo(i-1,j))-..
                       (0.5*dt*v/dy)*(fo(i,j+1)-fo(i,j-1))+...
              D*dt*( ((fo(i+1,j)-2*fo(i,j)+fo(i-1,j))/dx^2)...
                     +((fo(i,j+1)-2*fo(i,j)+fo(i,j-1))/dy^2) );
    end,end
    for i=1:Nx, f(i,1)=f(i,2);end;
    for j=1:Ny,f(1,j)=f(2,j);f(Nx,j)=f(Nx-1,j);end;
    time=time+dt;
end;
```

Fig. 15.11 shows the solution at time 1.4 and $D = 0.025$ as computed on a 32×32 grid with $\Delta t = 0.02$. If there was no diffusion, we would see a rectangular slug, where $f = 1$, move into the domain and across it, but because we include diffusion f spreads out as it is advected across the domain.

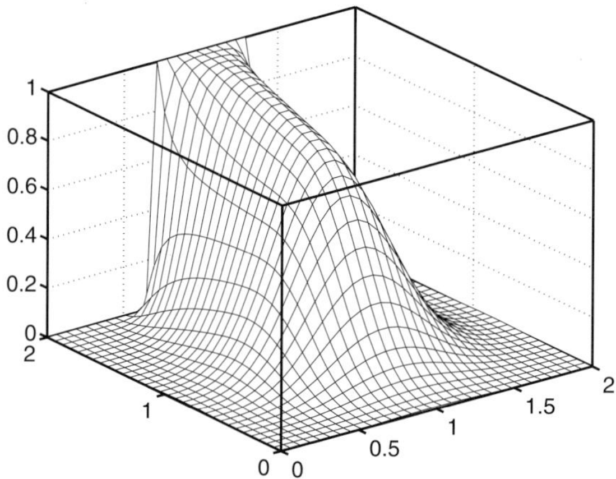

FIGURE 15.11 The solution of the unsteady two-dimensional advection-diffusion equation at time 1.4. Initially, $f = 0$ in the whole domain and then a strip with $f = 1$ is carried with the flow into the domain.

Multidimensional steady-state and boundary-value problems

For steady-state multidimensional problems we generally end up with a system of equations that must be solved simultaneously, just as we did when we examined the one-dimensional (in space) advection-diffusion equation. For multidimensional problems it is generally not possible to solve the system analytically and we must use either direct elimination or an iterative method to find the solution. Solving a linear set of equations is a vast subject where enormous progress has been made during the last several decades. Many excellent algorithms are available, both as part of computational environments such as Matlab or Octave or as separate packages. Any serious student of computational fluid dynamics, particularly those dealing with incompressible flows, should have a thorough understanding of how advanced methods work. Here, however, we will only examine elementary iterative schemes. These schemes allow us to obtain the solution to relatively small systems of equations in just a few lines of code.

To simplify the treatment as much as possible, we will focus only on the highest derivatives and lump the lower-order terms into a source $s(x, y)$, which could also include other effects such as the creation or destruction of f. We then have a Poisson equation, which is found in a wide range of situations in many branches of physics. We note that when the source terms consist of the advection terms of the advection-diffusion equation then the solution exhibits the same behavior for low diffusion and coarse resolution as we found for one-dimensional flow, and is therefore more difficult to solve than the pure diffusion equation.

The two-dimensional Poisson equation is:

$$\frac{\partial^2 f}{\partial x^2} + \frac{\partial^2 f}{\partial y^2} = s(x, y). \tag{15.53}$$

Using centered approximations for all derivatives, as before, we have:

$$\frac{f_{i+1,j} - 2f_{i,j} + f_{i-1,j}}{\Delta x^2} + \frac{f_{i,j+1} - 2f_{i,j} + f_{i,j-1}}{\Delta y^2} = s(i, j). \tag{15.54}$$

If $\Delta x = \Delta y = h$, as is often the case, the discrete equation is even simpler, or:

$$\frac{f_{i+1,j} + f_{i-1,j} + f_{i,j+1} + f_{i,j-1} - 4f_{i,j}}{h^2} = s_{i,j}. \tag{15.55}$$

To solve this equation iteratively, we first isolate $f_{i,j}$ on the left-hand side:

$$f_{i,j} = \frac{1}{4}\left(f_{i+1,j} + f_{i-1,j} + f_{i,j+1} + f_{i,j-1} - h^2 s_{i,j}\right). \tag{15.56}$$

We can use (15.56) as it stands to find f iteratively. When we use a Jacobi iteration we compute a new value of $f_{i,j}$ by using the values of f from the last iteration on the right-hand side. A slightly better approach is to use the most recent value of f on the right-hand side, leading to the so-called Gauss-Seidel iteration, which converges about twice as fast as the Jacobi iteration. An even better approach, and just as simple, is to *accelerate* the iteration by extrapolating the new value from the new one and the old one. This is called successive over relaxation (SOR) iterations and can be written as:

$$f_{i,j}^{\alpha+1} = (1-\beta) f_{i,j}^{\alpha} + \frac{\beta}{4}\left(f_{i+1,j}^{\alpha} + f_{i-1,j}^{\alpha+1} + f_{i,j+1}^{\alpha} + f_{i,j-1}^{\alpha+1} - h^2 s_{i,j}^{n}\right). \tag{15.57}$$

The parameter β is an *acceleration* parameter that must satisfy $1 \geq \beta \geq 2$. Usually $\beta = 1.5$ is a good starting choice. Although the notation used here, with subscripts α and $\alpha + 1$ on the right-hand side, suggests some complexity, the computational implementation of (15.57) is very simple. We have one array for $f_{i,j}$ that we update, and when computing the right-hand side some of the values have already been updated, while others have not.

15.3 Incompressible flows in rectangular domains

To show how the concepts described above work for the full fluid equations, here we describe codes to solve the so-called driven cavity problem. The driven cavity problem, sketched in Fig. 15.12, is a very popular test problem for methods to solve the Navier-Stokes equations for incompressible flows. It consists of a square or rectangular domain (a cavity) surrounded by stationary walls on three sides and a moving wall on the top. The motion of the top wall drives a circular motion in the cavity (dashed circle). This domain is particularly well suited for methods that use regular square grids and the absence of in- and out-flow boundaries simplifies the setup. The top wall is usually given a constant velocity, resulting in a velocity discontinuity at the top corners. This leads to a singularity where the gradient of the velocity depends on the fineness of the grid but does not, however, prevent the convergence of the solution in the rest of the domain. Although this problem is often used to test methods for the steady-state solution, it can also be used for unsteady flow, in which case the initial fluid velocity is usually taken to be zero.

We describe two methods to solve the incompressible fluid equations. The first one, where we use the vorticity-stream function form of the equations, is a straightforward application of what has already been discussed, but the second one, where we work with the velocity pressure formulation, requires additional considerations, such as for the time integration and the structure of the grid. The second approach, although more complex, is much more general and is, in particular, easily extended to fully three-dimensional and more complex flows, such as those with variable material properties.

Vorticity-stream function methods

The simplest way to solve the Navier-Stokes equations in two dimensions is to work with the vorticity-stream function form of the equations. Those are derived in Chapter 5 in vorticity-velocity form (5.13), but we give them here for two-dimensional flow, where the system of equations consists of one advection-diffusion equation for the vorticity:

$$\frac{\partial \omega}{\partial t} = -\frac{\partial \psi}{\partial y}\frac{\partial \omega}{\partial x} + \frac{\partial \psi}{\partial x}\frac{\partial \omega}{\partial y} + \nu\left(\frac{\partial^2 \omega}{\partial x^2} + \frac{\partial^2 \omega}{\partial y^2}\right), \tag{15.58}$$

and a Poisson equation that relates the stream function to the vorticity:

$$\frac{\partial^2 \psi}{\partial x^2} + \frac{\partial^2 \psi}{\partial y^2} = -\omega. \tag{15.59}$$

Once the stream function is known, the velocities are found by:

$$u = \frac{\partial \psi}{\partial y}; \quad v = -\frac{\partial \psi}{\partial x}. \tag{15.60}$$

Eqs. (15.58) and (15.59) form a consistent and self-contained system of equations that allow us to solve for the flow, given the appropriate boundary and initial conditions.

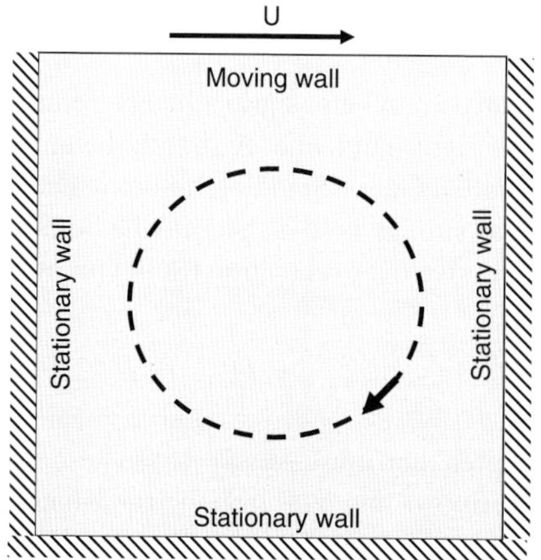

FIGURE 15.12 The driven cavity problem is often used to test methods to solve the Navier-Stokes equations. The cavity is a rectangular domain bounded by three stationary walls and a moving wall at the top, which drives a circulatory motion.

The boundary conditions are that the velocities of the fluid at a wall are equal to the wall velocity, or $u = v = 0$ on the stationary walls, and $v = 0$ and $u = U$ on the top wall. From the definition of the stream function (15.60), it is constant if the normal velocity is zero. For inflow boundary conditions, where the normal velocity is given, the rate of change of the stream function along the wall is given by (15.60).

Numerical approximations

We will solve the vorticity-stream function equations numerically on a discrete grid, where the grid points are labeled (i, j), with $i = 1...N_x$ changing in the x-direction and $j = 1...N_y$ in the y-direction. To simplify the situation we take the spacing between the grid points to be the same in both directions so $\Delta x = \Delta y = h$. The first and last grid points are on the opposite boundaries. The notation is the same as used earlier and shown in Fig. 15.10.

We start by writing (15.58) and (15.59) at grid point (i, j), at time n:

$$\left.\frac{\partial \omega}{\partial t}\right)_{i,j}^n = -\left.\frac{\partial \psi}{\partial y}\frac{\partial \omega}{\partial x}\right)_{i,j}^n + \left.\frac{\partial \psi}{\partial x}\frac{\partial \omega}{\partial y}\right)_{i,j}^n + \nu\left(\frac{\partial^2 \omega}{\partial x^2} + \frac{\partial^2 \omega}{\partial y^2}\right)_{i,j}^n, \tag{15.61}$$

$$\left(\frac{\partial^2 \psi}{\partial x^2} + \frac{\partial^2 \psi}{\partial y^2}\right)_{i,j}^n = -\omega_{i,j}^n, \tag{15.62}$$

using the same numerical approximation as we used for the two-dimensional advection-diffusion equation, approximating the time derivative by one-sided forward differences (15.15), and the spatial derivatives by centered difference approximations ((15.17) for the first derivatives and (15.18) for the second derivatives). When the grid spacing in x and y is the same, as we assume here, the sum of the second derivatives (the Laplacian) can be simplified slightly:

$$\frac{\partial^2 f}{\partial x^2} + \frac{\partial^2 f}{\partial y^2} \approx \frac{f_{i+1,j} - 2f_{i,j} + f_{i-1,j}}{h^2} + \frac{f_{i,j+1} - 2f_{i,j} + f_{i,j-1}}{h^2} = \frac{f_{i+1,j} + f_{i-1,j} + f_{i,j+1} + f_{i,j-1} - 4f_{i,j}}{h^2}.$$

Using those approximations the vorticity advection equation (15.61) becomes:

$$\frac{\omega_{i,j}^{n+1}-\omega_{i,j}^{n}}{\Delta t}=-\left(\frac{\psi_{i,j+1}^{n}-\psi_{i,j-1}^{n}}{2h}\right)\left(\frac{\omega_{i+1,j}^{n}-\omega_{i-1,j}^{n}}{2h}\right)+\left(\frac{\psi_{i+1,j}^{n}-\psi_{i-1,j}^{n}}{2h}\right)\left(\frac{\omega_{i,j+1}^{n}-\omega_{i,j-1}^{n}}{2h}\right)$$
$$+\nu\left(\frac{\omega_{i+1,j}^{n}+\omega_{i-1,j}^{n}+\omega_{i,j+1}^{n}+\omega_{i,j-1}^{n}-4\omega_{i,j}^{n}}{h^2}\right). \tag{15.63}$$

Thus, given the vorticity and stream function everywhere at time n, we can compute the vorticity at all *interior* points directly, by (15.63).

For (15.62), we have:

$$\frac{\psi_{i+1,j}^{n}+\psi_{i-1,j}^{n}+\psi_{i,j+1}^{n}+\psi_{i,j-1}^{n}-4\psi_{i,j}^{n}}{h^2}=-\omega_{i,j}^{n}, \tag{15.64}$$

and since ψ on the boundary is known, (15.64) allows us to find $\psi_{i,j}$ for all interior points. This means, in particular, that we do not need to know ω on the boundary to solve (15.64). Eq. (15.64) can be solved in the same way as discussed for multidimensional boundary-value problems, and we will use the SOR method introduced earlier (15.57) when we write our numerical code.

Boundary conditions

While the boundary conditions for the stream function are generally given at the outset and do not change as the computations proceed, the vorticity on the boundary must be determined at every time step. Consider a flat, rigid wall that coincides with the x-axis, and has fluid on the $y > 0$ side. The stream function, one mesh block away, can be written using a Taylor series expansion around the boundary point:

$$\psi_{i,2}=\psi_{i,1}+\frac{\partial\psi_{i,1}}{\partial y}h+\frac{\partial^2\psi_{i,1}}{\partial y^2}\frac{h^2}{2}+O(h^3). \tag{15.65}$$

Using that $\partial\psi_{i,1}/\partial y = U_{wall}$ and that $\omega_{wall} = -\partial^2\psi_{i,1}/\partial y^2$ this becomes:

$$\psi_{i,2}=\psi_{i,1}+U_{wall}h-\omega_{wall}\frac{h^2}{2}+O(h^3), \tag{15.66}$$

which can be solved for the wall vorticity:

$$\omega_{wall}=(\psi_{i,1}-\psi_{i,2})\frac{2}{h^2}+U_{wall}\cdot\frac{2}{h}+O(h). \tag{15.67}$$

Thus, given the vorticity in the interior and the stream function everywhere at time level n, we can find the vorticity at the wall and thus $\omega_{i,j}^{n+1}$. Although (15.67) is only first order accurate, it is fairly robust and usually does not lead to significant loss of overall accuracy when used with a second order interior scheme.

Simulations using ψ and ω

We now have all the elements to construct a numerical code. However, since some quantities depend on other quantities that are found as a part of the solution, it is important to solve our equations in the correct order. We generally start with a given vorticity distribution, find the stream function, determine the boundary vorticity, and then update the vorticity for the next time level. Notice that we do not need the vorticity on the boundary to solve the vorticity-stream function equation. Notice also that in two dimensions any vorticity distribution works as initial conditions. More formally, our algorithm consists of the following steps:

1. Given $\omega_{i,j}$ at all interior points, solve (15.64) for $\psi_{i,j}$.
2. Find the boundary vorticity, w_{wall}, by (15.67).
3. Calculate the vorticity at the new time, $w_{i,j}^{n+1}$ using (15.63) for all the interior points.
4. Set $t = t + \Delta t$ and go back to the first step.

In the full code, listed below, we have put a few statements on the same line, as allowed in Matlab, but only those statements that clearly belong together. The parallelization feature of Matlab, where the range of an index can be put in the argument list of an array is also used, but only when the readability of the program is not affected.

CODE 3: SOLUTION OF THE UNSTEADY TWO-DIMENSIONAL NAVIER-STOKES EQUATIONS IN VORTICITY-STREAM FUNCTION FORM, USING AN EXPLICIT FORWARD IN TIME, CENTERED IN SPACE SCHEME

```
% Driven Cavity by Vorticity-Stream Function Method
Nx=17; Ny=17; MaxStep=200; Visc=0.1; dt=0.005; time=0.0;
MaxIt=100; Beta=1.5; MaxErr=0.001; % parameters for SOR
sf=zeros(Nx,Ny); vt=zeros(Nx,Ny); vto=zeros(Nx,Ny);
x=zeros(Nx,Ny); y=zeros(Nx,Ny); h=1.0/(Nx-1);
for i=1:Nx,for j=1:Ny,x(i,j)=h*(i-1);y(i,j)=h*(j-1);end,end;

for istep=1:MaxStep,                        % Time loop
  for iter=1:MaxIt,                         % solve for the stream function
    vto=sf;                                 % by SOR iteration
    for i=2:Nx-1; for j=2:Ny-1
      sf(i,j)=0.25*Beta*(sf(i+1,j)+sf(i-1,j)...
        +sf(i,j+1)+sf(i,j-1)+h*h*vt(i,j))+(1.0-Beta)*sf(i,j);
    end; end;
    Err=0.0; for i=1:Nx; for j=1:Ny,         % check error
        Err=Err+abs(vto(i,j)-sf(i,j))       ; end; end;
    if Err ≤ MaxErr, break, end             % stop if converged
  end;
  vt(2:Nx-1,1)=-2.0*sf(2:Nx-1,2)/(h*h);      % vorticity on bdrys
  vt(2:Nx-1,Ny)=-2.0*sf(2:Nx-1,Ny-1)/(h*h)-2.0/h; % top wall
  vt(1,2:Ny-1)=-2.0*sf(2,2:Ny-1)/(h*h);            % right wall
  vt(Nx,2:Ny-1)=-2.0*sf(Nx-1,2:Ny-1)/(h*h);        % left wall
  vto=vt;
  for i=2:Nx-1; for j=2:Ny-1
    vt(i,j)=vt(i,j)+dt*(- 0.25*((sf(i,j+1)-sf(i,j-1))*..
                (vto(i+1,j)-vto(i-1,j))-(sf(i+1,j)-sf(i-1,j))*..
                (vto(i,j+1)-vto(i,j-1)))/(h*h)...
          +Visc*(vto(i+1,j)+vto(i-1,j)+vto(i,j+1)+...
                 vto(i,j-1)-4.0*vto(i,j))/(h^2) );
  end; end;
  time=time+dt
  subplot(121), contour(x,y,vt,40), axis('square');
  subplot(122), contour(x,y,sf), axis('square');
  pause(0.01)
  end;
```

For the problem that we solve here we take the size of the domain to be 1 by 1 in the arbitrary computational units used here, the velocity at the top is 1.0, and the viscosity is 0.1. This is a relatively high viscosity so we expect the solution to reach a steady state quickly. The time step needs to be selected so that the method is stable and it must satisfy the conditions used for the unsteady advection diffusion equation in two dimensions. While we could have the code compute the permissible time step size automatically, we have not done so here and Δt must be determined before the code is run. The code can be run interactively to see how the flow evolves, but in Fig. 15.13 we only show the solution at time 1.0, after the flow has almost reached a steady state.

To examine if the grid resolution is fine enough to give a converged solution, we have run the code using a grid that has half the resolution and another one that has twice the resolution. For the finest resolution we used $\Delta t = 0.00125$ and 800 time steps to reach $t = 1.0$. One measure of the flow rate in the cavity is to compute the total volume of fluid moving to the right (or left, or up, or down, they must all be the same), given by $Q = \psi_{top} - \psi_{min}$, where ψ_{min} is the minimum value of the stream function. In Fig. 15.14, we plot Q versus time for all three runs and it is clear that the two finer grids give very similar results. This figure also shows that at time equal to 1.0 the solution has nearly reached a steady state.

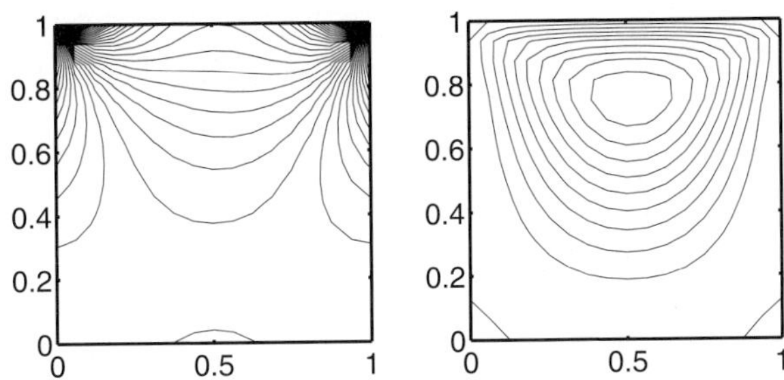

FIGURE 15.13 The vorticity (left frame) and stream function (right frame) after the flow has almost reached a steady state, as computed on a 17×17 grid, using $\Delta t = 0.005$.

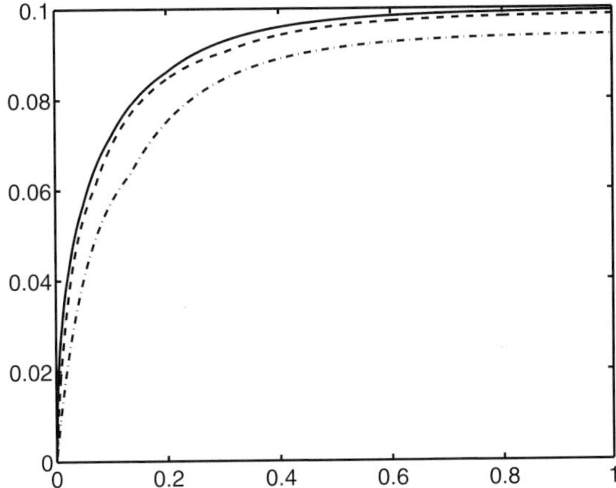

FIGURE 15.14 The total volume flow to the right in the driven cavity versus time, computed as the difference between the value of the stream function on walls and its minimum value, for three resolutions.

Often it is the velocity field that is of main interest and to find it we differentiate the stream function, using centered differences:

$$u_{i,j} = \frac{\psi_{i,j+1} - \psi_{i,j-1}}{2h} \tag{15.68}$$

$$v_{i,j} = -\frac{\psi_{i+1,j} - \psi_{i-1,j}}{2h}. \tag{15.69}$$

Those are plotted in Fig. 15.15 for the 32×32 grid. To examine how well the velocities have converged, we also plot the u velocity along a vertical line through the middle of the domain and the v-velocity along a horizontal line through the middle of the domain for all three resolutions and we see that while the results on the coarsest grid differ slightly, the results on the two finer grids are essentially identical. The velocity plot also shows clearly that the flow is as we expect. The fluid is dragged to the right with the moving top wall and forced down as it encounters the right boundary. It then returns to the left side in a leisurely way in the lower part of the domain and moves up along the left wall, to fill the void in the top left corner left by the fluid dragged away by the moving wall.

The code presented here is obviously extremely simple and mainly intended to show one possible way to solve the Navier-Stokes equations. Nevertheless, it is a complete code, capable of producing accurate solutions for flow in a driven cavity. It is only first order in time, but if we are only interested in the steady-state solution that is irrelevant. Furthermore, it is easily extended to second order, as discussed earlier.

Velocity-pressure method for incompressible flows

Most numerical codes for the Navier-Stokes equations work with the original equations (often referred to as the *primitive form*), and here we describe a method that solves for the flow directly in terms of the velocity and the pressure, for two-dimensional flows. We take the density and viscosity to be constant and use a simple first-order time integration method and a finite volume approach to derive the spatial discretization. By using separate control volumes for each velocity component

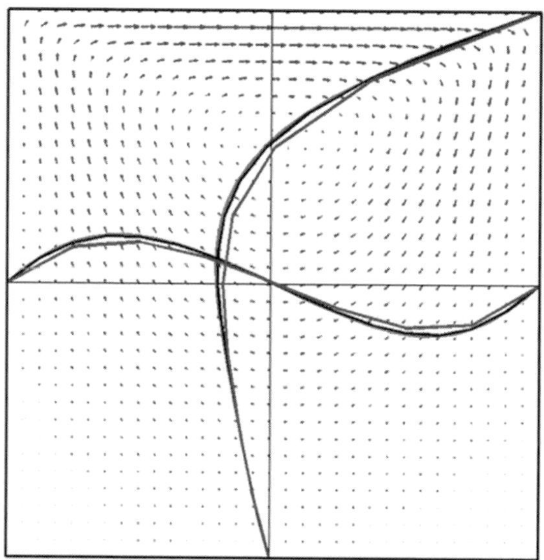

FIGURE 15.15 The velocity in the driven cavity, found by differentiating the stream function, along with the horizontal velocity on a vertical axis through the center of the domain and the vertical velocity on a horizontal axis through the center of the domain, for three resolutions.

and the pressure, we end up with a *staggered* grid. While it is possible to use a *colocated* grid, where the different variables are stored at the same location, the staggered grid results in a particularly simple and robust method.

In a finite volume method we work with the average velocity in the control volume, defined by:

$$\mathbf{u} = \frac{1}{V}\int_V \mathbf{u}(\mathbf{x})dv, \tag{15.70}$$

where V is the control volume. To derive an equation for the evolution of $\mathbf{u}$ in time, we integrate the Navier-Stokes equations over V, and by approximating the time derivative by a first-order, forward in time approximation and evaluating all spatial derivatives at the current time, we have:

$$\frac{\mathbf{u}^{n+1} - \mathbf{u}^n}{\Delta t} = -\frac{1}{\rho}\nabla_h p - \nabla_h \cdot \mathbf{u}^n\mathbf{u}^n + \frac{\mu}{\rho}\nabla_h^2 \mathbf{u}^n, \tag{15.71}$$

where the subscript h on the nable denotes a discrete approximation. The pressure, advection, and diffusion terms are given by:

$$\nabla_h p = \frac{1}{V}\int_V \nabla p\, dv = \frac{1}{V}\oint_S p\mathbf{n}\, ds, \tag{15.72}$$

$$\nabla_h \cdot \mathbf{u}^n\mathbf{u}^n = \frac{1}{V}\int_V \nabla \cdot \mathbf{u}\mathbf{u}\, dv = \frac{1}{V}\oint_S \mathbf{u}(\mathbf{u}\cdot\mathbf{n})ds, \tag{15.73}$$

$$\nabla_h^2 \mathbf{u}^n = \frac{1}{V}\int_V \nabla^2 \mathbf{u}\, dv = \frac{1}{V}\int_V \nabla \cdot \nabla \mathbf{u}\, dv = \frac{1}{V}\oint_S \nabla\mathbf{u}\cdot\mathbf{n}\, ds. \tag{15.74}$$

In all cases we have converted the volume integral to an integral over the control surface. Here, as before, the superscript n denotes a variable evaluated at the current time (time t) and $n+1$ stands for a variable at the end of the time step (time $t+\Delta t$).

The incompressibility condition states that the velocity field must be divergence free. Integrating over a control volume, we write:

$$\nabla_h \cdot \mathbf{u} = \int_V \nabla \cdot \mathbf{u} dv = \oint_S \mathbf{u}\cdot\mathbf{n} dv = 0. \tag{15.75}$$

Applying (15.75) to the new velocity at time step $n+1$ yields:

$$\nabla_h \cdot \mathbf{u}^{n+1} = 0. \tag{15.76}$$

Notice that we require the velocity at the new time step to be incompressible, but do not explicitly assume that $\nabla_h \cdot \mathbf{u}^n = 0$. Usually, the velocity field at t^n is not exactly divergence free but even if that is the case, we strive to make the divergence of the new velocity field zero.

The main challenge in integrating the Navier-Stokes equations in time in the velocity-pressure form is that we have two equations for the velocity and no explicit equation for the pressure. Rather, the pressure in the first equation (15.71) must be whatever is needed to satisfy the second one (15.76). To get around this problem, we split the momentum equation (15.71) by first computing the velocity field without considering the pressure:

$$\frac{\mathbf{u}^* - \mathbf{u}^n}{\Delta t} = -\nabla_h \cdot \mathbf{u}^n \mathbf{u}^n + \frac{\mu}{\rho} \nabla_h^2 \mathbf{u}^n, \tag{15.77}$$

and then adding the pressure:

$$\frac{\mathbf{u}^{n+1} - \mathbf{u}^*}{\Delta t} = -\frac{\nabla_h p}{\rho}. \tag{15.78}$$

Here, we have introduced an extra variable $\mathbf{u}^*$, which is the new velocity if the effect of pressure is ignored. Adding (15.77) and (15.78) cancels $\mathbf{u}^*$ and gives (15.71).

The pressure is found by taking the divergence of (15.78) and using the incompressibility conditions (15.76). Symbolically, we have:

$$\nabla_h \cdot \nabla_h p = \nabla_h^2 p = \frac{\rho}{\Delta t} \nabla_h \cdot \mathbf{u}^*, \tag{15.79}$$

which is an elliptic equation for the pressure. After the pressure has been found by solving (15.79), the final velocity is found by (15.78). By solving (15.77), (15.79), and (15.78) in that order, the velocity field at the new time level, $\mathbf{u}^{n+1}$ can be computed. This two-step time integration is usually referred to as a projection method.

Spatial discretization

For the numerical code we are developing here, we divide the computational domain into rectangular equal-sized control volumes, using structured Cartesian grids where the various variables are stored at points defined by the intersection of orthogonal grid lines, as shown in Fig. 15.10. For incompressible flows, experience shows that it is simplest to use one control volume for the pressure and different control volumes for each of the velocity components. The motivation for doing so comes from considering the incompressibility condition (15.76), which states that flow into the control volume must be equal to the outflow. If the in- and outflows do not balance, then the pressure in the control volume must be increased or decreased to increase or decrease the flow in or out of the control volume. The control volume for the pressure is shown in Fig. 15.16. In two dimensions, the volume (actually the area, but we will speak of a volume to make the transition to three-dimensional flows easier) is given by $V = \Delta x \Delta y$, where Δx and Δy are the dimensions of the control volume in the x and y direction, respectively.

Approximating (15.76) by integrating over the edges of the control volume in Fig. 15.16, using the midpoint rule yields:

$$\Delta y \left(u_{i+1/2,j} - u_{i-1/2,j}\right) + \Delta x \left(v_{i,j+1/2} - v_{i,j-1/2}\right) = 0, \tag{15.80}$$

where it is clear that we need the horizontal velocity components (u) at the vertical boundaries and the vertical velocity components (v) on the horizontal boundaries. Thus, it is natural to define new control volumes for each component, with the center of the control volume for the u-velocity component located at the middle of the vertical boundary of the pressure control volume and the control volume for the v-velocity component centered at the middle of the horizontal boundary of the pressure control volume. Thus, we can think of those control volumes as being displaced half a mesh to the right from the pressure node for the horizontal velocity and half a mesh upward for the vertical velocity (see Fig. 15.17). It is customary to identify the pressure nodes by the indices (i, j) and to refer to the location of the u-velocity component by $(i + 1/2, j)$ and the location of the u-velocity component by $(i, j + 1/2)$. In an actual computer code, the grids are, of course, simply shifted and each component referenced by an integer.

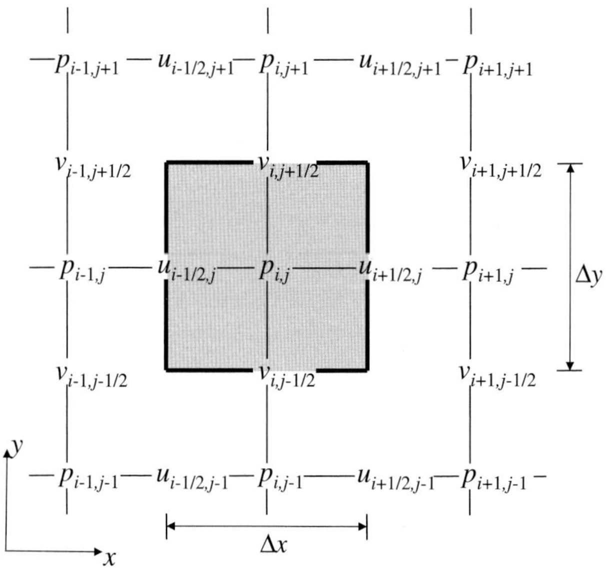

FIGURE 15.16 The notation used for a standard staggered mesh. The pressure is assumed to be known at the center of the control volume outlined by a thick solid line.

For the advection terms (15.73), we approximate the fluxes through each boundary by the value at the center of the boundary and multiply by the edge length. For the u velocity at point $(i+1/2, j)$, we have:

$$\begin{aligned}&\left(\frac{1}{V}\oint_S \mathbf{u}(\mathbf{u}\cdot\mathbf{n})ds\right)_x\Bigg)_{i+1/2,j}\\&\quad\approx\frac{1}{\Delta x\Delta y}\left\{\left[(uu)_{i+1,j}-(uu)_{i,j}\right]\Delta y+\left[(uv)_{i+1/2,j+1/2}-(uv)_{i+1/2,j-1/2}\right]\Delta x\right\},\end{aligned}\tag{15.81}$$

and the advection term for the v velocity, at point $(i, j+1/2)$, is:

$$\begin{aligned}&\left(\frac{1}{V}\oint_S \mathbf{u}(\mathbf{u}\cdot\mathbf{n})ds\right)_y\Bigg)_{i,j+1/2}\\&\quad\approx\frac{1}{\Delta x\Delta y}\left\{\left[(uv)_{i+1/2,j+1/2}-(uv)_{i-1/2,j+1/2}\right]\Delta y+\left[(vv)_{i,j+1}-(vv)_{i,j}\right]\Delta x\right\}.\end{aligned}\tag{15.82}$$

Approximating the integral of the viscous fluxes around the boundaries of the velocity control volumes (15.74) by the value at the midpoint of each edge times the length of the edge, and then finding the derivatives by standard second-order centered differences, results in:

$$\begin{aligned}&\left(\frac{1}{V}\oint_S \nabla\mathbf{u}\cdot\mathbf{n}ds\right)_x\Bigg)_{i+1/2,j}\\&\quad\approx\frac{1}{\Delta x\Delta y}\left\{\left(\left(\frac{\partial u}{\partial x}\right)_{i+1,j}-\left(\frac{\partial u}{\partial x}\right)_{i,j}\right)\Delta y+\left(\left(\frac{\partial u}{\partial y}\right)_{i+1/2,j+1/2}-\left(\frac{\partial u}{\partial y}\right)_{i+1/2,j-1/2}\right)\Delta x\right\},\end{aligned}\tag{15.83}$$

and

$$\begin{aligned}&\left(\frac{1}{V}\oint_S \nabla\mathbf{u}\cdot\mathbf{n}ds\right)_y\Bigg)_{i,j+1/2}\\&\quad\approx\frac{1}{\Delta x\Delta y}\left\{\left(\left(\frac{\partial v}{\partial x}\right)_{i+1/2,j+1/2}-\left(\frac{\partial v}{\partial x}\right)_{i-1/2,j+1/2}\right)\Delta y+\left(\left(\frac{\partial v}{\partial y}\right)_{i,j+1}-\left(\frac{\partial v}{\partial y}\right)_{i,j}\right)\Delta x\right\}.\end{aligned}\tag{15.84}$$

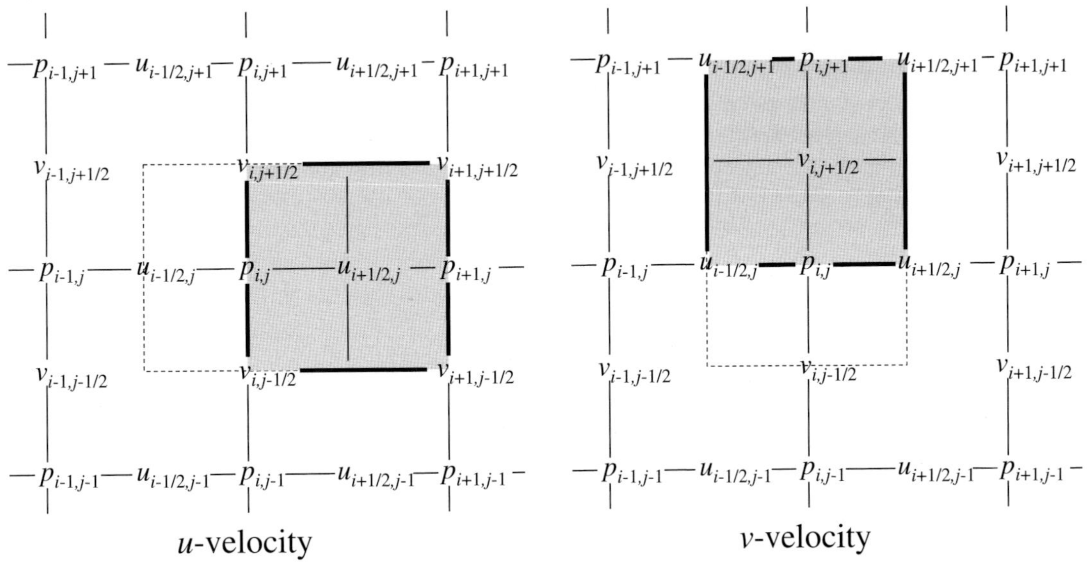

FIGURE 15.17 The notation used for a standard staggered mesh. The horizontal velocity components (u) are stored at the middle of the left and right edges of the pressure control volume and the vertical velocity components (v) are stored at the middle of the top and bottom edges.

The velocities on the boundaries in the approximations for the advection terms are generally not defined where we need them, so we interpolate those by taking the average of the neighboring components. The derivatives on the boundaries in the approximations for the diffusion terms are found by centered differences and the velocities that we need for those are defined exactly where we need them. Replacing the velocities in (15.81) and (15.82) by their interpolated values and the derivatives in (15.83) and (15.84) by the centered approximations, and using the grid in Figs. 15.16 and 15.17 along with the notation introduced above, the discrete approximations for the u and the v component of the predicted velocities (15.77) are:

$$
\begin{aligned}
u^*_{i+1/2,j} = u^n_{i+1/2,j} + \Delta t \Bigg\{ & -\frac{1}{\Delta x}\left[\left(\frac{u^n_{i+3/2,j}+u^n_{i+1/2,j}}{2}\right)^2 - \left(\frac{u^n_{i+1/2,j}+u^n_{i-1/2,j}}{2}\right)^2\right] \\
& -\frac{1}{\Delta y}\left[\left(\frac{u^n_{i+1/2,j+1}+u^n_{i+1/2,j}}{2}\right)\left(\frac{v^n_{i+1,j+1/2}+v^n_{i,j+1/2}}{2}\right)\right. \\
& \qquad \left. -\left(\frac{u^n_{i+1/2,j}+u^n_{i+1/2,j-1}}{2}\right)\left(\frac{v^n_{i+1,j-1/2}+v^n_{i,j-1/2}}{2}\right)\right] \\
& +\frac{\mu}{\rho}\left(\frac{u^n_{i+3/2,j}-2u^n_{i+1/2,j}+u^n_{i-1/2,j}}{\Delta x^2}+\frac{u^n_{i+1/2,j+1}-2u^n_{i+1/2,j}+u^n_{i+1/2,j-1}}{\Delta y^2}\right)\Bigg\},
\end{aligned}
\tag{15.85}
$$

and

$$
\begin{aligned}
v^*_{i,j+1/2} = v^n_{i,j+1/2} + \Delta t \Bigg\{ & -\frac{1}{\Delta x}\left[\left(\frac{u^n_{i+1/2,j}+u^n_{i+1/2,j+1}}{2}\right)\left(\frac{v^n_{i,j+1/2}+v^n_{i+1,j+1/2}}{2}\right)\right. \\
& \qquad \left. -\left(\frac{u^n_{i-1/2,j+1}+u^n_{i-1/2,j}}{2}\right)\left(\frac{v^n_{i,j+1/2}+v^n_{i-1,j+1/2}}{2}\right)\right] \\
& -\frac{1}{\Delta y}\left[\left(\frac{v^n_{i,j+3/2}+v^n_{i,j+1/2}}{2}\right)^2 - \left(\frac{v^n_{i,j+1/2}+v^n_{i,j-1/2}}{2}\right)^2\right] \\
& +\frac{\mu}{\rho}\left(\frac{v^n_{i+1,j+1/2}-2v^n_{i,j+1/2}+v^n_{i-1,j+1/2}}{\Delta x^2}+\frac{v^n_{i,j+3/2}-2v^n_{i,j+1/2}+v^n_{i,j-1/2}}{\Delta y^2}\right)\Bigg\}.
\end{aligned}
\tag{15.86}
$$

The pressure terms in the correction equations (15.78) are found in the same way by approximating the integrals using the mid-point rule. For the u velocity, at point $(i+1/2, j)$, we have:

$$\left(\frac{1}{V}\oint_S p\mathbf{n}\,ds\right)_x\Bigg)_{i,j+1/2,j} \approx \frac{1}{\Delta x\Delta y}\left(p_{i+1,j}-p_{i,j}\right)\Delta y, \tag{15.87}$$

and for the v velocity, at point $(i, j+1/2)$, we have:

$$\left(\frac{1}{V}\oint_S p\mathbf{n}\,ds\right)_y\Bigg)_{i,j+1/2} \approx \frac{1}{\Delta x\Delta y}\left(p_{i,j+1}-p_{i,j}\right)\Delta x. \tag{15.88}$$

The equations for the corrected velocities (15.78) therefore are:

$$u^{n+1}_{i+1/2,j} = u^{*}_{i+1/2,j} - \frac{\Delta t}{\rho}\left(\frac{p_{i+1,j}-p_{i,j}}{\Delta x}\right), \tag{15.89}$$

and

$$v^{n+1}_{i,j+1/2} = v^{*}_{i,j+1/2} - \frac{\Delta t}{\rho}\left(\frac{p_{i,j+1}-p_{i,j}}{\Delta y}\right). \tag{15.90}$$

The linear interpolations used here result in a second-order centered scheme for the advection terms. Other alternatives can, of course, be used but the centered second-order scheme is more accurate than any non-centered second-order scheme and usually gives the best results for fully resolved flows. It has, however, two serious shortcomings. The first is that for flows that are not fully resolved it can produce unphysical oscillations that can degrade the quality of the results. The second problem is that the centered second-order scheme is unconditionally unstable for inviscid flows when used in combination with the explicit forward-in-time integration given by (15.77) and (15.78). It is only the addition of the viscosity terms that makes the scheme stable and if the viscosity is small, the time step must be small. At high Reynolds numbers, this usually results in excessively small time steps and more sophisticated schemes are required.

The pressure equation is derived by substituting the expression for the corrected velocity, (15.89) and (15.90), into the discrete mass conservation equation (15.80). The result is:

$$\begin{aligned}\frac{1}{\Delta x^2}\left(p_{i+1,j}-2p_{i,j}+p_{i-1,j}\right)+\frac{1}{\Delta y^2}\left(p_{i,j+1}-2p_{i,j}+p_{i,j-1}\right)\\ =\frac{\rho}{\Delta t}\left(\frac{u^{*}_{i+1/2,j}-u^{*}_{i-1/2,j}}{\Delta x}+\frac{v^{*}_{i,j+1/2}-v^{*}_{i,j-1/2}}{\Delta y}\right).\end{aligned} \tag{15.91}$$

Solving the pressure equation is usually the most expensive part of any simulation of incompressible flows and usually it is necessary to use an advanced pressure solver to achieve reasonable computational times. Here, however, we will use the simple SOR method introduced earlier and used already for the stream function vorticity equation. To do so, we rearrange (15.91) and isolate $p_{i,j}$ on the left-hand side. The pressure is then updated iteratively by substituting for pressure values on the right-hand side the approximate values $p^{\alpha}_{i,j}$ of the pressure from the previous iteration, and extrapolating by taking the weighted average of the updated value and the pressure $p^{\alpha}_{i,j}$ from the last iteration. The new value is:

$$\begin{aligned}p^{\alpha+1}_{i,j}=\beta\left[\frac{2}{\Delta x^2}+\frac{2}{\Delta y^2}\right]^{-1}\Bigg[\frac{1}{\Delta x^2}\left(p^{\alpha}_{i+1,j}+p^{\alpha+1}_{i-1,j}\right)+\frac{1}{\Delta y^2}\left(p^{\alpha}_{i,j+1}+p^{\alpha+1}_{i,j-1}\right)\\ -\frac{\rho}{\Delta t}\left(\frac{u^{*}_{i+1/2,j}-u^{*}_{i-1/2,j}}{\Delta x}+\frac{v^{*}_{i,j+1/2}-v^{*}_{i,j-1/2}}{\Delta y}\right)\Bigg]+(1-\beta)p^{\alpha}_{i,j}.\end{aligned} \tag{15.92}$$

The relaxation parameter β must be in the range $1<\beta<2$, and taking $\beta=1.5$ is usually a good compromise between stability and accelerated convergence. To assess how well the iteration has converged, we monitor the maximum difference between the pressure at successive iterations and stop, once this difference is small enough. A more sophisticated approach would be to monitor the *residual,* i.e., the difference between the left- and the right-hand side of (15.91), but this generally requires extra computational effort. While the SOR method works well for our purpose, for serious computations more advanced pressure solvers are generally used.

The computational domain and boundary condition

The computational domain is a rectangle of size L_x by L_y, divided into N_x by N_y pressure control volumes. The pressure control volumes are placed inside the computational domain, such that their boundaries coincide with the domain boundary, at the edge of the domain. In addition to the control volumes inside the domain we usually provide one row of *ghost cells* outside the domain to help implementing boundary conditions. Thus, the pressure array needs be dimensioned $p(N_x + 2, N_y + 2)$. Similarly, we will need ghost points for the tangential velocity, but not the normal velocity, which is given. The velocity arrays thus have dimensions: $u(N_x + 1, N_y + 2)$ and $v(N_x + 2, N_y + 1)$. Fig. 15.18 shows the layout of the full grid.

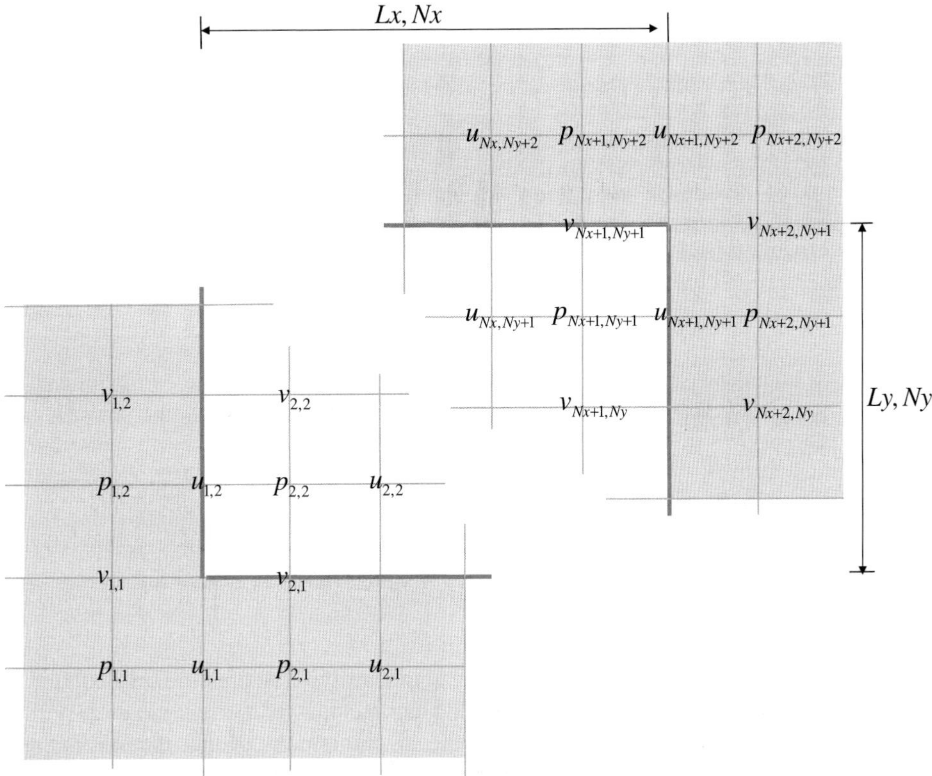

FIGURE 15.18 The full grid, with ghost cells. The lower-left corner and the upper-right corner are shown. There are $N_x \times N_y$ pressure control volumes inside the domain, so the size of the pressure array is $p(N_x + 2, N_y + 2)$. The size of the velocity arrays are $u(N_x + 1, N_y + 2)$ and $v(N_x + 2, N_y + 1)$.

Before attempting to evolve the solution using the discrete equations that we just derived, we must establish the appropriate boundary conditions. For the normal velocities the specification of the boundary values is very simple. Since the location of the center of the control volume coincides with the boundary, we can set the velocity equal to what it should be. For a rigid wall, the normal velocity is usually zero and for inflow boundaries the normal velocity is generally given. For viscous flows we must also specify the tangential velocity and this is slightly more complicated, since we do not store that velocity on the boundary. We do, however, have the tangent velocity half a grid spacing inside the domain and using this value, along with the known value on the boundary, we can specify the *ghost value* at the center of the ghost cell outside the boundary. To do so we assume that we know the ghost value and use that the wall velocity is given by a linear interpolation between the ghost velocity and the velocity just inside the domain. For the u-velocity on the bottom boundary, for example:

$$u_{wall} = (1/2)(u_{i,1} + u_{i,2}), \tag{15.93}$$

where u_{wall} is the tangent velocity on the wall and $u_{i,1}$ is the ghost velocity. Since the wall velocity and the velocity inside the domain, $u_{i,2}$, are known, we can easily find the ghost velocity:

$$u_{i,1} = 2u_{wall} - u_{i,2}. \tag{15.94}$$

Similar equations are derived for the other boundaries. Notice that if the wall velocity is zero, the ghost velocity is simply a reflection of the velocity inside.

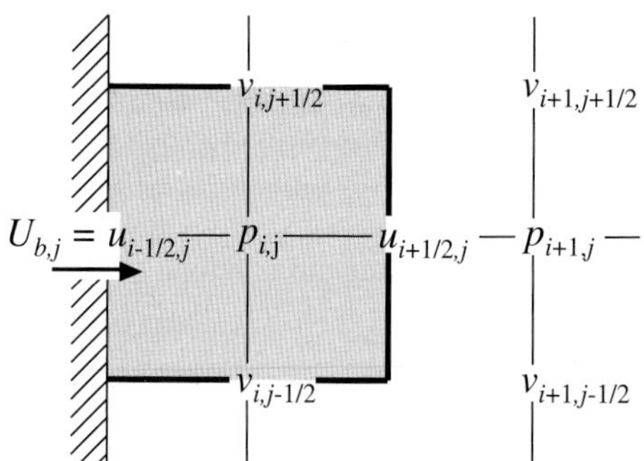

FIGURE 15.19 A control volume next to a boundary where the normal velocity is known.

The boundary conditions for the pressure become particularly simple on a staggered grid. Consider the vertical boundary in Fig. 15.19, on the left side of the domain and passing through node $(i - 1/2, j)$, where the horizontal velocity is $U_{b,j}$. The continuity equation for the cell next to the boundary, surrounding the pressure node $p(i, j)$, is:

$$\frac{u_{i+1/2,j}^{n+1} - U_{b,j}}{\Delta x} + \frac{v_{i,j+1/2}^{n+1} - v_{i,j-1/2}^{n+1}}{\Delta y} = 0. \tag{15.95}$$

Since the velocity at the left boundary is known, we only substitute the equations for the correction velocities, (15.89) and (15.90), for the three unknown velocities, through the top, bottom, and right edge. Thus, the pressure equation for a point next to the left boundary becomes:

$$\begin{aligned} &\frac{p_{i+1,j} - p_{i,j}}{\Delta x^2} + \frac{p_{i,j+1} + p_{i,j-1} - 2p_{i,j}}{\Delta y^2} \\ &= \frac{\rho}{\Delta t}\left(\frac{u_{i+1/2,j}^{*} - U_{b,j}}{\Delta x} + \frac{v_{i,j+1/2}^{*} - v_{i,j-1/2}^{*}}{\Delta y}\right). \end{aligned} \tag{15.96}$$

Solving this equation for $p_{i,j}$ shows that it differs from (15.92) in three ways. The pressure on the left is absent, the normal velocity on the left boundary is specified, and the coefficient in front includes $1/\Delta x^2$ instead of $2/\Delta x^2$. Similar equations are derived for the pressure next to the other boundaries and for each corner point. Notice that it is not necessary to impose any new conditions on the pressure at the boundaries and that simply using incompressibility yields the correct boundary equations. Sometimes it is, however, necessary to either fix the pressure at one point in the domain or specify the average pressure.

Numerical code and results

A `MATLAB` code implementing the algorithm described in this section is shown below. The implementation follows the description above as closely as possible. To simplify the programming we use only one set of equations for all the pressure points, putting the pressure at the ghost points to zero and defining a coefficient array, `c(i, j)`, which is different for the interior points, boundary points, and corner points. In this way, all the pressure nodes can be updated in one loop.

We note that as for the vorticity-stream function code, we have put a few statements on the same line, as allowed in `MATLAB`, but only those statements that clearly belong together. The parallelization feature of `MATLAB`, where the range of an index can be put in the argument list of an array is also used, but only when the readability of the program is not affected. The computation of the advection and diffusion terms as well as the pressure in unparalellized format is therefore left as an explicit for-loop.

The results from the code are essentially the same as for the vorticity-stream function method. In Fig. 15.20, the velocity vectors and pressure contours are shown at time 1.0 computed on a 32 × 32 grid (pressure control volumes inside the domain), along with the velocities along a horizontal and vertical axis through the center of the domain. The convergence properties of the scheme are similar to those of the vorticity-stream function method, as seen in Fig. 15.21, where we plot the net volume flow to the right versus time, for three different resolutions.

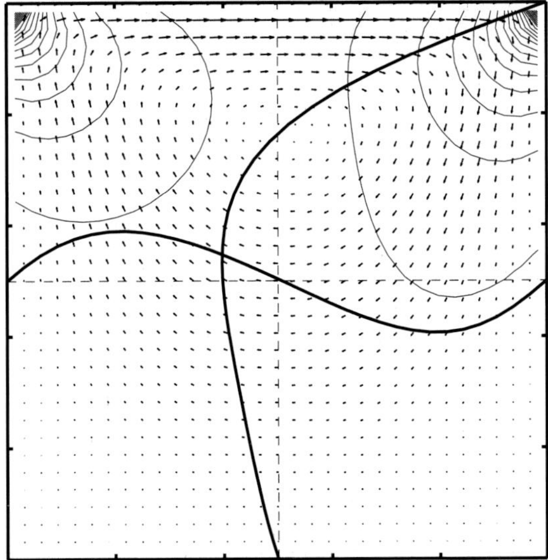

FIGURE 15.20 The velocity and pressure at steady state computed by solving the Navier-Stokes equations in the pressure-velocity form using a staggered grid on a 32×32 grid (pressure control volumes inside the domain). The horizontal and the vertical velocities are also plotted along lines through the middle of the domain.

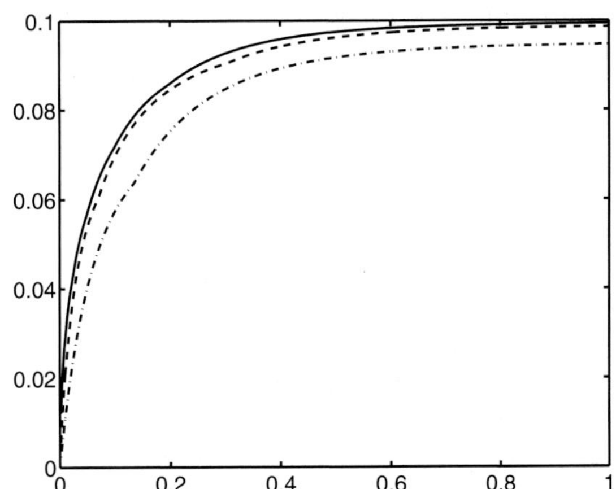

FIGURE 15.21 The net volume flow to the right versus time, as computed on a 8×8, 16×16, and 32×32 grid.

CODE 4: SOLUTION OF THE UNSTEADY TWO-DIMENSIONAL NAVIER-STOKES EQUATIONS IN PRESSURE-VELOCITY FORM, USING AN EXPLICIT FORWARD IN TIME, CENTERED IN SPACE SCHEME ON A STAGGERED GRID

```
% Driven Cavity by the MAC Method
Nx=32; Ny=32; Lx=1; Ly=1; MaxStep=500; Visc=0.1; rho=1.0;
MaxIt=100; Beta=1.5; MaxErr=0.001; % parameters for SOR
dx=Lx/Nx;dy=Ly/Ny; un=1;us=0;ve=0;vw=0;time=0.0; dt=0.002;
u=zeros(Nx+1,Ny+2);v=zeros(Nx+2,Ny+1);p=zeros(Nx+2,Ny+2);
ut=zeros(Nx+1,Ny+2);vt=zeros(Nx+2,Ny+1); pold=zeros(Nx+2,Ny+2);
c=zeros(Nx+1,Ny+1)+1/(2/dx^2+2/dy^2);
c(2,3:Ny)=1/(1/dx^2+2/dy^2); c(Nx+1,3:Ny)=1/(1/dx^2+2/dy^2);
c(3:Nx,2)=1/(1/dx^2+2/dy^2); c(3:Nx,Ny+1)=1/(1/dx^2+2/dy^2);
c(2,2)=1/(1/dx^2+1/dy^2); c(2,Ny+1)=1/(1/dx^2+1/dy^2);
c(Nx+1,2)=1/(1/dx^2+1/dy^2); c(Nx+1,Ny+1)=1/(1/dx^2+1/dy^2);
for i=1:Nx+1,for j=1:Ny+1,
  x(i,j)=dx*(i-1);y(i,j)=dy*(j-1);
```

```
end,end;

for is=1:MaxStep
  u(1:Nx+1,1)=2*us-u(1:Nx+1,2);
  u(1:Nx+1,Ny+2)=2*un-u(1:Nx+1,Ny+1);
  v(1,1:Ny+1)=2*vw-v(2,1:Ny+1);
  v(Nx+2,1:Ny+1)=2*ve-v(Nx+1,1:Ny+1);

  for i=2:Nx,for j=2:Ny+1      % temporary u-velocity
    ut(i,j)=u(i,j)+dt*(-0.25*(...
      ( (u(i+1,j)+u(i,j))^2-(u(i,j)+u(i-1,j))^2 )/dx+...
      ( (u(i,j+1)+u(i,j))*(v(i+1,j)+v(i,j))-...
        (u(i,j)+u(i,j-1))*(v(i+1,j-1)+v(i,j-1)))/dy )+...
        Visc*((u(i+1,j)+u(i-1,j)-2*u(i,j))/dx^2+...
              (u(i,j+1)+u(i,j-1)-2*u(i,j))/dy^2));

  end,end

  for i=2:Nx+1,for j=2:Ny      % temporary v-velocity
    vt(i,j)=v(i,j)+dt*( - 0.25*(...
         ( (u(i,j+1)+u(i,j))*(v(i+1,j)+v(i,j))-...
           (u(i-1,j+1)+u(i-1,j))*(v(i,j)+v(i-1,j)) )/dx+...
         ( (v(i,j+1)+v(i,j))^2-(v(i,j)+v(i,j-1))^2 )/dy )+...
           Visc*((v(i+1,j)+v(i-1,j)-2*v(i,j))/dx^2+...
                 (v(i,j+1)+v(i,j-1)-2*v(i,j))/dy^2));

  end,end

  for it=1:MaxIt                       % solve for pressure
    pold=p;
    for i=2:Nx+1,for j=2:Ny+1
      p(i,j)=Beta*c(i,j)*....
      ( (p(i+1,j)+p(i-1,j))/dx^2+(p(i,j+1)+p(i,j-1))/dy^2 -...
        (rho/dt)*( (ut(i,j)-ut(i-1,j))/dx+...
        (vt(i,j)-vt(i,j-1))/dy ) ) + (1-Beta)*p(i,j);
    end,end
    Err=0.0; for i=2:Nx+1; for j=2:Ny+1,    % check error
          Err=Err+abs(pold(i,j)-p(i,j)); end; end;
    if Err ≤ MaxErr, break, end          % stop if converged
  end
                                          % correct the velocity
  u(2:Nx,2:Ny+1)=...
    ut(2:Nx,2:Ny+1)-(dt/dx)*(p(3:Nx+1,2:Ny+1)-p(2:Nx,2:Ny+1));
  v(2:Nx+1,2:Ny)=...
    vt(2:Nx+1,2:Ny)-(dt/dy)*(p(2:Nx+1,3:Ny+1)-p(2:Nx+1,2:Ny));

  time=time+dt                          % plot the results
  uu(1:Nx+1,1:Ny+1)=0.5*(u(1:Nx+1,2:Ny+2)+u(1:Nx+1,1:Ny+1));
  vv(1:Nx+1,1:Ny+1)=0.5*(v(2:Nx+2,1:Ny+1)+v(1:Nx+1,1:Ny+1));
  quiver(x,y,uu,vv,'linewidth',1);hold on
  axis equal; axis([0,1,0,1]), hold off,pause(0.01)
end
```

15.4 Flow in complex domains

Most fluid problems of practical interest involve complex geometries. Thus, while rectangular geometries are generally used to develop numerical methods, and can often be used to study problems of theoretical interest, adapting numerical methods to non-rectangular geometries has remained a major thrust of research in computational fluid dynamics. Managing complex geometries is one of the driving considerations for commercial codes, and often the ease by which the flow domain can be represented dictates the numerical method that is used. This has led to elaborate schemes with arbitrarily shaped control volumes and sophisticated ways of identifying how the control volumes are connected. As such schemes have been developed and applied to complex problems it has, however, become clear that no methods can compete with schemes that use regular structured grids in terms of accuracy, efficiency, and simplicity *for those problems where they can be used.* Thus, the methods introduced earlier in this chapter continue to enjoy widespread use.

There are two main ways in which flows in complex domains can be computed using schemes employing regular structured grids: immersed boundary methods and mapped grid methods. Below, we describe both of them briefly, although for the mapped grids we only discuss methods based on the vorticity-stream function formulation. Mapped grid methods can lead to very accurate solutions since, in addition to aligning boundaries with grid lines, they allow us to make the grid finer in some parts of the domain. They are, however, limited in how complex the geometry can be. Immersed boundary methods can cope with essentially arbitrarily complex geometries but their accuracy is usually not as good. Since immersed boundary methods are simpler, at least in their elementary form, we start with them.

Immersed boundary methods

When applying an immersed boundary method to solve for fluid flow in a complex geometry, we select a (usually) rectangular domain that contains both the fluid region and also the solid. The resulting rectangular domain now contains fluid regions where the fluid velocity is governed by the Navier-Stokes equations and solid regions where the velocity is zero, or given. The domain is resolved by a regular structured grid and some grid points (or control volumes) are in the fluid region and some are in the solid region. If we use a marker function H, defined by:

$$H(x) = \begin{cases} 1 \text{ in the fluid} \\ 0 \text{ in the solid,} \end{cases} \tag{15.97}$$

then the velocity in the whole domain can be written as:

$$\mathbf{u}(x) = H(x)\mathbf{u}_f(x) + (1 - H(x))\mathbf{u}_S(x), \tag{15.98}$$

where $\mathbf{u}_f$ is the velocity in the fluid and $\mathbf{u}_S$ is the velocity in the solid (often zero).

If we solve for the flow field using a first-order, forward in time projection method, then the predicted fluid velocity is given by:

$$\mathbf{u}_f^* = \mathbf{u}^n + \Delta t(-\nabla_h \cdot \mathbf{u}^n\mathbf{u}^n + \nu\nabla_h^2\mathbf{u}^n), \tag{15.99}$$

using the notation introduced earlier. Here we use $\mathbf{u}^n$ instead of $\mathbf{u}_f^n$ on the right-hand side, since $\mathbf{u}^n = \mathbf{u}_f^n$ in the fluid region. This velocity is then corrected by adding a pressure gradient:

$$\mathbf{u}_f^{n+1} = \mathbf{u}_f^* - \frac{\Delta t}{\rho}\nabla_h p. \tag{15.100}$$

The pressure gradient should make the final velocity:

$$\mathbf{u}^{n+1} = H\mathbf{u}_f^{n+1} + (1 - H)\mathbf{u}_S^{n+1}, \tag{15.101}$$

divergence-free, or $\nabla_h \cdot \mathbf{u}^{n+1} = 0$. However:

$$\begin{aligned} \nabla_h \cdot \mathbf{u}^{n+1} &= \nabla_h \cdot H\mathbf{u}_f^{n+1} + \nabla_h \cdot (1 - H)\mathbf{u}_S^{n+1} \\ &= H\nabla_h \cdot \mathbf{u}_f^{n+1} + H\nabla_h \cdot \mathbf{u}_S^{n+1} + (\mathbf{u}_f^{n+1} - \mathbf{u}_S^{n+1}) \cdot \nabla_h H, \end{aligned} \tag{15.102}$$

where the last two terms are zero since the velocity in the solid is constant–often zero–and the normal velocity is continuous at the solid surface. Thus, we can find the pressure by considering only $\nabla_h \cdot \mathbf{u}_f^{n+1} = 0$, which gives the same pressure equation as for a flow without a solid body:

$$\nabla_h^2 p = \frac{\Delta t}{\rho}\nabla_h \cdot \mathbf{u}^*, \tag{15.103}$$

where $\mathbf{u}^*$ is the predicted velocity in the whole domain, computed using (15.99).

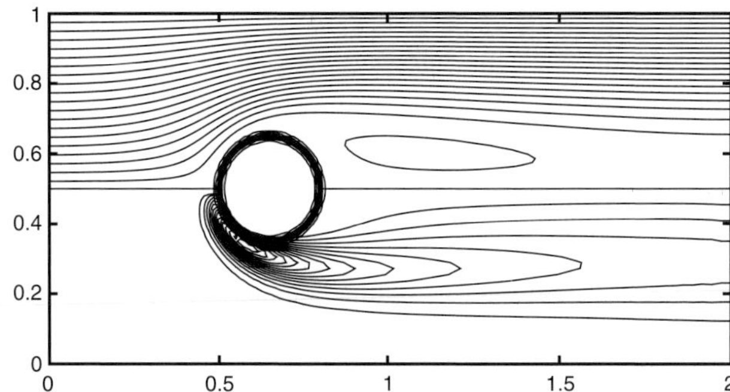

FIGURE 15.22 The steady-state flow around a circular cylinder at a Reynolds number of 75 computed using a simple immersed boundary method on a 60 by 120 stretched grid. The cylinder is resolved by about 25 grid points across its diameter. The stream function is shown in the top of the figure and the vorticity in the bottom half.

This makes it particularly simple to extend a code written for fluid flow to include solids. We simply need to put the velocity inside the solids equal to $\mathbf{u}_S^{n+1}$ at the end of the time step. Often, H is taken to have a smooth transition from 0 to 1 where the surface of the solid is, and this is the case in Fig. 15.22, where an example of flow around a stationary cylinder computed using this approach is shown. More recent implementations of this idea have focused on representing the velocity field near the surface as accurately as possible.

In addition to accommodating complex boundaries, grid points are sometimes distributed unevenly to allow fine resolution of part of the domain, while using a coarser resolution in other parts. The methods described so far can be easily extended to space the gridlines unevenly, providing a useful but rather limited ability for local refinement. Immersed boundaries can be used with flow solvers using unevenly spaced gridlines, as well as with solvers based on using rectangular patches of grids with finer resolution. Such patches of fine grids aligned with the coarser grid preserve many of the properties of the regular structured methods, although at the cost of some overhead.

Mapped grid methods

By *coordinate mapping,* we transform the complex domain that we want to compute into a simpler—usually a rectangular—domain. This leads to more complex equations than we are used to, but, on the other hand, allows us to use all the machinery that we have already developed. The use of coordinate mapping is a relatively standard trick in applied mathematics, and we have all dealt with problems in axial or cylindrical geometry.

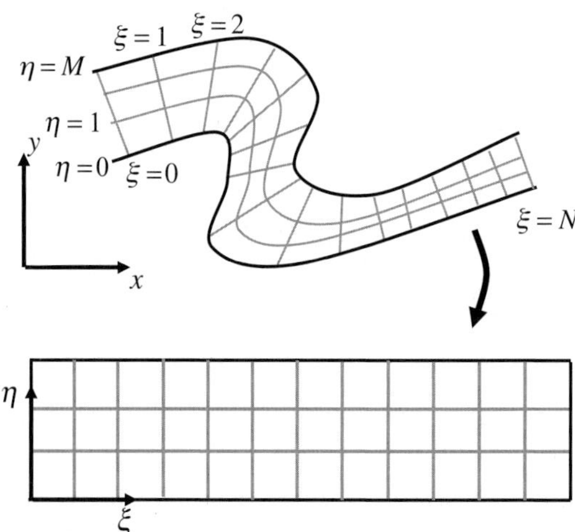

FIGURE 15.23 The mapping of an irregular domain into a rectangular one.

The basic idea of coordinate mapping is introduced in Fig. 15.23. Here, our region is bounded by curved walls in the x, y plane. Imagine now that we draw two set of lines on the domain, denoted by $\xi_1, \xi_2, \xi_3, \ldots, \xi_N$, and $\eta_1, \eta_2, \eta_3, \ldots, \eta_M$, such that the first and last line of each set coincides with a boundary, and the markers increase in value from one boundary to the other. In the figure we have drawn one such set, but notice that except for the requirement that the boundary fall on a constant ξ or η line, other lines would do just as well. We now assume that every point in the domain has a unique value of ξ and η (interpolated from the lines we drew). Therefore, instead of specifying the coordinates (x, y), we could just as well specify (ξ, η). Furthermore, since we are talking about a specific point, the new values depend in a unique way on the original values, or $\xi = \xi(x, y)$ and $\eta = \eta(x, y)$. Since both coordinate values are monotonically increasing, these relationships are invertible, and we can also write $x = x(\xi, \eta)$ and $y = y(\xi, \eta)$. Any function, say $f(x, y)$, can therefore be written as a function of ξ and η as[1]:

$$f(x, y) = f(x(\xi, \eta), y(\xi, \eta)) = f(\xi, \eta). \tag{15.104}$$

The main purpose of introducing the new coordinates, ξ and η, is to simplify the treatment of the boundaries. If we plot our computational domain in the ξ, η plane it is obvious that we have achieved our goal. The complex region in the x, y plane is now a simple rectangle, with boundaries that coincide with coordinate lines. However, there is a price to be paid. While the domain became simpler, our equations became more complicated.

To find expressions for the derivatives in the new coordinate system, we first introduce the following short-hand notation:

$$f_\xi = \frac{\partial f}{\partial \xi}; \quad f_\eta = \frac{\partial f}{\partial \eta}; \tag{15.105}$$

$$f_x = \frac{\partial f}{\partial x}; \quad f_y = \frac{\partial f}{\partial y}. \tag{15.106}$$

Since $f = f(x(\xi, \eta), y(\xi, \eta))$ we write, by the chain rule:

$$f_\xi = f_x x_\xi + f_y y_\xi, \tag{15.107}$$

$$f_\eta = f_x x_\eta + f_y y_\eta. \tag{15.108}$$

Notice that we will discretize the equations in the new system and it is important to end up with terms like x_ξ but not ξ_x. Therefore, we expand f_ξ by the chain rule but not f_x. Solving these two equations for f_x and f_y yields:

$$f_x = \frac{f_\xi y_\eta - f_\eta y_\xi}{x_\xi y_\eta - x_\eta y_\xi} \quad f_y = \frac{f_\eta x_\xi - f_\xi x_\eta}{x_\xi y_\eta - x_\eta y_\xi}. \tag{15.109}$$

The quantity:

$$J = x_\xi y_\eta - x_\eta y_\xi, \tag{15.110}$$

which involves only the mapping, is called the Jacobian of the transformation, and usually we write:

$$f_x = \frac{1}{J}\left(f_\xi y_\eta - f_\eta y_\xi\right); \quad f_y = \frac{1}{J}\left(f_\eta x_\xi - f_\xi x_\eta\right). \tag{15.111}$$

The second derivative is found by repeated application of the rule for the first derivative:

$$\begin{aligned}\frac{\partial^2 f}{\partial x^2} &= \frac{\partial}{\partial x}\left(\frac{\partial f}{\partial x}\right) = \frac{1}{J}\left[\left(\frac{\partial f}{\partial x}\right)_\xi y_\eta - \left(\frac{\partial f}{\partial x}\right)_\eta y_\xi\right] \\ &= \frac{1}{J}\left[\left(\frac{1}{J}\left(f_\xi y_\eta - f_\eta y_\xi\right)\right)_\xi y_\eta - \left(\frac{1}{J}\left(f_\xi y_\eta - f_\eta y_\xi\right)\right)_\eta y_\xi\right].\end{aligned} \tag{15.112}$$

Similarly:

$$\frac{\partial^2 f}{\partial y^2} = \frac{1}{J}\left[\left(\frac{1}{J}\left(f_\eta x_\xi - f_\xi x_\eta\right)\right)_\eta x_\xi - \left(\frac{1}{J}\left(f_\eta x_\xi - f_\xi x_\eta\right)\right)_\xi x_\eta\right]. \tag{15.113}$$

---

1. We are being a little bit cavalier here with our notation. Strictly speaking we should not write $f(\xi, \eta)$ since it is different from $f(x, y)$. However, the content should make the usage clear, and there is a certain economy of not introducing an excessive number of new notations.

Adding these equations yields:

$$\frac{\partial^2 f}{\partial x^2}+\frac{\partial^2 f}{\partial y^2}=\frac{1}{J^2}\left[\frac{\partial}{\partial \xi}\left(q_1 f_\xi - q_2 f_\eta\right)+\frac{\partial}{\partial \eta}\left(q_3 f_\eta - q_2 f_\xi\right)\right], \tag{15.114}$$

where:

$$q_1 = x_\eta^2 + y_\eta^2, \tag{15.115}$$

$$q_2 = x_\xi x_\eta + y_\xi y_\eta, \tag{15.116}$$

$$q_3 = x_\xi^2 + y_\xi^2. \tag{15.117}$$

Expanding the derivatives allows the Laplacian to be written as:

$$\nabla^2 f = \frac{1}{J^2}(q_1 f_{\xi\xi} - 2q_2 f_{\xi\eta} + q_3 f_{\eta\eta}) + (\nabla^2 \xi) f_\xi + (\nabla^2 \eta) f_\eta, \tag{15.118}$$

where:

$$\nabla^2 \xi = \frac{1}{J^3}(q_1(x_\eta y_{\xi\xi} - y_\eta x_{\xi\xi}) - 2q_2(x_\eta y_{\xi\eta} - y_\eta x_{\xi\eta}) + q_3(x_\eta y_{\eta\eta} - y_\eta x_{\eta\eta})), \tag{15.119}$$

and:

$$\nabla^2 \eta = \frac{1}{J^3}(q_1(y_\xi x_{\xi\xi} - x_\eta y_{\eta\eta}) - 2q_2(y_\xi x_{\xi\eta} - x_\xi y_{\xi\eta}) + q_3(y_\xi x_{\eta\eta} - x_\xi y_{\eta\eta})). \tag{15.120}$$

We also have, for any two functions f and g:

$$g_y f_x - g_x f_y = \frac{1}{J}\left(g_\eta f_\xi - g_\xi f_\eta\right). \tag{15.121}$$

The expressions developed above can be used to rewrite the vorticity-steam function equations for two-dimensional flow. Those are given by Eqs. (15.58) and (15.59), but we repeat them here for completeness:

$$\frac{\partial \omega}{\partial t} + \psi_y \omega_x - \psi_x \omega_y = \nu \nabla^2 \omega, \quad \text{and} \quad \nabla^2 \psi = -\omega.$$

Substituting for the derivatives results in:

$$\frac{\partial \omega}{\partial t} + \frac{1}{J}(\psi_\eta \omega_\xi - \psi_\xi \omega_\eta) = \nu\left(\frac{1}{J^2}(q_1 \omega_{\xi\xi} - 2q_2 \omega_{\xi\eta} + q_3 \omega_{\eta\eta}) + (\nabla^2 \xi)\omega_\xi + (\nabla^2 \eta)\omega_\eta\right) \tag{15.122}$$

and:

$$\frac{1}{J^2}(q_1 \psi_{\xi\xi} - 2q_2 \psi_{\xi\eta} + q_3 \psi_{\eta\eta}) + (\nabla^2 \xi)\psi_\xi + (\nabla^2 \eta)\psi_\eta = -\omega. \tag{15.123}$$

In addition to the equations themselves, we need boundary conditions. Let us assume that we intend to compute the flow in the channel in Fig. 15.23 and that there is an inflow at $\xi = 0$ and outflow at $\xi = N$. At the walls given by $\eta = 0$ and $\eta = M$, the no-slip boundary conditions must be enforced. Since the walls are streamlines, ψ = constant there, as well as any derivatives of the stream function along the wall. In particular, $\psi_\xi = \psi_{\xi\xi} = 0$. At the lower wall we can set $\psi = 0$, but at the upper wall the constant is determined by the total volume flow through the channel, since the volume flux between any two streamlines is determined by:

$$Q = \int_1^2 (u dy - v dx) = \int_1^2 d\psi = \psi_2 - \psi_1. \tag{15.124}$$

Thus, at the top wall, $\psi = Q$.

To find the vorticity on the bottom wall, we first write the stream function at the first grid point away from the wall as an expansion around the value at the wall:

$$\psi(1) = \psi(0) + \psi_\eta(1) + \psi_{\eta\eta}\frac{1^2}{2} + \cdots. \tag{15.125}$$

Using that the velocity at the wall is zero, we have $\psi_\xi = \psi_\eta = 0$ and using that:

$$\omega(0) = -\frac{q_3}{J^2}\psi_{\eta\eta}, \tag{15.126}$$

we find that the boundary vorticity can be computed by:

$$\omega(0) = 2\frac{q_3}{J^2}(\psi(0) - \psi(1)). \tag{15.127}$$

A similar expression gives the vorticity on other walls.

In general we would need to specify the inflow at one end of the channel and outflow conditions where we allow the flow to exit as smoothly as possible at the other. The inflow velocity can be specified by giving the stream function and the vorticity and for the outflow the trick is to disrupt the flow as little as possible, so the conditions there have no significant effect on the upstream flow. We can, for example, assume that $\partial\psi/\partial n = 0$, which requires the streamlines at the outlet to be straight, but allows the flow to be otherwise non-uniform. If the channel can be assumed to be periodic, so that whatever flows out at one end enter through the other, then the boundary conditions are even simpler. This is the approach taken for the channel in Fig. 15.24.

The generation of body-fitted grids

Before we proceed to solve the transformed equations for an irregular domain we must set up the grid. This can be done in several ways, and elaborate techniques to automate the generation of body-fitted grids have been developed (see, for example, Thompson et al., 1998). For simple problems we can, however, do this by straightforward interpolations. For domains that have two boundaries that are clearly opposite each other, such as the domain in Fig. 15.23, we can often put an equal number of points on those boundaries that define the $\eta = 0$ and $\eta = M$ grid lines and then draw a straight line between those, defining the ξ grid lines. The remaining η grid lines can then be defined by dividing the ξ grid lines evenly.

In Fig. 15.24, we show one example of the use of the mapped Navier-Stokes equations in vorticity-stream function form. The domain is a curved channel, with periodic boundaries, so the flow leaving through the right boundary enters through the left one. The computations were done using a 30×60 grid but in the top frame we show only a few grid lines. The viscosity is $\nu = 0.01$ and the width of the channel at the entrance is 0.24 computational units. The solution was advanced with $\Delta t = 0.0005$ up to time 0.1. The stream function at that time is shown in the middle frame and the vorticity in the bottom frame. The solution is unsteady and once the vortices grow large enough they are swept downstream and re-enter through the left boundary. The computer code used for the simulation is not included here, but it has essentially the same structure as Code 3, for the rectangular domain, except that the grid must be generated and the equations are more complicated.

We note that even though any mapping should in principle work, for accurate results it is better to avoid sudden changes in the grid spacing and to keep the coordinate lines as orthogonal as practical. Furthermore, we note that it can be shown that it is more accurate to evaluate the geometric quantities x_η, x_ξ, and so on by the same numerical scheme as is used for the partial differential equation, even though the analytical form of the mapping may be known. For more complicated domains, we can sometimes divide the domain into more easily gridded patches and interpolate the grid points linearly within each patch. Usually the resulting grid can then by smoothed a little bit to make the transition between different patches smoother.

In addition to mapping the grid to align the grid lines with the boundary of the domain, mapping can also be used to refine the grid locally. In the simple interpolation strategy described above, we could, for example, use nonlinear interpolation to cluster grid lines near one wall. More sophisticated techniques allow for the grid to adapt to the solution, but those are beyond the scope of this brief introduction.

Unstructured grids

In spite of their accuracy and simplicity, structured grids (including body-fitted ones) are not well suited for complex geometries and essentially all commercial solvers use *unstructured grids*. In principle, the control volumes can have arbitrary

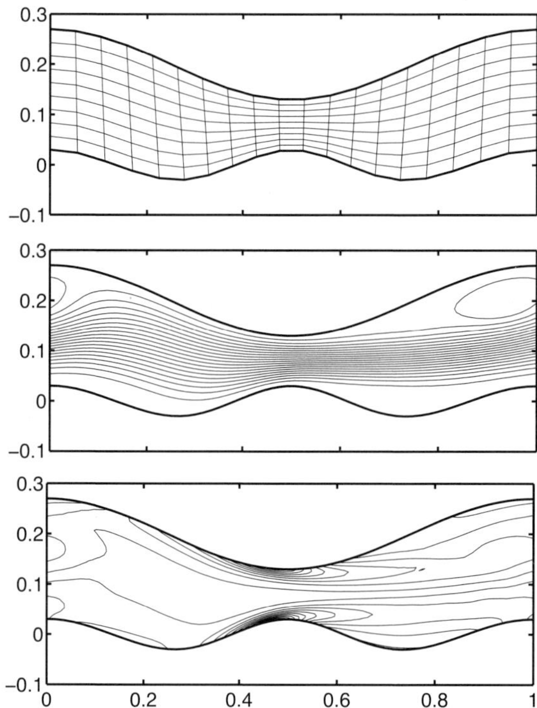

FIGURE 15.24 The flow in a channel with curved walls. The top frame shows every fifth grid line, the middle frame shows the stream function at time 0.1, and the vorticity at the same time is plotted in the bottom frame.

shapes, although tetrahedrons (or triangular pyramids) are common. Their main advantage is the possibility of adopting the grid to a complex boundary and thus discretizes complex geometries better and more easily. The flexibility comes at the cost of significant complexity, since the connectivity, or what control volumes share faces with each other, must be specified, the execution times are usually longer, and the accuracy less, than for structured grid of the same grid density. Extending codes based on unstructured grids from two-dimension to three-dimensions generally requires more effort than for codes based on structured grids. Nevertheless, such grids allow solutions of problems that would be difficult or impossible to compute in any other way. Although numerical algorithms used to solve for fluid flow using unstructured grids are much more complex than those described here, the basic concepts and principles are the same.

15.5 Velocity-pressure method for compressible flow

The equations governing compressible flows are, in principle, solved in the same way as those for incompressible flows. There are, however, some major differences. First of all, pressure is now a thermodynamic variable that is found by solving the energy equation. While this adds one more equation that needs to be integrated in time, the absence of an elliptic equation that connects all the grid points together at every time step is a significant simplification. This does, in particular, allow the equations to be solved using *splitting* where multidimensional problems are solved as a sequence of one-dimensional problems. Furthermore, viscosity is often ignored, which reduces the number of boundary conditions that must be specified, and frees us from worrying about the resolution of boundary layers and internal strained-diffusion layers. However, compressible flows often have shocks that can form spontaneously as the flow evolves and where the flow state changes abruptly. The history of the development of solution methods for compressible flows is, to a large extent, the story of capturing shocks accurately.

Before moving to the full Euler equations, it is worth introducing the essential problem with numerical computations of shocks by a simple example. To do so we solve the linear advection equation ((15.11), with $D = 0$), using discontinuous initial conditions where the solution is unity left of the discontinuity and zero to the right. Here we take the advection velocity to be positive and equal to unity. We show the solution in Fig. 15.25 as obtained by two methods, using the same spatial and temporal resolution in both cases. In one case we use a simple first-order upwind method and in the other case we use a second-order centered difference method. Since the simple forward-in-time, centered-in-space method used for the advection-diffusion equation is unstable when there is no diffusion, here we have used the so-called two-step Lax-Wendroff method, where the solution is found by:

$$
\begin{aligned}
f_{j+1/2}^{n+1/2} &= \frac{1}{2}\left(f_{j+1}^n + f_j^n\right) - \frac{\Delta t U}{2\Delta x}\left(f_{j+1}^n - f_j^n\right) \\
f_j^{n+1} &= f_j^n - \frac{\Delta t U}{\Delta x}\left(f_{j+1/2}^{n+1/2} - f_{j-1/2}^{n+1/2}\right).
\end{aligned}
\tag{15.128}
$$

The subscript $j \pm 1/2$ and the superscript $n + 1/2$ denote intermediate values that can be interpreted to lie in between the original grid points. Clearly, both methods do a miserable job, but the errors are very different. For the low-order method (upwind) the solution is smeared out so the discontinuity is hardly recognizable but for the higher-order scheme the solution oscillates around the shock, again making it non-obvious that the true solution is a simple discontinuity that propagates to the right with unit velocity. For linear schemes (namely those that we have been dealing with) this is the general case: first-order schemes give a monotonic (no oscillation) solution that is smeared, but higher-order schemes produce oscillations around shocks.

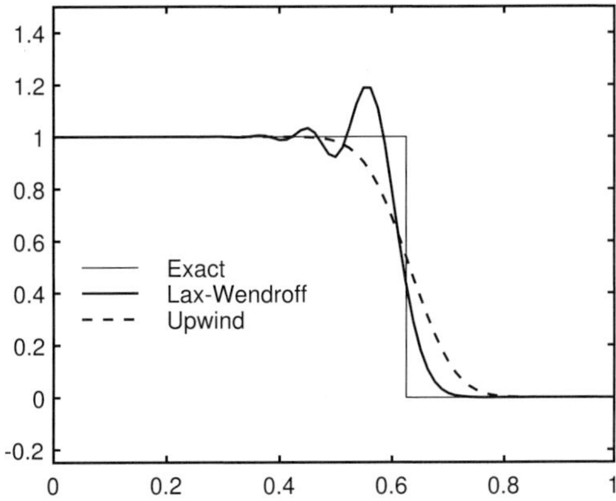

FIGURE 15.25 Advection of a shock by linear schemes. Here, $U = 1$, and a domain of length 1 is resolved by 81 evenly spaced grid points. $\Delta t = \Delta x/2$.

Modern methods are capable of producing accurate solutions for flows with sharp shocks and are high order almost everywhere, but here we will only present a simple low-order method. Low-order methods are usually very robust and are often preferable during initial exploration of a new problem. Relatively few things can go wrong and we can assure ourselves that we have the problem setup and the basic aspects correct before moving on to more sophisticated approaches. The simplest approach is clearly to use a centered scheme where all parts of the domain are treated in the same way for all governing parameters and initial and boundary conditions. Unfortunately, the forward in time, centered in space scheme used earlier is not only second order (which is usually good, but not if we are looking specifically for a first-order scheme), but also unstable for the advection only problem. The simplest first-order centered difference scheme is the Lax-Fredrich scheme, obtained by replacing f_j^n by $\frac{1}{2}(f_{j+1}^n + f_{j-1}^n)$ in the forward in time, centered in space scheme:

$$
f_j^{n+1} = \frac{1}{2}\left(f_{j+1}^n + f_{j-1}^n\right) - \frac{\Delta t U}{2\Delta x}\left(f_{j+1}^n - f_{j-1}^n\right).
\tag{15.129}
$$

Unfortunately it is so diffusive that it is rarely used directly, but sometimes it is the starting point for more sophisticated schemes.

The next alternative is the first-order upwind scheme (15.37). For a single advection equation it is straightforward but for the Euler equations we have signal propagation not only by the flow velocity u but also by the speed of sound $(u \pm c)$. If the flow velocity exceeds the speed of sound then all three signal speeds have the same sign, but if the flow velocity is lower, then the "upstream" direction can depend on what signal speed we are looking at. The applications of upwind methods to systems of equations thus require us to determine what part of the flux goes where. The standard way of dealing with this is to split the fluxes into two, accounting for the different directions.

One-dimensional flow

The Euler equations can be written as one vector equation:

$$\frac{\partial \mathbf{f}}{\partial t}+\frac{\partial \mathbf{F}}{\partial x}=0, \tag{15.130}$$

where:

$$\mathbf{f}=\begin{bmatrix}\rho\\ \rho u\\ \rho e\end{bmatrix} \quad \text{and} \quad \mathbf{F}=\begin{bmatrix}\rho u\\ \rho u^2+p\\ \rho e u+pu\end{bmatrix}. \tag{15.131}$$

To apply upwinding we split the fluxes into two parts, $\mathbf{F}=\mathbf{F}^{+}+\mathbf{F}^{-}$, where the first part accounts for right running waves and can be discretized using points j and $j-1$ and the second part includes left running waves and can be discretized using points j and $j+1$. The fluxes can be split in different ways and it is not obvious which is the best one. Here we will use the Zha-Bilgen flux vector splitting, which is one of the simplest ones and has a clear physical interpretation. We start by splitting the fluxes into two parts:

$$\mathbf{F}=\begin{bmatrix}\rho u\\ \rho u^2\\ \rho e u\end{bmatrix}+\begin{bmatrix}0\\ p\\ pu\end{bmatrix}. \tag{15.132}$$

The first part can be written as $u\mathbf{f}$ and it can be shown that it represents information carried by the flow velocity u, so the upwind direction depends only on the sign of u. The second part of the fluxes represents information carried by $u \pm c$ and for supersonic flow, where the Mach number is either larger than 1 or smaller than -1, the upwind direction again depends on the sign of the velocity. For $-1 < M < 1$, the second term is split into two and allocated to the upwind and downwind directions as a linear function of the Mach number. Thus, following Laney (1998), the scheme is:

$$\mathbf{F}^{+}=\max(u,0)\begin{bmatrix}\rho\\ \rho u\\ \rho e\end{bmatrix}+\begin{bmatrix}0\\ p^{+}\\ (pu)^{+}\end{bmatrix} \tag{15.133}$$

and

$$\mathbf{F}^{-}=\min(u,0)\begin{bmatrix}\rho\\ \rho u\\ \rho e\end{bmatrix}+\begin{bmatrix}0\\ p^{-}\\ (pu)^{-}\end{bmatrix}, \tag{15.134}$$

where we note that the first part depends only on the sign of the velocity. The variables in the second part are found by:

$$p^{+}=p\begin{cases}0, & M\le +1\\ \frac{1}{2}(1+M), & -1<M<1,\\ 1, & M\ge 1\end{cases} \qquad p^{-}=p\begin{cases}1, & M\le -1\\ \frac{1}{2}(1-M), & -1<M<1,\\ 0, & M\ge 1\end{cases} \tag{15.135}$$

and

$$(pu)^{+}=p\begin{cases}0, & M\le -1\\ \frac{1}{2}(u+c), & -1<M<1\\ u, & M\ge 1\end{cases} \qquad (pu)^{-}=p\begin{cases}u, & M\le -1\\ \frac{1}{2}(u-c), & -1<M<1.\\ 0, & M\ge 1\end{cases} \tag{15.136}$$

For $M \ge 1$, $\mathbf{F}=\mathbf{F}^{+}$ and for $M \le 1$, $\mathbf{F}=\mathbf{F}^{-}$, as we intended.

To discretize Eq. (15.130), we first write:

$$\frac{\partial \mathbf{f}}{\partial t}+\frac{\partial \mathbf{F}^{+}}{\partial x}+\frac{\partial \mathbf{F}^{-}}{\partial x}=0, \tag{15.137}$$

and then approximate the fluxes using the first-order upwind scheme:

$$\mathbf{f}_j^{n+1} = \mathbf{f}_j^n - \frac{\Delta t}{\Delta x}\left(F^+\left(\mathbf{f}_j^n\right) - \mathbf{F}^+\left(\mathbf{f}_{j-1}^n\right) + \mathbf{F}^-\left(\mathbf{f}_{j+1}^n\right) - \mathbf{F}^-\left(\mathbf{f}_j^n\right)\right). \tag{15.138}$$

For a numerical implementation it is possible to rewrite (15.138) as:

$$\mathbf{f}_j^{n+1} = \mathbf{f}_j^n - \frac{\Delta t}{\Delta x}\left(\widehat{\mathbf{F}}_{j+1/2}^n - \widehat{\mathbf{F}}_{j-1/2}^n\right), \tag{15.139}$$

where the fluxes at the half-points are found by:

$$\widehat{\mathbf{F}}_{j+1/2}^n = \mathbf{F}^+\left(\mathbf{f}_j^n\right) + \mathbf{F}^-\left(f_{j+1}^n\right). \tag{15.140}$$

Using that $\max(f, 0) = 0.5(f + |f|)$ and $\min(f, 0) = 0.5(f - |f|)$ for any f, we can write the fluxes for $-1 < M < 1$ as:

$$\begin{aligned}\widehat{\mathbf{F}}_{j+1/2}^n = {} & \mathbf{F}\left(f_{j+1}^n\right) + \mathbf{F}\left(f_j^n\right) - \frac{1}{2}\left(|u_{j+1}^n|\begin{bmatrix}\rho_{j+1}^n \\ \rho_{j+1}^n u_{j+1}^n \\ \rho_{j+1}^n e_{j+1}^n\end{bmatrix} - |u_j^n|\begin{bmatrix}\rho_j^n \\ \rho_j^n u_j^n \\ \rho_j^n e_j^n\end{bmatrix}\right) \\ & - \frac{1}{2}\left(\begin{bmatrix}0 \\ p_{j+1}^n M_{j+1}^n \\ p_{j+1}^n c_{j+1}^n\end{bmatrix} - \begin{bmatrix}0 \\ p_j^n M_j^n \\ p_j^n c_j^n\end{bmatrix}\right),\end{aligned} \tag{15.141}$$

and we use this form in the code shown below. Once the fluxes have been found by (15.141), the variables are updated using (15.139).

As methods for the Euler equations are developed, new methods are often tested using the problems used to test earlier methods. Several of these tests have been used by so many researchers that they have become de facto standards. One such test was introduced by Sod (1978) and consists of discontinuous initial conditions with high-pressure gas occupying the left half of a domain and a low pressure gas in the right half. The solution consists of the high pressure gas expanding into the low pressure gas, sending a shock wave ahead and a rarifaction wave in the other direction. Thus, the solution consists of (moving from right to left): undisturbed gas at low pressure on the right; compressed low-pressure gas (separated by a shock from the undisturbed gas); expanded high-pressure gas moving into the right hand side of the domain; a rarifaction fan where the initially compressed gas expands smoothly; and finally the undisturbed high-pressure gas on the left. The exact solution is easily found, as explained in Chapter 14, and once the conditions in each region have been found, the time evolution is simply given by expanding the time axis at a constant rate. A code for this problem, using (15.141) and (15.139), is given below and sample results are given in Fig. 15.26, where we plot the density, the pressure, the velocity, and the Mach number at time 0.005. In all cases, we plot results for three different resolutions. The time step must be limited by the Courant conditions, which say that the signal cannot travel more than one grid space in each time step, or $(u + c)\Delta t < \Delta t$, and we take the time step to be 0.45 times the maximum step. The shock (the rightmost discontinuity) is captured reasonably well for the finer resolutions, but the contact discontinuity (the second density discontinuity from the right) is relatively smooth for all three resolutions. It is generally found that maintaining sharp contacts is more difficult than keeping shocks sharp.

CODE 5: SOLUTION OF THE UNSTEADY ONE-DIMENSIONAL EULER EQUATIONS

```
% Upwind-Zha-Bilgen flux splitting
nx=40*256; tfinal=0.005; xl=10.0; time=0; gg=1.4;
h=xl/(nx-1);for i=1:nx,x(i)=h*(i-1);end
p_left=100000;p_right=10000;r_left=1;r_right=0.125;u_left=0;

r=zeros(1,nx);ru=zeros(1,nx);rE=zeros(1,nx);
p=zeros(1,nx);c=zeros(1,nx);u=zeros(1,nx);m=zeros(1,nx);
F1=zeros(1,nx);F2=zeros(1,nx);F3=zeros(1,nx);
```

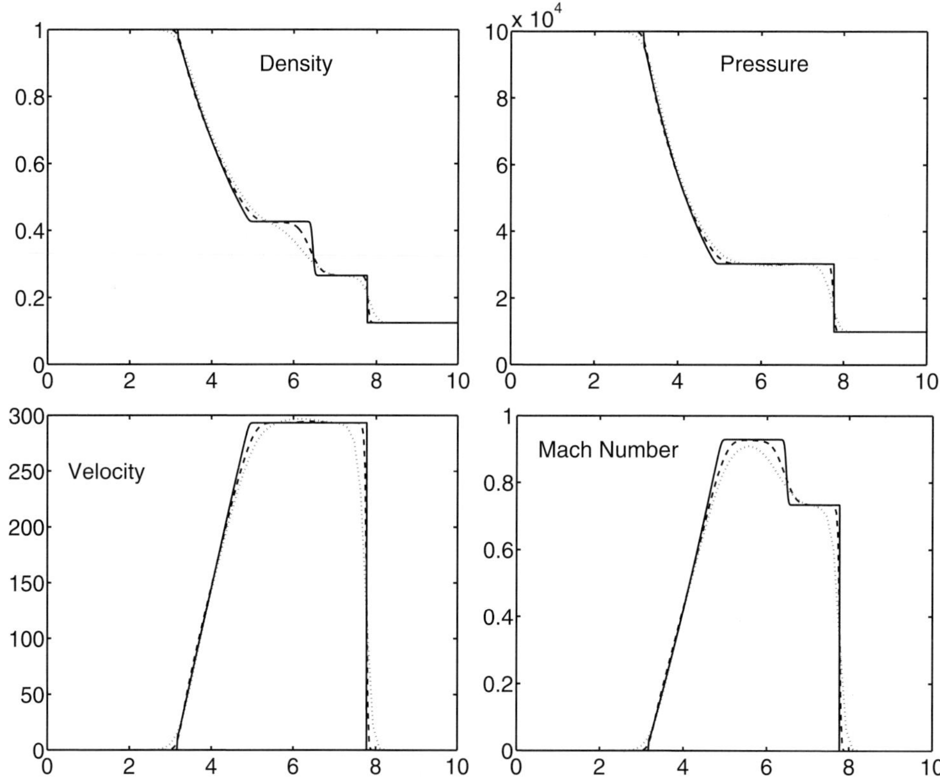

FIGURE 15.26 The solution of a shock tube problem at time 0.005 computed using a first-order upwind method with the Zha-Bilgen splitting, with grids using 64 (dotted line), 256 (dashed line), and 1024 (solid line) grid points.

```
for i=1:nx, r(i)=r_right;ru(i)=0.0;rE(i)=p_right/(gg-1);end
for i=1:nx/2; r(i)=r_left; rE(i)=p_left/(gg-1); end
cmax=sqrt( max(gg*p_right/r_right,gg*p_left/r_left) );
dt=0.45*h/cmax; maxstep=tfinal/dt;

for istep=1:maxstep
  for i=1:nx,p(i)=(gg-1)*(rE(i)-0.5*(ru(i)*ru(i)/r(i)));end
  for i=1:nx,c(i)=sqrt( gg*p(i)/r(i) );end
  for i=1:nx,u(i)=ru(i)/r(i);end;
  for i=1:nx,m(i)=u(i)/c(i);end

  for i=1:nx-1 % Find fluxes
    F1(i)=0.5*(ru(i+1)+ru(i))-0.5*(abs(ru(i+1))-abs(ru(i)));
    F2(i)=0.5*(u(i+1)*ru(i+1)+p(i+1)+u(i)*ru(i)+p(i))...
         -0.5*(abs(u(i+1))*ru(i+1)-abs(u(i))*ru(i))...
         -0.5*(p(i+1)*m(i+1)-p(i)*m(i));
    F3(i)=0.5*(u(i+1)*(rE(i+1)+p(i+1))+u(i)*(rE(i)+p(i)))...
         -0.5*(abs(u(i+1))*rE(i+1)-abs(u(i))*rE(i))...
         -0.5*(p(i+1)*c(i+1)-p(i)*c(i));
    if m(i) > 1, F2(i)=ru(i)*u(i)+p(i);
                         F3(i)=(rE(i)+p(i))*u(i);end
    if m(i) < -1, F2(i)=ru(i+1)*u(i+1)+p(i+1);
                       F3(i)=(rE(i+1)+p(i+1))*u(i+1);end
  end

  for i=2:nx-2 % Update solution
    r(i)=r(i)-(dt/h)*(F1(i)-F1(i-1));
    ru(i)=ru(i)-(dt/h)*(F2(i)-F2(i-1));
    rE(i)=rE(i)-(dt/h)*(F3(i)-F3(i-1));
```

```
    end
    time=time+dt,istep
  end
  plot(x,r,'k','linewidth',2);
```

We iterate again that the Zha-Bilgen scheme is just one of many splitting techniques that have been proposed.

Two-dimensional flow

Methods for compressible flows are usually extended to multidimensional flows by simply applying the one-dimensional method to each direction separately. There are at least a couple of ways to do this. We can do both directions simultaneously or we can do the advection in different directions in sequence. Here we take the first approach. Just as for one-dimensional problems, in two dimensions there are now a few test cases that have become standard and are used for essentially every new method. The one we use here was introduced by Woodward and Colella (1984) and consists of the collision of an oblique shock with a solid wedge, computed in a rectangular domain. Fig. 15.27 shows the setup. A strong shock propagates from the left and collides with the wedge. The domain is aligned with the sloping wall and is taken to be a 1×4 box, with the shock initially at a 60° angle near the left boundary. The shock is pinned at to the bottom boundary at $x = 1/6$, by specifying the conditions behind the shock to the left of this point and symmetry boundary conditions are applied on the rest of the $y = 0$ axis. The conditions behind the shock are propagated along the upper wall with the shock velocity and the rest of the wall is taken to be at the rest conditions ahead of the shock. When the shock hits the wedge, it is reflected, forming a complex pattern of shocks and contact discontinuities. The density contours at time 0.2, computed using a straightforward extension of the one-dimensional code to two dimensions, are shown in Fig. 15.28, computed using three different grid resolutions. Even on the finest grid, the contacts in the region where the original shock hits the wall are not resolved. The overall shock structure, however, is.

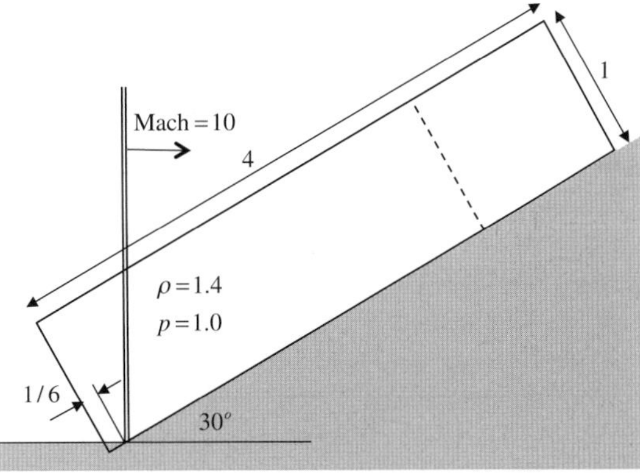

FIGURE 15.27 The collision of an oblique shock with a wedge. The schematic is based on: http://folk.uio.no/kalie/fronttrack/gas/wedge/index.html.

The first-order upwind method presented here requires a fairly large number of grid points to produce reasonably accurate solution. Obviously, more accurate higher-order schemes would be much better. Many high-order schemes, such as the Lax-Wendroff scheme (15.128), have been developed for the Euler equations and generally they work well for smooth flows but produce unsatisfactory oscillatory solutions around shocks and discontinuities. The simplest remedy is to smooth the oscillations by adding viscous terms to the equations. The physical viscosity does, of course, smooth shocks but this smoothing is far too small for the grid resolutions used in practice. However, an *artificial viscosity* constructed in such a way that it is essentially zero where the solution is smooth but takes on large enough values to modify the solution at discontinuities was used early on to allow the use of higher-order scheme for flows with shocks. Modern approaches to capturing shocks approach the problem somewhat differently, although they also selectively add dissipation around shocks to damp out oscillations. This is usually achieved by modifying the fluxes in such a way that the solution at the new time step is free of oscillations. For higher-order methods it is generally necessary to find the value of the variable in each cell at the cell boundary (only for first-order methods can this value be assumed to be equal to the average value) and modern methods are generally based on adjusting the value (or the flux based on this value) in such a way that no oscillations appear. For an introduction to the large body of work describing modern shock capturing methods see, for example, Laney (1998).

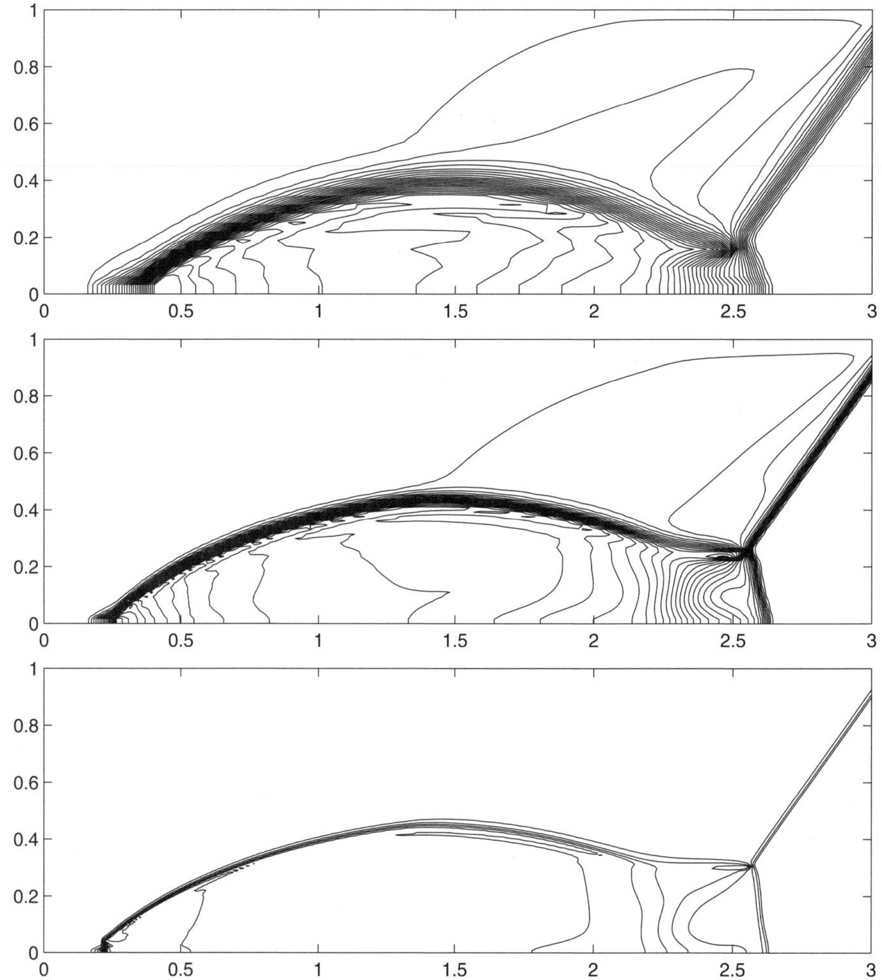

FIGURE 15.28 The collision of a strong oblique shock with a 30° wedge. Density contours are shown for three grids at time 0.02, with $\Delta x = \Delta y = 1/32$ at the top, $\Delta x = \Delta y = 1/64$ in the middle, and $\Delta x = \Delta y = 1/128$ at the bottom. The computational domain is 1×4 but only 1×3 is plotted.

15.6 More to explore

Computational fluid dynamics is a complicated subject but we have hopefully been able to show here that the foundation rests on relatively simple applications of the fundamental principles of fluid dynamics and numerical approximations. As we have only covered the very elementary aspects of computational fluid dynamics, there are many topics that we have left out. We focused on two-dimensional flows to keep the presentation and the visualization of the results simple but extensions to three-dimensional flows are relatively straightforward, particularly for the pressure-velocity formulation. Here we discuss briefly a few of the other topics that an expert in CFD needs to know.

Other methods

The computational approaches described here—finite volume and finite difference for simple domains—are the most straightforward discretizations of the Navier-Stokes and Euler equations. They are, however, far from the only ones and more complicated methods fall into several categories. First of all, the methods described here can be improved by higher-order and/or more robust discretization schemes, while staying with a regular structured grid. There is a large number of possibilities available to improve their stability, accuracy, and robustness and those have not been all been explored yet. For unstructured grids and control volumes of arbitrary shapes this is even more the case.

Other discretization strategies are also possible. The most common of those are finite element methods, spectral methods, Lattice Boltzmann method, vortex methods, and smoothed particle hydrodynamics methods. All of those come in several different versions.

Finite element methods dominate computations of stresses and strains in solid mechanics and considerable effort has been devoted to develop finite element methods for fluid dynamics. As the name implies, the computational domain is divided into finite elements that connect nodal points. In solid mechanics those are often material points, but in fluid dynamics the finite elements are usually stationary and thus resemble the control volumes used in the finite volume method. The governing equations are approximated by low-order functions in each element and discrete approximations are obtained by taking the weighted average over each element. In a widely used variant the governing equations are weighted by the functions describing the solution in each element, resulting in Galerkin methods. In traditional finite element methods the functions are continuous at element boundaries so the nodal points are all connected and for unsteady problems we need to solve a set of linear equations, even when using an explicit time integration. More recently, this requirement has been abandoned in discontinuous Galerkin (DG) methods where the solution is approximated within each element by a smooth function, but allowed to be discontinuous across element boundary. This is, of course, exactly what happens in finite volume methods where the solution is approximated as a constant within each control volume. Indeed, higher-order finite volume methods can be constructed using similar ideas. Finite element methods have been used successfully to solve a wide range of fluid mechanics problems, but they are not as intuitive as finite volume methods and many new ideas, such as for shock capturing, were developed first for finite volume methods and only later implemented in finite element methods. Furthermore, a large fraction of the finite element terminology has been developed in the context of solid mechanic applications and is therefore not immediately intuitive in a fluid mechanics context.

Traditional finite elements assume that the solution in each element can be approximated by a series of known elementary functions, whose coefficients we need to find. Usually we only keep the first few terms in the series and if a more accurate solution is needed, we reduce the size of the elements and increase their numbers. Alternatively, we could increase the order of the series and use more terms. In the limit, we can take the whole domain as one big element and use a series with a large number of terms. Such methods are called spectral methods, partly because Fourier series are most commonly used to approximate the solution. Spectral methods work extremely well for simple domains with simple boundary conditions and are generally more accurate than any other numerical method for comparable resolution. Thus, early fully resolved simulations of turbulent flows relied almost exclusively on spectral methods and domains with periodic boundary conditions. While it is possible to use other functions (such as Chebyshev polynomials) to accommodate more complicated boundaries and spectral-element methods attempt to combine the accuracy of spectral methods with the versatility of finite element methods, spectral methods remain a rather specialized branch of CFD, used mostly for fundamental studies of turbulent flows in simple geometries.

While finite volume, finite element, and spectral methods are usually based on the Eulerian formulation of the governing equations, a number of methods have been developed that use the Lagrangian formulation so that the computational elements follow material particles. The best-known and most successful of those are vortex methods for inviscid incompressible flows. Vortex methods use the fact that vorticity moves with material particles and that for unbounded domain the fluid velocity is determined by the vorticity, as discussed in Chapter 5. Thus, it is necessary to follow only the fluid that has non-zero vorticity and often the vortical fluid occupies only a small fraction of the domain. Thus, by representing the vorticity by smooth vortex "blobs," we only need to follow the motion of the particles representing the "blobs." Vortex methods have been extended to include viscous diffusion of vorticity and vorticity generation at boundaries, but in general they are most suitable for high Reynolds number unbounded flows where vorticity regions are compact and diffusion is small.

For inviscid incompressible flows over boundaries, the flow is irrotational in the absence of vorticity and determined completely by the boundary conditions. The velocity potential can be found by solving a Laplace equation by a boundary integral method. Such *panel* methods once enjoyed considerable popularity in the aerospace industry (since flow over a well-designed wing will shed very little vorticity) but have now largely been replaced with full solutions of the Euler and the Navier-Stokes equations. Boundary integral methods have also been used to compute the evolution of free surfaces and internal waves in inviscid irrotational flows, and for some applications boundary integral and vortex methods have been combined.

Other particle methods are based on representing fluid masses by particles directly. In the smoothed particle hydrodynamics (SPH) method the particles are smoothed so that they overlap slightly and, as in vortex methods, the problem is reduced to follow the motion of particles representing the smoothed blobs. However, unlike vortex methods for inviscid flows, we generally need to account for interactions of particles. SPH are particularly well suited for large amplitude-free surface motion, where the fluid only occupies part of the computational domain, but the quality of the results generally are not compatible with what can be obtained by finite volume methods specialized to handle free-surface flows.

Other alternative methods include the use of physical models whose connection to the Navier-Stokes equations is more tenuous, such as the Lattice Boltzmann method (LBM), where discrete "particles" are moved on a regular grid in such a way that the average motion approximates solutions to the Navier-Stokes equations. While considerable effort has been devoted to develop the various versions of the LBM and it is capable of producing solutions comparable to those obtained by other methods, it is not obvious that it offers much advantage over comparable finite volume methods. Extending LBM to problems with complex physics is, in particular, not as straightforward. LBM methods are sometimes used to model "mesoscopic" physics not captured by the Navier-Stokes equations but are not quite suitable for molecular simulations either. We will not discuss methods for flows where the continuum assumption breaks down (as in rarified gases, for example), but note that modeling of such flows is an active research area.

More complex physics

In this chapter, we focused exclusively on the fluid flow. In many practical applications we are interested in the fluid flow because it affects other physical processes, such as heat or mass transfer or combustion. In principle, additional physics usually results in additional equations that can be solved by the methods already described. Heat and mass transfer, for example, lead to additional advection-diffusion equations, often with source terms such as those that arise due to combustion. In practice, however, the new physical processes often take place at time and length scales that are very different from the fluid flow and this disparity may make it difficult to solve the equations together. The diffusion of mass in liquid, for example, is many times slower than the diffusion of momentum, resulting in mass boundary layers that are many times thinner than the fluid boundary layers and thus require much finer grids. The same thing is true for combustion when the reaction zones are often orders of magnitude thinner than any flow scale.

Another class of problems that we have not addressed here include free-surface and interface and multiphase flows. Such problems are fairly common in both industry and nature and include, for example, ocean waves, rain, boiling heat transfer, atomization and sprays for many applications, condensation, and refrigeration. For such problems, it is generally necessary to track the interface accurately since not only does the interface separate regions of very different material properties, such as density and viscosity, but it can also introduce additional physical effects like surface tension. Often, just as for homogeneous flow problems, free-surface and interface problems also require us to account for other physical effects such as heat and mass transfer, electrical and magnetic forces, and so on. In addition, fluid problems must often be solved along with equations governing stresses and strains and heat transfer in adjacent solids.

Direct numerical simulations

For many problems of practical interest, we find that the fluid flow is turbulent and consists of whorls and swirls of many shapes and sizes. For most industrial problems the detailed motion of each whorl is not of much interest and we are interested only in the average motion. However, as discussed in Chapter 10, the derivation of equations for the average motion, by averaging the Navier-Stokes equations, results in unknown terms that describe the mixing and momentum transfer by the unsteady small-scale motion. Finding how these terms depend on the average motion, and sometimes on quantities that describe the average state of the small-scale motion, is referred to as the closure problem and is widely regarded as one of the great unsolved problems in physics.

It is important, however, to remember that the fluid motion is described by the Navier-Stokes equations and that turbulent motion is simply a complicated and unsteady solution of the Navier-Stokes equations. Thus, if we had an infinitely large and fast computer, we could solve for the flow, turbulence, and all! Although we do not have infinitely large computers, we have very large ones and it *is* possible to solve the Navier-Stokes equations for simple turbulent flows, where the range of resolved scales is modest, and then average the results to find what the closure terms look like and how they depend on the average motion. Such simulations are referred to as direct numerical simulations, or DNS, and although the term is sometimes used more broadly, generally implying a fully resolved motion of unsteady flow that contains a reasonably large range of scales. DNS studies of turbulent flows have been a very active research area for the last few decades and have had a profound impact on our understanding of turbulent flows (see, for example, Pope, 2000). Although DNS were originally limited to constant density flows in simple geometries, more complex situations, including multiphase and reacting flows, are now being studied in this way (as discussed, for example, in Tryggvason et al., 2011). The methods described earlier in this chapter, extended to three-dimensional flows, allow us to carry out DNS, although in practice higher-order versions are usually preferred.

Multiscale modeling

For the vast majority of fluid dynamics problems of practical interest the flow is turbulent and we are only after the average or large-scale behavior. Even for those cases where DNS could be done, they are generally too expensive for routine engineering work. Furthermore, a large number of experimental studies suggest that to a large degree the smaller scales exhibit a significant universality. That is, given the large-scale flow—and perhaps a few characteristics of the small-scale motion—the average effect of the small scales is determined. Thus, there are reasons to believe that in many cases we can model the effect of the small scales on the large or average scales without resolving them in detail. As discussed in Chapter 10, the Reynolds-average Navier-Stokes (RANS) equations contain unknown terms, the Reynolds stresses, that result from averaging the nonlinear advection terms. These terms are generally modeled by introducing a turbulent viscosity that multiplies the average deformation tensor. Unlike the regular viscosity, which is a material constant, the turbulent viscosity depends on the flow, and several proposals have been forwarded to account for that relationship. Current industrial practice is to assume that the turbulent viscosity depends on two quantities that describe the average properties of the unresolved motion. The first of these is usually taken to be the turbulent kinetic energy, k, which roughly describes the intensity of the turbulent motion. The second variable describes the length scale of the motion, and while several variables are in use, the turbulent dissipation rate, ε, is most commonly used. k and ε at each point in the flow are found by solving advection-diffusion equations with source and sink terms. These equations can be constructed from the Navier-Stokes equations but contain their own closure terms. Thus, from a computational point of view the inclusion of a turbulence model requires the solution of additional equations somewhat similar to the Navier-Stokes equations. The new equations, however, introduce two new challenges: stiffness and what to do at walls. The first problem comes from the different effective diffusivity and the source terms, and the second comes from the fact that the flow goes to zero in a very thin layer near the wall. The first problem may call for implicit methods and the second is often solved by additional modeling where a "wall-function" is introduced to account for the thin layer near the wall. Several specialized books are available that cover turbulence modeling and computations in great detail (see Wilcox, 2006 and Pope, 2000, for example).

In addition to sub-grid models for unresolved turbulent scales, sub-grid models are also required for a large range of other physical phenomena encountered in flows of interest for practical applications. Those include multiphase flows, gas-liquid, and gas/liquid-solid flows where the phases are either mixed together or one phase is dispersed in another phase, as well as combustion and other flows with mass transfer and chemical reactions. A large range of sub-models have been developed, some relying on solving another advection/diffusion like equation or equations and others relying on tracking either individual particles (or drops or bubbles) or groups of particles.

Modeling of the effect of small scales on larger scales is generally most successful when the physical problem possesses large scale-separation (as is the case for gas molecules and continuum scales, for example). For turbulent flow this is not always the case and often the flow contains large-scale structures of the order of the domain in addition to a continuous range of smaller scales. Large scales can exhibit significant unsteady motion on relatively large time scales and that poses considerable challenges for averaged models. As computer power has increased it has become increasingly feasible to resolve part of the unsteady flow and use modeling only for small scales. Such large eddy simulations (LES) are increasingly common (see Sagaut, 2006, for an introduction). In the DNS community such efforts were initially focused on modeling only the energy dissipative scales and resolving all other scales and thus "true" LES require us to resolve a very large range of scales. More recently, the definition has been applied more loosely, and industry practice spans the range from unsteady RANS (URANS) and very large eddy simulations (VLES) to true LES. While the theoretical justification may not always be strong, the reality is that in the hands of an expert such simulations can have incredible predictive power.

Computational modeling of complicated flows is a rapidly evolving field where new progress is being made essentially every day. Increasingly, such simulations play a major role in the design of complex artifacts, and national laboratories and academic institution are hard at work to enable the modeling of even more challenging problems. As our abilities to conduct simulations of problems of increasing complexity increase and we desire to use computational modeling for the design of ever more complex systems, the question of how reliable the solution is, is becoming increasingly important. Thus, considerable effort has been devoted to verification and validation, often referred to as V&V. Those are fundamentally different tasks. Verification, as used in the context of computational predictions, is intended to confirm that a numerical code solves correctly a specific set of equations. Validation addressed whether the equations describe the physical process that we are examining sufficiently accurately for our purpose. In many cases our problems specification is inherently uncertain or incomplete (boundaries may not be as smooth as we assume, our material properties may vary, the system may be subject to disturbances that we have left out, often due to our inability to accurately describe them, and so on). In those cases, we often seek to predict how a given uncertainty in our problem specification translates into an uncertainty in the solution. *Uncertainty quantification*, or UQ, is a vast topic and we will not address in here, but a serious user of CFD should be aware of it. Finally, we note that many new developments are taking place in computational modeling in general and CFD in particular that attempt to use machine learning to greatly extend what we can do and how we do it. Those include the development of *surrogate models* which attempt to correlate results of expensive computations to allow fast predictions, the reconstruction of flow field from limited observations, development of subgrid closure models for multiscale simulations, and many other tasks. It is difficult at this time to predict exactly how machine learning, or artificial intelligence in general, will affect our abilities to predict, but the probability is that it will be significant.

We end this chapter by noting that CFD, like any other tool, must be used with care. The fact that a code produces a solution does not mean that it is the correct solution and a user of a numerical code is responsible for assessing the accuracy of his or her results. In addition to the possibility of the code producing an inaccurate—or even a wrong—solution, making a mistake in the problem specification, such as when specifying material properties and boundary conditions, is fairly easy. The computer is not a substitute for sound judgment and if the solution looks wrong, it is probably because it is.

Exercises

15.1 Show, by Taylor expansion, that:

$$\frac{d^3 f}{dx^3} \approx \frac{f_{j+2} - 2f_{j+1} + 2f_{j-1} - f_{j-2}}{2\Delta x^3}.$$

What is the order of this approximation?

15.2 Consider the following "backward in time" approximation for the diffusion equation:

$$f_j^{n+1} = f_j^n + \frac{\Delta t D}{\Delta x^2}\left(f_{j+1}^{n+1} + f_{j-1}^{n+1} - 2f_j^{n+1}\right)$$

(a) Determine the accuracy of this scheme.

(b) Find its stability properties by von Neumann's method. How does it compare with the forward in time, centered in space approximation considered earlier?

15.3 Approximate the linear advection equation:

$$\frac{\partial f}{\partial t} + U\frac{\partial f}{\partial x} = 0 \quad U > 0$$

by the backward in time method from problem 2. Use the standard second-order centered difference approximation for the spatial derivative.

(a) Write down the finite difference equation.

(b) Write down the modified equation.

(c) Find the accuracy of the scheme.

(d) Use von Neuman's method to determine the stability of the scheme.

15.4 Consider the following finite difference approximation to the diffusion equation:

$$f_j^{n+1} = f_j^n + 2\frac{\Delta t D}{\Delta x^2}\left(f_{j+1}^n - f_j^{n+1} - f_j^{n-1} + f_j^n\right).$$

This is the so-called Dufort-Frankel scheme, where the time integration is the "Leapfrog" method, and the spatial derivative is the usual center difference approximation, except that we have replaced f_j^n by (1/2) $(f_j^{n+1} + f_j^{n-1})$. Derive the modified equation and determine the accuracy of the scheme. Are there any surprises?

15.5 The following finite difference approximation is given:

$$f_j^{n+1} = \frac{1}{2}\left(f_{j+1}^n + f_{j-1}^n\right) - \frac{\Delta t U}{2\Delta x}\left(f_{j+1}^n - f_{j-1}^n\right).$$

(a) Write down the modified equation.

(b) What equation is being approximated?

(c) Determine the accuracy of the scheme.

(d) Use the von Neuman's method to examine under which conditions this scheme is stable.

15.6 Consider the equation:

$$\frac{\partial f}{\partial t} = g(f),$$

and the second-order predictor-corrector method:

$$f_j^* = f_j^n + \Delta t g(f^n)$$
$$f_j^{n+1} = f_j^n + \frac{\Delta t}{2}(g(f^n) + g(f^*).$$

Show that this method can also be written as:

$$f_j^* = f_j^n + \Delta t g(f^n)$$
$$f_j^{**} = f_j^* + \Delta t g(f^*)$$
$$f_j^{n+1} = (1/2)(f^n + f^{**}).$$

That is, you simply take two explicit Euler steps and then average the solution at the beginning of the time step and the end. This makes it particularly simple to extend a first-order explicit time integration scheme to second order.

15.7 Modify the code used to solve the one-dimensional linear advection equation (Code 1) to solve the Burgers equation:

$$\frac{\partial f}{\partial t} + \frac{\partial}{\partial x}\left(\frac{f^2}{2}\right) = D\frac{\partial^2 f}{\partial x^2}$$

using the same initial conditions. What happens? Refine the grid. How does the solution change if we add a constant (say 1) to the initial conditions?

15.8 Modify the code used to solve the two-dimensional linear advection equation (Code 2) to simulate the advection of an initially square blob with $f = 1$ diagonally across a square domain by setting $u = v = 1$. The dimension of the blob is 0.2×0.2, and it is initially located near the origin. Refine the grid and show that the solution converges by comparing the results before the blob flows out of the domain.

15.9 Derive a second-order expression for the boundary vorticity by writing the stream function at $j = 2$ and $j = 3$ as a Taylor series expansion around the value at the wall ($j = 1$). How does the expression compare with Eq. (15.67)?

15.10 Modify the vorticity-stream function code used to simulate the two-dimensional driven cavity problem (Code 3) to simulate the flow in a rectangular 2×1 channel with periodic boundaries. Set the value of the stream function at the top equal to $\psi = 1$. As initial conditions place two circular blobs with radius $r = 0.25$ and $\omega = 10$ on the centerline of the channel at $y = 0.5$ and $x = 0.6$ and $x = 1.4$. Refine the grid to ensure that the solution converges. Describe the evolution of the flow.

15.11 Derive the discrete pressure equation for a corner point for the velocity-pressure method described in Section 15.3. How does it compare with the equation for an interior point, (15.91) and a point next to a straight boundary (15.96)?

15.12 Modify the velocity-pressure code used to simulate the two-dimensional driven cavity problem (Code 4) to simulate the mixing of a jet of fast fluid with slower fluid. Change the length of the domain (L_x) to 3 and specify an inflow velocity of 1 in the middle third of the left boundary and an inflow velocity of 0.25 for the rest of the boundary. For the right boundary specify a uniform outflow velocity of 0.5. Keep other parameters the same. Refine the grid and check the convergence of the solution.

15.13 Extend the velocity-pressure code used to simulate the two-dimensional driven cavity problem (Code 4) to three dimensions. Assume that the third dimension is unity (as the current dimensions) and take the velocity of the top wall and the material properties to be the same. Compute the flow on a 9^3 and 17^3 grids and compare the results by plotting the velocities along lines through the center of the domain. How do the velocities in the center compare with the two-dimensional results?

15.14 Derive Eq. (15.121):

$$g_y f_x - g_x f_y = \frac{1}{J}\left(g_\eta f_\xi - g_\xi f_\eta\right).$$

15.15 Show that the equations for the first derivatives in the mapped coordinates (15.111) can be written in the so-called conservative form:

$$f_x = \frac{1}{J}\left((f y_\eta)_\xi - (f y_\xi)_\eta\right) \quad \text{and} \quad f_y = \frac{1}{J}\left((f x_\xi)_\eta - (f x_\eta)_\xi\right).$$

15.16 Derive Eq. (15.120):

$$\nabla^2 \xi = \frac{1}{J^3}\left(q_1(x_\eta y_{\xi\xi} - y_\eta x_{\xi\xi}) - 2q_2(x_\eta y_{\xi\eta} - y_\eta x_{\xi\eta}) + q_3(x_\eta y_{\eta\eta} - y_\eta x_{\eta\eta})\right).$$

Take $f = \xi$ and use that $\xi_\eta = 0$ and so on.

15.17 Derive numerical approximations for the velocity-pressure equations for a mapping where the grid lines are straight and orthogonal, but unevenly spaced. That is, $x = x(\xi)$ and $y = y(\eta)$ only. Assume that $\Delta\xi = \Delta\eta = 1$. How do these equations compare with (15.85), (15.86), (15.89), (15.90), and (15.91)?

15.18 When the velocity is high and diffusion is small, the linear advection-diffusion equation can exhibit boundary layer behavior. Assume that you want to solve (15.11) in a domain given by $0 \le x \le 1$, that $U > 0$, and that the boundary conditions are $f(0) = 0$ and $f(1) = 1$. The velocity U is high and the diffusion D is small so we expect a boundary layer near $x = 1$.

(a) Sketch the solution for high U and low D.

(b) Propose a mapping function that will cluster the grid points near the x = 1 boundary.

(c) Write the equation in the mapped coordinates.

15.19 Derive Eq. (15.141).

15.20 Propose a numerical scheme to solve for the unsteady flow over a rectangular cube in an unbounded domain. The Reynolds number is relatively low, 500-1000. Identify the key issues that must be addressed and propose a solution. Limit your discussion to one page and do NOT write down the detailed numerical approximations, but state clearly what kind of spatial and temporal discretization you would use.

15.21 You compare an analytic solution of the Navier-Stokes equations for a laminar planar jet to the numerical solution obtained with your Navier-Stokes code. Is this an exercise in validation or verification?

15.22 After implementing an immersed boundary method to simulate flow past a spherical object you compare the resulting drag coefficient against an empirical correlation obtained from Fig. 8.24. Is this an exercise in validation or verification?

References

Laney, C.B., 1998. Computational Gasdynamics. Cambridge University Press.

Pope, S.B., 2000. Turbulent Flows. Cambridge University Press.

Sagaut, P., 2006. Large Eddy Simulation for Incompressible Flows: An Introduction. Springer.

Sod, G.A., 1978. A survey of several finite difference methods for systems of nonlinear hyperbolic conservation laws. Journal of Computational Physics 27, 1–31.

Thompson, J.F., Soni, B.K., Weatherill, N.P. (Eds.), 1998. Handbook of Grid Generation. CRC Press.

Tryggvason, G., Scardovelli, R., Zaleski, S., 2011. Direct Numerical Simulations of Gas-Liquid Multiphase Flows. Cambridge University Press.

Wilcox, D.C., 2006. Turbulence Modeling for CFD, 3rd ed. DCW Industries.

Woodward, P., Colella, P., 1984. The numerical solution of two-dimensional fluid flow with strong shocks. Journal of Computational Physics 54, 115–173.

Further reading

Fletcher, C.A.J., 1988. Computational Techniques for Fluid Dynamics. Volumes I and II. Springer.

Ferziger, J.H., Peric, M., 2013. Computational Methods for Fluid Dynamics, 3rd ed. Springer.

Pletcher, R.H., Tannehill, J.C., Anderson, D., 2011. Computational Fluid Mechanics and Heat Transfer, 3rd ed. CRC Press.

Wesseling, P., 2000. Principles of Computational Fluid Dynamics. Springer.

Appendix A

Conversion factors, constants, and fluid properties

A.1 Conversion factors

Length:	1 m = 3.2808 ft 1 inch = 2.540 cm 1 mile = 1.609 km 1 nautical mile = 1.852 km
Mass:[a]	1 kg = 0.06854 slug = 1000 g ↔ 2.205 lbs 1 metric ton = 1000 kg
Time:	1 day = 86,400 s
Density:[a]	1 kg m^{-3} = 1.941 × 10^{-3} slugs ft^{-3} ↔ 0.06244 lbs/ft^3
Velocity:	1 knot = 0.5144 m/s
Force:	1 N = 10^5 dyn = 0.2248 lbs
Pressure:	1 dyn cm^{-2} = 0.1 N/m^2 = 0.1 Pa 1 bar = 10^5 Pa
Temperature:	°C = K − 273.15 = (5/9)(°F − 32)
Energy:	1 J = 10^7 erg = 0.2390 cal 1 cal = 4.184 J
Energy flux:	1 W m^{-2} = 2.39 × 10^{-5} cal cm^{-2} s^{-1}

[a] *At the earth's surface, the weight of a 1 kg mass is 2.205 lbs.*

A.2 Physical constants

Atmospheric Pressure:	101.3 kPa
Avogadro's Number:	6.023 × 10^{23} $gmole^{-1}$
Boltzmann's Constant:	1.381 × 10^{-23} J K^{-1}
Gravitational Acceleration:	9.807 m s^{-2} = 32.17 ft s^{-2} (at the surface of the earth)
Gravitational Constant:	6.67 × 10^{-11} m^3 kg^{-1} s^{-2}
Planck's Constant:	6.626 × 10^{-34} J s
Speed of Light in Vacuum:	2.998 × 10^8 m s^{-1}
Universal Gas Constant:	8.314 J $gmole^{-1}$ K^{-1}

A.3 Properties of pure water at atmospheric pressure

Here, ρ = density, α = coefficient of thermal expansion, μ = shear viscosity, v = kinematic viscosity = μ/ρ, κ = thermal diffusivity = $k/(\rho c_p)$, (k is first defined in Section 1.5) Pr = Prandtl number, and 1.0×10^{-n} is written as 1.0E − n.

T °C	ρ kg/m^3	α K^{-1}	μ kg m^{-1} s^{-1}	v m^2/s	κ m^2/s	c_p J kg^{-1} K^{-1}	Prv/κ
0	1000	−0.6E − 4	1.787E − 3	1.787E − 6	1.33E − 7	4217	13.4
10	1000	+0.9E − 4	1.307E − 3	1.307E − 6	1.38E − 7	4192	9.5
20	998	2.1E − 4	1.002E − 3	1.004E − 6	1.42E − 7	4182	7.1
30	996	3.0E − 4	0.799E − 3	0.802E − 6	1.46E − 7	4178	5.5
40	992	3.8E − 4	0.653E − 3	0.658E − 6	1.52E − 7	4178	4.3
50	988	4.5E − 4	0.548E − 3	0.555E − 6	1.58E − 7	4180	3.5

Latent heat of vaporization at 100 °C = 2.257×10^6 J/kg.
Latent heat of melting of ice at 0 °C = 0.334×10^6 J/kg.
Density of ice = 920 kg/m^3.
Surface tension between water and air at 20 °C = 0.0728 N/m.
Sound speed at 20 °C = 1481 m/s.

A.4 Properties of dry air at atmospheric pressure

T °C	ρ kg/m^3	μ kg m^{-1} s^{-1}	v m^2/s	κ m^2/s	Prv/κ
−20	1.394	1.61E − 5	1.16E − 5	1.61E − 5	0.72
−10	1.341	1.66E − 5	1.24E − 5	1.73E − 5	0.72
0	1.292	1.71E − 5	1.33E − 5	1.84E − 5	0.72
10	1.246	1.76E − 5	1.41E − 5	1.96E − 5	0.72
20	1.204	1.81E − 5	1.50E − 5	2.08E − 5	0.72
30	1.164	1.86E − 5	1.60E − 5	2.25E − 5	0.71
40	1.127	1.87E − 5	1.66E − 5	2.38E − 5	0.71
60	1.060	1.97E − 5	1.86E − 5	2.65E − 5	0.71
80	1.000	2.07E − 5	2.07E − 5	2.99E − 5	0.70
100	0.946	2.17E − 5	2.29E − 5	3.28E − 5	0.70

At 20 °C and 1 atm,

Specific heat capacity at constant pressure:	c_p = 1004 J kg^{-1} K^{-1}
Specific heat capacity at constant volume:	c_v = 717 J kg^{-1} K^{-1}
Ratio of specific heat capacities:	γ = 1.40
Coefficient of thermal expansion:	$\alpha = 3.41 \times 10^{-3}$ K^{-1}
Speed of sound:	c = 343 m s^{-1}

Constants for dry air:

Gas constant:	R = 287 J kg^{-1} K^{-1}
Molecular mass:	28.97 g gmole^{-1} or kg kmole^{-1}

A.5 Properties of standard atmosphere

The following average values are accepted by international agreement. Here, z is the height above sea level.

z km	T °C	p kPa	ρ kg/m^3
0	15.0	101.3	1.225
0.5	11.5	95.5	1.168
1	8.5	89.9	1.112
2	2.0	79.5	1.007
3	−4.5	70.1	0.909
4	−11.0	61.6	0.819
5	−17.5	54.0	0.736
6	−24.0	47.2	0.660
8	−37.0	35.6	0.525
10	−50.0	26.4	0.413
12	−56.5	19.3	0.311
14	−56.5	14.1	0.226
16	−56.5	10.3	0.165
18	−56.5	7.5	0.120
20	−56.5	5.5	0.088

Appendix B

Mathematical tools and resources

B.1 Partial and total differentiation

In fluid mechanics, the field quantities like fluid velocity, fluid density, pressure, etc., may vary in time, t, and across three-dimensional space, herein specified by three coordinates as a vector $\mathbf{x} = (x, y, z)$ or (x_1, x_2, x_3). For multivariable functions, such as $f(x_1, x_2, x_3, t)$, there are important differences between partial and total derivatives, for example between $\partial f/\partial t$ and df/dt.

Partial differentiation

$(\partial/\partial t) f(x_1, x_2, x_3, t)$ means differentiate the function $f(x_1, x_2, x_3, t)$ with respect to time, t, treating all other independent variables as constants. Additional information and specifications are not needed. And, multiple partial derivatives that operate on different variables can be applied in either order, that is, $(\partial/\partial t)(\partial f/\partial x_i) = (\partial/\partial x_i)(\partial f/\partial t)$ and $(\partial/\partial x_i)(\partial f/\partial x_j) = (\partial/\partial x_j)(\partial f/\partial x_i)$.

Total differentiation

$(d/dt) f(x_1, x_2, x_3, t)$ means differentiate the function $f(x_1, x_2, x_3, t)$ with respect to time, t, including the time variation of the spatial coordinates. This total time derivative has meaning along a space-time path specified through the three-dimensional domain. Such a path specification may be given as a vector function of time, for example $\mathbf{x} = (X_1(t), X_2(t), X_3(t))$. Without such a path specification, the total time derivative of f is not fully defined; however, when the path is specified, then:

$$\frac{d}{dt} f(x_1, x_2, x_3, t) = \frac{\partial f}{\partial x_1}\frac{dX_1}{dt} + \frac{\partial f}{\partial x_2}\frac{dX_2}{dt} + \frac{\partial f}{\partial x_3}\frac{dX_3}{dt} + \frac{\partial f}{\partial t}.$$

When studying fluid mechanics, the space-time path, $\mathbf{x}(t)$, most commonly chosen is that of a fluid particle. This path specification is commonly denoted by use of capital Ds:

$$\begin{aligned}\frac{D}{Dt} f(x, t) &\equiv \left[\frac{d}{dt} f(x, t)\right]_{\text{following a fluid particle}} \\ &= \left[\frac{\partial f}{\partial x_1}\frac{dX_1}{dt} + \frac{\partial f}{\partial x_2}\frac{dX_2}{dt} + \frac{\partial f}{\partial x_3}\frac{dX_3}{dt} + \frac{\partial f}{\partial t}\right]_{\text{following a fluid particle}}.\end{aligned} \tag{B.1.1}$$

Here, the evaluation of the total derivative *following a fluid particle* can be formally completed by using the fluid-particle velocity matching condition:

$$\textit{fluid particle velocity} \equiv \frac{d}{dt} x(t) = \left(\frac{dX_1(t)}{dt}, \frac{dX_2(t)}{dt}, \frac{dX_3(t)}{dt}\right) = (u_1, u_2, u_3)|_{x(t)} = \mathbf{u}(\mathbf{x}, t), \tag{B.1.2}$$

where $\mathbf{u}(\mathbf{x}, t)$ is the fluid velocity at the particle location, and u_1, u_2, and u_3 are the Cartesian components of the fluid velocity. The third equality in (B.1.2) provides three velocity-component matching conditions:

$$dX_1/dt = u_1, \qquad dX_2/dt = u_2, \quad \text{and} \quad dX_3/dt = u_3. \tag{B.1.3}$$

When the various parts of (B.1.3) are substituted into (B.1.1), a final form for Df/Dt emerges:

$$\frac{D}{Dt} f(\mathbf{x}, t) = \frac{\partial f}{\partial t} + u_1\frac{\partial f}{\partial x_1} + u_2\frac{\partial f}{\partial x_2} + u_3\frac{\partial f}{\partial x_3} = \frac{\partial f}{\partial t} + \mathbf{u}\cdot\nabla f = \frac{\partial f}{\partial t} + u_i\frac{\partial f}{\partial x_i}, \tag{B.1.4}$$

which is the same as (3.4). Here the final two equalities involve vector and index notation, respectively. These notations are described in Chapter 2. All three forms of Df/Dt are used in this text. *Total* and *partial* differentiation are the same when they operate on the same independent variable and this independent variable is the only independent variable.

Uses of partial and total derivatives

There are situations in the study of fluid mechanics where a first-order partial differential equation, involving both time and space derivatives, like:

$$A(x,t)\frac{\partial f(x,t)}{\partial t} + B(x,t)\frac{\partial f(x,t)}{\partial x} = g(x,t,f) \quad \text{(B.1.5)}$$

needs to be solved to find $f(x,t)$. To accomplish this task, assume there exists a curve C in x-t space described by equations $x = X(s)$ and $t = T(s)$ that allows (B.1.5) to be recast as a total derivative with respect to s. Here s is the *arc length* in x-t space along the curve C. The total derivative of f along s is:

$$\frac{df}{ds} = \frac{\partial f(x,t)}{\partial t}\frac{dT(s)}{ds} + \frac{\partial f(x,t)}{\partial x}\frac{dX(s)}{ds}. \quad \text{(B.1.6)}$$

Thus, (B.1.5) can be simplified to:

$$df/ds = g \quad \text{when} \quad dT/ds = A \quad \text{and} \quad dX/ds = B. \quad \text{(B.1.7)}$$

Taking a ratio of the last two equations produces:

$$dX/dT = B(X,T)/A(X,T), \quad \text{(B.1.8)}$$

which parametrically specifies a set of curves C. Along any such curve, $df/ds = g$ and this equation can be integrated starting from an initial condition or boundary condition to determine f.

Example B.1

Consider one-dimensional unidirectional wave propagation as specified by:

$$\frac{\partial f(x,t)}{\partial t} + U(t)\frac{\partial f(x,t)}{\partial x} = 0 \quad \text{where} \quad f(x,0) = \phi(x); \quad \text{(B.1.9, B.1.10)}$$

f represents a propagating disturbance of some type, and U is the propagation velocity. In this case, $A = 1$ and $B = U$; thus, (B.1.8) specifies the C curves via

$$\frac{dX}{dT} = U(T), \quad \text{or} \quad X(T) = X_o + \int_0^T U(\tau)d\tau. \quad \text{(B.1.11)}$$

With $A = 1$, the middle equation of (B.1.7) implies $T = T_o + s$, so (B.1.11) leads to:

$$x = X(s) = X_o + \int_o^{T_o+s} U(\tau)d\tau, \quad \text{and} \quad t = T(s) = T_o + s. \quad \text{(B.1.12, B.1.13)}$$

These two equations define the set of C curves in x-t space along which the behavior of f is easily determined from the first equation of (B.1.7) with $g = 0$:

$$\frac{df}{ds} = 0, \quad \text{or} \quad f_o = f(x,t) = f(X(s),T(s)) = f\left(X_o + \int_0^{T_o+s} U(\tau)d\tau, T_o + s\right). \quad \text{(B.1.14)}$$

Here f_o is the constant value of $f(x,t)$ that is found when s varies along a particular C curve, and X_o and T_o are constants of integration that specify the x-t location of $s = 0$ on this C curve. These constants can be evaluated using the initial condition specified in (B.1.10) in terms of ϕ at $T = T_o + s = 0$, and the last form for f in (B.1.14):

$$f_o = f(X_o, 0) = \phi(X_o). \quad \text{(B.1.15)}$$

Here it is important to note that the constant f_o may be different for the various C curves that start from different x-t locations. To reach the final solution of (B.1.9), eliminate f_o and X_o from (B.1.15) using (B.1.12) through (B.1.14) in favor of x, t, and $f(x,t)$:

$$f(x,t) = \phi\left(x - \int_o^t U(\tau)d\tau\right). \tag{B.1.16}$$

This approach to differential equation solving where special paths are found that simplify the governing equation (or equations) can be formalized and generalized; it is called the *method of characteristics*. But, independent of this and perhaps more important, the two fundamental and enduring features of partial differential equation solving are displayed here.

a) Partial differential equations are solved by rearrangement and *integration*. Extra differentiation is typically not useful; first look for ways to *integrate* to find a solution.

b) Difficulty is not entirely eliminated by changing from partial to total derivatives or vice versa. In the above example, there is initially one unknown function, f, and two independent coordinates, x and t, but this is transformed (via the method of characteristics) into a problem with two unknown functions, f and X, and one independent variable, s or t.

Integration of partial derivatives

There is really nothing special here except to note that constants of integration turn into functions that may depend on all the not-integrated-over independent variables. For example, consider $f(x, y, z, t)$ that solves the partial differential equation: $\partial f/\partial x = Ax + By$. Direct integration with y, z, and t treated as constants produces:

$$f = \int (Ax + By)dx = Ax^2/2 + Byx + C(y, z, t),$$

where $C(y, z, t)$ is an unknown function that does not depend on x; it replaces the usual constant of integration in one-variable indefinite integration.

B.2 Changing independent variables

Two situations commonly arise in the study of fluid mechanics where changing the independent variable(s) is advantageous. The first situation is changing coordinate systems. Here the number of new and old independent variables will usually be the same. Consider the situation where a partial differential equation is known in Cartesian-time coordinates (x, y, z, t), but it will be easier to solve in another coordinate system (ξ, ψ, ζ, τ). Assume the trans-formation between the two coordinate systems is given by: $\xi = X(x, y, z, t)$, $\psi = Y(x, y, z, t)$, $\zeta = Z(x, y, z, t)$, and $\tau = T(x, y, z, t)$. Cartesian and temporal partial derivatives can be transformed as follows:

$$\frac{\partial}{\partial x} = \frac{\partial X}{\partial x}\frac{\partial}{\partial \xi} + \frac{\partial Y}{\partial x}\frac{\partial}{\partial \psi} + \frac{\partial Z}{\partial x}\frac{\partial}{\partial \zeta} + \frac{\partial T}{\partial x}\frac{\partial}{\partial \tau}, \quad \frac{\partial}{\partial y} = \frac{\partial X}{\partial y}\frac{\partial}{\partial \xi} + \frac{\partial Y}{\partial y}\frac{\partial}{\partial \psi} + \frac{\partial Z}{\partial y}\frac{\partial}{\partial \zeta} + \frac{\partial T}{\partial y}\frac{\partial}{\partial \tau},$$
$$\frac{\partial}{\partial z} = \frac{\partial X}{\partial z}\frac{\partial}{\partial \xi} + \frac{\partial Y}{\partial z}\frac{\partial}{\partial \psi} + \frac{\partial Z}{\partial z}\frac{\partial}{\partial \zeta} + \frac{\partial T}{\partial z}\frac{\partial}{\partial \tau}, \quad \text{and} \quad \frac{\partial}{\partial t} = \frac{\partial X}{\partial t}\frac{\partial}{\partial \xi} + \frac{\partial Y}{\partial t}\frac{\partial}{\partial \psi} + \frac{\partial Z}{\partial t}\frac{\partial}{\partial \zeta} + \frac{\partial T}{\partial t}\frac{\partial}{\partial \tau}. \tag{B.2.1}$$

Example B.2

Consider the case where (x, y, z, t) and (ξ, ψ, ζ, τ) represent Cartesian systems with parallel axes that are moving with respect to each other at a constant velocity (U, V, W) when observed in (x, y, z, t), so that $\xi = x - Ut$, $\psi = y - Vt$, $\zeta = z - Wt$, and $\tau = t$. Application of the above derivative transformations (B.2.1) produces:

$$\frac{\partial}{\partial x} = \frac{\partial}{\partial \xi}, \quad \frac{\partial}{\partial y} = \frac{\partial}{\partial \psi}, \quad \frac{\partial}{\partial z} = \frac{\partial}{\partial \zeta}, \quad \text{and} \quad \frac{\partial}{\partial t} = -U\frac{\partial}{\partial \xi} - V\frac{\partial}{\partial \psi} - W\frac{\partial}{\partial \zeta} + \frac{\partial}{\partial \tau}. \tag{B.2.2}$$

Perhaps unexpectedly, extra differentiations only appear in the transformed time derivative, even though the time variable transformation equation was simplest.

The second situation that requires changing independent variables occurs when a combination of independent variables (and parameters) is found that might simplify a partial differential equation. Here the usual goal is to convert a partial differential equation having multiple independent variables into a total differential equation with one independent variable.

If $\eta = H(x, y, z, t)$ is the combination variable, then a straightforward application of the chain rule for partial differentiation produces:

$$\frac{\partial}{\partial x} = \frac{\partial H}{\partial x}\frac{d}{d\eta}, \quad \frac{\partial}{\partial y} = \frac{\partial H}{\partial y}\frac{d}{d\eta}, \quad \frac{\partial}{\partial z} = \frac{\partial H}{\partial z}\frac{d}{d\eta}, \quad \text{and} \quad \frac{\partial}{\partial t} = \frac{\partial H}{\partial t}\frac{d}{d\eta}. \tag{B.2.3}$$

Example B.3

Consider a function with two independent variables, $f(x, t)$, for which we hypothesize the existence of a special combination (or similarity) variable $\eta = xt^{\alpha}$, where α is a real number, that facilitates the solution of the partial differential equation for $f(x, t)$. Mathematically, this hypothesis can be stated as: $f(x, t) = f(\eta) = f(xt^{\alpha})$, and partial derivatives of f can be obtained from the first and last equations of (B.2.3) with $H = xt^{\alpha}$:

$$\frac{\partial}{\partial x} f(x, t) = \frac{\partial(xt^{\alpha})}{\partial x}\frac{d}{d\eta} f(\eta) = t^{\alpha}\frac{df}{d\eta}, \quad \text{and} \quad \frac{\partial}{\partial t} f(x, t) = \frac{\partial(xt^{\alpha})}{\partial t}\frac{d}{d\eta} f(\eta) = \alpha x t^{\alpha-1}\frac{df}{d\eta} = \frac{\alpha}{t}\eta\frac{df}{d\eta}.$$

Second-order derivatives are generated by appropriately differentiating these first-order results.

B.3 Basic vector calculus

The gradient operator, ∇, is the general-purpose directional derivative for multiple spatial coordinates. It is a vector operator, and it exists in all suitably defined coordinate systems. Its properties are a combination of those of ordinary partial derivatives and ordinary vectors. It has components and its position and operation character (multiply, dot product, cross product, etc.) matter within a set or grouping of functions or variables. For example, $(\mathbf{u} \cdot \nabla)\mathbf{v} \neq \mathbf{v}(\nabla \cdot \mathbf{u})$ in general, even though these two expressions would be an equal if ∇ were replaced by an ordinary vector. Some properties of ∇ are listed here:

- In Cartesian coordinates, $\mathbf{x} = (x, y, z)$: $\nabla = \mathbf{e}_x \frac{\partial}{\partial x} + \mathbf{e}_y \frac{\partial}{\partial y} + \mathbf{e}_z \frac{\partial}{\partial z}$ where the $\mathbf{e}$s are unit vectors
- The *gradient* of the scalar field ρ is: $\nabla\rho = \mathbf{e}_x \frac{\partial \rho}{\partial x} + \mathbf{e}_y \frac{\partial \rho}{\partial y} + \mathbf{e}_z \frac{\partial \rho}{\partial z}$
- The *divergence* of a vector field $\mathbf{u} = (u, v, w)$ is: $\nabla \cdot \mathbf{u} = \frac{\partial u}{\partial x} + \frac{\partial v}{\partial y} + \frac{\partial w}{\partial z}$
- The *curl* of a vector field $\mathbf{u} = (u, v, w)$ is: $\nabla \times \mathbf{u} = \det \begin{vmatrix} \mathbf{e}_x & \mathbf{e}_y & \mathbf{e}_z \\ \partial/\partial x & \partial/\partial y & \partial/\partial z \\ u & v & w \end{vmatrix}$.

Vector identities involving ∇

Here ρ and ϕ are scalar functions, $\mathbf{u}$ and $\mathbf{F}$ are vector functions, and $\mathbf{x}$ is the position vector.

$$\nabla \cdot \mathbf{x} = 3 \tag{B.3.1}$$

$$\nabla \times \mathbf{x} = 0 \tag{B.3.2}$$

$$\nabla \cdot \left(\mathbf{x}/|\mathbf{x}|^3\right) = 0 \tag{B.3.3}$$

$$(\mathbf{u} \cdot \nabla)\mathbf{x} = \mathbf{u} \tag{B.3.4}$$

$$\nabla(\rho\phi) = \rho\nabla\phi + \phi\nabla\rho \tag{B.3.5}$$

$$\nabla \cdot (\rho\mathbf{u}) = \rho\nabla \cdot \mathbf{u} + (\mathbf{u} \cdot \nabla)\rho \tag{B.3.6}$$

$$\nabla \times (\rho\mathbf{u}) = \rho\nabla \times \mathbf{u} + (\nabla\rho) \times \mathbf{u} \tag{B.3.7}$$

$$\nabla \cdot (\mathbf{u} \times \mathbf{F}) = (\nabla \times \mathbf{u}) \cdot \mathbf{F} - \mathbf{u} \cdot (\nabla \times \mathbf{F}) \tag{B.3.8}$$

$$\nabla(\mathbf{u} \cdot \mathbf{F}) = \mathbf{u} \times (\nabla \times \mathbf{F}) + \mathbf{F} \times (\nabla \times \mathbf{u}) + (\mathbf{u} \cdot \nabla)\mathbf{F} + (\mathbf{F} \cdot \nabla)\mathbf{u} \tag{B.3.9}$$

$$\nabla \times (\mathbf{u} \times \mathbf{F}) = (\mathbf{F} \cdot \nabla)\mathbf{u} - \mathbf{F}(\nabla \cdot \mathbf{u}) + \mathbf{u}(\nabla \cdot \mathbf{F}) - (\mathbf{u} \cdot \nabla)\mathbf{F} \tag{B.3.10}$$

$$\nabla \times (\nabla\rho) = 0 \tag{B.3.11}$$

$$\nabla \cdot (\nabla \times \mathbf{u}) = 0 \quad \text{(B.3.12)}$$

$$\nabla \times (\nabla \times \mathbf{u}) = \nabla(\nabla \cdot \mathbf{u}) - \nabla^2 \mathbf{u} \quad \text{(B.3.13)}$$

Integral theorems involving ∇

These are discussed in Sections 2.12–2.13.

- For a closed surface A that contains volume V with $\mathbf{n}$ = the outward normal on A, Gauss' Theorem is:

$$\int_A \rho \mathbf{n} dA = \int_V \nabla \rho dV \quad \text{for scalars,} \quad \text{and} \quad \int_A \mathbf{u} \cdot \mathbf{n} dA = \int_V \nabla \cdot \mathbf{u} dV \quad \text{for vectors.}$$

- For a closed curve C that bounds surface A with $\mathbf{n}$ = the normal to A and $\boldsymbol{t}$ the tangent to C, Stokes' Theorem is: $\oint_C \mathbf{u} \cdot \mathbf{t} ds = \int_A (\nabla \times \mathbf{u}) \cdot \mathbf{n} dA$, where s is the arc length along C.

B.4 The Dirac delta function

The Dirac delta function is commonly denoted $\delta(x)$, where x is a real variable. It is a unit-area impulse that exists at only one point in space; it is zero everywhere except where its argument is zero. The Dirac delta-function can be defined as a limit of a smooth function, such as:

$$\delta(x) = \lim_{\sigma \to 0} \left(\sqrt{2\pi}\sigma\right)^{-1} \exp\left\{-x^2/2\sigma^2\right\}. \quad \text{(B.4.1)}$$

The value of $\delta(x)$ is infinite at $x = 0$ but its integral is unity. Here are a few properties of $\delta(x)$ for a, b, and x_o real constants and $b > a$:

$$x\delta(x - a) = a\delta(x - a), \quad \text{(B.4.2)}$$

$$\int_a^b \delta(x - x_o) dx = \left\{ \begin{array}{l} 1 \text{ for } a \le x_o \le b \\ 0 \text{ for } x_o < a \text{ or } b < x_o \end{array} \right\}, \quad \text{(B.4.3)}$$

$$\int_{-\infty}^{+\infty} f(x)\delta(x - x_o) dx = f(x_o). \quad \text{(B.4.4)}$$

These properties ease the evaluation of complicated integrals when a Dirac delta function appears in the integrand. In more dimensions where $\mathbf{x} = (x, y, z)$, the following notation is common:

$$\delta(\mathbf{x} - \mathbf{x}_o) = \delta(x - x_o)\delta(y - y_o)\delta(z - z_o).$$

In the study of fluid mechanics, the usual notation for the Dirac delta-function is potentially confusing because δ is also commonly used to denote a length scale of interest in the flow field, such as a boundary-layer thickness or the length scale of a similarity variable. Thus, specific mention of the Dirac delta function is made where it is used in the text.

Example B.4

Evaluate the integral: $I = \int_{-\infty}^{+\infty} F(x)[(x_o - x)^2 + r_o^2]^{-1/2} e^{ikx} \delta(x - ct) dx$. Here the limits of integration ensure that x will equal ct somewhere in the integration. Eq. (B.4.4) implies that the value of this integral is determined by replacing x with ct in the integrand; therefore: $I = F(ct)[(x_o - ct)^2 + r_o^2]^{-1/2} e^{ikct}$.

B.5 Common three-dimensional coordinate systems

In all cases that follow, ξ, ψ, and ζ are constants.

Cartesian coordinates (Fig. B.1)

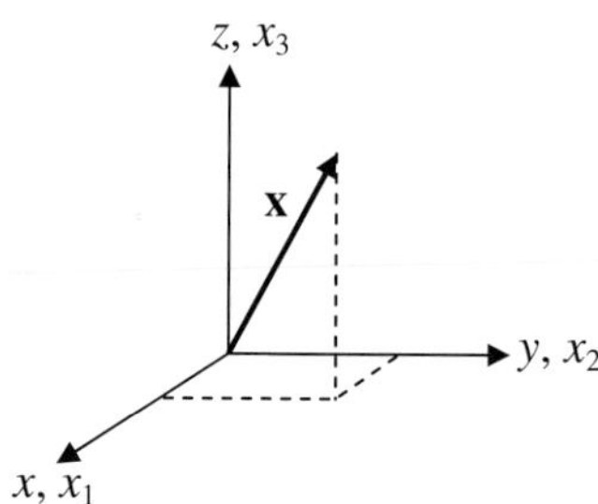

FIGURE B.1 Cartesian coordinate system.

Position: $\mathbf{x} = (x, y, z) = (x_1, x_2, x_3) = x_1\mathbf{e}_1 + x_2\mathbf{e}_2 + x_3\mathbf{e}_3$

Unit vectors: $\mathbf{e}_x$, $\mathbf{e}_y$, and $\mathbf{e}_z$, or $\mathbf{e}_1$, $\mathbf{e}_2$, and $\mathbf{e}_3$

Unit vector dependencies: $\partial\mathbf{e}_i/\partial x_j = 0$ for i and $j = 1$, 2, or 3; that is, Cartesian unit vectors are independent of the coordinate values

Gradient operator: $\nabla = \mathbf{e}_x\frac{\partial}{\partial x} + \mathbf{e}_y\frac{\partial}{\partial y} + \mathbf{e}_z\frac{\partial}{\partial z} = \mathbf{e}_1\frac{\partial}{\partial x_1} + \mathbf{e}_2\frac{\partial}{\partial x_2} + \mathbf{e}_3\frac{\partial}{\partial x_3}$

Surface integral, S, of $f(x, y, z)$ over the plane defined by $x = \xi{:}S = \int_{y=-\infty}^{+\infty}\int_{z=-\infty}^{+\infty} f(\xi, y, z)dzdy$

Surface integral, S, of $f(x, y, z)$ over the plane defined by $y = \psi{:}S = \int_{x=-\infty}^{+\infty}\int_{z=-\infty}^{+\infty} f(x, \psi, z)dzdx$

Surface integral, S, of $f(x, y, z)$ over the plane defined by $z = \zeta{:}S = \int_{x=-\infty}^{+\infty}\int_{y=-\infty}^{+\infty} f(x, y, \zeta)dydx$

Volume integral, V, of $f(x, y, z)$ over all space: $V = \int_{x=-\infty}^{+\infty}\int_{y=-\infty}^{+\infty}\int_{z=-\infty}^{+\infty} f(x, y, z)dzdydx$.

Cylindrical coordinates (Fig. B.2)

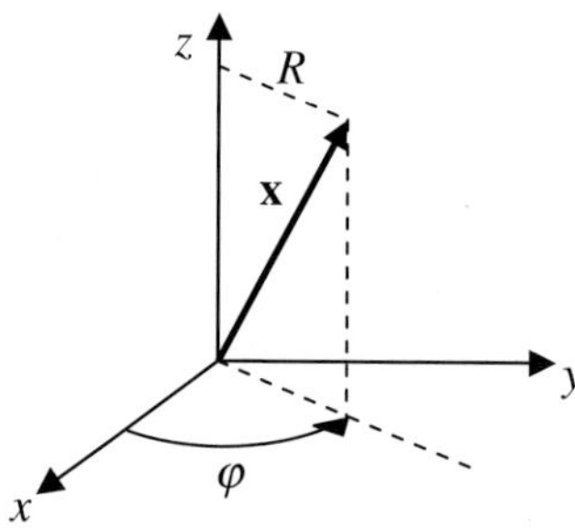

FIGURE B.2 Cylindrical coordinate system.

Position: $\mathbf{x} = (R, \varphi, z) = R\mathbf{e}_R + z\mathbf{e}_z$; $x = R\cos\varphi$, $y = R\sin\varphi$, $z = z$; or $R = \sqrt{x^2 + y^2}$, $\varphi = \tan^{-1}(y/x)$

Unit vectors: $\mathbf{e}_R = \mathbf{e}_x\cos\varphi + \mathbf{e}_y\sin\varphi$, $\mathbf{e}_\varphi = -\mathbf{e}_x\sin\varphi + \mathbf{e}_y\cos\varphi$, $\mathbf{e}_z$ = same as Cartesian

Unit vector dependencies: $\partial\mathbf{e}_R/\partial R = 0,\ \partial\mathbf{e}_R/\partial\varphi = \mathbf{e}_\varphi,\ \partial\mathbf{e}_R/\partial z = 0$

$$\partial\mathbf{e}_\varphi/\partial R = 0,\ \partial\mathbf{e}_\varphi/\partial\varphi = -\mathbf{e}_R,\ \partial\mathbf{e}_\varphi/\partial z = 0$$

$$\partial\mathbf{e}_z/\partial R = 0,\ \partial\mathbf{e}_z/\partial\varphi = 0,\ \partial\mathbf{e}_z/\partial z = 0$$

Gradient Operator: $\nabla = \mathbf{e}_R\frac{\partial}{\partial R} + \mathbf{e}_\varphi\frac{1}{R}\frac{\partial}{\partial\varphi} + \mathbf{e}_z\frac{\partial}{\partial z}$

Surface integral, S, of $f(R, \theta, z)$ over the cylinder defined by $R = \xi{:}S = \int_{\varphi=0}^{2\pi}\int_{z=-\infty}^{+\infty} f(\xi, \varphi, z)\xi dzd\varphi$

Surface integral, S, of $f(R, \theta, z)$ over the half plane defined by $\varphi = \psi{:}S = \int_{R=0}^{+\infty}\int_{z=-\infty}^{+\infty} f(R, \psi, z)dzdR$

Surface integral, S, of $f(R, \theta, z)$ over the plane defined by $z = \zeta{:}S = \int_{R=0}^{+\infty}\int_{\varphi=0}^{2\pi} f(R, \varphi, \zeta)Rd\varphi dR$

Volume integral, V, of $f(R, \theta, z)$ over all space: $V = \int_{z=-\infty}^{+\infty}\int_{R=0}^{+\infty}\int_{\varphi=0}^{2\pi} f(R, \varphi, z)Rd\varphi dRdz$

Spherical coordinates (Fig. B.3)

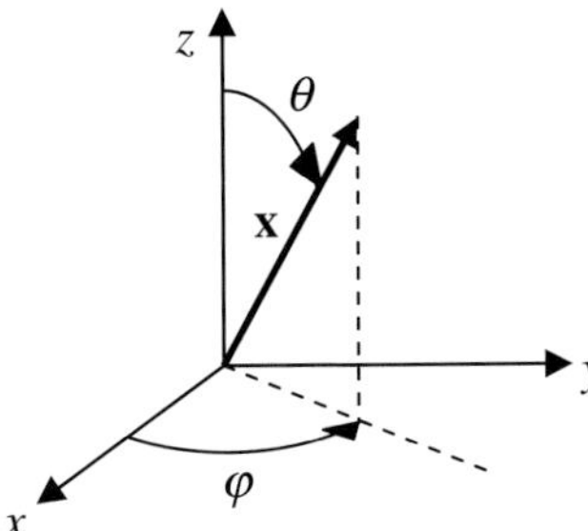

FIGURE B.3 Spherical coordinate system.

Position: $\mathbf{x} = (r, \theta, \varphi) = r\mathbf{e}_r$; $x = r\cos\varphi\sin\theta$, $y = r\sin\varphi\sin\theta$, $z = r\cos\theta$; or $r = \sqrt{x^2 + y^2 + z^2}$, $\theta = \tan^{-1}(\sqrt{x^2 + y^2}/z)$, and $\varphi = \tan^{-1}(y/x)$

Unit vectors: $\mathbf{e}_r = \mathbf{e}_x \sin\theta\cos\varphi + \mathbf{e}_y \sin\theta\sin\varphi + \mathbf{e}_z\cos\theta$,

$\mathbf{e}_\theta = \mathbf{e}_x\cos\theta\cos\varphi + \mathbf{e}_y\cos\theta\sin\varphi - \mathbf{e}_z\sin\theta$, $\mathbf{e}_\varphi = -\mathbf{e}_x\sin\varphi + \mathbf{e}_y\cos\varphi$

Unit vector dependencies: $\partial\mathbf{e}_r/\partial r = 0$, $\partial\mathbf{e}_r/\partial\theta = \mathbf{e}_\theta$, $\partial\mathbf{e}_r/\partial\varphi = \mathbf{e}_\varphi\sin\theta$

$\partial\mathbf{e}_\theta/\partial r = 0$, $\partial\mathbf{e}_\theta/\partial\theta = -\mathbf{e}_r$, $\partial\mathbf{e}_\theta/\partial\varphi = \mathbf{e}_\varphi\cos\theta$

$\partial\mathbf{e}_\varphi/\partial r = 0$, $\partial\mathbf{e}_\varphi/\partial\theta = 0$, $\partial\mathbf{e}_\varphi/\partial\varphi = -\mathbf{e}_r\sin\theta - \mathbf{e}_\theta\cos\theta$

Gradient Operator: $\nabla = \mathbf{e}_r\frac{\partial}{\partial r} + \mathbf{e}_\theta\frac{1}{r}\frac{\partial}{\partial\theta} + \mathbf{e}_\varphi\frac{1}{r\sin\theta}\frac{\partial}{\partial\varphi}$

Surface integral, S, of $f(r,\theta,\varphi)$ over the sphere defined by $r = \xi{:}S = \int_{\theta=0}^{\pi}\int_{\varphi=0}^{2\pi} f(\xi,\theta,\varphi)\xi^2\sin\theta d\varphi d\theta$

Surface integral, S, of $f(r,\theta,\varphi)$ over the cone defined by $\theta = \psi{:}S = \int_{r=0}^{+\infty}\int_{\varphi=0}^{2\pi} f(r,\psi,\varphi)r\sin\psi d\varphi dr$

Surface integral, S, of $f(r,\theta,\varphi)$ over the half plane defined by $\varphi = \zeta{:}S = \int_{r=0}^{+\infty}\int_{\theta=0}^{\pi} f(r,\theta,\zeta)rd\theta dr$

Volume integral, V, of $f(r,\theta,\varphi)$ over all space: $V = \int_{r=0}^{+\infty}\int_{\theta=0}^{\pi}\int_{\varphi=0}^{2\pi} f(r,\theta,\varphi)r^2\sin\theta d\varphi d\theta dr$

B.6 Equations in curvilinear coordinates

Plane polar coordinates (Fig. 3.3a)

Position and velocity vectors $\mathbf{x} = (r,\theta) = r\mathbf{e}_r$; $\mathbf{u} = (u_r, u_\theta) = u_r\mathbf{e}_r + u_\theta\mathbf{e}_\theta$

Gradient of a scalar ψ: $\nabla\psi = \mathbf{e}_r\frac{\partial\psi}{\partial r} + \mathbf{e}_\theta\frac{1}{r}\frac{\partial\psi}{\partial\theta}$

Laplacian of a scalar ψ: $\nabla^2\psi = \frac{1}{r}\frac{\partial}{\partial r}\left(r\frac{\partial\psi}{\partial r}\right) + \frac{1}{r^2}\frac{\partial^2\psi}{\partial\theta^2}$

Divergence of a vector: $\nabla\cdot\mathbf{u} = \frac{1}{r}\frac{\partial}{\partial r}(ru_r) + \frac{1}{r}\frac{\partial u_\theta}{\partial\theta}$

Curl of a vector, vorticity: $\boldsymbol{\omega} = \nabla\times\mathbf{u} = \mathbf{e}_z\left(\frac{1}{r}\frac{\partial(ru_\theta)}{\partial r} - \frac{1}{r}\frac{\partial u_r}{\partial\theta}\right)$

Laplacian of a vector: $\nabla^2\mathbf{u} = \mathbf{e}_r\left(\nabla^2 u_r - \frac{u_r}{r^2} - \frac{2}{r^2}\frac{\partial u_\theta}{\partial\theta}\right) + \mathbf{e}_\theta\left(\nabla^2 u_\theta + \frac{2}{r^2}\frac{\partial u_r}{\partial\theta} - \frac{u_\theta}{r^2}\right)$

Strain rate S_{ij} and viscous stress τ_{ij} for an incompressible fluid where $\tau_{ij} = 2\mu S_{ij}$:

$$S_{rr} = \frac{\partial u_r}{\partial r} = \frac{1}{2\mu}\tau_{rr}, \qquad S_{\theta\theta} = \frac{1}{r}\frac{\partial u_\theta}{\partial\theta} + \frac{u_r}{r} = \frac{1}{2\mu}\tau_{\theta\theta}, \qquad S_{r\theta} = \frac{r}{2}\frac{\partial}{\partial r}\left(\frac{u_\theta}{r}\right) + \frac{1}{2r}\frac{\partial u_r}{\partial\theta} = \frac{1}{2\mu}\tau_{r\theta}$$

Equation of continuity: $\frac{\partial\rho}{\partial t} + \frac{1}{r}\frac{\partial}{\partial r}(r\rho u_r) + \frac{1}{r}\frac{\partial}{\partial\theta}(\rho u_\theta) = 0$

Navier-Stokes equations with constant ρ, constant ν, and no body force:

$$\frac{\partial u_r}{\partial t} + u_r\frac{\partial u_r}{\partial r} + \frac{u_\theta}{r}\frac{\partial u_r}{\partial\theta} - \frac{u_\theta^2}{r} = -\frac{1}{\rho}\frac{\partial p}{\partial r} + \nu\left(\nabla^2 u_r - \frac{u_r}{r^2} - \frac{2}{r^2}\frac{\partial u_\theta}{\partial\theta}\right),$$

$$\frac{\partial u_\theta}{\partial t} + u_r\frac{\partial u_\theta}{\partial r} + \frac{u_\theta}{r}\frac{\partial u_\theta}{\partial\theta} + \frac{u_r u_\theta}{r} = -\frac{1}{\rho r}\frac{\partial p}{\partial\theta} + \nu\left(\nabla^2 u_\theta + \frac{2}{r^2}\frac{\partial u_r}{\partial\theta} - \frac{u_\theta}{r^2}\right),$$

where $\nabla^2 = \frac{1}{r}\frac{\partial}{\partial r}\left(r\frac{\partial}{\partial r}\right) + \frac{1}{r^2}\frac{\partial^2}{\partial\theta^2}$.

Cylindrical coordinates (Fig. B.2)

Position and velocity vectors: $\mathbf{x} = (R, \varphi, z) = R\mathbf{e}_R + z\mathbf{e}_z$; $\mathbf{u} = (u_R, u_\varphi, u_z) = u_R\mathbf{e}_R + u_\varphi\mathbf{e}_\varphi + u_z\mathbf{e}_z$

Gradient of a scalar ψ: $\nabla\psi = \mathbf{e}_R\frac{\partial\psi}{\partial R} + \mathbf{e}_\varphi\frac{1}{R}\frac{\partial\psi}{\partial\varphi} + \mathbf{e}_z\frac{\partial\psi}{\partial z}$

Laplacian of a scalar ψ: $\nabla^2\psi = \frac{1}{R}\frac{\partial}{\partial R}\left(R\frac{\partial\psi}{\partial R}\right) + \frac{1}{R^2}\frac{\partial^2\psi}{\partial\varphi^2} + \frac{\partial^2\psi}{\partial z^2}$

Divergence of a vector: $\nabla\cdot\mathbf{u} = \frac{1}{R}\frac{\partial}{\partial R}(Ru_R) + \frac{1}{R}\frac{\partial u_\varphi}{\partial\varphi} + \frac{\partial u_z}{\partial z}$

Curl of a vector, vorticity: $\boldsymbol{\omega} = \nabla\times\mathbf{u} = \mathbf{e}_R\left(\frac{1}{R}\frac{\partial u_z}{\partial\varphi} - \frac{\partial u_\varphi}{\partial z}\right) + \mathbf{e}_\varphi\left(\frac{\partial u_R}{\partial z} - \frac{\partial u_z}{\partial R}\right) + \mathbf{e}_z\left(\frac{1}{R}\frac{\partial(Ru_\varphi)}{\partial R} - \frac{1}{R}\frac{\partial u_R}{\partial\varphi}\right)$

Laplacian of a vector: $\nabla^2\mathbf{u} = \mathbf{e}_R\left(\nabla^2u_R - \frac{u_R}{R^2} - \frac{2}{R^2}\frac{\partial u_\varphi}{\partial\varphi}\right) + \mathbf{e}_\varphi\left(\nabla^2u_\varphi + \frac{2}{R^2}\frac{\partial u_R}{\partial\varphi} - \frac{u_\varphi}{R^2}\right) + \mathbf{e}_z\nabla^2u_z$

Strain rate S_{ij} and viscous stress τ_{ij} for an incompressible fluid where $\tau_{ij} = 2\mu S_{ij}$:

$$S_{RR} = \frac{\partial u_R}{\partial R} = \frac{1}{2\mu}\tau_{RR}, \qquad S_{\varphi\varphi} = \frac{1}{R}\frac{\partial u_\varphi}{\partial\varphi} + \frac{u_R}{R} = \frac{1}{2\mu}\tau_{\varphi\varphi}, \qquad S_{zz} = \frac{\partial u_z}{\partial z} = \frac{1}{2\mu}\tau_{zz}$$

$$S_{R\varphi} = \frac{R}{2}\frac{\partial}{\partial R}\left(\frac{u_\varphi}{R}\right) + \frac{1}{2R}\frac{\partial u_R}{\partial\varphi} = \frac{1}{2\mu}\tau_{R\varphi}, \qquad S_{\varphi z} = \frac{1}{2R}\frac{\partial u_z}{\partial\varphi} + \frac{1}{2}\frac{\partial u_\varphi}{\partial z} = \frac{1}{2\mu}\tau_{\varphi z},$$

$$S_{zR} = \frac{1}{2}\left(\frac{\partial u_R}{\partial z} + \frac{\partial u_z}{\partial R}\right) = \frac{1}{2\mu}\tau_{zR}$$

Equation of continuity: $\frac{\partial\rho}{\partial t} + \frac{1}{R}\frac{\partial}{\partial R}(R\rho u_R) + \frac{1}{R}\frac{\partial}{\partial\varphi}(\rho u_\varphi) + \frac{\partial}{\partial z}(\rho u_z) = 0$

Navier-Stokes equations with constant ρ, constant ν, and no body force:

$$\frac{\partial u_R}{\partial t} + (\mathbf{u}\cdot\nabla)u_R - \frac{u_\varphi^2}{R} = -\frac{1}{\rho}\frac{\partial p}{\partial R} + \nu\left(\nabla^2u_R - \frac{u_R}{R^2} - \frac{2}{R^2}\frac{\partial u_\varphi}{\partial\varphi}\right),$$

$$\frac{\partial u_\varphi}{\partial t} + (\mathbf{u}\cdot\nabla)u_\varphi + \frac{u_Ru_\varphi}{R} = -\frac{1}{\rho R}\frac{\partial p}{\partial\varphi} + \nu\left(\nabla^2u_\varphi + \frac{2}{R^2}\frac{\partial u_R}{\partial\varphi} - \frac{u_\varphi}{R^2}\right),$$

$$\frac{\partial u_z}{\partial t} + (\mathbf{u}\cdot\nabla)u_z = -\frac{1}{\rho}\frac{\partial p}{\partial z} + \nu\nabla^2u_z$$

where: $\mathbf{u}\cdot\nabla = u_R\frac{\partial}{\partial R} + \frac{u_\varphi}{R}\frac{\partial}{\partial\varphi} + u_z\frac{\partial}{\partial z}$ and $\nabla^2 = \frac{1}{R}\frac{\partial}{\partial R}\left(R\frac{\partial}{\partial R}\right) + \frac{1}{R^2}\frac{\partial^2}{\partial\varphi^2} + \frac{\partial^2}{\partial z^2}$.

Spherical coordinates (Fig. B.3)

Position and velocity vectors: $\mathbf{x} = (r, \theta, \varphi) = r\mathbf{e}_r$; $\mathbf{u} = (u_r, u_\theta, u_\varphi) = u_r\mathbf{e}_r + u_\theta\mathbf{e}_\theta + u_\varphi\mathbf{e}_\varphi$

Gradient of a scalar ψ: $\nabla\psi = \mathbf{e}_r\frac{\partial\psi}{\partial r} + \mathbf{e}_\theta\frac{1}{r}\frac{\partial\psi}{\partial\theta} + \mathbf{e}_\varphi\frac{1}{r\sin\theta}\frac{\partial\psi}{\partial\varphi}$

Laplacian of a scalar ψ: $\nabla^2\psi = \frac{1}{r^2}\frac{\partial}{\partial r}\left(r^2\frac{\partial\psi}{\partial r}\right) + \frac{1}{r^2\sin\theta}\frac{\partial}{\partial\theta}\left(\sin\theta\frac{\partial\psi}{\partial\theta}\right) + \frac{1}{r^2\sin^2\theta}\frac{\partial^2\psi}{\partial\varphi^2}$

Divergence of a vector: $\nabla\cdot\mathbf{u} = \frac{1}{r^2}\frac{\partial}{\partial r}(r^2u_r) + \frac{1}{r\sin\theta}\frac{\partial(u_\theta\sin\theta)}{\partial\theta} + \frac{1}{r\sin\theta}\frac{\partial u_\varphi}{\partial\varphi}$

Curl of a vector, vorticity: $\boldsymbol{\omega} = \nabla\times\mathbf{u} = \frac{\mathbf{e}_r}{r\sin\theta}\left(\frac{\partial(u_\varphi\sin\theta)}{\partial\theta} - \frac{\partial u_\theta}{\partial\varphi}\right) + \frac{\mathbf{e}_\theta}{r}\left(\frac{1}{\sin\theta}\frac{\partial u_r}{\partial\varphi} - \frac{\partial(ru_\varphi)}{\partial r}\right) + \frac{\mathbf{e}_\varphi}{r}\left(\frac{\partial(ru_\theta)}{\partial r} - \frac{\partial u_r}{\partial\theta}\right)$

Laplacian of a vector:

$$\nabla^2\mathbf{u} = \mathbf{e}_r\left(\nabla^2u_r - \frac{2u_r}{r^2} - \frac{2}{r^2\sin\theta}\frac{\partial(u_\theta\sin\theta)}{\partial\theta} - \frac{2}{r^2\sin\theta}\frac{\partial u_\varphi}{\partial\varphi}\right)$$
$$+\mathbf{e}_\theta\left(\nabla^2u_\theta + \frac{2}{r^2}\frac{\partial u_r}{\partial\theta} - \frac{u_\theta}{r^2\sin^2\theta} - \frac{2\cos\theta}{r^2\sin^2\theta}\frac{\partial u_\varphi}{\partial\varphi}\right)$$
$$+\mathbf{e}_\varphi\left(\nabla^2u_\varphi + \frac{2}{r^2\sin\theta}\frac{\partial u_r}{\partial\varphi} + \frac{2\cos\theta}{r^2\sin^2\theta}\frac{\partial u_\theta}{\partial\varphi} - \frac{u_\varphi}{r^2\sin^2\theta}\right)$$

Strain rate S_{ij} and viscous stress τ_{ij} for an incompressible fluid where $\tau_{ij} = 2\mu S_{ij}$:

$$S_{rr} = \frac{\partial u_r}{\partial r} = \frac{1}{2\mu}\tau_{rr}, \qquad S_{\theta\theta} = \frac{1}{r}\frac{\partial u_\theta}{\partial\theta} + \frac{u_r}{r} = \frac{1}{2\mu}\tau_{\theta\theta}, \qquad S_{\varphi\varphi} = \frac{1}{r\sin\theta}\frac{\partial u_\varphi}{\partial\varphi} + \frac{u_r}{r} + \frac{u_\theta\cot\theta}{r} = \frac{1}{2\mu}\tau_{\varphi\varphi},$$

$$S_{\theta\varphi} = \frac{\sin\theta}{2r}\frac{\partial}{\partial\theta}\left(\frac{u_\varphi}{\sin\theta}\right) + \frac{1}{2r\sin\theta}\frac{\partial u_\theta}{\partial\varphi} = \frac{1}{2\mu}\tau_{\theta\varphi}, \qquad S_{\varphi r} = \frac{1}{2r\sin\theta}\frac{\partial u_r}{\partial\varphi} + \frac{r}{2}\frac{\partial}{\partial r}\left(\frac{u_\varphi}{r}\right) = \frac{1}{2\mu}\tau_{\varphi r},$$

$$S_{r\theta} = \frac{r}{2}\frac{\partial}{\partial r}\left(\frac{u_\theta}{r}\right) + \frac{1}{2r}\frac{\partial u_r}{\partial\theta} = \frac{1}{2\mu}\tau_{r\theta}$$

Equation of continuity:

$$\frac{\partial\rho}{\partial t} + \frac{1}{r^2}\frac{\partial}{\partial r}\left(\rho r^2 u_r\right) + \frac{1}{r\sin\theta}\frac{\partial}{\partial\theta}(\rho u_\theta\sin\theta) + \frac{1}{r\sin\theta}\frac{\partial}{\partial\varphi}(\rho u_\varphi) = 0$$

Navier-Stokes equations with constant ρ, constant ν, and no body force:

$$\frac{\partial u_r}{\partial t} + (\mathbf{u}\cdot\nabla)u_r - \frac{u_\theta^2 + u_\varphi^2}{r} = -\frac{1}{\rho}\frac{\partial p}{\partial r} + \nu\left[\nabla^2 u_r - \frac{2u_r}{r^2} - \frac{2}{r^2\sin\theta}\frac{\partial(u_\theta\sin\theta)}{\partial\theta} - \frac{2}{r^2\sin\theta}\frac{\partial u_\varphi}{\partial\varphi}\right],$$

$$\frac{\partial u_\theta}{\partial t} + (\mathbf{u}\cdot\nabla)u_\theta + \frac{u_r u_\theta}{r} - \frac{u_\varphi^2\cot\theta}{r} = -\frac{1}{\rho r}\frac{\partial p}{\partial\theta} + \nu\left[\nabla^2 u_\theta + \frac{2}{r^2}\frac{\partial u_r}{\partial\theta} - \frac{u_\theta}{r^2\sin^2\theta} - \frac{2\cos\theta}{r^2\sin^2\theta}\frac{\partial u_\varphi}{\partial\varphi}\right],$$

$$\frac{\partial u_\varphi}{\partial t} + (\mathbf{u}\cdot\nabla)u_\varphi + \frac{u_\varphi u_r}{r} + \frac{u_\theta u_\varphi\cot\theta}{r} = -\frac{1}{\rho r\sin\theta}\frac{\partial p}{\partial\varphi} + \nu\left[\nabla^2 u_\varphi + \frac{2}{r^2\sin\theta}\frac{\partial u_r}{\partial\varphi} + \frac{2\cos\theta}{r^2\sin^2\theta}\frac{\partial u_\theta}{\partial\varphi} - \frac{u_\varphi}{r^2\sin^2\theta}\right]$$

where

$$\mathbf{u}\cdot\nabla = u_r\frac{\partial}{\partial r} + \frac{u_\theta}{r}\frac{\partial}{\partial\theta} + \frac{u_\varphi}{r\sin\theta}\frac{\partial}{\partial\varphi},$$

$$\nabla^2 = \frac{1}{r^2}\frac{\partial}{\partial r}\left(r^2\frac{\partial}{\partial r}\right) + \frac{1}{r^2\sin\theta}\frac{\partial}{\partial\theta}\left(\sin\theta\frac{\partial}{\partial\theta}\right) + \frac{1}{r^2\sin^2\theta}\frac{\partial^2}{\partial\varphi^2}.$$

Appendix C

Founders of modern fluid dynamics

Ludwig Prandtl (1875–1953)

Ludwig Prandtl was born in Freising, Germany, in 1875. He studied mechanical engineering in Munich. For his doctoral thesis, he worked on a problem on elasticity under August Föppl, who himself did pioneering work in bringing together applied and theoretical mechanics. Later, Prandtl became Föppl's son-in-law, following the good German academic tradition in those days. In 1901, he became Professor of Mechanics at the University of Hanover, where he continued his earlier efforts to provide a sound theoretical basis for fluid mechanics. The famous mathematician Felix Klein, who stressed the use of mathematics in engineering education, became interested in Prandtl and enticed him to come to the University of Göttingen. Prandtl was a great admirer of Klein and kept a large portrait of him in his office. He served as Professor of Applied Mechanics at Göttingen from 1904 to 1953; the quiet university town of Göttingen became an international center of aerodynamic research.

In 1904, Prandtl conceived the idea of a boundary layer, which adjoins the surface of a body moving through a fluid, and is perhaps the greatest single discovery in the history of fluid mechanics. He showed that frictional effects in a slightly viscous fluid are confined to a thin layer near the surface of the body; the rest of the flow can be considered inviscid. The idea led to a rational way of simplifying the equations of motion in the different regions of the flow field. Since then the boundary-layer technique has been generalized and has become a most useful tool in many branches of science.

Prandtl's work on wings of finite span (the Prandtl-Lanchester wing theory) elucidated the generation of induced drag. In compressible fluid motions he contributed the Prandtl-Glauert rule of subsonic flow, the Prandtl-Meyer expansion fan in supersonic flow around a corner, and published the first estimate of the thickness of a shock wave. He made notable innovations in the design of wind tunnels and other aerodynamic equipment. His advocacy of monoplanes greatly advanced heavier-than-air aviation. In experimental fluid mechanics, he designed the Pitot-static tube for measuring velocity. In turbulence theory he contributed the mixing length theory.

Prandtl liked to describe himself as a plain mechanical engineer. So naturally he was also interested in solid mechanics; for example, he devised a soap-film analogy for analyzing the torsion stresses of structures with noncircular cross sections. In this respect he was like G.I. Taylor, and his famous student von Karman; all three of them did a considerable amount of work on solid mechanics. Toward the end of his career Prandtl became interested in dynamic meteorology and published a paper generalizing the Ekman spiral for turbulent flows.

Prandtl was endowed with rare vision for understanding physical phenomena. His mastery of mathematical tricks was limited; indeed many of his collaborators were better mathematicians. However, Prandtl had an unusual ability for putting ideas in simple mathematical forms. In 1948, Prandtl published a simple and popular textbook on fluid mechanics, which has been referred to in several places here. His varied interest and simplicity of analysis is evident throughout this book. Prandtl died in Göttingen in 1953.

Geoffrey Ingram Taylor (1886–1975)

Geoffrey Ingram Taylor's name almost always includes his initials G.I. in references, and his associates and friends simply refer to him as "G.I." He was born in 1886 in London. He apparently inherited a bent toward mathematics from his mother, who was the daughter of George Boole, the originator of "Boolean algebra." After graduating from the University of Cambridge, Taylor started to work with J.J. Thomson in pure physics.

He soon gave up pure physics and changed his interest to the mechanics of fluids and solids. At this time a research position in dynamic meteorology was created at Cambridge and it was awarded to Taylor, although he had no knowledge of meteorology! At the age of 27, he was invited to serve as meteorologist on a British ship that sailed to Newfoundland to investigate the sinking of the *Titanic*. He took the opportunity to make measurements of velocity, temperature, and humidity profiles up to 2000 m by flying kites and releasing balloons from the ship. These were the very first measurements on the turbulent transfers of momentum and heat in the frictional layer of the atmosphere. This activity started his lifelong interest in turbulent flows.

During World War I, he was commissioned as a meteorologist by the British Air Force. He learned to fly and became interested in aeronautics. He made the first measurements of the pressure distribution over a wing in full-scale flight. Involvement in aeronautics led him to an analysis of the stress distribution in propeller shafts. This work finally resulted in a fundamental advance in solid mechanics, the "Taylor dislocation theory."

Taylor had an extraordinarily long and productive research career (1909–1972). The amount and versatility of his work can be illustrated by the size and range of his *Collected Works* published in 1954: Volume I contains "Mechanics of Solids" (41 papers, 593 pages); Volume II contains "Meteorology, Oceanography, and Turbulent Flow" (45 papers, 515 pages); Volume III contains "Aerodynamics and the Mechanics of Projectiles and Explosions" (58 papers, 559 pages); and Volume IV contains "Miscellaneous Papers on Mechanics of Fluids" (49 papers, 579 pages). Perhaps G.I. Taylor is best known for his work on turbulence. When asked, however, what gave him maximum *satisfaction*, Taylor singled out his work on the stability of Couette flow.

Professor George Batchelor, who has encountered many great physicists at Cambridge, described G.I. Taylor as one of the greatest physicists of the century. He combined a remarkable capacity for analytical thought with physical insight by which he knew "how things worked." He loved to conduct simple experiments, not to gather data to understand a phenomenon, but to demonstrate his theoretical calculations; in most cases he already knew what the experiment would show. Professor Batchelor has stated that Taylor was a thoroughly lovable man who did not suffer from the maladjustment and self-concern that many of today's institutional scientists seem to suffer (because of pressure!), and this allowed his creative energy to be used to the fullest extent.

He thought of himself as an amateur, and worked for pleasure alone. He did not take up a regular faculty position at Cambridge, had no teaching responsibilities, and did not visit another institution to pursue his research. He never had a secretary or applied for a research grant; the only facility he needed was a one-room laboratory and one technical assistant. He did not "keep up with the literature," tended to take up problems that were entirely new, and chose to work alone. Instead of mastering tensor notation, electronics, or numerical computations, G.I. Taylor chose to do things his own way, and did them better than anybody else.

Further reading

Batchelor, G.K., 1976. Geoffrey Ingram Taylor, 1886–1975. Biographical Memoirs of Fellows of the Royal Society 22, 565–633.

Batchelor, G.K., 1986. Geoffrey Ingram Taylor, 7 March 1886–27 June 1975. Journal of Fluid Mechanics 173, 1–14.

Oswatitsch, K., Wieghardt, K., 1987. Ludwig Prandtl and his Kaiser-Wilhelm-Institute. Annual Review of Fluid Mechanics 19, 1–25.

Von Karman, T., 1954. Aerodynamics. McGraw-Hill, New York.

Appendix D

Visual resources

Following is a list of films, all but the first by the National Committee for Fluid Mechanics Films (NCFMF), founded in 1961 by the late Ascher H. Shapiro, then Professor of Mechanical Engineering at the Massachusetts Institute of Technology. Descriptive text for the films was published separately as described below.

The Fluid Dynamics of Drag, Parts I, II, III, IV (1960)
Text: Ascher H. Shapiro, Shape and Flow: The Fluid Dynamics of Drag, Doubleday and Co., New York (1961).
Vorticity, Parts I, II (1961)
The text for this and all following films is: NCFMF, Illustrated Experiments in Fluid Mechanics, MIT Press, Cambridge, MA (1972).
Deformation of Continuous Media (1963)
Flow Visualization (1963)
Pressure Fields and Fluid Acceleration (1963)
Surface Tension in Fluid Mechanics (1964)
Waves in Fluids (1964)
*Boundary Layer Control (1965)
Rheological Behavior of Fluids (1965)
Secondary Flow (1965)
Channel Flow of a Compressible Fluid (1967)
Low-Reynolds-Number Flows (1967)
Magnetohydrodynamics (1967)
Cavitation (1968)
Eulerian and Lagrangian Descriptions in Fluid Mechanics (1968)
Flow Instabilities (1968)
Fundamentals of Boundary Layers (1968)
Rarefied Gas Dynamics (1968)
Stratified Flow (1968)
Aerodynamic Generation of Sound (1969)
Rotating Flows (1969)
Turbulence (1969)

Although these films are decades old, they remain excellent visualizations of the principles of fluid mechanics. All but the one marked with an asterisk are available for viewing on the MIT website: http://web.mit.edu/fluids/www/Shapiro/ncfmf.html. It would be very beneficial to view the film appropriate to the corresponding section of the text.

Index

Symbols

β-plane model, 495, 496
Π-group, 19

A

Absolute angle of attack, 556
Absolute pressure, 8
Absolute vorticity, 162, 519
Acceleration
  centripetal, 103
  Coriolis, 103
  fluid, 64
  gravitational, 669
  parameter, 637
Accumulation point, 364
Accuracy, 627
Acoustics, 577
Added mass, 207
Adiabatic density gradient, 422, 491
Adiabatic lapse rate, 17
Adiabatic process, 13
Adiabatic temperature gradient, 17, 491
Adjacent index, 41
Advection, 63
Advection-diffusion equation, 623
Adverse, 291
Aerodynamics, 543
  incompressible, 543
Aerofoil, 547
Aileron, 544
Aircraft, 544
Airfoil, 547
  Zhukhovsky, 555
Algebraic model, 419
Alternating tensor, 43, 46, 47
Amplitude, 446
Angle of attack, 547
  absolute, 556
Angle of incidence, 547
Angular momentum principle, 110
Antisymmetric tensor, 50
Aperiodic solution, 361
Apparent mass, 206, 207
Approximation
  Boussinesq, 117, 493
  finite difference, 625
  finite volume, 629
  first-order, 626
  long-wave, 476
  narrow-gap, 341
  Oseen's, 250
  rigid lid, 509
  second-order, 626
  shallow-water, 476
  WKB, 481, 521
  zeroth-order, 232
Arbitrarily moving control volume, 86
Arbitrary process, 11
Arc length, 674
Artificial viscosity, 661
Aspect ratio, 544
Atmospheric pressure, 669
Attractors, 361
Autocorrelation function, 380
Average, 349, 379
  ensemble, 376
  manifestation, 3
  phase, 435
Averaged equations of motion, 383
Avogadro's number, 669
Axisymmetric
  flow, 199, 202, 203, 216, 247, 371
  ideal flow, 202, 204

B

Background density distribution, 478
Background state, 321
Baroclinic
  flow, 156
  instability, 320, 532, 533
  mode, 475, 509
  torque, 162
Barotropic
  fluid, 102
  instability, 531
  mode, 475, 509
Basic state, 321
Basic vector calculus, 676
Bathtub vortex, 156
Bearing number, 231
Bénard convection, 330
Bénard problem, 329
Bernoulli equation, 111, 112
Bifurcation, 361
Blasius solution, 277
Blasius theorem, 193, 194
Blocking, 480
Blunt body, 281
Boltzmann's constant, 669
Borderline, 321
Bound vortex, 559
Boundary, 10
  condition, 119, 639, 647
  layer, 113, 177, 246, 277, 356
Boundary-layer
  instability, 357
  separation, 289
  thickness, 275
Boundary-value problem, 633, 636
Boussinesq approximation, 117, 493
Brunt–Väisälä frequency, 16, 478
Buffer layer, 409
Bulk strain rate, 70
Buoyancy frequency, 478, 491
Buoyant
  destruction, 394
  generation, 394
  production, 394, 422

C

Caloric equations of state, 11
Camber, 544, 547, 554
  line, 547
Capillarity, 7
Capillary tubes, 7
Capillary waves, 457
Capillus, 7
Cartesian coordinates, 678
Cauchy-Riemann conditions, 191
Cauchy's equation of motion, 98
Cavitation number, 127
Cell Reynolds number, 633
Cellular convection, 321, 333
Centrifugal instability, 339
Centripetal acceleration, 103
Chaotic saddle in state space, 355
Chaotic system, 360
Characteristic curve, 604
Chord, 547
  length, 544
Circular
  Couette flow, 228
  cylinder, 294
  Poiseuille flow, 227
Circulation, 55, 71
  clockwise, 192
  per unit area, 71
Clam-shell control volume, 215
Classical thermodynamics, 10
Clockwise
  circulation, 192
  velocity, 185
Closure problem, 387

Cnoidal wave, 471
Cockpit, 544
Coefficient
correlation, 380
drag, 281
profile, 566
of bulk viscosity, 100
pressure, 126, 184
skin friction, 275
thermal expansion, 12
Colocated grid, 642
Common three-dimensional coordinate systems, 677
Complex
notation, 472
potential, 191
velocity, 191, 500
Compressible flow, 575, 656
Compression, 580
corner, 609
region, 467
side, 547
Computational domain, 647
Computational fluid dynamics (CFD), 199
Concentration, 5
Condition
boundary, 119, 639, 647
constant surface tension, 120
kinematic, 448
Cauchy-Riemann, 191
Courant, 435
critical, 585
dynamic, 449
Kutta, 549
no-slip, 123
sonic, 585
Conformal mapping, 195
Conformal transformation, 552
Conservation of energy, 106
boundary conditions, 123
Conservation of mass, 86
boundary condition, 120
Conservation of momentum, 89
boundary conditions, 122
Conservative
body forces, 90
force, 90
form, 668
Conserved scalars, 387
Consistent, 629
Constant, 397, 533
density flow, 87
diffusivity, 429
gas, 13
universal, 13, 669
gravitational, 669
shear stress, 313
Constant surface tension boundary condition, 120
Constant-area ducts, 600, 603
Constitutive equation, 98, 99
Contact angle, 9
Continuity equation, 87, 477
Continuously stratified fluid, 477
Continuously stratified layer, 507
Continuum fluid, 3
Continuum hypothesis, 3
Contraction, 43
Control surface, 77, 544, 623
Control volume, 77, 623, 629
Convecting layers, 339
Convection, 63
Bénard, 330
cellular, 321, 333
Convergent nozzle, 597
Convergent–divergent nozzle, 589, 597
Conversion factors, 669
Coordinate mapping, 652
Coordinate systems, 59
Coriolis acceleration, 103
Coriolis frequency, 495
Coriolis parameter, 495
Correlation, 380
coefficients, 380
function, 380
Couette flow
circular, 228
plane, 226, 355
Courant condition, 435
Creeping flow, 246
Critical conditions, 585
Critical layer, 352
Cross products, 47
Cross stresses, 434
Cross-correlation function, 380
Curl, 48, 49, 676
free vector field, 49
Cusp, 555
Cyclic frequency, 446
Cyclone, 504
Cylindrical coordinates, 678, 680

D

d'Alembert's paradox, 185
d'Alembert's solution, 580
Dead water, 474
Decay, 242
Deep-water waves, 454
Definitions, 10
Deflection angle, 608
Deformation rates, 69
Degree of freedom, 361
Delta wings, 566
Density, 669
equation, 477
interface, 472
Departure, 529
Depth, 523
Derivative, 102
material, 63
partial, 674, 675
particle, 63
substantial, 63
total, 674
Deterministic chaos, 360
Deviatoric stress tensor, 99
Difference, 453
Differential form, 85
Differentiation
partial, 673
total, 673
Diffusers, 585
Diffusion, 156
Diffusivity, 374
Dilatant, 101
Dimensional analysis, 19
Dimensional homogeneity, 19
Dimensionless forms, 124
Dimensionless group, 19
Dirac delta function, 677
Direct numerical simulation (DNS), 388, 435
Direct solution, 552
Direct viscous dissipation, 393
Direction, 134
Dispersion, 450, 461, 481
relation, 450, 523, 529
Displacement thickness, 275
Dissipation, 374
Distribution
background density, 478
elliptical lift, 562
spatial, 8
Disturbances, 351
Divergence, 48, 676
free vector field, 49
theorem, 53
Doppler shifted wave, 447
Double contraction, 44
Double dot product, 44
Double-diffusive instability, 336
Double-sided normalization, 397
Doublet, 182
Downstream direction, 274
Downwash, 560
Drag, 565
coefficient, 281
force, 543
form, 292, 547, 566
friction, 547, 566
induced, 559, 561, 566
polar, 566
pressure, 547, 566
profile, 566
vortex, 561
wave, 130, 561, 595, 613
Driven cavity problem, 637
Dufort-Frankel scheme, 666
Dynamic
condition, 449
pressure, 115
similarity, 124
viscosity, 5
Dynamics, 59
gas, 543, 575

E

Eddy, 374
diffusivities, 418
viscosity, 418
Effect of friction, 600

Effect of heat transfer, 601
Effective
  angle of attack, 560
  gravity, 105
  viscosity, 418
Eigenvalues, 51
Eigenvectors, 51
Einstein summation convention, 38
Ekman layer, 499, 500
  at a free surface, 500
  on a rigid surface, 503
Ekman spiral, 500
Elementary lifting line theory, 558
Elementary lubrication theory, 230
Elevator, 544
Elliptical lift distribution, 562
Energetics, 536
Energy, 669
  considerations, 482
  flux, 453, 461, 482, 669
  in transition, 10
  internal, 10
  kinetic, 394
  potential, 394, 453, 536
  spectrum, 382
Energy-balance equation, 393
Energy-budget equation, 393
Ensemble, 376
  average, 376
Enstrophy, 538
Enthalpy, 11
Entire flow, 223
Entrainment, 303, 399
Entropy, 10, 11
  production, 109
Epsilon delta relation, 47
Equation
  advection-diffusion, 623
  Bernoulli, 111, 112
  constitutive, 98, 99
  continuity, 87, 477
  density, 477
  energy-balance, 393
  energy-budget, 393
  Euler, 102
  in curvilinear coordinates, 679
  Korteweg–de Vries, 471
  laminar boundary-layer, 282
  modified, 629
  Navier-Stokes momentum, 101
  Orr-Sommerfeld, 349, 351
  Oseen's, 250
  perturbation vorticity, 533
  quasi-geostrophic vorticity, 528
  RANS, 387
  Rayleigh, 351
  shallow-water, 505, 506, 510
  Taylor-Goldstein, 345
  two-dimensional advection-diffusion, 634
  vorticity, 158, 160, 162
Equation of motion
  averaged, 383
  Cauchy's, 98
  for geophysical flows, 493
Equation of state, 11
  caloric, 11
  thermal, 11
Equilibrium state, 321
Equivalent depth, 506, 508
Euler equation, 102
Euler number, 126
Eulerian description, 61
Evolution, 528
Expansion, 580
  waves, 610
Expected value, 376
Explicit schemes, 632
Extensive variables, 10, 33
External flows, 575

F

f-plane model, 495
Falkner-Skan similarity solutions, 282
Far field, 402
Favorable pressure gradient, 278, 291
Fick's law, 5
Fictitious body forces, 90
Fineness ratio, 231
Finite amplitude, 466
  waves, 582
Finite difference approximation, 625
Finite volume approach, 629
Finite volume approximation, 629
First law of thermodynamics, 10
First moment, 379
First normal-stress difference, 101
First-order approximation, 626
Fixed point, 361
Flap, 545
Flow
  axisymmetric, 199, 202, 203, 216, 247, 371
  baroclinic, 156
  blockage effect, 299
  compressible, 575, 656
  creeping, 246
  entire, 223
  external, 575
  geostrophic, 496
  hypersonic, 576
  ideal, 177
  in complex domains, 650
  incompressible, 87, 576, 637, 641
  internal, 575
  lines, 64
  Marangoni, 121
  one-dimensional, 59, 658
  parallel, 351
  pipe, 355
  plane, 59
  potential, 180, 204
  stagnation, 192
  steady, 59, 226–228
  subcritical, 467
  subsonic, 128, 576
  supercritical, 467
  supersonic, 128, 576, 612
  Taylor-Couette, 228
  three-dimensional, 60
    axisymmetric, 59
  transonic, 576
  turbulent, 376
  two-dimensional, 59, 661
  unsteady, 59
  with oscillations, 243
  work, 583
Fluctuations, 374
Fluid
  acceleration, 64
  barotropic, 102
  continuum, 3
  density, 158
  homogeneous, 497, 506
  incompressible, 87, 100
  mechanics, 1
  Newtonian, 98–100
  particles, 10
  statics, 8
Flux, 71
  divergence, 87
  of vorticity, 71
  Richardson number, 423
Force, 669
  fields, 90
  on a surface, 44
  on a two-dimensional body, 193
  per unit area, 41
  potential, 90
Forced convection region, 423
Form drag, 292, 547, 566
Fourier's law, 5
Free shear flows, 303
Free surface, 449
Free turbulent shear flows, 399
Frequency
  Brunt–Väisälä, 16, 478
  buoyancy, 478, 491
  Coriolis, 495
  cyclic, 446
  intrinsic, 447, 526
  observed, 447, 526
  radian, 446
Friction
  drag, 547, 566
  effect, 600
  velocity, 317, 407
Function
  autocorrelation, 380
  correlation, 380
  cross-correlation, 380
  Dirac delta, 677
  harmonic, 180
  nonlinear, 101
  of time, 23
  path, 10
  Prandtl-Meyer, 612
  state, 10, 12
  Stokes stream, 202
  stream, 88
  velocity potential, 180
  wall, 421
Fuselage, 544

G

Galilean transformation, 64
Gas, 2
  constant, 13
    universal, 13, 669
  dynamics, 543, 575
Gauge pressure, 8
Gauss' theorem, 52
Gauss-Seidel iteration, 637
Geoffrey Ingram Taylor, 683
Geostrophic
  balance, 496
  flow, 496
  turbulence, 374, 537
Ghost cells, 647
Görtler vortices, 343
Gradient, 48, 100, 676
  diffusion hypothesis, 418
  Richardson number, 346, 423
Gravitational acceleration, 669
Gravitational constant, 669
Gravity waves with rotation, 512
Grid points, 199
Ground effect, 552
Group
  speed, 454
  velocity, 461

H

Half-body, 184
Harmonic functions, 180
Harmonic pressure fluctuations, 422
Heat transfer, 601
Helmholtz's theorem, 155
High Reynolds numbers, 296
High-frequency regime, 510, 525
Higher-order central moments, 379
Hill's spherical vortex, 174
History term, 252
Hodograph, 500
Homogeneous
  fluid, 497, 506
  isotropic turbulence, 389
  layer, 508, 510
Horizontal
  density surfaces, 537
  gradient of density, 497
  Laplacian operator, 478
Howard semicircle theorem, 348
Hydraulic diameter, 412
Hydraulic jump, 467
Hydrodynamically smooth surfaces, 413
Hydrostatic waves, 455
Hydrostatics, 9
Hypersonic flow, 576
Hypothesis
  continuum, 3
  gradient diffusion, 418
  Taylor's, 383
  turbulent viscosity, 418
  Zhukhovsky, 549

I

Ideal flow, 177
Ideal line vortex, 151
Immersed boundary methods, 651
Implicit schemes, 632
Imposed time scale, 127
Incompressible
  aerodynamics, 543
  flow, 87, 576, 637, 641
  fluid, 87, 100
Index-based notation, 38
Indicial notation, 38
Induced drag, 559, 561, 566
Inertial
  circles, 514
  frame of reference, 158, 160
  motion, 514
  period, 495, 514
  regions, 538
  sub-range, 397
Inflectional profiles, 354
Initial
  behavior, 320
  state, 321
  vorticity, 519
Inner
  layer, 407
  problem, 272
  region, 178
Instability, 533, 629
  baroclinic, 320, 532, 533
  barotropic, 531
  boundary-layer, 357
  centrifugal, 339
  criterion, 535
  double-diffusive, 336
  Kelvin-Helmholtz, 322, 344
  Rayleigh-Plateau, 326
  Rayleigh-Taylor, 324, 366
  sausage, 371
  secondary, 359
  thermal, 329
Integral form, 85
Integral theorems, 677
Integral time scale, 381
Integration, 675
Intensive variables, 10
Interaction of vortices, 165
Interior points, 639
Internal
  energy, 10
  flows, 575
  gravity waves, 506, 511
  Kelvin waves, 516
  mode, 475
  Rossby radius of deformation, 516
  value, 529
  waves, 477, 479, 520
Intrinsic frequency, 447, 526
Invariant, 43
Inversion, 17
Inviscid stability, 351
Irrotational
  constant-density flow theory, 177
  fluid motion, 71
  vector field, 49
  vortex, 151
Isentropic
  process, 13
  table, 599
Isotropic
  stress, 99
  tensor, 47

J

Jacobian, 653
Jet, 31

K

k-ε closure model, 417
Karman vortex streets, 480
Kelvin, 2
  cat's eye pattern, 354
  wave, 514, 515
Kelvin-Helmholtz instability, 322, 344
Kelvin's theorem, 155
Kilogram, 2
Kinematic boundary condition, 448
Kinematic viscosity, 6
Kinematics, 59
  of simple plane flows, 73
Kinetic energy, 394
Kinetics, 59
kmole, 22
Kolmogorov microscale, 396
Kolmogorov's $k^{-5/3}$ law, 397
Korteweg–de Vries equation, 471
Kronecker delta, 43, 46
Kutta condition, 549
Kutta-Zhukhovsky lift theorem, 188, 194, 195, 547

L

Laboratory frame, 103
Lagrangian description, 61
Lamb surfaces, 112
Laminar boundary-layer equations, 282
Laminar viscous flow, 223
Lanchester, 564
Laplace pressure, 7
Lapse rate, 16
  adiabatic, 17
Law
  Fick's, 5
  Fourier's, 5
  Kolmogorov's $k^{-5/3}$, 397
  logarithmic, 408
  Newton's, 5
  of Biot and Savart, 163
  of the wall, 407
  Pascal's, 8
  phenomenological, 5
  Richardson's four-thirds, 376
  velocity defect, 407, 408
Layer
  boundary, 113, 177, 246, 277, 356
  buffer, 409
  convecting, 339
  critical, 352
  Ekman, 499, 500

homogeneous, 508, 510
inner, 407
logarithmic, 408
outer, 407, 408
overlap, 408
two-stream shear, 354
Leading-edge suction, 571
Leap-Frog scheme, 632
Lee wave, 526
Leibniz's theorem, 77
Length, 2, 669
Monin-Obukhov, 423
Leonard stresses, 434
Lift, 565
force, 188, 543
of a Zhukhovsky airfoil, 555
Lifting line, 559
Lighthill stress tensor, 579
Limit cycle, 361
Linear liquid-surface gravity waves, 447
Linear strain rates, 69
Liquid, 2
Local Cartesian system, 494
Logarithmic law, 408
Logarithmic layer, 408
Long-wave approximation, 476
Loop, 156
Low Reynolds numbers, 294, 350
Low-frequency regimes, 510, 525
Ludwig Prandtl, 683

M

Mach
angle, 607
cone, 607
lines, 607
number, 467, 576
Magnitude, 134, 166
Magnus effect, 188, 300
negative, 301
Manometer, 26
Mapped grid methods, 652
Marangoni flows, 121
Marginal state, 321
Mass, 2, 669
apparent, 206, 207
diffusivity, 5
transport velocity, 470
Material
derivative, 63
lines, 162
volume, 86
Mean pressure, 100
Memory of the turbulence, 381
Meter, 2
Method
immersed boundary, 651
mapped grid, 652
normal mode, 321
of characteristics, 466, 604, 675
of images, 166
of normal modes, 321
panel, 663
Schlieren, 576
spectral, 663
Theodorsen's, 552
Thwaites', 285, 286
two-step Lax-Wendroff, 657
velocity-pressure, 641, 656
vortex blob, 170
vorticity-stream function, 637
Mid-frequency regime, 526
Mixing length, 418
model, 418
theory, 408
Mode
baroclinic, 475, 509
barotropic, 475, 509
normal, 507
oscillatory, 321
Model
β-plane, 495, 496
algebraic, 419
mixing length, 418
one-equation, 420
Reynolds stress, 421
surrogate, 666
two-equation, 420
zero-equation, 419
Moderate Reynolds numbers, 294
Modified equation, 629
Molecular transport phenomena, 5
Molecular volume, 25
Moment of interest, 87
Moments, 376
Momentum diffusivity, 224
Monin-Obukhov length, 423
Moving equilibrium, 400
Moving medium, 591
Multidimensional steady-state problem, 636
Multiplication, 43
of matrices, 40
Multiscale modeling, 665

N

Nabla, 48
Narrow-gap approximation, 341
Navier-Stokes momentum equation, 101, 102
Neutrally stable atmosphere, 16
Neutrally stable state, 321
Newtonian fluid, 98–100
Newton's law, 5
No-slip condition, 123
Nodes, 459
Noninertial frame of reference, 103
Nonlinear
effects, 359
function, 101
waves, 466
Nonlinearity, 374
Normal mode, 507
method, 321
Normal shock, 591
structure, 595
waves, 591
Notation, 37
Nozzle, 585
operation, 597
Number
Avogadro's, 669
bearing, 231
cavitation, 127
Euler, 126
Mach, 467, 576
Rayleigh, 330
Reynolds, 125
cell, 633
high, 296
low, 294, 350
moderate, 294
Richardson, 422
flux, 423
gradient, 346, 423
Rossby, 496
Strouhal, 295
turbulent Prandtl, 423
wave, 446

O

Observed frequency, 447, 526
Ocean, 18
One-dimensional flow, 59, 658
One-equation model, 420
Operation of nozzles, 597
Orientation, 567
Orr-Sommerfeld equation, 349, 351
Orthogonal matrix, 57
Oscillatory mode, 321
Oseen's approximation, 250
Oseen's equation, 250
Outer
layer, 407, 408
length scale, 395
problem, 272
region, 178
Overlap layer, 408
Overstability, 321

P

Panel methods, 663
Parallel flows, 351
Partial derivatives, 674, 675
Partial differentiation, 673
Particle
derivative, 63
fluid, 10
orbit, 513, 523
Particle image velocimetry (PIV), 65, 382
Pascal's law, 8
Passive scalars, 387
Path, 10
functions, 10
line, 64
Perfect gas, 13
thermodynamic relations, 576
Period, 446
Permutation symbol, 47
Persistent turbulence, 399
Perturbation vorticity equation, 533
Perturbations, 495
Phase, 446

average, 435
space, 361
speed, 446
Phenomenological laws, 5
Physical constants, 669
Pipe flow, 355
Pitch axis, 544
Pitot tube, 114
Planck's constant, 669
Plane
flow, 59
Couette, 226, 355
Poiseuille, 226, 254, 355, 356
polar coordinates, 679
Planetary vorticity, 162, 495
Planetary waves, 528
Planform area, 127, 544
Plastic, 3
Plume, 31
Poincaré waves, 512
Point of inflection, 290
Polar plots, 566
Positive stresses, 98
Potential
density, 17, 422, 478
energy, 394, 453, 536
flow, 180, 204
temperature, 17
velocity, 179
vorticity, 519
conservation, 517
Prandtl, Ludwig, 564, 683
Prandtl-Meyer expansion fan, 611
Prandtl-Meyer function, 612
Pressure, 3, 8, 669
absolute, 8
coefficient, 126, 184
drag, 547, 566
dynamic, 115
gauge, 8
gradient, 226, 289, 356
favorable, 278, 291
mean, 100
side, 547
stagnation, 115, 584
total, 115
Primitive form, 641
Principal axes, 51, 72, 73
Principle
angular momentum, 110
of equipartition of energy, 453
of exchange of stabilities, 333
Problem
Bénard, 329
boundary-value, 633, 636
closure, 387
driven cavity, 637
inner, 272
multidimensional steady-state, 636
outer, 272
steady-state, 633
Stokes' first, 236
Stokes' second, 243
Taylor, 339
viscous, 355
Process, 10
adiabatic, 13
arbitrary, 11
isentropic, 13
reversible, 10
stochastic, 376
Profile drag, 566
coefficient, 566
Pseudo-turbulence, 434
Pseudoplastic, 101
Pure capillary waves, 458

Q

Q-criterion, 81
Quasi-geostrophic motions, 528
Quasi-geostrophic vorticity equation, 528
Quonset hut, 214

R

Radian frequency, 446
Random variables, 376
Random walk, 428, 429
Rankine-Hugoniot relations, 592
RANS equations, 387
Rapid pressure fluctuations, 422
Rarefaction, 580
Rayleigh equation, 351
Rayleigh number, 330
Rayleigh-Plateau instability, 326
Rayleigh-Taylor instability, 324, 366
Rayleigh's inflection point criterion, 531
Reduced gravity, 126
Refraction, 455
Regime
high-frequency, 510, 525
low-frequency, 510, 525
mid-frequency, 526
Region
compression, 467
forced convection, 423
inertial, 538
inner, 178
outer, 178
wake, 178
Regular perturbation, 354
Relation
dispersion, 450, 523, 529
epsilon delta, 47
perfect gas thermodynamic, 576
Rankine-Hugoniot, 592
Relative
speed, 25
velocity, 4
vorticity, 518
Relaxation time, 10
Repeller, 362
Residual, 646
stress tensor, 433
Residue, 195
Reversible processes, 10
Revolution, 74
Reynolds
analogy, 423
averaging, 387
decomposition, 376, 383
number, 125
cell, 633
high, 296
low, 294, 350
moderate, 294
stress, 434
models, 421
tensor, 385
transport theorem, 77
Richardson number, 422
flux, 423
gradient, 346, 423
Richardson's four-thirds law, 376
Riemann invariants, 604
Rigid lid approximation, 509
Ripples, 458
Robins effect, 301
Roll axis, 544
Root-mean-square, 379
Rossby
number, 496
radius of deformation, 516
wave, 528, 529
Rotation, 74, 493
rate, 69
tensor, 50, 69
Rotational gravity waves, 512
Rotor ships, 300
Rough surfaces, 413
Round tube, 227
Rudder, 544

S

Saffman lift, 300
Salinity, 18
Salt fingers, 337
Sausage instability, 371
Scalars, 37, 38
Scale height, 18
Scatter plot, 389
Scheme
Dufort-Frankel, 666
explicit, 632
implicit, 632
Leap-Frog, 632
upwind, 632
Schlieren method, 576
Schwartz inequality, 380
Second, 2
Second law of thermodynamics, 11
Second normal-stress difference, 101
Second-order
approximation, 626
closures, 421
tensor, 37, 41, 42
Secondary flow, 308, 321, 343
Secondary instability, 359
Seiche, 460
Self-preservation, 400
Self-preserving, 399
Separated flow region, 178
Separation bubble, 178, 294

Separation point, 291
Settling velocity, 253
Shallow-water
approximation, 476
equations, 505, 506, 510
theory, 517
waves, 455
Shape factor, 285
Shear
correlation, 285
production, 393, 394
thickening, 101
thinning, 101
velocity, 407
waves, 480
Shock angle, 608
Shock wave, 582
Similarity solution, 236, 239
Singularities, 191
Sinuous, 371
Skin friction coefficient, 275
Sloping convection, 537
Small-scale geophysical phenomena, 490
Solenoidal vector field, 49
Solid body rotation, 74
Solids, 2
Solitary wave, 471
Soliton, 471
Solution
aperiodic, 361
Blasius, 277
d'Alembert's, 580
direct, 552
similarity, 236, 239
Falkner-Skan, 282
variable, 19
vertical mode, 508
wave, 535
WKB, 521
Sonic condition, 585
Span, 127
Span-wise direction, 359
Spatial discretization, 643
Spatial distribution, 8
Species, 5
Specific
heats, 11
impulse, 95
volume, 10
Spectral methods, 663
Speed of light in vacuum, 669
Speed of sound, 12
Spherical coordinates, 679, 680
Spin parameter, 301
Splitting, 656
Spoilers, 545
Squire transformation, 350
Squire's theorem, 344, 349
Stability, 629
Staggered grid, 642
Stagnation
density, 584
enthalpy, 584
flow, 192
pressure, 115, 584
temperature, 584
Stall, 549, 565
Standard deviation, 379
Standing waves, 459
Starting vortex, 550
State functions, 10, 12
Stationary in space, 377
Stationary normal shock wave, 591
Stationary pattern, 321
Statistics, 376, 377
Steady flow, 59, 226–228
Steady-state problem, 633
Stochastic processes, 376
Stokes'
first problem, 236
law of resistance, 248
second problem, 243
theorem, 55
Stokes assumption, 100
Stokes drift, 470
Stokes stream function, 202
Stokes waves, 469
Strain, 69
rate, 69
bulk, 70
linear, 69
tensor, 69
volumetric, 70
Strange attractor, 363
Stratopause, 491
Stratosphere, 491
Streak line, 65
Stream functions, 88
Stream tube, 64
Streamline, 64
Strength of a vortex sheet, 168
Strength of a vortex tube, 152
Stress
isotropic, 99
Leonard, 434
positive, 98
Reynolds, 434
tensor
deviatoric, 99
lighthill, 579
residual, 433
Reynolds, 385
Strong shock, 609
Strouhal number, 295
Sturm-Liouville form, 507
Subcritical flow, 467
Subharmonics, 364
Subsonic flow, 128, 576
Substantial derivative, 63
Successive over relaxation (SOR), 637
Suction side, 547
Supercritical flow, 467
Supersonic flow, 128, 576, 612
Surface, 41
capillary-gravity waves, 445
Lamb, 112
tension, 7, 120
Surrogate models, 666
Sverdrup waves, 512
Sweepback angle, 566
Symmetric tensor, 50, 51
System
chaotic, 360
coordinate, 59
common three-dimensional, 677
local Cartesian, 494
thermodynamic, 10

T

Tangent, 461
plane, 494
Taylor, 683
column, 498
problem, 339
vortices, 342
Taylor-Couette flow, 228
Taylor-Goldstein equation, 345
Taylor-Proudman theorem, 497, 498
Taylor's hypothesis, 383
Taylor's theory of turbulent dispersion, 426
Temperature, 2, 669
potential, 17
Tensor, 37
alternating, 43, 46, 47
antisymmetric, 50
isotropic, 47
rotation, 50, 69
second-order, 37, 41, 42
strain rate, 69
symmetric, 50, 51
velocity gradient, 69
Theodorsen's method, 552
Theorem
Blasius, 193, 194
divergence, 53
Gauss', 52
Helmholtz's, 155
Howard semicircle, 348
integral, 677
Kelvin's, 155
Kutta-Zhukhovsky lift, 188, 194, 195, 547
Leibniz's, 77
Reynolds transport, 77
Squire's, 344, 349
Stokes', 55
Taylor-Proudman, 497, 498
Thermal diffusivity, 224
Thermal equations of state, 11
Thermal expansion coefficient, 12
Thermal instability, 329
Thermal wind, 497
Thermocline, 488, 491
Thermodynamic
potential, 120
pressure, 99
system, 10
Thin-airfoil theory, 552, 612
Three-dimensional
flow, 60
axisymmetric, 59
potential, 206
waves, 359

Throat, 589
Thrust, 544
Thrust-specific fuel consumption, 574
Thwaites' method, 285, 286
Tilting, 519
Time, 2, 669
    lag, 380
Tip vortices, 559
Tollmien-Schlichting waves, 355, 358
Total
    derivatives, 674
    differentiation, 673
    pressure, 115
Trace velocity, 446
Trailing vortices, 558
Trajectory, 361
Transformation
    conformal, 552
    Galilean, 64
    Squire, 350
    Zhukhovsky, 552
Transition, 289, 359
Transitional viscous flow, 223
Transonic flow, 576
Transport, 87, 393
Transpose of a matrix, 38
Traveling oscillations, 321
Tropopause, 491
Troposphere, 491
Turbulence, 223, 319, 422
    geostrophic, 374, 537
    modeling, 417
    persistent, 399
Turbulent
    diffusivity, 429
    flow, 376
        viscous, 223
    intensities, 404
    kinetic energy, 393, 436
    Prandtl number, 423
    viscosity hypothesis, 418
Two-dimensional
    advection-diffusion equation, 634
    flow, 59, 661
    jets, 303
    steady compressible flow, 606
    stream function, 179
Two-equation models, 420
Two-step Lax-Wendroff method, 657
Two-stream shear layer, 354

U

Uncertainty quantification, 665
Universal gas constant, 13, 669
Unsteady flow, 59
Unstructured grids, 655
Upwelling, 516
Upwind scheme, 632

V

Vapor trails, 559
Vector, 37, 39
    dot, 47
    vorticity, 50
Vector field
    curl free, 49
    divergence free, 49
    irrotational, 49
    solenoidal, 49
Velocity, 669
    clockwise, 185
    complex, 191, 500
    defect, 408
        law, 407, 408
    friction, 317, 407
    gradient tensor, 69
    group, 461
    potential, 112, 179
        function, 180
    relative, 4
    scale, 396
    shear, 407
    trace, 446
Velocity-pressure method, 641, 656
Venturi tube, 590
Viscosity
    artificial, 661
    dynamic, 5
    eddy, 418
    effective, 418
    kinematic, 6
Viscous
    convective sub-range, 425
    problem, 355
    sub-layer, 407
    torque, 162
    wall unit, 407
Volume per molecule, 25
Volumetric strain rate, 70
von Karman
    boundary-layer momentum integral
        equation, 284
    constant, 408
    momentum integral equation, 283
    vortex street, 294, 298
Vortex, 151
    bathtub, 156
    blob method, 170
    bound, 559
    drag, 561
    Görtler, 343
    Hill's spherical, 174
    ideal line, 151
    irrotational, 151
    motion, 151
    ring, 156
    sheet, 168, 474, 558
        strength, 168
    starting, 550
    stretching, 519
    Taylor, 342
    tip, 559
    trailing, 558
    tube, 152
        strength, 152
Vorticity, 70, 374
    absolute, 162, 519
    equation, 158, 160, 162
    initial, 519
    planetary, 162, 495
    potential, 519
    relative, 518
    vector, 50
Vorticity-stream function methods, 637

W

Wake region, 178
Wall functions, 421
Wall jet, 306
Wall-bounded turbulent shear flows, 399, 405
Water waves, 445
Wave
    angle, 608
    capillary, 457
    cnoidal, 471
    deep-water, 454
    Doppler shifted, 447
    drag, 130, 561, 595, 613
    expansion, 610
    finite amplitude, 582
    hydrostatic, 455
    internal, 477, 479, 520
        gravity, 506, 511
        Kelvin, 516
    Kelvin, 514, 515
    Lee, 526
    nonlinear, 466
    number, 446
        vector, 446
    packet, 461
    planetary, 528
    Poincaré, 512
    pure capillary, 458
    refraction, 456
    Rossby, 528, 529
    shallow-water, 455
    shear, 480
    shock, 582
        normal, 591
    solitary, 471
    solution, 535
    standing, 459
    stationary normal shock, 591
    Stokes, 469
    surface capillary-gravity, 445
    Sverdrup, 512
    three-dimensional, 359
    Tollmien-Schlichting, 355, 358
    water, 445
Waveform, 446
Wavelength, 446
Weak shock, 609
Wedge of instability, 537
Wing root, 544
Wing tip, 544
Winglets, 574
Wingspan, 544

WKB approximation, 481, 521
WKB solution, 521

Y

Yaw axis, 544
Yield point, 3

Z

Zero-equation model, 419
Zeroth-order approximation, 232
Zhukhovsky
  airfoils, 555
  hypothesis, 549
  transformation, 552
Zone of
  action, 607
  free convection, 424
  silence, 607